Linear Algebra and Matrix Analysis for Statistics

CHAPMAN & HALL/CRC
Texts in Statistical Science Series

Series Editors
Francesca Dominici, *Harvard School of Public Health, USA*
Julian J. Faraway, *University of Bath, UK*
Martin Tanner, *Northwestern University, USA*
Jim Zidek, *University of British Columbia, Canada*

Analysis of Failure and Survival Data
P. J. Smith

The Analysis of Time Series: An Introduction, Sixth Edition
C. Chatfield

Applied Bayesian Forecasting and Time Series Analysis
A. Pole, M. West, and J. Harrison

Applied Categorical and Count Data Analysis
W. Tang, H. He, and X.M. Tu

Applied Nonparametric Statistical Methods, Fourth Edition
P. Sprent and N.C. Smeeton

Applied Statistics: Handbook of GENSTAT Analyses
E.J. Snell and H. Simpson

Applied Statistics: Principles and Examples
D.R. Cox and E.J. Snell

Applied Stochastic Modelling, Second Edition
B.J.T. Morgan

Bayesian Data Analysis, Third Edition
A. Gelman, J.B. Carlin, H.S. Stern, D.B. Dunson, A. Vehtari, and D.B. Rubin

Bayesian Ideas and Data Analysis: An Introduction for Scientists and Statisticians
R. Christensen, W. Johnson, A. Branscum, and T.E. Hanson

Bayesian Methods for Data Analysis, Third Edition
B.P. Carlin and T.A. Louis

Beyond ANOVA: Basics of Applied Statistics
R.G. Miller, Jr.

The BUGS Book: A Practical Introduction to Bayesian Analysis
D. Lunn, C. Jackson, N. Best, A. Thomas, and D. Spiegelhalter

A Course in Categorical Data Analysis
T. Leonard

A Course in Large Sample Theory
T.S. Ferguson

Data Driven Statistical Methods
P. Sprent

Decision Analysis: A Bayesian Approach
J.Q. Smith

Design and Analysis of Experiments with SAS
J. Lawson

Elementary Applications of Probability Theory, Second Edition
H.C. Tuckwell

Elements of Simulation
B.J.T. Morgan

Epidemiology: Study Design and Data Analysis, Third Edition
M. Woodward

Essential Statistics, Fourth Edition
D.A.G. Rees

Exercises and Solutions in Statistical Theory
L.L. Kupper, B.H. Neelon, and S.M. O'Brien

Exercises and Solutions in Biostatistical Theory
L.L. Kupper, B.H. Neelon, and S.M. O'Brien

Extending the Linear Model with R: Generalized Linear, Mixed Effects and Nonparametric Regression Models
J.J. Faraway

A First Course in Linear Model Theory
N. Ravishanker and D.K. Dey

Generalized Additive Models: An Introduction with R
S. Wood

Generalized Linear Mixed Models: Modern Concepts, Methods and Applications
W. W. Stroup

Graphics for Statistics and Data Analysis with R
K.J. Keen

Interpreting Data: A First Course in Statistics
A.J.B. Anderson

Introduction to General and Generalized Linear Models
H. Madsen and P. Thyregod

An Introduction to Generalized Linear Models, Third Edition
A.J. Dobson and A.G. Barnett

Introduction to Multivariate Analysis
C. Chatfield and A.J. Collins

Introduction to Optimization Methods and Their Applications in Statistics
B.S. Everitt

Introduction to Probability with R
K. Baclawski

Introduction to Randomized Controlled Clinical Trials, Second Edition
J.N.S. Matthews

Introduction to Statistical Inference and Its Applications with R
M.W. Trosset

Introduction to Statistical Limit Theory
A.M. Polansky

Introduction to Statistical Methods for Clinical Trials
T.D. Cook and D.L. DeMets

Introduction to Statistical Process Control
P. Qiu

Introduction to the Theory of Statistical Inference
H. Liero and S. Zwanzig

Large Sample Methods in Statistics
P.K. Sen and J. da Motta Singer

Linear Algebra and Matrix Analysis for Statistics
S. Banerjee and A. Roy

Logistic Regression Models
J.M. Hilbe

Markov Chain Monte Carlo: Stochastic Simulation for Bayesian Inference, Second Edition
D. Gamerman and H.F. Lopes

Mathematical Statistics
K. Knight

Modeling and Analysis of Stochastic Systems, Second Edition
V.G. Kulkarni

Modelling Binary Data, Second Edition
D. Collett

Modelling Survival Data in Medical Research, Second Edition
D. Collett

Multivariate Analysis of Variance and Repeated Measures: A Practical Approach for Behavioural Scientists
D.J. Hand and C.C. Taylor

Multivariate Statistics: A Practical Approach
B. Flury and H. Riedwyl

Multivariate Survival Analysis and Competing Risks
M. Crowder

Nonparametric Methods in Statistics with SAS Applications
O. Korosteleva

Pólya Urn Models
H. Mahmoud

Practical Data Analysis for Designed Experiments
B.S. Yandell

Practical Longitudinal Data Analysis
D.J. Hand and M. Crowder

Practical Multivariate Analysis, Fifth Edition
A. Afifi, S. May, and V.A. Clark

Practical Statistics for Medical Research
D.G. Altman

A Primer on Linear Models
J.F. Monahan

Principles of Uncertainty
J.B. Kadane

Probability: Methods and Measurement
A. O'Hagan

Problem Solving: A Statistician's Guide, Second Edition
C. Chatfield

Randomization, Bootstrap and Monte Carlo Methods in Biology, Third Edition
B.F.J. Manly

Readings in Decision Analysis
S. French

Sampling Methodologies with Applications
P.S.R.S. Rao

Stationary Stochastic Processes: Theory and Applications
G. Lindgren

Statistical Analysis of Reliability Data
M.J. Crowder, A.C. Kimber, T.J. Sweeting, and R.L. Smith

Statistical Methods for Spatial Data Analysis
O. Schabenberger and C.A. Gotway

Statistical Methods for SPC and TQM
D. Bissell

Statistical Methods in Agriculture and Experimental Biology, Second Edition
R. Mead, R.N. Curnow, and A.M. Hasted

Statistical Process Control: Theory and Practice, Third Edition
G.B. Wetherill and D.W. Brown

Statistical Theory: A Concise Introduction
F. Abramovich and Y. Ritov

Statistical Theory, Fourth Edition
B.W. Lindgren

Statistics for Accountants
S. Letchford

Statistics for Epidemiology
N.P. Jewell

Statistics for Technology: A Course in Applied Statistics, Third Edition
C. Chatfield

Statistics in Engineering: A Practical Approach
A.V. Metcalfe

Statistics in Research and Development, Second Edition
R. Caulcutt

Stochastic Processes: An Introduction, Second Edition
P.W. Jones and P. Smith

Survival Analysis Using S: Analysis of Time-to-Event Data
M. Tableman and J.S. Kim

The Theory of Linear Models
B. Jørgensen

Time Series Analysis
H. Madsen

Time Series: Modeling, Computation, and Inference
R. Prado and M. West

Understanding Advanced Statistical Methods
P.H. Westfall and K.S.S. Henning

Texts in Statistical Science

Linear Algebra and Matrix Analysis for Statistics

Sudipto Banerjee
Professor of Biostatistics
School of Public Health
University of Minnesota, U.S.A.

Anindya Roy
Professor of Statistics
Department of Mathematics and Statistics
University of Maryland, Baltimore County, U.S.A.

CRC Press
Taylor & Francis Group
Boca Raton London New York

CRC Press is an imprint of the
Taylor & Francis Group an **informa** business
A CHAPMAN & HALL BOOK

CRC Press
Taylor & Francis Group
6000 Broken Sound Parkway NW, Suite 300
Boca Raton, FL 33487-2742

Printed on acid-free paper
Version Date: 20140407

International Standard Book Number-13: 978-1-4200-9538-8 (Hardback)

Library of Congress Cataloging-in-Publication Data

Banerjee, Sudipto, author.
Linear algebra and matrix analysis for statistics / Sudipto Banerjee, professor of biostatistics, School of Public Health, University of Minnesota, U.S.A., Anindya Roy, professor of statistics, Department of Mathematics and Statistics, University of Maryland, Baltimore County, U.S.A.
pages cm. -- (Chapman & Hall/CRC texts in statistical science)
Summary: "Linear algebra and the study of matrix algorithms have become fundamental to the development of statistical models. Using a vector-space approach, this book provides an understanding of the major concepts that underlie linear algebra and matrix analysis. Each chapter introduces a key topic, such as infinite-dimensional spaces, and provides illustrative examples. The authors examine recent developments in diverse fields such as spatial statistics, machine learning, data mining, and social network analysis. Complete in its coverage and accessible to students without prior knowledge of linear algebra, the text also includes results that are useful for traditional statistical applications."-- Provided by publisher.
Includes bibliographical references and index.
ISBN 978-1-4200-9538-8 (hardback)
1. Algebras, Linear. 2. Matrices. 3. Mathematical statistics. I. Roy, Anindya, 1970- author. II. Title.

QA184.2.B36 2014
512'.5--dc23 2014004884

Visit the Taylor & Francis Web site at
http://www.taylorandfrancis.com

and the CRC Press Web site at
http://www.crcpress.com

To my parents, Shyamali and Sunit, my wife, Sharbani, and my son, Shubho.

—Sudipto Banerjee

To my wife, Ishani.

—Anindya Roy

Contents

Preface

Linear algebra constitutes one of the core mathematical components in any modern curriculum involving statistics. Usually students studying statistics are expected to have seen at least one semester of linear algebra (or applied linear algebra) at the undergraduate level. In particular, students pursuing graduate studies in statistics or biostatistics are expected to have a sound conceptual grasp of vector spaces and subspaces associated with matrices, orthogonality, projections, quadratic forms and so on.

As the relevance and attraction of statistics as a discipline for graduate studies continues to increase for students with more diverse academic preparations, the need to accommodate their mathematical needs also keeps growing. In particular, many students find their undergraduate preparation in linear algebra rather different from what is required in graduate school. There are several excellent texts on the subject that provide as comprehensive a coverage of the subject as possible at the undergraduate level. However, some of these texts cater to a broader audience (e.g., scientists and engineers) and several formal concepts that are important in theoretical statistics are not emphasized.

There are several excellent texts on linear algebra. For example, there are classics by Halmos (1974), Hoffman and Kunze (1984) and Axler (1997) that make heavy use of vector spaces and linear transformations to provide a coordinate-free approach. A remarkable feature of the latter is that it develops the subject without using determinants at all. Then, there are the books by Strang (2005, 2009) and Meyer (2001) that make heavy use of echelon forms and canonical forms to reveal the properties of subspaces associated with a matrix. This approach is tangible, but may not turn out to be the most convenient to derive and prove results often encountered in statistical modeling. Among texts geared toward statistics, Rao and Bhimsankaram (2000), Searle (1982) and Graybill (2001) have stood the test of time. The book by Harville (1997) stands out in its breadth of coverage and is already considered a modern classic. Several other excellent texts exist for statisticians including Healy (2000), Abadir and Magnus (2005), Schott (2005) and Gentle (2010). The concise text by Bapat (2012) is a delightful blend of linear algebra and statistical linear models.

While the above texts offer excellent coverage, some expect substantial mathematical maturity from the reader. Our attempt here has been to offer a more gradual exposition to linear algebra without really dumbing down the subject. The book tries to be as self-contained as possible and does not assume any prior knowledge of linear algebra. However, those who have seen some elementary linear algebra will be able to move more quickly through the early chapters. We have attempted to present both the vector space approach as well as the canonical forms in matrix theory.

Although we adopt the vector space approach for much of the later development, the book does not begin with vector spaces. Instead, it addresses the rudimentary mechanics of linear systems using Gaussian elimination and the resultant decompositions (Chapters 1–3). Chapter 4 introduces Euclidean vector spaces using less abstract concepts and makes connections to systems of linear equations wherever possible. Chapter 5 is on the rank of a matrix. Why devote an entire chapter to rank? We believe that the concept of rank is *that* important for a thorough understanding. In several cases we show how the same result may be derived using multiple techniques, which, we hope, will offer insight into the subject and ensure a better conceptual grasp of the material. Chapter 6 introduces complementary subspaces and oblique projectors. Chapter 7 introduces orthogonality and orthogonal projections and leads us to the Fundamental Theorem of Linear Algebra, which connects the four fundamental subspaces associated with a matrix. Chapter 8 builds upon the previous chapter and focuses on orthogonal projectors, which is fundamental to linear statistical models, and also introduces several computational techniques for orthogonal reduction. Chapter 9 revisits linear equations from a more mature perspective and shows how the theoretical concepts developed thus far can be handy in analyzing solutions for linear systems. The reader, at this point, will have realized that there is much more to linear equations than Gaussian elimination and echelon forms. Chapter 10 discusses determinants. Unlike some classical texts, we introduce determinants a bit late in the game and present it as a useful tool for characterizing and obtaining certain useful results. Chapter 11 introduces eigenvalues and eigenvectors and is the first time complex numbers make an appearance. Results on general real matrices are followed by those of real symmetric matrices. The popular algorithms for eigenvalues and eigenvectors are outlined both for symmetric and unsymmetric matrices. Chapter 12 derives the Singular value decomposition and the Jordan Canonical Form and presents an accessible proof of the latter. Chapter 13 is devoted to quadratic forms, another topic of fundamental importance to statistical theory and methods. Chapter 14 presents Kronecker and Hadamard products and other related materials that have become conspicuous in multivariate statistics and econometrics. Chapters 15 and 16 provide a taste of some more advanced topics but, hopefully, in a more accessible manner than more advanced texts. The former presents some aspects of linear iterative systems and convergence of matrices, while the latter introduces more general vector spaces, linear transformations and Hilbert spaces.

We remark that this is not a book on matrix computations, although we describe several numerical procedures in some detail. We have refrained from undertaking a thorough exploration of the most numerically stable algorithms as they would require a lot more theory and be too much of a digression. However, readers who grasp the material provided here should find it easier to study more specialized texts on matrix computations (e.g., Golub and Van Loan, 2013; Trefthen and Bau III, 1997). Also, while we have included many exercises that can be solved using languages such as

`MATLAB` and `R`, we decided not to marry the text to a specific language or platform. We have also not included statistical theory and applications here. This decision was taken neither in haste nor without deliberation. There are plenty of excellent texts on the theory of linear models, regression and modeling that make abundant use of linear algebra. Our hope is that readers of this text will find it easier to grasp the material in such texts. In fact, we believe that this book can be used as a companion text in the more theoretical courses on linear regression or, perhaps, stand alone as a one-semester course devoted to linear algebra for statistics and econometrics.

Finally, we have plenty of people to thank for this book. We have been greatly influenced by our teachers at the Indian Statistical Institute, Kolkata. The book by Rao and Bhimsankaram (2000), written by two of our former teachers whose lectures and notes are still vivid in our minds, certainly shaped our preparation. Sudipto Banerjee would also like to acknowledge Professor Alan Gelfand of Duke University with whom he has had several discussions regarding the role of linear algebra in Bayesian hierarchical models and spatial statistics. The first author also thanks Dr. Govindan Rangarajan of the Indian Institute of Science, Bangalore, India, Dr. Anjana Narayan of California Polytechnic State University, Pomona, and Dr. Mohan Delampady of Indian Statistical Institute, Bangalore, for allowing the author to work on this manuscript as a visitor in their respective institutes. We thank the Division of Biostatistics at the University of Minnesota, Twin Cities, and the Department of Statistics at the University of Maryland, Baltimore, for providing us with an ambience most conducive to this project. Special mention must be made of Dr. Rajarshi Guhaniyogi, Dr. Joao Monteiro and Dr. Qian Ren, former graduate students at the University of Minnesota, who have painstakingly helped with proof-reading and typesetting parts of the text. This book would also not have happened without the incredible patience and cooperation of Rob Calver, Rachel Holt, Sarah Gelson, Kate Gallo, Charlotte Byrnes and Shashi Kumar at CRC Press/Chapman and Hall. Finally, we thank our families, whose ongoing love and support made all of this possible.

SUDIPTO BANERJEE — Minneapolis, Minnesota
ANINDYA ROY — Baltimore, Maryland
August, 2013.

CHAPTER 1

Matrices, Vectors and Their Operations

Linear algebra usually starts with the analysis and solutions for systems of linear equations such as

$$\begin{array}{ccccccccc} a_{11}x_1 & + & a_{12}x_2 & + & \dots & + & a_{1n}x_n & = & b_1, \\ a_{21}x_1 & + & a_{22}x_2 & + & \dots & + & a_{2n}x_n & = & b_2, \\ & & & & \vdots & & & & \\ a_{m1}x_1 & + & a_{m2}x_2 & + & \dots & + & a_{mn}x_n & = & b_m. \end{array}$$

Such systems are of fundamental importance because they arise in diverse mathematical and scientific disciplines. The a_{ij}'s and b_i's are usually known from the manner in which these equations arise. The $x_i's$ are unknowns that satisfy the above set of equations and need to be found. The solution to such a system, depends on the a_{ij}'s and b_i's. They contain all the information we need about the system. It is, therefore, natural to store these numbers in an array and develop mathematical operations for these arrays that will lead us to the x_i's.

Example 1.1 Consider the following system of three equations in four unknowns:

$$\begin{array}{ccccccccc} 4x_1 & + & 7x_2 & + & 2x_3 & & & = & 2, \\ -6x_1 & - & 10x_2 & & & + & x_4 & = & 1, \\ 4x_1 & + & 6x_2 & + & 4x_3 & + & 5x_4 & = & 0. \end{array} \tag{1.1}$$

All the information contained in the above system can be stored in a rectangular array with three rows and four columns containing the coefficients of the unknowns and another single column comprising the entries in the right-hand side of the equation. Thus,

$$\begin{bmatrix} 4 & 7 & 2 & 0 \\ -6 & -10 & 0 & 1 \\ 4 & 6 & 4 & 5 \end{bmatrix} \text{ and } \begin{bmatrix} 2 \\ 1 \\ 0 \end{bmatrix}$$

are the two arrays that represent the linear system. We use two different arrays to distinguish between the coefficients on the left-hand side and the right-hand side. Alternatively, one could create one *augmented* array

$$\left[\begin{array}{cccc|c} 4 & 7 & 2 & 0 & 2 \\ -6 & -10 & 0 & 1 & 1 \\ 4 & 6 & 4 & 5 & 0 \end{array}\right]$$

with a "|" to distinguish the right-hand side of the linear system. ■

We will return to solving linear equations using matrices in Chapter 2. More generally, rectangular arrays are often used as data structures to store information in computers. If we can define algebraic operations on such arrays in a meaningful way, then we can not only use arrays as mere storage devices but also for solving linear systems of equations in computers. Rectangular arrays of numbers are called ***matrices***. When the array is a single row or column it is called a ***vector***. In this chapter we introduce notations and develop some algebraic operations involving matrices and vectors that will be used extensively in subsequent chapters.

1.1 Basic definitions and notations

Definition 1.1 *A* **matrix** *of order (or dimension) $m \times n$ is a collection of mn items arranged in a rectangular array with m rows and n columns as below:*

$$\boldsymbol{A} = \begin{pmatrix} a_{11} & a_{12} & \dots & a_{1n} \\ a_{21} & a_{22} & \dots & a_{2n} \\ \vdots & \vdots & \vdots & \vdots \\ a_{m1} & a_{m2} & \dots & a_{mn} \end{pmatrix} \quad or \quad \begin{bmatrix} a_{11} & a_{12} & \dots & a_{1n} \\ a_{21} & a_{22} & \dots & a_{2n} \\ \vdots & \vdots & \vdots & \vdots \\ a_{m1} & a_{m2} & \dots & a_{mn} \end{bmatrix}.$$

The individual items in a matrix are called its **elements** *or* **entries.**

The elements of a matrix are often called ***scalars***. *In this book, unless explicitly stated otherwise, the elements of a matrix will be assumed to be real numbers, i.e., $a_{ij} \in \Re$ for $i = 1, \dots, m$ and $j = 1, \dots, n$.*

A matrix is often written in short form as $\boldsymbol{A} = \{a_{ij}\}_{i,j=1}^{m,n}$, or simply as $\{a_{ij}\}$ when the dimensions are obvious. We also write $[\boldsymbol{A}]_{ij}$ or $(\boldsymbol{A})_{ij}$ to denote the (i, j)-th element of $\boldsymbol{A}$. When the order of a matrix needs to be highlighted, we will often write $\boldsymbol{A}_{m\times n}$ to signify an $m \times n$ matrix.

Example 1.2 Suppose we have collected data on seven rats, each of whom had its heights measured weekly for five weeks. We present the data as a 7×5 matrix, where the entry in each row corresponds to the five weekly measurements:

$$\boldsymbol{A} = \begin{bmatrix} 151 & 199 & 246 & 283 & 320 \\ 145 & 199 & 249 & 293 & 354 \\ 155 & 200 & 237 & 272 & 297 \\ 135 & 188 & 230 & 280 & 323 \\ 159 & 210 & 252 & 298 & 331 \\ 141 & 189 & 231 & 275 & 305 \\ 159 & 201 & 248 & 297 & 338 \end{bmatrix}. \tag{1.2}$$

The entry at the intersection of the i-th row and j-th column is the (i, j)-th element and denoted by a_{ij}. For example, the $(2, 3)$-th element is located at the intersection of the second row and third column: $a_{23} = 249$. ■

Two matrices $\boldsymbol{A} = \{a_{ij}\}$ and $\boldsymbol{B} = \{b_{ij}\}$ are equal if (a) they both have the same order, i.e., they have the same number of rows and columns, and (b) if $a_{ij} = b_{ij}$ for all i and j. We then write $\boldsymbol{A} = \boldsymbol{B}$.

If the number of rows equals the number of columns in $\boldsymbol{A}$, i.e., $m = n$, then we call $\boldsymbol{A}$ a ***square*** matrix of order m. In this text we will denote matrices by uppercase bold italics (e.g., $\boldsymbol{A}$, $\boldsymbol{B}$, $\boldsymbol{X}$, $\boldsymbol{Y}$, $\boldsymbol{\Gamma}$, $\boldsymbol{\Theta}$ etc).

Definition 1.2 *The* **transpose** *of the $m \times n$ matrix $\boldsymbol{A}$, denoted as $\boldsymbol{A}'$, is the $n \times m$ matrix formed by placing the columns of $\boldsymbol{A}$ as its rows. Thus, the (i, j)-th element of $\boldsymbol{A}'$ is a_{ji}, where a_{ji} is the (j, i)-th element of $\boldsymbol{A}$.*

Example 1.3 The transpose of the matrix $\boldsymbol{A}$ in (1.2) is given by

$$\boldsymbol{A}' = \begin{bmatrix} 151 & 145 & 155 & 135 & 159 & 141 & 159 \\ 199 & 199 & 200 & 188 & 210 & 189 & 201 \\ 246 & 249 & 237 & 230 & 252 & 231 & 248 \\ 283 & 293 & 272 & 280 & 298 & 275 & 297 \\ 320 & 354 & 297 & 323 & 331 & 305 & 338 \end{bmatrix} .$$

Note that $(\boldsymbol{A}')' = \boldsymbol{A}$; thus, transposing the transpose of a matrix yields the original matrix.

Definition 1.3 **Symmetric matrices.** *A square matrix $\boldsymbol{A}$ is called* **symmetric** *if $a_{ij} = a_{ji}$ for all $i, j = 1, \ldots, n$ or, equivalently, if $\boldsymbol{A} = \boldsymbol{A}'$.*

Note that symmetric matrices must be square matrices for the preceding definition to make sense.

Example 1.4 The following 3×3 matrix is symmetric:

$$\boldsymbol{A} = \begin{bmatrix} 1 & 2 & 3 \\ 2 & 5 & -6 \\ 3 & -6 & -1 \end{bmatrix} = \boldsymbol{A}' . \blacksquare$$

The matrix with all its entries equal to 0 is called the ***null matrix*** and is denoted by $\boldsymbol{O}$. When the order is important, we denote it as $\boldsymbol{O}_{m \times n}$. A square null matrix is an example of a symmetric matrix. Another class of symmetric matrices are *diagonal* matrices defined below.

Definition 1.4 **Diagonal matrices.** *A* **diagonal** *matrix is a square matrix in which the entries outside the main diagonal are all zero. The diagonal entries themselves may or may not be zero. Thus, the matrix $\boldsymbol{D} = \{d_{ij}\}$ with n columns and n rows is diagonal if $d_{ij} = 0$ whenever $i \neq j$, $i, j = 1, 2, \ldots, n$.*

An extremely important diagonal matrix is the ***identity*** matrix whose diagonal elements are all 1 and all other elements are 0. We will denote an identity matrix of order n to be $\boldsymbol{I}_n$, or simply as $\boldsymbol{I}$ when the dimension is obvious from the context.

Definition 1.5 *An* $m \times 1$ *matrix is called a* **vector**, *or a* **column vector** *and is written as*

$$a = \begin{pmatrix} a_1 \\ a_2 \\ \vdots \\ a_m \end{pmatrix} \quad or \quad \begin{bmatrix} a_1 \\ a_2 \\ \vdots \\ a_m \end{bmatrix}.$$

A $1 \times n$ *matrix is called a* **row vector** *and will be written as* $\boldsymbol{b}' = (b_1, b_2, \ldots, b_n)$.

Note that $m \times 1$ column vectors, when transposed become $1 \times m$ row vectors; likewise row vectors become column vectors upon transposition. We will denote column vectors by by lowercase bold italics such as $\boldsymbol{a}$, $\boldsymbol{b}$, $\boldsymbol{x}$, $\boldsymbol{y}$, $\boldsymbol{\beta}$, $\boldsymbol{\mu}$ etc. while $\boldsymbol{a}'$, $\boldsymbol{b}'$, $\boldsymbol{x}'$, $\boldsymbol{y}'$, $\boldsymbol{\beta}'$, $\boldsymbol{\mu}'$, etc., denote the corresponding row vectors. Since writing row vectors requires less space than column vectors, we will often write $\boldsymbol{x} = \{x_i\}_{i=1}^m$ to denote an $m \times 1$ column vector, while the corresponding row vector will be denoted by $\boldsymbol{x}' = (x_1, \ldots, x_m)$ or $\boldsymbol{x}' = [x_1, x_2, \ldots, x_m]$ or sometimes by $\boldsymbol{x}' = [x_1 : x_2 : \ldots : x_m]$. By this convention, $\boldsymbol{x} = [x_1, x_2, \ldots, x_m]'$ is again the $m \times 1$ column vector.

Matrices are often written as a collection of its row or column vectors, such as

$$\boldsymbol{A} = [\boldsymbol{a}_{*1}, \ldots, \boldsymbol{a}_{*n}] = \begin{bmatrix} \boldsymbol{a}'_{1*} \\ \boldsymbol{a}'_{2*} \\ \vdots \\ \boldsymbol{a}'_{m*} \end{bmatrix},$$

where $\boldsymbol{A}$ is an $m \times n$ matrix. The vectors $\boldsymbol{a}_{*1}, \ldots, \boldsymbol{a}_{*n}$ are each $m \times 1$ and referred to as the column vectors of $\boldsymbol{A}$. The vectors $\boldsymbol{a}'_{1*}, \ldots, \boldsymbol{a}'_{m*}$ are each $1 \times n$ and are called the row vectors of $\boldsymbol{A}$. Note that $\boldsymbol{a}_{1*}, \ldots, \boldsymbol{a}_{m*}$ are each $n \times 1$ vectors obtained by transposing the row vectors of $\boldsymbol{A}$. We sometimes separate the vectors by a "colon" instead of a "comma": $\boldsymbol{A} = [\boldsymbol{a}_{*1} : \boldsymbol{a}_{*2} : \ldots : \boldsymbol{a}_{*n}]$.

The transpose of the above matrix is an $n \times m$ matrix $\boldsymbol{A}'$ and can be written in terms of the column and row vectors of $\boldsymbol{A}$ as

$$\boldsymbol{A}' = \begin{bmatrix} \boldsymbol{a}'_{*1} \\ \boldsymbol{a}'_{*2} \\ \vdots \\ \boldsymbol{a}'_{*n} \end{bmatrix} = [\boldsymbol{a}_{1*}, \ldots, \boldsymbol{a}_{m*}].$$

We again draw attention to our notation here: each $\boldsymbol{a}'_{*j}$ is the $1 \times m$ row vector obtained by transposing the $m \times 1$ j-th column vector of $\boldsymbol{A}$, viz. $\boldsymbol{a}_{*j}$, while $\boldsymbol{a}_{i*}$'s are $n \times 1$ column vectors corresponding to the row vectors of $\boldsymbol{A}$. When there is no scope for confusion, we will drop the $*$ from the index in $\boldsymbol{a}_{*j}$ and simply write $\boldsymbol{A} = [\boldsymbol{a}_1, \ldots, \boldsymbol{a}_n]$ in terms of its $m \times 1$ column vectors.

Example 1.5 Consider the matrix $\boldsymbol{A}'$ in Example 1.3. The fourth row vector of this matrix is written as $\boldsymbol{a}'_{4*} = [135, 188, 230, 280, 323]$, while the fifth column vector, written in transpose form (to save space) is $\boldsymbol{a}'_{*5} = [320, 354, 297, 323, 331, 305, 328]$ or as $\boldsymbol{a}_{*5} = [320, 354, 297, 323, 331, 305, 328]'$. ■

For a symmetric matrix, $\boldsymbol{A} = \boldsymbol{A}'$, hence the column and row vectors of $\boldsymbol{A}$ have the same elements. The column (or row) vectors of the identity matrix have a very special role to play in linear algebra. We have a special definition for them.

Definition 1.6 *The column or row vectors of an identity matrix of order n are called the* **standard unit vectors** *or the* **canonical vectors** *in $\Re^n$. Letting $\boldsymbol{I} = [\boldsymbol{e}_1, \ldots, \boldsymbol{e}_n]$, we have*

$$\boldsymbol{e}_1 = \begin{bmatrix} 1 \\ 0 \\ \vdots \\ 0 \end{bmatrix}, \; \boldsymbol{e}_2 = \begin{bmatrix} 0 \\ 1 \\ \vdots \\ 0 \end{bmatrix}, \; \ldots, \boldsymbol{e}_n = \begin{bmatrix} 0 \\ 0 \\ \vdots \\ 1 \end{bmatrix}.$$

The vector $\boldsymbol{e}_i$ consists of all zeroes except in its i-th position, which is occupied by a 1.

The $\boldsymbol{e}_i$'s point in the direction of each axis of the Cartesian or Euclidean coordinate system. Usually the length of these vectors is clear from the context and we do not include it in the notation. For example, we use $\boldsymbol{e}_1$ and $\boldsymbol{e}_2$ to denote the first two columns of $\boldsymbol{I}_2$ as well as $\boldsymbol{I}_3$. In the former, $\boldsymbol{e}_1 = (1, 0)'$ and $\boldsymbol{e}_2 = (0, 1)'$ have only two elements, while in the latter $\boldsymbol{e}_1 = (1, 0, 0)'$ and $\boldsymbol{e}_2 = (0, 1, 0)'$ have three elements. While the $\boldsymbol{e}_1$ and $\boldsymbol{e}_2$ in $\Re^2$ are not the same as the $\boldsymbol{e}_1$ and $\boldsymbol{e}_2$ in $\Re^3$, we do not use different notations. The number of elements these vectors will have are usually clear from the context and we do not foresee any confusion.

1.2 Matrix addition and scalar-matrix multiplication

Basic operations on matrices include ***addition*** and ***subtraction*** defined as element-wise addition and subtraction respectively. Below is a formal definition.

Definition 1.7 Addition and subtraction of matrices. *Let $\boldsymbol{A} = \{a_{ij}\}$ and $\boldsymbol{B} = \{b_{ij}\}$ be two $m \times n$ matrices. Then the* **sum** *of $\boldsymbol{A}$ and $\boldsymbol{B}$ is the $m \times n$ matrix $\boldsymbol{C}$, written as $\boldsymbol{C} = \boldsymbol{A} + \boldsymbol{B}$, whose (i, j)-th element is given by $c_{ij} = a_{ij} + b_{ij}$. The* **difference** *of $\boldsymbol{A}$ and $\boldsymbol{B}$ is the $m \times n$ matrix $\boldsymbol{C}$, written as $\boldsymbol{C} = \boldsymbol{A} - \boldsymbol{B}$, whose (i, j)-th element is given by $c_{ij} = a_{ij} - b_{ij}$.*

In other words, the sum of two $m \times n$ matrices $\boldsymbol{A}$ and $\boldsymbol{B}$, denoted by $\boldsymbol{A} + \boldsymbol{B}$, is again an $m \times n$ matrix computed by adding their corresponding elements. Note that two matrices must be of the same order to be added. This definition implies that $\boldsymbol{A} + \boldsymbol{B} = \boldsymbol{B} + \boldsymbol{A}$ since the (i, j)-th element of $\boldsymbol{A} + \boldsymbol{B}$ is the same as the (i, j)-th element of $\boldsymbol{B} + \boldsymbol{A}$. That is,

$$[\boldsymbol{A} + \boldsymbol{B}]_{ij} = a_{ij} + b_{ij} = b_{ij} + a_{ij} = [\boldsymbol{B} + \boldsymbol{A}]_{ij}.$$

Example 1.6 Consider the two 3×2 matrices:

$$\boldsymbol{A} = \begin{bmatrix} 1 & -3 \\ 2 & 7 \\ 5 & 9 \end{bmatrix} \quad \text{and} \quad \boldsymbol{B} = \begin{bmatrix} 0 & 5 \\ 7 & -8 \\ 2 & 4 \end{bmatrix}.$$

Then,

$$\boldsymbol{A}+\boldsymbol{B} = \begin{bmatrix} 1 & -3 \\ 2 & 7 \\ 5 & 9 \end{bmatrix} + \begin{bmatrix} 0 & 5 \\ 7 & -8 \\ 2 & 4 \end{bmatrix} = \begin{bmatrix} 1+0 & (-3)+5 \\ 2+7 & 7+(-8) \\ 5+2 & 9+4 \end{bmatrix} = \begin{bmatrix} 1 & 2 \\ 9 & -1 \\ 7 & 13 \end{bmatrix};$$

$$\boldsymbol{A}-\boldsymbol{B} = \begin{bmatrix} 1 & -3 \\ 2 & 7 \\ 5 & 9 \end{bmatrix} - \begin{bmatrix} 0 & 5 \\ 7 & -8 \\ 2 & 4 \end{bmatrix} = \begin{bmatrix} 1-0 & (-3)-5 \\ 2-7 & 7-(-8) \\ 5-2 & 9-4 \end{bmatrix} = \begin{bmatrix} 1 & -8 \\ -5 & 15 \\ 3 & 5 \end{bmatrix}.$$

■

Definition 1.8 *Let $\boldsymbol{A}$ be an $m \times n$ matrix and α a real number (scalar).* **Scalar multiplication** *between $\boldsymbol{A}$ and α results in the $m \times n$ matrix $\boldsymbol{C}$, denoted $\boldsymbol{C} = \alpha\boldsymbol{A} = \boldsymbol{A}\alpha$, whose (i,j)-th element is given by $c_{ij} = \alpha a_{ij}$.*

Addition and scalar multiplication can be combined to meaningfully construct $\alpha\boldsymbol{A} + \beta\boldsymbol{B}$, where α and β are scalars, in the following way:

$$\alpha\boldsymbol{A}+\beta\boldsymbol{B} = \begin{bmatrix} \alpha a_{11}+\beta b_{11} & \alpha a_{12}+\beta b_{12} & \dots & \alpha a_{1n}+\beta b_{1n} \\ \alpha a_{21}+\beta b_{21} & \alpha a_{22}+\beta b_{22} & \dots & \alpha a_{2n}+\beta b_{2n} \\ \vdots & \vdots & \vdots & \vdots \\ \alpha a_{m1}+\beta a_{m1} & \alpha a_{m2}+\beta b_{m2} & \dots & \alpha a_{mn}+\beta b_{mn} \end{bmatrix}.$$

In particular, we have $\boldsymbol{A}-\boldsymbol{B} = \boldsymbol{A}+(-1)\boldsymbol{B} = \{a_{ij}-b_{ij}\}_{i,j=1}^{m,n}$. Expressions such as $\alpha\boldsymbol{A}+\beta\boldsymbol{B}$ are examples of ***linear combinations*** of matrices. More generally, linear combinations of p matrices are defined as $\sum_{i-1}^{p} \alpha_i \boldsymbol{A}_i$, where $\boldsymbol{A}_i$'s are matrices of the same dimensions and α_i's are scalars.

Theorem 1.1 *Let $\boldsymbol{A}$, $\boldsymbol{B}$ and $\boldsymbol{C}$ denote matrices each of order $m \times n$, let α and β denote any two scalars. The following properties are immediate consequences of the above definitions.*

(i) Associative property of matrix addition: $\boldsymbol{A}+(\boldsymbol{B}+\boldsymbol{C}) = (\boldsymbol{A}+\boldsymbol{B})+\boldsymbol{C} = \boldsymbol{A}+\boldsymbol{B}+\boldsymbol{C}$.

(ii) Commutative property of matrix addition: $\boldsymbol{A}+\boldsymbol{B} = \boldsymbol{B}+\boldsymbol{A}$.

(iii) The null matrix has the property: $\boldsymbol{A}+\boldsymbol{O} = \boldsymbol{A}$.

(iv) The matrix $(-\boldsymbol{A})$ has the property $\boldsymbol{A}+(-\boldsymbol{A}) = \boldsymbol{O}$.

(v) Distributive laws: $(\alpha+\beta)\boldsymbol{A} = \alpha\boldsymbol{A}+\beta\boldsymbol{A}$ *and* $\alpha(\boldsymbol{A}+\boldsymbol{B}) = \alpha\boldsymbol{A}+\alpha\boldsymbol{B}$.

(vi) $(\alpha\boldsymbol{A}+\beta\boldsymbol{B})' = \alpha\boldsymbol{A}'+\beta\boldsymbol{B}'$. *In particular* $(\alpha\boldsymbol{A})' = \alpha\boldsymbol{A}'$.

Proof. Parts (i) – (v) follow easily from the definitions of matrix addition and scalar-matrix multiplication and are left to the reader as exercises. For part (vi), note that the (i, j)-th element of $\alpha\boldsymbol{A} + \beta\boldsymbol{B}$ is given by $\alpha a_{ij} + \beta b_{ij}$. Therefore, the (i, j)-th element of $(\alpha\boldsymbol{A} + \beta\boldsymbol{B})'$ is given by $\alpha a_{ji} + \beta b_{ji}$. Note that the (i, j)-th element of $\alpha\boldsymbol{A}'$ is αa_{ji} and that of $\beta\boldsymbol{B}'$ is βb_{ij}, so the (i, j)-th element of $\alpha\boldsymbol{A}' + \beta\boldsymbol{B}'$ is $\alpha a_{ji} + \beta b_{ji}$. This proves that the (i, j)-th element of $(\alpha\boldsymbol{A} + \beta\boldsymbol{B})'$ is the same as that of $\alpha\boldsymbol{A}' + \beta\boldsymbol{B}'$. That $(\alpha\boldsymbol{A})' = \alpha\boldsymbol{A}'$ is true follows by simply taking $\beta = 0$. This completes the proof. □

Part (vi) of the above theorem states that the transpose of linear combinations of matrices is the linear combination of the transposes. More generally, we have $(\sum_{i=1}^{p} \alpha_i \boldsymbol{A}_i)' = \sum_{i=1}^{p} \alpha_i \boldsymbol{A}_i'$. The idea of the proof is no different from the case with $p = 2$ and easy to fill in.

1.3 Matrix multiplication

We first define ***matrix-vector*** multiplication between an $m \times n$ matrix $\boldsymbol{A}$ and an $n \times 1$ column vector $\boldsymbol{x}$ to be

$$\boldsymbol{A}\boldsymbol{x} = \begin{bmatrix} a_{11} & a_{12} & \dots & a_{1n} \\ a_{21} & a_{22} & \dots & a_{2n} \\ \vdots & \vdots & \vdots & \vdots \\ a_{m1} & a_{m2} & \dots & a_{mn} \end{bmatrix} \begin{bmatrix} x_1 \\ x_2 \\ \vdots \\ x_n \end{bmatrix} = \begin{bmatrix} \sum_{j=1}^{n} a_{1j}x_j \\ \sum_{j=1}^{n} a_{2j}x_j \\ \vdots \\ \sum_{j=1}^{n} a_{mj}x_j \end{bmatrix}. \tag{1.3}$$

Note that we must have the column vector's dimension to be the same as the number of columns in $\boldsymbol{A}$ for this multiplication to be meaningful. In short, we write $\boldsymbol{A}\boldsymbol{x} = \boldsymbol{y}$, where $\boldsymbol{y} = (y_1, y_2, \dots, y_m)'$ is an $m \times 1$ column vector with $y_i = \sum_{j=1}^{m} a_{ij}x_j$. Among other applications, we will see that the matrix-vector product is very useful in studying systems of linear equations.

Example 1.7 Suppose we want to find the product $\boldsymbol{A}\boldsymbol{x}$ where

$$\boldsymbol{A} = \begin{bmatrix} 1 & 0 & -3 & 0 \\ 0 & 1 & 4 & 0 \end{bmatrix} \text{ and } \boldsymbol{x} = \begin{bmatrix} 6 \\ 3 \\ 0 \\ -5 \end{bmatrix}.$$

The product $\boldsymbol{A}\boldsymbol{x}$ is obtained from (1.3) as

$$\boldsymbol{A}\boldsymbol{x} = \begin{bmatrix} 1 & 0 & -3 & 0 \\ 0 & 1 & 4 & 0 \end{bmatrix} \begin{bmatrix} 6 \\ 3 \\ 0 \\ -5 \end{bmatrix} = \begin{bmatrix} 1 \times 6 + 0 \times 3 + (-3) \times 0 + 0 \times (-5) \\ 0 \times 6 + 1 \times 3 + 4 \times (0) + 0 \times (-5) \end{bmatrix}$$

$$= \begin{bmatrix} 6 \\ 3 \end{bmatrix}. \blacksquare$$

Note that this matrix-vector product can also be written as

$$\boldsymbol{A}\boldsymbol{x} = x_1 \begin{bmatrix} a_{11} \\ a_{21} \\ \vdots \\ a_{m1} \end{bmatrix} + x_2 \begin{bmatrix} a_{12} \\ a_{22} \\ \vdots \\ a_{m2} \end{bmatrix} + \ldots + x_n \begin{bmatrix} a_{1n} \\ a_{2n} \\ \vdots \\ a_{mn} \end{bmatrix} = \sum_{j=1}^{n} x_i \boldsymbol{a}_j, \qquad (1.4)$$

where $\boldsymbol{a}_j$ is the j-th column vector of $\boldsymbol{A}$. This shows that the above matrix-vector product may be looked upon as a ***linear combination***, or weighted average, of the columns of the matrix—the weights being the elements of the vector $\boldsymbol{x}$.

Example 1.8 In Example 1.2, suppose we want to compute the average height over the five week period for each rat. This can be easily computed as $\boldsymbol{A}\boldsymbol{x}$, where $\boldsymbol{A}$ is as in (1.2) and $\boldsymbol{x} = (1/5)\boldsymbol{1} = (1/5, 1/5, 1/5, 1/5, 1/5)'$:

$$\begin{bmatrix} 151 & 199 & 246 & 283 & 320 \\ 145 & 199 & 249 & 293 & 354 \\ 155 & 200 & 237 & 272 & 297 \\ 135 & 188 & 230 & 280 & 323 \\ 159 & 210 & 252 & 298 & 331 \\ 141 & 189 & 231 & 275 & 305 \\ 159 & 201 & 248 & 297 & 338 \end{bmatrix} \begin{bmatrix} 1/5 \\ 1/5 \\ 1/5 \\ 1/5 \\ 1/5 \end{bmatrix} = \begin{bmatrix} 239.8 \\ 248.0 \\ 252.8 \\ 232.2 \\ 231.2 \\ 250.0 \\ 228.2 \end{bmatrix}. \quad \blacksquare$$

A special case of the matrix-vector product is when we take the matrix to be a $1 \times n$ row vector, say $\boldsymbol{y}' = (y_1, \ldots, y_n)$. Then, applying the above matrix-vector product yields a scalar: $\boldsymbol{y}'\boldsymbol{x} = \sum_{i=1}^{n} x_i y_i$. Note that this same scalar would have been obtained had we interchanged the roles of $\boldsymbol{x}$ and $\boldsymbol{y}$ to obtain $\boldsymbol{x}'\boldsymbol{y}$. This leads to the following definition.

Definition 1.9 *The* **standard inner product** *or simply the* **inner product** *of two $n \times 1$ vectors $\boldsymbol{x}$ and $\boldsymbol{y}$, denoted by $\langle \boldsymbol{x}, \boldsymbol{y} \rangle$, and defined as*

$$\langle \boldsymbol{x}, \boldsymbol{y} \rangle = \boldsymbol{y}'\boldsymbol{x} = \sum_{i=1}^{n} x_i y_i = \boldsymbol{x}'\boldsymbol{y} = \langle \boldsymbol{y}, \boldsymbol{x} \rangle , \qquad (1.5)$$

where x_i and y_i are the i-th elements of $\boldsymbol{x}$ and $\boldsymbol{y}$, respectively. The inner product of a real vector $\boldsymbol{x}$ with itself, i.e. $\langle \boldsymbol{x}, \boldsymbol{x} \rangle$, is of particular importance and represents the square of the **length** *or* **norm** *of a vector. We denote the norm or length of a real vector $\boldsymbol{x}$ as $\|\boldsymbol{x}\| = \sqrt{\langle \boldsymbol{x}, \boldsymbol{x} \rangle} = \sqrt{\sum_{i=1}^{n} x_i^2}$.*

Example 1.9 Let $\boldsymbol{x} = (1, 0, -3, 0)'$ and $\boldsymbol{y} = (4, 3, 0, 4)'$. Then

$$\langle \boldsymbol{x}, \boldsymbol{y} \rangle = \boldsymbol{y}'\boldsymbol{x} = [4 : 3 : 0 : 4] \begin{bmatrix} 1 \\ 0 \\ -3 \\ 0 \end{bmatrix} = 4 \times 1 + 3 \times 0 + 0 \times (-3) + 4 \times 0 = 4 .$$

Also, $\|\boldsymbol{x}\| = \sqrt{\sum_{i=1}^4 x_i^2} = \sqrt{1^2+0^2+(-3)^2+0^2} = \sqrt{10}$ and $\|\boldsymbol{y}\| = \sqrt{\sum_{i=1}^4 y_i^2} = \sqrt{4^2+3^2+0^2+4^2} = \sqrt{41}$. ■

We will study inner products and norms in much greater detail later. For now, we point out the simple and useful observation that the only real vector of length 0 is the null vector:

Lemma 1.1 *Let $\boldsymbol{x}$ be an $n \times 1$ vector with elements that are real numbers. Then $\|\boldsymbol{x}\| = 0$ if and only if $\boldsymbol{x} = \boldsymbol{0}$.*

Proof. Let x_i denote the i-th element of $\boldsymbol{x}$. Then $\|\boldsymbol{x}\|^2 = \boldsymbol{x}'\boldsymbol{x} = \sum_{i=1}^n x_i^2$ is the sum of m non-negative numbers which is equal to zero if and only if $x_i = 0$ for each i, i.e., if and only $\boldsymbol{x} = \boldsymbol{0}$. □

We now use matrix-vector products to define ***matrix multiplication***, or the product of two matrices. Let $\boldsymbol{A}$ be an $m \times p$ matrix and let $\boldsymbol{B}$ be a $p \times n$ matrix. We write $\boldsymbol{B} = [\boldsymbol{b}_{*1}, \ldots, \boldsymbol{b}_{*n}]$ in terms of its $p \times 1$ column vectors. The matrix product $\boldsymbol{AB}$ is then defined as

$$\boldsymbol{AB} = [\boldsymbol{Ab}_{*1}, \ldots, \boldsymbol{Ab}_{*n}],$$

where, $\boldsymbol{AB}$ is the $m \times n$ matrix whose j-th column vector is given by the matrix-vector product $\boldsymbol{Ab}_{*j}$. Note that each $\boldsymbol{Ab}_{*j}$ is an $m \times 1$ column vector, hence the dimension of $\boldsymbol{AB}$ is $m \times n$.

Example 1.10 Suppose we want to find the product $\boldsymbol{AB}$ where

$$\boldsymbol{A} = \begin{bmatrix} 1 & 0 & -3 & 0 \\ 0 & 1 & 4 & 0 \end{bmatrix} \text{ and } \boldsymbol{B} = \begin{bmatrix} 4 & -1 & 6 \\ 3 & 8 & 3 \\ 0 & 0 & 0 \\ 4 & 3 & -5 \end{bmatrix}.$$

The product $\boldsymbol{AB}$ is obtained by post-multiplying the matrix $\boldsymbol{A}$ with each of the columns of $\boldsymbol{B}$ and placing the resulting column vectors in a new matrix $\boldsymbol{C} = \boldsymbol{AB}$. Thus,

$$\boldsymbol{Ab}_{*1} = \begin{bmatrix} 1 & 0 & -3 & 0 \\ 0 & 1 & 4 & 0 \end{bmatrix} \begin{bmatrix} 4 \\ 3 \\ 0 \\ 4 \end{bmatrix} = \begin{bmatrix} 4 \\ 3 \end{bmatrix};$$

$$\boldsymbol{Ab}_{*2} = \begin{bmatrix} 1 & 0 & -3 & 0 \\ 0 & 1 & 4 & 0 \end{bmatrix} \begin{bmatrix} -1 \\ 8 \\ 0 \\ 3 \end{bmatrix} = \begin{bmatrix} -1 \\ 8 \end{bmatrix};$$

$$\boldsymbol{Ab}_{*3} = \begin{bmatrix} 1 & 0 & -3 & 0 \\ 0 & 1 & 4 & 0 \end{bmatrix} \begin{bmatrix} 6 \\ 3 \\ 0 \\ -5 \end{bmatrix} = \begin{bmatrix} 6 \\ 3 \end{bmatrix}.$$

Therefore,

$$C = AB = [Ab_{*1} : Ab_{*2} : Ab_{*3}] = \begin{bmatrix} 4 & -1 & 6 \\ 3 & 8 & 3 \end{bmatrix}. \blacksquare$$

This definition implies that if we write $C = AB$, then C is the $m \times n$ matrix whose (i,j)-th element is given by $c_{ij} = \sum_{k=1}^{p} a_{ik}b_{kj}$. More explicitly, we write the matrix product in terms of its elements as

$$AB = \begin{bmatrix} \sum_{k=1}^{n} a_{1k}b_{k1} & \sum_{k=1}^{n} a_{1k}b_{k2} & \dots & \sum_{k=1}^{n} a_{1k}b_{kp} \\ \sum_{k=1}^{n} a_{2k}b_{k1} & \sum_{k=1}^{n} a_{2k}b_{k2} & \dots & \sum_{k=1}^{n} a_{2k}b_{kp} \\ \vdots & \vdots & \vdots & \vdots \\ \sum_{k=1}^{n} a_{mk}b_{k1} & \sum_{k=1}^{n} a_{mk}b_{k2} & \dots & \sum_{k=1}^{n} a_{mk}b_{kp} \end{bmatrix}. \tag{1.6}$$

It is also worth noting that each element of $C = AB$ can also be expressed as the inner product between the i-th row vector of A and the j-th column vector of B, i.e., $c_{ij} = a'_{i*}b_{*j}$. More explicitly,

$$AB = \begin{bmatrix} a'_{1*}b_{*1} & a'_{1*}b_{*2} & \dots & a'_{1*}b_{*n} \\ a'_{2*}b_{*1} & a'_{2*}b_{*2} & \dots & a'_{2*}b_{*n} \\ \vdots & \vdots & \vdots & \vdots \\ a'_{m*}b_{*1} & a'_{m*}b_{*2} & \dots & a'_{m*}b_{*n}. \end{bmatrix}. \tag{1.7}$$

From the above definitions we note that for two matrices to be ***conformable*** for multiplication we must have the number of columns of the first matrix must equal to the number of rows of the second. Thus if AB and BA are both well-defined, A and B are necessarily both square matrices of the same order. Furthermore, matrix multiplication is not commutative in general: AB is not necessarily equal to BA even if both the products are well-defined.

A $1 \times m$ row-vector y' is conformable for multiplication with an $m \times n$ matrix A to form the $1 \times n$ row vector as

$$y'A = [y_1 : y_2 : \dots : y_m] \begin{bmatrix} a_{11} & a_{12} & \dots & a_{1n} \\ a_{21} & a_{22} & \dots & a_{2n} \\ \vdots & \vdots & \vdots & \vdots \\ a_{m1} & a_{m2} & \dots & a_{mn} \end{bmatrix} = \left[\sum_{i=1}^{m} y_i a_{i1}, \dots, \sum_{i=1}^{m} y_i a_{in} \right]$$

$$= y_1[a_{11}, \dots, a_{1n}] + \dots + y_m[a_{m1}, \dots, a_{mn}] = \sum_{j=1}^{m} y_i a'_{i*},$$

where a'_{i*}'s are the row vectors of the matrix A. This shows the above vector-matrix multiplication results in a ***linear combination*** of the row vectors.

Example 1.11 In Example 1.2, suppose we want to compute the average weekly heights for the rats. This can be easily computed as $y'A$, where A is as in (1.2) and

$\boldsymbol{y}' = (1/7)\mathbf{1}' = [1/7, 1/7, 1/7, 1/7, 1/7, 1/7, 1/7]$:

$$(1/7)\mathbf{1}' \begin{bmatrix} 151 & 199 & 246 & 283 & 320 \\ 145 & 199 & 249 & 293 & 354 \\ 155 & 200 & 237 & 272 & 297 \\ 135 & 188 & 230 & 280 & 323 \\ 159 & 210 & 252 & 298 & 331 \\ 141 & 189 & 231 & 275 & 305 \\ 159 & 201 & 248 & 297 & 338 \end{bmatrix} = [147.57, 199.86, 244287.57, 322.57].$$

■

These definitions imply that if we want to *postmultiply* a matrix $\boldsymbol{A}$ with a vector $\boldsymbol{x}$ to form $\boldsymbol{Ax}$, then $\boldsymbol{x}$ must be a column vector with the same number of entries as the number of columns in $\boldsymbol{A}$. If we want to *premultiply* $\boldsymbol{A}$ with a vector $\boldsymbol{y}'$ to form $\boldsymbol{y}'\boldsymbol{A}$, then $\boldsymbol{y}'$ must be a row vector with the same number of entries as the number of rows in $\boldsymbol{A}$. It is also easy to check that the row vector $\boldsymbol{y}'\boldsymbol{A}$ is the transpose of the column vector $\boldsymbol{A}'\boldsymbol{y}$. That is, $(\boldsymbol{y}'\boldsymbol{A})' = \boldsymbol{A}'\boldsymbol{y}$, which is a special case of a more general result on transposition in Theorem 1.3.

We can also legally multiply an $m \times 1$ column vector $\boldsymbol{u}$ with a $1 \times n$ row vector $\boldsymbol{v}'$ to yield a $m \times n$ ***matrix***:

$$\boldsymbol{uv}' = \begin{bmatrix} u_1 \\ u_2 \\ \vdots \\ u_m \end{bmatrix} [v_1, v_2, \ldots, v_n] = \begin{bmatrix} u_1v_1 & u_1v_2 & \ldots & u_1v_n \\ u_2v_1 & u_2v_2 & \ldots & u_2v_n \\ \vdots & \vdots & \vdots & \vdots \\ u_mv_1 & u_mv_2 & \ldots & u_mv_n \end{bmatrix}. \quad (1.8)$$

This product of two vectors is often called the ***outer product***. In short, we write $\boldsymbol{uv}' = [u_iv_j]_{i,j=1}^{m,n}$. We point out that the outer product is defined for two vectors of any dimensions, and produces a matrix as a result. This is in contrast with the inner product or inner product, which is defined only for two vectors of the same dimension and produces a scalar. Also, the outer product is certainly not symmetric (i.e. $\boldsymbol{uv}' \neq \boldsymbol{vu}'$), but the inner product is symmetric (i.e. $\boldsymbol{y}'\boldsymbol{x} = \boldsymbol{x}'\boldsymbol{y}$).

Recall that the elements of the product of two matrices can be written as the inner product between the row vectors of the first matrix and the column vectors of the second matrix. We can also represent matrix multiplication as a sum of outer products of the column vectors of the first matrix and the row vectors of the second matrix:

$$\boldsymbol{AB} = [\boldsymbol{a}_{*1}, \ldots, \boldsymbol{a}_{*p}] \begin{bmatrix} \boldsymbol{b}'_{1*} \\ \boldsymbol{b}'_{2*} \\ \vdots \\ \boldsymbol{b}'_{m*} \end{bmatrix} = \sum_{j=1}^{p} \boldsymbol{a}_{*j}\boldsymbol{b}'_{j*}. \quad (1.9)$$

Several elementary properties of matrix operations are easily derived from the above definitions. For example, an easy way to extract the j-th column of a matrix is to postmultiply it with the j-th canonical vector (i.e. the j-th column of the identity matrix): $\boldsymbol{A}\boldsymbol{e}_j = \boldsymbol{a}_{*j}$. Premultiplication by the i-th canonical vector yields the i-th row vector of $\boldsymbol{A}$, i.e. $\boldsymbol{e}_i'\boldsymbol{A} = \boldsymbol{a}_{i*}'$. It is also easy to see that if $\boldsymbol{A}$ is an $n \times n$ matrix then $\boldsymbol{A}\boldsymbol{I}_n = \boldsymbol{I}_n\boldsymbol{A} = \boldsymbol{A}$.

The following theorem lists some fundamental properties of matrices and operations that are quite easily derived.

Theorem 1.2

(i) *Let $\boldsymbol{A}$ be an $m \times n$ matrix, $\boldsymbol{x}$ and $\boldsymbol{y}$ be two $n \times 1$ vectors, and α and β be two scalars. Then $\boldsymbol{A}(\alpha\boldsymbol{x} + \beta\boldsymbol{y}) = \alpha\boldsymbol{A}\boldsymbol{x} + \beta\boldsymbol{A}\boldsymbol{y}$.*

(ii) *Let $\boldsymbol{A}$ be an $m \times p$ and $\boldsymbol{B}$ a $p \times n$ matrix. Let α and β be scalars. Then $\alpha(\beta\boldsymbol{A}) = (\alpha\beta)\boldsymbol{A}$ and $(\alpha\boldsymbol{A})(\beta\boldsymbol{B}) = (\alpha\beta)(\boldsymbol{A}\boldsymbol{B})$*

(iii) *Let $\boldsymbol{C} = \boldsymbol{A}\boldsymbol{B}$. Then the i-th row vector of $\boldsymbol{C}$ is given by $\boldsymbol{c}_{i*}' = \boldsymbol{a}_{i*}'\boldsymbol{B}$ and the j-th column vector of $\boldsymbol{C}$ is given by $\boldsymbol{c}_{*j} = \boldsymbol{A}\boldsymbol{b}_{*j}$, where $\boldsymbol{a}_{i*}'$ is the i-th row vector of $\boldsymbol{A}$ and $\boldsymbol{b}_{*j}$ is the j-th column vector of $\boldsymbol{B}$.*

(iv) Left-hand distributive law: *Let $\boldsymbol{A}$ be an $m \times p$ matrix, $\boldsymbol{B}$ and $\boldsymbol{C}$ both be $p \times n$ matrices. Then, $\boldsymbol{A}(\boldsymbol{B} + \boldsymbol{C}) = \boldsymbol{A}\boldsymbol{B} + \boldsymbol{B}\boldsymbol{C}$.*

(v) Right-hand distributive law: *Let $\boldsymbol{A}$ and $\boldsymbol{B}$ both be $m \times p$ matrices and let $\boldsymbol{C}$ be a $p \times n$ matrix. Then, $(\boldsymbol{A} + \boldsymbol{B})\boldsymbol{C} = \boldsymbol{A}\boldsymbol{C} + \boldsymbol{B}\boldsymbol{C}$.*

(vi) Associative law of matrix multiplication: *Let $\boldsymbol{A}$ be $m \times k$, $\boldsymbol{B}$ be $k \times p$ and $\boldsymbol{C}$ be $p \times n$. Then $\boldsymbol{A}(\boldsymbol{B}\boldsymbol{C}) = (\boldsymbol{A}\boldsymbol{B})\boldsymbol{C}$.*

(vii) *Let $\boldsymbol{D} = \boldsymbol{A}\boldsymbol{B}\boldsymbol{C}$. Then the (i,j)-th element of $\boldsymbol{D}$ is given by $[\boldsymbol{D}]_{ij} = \boldsymbol{a}_{i*}'\boldsymbol{B}\boldsymbol{c}_{*j}$, where $\boldsymbol{a}_{i*}'$ is the i-th row vector of $\boldsymbol{A}$ and $\boldsymbol{c}_{*j}$ is the j-th column vector of the matrix $\boldsymbol{C}$.*

Proof. Parts (i)–(v) are very easily verified from the definitions. We leave them to the reader. For part (vi), it is convenient to treat some of the matrices in terms of their column vectors. Therefore, we write $\boldsymbol{B}\boldsymbol{C} = [\boldsymbol{B}\boldsymbol{c}_{*1}, \ldots, \boldsymbol{B}\boldsymbol{c}_{*n}]$, where $\boldsymbol{c}_{*j}$ is the j-th column of $\boldsymbol{C}$. Then,

$$\begin{aligned}\boldsymbol{A}(\boldsymbol{B}\boldsymbol{C}) &= \boldsymbol{A}[\boldsymbol{B}\boldsymbol{c}_{*1}, \ldots, \boldsymbol{B}\boldsymbol{c}_{*n}] = [\boldsymbol{A}\boldsymbol{B}\boldsymbol{c}_{*1}, \ldots, \boldsymbol{A}\boldsymbol{B}\boldsymbol{c}_{*n}] \\ &= (\boldsymbol{A}\boldsymbol{B})[\boldsymbol{c}_{*1}, \ldots, \boldsymbol{c}_{*n}] = (\boldsymbol{A}\boldsymbol{B})\boldsymbol{C}.\end{aligned}$$

For part (vii), we first note that the (i,j)-th element of $\boldsymbol{D}$ is the (i,j)-th element of $\boldsymbol{A}(\boldsymbol{B}\boldsymbol{C})$ (using the associative law in part (v)). Hence, it can be expressed as the inner product between the i-th row vector of $\boldsymbol{A}$ and the j-th column vector of $\boldsymbol{B}\boldsymbol{C}$. The j-th column of $\boldsymbol{B}\boldsymbol{C}$ is precisely $\boldsymbol{B}\boldsymbol{c}_{*j}$ (using part (i)). Thus, $[\boldsymbol{D}]_{ij} = \boldsymbol{a}_{i*}'\boldsymbol{B}\boldsymbol{c}_{*j}$. □

A function that satisfies the property $f(\sum_{i=1}^n \alpha_i\boldsymbol{x}_i) = \sum_{i=1}^n \alpha_i f(\boldsymbol{x}_i)$, i.e.,

$$f(\alpha_1\boldsymbol{x}_1 + \alpha_2\boldsymbol{x}_2 + \cdots + \alpha_n\boldsymbol{x}_n) = \alpha_1 f(\boldsymbol{x}_1) + \alpha_2 f(\boldsymbol{x}_2) + \cdots + \alpha_n f(\boldsymbol{x}_n) ,$$

is said to be ***linear*** in its arguments and is called a ***linear function***. These functions lie at the very heart of linear algebra. Repeated application of part (i) of Theorem 1.2 shows that $f(\boldsymbol{x}) = \boldsymbol{A}\boldsymbol{x}$ is a well-defined linear function from $\Re^n$ to $\Re^m$.

We next arrive at an extremely important property of matrix products and transposes.

Theorem 1.3 *If $\boldsymbol{A}$ and $\boldsymbol{B}$ are such that $\boldsymbol{AB}$ exists, then* $(\boldsymbol{AB})' = \boldsymbol{B}'\boldsymbol{A}'$.

Proof. Let $\boldsymbol{A}$ be an $m \times p$ matrix and let $\boldsymbol{B}$ be a $p \times n$ matrix. Then, $(\boldsymbol{AB})'$ and $\boldsymbol{B}'\boldsymbol{A}'$ are both $n \times m$. Therefore, if we can prove that the (i, j)-th element of $(\boldsymbol{AB})'$ is the same as that of $\boldsymbol{B}'\boldsymbol{A}'$, we would be done. We will prove this result by using (1.7). The (j, i)-th element of $\boldsymbol{AB}$ is given by $\boldsymbol{a}'_{j*}\boldsymbol{b}_{*i}$ and is precisely the (i, j)-th element of $(\boldsymbol{AB})'$. Also note from the definition of the scalar prroduct that $\boldsymbol{a}'_{j*}\boldsymbol{b}_{*i} = \boldsymbol{b}'_{*i}\boldsymbol{a}_{j*}$. We now make the observation that $\boldsymbol{b}'_{*i}$ is the i-th row of $\boldsymbol{B}'$ and $\boldsymbol{a}_{j*}$ is the j-th column of $\boldsymbol{A}'$. Therefore, $\boldsymbol{b}'_{*i}\boldsymbol{a}_{j*}$ is also the i, j-th element of $\boldsymbol{B}'\boldsymbol{A}'$. □

Example 1.12 Consider the matrices $\boldsymbol{A}$ and $\boldsymbol{B}$ in Example 1.10. The transpose of $\boldsymbol{AB}$ is

$$(\boldsymbol{AB})' = \begin{bmatrix} 4 & -1 & 6 \\ 3 & 8 & 3 \end{bmatrix}' = \begin{bmatrix} 4 & 3 \\ -1 & 8 \\ 6 & 3 \end{bmatrix}.$$

Also,

$$\boldsymbol{B}'\boldsymbol{A}' = \begin{bmatrix} 4 & 3 & 0 & 4 \\ -1 & 8 & 0 & 3 \\ 6 & 3 & 0 & -5 \end{bmatrix} \begin{bmatrix} 1 & 0 \\ 0 & 1 \\ -3 & 4 \\ 0 & 0 \end{bmatrix} = \begin{bmatrix} 4 & 3 \\ -1 & 8 \\ 6 & 3 \end{bmatrix},$$

which is equal to $(\boldsymbol{AB})'$. ■

In deriving results involving matrix-matrix products, it is sometimes useful to first consider the result for matrix-vector products. For example, we can derive Theorem 1.3 by first deriving the result for products such as $\boldsymbol{Ax}$. The matrix-vector product $\boldsymbol{Ax}$, where $\boldsymbol{x}$ is a $p \times 1$ column vector, can be written as $\sum_{j=1}^{p} x_j \boldsymbol{a}_{*j}$. Therefore,

$$(\boldsymbol{Ax})' = (\sum_{j=1}^{p} x_j \boldsymbol{a}_{*j})' = \sum_{j=1}^{p} x_j \boldsymbol{a}'_{*j} = (x_1, \ldots, x_n) \begin{bmatrix} \boldsymbol{a}'_{*1} \\ \boldsymbol{a}'_{*2} \\ \vdots \\ \boldsymbol{a}'_{*n} \end{bmatrix} = \boldsymbol{x}'\boldsymbol{A}' .$$

In the above we have used $(\sum_{j=1}^{p} x_j \boldsymbol{a}_{*j})' = \sum_{j=1}^{p} x_j \boldsymbol{a}'_{*j}$ (see part (vi) in Theorem 1.1). The result, $\boldsymbol{x}'\boldsymbol{A}'$ is the $1 \times m$ transpose of the $m \times 1$ vector $\boldsymbol{Ax}$. We have, therefore, proved Theorem 1.3 for the special case of matrix-vector products. Now we write $\boldsymbol{AB} = [\boldsymbol{Ab}_{*1}, \ldots, \boldsymbol{Ab}_{*n}]$ and use the result for matrix-vector products to obtain

$$(\boldsymbol{AB})' = \begin{bmatrix} (\boldsymbol{Ab}_{*1})' \\ (\boldsymbol{Ab}_{*2})' \\ \vdots \\ (\boldsymbol{Ab}_{*n})' \end{bmatrix} = \begin{bmatrix} \boldsymbol{b}'_{*1}\boldsymbol{A}' \\ \boldsymbol{a}'_{*2}\boldsymbol{A}' \\ \vdots \\ \boldsymbol{a}'_{*n}\boldsymbol{A}' \end{bmatrix} = \begin{bmatrix} \boldsymbol{b}'_{*1} \\ \boldsymbol{a}'_{*2} \\ \vdots \\ \boldsymbol{a}'_{*n} \end{bmatrix} \boldsymbol{A}' = \boldsymbol{B}'\boldsymbol{A}'.$$

1.4 Partitioned matrices

Example 1.13 Let us revisit the problem of finding the product $\boldsymbol{AB}$ where

$$\boldsymbol{A} = \begin{bmatrix} 1 & 0 & -3 & 0 \\ 0 & 1 & 4 & 0 \end{bmatrix} \text{ and } \boldsymbol{B} = \begin{bmatrix} 4 & -1 & 6 \\ 3 & 8 & 3 \\ 0 & 0 & 0 \\ 4 & 3 & -5 \end{bmatrix}.$$

From Example 1.10, we found that

$$\boldsymbol{AB} = \begin{bmatrix} 4 & -1 & 6 \\ 3 & 8 & 3 \end{bmatrix}. \tag{1.10}$$

There is, however, a simpler way of computing the product here. Note that the first two columns of $\boldsymbol{A}$ form the 2×2 identity matrix $\boldsymbol{I}_2$ and the fourth column is $\boldsymbol{0}$, while the third row is $\boldsymbol{0}'$. Let us write $\boldsymbol{A}$ and $\boldsymbol{B}$ as

$$\boldsymbol{A} = \left[\begin{array}{c|c|c} \boldsymbol{I}_2 & \begin{matrix} -3 \\ 4 \end{matrix} & \boldsymbol{0} \end{array}\right] \text{ and } \boldsymbol{B} = \left[\begin{array}{ccc} 4 & -1 & 6 \\ 3 & 8 & 3 \\ \hline & \boldsymbol{0}' & \\ \hline 4 & 3 & -5 \end{array}\right]$$

and now multiply $\boldsymbol{A}$ and $\boldsymbol{B}$ as if $\boldsymbol{A}$ were a row and $\boldsymbol{B}$ a column. This yields

$$\boldsymbol{AB} = \boldsymbol{I}_2 \begin{bmatrix} 4 & -1 & 6 \\ 3 & 8 & 3 \end{bmatrix} + \begin{bmatrix} -3 \\ 4 \end{bmatrix} \boldsymbol{0}' + \boldsymbol{0} \begin{bmatrix} 4 & 3 & -5 \end{bmatrix}$$

and the result is clearly the matrix in (1.10). ■

This type of multiplication is valid provided the columns of $\boldsymbol{A}$ and $\boldsymbol{B}$ are *partitioned* the same way. Partitioning essentially amounts to drawing horizontal lines to form row partitions and vertical lines to form column partitions. With proper partitioning of the matrices, we can often achieve economy in matrix operations. We will be dealing with partitioned matrices throughout this book. In the next two subsections we outline some definitions, notations and basic results for partitioned matrices.

1.4.1 2×2 partitioned matrices

In the preceding section, we have seen how we can represent matrices in terms of their column vectors or row vectors. For example, suppose we write the $m \times n$ matrix $\boldsymbol{A}$ as $\boldsymbol{A} = [\boldsymbol{a}_{*1}, \ldots, \boldsymbol{a}_{*n}]$. Since each column vector $\boldsymbol{a}_{*j}$ is itself an $m \times 1$ matrix, we can say that we have ***partitioned***, or ***blocked*** the $m \times n$ matrix $\boldsymbol{A}$ to form a $1 \times n$ matrix in terms of $m \times 1$ ***submatrices*** . Note that this partitioning makes sense only because each of the column vectors has the same dimension. Analogously, we could partition $\boldsymbol{A}$ in terms of its rows: we treat $\boldsymbol{A}$ as an $m \times 1$ matrix blocked in terms of $1 \times n$ row vectors.

Let A be an $m \times n$ matrix partitioned into 2×2 blocks:

$$A = \begin{bmatrix} A_{11} & A_{12} \\ A_{21} & A_{22} \end{bmatrix}, \tag{1.11}$$

where A_{11} is $m_1 \times n_1$, A_{12} is $m_1 \times n_2$, A_{21} is $m_2 \times n_1$ and A_{22} is $m_2 \times n_2$. Here $m = m_1 + m_2$ and $n = n_1 + n_2$. For this partitioning to make sense, the submatrices forming the rows must have the same number of rows, while the submatrices forming the columns must have the same number of columns. More precisely, A_{11} and A_{12} have the same number of rows (m_1) as do A_{21} and A_{22} (m_2). On the other hand, A_{11} and A_{21} have the same number of columns (n_1), as do A_{12} and A_{22} (n_2).

How can we express the transpose of the partitioned matrix A in (1.11) in terms of its submatrices? To visualize this, consider a row in the upper part of A, i.e., a'_{i*}, where $i \in \{1, \ldots, m_1\}$, and another, say a'_{j*} in the lower half of A, i.e. where $j \in \{m_1+1, \ldots, m\}$. These rows become the i-th and j-th columns of A' as below:

$$A = \begin{bmatrix} \cdots & \cdots & \cdots & \cdots & \cdots & \cdots \\ a_{i1} & \cdots & a_{in_1} & a_{in_1+1} & \cdots & a_{in} \\ \cdots & \cdots & \cdots & \cdots & \cdots & \cdots \\ & & & & & \\ \cdots & \cdots & \cdots & \cdots & \cdots & \cdots \\ a_{j1} & \cdots & a_{jn_1} & a_{jn_1+1} & \cdots & a_{jn} \\ \cdots & \cdots & \cdots & \cdots & \cdots & \cdots \end{bmatrix} \text{ and}$$

$$A' = \begin{bmatrix} \vdots & a_{i1} & \vdots & \vdots & a_{j1} & \vdots \\ \vdots & \vdots & \vdots & \vdots & \vdots & \vdots \\ \vdots & a_{in_1} & \vdots & \vdots & a_{jn_1} & \vdots \\ & & & & & \\ \vdots & a_{in_1+1} & \vdots & \vdots & a_{jn_1+1} & \vdots \\ \vdots & \vdots & \vdots & \vdots & \vdots & \vdots \\ \vdots & a_{in} & \vdots & \vdots & a_{jn} & \vdots \end{bmatrix}.$$

In A' above, note that the column vector $(a_{i1}, \ldots, a_{in_1})'$ is the i-th column of A'_{11}, $(a_{in_1+1}, \ldots, a_{in})'$ is the i-th column of A'_{12}, $(a_{j1}, \ldots, a_{jn_1})'$ is the j-th column of A'_{21} and $(a_{jn_1+1}, \ldots, a_{jn})'$ is the j-th column of A'_{22}. Therefore,

$$A = \begin{bmatrix} A_{11} & A_{12} \\ A_{21} & A_{22} \end{bmatrix} \text{ implies that } A' = \begin{bmatrix} A'_{11} & A'_{21} \\ A'_{12} & A'_{22} \end{bmatrix}. \tag{1.12}$$

The matrix A' is now $n \times m$ with A'_{11} being $n_1 \times m_1$, A'_{12} being $n_2 \times m_1$, A'_{21} being $n_1 \times m_2$ and A'_{22} being $n_2 \times m_2$. Note that A' is formed by transposing each of the four block matrices and, in addition, *switching* the positions of A_{12} and A_{21}.

We say that two partitioned matrices A and B are conformably partitioned for addition when their corresponding blocks have the same order. The linear combination of two such conformable matrices can be written as

$$\alpha \boldsymbol{A} + \beta \boldsymbol{B} = \alpha \begin{bmatrix} \boldsymbol{A}_{11} & \boldsymbol{A}_{12} \\ \boldsymbol{A}_{21} & \boldsymbol{A}_{22} \end{bmatrix} + \beta \begin{bmatrix} \boldsymbol{B}_{11} & \boldsymbol{B}_{12} \\ \boldsymbol{B}_{21} & \boldsymbol{B}_{22} \end{bmatrix}$$
$$= \begin{bmatrix} \alpha\boldsymbol{A}_{11} + \beta\boldsymbol{B}_{11} & \alpha\boldsymbol{A}_{12} + \beta\boldsymbol{B}_{12} \\ \alpha\boldsymbol{A}_{21} + \beta\boldsymbol{B}_{21} & \alpha\boldsymbol{A}_{22} + \beta\boldsymbol{B}_{22} \end{bmatrix}.$$

The product of two partitioned matrices can also be expressed in terms of the submatrices, provided they are conformable for multiplication. Consider, for example, the $m \times n$ matrix $\boldsymbol{A}$ in (1.11) and let $\boldsymbol{B}$ be an $n \times p$ matrix, partitioned into a 2×2 block matrix with $\boldsymbol{B}_{11}$ being $n_1 \times p_1$, $\boldsymbol{B}_{12}$ being $n_1 \times p_2$, $\boldsymbol{B}_{21}$ being $n_2 \times p_1$ and $\boldsymbol{B}_{22}$ being $n_2 \times p_2$. In that case, the products $\boldsymbol{A}_{il}\boldsymbol{B}_{lj}$ are defined for $i, l, j = 1, 2$. In fact, consider the (i,j)-th element of $\boldsymbol{AB}$, i.e., $\boldsymbol{a}'_{i*}\boldsymbol{b}_{*j}$, which we can write as

$$\boldsymbol{a}'_{i*}\boldsymbol{b}_{*j} = \sum_{k=1}^{n} a_{ik}b_{kj} = \sum_{k=1}^{n_1} a_{ik}b_{kj} + \sum_{k=n_1+1}^{n} a_{ik}b_{kj}.$$

Suppose, for example, that $i \in \{1, \ldots, m_1\}$ and $j \in \{1, \ldots, n_1\}$, so that the (i,j)-th element is a member of the $(1,1)$-th block. Then the first sum on the right-hand side of the above equation gives precisely the (i,j)-th element of $\boldsymbol{A}_{11}\boldsymbol{B}_{11}$ and the second sum gives the (i,j)-th element of $\boldsymbol{A}_{12}\boldsymbol{B}_{21}$. Thus, we have that $\boldsymbol{a}'_{i*}\boldsymbol{b}_{*j}$ is the (i,j)-th entry in $\boldsymbol{A}_{11}\boldsymbol{B}_{11} + \boldsymbol{A}_{12}\boldsymbol{B}_{21}$. Analogous arguments hold for elements in each of the other blocks and we obtain

$$\boldsymbol{AB} = \begin{bmatrix} \boldsymbol{A}_{11} & \boldsymbol{A}_{12} \\ \boldsymbol{A}_{21} & \boldsymbol{A}_{22} \end{bmatrix} \begin{bmatrix} \boldsymbol{B}_{11} & \boldsymbol{B}_{12} \\ \boldsymbol{B}_{21} & \boldsymbol{B}_{22} \end{bmatrix}$$
$$= \begin{bmatrix} \boldsymbol{A}_{11}\boldsymbol{B}_{11} + \boldsymbol{A}_{12}\boldsymbol{B}_{21} & \boldsymbol{A}_{11}\boldsymbol{B}_{12} + \boldsymbol{A}_{12}\boldsymbol{B}_{22} \\ \boldsymbol{A}_{21}\boldsymbol{B}_{11} + \boldsymbol{A}_{22}\boldsymbol{B}_{21} & \boldsymbol{A}_{21}\boldsymbol{B}_{12} + \boldsymbol{A}_{22}\boldsymbol{B}_{22} \end{bmatrix}.$$

1.4.2 General partitioned matrices

The 2×2 partitioned matrices reveal much insight about operations with submatrices. For a more general treatment of partitioned matrices, we can form ***submatrices*** formed by striking out rows and columns of a matrix. Consider $\boldsymbol{A}$ as an $m \times n$ matrix and let $I_r \subseteq \{1, \ldots, m\}$ and $I_c \subseteq \{1, \ldots, n\}$. Then the matrix formed by retaining only the rows of $\boldsymbol{A}$ indexed in I_r is denoted $\boldsymbol{A}_{I_r,\cdot}$. Similarly the matrix formed by retaining only the columns of $\boldsymbol{A}$ indexed in I_c is denoted $\boldsymbol{A}_{\cdot,I_c}$. In general, retaining the rows in I_r and columns in I_c result in a sub-matrix $\boldsymbol{A}_{I_r,I_c}$.

Definition 1.10 *A submatrix of an $n \times n$ matrix is called a* **principal submatrix** *if it can be obtained by striking out the same rows as columns; in other words if $I_r = I_c$. It is called a* **leading principal submatrix** *when $I_c = I_r = \{1, 2, \ldots, k\}$ for some $k < n$.*

More generally, partitioned matrices look like:

$$A = \begin{bmatrix} A_{11} & A_{12} & \dots & A_{1c} \\ A_{21} & A_{22} & \dots & A_{2c} \\ \vdots & \vdots & \vdots & \vdots \\ A_{r1} & A_{r2} & \dots & A_{rc} \end{bmatrix}, \tag{1.13}$$

where $\boldsymbol{A}$ is $m \times n$ and each $\boldsymbol{A}_{ij}$ is an $m_i \times n_j$ matrix, $i = 1, \dots, r$; $j = 1, \dots, c$ with $\sum_{i=1}^{r} m_i = m$ and $\sum_{j=1}^{c} n_j = n$. This is called a ***conformable partition*** and is often written more concisely as $\boldsymbol{A} = \{\boldsymbol{A}_{ij}\}_{i,j=1}^{r,c}$. Note that the individual dimensions of the $\boldsymbol{A}_{ij}$'s are suppressed in this concise notation. A matrix can be partitioned conformably by drawing horizontal or vertical lines between the various rows and columns. A matrix that is subdivided irregularly using scattered lines is not conformably partitioned and will not be considered a partitioned matrix.

Using an argument analogous to the one used for 2×2 partitioned matrices, we immediately see that transposition of a conformably partitioned matrix yields another conformably partitioned matrix:

$$A' = \begin{bmatrix} A'_{11} & A'_{21} & \dots & A'_{r1} \\ A'_{12} & A'_{22} & \dots & A'_{r2} \\ \vdots & \vdots & \vdots & \vdots \\ A'_{1c} & A'_{2c} & \dots & A'_{rc} \end{bmatrix},$$

or, more concisely, $\boldsymbol{A}' = \{\boldsymbol{A}'_{ji}\}_{i,j=1}^{r,c}$.

Now consider another partitioned matrix

$$B = \begin{bmatrix} B_{11} & B_{12} & \dots & B_{1v} \\ B_{21} & B_{22} & \dots & B_{2v} \\ \vdots & \vdots & \vdots & \vdots \\ B_{u1} & B_{r2} & \dots & B_{uv} \end{bmatrix}.$$

This represents a $p \times q$ matrix whose (i, j)-th block is the $p_i \times q_j$ matrix $\boldsymbol{B}_{ij}$ with $\sum_{i=1}^{u} p_i = p$ and $\sum_{j=1}^{v} q_j = q$.

The matrices $\boldsymbol{A}$ and $\boldsymbol{B}$ are conformable for addition if $p = m$ and $q = n$. However, for the partitioning to be conformable for addition as well, we must have $u = r$, $v = c$, $m_i = p_i$, $i = 1, \dots, r$ and $n_j = q_j$, $j = 1, \dots, c$. In that case, operations on (conformably) partitioned matrices carry over immediately to yield matrix addition as

$$A + B = \begin{bmatrix} A_{11} + B_{11} & A_{12} + B_{12} & \dots & A_{1c} + B_{1c} \\ A_{21} + B_{21} & A_{22} + B_{22} & \dots & A_{2c} + B_{2c} \\ \vdots & \vdots & \vdots & \vdots \\ A_{r1} + A_{r1} & A_{r2} + B_{r2} & \dots & A_{rc} + B_{rc} \end{bmatrix}.$$

The matrix product $\boldsymbol{AB}$ is defined provided that $n = p$. If $c = u$ and $n_k = p_k$, $k = 1, \dots, c$, whereupon all the products $\boldsymbol{A}_{ik}\boldsymbol{B}_{kj}$, $i = 1, \dots, r$; $j = 1, \dots, v$; $k =$

$1, \ldots, c$ are well-defined, along with $\boldsymbol{AB}$. Then,

$$\boldsymbol{AB} = \begin{bmatrix} \boldsymbol{C}_{11} & \boldsymbol{C}_{12} & \ldots & \boldsymbol{C}_{1v} \\ \boldsymbol{C}_{21} & \boldsymbol{C}_{22} & \ldots & \boldsymbol{C}_{2v} \\ \vdots & \vdots & \vdots & \vdots \\ \boldsymbol{C}_{r1} & \boldsymbol{C}_{r2} & \ldots & \boldsymbol{C}_{rv} \end{bmatrix},$$

where

$$\boldsymbol{C}_{ij} = \sum_{k=1}^{c} \boldsymbol{A}_{ik}\boldsymbol{B}_{kj} = \boldsymbol{A}_{i1}\boldsymbol{B}_{1j} + \boldsymbol{A}_{i2}\boldsymbol{B}_{2j} + \cdots + \boldsymbol{A}_{ic}\boldsymbol{B}_{cj}.$$

1.5 The "trace" of a square matrix

Definition 1.11 *The* **trace** *of a square matrix* $\boldsymbol{A} = [a_{ij}]_{i,j=1}^{n}$ *of order* n *is defined as the sum of its diagonal elements:*

$$tr(\boldsymbol{A}) = a_{11} + \ldots + a_{nn} . \tag{1.14}$$

For example, $\text{tr}(\boldsymbol{O}) = 0$ and $\text{tr}(\boldsymbol{I}_n) = n$. Thus, the trace is a function that takes a square matrix as its input and yields a real number as output. The trace is an extremely useful function for matrix manipulations and has found wide applicability in statistics, especially in multivariate analysis and design of experiments.

Since a scalar α is a 1×1 matrix whose only element is α, the trace of a scalar is the scalar itself. That is $\text{tr}(\alpha) = \text{tr}([\alpha]) = \alpha$. The following properties of the trace function are easily verified and used often, so we list them in a theorem.

Theorem 1.4 *Let* $\boldsymbol{A}$ *and* $\boldsymbol{B}$ *be two square matrices of order* n*, and let* α *be a scalar.*

(i) $tr(\boldsymbol{A}) = tr(\boldsymbol{A}')$.

(ii) $tr(\alpha\boldsymbol{A}) = \alpha tr(\boldsymbol{A})$.

(iii) $tr(\boldsymbol{A} + \boldsymbol{B}) = tr(\boldsymbol{A}) + tr(\boldsymbol{B})$.

(iv) Let $\boldsymbol{A}_1, \ldots, \boldsymbol{A}_k$ *be a sequence of* $n \times n$ *square matrices and let* $\alpha_1, \ldots, \alpha_k$ *denote a sequence of* k *scalars. Then,*

$$tr\left(\sum_{i=1}^{k} \alpha_i \boldsymbol{A}_i\right) = \alpha_1 tr(\boldsymbol{A}_1) + \ldots + \alpha_k tr(\boldsymbol{A}_k) = \sum_{i=1}^{k} \alpha_i tr(\boldsymbol{A}_i) .$$

Proof. Parts (i)–(iii) are easily verified using the definition of trace. Part (iv) follows from repeated application of parts (ii) and (iii). □

Part (iv) shows that the trace function is a particular example of a linear function.

Consider again the partitioned matrix $\boldsymbol{A}$ in (1.11) where each diagonal block, $\boldsymbol{A}_{ii}$, is a square matrix of order m_i. Note that this implies that $\boldsymbol{A}$ is a square matrix of order

$m = \sum_{i=k} m_i$. It is easily verified that the trace of $\boldsymbol{A}$ is the sum of the traces of its diagonal blocks, i.e., $\text{tr}(\boldsymbol{A}) = \text{tr}(\boldsymbol{A}_{11}) + \ldots + \text{tr}(\boldsymbol{A}_{kk})$.

The trace function is defined only for square matrices. Therefore, it is not defined for an $m \times n$ matrix $\boldsymbol{A}$ or an $n \times m$ matrix $\boldsymbol{B}$, but it *is defined* for the product $\boldsymbol{AB}$, which is a square matrix of order m, and also for $\boldsymbol{BA}$ which are. The following is another extremely important result concerning the trace function and matrix products.

Theorem 1.5

(i) Let $\boldsymbol{u}$ and $\boldsymbol{v}$ both be $m \times 1$ vectors. Then tr$(\boldsymbol{uv}') = \boldsymbol{u}'\boldsymbol{v} = \boldsymbol{v}'\boldsymbol{u}$.

(ii) Let $\boldsymbol{A}$ be an $m \times n$ matrix and $\boldsymbol{B}$ be an $n \times m$ matrix. Then tr$(\boldsymbol{AB})$ = tr$(\boldsymbol{BA})$.

Proof. Part (i) follows from the simple observation that the diagonal elements of the $m \times m$ outer-product matrix $\boldsymbol{uv}'$ is given by $u_i v_i$. Therefore, $\text{tr}(\boldsymbol{uv}') = \sum_{i=1}^m u_i v_i$, but $\sum_{i=1}^m u_i v_i = \boldsymbol{v}'\boldsymbol{u} = \boldsymbol{u}'\boldsymbol{v}$.

For part (ii), note that the i-th diagonal element of $\boldsymbol{AB}$ is given by $\boldsymbol{a}_{i*}'\boldsymbol{b}_{*i} = \sum_{k=1}^n a_{ik}b_{ki}$, while the j-th diagonal element of $\boldsymbol{BA}$ is given by $\boldsymbol{b}_{j*}'\boldsymbol{a}_{*j} = \sum_{k=1}^m b_{jk}a_{kj}$. Now,

$$\text{tr}(\boldsymbol{AB}) = \sum_{i=1}^m \sum_{j=1}^n a_{ij}b_{ji} = \sum_{i=1}^m \sum_{j=1}^n b_{ji}a_{ij} = \sum_{j=1}^n \sum_{i=1}^m b_{ji}a_{ij} = \text{tr}(\boldsymbol{BA}).$$

This proves the result. □

Instead of using the elements of the matrix in part (ii) of the above theorem, we could have carried out the derivation at the "vector level" as follows:

$$\text{tr}(\boldsymbol{AB}) = \sum_{i=1}^m \boldsymbol{a}_{i*}'\boldsymbol{b}_{*i} = \sum_{i=1}^m \text{tr}(\boldsymbol{b}_{*i}\boldsymbol{a}_{i*}') = \text{tr}\left(\sum_{i=1}^m \boldsymbol{b}_{*i}\boldsymbol{a}_{i*}'\right) = \text{tr}(\boldsymbol{BA}). \tag{1.15}$$

Here, the second equality follows from part (i), the third equality follows from the linearity of the trace function and the last equality follows from the outer-product matrix representation $\boldsymbol{BA} = \sum_{i=1}^m \boldsymbol{b}_{*i}\boldsymbol{a}_{i*}'$ in (1.9).

Part (i) of Theorem 1.4 extends (1.15) to

$$\text{tr}(\boldsymbol{AB}) = \text{tr}(\boldsymbol{B}'\boldsymbol{A}') = \text{tr}(\boldsymbol{A}'\boldsymbol{B}') = \text{tr}(\boldsymbol{BA}) \,. \tag{1.16}$$

Also, (1.15) can be extended to derive relationships involving more than two matrices. For example, with three matrices, we obtain

$$\text{tr}(\boldsymbol{ABC}) = \text{tr}(\boldsymbol{CAB}) = \text{tr}(\boldsymbol{BCA}) \,, \tag{1.17}$$

where the products are well-defined. This can be extended to products involving several (more than three) matrices.

A particularly useful version of (1.15) is the following:

$$\text{tr}(\boldsymbol{AA}') = \text{tr}(\boldsymbol{A}'\boldsymbol{A}) = \sum_{i=1}^n \boldsymbol{a}_{*i}'\boldsymbol{a}_{*i} = \sum_{i=1}^n \|\boldsymbol{a}_{*i}\|^2 = \sum_{i=1}^n \sum_{j=1}^n a_{ij}^2 \geq 0 \,, \tag{1.18}$$

with equality holding if and only if $\boldsymbol{A} = \boldsymbol{O}$.

The trace function has several applications in linear algebra. It is useful in deriving several interesting results and we will encounter many of them in subsequent chapters. Here is one example.

Example 1.14 If we are given a square matrix $\boldsymbol{A}$, then can we find another square matrix $\boldsymbol{X}$ that satisfies the equation $\boldsymbol{AX} - \boldsymbol{XA} = \boldsymbol{I}$? The answer is ***no***. To see why, simply apply the trace function to both sides of the equation and use Part (ii) of Theorem 1.5 to conclude that the left-hand side would yield 0 and the right-hand side would yield n. ■

1.6 Some special matrices

We will conclude this chapter with two special types of matrices, *permutation matrices* and *triangular matrices*, that we will encounter in the next chapter.

1.6.1 Permutation matrices

Definition 1.12 *An $n \times n$* **permutation matrix** *is a matrix obtained by permuting the columns or rows of an $n \times n$ identity matrix.*

We can express a permutation matrix in terms of its column and row vectors as

$$\boldsymbol{P} = [\boldsymbol{e}_{j_1} : \boldsymbol{e}_{j_2} : \ldots : \boldsymbol{e}_{j_n}] = \begin{bmatrix} \boldsymbol{e}'_{i_1} \\ \boldsymbol{e}'_{i_2} \\ \vdots \\ \boldsymbol{e}'_{i_n} \end{bmatrix},$$

where $(i_1, \ldots, i_n)$ and $(j_1, \ldots, j_n)$ are both permutations of $(1, \ldots, n)$, but *not* the same permutation. For example, consider the following 4×4 matrix:

$$\boldsymbol{P} = \begin{bmatrix} 0 & 0 & 1 & 0 \\ 1 & 0 & 0 & 0 \\ 0 & 0 & 0 & 1 \\ 0 & 1 & 0 & 0 \end{bmatrix} = [\boldsymbol{e}_2 : \boldsymbol{e}_4 : \boldsymbol{e}_1 : \boldsymbol{e}_3] = \begin{bmatrix} \boldsymbol{e}'_3 \\ \boldsymbol{e}'_1 \\ \boldsymbol{e}'_4 \\ \boldsymbol{e}'_2 \end{bmatrix}.$$

Here $\boldsymbol{P}$ is derived from $\boldsymbol{I}_4$ by permuting the latter's columns as $(j_1, j_2, j_3, j_4) = (2, 4, 1, 3)$ or, equivalently, by permuting the rows as $(i_1, i_2, i_3, i_4) = (3, 1, 4, 2)$. An alternative, but equivalent, definition of a permutation matrix that avoids "permutation" of rows or columns can be given as below:

Definition 1.13 *A* **permutation matrix** *is a square matrix whose every row and column contains precisely a single 1 with 0's everywhere else.*

What does a permutation matrix do to a general matrix? If $\boldsymbol{A}$ is an $m \times n$ matrix and $\boldsymbol{P}$ is a $m \times m$ permutation matrix obtained from $\boldsymbol{I}_m$ by permuting its rows as $(i_1, \ldots, i_m)$, then we can write

$$\boldsymbol{P}\boldsymbol{A} = \begin{bmatrix} \boldsymbol{e}'_{i_1} \\ \boldsymbol{e}'_{i_2} \\ \vdots \\ \boldsymbol{e}'_{i_m} \end{bmatrix} \boldsymbol{A} = \begin{bmatrix} \boldsymbol{e}'_{i_1}\boldsymbol{A} \\ \boldsymbol{e}'_{i_2}\boldsymbol{A} \\ \vdots \\ \boldsymbol{e}'_{i_m}\boldsymbol{A} \end{bmatrix} = \begin{bmatrix} \boldsymbol{a}'_{i_1 *} \\ \boldsymbol{a}'_{i_2 *} \\ \vdots \\ \boldsymbol{a}'_{i_m *} \end{bmatrix}$$

to see how *premultiplication* by $\boldsymbol{P}$ has permuted the *rows* of $\boldsymbol{A}$ through $(i_1, \ldots, i_m)$. On the other hand, if $\boldsymbol{P}$ is an $n \times n$ permutation matrix whose columns arise from permuting the columns of $\boldsymbol{I}_n$ through $(j_1, \ldots, j_n)$, then we have

$$\begin{aligned} \boldsymbol{A}\boldsymbol{P} &= \boldsymbol{A}[\boldsymbol{e}_{j_1} : \boldsymbol{e}_{j_2} : \ldots : \boldsymbol{e}_{j_n}] = [\boldsymbol{A}\boldsymbol{e}_{j_1} : \boldsymbol{e}_{j_2} : \ldots : \boldsymbol{A}\boldsymbol{e}_{j_n}] \\ &= [\boldsymbol{a}_{*j_1} : \cdots : \boldsymbol{a}_{*j_n}]. \end{aligned}$$

This shows how *post-multiplication* by $\boldsymbol{P}$ permutes the *columns* of $\boldsymbol{A}$ according to $(j_1, \ldots, j_n)$.

Example 1.15 Suppose we premultiply a 4×4 matrix $\boldsymbol{A}$ with the 4×4 permutation matrix $\boldsymbol{P}$ above to form

$$\boldsymbol{P}\boldsymbol{A} = \begin{bmatrix} 0 & 0 & 1 & 0 \\ 1 & 0 & 0 & 0 \\ 0 & 0 & 0 & 1 \\ 0 & 1 & 0 & 0 \end{bmatrix} \begin{bmatrix} 5 & 3 & 7 & 6 \\ 2 & 9 & 4 & 1 \\ 6 & 4 & 5 & 2 \\ 7 & 7 & 8 & 2 \end{bmatrix} = \begin{bmatrix} 6 & 4 & 5 & 2 \\ 5 & 3 & 7 & 6 \\ 7 & 7 & 8 & 2 \\ 2 & 9 & 4 & 1 \end{bmatrix}.$$

Note that premultiplication by $\boldsymbol{P}$ has simply permuted the rows of $\boldsymbol{A}$ according to the same permutation that led to $\boldsymbol{P}$ from the rows of $\boldsymbol{I}$. Next, suppose that we post-multiply $\boldsymbol{A}$ with $\boldsymbol{P}$. Then, we obtain

$$\boldsymbol{A}\boldsymbol{P} = \begin{bmatrix} 5 & 3 & 7 & 6 \\ 2 & 9 & 4 & 1 \\ 6 & 4 & 5 & 2 \\ 7 & 7 & 8 & 2 \end{bmatrix} \begin{bmatrix} 0 & 0 & 1 & 0 \\ 1 & 0 & 0 & 0 \\ 0 & 0 & 0 & 1 \\ 0 & 1 & 0 & 0 \end{bmatrix} = \begin{bmatrix} 3 & 6 & 5 & 7 \\ 9 & 1 & 2 & 4 \\ 4 & 2 & 6 & 5 \\ 7 & 2 & 7 & 8 \end{bmatrix}.$$

Postmultiplication by $\boldsymbol{P}$ has permuted the columns of $\boldsymbol{A}$ according to the same permutation that delivered $\boldsymbol{P}$ from the columns of $\boldsymbol{I}$.

From the above property, we can easily deduce that the product of two permutation matrices is again a permutation matrix. We state this as a theorem.

Theorem 1.6 *If $\boldsymbol{P}$ and $\boldsymbol{Q}$ are two $n \times n$ permutation matrices, then their product $\boldsymbol{P}\boldsymbol{Q}$ is also an $n \times n$ permutation matrix.*

Proof. Since $\boldsymbol{Q}$ is a permutation matrix, we can write $\boldsymbol{Q} = [\boldsymbol{e}_{*j_1} : \cdots : \boldsymbol{e}_{*j_n}]$. Post-multiplication by $\boldsymbol{Q}$ permutes the columns of $\boldsymbol{P}$ as $\boldsymbol{P}\boldsymbol{Q} = [\boldsymbol{P}_{*j_1} : \cdots : \boldsymbol{P}_{*j_n}]$. As the columns of $\boldsymbol{P}$ are a permutation of the identity matrix, the columns of $\boldsymbol{P}\boldsymbol{Q}$ are also a permutation of the identity matrix. Therefore, $\boldsymbol{P}\boldsymbol{Q}$ is a permutation matrix.
□

If we have permuted the rows (columns) of $\boldsymbol{A}$ by pre- (post-) multiplication with a permutation matrix $\boldsymbol{P}$, then how can we "undo" the operation and recover the original $\boldsymbol{A}$ matrix? The answer lies in the transpose of the permutation matrix as we see below.

Theorem 1.7 *Let $\boldsymbol{P}$ be an $n \times n$ permutation matrix. Then, $\boldsymbol{P}\boldsymbol{P}' = \boldsymbol{I} = \boldsymbol{P}'\boldsymbol{P}$.*

Proof. Writing $\boldsymbol{P}$ in terms of its row vectors and multiplying with the transpose, we have

$$\boldsymbol{P}\boldsymbol{P}' = \begin{bmatrix} \boldsymbol{e}'_{i_1} \\ \boldsymbol{e}'_{i_2} \\ \vdots \\ \boldsymbol{e}'_{i_n} \end{bmatrix} [\boldsymbol{e}_{i_1} : \boldsymbol{e}_{i_2} : \ldots : \boldsymbol{e}_{i_n}] = \begin{bmatrix} \boldsymbol{e}'_{i_1}\boldsymbol{e}_{i_1} & \boldsymbol{e}'_{i_1}\boldsymbol{e}_{i_2} & \ldots & \boldsymbol{e}'_{i_1}\boldsymbol{e}_{i_n} \\ \boldsymbol{e}'_{i_2}\boldsymbol{e}_{i_1} & \boldsymbol{e}'_{i_2}\boldsymbol{e}_{i_2} & \ldots & \boldsymbol{e}'_{i_2}\boldsymbol{e}_{i_n} \\ \vdots & \vdots & \ddots & \vdots \\ \boldsymbol{e}'_{i_n}\boldsymbol{e}_{i_1} & \boldsymbol{e}'_{i_n}\boldsymbol{e}_{i_2} & \ldots & \boldsymbol{e}'_{i_n}\boldsymbol{e}_{i_n} \end{bmatrix},$$

which is equal to the identity matrix $\boldsymbol{I}_n$ since $\boldsymbol{e}'_k\boldsymbol{e}_l$ equals 1 if $k = l$ and 0 otherwise.

The identity $\boldsymbol{P}'\boldsymbol{P} = \boldsymbol{I}$ follows in analogous fashion by expressing $\boldsymbol{P}$ in terms of its column vectors. □

The above result shows that if we have permuted the rows (columns) of a matrix $\boldsymbol{A}$ by pre- (post-) multiplication with a permutation matrix $\boldsymbol{P}$, then pre- (post-) multiplication with $\boldsymbol{P}'$ will again yield $\boldsymbol{A}$. We say that $\boldsymbol{P}'$ is the *inverse* of $\boldsymbol{P}$. We will learn more about inverse matrices in the subsequent chapters.

1.6.2 *Triangular matrices*

Definition 1.14 *A* **triangular matrix** *is a square matrix where the entries either below or above the main diagonal are zero. Triangular matrices look like*

$$\boldsymbol{L} = \begin{bmatrix} l_{11} & 0 & 0 & \cdots & 0 \\ l_{21} & l_{22} & 0 & \cdots & 0 \\ l_{31} & l_{32} & l_{33} & \cdots & 0 \\ \vdots & \vdots & \vdots & \ddots & \vdots \\ l_{n1} & l_{n2} & l_{n3} & \cdots & l_{nn} \end{bmatrix} \text{ and } \boldsymbol{U} = \begin{bmatrix} u_{11} & u_{12} & u_{13} & \cdots & u_{1n} \\ 0 & u_{22} & u_{23} & \cdots & u_{2n} \\ 0 & 0 & u_{33} & \cdots & u_{3n} \\ \vdots & \vdots & \vdots & \ddots & \vdots \\ 0 & 0 & 0 & \cdots & u_{nn} \end{bmatrix}.$$

A square matrix $\boldsymbol{L} = \{l_{ij}\}$ is called **lower-triangular** *if all its elements above the diagonal are zero, i.e., $l_{ij} = 0$ for all $j > i$. A square matrix $\boldsymbol{U} = \{u_{ij}\}$ is called* **upper-triangular** *if all its elements below the diagonal are zero, i.e., $u_{ij} = 0$ for all $i > j$.*

Clearly, if $\boldsymbol{L}$ is lower-triangular, then its transpose, $\boldsymbol{L}'$ is upper-triangular. Analogously, if $\boldsymbol{U}$ is upper-triangular, then its transpose, $\boldsymbol{U}'$, is lower-triangular. The above structures immediately reveal that the sum of two upper-triangular matrices is an upper-triangular matrix and that of two lower-triangular matrices is another lower-triangular matrix.

Products of triangular matrices are again triangular matrices.

Theorem 1.8 *Let $\boldsymbol{A}$ and $\boldsymbol{B}$ be square matrices of the same order.*

(i) If $\boldsymbol{A}$ and $\boldsymbol{B}$ are lower-triangular then $\boldsymbol{AB}$ is lower-triangular.
(ii) If $\boldsymbol{A}$ and $\boldsymbol{B}$ are upper-triangular then $\boldsymbol{AB}$ is upper-triangular.

Proof. Let $\boldsymbol{A}$ and $\boldsymbol{B}$ be $n \times n$ lower-triangular matrices. Then, the (i, j)-th element of $\boldsymbol{AB}$ is given by

$$(\boldsymbol{AB})_{ij} = \sum_{k=1}^{n} a_{ik}b_{kj} = \sum_{k=1}^{i} a_{ik}b_{kj} + \sum_{k=i+1}^{n} a_{ik}b_{kj}.$$

Suppose $i < j$. Then, since $\boldsymbol{B}$ is lower-triangular, $b_{kj} = 0$ for each k in the first sum as $k \leq i < j$. Hence, the first term is zero. Now look at the second term. Note that $a_{ik} = 0$ for all $k > i$ because $\boldsymbol{A}$ is lower-triangular. This means that the second term is also zero. Therefore, the (i, j)-th element of $\boldsymbol{AB}$ is zero for all $i < j$; hence $\boldsymbol{AB}$ is lower-triangular.

The proof for upper-triangular matrices can be constructed on similar lines or, alternatively, can be deduced from Part (i) using transposes. □

Sometimes operations with structured matrices, such as triangular matrices, are best revealed using a symbolic or schematic representation, where a $*$ denotes any entry that is not necessarily zero. Such schematic representations are very useful when interest resides with what entries are necessarily zero, rather than what their actual values are. They are often useful in deriving results involving operations with triangular matrices. For example, Part (ii) of Theorem 1.8 can be easily visualized using 4×4 matrices as below:

$$\boldsymbol{AB} = \begin{bmatrix} * & * & * & * \\ 0 & * & * & * \\ 0 & 0 & * & * \\ 0 & 0 & 0 & * \end{bmatrix} \begin{bmatrix} * & * & * & * \\ 0 & * & * & * \\ 0 & 0 & * & * \\ 0 & 0 & 0 & * \end{bmatrix} = \begin{bmatrix} * & * & * & * \\ 0 & * & * & * \\ 0 & 0 & * & * \\ 0 & 0 & 0 & * \end{bmatrix}.$$

The pattern of the 0's and the $*$'s in the matrices in the left-hand side results in the upper-triangular structure on the right-hand side. For example, to check if the $(3, 2)$-th element of $\boldsymbol{AB}$ is 0 or $*$ (some nonzero entry), we multiply the third row vector of $\boldsymbol{A}$ with the second column vector of $\boldsymbol{B}$. Since the first two entries in the third row of $\boldsymbol{A}$ are zero and the last two entries in the second column of $\boldsymbol{B}$ are zero, the product is 0. These schematic presentations are often more useful than constructing formal proofs such as in Theorem 1.8.

If the diagonal elements of a triangular matrix are all zero, we call it *strictly* triangular. The following is an interesting property of strictly triangular matrices.

Lemma 1.2 *Let $\boldsymbol{A}$ be an $n \times n$ triangular matrix with all its diagonal entries equal to zero. Then $\boldsymbol{A}^n = \boldsymbol{O}$.*

Proof. This can be established by direct verification, but here is an argument using induction. Consider upper-triangular matrices. The result is trivially true for $n = 1$. For $n > 1$, assume that the result holds for all $(n-1) \times (n-1)$ upper-triangular matrices with zeroes along the diagonal. Let $\boldsymbol{A}$ be an $n \times n$ upper-triangular matrix partitioned as

$$\boldsymbol{A} = \begin{bmatrix} 0 & \boldsymbol{u}' \\ \boldsymbol{0} & \boldsymbol{B} \end{bmatrix},$$

where $\boldsymbol{B}$ is an $(n-1) \times (n-1)$ triangular matrix with all its diagonal entries equal to zero. It is easy to see that

$$\boldsymbol{A}^n = \begin{bmatrix} 0 & \boldsymbol{u}'\boldsymbol{B}^{n-1} \\ \boldsymbol{0} & \boldsymbol{B}^n \end{bmatrix} = \begin{bmatrix} 0 & \boldsymbol{0}' \\ \boldsymbol{0} & \boldsymbol{O} \end{bmatrix},$$

where the last equality holds because $\boldsymbol{B}^{n-1} = \boldsymbol{O}$ by the induction hypothesis. □

Any square matrix $\boldsymbol{A}$ (not necessarily triangular) that satisfies $\boldsymbol{A}^k = \boldsymbol{O}$ for some positive integer k is said to be ***nilpotent***. Lemma 1.2 reveals that triangular matrices whose diagonal entries are all zero are nilpotent.

1.6.3 Hessenberg matrices

Triangular matrices play a very important role in numerical linear algebra. Most algorithms for solving general systems of linear equations rely upon reducing them to triangular systems. We will see more of these in the subsequent chapters. If, however, the constraints of an algorithm do not allow a general matrix to be conveniently reduced to a triangular form, reduction to matrices that are *almost* triangular often prove to be the next best thing. A particular type, known as *Hessenberg matrices*, is especially important in numerical linear algebra.

Definition 1.15 *A* **Hessenberg matrix** *is a square matrix that is "almost" triangular in that it has zero entries either above the super-diagonal (lower-Hessenberg) or below the sub-diagonal (upper-Hessenberg). A square matrix* $\boldsymbol{H} = \{h_{ij}\}$ *is* **lower-Hessenberg** *matrix if* $h_{ij} = 0$ *for all* $j > i + 1$. *It is* **upper-Hessenberg** *if* $h_{ij} = 0$ *for all* $i > j + 1$.

Below are symbolic displays of a lower- and an upper-Hessenberg matrix (5×5)

$$\begin{bmatrix} * & * & 0 & 0 & 0 \\ * & * & * & 0 & 0 \\ * & * & * & * & 0 \\ * & * & * & * & * \\ * & * & * & * & * \end{bmatrix} \text{ and } \begin{bmatrix} * & * & * & * & * \\ * & * & * & * & * \\ 0 & * & * & * & * \\ 0 & 0 & * & * & * \\ 0 & 0 & 0 & * & * \end{bmatrix}, \text{ respectively.}$$

If $\boldsymbol{H}$ is lower-Hessenberg, then $\boldsymbol{H}'$ is upper-Hessenberg. If $\boldsymbol{H}$ is upper-Hessenberg, then $\boldsymbol{H}'$ is lower-Hessenberg. If $\boldsymbol{H}$ is symmetric, so that $\boldsymbol{H} = \boldsymbol{H}'$, then $h_{ij} = 0$

whenever $j > i+1$ or $i > j+1$. Such a matrix is called ***tridiagonal***. Below is the structure of a 5×5 tridiagonal matrix:

$$\begin{bmatrix} * & * & 0 & 0 & 0 \\ * & * & * & 0 & 0 \\ 0 & * & * & * & 0 \\ 0 & 0 & * & * & * \\ 0 & 0 & 0 & * & * \end{bmatrix}.$$

A tridiagonal matrix that has its subdiagonal entries equal to zero is called an ***upper-bidiagonal matrix***. A tridiagonal matrix that has its superdiagonal entries equal to zero is called a ***lower bidiagonal matrix***. These look like

$$\begin{bmatrix} * & * & 0 & 0 & 0 \\ 0 & * & * & 0 & 0 \\ 0 & 0 & * & * & 0 \\ 0 & 0 & 0 & * & * \\ 0 & 0 & 0 & 0 & * \end{bmatrix} \text{ and } \begin{bmatrix} * & 0 & 0 & 0 & 0 \\ * & * & 0 & 0 & 0 \\ 0 & * & * & 0 & 0 \\ 0 & 0 & * & * & 0 \\ 0 & 0 & 0 & * & * \end{bmatrix}, \text{ respectively.}$$

A triangular matrix is also a Hessenberg matrix because it satisfies the conditions in Definition 1.15. An upper-triangular matrix has all entries below the diagonal equal to zero, so all its entries below the subdiagonal are zero. In a lower-triangular matrix all entries above the diagonal, hence above the superdiagonal as well, are zero.

The product of two Hessenberg matrices need not be a Hessenberg matrix, as seen below with upper-Hessenberg matrices:

$$\begin{bmatrix} * & * & * & * \\ * & * & * & * \\ 0 & * & * & * \\ 0 & 0 & * & * \end{bmatrix} \begin{bmatrix} * & * & * & * \\ * & * & * & * \\ 0 & * & * & * \\ 0 & 0 & * & * \end{bmatrix} = \begin{bmatrix} * & * & * & * \\ * & * & * & * \\ * & * & * & * \\ 0 & * & * & * \end{bmatrix}.$$

However, multiplication by an upper- (lower-) triangular matrix, either from the left or from the right, does not alter the upper (lower) Hessenberg form. We have the following theorem.

Theorem 1.9 *If $\boldsymbol{H}$ is upper (lower) Hessenberg and $\boldsymbol{T}$ is upper (lower) triangular, then $\boldsymbol{TH}$ and $\boldsymbol{HT}$ are both upper (lower) Hessenberg whenever these products are well-defined.*

Proof. Let $\boldsymbol{T} = \{t_{ij}\}$ and $\boldsymbol{H} = \{h_{ij}\}$ be $n \times n$ upper-triangular and upper-Hessenberg, respectively.

Let i and j be integers between 1 and n, such that $i > j+1 \geq 2$. The (i,j)-th element of $\boldsymbol{TH}$ is

$$(\boldsymbol{TH})_{ij} = \sum_{k=1}^{n} t_{ik}h_{kj} = \sum_{k=1}^{i-1} t_{ik}h_{kj} + \sum_{k=i}^{n} t_{ik}h_{kj}.$$

Since $\boldsymbol{T}$ is upper-triangular, $t_{ik} = 0$ for every $k < i$. Hence, the first sum is zero.

Now look at the second sum. Since $\boldsymbol{H}$ is upper-Hessenberg, $h_{kj} = 0$ for every $k \geq i > j+1$. This means that the second sum is also zero. Therefore, the (i,j)-th element of $\boldsymbol{TH}$ is zero for all $j > i+1$, so $\boldsymbol{TH}$ is upper-Hessenberg.

Next, consider the (i,j)-th element of the product $\boldsymbol{HT}$, where $i > j+1 \geq 2$:

$$(\boldsymbol{HT})_{ij} = \sum_{k=1}^{n} h_{ik}t_{kj} = \sum_{k=1}^{i-2} h_{ik}t_{kj} + \sum_{k=i-1}^{n} h_{ik}t_{kj}.$$

The first sum is zero because $h_{ik} = 0$ for every $k \leq i-2$ (or, equivalently, $i > k+1$) since $\boldsymbol{H}$ is upper-Hessenberg. Since $\boldsymbol{T}$ is upper-triangular, $t_{kj} = 0$ for every $k > j$, which happens throughout the second sum because $k \geq i-1 > j$. Hence, the second sum is also zero and $\boldsymbol{HT}$ is upper-Hessenberg. □

If $\boldsymbol{H}$ is upper- (lower-) Hessenberg and $\boldsymbol{T}_1$ and $\boldsymbol{T}_2$ are any two upper- (lower-) triangular matrices, then Theorem 1.9 implies that $\boldsymbol{T}_1\boldsymbol{H}\boldsymbol{T}_2$ is upper (lower) Hessenberg. Symbolically, this is easily seen for 4×4 structures:

$$\begin{bmatrix} * & * & * & * \\ 0 & * & * & * \\ 0 & 0 & * & * \\ 0 & 0 & 0 & * \end{bmatrix} \begin{bmatrix} * & * & * & * \\ * & * & * & * \\ 0 & * & * & * \\ 0 & 0 & * & * \end{bmatrix} \begin{bmatrix} * & * & * & * \\ 0 & * & * & * \\ 0 & 0 & * & * \\ 0 & 0 & 0 & * \end{bmatrix} = \begin{bmatrix} * & * & * & * \\ * & * & * & * \\ 0 & * & * & * \\ 0 & 0 & * & * \end{bmatrix}.$$

1.6.4 Sparse matrices

Diagonal, triangular and Hessenberg matrices are examples of matrices with *structural zeros*. This means that certain entries are stipulated to be 0 by their very definition. In general, we say that a matrix is ***sparse*** if its entries are populated primarily with zeros. If the majority of entries are not necessarily zero, then it is common to refer to the matrix as ***dense***.

There is usually no exact percentage (or number) of zero entries that qualifies a matrix to be treated as sparse. This usually depends upon the specific application and whether treating a matrix as sparse will in fact deliver computational benefits. Specialized algorithms and data structures that exploit the sparse structure of the matrix are often useful, and sometimes even necessary, for storage and numerical computation. These algorithms are different from those used for dense matrices, which do not attempt to exploit structural zeroes. For example, consider storing a sparse matrix in a computer. While an $n \times n$ unstructured dense matrix is usually requires storage of all its n^2 entries, one only needs to store the nonzero entries of a sparse matrix.

There are several storage mechanism for sparse matrices that are designed based upon the specific operations that require the elements in the matrix need to be accessed. We describe one such storage mechanism, called the ***compressed row storage***

(CRS). Consider the following 4×4 sparse matrix:

$$A = \begin{bmatrix} 1 & 0 & 0 & 2 \\ 0 & 3 & 4 & 5 \\ 6 & 0 & 7 & 0 \\ 0 & 0 & 0 & 8 \end{bmatrix} .$$

There are 8 nonzero elements in the matrix. CRS uses three "lists" to store $\boldsymbol{A}$. The first is list collects all the nonzero values in $\boldsymbol{A}$, while the second list stores the corresponding columns where the nonzero values arise. These are called `values` and `column index`, respectively, and are displayed below for $\boldsymbol{A}$:

values	1	2	3	4	5	6	7	8
column index	1	4	2	3	4	1	3	4

Then, a third list, often called `row pointer`, lists the positions in `values` that are occupied by the first element in each row. For $\boldsymbol{A}$, this is given by

row pointer	1	3	6	8

Thus, the first, third, sixth and eighth elements in `values` are the elements appearing as the first nonzero entry in a row. It is easy to see that the matrix $\boldsymbol{A}$ can be reconstructed from the three lists `values`, `column index` and `row pointer`.

CRS is efficient if we want to compute matrix-vector multiplications of the form $\boldsymbol{u} = \boldsymbol{A}\boldsymbol{x}$ when $\boldsymbol{A}$ is sparse because the required nonzero entries in the rows that contribute to the product can be effectively extracted from these three lists. On the other hand, if one wanted to compute $\boldsymbol{v} = \boldsymbol{A}'\boldsymbol{x}$, then CRS would be less efficient and one could construct a *compressed column storage* (CCS) mechanism, which is built by essentially changing the roles of the columns and rows.

1.6.5 Banded matrices

Definition 1.16 *A* **band matrix** *or* **banded matrix** *is a particular type of sparse matrix, where all its entries outside a diagonally bordered band are zero. More formally, an $n \times n$ matrix $\boldsymbol{A} = \{a_{ij}\}$ is a band matrix if there are integers $p_1 \geq 0$ and $p_2 \geq 0$ such that*

$$a_{ij} = 0 \text{ whenever } i < j + p_1 \text{ or } j > i + p_2 .$$

The integers p_1 and p_2 are called the **lower bandwidth** *and* **upper bandwidth** *respectively. The* **bandwidth** *of the matrix is $p_1 + p_2 + 1$. Put differently, the bandwidth is the smallest number of adjacent diagonals to which the nonzero elements are confined.*

Band matrices need not be square matrices. A 6×5 band matrix with lower bandwidth $p_1 = 1$ and upper bandwidth $p_2 = 2$ has the following structure:

$$\begin{bmatrix} * & * & * & 0 & 0 \\ * & * & * & * & 0 \\ 0 & * & * & * & * \\ 0 & 0 & * & * & * \\ 0 & 0 & 0 & * & * \\ 0 & 0 & 0 & 0 & * \end{bmatrix}.$$

The bandwidth is $p_1 + p_2 + 1 = 4$, which can be regarded as the length of the "band" beyond which all entries are zero.

Examples of banded matrices abound in linear algebra. The following are special cases of band matrices:

1. A diagonal matrix has lower bandwidth and upper bandwidth both equal to zero. Its bandwidth is 1.
2. An $n \times n$ upper-triangular matrix has lower bandwidth 0 and upper bandwidth equal to $n - 1$.
3. An $n\times n$ lower-triangular matrix has lower bandwidth $n-1$ and upper bandwidth 0.
4. An $n \times n$ tridiagonal matrix has lower and upper bandwidth both equal to 1.
5. An $n \times n$ upper-Hessenberg matrix has lower bandwidth 1 and upper bandwidth $n - 1$.
6. An $n \times n$ lower-Hessenberg matrix has lower bandwidth $n - 1$ and upper bandwidth 1.
7. An $n\times n$ upper-bidiagonal matrix has lower bandwidth 0 and upper bandwidth 1.
8. An $n\times n$ lower-bidiagonal matrix has lower bandwidth 1 and upper bandwidth 0.

Band matrices are also convenient to characterize *rectangular* versions of some of the matrices we have seen before. For example, an $m \times n$ upper-triangular matrix is defined as a band matrix with lower bandwidth 0 and upper bandwidth $n - 1$. Here is what a 4×5 upper-triangular matrix would look like:

$$\begin{bmatrix} * & * & * & * & * \\ 0 & * & * & * & * \\ 0 & 0 & * & * & * \\ 0 & 0 & 0 & * & * \end{bmatrix}.$$

Such matrices are often called ***trapezoidal matrices***. Table 1.6.5 shows how some of the more widely used sparse and structured matrices are defined using band terminology, assuming the matrices are $m \times n$.

Type of matrix	Lower bandwidth	Upper bandwidth
Diagonal	0	0
Upper triangular	0	$n-1$
Lower triangular	$m-1$	0
Tridiagonal	1	1
Upper Hessenberg	1	$n-1$
Lower Hessenberg	$m-1$	1
Upper bidiagonal	0	1
Lower bidiagonal	1	0

Table 1.1 *Characterization of some common matrices using bandwidths.*

1.7 Exercises

1. Consider the following matrices:

$$\boldsymbol{A} = \begin{bmatrix} 1 & -9 \\ 2 & 8 \\ 3 & -7 \\ 4 & 6 \end{bmatrix}, \quad \boldsymbol{B} = \begin{bmatrix} 1 & 2 \\ 1 & 1 \\ 0 & 4 \\ 1 & -1 \end{bmatrix} \text{ and } \boldsymbol{u} = \begin{bmatrix} -2 \\ 3 \\ -4 \\ 1 \end{bmatrix}.$$

Which of the following operations are well-defined: $\boldsymbol{A}+\boldsymbol{B}$, $\boldsymbol{AB}$, $\boldsymbol{A}'\boldsymbol{B}$, $\boldsymbol{B}'\boldsymbol{A}$, $\boldsymbol{Au}$, $\boldsymbol{Bu}$, $\boldsymbol{u}'\boldsymbol{A}$ and $\boldsymbol{u}'\boldsymbol{B}$? Find the resulting matrices when they are well-defined.

2. Find w, x, y and z in the following:

$$2\begin{bmatrix} x+5 & y-3 \\ 5 & 7+2z \end{bmatrix} = \begin{bmatrix} 2 & 3-y \\ w+7 & 14+z \end{bmatrix}.$$

3. For each of the following pairs of matrices, find $\boldsymbol{AB}$ and $\boldsymbol{BA}$:

(a) $\boldsymbol{A} = \begin{bmatrix} 0 & 0 \\ 0 & 1 \end{bmatrix}$ and $\boldsymbol{B} = \begin{bmatrix} 0 & 1 \\ 0 & 0 \end{bmatrix}$.

(b) $\boldsymbol{A} = \begin{bmatrix} 1 & 0 & 0 \\ 0 & 2 & 0 \\ 0 & 0 & 3 \end{bmatrix}$ and $\boldsymbol{B} = \begin{bmatrix} 1 & -1 & 0 \\ 2 & 3 & -2 \\ 7 & 5 & 3 \end{bmatrix}$.

(c) $\boldsymbol{A} = \begin{bmatrix} 1 & 2 & 3 \\ 0 & 4 & 5 \\ 0 & 0 & 6 \end{bmatrix}$ and $\boldsymbol{B} = \begin{bmatrix} 7 & 8 & 9 \\ 0 & 1 & 2 \\ 0 & 0 & 3 \end{bmatrix}$.

(d) $\boldsymbol{A} = \begin{bmatrix} 0 & 0 & 0 & a_{14} \\ 0 & 0 & a_{23} & 0 \\ 0 & a_{32} & 0 & 0 \\ a_{41} & 0 & 0 & 0 \end{bmatrix}$ and $\boldsymbol{B} = \begin{bmatrix} 0 & 0 & 0 & b_{14} \\ 0 & 0 & b_{23} & 0 \\ 0 & b_{32} & 0 & 0 \\ b_{41} & 0 & 0 & 0 \end{bmatrix}$.

4. For each of the matrices in the above exercise, verify that $(\boldsymbol{AB})' = \boldsymbol{B}'\boldsymbol{A}'$.

5. True or false: If $\boldsymbol{A}$ is a nonzero real matrix, then $\boldsymbol{AA}' \neq \boldsymbol{O}$.

6. True or false: $(\boldsymbol{A}'\boldsymbol{A})' = \boldsymbol{A}'\boldsymbol{A}$.

7. True or false: If $\boldsymbol{AB} = \boldsymbol{O}$, then either $\boldsymbol{A} = \boldsymbol{O}$ or $\boldsymbol{B} = \boldsymbol{O}$.

8. Find $\boldsymbol{uv}'$ and $\boldsymbol{u}'\boldsymbol{v}$, where

$$\boldsymbol{u} = \begin{bmatrix} 1 \\ -2 \\ 5 \end{bmatrix} \text{ and } \boldsymbol{v} = \begin{bmatrix} 3 \\ -6 \\ 4 \end{bmatrix}.$$

9. Let $\boldsymbol{A}$ be $m \times n$ and $\boldsymbol{B}$ be $n \times p$. How many floating point operations (i.e., scalar multiplications and additions) are needed to compute $\boldsymbol{AB}$?

10. Suppose we want to compute $\boldsymbol{xy}'\boldsymbol{z}$, where

$$\boldsymbol{x} = \begin{bmatrix} 1 \\ 2 \\ 0 \end{bmatrix}, \quad \boldsymbol{y} = \begin{bmatrix} 3 \\ 0 \\ 4 \\ 5 \end{bmatrix} \text{ and } \boldsymbol{z} = \begin{bmatrix} 0 \\ 6 \\ 7 \\ 8 \end{bmatrix}.$$

We can do this in two ways. The first method computes the matrix $\boldsymbol{xy}'$ first and then multiplies it with the vector $\boldsymbol{z}$. The second method computes the scalar $\boldsymbol{y}'\boldsymbol{z}$ and multiplies it with the vector $\boldsymbol{x}$. Which one requires fewer floating point operations (i.e., scalar multiplications and additions)?

11. Let $\boldsymbol{A}$ be $m \times n$ and $\boldsymbol{B}$ be $n \times p$. How many different ways can we compute the scalar $\boldsymbol{x}'\boldsymbol{ABy}$, where $\boldsymbol{x}$ and $\boldsymbol{y}$ are $m \times 1$ and $p \times 1$, respectively? What is the cheapest (in terms of floating point operations) way to evaluate this?

12. Let $\boldsymbol{A}$ be $m \times n$, $\boldsymbol{x}$ is $n \times 1$, $\boldsymbol{y}$ is $p \times 1$ and $\boldsymbol{B}$ is $p \times q$. What is the simplest way to compute $\boldsymbol{Axy}'\boldsymbol{B}$?

13. Let $\boldsymbol{A} = \{a_{ij}\}$ be an $m \times n$ matrix. Suppose we stack the column vectors of $\boldsymbol{A}$ one above the other to form the $mn \times 1$ vector $\boldsymbol{x}$ and we stack the rows of $\boldsymbol{A}$ one above the other to form the $mn \times 1$ vector $\boldsymbol{y}$. What position does a_{ij} occupy in $\boldsymbol{x}$ and $\boldsymbol{y}$?

14. Let $\boldsymbol{C} = \{c_{ij}\}$ be an $n \times n$ matrix, where each $c_{ij} = r_{ij}d_i d_j$. Show how you can write $\boldsymbol{C} = \boldsymbol{DRD}$, where $\boldsymbol{D}$ is an $n \times n$ diagonal matrix.

15. Find all matrices that commute with the diagonal matrix

$$\boldsymbol{D} = \begin{bmatrix} a_{11} & 0 & 0 \\ 0 & a_{22} & 0 \\ 0 & 0 & a_{33} \end{bmatrix},$$

where a_{11}, a_{22} and a_{33} are all different nonzero real numbers.

16. Show that $(\boldsymbol{A} + \boldsymbol{B})^2 = \boldsymbol{A}^2 + 2\boldsymbol{AB} + \boldsymbol{B}^2$ if and only if $\boldsymbol{AB} = \boldsymbol{BA}$.

17. Show that $(\boldsymbol{A} - \boldsymbol{B})(\boldsymbol{A} + \boldsymbol{B}) = \boldsymbol{A}^2 - \boldsymbol{B}^2$ if and only if $\boldsymbol{AB} = \boldsymbol{BA}$.

18. If $\boldsymbol{AB} = \boldsymbol{BA}$, prove that $\boldsymbol{A}^k\boldsymbol{B}^m = \boldsymbol{B}^m\boldsymbol{A}^k$ for all integers k and m.

19. If $\boldsymbol{AB} = \boldsymbol{BA}$, prove that for any positive integer k, there exists a matrix $\boldsymbol{C}$ such that $\boldsymbol{A}^k - \boldsymbol{B}^k = (\boldsymbol{A} - \boldsymbol{B})\boldsymbol{C}$.

20. If $\boldsymbol{A}$ is a square matrix, then $\boldsymbol{A} + \boldsymbol{A}'$ is a symmetric matrix.

21. A square matrix is said to be ***skew-symmetric*** if $\boldsymbol{A} = -\boldsymbol{A}'$. Show that $\boldsymbol{A} - \boldsymbol{A}'$ is skew-symmetric.

22. Prove that there is one and only one way to express a square matrix as the sum of a symmetric matrix and a skew-symmetric matrix.

23. Use a convenient partitioning and block multiplication to evaluate $\boldsymbol{AB}$, where

$$\boldsymbol{A} = \begin{bmatrix} 2 & 1 & -1 & 0 & 0 & 0 \\ 1 & 0 & 0 & 1 & 0 & 5 \\ 0 & 1 & 1 & 0 & 1 & -2 \end{bmatrix} \text{ and } \boldsymbol{B} = \begin{bmatrix} 1 & 0 & 0 & 0 \\ 0 & 1 & 0 & 0 \\ 1 & 2 & 3 & 4 \\ 2 & -1 & 3 & 0 \\ -1 & 3 & 0 & 1 \\ 0 & 1 & 0 & 0 \end{bmatrix}.$$

24. Let $\boldsymbol{M} = \begin{bmatrix} \boldsymbol{A} & \boldsymbol{B} \\ \boldsymbol{C} & \boldsymbol{D} \end{bmatrix}$ be a block matrix of order $m \times n$, where $\boldsymbol{A}$ is $k \times l$. What are the sizes of $\boldsymbol{B}$, $\boldsymbol{C}$ and $\boldsymbol{D}$?

25. True or false: $\boldsymbol{AB} + \boldsymbol{CD} = [\boldsymbol{A} : \boldsymbol{B}] \begin{bmatrix} \boldsymbol{C} \\ \boldsymbol{D} \end{bmatrix}$.

26. Find the following products for conformably partitioned matrices:

(a) $\begin{bmatrix} \boldsymbol{A} & \boldsymbol{B} \\ \boldsymbol{C} & \boldsymbol{D} \end{bmatrix} \begin{bmatrix} \boldsymbol{M} & \boldsymbol{O} \\ \boldsymbol{O} & \boldsymbol{O} \end{bmatrix} \begin{bmatrix} \boldsymbol{P} & \boldsymbol{Q} \\ \boldsymbol{R} & \boldsymbol{S} \end{bmatrix}$;

(b) $\begin{bmatrix} \boldsymbol{A} \\ \boldsymbol{C} \end{bmatrix} \boldsymbol{D} [\boldsymbol{P} : \boldsymbol{Q}]$;

(c) $\begin{bmatrix} \boldsymbol{A} & \boldsymbol{B} \\ \boldsymbol{O} & \boldsymbol{D} \end{bmatrix} \begin{bmatrix} \boldsymbol{P} & \boldsymbol{Q} \\ \boldsymbol{O} & \boldsymbol{S} \end{bmatrix}$.

27. What would be the best way to partition $\boldsymbol{A}$ and $\boldsymbol{C}$ to evaluate $\boldsymbol{ABC}$, where $\boldsymbol{B} = \begin{bmatrix} \boldsymbol{B}_{11} & \boldsymbol{O} \\ \boldsymbol{O} & \boldsymbol{B}_{22} \end{bmatrix}$?

28. Consider the conformably partitioned matrices

$$\boldsymbol{A} = \begin{bmatrix} \boldsymbol{I} & \boldsymbol{O} \\ \boldsymbol{C} & \boldsymbol{I} \end{bmatrix} \text{ and } \boldsymbol{B} = \begin{bmatrix} \boldsymbol{I} & \boldsymbol{O} \\ -\boldsymbol{C} & \boldsymbol{I} \end{bmatrix},$$

so that $\boldsymbol{AB}$ and $\boldsymbol{BA}$ are well-defined. Verify that $\boldsymbol{AB} = \boldsymbol{BA} = \boldsymbol{I}$ for every $\boldsymbol{C}$.

29. A square matrix $\boldsymbol{P}$ is called ***idempotent*** if $\boldsymbol{P}^2 = \boldsymbol{P}$. Find $\boldsymbol{A}^{500}$, where

$$\boldsymbol{A} = \begin{bmatrix} \boldsymbol{I} & \boldsymbol{P} \\ \boldsymbol{O} & \boldsymbol{P} \end{bmatrix} \text{ and } \boldsymbol{P} \text{ is idempotent.}$$

30. Find the trace of the matrix

$$\boldsymbol{A} = \begin{bmatrix} 8 & 1 & 6 \\ 3 & 5 & 7 \\ 4 & 9 & 2 \end{bmatrix}.$$

Verify that $\boldsymbol{a}_{i*}'\boldsymbol{1} = \boldsymbol{a}_{*i}'\boldsymbol{1} = \text{tr}(\boldsymbol{A})$ for each $i = 1, 2, 3$ and that the sum of the elements in the "backward diagonal," i.e., $a_{13} + a_{22} + a_{31}$, is also equal to $\text{tr}(\boldsymbol{A})$.

31. Let $A = \begin{bmatrix} a & b \\ c & d \end{bmatrix}$. The ***determinant*** of A is defined to be the scalar $\det(A) = ad - bc$. Show that $A^2 - \text{tr}(A)A + \det(A)I_2 = O$.
32. Prove that $\text{tr}(A'B) = \text{tr}(AB')$.
33. Find matrices A, B and C such that $\text{tr}(ABC) \neq \text{tr}(BAC)$.
34. True or false: $x'Ax = \text{tr}(Axx')$, where x is $n \times 1$ and A is $n \times n$.
35. Find PA, AP and PAP without any explicit matrix multiplication, where

$$P = \begin{bmatrix} 0 & 1 & 0 \\ 0 & 0 & 1 \\ 1 & 0 & 0 \end{bmatrix} \text{ and } A = \begin{bmatrix} 1 & 2 & 3 \\ 4 & 5 & 6 \\ 7 & 8 & 9 \end{bmatrix}.$$

Hint: Note that P is a permutation matrix.

36. Verify that

$$\begin{bmatrix} l_{11} & 0 \\ l_{21} & l_{22} \end{bmatrix} = \begin{bmatrix} l_{11} & 0 \\ l_{21} & 1 \end{bmatrix} \begin{bmatrix} 1 & 0 \\ 0 & l_{22} \end{bmatrix}.$$

Also verify that

$$\begin{bmatrix} l_{11} & 0 & 0 \\ l_{21} & l_{22} & 0 \\ l_{31} & l_{32} & l_{33} \end{bmatrix} = \begin{bmatrix} l_{11} & 0 & 0 \\ l_{21} & 1 & 0 \\ l_{31} & 0 & 1 \end{bmatrix} \begin{bmatrix} 1 & 0 & 0 \\ 0 & l_{22} & 0 \\ 0 & l_{32} & 1 \end{bmatrix} \begin{bmatrix} 1 & 0 & 0 \\ 0 & 1 & 0 \\ 0 & 0 & l_{33} \end{bmatrix}.$$

In general verify that if L is an $n \times n$ lower-triangular matrix, then $L = L_1 L_2 \cdots L_n$, where L_i is the $n \times n$ matrix obtained by replacing the i-th column of I_n by the i-th column of L. Derive an analogous result for upper-triangular matrices.

37. A matrix $T = \{T_{ij}\}$ is said to be ***block upper-triangular*** if the row and column partition have the same number of blocks, each diagonal block T_{ii} is upper-triangular and each block below the diagonal is the zero matrix, i.e., $T_{ij} = O$ whenever $i > j$. Here is an example of a block upper-triangular matrix

$$T = \begin{bmatrix} T_{11} & T_{12} & T_{13} \\ O & T_{22} & T_{23} \\ O & O & T_{33} \end{bmatrix},$$

where each T_{ii} is a square matrix for $i = 1, 2, 3$. Prove the following statements for block upper-triangular matrices A and B:

(a) $A + B$ and AB are block upper-triangular, whenever the sum and product are well-defined.

(b) αA is also block-triangular for all scalars α.

(c) If every diagonal block A_{ii} in A is zero, then $A^m = O$, where m is the number of blocks in the row (and column) partitions.

CHAPTER 2

Systems of Linear Equations

2.1 Introduction

Matrices play an indispensable role in the study of systems of linear equations. A system of linear equations is a system of the form:

$$\begin{array}{ccccccccc} a_{11}x_1 & + & a_{12}x_2 & + & \dots & + & a_{1n}x_n & = & b_1, \\ a_{21}x_1 & + & a_{22}x_2 & + & \dots & + & a_{2n}x_n & = & b_2, \\ & & & & \vdots & & & & \\ a_{m1}x_1 & + & a_{m2}x_2 & + & \dots & + & a_{mn}x_n & = & b_m. \end{array} \tag{2.1}$$

Here the a_{ij}'s and the b_i's are known scalars (constants), while the x_j's are unknown scalars (variables). The a_{ij}'s are called the ***coefficients*** of the system, and the set of b_i's is often referred to as the right-hand side of the system. When $m = n$, i.e., there are as many equations as there are unknowns, we call (2.1) a ***square system***. When $m \neq n$, we call it a ***rectangular system***.

The system in (2.1) can be written as $\boldsymbol{Ax} = \boldsymbol{b}$, where $\boldsymbol{A} = [a_{ij}]$ is the $m \times n$ matrix of the coefficients a_{ij}, $\boldsymbol{x} = (x_1, \dots, x_n)'$ is the $n \times 1$ (column) vector of unknowns and $\boldsymbol{b} = (b_1, \dots, b_n)'$ is the $m \times 1$ (column) vector of constants. Any vector $\boldsymbol{u}$ such that $\boldsymbol{Au} = \boldsymbol{b}$ is said to be a solution of $\boldsymbol{Ax} = \boldsymbol{b}$. For any such system, there are exactly three possibilities for the set of solutions:

(i) **Unique solution:** There is one and only one vector $\boldsymbol{x}$ that satisfies $\boldsymbol{Ax} = \boldsymbol{b}$;

(ii) **No solution:** There is no $\boldsymbol{x}$ that satisfies $\boldsymbol{Ax} = \boldsymbol{b}$;

(iii) **Infinitely many solutions:** There are infinitely many different vectors $\boldsymbol{x}$ that satisfy $\boldsymbol{Ax} = \boldsymbol{b}$. In fact, if a system $\boldsymbol{Ax} = \boldsymbol{b}$ has two solutions $\boldsymbol{u}_1$ and $\boldsymbol{u}_2$, then every convex combination $\beta\boldsymbol{u}_1 + (1-\beta)\boldsymbol{u}_2$, $\beta \in [0, 1]$, is also a solution. *In other words, if a linear system has more than one solution, it must have an infinite number of solutions.*

It is easy to construct examples of linear systems that have a unique (i.e., exactly one) solution, have no solution and have more than one solutions. For instance, the system

$$4x_1 + 3x_2 = 11,$$
$$4x_1 - 3x_2 = 5$$

has a unique solution, viz. $x_1 = 2$ and $x_2 = 1$. The system

$$4x_1 + 3x_2 = 11,$$
$$8x_1 + 6x_2 = 22$$

has more than one solution. Indeed the vector $(\alpha, (11-4\alpha)/3)$ is a solution for every real number α. Finally, note that the system

$$4x_1 + 3x_2 = 11,$$
$$8x_1 + 6x_2 = 20$$

has no solutions.

Definition 2.1 Consistency of a linear system. *A linear system* $\boldsymbol{Ax} = \boldsymbol{b}$ *is said to be* **consistent** *if it has at least one solution and is said to be* **inconsistent** *otherwise.*

In this chapter we will see how matrix operations help us analyze and solve systems of linear equations. Matrices offer a nice way to store a linear system, such as in (2.1), in computers through an *augmented matrix* associated with the linear system:

$$\left[\begin{array}{cccc|c} a_{11} & a_{12} & \cdots & a_{1n} & b_1 \\ a_{21} & a_{22} & \cdots & a_{2n} & b_1 \\ \vdots & \vdots & \ddots & \vdots & \vdots \\ a_{m1} & a_{m2} & \cdots & a_{mn} & b_n \end{array}\right] . \tag{2.2}$$

Each row of the augmented matrix represents an equation, while the vertical line emphasizes where "=" appeared. We write an augmented matrix associated with the linear system $\boldsymbol{Ax} = \boldsymbol{b}$ as $[\boldsymbol{A} : \boldsymbol{b}]$.

2.2 Gaussian elimination

Consider the following linear system involving n linear algebraic equations in n unknowns.

$$\begin{array}{ccccccccc} a_{11}x_1 & + & a_{12}x_2 & + & \ldots & + & a_{1n}x_n & = & b_1, \\ a_{21}x_1 & + & a_{22}x_2 & + & \ldots & + & a_{2n}x_n & = & b_2, \\ & & & & \vdots & & & & \\ a_{n1}x_1 & + & a_{m2}x_2 & + & \ldots & + & a_{nn}x_n & = & b_n. \end{array} \tag{2.3}$$

Thus, we have $\boldsymbol{Ax} = \boldsymbol{b}$, where $\boldsymbol{A}$ is now an $n \times n$ coefficient matrix, $\boldsymbol{x}$ is $n \times 1$ and $\boldsymbol{b}$ is also $n \times 1$. Gaussian elimination is a sequential procedure of transforming a linear system of n linear algebraic equations in n unknowns, into another simpler system having the same solution set. Gaussian elimination proceeds by successively eliminating unknowns and eventually arriving at a system that is easily solvable. The elimination process relies on three simple operations by which to transform one system to another equivalent system.

The Gaussian elimination process relies on three simple operations which to transform the system in (2.3) into a triangular system. These are known as *elementary row operations.*

Definition 2.2 *An* **elementary row operation** *on a matrix is any one of the following:*

(i) **Type-I:** *interchange two rows of the matrix,*

(ii) **Type-II:** *multiply a row by a nonzero scalar, and*

(iii) **Type-III:** *replace a row by the sum of that row and a scalar multiple of another row.*

We will use the notation E_{ik} for interchanging the i-th and k-th rows, $E_i(\alpha)$ for multiplying the i-th row by α, and $E_{ik}(\beta)$, with $i \neq k$ for replacing the i-th row with the sum of the i-th row and β times the k-th row.

We now demonstrate how these operations can be used to solve linear equations. The key observation is that the solution vector $\boldsymbol{x}$ of a linear system $\boldsymbol{Ax} = \boldsymbol{b}$, as in (2.1), will remain unaltered as long as we apply the same elementary operations on both $\boldsymbol{A}$ and $\boldsymbol{b}$, i.e., we apply the elementary operations to *both* sides of the equation. Note that this will be automatically ensured by writing the linear system as the augmented matrix $[\boldsymbol{A} : \boldsymbol{b}]$ and performing the elementary row operations on the augmented matrix.

We now illustrate Gaussian elimination with the following example.

Example 2.1 Consider the following system of four equations in four variables:

$$\begin{array}{rcrcrcrcl} 2x_1 & + & 3x_2 & & & & & = & 1, \\ 4x_1 & + & 7x_2 & + & 2x_3 & & & = & 2, \\ -6x_1 & - & 10x_2 & & & + & x_4 & = & 1, \\ 4x_1 & + & 6x_2 & + & 4x_3 & + & 5x_4 & = & 0. \end{array} \tag{2.4}$$

The augmented matrix associated with this system is

$$\left[\begin{array}{rrrr|r} 2 & 3 & 0 & 0 & 1 \\ 4 & 7 & 2 & 0 & 2 \\ -6 & -10 & 0 & 1 & 1 \\ 4 & 6 & 4 & 5 & 0 \end{array}\right]. \tag{2.5}$$

At each step, Gaussian elimination focuses on one position, called the ***pivot position*** and eliminates all terms below this position using the elementary operations. The coefficient in the pivot position is called a ***pivot*** and the row containing the pivot is called the ***pivotal row***. Only nonzero numbers are allowed to be pivots. If a coefficient in a pivot position is 0, then that row is interchanged with some row below the pivotal below it to produce a nonzero pivot. (This is always possible for square systems possessing a unique solution.)

Unless it is 0, we take the first element of the first row in the augmented matrix as the first pivot. Otherwise, if it is 0, we can always rearrange the system of equations so that an equation with a nonzero coefficient of x_1 appears as the first equation. This will ensure that the first element in the first row of the augmented matrix is nonzero. Thus, in (2.5) 2 is the pivot in the first row. We now conduct Gaussian elimination in the following steps.

Step 1: Eliminate all terms below the first pivot. The pivot is circled and elementary operation is indicated to the side of the matrix:

$$\begin{array}{l} \\ E_{21}(-2) \\ E_{31}(3) \\ E_{41}(-2) \end{array} \left[\begin{array}{rrrr|r} \textcircled{2} & 3 & 0 & 0 & 1 \\ 4 & 7 & 2 & 0 & 2 \\ -6 & -10 & 0 & 1 & 1 \\ 4 & 6 & 4 & 5 & 0 \end{array}\right] \rightarrow \left[\begin{array}{rrrr|r} 2 & 3 & 0 & 0 & 1 \\ 0 & 1 & 2 & 0 & 0 \\ 0 & -1 & 0 & 1 & 4 \\ 0 & 0 & 4 & 5 & -2 \end{array}\right].$$

Step 2: Select a new pivot. Initially, attempt to select a new pivot by moving down and to the right (i.e., try the next diagonal element, or, the $(2, 2)$-th position). If this coefficient is not 0, then it is the next pivot. Otherwise, interchange with a row below this position so as to bring a nonzero number into this pivotal position. Once the pivot has been selected, proceed to eliminate the elements below the pivot. In our example, this strategy works and we find the new pivot to be 1 (circled) and we eliminate all the numbers below it:

$$\begin{array}{l} \\ \\ E_{32}(1) \\ \\ \end{array} \left[\begin{array}{rrrr|r} 2 & 3 & 0 & 0 & 1 \\ 0 & \textcircled{1} & 2 & 0 & 0 \\ 0 & -1 & 0 & 1 & 4 \\ 0 & 0 & 4 & 5 & -2 \end{array}\right] \rightarrow \left[\begin{array}{rrrr|r} 2 & 3 & 0 & 0 & 1 \\ 0 & 1 & 2 & 0 & 0 \\ 0 & 0 & 2 & 1 & 4 \\ 0 & 0 & 4 & 5 & -2 \end{array}\right].$$

Step 3: Select the third pivot using the same strategy as in Step 2, i.e., move to the next diagonal element which is in the $(3, 3)$-th position. We find this to be 2 (circled) below. Eliminate numbers below it:

$$\begin{array}{l} \\ \\ \\ E_{43}(-2) \end{array} \left[\begin{array}{rrrr|r} 2 & 3 & 0 & 0 & 1 \\ 0 & 1 & 2 & 0 & 0 \\ 0 & 0 & \textcircled{2} & 1 & 4 \\ 0 & 0 & 4 & 5 & -2 \end{array}\right] \rightarrow \left[\begin{array}{rrrr|r} 2 & 3 & 0 & 0 & 1 \\ 0 & 1 & 2 & 0 & 0 \\ 0 & 0 & 2 & 1 & 4 \\ 0 & 0 & 0 & 3 & -10 \end{array}\right] = [\boldsymbol{U} : \boldsymbol{b}^*].$$

Step 4: Observe that the corresponding matrix obtained at the end of Step 3 corresponds to a triangular system:

$$\begin{array}{rcrcrcrcr} 2x_1 & + & 3x_2 & & & & & = & 1, \\ & & x_2 & + & 2x_3 & & & = & 0, \\ & & & & 2x_3 & + & x_4 & = & 4, \\ & & & & & & 3x_4 & = & -10. \end{array} \tag{2.6}$$

This yields $x_4 = -10/3$ from the fourth equation, $x_3 = 11/3$ from the third, $x_2 = -22/3$ from the second and, finally, $x_1 = 23/2$ from the first. Thus,

$\boldsymbol{x} = (23/2, -22/3, 11/3, -10/3)'$. Many computer programs replace the $\boldsymbol{b}$ in the augmented system with the solution vector. That is, they provide $[\boldsymbol{U} : \boldsymbol{x}]$ as the output:

$$\left[\begin{array}{cccc|c} 2 & 3 & 0 & 0 & 23/2 \\ 0 & 1 & 2 & 0 & -22/3 \\ 0 & 0 & 2 & 1 & 11/3 \\ 0 & 0 & 0 & 3 & -10/3 \end{array}\right].$$

It is easily verified that $\boldsymbol{U}\boldsymbol{x} = \boldsymbol{b}^*$ and $\boldsymbol{A}\boldsymbol{x} = \boldsymbol{b}$. ■

In general, if an $n \times n$ system has been *triangularized* (i.e., the coefficient matrix is triangular) to yield the augmented matrix:

$$\left[\begin{array}{cccc|c} u_{11} & u_{12} & \cdots & u_{1n} & b_1 \\ 0 & u_{22} & \cdots & u_{2n} & b_2 \\ \vdots & \vdots & \ddots & \vdots & \vdots \\ 0 & 0 & \cdots & u_{nn} & b_n \end{array}\right],$$

in which each $u_{ii} \neq 0$, then the general algorithm for back substitution is given as:

Step 1: Solve for $x_n = b_n/u_{ii}$;

Step 2: Recursively compute

$$x_i = \frac{1}{u_{ii}}\left(b_i - \sum_{j=i+1}^{n} u_{ij}x_j\right), \text{ for } i = n, n-1\ldots, 2, 1.$$

In many practical settings, one wishes to solve several linear systems but with the same coefficient matrix. For instance, suppose we have p linear systems $\boldsymbol{A}\boldsymbol{x}_j = \boldsymbol{b}_j$ for $j = 1, \ldots p$. Such a system can be written compactly as a matrix equation $\boldsymbol{A}\boldsymbol{X} = \boldsymbol{B}$, where $\boldsymbol{A}$ is $n \times n$, $\boldsymbol{X} = [\boldsymbol{x}_1 : \ldots, \boldsymbol{x}_p]$ is an $n \times p$ matrix whose j-th column is the unknown vector corresponding to the j-th linear system and $\boldsymbol{B} = [\boldsymbol{b}_1 : \ldots : \boldsymbol{b}_p]$. Note that we will need to perform Gaussian elimination only once to transform the matrix $\boldsymbol{A}$ to an upper-triangular matrix $\boldsymbol{U}$. More precisely, we use elementary operations on the augmented matrix $[\boldsymbol{A} : \boldsymbol{b}_1 : \boldsymbol{b}_2 : \ldots : \boldsymbol{b}_p]$ to convert to $[\boldsymbol{U} : \boldsymbol{b}_1^* : \ldots : \boldsymbol{b}_p^*]$.

Example 2.2 Let us now consider a *rectangular* system of equations, $\boldsymbol{A}\boldsymbol{x} = \boldsymbol{b}$, where

$$\boldsymbol{A} = \begin{bmatrix} 2 & -2 & -5 & -3 & -1 & 2 \\ 2 & -1 & -3 & 2 & 3 & 2 \\ 4 & -1 & -4 & 10 & 11 & 4 \\ 0 & 1 & 2 & 5 & 4 & 0 \end{bmatrix} \quad \text{and} \quad \boldsymbol{b} = \begin{bmatrix} 9 \\ 4 \\ 4 \\ -5 \end{bmatrix}.$$

Note that the above matrix has more columns than rows. Thus, there are more variables than there are equations. We construct the augmented matrix $[\boldsymbol{A} : \boldsymbol{b}]$, i.e.,

$$\left[\begin{array}{cccccc|c} 2 & -2 & -5 & -3 & -1 & 2 & 9 \\ 2 & -1 & -3 & 2 & 3 & 2 & 4 \\ 4 & -1 & -4 & 10 & 11 & 4 & 4 \\ 0 & 1 & 2 & 5 & 4 & 0 & -5 \end{array}\right],$$

by adding $\boldsymbol{b}$ as an extra column to the matrix $\boldsymbol{A}$ on the far right. What happens when we apply Gaussian elimination to solve the above system? The steps are analogous to Example 2.1.

Step 1: Eliminate all terms below the first pivot (circled):

$$\begin{array}{c} \\ E_{21}(-1) \\ E_{31}(-2) \\ \\ \end{array} \left[\begin{array}{cccccc|c} \textcircled{2} & -2 & -5 & -3 & -1 & 2 & 9 \\ 2 & -1 & -3 & 2 & 3 & 2 & 4 \\ 4 & -1 & -4 & 10 & 11 & 4 & 4 \\ 0 & 1 & 2 & 5 & 4 & 0 & -5 \end{array}\right]$$

$$\rightarrow \left[\begin{array}{cccccc|c} 2 & -2 & -5 & -3 & -1 & 2 & 9 \\ 0 & 1 & 2 & 5 & 4 & 0 & -5 \\ 0 & 3 & 6 & 16 & 13 & 0 & -14 \\ 0 & 1 & 2 & 5 & 4 & 0 & -5 \end{array}\right].$$

Step 2: Go to the second row. Since the $(2, 2)$-th element is nonzero, it acts as the pivot. Sweep out the elements below it:

$$\begin{array}{c} \\ \\ E_{32}(-3) \\ E_{42}(-1) \end{array} \left[\begin{array}{cccccc|c} 2 & -2 & -5 & -3 & -1 & 2 & 9 \\ 0 & \textcircled{1} & 2 & 5 & 4 & 0 & -5 \\ 0 & 3 & 6 & 16 & 13 & 0 & -14 \\ 0 & 1 & 2 & 5 & 4 & 0 & -5 \end{array}\right]$$

$$\rightarrow \left[\begin{array}{cccccc|c} 2 & -2 & -5 & -3 & -1 & 2 & 9 \\ 0 & 1 & 2 & 5 & 4 & 0 & -5 \\ 0 & 0 & 0 & 1 & 1 & 0 & 1 \\ 0 & 0 & 0 & 0 & 0 & 0 & 0 \end{array}\right].$$

Step 3: Next we try to solve the system corresponding to the augmented matrix obtained at the end of Step 2. The associated *reduced system* is

$$\begin{array}{ccccccccccccc} 2x_1 & - & 2x_2 & - & 5x_3 & - & 3x_4 & - & x_5 & + & 2x_6 & = & 9, \\ & & x_2 & + & 2x_3 & + & 5x_4 & + & 4x_5 & & & = & -5, \\ & & & & & & x_4 & + & x_5 & & & = & 1. \end{array}$$

The reduced system contains six unknowns but only three equations. It is, therefore, impossible to obtain a unique solution. Customarily we pick three *basic* unknowns which are called the ***basic variables*** and solve for these in terms of the other three unknowns that are referred to as the ***free variables***. There are several possibilities for selecting a set of basic variables, but the convention is to always solve for the unknowns corresponding to the pivotal positions. In this example the pivot variables lie in the first, second and fourth positions, so we apply back substitution to solve the reduced system for the basic variables x_1, x_2 and x_4 in terms of the free variables

x_3, x_5 and x_6. From the third equation in the reduced system, we obtain

$$x_4 = 1 - x_5 .$$

Substituting the above value of x_4 in the second equation, we find

$$x_2 = -5 - 2x_3 - 5(1 - x_5) - 4x_5 = -10 - 2x_3 + x_5 .$$

Finally, we solve for x_1 from the first equation

$$x_1 = \frac{9}{2} + (-10 - 2x_3 + x_5) + \frac{5}{2}x_3 + \frac{3}{2}(1 - x_5) + \frac{1}{2}x_5 - x_6 = -4 + \frac{1}{2}x_3 - x_6.$$

The solution set can be described more concisely as

$$\boldsymbol{x} = \begin{bmatrix} x_1 \\ x_2 \\ x_3 \\ x_4 \\ x_5 \\ x_6 \end{bmatrix} = \begin{bmatrix} -4 \\ -10 \\ 0 \\ 1 \\ 0 \\ 0 \end{bmatrix} + x_3 \begin{bmatrix} 1/2 \\ -2 \\ 1 \\ 0 \\ 0 \\ 0 \end{bmatrix} + x_5 \begin{bmatrix} 0 \\ 1 \\ 0 \\ -1 \\ 1 \\ 0 \end{bmatrix} + x_6 \begin{bmatrix} -1 \\ 0 \\ 0 \\ 0 \\ 0 \\ 1 \end{bmatrix} .$$

This representation is referred to as the ***general solution*** of a linear system. It represents a solution set with x_3, x_5 and x_6 being free variables that are "free" to take any possible number. ■

It is worth pointing out that the final result of applying Gaussian elimination in the above example is not a purely triangular form but rather a "stair-step" type of triangular form. Such forms are known as *row echelon* forms.

Definition 2.3 *An $m \times n$ matrix $\boldsymbol{U}$ is said to be in* **row echelon** *form if the following two conditions hold:*

- *If a row, say $\boldsymbol{u}'_{i*}$, consists of all zeroes, i.e., $\boldsymbol{u}'_{i*} = \boldsymbol{0}'$, then all rows below $\boldsymbol{u}'_{i*}$ are also entirely zeroes. This implies that all zero rows are at the bottom.*
- *If the first nonzero entry in $\boldsymbol{u}'_{i*}$ lies in the j-th position, then* all *entries below the i-th position in each of the columns $\boldsymbol{u}_{*1}, \ldots, \boldsymbol{u}_{*j}$ are* 0.

The above two conditions imply that the nonzero entries in an echelon form must lie on or above a stair-step line. The pivots are the first nonzero entries in each row. A typical structure for a matrix in row echelon form is illustrated below with the pivots circled.

$$\begin{bmatrix} \circledast & * & * & * & * & * & * & * \\ 0 & \circledast & * & * & * & * & * & * \\ 0 & 0 & 0 & \circledast & * & * & * & * \\ 0 & 0 & 0 & 0 & 0 & 0 & \circledast & * \\ 0 & 0 & 0 & 0 & 0 & 0 & 0 & 0 \\ 0 & 0 & 0 & 0 & 0 & 0 & 0 & 0 \end{bmatrix} .$$

The number of pivots equals the number of nonzero rows in a reduced echelon matrix. The columns containing the pivots are known as ***basic columns***. Thus the number of pivots also equals the number of basic columns.

Here, an important point needs to be clarified. Because there are several different sequences of elementary operations that will reduce a general $m \times n$ matrix $\boldsymbol{A}$ to an echelon form $\boldsymbol{U}$, the final echelon form is not unique. How, then, can we be sure that no matter what sequence of elementary operations we employ to arrive at an echelon form, these will still yield the same number of pivots? Fortunately, the answer is in the affirmative. *It is a fact that no matter what sequence of elementary operations have been used to obtain the echelon form, the number of pivots will remain the same. In fact, the very definition of row echelon forms ensures that the* positions *of the pivots will be the same.* And, if the number of pivots remain the same, so would the number of basic and free variables. These facts allow us to refer to these pivots simply as the "pivots of $\boldsymbol{A}$."

It will be worthwhile to pause for a while and consider what would happen if the number of pivots did not remain the same. Consider a linear system where the right-hand side is $\boldsymbol{0}$, i.e., $\boldsymbol{A}\boldsymbol{x} = \boldsymbol{0}$. Suppose we employed two different sequences of row operations to arrive at two echelon forms $\boldsymbol{U}_1$ and $\boldsymbol{U}_2$. Since the elementary operations do not affect the $\boldsymbol{0}$ vector, we have two reduced systems $\boldsymbol{U}_1\boldsymbol{x} = \boldsymbol{0}$ and $\boldsymbol{U}_2\boldsymbol{x} = \boldsymbol{0}$. If the number of pivots are different in $\boldsymbol{U}_1$ and $\boldsymbol{U}_2$, then the number of basic and free variables in $\boldsymbol{U}_1$ and $\boldsymbol{U}_2$ will also be different. But this means that we would be able to find an $\boldsymbol{x}_p$ such that $\boldsymbol{U}_1\boldsymbol{x}_p = \boldsymbol{0}$ while $\boldsymbol{U}_2\boldsymbol{x}_p \neq \boldsymbol{0}$. But this would be a contradiction: $\boldsymbol{U}_1\boldsymbol{x}_p = \boldsymbol{0}$ would imply that $\boldsymbol{x}_p$ is a solution to $\boldsymbol{A}\boldsymbol{x} = \boldsymbol{0}$, while $\boldsymbol{U}_2\boldsymbol{x}_p \neq \boldsymbol{0}$ implies that it is not!

Therefore, even if two different echelon matrices are derived from $\boldsymbol{A}$, the number of pivots or basic columns (and hence the number of basic and free variables) remain the same. This number, therefore, is a characteristic unique to the matrix $\boldsymbol{A}$. Therefore, if $\boldsymbol{U}$ is *any* row echelon matrix obtained by row operations on $\boldsymbol{A}$, then we can unambiguously define:

$$\begin{aligned}\text{number of pivots of } \boldsymbol{A} &= \text{number of pivots in } \boldsymbol{U}\\ &= \text{number of nonzero rows in } \boldsymbol{U}\\ &= \text{number of basic columns in } \boldsymbol{U}\ .\end{aligned}$$

The number of pivots of $\boldsymbol{A}$ is often referred to as the ***rank*** of $\boldsymbol{A}$ and plays an extremely important role in matrix analysis. In summary, the rank of a matrix is the number of pivots which equals the number of nonzero rows in any row echelon form derived from $\boldsymbol{A}$ and is also the same as the number of basic columns therein. It is, however, rather cumbersome to develop the concept of rank from row echelon forms. A more elegant development employs the concept of vector spaces and gives an alternative, but equivalent, definition of rank. We will pursue this approach later. For now, we consider echelon forms as useful structures associated with systems of linear equations that arise naturally from Gaussian elimination.

Gaussian elimination also reveals when a linear system is *consistent.* A system of m linear equations in n unknowns is said to be a consistent system if it possesses at least one solution. If there are no solutions, then the system is called *inconsistent.*

We have already seen an example of an inconsistent system in Section 2.1:

$$4x_1 + 3x_2 = 11,$$
$$8x_1 + 6x_2 = 20.$$

What happens if we apply Gaussian elimination to the above system? We have

$$\begin{array}{c} \\ E_{21}(-2) \end{array} \left[\begin{array}{cc|c} \textcircled{4} & 3 & 11 \\ 8 & 6 & 20 \end{array}\right] \rightarrow \left[\begin{array}{cc|c} 4 & 3 & 11 \\ 0 & 0 & -2 \end{array}\right].$$

Note that the last row corresponds to the equation

$$0x_1 + 0x_2 = -2,$$

which clearly does not have a solution. Geometrically, the two equations above represent two straight lines that are parallel to each other and hence do not intersect. A linear equation in two unknowns represents a line in 2-space. Therefore, a linear system of m equations in two unknowns is consistent if and only if the m lines defined by the m equations intersect at a common point. Similarly, a linear equation in three unknowns is a plane in 3-space. A linear system of m equations in three unknowns is consistent if and only if the m planes have at least one common point of intersection.

For larger m and n the geometric visualizations of intersecting lines or planes become difficult. Here Gaussian elimination helps. Suppose we apply Gaussian elimination to reduce the augmented matrix $[\boldsymbol{A}|\boldsymbol{b}]$ associated with the system $\boldsymbol{Ax} = \boldsymbol{b}$ to a row echelon form. Suppose, in the process of Gaussian elimination, we arrive at a row where the only nonzero entry appears on the right-hand side as

$$\left[\begin{array}{cccccc|c} * & * & * & * & * & * & * \\ 0 & * & * & * & * & * & * \\ 0 & 0 & 0 & * & * & * & * \\ 0 & 0 & 0 & 0 & 0 & 0 & \alpha \\ \bullet & \bullet & \bullet & \bullet & \bullet & \bullet & \bullet \\ \bullet & \bullet & \bullet & \bullet & \bullet & \bullet & \bullet \end{array}\right].$$

If $\alpha \neq 0$, then the system is inconsistent. In fact, there is no need to proceed further with the Gaussian elimination algorithm below that row. For $\alpha \neq 0$, the corresponding equation, $0x_1 + 0x_2 + \cdots + 0x_n = \alpha$, will have no solution. Because row operations do not affect the solution of the original system, this implies that $\boldsymbol{Ax} = \boldsymbol{b}$ is an inconsistent system.

The converse is also true: If a system is inconsistent, then Gaussian elimination must produce a row of the form

$$\left[\begin{array}{cccc|c} 0 & 0 & \cdots & 0 & \alpha \end{array}\right],$$

where $\alpha \neq 0$. If such a row does not emerge, we can solve the system by completing the back substitutions. Note that when $\alpha = 0$, the left-hand and right-hand sides of the equation are both zero. This does not contradict the existence of a solution although it is unhelpful in finding a solution.

From the above, it is clear that the system $\boldsymbol{Ax} = \boldsymbol{b}$ is consistent if and only if the number of pivots in $\boldsymbol{A}$ is the same as that in $[\boldsymbol{A} : \boldsymbol{b}]$.

2.3 Gauss-Jordan elimination

The ***Gauss-Jordan*** algorithm is a variation of Gaussian elimination. The two features that distinguish the Gauss-Jordan method from standard Gaussian elimination are: (i) each pivot element is forced to be 1; and (ii) the elements below *and above* the pivot are eliminated. In other words, if

$$\left[\begin{array}{cccc|c} a_{11} & a_{12} & \cdots & a_{1n} & b_1 \\ a_{21} & a_{22} & \cdots & a_{2n} & b_2 \\ \vdots & \vdots & \ddots & \vdots & \vdots \\ a_{n1} & a_{n2} & \cdots & a_{nn} & b_n \end{array}\right]$$

is the augmented matrix associated with $\boldsymbol{Ax} = \boldsymbol{b}$, then elementary row operations are used to reduce this matrix to

$$\left[\begin{array}{cccc|c} 1 & 0 & \cdots & 0 & x_1 \\ 0 & 1 & \cdots & 0 & x_2 \\ \vdots & \vdots & \ddots & \vdots & \vdots \\ 0 & 0 & \cdots & 1 & x_n \end{array}\right],$$

where $\boldsymbol{x} = (x_1, \ldots, x_n)'$ is the solution to $\boldsymbol{Ax} = \boldsymbol{b}$. Upon successful completion of the algorithm, the last column holds the solution vector and, unlike in Gaussian elimination, there is no need to perform back substitutions.

Example 2.3 To illustrate, let us revisit the system in Example 2.1 in Section 2.2. The Gauss-Jordan algorithm can proceed exactly like Gaussian elimination until Step 3. Using Gaussian elimination, at the end of Step 3 in Example 2.1, we finally arrived at the augmented system:

$$\left[\begin{array}{cccc|c} 2 & 3 & 0 & 0 & 1 \\ 0 & 1 & 2 & 0 & 0 \\ 0 & 0 & 2 & 1 & 4 \\ 0 & 0 & 0 & 3 & -10 \end{array}\right].$$

In Step 4 we depart from Gaussian elimination. Instead of proceeding with back substitution to solve the system, Gauss-Jordan will proceed with the following steps:

Step 4: Using elementary operations, force all the pivots that are not equal to 1 (circled) to equal 1:

$$\begin{array}{c} E_1(1/2) \\ \\ E_3(1/2) \\ E_4(1/3) \end{array} \left[\begin{array}{cccc|c} \textcircled{2} & 3 & 0 & 0 & 1 \\ 0 & 1 & 2 & 0 & 0 \\ 0 & 0 & \textcircled{2} & 1 & 4 \\ 0 & 0 & 0 & \textcircled{3} & -10 \end{array}\right] \rightarrow \left[\begin{array}{cccc|c} 1 & 3/2 & 0 & 0 & 1/2 \\ 0 & 1 & 2 & 0 & 0 \\ 0 & 0 & 1 & 1/2 & 2 \\ 0 & 0 & 0 & 1 & -10/3 \end{array}\right].$$

Step 5: "Sweep out" each column by eliminating the entries above the pivot. Do this column by column, starting with the second column. Each row indicates the elementary operation to sweep out a column. The corresponding pivots are circled.

$$\begin{matrix} E_{12}(-3/2) \end{matrix} \left[\begin{array}{cccc|c} 1 & 3/2 & 0 & 0 & 1/2 \\ 0 & \textcircled{1} & 2 & 0 & 0 \\ 0 & 0 & 1 & 1/2 & 2 \\ 0 & 0 & 0 & 1 & -10/3 \end{array}\right] \rightarrow \left[\begin{array}{cccc|c} 1 & 0 & -3 & 0 & 1/2 \\ 0 & 1 & 2 & 0 & 0 \\ 0 & 0 & 1 & 1/2 & 2 \\ 0 & 0 & 0 & 1 & -10/3 \end{array}\right];$$

$$\begin{matrix} E_{13}(3) \\ E_{23}(-2) \end{matrix} \left[\begin{array}{cccc|c} 1 & 0 & -3 & 0 & 1/2 \\ 0 & 1 & 2 & 0 & 0 \\ 0 & 0 & \textcircled{1} & 1/2 & 2 \\ 0 & 0 & 0 & 1 & -10/3 \end{array}\right] \rightarrow \left[\begin{array}{cccc|c} 1 & 0 & 0 & 3/2 & 13/2 \\ 0 & 1 & 0 & -1 & -4 \\ 0 & 0 & \textcircled{1} & 1/2 & 2 \\ 0 & 0 & 0 & 1 & -10/3 \end{array}\right];$$

$$\begin{matrix} E_{14}(-3/2) \\ E_{24}(1) \\ E_{34}(-1/2) \end{matrix} \left[\begin{array}{cccc|c} 1 & 0 & 0 & 3/2 & 13/2 \\ 0 & 1 & 0 & -1 & -4 \\ 0 & 0 & 1 & 1/2 & 2 \\ 0 & 0 & 0 & \textcircled{1} & -10/3 \end{array}\right] \rightarrow \left[\begin{array}{cccc|c} 1 & 0 & 0 & 0 & 23/2 \\ 0 & 1 & 0 & 0 & -22/3 \\ 0 & 0 & 1 & 0 & 11/3 \\ 0 & 0 & 0 & 1 & -10/3 \end{array}\right].$$

Upon successful termination, Gauss-Jordan elimination produces an augmented matrix of the form $[\boldsymbol{I} : \boldsymbol{b}^*]$ and the solution vector is immediately seen to be $\boldsymbol{x} = \boldsymbol{b}^*$, which is precisely the last column. ■

Suppose we apply the Gauss-Jordan elimination to a general $m \times n$ matrix. Each step of the Gauss-Jordan method forces the pivot to be equal to 1, and then annihilates all entries above and below the pivot are annihilated. The final outcome is a special echelon matrix with elements *below and above* the pivot being 0. Such matrices are known as ***reduced row echelon*** matrices. A formal definition is given below.

Definition 2.4 *An $m \times n$ matrix $\boldsymbol{E}$ is said to be in* **reduced row echelon form** *if it satisfies the following three conditions:*

(i) $\boldsymbol{E}$ is in echelon form.

(ii) The first nonzero entry in each row (i.e., each pivot) is equal to 1.

(iii) All entries above each pivot are equal to 0.

Because $\boldsymbol{E}$ is in echelon form, note that (iii) implies that the elements below and above each pivot are zero. A typical structure for a matrix in reduced row echelon form is illustrated below with the pivots circled:

$$\begin{bmatrix} 1 & 0 & * & 0 & * & * & 0 & * \\ 0 & 1 & * & 0 & * & * & 0 & * \\ 0 & 0 & 0 & 1 & * & * & 0 & * \\ 0 & 0 & 0 & 0 & 0 & 0 & 1 & * \\ 0 & 0 & 0 & 0 & 0 & 0 & 0 & 0 \\ 0 & 0 & 0 & 0 & 0 & 0 & 0 & 0 \end{bmatrix}.$$

Suppose $\boldsymbol{E}$ is an $m \times n$ matrix with r pivots that is in reduced row echelon form. The basic columns of $\boldsymbol{E}$ are in fact the canonical vectors $\boldsymbol{e}_1, \boldsymbol{e}_2, \ldots, \boldsymbol{e}_r$ in $\Re^m$. We can permute the columns to bring the basic columns to the first r positions. Therefore, there is a permutation matrix $\boldsymbol{P}$ such that $\boldsymbol{EP} = \begin{bmatrix} \boldsymbol{I}_r & \boldsymbol{J} \\ \boldsymbol{O} & \boldsymbol{O} \end{bmatrix}$. This reveals that each non-basic column is a linear combinations of the basic columns in a reduced row echelon form.

Gaussian or Gauss-Jordan elimination: Which is faster?

For assessing the efficiency of a matrix algorithm we need to count the number of arithmetical operations required. For matrix algorithms, customarily, additions and subtractions count as one operation, as do multiplications and divisions. However, multiplications or divisions are usually counted separately from additions or subtractions.

Counting operations in Gaussian elimination for an $n \times n$ matrix formally yields $n^3/3 + n^2 - n/3$ multiplications or divisions, and $n^3/3 + n^2/2 - 5n/6$ additions or subtractions. The corresponding numbers for Gauss-Jordan elimination applied to an $n \times n$ matrix are $n^3/2 + n^2/2$ and $n^3/2 - n/2$, respectively. Since computational considerations are more relevant for large matrices, we often describe the computational complexity of these algorithms in terms of the highest power of n as that is what dominates the remaining terms. Using this norm, we say that Gaussian elimination with back substitution has complexity of the "order of" $n^3/3$ (written as $O(n^3/3)$) and Gauss-Jordan has complexity of $n^3/2$ (i.e., $O(n^3/2)$).

The above numbers show that Gauss-Jordan requires more arithmetic (approximately by a factor of $3/2$) than Gaussian elimination with back substitution. It is, therefore, incorrect to conclude that Gauss-Jordan and Gaussian elimination with back substitution are "equivalent" algorithms. In fact, eliminating terms above the pivot with Gauss-Jordan will be slightly more expensive than performing back substitution. While in small and medium size linear systems, this difference will not manifest itself, for large systems every saving of arithmetic operations counts and Gaussian elimination with back substitution will be preferred.

2.4 Elementary matrices

Associated with each elementary operation is an elementary matrix that is obtained by applying the elementary operation on the identity matrix. For example, in a 4×4 system (i.e., four equations and four unknowns) the elementary matrix associated with E_{23} (i.e., interchanging the second and third equations) is denoted by $\boldsymbol{E}_{23}$ and obtained by interchanging the second and third rows of the identity matrix:

$$\boldsymbol{E}_{23} = \begin{bmatrix} 1 & 0 & 0 & 0 \\ 0 & 0 & 1 & 0 \\ 0 & 1 & 0 & 0 \\ 0 & 0 & 0 & 1 \end{bmatrix}.$$

Similarly, the matrix associated with, say, $E_2(\alpha)$ is

$$\boldsymbol{E}_2(\alpha) = \begin{bmatrix} 1 & 0 & 0 & 0 \\ 0 & \alpha & 0 & 0 \\ 0 & 0 & 1 & 0 \\ 0 & 0 & 0 & 1 \end{bmatrix}$$

and that with, say, $E_{31}(\beta)$ is

$$\boldsymbol{E}_{31}(\beta) = \begin{bmatrix} 1 & 0 & 0 & 0 \\ 0 & 1 & 0 & 0 \\ \beta & 0 & 1 & 0 \\ 0 & 0 & 0 & 1 \end{bmatrix}.$$

Premultiplying any given matrix, say $\boldsymbol{A}$, with these elementary matrices achieves the corresponding elementary row operation on $\boldsymbol{A}$. We illustrate this with an example, using the same system of equations as in (2.4). This gives us a way to use matrix operations reduce a general system of equations into an upper-triangular system.

Example 2.4 We write the system in (2.4) as an *augmented matrix* where the vector of constants on the right-hand side is appended to the coefficient matrix:

$$[\boldsymbol{A} : \boldsymbol{b}] = \left[\begin{array}{cccc|c} 2 & 3 & 0 & 0 & 1 \\ 4 & 7 & 2 & 0 & 2 \\ -6 & -10 & 0 & 1 & 1 \\ 4 & 6 & 4 & 5 & 0 \end{array}\right]. \tag{2.7}$$

The augmented matrix is a 4×5 rectangular matrix. We follow exactly same sequence of elementary operations as in Example 2.1. As can be easily verified, we obtain

$$\boldsymbol{E}_{43}(-2)\boldsymbol{E}_{32}(1)\boldsymbol{E}_{41}(-2)\boldsymbol{E}_{31}(3)\boldsymbol{E}_{21}(-2)[\boldsymbol{A} : \boldsymbol{b}] = \left[\begin{array}{cccc|c} 2 & 3 & 0 & 0 & 1 \\ 0 & 1 & 2 & 0 & 0 \\ 0 & 0 & 2 & 1 & 4 \\ 0 & 0 & 0 & 3 & -10 \end{array}\right],$$

where the right-hand side above is the augmented matrix corresponding to the upper-triangular system in (2.6). In analyzing linear systems, some of the elements in the upper-triangular matrix play an especially important role. These are the first nonzero element in each row. They are known as the ***pivots*** and are circled above. The columns containing the pivots are called ***basic columns***.

Note that each of the elementary matrices in the above sequence is a unit lower-triangular matrix (i.e., with diagonal elements being equal to one). Let $\boldsymbol{E} = \boldsymbol{E}_{43}(-2)\boldsymbol{E}_{32}(1)\boldsymbol{E}_{41}(-2)\boldsymbol{E}_{31}(3)\boldsymbol{E}_{21}(-2)$. We find that

$$\boldsymbol{E} = \begin{bmatrix} 1 & 0 & 0 & 0 \\ -2 & 1 & 0 & 0 \\ 1 & 1 & 1 & 0 \\ -4 & -2 & -2 & 1 \end{bmatrix}$$

is also a unit lower-triangular matrix. This is no coincidence and is true in general. We will see this later when we explore the relationship between these triangular matrices and Gaussian elimination. A formal definition follows.

Definition 2.5 *An* **elementary matrix** *of order $n \times n$ is a square matrix of the form $\boldsymbol{I}_n + \boldsymbol{u}\boldsymbol{v}'$, where $\boldsymbol{u}$ and $\boldsymbol{v}$ are $n \times 1$ vectors such that $\boldsymbol{v}'\boldsymbol{u} \neq -1$. Three special types of elementary matrices that can be expressed in terms of the identity matrix $\boldsymbol{I}$ and a judicious choice of canonical vectors.are particularly important:*

(i) A **Type I elementary matrix** *is denoted by $\boldsymbol{E}_{ij}$ and is defined as*

$$\boldsymbol{E}_{ij} = \boldsymbol{I} - (\boldsymbol{e}_j - \boldsymbol{e}_i)(\boldsymbol{e}_j - \boldsymbol{e}_i)' . \tag{2.8}$$

(ii) A **Type II elementary matrix** *is denoted by $\boldsymbol{E}_i(\alpha)$ and is defined as*

$$\boldsymbol{E}_i(\alpha) = \boldsymbol{I} + (\alpha - 1)\boldsymbol{e}_i\boldsymbol{e}_i' . \tag{2.9}$$

(iii) A **Type III elementary matrix** *is denoted by $\boldsymbol{E}_{ij}(\beta)$ for $i \neq j$ and is defined as*

$$\boldsymbol{E}_{ij}(\beta) = \boldsymbol{I} + \beta\boldsymbol{e}_j\boldsymbol{e}_i' . \tag{2.10}$$

We now summarize the effect of premultiplication by an elementary matrix.

Theorem 2.1 *Let $\boldsymbol{A}$ be an $m \times n$ matrix. An elementary row operation on $\boldsymbol{A}$ is achieved by multiplying $\boldsymbol{A}$ on the left with the corresponding elementary matrix.*

Proof. We carry out straightforward verifications.

Type I operations are achieved by $\boldsymbol{E}_{ik}\boldsymbol{A}$:

$$(\boldsymbol{I} - (\boldsymbol{e}_k - \boldsymbol{e}_i)(\boldsymbol{e}_k - \boldsymbol{e}_i)')\boldsymbol{A} = \boldsymbol{A} - (\boldsymbol{e}_k - \boldsymbol{e}_i)(\boldsymbol{a}_{k*}' - \boldsymbol{a}_{i*}') = \begin{bmatrix} \boldsymbol{a}_{1*}' \\ \vdots \\ \boldsymbol{a}_{k*}' \\ \vdots \\ \boldsymbol{a}_{i*}' \\ \vdots \\ \boldsymbol{a}_{m*}' \end{bmatrix} ;$$

Type-II operations is achieved by premultiplying with $\boldsymbol{E}_i(\alpha)$

$$\boldsymbol{E}_i(\alpha)\boldsymbol{A} = (\boldsymbol{I} + (\alpha - 1)\boldsymbol{e}_i\boldsymbol{e}_i')\boldsymbol{A} = \boldsymbol{A} + (\alpha - 1)\boldsymbol{e}_i\boldsymbol{a}_{i*}' = \begin{bmatrix} \boldsymbol{a}_{1*}' \\ \vdots \\ \alpha\boldsymbol{a}_{i*}' \\ \vdots \\ \boldsymbol{a}_{m*}' \end{bmatrix} ;$$

and Type-III operations is obtained by premultiplying with $\boldsymbol{E}_{ik}(\alpha)$

$$\boldsymbol{E}_{ik}(\beta)\boldsymbol{A} = (\boldsymbol{I} + \beta \boldsymbol{e}_k \boldsymbol{e}_i')\boldsymbol{A} = \boldsymbol{A} + \beta \boldsymbol{e}_k \boldsymbol{a}_{i*}' = \begin{bmatrix} \boldsymbol{a}_{1*}' \\ \vdots \\ \boldsymbol{a}_{i*}' + \beta \boldsymbol{a}_{k*}' \\ \vdots \\ \boldsymbol{a}_{k*}' \\ \vdots \\ \boldsymbol{a}_{m*}' \end{bmatrix} .$$

□

Analogous to row operations, we can also define *elementary column operations* on matrices.

Definition 2.6 *An* **elementary column operation** *on a matrix is any one of the following:*

(i) **Type-I**: *interchange two columns of the matrix,*

(ii) **Type-II**: *multiply a column by a nonzero scalar, and*

(iii) **Type-III**: *replace a column by the sum of that column and a scalar multiple of another column.*

We will use the notation E_{jk}' for interchanging the j-th and k-th columns, $E_j(\alpha)'$ for multiplying the j-th column by α, and $E_{jk}(\beta)'$ for adding β times the k-th column to the j-th column. The following theorem is the analogue of Theorem 2.1 for column operations.

Theorem 2.2 *Let $\boldsymbol{A}$ be an $m \times n$ matrix. Making the elementary column operations E_{jk}', $E_j(\alpha)'$ and $E_{kj}(\beta)'$ on the matrix $\boldsymbol{A}$ is equivalent to* postmultiplying $\boldsymbol{A}$ *by the elementary matrices $\boldsymbol{E}_{jk}$, $\boldsymbol{E}_j(\alpha)$ and $\boldsymbol{E}_{kj}(\beta)$, respectively.*

Proof. The proof of this theorem is by direct verification and is left to the reader. *It is worth pointing out that the elementary matrix postmultiplying $\boldsymbol{A}$ to describe the operation $E_{jk}(\beta)'$ on $\boldsymbol{A}$ is $\boldsymbol{E}_{kj}(\beta)$ and not $\boldsymbol{E}_{jk}(\beta)$.* □

Theorem 2.3 *Let $\boldsymbol{E}$ be the elementary matrix required to premultiply a square matrix $\boldsymbol{A}$ for making an elementary row operation on $\boldsymbol{A}$. Then its transpose $\boldsymbol{E}'$ is the elementary matrix postmultiplying $\boldsymbol{A}$ for making the corresponding elementary column operation.*

Proof. Clearly, $\boldsymbol{E}_{ik}$ and $\boldsymbol{E}_i(\alpha)$ are both symmetric and $\boldsymbol{E}_{ik}(\beta)' = \boldsymbol{E}_{ki}(\beta)$. Hence, the theorem follows directly from Theorem 2.2. □

We conclude this section with another relevant observation. The order in which a single elementary row operation and a single elementary column operation on $\boldsymbol{A}$ is immaterial since $(\boldsymbol{E}_1\boldsymbol{A})\boldsymbol{E}_2' = \boldsymbol{E}_1(\boldsymbol{A}\boldsymbol{E}_2')$, where $\boldsymbol{E}_1$ and $\boldsymbol{E}_2'$ are the elementary matrices corresponding to the row and column operation. However, when several row (or several column) operations are performed, the order in which they are performed is important.

2.5 Homogeneous linear systems

Definition 2.7 *A linear system* $\boldsymbol{A}\boldsymbol{x} = \boldsymbol{0}$ *is called a* **homogeneous linear system**, *where* $\boldsymbol{A}$ *is an* $m \times n$ *matrix and* $\boldsymbol{x}$ *is an* $n \times 1$ *vector.*

Homogeneous linear systems are *always* consistent because $\boldsymbol{x} = \boldsymbol{0}$ is always one solution regardless of the values of the coefficients. The solution $\boldsymbol{x} = \boldsymbol{0}$ is, therefore, often referred to as the *trivial* solution. To find out whether there exists any other solution besides the trivial solution, we resort to Gaussian elimination. Here is an example.

Example 2.5 Consider the homogeneous system $\boldsymbol{A}\boldsymbol{x} = \boldsymbol{0}$ with the coefficient matrix given by

$$\boldsymbol{A} = \begin{bmatrix} 1 & 2 & 3 \\ 4 & 5 & 6 \\ 7 & 8 & 9 \end{bmatrix}. \tag{2.11}$$

Here, it is worth noting that while reducing the augmented matrix $[\boldsymbol{A} : \boldsymbol{0}]$ of a homogeneous system to a row echelon form using Gaussian elimination, the zero column on the right-hand side is not altered by the elementary row operations. Hence, applying elementary row operations to $[\boldsymbol{A} : \boldsymbol{0}]$ will yield the form $[\boldsymbol{U} : \boldsymbol{0}]$. Thus, it is not needed to form an augmented matrix for homogeneous systems—we simply reduce the coefficient matrix $\boldsymbol{A}$ to a row echelon form $\boldsymbol{U}$, keeping in mind that the right-hand side is entirely zero during back substitution. More precisely, we have

$$\boldsymbol{E}_{32}(-2)\boldsymbol{E}_{31}(-7)\boldsymbol{E}_{21}(-4)\boldsymbol{A} = \begin{bmatrix} 1 & 2 & 3 \\ 0 & -3 & -6 \\ 0 & 0 & 0 \end{bmatrix}.$$

The last row is $\boldsymbol{0}'$. In other words, letting $x_3 = \alpha$ be any real number, the second equation yields $x_2 = -2\alpha$ and the first equation, upon substitution, yields $x_1 = \alpha$. Thus, any vector of the form $\boldsymbol{x} = (\alpha, -2\alpha, \alpha)'$ is a solution to the homogeneous system $\boldsymbol{A}\boldsymbol{x} = \boldsymbol{0}$. ■

Consider now a rectangular homogeneous system, $\boldsymbol{A}\boldsymbol{x} = \boldsymbol{0}$, where $\boldsymbol{A}$ is now an $m \times n$ matrix with $m < n$, $\boldsymbol{x}$ is $n \times 1$ and $\boldsymbol{0}$ is $m \times 1$. It is worth noting that such a system will have at least one nonzero solution. This is easily seen by considering an echelon reduction of $\boldsymbol{A}$ as we demonstrate in the following example.

Example 2.6 Let us consider the homogeneous system $\boldsymbol{Ax} = \mathbf{0}$ with $\boldsymbol{A}$ as the same rectangular matrix in Example 2.2:

$$\boldsymbol{A} = \begin{bmatrix} 2 & -2 & -5 & -3 & -1 & 2 \\ 2 & -1 & -3 & 2 & 3 & 2 \\ 4 & -1 & -4 & 10 & 11 & 4 \\ 0 & 1 & 2 & 5 & 4 & 0 \end{bmatrix}.$$

As we saw in Example 2.2, applying Gaussian elimination to $\boldsymbol{A}$ eventually yields

$$\boldsymbol{U} = \begin{bmatrix} 2 & -2 & -5 & -3 & -1 & 2 \\ 0 & 1 & 2 & 5 & 4 & 0 \\ 0 & 0 & 0 & 1 & 1 & 0 \\ 0 & 0 & 0 & 0 & 0 & 0 \end{bmatrix}.$$

This corresponds to the following homogeneous system:

$$\begin{array}{rcrcrcrcrcrcl} 2x_1 & - & 2x_2 & - & 5x_3 & - & 3x_4 & - & x_5 & + & 2x_6 & = & 0, \\ & & x_2 & + & 2x_3 & + & 5x_4 & + & 4x_5 & & & = & 0, \\ & & & & & & x_4 & + & x_5 & & & = & 0. \end{array}$$

As in Example 2.2, we choose the *basic variables* to be those corresponding to the pivots, i.e., x_1, x_2 and x_4 and solve them in terms of the *free variables* x_3, x_5 and x_6. From the third equation in the reduced system,

$$x_4 = -x_5 \,.$$

Substituting the above value of x_4 in the second equation, we find

$$x_2 = -2x_3 - 5(-x_5) - 4x_5 = -2x_3 + x_5 \,,$$

and, finally, we solve for x_1 from the first equation

$$x_1 = (-2x_3 + x_5) + \frac{5}{2}x_3 + \frac{3}{2}(-x_5) + \frac{1}{2}x_5 - x_6 = \frac{1}{2}x_3 - x_6 \,.$$

Thus, the basic variables are a linear combination of the free variables:

$$\begin{array}{rcrcrcr} x_1 & = & (1/2)x_3 & + & 0x_5 & + & (-1)x_6, \\ x_2 & = & (-2)x_3 & + & 1x_5 & + & 0x_6, \\ x_4 & = & 0x_3 & + & (-1)x_5 & + & 0x_6. \end{array}$$

The free variables are "free" to be any real number. Each such choice leads to a *special solution* or a *particular solution.*

A common particular solution of a homogeneous linear system is obtained by setting exactly one of the free variables to one and all the others to zero. For example, let $\boldsymbol{h}_1$ be the solution when $x_3 = 1$, $x_5 = x_6 = 0$, $\boldsymbol{h}_2$ be the solution when $x_5 = 1$,

$x_3 = x_6 = 0$, and $\boldsymbol{h}_3$ be the solution when $x_6 = 1$ and $x_3 = x_5 = 0$. Then, we have

$$\boldsymbol{h}_1 = \begin{bmatrix} 1/2 \\ -2 \\ 1 \\ 0 \\ 0 \\ 0 \end{bmatrix}, \quad \boldsymbol{h}_2 = \begin{bmatrix} 0 \\ 1 \\ 0 \\ -1 \\ 1 \\ 0 \end{bmatrix} \quad \text{and} \quad \boldsymbol{h}_3 = \begin{bmatrix} -1 \\ 0 \\ 0 \\ 0 \\ 0 \\ 1 \end{bmatrix}.$$

The *general solution* of a homogeneous linear system is now described by a linear combination of the particular solutions with the free variables as the coefficients:

$$\boldsymbol{x} = x_3\boldsymbol{h}_1 + x_5\boldsymbol{h}_2 + x_6\boldsymbol{h}_3 \ . \ \blacksquare$$

The above example is symptomatic of what happens in general. Consider a general homogeneous system $\boldsymbol{A}\boldsymbol{x} = \boldsymbol{0}$ of m linear equations in n unknowns. If the $m \times n$ coefficient matrix $\boldsymbol{A}$ yields an echelon form with r pivots after Gaussian elimination, then it should be apparent from the preceding discussion that there will be exactly r basic variables (corresponding to the pivots) and, therefore, exactly $n - r$ free variables. Reducing $\boldsymbol{A}$ to a row echelon form using Gaussian elimination and then using back substitution to solve for the basic variables in terms of the free variables produces the general solution,

$$\boldsymbol{x} = x_{f_1}\boldsymbol{h}_1 + x_{f_2}\boldsymbol{h}_2 + \cdots + x_{f_{n-r}}\boldsymbol{h}_{n-r}, \tag{2.12}$$

where $x_{f_1}, \cdots, x_{f_{n-r}}$ are the free variables and where $\boldsymbol{h}_1, \boldsymbol{h}_2, \ldots, \boldsymbol{h}_{n-r}$ are $n \times 1$ columns that represent particular solutions of the system. As the free variables range over all possible values, the general solution generates all possible solutions.

The form of the general solution (2.12) also makes transparent when the trivial solution $\boldsymbol{x} = \boldsymbol{0}$ is the *only* solution to a homogeneous system. As long as there is at least one free variable, then it is clear from (2.12) that there will be an infinite number of solutions. *Consequently, the trivial solution is the only solution if and only if there are no free variables.*

Because there are $n - r$ free variables, where r is the number of pivots, the above condition is equivalent to the statement that the trivial solution is the only solution to $\boldsymbol{A}\boldsymbol{x} = \boldsymbol{0}$ if and only if the number of pivots equals the number of columns. Because each column can have at most one pivot, this implies that there must be a pivot in every column. Also, each row can have at most one pivot. This implies that the trivial solution can be the only solution only for systems that have at least as many rows as columns. These observations yield the following important theorem.

Theorem 2.4 *If $\boldsymbol{A}$ is an $m \times n$ matrix with $n > m$, then $\boldsymbol{A}\boldsymbol{x} = \boldsymbol{0}$ has non-trivial (i.e., $\boldsymbol{x} \neq \boldsymbol{0}$) solutions.*

Proof. Since $\boldsymbol{A}$ has m rows, there are at most m pivots. With $n > m$, the system $\boldsymbol{A}\boldsymbol{x} = \boldsymbol{0}$ must have at least $n - m > 0$ free variables. This clearly shows the existence of nonzero solutions. In fact, there are an infinite number of nonzero solutions because any multiple $c\boldsymbol{x}$ is a solution if $\boldsymbol{x}$ is. $\square$

The general solution of homogeneous systems also help us describe the general solution of a non-homogeneous system $\boldsymbol{Ax} = \boldsymbol{b}$ with $\boldsymbol{b} \neq \boldsymbol{0}$. Let us revisit the solution set obtained for the non-homogeneous system $\boldsymbol{Ax} = \boldsymbol{b}$ in Example 2.2. Recall that we have the same coefficient matrix $\boldsymbol{A}$ as above, but $\boldsymbol{b} = [9, 4, 4, -5]'$. The general solution we obtained there can be expressed as

$$\boldsymbol{x} = \boldsymbol{x}_p + x_3\boldsymbol{h}_1 + x_5\boldsymbol{h}_2 + x_6\boldsymbol{h}_3,$$

where $\boldsymbol{x}_p = [-4, -10, 0, 1, 0, 0]'$ and where $\boldsymbol{h}_1$, $\boldsymbol{h}_2$ and $\boldsymbol{h}_3$ are as in Example 2.6. How does this solution relate to that of the corresponding homogeneous system $\boldsymbol{Ax} = \boldsymbol{0}$? Notice that $\boldsymbol{x}_p$ is a particular solution—it corresponds to setting all the free variables to zero. Thus, the general solution of the non-homogeneous system $\boldsymbol{Ax} = \boldsymbol{b}$ is given by the sum of a particular solution of $\boldsymbol{Ax} = \boldsymbol{b}$, say $\boldsymbol{x}_p$, and the general solution of the corresponding homogeneous system $\boldsymbol{Ax} = \boldsymbol{0}$.

More generally, for linear systems $\boldsymbol{Ax} = \boldsymbol{b}$ having m equations in n unknowns, the complete solution is $\boldsymbol{x} = \boldsymbol{x}_p + \boldsymbol{x}_h$. To a particular solution, $\boldsymbol{x}_p$, we add all solutions of $\boldsymbol{Ax}_h = \boldsymbol{0}$. The number of pivots and free variables in a linear system are characteristics of the coefficient matrix $\boldsymbol{A}$ and is the same irrespective of whether the system is homogeneous or not.

2.6 The inverse of a matrix

Definition 2.8 *An $n \times n$ square matrix $\boldsymbol{A}$ is said to be* **invertible** *or* **nonsingular** *if there exists an $n \times n$ matrix $\boldsymbol{B}$ such that*

$$\boldsymbol{AB} = \boldsymbol{I}_n \quad \text{and} \quad \boldsymbol{BA} = \boldsymbol{I}_n.$$

The matrix $\boldsymbol{B}$, when it exists, is called the **inverse** *of $\boldsymbol{A}$ and is denoted as $\boldsymbol{A}^{-1}$. A matrix that does not have an inverse is called a* **singular** *matrix.*

Note that an invertible matrix $\boldsymbol{A}$ *must* be a square matrix for the above definition to make sense and $\boldsymbol{A}^{-1}$ must have the same order as $\boldsymbol{A}$.

Example 2.7 Let $\boldsymbol{A} = \begin{bmatrix} a & b \\ c & d \end{bmatrix}$ be a 2×2 matrix with real entries such that $ad-bc \neq 0$. Then

$$\boldsymbol{B} = \frac{1}{ad-bc}\begin{bmatrix} d & -b \\ -c & a \end{bmatrix}$$

is the inverse of $\boldsymbol{A}$ because it is easily verified that $\boldsymbol{AB} = \boldsymbol{I} = \boldsymbol{BA}$. ■

Not all matrices have an inverse. For example, try solving $\boldsymbol{AX} = \boldsymbol{I}_2$ where $\boldsymbol{A}$ is 2×2 as in the above example but with $ad - bc = 0$ and see what happens. You can also verify that a square matrix with a row or a column of only zeroes does not have an inverse. In particular, the null matrix (with all zero entries) is an example (a rather extreme example!) of a singular matrix.

A linear system $\boldsymbol{Ax} = \boldsymbol{b}$, where $\boldsymbol{A}$ is $n \times n$, and $\boldsymbol{x}$ and $\boldsymbol{b}$ are $n \times 1$ is said to be a ***nonsingular system*** when $\boldsymbol{A}$ is nonsingular. For a nonsingular system, we note

$$\boldsymbol{Ax} = \boldsymbol{b} \Longrightarrow \boldsymbol{A}^{-1}\boldsymbol{Ax} = \boldsymbol{A}^{-1}\boldsymbol{b} \Longrightarrow \boldsymbol{x} = \boldsymbol{A}^{-1}\boldsymbol{b}\,.$$

Therefore, $\boldsymbol{x} = \boldsymbol{A}^{-1}\boldsymbol{b}$ is a solution. Matters are no more complicated when we have a collection of p nonsingular linear systems, $\boldsymbol{Ax}_i = \boldsymbol{b}_i$ for $i = 1, 2, \ldots, p$. We write this as $\boldsymbol{AX} = \boldsymbol{B}$, where $\boldsymbol{X}$ and $\boldsymbol{B}$ are $n \times p$ with columns $\boldsymbol{x}_i$ and $\boldsymbol{b}_i$, respectively. Multiplying both sides from the left by $\boldsymbol{A}^{-1}$ yields $\boldsymbol{X} = \boldsymbol{A}^{-1}\boldsymbol{B}$ as a solution.

Let $\boldsymbol{A}$ be an $n \times n$ matrix. From Definition 2.8 it is clear that for $\boldsymbol{A}^{-1}$ to exist we must be able to find a solution for $\boldsymbol{AX} = \boldsymbol{I}_n$. But is that enough? Should we not verify the second condition: that the solution $\boldsymbol{X}$ also satisfies $\boldsymbol{XA} = \boldsymbol{I}$? The following lemma shows that this second condition is implied by the first for square matrices and there is no need for such a verification.

Lemma 2.1 *Let $\boldsymbol{A}$ and $\boldsymbol{B}$ be square matrices of the same order. Then $\boldsymbol{BA} = \boldsymbol{I}$ if and only if $\boldsymbol{AB} = \boldsymbol{I}$.*

Proof. Suppose $\boldsymbol{A}$ and $\boldsymbol{B}$ are both $n \times n$ matrices such that $\boldsymbol{AB} = \boldsymbol{I}$. We show that $\boldsymbol{X} = \boldsymbol{A}$ is a solution for the system $\boldsymbol{BX} = \boldsymbol{I}$:

$$\boldsymbol{BX} = \boldsymbol{I} \Longrightarrow \boldsymbol{A}(\boldsymbol{BX}) = \boldsymbol{AI} = \boldsymbol{A} \Longrightarrow (\boldsymbol{AB})\boldsymbol{X} = \boldsymbol{A} \Longrightarrow \boldsymbol{X} = \boldsymbol{A}\,.$$

This proves the "if" part: $\boldsymbol{AB} = \boldsymbol{I} \Longrightarrow \boldsymbol{BA} = \boldsymbol{I}$. The only if part follows by simply interchanging the roles of $\boldsymbol{A}$ and $\boldsymbol{B}$. □

It follows immediately from Definition 2.8 and Lemma 2.1 that $\boldsymbol{B} = \boldsymbol{A}^{-1}$ if and only if $\boldsymbol{A} = \boldsymbol{B}^{-1}$.

Lemma 2.1 shows that it suffices to use *only one* of the equations, $\boldsymbol{AB} = \boldsymbol{I}$ or $\boldsymbol{BA} = \boldsymbol{I}$, in Definition 2.8 to unambiguously define the inverse of a square matrix. If, however, $\boldsymbol{A}$ is not square then Lemma 2.1 is no longer true. To be precise, if $\boldsymbol{A}$ is an $m \times n$ matrix, then there may exist an $n \times m$ matrix $\boldsymbol{B}$ such that $\boldsymbol{AB} = \boldsymbol{I}_m$ but $\boldsymbol{BA} \neq \boldsymbol{I}_n$. We say that $\boldsymbol{B}$ is a ***right inverse*** of $\boldsymbol{A}$. The reverse could also be true: there may exist an $n \times m$ matrix $\boldsymbol{C}$ such that $\boldsymbol{CA} = \boldsymbol{I}_n$ but $\boldsymbol{AC} \neq \boldsymbol{I}_m$. In that case we call $\boldsymbol{C}$ a ***left inverse*** of $\boldsymbol{A}$. And, in general, $\boldsymbol{B} \neq \boldsymbol{C}$. Lemma 2.1 tells us that if $\boldsymbol{A}$ is a square matrix then any right inverse is also a left inverse and vice-versa. The use of both $\boldsymbol{AB} = \boldsymbol{I}$ and $\boldsymbol{BA} = \boldsymbol{I}$ in Definition 2.8, where any one of them would have sufficed, emphasizes that when we say $\boldsymbol{B}$ is simply an inverse of a square matrix $\boldsymbol{A}$ we mean that $\boldsymbol{B}$ is both a left-inverse and a right inverse. We will study left-inverses and right inverses in Section 5.3. In this chapter we will consider only inverses of square matrices.

The following important consequences of Definition 2.8 are widely used. The first of these states that when the inverse of a square matrix exists it must be unique.

Theorem 2.5

(i) A nonsingular matrix cannot have two different inverses.

(ii) *If $\boldsymbol{A}$ and $\boldsymbol{B}$ are invertible matrices of the same order. Then the product $\boldsymbol{AB}$ is invertible and is given by*

$$(\boldsymbol{AB})^{-1} = \boldsymbol{B}^{-1}\boldsymbol{A}^{-1}.$$

More generally, if $\boldsymbol{A}_1, \boldsymbol{A}_2, \ldots, \boldsymbol{A}_n$ be a set of n invertible matrices of the same order, then their product is invertible and is given by

$$(\boldsymbol{A}_1 \cdots \boldsymbol{A}_n)^{-1} = \boldsymbol{A}_n^{-1}\boldsymbol{A}_{n-1}^{-1} \cdots \boldsymbol{A}_1^{-1}. \tag{2.13}$$

(iii) *Let $\boldsymbol{A}$ be any $n \times n$ invertible matrix. Then $(\boldsymbol{A}')^{-1} = (\boldsymbol{A}^{-1})'$.*

Proof. **Proof of (i):** Suppose $\boldsymbol{B}$ and $\boldsymbol{C}$ are two different inverses for $\boldsymbol{A}$. Then, $\boldsymbol{B}$ satisfies $\boldsymbol{BA} = \boldsymbol{I}$ and $\boldsymbol{C}$ satisfies $\boldsymbol{AC} = \boldsymbol{I}$. The associative law of matrix multiplication now gives $\boldsymbol{B}(\boldsymbol{AC}) = (\boldsymbol{BA})\boldsymbol{C}$ which immediately gives $\boldsymbol{BI} = \boldsymbol{IC}$ or $\boldsymbol{B} = \boldsymbol{C}$.

Proof of (ii): This follows from multiplying $\boldsymbol{AB}$ on the left with $\boldsymbol{A}^{-1}$ followed by $\boldsymbol{B}^{-1}$ to get

$$\boldsymbol{B}^{-1}\boldsymbol{A}^{-1}(\boldsymbol{AB}) = \boldsymbol{B}^{-1}(\boldsymbol{A}^{-1}\boldsymbol{A})(\boldsymbol{B}) = \boldsymbol{B}^{-1}\boldsymbol{IB} = \boldsymbol{I}.$$

Similarly, multiplying $\boldsymbol{AB}$ on the right with $\boldsymbol{B}^{-1}$ followed by $\boldsymbol{A}^{-1}$ equals $\boldsymbol{AIA}^{-1} = \boldsymbol{AA}^{-1} = \boldsymbol{I}$. This proves that the inverse of the product of two matrices comes in reverse order. The same idea applies to a general set of n invertible matrices and we have

$$\begin{aligned}
\boldsymbol{A}_n^{-1}\boldsymbol{A}_{n-1}^{-1} \cdots \boldsymbol{A}_1^{-1}(\boldsymbol{A}_1\boldsymbol{A}_2 \cdots \boldsymbol{A}_n) &= \boldsymbol{A}_n^{-1}\boldsymbol{A}_{n-1}^{-1} \cdots \boldsymbol{A}_2(\boldsymbol{A}_1^{-1}\boldsymbol{A}_1)(\boldsymbol{A}_2 \cdots \boldsymbol{A}_n) \\
&= \boldsymbol{A}_n^{-1}\boldsymbol{A}_{n-1}^{-1} \cdots \boldsymbol{A}_3(\boldsymbol{A}_2^{-1}\boldsymbol{A}_2)\boldsymbol{A}_3 \cdots \boldsymbol{A}_n) \\
&= \boldsymbol{A}_n^{-1}\boldsymbol{A}_{n-1}^{-1} \cdots \boldsymbol{A}_4(\boldsymbol{A}_3^{-1}\boldsymbol{A}_3)\boldsymbol{A}_4 \cdots \boldsymbol{A}_n) \\
&= \cdots = \cdots \\
&= \boldsymbol{A}_n^{-1}(\boldsymbol{A}_{n-1}^{-1}\boldsymbol{A}_{n-1})\boldsymbol{A}_n = \boldsymbol{A}_n^{-1}\boldsymbol{A}_n = \boldsymbol{I}\,.
\end{aligned}$$

Similarly, multiplying $\boldsymbol{A}_1\boldsymbol{A}_2 \cdots \boldsymbol{A}_n$ on the right with $\boldsymbol{A}_n^{-1}\boldsymbol{A}_{n-1}^{-1} \cdots \boldsymbol{A}_1^{-1}$ again produces the identity matrix. This proves (2.13).

Proof of (iii): Let $\boldsymbol{X} = (\boldsymbol{A}^{-1})'$. Now, using the reverse order law for transposition of matrix products (Lemma 1.3) to write

$$\boldsymbol{A}'\boldsymbol{X} = \boldsymbol{A}(\boldsymbol{A}^{-1})' = (\boldsymbol{A}^{-1}\boldsymbol{A})' = \boldsymbol{I}' = \boldsymbol{I}.$$

This proves $\boldsymbol{X} = (\boldsymbol{A}')^{-1}$. □

Part (i) of Theorem 2.5 ensures that $\boldsymbol{x} = \boldsymbol{A}^{-1}\boldsymbol{b}$ and $\boldsymbol{X} = \boldsymbol{A}^{-1}\boldsymbol{B}$ are the ***unique*** solutions for nonsingular systems $\boldsymbol{Ax} = \boldsymbol{b}$ and $\boldsymbol{AX} = \boldsymbol{B}$, respectively. We emphasize, however, that in practice it is inefficient to solve $\boldsymbol{Ax} = \boldsymbol{b}$ by first computing $\boldsymbol{A}^{-1}$ and then evaluating $\boldsymbol{x} = \boldsymbol{A}^{-1}\boldsymbol{b}$. In fact, in the rare cases where an inverse may be needed, it is obtained by solving linear systems.

So, given an $n \times n$ matrix $\boldsymbol{A}$, how do we compute its inverse? Since the inverse must satisfy the system $\boldsymbol{AX} = \boldsymbol{I}$, we can solve $\boldsymbol{Ax}_j = \boldsymbol{e}_j$ for $j = 1, \ldots, n$ using Gaussian elimination. Lemma 2.1 guarantees that the solution $\boldsymbol{X} = [\boldsymbol{x}_1 : \ldots : \boldsymbol{x}_n]$ will also satisfy $\boldsymbol{XA} = \boldsymbol{I}$ and is precisely the inverse of $\boldsymbol{A}$, i.e., $\boldsymbol{X} = \boldsymbol{A}^{-1}$. Also, part (i) of Theorem 2.5 tells us that such an $\boldsymbol{X}$ must be the unique solution. We now illustrate with some examples.

Example 2.8 Computing an inverse using Gaussian elimination. Let us return to the coefficient matrix in Example 2.4, but replace $\boldsymbol{b}$ in that example with $\boldsymbol{e}_1, \ldots, \boldsymbol{e}_4$. We write down the augmented matrix:

$$[\boldsymbol{A} : \boldsymbol{I}] = \left[\begin{array}{rrrr|rrrr} 2 & 3 & 0 & 0 & 1 & 0 & 0 & 0 \\ 4 & 7 & 2 & 0 & 0 & 1 & 0 & 0 \\ -6 & -10 & 0 & 1 & 0 & 0 & 1 & 0 \\ 4 & 6 & 4 & 5 & 0 & 0 & 0 & 1 \end{array}\right]. \tag{2.14}$$

Using same sequence of elementary operations as in Example 2.4, we can reduce the system to the upper-triangular system:

$$\boldsymbol{E}[\boldsymbol{A} : \boldsymbol{I}] = \left[\begin{array}{rrrr|rrrr} 2 & 3 & 0 & 0 & 1 & 0 & 0 & 0 \\ 0 & 1 & 2 & 0 & -2 & 1 & 0 & 0 \\ 0 & 0 & 2 & 1 & 1 & 1 & 1 & 0 \\ 0 & 0 & 0 & 3 & -4 & -2 & -2 & 1 \end{array}\right] = [\boldsymbol{U} : \boldsymbol{B}], \tag{2.15}$$

where $\boldsymbol{E} = \boldsymbol{E}_{43}(-2)\boldsymbol{E}_{32}(1)\boldsymbol{E}_{41}(-2)\boldsymbol{E}_{31}(3)\boldsymbol{E}_{21}(-2)$. One could now solve the upper-triangular system $\boldsymbol{Ux}_j = \boldsymbol{B}_{*j}$ for each $j = 1, \ldots, 4$. The resulting inverse turns out to be

$$\boldsymbol{A}^{-1} = \boldsymbol{X} = [\boldsymbol{x}_1 : \ldots : \boldsymbol{x}_4] = \begin{bmatrix} 7 & 1 & 5/2 & 1/2 \\ -13/3 & -2/3 & -5/3 & 1/3 \\ 7/6 & 5/6 & 5/6 & -1/6 \\ -4/3 & -2/3 & -2/3 & 1/3 \end{bmatrix}. \blacksquare \tag{2.16}$$

We now illustrate the ***Gauss-Jordan elimination*** using the same example. Let us start where we left off with the augmented system in (2.15). To reduce $\boldsymbol{U}$ to $\boldsymbol{I}$, we proceed as in Example 2.3 by first sweeping out the first column:

$$\begin{array}{c} E_1(1/2) \\ \\ E_3(1/2) \\ E_4(1/3) \end{array} \left[\begin{array}{rrrr|rrrr} 2 & 3 & 0 & 0 & 1 & 0 & 0 & 0 \\ 0 & 1 & 2 & 0 & -2 & 1 & 0 & 0 \\ 0 & 0 & 2 & 1 & 1 & 1 & 1 & 0 \\ 0 & 0 & 0 & 3 & -4 & -2 & -2 & 1 \end{array}\right]$$

$$\rightarrow \left[\begin{array}{rrrr|rrrr} 1 & 3/2 & 0 & 0 & 1/2 & 0 & 0 & 0 \\ 0 & 1 & 2 & 0 & -2 & 1 & 0 & 0 \\ 0 & 0 & 1 & 1/2 & 1/2 & 1/2 & 1/2 & 0 \\ 0 & 0 & 0 & 1 & -4/3 & -2/3 & -2/3 & 1/3 \end{array}\right].$$

Now we proceed to sweep out the second column (elements both above and below the pivot):

$$E_{12}(-3/2)\quad \left[\begin{array}{cccc|cccc} 1 & 3/2 & 0 & 0 & 1/2 & 0 & 0 & 0 \\ 0 & 1 & 2 & 0 & -2 & 1 & 0 & 0 \\ 0 & 0 & 1 & 1/2 & 1/2 & 1/2 & 1/2 & 0 \\ 0 & 0 & 0 & 1 & -4/3 & -2/3 & -2/3 & 1/3 \end{array}\right]$$

$$\rightarrow \left[\begin{array}{cccc|cccc} 1 & 0 & -3 & 0 & 7/2 & -3/2 & 0 & 0 \\ 0 & 1 & 2 & 0 & -2 & 1 & 0 & 0 \\ 0 & 0 & 1 & 1/2 & 1/2 & 1/2 & 1/2 & 0 \\ 0 & 0 & 0 & 1 & -4/3 & -2/3 & -2/3 & 1/3 \end{array}\right].$$

Then onto the third and fourth columns:

$$\begin{array}{l} E_{13}(3) \\ E_{23}(-2) \\ \\ \\ \end{array} \left[\begin{array}{cccc|cccc} 1 & 0 & -3 & 0 & 7/2 & -3/2 & 0 & 0 \\ 0 & 1 & 2 & 0 & -2 & 1 & 0 & 0 \\ 0 & 0 & 1 & 1/2 & 1/2 & 1/2 & 1/2 & 0 \\ 0 & 0 & 0 & 1 & -4/3 & -2/3 & -2/3 & 1/3 \end{array}\right]$$

$$\rightarrow \left[\begin{array}{cccc|cccc} 1 & 0 & 0 & 3/2 & 5 & 0 & 3/2 & 0 \\ 0 & 1 & 0 & -1 & -3 & 0 & -1 & 0 \\ 0 & 0 & 1 & 1/2 & 1/2 & 1/2 & 1/2 & 0 \\ 0 & 0 & 0 & 1 & -4/3 & -2/3 & -2/3 & 1/3 \end{array}\right];$$

$$\begin{array}{l} E_{14}(-3/2) \\ E_{24}(1) \\ E_{34}(-1/2) \\ \\ \end{array} \left[\begin{array}{cccc|cccc} 1 & 0 & 0 & 3/2 & 5 & 0 & 3/2 & 0 \\ 0 & 1 & 0 & -1 & -3 & 0 & -1 & 0 \\ 0 & 0 & 1 & 1/2 & 1/2 & 1/2 & 1/2 & 0 \\ 0 & 0 & 0 & 1 & -4/3 & -2/3 & -2/3 & 1/3 \end{array}\right]$$

$$\rightarrow \left[\begin{array}{cccc|cccc} 1 & 0 & 0 & 0 & 7 & 1 & 5/2 & -1/2 \\ 0 & 1 & 0 & 0 & -13/3 & -2/3 & -5/3 & 1/3 \\ 0 & 0 & 1 & 0 & 7/6 & 5/6 & 5/6 & -1/6 \\ 0 & 0 & 0 & 1 & -4/3 & -2/3 & -2/3 & 1/3 \end{array}\right].$$

Therefore, Gauss-Jordan reduces the augmented system to $[\boldsymbol{I} : \boldsymbol{A}^{-1}]$. ■

In fact, Gauss-Jordan elimination amounts to premultiplying the matrix $[\boldsymbol{A} : \boldsymbol{I}]$ by an invertible matrix $\boldsymbol{L}$ (formed by a composition of elementary operations) so that

$$\boldsymbol{L}[\boldsymbol{A} : \boldsymbol{I}] = [\boldsymbol{LA} : \boldsymbol{L}] = [\boldsymbol{I} : \boldsymbol{L}] \Rightarrow \boldsymbol{L} = \boldsymbol{A}^{-1}.$$

The inverse of a diagonal matrix none of whose diagonal elements are zero is especially easy to find:

$$\text{If } \boldsymbol{A} = \begin{bmatrix} a_{11} & 0 & \dots & 0 \\ 0 & a_{22} & \dots & 0 \\ 0 & 0 & \ddots & 0 \\ 0 & 0 & \dots & a_{nn} \end{bmatrix} \text{ then } \boldsymbol{A}^{-1} = \begin{bmatrix} \frac{1}{a_{11}} & 0 & \dots & 0 \\ 0 & \frac{1}{a_{22}} & \dots & 0 \\ 0 & 0 & \ddots & 0 \\ 0 & 0 & \dots & \frac{1}{a_{nn}} \end{bmatrix}.$$

Elementary matrices as defined in Definition 2.5, are nonsingular. In particular, the inverses of elementary matrices of Types I, II and III are again elementary matrices of the same type. We prove this in the following lemma.

Lemma 2.2 *Any elementary matrix is nonsingular and has an inverse given by*

$$(\boldsymbol{I} + \boldsymbol{u}\boldsymbol{v}')^{-1} = \boldsymbol{I} - \frac{\boldsymbol{u}\boldsymbol{v}'}{1 - \boldsymbol{v}'\boldsymbol{u}} \,. \tag{2.17}$$

Furthermore, if $\boldsymbol{E}$ is an elementary matrix of Type I, II or III, then $\boldsymbol{E}$ is nonsingular and $\boldsymbol{E}^{-1}$ is an elementary matrix of the same type as $\boldsymbol{E}$.

Proof. First of all, recall from Definition 2.5 that $\boldsymbol{v}'\boldsymbol{u} \neq -1$, so the expression on the right-hand side of (2.17) is well-defined. To prove (2.17), we simply multiply $\boldsymbol{I} + \boldsymbol{u}\boldsymbol{v}'$ by the matrix on the right-hand side of (2.17) and show that it equals $\boldsymbol{I}$. More precisely,

$$\begin{aligned}(\boldsymbol{I} + \boldsymbol{u}\boldsymbol{v}')\left(\boldsymbol{I} - \frac{\boldsymbol{u}\boldsymbol{v}'}{1+\boldsymbol{v}'\boldsymbol{u}}\right) &= \boldsymbol{I} + \boldsymbol{u}\boldsymbol{v}' - \frac{\boldsymbol{u}\boldsymbol{v}'}{1+\boldsymbol{v}'\boldsymbol{u}} - \frac{\boldsymbol{u}\boldsymbol{v}'\boldsymbol{u}\boldsymbol{v}'}{1+\boldsymbol{v}'\boldsymbol{u}} \\ &= \boldsymbol{I} + \boldsymbol{u}\left(1 - \frac{1}{1+\boldsymbol{v}'\boldsymbol{u}} - \frac{\boldsymbol{v}'\boldsymbol{u}}{1+\boldsymbol{v}'\boldsymbol{u}}\right)\boldsymbol{v}' = \boldsymbol{I}\,,\end{aligned}$$

where the last equality follows because

$$1 - \frac{1}{1+\boldsymbol{v}'\boldsymbol{u}} - \frac{\boldsymbol{v}'\boldsymbol{u}}{1+\boldsymbol{v}'\boldsymbol{u}} = \frac{1+\boldsymbol{v}'\boldsymbol{u} - 1 - \boldsymbol{v}'\boldsymbol{u}}{1+\boldsymbol{v}'\boldsymbol{u}} = 0\,.$$

Similarly, one can prove that $\left(\boldsymbol{I} - \frac{\boldsymbol{u}\boldsymbol{v}'}{1+\boldsymbol{v}'\boldsymbol{u}}\right)(\boldsymbol{I} + \boldsymbol{u}\boldsymbol{v}') = \boldsymbol{I}$, thereby completing the verification of (2.17). (A derivation of (2.17) when one is not given the right-hand side is provided by the Sherman-Woodbury-Morrison formula that we discuss later in Section 3.7.)

We next prove that the inverses of elementary matrices of Types I, II and III are again elementary matrices of the same type.

Let $\boldsymbol{E} = \boldsymbol{E}_{ij}$ be an elementary matrix of Type I. From (2.8), we have that $\boldsymbol{E}_{ij} = \boldsymbol{I} - \boldsymbol{u}_{ij}\boldsymbol{u}_{ij}'$, where $\boldsymbol{u}_{ij} = \boldsymbol{e}_j - \boldsymbol{e}_i$. Notice that $\boldsymbol{u}_{ij}'\boldsymbol{u}_{ij} = 2$. Then,

$$\boldsymbol{E}_{ij}^2 = (\boldsymbol{I} - \boldsymbol{u}_{ij}\boldsymbol{u}_{ij}')(\boldsymbol{I} - \boldsymbol{u}_{ij}\boldsymbol{u}_{ij}') = \boldsymbol{I} - 2\boldsymbol{u}_{ij}\boldsymbol{u}_{ij}' + \boldsymbol{u}_{ij}(\boldsymbol{u}_{ij}'\boldsymbol{u}_{ij})\boldsymbol{u}_{ij}' = \boldsymbol{I}.$$

Therefore $\boldsymbol{E}_{ij}^{-1} = \boldsymbol{E}_{ij}$, i.e., $\boldsymbol{E}_{ij}$ is its own inverse. This is also obvious because $\boldsymbol{E}_{ij}$ is obtained by interchanging the i-th and j-th rows of the identity matrix. Therefore, applying $\boldsymbol{E}_{ij}$ to itself will restore the interchanged rows to their original positions producing the identity matrix again.

Next suppose $\boldsymbol{E} = \boldsymbol{E}_i(\alpha)$ is an elementary matrix of Type II. $\boldsymbol{E}_i(\alpha)$ is obtained by multiplying the i-th row of the identity matrix by α. Therefore, multiplying that same row by $1/\alpha$ will restore the orginal matrix. This means that $\boldsymbol{E}_i(1/\alpha)$ is the inverse of $\boldsymbol{E}_i(\alpha)$. This can also be shown by using $\boldsymbol{E}_i(\alpha) = \boldsymbol{I} + (\alpha - 1)\boldsymbol{e}_i\boldsymbol{e}_i'$ (see (2.9)) and noting that

$$\begin{aligned}E_i(\alpha)E_i(1/\alpha) &= (I + (\alpha-1)e_ie_i')(I + (1/\alpha-1)e_ie_i')\\&= I + (\alpha-1)e_ie_i' + (1/\alpha-1)e_ie_i' + (\alpha-1)(1/\alpha-1)e_i(e_i'e_i)e_i'\\&= I + (\alpha-1)e_ie_i' + (1/\alpha-1)e_ie_i' + (\alpha-1)(1/\alpha-1)e_ie_i'\\&= I + \{\alpha-1+1/\alpha-1+(\alpha-1)(1/\alpha-1)\}e_ie_i'\\&= I + \frac{1}{\alpha}(\alpha^2-2\alpha+1-(\alpha-1)^2)e_ie_i' = I + \{0\}e_ie_i' = I\,.\end{aligned}$$

Similarly, we can show that $E_i(1/\alpha)E_i(\alpha) = I$. Therefore $E_i(\alpha)^{-1} = E_i(1/\alpha)$, which is a Type II elementary matrix.

Finally, let $E = E_{ij}(\beta)$ be a Type III elementary matrix. Then, from (2.10) we know that $E_{ij}(\beta) = I + \beta e_je_i'$. Now write

$$\begin{aligned}E_{ij}(\beta)E_{ij}(-\beta) &= (I + \beta e_je_i')(I - \beta e_je_i') = I - \beta^2 e_je_i'e_je_i'\\&= I - \beta^2 e_j(e_i'e_j)e_i' = I,\end{aligned}$$

where we have used the fact that $e_i'e_j = 0$. Similarly, we can verify that $E_{ij}(-\beta)E_{ij}(\beta) = I$. This proves that $E_{ij}(\beta)^{-1}E_{ij}(-\beta)$, which is again of Type III. This concludes the proof. □

The invertibility (or nonsingularity) of the elementary matrices immediately leads to the following two useful results. The first is a direct consequence of Gaussian elimination algorithm, while the second emerges from the Gauss-Jordan algorithm.

Lemma 2.3 *Let A be an $m \times n$ matrix and assume $A \neq O$ (so at least one of its elements is nonzero). Then the following statements are true:*

(i) There exists an invertible matrix G such that $GA = U$, where U is in row echelon form.

(ii) There exists an invertible matrix H such that $HA = E_A$, where E_A is in reduced row echelon form (RREF).

Proof. **Proof of (i):** Gaussian elimination using elementary row operations will reduce A to an echelon matrix U. Each such row operation amounts to multiplication on the left by an $m \times m$ elementary matrix of Type I or III. Let $E_1, E_2, \ldots, E_k$ be the set of elementary matrices. Then setting $G = E_1E_2\cdots E_k$, we have that $GA = U$. Since G is the product of invertible matrices, G itself is invertible (Part (ii) of Theorem 2.5). This proves part (i).

Proof of (ii): This follows by simply using the Gauss-Jordan elimination rather than Gaussian elimination in the proof for part (i). □

The Gauss-Jordan algorithm immediately leads to the following characterization of nonsingular matrices.

Theorem 2.6 *A is a nonsingular matrix if and only if A is the product of elementary matrices of Type I, II, or III.*

Proof. If $\boldsymbol{A}$ is nonsingular then Gauss-Jordan elimination reduces $\boldsymbol{A}$ to $\boldsymbol{I}$ using elementary row operations. In other words, let $\boldsymbol{G}_1, \ldots, \boldsymbol{G}_k$ denote the elementary matrices of Type I, II, or III that reduce $\boldsymbol{A}$ to $\boldsymbol{I}$. Then $\boldsymbol{G}_k \boldsymbol{G}_{k-1} \cdots \boldsymbol{G}_1 \boldsymbol{A} = \boldsymbol{I}$, which implies that $\boldsymbol{A} = \boldsymbol{G}_1^{-1} \cdots \boldsymbol{G}_{k-1}^{-1} \boldsymbol{G}_k^{-1}$. Note that the inverse of an elementary matrix is again an elementary matrix of the same type, i.e., each $\boldsymbol{G}_j^{-1}$ is an elementary matrix of the same type as $\boldsymbol{G}_j$. Therefore, $\boldsymbol{A}$ is the product of elementary matrices of Type I, II, or III.

Conversely, suppose $\boldsymbol{A} = \boldsymbol{E}_1 \cdots \boldsymbol{E}_k$, where each $\boldsymbol{E}_j$ is an elementary matrix. Then $\boldsymbol{A}$ must be nonsingular because the $\boldsymbol{E}_i$'s are nonsingular and a product of nonsingular matrices is also nonsingular. □

Not all square matrices have an inverse. One immediate example is the null matrix. Another simple example is a diagonal matrix at least one of whose diagonal elements is zero. An easy way to verify if a matrix is nonsingular is to check if $\boldsymbol{Ax} = \boldsymbol{0}$ implies that $\boldsymbol{x} = \boldsymbol{0}$. In other words, is the trivial solution the *only* solution? The following lemma throws light on this matter.

Lemma 2.4 *Let $\boldsymbol{A}$ be an $n \times n$ matrix. The homogeneous system $\boldsymbol{Ax} = \boldsymbol{0}$ will have the trivial solution $\boldsymbol{x} = \boldsymbol{0}$ as its only solution if and only if $\boldsymbol{A}$ is nonsingular.*

Proof. Suppose $\boldsymbol{A}$ is nonsingular. Then we would be able to premultiply both sides of $\boldsymbol{Ax} = \boldsymbol{0}$ by its inverse to obtain

$$\boldsymbol{Ax} = \boldsymbol{0} \Rightarrow \boldsymbol{x} = \boldsymbol{A}^{-1}\boldsymbol{0} = \boldsymbol{0} \, .$$

This proves the "if" part. Now suppose $\boldsymbol{Ax} = \boldsymbol{0} \Rightarrow \boldsymbol{x} = \boldsymbol{0}$. This implies that $\boldsymbol{A}$ will have n pivots and one will be able to solve for $\boldsymbol{x}_j$ in each of the systems $\boldsymbol{Ax}_j = \boldsymbol{e}_j$ for $j = 1, 2, \ldots, n$. This proves the "only if" part. □

If we apply Gaussian elimination to reduce a matrix $\boldsymbol{A}$ to its upper-triangular form which has at least one zero entry in its diagonal, then $\boldsymbol{A}$ will not have an inverse because it will not lead to a unique solution for $\boldsymbol{Ax} = \boldsymbol{0}$. For example, we saw in Example 2.5 that the system $\boldsymbol{Ax} = \boldsymbol{0}$, where

$$\boldsymbol{A} = \begin{bmatrix} 1 & 2 & 3 \\ 4 & 5 & 6 \\ 7 & 8 & 9 \end{bmatrix} ,$$

had non-trivial solutions. Therefore, $\boldsymbol{A}$ is singular.

Example 2.9 Polynomial interpolation and the Vandermonde matrix. Suppose we are given a set of n observations on two variables x and y, say

$$\{(x_1, y_1), (x_2, y_2), \ldots, (x_n, y_n)\} ,$$

where $n \geq 2$. We want to fit a polynomial $p(t)$ to the data so that $y_i = p(x_i)$ for $i = 1, 2, \ldots, n$. This is an exercise in curve fitting. A polynomial with degree $n - 1$

is the function $p(t) = \beta_0 + \beta_1 t + \beta_2 t^2 + \cdots + \beta_{n-1} t^{n-1}$. The conditions $y_i = p(x_i)$ for $i = 1, 2, \ldots, n$ implies the following linear system

$$\begin{bmatrix} 1 & x_1 & x_1^2 & \ldots & x_1^{n-1} \\ 1 & x_2 & x_2^2 & \ldots & x_2^{n-1} \\ 1 & x_3 & x_3^2 & \ldots & x_3^{n-1} \\ \vdots & \vdots & \vdots & \ddots & \vdots \\ 1 & x_n & x_n^2 & \ldots & x_n^{n-1} \end{bmatrix} \begin{bmatrix} \beta_0 \\ \beta_1 \\ \beta_2 \\ \vdots \\ \beta_{n-1} \end{bmatrix} = \begin{bmatrix} y_1 \\ y_2 \\ y_3 \\ \vdots \\ y_n \end{bmatrix}, \tag{2.18}$$

which we write as $\boldsymbol{V\beta} = \boldsymbol{y}$, where $\boldsymbol{V}$ is the $n \times n$ coefficient matrix, and $\boldsymbol{\beta}$ and $\boldsymbol{y}$ are the $n \times 1$ vectors of β_i's and y_i's in (2.18). The matrix $\boldsymbol{V}$ is named after the French mathematician Alexandre-Théophile Vandermonde and called the ***Vandermonde matrix*** and we refer to (2.18) as the Vandermonde linear system.

Is the above system nonsingular? Clearly, if any two x_i's are equal then $\boldsymbol{V}$ has identical rows and will be singular. Let us assume that all the x_i's are distinct. To see if $\boldsymbol{V}$ is nonsingular, we will use Lemma 2.4. Consider the homogeneous system,

$$\begin{bmatrix} 1 & x_1 & x_1^2 & \ldots & x_1^{n-1} \\ 1 & x_2 & x_2^2 & \ldots & x_2^{n-1} \\ 1 & x_3 & x_3^2 & \ldots & x_3^{n-1} \\ \vdots & \vdots & \vdots & \ddots & \vdots \\ 1 & x_n & x_n^2 & \ldots & x_n^{n-1} \end{bmatrix} \begin{bmatrix} \beta_0 \\ \beta_1 \\ \beta_2 \\ \vdots \\ \beta_{n-1} \end{bmatrix} = \begin{bmatrix} 0 \\ 0 \\ 0 \\ \vdots \\ 0 \end{bmatrix}.$$

Then, for each $i = 1, 2, \ldots, n$, we have

$$p(x_i) = \beta_0 + \beta_1 x_i + \beta_2 x_i^2 + \cdots + \beta_{n-1} x_i^{n-1} = 0 .$$

This implies that each x_i is a root of the polynomial $p(x)$. Thus, $p(x)$ has n roots. But this contradicts the fundamental theorem of algebra, which states that a polynomial of degree $n-1$ can have at most $n-1$ roots *unless* it is the zero polynomial. Therefore, $p(x)$ must be the zero polynomial, which implies that $\beta_j = 0$ for $j = 0, 1, 2, \ldots, n-1$. Thus, $\boldsymbol{V\beta} = \boldsymbol{0}$ implies that $\boldsymbol{\beta} = \boldsymbol{0}$. Lemma 2.4 ensures that $\boldsymbol{V}$ is nonsingular.

We mentioned earlier that it is computationally inefficient to solve nonsingular systems by actually inverting the coefficient matrix. Let us elaborate with the Vandermonde linear system in (2.18). Suppose we wanted to compute the inverse of the Vandermonde matrix. This would, however, be tedious. Fortunately, we can solve (2.18) very cheaply by "deriving" the solution from a well-known result in polynomial interpolation. Consider the ***Lagrange interpolating polynomial*** of degree $n-1$

$$l(t) = \gamma_1(t) y_1 + \gamma_2(t) y_2 + \cdots + \gamma_n(t) y_n , \quad \text{where } \gamma_i(t) = \frac{\prod_{j=1 \neq i}^{n} (t - x_j)}{\prod_{j=1 \neq i}^{n} (x_i - x_j)} . \tag{2.19}$$

Clearly, each $\gamma_i(t)$ is a polynomial in t of degree $n-1$, so $l(t)$ is also a polynomial in t of degree $n-1$. Also, it is easy to verify that each $\gamma_i(t) = 1$ if $t = x_i$ and 0 if $t = x_j$ for $j \neq i$. Therefore, $l(x_i) = y_i$ for $i = 1, 2, \ldots, n$ so it is an interpolating polynomial. If we rewrite $l(t)$ as

$$l(t) = \beta_0 + \beta_1 t + \beta_2 t^2 + \cdots + \beta_{n-1} t^{n-1} ,$$

it is clear that $\boldsymbol{\beta} = [\beta_1 : \beta_2 : \ldots : \beta_{n-1}]'$ satisfies $\boldsymbol{V\beta} = \boldsymbol{y}$. Since the system in (2.18) is nonsingular $\boldsymbol{\beta}$ is the unique solution. ■

Testing non-singularity of matrices by forming homogeneous linear systems can help derive several important properties. The following lemma is an example. An important property of triangular matrices is that their inverses, when they exist, are also triangular. We state and prove this in the next theorem.

Theorem 2.7 *An $n \times n$ lower (upper) triangular matrix with all diagonal entries nonzero is invertible and its inverse is lower (upper) triangular. Also, each diagonal element of the inverse (when it exists) is the reciprocal of the corresponding diagonal element of the original matrix.*

Proof. Let $\boldsymbol{L}$ be an $n \times n$ lower-triangular matrix given by

$$\begin{bmatrix} l_{11} & 0 & 0 & \ldots & 0 \\ l_{21} & l_{22} & 0 & \ldots & 0 \\ l_{31} & l_{32} & l_{33} & \ldots & 0 \\ \vdots & \vdots & \vdots & \ddots & \vdots \\ l_{n1} & l_{n2} & l_{n3} & \ldots & l_{nn} \end{bmatrix},$$

where $l_{ii} \neq 0$ for $i = 1, 2, \ldots, n$. To see whether the inverse of $\boldsymbol{L}$ exists, we check whether $\boldsymbol{x} = \boldsymbol{0}$ is the only solution of the homogeneous system $\boldsymbol{Lx} = \boldsymbol{0}$. We write the system as

$$\begin{array}{lllllll} l_{11}x_1 & & & & & = & 0, \\ l_{21}x_1 & + & l_{22}x_2 & & & = & 0, \\ l_{31}x_1 & + & l_{32}x_2 & + & l_{33}x_3 & = & 0, \\ & & & & \vdots & & \\ l_{n1}x_1 & + & l_{n2}x_2 & + & \cdots \quad + \quad \cdots \quad l_{nn}x_n & = & 0. \end{array}$$

Since $l_{11} \neq 0$, the first equation yields $x_1 = 0$. Substituting $x_1 = 0$ in the second equation yields $x_2 = 0$ because $l_{22} \neq 0$. With $x_1 = x_2 = 0$, the third equation simplifies to $l_{33}x_3 = 0$, which implies that $x_3 = 0$. Proceeding in this manner, we eventually arrive at $x_n = 0$ at the n-th equation. Therefore, $\boldsymbol{Lx} = \boldsymbol{0} \Rightarrow \boldsymbol{x} = \boldsymbol{0}$ which ensures the existence of the inverse of $\boldsymbol{L}$.

It remains to show that the inverse is also a lower-triangular matrix. The inverse will be obtained by solving for $\boldsymbol{X}$ in $\boldsymbol{LX} = \boldsymbol{I}$. In other words, we solve the n systems:

$$\boldsymbol{L}[\boldsymbol{x}_{*1} : \boldsymbol{x}_{*2} : \ldots : \boldsymbol{x}_{*n}] = [\boldsymbol{e}_1 : \boldsymbol{e}_2 : \ldots : \boldsymbol{e}_n].$$

We need to prove that the first $j-1$ elements in $\boldsymbol{x}_{*j}$ are zero for each $j = 2, \ldots, n$. But notice that the first $j-1$ equations of $\boldsymbol{Lx}_{*j} = \boldsymbol{e}_j$ form a homogeneous system by themselves: $\boldsymbol{L}_{11}\boldsymbol{x}_1 = \boldsymbol{0}$, where $\boldsymbol{L}_{11}$ is the $(j-1) \times (j-1)$ lower-triangular matrix formed from the first $(j-1)$ rows and columns of $\boldsymbol{L}$ and $\boldsymbol{x}_1 =$

$(x_{1j}, x_{2j}, \ldots, x_{(j-1),j})'$. Since the diagonal elements of $\boldsymbol{L}_{11}$ are all nonzero, this homogeneous system would also imply that $\boldsymbol{x}_1 = \boldsymbol{0}$.

Applying the above argument to $\boldsymbol{L}\boldsymbol{x}_{*j} = \boldsymbol{e}_j$ for $j = 2, , \ldots, n$ reveals that $x_{ij} = 0$ whenever $i < j$ in the matrix $\boldsymbol{X} = [x_{ij}]$ that solves $\boldsymbol{L}\boldsymbol{X} = \boldsymbol{I}$. Hence $\boldsymbol{L}^{-1}$ is lower-triangular. The proof for upper-triangular matrices proceeds analogous to the preceding arguments or, even simpler, by taking transposes and using Part (ii) of Theorem 2.5.

It remains to prove that each diagonal element of $\boldsymbol{T}^{-1}$, where $\boldsymbol{T}$ is triangular (lower or upper), is the reciprocal of the corresponding diagonal element of $\boldsymbol{T}$. This is a straightforward verification. Observe that the i-th diagonal element of $\boldsymbol{T}\boldsymbol{T}^{-1}$ is simply the product of the i-th diagonal elements of $\boldsymbol{T}$ and $\boldsymbol{T}^{-1}$. Since $\boldsymbol{T}\boldsymbol{T}^{-1} = \boldsymbol{I}$, the result follows. □

It is worth pointing out that if a triangular matrix is invertible, then its diagonal elements must all be nonzero. This is the converse of the above theorem. We leave this to the reader to verify (show that if any one of the diagonal elements is indeed zero then the triangular homogeneous system will have a non-trivial solution).

2.7 Exercises

1. Using Gaussian elimination solve the linear equation

$$\begin{array}{rcrcrcr} 2x_1 & - & x_1 & & & = & 0, \\ -x_1 & + & 2x_2 & + & x_3 & = & 6, \\ -x_1 & & & - & x_3 & = & -4. \end{array}$$

   Also use Gaussian-Jordan elimination to obtain the solution. Verify that the two solutions coincide. How many more floating point operations (i.e., scalar multiplications and additions) were needed by the Gauss-Jordan elimination over Gaussian elimination?

2. Using Gaussian elimination solve $\boldsymbol{A}\boldsymbol{x} = \boldsymbol{b}$ for the following choices of $\boldsymbol{A}$ and $\boldsymbol{b}$:

   (a) $\boldsymbol{A} = \begin{bmatrix} 2 & 6 & 1 \\ 3 & 9 & 2 \\ 0 & -1 & 3 \end{bmatrix}$ and $\boldsymbol{b} = \begin{bmatrix} 5 \\ 8 \\ 3 \end{bmatrix}$; repeat for $\boldsymbol{b} = \begin{bmatrix} -1 \\ -1 \\ 4 \end{bmatrix}$;

   (b) $\boldsymbol{A} = \begin{bmatrix} 0 & 2 & 5 \\ -1 & 7 & -5 \\ -1 & 8 & 3 \end{bmatrix}$ and $\boldsymbol{b} = \begin{bmatrix} 15 \\ -20 \\ 4 \end{bmatrix}$;

   (c) $\boldsymbol{A} = \begin{bmatrix} 1 & -4 & 0 & -3 \\ 0 & 0 & 1 & 0 \\ 0 & 0 & 1 & 4 \\ 0 & 1 & 0 & 0 \end{bmatrix}$ and $\boldsymbol{b} = \begin{bmatrix} 11 \\ 2 \\ -6 \\ -1 \end{bmatrix}$;

   (d) $\boldsymbol{A} = \begin{bmatrix} 1 & 1/2 & 1/3 \\ 1/2 & 1/3 & 1/4 \\ 1/3 & 1/4 & 1/5 \end{bmatrix}$ and $\boldsymbol{b} = \begin{bmatrix} 1/3 \\ 1/4 \\ 1/5 \end{bmatrix}$.

Solve each of the above linear systems using Gauss-Jordan elimination.

3. For each matrix in Exercise 2, find $\boldsymbol{A}^{-1}$, when it exists, by solving $\boldsymbol{AX} = \boldsymbol{I}$.

4. Find the sequences of elementary operations that will reduce $\boldsymbol{A}$ is an echelon matrix, where $\boldsymbol{A}$ is the following:

$$\text{(a)}\quad \boldsymbol{A} = \begin{bmatrix} 3 & 0 & 2 & -1 \\ 2 & 3 & -1 & 0 \\ 1 & 5 & 0 & 8 \end{bmatrix}, \quad \text{and} \quad \text{(b)}\quad \boldsymbol{A} = \begin{bmatrix} 1 & 0 & 0 & 0 \\ -1 & 1 & 0 & 0 \\ -1 & -1 & 1 & 0 \\ -1 & -1 & -1 & 1 \end{bmatrix}.$$

5. True or false: Interchanging two rows (Type-I operation) of a matrix can be achieved using Type-II and Type-III operations.

6. By setting up a 3×3 linear system, find the coefficients α, β and γ in the parabola $y = \alpha + \beta x + \gamma x^2$ that passes through the points $(1, 6)$, $(2, 17)$ and $(3, 34)$.

7. Verify whether $\boldsymbol{Ax} = \boldsymbol{b}$ is consistent for the following choices of $\boldsymbol{A}$ and $\boldsymbol{b}$ and find a solution when it exists:

$$\text{(a)}\ \boldsymbol{A} = \begin{bmatrix} 2 & 1 & 1 \\ 1 & 2 & -4 \\ 0 & 1 & -3 \\ -1 & 0 & -2 \end{bmatrix} \text{ and } \boldsymbol{b} - \begin{bmatrix} 1 \\ -1 \\ -1 \\ -1 \end{bmatrix}.$$

$$\text{(b)}\ \boldsymbol{A} \text{ as above and } \boldsymbol{b} = \begin{bmatrix} 3 \\ 3 \\ 1 \\ 1 \end{bmatrix}.$$

$$\text{(c)}\ \boldsymbol{A} = \begin{bmatrix} 1 & 2 & 3 \\ 4 & 5 & 6 \\ 7 & 8 & 9 \end{bmatrix} \text{ and } \boldsymbol{b} = \begin{bmatrix} 1 \\ 2 \\ 3 \end{bmatrix}.$$

8. Consider the linear system $\boldsymbol{Ax} = \boldsymbol{b}$, where

$$\boldsymbol{A} = \begin{bmatrix} 2 & 4 & (\alpha+3) \\ 1 & 3 & 1 \\ (\alpha-2) & 2 & 3 \end{bmatrix} \quad \text{and} \quad \boldsymbol{b} = \begin{bmatrix} 2 \\ 2 \\ \beta \end{bmatrix}.$$

For what values of α and β is the above system consistent?

9. Find a general solution of the homogeneous linear system $\boldsymbol{Ax} = \boldsymbol{0}$ for the following choices of $\boldsymbol{A}$:

$$(a)\quad \boldsymbol{A} = \begin{bmatrix} 2 & 6 & 1 & 5 \\ 1 & 3 & 0 & 2 \\ 1 & 3 & -1 & 1 \end{bmatrix}, \text{ and (b)} \quad \boldsymbol{A} = \begin{bmatrix} 1 & 1 & 2 & 3 \\ 2 & 0 & 4 & 4 \\ 1 & -1 & 2 & 1 \\ 1 & -2 & 2 & 0 \end{bmatrix}.$$

10. True or false: Every system $\boldsymbol{Ax} = \boldsymbol{0}$, where $\boldsymbol{A}$ is 2×3, has a solution $\boldsymbol{x}$ with the first entry $x_1 \neq 0$.

CHAPTER 3

More on Linear Equations

Consider the linear system $\boldsymbol{Ax} = \boldsymbol{b}$, where $\boldsymbol{A}$ is an $n \times n$ nonsingular matrix and $\boldsymbol{b}$ is a known vector. In Chapter 2, we discussed how Gaussian (and Gauss-Jordan) elimination can be used to solve such nonsingular systems. In this chapter, we discuss the connection between Gaussian elimination and the factorization of a nonsingular matrix into a lower-triangular and upper-triangular matrix. Such decompositions are extremely useful in solving nonsingular linear systems.

3.1 The LU decomposition

Definition 3.1 The $\boldsymbol{LU}$ decomposition. *An $n \times n$ matrix $\boldsymbol{A}$ is said to have a* $\boldsymbol{LU}$ *decomposition if it can be factored as the product $\boldsymbol{A} = \boldsymbol{LU}$, where*

(i) $\boldsymbol{L}$ is an $n \times n$ lower-triangular matrix with $l_{ii} = 1$ for each $i = 1, \ldots, n$, and

(ii) $\boldsymbol{U}$ is an $n \times n$ upper-triangular matrix with $u_{ii} \neq 0$ for each $i = 1, \ldots, n$.

In the above definition, $\boldsymbol{L}$ is restricted to be ***unit lower-triangular***, i.e., it has 1's along its diagonal. No such restriction applies to $\boldsymbol{U}$ but we do require that its diagonal elements are nonzero. When we say that $\boldsymbol{A}$ has an $\boldsymbol{LU}$ decomposition, we will mean that $\boldsymbol{L}$ and $\boldsymbol{U}$ have these forms. When such a factorization exists, we call $\boldsymbol{L}$ the ***(unit) lower factor*** (or, sometimes, the "unit" lower factor) and $\boldsymbol{U}$ the ***upper factor***.

We will demonstrate how $\boldsymbol{A}$ can be factorized as $\boldsymbol{A} = \boldsymbol{LU}$ when Gaussian elimination can be executed on $\boldsymbol{A}$ using only Type-III elementary row operations. In other words, no zero pivots are encountered in the sequence of Gaussian elimination, so no row interchanges are necessary.

To facilitate subsequent discussions, we introduce the concept of an $n \times n$ ***elementary lower-triangular matrix***, defined as

$$\boldsymbol{T}_k = \boldsymbol{I}_n - \boldsymbol{\tau}_k \boldsymbol{e}_k', \tag{3.1}$$

where $\boldsymbol{\tau}_k$ is a column vector with zeros in the first k positions. More explicitly,

$$\text{if } \boldsymbol{\tau}_k = \begin{bmatrix} 0 \\ \vdots \\ 0 \\ \tau_{k+1,k} \\ \vdots \\ \tau_{n,k} \end{bmatrix}, \text{ then } \boldsymbol{T}_k = \begin{bmatrix} 1 & 0 & \cdots & 0 & 0 & \cdots & 0 \\ 0 & 1 & \cdots & 0 & 0 & \cdots & 0 \\ \vdots & \vdots & \ddots & \vdots & \vdots & \vdots & \vdots \\ 0 & 0 & \cdots & 1 & 0 & \cdots & 0 \\ 0 & 0 & \cdots & -\tau_{k+1,k} & 1 & \cdots & 0 \\ \vdots & \vdots & \vdots & \vdots & \vdots & \ddots & \vdots \\ 0 & 0 & \cdots & -\tau_{n,k} & 0 & \cdots & 1 \end{bmatrix}.$$

Clearly $\boldsymbol{e}_k'\boldsymbol{\tau}_k = 0$, which implies

$$(\boldsymbol{I} - \boldsymbol{\tau}_k\boldsymbol{e}_k')(\boldsymbol{I} + \boldsymbol{\tau}_k\boldsymbol{e}_k') = \boldsymbol{I} - (\boldsymbol{\tau}_k\boldsymbol{e}_k')(\boldsymbol{\tau}_k\boldsymbol{e}_k') = \boldsymbol{I} - \boldsymbol{u}_k(\boldsymbol{e}_k'\boldsymbol{\tau}_k)\boldsymbol{e}_k' = \boldsymbol{I}.$$

This shows that the inverse of $\boldsymbol{T}_k$ is given by $\boldsymbol{T}_k^{-1} = \boldsymbol{I} + \boldsymbol{\tau}_k\boldsymbol{e}_k'$ or, more explicitly,

$$\boldsymbol{T}_k^{-1} = \begin{bmatrix} 1 & 0 & \cdots & 0 & 0 & \cdots & 0 \\ 0 & 1 & \cdots & 0 & 0 & \cdots & 0 \\ \vdots & \vdots & \ddots & \vdots & \vdots & \vdots & \vdots \\ 0 & 0 & \cdots & 1 & 0 & \cdots & 0 \\ 0 & 0 & \cdots & \tau_{k+1,k} & 1 & \cdots & 0 \\ \vdots & \vdots & \vdots & \vdots & \vdots & \ddots & \vdots \\ 0 & 0 & \cdots & \tau_{n,k} & 0 & \cdots & 1 \end{bmatrix},$$

which is again an elementary lower-triangular matrix. Pre-multiplication by $\boldsymbol{T}_k$ annihilates the entries below the k-th pivot. To see why this is true, let

$$\boldsymbol{A}^{(k-1)} = \begin{bmatrix} * & * & \cdots & a_{1,k}^{(k-1)} & * & \cdots & * \\ 0 & * & \cdots & a_{2,k}^{(k-1)} & * & \cdots & * \\ \vdots & \vdots & \ddots & \vdots & \vdots & \vdots & \vdots \\ 0 & 0 & \cdots & a_{k,k}^{(k-1)} & * & \cdots & * \\ 0 & 0 & \cdots & a_{k+1,k}^{(k-1)} & * & \cdots & * \\ \vdots & \vdots & \vdots & \vdots & \vdots & \ddots & \vdots \\ 0 & 0 & \cdots & a_{n,k}^{(k-1)} & * & \cdots & * \end{bmatrix}$$

be partially triangularized after $k-1$ elimination steps with $a_{k,k}^{(k-1)} \neq 0$. Then

$$\boldsymbol{T}_k\boldsymbol{A}^{(k-1)} = (\boldsymbol{I} - \boldsymbol{\tau}_k\boldsymbol{e}_k')\boldsymbol{A}^{(k-1)} = \boldsymbol{A}^{(k-1)} - \boldsymbol{\tau}_k\boldsymbol{e}_k'\boldsymbol{A}^{(k-1)}$$

$$= \begin{bmatrix} * & * & \cdots & a_{1,k}^{(k-1)} & * & \cdots & * \\ 0 & * & \cdots & a_{2,k}^{(k-1)} & * & \cdots & * \\ \vdots & \vdots & \ddots & \vdots & \vdots & \vdots & \vdots \\ 0 & 0 & \cdots & a_{k,k}^{(k-1)} & * & \cdots & * \\ 0 & 0 & \cdots & 0 & * & \cdots & * \\ \vdots & \vdots & \vdots & \vdots & \vdots & \ddots & \vdots \\ 0 & 0 & \cdots & 0 & * & \cdots & * \end{bmatrix}, \text{ where } \boldsymbol{\tau}_k = \begin{bmatrix} 0 \\ \vdots \\ 0 \\ l_{k+1,k} \\ \vdots \\ l_{n,k} \end{bmatrix}$$

with $l_{ik} = a_{i,k}^{(k-1)}/a_{k,k}^{(k-1)}$ (for $i = k+1, \ldots, n$) being nonzero elements of $\boldsymbol{\tau}_k$. In fact, the l_{ik}'s are precisely the premultipliers for annihilating the entries below the k-th pivot. Furthermore, $\boldsymbol{T}_k$ does not alter the first $k-1$ columns of $\boldsymbol{A}^{(k-1)}$ because $\boldsymbol{e}_k'\boldsymbol{A}_{*j}^{(k-1)} = 0$ whenever $j \leq k-1$. Therefore, if no row interchanges are required, then reducing $\boldsymbol{A}$ to an upper-triangular matrix $\boldsymbol{U}$ by Gaussian elimination is equivalent to executing a sequence of $n-1$ left-hand multiplications with elementary lower-triangular matrices. More precisely, $\boldsymbol{T}_{n-1}\cdots\boldsymbol{T}_2\boldsymbol{T}_1\boldsymbol{A} = \boldsymbol{U}$, and hence

$$\boldsymbol{A} = \boldsymbol{T}_1^{-1}\boldsymbol{T}_2^{-1}\cdots\boldsymbol{T}_{n-1}^{-1}\boldsymbol{U}. \tag{3.2}$$

Now, using the expression for the inverse of $\boldsymbol{T}_k^{-1}$, we obtain

$$\begin{aligned}\boldsymbol{T}_1^{-1}\boldsymbol{T}_2^{-1}\cdots\boldsymbol{T}_{n-1}^{-1} &= (\boldsymbol{I} + \boldsymbol{\tau}_1\boldsymbol{e}_1')(\boldsymbol{I} + \boldsymbol{\tau}_2\boldsymbol{e}_2')\cdots(\boldsymbol{I} + \boldsymbol{\tau}_{n-1}\boldsymbol{e}_{n-1}') \\ &= \boldsymbol{I} + \boldsymbol{\tau}_1\boldsymbol{e}_1' + \boldsymbol{\tau}_2\boldsymbol{e}_2' + \ldots + \boldsymbol{\tau}_{n-1}\boldsymbol{e}_{n-1}',\end{aligned}$$

where we have used the fact that that $\boldsymbol{e}_j'\boldsymbol{\tau}_k = 0$ whenever $j < k$. Furthermore,

$$\boldsymbol{\tau}_k\boldsymbol{e}_k' = \begin{bmatrix} 0 & 0 & \cdots & 0 & 0 & \cdots & 0 \\ 0 & 0 & \cdots & 0 & 0 & \cdots & 0 \\ \vdots & \vdots & \ddots & \vdots & \vdots & \vdots & \vdots \\ 0 & 0 & \cdots & 0 & * & \cdots & * \\ 0 & 0 & \cdots & l_{k+1,k} & 0 & \cdots & 0 \\ \vdots & \vdots & \vdots & \vdots & \vdots & \ddots & \vdots \\ 0 & 0 & \cdots & l_{nk} & * & \cdots & * \end{bmatrix},$$

which yields

$$\boldsymbol{I} + \boldsymbol{\tau}_1\boldsymbol{e}_1' + \boldsymbol{\tau}_2\boldsymbol{e}_2' + \ldots + \boldsymbol{\tau}_{n-1}\boldsymbol{e}_{n-1}' = \begin{bmatrix} 1 & 0 & 0 & \cdots & 0 \\ l_{21} & 1 & 0 & \cdots & 0 \\ l_{31} & l_{32} & 1 & \cdots & 0 \\ \vdots & \vdots & \vdots & \ddots & \vdots \\ l_{n1} & l_{n2} & l_{n3} & \cdots & 1 \end{bmatrix}. \tag{3.3}$$

We denote the matrix in (3.3) by $\boldsymbol{L}$. It is a lower-triangular matrix with 1's on the diagonal. The elements below the diagonal, i.e., the l_{ij}'s with $i \neq j$, are precisely the multipliers used to annihilate the (i,j)-th position during Gaussian elimination.

From (3.2), we immediately obtain $\boldsymbol{A} = \boldsymbol{L}\boldsymbol{U}$, where $\boldsymbol{L}$ is as above and $\boldsymbol{U}$ is the matrix obtained in the last step of the Gaussian elimination algorithm. In other words, the $\boldsymbol{L}\boldsymbol{U}$ factorization is the matrix formulation of Gaussian elimination, with the understanding that no row interchanges are used.

Once $\boldsymbol{L}$ and $\boldsymbol{U}$ have been obtained such that $\boldsymbol{A} = \boldsymbol{L}\boldsymbol{U}$, we solve the linear system $\boldsymbol{A}\boldsymbol{x} = \boldsymbol{b}$ by rewriting $\boldsymbol{A}\boldsymbol{x} = \boldsymbol{b}$ as $\boldsymbol{L}(\boldsymbol{U}\boldsymbol{x}) = \boldsymbol{b}$. Now letting $\boldsymbol{y} = \boldsymbol{U}\boldsymbol{x}$, solving $\boldsymbol{A}\boldsymbol{x} = \boldsymbol{b}$ is equivalent to solving the two triangular systems $\boldsymbol{L}\boldsymbol{y} = \boldsymbol{b}$ and $\boldsymbol{U}\boldsymbol{x} = \boldsymbol{y}$.

First, the lower-triangular system $\boldsymbol{L}\boldsymbol{y} = \boldsymbol{b}$ is solved for $\boldsymbol{y}$ by ***forward substitution***. More precisely, we write the system as

$$\begin{bmatrix} 1 & 0 & 0 & \dots & 0 \\ l_{21} & 1 & 0 & \dots & 0 \\ l_{31} & l_{32} & 1 & \dots & 0 \\ \vdots & \vdots & \vdots & \ddots & 0 \\ l_{n1} & l_{n2} & l_{n3} & \dots & 1 \end{bmatrix} \begin{bmatrix} y_1 \\ y_2 \\ y_3 \\ \vdots \\ y_n \end{bmatrix} = \begin{bmatrix} b_1 \\ b_2 \\ b_3 \\ \vdots \\ b_n \end{bmatrix}.$$

We can now solve the system as

$$y_1 = b_1 \quad \text{and} \quad y_i = b_i - \sum_{j=1}^{i-1} l_{ij} y_i, \text{ for } i = 2, 3 \dots, n. \tag{3.4}$$

After $\boldsymbol{y}$ has been obtained, the upper-triangular system $\boldsymbol{U}\boldsymbol{x} = \boldsymbol{y}$ is solved using the ***back substitution*** procedure for upper-triangular systems (as used in the last step of Gaussian elimination). Here, we have

$$\begin{bmatrix} u_{11} & u_{12} & u_{13} & \dots & u_{1n} \\ 0 & u_{22} & u_{23} & \dots & u_{2n} \\ 0 & 0 & u_{33} & \dots & 0 \\ \vdots & \vdots & \vdots & \ddots & 0 \\ 0 & 0 & 0 & \dots & u_{nn} \end{bmatrix} \begin{bmatrix} x_1 \\ x_2 \\ x_3 \\ \vdots \\ x_n \end{bmatrix} = \begin{bmatrix} y_1 \\ y_2 \\ y_3 \\ \vdots \\ y_n \end{bmatrix}.$$

We now solve for $\boldsymbol{x}$ using the back substitution:

$$x_n = \frac{y_n}{u_{nn}} \quad \text{and} \quad x_i = \frac{1}{u_{ii}} \left(y_i - \sum_{j=1}^{i+1} u_{ij} x_j \right) \quad \text{for } i = n-1, n-2, \dots, 1. \tag{3.5}$$

The connection between the $\boldsymbol{L}\boldsymbol{U}$ decomposition of $\boldsymbol{A}$ and Gaussian elimination is beautiful. The upper-triangular matrix $\boldsymbol{U}$ is precisely that obtained as a result of Gaussian elimination. The diagonal elements of $\boldsymbol{U}$ are, therefore, precisely the pivots of $\boldsymbol{A}$. These are sometimes referred to as the ***pivots*** of an $\boldsymbol{L}\boldsymbol{U}$ decomposition. It is the elementary row operations that hide $\boldsymbol{L}$. For each set of row operations used to sweep out columns, the multipliers used in Gaussian elimination occupy their proper places in a lower-triangular matrix. We demonstrate this below with a simple numerical example.

Example 3.1 Let us return to the coefficient matrix associated with the linear system in Example 2.1:

$$\boldsymbol{A} = \begin{bmatrix} 2 & 3 & 0 & 0 \\ 4 & 7 & 2 & 0 \\ -6 & -10 & 0 & 1 \\ 4 & 6 & 4 & 5 \end{bmatrix}.$$

We start by writing

$$\boldsymbol{A} = \begin{bmatrix} 1 & 0 & 0 & 0 \\ 0 & 1 & 0 & 0 \\ 0 & 0 & 1 & 0 \\ 0 & 0 & 0 & 1 \end{bmatrix} \begin{bmatrix} 2 & 3 & 0 & 0 \\ 4 & 7 & 2 & 0 \\ -6 & -10 & 0 & 1 \\ 4 & 6 & 4 & 5 \end{bmatrix}.$$

Recall the elementary row operations used to sweep out the first column: we used $E_{21}(-2)$, $E_{31}(3)$ and $E_{41}(-2)$. These multipliers (in fact their "negatives") occupy positions below the first diagonal of the identity matrix at the left and the result, at the end of the first step, can be expressed as

$$\boldsymbol{A} = \begin{bmatrix} 1 & 0 & 0 & 0 \\ 2 & 1 & 0 & 0 \\ -3 & 0 & 1 & 0 \\ 2 & 0 & 0 & 1 \end{bmatrix} \begin{bmatrix} 2 & 3 & 0 & 0 \\ 0 & 1 & 2 & 0 \\ 0 & -1 & 0 & 1 \\ 0 & 0 & 4 & 5 \end{bmatrix}.$$

Note that the matrix on the right is precisely the matrix obtained at the end of Step 1 in Example 2.1.

Next recall that we used $E_{32}(1)$ and $E_{42}(0)$ to annihilate the nonzero elements below the diagonal in the second column. Thus -1 and 0 occupy positions below the second diagonal and this results in

$$\boldsymbol{A} = \begin{bmatrix} 1 & 0 & 0 & 0 \\ 2 & 1 & 0 & 0 \\ -3 & -1 & 1 & 0 \\ 2 & 0 & 0 & 1 \end{bmatrix} \begin{bmatrix} 2 & 3 & 0 & 0 \\ 0 & 1 & 2 & 0 \\ 0 & 0 & 2 & 1 \\ 0 & 0 & 4 & 5 \end{bmatrix},$$

where the matrix on the right is that obtained at the end of Step 2 in Example 2.1.

Finally, we used $E_{43}(-2)$ to sweep out the third column and this results in 2 occupying the position below the third diagonal. Indeed, we have arrived at the ***LU*** decomposition:

$$\boldsymbol{A} = \begin{bmatrix} 1 & 0 & 0 & 0 \\ 2 & 1 & 0 & 0 \\ -3 & -1 & 1 & 0 \\ 2 & 0 & 2 & 1 \end{bmatrix} \begin{bmatrix} 2 & 3 & 0 & 0 \\ 0 & 1 & 2 & 0 \\ 0 & 0 & 2 & 1 \\ 0 & 0 & 0 & 3 \end{bmatrix} = \boldsymbol{LU},$$

where $\boldsymbol{L}$'s lower-triangular elements are entirely filled up using the multipliers in Gaussian elimination and $\boldsymbol{U}$ is indeed the coefficient matrix obtained at the end of Step 3 in Example 2.1.

Now suppose we want to solve the system $\boldsymbol{Ax} = \boldsymbol{b}$ in Example 2.1. We first solve $\boldsymbol{Ly} = \boldsymbol{b}$ using the forward substitution algorithm (3.4) and then we solve $\boldsymbol{Ux} = \boldsymbol{y}$ using the back substitution algorithm (3.5).

The system $\boldsymbol{Ly} = \boldsymbol{b}$ is

$$\begin{array}{rcrcrcrcr} y_1 & + & & & & & & = & 1, \\ 2y_1 & + & y_2 & & & & & = & 2, \\ -3y_1 & - & y_2 & + & y_3 & & & = & 1, \\ 2y_1 & & & + & 2y_3 & + & y_4 & = & 0. \end{array} \tag{3.6}$$

We start with the first row and proceed with successive substitutions. From the first row we immediately obtain $y_1 = 1$. The second row then yields $y_2 = 2 - 2y_1 = 0$. We solve for y_3 from the third row as $y_3 = 1 + 3y_1 + y_2 = 4$ and finally substitution in the last equation yields $y_4 = -10$. Thus, the solution vector is $\boldsymbol{y} = (1, 0, 4, -10)'$.

Next we solve the system $\boldsymbol{Ux} = \boldsymbol{y}$:

$$\begin{array}{rcrcrcrcr} 2x_1 & + & 3x_2 & & & & & = & 1, \\ & & x_2 & + & 2x_3 & & & = & 0, \\ & & & & 2x_3 & + & x_4 & = & 4, \\ & & & & & & 3x_4 & = & -10. \end{array} \tag{3.7}$$

We now solve this system using successive substitutions but beginning with the last equation. This yields $x_4 = -10/3$ from the fourth equation, $x_3 = 11/3$ from the third, $x_2 = -22/3$ from the second and, finally, $x_1 = 23/2$ from the first.

We arrive at $\boldsymbol{x} = (23/2, -22/3, 11/3, -10/3)'$ as the solution for $\boldsymbol{Ax} = \boldsymbol{b}$ – the same solution we obtained in Example 2.1 using Gaussian elimination. ■

Once the matrices $\boldsymbol{L}$ and $\boldsymbol{U}$ have been obtained, solving $\boldsymbol{Ax} = \boldsymbol{b}$ for $\boldsymbol{x}$ is relatively cheap. In fact, only n^2 multiplications/divisions and $n^2 - n$ additions/subtractions are required to solve the two triangular systems $\boldsymbol{Ly} = \boldsymbol{b}$ and $\boldsymbol{Ux} = \boldsymbol{y}$. For solving a single linear system $\boldsymbol{Ax} = \boldsymbol{b}$, there is no significant difference between the technique of Gaussian elimination and the $\boldsymbol{LU}$ factorization algorithms. However, when we want to solve several systems with the same coefficient matrix but with different right-hand sides (as is needed when we want to invert matrices), the $\boldsymbol{LU}$ factorization has merits. If $\boldsymbol{L}$ and $\boldsymbol{U}$ were already computed and saved for $\boldsymbol{A}$ when one system was solved, then they need not be recomputed, and the solutions to all subsequent systems $\boldsymbol{Ax} = \tilde{\boldsymbol{b}}$ are therefore relatively cheap to obtain. In fact, the operation count for each subsequent system is in the order of n^2, whereas this count would be on the order of $n^3/3$ were we to start from scratch each time for every new right-hand side.

From the description of how Gaussian elimination yields $\boldsymbol{L}$ and $\boldsymbol{U}$, it is clear that we cannot drop the assumption that no zero pivots are encountered. Otherwise, the process breaks down and we will not be able to factorize $\boldsymbol{A} = \boldsymbol{LU}$. While Gaussian elimination provides a nice way to reveal the $\boldsymbol{LU}$ factorization, it may be desirable to characterize when it will be possible to factorize $\boldsymbol{A} = \boldsymbol{LU}$ in terms of the structure of $\boldsymbol{A}$. The following theorem provides such a characterization in terms of principal submatrices (see Definition 1.10).

Theorem 3.1 *A nonsingular matrix $\boldsymbol{A}$ possesses an $\boldsymbol{LU}$ factorization if and only if all its leading principal submatrices are nonsingular.*

Proof. Let $\boldsymbol{A}$ be an $n \times n$ matrix. First assume that each leading principal submatrix of $\boldsymbol{A}$ is nonsingular. We will use induction to prove that each leading principal submatrix, say $\boldsymbol{A}_k$ (of order $k \times k$), has an $\boldsymbol{LU}$ factorization.

For $k = 1$, the leading principal submatrix is $\boldsymbol{A}_1 = (1)(a_{11})$ and the $\boldsymbol{LU}$ factorization is obvious (note: $a_{11} \neq 0$ since $\boldsymbol{A}_1$ is nonsingular). Now assume that the k-th leading principal submatrix has an $\boldsymbol{LU}$ factorization, viz. $\boldsymbol{A}_k = \boldsymbol{L}_k\boldsymbol{U}_k$. We partition the $k+1$-th principal submatrix as

$$\boldsymbol{A}_{k+1} = \begin{bmatrix} \boldsymbol{A}_k & \boldsymbol{u} \\ \boldsymbol{v}' & \alpha_{k+1} \end{bmatrix},$$

where $\boldsymbol{u}$ and $\boldsymbol{v}'$ contain the first k components of the $k+1$-th column and $k+1$-th row of $\boldsymbol{A}_{k+1}$. We now want to find a unit lower-triangular matrix $\boldsymbol{L}_{k+1}$ and an upper-triangular matrix $\boldsymbol{U}_{k+1}$ such that $\boldsymbol{A}_{k+1} = \boldsymbol{L}_{k+1}\boldsymbol{U}_{k+1}$. Suppose we write

$$\boldsymbol{L}_{k+1} = \begin{bmatrix} \boldsymbol{L}_k & \boldsymbol{0} \\ \boldsymbol{x}' & 1 \end{bmatrix} \text{ and } \boldsymbol{U}_{k+1} = \begin{bmatrix} \boldsymbol{U}_k & \boldsymbol{y} \\ \boldsymbol{0} & z \end{bmatrix},$$

where $\boldsymbol{x}$ and $\boldsymbol{y}$ are unknown vectors and z is an unknown scalar. We now attempt to solve for these unknowns by writing $\boldsymbol{A}_{k+1} = \boldsymbol{L}_{k+1}\boldsymbol{U}_{k+1}$. More explicitly, we write $\boldsymbol{A}_{k+1}$ as

$$\boldsymbol{A}_{k+1} = \begin{bmatrix} \boldsymbol{A}_k & \boldsymbol{u} \\ \boldsymbol{v}' & \alpha_{k+1} \end{bmatrix} = \begin{bmatrix} \boldsymbol{L}_k & \boldsymbol{0} \\ \boldsymbol{x}' & 1 \end{bmatrix} \begin{bmatrix} \boldsymbol{U}_k & \boldsymbol{y} \\ \boldsymbol{0} & z \end{bmatrix} = \boldsymbol{L}_{k+1}\boldsymbol{U}_{k+1}.$$

Note that the above yields $\boldsymbol{A}_k = \boldsymbol{L}_k\boldsymbol{U}_k$. Also, since $\boldsymbol{L}_k$ and $\boldsymbol{U}_k$ are both nonsingular, we have

$$\boldsymbol{v}' = \boldsymbol{x}'\boldsymbol{U}_k \Rightarrow \boldsymbol{x}' = \boldsymbol{v}'\boldsymbol{U}_k^{-1}; \qquad \boldsymbol{u} = \boldsymbol{L}_k\boldsymbol{y} \Rightarrow \boldsymbol{y} = \boldsymbol{L}_k^{-1}\boldsymbol{u};$$
$$\text{and } \alpha_{k+1} = \boldsymbol{x}'\boldsymbol{y} + z \Rightarrow z = \alpha_{k+1} - \boldsymbol{x}'\boldsymbol{y} = \alpha_{k+1} - \boldsymbol{v}'\boldsymbol{A}_k^{-1}\boldsymbol{u}.$$

In solving for z, we use the nonsingularity of $\boldsymbol{A}_k$, $\boldsymbol{L}_k$ and $\boldsymbol{U}_k$ to write $\boldsymbol{A}_k^{-1} = \boldsymbol{U}_k^{-1}\boldsymbol{L}_k^{-1}$ and, hence, $\boldsymbol{x}'\boldsymbol{y} = \boldsymbol{v}'\boldsymbol{U}_k^{-1}\boldsymbol{L}_k^{-1}\boldsymbol{u} = \boldsymbol{v}'\boldsymbol{A}_k^{-1}\boldsymbol{u}$. We now have the explicit factorization $\boldsymbol{A}_{k+1} = \boldsymbol{L}_{k+1}\boldsymbol{U}_{k+1}$, where

$$\boldsymbol{L}_{k+1} = \begin{bmatrix} \boldsymbol{L}_k & 0 \\ \boldsymbol{v}'\boldsymbol{U}_k^{-1} & 1 \end{bmatrix} \text{ and } \boldsymbol{U}_{k+1} = \begin{bmatrix} \boldsymbol{U}_k & \boldsymbol{L}_k^{-1}\boldsymbol{u} \\ \boldsymbol{0} & \alpha_{k+1} - \boldsymbol{v}'\boldsymbol{A}_k^{-1}\boldsymbol{u} \end{bmatrix}$$

are lower- and upper-triangular matrices, respectively, and $\boldsymbol{L}_{k+1}$ has 1's on its diagonals ($\boldsymbol{L}_k$ has 1's on its diagonals by induction hypothesis). That $\boldsymbol{L}_{k+1}$ is nonsingular, follows immediately from the induction hypothesis that $\boldsymbol{L}_k$ is nonsingular.

Furthermore, since $\boldsymbol{A}_{k+1}$ is nonsingular (recall that we assumed all leading principal submatrices are nonsingular), we have that $\boldsymbol{U}_{k+1} = \boldsymbol{L}_{k+1}^{-1}\boldsymbol{A}_{k+1}$ is nonsingular as well. (This proves that $\alpha_{k+1} - \boldsymbol{v}'\boldsymbol{A}_k^{-1}\boldsymbol{u} \neq 0$.) Therefore, the nonsingularity of the leading principal submatrices implies that each $\boldsymbol{A}_k$ possesses an $\boldsymbol{LU}$ factorization, and hence $\boldsymbol{A}_n = \boldsymbol{A}$ must have an $\boldsymbol{LU}$ factorization. This proves the "if" part of the theorem.

Let us now assume that $\boldsymbol{A}$ is nonsingular and possesses an $\boldsymbol{LU}$ factorization. We can then write

$$\boldsymbol{A} = \boldsymbol{LU} = \begin{bmatrix} \boldsymbol{L}_{11} & \boldsymbol{O} \\ \boldsymbol{L}_{21} & \boldsymbol{L}_{22} \end{bmatrix} \begin{bmatrix} \boldsymbol{U}_{11} & \boldsymbol{U}_{12} \\ \boldsymbol{O} & \boldsymbol{U}_{22} \end{bmatrix} = \begin{bmatrix} \boldsymbol{L}_{11}\boldsymbol{U}_{11} & * \\ * & * \end{bmatrix},$$

where $\boldsymbol{L}_{11}$ and $\boldsymbol{U}_{11}$ are each $k \times k$ and all submatrices are conformably partitioned. We do not need to keep track of the blocks represented by $*$. Note that $\boldsymbol{L}_{11}$ is a lower-triangular matrix with each diagonal element equal to 1 and $\boldsymbol{U}_{11}$ is upper-triangular with nonzero diagonal elements. Therefore, $\boldsymbol{L}_{11}$ and $\boldsymbol{U}_{11}$ are both nonsingular and so is the leading principal submatrix $\boldsymbol{A}_k = \boldsymbol{L}_{11}\boldsymbol{U}_{11}$. This completes the "only if" part of the proof. □

The next theorem says that if $\boldsymbol{A}$ has an $\boldsymbol{LU}$ decomposition, then that decomposition must be unique.

Theorem 3.2 *If $\boldsymbol{A} = \boldsymbol{LU}$ is an $\boldsymbol{LU}$ decomposition of the nonsingular matrix $\boldsymbol{A}$, where $\mathbf{L}$ is unit lower-triangular (i.e., each diagonal element is 1) and $\mathbf{U}$ is upper-triangular with nonzero diagonal elements, then $\mathbf{L}$ and $\mathbf{U}$ are unique.*

Proof. We will prove that if $\boldsymbol{A} = \boldsymbol{L}_1\boldsymbol{U}_1 = \boldsymbol{L}_2\boldsymbol{U}_2$ are two $\boldsymbol{LU}$ decompositions of $\boldsymbol{A}$, then $\boldsymbol{L}_1 = \boldsymbol{L}_2$ and $\boldsymbol{U}_1 = \boldsymbol{U}_2$.

The $\boldsymbol{L}_i$'s and $\boldsymbol{U}_i$'s are nonsingular because they are triangular matrices with none of their diagonal elements equal to zero. Therefore,

$$\boldsymbol{L}_1\boldsymbol{U}_1 = \boldsymbol{L}_2\boldsymbol{U}_2 \Longrightarrow \boldsymbol{L}_2^{-1}\boldsymbol{L}_1 = \boldsymbol{U}_2\boldsymbol{U}_1^{-1}\,.$$

Now we make a few observations about lower-triangular matrices that are easily verified:

(i) Since $\boldsymbol{L}_2$ is lower-triangular with 1's along its diagonal, $\boldsymbol{L}_2^{-1}$ is also lower-triangular with 1's along its diagonal. This follows from Theorem 2.7.

(ii) $\boldsymbol{L}_2^{-1}\boldsymbol{L}_1$ is also lower-triangular. We have already verified this in Theorem 1.8.

(iii) $\boldsymbol{U}_1^{-1}$ is also upper-triangular (Theorem 2.7) and so is $\boldsymbol{U}_2\boldsymbol{U}_1^{-1}$ (Theorem 1.8).

Since $\boldsymbol{L}_2^{-1}\boldsymbol{L}_1$ is lower-triangular and $\boldsymbol{U}_2\boldsymbol{U}_1^{-1}$ is upper-triangular, their equality implies that they must both be diagonal. After all, the only matrix that is both lower and upper-triangular is the diagonal matrix. We also know that all the diagonal elements in $\boldsymbol{L}_2^{-1}\boldsymbol{L}_1$ is equal to one. So the diagonal matrix must be the identity matrix and we find

$$\boldsymbol{L}_2^{-1}\boldsymbol{L}_1 = \boldsymbol{U}_2\boldsymbol{U}_1^{-1} = \boldsymbol{I} \Longrightarrow \boldsymbol{L}_1 = \boldsymbol{L}_2 \text{ for } \boldsymbol{U}_1 = \boldsymbol{U}_2\,.$$

This proves the uniqueness. □

3.2 Crout's Algorithm

Earlier we saw how Gaussian elimination yields the $\boldsymbol{LU}$ factors of a matrix $\boldsymbol{A}$. However, when programming an $\boldsymbol{LU}$ factorization, alternative strategies can be more efficient. One alternative is to assume the existence of an $\boldsymbol{LU}$ decomposition (with an unit lower-triangular $\boldsymbol{L}$) and actually solve for the other elements of $\boldsymbol{L}$ and $\boldsymbol{U}$.

What we discuss below is known as ***Crout's algorithm*** (see, e.g., Wilkinson, 1965; Stewart, 1998) and forms the basis of several existing computer algorithms for solving linear systems. Here we write

$$\begin{bmatrix} a_{11} & a_{12} & \cdots & a_{1n} \\ a_{21} & a_{22} & \cdots & a_{2n} \\ \vdots & \vdots & \ddots & \vdots \\ a_{n1} & a_{n2} & \cdots & a_{nn} \end{bmatrix} = \begin{bmatrix} 1 & 0 & \cdots & 0 \\ l_{21} & 1 & \cdots & 0 \\ \vdots & \vdots & \ddots & \vdots \\ l_{n1} & l_{n2} & \cdots & 1 \end{bmatrix} \begin{bmatrix} u_{11} & u_{12} & \cdots & u_{1n} \\ 0 & u_{22} & \cdots & u_{2n} \\ \vdots & \vdots & \ddots & \vdots \\ 0 & 0 & \cdots & u_{nn} \end{bmatrix}.$$

Crout's algorithm proceeds by now solving for the elements of L and U in a specific order. We first solve for the first row of $\boldsymbol{U}$:

$$u_{1j} = a_{1j} \text{ for } j = 1, 2, ..., n .$$

Next we solve for the first column of L:

$$l_{i1} = a_{i1}/u_{11} , \quad i = 2, 3, ..., n.$$

Then we return to solving for the second row of $\boldsymbol{U}$:

$$l_{21}u_{1j} + u_{2j} = a_{2j} , \quad j = 2, 3, ...n;$$
$$\text{so, } u_{2j} = a_{2j} - l_{21}u_{1j} ; \quad j = 2, 3, ..., n .$$

Then we solve for the second column of L and so on. Crout's algorithm may be written down in terms of the following general equations:

For each $i = 1, 2, ...n$,

$$u_{ij} = a_{ij} - \sum_{k=1}^{i-1} l_{ik}u_{kj} , \quad j = i, i+1, ..., n ;$$
$$l_{ji} = \frac{a_{ji} - \sum_{k=1}^{i-1} l_{jk}u_{ki}}{u_{ii}} , \quad j = i+1, ..., n.$$

In the above equations, all the a_{ij}'s are known and $l_{ii} = 1$ for $i = 1, 2, ..., n$. It is important to note the specific order in which the above equations are solved, alternating between the rows of $\boldsymbol{U}$ and the columns of $\boldsymbol{L}$. To be precise, first we solve for the elements of the first row of $\boldsymbol{U}$. This gives us u_{1j} for $j = 1, 2, \ldots, n$. Then, we solve for all the elements in the first column of $\boldsymbol{L}$ to obtain l_{j1}'s. Next, we move to the second row of $\boldsymbol{U}$ and obtain the u_{2j}'s. Note that the u_{2j}'s only involve elements in the first row of $\boldsymbol{U}$ and the first column of $\boldsymbol{L}$, which have already been obtained. Then, we move to the second column of $\boldsymbol{L}$ and obtain l_{j2}'s, which involves only elements in the first two rows of $\boldsymbol{U}$ and the first column of $\boldsymbol{L}$.

Proceeding in this manner, we will obtain $\boldsymbol{L}$ and $\boldsymbol{U}$ in the end, provided we do not encounter an $u_{ii} = 0$ at any stage. If we do find $u_{ii} = 0$ for some i, we can conclude that $\boldsymbol{A} = \boldsymbol{LU}$ does not exist. This does not necessarily mean that $\boldsymbol{A}$ is singular. It means that some row interchanges may be needed in $\boldsymbol{A}$ to obtain $\boldsymbol{PA} = \boldsymbol{LU}$, where $\boldsymbol{P}$ is a permutation matrix that brings about row interchanges in $\boldsymbol{A}$. Details on this are provided in the next section.

Crout's algorithm has several benefits. First of all, it offers a constructive "proof" that $\boldsymbol{A} = \boldsymbol{LU}$, where $\boldsymbol{L}$ is unit lower-triangular and $\boldsymbol{U}$ is upper-triangular. It also demonstrates an efficient way to compute the $\boldsymbol{LU}$ decomposition and also demonstrates that the decomposition is unique. This is because it is clear that the above system leads to unique solutions for $\boldsymbol{L}$ and $\boldsymbol{U}$, provided we restrict the diagonal elements in $\boldsymbol{L}$ to one. This was also proved in Theorem 3.2.

Gaussian elimination and Crout's algorithm differ only in the ordering of operations. Both algorithms are theoretically and numerically equivalent with complexity $O\left(n^3\right)$ (actually, the number of operations is approximately $n^3/3$, where 1 operation = 1 multiplication + 1 addition).

3.3 LU decomposition with row interchanges

Not all nonsingular matrices possess an LU factorization. For example, try finding the following $\boldsymbol{LU}$ decomposition:

$$\boldsymbol{A} = \begin{bmatrix} 0 & 1 \\ 1 & a_{22} \end{bmatrix} = \begin{bmatrix} 1 & 0 \\ l_{21} & 1 \end{bmatrix} \begin{bmatrix} u_{11} & u_{12} \\ 0 & u_{22} \end{bmatrix}.$$

Irrespective of the value of a_{22}, there is no nonzero value of u_{11} that will satisfy the above equation, thus precluding an $\boldsymbol{LU}$ decomposition. The problem here is that a zero pivot occurs in the $(1,1)$-th position. In other words, the first principal submatrix $[0]$ is singular, thus violating the necessary condition in Theorem 3.1.

We can, nevertheless, pre-multiply the above matrix by a permutation matrix $\boldsymbol{P}$ that will interchange the rows and obtain the following decomposition:

$$\boldsymbol{PA} = \begin{bmatrix} 0 & 1 \\ 1 & 0 \end{bmatrix} \begin{bmatrix} 0 & 1 \\ 1 & a_{22} \end{bmatrix} = \begin{bmatrix} 1 & a_{22} \\ 0 & 1 \end{bmatrix} = \begin{bmatrix} 1 & 0 \\ 0 & 1 \end{bmatrix} \begin{bmatrix} 1 & a_{22} \\ 0 & 1 \end{bmatrix} = \boldsymbol{LU}.$$

When row interchanges are allowed, zero pivots can always be avoided when the original matrix $\boldsymbol{A}$ is nonsingular. *In general, it is true that for every nonsingular matrix* $\boldsymbol{A}$*, there exists a permutation matrix* $\boldsymbol{P}$ *(a product of elementary interchange matrices) such that* $\boldsymbol{PA}$ *has an* $\boldsymbol{LU}$ *factorization.*

We now describe the effect of row interchanges and how the permutation matrix appears explicitly in Gaussian elimination. As in Section 3.1, assume we are row reducing an $n \times n$ nonsingular matrix $\boldsymbol{A}$, but, at the k-stage, require an interchange of rows, say between $k+i$ and $k+j$. Let $\boldsymbol{E}$ denote the elementary matrix of Type I that interchanges rows $k+i$ and $k+j$. Therefore, the sequence of pre-multiplications $\boldsymbol{ET}_k\boldsymbol{T}_{k-1}\cdots\boldsymbol{T}_1$ is applied to $\boldsymbol{A}$, where $\boldsymbol{T}_k$'s are elementary triangular matrices as defined in Section 3.1.

From Lemma 2.2, we know that $\boldsymbol{E}^2 = \boldsymbol{I}$. Therefore,

$$\begin{aligned}\boldsymbol{E}\boldsymbol{T}_k\boldsymbol{T}_{k-1}\cdots\boldsymbol{T}_1\boldsymbol{A} &= \boldsymbol{E}\boldsymbol{T}_k\boldsymbol{E}^2\boldsymbol{T}_{k-1}\boldsymbol{E}^2\cdots\boldsymbol{E}^2\boldsymbol{T}_1\boldsymbol{E}^2\boldsymbol{A}\\ &= (\boldsymbol{E}\boldsymbol{T}_k\boldsymbol{E})(\boldsymbol{E}\boldsymbol{T}_{k-1}\boldsymbol{E})\cdots(\boldsymbol{E}\boldsymbol{T}_1\boldsymbol{E})\boldsymbol{E}\boldsymbol{A}\\ &= \tilde{\boldsymbol{T}}_k\tilde{\boldsymbol{T}}_{k-1}\cdots\tilde{\boldsymbol{T}}_1\boldsymbol{E}\boldsymbol{A},\end{aligned}$$

where $\tilde{\boldsymbol{T}}_l = \boldsymbol{E}\boldsymbol{T}_l\boldsymbol{E}$. The above sequence of operations has moved the matrix $\boldsymbol{E}$ from the far-left position to that immediately preceding $\boldsymbol{A}$. Furthermore, each of the $\tilde{\boldsymbol{T}}_l$'s are themselves lower-triangular matrices. To see this, write $\boldsymbol{T}_l = \boldsymbol{I} - \boldsymbol{\tau}_l\boldsymbol{e}_l'$ (as in Section 3.1) and $\boldsymbol{E} = \boldsymbol{I} - \boldsymbol{u}\boldsymbol{u}'$ with $\boldsymbol{u} = \boldsymbol{e}_{k+i} - \boldsymbol{e}_{k+j}$. It now follows that

$$\boldsymbol{E}\boldsymbol{T}_l\boldsymbol{E} = \boldsymbol{E}(\boldsymbol{I} - \boldsymbol{\tau}_l\boldsymbol{e}_l')\boldsymbol{E} = \boldsymbol{E}^2 - \boldsymbol{E}\boldsymbol{\tau}_l\boldsymbol{e}_l'\boldsymbol{E} = \boldsymbol{I} - \tilde{\boldsymbol{\tau}}_l\boldsymbol{e}_l',$$

where $\tilde{\boldsymbol{\tau}}_l = \boldsymbol{E}\boldsymbol{\tau}_l$. From this structure, it is evident $\tilde{\boldsymbol{T}}_l$ is itself an elementary lower-triangular matrix which agrees with $\boldsymbol{T}_l$ in all positions except that the multipliers in the $k+i$-th and $k+j$-th positions have traded places.

Why the general $\boldsymbol{LU}$ factorization with row interchanges works can be summarized as follows. For the steps requiring row-interchanges, the necessary interchange matrices $\boldsymbol{E}$ can all be "factored" to the far right-hand side, and the matrices $\tilde{\boldsymbol{T}}_l$'s retain the desirable feature of being elementary lower-triangular matrices. Furthermore, it can be directly verified that $\boldsymbol{T}_k\boldsymbol{T}_{k-1}\cdots\boldsymbol{T}_1$ differs from $\boldsymbol{T}_k\boldsymbol{T}_{k-1}\cdots\boldsymbol{T}_1$ only in the sense that the multipliers in rows $k+i$ and $k+j$ have traded places. We can now write

$$\tilde{\boldsymbol{T}}_{n-1}\cdots\tilde{\boldsymbol{T}}_1\boldsymbol{P}\boldsymbol{A} = \boldsymbol{U},$$

where $\boldsymbol{P}$ is the product of all elementary interchange matrices used during the reduction. Since all of the $\tilde{\boldsymbol{T}}_k$'s are elementary lower-triangular matrices, we have (analogous to Section 3.1) that $\boldsymbol{P}\boldsymbol{A} = \boldsymbol{L}\boldsymbol{U}$, where $\boldsymbol{L} = \tilde{\boldsymbol{T}}_1^{-1}\cdots\tilde{\boldsymbol{T}}_{n-1}^{-1}$.

The $\boldsymbol{LU}$ decomposition with row interchanges is widely adopted in numerical linear algebra to solve linear systems even when we do not encounter zero pivots. For ensuring numerical stability, at each step we search the positions on and below the pivotal position for the entry with largest magnitude. If necessary, we perform the appropriate row interchanges to bring this largest entry into the pivotal position. Below is an illustration in a typical setting, where x denotes the largest entry in the third column after the entries below the pivots in the first two columns have been eliminated:

$$\begin{bmatrix} * & * & * & * & * & * \\ 0 & * & * & * & * & * \\ 0 & 0 & * & * & * & * \\ 0 & 0 & * & * & * & * \\ 0 & 0 & x & * & * & * \\ 0 & 0 & * & * & * & * \end{bmatrix} \xrightarrow{\boldsymbol{P}_1} \begin{bmatrix} * & * & * & * & * & * \\ 0 & * & * & * & * & * \\ 0 & 0 & x & * & * & * \\ 0 & 0 & * & * & * & * \\ 0 & 0 & * & * & * & * \\ 0 & 0 & * & * & * & * \end{bmatrix} \xrightarrow{\boldsymbol{T}_3} \begin{bmatrix} * & * & * & * & * & * \\ 0 & * & * & * & * & * \\ 0 & 0 & x & * & * & * \\ 0 & 0 & 0 & * & * & * \\ 0 & 0 & 0 & * & * & * \\ 0 & 0 & 0 & * & * & * \end{bmatrix}.$$

Here $\boldsymbol{P}_1 = \boldsymbol{E}_{35}$ is the permutation matrix that interchanges rows 3 and 5, and $\boldsymbol{T}_3$ is the elementary lower-triangular matrix that annihilates all the entries below the third pivot. This method is known as the ***partial pivoting*** algorithm.

One can also apply the same line of reasoning used to arrive at (3.3) to conclude that the multipliers in the matrix $\boldsymbol{L}$, where $\boldsymbol{PA} = \boldsymbol{LU}$, are permuted according to the row interchanges executed. More specifically, if rows i and j with $i < j$ are interchanged to create the i-th pivot, then the multipliers $l_{i1}, l_{i2}, \ldots, l_{i,(i-1)}$ and $l_{j1}, l_{j2}, \ldots, l_{j,(i-1)}$ trade places in the formation of $\boldsymbol{L}$. Efficient computer programs that implement $\boldsymbol{LU}$ factorizations save storage space by successively overwriting the entries in $\boldsymbol{A}$ with entries from $\boldsymbol{L}$ and $\boldsymbol{U}$. Information on the row interchanges are stored in a permutation vector. Upon termination of the algorithm, the input matrix $\boldsymbol{A}$ is overwritten as follows:

$$\begin{bmatrix} a_{11} & a_{12} & \cdots & a_{1n} \\ a_{21} & a_{22} & \cdots & a_{2n} \\ \vdots & \vdots & \ddots & \vdots \\ a_{n1} & a_{n2} & \cdots & a_{nn} \end{bmatrix} \xrightarrow{\boldsymbol{PA}=\boldsymbol{LU}} \begin{bmatrix} u_{11} & u_{12} & \cdots & u_{1n} \\ l_{21} & u_{22} & \cdots & u_{2n} \\ \vdots & \vdots & \ddots & \vdots \\ l_{n1} & l_{n2} & \cdots & u_{nn} \end{bmatrix} ; \boldsymbol{p} = \begin{bmatrix} i_1 \\ i_2 \\ \vdots \\ i_n \end{bmatrix}. \tag{3.8}$$

We illustrate with the following example.

Example 3.2 $\boldsymbol{PA} = \boldsymbol{LU}$**.** Let us again consider the matrix in Examples 2.1 and 3.1. We also keep track of a permutation vector $\boldsymbol{p}$ that is initially set to the natural order $\boldsymbol{p} = (1, 2, 3, 4)'$. We start with

$$\boldsymbol{A} = \begin{bmatrix} 2 & 3 & 0 & 0 \\ 4 & 7 & 2 & 0 \\ -6 & -10 & 0 & 1 \\ 4 & 6 & 4 & 5 \end{bmatrix} ; \quad \boldsymbol{p} = \begin{bmatrix} 1 \\ 2 \\ 3 \\ 4 \end{bmatrix}.$$

The entry with maximum magnitude in the first column is -6 which is the first element of the third row. Therefore, we interchange the first and third rows to obtain

$$\begin{bmatrix} 2 & 3 & 0 & 0 \\ 4 & 7 & 2 & 0 \\ -6 & -10 & 0 & 1 \\ 4 & 6 & 4 & 5 \end{bmatrix} \longrightarrow \begin{bmatrix} -6 & -10 & 0 & 1 \\ 4 & 7 & 2 & 0 \\ 2 & 3 & 0 & 0 \\ 4 & 6 & 4 & 5 \end{bmatrix} ; \quad \boldsymbol{p} = \begin{bmatrix} 3 \\ 2 \\ 1 \\ 4 \end{bmatrix}.$$

We now eliminate the entries below the diagonal in the first column. In the process, we successively overwrite the zero entries with entries from $\boldsymbol{L}$. For purposes of clarity, the l_{ij}'s are shown in boldface type.

$$\begin{bmatrix} -6 & -10 & 0 & 1 \\ 4 & 7 & 2 & 0 \\ 2 & 3 & 0 & 0 \\ 4 & 6 & 4 & 5 \end{bmatrix} \longrightarrow \begin{bmatrix} -6 & -10 & 0 & 1 \\ \mathbf{-2/3} & 1/3 & 2 & 2/3 \\ \mathbf{-1/3} & -1/3 & 0 & 1/3 \\ \mathbf{-2/3} & -2/3 & 4 & 17/3 \end{bmatrix} ; \quad \boldsymbol{p} = \begin{bmatrix} 3 \\ 2 \\ 1 \\ 4 \end{bmatrix}.$$

Next we look for the entry below the diagonal in the second column that has the largest magnitude. We find this to be the $(4, 2)$-th element. Therefore, we interchange the second and fourth columns and apply this change to $\boldsymbol{p}$ as well. *Note that the* l_{ij}*'s in boldface are also interchanged.*

$$\begin{bmatrix} -6 & -10 & 0 & 1 \\ \mathbf{-2/3} & 1/3 & 2 & 2/3 \\ \mathbf{-1/3} & -1/3 & 0 & 1/3 \\ \mathbf{-2/3} & -2/3 & 4 & 17/3 \end{bmatrix} \rightarrow \begin{bmatrix} -6 & -10 & 0 & 1 \\ \mathbf{-2/3} & -2/3 & 4 & 17/3 \\ \mathbf{-1/3} & -1/3 & 0 & 1/3 \\ \mathbf{-2/3} & 1/3 & 2 & 2/3 \end{bmatrix}; \quad \boldsymbol{p} = \begin{bmatrix} 3 \\ 4 \\ 1 \\ 2 \end{bmatrix}.$$

We now proceed to sweep out the entries below the diagonal in the second column, storing the multipliers in those positions. *Note that the l_{ij}'s (boldface) in the preceding steps are not changed as they are simply stored in place of zeros to save space.*

$$\begin{bmatrix} -6 & -10 & 0 & 1 \\ \mathbf{-2/3} & -2/3 & 4 & 17/3 \\ \mathbf{-1/3} & -1/3 & 0 & 1/3 \\ \mathbf{-2/3} & 1/3 & 2 & 2/3 \end{bmatrix} \rightarrow \begin{bmatrix} -6 & -10 & 0 & 1 \\ \mathbf{-2/3} & -2/3 & 4 & 17/3 \\ \mathbf{-1/3} & \mathbf{1/2} & -2 & -5/2 \\ \mathbf{-2/3} & \mathbf{-1/2} & 4 & 7/2 \end{bmatrix}; \boldsymbol{p} = \begin{bmatrix} 3 \\ 4 \\ 1 \\ 2 \end{bmatrix}.$$

Next, we move to the third column. Here the diagonal entry is -2 and the entry below the diagonal is 4. So, we interchange the third and fourth rows (including the l_{ij}'s):

$$\begin{bmatrix} -6 & -10 & 0 & 1 \\ \mathbf{-2/3} & -2/3 & 4 & 17/3 \\ \mathbf{-1/3} & \mathbf{1/2} & -2 & -5/2 \\ \mathbf{-2/3} & \mathbf{-1/2} & 4 & 7/2 \end{bmatrix} \rightarrow \begin{bmatrix} -6 & -10 & 0 & 1 \\ \mathbf{-2/3} & -2/3 & 4 & 17/3 \\ \mathbf{-2/3} & \mathbf{-1/2} & 4 & 7/2 \\ \mathbf{-1/3} & \mathbf{1/2} & -2 & -5/2 \end{bmatrix}; \boldsymbol{p} = \begin{bmatrix} 3 \\ 4 \\ 2 \\ 1 \end{bmatrix}.$$

Our final step annihilates the entry below the diagonal in the third column and we arrive at our final output in the form given by (3.8):

$$\begin{bmatrix} -6 & -10 & 0 & 1 \\ \mathbf{-2/3} & -2/3 & 4 & 17/3 \\ \mathbf{-2/3} & \mathbf{-1/2} & 4 & 7/2 \\ \mathbf{-1/3} & \mathbf{1/2} & \mathbf{-1/2} & -3/4 \end{bmatrix}; \quad \boldsymbol{p} = \begin{bmatrix} 3 \\ 4 \\ 2 \\ 1 \end{bmatrix}.$$

The matrix $\boldsymbol{P}$ is obtained by applying the final permutation given by $\boldsymbol{p}$ to the rows of $\boldsymbol{I}_4$. Therefore,

$$\boldsymbol{L} = \begin{bmatrix} 1 & 0 & 0 & 0 \\ -2/3 & 1 & 0 & 0 \\ -2/3 & -1/2 & 1 & 0 \\ -1/3 & 1/2 & -1/2 & 1 \end{bmatrix}; \quad \boldsymbol{U} = \begin{bmatrix} -6 & -10 & 0 & 1 \\ 0 & -2/3 & 4 & 17/3 \\ 0 & 0 & 4 & 7/2 \\ 0 & 0 & 0 & -3/4 \end{bmatrix};$$

$$\boldsymbol{P} = \begin{bmatrix} 0 & 0 & 1 & 0 \\ 0 & 0 & 0 & 1 \\ 0 & 1 & 0 & 0 \\ 1 & 0 & 0 & 0 \end{bmatrix}. \qquad \blacksquare$$

The LU decomposition with partial pivoting can be an effective tool for solving a nonsingular system $\boldsymbol{Ax} = \boldsymbol{b}$. Observe that

$$\boldsymbol{Ax} = \boldsymbol{b} \Longleftrightarrow \boldsymbol{PAx} = \boldsymbol{Pb},$$

where $\boldsymbol{P}$ is a permutation matrix. The two systems above are equivalent because permutation matrices are nonsingular, which implies that we can multiply both sides of an equation by P as well as P^{-1}. Once we have computed the factorization $PA = LU$, we first solve the lower-triangular system $Ly = Pb$ using forward substitution to obtain y and then we solve the upper-triangular system $Ux = y$ using backward substitution to obtain the solution x.

The preceding developments reveal how Gaussian elimination with partial pivoting leads to the decomposition $PA = LU$. It shows, in particular, that every nonsingular matrix has such a decomposition. It is instructive to provide a more direct proof of this result using induction.

Theorem 3.3 *If A is any nonsingular matrix, then there exists a permutation matrix P such that PA has an LU decomposition,*

$$PA = LU ,$$

where L is unit lower-triangular (with ones along the diagonal) and U is upper-triangular.

Proof. The 1×1 case is trivial and it is not difficult to prove the result for 2×2 matrices (by swapping rows if necessary). We leave the details to the reader and prove the result for $n \times n$ matrices with $n \geq 3$ using induction.

Suppose the result is true for all $(n-1) \times (n-1)$ matrices, where $n \geq 3$. Let A be an $n \times n$ nonsingular matrix. This means that the first column of A must have at least one nonzero entry and by permuting the rows we can always bring the entry with the largest magnitude in the first column to become the $(1,1)$-th entry. Therefore, there exists a permutation matrix P_1 such that $P_1 A$ can be partitioned as

$$P_1 A = \begin{bmatrix} a_{11} & a_{12}' \\ a_{21} & A_{22} \end{bmatrix} ,$$

where $a_{11} \neq 0$ and A_{22} is $(n-1) \times (n-1)$. Let $L_1 = \begin{bmatrix} 1 & 0' \\ l & I_{n-1} \end{bmatrix}$ be an $n \times n$ lower-triangular matrix, where $l = (1/a_{11}) a_{21}$. This is well-defined since $a_{11} \neq 0$. It is easy to check that $L_1^{-1} = \begin{bmatrix} 1 & 0' \\ -l & I_{n-1} \end{bmatrix}$. Therefore,

$$L_1^{-1} P_1 A = \begin{bmatrix} 1 & 0' \\ -l & I_{n-1} \end{bmatrix} \begin{bmatrix} a_{11} & a_{12}' \\ a_{21} & A_{22} \end{bmatrix} = \begin{bmatrix} a_{11} & a_{12}' \\ 0 & B \end{bmatrix} = \begin{bmatrix} a_{11} & a_{12}' \\ 0 & P_B^{-1} L_B U_B \end{bmatrix} ,$$

where $B = A_{22} - l a_{12}'$ is $(n-1) \times (n-1)$ and the induction hypothesis ensures that there exists an $(n-1) \times (n-1)$ permutation matrix P_B such that $P_B B = L_B U_B$ is an LU decomposition. Therefore,

$$\begin{aligned} P_1 A &= \begin{bmatrix} 1 & 0' \\ l & I_{n-1} \end{bmatrix} \begin{bmatrix} a_{11} & a_{12}' \\ 0 & P_B^{-1} L_B U_B \end{bmatrix} = \begin{bmatrix} 1 & 0' \\ l & I_{n-1} \end{bmatrix} \begin{bmatrix} 1 & 0' \\ 0 & P_B^{-1} \end{bmatrix} \begin{bmatrix} a_{11} & a_{12}' \\ 0 & L_B U_B \end{bmatrix} \\ &= \begin{bmatrix} 1 & 0' \\ l & P_B^{-1} \end{bmatrix} \begin{bmatrix} a_{11} & a_{12}' \\ 0 & L_B U_B \end{bmatrix} = \begin{bmatrix} 1 & 0' \\ 0 & P_B^{-1} \end{bmatrix} \begin{bmatrix} 1 & 0' \\ P_B l & L_B \end{bmatrix} \begin{bmatrix} a_{11} & a_{12}' \\ 0 & U_B \end{bmatrix} . \end{aligned}$$

From the above, it follows that $PA = LU$, where

$$P = \begin{bmatrix} 1 & 0 \\ 0 & P_B \end{bmatrix} P_1 \,, \quad L = \begin{bmatrix} 1 & 0' \\ P_B l & L_B \end{bmatrix} \quad \text{and } U = \begin{bmatrix} a_{11} & a'_{12} \\ 0 & U_B \end{bmatrix} .$$

It is easily seen that P is a permutation matrix, L is unit lower-triangular and U is upper-triangular, which shows that $PA = LU$ is indeed an LU decomposition. □

The decomposition $PA = LU$ is sometimes call the PLU decomposition. Note that $PA = LU$ implies that $A = P'LU$ because P is a permutation matrix and $P'P = I$. So it may seem more natural to call this a $P'LU$ decomposition. This is just a matter of convention. We could have simply represented the row interchanges as a permutation matrix P', which would yield $P'A = LU$ and $A = PLU$.

3.4 The LDU and Cholesky factorizations

Let $PA = LU$ be the LU decomposition for a nonsingular matrix A with possible row permutations. Recall that the lower factor L is a unit lower-triangular matrix (with 1's along the diagonal), while the upper factor U is an upper-triangular matrix with nonzero u_{ii}'s along the diagonal. This apparent "asymmetry" can be easily rectified by a LDU factorization, where L and U are now unit lower and unit upper-triangular, respectively, and D is diagonal with the u_{ii}'s as its diagonal elements. In other words, the LDU decomposition results from factoring the diagonal entries out of the upper factor in an LU decomposition.

Theorem 3.4 *Let A be an $n \times n$ nonsingular matrix. Then, $PA = LDU$, where*

(i) P is some $n \times n$ permutation matrix;

(ii) L is an $n \times n$ unit lower-triangular matrix with $l_{ii} = 1$, $i = 1, 2, \ldots, n$;

(iii) D is an $n \times n$ diagonal matrix with nonzero diagonal elements; and

(iv) U is an $n \times n$ unit upper-triangular matrix with $u_{ii} = 1$, $i = 1, 2, \ldots, n$.

Proof. Since A is nonsingular, there exists some permutation matrix such that $PA = LU$ is the LU decomposition of A with possible row interchanges as discussed in Section 3.3. Write U as

$$\begin{bmatrix} u_{11} & u_{12} & \ldots & u_{1n} \\ 0 & u_{22} & \ldots & u_{2n} \\ \vdots & \vdots & \ddots & \vdots \\ 0 & 0 & \ldots & u_{nn} \end{bmatrix} = \begin{bmatrix} u_{11} & 0 & \ldots & 0 \\ 0 & u_{22} & \ldots & 0 \\ \vdots & \vdots & \ddots & \vdots \\ 0 & 0 & \ldots & u_{nn} \end{bmatrix} \begin{bmatrix} 1 & u_{12}/u_{11} & \ldots & u_{1n}/u_{11} \\ 0 & 1 & \ldots & u_{2n}/u_{22} \\ \vdots & \vdots & \ddots & \vdots \\ 0 & 0 & \ldots & 1 \end{bmatrix} .$$

In other words, we have $U = D\tilde{U}$, where D is diagonal with u_{ii}'s along its diagonal and $\tilde{U} = D^{-1}U$ is an upper-triangular matrix with with 1's along the diagonal and u_{ij}/u_{ii} as its (i, j)-th element. Note that this is possible only because $u_{ii} \neq 0$, which is ensured by the nonsingularity of A. Redefining $\tilde{U}$ as U yields $PA = LDU$ with L and U as unit triangular. □

Recall from Theorem 3.1 that if the leading principal submatrices of a matrix are nonsingular, then $\boldsymbol{A} = \boldsymbol{LU}$ without the need for $\boldsymbol{P}$. This means that if $\boldsymbol{A}$ has nonsingular leading principal submatrices, then $\boldsymbol{A} = \boldsymbol{LDU}$ decomposition and the permutation matrix $\boldsymbol{P}$ in Theorem 3.4 is not needed (or $\boldsymbol{P} = \boldsymbol{I}$).

If $\boldsymbol{A} = \boldsymbol{LU}$, or $\boldsymbol{PA} = \boldsymbol{LU}$ then the diagonal elements of $\boldsymbol{U}$ are the pivots of $\boldsymbol{A}$. These pivots appear as the diagonal elements of $\boldsymbol{D}$ in the corresponding $\boldsymbol{LDU}$ decomposition. Since every nonsingular matrix is guaranteed to have all its pivots as nonzero, the diagonal matrix $\boldsymbol{D}$ in the $\boldsymbol{LDU}$ decomposition has all its diagonal elements to be nonzero.

Observe that if $\boldsymbol{A} = \boldsymbol{LU}$, then $\boldsymbol{L}$ and $\boldsymbol{U}$ are unique (Theorem 3.2). This means that the corresponding $\boldsymbol{LDU}$ decomposition will also be unique as long as we restrict both $\boldsymbol{L}$ and $\boldsymbol{U}$ to have unit diagonal elements. This uniqueness yields the following useful result.

Lemma 3.1 *Let $\boldsymbol{A}$ be an $n \times n$ symmetric matrix such that $\boldsymbol{A} = \boldsymbol{LDU}$, where $\boldsymbol{L}$ and $\boldsymbol{U}$ are unit lower-triangular and unit upper-triangular matrices, respectively, and $\boldsymbol{D}$ is a diagonal matrix with nonzero diagonal elements. Then, $\boldsymbol{U} = \boldsymbol{L}'$ and $\boldsymbol{A} = \boldsymbol{LDL}'$.*

Proof. Since $\boldsymbol{A} = \boldsymbol{LDU}$ is symmetric, and $\boldsymbol{D}$ is symmetric, we have

$$\boldsymbol{LDU} = \boldsymbol{A} = \boldsymbol{A}' = \boldsymbol{U}'\boldsymbol{D}'\boldsymbol{L}' = \boldsymbol{U}'\boldsymbol{D}\boldsymbol{L}' \ .$$

Therefore, $\boldsymbol{LDU}$ and $\boldsymbol{U}'\boldsymbol{DL}'$ are both $\boldsymbol{LDU}$ decompositions of $\boldsymbol{A}$. The uniqueness of the $\boldsymbol{LU}$ decomposition (Theorem 3.2) also implies the uniqueness of $\boldsymbol{L}$, $\boldsymbol{D}$ and $\boldsymbol{U}$ in Theorem 3.4. Therefore, $\boldsymbol{U} = \boldsymbol{L}'$ and $\boldsymbol{A} = \boldsymbol{LDL}'$. □

The following factorization of some special symmetric matrices is extremely useful.

Theorem 3.5 *Let $\boldsymbol{A}$ be an $n \times n$ symmetric matrix that has an $\boldsymbol{LU}$ decomposition with strictly positive pivots. Then, there exists a lower-triangular matrix $\boldsymbol{T}$ such that $\boldsymbol{A} = \boldsymbol{TT}'$. Also, each diagonal element of $\boldsymbol{T}$ is positive.*

Proof. If $\boldsymbol{A}$ is symmetric and has an $\boldsymbol{LU}$ decomposition, then $\boldsymbol{A} = \boldsymbol{LDL}'$, where $\boldsymbol{L}$ is unit lower-triangular and $\boldsymbol{D}$ is diagonal (Lemma 3.1). Because the pivots of $\boldsymbol{A}$ are strictly positive, it follows from the construction of $\boldsymbol{D} = \text{diag}(u_{11}, u_{22}, \ldots, u_{nn})$ that each of its diagonal elements is positive. This means that we can define a "square root" matrix of $\boldsymbol{D}$, say $\boldsymbol{D}^{1/2} = \text{diag}(\sqrt{u_{11}}, \sqrt{u_{22}}, \ldots, \sqrt{u_{nn}})$. Therefore,

$$\boldsymbol{A} = \boldsymbol{LDL}' = \boldsymbol{LD}^{1/2}\boldsymbol{D}^{1/2}\boldsymbol{L}' = (\boldsymbol{LD}^{1/2})(\boldsymbol{D}^{1/2})' = \boldsymbol{TT}' \ ,$$

where $\boldsymbol{T} = \boldsymbol{LD}^{1/2}$ is lower-triangular. Each diagonal element of $\boldsymbol{T}$ is the same as the corresponding diagonal element of $\boldsymbol{D}^{1/2}$ and is, therefore, positive. □

If all the pivots of a matrix $\boldsymbol{A}$ are strictly positive, then Gaussian elimination can always be accomplished without the need for row interchanges. This means that $\boldsymbol{A}$

has a $\boldsymbol{LU}$ decomposition with positive diagonal elements for $\boldsymbol{U}$ or an $\boldsymbol{LDU}$ decomposition with strictly positive diagonal elements of $\boldsymbol{D}$.

Symmetric square matrices that have strictly positive pivots are called ***positive definite matrices***. These play an extremely important role in statistics. From what we have seen so far, we can equivalently define positive definite matrices as those that can be factored as $\boldsymbol{LDL}'$ with strictly positive diagonal elements in $\boldsymbol{D}$. Theorem 3.5 shows a positive definite matrix $\boldsymbol{A}$ can be expressed as $\boldsymbol{A} = \boldsymbol{TT}'$, where $\boldsymbol{T}$ is lower-triangular with positive diagonal elements. This decomposition is often referred to as the ***Cholesky decomposition*** and is an important characteristic for positive definite matrices. We will see more of positive definite matrices later.

3.5 Inverse of partitioned matrices

Let $\boldsymbol{A}$ be an $n \times n$ matrix, partitioned as

$$\boldsymbol{A} = \begin{bmatrix} \boldsymbol{A}_{11} & \boldsymbol{A}_{12} \\ \boldsymbol{A}_{21} & \boldsymbol{A}_{22} \end{bmatrix},$$

where $\boldsymbol{A}_{11}$ is $n_1 \times n_1$, $\boldsymbol{A}_{12}$ is $n_1 \times n_2$, $\boldsymbol{A}_{21}$ is $n_2 \times n_1$ and $\boldsymbol{A}_{22}$ is $n_2 \times n_2$. Here $n = n_1 + n_2$. When $\boldsymbol{A}_{11}$ and $\boldsymbol{A}_{22}$ are nonsingular, the matrices defined as

$$\boldsymbol{F} = \boldsymbol{A}_{22} - \boldsymbol{A}_{21}\boldsymbol{A}_{11}^{-1}\boldsymbol{A}_{12} \quad \text{and} \quad \boldsymbol{G} = \boldsymbol{A}_{11} - \boldsymbol{A}_{12}\boldsymbol{A}_{22}^{-1}\boldsymbol{A}_{21} \tag{3.9}$$

are called the ***Schur's complements*** for $\boldsymbol{A}_{11}$ and $\boldsymbol{A}_{22}$, respectively. Note that $\boldsymbol{F}$ is $n_2 \times n_2$ and $\boldsymbol{G}$ is $n_1 \times n_1$. As we will see below, these matrices arise naturally in the elimination of blocks to reduce $\boldsymbol{A}$ to a block-triangular form.

To find $\boldsymbol{A}^{-1}$ in terms of its submatrices, we solve the system

$$\begin{bmatrix} \boldsymbol{A}_{11} & \boldsymbol{A}_{12} \\ \boldsymbol{A}_{21} & \boldsymbol{A}_{22} \end{bmatrix} \begin{bmatrix} \boldsymbol{X}_{11} & \boldsymbol{X}_{12} \\ \boldsymbol{X}_{21} & \boldsymbol{X}_{22} \end{bmatrix} = \begin{bmatrix} \boldsymbol{I}_{n_1} & \boldsymbol{O} \\ \boldsymbol{O} & \boldsymbol{I}_{n_2} \end{bmatrix}. \tag{3.10}$$

A natural way to proceed is to use block-wise elimination. Block-wise elimination uses elementary block operations to reduce general partitioned matrices into block-triangular forms. We demonstrate below.

Let us first assume that $\boldsymbol{A}_{11}$ and its Schur's complement $\boldsymbol{F}$ are both invertible. We can then eliminate the block below $\boldsymbol{A}_{11}$ by premultiplying both sides of (3.10) with an "elementary block matrix." Elementary block operations are analogous to elementary operations but with scalars replaced by matrices. Here one needs to be careful about the dimension of the matrices and the order in which they are multiplied. The elementary block operation we use is to subtract $-\boldsymbol{A}_{21}\boldsymbol{A}_{11}^{-1}$ times the first row from the second row. To be precise, note that

$$\begin{bmatrix} \boldsymbol{I}_{n_1} & \boldsymbol{O} \\ -\boldsymbol{A}_{21}\boldsymbol{A}_{11}^{-1} & \boldsymbol{I}_{n_2} \end{bmatrix} \begin{bmatrix} \boldsymbol{A}_{11} & \boldsymbol{A}_{12} \\ \boldsymbol{A}_{21} & \boldsymbol{A}_{22} \end{bmatrix} = \begin{bmatrix} \boldsymbol{A}_{11} & \boldsymbol{A}_{12} \\ \boldsymbol{O} & \boldsymbol{F} \end{bmatrix}.$$

Applying this elementary block operation to both sides of (3.10) yields

$$\begin{bmatrix} \boldsymbol{A}_{11} & \boldsymbol{A}_{12} \\ \boldsymbol{O} & \boldsymbol{F} \end{bmatrix} \begin{bmatrix} \boldsymbol{X}_{11} & \boldsymbol{X}_{12} \\ \boldsymbol{X}_{21} & \boldsymbol{X}_{22} \end{bmatrix} = \begin{bmatrix} \boldsymbol{I}_{n_1} & \boldsymbol{O} \\ -\boldsymbol{A}_{21}\boldsymbol{A}_{11}^{-1} & \boldsymbol{I}_{n_2} \end{bmatrix},$$

which is a "block-triangular" system. The second row block yields

$$\boldsymbol{F}\boldsymbol{X}_{21} = -\boldsymbol{A}_{21}\boldsymbol{A}_{11}^{-1} \text{ and } \boldsymbol{F}\boldsymbol{X}_{22} = \boldsymbol{I}_{n_2} .$$

Therefore, $\boldsymbol{X}_{21} = -\boldsymbol{F}^{-1}\boldsymbol{A}_{21}\boldsymbol{A}_{11}^{-1}$ and $\boldsymbol{X}_{22} = \boldsymbol{F}^{-1}$.

We then solve for $\boldsymbol{X}_{11}$ and $\boldsymbol{X}_{12}$ by substituting the expressions for $\boldsymbol{X}_{21}$ and $\boldsymbol{X}_{22}$:

$$\boldsymbol{A}_{11}\boldsymbol{X}_{11} + \boldsymbol{A}_{12}\boldsymbol{X}_{21} = \boldsymbol{I}_{n_1} \Longrightarrow \boldsymbol{X}_{11} = \boldsymbol{A}_{11}^{-1} + \boldsymbol{A}_{11}^{-1}\boldsymbol{A}_{12}\boldsymbol{F}^{-1}\boldsymbol{A}_{21}\boldsymbol{A}_{11}^{-1};$$
$$\boldsymbol{A}_{11}\boldsymbol{X}_{12} + \boldsymbol{A}_{12}\boldsymbol{X}_{22} = \boldsymbol{O} \Longrightarrow \boldsymbol{X}_{22} = -\boldsymbol{A}_{11}^{-1}\boldsymbol{A}_{12}\boldsymbol{F}^{-1}.$$

Therefore, assuming that $\boldsymbol{A}_{11}$ and its Schur's complement $\boldsymbol{F}$ are both invertible, we obtain the following expression for the inverse of $\boldsymbol{A}$:

$$\boldsymbol{A}^{-1} = \begin{bmatrix} \boldsymbol{A}_{11}^{-1} + \boldsymbol{A}_{11}^{-1}\boldsymbol{A}_{12}\boldsymbol{F}^{-1}\boldsymbol{A}_{21}\boldsymbol{A}_{11}^{-1} & -\boldsymbol{A}_{11}^{-1}\boldsymbol{A}_{12}\boldsymbol{F}^{-1} \\ -\boldsymbol{F}^{-1}\boldsymbol{A}_{21}\boldsymbol{A}_{11}^{-1} & \boldsymbol{F}^{-1} \end{bmatrix}. \tag{3.11}$$

This is sometimes expressed as

$$\boldsymbol{A}^{-1} = \begin{bmatrix} \boldsymbol{A}_{11}^{-1} & \boldsymbol{O} \\ \boldsymbol{O} & \boldsymbol{O} \end{bmatrix} + \begin{bmatrix} -\boldsymbol{A}_{11}^{-1}\boldsymbol{A}_{12} \\ \boldsymbol{I}_{n_2} \end{bmatrix} \boldsymbol{F}^{-1} \begin{bmatrix} -\boldsymbol{A}_{21}\boldsymbol{A}_{11}^{-1} & \boldsymbol{I}_{n_2} \end{bmatrix}.$$

Equation (3.11) gives the inverse of $\boldsymbol{A}$ assuming $\boldsymbol{A}_{11}^{-1}$ and $\boldsymbol{F}^{-1}$ exist. We can also find an analogous expression for $\boldsymbol{A}^{-1}$ when $\boldsymbol{A}_{22}$ and its Schur's complement $\boldsymbol{G}$ are both nonsingular. In that case, we can we can eliminate the block above $\boldsymbol{A}_{22}$ using

$$\begin{bmatrix} \boldsymbol{I}_{n_1} & -\boldsymbol{A}_{12}\boldsymbol{A}_{22}^{-1} \\ \boldsymbol{O} & \boldsymbol{I}_{n_2} \end{bmatrix} \begin{bmatrix} \boldsymbol{A}_{11} & \boldsymbol{A}_{12} \\ \boldsymbol{A}_{21} & \boldsymbol{A}_{22} \end{bmatrix} = \begin{bmatrix} \boldsymbol{G} & \boldsymbol{O} \\ \boldsymbol{A}_{21} & \boldsymbol{A}_{22} \end{bmatrix}.$$

This yields the block lower-triangular system from (3.10)

$$\begin{bmatrix} \boldsymbol{G} & \boldsymbol{O} \\ \boldsymbol{A}_{21} & \boldsymbol{A}_{22} \end{bmatrix} \begin{bmatrix} \boldsymbol{X}_{11} & \boldsymbol{X}_{12} \\ \boldsymbol{X}_{21} & \boldsymbol{X}_{22} \end{bmatrix} = \begin{bmatrix} \boldsymbol{I}_{n_1} & -\boldsymbol{A}_{12}\boldsymbol{A}_{22}^{-1} \\ \boldsymbol{O} & \boldsymbol{I}_{n_2} \end{bmatrix}.$$

From the first row block, we obtain

$$\boldsymbol{G}\boldsymbol{X}_{11} = \boldsymbol{I} \text{ and } \boldsymbol{G}\boldsymbol{X}_{12} = -\boldsymbol{A}_{12}\boldsymbol{A}_{22}^{-1} .$$

These yield $\boldsymbol{X}_{11} = \boldsymbol{G}^{-1}$ and $\boldsymbol{X}_{12} = -\boldsymbol{G}^{-1}\boldsymbol{A}_{12}\boldsymbol{A}_{22}^{-1}$.

We then solve for $\boldsymbol{X}_{21}$ and $\boldsymbol{X}_{22}$ by substituting the expressions for $\boldsymbol{X}_{11}$ and $\boldsymbol{X}_{12}$:

$$\boldsymbol{A}_{21}\boldsymbol{X}_{11} + \boldsymbol{A}_{22}\boldsymbol{X}_{21} = \boldsymbol{O} \Longrightarrow \boldsymbol{X}_{21} = -\boldsymbol{A}_{22}^{-1}\boldsymbol{A}_{21}\boldsymbol{G}^{-1};$$
$$\boldsymbol{A}_{21}\boldsymbol{X}_{12} + \boldsymbol{A}_{22}\boldsymbol{X}_{22} = \boldsymbol{I} \Longrightarrow \boldsymbol{X}_{22} = \boldsymbol{A}_{22}^{-1} + \boldsymbol{A}_{22}^{-1}\boldsymbol{A}_{21}\boldsymbol{G}^{-1}\boldsymbol{A}_{12}\boldsymbol{A}_{22}^{-1} ,$$

which yields

$$\boldsymbol{A}^{-1} = \begin{bmatrix} \boldsymbol{G}^{-1} & -\boldsymbol{G}^{-1}\boldsymbol{A}_{22}^{-1}\boldsymbol{A}_{21} \\ -\boldsymbol{A}_{12}\boldsymbol{A}_{22}^{-1}\boldsymbol{G}^{-1} & \boldsymbol{A}_{22}^{-1} + \boldsymbol{A}_{22}^{-1}\boldsymbol{A}_{21}\boldsymbol{G}^{-1}\boldsymbol{A}_{12}\boldsymbol{A}_{22}^{-1} \end{bmatrix}. \tag{3.12}$$

This is also written as

$$\boldsymbol{A}^{-1} = \begin{bmatrix} \boldsymbol{O} & \boldsymbol{O} \\ \boldsymbol{O} & \boldsymbol{A}_{22}^{-1} \end{bmatrix} + \begin{bmatrix} \boldsymbol{I}_{n_1} \\ -\boldsymbol{A}_{22}^{-1}\boldsymbol{A}_{21} \end{bmatrix} \boldsymbol{G}^{-1} \begin{bmatrix} \boldsymbol{I}_{n_1} & -\boldsymbol{A}_{12}\boldsymbol{A}_{22}^{-1} \end{bmatrix}.$$

3.6 The LDU decomposition for partitioned matrices

Consider the 2×2 partitioned matrix A from Section 3.5. We can eliminate the block below A_{11} using pre-multiplication:

$$\begin{bmatrix} I_{n_1} & O \\ -A_{21}A_{11}^{-1} & I_{n_2} \end{bmatrix} \begin{bmatrix} A_{11} & A_{12} \\ A_{21} & A_{22} \end{bmatrix} = \begin{bmatrix} A_{11} & A_{12} \\ O & F \end{bmatrix}.$$

And we can eliminate the block above A_{22} using post-multiplication:

$$\begin{bmatrix} A_{11} & A_{12} \\ A_{21} & A_{22} \end{bmatrix} \begin{bmatrix} I_{n_1} & -A_{11}^{-1}A_{12} \\ O & I_{n_2} \end{bmatrix} = \begin{bmatrix} A_{11} & O \\ A_{21} & F \end{bmatrix}.$$

If we use both the above operations, we obtain

$$\begin{bmatrix} I_{n_1} & O \\ -A_{21}A_{11}^{-1} & I_{n_2} \end{bmatrix} \begin{bmatrix} A_{11} & A_{12} \\ A_{21} & A_{22} \end{bmatrix} \begin{bmatrix} I_{n_1} & -A_{11}^{-1}A_{12} \\ O & I_{n_2} \end{bmatrix} = \begin{bmatrix} A_{11} & O \\ O & F \end{bmatrix}.$$

It is easy to verify that

$$\begin{bmatrix} I_{n_1} & O \\ -A_{21}A_{11}^{-1} & I_{n_2} \end{bmatrix}^{-1} = \begin{bmatrix} I_{n_1} & O \\ A_{21}A_{11}^{-1} & I_{n_2} \end{bmatrix} \text{ and}$$

$$\begin{bmatrix} I_{n_1} & -A_{11}^{-1}A_{12} \\ O & I_{n_2} \end{bmatrix}^{-1} = \begin{bmatrix} I_{n_1} & A_{11}^{-1}A_{12} \\ O & I_{n_2} \end{bmatrix}.$$

This produces the ***block LDU decomposition***:

$$A = \begin{bmatrix} I_{n_1} & O \\ A_{21}A_{11}^{-1} & I_{n_2} \end{bmatrix} \begin{bmatrix} A_{11} & O \\ O & F \end{bmatrix} \begin{bmatrix} I_{n_1} & A_{11}^{-1}A_{12} \\ O & I_{n_2} \end{bmatrix}, \tag{3.13}$$

which is a decomposition of a partitioned matrix into a lower-triangular matrix, a block diagonal matrix and an upper-triangular matrix.

The block LDU decomposition makes evaluating A^{-1} easy:

$$\begin{aligned} A^{-1} &= \begin{bmatrix} I_{n_1} & A_{11}^{-1}A_{12} \\ O & I_{n_2} \end{bmatrix}^{-1} \begin{bmatrix} A_{11}^{-1} & O \\ O & F^{-1} \end{bmatrix} \begin{bmatrix} I_{n_1} & O \\ A_{21}A_{11}^{-1} & I_{n_2} \end{bmatrix}^{-1} \\ &= \begin{bmatrix} I_{n_1} & -A_{11}^{-1}A_{12} \\ O & I_{n_2} \end{bmatrix} \begin{bmatrix} A_{11}^{-1} & O \\ O & F^{-1} \end{bmatrix} \begin{bmatrix} I_{n_1} & O \\ -A_{21}A_{11}^{-1} & I_{n_2} \end{bmatrix} \\ &= \begin{bmatrix} A_{11}^{-1} + A_{11}^{-1}A_{12}F^{-1}A_{21}A_{11}^{-1} & -A_{11}^{-1}A_{12}F^{-1} \\ -F^{-1}A_{21}A_{11}^{-1} & F^{-1} \end{bmatrix}. \end{aligned}$$

This affords an alternative derivation of (3.11). Assuming that A_{22}^{-1} and G^{-1} exist, we can obtain an analogous block LDU decomposition to (3.13) and, subsequently, an alternative derivation of (3.12). We leave the details to the reader.

3.7 The Sherman-Woodbury-Morrison formula

Consider the partitioned matrix $A = \begin{bmatrix} A_{11} & A_{12} \\ A_{21} & A_{22} \end{bmatrix}$ and suppose that A_{11}, A_{22} are both invertible and their Schur's complements, F and G as defined in (3.9) are well-defined. Then, the inverse of $\boldsymbol{A}$ can be expressed either as in (3.11) or as in (3.12). These two expressions must be equal. In particular, the $(1,1)$-th block yields

$$\boldsymbol{G}^{-1} = \boldsymbol{A}_{11}^{-1} + \boldsymbol{A}_{11}^{-1}\boldsymbol{A}_{12}\boldsymbol{F}^{-1}\boldsymbol{A}_{21}\boldsymbol{A}_{11}^{-1}.$$

More explicitly, writing out the expressions for $\boldsymbol{G}$ and $\boldsymbol{F}$, we obtain

$$\left(\boldsymbol{A}_{11} - \boldsymbol{A}_{12}\boldsymbol{A}_{22}^{-1}\boldsymbol{A}_{21}\right)^{-1} = \boldsymbol{A}_{11}^{-1} + \boldsymbol{A}_{11}^{-1}\boldsymbol{A}_{12}\left(\boldsymbol{A}_{22} - \boldsymbol{A}_{21}\boldsymbol{A}_{11}^{-1}\boldsymbol{A}_{12}\right)^{-1}\boldsymbol{A}_{21}\boldsymbol{A}_{11}^{-1}. \tag{3.14}$$

This is the ***Sherman-Woodbury-Morrison*** formula. Often this is written with $-\boldsymbol{A}_{12}$ instead of $\boldsymbol{A}_{12}$ in (3.14), yielding

$$\left(\boldsymbol{A}_{11} + \boldsymbol{A}_{12}\boldsymbol{A}_{22}^{-1}\boldsymbol{A}_{21}\right)^{-1} = \boldsymbol{A}_{11}^{-1} - \boldsymbol{A}_{11}^{-1}\boldsymbol{A}_{12}\left(\boldsymbol{A}_{22} + \boldsymbol{A}_{21}\boldsymbol{A}_{11}^{-1}\boldsymbol{A}_{12}\right)^{-1}\boldsymbol{A}_{21}\boldsymbol{A}_{11}^{-1}. \tag{3.15}$$

Some further insight into the above formulas may be helpful. Suppose we want to compute $(\boldsymbol{D} + \boldsymbol{U}\boldsymbol{W}\boldsymbol{V})^{-1}$, where $\boldsymbol{D}$ is $n \times n$, $\boldsymbol{U}$ is $n \times p$, $\boldsymbol{W}$ is $p \times p$ and $\boldsymbol{V}$ is $p \times n$. Also, $\boldsymbol{D}$ and $\boldsymbol{W}$ are nonsingular. Note that $\boldsymbol{D}+\boldsymbol{U}\boldsymbol{W}\boldsymbol{V}$ is $n \times n$. If n is large, then directly computing the inverse of $\boldsymbol{D} + \boldsymbol{U}\boldsymbol{W}\boldsymbol{V}$ using Gaussian elimination or the $\boldsymbol{L}\boldsymbol{U}$ decomposition will be expensive.

The Sherman-Woodbury-Morrison formula can considerably improve matters in certain situations. Consider the setting where $\boldsymbol{D}^{-1}$ is easily available or inexpensive to compute (e.g., $\boldsymbol{D}$ is diagonal) and p is much smaller than n so that solving systems involving $p \times p$ matrices (and hence inverting $p \times p$ matrices) is also inexpensive. In that case, (3.15) can be used by setting $\boldsymbol{A}_{11} = \boldsymbol{D}$, $\boldsymbol{A}_{12} = \boldsymbol{U}$, $\boldsymbol{A}_{22}^{-1} = \boldsymbol{W}$ and $\boldsymbol{A}_{21} = \boldsymbol{V}$. Note that the expression on the right-hand side of (3.12) involves only the inverse of $\boldsymbol{D}$ and that of a $p \times p$ matrix.

In practice, however, directly evaluating the right-hand side of (3.15) is not the most efficient strategy. As we have seen whenever we want to find inverses, we usually solve a system of equations. The Sherman-Woodbury-Morrison formulas in (3.14) and (3.15) were derived from expressions for the inverse of a partitioned matrix. One should, therefore, compute $(\boldsymbol{D} + \boldsymbol{U}\boldsymbol{W}\boldsymbol{V})^{-1}$ using a system of equations.

Consider the following system of block equations:

$$\begin{bmatrix} \boldsymbol{D} & -\boldsymbol{U} \\ \boldsymbol{V} & \boldsymbol{W}^{-1} \end{bmatrix} \begin{bmatrix} \boldsymbol{X}_1 \\ \boldsymbol{X}_2 \end{bmatrix} = \begin{bmatrix} \boldsymbol{I} \\ \boldsymbol{O} \end{bmatrix}. \tag{3.16}$$

The above is an $(n+p) \times (n+p)$ system. The first block of rows in (3.16) represents n equations, while the second block of rows represents p equations. At the outset make the following observation. Suppose we eliminate $\boldsymbol{X}_2$ from the first equation of (3.16) as below:

$$\begin{bmatrix} \boldsymbol{I} & \boldsymbol{U}\boldsymbol{W} \\ \boldsymbol{O} & \boldsymbol{I} \end{bmatrix} \begin{bmatrix} \boldsymbol{D} & -\boldsymbol{U} \\ \boldsymbol{V} & \boldsymbol{W}^{-1} \end{bmatrix} \begin{bmatrix} \boldsymbol{X}_1 \\ \boldsymbol{X}_2 \end{bmatrix} = \begin{bmatrix} \boldsymbol{I} & \boldsymbol{U}\boldsymbol{W} \\ \boldsymbol{O} & \boldsymbol{I} \end{bmatrix} \begin{bmatrix} \boldsymbol{I} \\ \boldsymbol{O} \end{bmatrix}$$

$$\Longrightarrow \begin{bmatrix} \boldsymbol{D}+\boldsymbol{U}\boldsymbol{W}\boldsymbol{V} & \boldsymbol{O} \\ \boldsymbol{V} & \boldsymbol{W}^{-1} \end{bmatrix} \begin{bmatrix} \boldsymbol{X}_1 \\ \boldsymbol{X}_2 \end{bmatrix} = \begin{bmatrix} \boldsymbol{I} \\ \boldsymbol{O} \end{bmatrix}. \tag{3.17}$$

The first block of equations in (3.17) reveals that $(\boldsymbol{D}+\boldsymbol{UWV})\boldsymbol{X}_1 = \boldsymbol{I}$. This means that solving for $\boldsymbol{X}_1$ in (3.16) yields $\boldsymbol{X}_1 = (\boldsymbol{D}+\boldsymbol{UWV})^{-1}$.

But, obviously, we want to avoid solving the $n \times n$ system involving $\boldsymbol{D}+\boldsymbol{UWV}$. Therefore, we will first eliminate $\boldsymbol{X}_1$ from the second block equation, solve for $\boldsymbol{X}_2$ and then solve back for $\boldsymbol{X}_1$ from the first block equation. We eliminate $\boldsymbol{X}_1$ from the second equation as below:

$$\begin{bmatrix} \boldsymbol{I} & \boldsymbol{O} \\ -\boldsymbol{V}\boldsymbol{D}^{-1} & \boldsymbol{I} \end{bmatrix} \begin{bmatrix} \boldsymbol{D} & -\boldsymbol{U} \\ \boldsymbol{V} & \boldsymbol{W}^{-1} \end{bmatrix} \begin{bmatrix} \boldsymbol{X}_1 \\ \boldsymbol{X}_2 \end{bmatrix} = \begin{bmatrix} \boldsymbol{I} & \boldsymbol{O} \\ -\boldsymbol{V}\boldsymbol{D}^{-1} & \boldsymbol{I} \end{bmatrix} \begin{bmatrix} \boldsymbol{I} \\ \boldsymbol{O} \end{bmatrix}$$

$$\Longrightarrow \begin{bmatrix} \boldsymbol{D} & -\boldsymbol{U} \\ \boldsymbol{O} & \boldsymbol{W}^{-1} + \boldsymbol{V}\boldsymbol{W}^{-1}\boldsymbol{U} \end{bmatrix} \begin{bmatrix} \boldsymbol{X}_1 \\ \boldsymbol{X}_2 \end{bmatrix} = \begin{bmatrix} \boldsymbol{I} \\ -\boldsymbol{V}\boldsymbol{D}^{-1} \end{bmatrix}. \tag{3.18}$$

From the second equation, we first solve $(\boldsymbol{W}^{-1} + \boldsymbol{V}\boldsymbol{D}^{-1}\boldsymbol{U})\boldsymbol{X}_2 = -\boldsymbol{V}\boldsymbol{D}^{-1}$. Note that this is feasible because the coefficient matrix $\boldsymbol{W}^{-1} + \boldsymbol{V}\boldsymbol{W}^{-1}\boldsymbol{U}$ is $p \times p$ (and not $n \times n$) and $\boldsymbol{D}^{-1}$ is easily available. Next we solve $\boldsymbol{X}_1$ from the first block of equations in $\boldsymbol{D}\boldsymbol{X}_1 = \boldsymbol{I}+\boldsymbol{U}\boldsymbol{X}_2$. This is very simple because $\boldsymbol{D}^{-1}$ is easily available. Thus, we have obtained $\boldsymbol{X}_1$, which, from (3.17), is $(\boldsymbol{D}+\boldsymbol{UWV})^{-1}$.

Finally, note that this strategy yields another derivation of the Sherman-Woodbury-Morrison formula by substituting the solutions for $\boldsymbol{X}_1$ and $\boldsymbol{X}_2$ obtained from the above. From (3.18), we have found $\boldsymbol{X}_2 = -(\boldsymbol{W}^{-1}+\boldsymbol{V}\boldsymbol{D}^{-1}\boldsymbol{U})^{-1}\boldsymbol{V}\boldsymbol{D}^{-1}$ and $\boldsymbol{X}_1 = \boldsymbol{D}^{-1}(\boldsymbol{I}+\boldsymbol{U}\boldsymbol{X}_2)$. This yields

$$\begin{aligned}(\boldsymbol{D}+\boldsymbol{UWV})^{-1} &= \boldsymbol{X}_1 = \boldsymbol{D}^{-1}(\boldsymbol{I}+\boldsymbol{U}\boldsymbol{X}_2) = \boldsymbol{D}^{-1} + \boldsymbol{D}^{-1}\boldsymbol{U}\boldsymbol{X}_2 \\ &= \boldsymbol{D}^{-1} - \boldsymbol{D}^{-1}\boldsymbol{U}\left(\boldsymbol{W}^{-1} + \boldsymbol{V}\boldsymbol{D}^{-1}\boldsymbol{U}\right)^{-1}\boldsymbol{V}\boldsymbol{D}^{-1}.\end{aligned}$$

Henderson and Searle (1981) offer an excellent review of similar identities.

3.8 Exercises

1. Use an $\boldsymbol{LU}$ decomposition to solve the linear system in Exercise 1 of Section 2.7.
2. Find the $\boldsymbol{LU}$ decomposition (with row interchanges if needed) of $\boldsymbol{A}$ and solve $\boldsymbol{A}\boldsymbol{x} = \boldsymbol{b}$ for each $\boldsymbol{A}$ and $\boldsymbol{b}$ in Exercise 2 of Section 2.7. Also find the $\boldsymbol{LDU}$ decomposition for each $\boldsymbol{A}$ in Exercise 2 of Section 2.7.
3. Explain how you can efficiently obtain $\boldsymbol{A}^{-1}$ by solving $\boldsymbol{AX} = \boldsymbol{I}$ using the $\boldsymbol{LU}$ decomposition of $\boldsymbol{A}$. Find $\boldsymbol{A}^{-1}$ using the $\boldsymbol{LU}$ decomposition of $\boldsymbol{A}$ for each matrix $\boldsymbol{A}$ in Exercise 2 of Section 2.7.
4. Find the $\boldsymbol{LU}$ decomposition of the transposed Vandermonde matrices

$$\text{(a)} \begin{bmatrix} 1 & 1 \\ x_1 & x_2 \end{bmatrix}, \quad \text{(b)} \begin{bmatrix} 1 & 1 & 1 \\ x_1 & x_2 & x_3 \\ x_1^2 & x_2^2 & x_3^2 \end{bmatrix} \quad \text{and (c)} \begin{bmatrix} 1 & 1 & 1 & 1 \\ x_1 & x_2 & x_3 & x_4 \\ x_1^2 & x_2^2 & x_3^2 & x_4^2 \\ x_1^3 & x_2^3 & x_3^3 & x_4^3 \end{bmatrix},$$

   where the x_i's are distinct real numbers for $i = 1, 2, 3, 4$. Can you guess the pattern for the general $n \times n$ Vandermonde matrix?

5. Find an $\boldsymbol{LU}$ decomposition of the *tridiagonal* matrix

$$\boldsymbol{A} = \begin{bmatrix} a_{11} & a_{12} & 0 & 0 \\ a_{21} & a_{22} & a_{23} & 0 \\ 0 & a_{32} & a_{33} & a_{34} \\ 0 & 0 & a_{43} & a_{44} \end{bmatrix}.$$

6. True or false: If a matrix has one of its diagonal entries equal to zero, then it does not have an $\boldsymbol{LU}$ decomposition without row interchanges.

7. Count the exact number of floating point operations (i.e., scalar addition, subtraction, multiplication and division) required to compute the $\boldsymbol{LU}$ decomposition of an $n \times n$ matrix if (a) row interchanges are not required, and (b) row interchanges are required.

8. If $\boldsymbol{A} = \{a_{ij}\}$ is a matrix such that each a_{ij} is an integer and all of the pivots of $\boldsymbol{A}$ are 1, then explain why all the entries in $\boldsymbol{A}^{-1}$ must also be integers.

9. Find an $\boldsymbol{LDL}'$ for the following matrix:

$$\boldsymbol{A} = \begin{bmatrix} 1 & 2 & 3 \\ 2 & 8 & 10 \\ 3 & 10 & 16 \end{bmatrix}.$$

   Is $\boldsymbol{A}$ positive definite? If so, what is the Cholesky decomposition of $\boldsymbol{A}$?

10. This exercise develops an "$\boldsymbol{LU}$" decomposition of an $m \times n$ rectangular matrix. Let $\boldsymbol{A} = \{a_{ij}\}$ be an $m \times n$ matrix such that $m \leq n$ and the m leading principal submatrices of $\boldsymbol{A}$ are nonsingular. Prove that there exist matrices $\boldsymbol{L}$ and $\boldsymbol{U}$ such that $\boldsymbol{A} = \boldsymbol{LU}$, where $\boldsymbol{L}$ is $m \times m$ lower-triangular and $\boldsymbol{U}$ is $m \times n$ such that the first m columns of $\boldsymbol{A}$ form an upper-triangular matrix with all its diagonal entries equal to 1. Find such an "$\boldsymbol{LU}$" decomposition for

$$\boldsymbol{A} = \begin{bmatrix} 3 & 0 & 2 & -1 \\ 2 & 3 & -1 & 0 \\ 1 & 5 & 0 & 8 \end{bmatrix}.$$

11. Let $\boldsymbol{A} = \{a_{ij}\}$ be an $m \times n$ matrix with real entries such that $m \leq n$ and the m leading principal submatrices of $\boldsymbol{A}$ are nonsingular. Prove that there exist matrices $\boldsymbol{L} = \{l_{ij}\}$ and $\boldsymbol{U} = \{u_{ij}\}$ such that $\boldsymbol{A} = \boldsymbol{LU}$, where $\boldsymbol{L}$ is $m \times m$ lower-triangular and $\boldsymbol{U}$ is $m \times n$ such that the first m columns of $\boldsymbol{A}$ form an upper-triangular matrix and *either* $l_{11}, l_{22}, \ldots, l_{mm}$ are preassigned nonzero real numbers *or* $u_{11}, u_{22}, \ldots, u_{mm}$ are preassigned nonzero real numbers.

12. Show that $\begin{bmatrix} \boldsymbol{A} & \boldsymbol{B} \\ \boldsymbol{O} & \boldsymbol{D} \end{bmatrix}^{-1}$ is of the form $\begin{bmatrix} \boldsymbol{A}^{-1} & \boldsymbol{X} \\ \boldsymbol{O} & \boldsymbol{D}^{-1} \end{bmatrix}$, where $\boldsymbol{A}$ and $\boldsymbol{D}$ are nonsingular, and find $\boldsymbol{X}$. Also show that $\begin{bmatrix} \boldsymbol{A} & \boldsymbol{O} \\ \boldsymbol{C} & \boldsymbol{D} \end{bmatrix}^{-1}$ is of the form $\begin{bmatrix} \boldsymbol{A}^{-1} & \boldsymbol{O} \\ \boldsymbol{Y} & \boldsymbol{D}^{-1} \end{bmatrix}$, where $\boldsymbol{A}$ and $\boldsymbol{D}$ are nonsingular, and find $\boldsymbol{Y}$.

13. Verify the following:

   (a) $\begin{bmatrix} \boldsymbol{I}_{n_1} & \boldsymbol{O} \\ \boldsymbol{C} & \boldsymbol{I}_{n_2} \end{bmatrix}^{-1} = \begin{bmatrix} \boldsymbol{I}_{n_1} & \boldsymbol{O} \\ -\boldsymbol{C} & \boldsymbol{I}_{n_2} \end{bmatrix}$, where $\boldsymbol{C}$ is $n_2 \times n_1$.

(b) $\begin{bmatrix} I_{n_1} & B \\ O & I_{n_2} \end{bmatrix}^{-1} = \begin{bmatrix} I_{n_1} & -B \\ O & I_{n_2} \end{bmatrix}$, where B is $n_1 \times n_2$.

14. Consider the following matrix:

$$A = \begin{bmatrix} 1 & 0 & 0 & 1 & 1 \\ 0 & 2 & 0 & 0 & -1 \\ 0 & 0 & 3 & 1 & 3 \\ 2 & -1 & 1 & 2 & -1 \\ 0 & 2 & 2 & 1 & 3 \end{bmatrix} .$$

Using a suitable partitioning of A, find A^{-1}.

15. Verify the Sherman-Woodbury-Morrison identity in Section 3.7:

$$(D + UWV)^{-1} = D^{-1} - D^{-1}U\left(W^{-1} + VD^{-1}U\right)^{-1}VD^{-1}$$

by multiplying the right-hand side with $D+UWV$ and showing that this product is equal to the identity matrix.

16. Let A be $n \times n$ and nonsingular. Let u and v be $n \times 1$ vectors. Show that $A + uv'$ is nonsingular if and only if $v'A^{-1}u$ and in that case derive the formula

$$(A + uv')^{-1} = A^{-1} - \frac{A^{-1}uv'A^{-1}}{(1 + v'A^{-1}u)} .$$

17. Let A be an $n \times n$ nonsingular matrix for which A^{-1} is already available. Let B be an $n \times n$ matrix obtained from A by changing exactly one row or exactly one column. Using the preceding exercise, explain how one can obtain B^{-1} cheaply without having to invert B.

18. Let A and $(A + uv')$ be $n \times n$ nonsingular matrices. If $Ax = b$ and $Ay = u$, then find the solution for $(A + uv')z = b$ in terms of the vectors x, y and v.

19. Let A be an $n \times n$ matrix such that all its diagonal entries are equal to α and all its other entries are equal to β:

$$A = \begin{bmatrix} \alpha & \beta & \dots & \beta \\ \beta & \alpha & \cdots & \beta \\ \vdots & \vdots & \ddots & \vdots \\ \beta & \beta & \dots & \alpha \end{bmatrix} .$$

Show that A can be written in the form $D + uv'$, where D is diagonal and u and v are $n \times 1$ vectors. Deduce that A is nonsingular if and only if

$$(\alpha - \beta)[\alpha + (n-1)\beta] \neq 0 .$$

Also find an explicit expression for A^{-1}.

20. For what values of the scalar θ is $A = \begin{bmatrix} I & \theta 1 \\ \theta 1' & 1 \end{bmatrix}$ nonsingular? Find A^{-1} when it exists.

CHAPTER 4

Euclidean Spaces

4.1 Introduction

Until now we explored matrix algebra as a set of operations on a rectangular array of numbers. Using these operations we learned how elementary operations can reduce matrices to triangular or echelon forms and help us solve systems of linear equations. To ascertain when such a system is consistent, we relied upon the use of elementary row operations that reduced the matrix to echelon or triangular matrices. Once this was done, the "pivot" and "free" variables determined the solutions of the system.

While computationally attractive and transparent, understanding linear systems using echelon matrices is awkward and sometimes too complicated. It turns out that matrices and their structures are often best understood by studying what they are made up of: their column vectors and row vectors. In fact, whether we are looking at $\boldsymbol{A}\boldsymbol{x}$ or the product $\boldsymbol{A}\boldsymbol{B}$, we are looking at linear combinations of the column vectors of $\boldsymbol{A}$. In this chapter, we study ***spaces*** of vectors and their characteristics. Understanding such spaces, and especially some of their special subsets called ***subspaces***, is critical to understanding everything about $\boldsymbol{A}\boldsymbol{x} = \boldsymbol{b}$.

4.2 Vector addition and scalar multiplication

We will denote the set of finite real numbers by $\Re^1$, or simply $\Re$.

Definition 4.1 *The m-**dimensional real Euclidean space**, denoted by $\Re^m$, is the Cartesian product of m sets, each equal to $\Re^1$:*

$$\Re^m = \Re^1 \times \Re^1 \cdots \Re^1 \times \Re^1 \quad m \text{ times.}$$

Thus $\Re^m$ consists of m-tuples of real numbers and is often called the **real coordinate space**.

We can write m-tuples of real numbers either as $1 \times m$ row vectors or as $m \times 1$ column vectors, yielding the real coordinate spaces:

$$\Re^{1\times m} = \{(x_1, x_2, \ldots, x_n) : x_i \in \Re^1\} \quad \text{and} \quad \Re^{m\times 1} = \left\{ \begin{bmatrix} x_1 \\ x_2 \\ \vdots \\ x_m \end{bmatrix} : x_i \in \Re^1 \right\}.$$

When studying vectors, it usually makes no difference whether we treat a coordinate vector as a row or as a column. When distinguishing between row or column vectors is irrelevant, or when it is clear from the context, we will use the common symbol $\Re^m$ to designate a real coordinate space. In situations where we want to distinguish between row and column vectors, we will explicitly write $\Re^{1\times m}$ and $\Re^{m\times 1}$.

Two important operations with vectors are ***vector addition*** and ***scalar multiplication***, given by

$$\boldsymbol{x} + \boldsymbol{y} = \begin{bmatrix} x_1 + y_1 \\ x_2 + y_2 \\ \vdots \\ x_m + y_m \end{bmatrix} \quad \text{and} \quad a\boldsymbol{x} = \begin{bmatrix} ax_1 \\ ax_2 \\ \vdots \\ ax_m \end{bmatrix}, \tag{4.1}$$

where $\boldsymbol{x}$ and $\boldsymbol{y}$ are two vectors in $\Re^m$ and a is a scalar in $\Re^1$. The difference of two vectors is given by $\boldsymbol{x} - \boldsymbol{y} = \boldsymbol{x} + (-\boldsymbol{y})$. Note that the above operations are the same as those defined for matrices (see Section 1.2) when we treat *both* of $\boldsymbol{x}$ and $\boldsymbol{y}$ as $m \times 1$ column vectors or as $1 \times m$ row vectors.

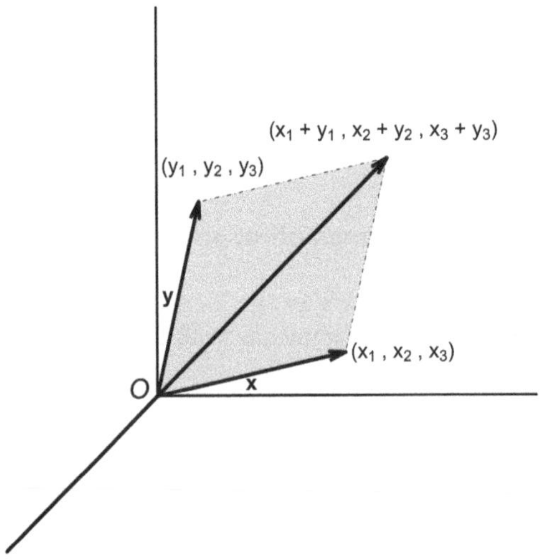

Figure 4.1 *Addition of vectors.*

The geometric interpretation of these operations can be easily visualized when $m = 2$ or $m = 3$, i.e., the vectors are simply points on the plane or in space. The sum $\boldsymbol{x}+\boldsymbol{y}$ is obtained by drawing the diagonal of the parallelogram obtained from $\boldsymbol{x}$, $\boldsymbol{y}$ and the origin (Figure 4.1). The scalar multiple $a\boldsymbol{x}$ is obtained by multiplying the magnitude of the vector by a (i.e., stretching it by a factor of a) and retaining the same direction if $a > 0$, while reversing the direction if $a < 0$ (Figure 4.2).

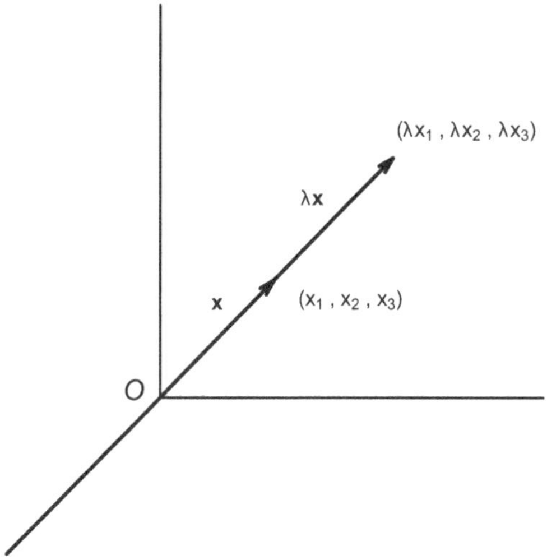

Figure 4.2 *Scalar multiplication on a vector.*

4.3 Linear spaces and subspaces

One can look upon $\Re^m$ as an example of a space where vectors reside. The definitions of vector addition and scalar multiplication in (4.1) along with basic properties of real numbers immediately yields the following important properties of $\Re^m$ as a set of vectors.

[A1] $\boldsymbol{x} + \boldsymbol{y} \in \Re^m$ for every $\boldsymbol{x}, \boldsymbol{y} \in \Re^m$. We say that $\Re^m$ is ***closed under vector addition***.

[A2] $\boldsymbol{x} + \boldsymbol{y} = \boldsymbol{y} + \boldsymbol{x}$ for every $\boldsymbol{x}, \boldsymbol{y} \in \Re^m$. This says that ***vector addition is commutative***.

[A3] $(\boldsymbol{x} + \boldsymbol{y}) + \boldsymbol{z} = \boldsymbol{x} + (\boldsymbol{y} + \boldsymbol{z})$ for every $\boldsymbol{x}, \boldsymbol{y}, \boldsymbol{z} \in \Re^m$. This says that ***vector addition is associative***.

[A4] $\boldsymbol{x} + \boldsymbol{0} = \boldsymbol{x}$ for every $\boldsymbol{x} \in \Re^m$. The null vector $\boldsymbol{0}$ is also called the ***additive identity*** element in $\Re^m$.

[A5] $\boldsymbol{x} + (-\boldsymbol{x}) = \boldsymbol{0}$ for every $\boldsymbol{x} \in \Re^m$. We call $(-\boldsymbol{x})$ the ***additive inverse*** of $\boldsymbol{x}$.

[M1] $ax \in \Re^m$ for all $a \in \Re^1$ and $x \in \Re^1$. This says that $\Re^m$ is ***closed under scalar multiplication.***

[M2] $a(bx) = (ab)x$ for every $a, b \in \Re^1$ and for every $x \in \Re^m$. This is called the ***compatibility property of scalar multiplication***.

[M3] $(a+b)x = ax + bx$ for every $a, b \in \Re^1$ and for every $x \in \Re^m$. This is called the ***distributive property of scalar multiplication with respect to scalar addition***

[M4] $a(x + y) = ax + ay$ for every $x, y \in \Re^m$ and every $a \in \Re^1$. This is called the ***distributive property of scalar multiplication with respect to vector addition***.

[M5] $1x = x$ for every $x \in \Re^m$. The element 1 is called the ***identity element of scalar multiplication***.

The above properties of vectors in $\Re^m$ motivates the study of more general mathematical structures known as ***vector spaces*** or ***linear spaces***. These can be looked upon as spaces where "vectors" reside, but the definition of "vectors" need not be restricted to m-tuples of real numbers as in $\Re^m$. For instance, they can be spaces of matrices or even functions. We will later see examples of more abstract vector spaces. For most of the book, we will mostly restrict ourselves to $\Re^m$.

Often, our interest will focus not upon the entire Euclidean real coordinate space $\Re^m$, but on certain subsets. The most interesting subsets of the Euclidean spaces are the non-empty sets whose members can be added or multiplied by real numbers without leaving the set. We call such a set a ***linear subspace*** or ***vector subspace*** or simply ***subspace*** of $\Re^m$. A formal definition follows.

Definition 4.2 *A non-empty set $\mathcal{S} \subseteq \Re^m$ is called a* **vector space** *in $\Re^m$ (or a* **subspace** *of $\Re^m$) if both of the following conditions are satisfied:*

- *If $x \in \mathcal{S}$ and $y \in \mathcal{S}$, then $x + y \in \mathcal{S}$. In other words, $\mathcal{S}$ is* **closed under vector addition.**
- *If $x \in \mathcal{S}$, then $\alpha x \in \mathcal{S}$ for all $\alpha \in \Re^1$. In other words, $\mathcal{S}$ is* **closed under scalar multiplication.**

The above two criteria can be combined to say that a non-empty set $\mathcal{S} \subseteq \Re^m$ is a subspace if $x + \alpha y \in \mathcal{S}$ for every $x, y \in \mathcal{S}$ and every $\alpha \in \Re^1$.

Remark: When we say that $\mathcal{S}$ is a vector space in $\Re^m$, what we precisely mean that $\mathcal{S}$ is a vector space comprising vectors in $\Re^m$. In other words, each member of $\mathcal{S}$ is an $m \times 1$ vector.

An immediate and important consequence of the above definition is the following result.

Theorem 4.1 *Let $\mathcal{S}$ be a vector space in $\Re^m$. Then $\mathbf{0} \in \mathcal{S}$.*

Proof. Since $\mathcal{S}$ is non-empty by definition, we can suppose $x \in \mathcal{S}$. If $x = \mathbf{0}$, we are done. Otherwise, since $-1 \in \Re^1$, closure under scalar multiplication implies that

$(-1)\boldsymbol{x} = -\boldsymbol{x} \in \mathcal{S}$. Since $\boldsymbol{x}$ and $-\boldsymbol{x}$ are both in $\mathcal{S}$, closure under addition implies that $\boldsymbol{x} - \boldsymbol{x} = \boldsymbol{0}$ is also in $\mathcal{S}$. □

Example 4.1 The set $\{\boldsymbol{0}\} \subset \Re^m$ is itself a subspace of $\Re^m$. We call it the trivial subspace.

Example 4.2 Consider the Euclidean plane $\Re^2$ and let $\mathcal{S}$ be the subset containing all points lying on a line through the origin. In other words,

$$\mathcal{S} = \{(x_1, x_2) : x_2 = \alpha x_1\}, \quad \alpha \in \Re^1 .$$

Let $\boldsymbol{x} = (x_1, x_2)$ and $\boldsymbol{y} = (y_1, y_2)$ be two members in $\mathcal{S}$. Therefore, $x_2 = \alpha x_1$ and $y_2 = \alpha y_1$. Let $\theta \in \Re^1$ be any given real number and let $\boldsymbol{z} = \boldsymbol{x} + \theta \boldsymbol{y}$. Then,

$$z_2 = x_2 + \theta y_2 = \alpha x_1 + \theta(\alpha y_1) = \alpha x_1 + \alpha \theta y_1 \alpha(x_1 + \theta y_1) = \alpha z_1 ,$$

which implies that $\boldsymbol{z} \in \mathcal{S}$. This shows that if $\boldsymbol{x}, \boldsymbol{y} \in \mathcal{S}$, then $\boldsymbol{x} + \theta \boldsymbol{y} \in \mathcal{S}$.

The above arguments hold in the special case when $\alpha = 0$. This shows that the x-axis of the Euclidean plane, described by $\{(x_1, 0) : x_1 \in \Re^1\}$ is itself a linear subspace.

Arguments analogous to the above can also be used to prove that the set

$$\mathcal{S} = \{(x_1, x_2) : x_1 = \alpha x_2\}, \quad \alpha \in \Re^1$$

is a linear subspace of $\Re^2$. Taking $\alpha = 0$ now shows that the y-axis of the Euclidean plane, described by $\{(0, x_2) : x_2 \in \Re^1\}$ is a linear subspace of $\Re^2$. ■

Example 4.2 shows that straight lines through the origin in $\Re^2$ are subspaces. However, straight lines that *do not* pass through the origin cannot be subspaces because subspaces must contain the null vector (Theorem 4.1). *Curves* (that are not straight lines) also do not form subspaces because they violate the closure property under vector addition. In other words, lines with a curvature have points $\boldsymbol{u}$ and $\boldsymbol{v}$ on the curve for which $\boldsymbol{u} + \boldsymbol{v}$ is not on the curve. In fact, the only proper subspaces of $\Re^2$ are the trivial subspace $\{\boldsymbol{0}\}$ and the lines through the origin.

Moving to $\Re^3$, the trivial subspace and straight lines passing through the origin are again subspaces. But now there is a new one—planes that contain the origin. We discuss this in the example below.

Example 4.3 A plane in $\Re^3$ is described parametrically as the set of all points of the form $\mathcal{P} = \{\boldsymbol{r} \in \Re^3 : \boldsymbol{r} = \boldsymbol{r}_0 + s\boldsymbol{u} + t\boldsymbol{v}\}$, where s and t range over $\Re^1$, $\boldsymbol{u}$ and $\boldsymbol{v}$ are given non-null vectors in $\Re^3$ defining the plane, and $\boldsymbol{r}_0 \in \Re^3$ is a vector representing the position of an arbitrary (but fixed) point on the plane. $\boldsymbol{r}_0$ is called the ***position vector*** of a point on the plane. The vectors $\boldsymbol{u}$ and $\boldsymbol{v}$ cannot be parallel, i.e., $\boldsymbol{u} \neq \alpha \boldsymbol{v}$ for any $\alpha \in \Re^1$. The vectors $\boldsymbol{u}$ and $\boldsymbol{v}$ can be visualized as originating from $\boldsymbol{r}_0$ and pointing in different directions along the plane.

A plane through the origin is described by taking $\boldsymbol{r}_0 = \boldsymbol{0}$, so that

$$\mathcal{P}_0 = \{\boldsymbol{r} \in \Re^3 : s\boldsymbol{u} + t\boldsymbol{v}\} .$$

Clearly $\boldsymbol{r} = \boldsymbol{0}$ is obtained when $s = t = 0$. Suppose $\boldsymbol{r}_1$ and $\boldsymbol{r}_2$ are two points in $\mathcal{P}_0$. Then, we can find real numbers s_1, t_1, s_2 and t_2 so that

$$\boldsymbol{r}_1 = s_1\boldsymbol{u} + t_1\boldsymbol{v} \quad \text{and} \quad \boldsymbol{r}_2 = s_2\boldsymbol{u} + t_2\boldsymbol{v} .$$

Let $\theta \in \Re^1$ be any given real number and consider the vector $\boldsymbol{r}_3 = \boldsymbol{r}_1 + \theta\boldsymbol{r}_2$. Then,

$$\boldsymbol{r}_3 = s_1\boldsymbol{u} + t_1\boldsymbol{v} + \theta(s_2\boldsymbol{u} + t_2\boldsymbol{v}) = (s_1 + \theta s_2)\boldsymbol{u} + (t_1 + \theta t_2)\boldsymbol{v} = s_3\boldsymbol{u} + t_3\boldsymbol{v} \in \mathcal{P}_0 ,$$

where $s_3 = s_1 + \theta s_2$ and $t_1 + \theta t_2$ are real numbers. This shows that $\mathcal{P}_0$ is a subspace of $\Re^3$. ■

Remark: Note that any subspace $\mathcal{V}$ of $\Re^m$ is a vector space and it is legitimate to consider further subspaces of $\mathcal{V}$. In fact, we will often consider situations where $\mathcal{V}$ is a vector space comprising vectors in the Euclidean space $\Re^m$ and $\mathcal{S}$ is a subspace of $\mathcal{V}$. Any result that we derive for a subspace $\mathcal{S}$ of $\mathcal{V}$ will hold for any subspace of $\Re^m$ by simply letting $\mathcal{V} = \Re^m$. Also, we will sometimes be informal and simply say "$\mathcal{S}$ is a subspace of $\mathcal{V}$." This will mean that $\mathcal{S}$ is a subspace of the vector space $\mathcal{V}$ which is itself some subspace of the Euclidean space.

4.4 Intersection and sum of subspaces

We can define the *intersection* of two subspaces in the same way as we define the intersection of two sets provided that the two subspaces reside within the same vector space.

Definition 4.3 *Let $\mathcal{S}_1$ and $\mathcal{S}_2$ be two subspaces of a vector space $\mathcal{V}$ in $\Re^n$. Their* **intersection** *is denoted by $\mathcal{S}_1 \cap \mathcal{S}_2$ and defined as*

$$\mathcal{S}_1 \cap \mathcal{S}_2 = \{\boldsymbol{x} \in \Re^n : \boldsymbol{x} \in \mathcal{S}_1 \quad \textit{AND} \quad \boldsymbol{x} \in \mathcal{S}_2\} .$$

It is true that the intersection of subspaces is always a subspace.

Theorem 4.2 *Let $\mathcal{S}_1$ and $\mathcal{S}_2$ be two subspaces of $\Re^m$. Then $\mathcal{S}_1 \cap \mathcal{S}_2$ is also a subspace.*

Proof. By definition both $\mathcal{S}_1$ and $\mathcal{S}_2$ contain $\boldsymbol{0}$, so $\boldsymbol{0} \in \mathcal{S}_1 \cap \mathcal{S}_2$. If $\boldsymbol{u}$, $\boldsymbol{v}$ are two vectors in $\mathcal{S}_1 \cap \mathcal{S}_2$, then any vector $\alpha\boldsymbol{u} + \boldsymbol{v}$ must belong to both $\mathcal{S}_1$ and $\mathcal{S}_2$ (as both are subspaces), and so $\alpha\boldsymbol{u} + \boldsymbol{v} \in \mathcal{S}_1 \cap \mathcal{S}_2$. □

Suppose $\mathcal{S}_1$ and $\mathcal{S}_2$ are two subspaces. Then, is $\mathcal{S}_1 \cup \mathcal{S}_2$ always a subspace? The following example gives the answer.

Example 4.4 The union of subspaces is not necessarily a subspace. Let $\mathcal{S}_1 = \{(x, 0) : x \in \Re^1\}$ and $\mathcal{S}_2 = \{(0, y) : y \in \Re^2\}$. In other words, $\mathcal{S}_1$ is the x-axis and $\mathcal{S}_2$ is the y-axis. Both are subspaces of $\mathcal{V} = \Re^2$, but $\mathcal{S}_1 \cup \mathcal{S}_2$ does not contain any non-trivial linear combination of one vector from $\mathcal{S}_1$ and another from $\mathcal{S}_2$. ■

If unions of subspaces are not subspaces, then can we somehow *extend* them to subspaces? The answer is *yes* and we show how. We first provide the following definition of a sum of subspaces.

Definition 4.4 *Let $\mathcal{S}_1$ and $\mathcal{S}_2$ be two subspaces of $\Re^m$. Then the* **sum** *of $\mathcal{S}_1$ and $\mathcal{S}_2$, written as $\mathcal{S}_1 + \mathcal{S}_2$, is defined to be*

$$\mathcal{S}_1 + \mathcal{S}_2 = \{\boldsymbol{x} + \boldsymbol{y} : \boldsymbol{x} \in \mathcal{S}_1,\ \boldsymbol{y} \in \mathcal{S}_2\}.$$

In other words, $\mathcal{S}_1 + \mathcal{S}_2$ is defined to be the sum of all possible vectors from $\mathcal{S}_1$ with vectors from $\mathcal{S}_2$.

It is clear from the above definition that $\boldsymbol{S}_1 + \boldsymbol{S}_2 = \boldsymbol{S}_2 + \boldsymbol{S}_1$. In other words, addition of subspaces is commutative. It is also easily verified that addition of subspaces is associative. In other words, we have $(\boldsymbol{S}_1 + \boldsymbol{S}_2) + \boldsymbol{S}_3 = \boldsymbol{S}_1 + (\boldsymbol{S}_2 + \boldsymbol{S}_3)$, which means that we can unambiguously define this sum as $\boldsymbol{S}_1 + \boldsymbol{S}_2 + \boldsymbol{S}_3$. Thus, the sum of several subspaces can be found by repeated addition of subspaces. Indeed, Definition 4.4 is easily generalized to the sum of any finite number of subspaces as follows:

$$\boldsymbol{S}_1 + \boldsymbol{S}_2 + \cdots + \boldsymbol{S}_k = \{\boldsymbol{x}_1 + \boldsymbol{x}_2 + \cdots + \boldsymbol{x}_k : \quad \boldsymbol{x}_i \in \boldsymbol{S}_i, \quad i = 1, 2, \ldots, k\}\,. \tag{4.2}$$

Notice that if $(i_1, i_2, \ldots, i_k)$ is a permutation of $(1, 2, \ldots, k)$, then $\boldsymbol{S}_{i_1} + \boldsymbol{S}_{i_2} + \cdots + \boldsymbol{S}_{i_k} = \boldsymbol{S}_1 + \boldsymbol{S}_2 + \cdots + \boldsymbol{S}_k$.

And now we have the following theorem.

Theorem 4.3 *Let $\mathcal{S}_1$ and $\mathcal{S}_2$ be two subspaces of $\Re^m$. Then their sum $\mathcal{S}_1 + \mathcal{S}_2$ is the smallest subspace of $\Re^m$ that contains $\mathcal{S}_1 \cup \mathcal{S}_2$.*

Proof. We first prove that $\mathcal{S}_1 + \mathcal{S}_2$ is a subspace. Since $\boldsymbol{0} \in \mathcal{S}_1$ and $\boldsymbol{0} \in \mathcal{S}_2$, we have $\boldsymbol{0} = \boldsymbol{0} + \boldsymbol{0} \in \mathcal{S}_1 + \mathcal{S}_2$. Also, let $\boldsymbol{u}$ and $\boldsymbol{v}$ both belong to $\mathcal{S}_1 + \mathcal{S}_2$. Then, $\boldsymbol{u} = \boldsymbol{x}_1 + \boldsymbol{x}_2$ for some $\boldsymbol{x}_1 \in \mathcal{S}_1$ and $\boldsymbol{x}_2 \in \mathcal{S}_2$, and $\boldsymbol{v} = \boldsymbol{y}_1 + \boldsymbol{y}_2$ for some $\boldsymbol{y}_1 \in \mathcal{S}_1$ and $\boldsymbol{y}_2 \in \mathcal{S}_2$. Then,

$$\alpha\boldsymbol{u} + \boldsymbol{v} = (\alpha\boldsymbol{x}_1 + \boldsymbol{y}_1) + (\alpha\boldsymbol{x}_2 + \boldsymbol{y}_2) \in \mathcal{S}_1 + \mathcal{S}_2$$

since $(\alpha\boldsymbol{x}_1 + \boldsymbol{y}_1) \in \mathcal{S}_1$ and $(\alpha\boldsymbol{x}_2 + \boldsymbol{y}_2) \in \mathcal{S}_2$.

We now prove that $\mathcal{S}_1 + \mathcal{S}_2$ is in fact the *smallest* subspace containing $\mathcal{S}_1 \cup \mathcal{S}_2$. We argue that if $\mathcal{S}_3$ is any subspace containing $\mathcal{S}_1 \cup \mathcal{S}_2$, then it *must* also contain $\mathcal{S}_1 + \mathcal{S}_2$. Toward this end, suppose $\boldsymbol{u} \in \mathcal{S}_1 + \mathcal{S}_2$. Then, $\boldsymbol{u} = \boldsymbol{x} + \boldsymbol{y}$ for some $\boldsymbol{x} \in \mathcal{S}_1$ and $\boldsymbol{y} \in \mathcal{S}_2$. But, $\boldsymbol{x}$ and $\boldsymbol{y}$ both belong to $\mathcal{S}_3$ which, being a subspace itself, must contain $\boldsymbol{x} + \boldsymbol{y}$. Thus, $\boldsymbol{u} \in \mathcal{S}_3$. □

Example 4.5 Suppose that $\mathcal{S}_1$ and $\mathcal{S}_2$ are two subspaces in $\Re^2$ that are defined by two lines through the origin that are not parallel to one another. Then $\mathcal{S}_1 + \mathcal{S}_2 = \Re^2$. This follows from the "parallelogram law" of vectors—see Figure 4.1.

More precisely, we can express lines through the origin in $\Re^2$ as $\mathcal{S}_1 = \{(x_1, x_2) : x_2 = \alpha x_1\}$ and $\mathcal{S}_2 = \{(y_1, y_2) : y_2 = \beta y_1\}$, where α and β are *fixed* real numbers. Therefore,

$$\mathcal{S}_1 + \mathcal{S}_2 = \{(x_1 + y_1, \alpha x_1 + \beta y_1) : x_1, y_1 \in \Re^1\}\,.$$

Since $\mathcal{S}_1 + \mathcal{S}_2$ is obviously a subset of $\Re^2$, we want to find out if every vector in $\Re^2$ lies in $\mathcal{S}_1 + \mathcal{S}_2$. Let $\boldsymbol{b} = (b_1, b_2)$ be any vector in $\Re^2$. To see if $\boldsymbol{b} \in \mathcal{S}_1 + \mathcal{S}_2$, we need to find x_1 and y_1 that will satisfy

$$\begin{bmatrix} 1 & 1 \\ \alpha & \beta \end{bmatrix} \begin{bmatrix} x_1 \\ y_1 \end{bmatrix} = \begin{bmatrix} b_1 \\ b_2 \end{bmatrix} \Longrightarrow \begin{bmatrix} 1 & 1 \\ 0 & \beta - \alpha \end{bmatrix} \begin{bmatrix} x_1 \\ y_1 \end{bmatrix} = \begin{bmatrix} b_1 \\ b_2 - \alpha b_1 \end{bmatrix}.$$

As long as the two lines are not parallel (i.e., $\beta \neq \alpha$), we will have two pivots and the above system will always have a solution. Therefore, every $\boldsymbol{b}$ in the $\Re^2$ plane belongs to $\mathcal{S}_1 + \mathcal{S}_2$ and so $\mathcal{S}_1 + \mathcal{S}_2 = \Re^2$. ■

4.5 Linear combinations and spans

Since subspaces are closed under vector addition and scalar multiplication, vectors of the form $\alpha_1 \boldsymbol{u}_1 + \alpha_2 \boldsymbol{u}_2$ will belong to $\mathcal{S}$ whenever $\boldsymbol{u}_1$ and $\boldsymbol{u}_2$ belong to $\mathcal{S}$ and $\alpha_1, \alpha_2 \in \Re^m$. Matters are no more complicated if we have three vectors. Let $\boldsymbol{u}_1, \boldsymbol{u}_2, \boldsymbol{u}_3 \in \mathcal{S}$ and let α_1, α_2, α_3 be any three real numbers Apply the closure properties to two vectors at a time. First note that the vector $\boldsymbol{x} = \alpha_1 \boldsymbol{u}_1 + \alpha_2 \boldsymbol{u}_2$ is in $\mathcal{S}$ and then note that $\boldsymbol{x} + \alpha_3 \boldsymbol{u}_3 \in \mathcal{S}$ to conclude that the vector $\alpha_1 \boldsymbol{u}_1 + \alpha_2 \boldsymbol{u}_2 + \alpha_3 \boldsymbol{u}_3$ will be in $\mathcal{S}$. This can be generalized to any finite number of vectors. If $\boldsymbol{u}_i$, $i = 1, 2, \ldots, n$ are n vectors residing in a subspace $\mathcal{S}$, then the vectors $\sum_{i=1}^{n} \alpha_i \boldsymbol{u}_i$, where α_i's are real numbers, will also remain in $\mathcal{S}$. This is an important observation and we proceed with a formal study of "linear combinations."

Definition 4.5 *Let $\mathcal{A} = \{\boldsymbol{a}_1, \ldots, \boldsymbol{a}_n\}$ be a set of n vectors with $\boldsymbol{a}_i \in \Re^m, i = 1, \ldots, n$ and $x_1, \ldots, x_n$ are real numbers. The vector*

$$\boldsymbol{b} = \sum_{i=1}^{n} x_i \boldsymbol{a}_i = x_1 \boldsymbol{a}_1 + x_2 \boldsymbol{a}_2 + \cdots + x_n \boldsymbol{a}_n \tag{4.3}$$

is called a **linear combination** *of $\boldsymbol{a}_1, \boldsymbol{a}_2, \ldots, \boldsymbol{a}_n$. The set of all possible linear combinations of the $\boldsymbol{a}_i$'s is called the* **span** *of $\mathcal{A}$ and written as*

$$Sp(\mathcal{A}) = \{\alpha_1 \boldsymbol{a}_1 + \alpha_2 \boldsymbol{a}_2 + \cdots + \alpha_n \boldsymbol{a}_n : \alpha_i \in \Re^1,\ i = 1, 2, \ldots, n\}. \tag{4.4}$$

If $\boldsymbol{b}$ is a linear combination of the vectors in a set $\mathcal{A}$, then we say that $\boldsymbol{b}$ belongs to the span of $\mathcal{A}$ and write $\boldsymbol{b} \in Sp(\mathcal{A})$ in set-theoretic notation.

Lines and planes in $\Re^3$ can be described in terms of spans. For example, if $\boldsymbol{u} \neq \boldsymbol{0}$ is a vector in $\Re^3$, then $Sp(\{\boldsymbol{u}\})$ is the straight line passing through the origin and $\boldsymbol{u}$. In Example 4.3, we saw that if $\boldsymbol{u}$ and $\boldsymbol{v}$ are two nonzero vectors in $\Re^3$ not lying on the same line, then $Sp(\{\boldsymbol{u}, \boldsymbol{v}\})$ describes the plane passing through the origin and the points $\boldsymbol{u}$ and $\boldsymbol{v}$.

The following result shows that the span of a non-empty set of vectors is a subspace.

Theorem 4.4 *Let $\mathcal{A}$ be a non-empty set of vectors in $\Re^m$. The span of $\mathcal{A}$ is a subspace of $\Re^m$.*

Proof. First of all note that if $\mathcal{A} = \{\mathbf{0}\}$. Then, the span of $\mathcal{A}$ is clearly $\{\mathbf{0}\}$ which, from Example 4.1, is a subspace.

Now suppose $\mathcal{A} = \{\boldsymbol{a}_1, \ldots, \boldsymbol{a}_n\}$ be a set of n vectors with $\boldsymbol{a}_i \in \Re^m$, $i = 1, \ldots, n$. Suppose $\boldsymbol{x}$ and $\boldsymbol{y}$ are two vectors in $\Re^m$ that belong to the span of $\mathcal{A}$. Then, both $\boldsymbol{x}$ and $\boldsymbol{y}$ can be expressed as linear combinations of the $\boldsymbol{a}_i$'s. Therefore, we can find real numbers $\alpha_1, \alpha_2, \ldots, \alpha_n$ and $\beta_1, \beta_2, \ldots, \beta_n$ such that $\boldsymbol{x} = \sum_{i=1}^n \alpha_i \boldsymbol{a}_i$ and $\boldsymbol{y} = \sum_{i=1}^n \beta_i \boldsymbol{a}_i$. Let $\boldsymbol{z} = \boldsymbol{x} + \eta \boldsymbol{y}$, where η is any real number. Then,

$$\boldsymbol{z} = \boldsymbol{x} + \eta \boldsymbol{y} = \sum_{i=1}^n \alpha_i \boldsymbol{a}_i + \eta \sum_{i=1}^n \beta_i \boldsymbol{a}_i = \sum_{i=1}^n (\alpha_i + \eta \beta_i) \boldsymbol{a}_i = \sum_{i=1}^n \zeta_i \boldsymbol{a}_i,$$

where $\zeta_i = \alpha_i + \eta \beta_i \in \Re$. This shows that $\boldsymbol{z}$ is also a linear combination of the vectors in $\mathcal{A}$ and, hence, belongs to the $Sp(\mathcal{A})$. Therefore, $Sp(\mathcal{A})$ is a subspace of $\Re^m$. □

Let $\mathcal{S}$ be a subspace of $\Re^m$. If every vector in $\mathcal{S}$ can be expressed as some linear combination of the vectors in $\mathcal{A}$, we say that $\mathcal{A}$ ***spans*** $\mathcal{S}$ or that $\mathcal{A}$ is a ***spanning set*** of $\mathcal{S}$. We write this as $\mathcal{S} = Sp(\mathcal{A})$. In fact, we will later see that all subspaces of $\Re^m$ are can be expressed as $Sp(\mathcal{B})$ for some finite set of vectors $\mathcal{B}$ in $\Re^m$.

If we are given a vector $\boldsymbol{b} \in \Re^m$ and a set of vectors $\mathcal{A} = \{\boldsymbol{a}_1, \ldots, \boldsymbol{a}_n\}$ with $\boldsymbol{a}_i \in \Re^m, i = 1, \ldots, n$, how do we verify if $\boldsymbol{b}$ belongs to $Sp(\mathcal{A})$? Such questions are most conveniently answered by expressing linear combinations of vectors as a linear system. If we form an $m \times n$ matrix $\boldsymbol{A} = [\boldsymbol{a}_1 : \boldsymbol{a}_2 : \ldots : \boldsymbol{a}_n]$ by placing $\boldsymbol{a}_i$ as the i-th column and let $\boldsymbol{x} = (x_1, x_2, \ldots, x_n)'$ denote an $n \times 1$ column vector, then the linear combination in (4.3) can be expressed as the vector $\boldsymbol{b} = \boldsymbol{A}\boldsymbol{x}$. Therefore, $\boldsymbol{b} \in Sp(\mathcal{A})$ if and only there exists an $n \times 1$ column vector $\boldsymbol{x}$ such that $\boldsymbol{A}\boldsymbol{x} = \boldsymbol{b}$. In other words, $\boldsymbol{b} \in Sp(\mathcal{A})$ if and only if the linear system $\boldsymbol{A}\boldsymbol{x} = \boldsymbol{b}$ is consistent. The following theorem uses the same principle to characterize when $\mathcal{A}$ will span a subspace of $\Re^m$ in terms of a system of linear equations.

Theorem 4.5 *Let $\mathcal{A} = \{\boldsymbol{a}_1, \ldots, \boldsymbol{a}_n\}$ be a set of vectors from some subspace $\mathcal{S} \subseteq \Re^m$. Let $\boldsymbol{A} = [\boldsymbol{a}_1, \boldsymbol{a}_2, \ldots, \boldsymbol{a}_n]$ be the $m \times n$ matrix formed by placing $\boldsymbol{a}_i$ as the i-th column. Then $\mathcal{A}$ spans $\mathcal{S}$ if and only if $\boldsymbol{A}\boldsymbol{x} = \boldsymbol{b}$ is a consistent system for every $\boldsymbol{b} \in \mathcal{S}$.*

Proof. From Definition 4.5, $\mathcal{A}$ will span $\mathcal{S}$ if and only if every $\boldsymbol{b}$ in $\mathcal{S}$ can be expressed as

$$\boldsymbol{b} = x_1 \boldsymbol{a}_1 + x_2 \boldsymbol{a}_2 + \cdots + x_n \boldsymbol{a}_n = [\boldsymbol{a}_1, \boldsymbol{a}_2, \ldots, \boldsymbol{a}_n] \begin{bmatrix} x_1 \\ x_2 \\ \vdots \\ x_n \end{bmatrix} = \boldsymbol{A}\boldsymbol{x}\,.$$

This simple observation proves the lemma. □

Here is an example.

Example 4.6 Let $\mathcal{A}\{(4,8),\ (3,6)\}$ be a set of two vectors in $\Re^2$. To see if $\mathcal{A}$ spans $\Re^2$ we need to verify whether the linear system

$$\begin{bmatrix} 4 & 3 \\ 8 & 6 \end{bmatrix} \begin{bmatrix} x_1 \\ x_2 \end{bmatrix} = \boldsymbol{b}$$

is consistent for *every* $\boldsymbol{b}$ in $\Re^2$. It is easy to see that the system is not consistent for $\boldsymbol{b} = (11, 20)'$. Therefore $\mathcal{A}$ does not span $\Re^2$.

On the other hand, the system is consistent whenever $\boldsymbol{b}$ is a vector of the form $\boldsymbol{b} = (\alpha, 2\alpha)'$. This shows that $\mathcal{A}$ spans the subspace $\mathcal{S} = \{(\alpha, 2\alpha) : \alpha \in \Re^1\}$, i.e., the subspace comprising of straight lines passing through the origin and having slope 2. ■

When a vector $\boldsymbol{b}$ belongs to the span of a set of vectors $\mathcal{S}$, we can write down $\boldsymbol{b}$ as a linear combination of the vectors in $\mathcal{S}$. Put another way, there exists linear dependencies in the set $\mathcal{S} \cup \{\boldsymbol{b}\}$. In fact, for any given set of vectors, say $\mathcal{A}$, linear dependencies among its members exist when it is possible to express one vector as a linear combination of the others or, put another way, when one vector belongs to the span of the remaining vectors. For example, consider the set of vectors

$$\mathcal{A} = \left\{ \boldsymbol{a}_1 = \begin{bmatrix} 1 \\ 2 \\ 3 \end{bmatrix}, \boldsymbol{a}_2 = \begin{bmatrix} 4 \\ 5 \\ 6 \end{bmatrix}, \boldsymbol{a}_3 = \begin{bmatrix} 14 \\ 19 \\ 24 \end{bmatrix} \right\}.$$

Then, it is easy to verify that $\boldsymbol{a}_3 = 2\boldsymbol{a}_1 + 3\boldsymbol{a}_3$. This can always be rewritten as a homogeneous system

$$2\boldsymbol{a}_1 + 3\boldsymbol{a}_3 - \boldsymbol{a}_3 = \boldsymbol{0}. \tag{4.5}$$

Linear dependence among the vectors in $\mathcal{A}$ implies that the homogeneous system

$$\boldsymbol{0} = x_1\boldsymbol{a}_1 + x_2\boldsymbol{a}_2 + x_3\boldsymbol{a}_3 = [\boldsymbol{a}_1, \boldsymbol{a}_2, \boldsymbol{a}_3] \begin{bmatrix} x_1 \\ x_2 \\ x_3 \end{bmatrix} = \boldsymbol{A}\boldsymbol{x}$$

will have some solution besides the trivial solution. Equation (4.5) confirms that $\boldsymbol{x} = (2, 3, -1)'$ is a non-trivial solution.

On the other hand, there are no dependency relationships in the set

$$\mathcal{E} = \left\{ \boldsymbol{e}_1 = \begin{bmatrix} 1 \\ 0 \\ 0 \end{bmatrix}, \boldsymbol{e}_2 = \begin{bmatrix} 0 \\ 1 \\ 0 \end{bmatrix}, \boldsymbol{e}_3 = \begin{bmatrix} 0 \\ 0 \\ 1 \end{bmatrix} \right\}$$

because no vector can be expressed as a combination of the others. Put another way, the *only* solution for x_1, x_2, and x_3 in the homogeneous equation

$$\boldsymbol{0} = x_1\boldsymbol{e}_1 + x_2\boldsymbol{e}_2 + x_3\boldsymbol{e}_3 = [\boldsymbol{e}_1, \boldsymbol{e}_2, \boldsymbol{e}_3] \begin{bmatrix} x_1 \\ x_2 \\ x_3 \end{bmatrix} = \boldsymbol{I}_3\boldsymbol{x} \tag{4.6}$$

is the trivial solution $x_1 = x_2 = x_3 = 0$.

4.6 Four fundamental subspaces

Central to the study of linear algebra are four fundamental subspaces associated with a matrix. These subspaces are intricately linked to the rows and columns of a matrix as well as the solution set of the homogeneous linear system associated with that matrix. These four subspaces will keep surfacing as we explore properties of matrices. Later on, after we study the concept of orthogonality, we will see how these fundamental subspaces lead to what is known as the fundamental theorem of linear algebra.

The first of the four fundamental subspaces is the *column space* of a matrix.

Definition 4.6 *Let $\boldsymbol{A}$ be an $m \times n$ matrix. The set*

$$\mathcal{C}(\boldsymbol{A}) = \{\boldsymbol{y} \in \Re^m : \boldsymbol{y} = \boldsymbol{A}\boldsymbol{x} \text{ for some } \boldsymbol{x} \in \Re^n\}$$

is called the **column space** *of $\boldsymbol{A}$. Any member of the set $\mathcal{C}(\boldsymbol{A})$ is an $m \times 1$ vector, so $\mathcal{C}(\boldsymbol{A})$ is a subset of $\Re^m$.*

The column space of a matrix is precisely the set obtained by taking all possible linear combinations of its columns. Put another way, $\mathcal{C}(\boldsymbol{A})$ is the span of the column vectors of $\boldsymbol{A}$. If $\boldsymbol{x} \in \mathcal{C}(\boldsymbol{A})$, then there must exist a vector $\boldsymbol{v} \in \Re^n$ such that $\boldsymbol{x} = \boldsymbol{A}\boldsymbol{v}$. Therefore, we can say that $\boldsymbol{x} \in \mathcal{C}(\boldsymbol{A})$ *if and only if* $\boldsymbol{x} = \boldsymbol{A}\boldsymbol{v}$ *for some* $\boldsymbol{v}$. In particular, every column vector of $\boldsymbol{A}$ is a member of the column space because the j-th column vector is given by $\boldsymbol{a}_{*j} = \boldsymbol{A}\boldsymbol{e}_{*j} \in \mathcal{C}(\boldsymbol{A})$, where $\boldsymbol{e}_{*j}$ is the j-th euclidean canonical vector. Also, $\boldsymbol{0}_{m\times 1} = \boldsymbol{A}\boldsymbol{0}_{n\times 1}$, so $\boldsymbol{0}_{m\times 1} \in \mathcal{C}(\boldsymbol{A})$.

Theorem 4.4 immediately tells us that $\mathcal{C}(\boldsymbol{A})$ is in fact a subspace of $\Re^m$, but here is a direct proof.

Lemma 4.1 *Let $\boldsymbol{A}$ be an $m \times n$ matrix in $\Re^{m\times n}$. Then $\mathcal{C}(\boldsymbol{A})$ is a subspace of $\Re^m$.*

Proof. Clearly $\mathcal{C}(\boldsymbol{A})$ is non-empty (it contains the column vectors of $\boldsymbol{A}$ and also the null vector). Let $\boldsymbol{x}$ and $\boldsymbol{y}$ be two vectors in $\mathcal{C}(\boldsymbol{A})$ and let α be a scalar. There exist vectors $\boldsymbol{u}$ and $\boldsymbol{v}$ in $\Re^n$ such that $\boldsymbol{x} = \boldsymbol{A}\boldsymbol{u}$ and $\boldsymbol{y} = \boldsymbol{A}\boldsymbol{v}$. But this means that

$$\boldsymbol{x} + \alpha\boldsymbol{y} = \boldsymbol{A}\boldsymbol{u} + \alpha\boldsymbol{A}\boldsymbol{v} = \boldsymbol{A}(\boldsymbol{x} + \alpha\boldsymbol{y}) \in \mathcal{C}(\boldsymbol{A}) .$$

From Definition 4.2, it follows that $\mathcal{C}(\boldsymbol{A})$ is a subspace. □

Let $\boldsymbol{C} = \boldsymbol{A}\boldsymbol{B}$. Then, each column vector in $\boldsymbol{C}$ is a linear combination of the column vectors of $\boldsymbol{A}$. It follows from the definition of matrix multiplication that the (i, j)-th element of $\boldsymbol{B}$ is the coefficient of $\boldsymbol{a}_{*i}$ when the j-th column of $\boldsymbol{C}$ is expressed as a linear combination of the n columns of $\boldsymbol{A}$. This means that the column space of $\boldsymbol{A}\boldsymbol{B}$ must be a subspace of the column space of $\boldsymbol{A}$. What is more interesting, though, is the converse: if $\boldsymbol{C}$ and $\boldsymbol{A}$ are two matrices such that the column space of $\boldsymbol{C}$ is a subspace of the column space of $\boldsymbol{A}$, then $\boldsymbol{C} = \boldsymbol{A}\boldsymbol{B}$ for some matrix $\boldsymbol{B}$. We formalize these notions in the following lemma.

Theorem 4.6 *Let A be an $m \times n$ matrix and C an $n \times p$ matrix. Then $\mathcal{C}(C) \subseteq \mathcal{C}(A)$ if and only if $C = AB$ for some $n \times p$ matrix B.*

Proof. We first prove the "if" part. Suppose $C = AB$ for some $n \times p$ matrix. The columns of A and those of $C = AB$ are $m \times 1$ vectors, so $\mathcal{C}(AB)$ and $\mathcal{C}(A)$ are both subspaces of $\Re^m$. Now we can argue as follows:

$$\begin{aligned} x \in \mathcal{C}(AB) &\Rightarrow x = ABv\,, \quad \text{where} \quad v \in \Re^p \\ &\Rightarrow x = Au \in \mathcal{C}(A)\,, \quad \text{where} \quad u = Bv\,. \end{aligned}$$

This proves that any member of $\mathcal{C}(AB)$ is also a member of $\mathcal{C}(A)$, i.e., $\mathcal{C}(AB) \subseteq \mathcal{C}(A)$. This proves the "if" part.

For the "only if" part, suppose that $\mathcal{C}(C) \subset \mathcal{C}(A)$. This means that every member of $\mathcal{C}(C)$ belongs to $\mathcal{C}(A)$. In particular, every column of C must belong to $\mathcal{C}(A)$. This means that the j-th column of C can be written as $c_{*j} = Av_j$, where v_j are vectors in $\Re^n$, for $j = 1, 2, \ldots, p$. If B is the $n \times p$ matrix formed by placing v_j as the j-th column, then we can write

$$\begin{aligned} C &= [c_{*1}, c_{*2}, \ldots, c_{*p}] = [Av_{*1}, Av_{*2}, \ldots, Av_{*p}] \\ &= A[v_{*1}, v_{*2}, \ldots, v_{*p}] = AB\,. \end{aligned}$$

This proves the "only if" part. □

The "if" part of the lemma is often concisely restated as: $\mathcal{C}(AB) \subseteq \mathcal{C}(A)$.

What is the column space of AB, where A is $m \times n$ and B is an $n \times n$ *nonsingular* matrix? We write $A = ABB^{-1}$ and show that $\mathcal{C}(A) = \mathcal{C}(AB)$ as below:

$$\mathcal{C}(A) = \mathcal{C}(ABB^{-1}) \subseteq \mathcal{C}(AB) \subseteq \mathcal{C}(A)\,. \tag{4.7}$$

In other words, we have proved that if B is nonsingular then $\mathcal{C}(A) \subseteq \mathcal{C}(AB)$. Since $\mathcal{C}(AB) \subseteq \mathcal{C}(A)$ is always true, we have $\mathcal{C}(A) = \mathcal{C}(AB)$ when B is nonsingular.

Note that $\mathcal{C}(A')$ is the space spanned by the columns of A'. But the column vectors of A' are precisely the row vectors of A, so $\mathcal{C}(A')$ is the subspace spanned by the row vectors of A. Consequently, $\mathcal{C}(A')$ is also known as the *row space* of A. We denote this by $\mathcal{R}(A)$ and provide a formal definition below.

Definition 4.7 *Let A be an $m \times n$ matrix in $\Re^{m \times n}$. The set*

$$\mathcal{R}(A) = \mathcal{C}(A') = \{y \in \Re^n : y = A'x \ \textit{for some} \ x \in \Re^m\}$$

is called the **row space** *of A. Any member of the set $\mathcal{R}(A)$ is an $n \times 1$ vector, so $\mathcal{R}(A)$ is a subset of $\Re^n$.*

The row space consists of all possible linear combinations of the row vectors of A or, equivalently, the column vectors of A'. All the individual row vectors of A are members of $\mathcal{R}(A)$; so is $\mathbf{0}_{n \times 1}$.

Properties for the row space of a matrix will follow from that of its column space by simply considering the column vectors of its transpose. For example, $\boldsymbol{x} \in \mathcal{R}(\boldsymbol{A})$ if and only if $\boldsymbol{x}' = \boldsymbol{u}'\boldsymbol{A}$ for some vector $\boldsymbol{u}$ because

$$\boldsymbol{x} \in \mathcal{R}(\boldsymbol{A}) \iff \boldsymbol{x} \in \mathcal{C}(\boldsymbol{A}') \iff \boldsymbol{x} = \boldsymbol{A}'\boldsymbol{u} \iff \boldsymbol{x}' = \boldsymbol{u}'\boldsymbol{A} .$$

An example is the following analogue of Theorem 4.6 for row spaces.

Theorem 4.7 *Let $\boldsymbol{A}$ be an $m \times n$ matrix and $\boldsymbol{C}$ a $p \times n$ matrix. Then $\mathcal{R}(\boldsymbol{C}) \subseteq \mathcal{R}(\boldsymbol{A})$ if and only if $\boldsymbol{C} = \boldsymbol{B}\boldsymbol{A}$ for some $p \times m$ matrix $\boldsymbol{B}$.*

Proof. This can be proved using a direct argument, but one can easily invoke Theorem 4.6 as follows:

$$\mathcal{R}(\boldsymbol{C}) \subseteq \mathcal{R}(\boldsymbol{A}) \iff \mathcal{C}(\boldsymbol{C}') \subseteq \mathcal{C}(\boldsymbol{A}') \iff \boldsymbol{C}' = \boldsymbol{A}'\boldsymbol{B}' \iff \boldsymbol{C} = \boldsymbol{B}\boldsymbol{A} ,$$

where $\boldsymbol{B}$ is some $p \times m$ matrix. The first implication ($\Leftrightarrow$) follows from the definition of row spaces and the second implication is a consequence of Theorem 4.6. □

If $\boldsymbol{A}$ is $m \times n$ and $\boldsymbol{B}$ is an $m \times m$ nonsingular matrix then we have $\mathcal{R}(\boldsymbol{B}\boldsymbol{A}) = \mathcal{R}(\boldsymbol{A})$ because

$$\mathcal{R}(\boldsymbol{A}) = \mathcal{C}(\boldsymbol{A}') = \mathcal{C}(\boldsymbol{A}'\boldsymbol{B}') = \mathcal{C}((\boldsymbol{B}\boldsymbol{A})') = \mathcal{R}(\boldsymbol{B}\boldsymbol{A}) , \tag{4.8}$$

where the second equality follows from (4.7) and the fact that $\boldsymbol{B}'$ is nonsingular since $\boldsymbol{B}$ is.

The column space and row space are two of the four fundamental subspaces associated with a matrix. The third fundamental subspace associated with the matrix $\boldsymbol{A}$ is the set of all solutions to the system of equations $\boldsymbol{A}\boldsymbol{x} = \boldsymbol{0}$. This is called the *null space*. We provide a formal definition below.

Definition 4.8 *Let $\boldsymbol{A}$ be an $m \times n$ matrix in $\Re^{m \times n}$. The set*

$$\mathcal{N}(\boldsymbol{A}) = \{\boldsymbol{x} \in \Re^n : \boldsymbol{A}\boldsymbol{x} = \boldsymbol{0}\}$$

is called the **null space** *of $\boldsymbol{A}$. Any member of the set $\mathcal{N}(\boldsymbol{A})$ is an $n \times 1$ vector, so $\mathcal{N}(\boldsymbol{A})$ is a subset of $\Re^n$.*

Saying that $\boldsymbol{x} \in \mathcal{N}(\boldsymbol{A})$ is the same as saying $\boldsymbol{x}$ satisfies $\boldsymbol{A}\boldsymbol{x} = \boldsymbol{0}$. It is easy to prove that the null space is a subspace. We do so in the following lemma.

Lemma 4.2 *Let $\boldsymbol{A}$ be any $m \times n$ matrix. Then $\mathcal{N}(\boldsymbol{A})$ is a subspace of $\Re^n$.*

Proof. First of all note that $\mathcal{N}(\boldsymbol{A})$ is non-empty—clearly $\boldsymbol{0}_{n \times 1} \in \mathcal{N}(\boldsymbol{A})$ (the *trivial* solution to $\boldsymbol{A}\boldsymbol{x} = \boldsymbol{0}$). Suppose $\boldsymbol{x}$ and $\boldsymbol{y}$ vectors in $\mathcal{N}(\boldsymbol{A})$ and let α be any scalar. Then $\boldsymbol{A}\boldsymbol{x} = \boldsymbol{A}\boldsymbol{y} = \boldsymbol{0}$ and hence

$$\boldsymbol{A}(\boldsymbol{x} + \alpha\boldsymbol{y}) = \boldsymbol{A}\boldsymbol{x} + \alpha\boldsymbol{A}\boldsymbol{y} = \boldsymbol{0} \Rightarrow \boldsymbol{x} + \alpha\boldsymbol{y} \in \mathcal{N}(\boldsymbol{A}) .$$

This proves that $\mathcal{N}(\boldsymbol{A})$ is a subspace. □

We have seen that $\mathcal{C}(\boldsymbol{AB})$ is a subspace of $\mathcal{C}(\boldsymbol{A})$. The following result says that $\mathcal{N}(\boldsymbol{B})$ is a subspace of $\mathcal{N}(\boldsymbol{AB})$.

Theorem 4.8 *Let $\boldsymbol{A}$ be an $m \times n$ matrix and $\boldsymbol{B}$ an $n \times p$ matrix. Then $\mathcal{N}(\boldsymbol{B}) \subseteq \mathcal{N}(\boldsymbol{AB})$. In particular, if $m = n$ and $\boldsymbol{A}$ is an $n \times n$ nonsingular matrix, then $\mathcal{N}(\boldsymbol{B}) = \mathcal{N}(\boldsymbol{AB})$.*

Proof. To prove $\mathcal{N}(\boldsymbol{B}) \subseteq \mathcal{N}(\boldsymbol{AB})$, we simply need to show that $\boldsymbol{x} \in \mathcal{N}(\boldsymbol{B}) \Rightarrow \boldsymbol{x} \in \mathcal{N}(\boldsymbol{AB})$. This is immediate:

$$\boldsymbol{x} \in \mathcal{N}(\boldsymbol{B}) \Rightarrow \boldsymbol{Bx} = \boldsymbol{0} \Rightarrow \boldsymbol{ABx} = \boldsymbol{0} \Rightarrow \boldsymbol{x} \in \mathcal{N}(\boldsymbol{AB}) . \tag{4.9}$$

When $\boldsymbol{A}$ is an $n \times n$ nonsingular matrix, $\boldsymbol{A}^{-1}$ exists and we can write

$$\boldsymbol{x} \in \mathcal{N}(\boldsymbol{AB}) \Rightarrow \boldsymbol{ABx} = \boldsymbol{0} \Rightarrow \boldsymbol{A}^{-1}\boldsymbol{ABx} = \boldsymbol{A}^{-1}\boldsymbol{0} \Rightarrow \boldsymbol{Bx} = \boldsymbol{0} \Rightarrow \boldsymbol{x} \in \mathcal{N}(\boldsymbol{B}) .$$

This proves that $\mathcal{N}(\boldsymbol{AB}) \subseteq \mathcal{N}(\boldsymbol{B})$ when $\boldsymbol{A}$ is nonsingular and, together with (4.9), proves $\mathcal{N}(\boldsymbol{B}) = \mathcal{N}(\boldsymbol{AB})$. $\square$

Recall from Section 4.4 that the sum and intersection of subspaces are again subspaces. The following lemma expresses the column space, row space and null space of partitioned matrices in terms of the sum and intersection of subspaces.

Theorem 4.9 *Each of the following statements are true:*

(i) If $\boldsymbol{A}$ and $\boldsymbol{B}$ are two matrices with the same number of rows, then

$$\mathcal{C}\left([\boldsymbol{A} : \boldsymbol{B}]\right) = \mathcal{C}(\boldsymbol{A}) + \mathcal{C}(\boldsymbol{B}) .$$

(ii) If $\boldsymbol{A}$ and $\boldsymbol{C}$ are two matrices with the same number of columns, then

$$\mathcal{R}\left(\begin{bmatrix} \boldsymbol{A} \\ \boldsymbol{C} \end{bmatrix}\right) = \mathcal{R}(\boldsymbol{A}) + \mathcal{R}(\boldsymbol{C}) \quad \textit{and} \quad \mathcal{N}\left(\begin{bmatrix} \boldsymbol{A} \\ \boldsymbol{C} \end{bmatrix}\right) = \mathcal{N}(\boldsymbol{A}) \cap \mathcal{N}(\boldsymbol{C}) .$$

Proof. **Proof of (i):** Let $\boldsymbol{A}$ and $\boldsymbol{B}$ be matrices of order $m \times n_1$ and $m \times n_2$, respectively. To prove part (i), we will prove that $\mathcal{C}\left([\boldsymbol{A} : \boldsymbol{B}]\right) \subseteq \mathcal{C}(\boldsymbol{A}) + \mathcal{C}(\boldsymbol{B})$ and $\mathcal{C}(\boldsymbol{A}) + \mathcal{C}(\boldsymbol{B}) \subseteq \mathcal{C}\left([\boldsymbol{A} : \boldsymbol{B}]\right)$. To prove this, we argue as below:

$$\begin{aligned} \boldsymbol{x} \in \mathcal{C}\left([\boldsymbol{A} : \boldsymbol{B}]\right) &\Longleftrightarrow \boldsymbol{x} = [\boldsymbol{A} : \boldsymbol{B}] \begin{bmatrix} \boldsymbol{u} \\ \boldsymbol{v} \end{bmatrix} = \boldsymbol{Au} + \boldsymbol{Bv} \text{ for some } \boldsymbol{u} \in \Re^{n_1},\ \boldsymbol{v} \in \Re^{n_2} \\ &\Longleftrightarrow \boldsymbol{x} \in \mathcal{C}(\boldsymbol{A}) + \mathcal{C}(\boldsymbol{B}) . \end{aligned}$$

This proves $\mathcal{C}\left([\boldsymbol{A} : \boldsymbol{B}]\right) = \mathcal{C}(\boldsymbol{A}) + \mathcal{C}(\boldsymbol{B})$.

Proof of (ii): The result for row spaces can be derived immediately from part (i) using transposes:

$$\mathcal{R}\left(\begin{bmatrix} \boldsymbol{A} \\ \boldsymbol{C} \end{bmatrix}\right) = \mathcal{C}\left([\boldsymbol{A}' : \boldsymbol{C}']\right) = \mathcal{C}(\boldsymbol{A}') + \mathcal{C}(\boldsymbol{C}') = \mathcal{R}(\boldsymbol{A}) + \mathcal{R}(\boldsymbol{C}) .$$

The result for the null spaces is proved easily by noting that

$$\boldsymbol{x} \in \mathcal{N}\left(\begin{bmatrix} \boldsymbol{A} \\ \boldsymbol{C} \end{bmatrix}\right) \Longleftrightarrow \begin{bmatrix} \boldsymbol{A} \\ \boldsymbol{C} \end{bmatrix} \boldsymbol{x} = \boldsymbol{0} \Longleftrightarrow \boldsymbol{A}\boldsymbol{x} = \boldsymbol{0} \text{ and } \boldsymbol{C}\boldsymbol{x} = \boldsymbol{0}$$
$$\Longleftrightarrow \boldsymbol{x} \in \mathcal{N}(\boldsymbol{A}) \cap \mathcal{N}(\boldsymbol{C}) .$$

This concludes the proof. □

If $\boldsymbol{A}$ is an $m \times n$ matrix and $\boldsymbol{B}$ is a $p \times n$ matrix, $\mathcal{N}(\boldsymbol{A})$ and $\mathcal{N}(\boldsymbol{B})$ are both subspaces of $\Re^n$. So, it is meaningful to discuss when the null space of $\boldsymbol{A}$ is contained in the null space of $\boldsymbol{B}$, or when they are equal. The following lemma tells us that when $\mathcal{N}(\boldsymbol{A})$ is contained in $\mathcal{N}(\boldsymbol{B})$, then every linear relationship that exists among the column vectors of $\boldsymbol{A}$ must also hold for the column vectors of $\boldsymbol{B}$. If the null spaces of $\boldsymbol{A}$ and $\boldsymbol{B}$ are equal, exactly the same linear relationships must exist among their columns.

Theorem 4.10 *Let $\boldsymbol{A}$ be an $m \times n$ matrix and let $\boldsymbol{B}$ be a $p \times n$ matrix and let $\boldsymbol{a}_{*j}$'s and $\boldsymbol{b}_{*j}$'s be the column vectors of $\boldsymbol{A}$ and $\boldsymbol{B}$, respectively.*

(i) Suppose $\mathcal{N}(\boldsymbol{A}) \subseteq \mathcal{N}(\boldsymbol{B})$ and assume that the columns of $\boldsymbol{A}$ satisfy the following:

$$\boldsymbol{a}_{*k} = \alpha_1 \boldsymbol{a}_{*1} + \alpha_2 \boldsymbol{a}_{*2} + \cdots + \alpha_{k-1} \boldsymbol{a}_{*k-1} + \alpha_{k+1} \boldsymbol{a}_{*k+1} + \cdots + \alpha_n \boldsymbol{a}_{*n} .$$

Then the columns of $\boldsymbol{B}$ must satisfy the following linear relationship:

$$\boldsymbol{b}_{*k} = \alpha_1 \boldsymbol{b}_{*1} + \alpha_2 \boldsymbol{b}_{*2} + \cdots + \alpha_{k-1} \boldsymbol{b}_{*k-1} + \alpha_{k+1} \boldsymbol{b}_{*k+1} + \cdots + \alpha_n \boldsymbol{b}_{*n} .$$

(ii) Suppose that $\mathcal{N}(\boldsymbol{A}) = \mathcal{N}(\boldsymbol{B})$. Then the columns of $\boldsymbol{A}$ satisfy

$$\boldsymbol{a}_{*k} = \alpha_1 \boldsymbol{a}_{*1} + \alpha_2 \boldsymbol{a}_{*2} + \cdots + \alpha_{k-1} \boldsymbol{a}_{*k-1} + \alpha_{k+1} \boldsymbol{a}_{*k+1} + \cdots + \alpha_n \boldsymbol{a}_{*n} ,$$

if and only if *the columns of $\boldsymbol{B}$ satisfy*

$$\boldsymbol{b}_{*k} = \alpha_1 \boldsymbol{b}_{*1} + \alpha_2 \boldsymbol{b}_{*2} + \cdots + \alpha_{k-1} \boldsymbol{b}_{*k-1} + \alpha_{k+1} \boldsymbol{b}_{*k+1} + \cdots + \alpha_n \boldsymbol{b}_{*n} .$$

Proof. **Proof of (i)**: Note that $\mathcal{N}(\boldsymbol{A})$ and $\mathcal{N}(\boldsymbol{B})$ are both subspaces in $\Re^n$. Since $\mathcal{N}(\boldsymbol{A}) \subseteq \mathcal{N}(\boldsymbol{B})$, every $n \times 1$ vector that belongs to $\mathcal{N}(\boldsymbol{A})$ is also a member of $\mathcal{N}(\boldsymbol{B})$. In oher words, $\boldsymbol{B}\boldsymbol{x} = \boldsymbol{0}$ whenever $\boldsymbol{A}\boldsymbol{x} = \boldsymbol{0}$. Suppose that

$$\boldsymbol{a}_{*k} = \alpha_1 \boldsymbol{a}_{*1} + \alpha_2 \boldsymbol{a}_{*2} + \cdots + \alpha_{k-1} \boldsymbol{a}_{k-1} + \alpha_{k+1} \boldsymbol{a}_{k+1} + \cdots + \alpha_n \boldsymbol{a}_{*n} .$$

This means that $\boldsymbol{A}\boldsymbol{x} = \boldsymbol{0}$, where $x_i = \alpha_i$ for $i = 1, 2, \ldots, k-1, k+1, \ldots, n$ and $x_k = -1$. This $\boldsymbol{x}$ must also satisfy $\boldsymbol{B}\boldsymbol{x} = \boldsymbol{0}$, which yields

$$\boldsymbol{b}_{*k} = \alpha_1 \boldsymbol{b}_{*1} + \alpha_2 \boldsymbol{b}_{*2} + \cdots + \alpha_{k-1} \boldsymbol{b}_{k-1} + \alpha_{k+1} \boldsymbol{b}_{k+1} + \cdots + \alpha_n \boldsymbol{b}_{*n} .$$

This proves part (i).

Proof of (ii): Since $\mathcal{N}(\boldsymbol{A}) = \mathcal{N}(\boldsymbol{B})$ means that $\mathcal{N}(\boldsymbol{A}) \subseteq \mathcal{N}(\boldsymbol{B})$ *and* $\mathcal{N}(\boldsymbol{B}) \subseteq \mathcal{N}(\boldsymbol{A})$, we have that $\boldsymbol{A}\boldsymbol{x} = \boldsymbol{0}$ *if and only if* $\boldsymbol{B}\boldsymbol{x} = \boldsymbol{0}$. So the result of part (i) can be applied in the other direction as well. This proves part (ii). □

A remarkable result is that the row space and the null space are *virtually disjoint.*

Theorem 4.11 *If $\boldsymbol{A}$ is an $m \times n$ matrix, then $\mathcal{R}(\boldsymbol{A}) \cap \mathcal{N}(\boldsymbol{A}) = \{\boldsymbol{0}\}$.*

Proof. Both $\mathcal{R}(\boldsymbol{A})$ and $\mathcal{N}(\boldsymbol{A})$ are subspaces in $\Re^n$. Therefore, their intersection $\mathcal{R}(\boldsymbol{A}) \cap \mathcal{N}(\boldsymbol{A})$ is a well-defined subspace in $\Re^n$. Clearly $\boldsymbol{0} \in \mathcal{R}(\boldsymbol{A}) \cap \mathcal{N}(\boldsymbol{A})$. If $\boldsymbol{x} \in \mathcal{R}(\boldsymbol{A}) \cap \mathcal{N}(\boldsymbol{A})$, then $\boldsymbol{x} \in \mathcal{N}(\boldsymbol{A})$ and, hence, $\boldsymbol{A}\boldsymbol{x} = \boldsymbol{0}$. Since $\boldsymbol{x} \in \mathcal{R}(\boldsymbol{A}) = \mathcal{C}(\boldsymbol{A}')$, we can write $\boldsymbol{x} = \boldsymbol{A}'\boldsymbol{u}$ for some $\boldsymbol{u} \in \Re^m$. Therefore,

$$\boldsymbol{x}'\boldsymbol{x} = \boldsymbol{x}'\boldsymbol{A}'\boldsymbol{u} = (\boldsymbol{A}\boldsymbol{x})'\boldsymbol{u} = \boldsymbol{0}'\boldsymbol{u} = 0 \Longrightarrow \boldsymbol{x} = \boldsymbol{0} .$$

This proves that $\mathcal{R}(\boldsymbol{A}) \cap \mathcal{N}(\boldsymbol{A}) = \{\boldsymbol{0}\}$. □

Two vectors $\boldsymbol{u}$ and $\boldsymbol{v}$ are said to be ***orthogonal*** or ***perpendicular*** to each other if $\boldsymbol{u}'\boldsymbol{v} = 0$. In analytical geometry of two or three dimensions, this is described by saying their "dot product" is zero. Suppose $\boldsymbol{x} \in \mathcal{N}(\boldsymbol{A})$. Then $\boldsymbol{a}_{i*'}\boldsymbol{x} = 0$ for each $i = 1, 2, \ldots, m$, where $\boldsymbol{a}'_{i*}$ is the i-th row vector of $\boldsymbol{A}$. This means that $\boldsymbol{x}$ is orthogonal to every row vector of $\boldsymbol{A}$, which implies that $\boldsymbol{x}$ is orthogonal to any linear combination of the row vectors of $\boldsymbol{A}$. This shows that any vector in the null space of $\boldsymbol{A}$ is orthogonal to any vector in the row space of $\boldsymbol{A}$.

Given that the second fundamental subspace (the row space) was obtained by considering a transpose (for the column space), you can guess that our fourth fundamental subspace is defined by another transposition! Indeed, our fourth fundamental subspace is $\mathcal{N}(\boldsymbol{A}')$, which is called the *left null space*. We give a formal definition.

Definition 4.9 *Let $\boldsymbol{A}$ be an $m \times n$ matrix in $\Re^{m \times n}$. The set*

$$\mathcal{N}(\boldsymbol{A}') = \{\boldsymbol{x} \in \Re^m : \boldsymbol{A}'\boldsymbol{x} = \boldsymbol{0}\}$$

is called the **left null space** *of $\boldsymbol{A}$. Any member of the set $\mathcal{N}(\boldsymbol{A}')$ is an $m \times 1$ vector, so $\mathcal{N}(\boldsymbol{A}')$ is a subset of $\Re^m$.*

The "left" is used because multiplying $\boldsymbol{A}$ from the "left" with $\boldsymbol{x}'$ also yields the null vector. In other words, the left null space can also be described as $\mathcal{N}(\boldsymbol{A}') = \{\boldsymbol{x} \in \Re^m : \boldsymbol{x}'\boldsymbol{A} = \boldsymbol{0}'\}$. Since null spaces are vector spaces, we conclude that $\mathcal{N}(\boldsymbol{A}')$ is a subspace of $\Re^m$.

Matrices of the form $\boldsymbol{A}'\boldsymbol{A}$ and $\boldsymbol{A}\boldsymbol{A}'$ arise frequently in statistical modeling and have the following important properties with regard to their null spaces.

Theorem 4.12 $\mathcal{N}(\boldsymbol{A}'\boldsymbol{A}) = \mathcal{N}(\boldsymbol{A})$ *and* $\mathcal{N}(\boldsymbol{A}\boldsymbol{A}') = \mathcal{N}(\boldsymbol{A}')$.

Proof. We first prove $\mathcal{N}(\boldsymbol{A}'\boldsymbol{A}) = \mathcal{N}(\boldsymbol{A})$. That $\mathcal{N}(\boldsymbol{A}) \subseteq \mathcal{N}(\boldsymbol{A}'\boldsymbol{A})$ is immediate:

$$\boldsymbol{x} \in \mathcal{N}(\boldsymbol{A}) \Rightarrow \boldsymbol{A}\boldsymbol{x} = \boldsymbol{0} \Rightarrow \boldsymbol{A}'\boldsymbol{A}\boldsymbol{x} = \boldsymbol{0} \Rightarrow \boldsymbol{x} \in \mathcal{N}(\boldsymbol{A}'\boldsymbol{A}) .$$

To prove that $\mathcal{N}(\boldsymbol{A}'\boldsymbol{A}) \subseteq \mathcal{N}(\boldsymbol{A})$, we argue as follows:

$$\begin{aligned} \boldsymbol{x} \in \mathcal{N}(\boldsymbol{A}'\boldsymbol{A}) &\Rightarrow \boldsymbol{A}'\boldsymbol{A}\boldsymbol{x} = \boldsymbol{0} \Rightarrow \boldsymbol{x}'\boldsymbol{A}'\boldsymbol{A}\boldsymbol{x} = 0 \Rightarrow \boldsymbol{u}'\boldsymbol{u} = \boldsymbol{0}\,, \text{ where } \boldsymbol{u} = \boldsymbol{A}\boldsymbol{x}\,, \\ &\Rightarrow \boldsymbol{u} = \boldsymbol{0} \Rightarrow \boldsymbol{A}\boldsymbol{x} = \boldsymbol{0} \Rightarrow \boldsymbol{x} \in \mathcal{N}(\boldsymbol{A})\,, \end{aligned} \tag{4.10}$$

where we have used the fact that $\boldsymbol{u}'\boldsymbol{u} = \|\boldsymbol{u}\|^2 = 0$ if and only if $\boldsymbol{u} = \boldsymbol{0}$ (Lemma 1.1). That $\mathcal{N}(\boldsymbol{A}\boldsymbol{A}') = \mathcal{N}(\boldsymbol{A}')$ follows by simply interchanging the roles of $\boldsymbol{A}$ and $\boldsymbol{A}'$ in the above argument. □

Let us share a few thoughts on identifying spanning sets for the fundamental subspaces. Obviously the columns of $\boldsymbol{A}$ span $\mathcal{C}(\boldsymbol{A})$, while the rows of $\boldsymbol{A}$ span $\mathcal{R}(\boldsymbol{A})$. But these are not necessarily the only spanning sets, and they need not be the spanning sets with the fewest number of vectors. A major aim of linear algebra is to find *minimal* spanning sets for these four fundamental subspaces, i.e., the fewest number of vectors that will still span the subspace.

4.7 Linear independence

Linear dependencies represent redundancies in spanning sets. Consider $Sp(\mathcal{A})$, the span of a finite set of vectors $\mathcal{A}$. If a vector in $\mathcal{A}$ that can be expressed as a linear combination of the other vectors in $\mathcal{A}$ is removed from $\mathcal{A}$, then the span of the resulting set is still $Sp(\mathcal{A})$. We formally state this as the following lemma.

Lemma 4.3 *Let $\mathcal{A} = \{\boldsymbol{a}_1, \ldots, \boldsymbol{a}_n\}$ be a set of vectors from some subspace $\mathcal{S} \subseteq \Re^m$. Suppose we find a vector $\boldsymbol{a}_j$ in $\mathcal{A}$ that can be expressed as a linear combination of the other vectors in $\mathcal{A}$. Then $Sp(\mathcal{A} \setminus \{\boldsymbol{a}_j\}) = Sp(\mathcal{A})$, where $\mathcal{A} \setminus \{\boldsymbol{a}_j\}$ denotes the set $\mathcal{A}$ without the element $\boldsymbol{a}_j$.*

Proof. We will prove that any vector belonging to $Sp(\mathcal{A} \setminus \{\boldsymbol{a}_j\})$ must belong to $Sp(\mathcal{A})$ and vice versa. One direction is obvious: if $\boldsymbol{v} \in Sp(\mathcal{A} \setminus \{\boldsymbol{a}_j\})$, then clearly $\boldsymbol{v} \in Sp(\mathcal{A})$ since $\mathcal{A} \setminus \{\boldsymbol{a}_j\}$ is a subset of $\mathcal{A}$.

Now suppose $\boldsymbol{v} \in Sp(\mathcal{A})$. Then we can find real numbers $x_1, x_2, \ldots, x_n$ such that

$$\boldsymbol{v} = x_1\boldsymbol{a}_1 + x_2\boldsymbol{a}_2 + \cdots + x_j\boldsymbol{a}_j + \cdots + x_n\boldsymbol{a}_n. \tag{4.11}$$

Since $\boldsymbol{a}_j$ is a linear combination of the other vectors in $\mathcal{A}$, we can find real numbers $\alpha_1, \alpha_2, \ldots, \alpha_{j-1}, \alpha_{j+1}, \ldots, \alpha_n$ such that

$$\boldsymbol{a}_j = \alpha_1\boldsymbol{a}_1 + \alpha_2\boldsymbol{a}_2 + \cdots + \alpha_{j-1}\boldsymbol{a}_{j-1} + \alpha_{j+1}\boldsymbol{a}_{j+1} + \alpha_n\boldsymbol{a}_n.$$

Substituting the above expression for $\boldsymbol{a}_j$ in (4.11) we get

$$\begin{aligned}\boldsymbol{v} &= x_1\boldsymbol{a}_1 + x_2\boldsymbol{a}_2 + \cdots \\ &\quad + x_{j-1}\boldsymbol{a}_{j-1} + x_j(\alpha_1\boldsymbol{a}_1 + \alpha_2\boldsymbol{a}_2 + \cdots + \alpha_{j-1}\boldsymbol{a}_{j-1} + \alpha_{j+1}\boldsymbol{a}_{j+1} + \alpha_n\boldsymbol{a}_n) \\ &\quad + x_{j+1}\boldsymbol{a}_{j+1} + \cdots + x_n\boldsymbol{a}_n \\ &= (x_1 + \alpha_1 x_j)\boldsymbol{a}_1 + (x_2 + \alpha_2 x_j)\boldsymbol{a}_2 + \cdots + (x_{j-1} + \alpha_{j-1}x_j)\boldsymbol{a}_{j-1} \\ &\quad + (x_{j+1} + \alpha_{j+1}x_j)\boldsymbol{a}_{j+1} + \cdots + x_n\boldsymbol{a}_n\,,\end{aligned}$$

which shows that $\boldsymbol{v}$ is a linear combination of the vectors in $\mathcal{A} \setminus \{\boldsymbol{a}_j\}$ and, therefore, $\boldsymbol{v} \in Sp(\mathcal{A} \setminus \{\boldsymbol{a}_j\})$. □

The objective is often to arrive at a "minimal" set of vectors, in some sense, that can be used to completely describe subspaces. To achieve this goal in a systematic manner, we will need to formally study the concepts of linear dependence and independence.

Definition 4.10 *Let $\mathcal{A} = \{\boldsymbol{a}_1, \boldsymbol{a}_2, \ldots, \boldsymbol{a}_n\}$ be a finite set of vectors with each $\boldsymbol{a}_i \in \Re^m$. The set $\mathcal{A}$ is said to be* **linearly independent** *if the following condition holds: whenever x_i's are real numbers such that*

$$x_1\boldsymbol{a}_1 + x_2\boldsymbol{a}_2 + \cdots + x_n\boldsymbol{a}_n = \boldsymbol{0},$$

we have $x_1 = x_2 = \cdots = x_n = 0$. On the other hand, whenever there exist real numbers $x_1, x_2, \ldots, x_n$, not all zero, such that $x_1\boldsymbol{a}_1 + x_2\boldsymbol{a}_2 + \cdots + x_n\boldsymbol{a}_n = \boldsymbol{0}$, we say that $\mathcal{A}$ is **linearly dependent**.

It is important to point out that the concepts of linear independence and dependence are defined only for sets of vectors. Individual vectors are neither linearly independent nor dependent. For example, consider the sets

$$\mathcal{A}_1 = \left\{ \begin{bmatrix} 1 \\ 0 \end{bmatrix}, \begin{bmatrix} 0 \\ 1 \end{bmatrix} \right\} \quad \text{and} \quad \mathcal{A}_2 = \left\{ \begin{bmatrix} 1 \\ 0 \end{bmatrix}, \begin{bmatrix} 0 \\ 1 \end{bmatrix}, \begin{bmatrix} 2 \\ 3 \end{bmatrix} \right\}.$$

Here $\mathcal{A}_1$ is clearly a linearly independent set, but $\mathcal{A}_2$ is clearly a linearly dependent set. Therefore, the individual vectors $(1, 0)$ and $(0, 1)$ can simultaneously belong to linearly independent sets as well as linearly dependent sets. Consequently, it only makes sense to talk about a "linearly independent set of vectors" or "linearly dependent set of vectors."

Definition 4.10 immediately connects linear independence with homogeneous linear systems, and hence with the null space of the coefficient matrix. We state this connection in the form of the following lemma.

Theorem 4.13 *Let $\mathcal{A} = \{\boldsymbol{a}_1, \ldots, \boldsymbol{a}_n\}$ be a set of vectors from some subspace $\mathcal{S} \subseteq \Re^m$. Let $\boldsymbol{A} = [\boldsymbol{a}_1, \boldsymbol{a}_2, \ldots, \boldsymbol{a}_n]$ be the $m \times n$ matrix formed by placing $\boldsymbol{a}_i$ as the i-th column. Then $\mathcal{A}$ is linearly independent if and only if $\mathcal{N}(\boldsymbol{A}) = \{\boldsymbol{0}\}$, which, in other words, means that the only solution to the homogeneous system $\boldsymbol{Ax} = \boldsymbol{0}$ is the trivial solution $\boldsymbol{x} = \boldsymbol{0}$.*

Proof. From Definition 4.10, we see that $\mathcal{A}$ is linearly independent if and only if the only solution to

$$\boldsymbol{0} = x_1\boldsymbol{a}_1 + x_2\boldsymbol{a}_2 + \cdots + x_n\boldsymbol{a}_n = [\boldsymbol{a}_1, \boldsymbol{a}_2, \ldots, \boldsymbol{a}_n] \begin{bmatrix} x_1 \\ x_2 \\ \vdots \\ x_n \end{bmatrix} = \boldsymbol{Ax}$$

is the trivial solution $\boldsymbol{x} = \boldsymbol{0}$. □

When $\mathcal{N}(\boldsymbol{A}) = \{\boldsymbol{0}\}$, we say that the columns of $\boldsymbol{A}$ are linearly independent or that $\boldsymbol{A}$ has linearly independent columns. When we say this we actually mean that the columns of $\boldsymbol{A}$ form a linearly independent set. It is clear that $\mathcal{N}(\boldsymbol{A}') = \{\boldsymbol{0}\}$ if and only if the rows of $\boldsymbol{A}$ are linearly independent. When $\boldsymbol{A}$ is a square matrix and $\mathcal{N}(\boldsymbol{A}) = \{\boldsymbol{0}\}$ then $\boldsymbol{A}$ is nonsingular.

Below we illustrate how the above lemma can be applied to find a set of linearly independent vectors in practice.

Example 4.7 Consider the set of vectors:

$$\mathcal{A} = \left\{ \boldsymbol{a}_1 = \begin{bmatrix} 2 \\ 4 \\ -6 \\ 4 \end{bmatrix}, \boldsymbol{a}_2 = \begin{bmatrix} 3 \\ 7 \\ -10 \\ 6 \end{bmatrix}, \boldsymbol{a}_3 = \begin{bmatrix} 0 \\ 2 \\ 0 \\ 4 \end{bmatrix}, \boldsymbol{a}_4 = \begin{bmatrix} 0 \\ 0 \\ 1 \\ 5 \end{bmatrix} \right\}. \tag{4.12}$$

To verify if $\mathcal{A}$ is a linearly independent set, we form the matrix $\boldsymbol{A}$ by placing $\boldsymbol{a}_i$ as the i-th column for $i = 1, 2, 3, 4$. From Theorem 4.13, this amounts to checking if $\boldsymbol{A}\boldsymbol{x} = \boldsymbol{0}$ implies $\boldsymbol{x} = \boldsymbol{0}$. Thus, we write

$$\boldsymbol{A}\boldsymbol{x} = \begin{bmatrix} 2 & 3 & 0 & 0 \\ 4 & 7 & 2 & 0 \\ -6 & -10 & 0 & 1 \\ 4 & 6 & 4 & 5 \end{bmatrix} \begin{bmatrix} x_1 \\ x_2 \\ x_3 \\ x_4 \end{bmatrix} = \begin{bmatrix} 0 \\ 0 \\ 0 \\ 0 \end{bmatrix}. \tag{4.13}$$

In Example 2.4 we used elementary operations to reduce $\boldsymbol{A}$ to an upper-triangular form. In fact, we have

$$\boldsymbol{E}_{43}(-2)\boldsymbol{E}_{32}(1)\boldsymbol{E}_{41}(-2)\boldsymbol{E}_{31}(3)\boldsymbol{E}_{21}(-2)\boldsymbol{A} = \begin{bmatrix} 2 & 3 & 0 & 0 \\ 0 & 1 & 2 & 0 \\ 0 & 0 & 2 & 1 \\ 0 & 0 & 0 & 3 \end{bmatrix}.$$

There are exactly four pivot variables in $\boldsymbol{A}$, which also has four columns. Therefore, the system $\boldsymbol{A}\boldsymbol{x} = \boldsymbol{0}$ has no free variables and, hence, $\boldsymbol{x} = \boldsymbol{0}$ is the only solution to (4.13). Consequently, the set $\mathcal{A}$ in (4.12) is linearly independent.

As another example, consider the set of vectors

$$\mathcal{A} = \left\{ \begin{bmatrix} 1 \\ 4 \\ 7 \end{bmatrix}, \begin{bmatrix} 2 \\ 5 \\ 8 \end{bmatrix}, \begin{bmatrix} 3 \\ 6 \\ 9 \end{bmatrix} \right\}. \tag{4.14}$$

Now the corresponding homogeneous system $\boldsymbol{A}\boldsymbol{x} = \boldsymbol{0}$ has the coefficient matrix

$$\boldsymbol{A} = \begin{bmatrix} 1 & 2 & 3 \\ 4 & 5 & 6 \\ 7 & 8 & 9 \end{bmatrix}. \tag{4.15}$$

In Example 2.5 we found that any vector of the form $\boldsymbol{x} = (\alpha, -2\alpha, \alpha)'$ is a solution to the homogeneous system $\boldsymbol{A}\boldsymbol{x} = \boldsymbol{0}$. Therefore the set in (4.14) is linearly dependent. ■

The above example looked at sets of m vectors in $\Re^m$—the set in (4.12) had $m = 4$, while the set in (4.14) had $m = 3$. This ensures that the coefficient matrices in (4.13) and (4.15) were square. In such cases, the columns of the coefficient matrix $\boldsymbol{A}$ are linearly independent if and only if $\boldsymbol{A}$ is non-singular (see Lemma 2.4).

Of course the same principles apply to any set of vectors and we may not always end up with square coefficient matrices. For instance, if we were to ascertain whether the set $\{\boldsymbol{a}_1, \boldsymbol{a}_2\}$ is linearly independent, where $\boldsymbol{a}_1$ and $\boldsymbol{a}_2$ are the first two vectors in (4.12), we would simply have formed the homogeneous system:

$$\boldsymbol{A}\boldsymbol{x} = \begin{bmatrix} 2 & 3 \\ 4 & 7 \\ -6 & -10 \\ 4 & 6 \end{bmatrix} \begin{bmatrix} x_1 \\ x_2 \end{bmatrix} = \begin{bmatrix} 0 \\ 0 \\ 0 \\ 0 \end{bmatrix} .$$

Now, using elementary operations we reduce the above system to

$$\boldsymbol{E}_{43}(-2)\boldsymbol{E}_{32}(1)\boldsymbol{E}_{41}(-2)\boldsymbol{E}_{31}(3)\boldsymbol{E}_{21}(-2)\boldsymbol{A} = \begin{bmatrix} 2 & 3 \\ 0 & 1 \\ 0 & 0 \\ 0 & 0 \end{bmatrix} .$$

Again, we have two pivots and there are two variables. Therefore, we have no free variables and indeed $x_1 = x_2 = 0$ is the *only* solution to the above homogeneous system. This shows that $\{\boldsymbol{a}_1, \boldsymbol{a}_2\}$ is linearly independent.

In fact, once we confirmed that the set $\mathcal{A}$ in (4.12) is linearly independent, we could immediately conclude that $\{\boldsymbol{a}_1, \boldsymbol{a}_2\}$ is also linearly independent. This is because any subset of a linearly independent set must be linearly independent. One way to explain this is from Definition 4.10 itself: it is immediate from the definition that any set containing a linearly dependent subset is itself linearly dependent. For those who prefer a more formal explanation, we present the following theorem.

Theorem 4.14 *Let $\mathcal{A} = \{\boldsymbol{a}_1, \ldots, \boldsymbol{a}_n\}$ be a nonempty set of vectors in some subspace $\mathcal{S}$. The following two statements are true:*

(i) If $\mathcal{A}$ contains a linearly dependent set, then $\mathcal{A}$ is itself linearly dependent.

(ii) If $\mathcal{A}$ is a linearly independent set and $\mathcal{A}_1$ is any subset of $\mathcal{A}$, then $\mathcal{A}_1$ is also a linearly independent set.

Proof. For the first statement, suppose that $\mathcal{A}$ contains a linearly dependent set which, for convenience, we can assume to be $\{\boldsymbol{a}_1, \boldsymbol{a}_2, \ldots, \boldsymbol{a}_k\}$ with $k < n$. From Definition 4.10, there must exist scalars $\alpha_1, \alpha_2, \ldots, \alpha_k$, not all of which are zero, such that $\alpha_1\boldsymbol{a}_1 + \alpha_2\boldsymbol{a}_2 + \cdots + \alpha_k\boldsymbol{a}_k = \boldsymbol{0}$. This means that we can write,

$$\alpha_1\boldsymbol{a}_1 + \alpha_2\boldsymbol{a}_2 + \cdots + \alpha_k\boldsymbol{a}_k + 0\boldsymbol{a}_{k+1} + 0\boldsymbol{a}_{k+2} + \cdots + 0\boldsymbol{a}_n = \boldsymbol{0} ,$$

which is a homogeneous linear combination of the vectors in $\mathcal{A}$ where all the scalars are not zero. Therefore, $\mathcal{A}$ is linearly dependent.

The second statement is a direct consequence of the first. Indeed, if $\mathcal{A}_1$ was linearly dependent, then the first statement would have forced $\mathcal{A}$ to be linearly dependent as well. Nevertheless, it is instructive to prove the second statement directly as it illustrates an application of Theorem 4.13. We give this proof below.

Assume that the $\boldsymbol{a}_i$'s are $m \times 1$ vectors and let $k < n$ be the number of vectors in $\mathcal{A}_1$. Let us form the $m \times n$ matrix $\boldsymbol{A} = [\boldsymbol{A}_1 : \boldsymbol{A}_2]$ so that the columns of the $m \times k$ submatrix $\boldsymbol{A}_1$ are the vectors in $\mathcal{A}_1$ and the remaining $n - k$ vectors in $\mathcal{A}$ are placed as the columns of $\boldsymbol{A}_2$. We want to show that $\boldsymbol{A}_1\boldsymbol{x}_1 = \boldsymbol{0} \Rightarrow \boldsymbol{x}_1 = \boldsymbol{0}$.

Since $\mathcal{A}$ is a linearly independent set, Theorem 4.13 tells us that $\boldsymbol{A}\boldsymbol{x} = \boldsymbol{0} \Rightarrow \boldsymbol{x} = \boldsymbol{0}$. We use this fact below to argue

$$\boldsymbol{0} = \boldsymbol{A}_1\boldsymbol{x}_1 = \boldsymbol{A}_1\boldsymbol{x}_1 + \boldsymbol{A}_2\boldsymbol{0} = [\boldsymbol{A}_1 : \boldsymbol{A}_2]\begin{bmatrix}\boldsymbol{x}_1\\ \boldsymbol{0}\end{bmatrix} = \boldsymbol{A}\boldsymbol{x}, \text{ where we write } \boldsymbol{x} = \begin{bmatrix}\boldsymbol{x}_1\\ \boldsymbol{0}\end{bmatrix},$$

$$\Rightarrow \boldsymbol{x} = \boldsymbol{0} \Rightarrow \boldsymbol{x}_1 = \boldsymbol{0}\,.$$

This proves the lemma. □

From Definition 4.10, it is trivially true that $\mathcal{A} = \{\boldsymbol{0}\}$ is a linearly dependent set. A consequence of this observation, in conjunction with Theorem 4.14, is that no linearly independent set can contain the $\boldsymbol{0}$ vector.

The next lemma shows that when we insert columns (rows) to a matrix whose columns (rows) are linearly independent, then the augmented matrix still has linearly independent columns (rows).

Lemma 4.4

(i) If $\boldsymbol{A}_1$ is an $m_1 \times r$ matrix with r linearly independent columns and $\boldsymbol{A}_2$ is any $m_2 \times r$ matrix, then $\boldsymbol{A} = \begin{bmatrix}\boldsymbol{A}_1\\ \boldsymbol{A}_2\end{bmatrix}$ has r linearly independent columns.

(ii) If $\boldsymbol{A}_1$ is an $r \times n_1$ matrix with r linearly independent rows and $\boldsymbol{A}_2$ is any $r \times n_2$ matrix, then, $\boldsymbol{A} = [\boldsymbol{A}_1 : \boldsymbol{A}_2]$ has r linearly independent rows.

Proof. **Proof of (i):** Since $\boldsymbol{A}_1$ has linearly independent columns, $\mathcal{N}(\boldsymbol{A}_1) = \{\boldsymbol{0}\}$. From Theorem 4.9, we have

$$\mathcal{N}(\boldsymbol{A}) = \mathcal{N}\left(\begin{bmatrix}\boldsymbol{A}_1\\ \boldsymbol{A}_2\end{bmatrix}\right) = \mathcal{N}(\boldsymbol{A}_1) \cap \mathcal{N}(\boldsymbol{A}_2) = \{\boldsymbol{0}\} \cap \mathcal{N}(\boldsymbol{A}_2) = \{\boldsymbol{0}\}\,.$$

Therefore, $\mathcal{N}(\boldsymbol{A}) = \{\boldsymbol{0}\}$, which means that the columns of $\boldsymbol{A}$ are linearly independent. This proves part (i).

Proof of (ii): We use transposes and part (i). Since $\boldsymbol{A}_1$ has linearly independent rows, $\boldsymbol{A}_1'$ has linearly independent columns. From part (i) we know that the matrix $\boldsymbol{A}' = \begin{bmatrix}\boldsymbol{A}_1'\\ \boldsymbol{A}_2'\end{bmatrix}$ also has linearly independent columns. But the columns of $\boldsymbol{A}'$ are the rows of $\boldsymbol{A}$. Therefore, $\boldsymbol{A} = [\boldsymbol{A}_1 : \boldsymbol{A}_2]$ has linearly independent rows. □

The above result really says something quite obvious: if you have a system of homogeneous equations and some of these equations yield the trivial solution as the only solution, then the remaining equations will not alter the solution. This does lead to interesting conclusions. For example, if you have a set of linearly independent vectors, say $\{a_1, a_2, \ldots, a_r\}$, then the augmented vectors $\begin{bmatrix} a_1 \\ b_1 \end{bmatrix}, \begin{bmatrix} a_2 \\ b_2 \end{bmatrix}, \ldots, \begin{bmatrix} a_r \\ b_r \end{bmatrix}$ will be linearly independent for any choice of b_i's. Another implication of Lemma 4.4 is that if an $m \times n$ matrix A has a $k \times r$ submatrix B with r (k) linearly independent columns (rows), then A must have r (k) linearly independent columns (rows).

The following lemma states a few other important facts about linear independence.

Theorem 4.15 *Let $\mathcal{A} = \{a_1, \ldots, a_n\}$ be a nonempty set of vectors in some subspace $\mathcal{S}$. The following statements are true:*

(i) If $\mathcal{A}$ is linearly independent and if $v \in \mathcal{S}$, then the extension set $\mathcal{A} \cup \{v\}$ is linearly independent if and only if $v \notin Sp(\mathcal{A})$.

(ii) Assume that $a_1 \neq 0$. The set $\mathcal{A}$ is linearly dependent if and only if a_j belongs to the span of $a_1, \ldots, a_{j-1}$ for some $2 \leq j \leq k$.

(iii) If $\mathcal{S} \subseteq \Re^m$ and if $n > m$, then $\mathcal{A}$ must be linearly dependent.

Proof. **Proof of (i):** First suppose $v \notin Sp(\mathcal{A})$. To prove that $\mathcal{A}$ is linearly independent, consider a homogeneous linear combination

$$x_1 a_1 + x_2 a_2 + \cdots + x_n a_n + x_{n+1} v = 0 \,. \tag{4.16}$$

Note that we must have $x_{n+1} = 0$—otherwise v would be a linear combination of the vectors in $\mathcal{A}$ and we would have $v \in Sp(\mathcal{A})$. With x_{n+1} gone, we have

$$x_1 a_1 + x_2 a_2 + \cdots + x_n a_n = 0 \,,$$

which implies $x_1 = x_2 = \ldots = x_n = 0$ because $\mathcal{A}$ is linearly independent. Therefore, the only solution for x_i's in (4.16) is the trivial solution, and hence $\mathcal{A} \cup \{v\}$ must be linearly independent. This proves the *if* part of (i).

Next suppose $\mathcal{A} \cup \{v\}$ is a linearly independent set. If v belonged to the span of $\mathcal{A}$, then v would be a linear combination of vectors from $\mathcal{A}$ thus forcing $\mathcal{A} \cup \{v\}$ to be a dependent set. Therefore, $\mathcal{A} \cup \{v\}$ being linearly independent would imply that $v \notin Sp(\mathcal{A})$. This proves the *only if* part.

Proof of (ii): The *if* part is immediate—if there exists $a_j \in Sp\{a_1, \ldots, a_{j-1}\}$, then $\{a_1, a_2, \ldots, a_{j-1}, a_j\}$ is linearly dependent (see Part (i)) and hence $\mathcal{A}$ is linearly dependent (ensured by Theorem 4.14).

To prove the *only if* part we argue as follows: if indeed $\mathcal{A}$ is linearly dependent then there exists α_i's, not all zero, such that

$$\alpha_1 a_1 + \alpha_2 a_2 + \cdots + \alpha_n a_n = 0 \,.$$

Let j be the largest suffix such that $\alpha_j \neq 0$. Therefore,

$$\begin{aligned} &\alpha_1 a_1 + \alpha_2 a_2 + \cdots + \alpha_j a_j = 0 \\ &\Rightarrow a_j = \left(-\frac{\alpha_1}{\alpha_j}\right) a_i + \left(-\frac{\alpha_2}{\alpha_j}\right) a_2 + \cdots + \left(-\frac{\alpha_{j-1}}{\alpha_j}\right) a_{j-1} \,, \end{aligned}$$

which, in turn, implies that $\boldsymbol{a}_j \in Sp\{\boldsymbol{a}_1, \ldots, \boldsymbol{a}_{j-1}\}$. This proves the *only if* part.

Proof of (iii): Place the $\boldsymbol{a}_i$'s as columns in the $m \times n$ matrix $\boldsymbol{A}$. This matrix is a "short wide" matrix with more columns than rows. Therefore, by Theorem 2.4, the homogeneous system $\boldsymbol{A}\boldsymbol{x} = \boldsymbol{0}$ has a non-trivial solution and hence the set $\mathcal{A}$ must be linearly dependent. This proves (iii). □

A central part of linear algebra is devoted to investigating the linear relationships that exist among the columns or rows of a matrix. In this regard, we will talk about linearly independent columns (or rows) of a matrix. There can be some confusion here. For example, suppose we say that $\boldsymbol{A}$ has two linearly independent columns. What do we exactly mean? After all, if $\boldsymbol{A}$ has, say five columns, there can be more than one way to select two columns that form linearly independent sets. Also, do we implicitly rule out the possibility that $\boldsymbol{A}$ can have three or more linearly independent columns? If there is a set of three linearly independent columns, then any of its subsets with vectors is also linearly independent. So the statement that $\boldsymbol{A}$ has two linearly independent columns is also true, is it not? To avoid such ambiguities, we should always refer to the *maximum* number of linearly independent rows or the maximum number of linearly independent columns. However, for brevity we adopt the following convention. In other words, when we refer to the "number" of linearly independent columns (rows) of a matrix, we will implicitly mean the *maximum* number of linearly independent columns in the matrix.

When we say that a matrix $\boldsymbol{A}$ *has* r *linearly independent columns (rows), we implicitly mean that the* **maximum** *number of linearly independent column (row) vectors one can find in* $\boldsymbol{A}$ *is* r.

If $\boldsymbol{A}$ has r columns (rows) then there is no ambiguity—the entire set of column (row) vectors is a linearly independent set and of course it has the maximum possible number of linearly independent columns (rows) in $\boldsymbol{A}$. If, however, $\boldsymbol{A}$ has more than r columns (rows), then it is implicit that every set of $r+1$ column (row) vectors in $\boldsymbol{A}$ is linearly dependent and there is at least one set of r column (row) vectors that is linearly independent. There can certainly be more than one selection of r linearly independent columns (rows), but that does not cause any ambiguity as far as the *number* r is concerned.

Quite remarkably, as we will see later, the maximum number of linearly independent rows is the same as the maximum number of linearly independent columns. This number—the maximum number of linearly independent columns or rows—plays a central role in linear algebra and is called the ***rank*** of the matrix.

Recall from Theorem 4.10 that, for two matrices $\boldsymbol{A}$ and $\boldsymbol{B}$ having the same number of columns, if the null space of $\boldsymbol{A}$ is contained in the null space of $\boldsymbol{B}$, then the columns in $\boldsymbol{B}$ must satisfy the same linear relationships that exist among the columns in $\boldsymbol{A}$. In other words, if a set of columns in $\boldsymbol{A}$ is *linearly dependent*, then so must be the set of corresponding columns in $\boldsymbol{B}$. And if a set of columns in $\boldsymbol{B}$ is *linearly independent*, then so must be the corresponding columns in $\boldsymbol{A}$—otherwise, the linear dependence in the columns of $\boldsymbol{A}$ would suggest the same for $\boldsymbol{B}$ as well. This means that if the maximum number of *linearly independent* columns in $\boldsymbol{B}$ is r, then we must be able to find a set of at least r linearly independent columns in $\boldsymbol{A}$. When the null spaces of $\boldsymbol{A}$ and $\boldsymbol{B}$ coincide, the maximum number of linearly independent columns in $\boldsymbol{A}$ and $\boldsymbol{B}$ are the same. The following theorem provides a formal statement.

Theorem 4.16 *Let $\boldsymbol{A}$ and $\boldsymbol{B}$ be matrices with the same number of columns.*

(i) If $\mathcal{N}(\boldsymbol{A}) \subseteq \mathcal{N}(\boldsymbol{B})$, then for any set of linearly independent columns in $\boldsymbol{B}$, the corresponding columns in $\boldsymbol{A}$ are linearly independent. In particular, $\boldsymbol{A}$ has at least as many linearly independent columns as there are in $\boldsymbol{B}$.

(ii) If $\mathcal{N}(\boldsymbol{A}) = \mathcal{N}(\boldsymbol{B})$, then any set of columns of $\boldsymbol{B}$ is linearly independent if and only if the corresponding columns of $\boldsymbol{A}$ are linearly independent. In particular, $\boldsymbol{A}$ and $\boldsymbol{B}$ have the same number of linearly independent columns.

Proof. Let $\boldsymbol{A}$ be $m \times n$ and $\boldsymbol{B}$ be $p \times n$. So $\mathcal{N}(\boldsymbol{A})$ and $\mathcal{N}(\boldsymbol{B})$ are both subspaces in $\Re^n$.

Proof of (i): Suppose the columns $\boldsymbol{b}_{*i_1}, \boldsymbol{b}_{*i_2}, \ldots, \boldsymbol{b}_{*i_r}$ of $\boldsymbol{B}$ are linearly independent. We want to prove that that the corresponding columns of $\boldsymbol{A}$, i.e., $\boldsymbol{a}_{*i_1}, \boldsymbol{a}_{*i_2}, \ldots, \boldsymbol{a}_{*i_r}$, are linearly independent.

Since $\mathcal{N}(\boldsymbol{A}) \subseteq \mathcal{N}(\boldsymbol{B})$, every vector in the null space of $\boldsymbol{A}$ is also a member of the null space of $\boldsymbol{B}$. In other words, $\boldsymbol{A}\boldsymbol{x} = \boldsymbol{0}$ implies that $\boldsymbol{B}\boldsymbol{x} = \boldsymbol{0}$. Therefore,

$$x_1 \boldsymbol{a}_{*1} + x_2 \boldsymbol{a}_{*2} + \cdots + x_n \boldsymbol{a}_{*n} = \boldsymbol{0} \Rightarrow x_1 \boldsymbol{b}_{*1} + x_2 \boldsymbol{b}_{*2} + \cdots + x_n \boldsymbol{b}_{*n} = \boldsymbol{0} \; .$$

Take $\boldsymbol{x}$ to be a vector such that $x_i = 0$ except, possibly, for $x_{i_1}, x_{i_2}, \ldots, x_{i_r}$. Then,

$$\begin{aligned} & x_{i_1} \boldsymbol{a}_{*i_1} + x_{i_2} \boldsymbol{a}_{*i_2} + \cdots + x_{i_r} \boldsymbol{a}_{*i_r} = \boldsymbol{0} \Rightarrow \boldsymbol{A}\boldsymbol{x} = \boldsymbol{0} \Rightarrow \boldsymbol{B}\boldsymbol{x} = \boldsymbol{0} \\ & \Rightarrow x_{i_1} \boldsymbol{b}_{*i_1} + x_{i_2} \boldsymbol{b}_{*i_2} + \cdots + x_{i_r} \boldsymbol{b}_{*i_r} = \boldsymbol{0} \Rightarrow x_{i_1} = x_{i_2} = \cdots = x_{i_r} = 0 \; , \end{aligned}$$

which proves that $\boldsymbol{a}_{*i_1}, \boldsymbol{a}_{*i_2}, \ldots, \boldsymbol{a}_{*i_r}$ are linearly independent. This also shows that if $\boldsymbol{B}$ has r linearly independent columns, then $\boldsymbol{A}$ must have at least r linearly independent columns. This proves part (i).

Proof of (ii): When $\mathcal{N}(\boldsymbol{A}) = \mathcal{N}(\boldsymbol{B})$, we have that $\mathcal{N}(\boldsymbol{A}) \subseteq \mathcal{N}(\boldsymbol{B})$ *and* $\mathcal{N}(\boldsymbol{B}) \subseteq \mathcal{N}(\boldsymbol{A})$. Applying the result proved in part (i), we have that $\boldsymbol{b}_{*i_1}, \boldsymbol{b}_{*i_2}, \ldots, \boldsymbol{b}_{*i_r}$ are linearly independent *if and only if* the columns $\boldsymbol{a}_{*i_1}, \boldsymbol{a}_{*i_2}, \ldots, \boldsymbol{a}_{*i_r}$ are linearly independent. Also, now the number of linearly independent columns in $\boldsymbol{B}$ cannot exceed that in $\boldsymbol{A}$ (because $\mathcal{N}(\boldsymbol{A}) \subseteq \mathcal{N}(\boldsymbol{B})$), and vice versa (because $\mathcal{N}(\boldsymbol{B}) \subseteq \mathcal{N}(\boldsymbol{A})$). Hence, they must have the same number of linearly independent columns. □

The next corollary shows that it is easy to identify the linearly independent columns of a matrix from its row echelon form.

Corollary 4.1 *Let $\boldsymbol{A}$ be an $m \times n$ matrix. A subset of columns from a matrix $\boldsymbol{A}$ is linearly independent if and only if the columns in the corresponding positions in a row echelon form of $\boldsymbol{A}$ form a linearly independent set.*

Proof. Let $\boldsymbol{U}$ be an $m \times n$ row echelon matrix obtained by applying elementary row operations on $\boldsymbol{A}$. Therefore, we can write $\boldsymbol{U} = \boldsymbol{G}\boldsymbol{A}$, where $\boldsymbol{G}$ is an $m \times m$ nonsingular matrix obtained as a product of the elementary matrices. Therefore $\boldsymbol{U}\boldsymbol{x} = \boldsymbol{0} \iff \boldsymbol{A}\boldsymbol{x} = \boldsymbol{G}^{-1}\boldsymbol{U}\boldsymbol{x} = \boldsymbol{0}$ and so $\mathcal{N}(\boldsymbol{A}) = \mathcal{N}(\boldsymbol{U})$ (also see Theorem 4.8). Part (ii) of Theorem 4.10 now tells us that any set of columns from $\boldsymbol{A}$ is linearly independent if and only if the columns in the corresponding positions in $\boldsymbol{U}$ are linearly independent. □

This gives us a simple way to find the linearly independent columns of a matrix. First reduce the matrix to a row echelon matrix using elementary row operations. The linearly independent columns in a row echelon matrix are easily seen to be precisely the basic columns (i.e., columns with pivots) (see Example 4.8 below). The columns in the corresponding positions in $\boldsymbol{A}$ are the linearly independent columns in $\boldsymbol{A}$. These columns of $\boldsymbol{A}$, because they correspond to the basic or pivot columns of its row echelon form, are often called the ***basic columns*** of $\boldsymbol{A}$.

Definition 4.11 *The linearly independent columns of a matrix* $\boldsymbol{A}$ *are called its* **basic columns.** *They occupy positions corresponding to the pivot columns in a row (or reduced row) echelon form of* $\boldsymbol{A}$.

The *non-basic* columns refer to the columns of $\boldsymbol{A}$ that are not basic. Clearly every non-basic column is a linear combination of the basic columns of $\boldsymbol{A}$.

Note that the positions of the linearly independent rows of $\boldsymbol{A}$ *do not* correspond to linearly independent rows of $\boldsymbol{U}$. This is because the elementary operations that helped reduce $\boldsymbol{A}$ to a row echelon form may include permutation of the rows, which changes the order of the rows. For instance, consider $\boldsymbol{A} = \begin{bmatrix} 0 & 0 & 0 \\ 1 & 0 & 0 \\ 0 & 0 & 1 \end{bmatrix}$. Clearly the second and third rows of $\boldsymbol{A}$ are linearly independent. But a row echelon form derived from $\boldsymbol{A}$ is $\boldsymbol{U} = \begin{bmatrix} 1 & 0 & 0 \\ 0 & 0 & 1 \\ 0 & 0 & 0 \end{bmatrix}$, which has its first and second rows as linearly independent. This is because the elementary operations yielding $\boldsymbol{U}$ from $\boldsymbol{A}$ include permuting the first and second rows and then the second and third rows. Because of this we do not have a definition for "basic rows" of a matrix.

The following example illustrates that it is fairly easy to find the linearly independent column vectors and row vectors in echelon matrices.

Example 4.8 Linearly independent rows and columns in echelon matrices. From the structure of echelon matrices it is fairly straightforward to show that the nonzero row vectors are linearly independent. Also, the pivot or basic columns are linearly independent. The matter is best illustrated with an example. Consider the matrix

$$U = \begin{bmatrix} 2 & -2 & -5 & -3 & -1 & 2 \\ 0 & 1 & 2 & 5 & 4 & 0 \\ 0 & 0 & 0 & 1 & 1 & 0 \\ 0 & 0 & 0 & 0 & 0 & 0 \end{bmatrix}. \tag{4.17}$$

Consider the nonzero rows of $\boldsymbol{U}$: $\boldsymbol{u}'_{1*}$, $\boldsymbol{u}'_{2*}$ and $\boldsymbol{u}'_{3*}$. We now form a homogeneous linear combination:

$$\begin{aligned} \mathbf{0}' &= x_1\boldsymbol{u}'_{1*} + x_2\boldsymbol{u}'_{2*} + x_3\boldsymbol{u}'_{3*} \\ &= (2x_1, -2x_1 + x_2, -5x_1 + 2x_2, -3x_1 + 5x_2 + x_3, -x_1 + 4x_2 + x_3, 2x_1) \,. \end{aligned}$$

We need to prove that $x_1 = x_2 = x_3 = 0$. Start with the first component. This yields $2x_1 = 0 \Rightarrow x_1 = 0$. With x_1 gone, the second component, $-2x_1 + x_2 = 0$, forces $x_2 = 0$. With $x_1 = x_2 = 0$, the third component $-5x_1 + 2x_2 = 0$ does not tell us anything new ($0 = 0$), but fourth component $-3x_1 + 5x_2 + x_3 = 0$ forces $x_3 = 0$. That is it: we have $x_1 = x_2 = x_3 = 0$. There is no need to examine the fifth and sixth components—they simply tell us $0 = 0$.

If we had a general echelon matrix with r nonzero rows, we would start with a homogeneous linear combination of these rows. We would look at the first pivot position. This would yield $x_1 = 0$. As we examine the subsequent positions, we will find $x_2 = 0$, then $x_3 = 0$, and so on. We would discover that all the x_i's would be zero. Thus, the r nonzero rows of an echelon matrix are linearly independent. In fact, these would be the only linearly independent rows in the echelon matrix as every other row is $\mathbf{0}$ and cannot be a part of any linearly independent set.

Let us now consider the columns with the pivots in (4.17). These occupy the first, second and fourth positions. Consider the homogeneous linear combination

$$\begin{aligned} \mathbf{0} &= x_1\boldsymbol{u}_{*1} + x_2\boldsymbol{u}_{*2} + x_3\boldsymbol{u}_{*4} = [\boldsymbol{u}_{*1} : \boldsymbol{u}_{*2} : \boldsymbol{u}_{*4}] \begin{bmatrix} x_1 \\ x_2 \\ x_3 \end{bmatrix} \\ &= \begin{bmatrix} 2 & -2 & -3 \\ 0 & 1 & 5 \\ 0 & 0 & 1 \\ 0 & 0 & 0 \end{bmatrix} \begin{bmatrix} x_1 \\ x_2 \\ x_3 \end{bmatrix} = \boldsymbol{U}_1\boldsymbol{x} = \begin{bmatrix} 2x_1 - 2x_2 - 3x_3 \\ x_2 + 5x_3 \\ x_3 \\ 0 \end{bmatrix}. \end{aligned}$$

We have a homogeneous system, $\boldsymbol{U}_1\boldsymbol{x} = \mathbf{0}$, where $\boldsymbol{U}_1 = [\boldsymbol{u}_{*1} : \boldsymbol{u}_{*2} : \boldsymbol{u}_{*4}]$ has three columns and three pivots. Therefore, there are no free variables which implies that the trivial solution $x_1 = x_2 = x_3 = 0$ is the only solution.

More explicitly, as for the rows, we can look at each component of $\boldsymbol{U}_1\boldsymbol{x}$, but this time we move backwards, i.e., from the bottom to the top. The fourth component provides no information ($0 = 0$). The third component gives $x_3 = 0$. With x_3 gone, the second component $x_2 + 5x_3 = 0$ forces $x_2 = 0$ and, finally, the first component $2x_1 - 2x_2 - 3x_3 = 0$ tells us $x_1 = 0$. Therefore, $x_1 = x_2 = x_3 = 0$.

Furthermore, the remaining columns in (4.17) are linear combinations of these three pivot columns. To see this, let us examine the consistency of $\boldsymbol{U}_1\boldsymbol{x} = \boldsymbol{b}$, where $\boldsymbol{b}$ is any 4×1 vector. We write

$$\begin{bmatrix} 2 & -2 & -3 \\ 0 & 1 & 5 \\ 0 & 0 & 1 \\ 0 & 0 & 0 \end{bmatrix} \begin{bmatrix} x_1 \\ x_2 \\ x_3 \end{bmatrix} = \begin{bmatrix} 2x_1 - 2x_2 - 3x_3 \\ x_2 + 5x_3 \\ x_3 \\ 0 \end{bmatrix} = \begin{bmatrix} b_1 \\ b_2 \\ b_3 \\ b_4 \end{bmatrix}.$$

Clearly we must have $b_4 = 0$ for the above system to be consistent. Since there are no free variables, we can use back substitution and solve for $\boldsymbol{x}$ in terms of any given b_1, b_2 and b_3. Therefore $\boldsymbol{U}_1\boldsymbol{x} = \boldsymbol{b}$ will be consistent for any $\boldsymbol{b}$ provided that $b_4 = 0$. Since every column in (4.17) has their fourth element as 0, $\boldsymbol{U}_1\boldsymbol{x} = \boldsymbol{a}_{*i}$ is consistent for $i = 1, 2, \ldots, 6$. This shows that any column in (4.17) can be expressed as a linear combination of the columns in $\boldsymbol{U}_1$, i.e., the pivot columns in (4.17).

The above arguments can be applied to general echelon matrices with r pivot columns to show that the r pivot columns are linearly independent and every other column is a linear combination of these r pivot columns. ■

The matrix $\boldsymbol{U}$ in (4.17) is obtained by applying elementary row operations to the matrix $\boldsymbol{A}$ in Example 2.2:

$$\boldsymbol{A} = \begin{bmatrix} 2 & -2 & -5 & -3 & -1 & 2 \\ 2 & -1 & -3 & 2 & 3 & 2 \\ 4 & -1 & -4 & 10 & 11 & 4 \\ 0 & 1 & 2 & 5 & 4 & 0 \end{bmatrix}.$$

From Example 4.8, we see that the first, second and fourth columns in $\boldsymbol{U}$ are linearly independent. By Corollary 4.1, the linearly independent columns of $\boldsymbol{A}$ are the first second and fourth columns of $\boldsymbol{A}$. In other words,

$$\begin{bmatrix} 2 \\ 2 \\ 4 \\ 0 \end{bmatrix}, \begin{bmatrix} -2 \\ -1 \\ -1 \\ 1 \end{bmatrix} \text{ and } \begin{bmatrix} -3 \\ 2 \\ 10 \\ 5 \end{bmatrix}$$

are the linearly independent columns (or basic columns) of $\boldsymbol{A}$.

The above example revealed that for any echelon matrix the number of linearly independent rows equals the number of linearly independent columns. This number is given by the number of pivots r. It is true, although not immediately apparent, that the number of linearly independent rows equals the number of linearly independent columns in *any* matrix $\boldsymbol{A}$. This number is called the ***rank*** of $\boldsymbol{A}$ and lies at the core of the four fundamental subspaces.

We will study rank in greater detail and also see, in later sections, a few elegant proofs of why the number of linearly independent rows and columns are the same. Below, we offer one proof that makes use of Theorem 4.16 and Part (iii) of Theorem 4.15.

Theorem 4.17 Number of linearly independent rows and columns of a matrix. *The number of linearly independent rows equals the number of linearly independent columns in any $m \times n$ matrix $\boldsymbol{A}$.*

Proof. Let $\boldsymbol{A}$ be an $m \times n$ matrix with exactly $r \leq m$ linearly independent rows and with $c \leq n$ linearly independent columns. We will prove that $r = c$.

Rearranging the order of the rows in $\boldsymbol{A}$ will clearly not change the number of linearly independent rows. Permuting the rows of $\boldsymbol{A}$ produces $\boldsymbol{PA}$, where $\boldsymbol{P}$ is a permutation matrix (hence nonsingular). This means that $\mathcal{N}(\boldsymbol{A}) = \mathcal{N}(\boldsymbol{PA})$ (Theorem 4.8) and so the number of linearly independent columns in $\boldsymbol{A}$ are the same as those in $\boldsymbol{PA}$ (part (ii) of Theorem 4.16). In other words, rearranging the rows of a matrix does not alter the number of linearly independent rows or columns. Therefore, without loss of generality, we can assume that the first r rows of the matrix are linearly independent.

We write $\boldsymbol{A} = \begin{bmatrix} \boldsymbol{A}_1 \\ \boldsymbol{A}_2 \end{bmatrix}$, where $\boldsymbol{A}_1$ is an $r \times n$ with r linearly independent rows and $\boldsymbol{A}_2$ is the $(m-r) \times n$ matrix each of whose rows is some linear combination of the rows of $\boldsymbol{A}_1$. This means that $\mathcal{R}(\boldsymbol{A}_2) \subseteq \mathcal{R}(\boldsymbol{A}_1)$ and, by Theorem 4.7, $\boldsymbol{A}_2 = \boldsymbol{B}\boldsymbol{A}_1$ for some $(m-r) \times r$ matrix $\boldsymbol{B}$. Therefore,

$$\boldsymbol{A}\boldsymbol{x} = \boldsymbol{0} \Leftrightarrow \begin{bmatrix} \boldsymbol{A}_1 \\ \boldsymbol{B}\boldsymbol{A}_1 \end{bmatrix} \boldsymbol{x} = \boldsymbol{0} \Leftrightarrow \begin{bmatrix} \boldsymbol{A}_1\boldsymbol{x} \\ \boldsymbol{B}\boldsymbol{A}_1\boldsymbol{x} \end{bmatrix} = \begin{bmatrix} \boldsymbol{0} \\ \boldsymbol{0} \end{bmatrix} \Leftrightarrow \boldsymbol{A}_1\boldsymbol{x} = \boldsymbol{0}\,.$$

In other words, $\mathcal{N}(\boldsymbol{A}) = \mathcal{N}(\boldsymbol{A}_1)$. From Theorems 4.10 and 4.16 we know that $\boldsymbol{A}$ and $\boldsymbol{A}_1$ must have the same number of linearly independent columns, viz. c. But note that each of the columns in $\boldsymbol{A}_1$ are $r \times 1$ vectors. In other words, they are members of $\Re^r$. Part (iii) of Theorem 4.15 tells us that $\boldsymbol{A}_1$ cannot have more than r linearly independent columns. Therefore $c \leq r$.

At this point we have proved that the number of linearly independent columns in *any* matrix cannot exceed the number of linearly independent rows in it. Now comes a neat trick: *Apply this result to the transpose of* $\boldsymbol{A}$. Since the matrix $\boldsymbol{A}'$ has c linearly independent *rows* and r linearly independent *columns*, we obtain $r \leq c$.

Thus, we obtained $c \leq r$ from $\boldsymbol{A}$ and $r \leq c$ from $\boldsymbol{A}'$. This proves $r = c$. □

Another central problem in linear algebra is to find the number of linearly independent solutions of the homogeneous system $\boldsymbol{A}\boldsymbol{x} = \boldsymbol{0}$, where $\boldsymbol{A}$ is an $m \times n$ matrix. In Section 2.5 we studied such systems and found the general solution to be a linear combination of particular solutions, as given in (2.12). These particular solutions can always be chosen to be linearly independent. For example, the solutions $\boldsymbol{h}_1$, $\boldsymbol{h}_2$ and $\boldsymbol{h}_3$ are easily verified to be linearly independent in Example 2.5.

Using the concepts developed up to this point, we now prove that if $\boldsymbol{A}$ has r linearly independent columns (which, by Theorem 4.17, is the number of linearly independent rows), then:

- we can find a set of $n - r$ linearly independent solutions for $\boldsymbol{A}\boldsymbol{x} = \boldsymbol{0}$ and
- every other solution can be expressed as a linear combination of these $n - r$ solutions.

Put another way, we prove that there is a set of $n - r$ linearly independent vectors that span $\mathcal{N}(\boldsymbol{A})$.

Theorem 4.18 Number of linearly independent solutions in a homogeneous linear system. *Let $\boldsymbol{A}$ be an $m \times n$ matrix with r linearly independent columns. There is a linearly independent set of $n-r$ solution vectors for the homogeneous system $\boldsymbol{Ax} = \boldsymbol{0}$ and any other solution vector can be expressed as a linear combination of those $n-r$ linearly independent solutions.*

Proof. Let us assume that the first r columns of $\boldsymbol{A}$ are linearly independent. (One can always rearrange the variables in the linear system so that this is the case, without changing the solution set.) Then we can write $\boldsymbol{A} = [\boldsymbol{A}_1 : \boldsymbol{A}_2]$, where $\boldsymbol{A}_1$ is $m \times r$ with r linearly independent column vectors and $\boldsymbol{A}_2$ is $m \times (n-r)$ each of whose $n-r$ columns are linear combinations of the columns of $\boldsymbol{A}_1$. Thus, $\mathcal{C}(\boldsymbol{A}_2) \subseteq \mathcal{C}(\boldsymbol{A}_1)$ and, by Theorem 4.6, $\boldsymbol{A}_2 = \boldsymbol{A}_1\boldsymbol{B}$ for some $r \times (n-r)$ matrix $\boldsymbol{B}$; thus, $\boldsymbol{A} = [\boldsymbol{A}_1 : \boldsymbol{A}_1\boldsymbol{B}]$.

Let $\boldsymbol{X} = \begin{bmatrix} -\boldsymbol{B} \\ \boldsymbol{I}_{n-r} \end{bmatrix}$. Then $\boldsymbol{X}$ is an $n \times (n-r)$ matrix that satisfies

$$\boldsymbol{AX} = [\boldsymbol{A}_1 : \boldsymbol{A}_1\boldsymbol{B}] \begin{bmatrix} -\boldsymbol{B} \\ \boldsymbol{I}_{n-r} \end{bmatrix} = -\boldsymbol{A}_1\boldsymbol{B} + \boldsymbol{A}_1\boldsymbol{B} = \boldsymbol{O}\,.$$

Therefore, each of the $(n-r)$ columns of $\boldsymbol{X}$ are particular solutions of $\boldsymbol{Ax} = \boldsymbol{0}$. Put another way, $\mathcal{C}(\boldsymbol{X}) \subseteq \mathcal{N}(\boldsymbol{A})$. Furthermore, the $n-r$ columns of $\boldsymbol{X}$ are linearly independent because

$$\boldsymbol{Xu} = \boldsymbol{0} \Rightarrow \begin{bmatrix} -\boldsymbol{B} \\ \boldsymbol{I}_{n-r} \end{bmatrix} \boldsymbol{u} = \boldsymbol{0} \Rightarrow \begin{bmatrix} -\boldsymbol{Bu} \\ \boldsymbol{u} \end{bmatrix} = \begin{bmatrix} \boldsymbol{0} \\ \boldsymbol{0} \end{bmatrix} \Rightarrow \boldsymbol{u} = \boldsymbol{0}\,.$$

Therefore, the column vectors of $\boldsymbol{X}$ constitute $n-r$ linearly independent solutions for $\boldsymbol{Ax} = \boldsymbol{0}$.

We now prove that *any* solution of $\boldsymbol{Ax} = \boldsymbol{0}$ must be a linear combination of the columns of $\boldsymbol{X}$, i.e., they must belong to $\mathcal{C}(\boldsymbol{X})$. For this, let $\boldsymbol{u} = \begin{bmatrix} \boldsymbol{u}_1 \\ \boldsymbol{u}_2 \end{bmatrix}$ be any vector such that $\boldsymbol{Au} = \boldsymbol{0}$. Note that since the columns of $\boldsymbol{A}_1$ are linearly independent, $\boldsymbol{A}_1\boldsymbol{x} = \boldsymbol{0} \Rightarrow \boldsymbol{x} = \boldsymbol{0}$ (Theorem 4.13). Then, we have

$$\begin{aligned} \boldsymbol{Au} = \boldsymbol{0} &\Rightarrow [\boldsymbol{A}_1 : \boldsymbol{A}_1\boldsymbol{B}] \begin{bmatrix} \boldsymbol{u}_1 \\ \boldsymbol{u}_2 \end{bmatrix} = \boldsymbol{0} \Rightarrow \boldsymbol{A}_1(\boldsymbol{u}_1 + \boldsymbol{Bu}_2) = \boldsymbol{0} \\ &\Rightarrow \boldsymbol{u}_1 + \boldsymbol{Bu}_2 = \boldsymbol{0} \Rightarrow \boldsymbol{u}_1 = -\boldsymbol{Bu}_2 \\ &\Rightarrow \boldsymbol{u} = \begin{bmatrix} \boldsymbol{u}_1 \\ \boldsymbol{u}_2 \end{bmatrix} = \begin{bmatrix} -\boldsymbol{B} \\ \boldsymbol{I}_{n-r} \end{bmatrix} \boldsymbol{u}_2 = \boldsymbol{Xu}_2 \Rightarrow \boldsymbol{u} \in \mathcal{C}(\boldsymbol{X})\,. \end{aligned}$$

This proves that any vector $\boldsymbol{u}$ that is a solution of $\boldsymbol{Ax} = \boldsymbol{0}$ must be a linear combination of the $n-r$ special solutions given by the columns of $\boldsymbol{X}$. In other words, we have proved that $\mathcal{N}(\boldsymbol{A}) \subseteq \mathcal{C}(\boldsymbol{X})$. Since we already saw that $\mathcal{C}(\boldsymbol{X}) \subseteq \mathcal{N}(\boldsymbol{A})$, this proves that $\mathcal{C}(\boldsymbol{X}) = \mathcal{N}(\boldsymbol{A})$. □

The column vectors of $\boldsymbol{X}$ in the preceding proof form a linearly independent spanning set for $\mathcal{N}(\boldsymbol{A})$. Such sets play a very special role in describing subspaces because there are no *redundancies* in them. We say that the column vectors of $\boldsymbol{X}$ form a *basis* of $\mathcal{N}(\boldsymbol{A})$. The next section studies these concepts in greater detail.

4.8 Basis and dimension

We have seen that a set of vectors, say $\mathcal{A}$, is a spanning set for a subspace $\mathcal{S}$ if and only if every vector in $\mathcal{S}$ is a linear combination of vectors in $\mathcal{A}$. However, if $\mathcal{A}$ is a linearly dependent set, then it contains redundant vectors. Consider a plane $\mathcal{P}_0$ in $\Re^3$ that passes through the origin. This plane can be spanned by many different set of vectors, but, as we saw in Example 4.3, two vectors that are not collinear (i.e., are linearly independent) form a minimal spanning set for the plane. Spanning sets that do not contain redundant vectors play an important role in linear algebra and motivate the following definition.

Definition 4.12 *A linearly independent subset of $\mathcal{S}$ which spans $\mathcal{S}$ is called a* **basis** *for $\mathcal{S}$.*

While a vector in $\mathcal{S}$ can be expressed as a linear combination of members in a spanning set for $\mathcal{S}$ in more than one way, linearly independent spanning vectors yield a unique representation. We present this as a lemma.

Lemma 4.5 *Let $\mathcal{A} = \{\boldsymbol{a}_1, \boldsymbol{a}_2, \ldots, \boldsymbol{a}_n\}$ be a linearly independent set and suppose $\boldsymbol{b} \in Sp(\mathcal{A})$. Then $\boldsymbol{b}$ is expressed* **uniquely** *as a linear combination of the vectors in $\mathcal{A}$.*

Proof. Since $\boldsymbol{b} \in Sp(\mathcal{A})$, there exists scalars $\{\theta_1, \theta_2, \ldots, \theta_n\}$ such that

$$\boldsymbol{b} = \theta_1\boldsymbol{a}_1 + \theta_2\boldsymbol{a}_2 + \cdots + \theta_n\boldsymbol{a}_n \ .$$

Suppose, we can find another set of scalars $\{\alpha_1, \alpha_2, \ldots, \alpha_n\}$ such that

$$\boldsymbol{b} = \alpha_1\boldsymbol{a}_1 + \alpha_2\boldsymbol{a}_2 + \cdots + \alpha_n\boldsymbol{a}_n \ .$$

Then, since subtracting the above two equations yields

$$\boldsymbol{0} = (\theta_1 - \alpha_1)\boldsymbol{a}_1 + (\theta_2 - \alpha_2)\boldsymbol{a}_2 + \cdots + (\theta_n - \alpha_n)\boldsymbol{a}_n \ ,$$

and the $\boldsymbol{a}_i$'s are linearly independent, we have $\theta_i = \alpha_i$. Therefore, the θ_i's are unique. □

The above lemma implies a formal definition for *coordinates* of a vector in terms of a basis.

Definition 4.13 *Let $\mathcal{A} = \{\boldsymbol{a}_1, \boldsymbol{a}_2, \ldots, \boldsymbol{a}_n\}$ be a basis for the subspace $\mathcal{S}$. The* **coordinates** *of a vector $\boldsymbol{b} \in \mathcal{S}$ are defined to be the unique set of scalars $\{\theta_1, \theta_2, \ldots, \theta_n\}$ such that $\boldsymbol{b} = \sum_{i=1}^{n} \theta_i\boldsymbol{a}_i$.*

The unit vectors $\mathcal{E} = \{\boldsymbol{e}_1, \boldsymbol{e}_2, \ldots, \boldsymbol{e}_m\}$ in $\Re^m$ are a basis for $\Re^m$. This is called the ***standard*** basis for $\Re^m$. When we usually refer to coordinates of a vector without referring to a basis, we refer to the standard basis. Thus, the coordinates of $\boldsymbol{v} = (1, 2, 6)'$ are what they are because of the representation $\boldsymbol{v} = 1\boldsymbol{e}_1 + 2\boldsymbol{e}_2 + 6\boldsymbol{e}_3$.

The set $\mathcal{E}$ provides a concrete example of the fact that a basis exists for $\Re^m$. In fact, a subspace can have several different sets of basis vectors. For example, in $\Re^3$, the sets $\mathcal{B}_1 = \{e_1, e_2, e_3\}$ and $\mathcal{B}_2 = \{(1,2,6)', (2,5,3)', (3,2,1)'\}$ are two different basis. However, a key fact is that these sets *must have the same number of vectors*. We call this the fundamental theorem of vector spaces. To prove this, we first prove an extremely useful result concerning the number of elements in linearly independent sets and spanning sets.

Theorem 4.19 *Let $\mathcal{S}$ be a subspace of $\Re^m$ and $\mathcal{B} = \{b_1, b_2, \ldots, b_l\}$ be a linearly independent subset of $\mathcal{S}$. Let $\mathcal{A} = \{a_1, a_2, \ldots, a_k\}$ be a spanning set of $\mathcal{S}$. Then $\mathcal{A}$ cannot have fewer elements than $\mathcal{B}$. In other words, we must have $k \geq l$.*

Proof. We will give two different proofs of this result. The first proof formulates the problem in terms of homogeneous linear systems and is easy to conceptualize. The second proof does not use linear systems or matrices and may be slightly abstract, but can be useful when studying vector spaces independently of matrices and linear systems. Both proofs use the *principle of contradiction*: we assume that the result is false and arrive at a contradiction.

Proof 1: Suppose, if possible, that $\mathcal{A}$ has fewer elements than $\mathcal{B}$. From $k < l$ we want to arrive at a contradiction. Since $\mathcal{A}$ is a spanning set, so every b_j is a linear combination of the a_i's. In other words, for $j = 1, 2, \ldots, l$ we have

$$b_j = u_{1j}a_1 + u_{2j}a_2 + \cdots + u_{kj}a_k = [a_1 : a_2 : \ldots : a_k] \begin{bmatrix} u_{1j} \\ u_{2j} \\ \vdots \\ u_{kj} \end{bmatrix} = Au_{*j} \,, \tag{4.18}$$

where $A = [a_1 : a_2 : \ldots : a_k]$ is $m \times k$ and $u_{*j} = (u_{1j}, u_{2j}, \ldots, u_{kj})'$ is a $k \times 1$ vector. Writing the vectors in $\mathcal{B}$ as columns in a matrix $B = [b_1 : b_2 : \cdots : b_l]$, we can write the system in (4.18) as

$$\begin{aligned} B = [b_1 : b_2 : \cdots : b_l] &= [Au_{*1} : Au_{*2} : \ldots : Au_{*l}] \\ &= A[u_{*1} : u_{*2} : \ldots : u_{*l}] = AU \,, \end{aligned}$$

where $U = [u_{*1} : u_{*2} : \ldots : u_{*l}]$ is $k \times l$. The key is that U has more columns than rows – it is a short, wide matrix. Theorem 2.4 ensures that there is a nonzero solution, say $x \neq 0$, for $Ux = 0$. But this non-trivial solution will also satisfy $AUx = 0$ and, hence, $Bx = 0$. But this means that the columns in B cannot be linearly independent (see Theorem 4.13). This is the contradiction we wanted. Hence we must have $k \geq l$. □

Proof 2: Let us assume $k < l$. Since $\mathcal{A}$ is a spanning set of $\mathcal{S}$, any vector in $\mathcal{S}$ can be expressed as a linear combination of the a_i's. In particular, b_1 can be expressed as a linear combination of the a_i's, which implies that the set $\{b_1\} \cup \mathcal{A} = \{b_1; a_1, a_2, \ldots, a_k\}$ (prepending b_1 to $\mathcal{A}$) is a linearly dependent set. Part (ii) of Theorem 4.15 ensures that there must be some vector $a_{j_1} \in \mathcal{A}$ that is a linear combination of its preceding vectors. Let us replace this $\boldsymbol{a}_{j_1}$ with $\boldsymbol{b}_2$ and form $\{\boldsymbol{b}_1, \boldsymbol{b}_2\} \cup \mathcal{A} \setminus \{\boldsymbol{a}_{j_1}\} = \{\boldsymbol{b}_2, \boldsymbol{b}_1; \boldsymbol{a}_1, \ldots, \boldsymbol{a}_{j_1-1}, \boldsymbol{a}_{j_1+1}, \ldots, \boldsymbol{a}_k\}$.

Since the $\boldsymbol{a}_{j_1}$ that we removed was a linear combination of $\boldsymbol{b}_1$ and the remaining $\boldsymbol{a}_i$'s, Lemma 4.3 tells us that $Sp(\{\boldsymbol{b}_1\} \cup \mathcal{A} \setminus \{\boldsymbol{a}_{j_1}\}) = Sp(\mathcal{A})$. Also, $\boldsymbol{b}_2 \in Sp(\mathcal{A})$, so it also belongs to $Sp(\{\boldsymbol{b}_1\} \cup \mathcal{A} \setminus \{\boldsymbol{a}_{j_1}\})$. This implies (see Part (i) of Theorem 4.15) that $\{\boldsymbol{b}_1, \boldsymbol{b}_2; \boldsymbol{a}_1, \ldots, \boldsymbol{a}_{j_1-1}, \boldsymbol{a}_{j_1+1}, \ldots, \boldsymbol{a}_k\}$ is a linearly dependent set.

Again, Part (ii) of Theorem 4.15 ensures that we can find some vector in $\{\boldsymbol{b}_1, \boldsymbol{b}_2\} \cup \mathcal{A} \setminus \{\boldsymbol{a}_{j_1}\}$ which will be a linear combination of its preceding vectors and replace it with $\boldsymbol{b}_3$. Furthermore, since the $\boldsymbol{b}_i$'s are independent (remember $\mathcal{B}$ is a basis), the vector to be removed must come from one of the $\boldsymbol{a}_i$'s. Let us denote this vector as $\boldsymbol{a}_{j_2}$. We now arrive at $\{\boldsymbol{b}_1, \boldsymbol{b}_2, \boldsymbol{b}_3\} \cup \mathcal{A} \setminus \{\boldsymbol{a}_{j_1}, \boldsymbol{a}_{j_2}\}$. Again, Lemma 4.3 ensures that $Sp(\{\boldsymbol{b}_1, \boldsymbol{b}_2\} \cup \mathcal{A} \setminus \{\boldsymbol{a}_{j_1}, \boldsymbol{a}_{j_2}\}) = Sp(\{\boldsymbol{b}_1\} \cup \mathcal{A} \setminus \{\boldsymbol{a}_{j_1}\}) = Sp(\mathcal{A})$. Since, $\boldsymbol{b}_3 \in Sp(\mathcal{A})$, the set $\{\boldsymbol{b}_1, \boldsymbol{b}_2, \boldsymbol{b}_3\} \cup \mathcal{A} \setminus \{\boldsymbol{a}_{j_1}, \boldsymbol{a}_{j_2}\}$ is linearly dependent.

We can proceed to find a $\boldsymbol{a}_{j_3}$ to be replaced by a $\boldsymbol{b}_4$. Continuing in this fashion, i.e., replacing some $\boldsymbol{a}_j$ with a $\boldsymbol{b}_i$ at each stage, we will eventually have replaced all the members of $\mathcal{A}$ and arrived at $\{\boldsymbol{b}_1, \ldots, \boldsymbol{b}_k\}$. Furthermore, at each stage of replacement, the resulting set is linearly dependent. But this means that the set $\{\boldsymbol{b}_1, \ldots, \boldsymbol{b}_k\}$ will be linearly dependent, which contradicts the linear independence of $\mathcal{B}$—recall that no subset of a linearly independent set can be linearly dependent. Therefore, our supposition $k < l$ cannot be true and we must have $k \geq l$. This concludes our proof. □

That every basis of a subspace must have the same number of elements is a straightforward corollary of Theorem 4.19, but given its importance we state it as a Theorem.

Theorem 4.20 *Let $\mathcal{S}$ be a subspace of $\Re^m$. Suppose $\mathcal{A} = \{\boldsymbol{a}_1, \boldsymbol{a}_2, \ldots, \boldsymbol{a}_k\}$ and $\mathcal{B} = \{\boldsymbol{b}_1, \boldsymbol{b}_2, \ldots, \boldsymbol{b}_l\}$ are two basis for $\mathcal{S}$. Then $k = l$.*

Proof. Since $\mathcal{A}$ and $\mathcal{B}$ are both linearly independent spanning sets, we can apply Theorem 4.19 twice—once we obtain $k \geq l$ and then we obtain $l \geq k$. Here is how.

First, we use the fact that $\mathcal{B}$ is linearly independent and $\mathcal{A}$ spans $\mathcal{S}$, Theorem 4.19 tells us that $\mathcal{A}$ must have at least as many vectors as $\mathcal{B}$. Therefore $k \geq l$.

Next, we reverse the roles of $\mathcal{A}$ and $\mathcal{B}$. Since $\mathcal{A}$ is linearly independent and $\mathcal{B}$ spans $\mathcal{S}$, Theorem 4.19 tells us that $\mathcal{B}$ must have at least as many vectors as $\mathcal{A}$. Therefore $l \geq k$.

Combining the above two results, we obtain $k = l$. □

Theorem 4.19 also yields the following important characterizations of a basis.

Corollary 4.2 *Let $\mathcal{S}$ be a subspace of $\Re^m$. The following statements about a subset $\mathcal{B}$ of $\mathcal{S}$ are equivalent.*

(i) $\mathcal{B}$ is a basis of $\mathcal{S}$.

(ii) *$\mathcal{B}$ is a* **minimal spanning set** *of $\mathcal{S}$. In other words, no set containing fewer than the number of vectors in $\mathcal{B}$ can span $\mathcal{S}$.*

(iii) *$\mathcal{B}$ is a* **maximal linear independent set** *of $\mathcal{S}$. In other words, every set containing more than the number of vectors in $\mathcal{B}$ must be linearly dependent.*

Proof. We will prove (i) $\Rightarrow$ (ii) $\Rightarrow$ (iii) $\Rightarrow$ (i).

Proof of (i) $\Rightarrow$ (ii): Since $\mathcal{B}$ is a basis, it is a linearly independent subset of $\mathcal{S}$ containing r vectors. By Theorem 4.19, no spanning set can have fewer than r vectors.

Proof of (ii) $\Rightarrow$ (iii): Let $\mathcal{B}$ be a minimal spanning set of $\mathcal{S}$ and suppose $\mathcal{B}$ contains r vectors. We want to argue that $\mathcal{S}$ cannot contain a linearly independent set with more than r vectors.

Suppose, if possible, there exists a linearly independent set with $k > r$ vectors. But Theorem 4.19 tells us that any spanning set must then contain at least as many as k ($> r$) vectors. This contradicts the existence of $\mathcal{B}$ (which is a spanning set of size r).

Proof of (iii) $\Rightarrow$ (i): Assume that $\mathcal{B}$ is a maximal linear independent set of r vectors in $\mathcal{S}$. If possible, suppose $\mathcal{B}$ is not a basis. Consider a basis of $\mathcal{S}$. Since a basis is also a spanning set of $\mathcal{S}$, Theorem 4.19 ensures that it must contain more than r vectors. But a basis is also a linearly independent set. This contradicts the maximal linear independence of $\mathcal{B}$. □

Now that we have proved that the number of vectors in a basis of a subspace is unique, we provide the following important definition.

Definition 4.14 *The* **dimension** *of a subspace $\mathcal{S}$ is defined to be the number of vectors in any basis for $\mathcal{S}$.*

Since the unit vectors $\mathcal{E} = \{e_1, e_2, \ldots, e_m\}$ in $\Re^m$ are a basis for $\Re^m$, the dimension of $\Re^m$ is m. Note that the set $\{\mathbf{0}\}$ is itself a vector space whose dimension is 0.

The concept of dimension plays an important role in understanding finite-dimensional vector spaces. In certain contexts it may be useful to regard dimension as a measure of the "size" of the space. For instance, a plane in $\Re^3$ is "larger" than a line in $\Re^3$, but is "smaller" than the space $\Re^3$ itself. In other contexts, dimension is often regarded as a measure of how much "movement" is possible in that space. This is also known as ***degrees of freedom***. In the trivial space $\{\mathbf{0}\}$, there are no degrees of freedom—we cannot move at all—whereas on a line there is one degree of freedom (length), in a plane there are two degrees of freedom (length and width) and in $\Re^3$ itself there are three degrees of freedom (length, width, and height).

The following theorem is useful in further understanding the roles of spanning sets and linearly independent sets with regard to a basis.

Theorem 4.21 *Let $\mathcal{S}$ be a subspace of $\Re^m$ and let $\dim(\mathcal{S}) = r$. Then the following are true.*

(i) Every spanning set of $\mathcal{S}$ can be reduced to a basis.

(ii) Every linearly independent subset in $\mathcal{S}$ can be extended to a basis of $\mathcal{S}$.

Proof. **Proof of (i)**: Let $\mathcal{B}$ be a set of k vectors spanning $\mathcal{S}$. Theorem 4.19 tells us that $k \geq r$. If $k = r$, then $\mathcal{B}$ is a minimal spanning set and, hence, a basis. There is nothing left to be done.

If $k > r$, then $\mathcal{B}$ is not a minimal spanning set. Therefore, we can remove some vector $\boldsymbol{v} \in \mathcal{B}$ and $\mathcal{B} \setminus \{\boldsymbol{v}\}$ would still span $\mathcal{S}$, but with one fewer vector than $\mathcal{B}$. If $k - 1 = r$, then we have a minimal spanning set and we have found our basis.

If $k - 1 > r$, then $\mathcal{B} \setminus \{\boldsymbol{v}\}$ is not minimal and we can remove one other element to arrive at a spanning set of size $k - 2$. Proceeding in this manner, we finally arrive at our spanning set of r vectors. This is minimal and, hence, our basis. □

Proof of (ii): Let $\mathcal{A}_k = \{\boldsymbol{a}_1, \boldsymbol{a}_2, \ldots, \boldsymbol{a}_k\}$ be a linearly independent set in $\mathcal{S}$. If $k = r$, then $\mathcal{A}_k$ is a maximal linearly independent set and, hence, is already a basis.

If $k < r$, then $\mathcal{A}_k$ cannot span $\mathcal{S}$ and there must exist some vector $\boldsymbol{a}_{k+1}$ in $\mathcal{S}$ such that $\boldsymbol{a}_{k+1} \notin Sp(\mathcal{A}_k)$. Part (i) of Theorem 4.15 tells us that $\mathcal{A}_{k+1} = \mathcal{A}_k \cup \{\boldsymbol{a}_{k+1}\}$ must be linearly independent. If $k + 1 = r$ then $\mathcal{A}_{k+1}$ is our basis.

If $k + 1 < r$, we find some $\boldsymbol{a}_{k+2} \in \mathcal{S}$ such that $\boldsymbol{a}_{k+2} \notin Sp(\mathcal{A}_{k+1})$ and form $\mathcal{A}_{k+2} = \mathcal{A}_{k+1} \cup \{\boldsymbol{a}_{k+2}\}$. Proceeding in this manner generates independent subsets $\mathcal{A}_{k+1} \subset \mathcal{A}_{k+2} \subset \cdots \subset \mathcal{A}_r$, eventually resulting in a maximal independent subset $\mathcal{A}_r$ containing r vectors and is a basis of $\mathcal{S}$. □

The following corollary, which is sometimes useful in matrix algebra, is a simple consequence of the above theorem. It says that any matrix with linearly independent columns (or rows) can be "extended" (or appended) to form a nonsingular matrix.

Corollary 4.3 *The following statements are true.*

(i) Let $\boldsymbol{A}_1$ be an $m \times r$ matrix (with $r < m$) whose r columns are linearly independent. Then, there exists an $m \times (m - r)$ matrix $\boldsymbol{A}_2$ such that the $m \times m$ matrix $\boldsymbol{A} = [\boldsymbol{A}_1 : \boldsymbol{A}_2]$ is nonsingular.

(ii) Let $\boldsymbol{A}_1$ be an $r \times n$ matrix (with $r < n$) whose r rows are linearly independent. Then, there exists an $(n-r)\times n$ matrix $\boldsymbol{A}_2$ such that the $n\times n$ matrix $\boldsymbol{A} = \begin{bmatrix} \boldsymbol{A}_1 \\ \boldsymbol{A}_2 \end{bmatrix}$ is nonsingular.

Proof. **Proof of (i):** The columns of $\boldsymbol{A}_1$ form a linearly independent set of $r < m$ vectors in $\Re^m$. Therefore, by part (ii) of Theorem 4.21, this set can be extended by appending $m - r$ vectors in $\Re^m$ to produce a basis of $\Re^m$. Place these additional $m - r$ vectors as the columns of $\boldsymbol{A}_2$. Now, $\boldsymbol{A} = [\boldsymbol{A}_1 : \boldsymbol{A}_2]$ is an $m \times m$ matrix whose m columns are linearly independent. This means that $\boldsymbol{A}$ is nonsingular (by Theorem 4.13).

Proof of (ii): The rows of $\boldsymbol{A}_1$ form a linearly independent set of $r < n$ vectors in

$\Re^n$. Therefore, by part (ii) of Theorem 4.21, this set can be extended by appending $n-r$ vectors in $\Re^n$ to produce a basis of $\Re^n$. Place these additional $n-r$ vectors as the rows of $\boldsymbol{A}_2$. Now, $\boldsymbol{A} = \begin{bmatrix} \boldsymbol{A}_1 \\ \boldsymbol{A}_2 \end{bmatrix}$ is an $n \times n$ matrix whose n rows are linearly independent. But by Theorem 4.17, this means that $\boldsymbol{A}$ has n linearly independent columns and, hence, $\boldsymbol{A}$ is nonsingular (by Theorem 4.13). □

Theorem 4.21 tell us that it is theoretically possible to find extend a set of linearly independent vectors to a basis, but the proofs given there do not offer much help in how we can actually find the extension vectors. We now discuss a practical strategy and also illustrate with a numerical example.

Let $\{\boldsymbol{a}_1, \boldsymbol{a}_2, \ldots, \boldsymbol{a}_r\}$ be a set of linearly independent vectors in $\Re^n$ such that $r < n$. We want to find extension vectors $\{\boldsymbol{a}_{r+1}, \boldsymbol{a}_{r+2}, \ldots, \boldsymbol{a}_n\}$ such that the set $\{\boldsymbol{a}_1, \boldsymbol{a}_2, \ldots, \boldsymbol{a}_n\}$ forms a basis for $\Re^n$. Let $\{\boldsymbol{e}_1, \boldsymbol{e}_2, \ldots, \boldsymbol{e}_n\}$ be the standard basis consisting of the unit vectors in $\Re^n$ (you can take any basis you like, but this standard basis is always readily available), and place the given $\boldsymbol{a}_i$'s along with the $\boldsymbol{e}_i$'s as columns in an $n \times (r+n)$ matrix $\boldsymbol{A}$:

$$\boldsymbol{A} = [\boldsymbol{a}_1 : \boldsymbol{a}_2 : \ldots : \boldsymbol{a}_r : \boldsymbol{e}_1 : \boldsymbol{e}_2 : \ldots : \boldsymbol{e}_n] \,.$$

Clearly, the columns of $\boldsymbol{A}$ span $\Re^n$, so they must contain n linearly independent column vectors. We want to find a maximal set of linearly independent vectors in $\boldsymbol{A}$—in other words, the basic columns of $\boldsymbol{A}$. Corollary 4.1 tells us that the basic columns of $\boldsymbol{A}$ will occupy the same positions in $\boldsymbol{A}$ as the pivot columns occupy in any row echelon form of $\boldsymbol{A}$. Let $\boldsymbol{U}$ be a row echelon form derived from $\boldsymbol{A}$ using elementary row operations. Since we know that the first r columns of $\boldsymbol{A}$ are linearly independent, Corollary 4.1 insures that the first r columns of $\boldsymbol{U}$ are pivot columns. Since $\boldsymbol{A}$ has n linearly independent columns, $\boldsymbol{U}$ must have n pivots. Let $\boldsymbol{u}_{*j_1}, \boldsymbol{u}_{*j_2}, \ldots, \boldsymbol{u}_{*j_{n-r}}$ be the remaining $n-r$ pivot columns of $\boldsymbol{U}$. This implies that $\boldsymbol{e}_{j_1}, \boldsymbol{e}_{j_2}, \ldots, \boldsymbol{e}_{j_{n-r}}$ are the remaining basic columns, hence the required extension vectors, so that $\{\boldsymbol{a}_1, \boldsymbol{a}_2, \ldots, \boldsymbol{a}_r, \boldsymbol{e}_{j_1}, \boldsymbol{e}_{j_2}, \ldots, \boldsymbol{e}_{j_{n-r}}\}$ is a basis of $\Re^n$. We illustrate this procedure with a numerical example.

Example 4.9 Extension to a basis. Consider the following two vectors in $\Re^4$:

$$\mathcal{A} = \left\{ \boldsymbol{a}_1 = \begin{bmatrix} 2 \\ 4 \\ -6 \\ 4 \end{bmatrix}, \boldsymbol{a}_2 = \begin{bmatrix} 3 \\ 7 \\ 0 \\ 0 \end{bmatrix} \right\} .$$

It is easily verified that $\boldsymbol{a}_1$ and $\boldsymbol{a}_2$ are linearly independent. We want to extend the set $\mathcal{A}$ to a basis of $\Re^4$. Following the discussion preceding this example, we first place $\boldsymbol{a}_1$, $\boldsymbol{a}_2$ as the first two columns and then the four unit vectors in $\Re^4$ to form

$$\boldsymbol{A} = \begin{bmatrix} 2 & 3 & 1 & 0 & 0 & 0 \\ 4 & 7 & 0 & 1 & 0 & 0 \\ -6 & 0 & 0 & 0 & 1 & 0 \\ 4 & 0 & 0 & 0 & 0 & 1 \end{bmatrix} .$$

With $G = E_{43}(2/3)E_{42}(6)E_{32}(-9)E_{21}(-2)E_{31}(3)E_{21}(-2)$ being the product of elementary matrices representing the elementary row operations, we arrive at the following row echelon form:

$$GA = U = \begin{bmatrix} 2 & 3 & 1 & 0 & 0 & 0 \\ 0 & 1 & -2 & 1 & 0 & 0 \\ 0 & 0 & 21 & -9 & 1 & 0 \\ 0 & 0 & 0 & 0 & 2/3 & 1 \end{bmatrix}.$$

We see that the first, second, third and fifth columns of U are the pivot columns. (**Note:** You may employ a different sequence of elementary operations to arrive at a row echelon form different from the one given above, but the positions of the pivot columns will be the same.) Therefore, the first, second, third and fifth columns of A form a basis for $\Re^4$. That is,

$$\mathcal{A}_{\text{ext}} = \left\{ a_1 = \begin{bmatrix} 2 \\ 4 \\ -6 \\ 4 \end{bmatrix}, a_2 = \begin{bmatrix} 3 \\ 7 \\ 0 \\ 0 \end{bmatrix}, a_3 = \begin{bmatrix} 1 \\ 0 \\ 0 \\ 0 \end{bmatrix}, a_4 = \begin{bmatrix} 0 \\ 0 \\ 1 \\ 0 \end{bmatrix} \right\}$$

is the set representing the extension of $\mathcal{A}$ to a basis of $\Re^4$. ■

It is important to distinguish between the dimension of a vector space $\mathcal{V}$ and the number of components contained in the individual vectors from $\mathcal{V}$. For instance, every point on a plane has three components (its coordinates), but the dimension of the plane is 2. In fact, Part (iii) of Theorem 4.15 (also the existence of the standard basis) showed us that that no linearly independent subset in $\Re^m$ can contain more than m vectors and, consequently, $\dim(\mathcal{S}) \leq m$ for every subspace $\mathcal{S} \subseteq \Re^m$. This observation generalizes to produce the following important result.

Theorem 4.22 *Let $\mathcal{S}$ and $\mathcal{V}$ be vector subspaces in $\Re^n$ such that $\mathcal{S} \subseteq \mathcal{V}$. The following statements are true:*

(i) $\dim(\mathcal{S}) \leq \dim(\mathcal{V})$.

(ii) If $\dim(\mathcal{S}) = \dim(\mathcal{V})$ *then we must have* $\mathcal{S} = \mathcal{V}$.

Proof. **Proof of (i)**: Let $\dim(\mathcal{S}) = r$ and $\dim(\mathcal{V}) = m$. We provide two different arguments to prove that $r \leq m$. The first recasts the problem in terms of column spaces and has a more constructive flavor to it, while the second uses the principle of contradiction.

First proof of (i): Let $S = [s_1, s_2, \ldots, s_r]$ be an $n \times r$ matrix whose columns form a basis for $\mathcal{S}$ and $V = [v_1, v_2, \ldots, v_m]$ be an $n \times m$ matrix whose columns form a basis for $\mathcal{V}$. Therefore, $\mathcal{S} = \mathcal{C}(S)$ and $\mathcal{V} = \mathcal{C}(V)$ and so $\mathcal{C}(S) \subset \mathcal{C}(V)$. From Theorem 4.6, we know that $S = VB$ for some $m \times r$ matrix B. Since $s_i = Vb_i$, where b_i is the i-th column of B, the vectors $Vb_1, Vb_2, \ldots, Vb_r$ form a linearly independent set of r vectors in $\mathcal{C}(V)$. Since $m = \dim(\mathcal{C}(V))$ is the size of any maximal linearly independent subset in $\mathcal{C}(V)$, we must have $r \leq m$. □

Second proof of (i): We use the principle of contradiction to prove $r \leq m$. If it were the case that $r > m$, a basis for $\mathcal{S}$ would constitute a linear independent subset of $\mathcal{V}$ containing more than m vectors. But this is impossible because m is the size of a maximal independent subset of $\mathcal{V}$. Therefore, we must have $r \leq m$. □

Proof of (ii): Denote $\dim(\mathcal{S}) = r$ and $\dim(\mathcal{V}) = m$ and assume that $r = m$. Suppose, if possible, that $\mathcal{S} \neq \mathcal{V}$ i.e., $\mathcal{S}$ is a proper subset of $\mathcal{V}$. Then there exists a vector $\boldsymbol{v}$ such that $\boldsymbol{v} \in \mathcal{V}$ but $\boldsymbol{v} \notin \mathcal{S}$. If $\mathcal{B}$ is a basis for $\mathcal{S}$, then $\boldsymbol{v} \notin Sp(\mathcal{B})$, and the extension set $\mathcal{B} \cup \{\boldsymbol{v}\}$ is a linearly independent subset of $\mathcal{V}$ (recall Part (i) of Theorem 4.15). But this implies that $\mathcal{B} \cup \{\boldsymbol{v}\}$ contains $r+1 = m+1$ vectors, which is impossible because $\dim(\mathcal{V}) = m$ is the size of a maximal independent subset of $\mathcal{V}$. Therefore, $\mathcal{S} = \mathcal{V}$. □

4.9 Change of basis and similar matrices

Every vector in $\Re^n$ can be expressed uniquely as a linear combination of the basis vectors. If we change the basis, the coordinates of the vector change. Sometimes it is useful to consider more than one basis for a vector space and see how the coordinates change with respect to a new basis. We study this by considering a new basis formed by the columns of a nonsingular matrix.

Let $\boldsymbol{A}$ be an $n \times n$ matrix and let $\boldsymbol{P}$ be an $n \times n$ nonsingular matrix. This means that the columns of $\boldsymbol{P}$ form a basis for $\Re^n$. If $\boldsymbol{p}_i$ is the i-th column of $\boldsymbol{P}$, $\boldsymbol{A}\boldsymbol{p}_i$ is a vector in $\Re^n$. Therefore, $\boldsymbol{A}\boldsymbol{p}_i$ can be expressed as a linear combination of the columns of $\boldsymbol{P}$ and there exists an $n \times 1$ vector $\boldsymbol{b}_i$ such that $\boldsymbol{A}\boldsymbol{p}_i = \boldsymbol{P}\boldsymbol{b}_i$. The elements are the coefficients of $\boldsymbol{A}\boldsymbol{p}_i$ with respect to the basis $\{\boldsymbol{p}_1, \boldsymbol{p}_2, \ldots, \boldsymbol{p}_n\}$. Therefore,

$$\boldsymbol{A}\boldsymbol{P} = \boldsymbol{A}\left[\boldsymbol{p}_1 : \boldsymbol{p}_2 : \ldots : \boldsymbol{p}_n\right] = \boldsymbol{P}\left[\boldsymbol{b}_1 : \boldsymbol{b}_2 : \ldots : \boldsymbol{b}_n\right] = \boldsymbol{P}\boldsymbol{B} ,$$

which implies that $\boldsymbol{P}^{-1}\boldsymbol{A}\boldsymbol{P} = \boldsymbol{B}$. This motivates the following definition.

Definition 4.15 Similar matrices. *Two $n \times n$ matrices $\boldsymbol{A}$ and $\boldsymbol{B}$ are said to be* **similar** *if there exists an $n \times n$ nonsingular matrix $\boldsymbol{P}$ such that $\boldsymbol{P}^{-1}\boldsymbol{A}\boldsymbol{P} = \boldsymbol{B}$. Transformations of the form $f(\boldsymbol{A}) = \boldsymbol{P}^{-1}\boldsymbol{A}\boldsymbol{P}$ are often called* **similarity transformations**.

Similar matrices arise naturally when expressing linear transformations with respect to a different basis. If $\boldsymbol{A}$ is used to transform a vector $\boldsymbol{x}$ in $\Re^n$, then the image $\boldsymbol{A}\boldsymbol{x}$ is a vector in $\Re^n$. If $\boldsymbol{P}$ is a nonsingular matrix, then its columns constitute a basis for $\Re^n$. We can express the "input" $\boldsymbol{x}$ and the "output" $\boldsymbol{A}\boldsymbol{x}$ in terms of the columns of $\boldsymbol{P}$. This means that there exists a vector $\boldsymbol{u} \in \Re^n$ such that $\boldsymbol{x} = \boldsymbol{P}\boldsymbol{u}$. And there is a vector $\boldsymbol{v} \in \Re^n$ such that $\boldsymbol{A}\boldsymbol{x} = \boldsymbol{P}\boldsymbol{v}$. In other words, the coordinates of $\boldsymbol{x}$ and $\boldsymbol{A}\boldsymbol{x}$ in terms of the columns of $\boldsymbol{P}$ are $\boldsymbol{u}$ and $\boldsymbol{v}$, respectively. Therefore,

$$\boldsymbol{v} = \boldsymbol{P}^{-1}\boldsymbol{A}\boldsymbol{x} = \boldsymbol{P}^{-1}\boldsymbol{A}\boldsymbol{P}(\boldsymbol{P}^{-1}\boldsymbol{x}) = \boldsymbol{B}\boldsymbol{u} , \quad \text{where } \boldsymbol{B} = \boldsymbol{P}^{-1}\boldsymbol{A}\boldsymbol{P} .$$

Therefore, $\boldsymbol{B}$ is the matrix that maps $\boldsymbol{u}$ to $\boldsymbol{v}$. This makes it clear that $\boldsymbol{B}$ and $\boldsymbol{A}$ are indeed "similar." Consider a linear map $f(\boldsymbol{x}) = \boldsymbol{A}\boldsymbol{x}$ with respect to the standard Euclidean basis. Express the input and output vectors in terms of the columns of a nonsingular matrix $\boldsymbol{P}$. Then, the corresponding transformation matrix changes from $\boldsymbol{A}$ to $\boldsymbol{B}$. The matrix $\boldsymbol{P}$ is sometimes referred to as the change of basis matrix.

Changing the basis is useful because it can yield simpler matrix representations for transformations. In fact, one of the fundamental goals of linear algebra is to find $\boldsymbol{P}$ such that $\boldsymbol{B} = \boldsymbol{P}^{-1}\boldsymbol{A}\boldsymbol{P}$ is really simple (e.g., a diagonal matrix). We will explore this problem in later chapters. For now, consider the following example of why simpler matrices may result for certain bases.

Let $\boldsymbol{A}$ and $\boldsymbol{P}$ be as above. Partition $\boldsymbol{P} = [\boldsymbol{P}_1 : \boldsymbol{P}_2]$, where $\boldsymbol{P}_1$ be an $n \times r$ matrix $(r < n)$ whose columns form a basis for a r-dimensional subspace $\mathcal{S}$ of $\Re^n$, and $\boldsymbol{P}_2$ is an $n \times (n - r)$ matrix whose columns are a basis for an $(n - r)$-dimensional subspace $\mathcal{T}$ of $\Re^n$. Since the columns of $\boldsymbol{P}$ form a basis for $\Re^n$, there exists a matrix $\boldsymbol{B}$ such that $\boldsymbol{A}\boldsymbol{P} = \boldsymbol{P}\boldsymbol{B}$, which we write as

$$\boldsymbol{A}\boldsymbol{P} = [\boldsymbol{A}\boldsymbol{P}_1 : \boldsymbol{A}\boldsymbol{P}_2] = [\boldsymbol{P}_1 : \boldsymbol{P}_2] \begin{bmatrix} \boldsymbol{B}_{11} & \boldsymbol{B}_{12} \\ \boldsymbol{B}_{21} & \boldsymbol{B}_{22} \end{bmatrix} .$$

This gives a nice geometric interpretation for submatrices in $\boldsymbol{B}$: $\boldsymbol{B}_{ij}$ holds the coordinates of $\boldsymbol{A}\boldsymbol{P}_i$ in terms of the basis vectors in $\boldsymbol{P}_j$.

Matters become a bit more interesting if one or both of $\mathcal{S}$ and $\mathcal{T}$ are ***invariant subspaces*** under $\boldsymbol{A}$. A subspace is invariant under $\boldsymbol{A}$ if any vector in that subspace remains in it after being transformed by $\boldsymbol{A}$. Assume that $\mathcal{S}$ is an invariant subspace under $\boldsymbol{A}$. Therefore, $\boldsymbol{A}\boldsymbol{x} \in \mathcal{S}$ for every $\boldsymbol{x} \in \mathcal{S}$. Because the columns of $\boldsymbol{P}_1$ are in $\mathcal{S}$, there exists a matrix $\boldsymbol{S}$ such that $\boldsymbol{A}\boldsymbol{P}_1 = \boldsymbol{P}_1\boldsymbol{S}$ so

$$\boldsymbol{A}\boldsymbol{P} = [\boldsymbol{A}\boldsymbol{P}_1 : \boldsymbol{A}\boldsymbol{P}_2] = [\boldsymbol{P}_1 : \boldsymbol{P}_2] \begin{bmatrix} \boldsymbol{S} & \boldsymbol{B}_{12} \\ \boldsymbol{O} & \boldsymbol{B}_{22} \end{bmatrix} .$$

Thus, irrespective of what $\boldsymbol{A}$ is, identifying an invariant subspace under $\boldsymbol{A}$ helps us arrive at a basis revealing that $\boldsymbol{A}$ is similar to a block upper-triangular matrix. What happens if *both* $\mathcal{S}$ and $\mathcal{T}$ are invariant under $\boldsymbol{A}$? Then, there will exist a matrix $\boldsymbol{T}$ such that $\boldsymbol{A}\boldsymbol{P}_2 = \boldsymbol{P}_2\boldsymbol{T}$. This will imply that

$$\boldsymbol{A}\boldsymbol{P} = [\boldsymbol{A}\boldsymbol{P}_1 : \boldsymbol{A}\boldsymbol{P}_2] = [\boldsymbol{P}_1 : \boldsymbol{P}_2] \begin{bmatrix} \boldsymbol{S} & \boldsymbol{O} \\ \boldsymbol{O} & \boldsymbol{T} \end{bmatrix} .$$

Now $\boldsymbol{P}$ reduces $\boldsymbol{A}$ to a block-diagonal form. Reducing a matrix to simpler structures using similarity and change of basis transformations is one of the major goals of linear algebra. Identifying invariant subspaces is one way of doing this.

4.10 Exercises

1. Show that $\mathcal{S}$ is a subspace in $\Re^3$, where

$$\mathcal{S} = \left\{ \boldsymbol{x} \in \Re^3 : \boldsymbol{x} = \alpha \begin{bmatrix} 1 \\ 2 \\ 3 \end{bmatrix} \text{ for some } \alpha \in \Re \right\} .$$

2. Prove that the set

$$\mathcal{S} = \left\{ \begin{bmatrix} x_1 \\ x_2 \\ x_3 \end{bmatrix} \in \Re^3 : \frac{x_1}{3} = \frac{x_2}{4} = \frac{x_3}{2} \right\}$$

is a subspace of $\Re^3$.

3. Which of the following are subspaces in $\Re^n$, where $n > 2$?

(a) $\{\boldsymbol{x} \in \Re^n : \boldsymbol{x} = \alpha \boldsymbol{u}$ for some $\alpha \in \Re$, where $\boldsymbol{u}$ is a fixed vector in $\Re^n\}$.

(b) $\left\{ \begin{bmatrix} x_1 \\ x_2 \\ \vdots \\ x_n \end{bmatrix} \in \Re^n : x_1 = 0 \right\}$.

(c) $\{\boldsymbol{x} \in \Re^n : \boldsymbol{1}'\boldsymbol{x} = 0\}$.

(d) $\{\boldsymbol{x} \in \Re^n : \boldsymbol{1}'\boldsymbol{x} = \alpha\}$, where $\alpha \neq 0$.

(e) $\{\boldsymbol{x} \in \Re^n : \boldsymbol{A}\boldsymbol{x} = \boldsymbol{b}\}$, where $\boldsymbol{A} \neq \boldsymbol{O}$ is $m \times n$ and $\boldsymbol{b} \neq \boldsymbol{0}$ is $m \times 1$.

4. Consider the following set of vectors in $\Re^3$:

$$\mathcal{X} = \left\{ \begin{bmatrix} 1 \\ 0 \\ -1 \end{bmatrix}, \begin{bmatrix} -2 \\ 1 \\ 1 \end{bmatrix} \right\}.$$

Show that one of the two vectors, $\boldsymbol{u} = \begin{bmatrix} -1 \\ 2 \\ 1 \end{bmatrix}$ and $\boldsymbol{v} = \begin{bmatrix} -1 \\ 1 \\ 1 \end{bmatrix}$, belongs to $Sp(\mathcal{X})$, while the other does not. Find a real number α such that $\begin{bmatrix} 1 \\ 1 \\ \alpha \end{bmatrix} \in Sp(\mathcal{X})$.

5. Consider the set

$$\mathcal{S} = \left\{ \begin{bmatrix} x_1 \\ x_2 \\ x_3 \end{bmatrix} \in \Re^3 : x_1 - 7x_2 + 10x_3 = 0 \right\}.$$

Show that $\mathcal{S}$ is a subspace of $\Re^3$ and find two vectors $\boldsymbol{x}_1$ and $\boldsymbol{x}_2$ that span $\mathcal{S}$. Show that $\mathcal{S}$ can also be expressed as

$$\left\{ \begin{bmatrix} a \\ b \\ \frac{7b-a}{10} \end{bmatrix} \in \Re^3 : a, b \in \Re \right\}.$$

6. Consider the set of vectors

$$\left\{ \begin{bmatrix} 1 \\ 2 \\ 1 \\ -1 \end{bmatrix}, \begin{bmatrix} 2 \\ 4 \\ 1 \\ 1 \end{bmatrix}, \begin{bmatrix} -1 \\ -2 \\ -2 \\ -4 \end{bmatrix}, \begin{bmatrix} 3 \\ 6 \\ 2 \\ 0 \end{bmatrix} \right\}.$$

Show that the span of the above set is the subspace:

$$\mathcal{N}(\boldsymbol{A}) = \{\boldsymbol{x} \in \Re^4 : \boldsymbol{A}\boldsymbol{x} = \boldsymbol{0}\} \ , \text{ where } \boldsymbol{A} = \begin{bmatrix} 2 & -1 & 0 & 0 \\ 2 & 0 & -3 & -1 \end{bmatrix} .$$

Show that $\mathcal{N}(\boldsymbol{A}) = \left\{ \begin{bmatrix} a \\ 2a \\ b \\ 2a - 3b \end{bmatrix} \in \Re^4 : a, b \in \Re \right\}$.

7. Let $\mathcal{X} = \{\boldsymbol{x}_1, \boldsymbol{x}_2, \ldots, \boldsymbol{x}_r\}$ be a set of vectors in $\Re^n$. Let $\mathcal{Y} = \mathcal{X} \cup \{\boldsymbol{y}\}$. Prove that $Sp(\mathcal{X}) = Sp(\mathcal{Y})$ if and only if $\boldsymbol{y} \in Sp(\mathcal{X})$.
8. True or false: If $\{\boldsymbol{x}_1, \boldsymbol{x}_2, \ldots, \boldsymbol{x}_r\}$ is a spanning set for $\mathcal{C}(\boldsymbol{X})$, then $\{\boldsymbol{A}\boldsymbol{x}_1, \boldsymbol{A}\boldsymbol{x}_2, \ldots, \boldsymbol{A}\boldsymbol{x}_r\}$ is a spanning set for $\mathcal{C}(\boldsymbol{A}\boldsymbol{X})$.
9. Verify whether each of the following sets are linearly independent:

$$\text{(a)} \left\{ \begin{bmatrix} 1 \\ 2 \\ 3 \end{bmatrix}, \begin{bmatrix} 1 \\ 0 \\ -1 \end{bmatrix}, \begin{bmatrix} -2 \\ 1 \\ 1 \end{bmatrix} \right\}; \quad \text{(b)} \left\{ \begin{bmatrix} 1 \\ 2 \\ 3 \end{bmatrix}, \begin{bmatrix} 4 \\ 5 \\ 6 \end{bmatrix}, \begin{bmatrix} 7 \\ 8 \\ 9 \end{bmatrix} \right\}$$

$$\text{(c)} \left\{ \begin{bmatrix} 1 \\ 2 \\ 3 \\ 4 \end{bmatrix}, \begin{bmatrix} 1 \\ 1 \\ 1 \\ 1 \end{bmatrix}, \begin{bmatrix} 1 \\ 3 \\ 1 \\ 2 \end{bmatrix}, \begin{bmatrix} 2 \\ 1 \\ 1 \\ 2 \end{bmatrix} \right\}; \quad \text{(d)} \left\{ \begin{bmatrix} 1 \\ 2 \\ -1 \\ 0 \end{bmatrix}, \begin{bmatrix} 0 \\ 1 \\ 2 \\ 1 \end{bmatrix}, \begin{bmatrix} 2 \\ 3 \\ -4 \\ -1 \end{bmatrix} \right\}.$$

10. Without doing any calculations, say why the following sets are linearly dependent:

$$\text{(a)} \left\{ \begin{bmatrix} 1 \\ 2 \end{bmatrix}, \begin{bmatrix} 3 \\ 4 \end{bmatrix}, \begin{bmatrix} 5 \\ 6 \end{bmatrix} \right\} \text{ and (b)} \left\{ \begin{bmatrix} 1 \\ 0 \\ 2 \end{bmatrix}, \begin{bmatrix} 1 \\ 1 \\ -1 \end{bmatrix}, \begin{bmatrix} 3 \\ 1 \\ 3 \end{bmatrix}, \begin{bmatrix} 1 \\ 3 \\ 4 \end{bmatrix} \right\}.$$

11. Find the values of a for which the following set is linearly independent:

$$\left\{ \begin{bmatrix} 0 \\ 1 \\ a \end{bmatrix}, \begin{bmatrix} a \\ 1 \\ 0 \end{bmatrix}, \begin{bmatrix} 1 \\ a \\ 1 \end{bmatrix} \right\}$$

12. Find the values of a and b for which the following set is linearly independent:

$$\left\{ \begin{bmatrix} a \\ b \\ b \\ b \end{bmatrix}, \begin{bmatrix} b \\ a \\ b \\ b \end{bmatrix}, \begin{bmatrix} b \\ b \\ a \\ b \end{bmatrix}, \begin{bmatrix} b \\ b \\ b \\ a \end{bmatrix} \right\}.$$

13. Show that the following set of vectors is linearly independent:

$$\left\{ \boldsymbol{x}_1 = \begin{bmatrix} 1 \\ -1 \\ 0 \\ 0 \\ \vdots \\ 0 \end{bmatrix}, \boldsymbol{x}_2 = \begin{bmatrix} 1 \\ 0 \\ -1 \\ 0 \\ \vdots \\ 0 \end{bmatrix}, \boldsymbol{x}_3 = \begin{bmatrix} 1 \\ 0 \\ 0 \\ -1 \\ \vdots \\ 0 \end{bmatrix}, \ldots, \boldsymbol{x}_{n-1} = \begin{bmatrix} 1 \\ 0 \\ 0 \\ 0 \\ \vdots \\ -1 \end{bmatrix} \right\}.$$

Show that $\boldsymbol{u} = [n-1, -1, -1, \ldots, -1]'$ is spanned by the above set of vectors.

14. Find a maximal set of linearly independent columns of the following matrix:

$$\begin{bmatrix} 1 & 1 & 1 & 0 & 2 \\ 1 & 2 & 0 & 1 & 0 \\ 0 & -1 & 1 & 1 & 2 \\ 1 & 0 & 2 & 1 & 4 \end{bmatrix}.$$

Also find a maximal set of linearly independent rows. ***Note:*** *By a maximal set of linearly independent columns (rows) we mean a largest possible set in the sense that adding any other column (row) would make the set linearly dependent.*

15. True or false: If $\boldsymbol{A}$ is an $m \times n$ matrix such that $\boldsymbol{A1} = \boldsymbol{0}$, then the columns of $\boldsymbol{A}$ must be linearly dependent.

16. Let $\{\boldsymbol{x}_1, \boldsymbol{x}_2, \ldots, \boldsymbol{x}_r\}$ be a linearly independent set and let $\boldsymbol{A}$ be an $n \times n$ nonsingular matrix. Prove that the set $\{\boldsymbol{Ax}_1, \boldsymbol{Ax}_2, \ldots, \boldsymbol{Ax}_r\}$ is also linearly independent.

17. Find a basis and the dimension of the subspaces $\mathcal{S}$ in Exercises 1, 2 and 5.

18. Verify that the following is a linearly independent set and extend it to a basis for $\Re^4$:

$$\left\{ \begin{bmatrix} 1 \\ 0 \\ -1 \\ 2 \end{bmatrix}, \begin{bmatrix} 0 \\ 0 \\ 1 \\ 2 \end{bmatrix} \right\}.$$

19. Let $\mathcal{X} = \{\boldsymbol{x}_1, \boldsymbol{x}_2, \ldots, \boldsymbol{x}_r\}$ be a basis of a subspace $\mathcal{S} \subseteq \Re^n$ and let $\boldsymbol{x} \in Sp(\mathcal{X})$. When can we *replace* a vector $\boldsymbol{x}_i$ in $\mathcal{X}$ with $\boldsymbol{x}$ and still obtain a basis for $\mathcal{S}$? More precisely, when is the set $(\mathcal{X} \setminus \{\boldsymbol{x}_i\}) \cup \{\boldsymbol{x}\}$ a basis for $\mathcal{S}$.

20. Let $\boldsymbol{A}$ be an $m \times n$ matrix and let $\mathcal{X} \subseteq \Re^n$ be an arbitrary subset of $\Re^n$. The ***image*** of $\mathcal{X}$ under the transformation $\boldsymbol{A}$ is defined to be the set

$$\boldsymbol{A}(\mathcal{X}) = \{\boldsymbol{Ax} : \boldsymbol{x} \in \mathcal{X}\}.$$

(a) Show that $\boldsymbol{A}(\mathcal{X}) \subseteq \mathcal{C}(\boldsymbol{A})$.

(b) If $\mathcal{X}$ is a subspace of $\Re^n$, then prove that $\boldsymbol{A}(\mathcal{X})$ is a subspace of $\Re^m$.

(c) Let $\mathcal{X}$ be a subspace of $\Re^n$ such that $\mathcal{X} \cap \mathcal{N}(\boldsymbol{A}) = \{\boldsymbol{0}\}$. If $\{\boldsymbol{x}_1, \boldsymbol{x}_2, \ldots, \boldsymbol{x}_r\}$ is a basis for $\mathcal{X}$, prove that $\{\boldsymbol{Ax}_1, \boldsymbol{Ax}_2, \ldots, \boldsymbol{Ax}_r\}$ is a basis for $\boldsymbol{A}(\mathcal{X})$. Hence, conclude that

$$\dim(\mathcal{X}) = \dim(\boldsymbol{A}(\mathcal{X})) \leq \dim(\mathcal{C}(\boldsymbol{A})).$$

(d) Consider the special case where $\mathcal{X} = \mathcal{R}(\boldsymbol{A}) = \mathcal{C}(\boldsymbol{A}')$. Recall from Theorem 4.11 that $\mathcal{R}(\boldsymbol{A}) \cap \mathcal{N}(\boldsymbol{A}) = \{\boldsymbol{0}\}$. Apply the above result to obtain the inequality:$\dim(\mathcal{R}(\boldsymbol{A})) \leq \dim(\mathcal{C}(\boldsymbol{A}))$. Now apply the above inequality to $\boldsymbol{A}'$ to conclude that

$$\dim(\mathcal{R}(\boldsymbol{A})) = \dim(\mathcal{C}(\boldsymbol{A})).$$

This remarkable result will be revisited several times in the text.

CHAPTER 5

The Rank of a Matrix

5.1 Rank and nullity of a matrix

At the core of linear algebra lies the four fundamental subspaces of a matrix. These were described in Section 4.6. The dimensions of the four fundamental subspaces can be characterized by two fundamentally important quantities—*rank* and *nullity*.

Definition 5.1 Rank and nullity. *The* **rank** *of a matrix $\boldsymbol{A}$ is the dimension of $\mathcal{C}(\boldsymbol{A})$. It is denoted by $\rho(\boldsymbol{A})$. The* **nullity** *of $\boldsymbol{A}$ refers to the dimension of $\mathcal{N}(\boldsymbol{A})$. It is denoted by $\nu(\boldsymbol{A})$.*

The rank of $\boldsymbol{A}$ is the number of vectors in any basis for the column space of $\boldsymbol{A}$ and, hence, is the maximum number of linearly independent columns in $\boldsymbol{A}$. The nullity of $\boldsymbol{A}$ is the number of vectors in any basis for the null space of $\boldsymbol{A}$. Equivalently, it is the maximum number of linearly independent solutions for the homogeneous system $\boldsymbol{A}\boldsymbol{x} = \boldsymbol{0}$. Based upon what we already know about linear independence of rows and columns of matrices in Chapter 4, the following properties are easily verified.

1. $\rho(\boldsymbol{A}) = \boldsymbol{O}$ if and only if $\boldsymbol{A} = \boldsymbol{O}$. Because of this, when we talk about an $m \times n$ matrix $\boldsymbol{A}$ with rank r, we will implicitly assume that $\boldsymbol{A}$ is non-null and r is a positive integer.
2. Let $\boldsymbol{T}$ be an $n \times n$ triangular (upper or lower) matrix with exactly k nonzero diagonal elements. Then $\rho(\boldsymbol{T}) = k$ and $\nu(\boldsymbol{T}) = n - k$. It is easy to verify that the k nonzero diagonal elements act as pivots and the columns containing the pivots are precisely the complete set of linearly independent columns in $\boldsymbol{T}$, thereby leading to $\rho(\boldsymbol{T}) = k$. There are also $n - k$ free variables that yield $n - k$ linearly independent solutions of $\boldsymbol{T}\boldsymbol{x} = \boldsymbol{0}$. This implies that $\nu(\boldsymbol{T}) = n - k$, which also follows from Theorem 4.18.
3. In particular, the above result is true when $\boldsymbol{T}$ is diagonal. Therefore $\rho(\boldsymbol{I}_n) = n$ and $\nu(\boldsymbol{I}_n) = 0$.
4. Part (i) of Theorem 4.16 tells us that if $\mathcal{N}(\boldsymbol{A}) \subseteq \mathcal{N}(\boldsymbol{B})$, then $\rho(\boldsymbol{A}) \geq \rho(\boldsymbol{B})$. Part (ii) of Theorem 4.16 tells us that if $\mathcal{N}(\boldsymbol{A}) = \mathcal{N}(\boldsymbol{B})$, then $\rho(\boldsymbol{B}) = \rho(\boldsymbol{A})$.
5. Theorem 4.17 tells us that the number of linearly independent rows and columns in a matrix are equal. Since the rows of $\boldsymbol{A}$ are precisely the columns of $\boldsymbol{A}'$, this

implies that $\rho(\boldsymbol{A}) = \rho(\boldsymbol{A}')$ and $\rho(\boldsymbol{A})$ could also be defined as the number of linearly independent rows in a matrix. We will provide a shorter and more elegant proof of this as a part of the *fundamental theorem of ranks* (see Theorem 5.2 and Corollary 5.2) in this section.

6. Let $\boldsymbol{A}$ be an $m \times n$ matrix. By the above argument, $\rho(\boldsymbol{A})$ is the number of linearly independent rows or columns of a matrix. Since the number of linearly independent rows (columns) cannot exceed m (n), we must have $\rho(\boldsymbol{A}) \leq \min\{m, n\}$.
7. Theorem 4.18 tells us that there are exactly $n - r$ linearly independent solutions of a homogeneous system $\boldsymbol{A}\boldsymbol{x} = \boldsymbol{0}$, where $\boldsymbol{A}$ is $m \times n$ with r linearly independent columns. In terms of rank and nullity, we can restate this as $\nu(\boldsymbol{A}) = n - \rho(\boldsymbol{A})$ for any $m \times n$ matrix $\boldsymbol{A}$. This result also emerges from a more general result (see Theorems 5.4 and 5.1) in this section.

Clearly, the rank of a matrix is an integer that lies between 0 and $\min\{m, n\}$. The only matrix with zero rank is the zero matrix. Also, given any integer r between 0 and $\min\{m, n\}$, we can always construct an $m \times n$ matrix of rank r. For example, take

$$\boldsymbol{A} = \begin{bmatrix} \boldsymbol{I}_r & \boldsymbol{O} \\ \boldsymbol{O} & \boldsymbol{O} \end{bmatrix} .$$

Matrices of the above form hold structural importance. We will see more about them later.

The following properties are straightforward consequences of Definition 5.1.

Lemma 5.1 *Let $\boldsymbol{A}$ and $\boldsymbol{B}$ be matrices such that $\boldsymbol{AB}$ is well-defined. Then $\rho(\boldsymbol{AB}) \leq \rho(\boldsymbol{A})$ and $\nu(\boldsymbol{AB}) \geq \nu(\boldsymbol{B})$.*

Proof. From Theorem 4.6 we know that $\mathcal{C}(\boldsymbol{AB}) \subseteq \mathcal{C}(\boldsymbol{B})$ and Theorem 4.8 tells us that $\mathcal{N}(\boldsymbol{B}) \subseteq \mathcal{N}(\boldsymbol{AB})$. Therefore, from part (i) of Theorem 4.22 we obtain

$$\begin{aligned} \rho(\boldsymbol{AB}) &= \dim(\mathcal{C}(\boldsymbol{AB})) \leq \dim(\mathcal{C}(\boldsymbol{A})) = \rho(\boldsymbol{A}) \\ \text{and} \quad \nu(\boldsymbol{AB}) &= \dim(\mathcal{N}(\boldsymbol{AB})) \geq \dim(\mathcal{N}(\boldsymbol{B})) = \nu(\boldsymbol{B}) . \end{aligned}$$

□

For matrices $\boldsymbol{A}'\boldsymbol{A}$ and $\boldsymbol{A}\boldsymbol{A}'$, we have the following result.

Lemma 5.2 *For any $m \times n$ matrix, $\nu(\boldsymbol{A}'\boldsymbol{A}) = \nu(\boldsymbol{A})$ and $\nu(\boldsymbol{A}\boldsymbol{A}') = \nu(\boldsymbol{A}')$.*

Proof. This follows immediately from $\mathcal{N}(\boldsymbol{A}'\boldsymbol{A}) = \mathcal{N}(\boldsymbol{A})$ and $\mathcal{N}(\boldsymbol{A}\boldsymbol{A}') = \mathcal{N}(\boldsymbol{A}')$—facts that were proved in Theorem 4.12. □

One of the most important results in linear algebra is the ***Rank-Nullity Theorem***. This states that the rank and the nullity of a matrix add up to the number of columns of the matrix. In terms of systems of linear equations this is equivalent to the fact that the number of pivots and the number of free variables add up to the number of columns of a matrix. We have also seen this result in Theorem 4.18. Nevertheless, below we state it as a theorem and also provide a rather elegant proof of this result. The idea of the proof is to start with a basis for $\mathcal{N}(\boldsymbol{A})$, extend it to a basis for $\Re^n$ and show that the image of the extension vectors after being transformed by $\boldsymbol{A}$ will in fact be a basis for $\mathcal{C}(\boldsymbol{A})$.

Theorem 5.1 The Rank-Nullity theorem. *Let A be an $m \times n$ matrix. Then,*

$$\rho(\boldsymbol{A}) + \nu(\boldsymbol{A}) = n \, .$$

Proof. If $\boldsymbol{A} = \boldsymbol{O}$, then every vector in $\Re^n$ belongs to its null space. So $\nu(\boldsymbol{A}) = n$ and $\rho(\boldsymbol{A}) = 0$ and the theorem is trivially true.

Assume that $\nu(\boldsymbol{A}) = k < n$ and let $\boldsymbol{X}_1 = [\boldsymbol{x}_1 : \boldsymbol{x}_2 : \ldots : \boldsymbol{x}_k]$ be an $n \times k$ matrix whose column vectors are a basis for $\mathcal{N}(\boldsymbol{A})$. Because the columns of $\boldsymbol{X}_1$ are a linearly independent set of k vectors in $\Re^n$, they can be extended to a basis for $\Re^n$ (part (ii) of Theorem 4.21). We place these $n - k$ extension vectors as columns of a matrix $\boldsymbol{X}_2 = [\boldsymbol{x}_{k+1} : \boldsymbol{x}_{k+2} : \ldots : \boldsymbol{x}_n]$. Thus, $\boldsymbol{X} = [\boldsymbol{X}_1 : \boldsymbol{X}_2]$ is an $n \times n$ nonsingular matrix whose columns form a basis for $\Re^n$ (recall Corollary 4.3).

We claim that the columns of the matrix $\boldsymbol{A}\boldsymbol{X}_2$ are a basis for $\mathcal{C}(\boldsymbol{A})$.

To prove our claim we need to show that (a) the columns of $\boldsymbol{A}\boldsymbol{X}_2$ are linearly independent and (b) they span $\mathcal{C}(\boldsymbol{A})$.

(a) *The columns of $\boldsymbol{A}\boldsymbol{X}_2$ are linearly independent:* Consider the homogeneous system $\boldsymbol{A}\boldsymbol{X}_2\boldsymbol{v} = \boldsymbol{0}$. Keeping in mind that $\mathcal{C}(\boldsymbol{X}_1) = \mathcal{N}(\boldsymbol{A})$, we can write

$$\begin{aligned} \boldsymbol{A}\boldsymbol{X}_2\boldsymbol{v} = \boldsymbol{0} &\Longrightarrow \boldsymbol{X}_2\boldsymbol{v} \in \mathcal{N}(\boldsymbol{A}) \Longrightarrow \boldsymbol{X}_2\boldsymbol{v} \in \mathcal{C}(\boldsymbol{X}_1) \\ &\Longrightarrow \boldsymbol{X}_2\boldsymbol{v} = \boldsymbol{X}_1\boldsymbol{u} \;\text{ for some vector }\; \boldsymbol{u} \in \Re^k \\ &\Longrightarrow [\boldsymbol{X}_1 : \boldsymbol{X}_2]\begin{bmatrix} -\boldsymbol{u} \\ \boldsymbol{v} \end{bmatrix} = \boldsymbol{0} \Longrightarrow \boldsymbol{u} = \boldsymbol{v} = \boldsymbol{0} \end{aligned}$$

because $\boldsymbol{X} = [\boldsymbol{X}_1 : \boldsymbol{X}_2]$ is nonsingular. This proves that $\boldsymbol{v} = \boldsymbol{0}$ whenever $\boldsymbol{A}\boldsymbol{X}_2\boldsymbol{v} = \boldsymbol{0}$ and so the columns of $\boldsymbol{A}\boldsymbol{X}_2$ are linearly independent.

(b) *The columns of $\boldsymbol{A}\boldsymbol{X}_2$ span $\mathcal{C}(\boldsymbol{A})$:* We need to prove that $\mathcal{C}(\boldsymbol{A}\boldsymbol{X}_2) = \mathcal{C}(\boldsymbol{A})$. One direction is immediate from Theorem 4.6: $\mathcal{C}(\boldsymbol{A}\boldsymbol{X}_2) \subseteq \mathcal{C}(\boldsymbol{A})$.
To prove $\mathcal{C}(\boldsymbol{A}) \subseteq \mathcal{C}(\boldsymbol{A}\boldsymbol{X}_2)$, consider a vector $\boldsymbol{w}$ in $\mathcal{C}(\boldsymbol{A})$. Then $\boldsymbol{w} = \boldsymbol{A}\boldsymbol{y}$ for some vector $\boldsymbol{y} \in \Re^n$. Because $\boldsymbol{y} \in \Re^n$, it can be expressed as a linear combination of the columns of $\boldsymbol{X}$ and so there are vectors $\boldsymbol{z}_1$ and $\boldsymbol{z}_2$ such that $\boldsymbol{y} = \boldsymbol{X}_1\boldsymbol{z}_1 + \boldsymbol{X}_2\boldsymbol{z}_2$. Therefore, $\boldsymbol{w} = \boldsymbol{A}\boldsymbol{y} = \boldsymbol{A}\boldsymbol{X}_1\boldsymbol{z}_1 + \boldsymbol{A}\boldsymbol{X}_2\boldsymbol{z}_2 = \boldsymbol{A}\boldsymbol{X}_2\boldsymbol{z}_2 \in \mathcal{C}(\boldsymbol{A}\boldsymbol{X}_2)$. This proves that $\mathcal{C}(\boldsymbol{A}) \subseteq \mathcal{C}(\boldsymbol{A}\boldsymbol{X}_2)$ and we conclude that $\mathcal{C}(\boldsymbol{A}\boldsymbol{X}_2) = \mathcal{C}(\boldsymbol{A})$.

We have now proved that the $n-k$ columns of $\boldsymbol{A}\boldsymbol{X}_2$ are indeed a basis for $\mathcal{C}(\boldsymbol{A})$. This means that the dimension of the column space of $\boldsymbol{A}$ is $n - k$. Thus, $\rho(\boldsymbol{A}) = n - k$. Recall that $k = \nu(\boldsymbol{A})$, so we obtain $\rho(\boldsymbol{A}) + \nu(\boldsymbol{A}) = n$. □

Theorem 5.1 immediately yields the following corollary.

Corollary 5.1 *If A and B are two matrices with the same number of columns, then $\rho(A) = \rho(B)$ if and only if $\nu(A) = \nu(B)$. In other words, they have the same rank if and only if they have the same nullity.*

Proof. Let A be an $m \times n$ matrix and B be an $p \times n$ matrix. Then, from Theorem 5.1 we have that $\rho(A) = \rho(B) \iff n - \nu(A) = n - \nu(B) \iff \nu(A) = \nu(B)$. □

It is worth comparing Corollary 5.1 with Theorem 4.16. While the latter assumes that the null spaces of A and B are the same, the former simply assumes that they have the same nullity (i.e., same dimension). In this sense Corollary 5.1 assumes less than Theorem 4.16. On the other hand, Theorem 4.16 assumes more but tells us more: not only do A and B have the same number of linearly independent columns (hence they have the same rank), their positions in A and B will correspond exactly.

We next derive another extremely important set of equalities concerning the rank and its transpose. The first of these equalities says that that the rank of $A'A$ equals the rank of A. This also leads to the fact that the rank of a matrix equals the rank of its transpose. We collect these into the following theorem, which we call the *fundamental theorem of ranks*.

Theorem 5.2 Fundamental theorem of ranks. *Let A be an $m \times n$ matrix. Then,*

$$\rho(A) = \rho(A'A) = \rho(A') = \rho(AA') . \tag{5.1}$$

Proof. Because $A'A$ and A have the same number of columns and $\mathcal{N}(A'A) = \mathcal{N}(A)$ (Theorem 4.12), Corollary 5.1 immediately tells us that $\rho(A'A) = \rho(A)$.

We next prove $\rho(A) = \rho(A')$. This is a restatement of Theorem 4.17, but we provide a much simpler and more elegant proof directly from the first equality in (5.1).

Lemma 5.1 tells us $\rho(A'A) \leq \rho(A')$, which implies $\rho(A) = \rho(A'A) \leq \rho(A')$. Thus, we have shown that the rank of a matrix cannot exceed the rank of its transpose. Now comes a neat trick: apply this result to the transposed matrix A'. Since the transpose of the transpose is the original matrix, this would yield $\rho(A') \leq \rho((A')') = \rho(A)$ and so $\rho(A') \leq \rho(A)$. Hence, $\rho(A) = \rho(A')$.

At this point we have proved $\rho(A) = \rho(A'A) = \rho(A')$.

To prove the last equality in (5.1), we apply the same arguments that proved $\rho(A) = \rho(A'A)$ to the transpose of A. More precisely, let $B = A'$ and use the first equality in (5.1) to write $\rho(A') = \rho(B) = \rho(B'B) = \rho(AA')$. We have now proved all the equalities in (5.1). □

A key result emerging from Theorem 5.2 is that *the rank of a matrix is equal to the rank of its transpose*, i.e., $\rho(A) = \rho(A')$. This says something extremely important about the dimensions of $\mathcal{C}(A)$ and $\mathcal{R}(A)$:

$$\dim(\mathcal{C}(A)) = \rho(A) = \rho(A') = \dim(\mathcal{C}(A')) = \dim(\mathcal{R}(A)) . \tag{5.2}$$

A different way of saying this is that the number of linearly independent columns in $\boldsymbol{A}$ is equal to the number of linearly independent rows of $\boldsymbol{A}$—a result that we have proved independently in Theorem 4.17. In fact, often the number of linearly independent rows is called the ***row rank*** and the number of linearly independent columns is called the ***column rank***. Their equality is immediate from Theorem 5.2.

Corollary 5.2 *The row rank of $\boldsymbol{A}$ equals the column rank of $\boldsymbol{A}$.*

Proof. This is simply a restatement of $\rho(\boldsymbol{A}) = \rho(\boldsymbol{A}')$, proved in Theorem 5.2. □

The dimension of the left null space is also easily connected to $\rho(\boldsymbol{A})$.

Corollary 5.3 *Let $\boldsymbol{A}$ be an $m \times n$ matrix. Then, $\nu(\boldsymbol{A}') = m - \rho(\boldsymbol{A})$.*

Proof. This follows immediately by applying Theorem 5.1 to $\boldsymbol{A}'$, which has m columns: $\nu(\boldsymbol{A}') = m - \rho(\boldsymbol{A}') = m - \rho(\boldsymbol{A})$ because $\rho(\boldsymbol{A}) = \rho(\boldsymbol{A}')$. □

Note that the rank of an $m \times n$ matrix $\boldsymbol{A}$ determines the dimensions of the four fundamental subspaces of $\boldsymbol{A}$. We summarize them below:

- $\dim(\mathcal{N}(\boldsymbol{A})) = n - \rho(\boldsymbol{A})$ (from Theorem 5.1);
- $\dim(\mathcal{C}(\boldsymbol{A})) = \rho(\boldsymbol{A}) = \rho(\boldsymbol{A}') = \dim(\mathcal{R}(\boldsymbol{A}))$ (from Theorem 5.2);
- $\dim(\mathcal{N}(\boldsymbol{A}')) = m - \rho(\boldsymbol{A})$ (from Corollary 5.3).

Indeed, $\rho(\boldsymbol{A})$ is the dimension of both the $\mathcal{C}(\boldsymbol{A})$ and $\mathcal{R}(\boldsymbol{A})$ (since $\rho(\boldsymbol{A}) = \rho(\boldsymbol{A}')$). From part (i) of Theorem 5.1, the dimension of $\mathcal{N}(\boldsymbol{A})$ is given by $n - \rho(\boldsymbol{A})$, while part (ii) tells us that the dimension of the left null space of $\boldsymbol{A}$ is obtained by subtracting the rank of $\boldsymbol{A}$ from the number of *rows* of $\boldsymbol{A}$.

Recall from Lemma 5.1 that the rank of $\boldsymbol{AB}$ cannot exceed the rank of $\boldsymbol{A}$. Because $\mathcal{N}(\boldsymbol{B}) \subseteq \mathcal{N}(\boldsymbol{AB})$, part (i) of Theorem 4.16 tells us $\rho(\boldsymbol{AB}) \leq \rho(\boldsymbol{B})$ also. So the rank of $\boldsymbol{AB}$ cannot exceed the rank of either $\boldsymbol{A}$ or $\boldsymbol{B}$. This important fact can also be proved as a corollary of the fundamental theorem of ranks.

Corollary 5.4 *Let $\boldsymbol{A}$ be an $m \times n$ matrix and $\boldsymbol{B}$ an $n \times p$ matrix. Then $\rho(\boldsymbol{AB}) \leq \min\{\rho(\boldsymbol{A}), \rho(\boldsymbol{B})\}$.*

Proof. This is another immediate consequence of the rank of a matrix being equal to the rank of its transpose. We already proved in Lemma 5.1 that $\rho(\boldsymbol{AB}) \leq \rho(\boldsymbol{A})$. Therefore, it remains to show that $\rho(\boldsymbol{AB}) \leq \rho(\boldsymbol{B})$.

$$\rho(\boldsymbol{AB}) = \rho((\boldsymbol{AB})') = \rho(\boldsymbol{B}'\boldsymbol{A}') \leq \rho(\boldsymbol{B}') = \rho(\boldsymbol{B}) , \tag{5.3}$$

where the inequality $\rho(\boldsymbol{B}'\boldsymbol{A}') \leq \rho(\boldsymbol{B}')$ follows from Lemma 5.1. □

The fact that the row rank and column rank of a matrix are the same can be used to construct a simple proof of the fact that the rank of a submatrix cannot exceed that of a matrix.

Corollary 5.5 *Let $\boldsymbol{A}$ be an $m \times n$ matrix of rank r and let $\boldsymbol{B}$ be a submatrix of $\boldsymbol{A}$. Then, $\rho(\boldsymbol{B}) \leq \rho(\boldsymbol{A})$.*

Proof. First consider the case where $\boldsymbol{B}$ has been obtained from $\boldsymbol{A}$ by selecting only some of the rows of $\boldsymbol{A}$. Clearly, the rows of $\boldsymbol{B}$ are a subset of the rows of $\boldsymbol{A}$ and so

$$\rho(\boldsymbol{B}) = \text{"row rank" of } \boldsymbol{B} \leq \text{"row rank" of } \boldsymbol{A} = \rho(\boldsymbol{A}) \; .$$

Next, suppose $\boldsymbol{B}$ has been obtained from $\boldsymbol{A}$ by selecting only some of the columns of $\boldsymbol{A}$. Now the columns of $\boldsymbol{B}$ are a subset of the columns of $\boldsymbol{A}$ and so

$$\rho(\boldsymbol{B}) = \text{"column rank" of } \boldsymbol{B} \leq \text{"column rank" of } \boldsymbol{A} = \rho(\boldsymbol{A}) \; .$$

Thus, when $\boldsymbol{B}$ has been obtained from $\boldsymbol{A}$ by omitting only some entire rows or only some entire columns, we find that $\rho(\boldsymbol{B}) \leq \rho(\boldsymbol{A})$.

Now, any submatrix $\boldsymbol{B}$ can be obtained from $\boldsymbol{A}$ by omitting some rows and then some columns. The moment one of these operations is performed we obtain a reduction in the rank. Therefore, $\rho(\boldsymbol{B}) \leq \rho(\boldsymbol{A})$ for any submatrix $\boldsymbol{B}$. □

The following theorem shows how the column and row spaces of $\boldsymbol{AB}$ are related to the ranks of $\boldsymbol{A}$ and $\boldsymbol{B}$.

Theorem 5.3 *Let $\boldsymbol{A}$ and $\boldsymbol{B}$ be matrices such that $\boldsymbol{AB}$ is defined. Then the following statements are true:*

(i) If $\rho(\boldsymbol{AB}) = \rho(\boldsymbol{A})$, then $\mathcal{C}(\boldsymbol{AB}) = \mathcal{C}(\boldsymbol{A})$ and $\boldsymbol{A} = \boldsymbol{ABC}$ for some matrix $\boldsymbol{C}$.

(ii) If $\rho(\boldsymbol{AB}) = \rho(\boldsymbol{B})$, then $\mathcal{R}(\boldsymbol{AB}) = \mathcal{R}(\boldsymbol{B})$ and $\boldsymbol{B} = \boldsymbol{DAB}$ for some matrix $\boldsymbol{D}$.

Proof. **Proof of part (i)**: Theorem 4.6 tells us that $\mathcal{C}(\boldsymbol{AB}) \subseteq \mathcal{C}(\boldsymbol{A})$. Now, $\rho(\boldsymbol{AB}) = \rho(\boldsymbol{A})$ means that $\dim(\mathcal{C}(\boldsymbol{AB})) = \dim(\mathcal{C}(\boldsymbol{A}))$. Part (ii) of Theorem 4.22 now yields $\mathcal{C}(\boldsymbol{AB}) = \mathcal{C}(\boldsymbol{A})$. To prove that $\boldsymbol{A} = \boldsymbol{ABC}$ for some matrix we use a neat trick: since $\mathcal{C}(\boldsymbol{AB}) = \mathcal{C}(\boldsymbol{A})$, it is true that $\mathcal{C}(\boldsymbol{A}) \subseteq \mathcal{C}(\boldsymbol{AB})$. Theorem 4.6 insures the existence of a matrix $\boldsymbol{C}$ such that $\boldsymbol{A} = \boldsymbol{ABC}$.

Proof of part (ii): This is the analogue of part (i) for row spaces and can be proved in similar fashion. Alternatively, we could apply part (i) to transposes (along with a neat application of the fact that $\rho(\boldsymbol{A}) = \rho(\boldsymbol{A}')$):

$$\begin{aligned} &\rho(\boldsymbol{AB}) = \rho(\boldsymbol{B}) \Rightarrow \rho((\boldsymbol{AB})') = \rho(\boldsymbol{B}') \Rightarrow \rho(\boldsymbol{B}'\boldsymbol{A}') = \rho(\boldsymbol{B}') \\ &\quad \Rightarrow \mathcal{C}(\boldsymbol{B}'\boldsymbol{A}') = \mathcal{C}(\boldsymbol{B}') \Rightarrow \mathcal{C}((\boldsymbol{AB})') = \mathcal{C}(\boldsymbol{B}') \Rightarrow \mathcal{R}(\boldsymbol{AB}) = \mathcal{R}(\boldsymbol{B}) \; . \end{aligned}$$

Since $\mathcal{R}(\boldsymbol{AB}) = \mathcal{R}(\boldsymbol{B})$, it is a fact that $\mathcal{R}(\boldsymbol{B}) \subseteq \mathcal{R}(\boldsymbol{AB})$ and Theorem 4.7 insures the existence of a matrix $\boldsymbol{C}$ such that $\boldsymbol{B} = \boldsymbol{DAB}$. □

The following lemma shows that the rank of a matrix remains unaltered by multiplication with a nonsingular matrix.

Lemma 5.3 *The following statements are true:*

(i) Let $\boldsymbol{A}$ be an $m \times n$ matrix and $\boldsymbol{B}$ an $n \times n$ nonsingular matrix. Then $\rho(\boldsymbol{AB}) = \rho(\boldsymbol{A})$ and $\mathcal{C}(\boldsymbol{AB}) = \mathcal{C}(\boldsymbol{A})$.

(ii) Let $\boldsymbol{A}$ be an $m \times m$ nonsingular matrix and $\boldsymbol{B}$ an $m \times p$ matrix. Then $\rho(\boldsymbol{AB}) = \rho(\boldsymbol{B})$ and $\mathcal{R}(\boldsymbol{AB}) = \mathcal{R}(\boldsymbol{B})$.

Proof. To prove part (i), note that we can write $\boldsymbol{A} = \boldsymbol{ABB}^{-1}$ (since $\boldsymbol{B}$ is nonsingular) and use Corollary 5.4 to show that

$$\rho(\boldsymbol{AB}) \leq \rho(\boldsymbol{A}) = \rho(\boldsymbol{ABB}^{-1}) \leq \rho(\boldsymbol{AB}) .$$

This proves $\rho(\boldsymbol{AB}) = \rho(\boldsymbol{A})$ when $\boldsymbol{B}$ is nonsingular. Part (i) of Theorem 5.3 tells us that $\mathcal{C}(\boldsymbol{AB}) = \mathcal{C}(\boldsymbol{A})$.

For part (ii), we use a similar argument. Now $\boldsymbol{A}$ is nonsingular, so $\boldsymbol{B} = \boldsymbol{A}^{-1}\boldsymbol{AB}$ and we use Corollary 5.4 to write

$$\rho(\boldsymbol{AB}) \leq \rho(\boldsymbol{B}) = \rho(\boldsymbol{A}^{-1}\boldsymbol{AB}) \leq \rho(\boldsymbol{AB}) .$$

This proves that $\rho(\boldsymbol{AB}) = \rho(\boldsymbol{B})$. That $\mathcal{R}(\boldsymbol{AB}) = \mathcal{R}(\boldsymbol{B})$ follows from part (ii) of Theorem 5.3. □

Note that the above lemma implies that $\rho(\boldsymbol{PAQ}) = \rho(\boldsymbol{A})$ as long as the matrix product $\boldsymbol{PAQ}$ is defined and *both* $\boldsymbol{P}$ and $\boldsymbol{Q}$ are nonsingular (hence square) matrices. However, multiplication by rectangular or singular matrices can alter the rank.

We have already seen that if $\boldsymbol{A}$ is a nonsingular matrix and we are given that $\boldsymbol{AB} = \boldsymbol{AD}$, then we can "cancel" $\boldsymbol{A}$ from both sides by premultiplying both sides with $\boldsymbol{A}^{-1}$. In other words,

$$\boldsymbol{AB} = \boldsymbol{AD} \Rightarrow \boldsymbol{A}^{-1}\boldsymbol{AB} = \boldsymbol{A}^{-1}\boldsymbol{AD} \Rightarrow \boldsymbol{B} = \boldsymbol{D} .$$

The following lemma shows that we can sometimes "cancel" matrices even when we do not assume that they are nonsingular, or even square. These are known as the *rank cancellation laws*.

Lemma 5.4 Rank cancellation laws. *Let $\boldsymbol{A}$, $\boldsymbol{B}$, $\boldsymbol{C}$ and $\boldsymbol{D}$ be matrices such that the products below are well-defined.*

(i) If $\boldsymbol{CAB} = \boldsymbol{DAB}$ and $\rho(\boldsymbol{AB}) = \rho(\boldsymbol{A})$, then we can "cancel" $\boldsymbol{B}$ and obtain $\boldsymbol{CA} = \boldsymbol{DA}$.

(ii) If $\boldsymbol{ABC} = \boldsymbol{ABD}$ and $\rho(\boldsymbol{AB}) = \rho(\boldsymbol{B})$, then we can "cancel" $\boldsymbol{A}$ and obtain $\boldsymbol{BC} = \boldsymbol{BD}$.

Proof. **Proof of (i)**: From part (i) of Theorem 5.3 we know that $\rho(\boldsymbol{AB}) = \rho(\boldsymbol{A})$ implies that $\mathcal{C}(\boldsymbol{AB}) = \mathcal{C}(\boldsymbol{A})$ and that there exists a matrix $\boldsymbol{X}$ such that $\boldsymbol{A} = \boldsymbol{ABX}$. Pre-multiplying both sides by $\boldsymbol{C}$ and $\boldsymbol{D}$ yield $\boldsymbol{CA} = \boldsymbol{CABX}$ and $\boldsymbol{DA} = \boldsymbol{DABX}$, respectively. This implies $\boldsymbol{CA} = \boldsymbol{DA}$ because

$$\boldsymbol{CA} = \boldsymbol{CABX} = \boldsymbol{DABX} = \boldsymbol{DA} ,$$

where the second equality follows because we are given that $\boldsymbol{CAB} = \boldsymbol{DAB}$.

Proof of (ii): This will follow using the result on row spaces in part (ii) of Theorem 5.3. To be precise, $\rho(\boldsymbol{AB}) = \rho(\boldsymbol{B})$ implies that $\mathcal{R}(\boldsymbol{AB}) = \mathcal{R}(\boldsymbol{B})$ and, therefore, there exists a matrix $\boldsymbol{Y}$ such that $\boldsymbol{B} = \boldsymbol{YAB}$. Post-multiplying both sides by $\boldsymbol{C}$ and $\boldsymbol{D}$ yield $\boldsymbol{BC} = \boldsymbol{YABC}$ and $\boldsymbol{BD} = \boldsymbol{YABD}$, respectively. This implies $\boldsymbol{BC} = \boldsymbol{BD}$ because

$$\boldsymbol{BC} = \boldsymbol{YABC} = \boldsymbol{YABD} = \boldsymbol{BD} ,$$

where the second equality follows because we are given that $\boldsymbol{ABC} = \boldsymbol{ABD}$. □

The following theorem is a generalization of the Rank-Nullity Theorem and gives a very fundamental result regarding the rank of a product of two matrices. Observe the similarity in the proof below and that for Theorem 5.1.

Theorem 5.4 *Let $\boldsymbol{A}$ be an $m \times n$ matrix and $\boldsymbol{B}$ an $n \times p$ matrix. Then,*

$$\rho(\boldsymbol{AB}) = \rho(\boldsymbol{B}) - \dim(\mathcal{N}(\boldsymbol{A}) \cap \mathcal{C}(\boldsymbol{B})) .$$

Proof. Note that $\mathcal{N}(\boldsymbol{A})$ and $\mathcal{C}(\boldsymbol{A})$ are both subspaces in $\Re^n$, so their intersection is well-defined. Since the intersection of two subspaces is a subspace (Lemma 4.2), $\mathcal{N}(\boldsymbol{A}) \cap \mathcal{C}(\boldsymbol{B})$ is a subspace in $\Re^n$. Let $\dim(\mathcal{N}(\boldsymbol{A}) \cap \mathcal{C}(\boldsymbol{B})) = s$ and let $\boldsymbol{X}$ be an $n \times s$ matrix whose column vectors are a basis of $\mathcal{N}(\boldsymbol{A}) \cap \mathcal{C}(\boldsymbol{B})$. Therefore,

$$\mathcal{C}(\boldsymbol{X}) = \mathcal{N}(\boldsymbol{A}) \cap \mathcal{C}(\boldsymbol{B}) \subseteq \mathcal{C}(\boldsymbol{B}) ,$$

which implies (Theorem 4.22) that $s \leq \dim(\mathcal{C}(\boldsymbol{B})) = \rho(\boldsymbol{B})$ and we can write $\rho(\boldsymbol{B}) = s + t$ for some integer $t \geq 0$. Also, $\mathcal{C}(\boldsymbol{X}) \subseteq \mathcal{N}(\boldsymbol{A}))$ and so $\boldsymbol{AX} = \boldsymbol{O}$.

Let us first consider the case $t = 0$. This means that $\dim(\mathcal{C}(\boldsymbol{B})) = \rho(\boldsymbol{B}) = s = \dim(\mathcal{C}(\boldsymbol{X}))$ and, since $\mathcal{C}(\boldsymbol{X}) \subseteq \mathcal{C}(\boldsymbol{B})$, we obtain $\mathcal{C}(\boldsymbol{X}) = \mathcal{C}(\boldsymbol{B})$ (part (ii) of Theorem 4.22). Therefore, the columns of $\boldsymbol{X}$ form a basis for $\mathcal{C}(\boldsymbol{B})$ and we can write $\boldsymbol{B} = \boldsymbol{XC}$ for some $s \times p$ matrix $\boldsymbol{C}$. Therefore, $\boldsymbol{AB} = \boldsymbol{AXC} = \boldsymbol{O}$, which implies that $\mathcal{C}(\boldsymbol{B}) \subseteq \mathcal{N}(\boldsymbol{A})$ and so $\mathcal{N}(\boldsymbol{A}) \cap \mathcal{C}(\boldsymbol{B}) = \mathcal{C}(\boldsymbol{B})$. Therefore,

$$\rho(\boldsymbol{AB}) = 0 = \rho(\boldsymbol{B}) - \rho(\boldsymbol{B}) = \rho(\boldsymbol{B}) - \dim(\mathcal{C}(\boldsymbol{B})) = \rho(\boldsymbol{B}) - \dim(\mathcal{N}(\boldsymbol{A}) \cap \mathcal{C}(\boldsymbol{B})) .$$

This shows that the theorem holds for $t = 0$.

Now assume that $t \geq 1$. The columns of $\boldsymbol{X}$ form a linearly independent set of s vectors in $\mathcal{C}(\boldsymbol{B})$ and, therefore, can be extended to a basis for $\mathcal{C}(\boldsymbol{B})$ (part (ii) of Lemma 4.21). Placing these extension vectors as columns of an $n \times t$ matrix $\boldsymbol{Y}$, we can construct an $n \times (s + t)$ matrix $\boldsymbol{Z} = [\boldsymbol{X} : \boldsymbol{Y}]$ whose columns form a basis for $\mathcal{C}(\boldsymbol{B})$. This means that the columns of $\boldsymbol{Z}$ are linearly independent.

We will prove that the columns of $\boldsymbol{AY}$ form a basis for $\mathcal{C}(\boldsymbol{AB})$.

We first prove that the columns of $\boldsymbol{AY}$ span $\mathcal{C}(\boldsymbol{AB})$. Observe that every column of $\boldsymbol{B}$ can be written as a linear combination of the columns of $\boldsymbol{Z}$ and, hence, we can write $\boldsymbol{B} = \boldsymbol{ZD}$ for some $(s + t) \times (s + t)$ matrix $\boldsymbol{D}$. Writing $\boldsymbol{D} = \begin{bmatrix} \boldsymbol{D}_1 \\ \boldsymbol{D}_2 \end{bmatrix}$ so that

$\boldsymbol{D}_1$ has s rows, we obtain

$$\boldsymbol{AB} = \boldsymbol{AZD} = \boldsymbol{A}[\boldsymbol{X} : \boldsymbol{Y}] \begin{bmatrix} \boldsymbol{D}_1 \\ \boldsymbol{D}_2 \end{bmatrix} = \boldsymbol{AXD}_1 + \boldsymbol{AYD}_2 = \boldsymbol{AYD}_2 \, .$$

This proves that $\mathcal{C}(\boldsymbol{AB}) \subseteq \mathcal{C}(\boldsymbol{AY})$, and so the columns of $\boldsymbol{AY}$ span $\mathcal{C}(\boldsymbol{AB})$.

It remains to prove that the columns of $\boldsymbol{AY}$ are linearly independent. Suppose that $\boldsymbol{AYv} = \boldsymbol{0}$ for some vector $\boldsymbol{v} \in \Re^t$. This means that $\boldsymbol{Yv} \in \mathcal{N}(\boldsymbol{A})$. Also, $\boldsymbol{Yv} \in \mathcal{C}(\boldsymbol{B})$ because the columns of $\boldsymbol{Y}$ are a part of a basis for $\mathcal{C}(\boldsymbol{B})$. Therefore, $\boldsymbol{Yv} \in \mathcal{N}(\boldsymbol{A}) \cap \mathcal{C}(\boldsymbol{B}) = \mathcal{C}(\boldsymbol{X})$ and there exists a vector $\boldsymbol{u} \in \Re^s$ such that $\boldsymbol{Yv} = \boldsymbol{Xu}$. Now we can argue:

$$\boldsymbol{Yv} = \boldsymbol{Xu} \Longrightarrow \boldsymbol{Xu} + \boldsymbol{Y}(-\boldsymbol{v}) = \boldsymbol{0} \Longrightarrow [\boldsymbol{X} : \boldsymbol{Y}] \begin{bmatrix} \boldsymbol{u} \\ -\boldsymbol{v} \end{bmatrix} = \boldsymbol{0} \Longrightarrow \boldsymbol{u} = \boldsymbol{v} = \boldsymbol{0} \, ,$$

where the last implication follows from the fact that the columns of $\boldsymbol{Z} = [\boldsymbol{X} : \boldsymbol{Y}]$ are linearly independent. This shows that the columns of $\boldsymbol{AY}$ are linearly independent.

From the above, the columns of $\boldsymbol{AY}$ form a basis for $\mathcal{C}(\boldsymbol{AB})$ and so $\dim(\mathcal{C}(\boldsymbol{AB})) = t$. This yields

$$\rho(\boldsymbol{AB}) = \dim(\mathcal{C}(\boldsymbol{AB})) = t = (s+t) - s = \rho(\boldsymbol{B}) - \dim(\mathcal{N}(\boldsymbol{A}) \cap \mathcal{C}(\boldsymbol{B})) \, ,$$

which concludes the proof of the theorem. □

The above proof produces the Rank-Nullity Theorem as a special case when $p = n$ and $\boldsymbol{B} = \boldsymbol{I}_n$. The theorem not only affirms that $\rho(\boldsymbol{AB}) \leq \rho(\boldsymbol{B})$, which we already proved in Corollary 5.4, but also expresses the difference between $\rho(\boldsymbol{B})$ and $\rho(\boldsymbol{AB})$ as the dimension of the subspace $\mathcal{N}(\boldsymbol{A}) \cap \mathcal{C}(\boldsymbol{B})$.

Corollary 5.4 yields an upper bound for the rank of $\boldsymbol{AB}$. Theorem 5.4 can be used to derive the following lower bound for $\rho(\boldsymbol{AB})$.

Corollary 5.6 *Let $\boldsymbol{A}$ be an $m \times n$ matrix and $\boldsymbol{B}$ an $n \times p$ matrix. Then,*

$$\rho(\boldsymbol{AB}) \geq \rho(\boldsymbol{A}) + \rho(\boldsymbol{B}) - n \, .$$

Proof. Note that

$$\dim(\mathcal{N}(\boldsymbol{A}) \cap \mathcal{C}(\boldsymbol{B})) \leq \dim(\mathcal{N}(\boldsymbol{A})) = n - \rho(\boldsymbol{A}) \, .$$

Therefore, using Theorem 5.4 we find

$$\rho(\boldsymbol{AB}) = \rho(\boldsymbol{B}) - \dim(\mathcal{N}(\boldsymbol{A}) \cap \mathcal{C}(\boldsymbol{B})) \geq \rho(\boldsymbol{B}) + \rho(\boldsymbol{A}) - n$$

and the lower bound is proved. □

The inequality in Corollary 5.6 is known as ***Sylvester's inequality***. Sylvester's inequality can also be expressed in terms of nullities using the Rank-Nullity Theorem. Note that

$$\begin{aligned} &\rho(\boldsymbol{AB}) \geq \rho(\boldsymbol{B}) + \rho(\boldsymbol{A}) - n = \rho(\boldsymbol{B}) - (n - \rho(\boldsymbol{A})) = \rho(\boldsymbol{B}) - \nu(\boldsymbol{A}) \\ &\Longrightarrow \rho(\boldsymbol{AB}) - p \geq \rho(\boldsymbol{B}) - p - \nu(\boldsymbol{A}) \Longrightarrow -\nu(\boldsymbol{AB}) \geq -\nu(\boldsymbol{B}) - \nu(\boldsymbol{A}) \\ &\Longrightarrow \nu(\boldsymbol{AB}) \leq \nu(\boldsymbol{A}) + \nu(\boldsymbol{B}) \, . \end{aligned}$$

Sylvester's inequality is easily generalized to several matrices as below:

$$\nu(\boldsymbol{A}_1\boldsymbol{A}_2\cdots\boldsymbol{A}_k) \leq \nu(\boldsymbol{A}_1) + \nu(\boldsymbol{A}_2) + \cdots \nu(\boldsymbol{A}_k) . \tag{5.4}$$

The Rank-Nullity Theorem is an extremely important theorem in linear algebra and helps in deriving a number of other results as we have already seen. We illustrate by deriving the second law in the rank-cancellation laws in Lemma 5.4. Let $\boldsymbol{A}$ and $\boldsymbol{B}$ be two matrices such that $\boldsymbol{AB}$ is well-defined. Assume that $\rho(\boldsymbol{AB}) = \rho(\boldsymbol{B})$. Since $\boldsymbol{AB}$ and $\boldsymbol{B}$ have the same number of columns, the Rank-Nullity Theorem implies that $\nu(\boldsymbol{AB}) = \nu(\boldsymbol{B})$. This means that $\mathcal{N}(\boldsymbol{AB}) = \mathcal{N}(\boldsymbol{B})$ because $\mathcal{N}(\boldsymbol{B}) \subseteq \mathcal{N}(\boldsymbol{AB})$ and they have the same dimension. Hence, $\boldsymbol{ABX} = \boldsymbol{O}$ if and only if $\boldsymbol{BX} = \boldsymbol{O}$. In particular, if $\boldsymbol{ABX} = \boldsymbol{O}$, we can cancel $\boldsymbol{A}$ from the left and obtain $\boldsymbol{BX} = \boldsymbol{O}$. Taking $\boldsymbol{X} = \boldsymbol{C} - \boldsymbol{D}$ yields the second law in Lemma 5.4. The first law in Lemma 5.4 can be proved by applying a similar argument to transposes.

5.2 Bases for the four fundamental subspaces

In this section, we see how a row echelon form for $\boldsymbol{A}$ reveals the basis vectors for the four fundamental subspaces. Let $\boldsymbol{A}$ be an $m \times n$ matrix of rank r and let $\boldsymbol{G}$ be the product of elementary matrices representing the elementary row operations that reduce $\boldsymbol{A}$ to an echelon matrix $\boldsymbol{U}$. Therefore, $\boldsymbol{GA} = \boldsymbol{U}$.

Since $\boldsymbol{G}$ is nonsingular, we have that $\rho(\boldsymbol{A}) = \rho(\boldsymbol{U})$ and $\mathcal{R}(\boldsymbol{A}) = \mathcal{R}(\boldsymbol{U})$ (recall part (ii) of Lemma 5.3). Therefore, the entire set of rows of $\boldsymbol{U}$ span $\mathcal{R}(\boldsymbol{A})$, but clearly the zero rows do not add anything, so the r nonzero rows of $\boldsymbol{U}$ form a spanning set for $\mathcal{R}(\boldsymbol{A})$. Also, the r nonzero rows of $\boldsymbol{U}$ are linearly independent (recall Example 4.8) and, hence, they form a basis of $\mathcal{R}(\boldsymbol{A})$.

To find a basis for the column space of $\boldsymbol{A}$, first note that the entire set of columns of $\boldsymbol{A}$ spans $\mathcal{C}(\boldsymbol{A})$, but they will not form a basis when linear dependencies exist among them. Corollary 4.1 insures that the column vectors in $\boldsymbol{A}$ and $\boldsymbol{U}$ that form maximal linearly independent sets must occupy the same positions in the respective matrices. Example 4.8 showed that the basic or pivot columns of $\boldsymbol{U}$ form a maximal linearly independent set. Therefore, the corresponding positions in $\boldsymbol{A}$ (i.e., the basic columns of $\boldsymbol{A}$) will form a basis of $\mathcal{C}(\boldsymbol{A})$. ***Note:*** *The column space of* $\boldsymbol{A}$ *and* $\boldsymbol{U}$ *need not be the same.*

We next turn to the null space $\mathcal{N}(\boldsymbol{A})$, which is the subspace formed by the solutions of the homogeneous linear system $\boldsymbol{Ax} = \boldsymbol{0}$. In Section 2.5 we studied such systems and found the general solution to be a linear combination of particular solutions. Indeed, the vectors $\mathcal{H} = \boldsymbol{h}_1, \boldsymbol{h}_2, \ldots, \boldsymbol{h}_{n-r}$ in (2.12) span $\mathcal{N}(\boldsymbol{A})$. These vectors are easily obtained from the row echelon form $\boldsymbol{U}$ by using back substitution to solve for the r basic variables in terms of the $n-r$ free variables. It is also easy to demonstrate that $\mathcal{H}$ is a linearly independent set and, therefore, form a basis of $\mathcal{N}(\boldsymbol{A})$. In fact, rather than explicitly demonstrating the linear independence of $\mathcal{H}$, we can argue as follows: since the dimension of $\mathcal{N}(\boldsymbol{A})$ is $n-r$ (Theorem 5.1) and $\mathcal{H}$ is a set of $n-r$ vectors that spans $\mathcal{N}(\boldsymbol{A})$, it follows that $\mathcal{H}$ is a basis of $\mathcal{N}(\boldsymbol{A})$.

Finally, we consider the left null space $\mathcal{N}(\boldsymbol{A}')$. One could of course apply the preceding argument to the homogeneous system $\boldsymbol{A}'\boldsymbol{x} = \boldsymbol{0}$ to derive a basis, but it would be more attractive if we could obtain a basis from the echelon form $\boldsymbol{U}$ itself and not repeat a row echelon reduction for $\boldsymbol{A}'$. The answer lies in the $\boldsymbol{G}$ matrix. We write

$$\begin{bmatrix} \boldsymbol{G}_1 \\ \boldsymbol{G}_2 \end{bmatrix} \boldsymbol{A} = \boldsymbol{G}\boldsymbol{A} = \boldsymbol{U} = \begin{bmatrix} \boldsymbol{U}_1 \\ \boldsymbol{O} \end{bmatrix} ,$$

where $\boldsymbol{G}_1$ is $r \times m$, $\boldsymbol{G}_2$ is $(m-r) \times m$ and $\boldsymbol{U}_1$ is $(m-r) \times n$. This implies that $\boldsymbol{G}_2\boldsymbol{A} = \boldsymbol{O}$. We claim that $\mathcal{R}(\boldsymbol{G}_2) = \mathcal{N}(\boldsymbol{A}')$. To see why this is true, observe that

$$\begin{aligned} &\boldsymbol{y}' \in \mathcal{R}(\boldsymbol{G}_2) = \mathcal{C}(\boldsymbol{G}_2') \Longrightarrow \boldsymbol{y} = \boldsymbol{G}_2'\boldsymbol{x} \\ &\Longrightarrow \boldsymbol{A}'\boldsymbol{y} = \boldsymbol{A}'\boldsymbol{G}_2'\boldsymbol{x} = (\boldsymbol{G}_2\boldsymbol{A})'\boldsymbol{x} = \boldsymbol{0} \Longrightarrow \boldsymbol{y} \in \mathcal{N}(\boldsymbol{A}') . \end{aligned}$$

This shows that $\mathcal{R}(\boldsymbol{G}_2) \subseteq \mathcal{N}(\boldsymbol{A}')$. To prove the inclusion in the other direction, we make use of a few elementary observations. We write $\boldsymbol{G}^{-1} = [\boldsymbol{H}_1 : \boldsymbol{H}_2]$, where $\boldsymbol{H}_1$ is $(m \times r)$, so that $\boldsymbol{A} = \boldsymbol{G}^{-1}\boldsymbol{U} = \boldsymbol{H}_1\boldsymbol{U}_1$. Also, since $\boldsymbol{G}\boldsymbol{G}^{-1} = \boldsymbol{G}^{-1}\boldsymbol{G} = \boldsymbol{I}_m$, it follows that $\boldsymbol{G}_1\boldsymbol{H}_1 = \boldsymbol{I}_r$ and $\boldsymbol{H}_1\boldsymbol{G}_1 = (\boldsymbol{I}_m - \boldsymbol{H}_2\boldsymbol{G}_2)$. We can now conclude that

$$\begin{aligned} &\boldsymbol{y} \in \mathcal{N}(\boldsymbol{A}') \Longrightarrow \boldsymbol{A}'\boldsymbol{y} = \boldsymbol{0} \Longrightarrow \boldsymbol{U}_1'\boldsymbol{H}_1'\boldsymbol{y} = \boldsymbol{0} \Longrightarrow \boldsymbol{H}_1'\boldsymbol{y} = \boldsymbol{0} \text{ since } \mathcal{N}(\boldsymbol{U}_1') = \{\boldsymbol{0}\} \\ &\Longrightarrow \boldsymbol{G}_1'\boldsymbol{H}_1'\boldsymbol{y} = \boldsymbol{0} \Longrightarrow (\boldsymbol{H}_1\boldsymbol{G}_1)'\boldsymbol{y} = \boldsymbol{0} \Longrightarrow (\boldsymbol{I}_m - \boldsymbol{H}_2\boldsymbol{G}_2)'\boldsymbol{y} = \boldsymbol{0} \\ &\Longrightarrow \boldsymbol{y} = \boldsymbol{G}_2'\boldsymbol{H}_2'\boldsymbol{y} \Longrightarrow \boldsymbol{y} \in \mathcal{C}(\boldsymbol{G}_2') = \mathcal{R}(\boldsymbol{G}_2) . \end{aligned}$$

Note: In the above argument we used the fact that $\mathcal{N}(\boldsymbol{U}_1') = \{\boldsymbol{0}\}$. This follows from $\nu(\boldsymbol{U}_1') = r - \rho(\boldsymbol{U}_1') = r - r = 0$.

Based upon the above argument, we conclude that $\mathcal{R}(\boldsymbol{G}_2) = \mathcal{N}(\boldsymbol{A}')$. Therefore, the $(m-r)$ rows of $\boldsymbol{G}_2$ span $\mathcal{N}(\boldsymbol{A}')$ and, since $\dim(\mathcal{N}(\boldsymbol{A}')) = \nu(\boldsymbol{A}') = m - r$, the rows of $\boldsymbol{G}_2$ form a basis for $\mathcal{N}(\boldsymbol{A}')$.

Example 5.1 Finding bases for the four fundamental subspaces of a matrix. Let us demonstrate the basis vectors corresponding to the four fundamental subspaces of the matrix:

$$\boldsymbol{A} = \begin{bmatrix} 2 & -2 & -5 & -3 & -1 & 2 \\ 2 & -1 & -3 & 2 & 3 & 2 \\ 4 & -1 & -4 & 10 & 11 & 4 \\ 0 & 1 & 2 & 5 & 4 & 0 \end{bmatrix} .$$

In Example 2.2 the row echelon form of $\boldsymbol{A}$ was found to be

$$\boldsymbol{U} = \begin{bmatrix} 2 & -2 & -5 & -3 & -1 & 2 \\ 0 & 1 & 2 & 5 & 4 & 0 \\ 0 & 0 & 0 & 1 & 1 & 0 \\ 0 & 0 & 0 & 0 & 0 & 0 \end{bmatrix} .$$

The first three nonzero row vectors of $\boldsymbol{U}$ form a basis for the row space of $\boldsymbol{U}$, which is the same as the row space of $\boldsymbol{A}$. The first, second and fourth columns are the pivot columns of $\boldsymbol{U}$. Therefore, the first, second and fourth columns of $\boldsymbol{A}$ are the basic columns of $\boldsymbol{A}$ and form a basis for the columns space of $\boldsymbol{A}$. Therefore,

$$\left\{ \begin{bmatrix} 2 \\ 2 \\ 4 \\ 0 \end{bmatrix}, \begin{bmatrix} -2 \\ -1 \\ -1 \\ 1 \end{bmatrix}, \begin{bmatrix} -3 \\ 2 \\ 10 \\ 5 \end{bmatrix} \right\}$$

form a basis for the column space of $\boldsymbol{A}$.

Note that the rank of $\boldsymbol{A}$ equals the rank of $\boldsymbol{U}$ and is easily counted as the number of pivots in $\boldsymbol{U}$.

A basis for the null space of $\boldsymbol{A}$ will contain $n - \rho(\boldsymbol{A}) = 6 - 3 = 3$ vectors and is given by any set of linearly independent vectors that are particular solutions of $\boldsymbol{Ax} = \boldsymbol{0}$. In Example 2.6 we analyzed the homogeneous system $\boldsymbol{Ax} = \boldsymbol{0}$ and found three particular solutions $\boldsymbol{h}_1$, $\boldsymbol{h}_2$ and $\boldsymbol{h}_3$. These form a basis of the null space of $\boldsymbol{A}$. Reading them off Example 2.6, we find

$$\mathcal{H} = \left\{ \begin{bmatrix} 1/2 \\ -2 \\ 1 \\ 0 \\ 0 \\ 0 \end{bmatrix}, \begin{bmatrix} 0 \\ 1 \\ 0 \\ -1 \\ 1 \\ 0 \end{bmatrix}, \begin{bmatrix} -1 \\ 0 \\ 0 \\ 0 \\ 0 \\ 1 \end{bmatrix} \right\}$$

form a basis of $\mathcal{N}(\boldsymbol{A})$.

Finally we turn to the left null space of $\boldsymbol{A}$, i.e., the null space of $\boldsymbol{A}'$. The dimension of this subspace is $m - \rho(\boldsymbol{A}) = 4 - 3 = 1$. From the discussions preceding this example, we know that the last row vector in $\boldsymbol{G}$ will form a basis for the left null space, where $\boldsymbol{GA} = \boldsymbol{U}$. Therefore, we need to find $\boldsymbol{G}$.

Let us take a closer look at Example 2.2 to see how we arrived at $\boldsymbol{U}$. We first sweep out the elements of $\boldsymbol{A}$ below the pivot in the first column by using the elementary row operations $E_{21}(-1)$ followed by $E_{31}(-2)$. Next we sweep out the elements below the pivot in the second column using $E_{32}(-3)$ followed by $E_{42}(-1)$. These resulted in the echelon form. In fact, recall from Section 3.1 that elementary row operations on $\boldsymbol{A}$ are equivalent to multiplying $\boldsymbol{A}$ on the left by lower-triangular matrices (see (3.1)). To be precise, we construct

$$\boldsymbol{L}_1 = \begin{bmatrix} 1 & 0 & 0 & 0 \\ -1 & 1 & 0 & 0 \\ -2 & 0 & 1 & 0 \\ 0 & 0 & 0 & 1 \end{bmatrix} \quad \text{and} \quad \boldsymbol{L}_2 = \begin{bmatrix} 1 & 0 & 0 & 0 \\ 0 & 1 & 0 & 0 \\ 0 & -3 & 1 & 0 \\ 0 & -1 & 0 & 1 \end{bmatrix},$$

where $\boldsymbol{L}_1$ and $\boldsymbol{L}_2$ sweep out the elements below the pivots in the first and second columns, respectively. Writing $\boldsymbol{G} = \boldsymbol{L}_1\boldsymbol{L}_2$, we have

$$\boldsymbol{G} = \begin{bmatrix} 1 & 0 & 0 & 0 \\ -1 & 1 & 0 & 0 \\ 1 & -3 & 1 & 0 \\ 1 & 1 & 0 & 1 \end{bmatrix}.$$

Therefore, the last row of G forms a basis for the left null space of A. ■

5.3 Rank and inverse

Basic properties of the inverse of (square) nonsingular matrices were discussed in Section 2.6. Here we extend the concept to *left* and *right* inverses for rectangular matrices. The inverse of a nonsingular matrix, as defined in Definition 2.8, will emerge as a special case.

Definition 5.2 Left and right inverses. *A* **left inverse** *of an* $m \times n$ *matrix* A *is any* $n \times m$ *matrix* B *such that* $BA = I_n$. *A* **right inverse** *of an* $m \times n$ *matrix* A *is any* $n \times m$ *matrix* C *such that* $AC = I_m$.

The rank of a matrix provides useful characterizations regarding when it has a left or right inverse.

Theorem 5.5 *Let* A *be an* $m \times n$ *matrix. Then the following statements are true:*

(i) A *has a right inverse if and only if* $\rho(A) = m$ *or, equivalently, if the all rows of* A *are linearly independent.*

(ii) A *has a left inverse if and only if* $\rho(A) = n$ *or, equivalently, if the all columns of* A *are linearly independent..*

Proof. Let us keep in mind that $\rho(A) \leq \min\{m, n\}$. Therefore, when $\rho(A) = m$ (as in part (i)), we must have $n \geq m$, and when $\rho(A) = n$ (as in part (ii)), we must have $m \geq n$.

Proof of (i): To prove the "if" part, note that $\rho(A) = m$ implies that there are m linearly independent column vectors in A and the dimension of $\mathcal{C}(A)$ is m. Note that $\mathcal{C}(A) \subseteq \Re^m$ and $\dim(\mathcal{C}(A)) = \rho(A) = m = \dim(\Re^m)$. Therefore, by part (ii) of Theorem 4.22, $\mathcal{C}(A) = \Re^m$.

Therefore, every vector in $\Re^m$ can be expressed as Ax for some $n \times 1$ vector x. In particular, each canonical vector e_j in $\Re^m$ can be expressed as $e_j = Ac_j$ for some $n \times 1$ vector c_j. Hence,

$$A[c_1 : c_2 : \cdots : c_m] = [e_1 : e_2 : \cdots : e_m] = I_m .$$

Therefore, $C = [c_1 : c_2 : \cdots : c_m]$ is a right inverse of A.

To prove the "only if" part, suppose that A has a right inverse C. Then, $AC = I_m = C'A'$ and we can write

$$\begin{aligned} A'x = 0 &\Longrightarrow C'A'x = x = 0 \Longrightarrow \mathcal{N}(A') = \{0\} \\ &\Longrightarrow \nu(A') = 0 \Longrightarrow \rho(A') = \rho(A) = m , \end{aligned}$$

where the last implication ($\Longrightarrow$) follows from the Rank-Nullity Theorem (Theorem 5.1). This proves the "only if" part.

Proof of (ii): This can be proved by a similar argument to the above, or simply by considering transposes. Note that $\boldsymbol{A}_L$ will be a left inverse of $\boldsymbol{A}$ if and only if $\boldsymbol{A}'_L$ is a right inverse of $\boldsymbol{A}'$: $\boldsymbol{A}_L\boldsymbol{A} = \boldsymbol{I}_n \iff \boldsymbol{A}'\boldsymbol{A}'_L = \boldsymbol{I}_m$. Therefore, $\boldsymbol{A}$ has a left inverse if and only $\boldsymbol{A}'$ has a right inverse, which happens if and only if all the rows of $\boldsymbol{A}'$ are linearly independent (from part (i)) and, since $\boldsymbol{A}'$ has n rows, if and only if $\rho(\boldsymbol{A}) = \rho(\boldsymbol{A}') = n$. $\square$

For an $m \times n$ matrix $\boldsymbol{A}$, when $\rho(\boldsymbol{A}) = m$ we say that $\boldsymbol{A}$ has ***full row rank*** and when $\rho(\boldsymbol{A}) = n$ we say that $\boldsymbol{A}$ has ***full column rank***. Theorem 5.5, in other words, tells us that every matrix with full row rank has a right inverse and every matrix with full column rank has a left inverse.

Theorem 5.5 can also be proved using Lemma 4.3. Let us consider the case where $\boldsymbol{A}$ is an $m \times n$ matrix with full column rank. Clearly $m \geq n$ and we can find an $m \times (m-n)$ matrix $\boldsymbol{B}$ such that $\boldsymbol{C} = [\boldsymbol{A} : \boldsymbol{B}]$ is an $m \times m$ nonsingular matrix (part (i) of Lemma 4.3). Therefore, $\boldsymbol{C}^{-1}$ exists and let us write it in partitioned form as

$$[\boldsymbol{A} : \boldsymbol{B}]^{-1} = \boldsymbol{C}^{-1} = \begin{bmatrix} \boldsymbol{G} \\ \boldsymbol{H} \end{bmatrix},$$

where $\boldsymbol{G}$ is $n \times m$. Then,

$$\boldsymbol{I}_m = \begin{bmatrix} \boldsymbol{I}_n & \boldsymbol{O} \\ \boldsymbol{O} & \boldsymbol{I}_{m-n} \end{bmatrix} = \boldsymbol{C}^{-1}[\boldsymbol{A} : \boldsymbol{B}] = \begin{bmatrix} \boldsymbol{G} \\ \boldsymbol{H} \end{bmatrix}[\boldsymbol{A} : \boldsymbol{B}] = \begin{bmatrix} \boldsymbol{GA} & \boldsymbol{GB} \\ \boldsymbol{HA} & \boldsymbol{HB} \end{bmatrix},$$

which implies that $\boldsymbol{GA} = \boldsymbol{I}_n$ and so $\boldsymbol{G}$ is a left inverse of $\boldsymbol{A}$. One can similarly prove part (ii) of Theorem 5.5 using part (ii) of Lemma 4.3.

What can we say when a matrix $\boldsymbol{A}$ has both a left inverse and a right inverse? The following theorem explains.

Theorem 5.6 *If a matrix $\boldsymbol{A}$ has a left inverse $\boldsymbol{B}$ and a right inverse $\boldsymbol{C}$, then the following statements are true:*

(i) $\boldsymbol{A}$ must be a square matrix and $\boldsymbol{B} = \boldsymbol{C}$.

(ii) $\boldsymbol{A}$ has a unique left inverse, a unique right inverse and a unique inverse of $\boldsymbol{A}$.

Proof. **Proof of (i)**: Suppose $\boldsymbol{A}$ is $m \times n$. From Theorem 5.5 we have that $\rho(\boldsymbol{A}) = m = n$. Therefore $\boldsymbol{A}$ is square.

Since $\boldsymbol{A}$ is $n \times n$, we have that $\boldsymbol{B}$ and $\boldsymbol{C}$ must be $n \times n$ as well. We can, therefore, write

$$\boldsymbol{B} = \boldsymbol{B}\boldsymbol{I}_n = \boldsymbol{B}(\boldsymbol{AC}) = (\boldsymbol{BA})\boldsymbol{C} = \boldsymbol{I}_n\boldsymbol{C} = \boldsymbol{C}.$$

This proves $\boldsymbol{B} = \boldsymbol{C}$.

Proof of (ii): Suppose $\boldsymbol{D}$ is another left inverse. Then, $\boldsymbol{DA} = \boldsymbol{I}_n = \boldsymbol{BA}$. Multiplying both sides by $\boldsymbol{C}$ on the right yields

$$\boldsymbol{D} = \boldsymbol{D}\boldsymbol{I}_n = \boldsymbol{D}(\boldsymbol{AC}) = (\boldsymbol{DA})\boldsymbol{C} = (\boldsymbol{BA})\boldsymbol{C} = \boldsymbol{B}(\boldsymbol{AC}) = \boldsymbol{B}\boldsymbol{I}_n = \boldsymbol{B}.$$

This proves that $\boldsymbol{D} = \boldsymbol{B}$ and the left inverse of a square matrix is unique. One can analogously prove that the right inverse of a square matrix is also equal.

Finally, note that $\boldsymbol{B} = \boldsymbol{C} = \boldsymbol{A}^{-1}$ satisfies the usual definition of the inverse of a nonsingular matrix (see Definition 2.8) and is the unique inverse of $\boldsymbol{A}$. □

The following is an immediate corollary of the preceding results.

Corollary 5.7 *An $n \times n$ matrix $\boldsymbol{A}$ is nonsingular if and only if $\rho(\boldsymbol{A}) = n$.*

Proof. Suppose $\boldsymbol{A}$ is an $n \times n$ matrix with $\rho(\boldsymbol{A}) = n$. Then, by Theorem 5.5, $\boldsymbol{A}$ has both a right and left inverse. Theorem 5.6 further tells us that the right and left inverses must be equal and is, in fact, the unique inverse. This proves the "if" part.

To prove the "only if" part, suppose $\boldsymbol{A}$ is nonsingular. Then $\boldsymbol{A}^{-1}$ exists and so

$$\boldsymbol{A}\boldsymbol{x} = \boldsymbol{0} \Rightarrow \boldsymbol{x} = \boldsymbol{A}^{-1}\boldsymbol{0} = \boldsymbol{0} \Rightarrow \mathcal{N}(\boldsymbol{A}) = \{\boldsymbol{0}\} \Rightarrow \nu(\boldsymbol{A}) = 0 \Rightarrow \rho(\boldsymbol{A}) = n \ ,$$

where the last equality follows from the Rank-Nullity Theorem (Theorem 5.1). This proves the "only if" part. □

The above corollary can also be proved without resorting to left and right inverses, but making use of Gaussian elimination as in Lemma 2.4. Recall from Lemma 2.4 that $\boldsymbol{A}$ is nonsingular if and only if $\boldsymbol{A}\boldsymbol{x} = \boldsymbol{0} \Rightarrow \boldsymbol{x} = \boldsymbol{0}$ or, equivalently, if and only if $\mathcal{N}(\boldsymbol{A}) = \{\boldsymbol{0}\}$. Therefore, $\boldsymbol{A}$ is nonsingular if and only if $\nu(\boldsymbol{A}) = 0$, which, from Theorem 5.1, is the same as $\rho(\boldsymbol{A}) = n$.

The following result is especially useful in the study of linear regression models.

Theorem 5.7 *Let $\boldsymbol{A}$ be an $m \times n$ matrix. Then*

(i) $\boldsymbol{A}$ is of full column rank (i.e., $\rho(\boldsymbol{A}) = n$) if and only if $\boldsymbol{A}'\boldsymbol{A}$ is nonsingular. Also, in this case $(\boldsymbol{A}'\boldsymbol{A})^{-1}\boldsymbol{A}'$ is a left-inverse of $\boldsymbol{A}$.

(ii) $\boldsymbol{A}$ is of full row rank (i.e., $\rho(\boldsymbol{A}) = m$) if and only if $\boldsymbol{A}\boldsymbol{A}'$ is nonsingular. Also, in this case $\boldsymbol{A}'(\boldsymbol{A}\boldsymbol{A}')^{-1}$ is a right-inverse of $\boldsymbol{A}$.

Proof. **Proof of (i):** Note that $\boldsymbol{A}'\boldsymbol{A}$ is $n \times n$. From the fundamental theorem of ranks (Theorem 5.2), we know that $\rho(\boldsymbol{A}'\boldsymbol{A}) = \rho(\boldsymbol{A})$. Therefore, $\rho(\boldsymbol{A}) = n$ if and only if $\rho(\boldsymbol{A}'\boldsymbol{A}) = n$, which is equivalent to $\boldsymbol{A}'\boldsymbol{A}$ being nonsingular (Corollary 5.7). When $\boldsymbol{A}'\boldsymbol{A}$ is nonsingular, the $n \times m$ matrix $(\boldsymbol{A}'\boldsymbol{A})^{-1}\boldsymbol{A}'$ is well-defined and is easily seen to be a left inverse of $\boldsymbol{A}$.

Proof of (ii): The proof is similar to (i) and left as an exercise. □

Consider the change of basis problem (recall Section 4.9) for subspaces of a vector space. Let $\boldsymbol{X} = [\boldsymbol{x}_1 : \boldsymbol{x}_2 : \ldots : \boldsymbol{x}_p]$ and $\boldsymbol{Y} = [\boldsymbol{y}_1 : \boldsymbol{y}_2 : \ldots : \boldsymbol{y}_p]$ be $n \times p$ matrices $(p < n)$ whose columns form bases for a p-dimensional subspace $\mathcal{S}$ of $\Re^n$. Let $\boldsymbol{u} \in \mathcal{S}$. Therefore, there exist $p \times 1$ vectors $\boldsymbol{\alpha}$ and $\boldsymbol{\beta}$ such that $\boldsymbol{u} = \boldsymbol{X}\boldsymbol{\alpha} = \boldsymbol{Y}\boldsymbol{\beta}$. Thus, $\boldsymbol{\alpha}$ and $\boldsymbol{\beta}$ are the coordinates of $\boldsymbol{u}$ in terms of the columns of $\boldsymbol{X}$ and $\boldsymbol{Y}$, respectively.

Since $\boldsymbol{\alpha}$ and $\boldsymbol{\beta}$ are coordinates of the same vector but with respect to different bases, there must be a matrix $\boldsymbol{P}$ such that $\boldsymbol{\beta} = \boldsymbol{P}\boldsymbol{\alpha}$. To find $\boldsymbol{P}$, we must solve $\boldsymbol{Y}\boldsymbol{\beta} = \boldsymbol{X}\boldsymbol{\alpha}$ for $\boldsymbol{\beta}$. Since $\boldsymbol{Y}$ has full column rank, $\boldsymbol{Y}'\boldsymbol{Y}$ is nonsingular. Therefore,

$$\boldsymbol{Y}\boldsymbol{\beta} = \boldsymbol{X}\boldsymbol{\alpha} \Longrightarrow \boldsymbol{Y}'\boldsymbol{Y}\boldsymbol{\beta} = \boldsymbol{Y}'\boldsymbol{X}\boldsymbol{\alpha} \Longrightarrow \boldsymbol{\beta} = (\boldsymbol{Y}'\boldsymbol{Y})^{-1}\boldsymbol{Y}'\boldsymbol{X}\boldsymbol{\alpha} = \boldsymbol{P}\boldsymbol{\alpha}\,,$$

where $\boldsymbol{P} = (\boldsymbol{Y}'\boldsymbol{Y})^{-1}\boldsymbol{Y}'\boldsymbol{X}$ is the change-of-basis matrix.

5.4 Rank factorization

How matrices can be "factorized" or "decomposed" constitutes a fundamentally important topic in linear algebra. For instance, from a computational standpoint, we have already seen in Chapter 2 when (and how) a square matrix admits the $\boldsymbol{LU}$ decomposition, which is essentially a factorization in terms of triangular matrices. Here we study factorizations based upon the rank of a matrix.

Definition 5.3 *Let $\boldsymbol{A}$ be an $m \times n$ matrix such that $\rho(\boldsymbol{A}) \geq 1$. A* **rank decomposition** *or* **rank factorization** *of $\boldsymbol{A}$ is a product $\boldsymbol{A} = \boldsymbol{C}\boldsymbol{R}$, where $\boldsymbol{C}$ is an $m \times r$ matrix and $\boldsymbol{R}$ is an $r \times n$ matrix.*

Every non-null matrix has a rank factorization. To see this, suppose $\boldsymbol{A}$ is an $m \times n$ matrix of rank r and let $\boldsymbol{c}_1, \boldsymbol{c}_2, \ldots, \boldsymbol{c}_r$ be a basis for $\mathcal{C}(\boldsymbol{A})$. Form the $m \times r$ matrix $\boldsymbol{C} = [\boldsymbol{c}_1 : \boldsymbol{c}_2 : \ldots : \boldsymbol{c}_r]$. Clearly $\mathcal{C}(\boldsymbol{A}) \subseteq \mathcal{C}(\boldsymbol{C})$ and, from Theorem 4.6, we know that there exists an $r \times n$ matrix $\boldsymbol{R}$ such that $\boldsymbol{A} = \boldsymbol{C}\boldsymbol{R}$.

Rank factorization leads to yet another proof of the fundamental result that the dimensions of the column space and row space of a matrix are equal. We have already seen how this arises from the fundamental theorem of ranks (see (5.2)), but here is another proof using rank factorizations.

Theorem 5.8 *Let $\boldsymbol{A}$ be an $m \times n$ matrix. Then*

$$\text{“column rank”} = \dim(\mathcal{C}(\boldsymbol{A})) = \dim(\mathcal{R}(\boldsymbol{A})) = \text{“row rank.”}$$

Proof. Let $\boldsymbol{A}$ be an $m \times n$ matrix whose "column rank" (i.e., the dimension of the column space of $\boldsymbol{A}$) is r. Then $\dim(\mathcal{C}(\boldsymbol{A})) = r$ and $\mathcal{C}(\boldsymbol{A})$ has a basis consisting of r vectors. Let $\{\boldsymbol{c}_1, \boldsymbol{c}_2, \ldots, \boldsymbol{c}_r\}$ be any such basis and let $\boldsymbol{C} = [\boldsymbol{c}_1 : \boldsymbol{c}_2 : \ldots : \boldsymbol{c}_r]$ be the $m \times r$ matrix whose columns are the basis vectors. Since every column of $\boldsymbol{A}$ is a linear combination of $\boldsymbol{c}_1, \boldsymbol{c}_2, \ldots, \boldsymbol{c}_r$, we can find an $r \times n$ matrix $\boldsymbol{R}$ such that $\boldsymbol{A} = \boldsymbol{C}\boldsymbol{R}$ (Theorem 4.6). Now, $\boldsymbol{A} = \boldsymbol{C}\boldsymbol{R}$ implies that the rows of $\boldsymbol{A}$ are linear combinations of the rows of $\boldsymbol{R}$; in other words, $\mathcal{R}(\boldsymbol{A}) \subset \mathcal{R}(\boldsymbol{R})$ (Theorem 4.7). This implies (see Theorem 4.22) that

$$\dim(\mathcal{R}(\boldsymbol{A})) \leq \dim(\mathcal{R}(\boldsymbol{R})) \leq r = \dim(\mathcal{C}(\boldsymbol{A}))\,. \tag{5.5}$$

(**Note:** $\dim(\mathcal{R}(\boldsymbol{R})) \leq r$ because $\boldsymbol{R}$ has r rows, so the dimension of the row space of

$\boldsymbol{R}$ cannot exceed r.) This proves that for any matrix $m \times n$ matrix $\boldsymbol{A}$, $\dim(\mathcal{R}(\boldsymbol{A})) \leq \dim(\mathcal{C}(\boldsymbol{A}))$. Now apply the above result to $\boldsymbol{A}'$. This yields

$$\dim(\mathcal{C}(\boldsymbol{A})) = \dim(\mathcal{R}(\boldsymbol{A}')) \leq \dim(\mathcal{C}(\boldsymbol{A}')) = \dim(\mathcal{R}(\boldsymbol{A})) . \tag{5.6}$$

Combining (5.5) and (5.6) yields the equality $\dim(\mathcal{C}(\boldsymbol{A})) = \dim(\mathcal{R}(\boldsymbol{A}))$. The dimension of the row space of $\boldsymbol{A}$ is called the ***row rank*** of $\boldsymbol{A}$ and we have just proved that the column rank and row rank of a matrix are equal. □

Since $\dim(\mathcal{R}(\boldsymbol{A})) = \rho(\boldsymbol{A}')$, the above theorem provides another proof of the fact that $\rho(\boldsymbol{A}) = \rho(\boldsymbol{A}')$. In fact, a direct argument is also easy:

$$\rho(\boldsymbol{A}') = \rho(\boldsymbol{R}'\boldsymbol{C}') \leq \rho(\boldsymbol{R}') \leq r = \rho(\boldsymbol{A}) = \rho((\boldsymbol{A}')') \leq \rho(\boldsymbol{A}') ,$$

where the first "$\leq$" follows because $\boldsymbol{R}'$ has r columns and the last "$\leq$" follows from applying $\rho(\boldsymbol{A}') \leq \rho(\boldsymbol{A})$ to $(\boldsymbol{A}')'$. This "sandwiches" the $\rho(\boldsymbol{A})$ between $\rho(\boldsymbol{A}')$'s and completes perhaps the shortest proof that $\rho(\boldsymbol{A}) = \rho(\boldsymbol{A}')$.

If an $m \times n$ matrix $\boldsymbol{A}$ can be factorized into $\boldsymbol{A} = \boldsymbol{U}\boldsymbol{V}$, where $\boldsymbol{U}$ is $m \times k$ and $\boldsymbol{V}$ is $k \times n$, then $k \geq \rho(\boldsymbol{A})$. This is because $\mathcal{C}(\boldsymbol{A}) \subseteq \mathcal{C}(\boldsymbol{U})$ and so

$$\rho(\boldsymbol{A}) = \dim(\mathcal{C}(\boldsymbol{A})) \leq \dim(\mathcal{C}(\boldsymbol{U})) \leq k .$$

Thus, the rank of an $m \times n$ matrix $\boldsymbol{A}$ is the *smallest* integer for which we can find an $m \times r$ matrix $\boldsymbol{C}$ and an $r \times n$ matrix $\boldsymbol{R}$ such that $\boldsymbol{A} = \boldsymbol{C}\boldsymbol{R}$. Sometimes it is convenient to use this as the definition of the rank and is referred to as ***decomposition rank***.

The following corollary tells us that if $\boldsymbol{A} = \boldsymbol{C}\boldsymbol{R}$ is a rank factorization then the ranks of $\boldsymbol{C}$ and $\boldsymbol{R}$ must equal that of $\boldsymbol{A}$.

Corollary 5.8 *Let $\boldsymbol{A}$ be an $m \times n$ matrix with $\rho(\boldsymbol{A}) = r$. If $\boldsymbol{A} = \boldsymbol{C}\boldsymbol{R}$, where $\boldsymbol{C}$ is $m \times r$ and $\boldsymbol{R}$ is $r \times n$, then $\rho(\boldsymbol{C}) = \rho(\boldsymbol{R}) = r$.*

Proof. Since $\mathcal{C}(\boldsymbol{A}) \subseteq \mathcal{C}(\boldsymbol{C})$, we have that $r = \rho(\boldsymbol{A}) \leq \rho(\boldsymbol{C})$. On the other hand, since $\boldsymbol{C}$ is $n \times r$, we have that $\rho(\boldsymbol{C}) \leq \min\{n, r\} \leq r$. Therefore, $\rho(\boldsymbol{C}) = r$.

Similarly, since $\mathcal{R}(\boldsymbol{A}) \subseteq \mathcal{R}(\boldsymbol{R})$ and since the dimensions of the row space and column space are equal (Theorem 5.8), we can write

$$r = \rho(\boldsymbol{A}) = \dim(\mathcal{C}(\boldsymbol{A})) = \dim(\mathcal{R}(\boldsymbol{A})) \leq \dim(\mathcal{R}(\boldsymbol{R})) = \dim(\mathcal{C}(\boldsymbol{R})) = \rho(\boldsymbol{R}) ,$$

which proves $r \leq \rho(\boldsymbol{R})$. Also, since $\boldsymbol{R}$ is $r \times n$, we have $\rho(\boldsymbol{R}) \leq \min\{r, n\} \leq r$. Therefore, $\rho(\boldsymbol{R}) = r$ and the proof is complete. □

The following corollary describes the column space, row space and null space of $\boldsymbol{A}$ in terms of its rank factorization.

Corollary 5.9 *Let $\boldsymbol{A} = \boldsymbol{C}\boldsymbol{R}$ be a rank factorization, where $\boldsymbol{A}$ is an $m \times n$ matrix with $\rho(\boldsymbol{A}) = r$. Then, $\mathcal{C}(\boldsymbol{A}) = \mathcal{C}(\boldsymbol{C})$, $\mathcal{R}(\boldsymbol{A}) = \mathcal{R}(\boldsymbol{R})$ and $\mathcal{N}(\boldsymbol{A}) = \mathcal{N}(\boldsymbol{R})$.*

Proof. Clearly $\mathcal{C}(\boldsymbol{A}) \subseteq \mathcal{C}(\boldsymbol{C})$ and $\mathcal{R}(\boldsymbol{A}) \subset \mathcal{R}(\boldsymbol{R})$. From the preceding lemma, $\rho(\boldsymbol{A}) = \rho(\boldsymbol{C}) = \rho(\boldsymbol{R})$. Therefore, Lemma 5.3 ensures $\mathcal{C}(\boldsymbol{A}) = \mathcal{C}(\boldsymbol{C})$ and $\mathcal{R}(\boldsymbol{A}) = \mathcal{R}(\boldsymbol{R})$.

Clearly $\mathcal{N}(\boldsymbol{R}) \subseteq \mathcal{N}(\boldsymbol{A})$. To prove the reverse inclusion, first of all note that all the columns of $\boldsymbol{C}$ are linearly independent, which ensures the existence of an $r \times m$ left inverse $\boldsymbol{C}_L$, such that $\boldsymbol{C}_L\boldsymbol{C} = \boldsymbol{I}_r$ (Theorem 5.5). Now, we can write

$$\boldsymbol{x} \in \mathcal{N}(\boldsymbol{A}) \Rightarrow \boldsymbol{CRx} = \boldsymbol{Ax} = \boldsymbol{0} \Rightarrow \boldsymbol{C}_L\boldsymbol{CRx} = \boldsymbol{0} \Rightarrow \boldsymbol{Rx} = \boldsymbol{0} \Rightarrow \boldsymbol{x} \in \mathcal{N}(\boldsymbol{R})\,.$$

This proves that $\mathcal{N}(\boldsymbol{A}) \subseteq \mathcal{N}(\boldsymbol{R})$ and, hence $\mathcal{N}(\boldsymbol{A}) = \mathcal{N}(\boldsymbol{R})$. □

Corollary 5.10 *Let $\boldsymbol{A} = \boldsymbol{CR}$ be a rank factorization for an $m \times n$ matrix with $\rho(\boldsymbol{A}) = r$. Then, the columns of $\boldsymbol{C}$ and rows of $\boldsymbol{R}$ constitute basis vectors for $\mathcal{C}(\boldsymbol{A})$ and $\mathcal{R}(\boldsymbol{A})$, respectively.*

Proof. This is straightforward from what we already know. The preceding corollary has shown that the columns of $\boldsymbol{C}$ span the column space of $\boldsymbol{A}$ and the rows of $\boldsymbol{R}$ span the row space of $\boldsymbol{R}$. Also, Corollary 5.8 ensures that $\boldsymbol{C}$ has full column rank and $\boldsymbol{R}$ has full row rank, which means that the columns of $\boldsymbol{C}$ are linearly independent as are the rows of $\boldsymbol{R}$. □

For a more constructive demonstration of why every matrix yields a rank factorization, one can proceed by simply permuting the rows and columns of the matrix to produce a nice form that will easily yield a rank factorization.

Theorem 5.9 *Let $\boldsymbol{A}$ be an $m \times n$ matrix with rank $r \geq 1$. Then there exist permutation matrices $\boldsymbol{P}$ and $\boldsymbol{Q}$, of order $m \times m$ and $n \times n$, respectively, such that*

$$\boldsymbol{PAQ} = \begin{bmatrix} \boldsymbol{B} & \boldsymbol{BC} \\ \boldsymbol{DB} & \boldsymbol{DBC} \end{bmatrix},$$

where $\boldsymbol{B}$ is an $r \times r$ nonsingular matrix, $\boldsymbol{C}$ is some $r \times (n - r)$ matrix and $\boldsymbol{D}$ is some $(m - r) \times r$ matrix.

Proof. We can permute the columns of $\boldsymbol{A}$ to bring the r linearly independent columns of $\boldsymbol{A}$ into the first r position. Therefore, there exists an $n \times n$ permutation matrix $\boldsymbol{Q}$ such that $\boldsymbol{AQ} = [\boldsymbol{F} : \boldsymbol{G}]$, where $\boldsymbol{F}$ is an $m \times r$ matrix whose r columns are linearly independent. So the $n - r$ columns of $\boldsymbol{G}$ are linear combinations of those of $\boldsymbol{F}$ and we can write $\boldsymbol{G} = \boldsymbol{FC}$ for some $r \times (n - r)$ matrix $\boldsymbol{C}$. Hence, $\boldsymbol{AQ} = [\boldsymbol{F} : \boldsymbol{FC}]$.

Now consider the $m \times r$ matrix $\boldsymbol{F}$. Since $\rho(\boldsymbol{F}) = r$. there are r linearly independent rows in $\boldsymbol{F}$. Use a permutation matrix $\boldsymbol{P}$ to bring the linearly independent rows into the first r positions. In other words, $\boldsymbol{PF} = \begin{bmatrix} \boldsymbol{B} \\ \boldsymbol{H} \end{bmatrix}$, where $\boldsymbol{B}$ is $r \times r$ and with rank r (hence, nonsingular). The rows of $\boldsymbol{H}$ must be linear combinations of the rows of $\boldsymbol{B}$ and we can write $\boldsymbol{H} = \boldsymbol{DB}$. Therefore, we can write

$$\boldsymbol{PAQ} = \boldsymbol{P}[\boldsymbol{F} : \boldsymbol{FC}] = \boldsymbol{PF}[\boldsymbol{I}_r : \boldsymbol{C}] = \begin{bmatrix} \boldsymbol{B} \\ \boldsymbol{DB} \end{bmatrix} [\boldsymbol{I}_r : \boldsymbol{C}] = \begin{bmatrix} \boldsymbol{B} & \boldsymbol{BC} \\ \boldsymbol{DB} & \boldsymbol{DBC} \end{bmatrix}.$$

This proves the theorem. □

The matrix in Theorem 5.9 immediately yields a rank factorization $\boldsymbol{A} = \boldsymbol{U}\boldsymbol{V}$, where

$$\boldsymbol{U} = \boldsymbol{P}^{-1}\begin{bmatrix} \boldsymbol{B} \\ \boldsymbol{D}\boldsymbol{B} \end{bmatrix} \quad \text{and} \quad \boldsymbol{V} = [\boldsymbol{I}_r : \boldsymbol{C}]\boldsymbol{Q}^{-1} .$$

The preceding discussions have shown that a full rank factorization always exists, but has done little to demonstrate how one can compute this in practice. Theorem 5.9 offers some insight, but we still need to know which columns and rows are linearly independent. One simple way to construct a rank factorization is to first compute a reduced row echelon form (RREF), say $\boldsymbol{E}_{\boldsymbol{A}}$, of the $m \times n$ matrix $\boldsymbol{A}$. This can be achieved using elementary row operations, which amount to multiplying $\boldsymbol{A}$ on the left by an $m \times m$ nonsingular matrix $\boldsymbol{G}$. Therefore,

$$\boldsymbol{G}\boldsymbol{A} = \boldsymbol{E}_{\boldsymbol{A}} = \begin{bmatrix} \boldsymbol{R} \\ \boldsymbol{O} \end{bmatrix} ,$$

where $\boldsymbol{R}$ is $r \times n$, $r = \rho(\boldsymbol{A}) = \rho(\boldsymbol{R})$ and $\boldsymbol{O}$ is an $(m-r) \times n$ matrix of zeroes. Then, $\boldsymbol{A} = \boldsymbol{G}^{-1}\begin{bmatrix} \boldsymbol{R} \\ \boldsymbol{O} \end{bmatrix}$. Let $\boldsymbol{F}_1$ be the $m \times r$ matrix consisting of the first r columns of $\boldsymbol{G}^{-1}$ and $\boldsymbol{F}_2$ be the $m \times (m - r)$ matrix consisting of the remaining $m - r$ columns of $\boldsymbol{G}^{-1}$. Then

$$\boldsymbol{A} = [\boldsymbol{F}_1 : \boldsymbol{F}_2]\begin{bmatrix} \boldsymbol{R} \\ \boldsymbol{O} \end{bmatrix} = \boldsymbol{F}_1\boldsymbol{R} + \boldsymbol{F}_2\boldsymbol{O} = \boldsymbol{F}_1\boldsymbol{R} .$$

Since $\boldsymbol{G}$ is nonsingular, $\boldsymbol{G}^{-1}$ has linearly independent columns. Therefore, $\boldsymbol{F}_1$ has r linearly independent columns and so has full column rank. Since $\boldsymbol{E}_{\boldsymbol{A}}$ is in RREF, $\boldsymbol{R}$ has r linearly independent rows. Therefore, $\boldsymbol{A} = \boldsymbol{F}_1\boldsymbol{R}$ is indeed a rank factorization for $\boldsymbol{A}$.

The above strategy is straightforward but it seems to involve computing $\boldsymbol{G}^{-1}$. Instead suppose we take $\boldsymbol{C}$ to be the $m \times r$ matrix whose columns are the r basic columns of $\boldsymbol{A}$ and take $\boldsymbol{R}$ to be $r \times n$ matrix obtained by keeping the r nonzero rows of $\boldsymbol{E}_{\boldsymbol{A}}$. Then $\boldsymbol{A} = \boldsymbol{C}\boldsymbol{R}$ is a rank factorization of $\boldsymbol{A}$.

To see why this is true, let $\boldsymbol{P}$ be an $n \times n$ permutation matrix that brings the basic (or pivot) columns of $\boldsymbol{A}$ into the first r positions, i.e., $\boldsymbol{A}\boldsymbol{P} = [\boldsymbol{C} : \boldsymbol{D}]$. Now, $\mathcal{C}(\boldsymbol{A})$ is spanned by the r columns of $\boldsymbol{C}$ and so $\mathcal{C}(\boldsymbol{D}) \subseteq \mathcal{C}(\boldsymbol{C})$. Therefore, each column of $\boldsymbol{D}$ is a linear combination of the columns of $\boldsymbol{C}$ and, hence, there exists an $r \times (n - r)$ matrix $\boldsymbol{F}$ such that $\boldsymbol{D} = \boldsymbol{C}\boldsymbol{F}$. Hence, we can write

$$\boldsymbol{A} = [\boldsymbol{C} : \boldsymbol{C}\boldsymbol{F}]\boldsymbol{P}^{-1} = \boldsymbol{C}[\boldsymbol{I}_r : \boldsymbol{F}]\boldsymbol{P}^{-1} , \tag{5.7}$$

where $\boldsymbol{I}_r$ is the $r \times r$ identity matrix.

Suppose we transform $\boldsymbol{A}$ into its reduced row echelon form using Gauss-Jordan elimination. This amounts to left-multiplying $\boldsymbol{A}$ by a product of elementary matrices, say $\boldsymbol{G}$. Note that the effect of these elementary transformations on the basic columns of

$\boldsymbol{A}$ is to annihilate everything above and below the pivot. Therefore, $\boldsymbol{GC} = \begin{bmatrix} \boldsymbol{I}_r \\ \boldsymbol{O} \end{bmatrix}$ and we can write

$$\boldsymbol{E_A} = \boldsymbol{GA} = \boldsymbol{GC}[\boldsymbol{I}_r : \boldsymbol{F}]\boldsymbol{P}^{-1} = \begin{bmatrix} \boldsymbol{I}_r \\ \boldsymbol{O} \end{bmatrix} [\boldsymbol{I}_r : \boldsymbol{F}]\boldsymbol{P}^{-1} = \begin{bmatrix} \boldsymbol{I}_r & \boldsymbol{F} \\ \boldsymbol{O} & \boldsymbol{O} \end{bmatrix} \boldsymbol{P}^{-1} = \begin{bmatrix} \boldsymbol{R} \\ \boldsymbol{O} \end{bmatrix},$$

where $\boldsymbol{R} = [\boldsymbol{I}_r : \boldsymbol{F}]\boldsymbol{P}^{-1}$ consists of the nonzero rows of $\boldsymbol{E_A}$. Also, from (5.7) we see that $\boldsymbol{A} = \boldsymbol{CR}$ and, hence, $\boldsymbol{CR}$ produces a rank factorization of $\boldsymbol{A}$.

5.5 The rank-normal form

We have already seen in Chapter 2 that using elementary row operations, we can reduce any rectangular matrix to echelon forms. In this section we derive the simplest form into which any rectangular matrix can be reduced when we use *both* row and column operations.

Performing elementary row operations on $\boldsymbol{A}$ is equivalent to multiplying $\boldsymbol{A}$ on the left by a sequence of elementary matrices and performing elementary column operations is equivalent to multiplying $\boldsymbol{A}$ on the right by a sequence of elementary matrices. Also, any nonsingular matrix is a product of elementary matrices (Theorem 2.6). In other words, we want to find the simplest form of a matrix $\boldsymbol{B}$ such that $\boldsymbol{B} = \boldsymbol{PAQ}$ for some $m \times m$ nonsingular matrix $\boldsymbol{P}$ and some $n \times n$ nonsingular matrix $\boldsymbol{Q}$.

The answer lies in the following definition and the theorem that follows.

Definition 5.4 *A matrix $\boldsymbol{A}$ is said to be in* **normal** *form if it is*

$$\boldsymbol{N}_r = \begin{bmatrix} \boldsymbol{I}_r & \boldsymbol{O} \\ \boldsymbol{O} & \boldsymbol{O} \end{bmatrix}$$

for some positive integer r.

Clearly the rank of a matrix in normal form is r. Sometimes $\boldsymbol{N}_0$ is defined to be the null matrix $\boldsymbol{O}$. We now prove that any matrix can be reduced to a normal form by pre-multiplication and post-multiplication with nonsingular matrices.

Theorem 5.10 *Let $\boldsymbol{A}$ be an $m \times n$ matrix with $\rho(\boldsymbol{A}) = r \geq 1$. Then there exists an $m \times m$ nonsingular matrix $\boldsymbol{P}$ and an $n \times n$ nonsingular matrix $\boldsymbol{Q}$ such that*

$$\boldsymbol{A} = \boldsymbol{P} \begin{bmatrix} \boldsymbol{I}_r & \boldsymbol{O} \\ \boldsymbol{O} & \boldsymbol{O} \end{bmatrix} \boldsymbol{Q}\,.$$

Proof. Since $\rho(\boldsymbol{A}) = r \geq 1$, $\boldsymbol{A}$ is non-null. Let $\boldsymbol{A} = \boldsymbol{P}_1\boldsymbol{Q}_1$ be a rank factorization (recall that every matrix has a rank factorization). This means that $\boldsymbol{P}_1$ is an $m \times r$ matrix with r linear independent columns and that $\boldsymbol{Q}_1$ is an $r \times n$ matrix with r linearly independent rows. By Corollary 4.3, we can find an $m \times (m-r)$ matrix $\boldsymbol{P}_2$

and an $(n-r) \times n$ matrix Q_2 such that $P = [P_1 : P_2]$ is an $m \times m$ nonsingular matrix and $Q = \begin{bmatrix} Q_1 \\ Q_2 \end{bmatrix}$ is an $n \times n$ nonsingular matrix. Therefore, we can write

$$A = P_1 Q_1 = [P_1 : P_2] \begin{bmatrix} I_r & O \\ O & O \end{bmatrix} \begin{bmatrix} Q_1 \\ Q_2 \end{bmatrix} = P \begin{bmatrix} I_r & O \\ O & O \end{bmatrix} Q \, .$$

This concludes our proof. □

Letting $G = P^{-1}$ and $H = Q^{-1}$, the above theorem tells us that GAH is a matrix in normal form, where G and H are obviously nonsingular. Because GA is the outcome of a sequence of elementary row operations on A and AH is the result of a sequence of elementary column operations on A, the above theorem tells us that every matrix can be reduced to a normal form using elementary row and column operations. This can also be derived in a constructive manner as described below.

Let G be the $m \times m$ nonsingular matrix such that $GA = E_A$, where E_A is in the reduced row echelon form (RREF) (Theorem 2.3). If $\rho(A) = r$, then the basic columns in E_A are the r unit columns. We can interchange the columns of E_A to move these r unit columns to occupy the first r positions. Let H_1 be the product of the elementary matrices corresponding to these column interchanges. Then, we have

$$GAH_1 = E_A H_1 = \begin{bmatrix} I_r & J \\ O & O \end{bmatrix} .$$

Let $H_2 = \begin{bmatrix} I_r & -J \\ O & I_{n-r} \end{bmatrix}$ and let $H = H_1 H_2$. Clearly H_2 is nonsingular and, therefore, so is $H = H_1 H_2$. Therefore,

$$GAH = GAH_1 H_2 = \begin{bmatrix} I_r & J \\ O & O \end{bmatrix} \begin{bmatrix} I_r & -J \\ O & I_{n-r} \end{bmatrix} = \begin{bmatrix} I_r & O \\ O & O \end{bmatrix} = N_r \, .$$

Taking $P = G^{-1}$ and $Q = H^{-1}$, we obtain $A = PN_rQ$.

We make a short remark here. For most of this book we deal with matrices whose entries are real numbers. For a matrix with complex entries we often study its *conjugate transpose*, which is obtained by first transposing the matrix and then taking the complex conjugate of each entry. The conjugate transpose of A is denoted by A^* and formally defined as below.

Definition 5.5 *The* **conjugate transpose** *or* **adjoint** *of an $m \times n$ matrix A, with possibly complex entries, is defined as the the matrix A^* whose (i,j)-th entry is the complex conjugate of the (j,i)-th element of A. In other words,*

$$A^* = \overline{A'} = (\overline{A})' \, ,$$

where $\overline{A}$ denotes the matrix whose entries are complex conjugates of the corresponding entries in A.

If A is a real matrix (i.e., all its elements are real numbers), then $A^* = A'$ because the complex conjugate of a real number is itself. The rank normal form of a matrix can be used to see that

$$\rho(\boldsymbol{A}) = \rho(\boldsymbol{A}') = \rho(\bar{\boldsymbol{A}}) = \rho(\bar{\boldsymbol{A}}') = \rho(\boldsymbol{A}^*) \,. \tag{5.8}$$

On more than one occasion we have proved that the important result that $\rho(\boldsymbol{A}) = \rho(\boldsymbol{A}')$ for matrices whose entries are all real numbers. The rank normal form provides another way of demonstrating this for a matrix $\boldsymbol{A}$, some or all of whose entries are complex numbers. The rank normal form for $\boldsymbol{A}'$ is immediately obtained from that of $\boldsymbol{A}$: if $\boldsymbol{GAH} = \boldsymbol{N}_r$, then $\boldsymbol{H}'\boldsymbol{A}'\boldsymbol{G}' = \boldsymbol{N}_r'$. Since $\boldsymbol{H}'$ and $\boldsymbol{G}'$ are nonsingular matrices, the rank normal form for $\boldsymbol{A}'$ is simply $\boldsymbol{N}_r'$, which implies that $\rho(\boldsymbol{A}) = \rho(\boldsymbol{A}')$. To see why $\rho(\boldsymbol{A}) = \rho(\bar{\boldsymbol{A}})$, let $\boldsymbol{GAH} = \boldsymbol{N}_r$ be the rank normal form of $\boldsymbol{A}$. Using elementary properties of complex conjugates we see that $\bar{\boldsymbol{G}}\bar{\boldsymbol{A}}\bar{\boldsymbol{H}} = \bar{\boldsymbol{N}}_r$. Note that $\bar{\boldsymbol{G}}$ and $\bar{\boldsymbol{H}}$ are nonsingular because $\boldsymbol{G}$ and $\boldsymbol{H}$ are nonsingular ($\bar{\boldsymbol{G}}^{-1} = \bar{\boldsymbol{G}}^{-1}$ and similarly for $\bar{\boldsymbol{H}}^{-1}$). Therefore, $\bar{\boldsymbol{N}}_r$ is the rank normal form of $\bar{\boldsymbol{A}}$ and hence $\rho(\bar{\boldsymbol{A}}) = r = \rho(\boldsymbol{A})$. Because the rank of a matrix is equal to that of its transpose, even for complex matrices, we have $\rho(\bar{\boldsymbol{A}}) = \rho(\bar{\boldsymbol{A}}') = \rho(\boldsymbol{A}^*)$. We will, for the most part, not deal with complex matrices unless we explicitly mention otherwise.

5.6 Rank of a partitioned matrix

The following results concerning ranks for partitioned matrices is useful can be easily derived using rank-normal forms.

Theorem 5.11 $\rho\begin{bmatrix} \boldsymbol{A}_{11} & \boldsymbol{O} \\ \boldsymbol{O} & \boldsymbol{A}_{22} \end{bmatrix} = \rho(\boldsymbol{A}_{11}) + \rho(\boldsymbol{A}_{22})$.

Proof. Let $\rho(\boldsymbol{A}_{11}) = r_1$ and $\rho(\boldsymbol{A}_{22}) = r_2$. Then, we can find nonsingular matrices $\boldsymbol{G}_1$, $\boldsymbol{G}_2$, $\boldsymbol{H}_1$ and $\boldsymbol{H}_2$ such that $\boldsymbol{G}_1\boldsymbol{A}_{11}\boldsymbol{H}_1 = \boldsymbol{N}_{r_1}$ and $\boldsymbol{G}_2\boldsymbol{A}_{22}\boldsymbol{H}_2 = \boldsymbol{N}_{r_2}$ are the normal forms for $\boldsymbol{A}_{11}$ and $\boldsymbol{A}_{22}$, respectively. Then

$$\begin{bmatrix} \boldsymbol{G}_1 & \boldsymbol{O} \\ \boldsymbol{O} & \boldsymbol{G}_2 \end{bmatrix}\begin{bmatrix} \boldsymbol{A}_{11} & \boldsymbol{O} \\ \boldsymbol{O} & \boldsymbol{A}_{22} \end{bmatrix}\begin{bmatrix} \boldsymbol{H}_1 & \boldsymbol{O} \\ \boldsymbol{O} & \boldsymbol{H}_2 \end{bmatrix} = \begin{bmatrix} \boldsymbol{N}_{r_1} & \boldsymbol{O} \\ \boldsymbol{O} & \boldsymbol{N}_{r_2} \end{bmatrix} .$$

The matrices $\begin{bmatrix} \boldsymbol{G}_1 & \boldsymbol{O} \\ \boldsymbol{O} & \boldsymbol{G}_2 \end{bmatrix}$ and $\begin{bmatrix} \boldsymbol{H}_1 & \boldsymbol{O} \\ \boldsymbol{O} & \boldsymbol{H}_2 \end{bmatrix}$ are clearly nonsingular—their inverses are the inverses of their diagonal blocks. Therefore, $\begin{bmatrix} \boldsymbol{N}_{r_1} & \boldsymbol{O} \\ \boldsymbol{O} & \boldsymbol{N}_{r_2} \end{bmatrix}$ is the normal form of $\begin{bmatrix} \boldsymbol{A}_{11} & \boldsymbol{O} \\ \boldsymbol{O} & \boldsymbol{A}_{22} \end{bmatrix}$ and $\rho\begin{bmatrix} \boldsymbol{A}_{11} & \boldsymbol{O} \\ \boldsymbol{O} & \boldsymbol{A}_{22} \end{bmatrix} = r_1 + r_2 = \rho(\boldsymbol{A}_{11}) + \rho(\boldsymbol{A}_{22})$. □

This easily generalizes for block-diagonal matrices of higher order: the rank of any block-diagonal matrix is the sum of the ranks of its diagonal blocks.

The rank of partitioned matrices can also be expressed in terms of its Schur's complement when at least one of the diagonal blocks is nonsingular. The following theorem summarizes.

Theorem 5.12 *Let* $\boldsymbol{A} = \begin{bmatrix} \boldsymbol{A}_{11} & \boldsymbol{A}_{12} \\ \boldsymbol{A}_{21} & \boldsymbol{A}_{22} \end{bmatrix}$ *be a* 2×2 *partitioned matrix.*

(i) If $\boldsymbol{A}_{11}$ *is nonsingular then* $\rho(\boldsymbol{A}) = \rho(\boldsymbol{A}_{11}) + \rho(\boldsymbol{F})$, *where* $\boldsymbol{F} = \boldsymbol{A}_{22} - \boldsymbol{A}_{21}\boldsymbol{A}_{11}^{-1}\boldsymbol{A}_{12}$.

(ii) If $\boldsymbol{A}_{22}$ *is nonsingular then* $\rho(\boldsymbol{A}) = \rho(\boldsymbol{A}_{22}) + \rho(\boldsymbol{G})$, *where* $\boldsymbol{G} = \boldsymbol{A}_{11} - \boldsymbol{A}_{12}\boldsymbol{A}_{22}^{-1}\boldsymbol{A}_{21}$.

Proof. **Proof of (i):** It is easily verified that (recall the *block* $\boldsymbol{LDU}$ *decomposition* in Section 3.6 and, in particular, Equation (3.13))

$$\begin{bmatrix} \boldsymbol{A}_{11} & \boldsymbol{A}_{12} \\ \boldsymbol{A}_{21} & \boldsymbol{A}_{22} \end{bmatrix} = \begin{bmatrix} \boldsymbol{I}_{n_1} & \boldsymbol{O} \\ \boldsymbol{A}_{21}\boldsymbol{A}_{11}^{-1} & \boldsymbol{I}_{n_2} \end{bmatrix} \begin{bmatrix} \boldsymbol{A}_{11} & \boldsymbol{O} \\ \boldsymbol{O} & \boldsymbol{F} \end{bmatrix} \begin{bmatrix} \boldsymbol{I}_{n_1} & \boldsymbol{A}_{11}^{-1}\boldsymbol{A}_{12} \\ \boldsymbol{O} & \boldsymbol{I}_{n_2} \end{bmatrix}.$$

Also, $\begin{bmatrix} \boldsymbol{I}_{n_1} & \boldsymbol{O} \\ -\boldsymbol{A}_{21}\boldsymbol{A}_{11}^{-1} & \boldsymbol{I}_{n_2} \end{bmatrix}$ and $\begin{bmatrix} \boldsymbol{I}_{n_1} & -\boldsymbol{A}_{11}^{-1}\boldsymbol{A}_{12} \\ \boldsymbol{O} & \boldsymbol{I}_{n_2} \end{bmatrix}$ are both triangular matrices with nonzero diagonal entries and, therefore, nonsingular. (In fact their inverses are obtained by simply changing the signs of their off-diagonal blocks—recall Section 3.6.) Since multiplication by nonsingular matrices does not alter rank (Lemma 5.3), we have

$$\rho(\boldsymbol{A}) = \rho \begin{bmatrix} \boldsymbol{A}_{11} & \boldsymbol{A}_{12} \\ \boldsymbol{A}_{21} & \boldsymbol{A}_{22} \end{bmatrix} = \rho \begin{bmatrix} \boldsymbol{A}_{11} & \boldsymbol{O} \\ \boldsymbol{O} & \boldsymbol{F} \end{bmatrix} = \rho(\boldsymbol{A}_{11}) + \rho(\boldsymbol{F}) ,$$

where the last equality follows from Theorem 5.11. This proves (i).

The proof of (ii) is similar and we leave the details to the reader. □

5.7 Bases for the fundamental subspaces using the rank normal form

In Section 5.2 we saw how row reduction to echelon forms yielded basis vectors for the four fundamental subspaces. The nonsingular matrices leading to the rank-normal form also reveal a set of basis vectors for each of the four fundamental subspaces. As in the preceding section, suppose $\boldsymbol{A}$ is $m \times n$ and $\boldsymbol{GAH} = \boldsymbol{N}_r$, where $\boldsymbol{G}$ is $m \times m$ and nonsingular and $\boldsymbol{H}$ is $n \times n$ and nonsingular. Let $\boldsymbol{P} = \boldsymbol{G}^{-1}$ and $\boldsymbol{Q} = \boldsymbol{H}^{-1}$.

If $\boldsymbol{P}_1$ is the $m \times r$ submatrix of $\boldsymbol{P}$ comprising the first r columns of $\boldsymbol{A}$ and $\boldsymbol{Q}_1$ is the submatrix comprising the first $r \times n$ rows of $\boldsymbol{Q}$, then clearly $\boldsymbol{A} = \boldsymbol{P}_1\boldsymbol{Q}_1$ is a rank factorization for $\boldsymbol{A}$. Therefore, from Corollary 5.9, we know that the columns of $\boldsymbol{P}_1$ are a basis for the column space of $\boldsymbol{A}$ and the rows of $\boldsymbol{R}$ are a basis for the row space of $\boldsymbol{A}$.

It is also not difficult to see that the last $n - r$ columns of $\boldsymbol{H}$ form a basis for $\mathcal{N}(\boldsymbol{A})$. Write $\boldsymbol{H} = [\boldsymbol{H}_1 : \boldsymbol{H}_2]$, where $\boldsymbol{H}_1$ is $n \times r$ and $\boldsymbol{H}_2$ is $n \times (n - r)$. From the rank-normal form it is clear that $\boldsymbol{AH}_2 = \boldsymbol{O}$, so the columns of $\boldsymbol{H}_2$ form a set of $(n - r)$ linearly independent columns in $\mathcal{N}(\boldsymbol{A})$. Now we can argue in two different ways:

(a) From Theorem 5.1, the dimension of $\mathcal{N}(\boldsymbol{A})$ is $n - r$. The columns of $\boldsymbol{H}_2$ form a set of $n - r$ linearly independent vectors in $\mathcal{N}(\boldsymbol{A})$ and, hence, is basis for $\mathcal{N}(\boldsymbol{A})$.

(b) Alternatively, we can construct a direct argument that shows $\mathcal{C}(\boldsymbol{H}_2) = \mathcal{N}(\boldsymbol{A})$. Since $\boldsymbol{A}\boldsymbol{H}_2 = \boldsymbol{O}$, we already know $\mathcal{C}(\boldsymbol{H}_2) \subseteq \mathcal{N}(\boldsymbol{A})$.
To prove the other direction, observe that

$$\boldsymbol{x} \in \mathcal{N}(\boldsymbol{A}) \Longrightarrow \boldsymbol{A}\boldsymbol{x} = \boldsymbol{0} \Longrightarrow \boldsymbol{P}\boldsymbol{N}_r\boldsymbol{Q}\boldsymbol{x} = \boldsymbol{0} \Longrightarrow \boldsymbol{N}_r\boldsymbol{Q}\boldsymbol{x} = \boldsymbol{0}$$
$$\Longrightarrow \begin{bmatrix} \boldsymbol{I}_r & \boldsymbol{O} \\ \boldsymbol{O} & \boldsymbol{O} \end{bmatrix} \begin{bmatrix} \boldsymbol{Q}_1 \\ \boldsymbol{Q}_2 \end{bmatrix} \boldsymbol{x} = \boldsymbol{0} \Longrightarrow \boldsymbol{Q}_1'\boldsymbol{x} = \boldsymbol{0} \Longrightarrow \boldsymbol{x} \in \mathcal{N}(\boldsymbol{Q}_1) ,$$

where $\boldsymbol{Q} = \begin{bmatrix} \boldsymbol{Q}_1 \\ \boldsymbol{Q}_2 \end{bmatrix}$ and $\boldsymbol{Q}_1$ is $r \times n$. Recall that $\boldsymbol{Q}$ is the inverse of $\boldsymbol{H}$, so $\boldsymbol{H}\boldsymbol{Q} = \boldsymbol{I}$, which implies that $[\boldsymbol{H}_1 : \boldsymbol{H}_2] \begin{bmatrix} \boldsymbol{Q}_1 \\ \boldsymbol{Q}_2 \end{bmatrix} = \boldsymbol{H}_1\boldsymbol{Q}_1 + \boldsymbol{H}_2\boldsymbol{Q}_2 = \boldsymbol{I}_n$. Therefore, $\boldsymbol{x} = \boldsymbol{I}\boldsymbol{x} = \boldsymbol{H}_1\boldsymbol{Q}_1\boldsymbol{x} + \boldsymbol{H}_2\boldsymbol{Q}_2\boldsymbol{x}$ for any $\boldsymbol{x} \in \Re^n$. In particular, if $\boldsymbol{x} \in \mathcal{N}(\boldsymbol{A})$, we saw that $\boldsymbol{Q}_1\boldsymbol{x} = \boldsymbol{0}$, so $\boldsymbol{x} = \boldsymbol{H}_2\boldsymbol{Q}_2\boldsymbol{x} \in \mathcal{C}(\boldsymbol{H}_2)$. This proves that every vector in $\mathcal{N}(\boldsymbol{A})$ belongs to $\mathcal{C}(\boldsymbol{H}_2)$ and, hence, $\mathcal{N}(\boldsymbol{A}) \subseteq \mathcal{C}(\boldsymbol{H}_2)$. Therefore, $\mathcal{C}(\boldsymbol{H}_2) = \mathcal{N}(\boldsymbol{A})$. This, in fact, constitutes an alternative proof of the Rank-Plus-Nullity Theorem.

Finally, since $\boldsymbol{H}'\boldsymbol{A}'\boldsymbol{G}' = \boldsymbol{N}_r$, we can apply the above result to conclude that the last $(m - r)$ columns of $\boldsymbol{G}'$ constitute a basis for the left null space $\mathcal{N}(\boldsymbol{A}')$.

5.8 Exercises

1. Without performing any calculations find the rank of $\boldsymbol{A}$ and $\boldsymbol{B}$, where

$$\boldsymbol{A} = \begin{bmatrix} 2 & -2 & -5 & -3 & -1 & 2 \\ 0 & 1 & 2 & 5 & 4 & 0 \\ 0 & 0 & 0 & 1 & 1 & 0 \\ 0 & 0 & 0 & 0 & 0 & 0 \end{bmatrix} \text{ and } \boldsymbol{B} = \begin{bmatrix} 2 & 3 & 1 & 0 & 0 & 0 \\ 4 & 7 & 0 & 1 & 0 & 0 \\ -6 & 0 & 0 & 0 & 1 & 0 \\ 4 & 0 & 0 & 0 & 0 & 1 \end{bmatrix}.$$

2. Compute the rank and nullity of

$$\boldsymbol{A} = \begin{bmatrix} 1 & 2 & 3 \\ 4 & 5 & 6 \\ 7 & 8 & 9 \end{bmatrix}.$$

3. Find a basis for each of the four fundamental subspaces of

$$\boldsymbol{A} = \begin{bmatrix} 2 & -2 & -5 & -3 & -1 & 2 \\ 2 & -1 & -3 & 2 & 3 & 2 \\ 4 & -1 & -4 & 10 & 11 & 4 \\ 0 & 1 & 2 & 5 & 4 & 0 \end{bmatrix} \text{ and } \boldsymbol{B} = \begin{bmatrix} 1 & 2 & -1 & 3 \\ 2 & 4 & -2 & 6 \\ 1 & 1 & -2 & 2 \\ -1 & 1 & -4 & 0 \end{bmatrix}.$$

What are the dimensions of each of the four fundamental subspaces of $\boldsymbol{A}$ and $\boldsymbol{B}$?

4. True or false: If $\rho(\boldsymbol{A}) = r$ and $\boldsymbol{B} = \begin{bmatrix} \boldsymbol{A} & \boldsymbol{O} \\ \boldsymbol{O} & \boldsymbol{O} \end{bmatrix}$, then $\rho(\boldsymbol{B}) = r$.

5. True or false: Adding a scalar multiple of one column of $\boldsymbol{A}$ to another column of $\boldsymbol{A}$ does not alter the rank of a matrix.

6. True or false: The rank of a submatrix of $\boldsymbol{A}$ cannot exceed the rank of $\boldsymbol{A}$.
7. Let $\boldsymbol{A}$ be an $m \times n$ matrix such that $\rho(\boldsymbol{A}) = 1$. Show that $\boldsymbol{A} = \boldsymbol{u}\boldsymbol{v}'$ for some $m \times 1$ vector $\boldsymbol{u}$ and $n \times 1$ vector $\boldsymbol{v}$.
8. Prove that if $\mathcal{R}(\boldsymbol{A}) \subseteq \mathcal{R}(\boldsymbol{B})$, then $\mathcal{R}(\boldsymbol{AC}) \subseteq \mathcal{R}(\boldsymbol{BC})$.
9. True or false: If $\mathcal{R}(\boldsymbol{A}) = \mathcal{R}(\boldsymbol{B})$, then $\rho(\boldsymbol{AC}) = \rho(\boldsymbol{BC})$.
10. Is it possible that $\rho(\boldsymbol{AB}) < \min\{\rho(\boldsymbol{A}), \rho(\boldsymbol{B})\}$?
11. True or false: $\rho(\boldsymbol{ABC}) \leq \rho(\boldsymbol{AC})$.
12. True or false: If $\boldsymbol{AB}$ and $\boldsymbol{BA}$ are both well-defined, then $\rho(\boldsymbol{AB}) = \rho(\boldsymbol{BA})$.
13. Let $\boldsymbol{B} = \boldsymbol{A}_1\boldsymbol{A}_2 \cdots \boldsymbol{A}_k$, where the product is well-defined. Prove that if at least one $\boldsymbol{A}_i$ is singular, then $\boldsymbol{B}$ is singular.
14. Prove that if $\boldsymbol{A}'\boldsymbol{A} = \boldsymbol{O}$, then $\boldsymbol{A} = \boldsymbol{O}$.
15. Prove that $\rho([\boldsymbol{A} : \boldsymbol{B}]) = \rho(\boldsymbol{A})$ if and only if $\boldsymbol{B} = \boldsymbol{AC}$ for some matrix $\boldsymbol{C}$.
16. Prove that $\rho\left(\begin{bmatrix} \boldsymbol{A} \\ \boldsymbol{D} \end{bmatrix}\right) = \rho(\boldsymbol{A})$ if and only if $\boldsymbol{D} = \boldsymbol{CA}$ for some matrix $\boldsymbol{C}$.
17. Let $\boldsymbol{A}$ and $\boldsymbol{B}$ be $n \times n$. Prove that $\rho\left(\begin{bmatrix} \boldsymbol{A} & \boldsymbol{I} \\ \boldsymbol{I} & \boldsymbol{B} \end{bmatrix}\right) = n$ if and only if $\boldsymbol{B} = \boldsymbol{A}^{-1}$.
18. Prove that $\mathcal{N}(\boldsymbol{A}) = \mathcal{C}(\boldsymbol{B})$ if and only if $\mathcal{C}(\boldsymbol{A}') = \mathcal{N}(\boldsymbol{B}')$.
19. Let $\mathcal{S} \subseteq \Re^n$ be a subspace and let $\boldsymbol{B}$ be the matrix whose columns are a basis of $\mathcal{S}$. Let $\boldsymbol{A}$ be the matrix such that the row vectors of $\boldsymbol{A}$ form a basis for $\mathcal{N}(\boldsymbol{B}')$. Prove that $\mathcal{S} = \mathcal{N}(\boldsymbol{A})$. *This shows that every subspace of $\Re^n$ corresponds to the null space of some matrix.*
20. Consider the matrix $\boldsymbol{A} = \begin{bmatrix} 4 & 1 & 6 & 0 \\ 0 & 1 & 2 & -4 \\ 1 & 0 & 1 & 1 \end{bmatrix}$. Find matrices $\boldsymbol{B}$ and $\boldsymbol{C}$ such that $\mathcal{C}(\boldsymbol{A}) = \mathcal{N}(\boldsymbol{B})$ and $\mathcal{R}(\boldsymbol{A}) = \mathcal{N}(\boldsymbol{C})$.
21. Prove that $\rho(\boldsymbol{AB}) = \rho(\boldsymbol{A}) - \dim(\mathcal{N}(\boldsymbol{B}') \cap \mathcal{C}(\boldsymbol{A}'))$.
22. True or false: If $\boldsymbol{AB}$ is well-defined, then $\nu(\boldsymbol{AB}) = \nu(\boldsymbol{B}) + \dim(\mathcal{C}(\boldsymbol{B}) \cap \mathcal{N}(\boldsymbol{A}))$.
23. Prove the following: If $\boldsymbol{AB} = \boldsymbol{O}$, then $\rho(\boldsymbol{A}) + \rho(\boldsymbol{B}) \leq n$, where n is the number of columns in $\boldsymbol{A}$.
24. If $\rho(\boldsymbol{A}^k) = \rho(\boldsymbol{A}^{k+1})$, then prove that $\rho(\boldsymbol{A}^{k+1}) = \rho(\boldsymbol{A}^{k+2})$.
25. For any $n \times n$ matrix $\boldsymbol{A}$, prove that there exists a positive integer $p < n$ such that

$$\rho(\boldsymbol{A}) > \rho(\boldsymbol{A}^2) > \cdots > \rho(\boldsymbol{A}^p) = \rho(\boldsymbol{A}^{p+1}) = \cdots$$

26. Let $\boldsymbol{A}$ be any square matrix and k be any positive integer. Prove that

$$\rho(\boldsymbol{A}^{k+1}) - 2\rho(\boldsymbol{A}^k) + \rho(\boldsymbol{A}^{k-1}) \geq 0\,.$$

27. Find a left inverse of $\boldsymbol{A}$ and a right inverse for $\boldsymbol{B}$, where

$$\boldsymbol{A} = \begin{bmatrix} 1 & 2 \\ 0 & 1 \\ 1 & 0 \end{bmatrix} \quad \text{and} \quad \boldsymbol{B} = \begin{bmatrix} 1 & 0 & 5 \\ 0 & 1 & 3 \end{bmatrix}.$$

28. True or false: Every nonzero $m \times 1$ matrix has a left inverse.
29. True or false: Every nonzero $1 \times n$ matrix has a right inverse.
30. Prove that if $\boldsymbol{G}_1$ and $\boldsymbol{G}_2$ are left inverses of $\boldsymbol{A}$, then $\alpha \boldsymbol{G}_1 + (1-\alpha)\boldsymbol{G}_2$ is also a left inverse of $\boldsymbol{A}$.
31. Let $\boldsymbol{A}$ be an $m \times n$ matrix which has a unique left inverse. Prove that $m = n$ and $\boldsymbol{A}$ has an inverse.
32. Let $\boldsymbol{M}$ be an $n \times n$ nonsingular matrix partitioned as follows:

$$\boldsymbol{M} = \begin{bmatrix} \boldsymbol{P} \\ \boldsymbol{Q} \end{bmatrix} \text{ and } \boldsymbol{M}^{-1} = [\boldsymbol{U} : \boldsymbol{V}] ,$$

where $\boldsymbol{P}$ is $k \times n$ and $\boldsymbol{U}$ is $n \times k$.

   (a) Prove that $\boldsymbol{U}$ is a right inverse of $\boldsymbol{P}$.
   (b) Prove that $\mathcal{N}(\boldsymbol{P}) = \mathcal{C}(\boldsymbol{V})$.
   (c) Can you derive the Rank-Nullity Theorem (Theorem 5.1) from this.

33. True or false: If $\boldsymbol{A}$ is an $m \times n$ matrix with rank m and $\boldsymbol{B}$ is an $n \times m$ matrix with rank m, then $\boldsymbol{AB}$ has an inverse.
34. Let $\boldsymbol{A}$ be $m \times n$ and let $\boldsymbol{b} \in \mathcal{C}(\boldsymbol{A})$. Prove that $\boldsymbol{Au} = \boldsymbol{b}$ if and only if $\boldsymbol{A}'\boldsymbol{Au} = \boldsymbol{b}$.
35. Prove that the linear system $\boldsymbol{A}'\boldsymbol{Ax} = \boldsymbol{b}$ always has a solution.
36. Prove that $\rho(\boldsymbol{A}'\boldsymbol{AB}) = \rho(\boldsymbol{AB}) = \rho(\boldsymbol{ABB}')$
37. True or false: $\rho(\boldsymbol{A}'\boldsymbol{AA}') = \rho(\boldsymbol{A})$.
38. Prove *Sylvester's law of nullity*: If $\boldsymbol{A}$ and $\boldsymbol{B}$ are square matrices, then

$$\max\{\nu(\boldsymbol{A}), \nu(\boldsymbol{B})\} \leq \nu(\boldsymbol{AB}) \leq \nu(\boldsymbol{A}) + \nu(\boldsymbol{B}) .$$

Provide an example to show that the above result does not hold for rectangular matrices.

39. Prove the *Frobenius inequality*:

$$\rho(\boldsymbol{ABC}) \geq \rho(\boldsymbol{AB}) + \rho(\boldsymbol{BC}) - \rho(\boldsymbol{B}) .$$

40. Find a rank factorization of the matrices in Exercises 1 and 2.
41. Let $\boldsymbol{A}$ be an $m \times n$ matrix and let $\boldsymbol{A} = \boldsymbol{PQ}_1 = \boldsymbol{PQ}_2$ be rank factorizations of $\boldsymbol{A}$. Prove that $\boldsymbol{Q}_1 = \boldsymbol{Q}_2$.
42. Let $\boldsymbol{A}$ be an $m \times n$ matrix and let $\boldsymbol{A} = \boldsymbol{P}_1\boldsymbol{Q} = \boldsymbol{P}_2\boldsymbol{Q}$ be rank factorizations of $\boldsymbol{A}$. Prove that $\boldsymbol{P}_1 = \boldsymbol{P}_2$.
43. If $\mathcal{C}(\boldsymbol{A}) \subseteq \mathcal{C}(\boldsymbol{B})$ and $\mathcal{R}(\boldsymbol{A}) \subseteq \mathcal{R}(\boldsymbol{D})$, then prove that $\boldsymbol{A} = \boldsymbol{BCD}$ for some matrix $\boldsymbol{C}$.
44. Prove that

$$\rho\left(\begin{bmatrix} \boldsymbol{A} & \boldsymbol{B} \\ \boldsymbol{O} & \boldsymbol{D} \end{bmatrix}\right) \geq \rho(\boldsymbol{A}) + \rho(\boldsymbol{D}) .$$

Give an example that strict inequality can occur in the above. Use the above inequality to conclude that the rank of a triangular matrix (upper or lower) cannot be less than the number of nonzero diagonal elements.

CHAPTER 6

Complementary Subspaces

6.1 Sum of subspaces

Recall Definition 4.4, where we defined the sum of two subspaces $\mathcal{S}_1$ and $\mathcal{S}_2$ of $\Re^m$. We also established that this sum, denoted by $\mathcal{S}_1 + \mathcal{S}_2$, is another subspace of $\Re^m$. For example, Figure 6.1 depicts $\mathcal{X}$ and $\mathcal{Y}$ in which $\mathcal{X}$ is a plane through the origin, and $\mathcal{Y}$ is a line through the origin. As seen in Section 4.3 (see Example 4.3 in particular), $\mathcal{X}$ and $\mathcal{Y}$ are both subspaces of $\Re^m$. It is worth pointing out that the intersection of $\mathcal{X}$

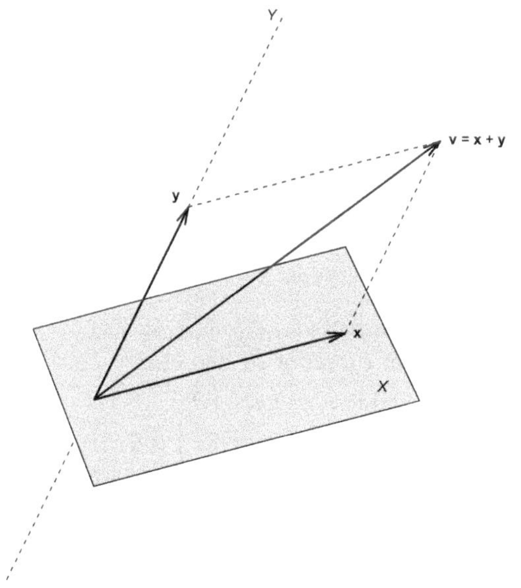

Figure 6.1 *A visual depiction of the sum of two subspaces, $\mathcal{X}$ and $\mathcal{Y}$, in $\Re^3$.*

and $\mathcal{Y}$ in Figure 6.1 comprises only the origin. In other words, these two subspaces are *essentially disjoint* in the sense that $\mathcal{X} \cap \mathcal{Y} = \mathbf{0}$. In fact, it follows easily from the parallelogram law for vector addition that $\mathcal{X} + \mathcal{Y} = \Re^3$ because each vector in $\Re^3$

can be written as the sum of a vector from $\mathcal{X}$ and a vector from $\mathcal{Y}$. In other words, the vector space $\Re^3$ is resolved into a pair of two disjoint subspaces $\mathcal{X}$ and $\mathcal{Y}$.

Why are we interested in studying the decomposition of vector spaces into disjoint subspaces? The concept arises in statistics and other areas of applied mathematics where vector spaces (and subspaces) represent information regarding a variable. For example, observations from three individuals may be collected into a 3×1 outcome vector that can be regarded as a point in $\Re^2$. Two explanatory variables, observed on the same three individuals, can be looked upon as two other points in $\Re^3$ that span the plane $\mathcal{X}$ in Figure 6.1. Statisticians often would like to understand how the space $\Re^3$, where the outcome resides, is geometrically related to the subspace spanned by the explanatory variables. And it is also of interest to understand what "remaining" subspace (or subspaces) are needed to capture the remainder of $\Re^3$, over and above what has been explained by $\mathcal{X}$. These ideas are formalized in a rigorous manner in this chapter. The chapter will also provide further insight into the four fundamental subspaces associated with a matrix and help us arrive at what is often called the "fundamental theorem of linear algebra."

6.2 The dimension of the sum of subspaces

The following theorem relates the dimension of $\mathcal{S}_1 + \mathcal{S}_2$ with respect to those of $\mathcal{S}_1$, $\mathcal{S}_2$ and their intersection.

Theorem 6.1 Dimension of a sum of subspaces. *Let $\mathcal{S}_1$ and $\mathcal{S}_2$ be two subspaces in $\Re^n$. Then,*

$$\dim(\mathcal{S}_1 + \mathcal{S}_2) = \dim(\mathcal{S}_1) + \dim(\mathcal{S}_2) - \dim(\mathcal{S}_1 \cap \mathcal{S}_2).$$

Proof. The idea is to extend a basis of $\mathcal{S}_1 \cap \mathcal{S}_2$ to ones for $\mathcal{S}_1$ and $\mathcal{S}_2$.

Let $\mathcal{B}_{12} = \{\boldsymbol{u}_1, \ldots, \boldsymbol{u}_k\}$ be a basis for $\mathcal{S}_1 \cap \mathcal{S}_2$, so that $k = \dim(\mathcal{S}_1 \cap \mathcal{S}_2)$. Since $\mathcal{B}_{12}$ is a linearly independent subset of $\mathcal{S}_1$ and also of $\mathcal{S}_2$, Part (ii) of Lemma 4.21 ensures that there are vectors $\{\boldsymbol{v}_{k+1}, \ldots, \boldsymbol{v}_p\}$ and $\{\boldsymbol{w}_{k+1}, \ldots, \boldsymbol{w}_m\}$ such that

$$\mathcal{B}_1 = \{\boldsymbol{u}_1, \ldots, \boldsymbol{u}_k, \boldsymbol{v}_{k+1}, \ldots, \boldsymbol{v}_p\} \text{ is a basis for } \mathcal{S}_1 \text{ and}$$
$$\mathcal{B}_2 = \{\boldsymbol{u}_1, \ldots, \boldsymbol{u}_k, \boldsymbol{w}_{k+1}, \ldots, \boldsymbol{w}_m\} \text{ is a basis for } \mathcal{S}_2\ ,$$

where $p = \dim(\mathcal{S}_1)$ and $m = \dim(\mathcal{S}_2)$.

Now we claim that the set

$$\mathcal{B} = \{\boldsymbol{u}_1, \ldots, \boldsymbol{u}_k, \boldsymbol{v}_{k+1}, \ldots, \boldsymbol{v}_p, \boldsymbol{w}_{k+1}, \ldots, \boldsymbol{w}_m\}$$

is a basis of $\mathcal{S}_1 + \mathcal{S}_2$. Since $\mathcal{B}$ has $p + m - k$ elements, this would imply $\dim(\mathcal{S}_1 + \mathcal{S}_2) = p + m - k$ and we will be done.

Clearly $\mathcal{B}$ spans $\mathcal{S}_1 + \mathcal{S}_2$ because $\mathcal{B}$ contains a basis of $\mathcal{S}_1$ as well as $\mathcal{S}_2$. It remains to verify whether $\mathcal{B}$ is a linearly independent set.

Consider the homogeneous linear combination:

$$\mathbf{0} = \sum_{i=1}^{k} \alpha_i \boldsymbol{u}_i + \sum_{i=k+1}^{p} \beta_i \boldsymbol{v}_i + \sum_{i=k+1}^{m} \gamma_i \boldsymbol{w}_i.$$

Rewriting the above sum as

$$-\sum_{i=k+1}^{m} \gamma_i \boldsymbol{w}_i = \sum_{i=1}^{k} \alpha_i \boldsymbol{u}_i + \sum_{i=k+1}^{p} \beta_i \boldsymbol{v}_i,$$

reveals that $\sum_{i=k+1}^{m} \gamma_i \boldsymbol{w}_i \in \mathcal{S}_1$. But $\sum_{i=k+1}^{m} \gamma_i \boldsymbol{w}_i \in \mathcal{S}_2$ already, so this means that $\sum_{i=k+1}^{m} \gamma_i \boldsymbol{w}_i \in \mathcal{S}_1 \cap \mathcal{S}_2$. But this would imply that $\sum_{i=k+1}^{m} \gamma_i \boldsymbol{w}_i \in Sp(\{\boldsymbol{u}_1, \ldots, \boldsymbol{u}_k\})$, contradicting the linear independence of the $\boldsymbol{w}_i$'s and $\boldsymbol{u}_i$'s. Therefore, we must have $\sum_{i=k+1}^{m} \gamma_i \boldsymbol{w}_i = \mathbf{0}$ implying that all the γ_i's are 0. But this leaves us with $\mathbf{0} = \sum_{i=1}^{k} \alpha_i \boldsymbol{u}_i + \sum_{i=k+1}^{p} \beta_i \boldsymbol{v}_i$, which implies that all the α_i's and β_i's are 0 because the $\boldsymbol{u}_i$'s and $\boldsymbol{v}_i$'s are linearly independent. This proves the linear independence of $\mathcal{B}$ and that $\mathcal{B}$ is indeed a basis of $\mathcal{S}_1 + \mathcal{S}_2$. Therefore, $\dim(\mathcal{S}_1 + \mathcal{S}_2) = p + m - k$ and we are done. □

The following important corollary is an immediate consequence of Theorem 6.1. It consists of two statements. The first says that the dimension of the sum of two subspaces cannot exceed the sum of their respective dimensions. The second statement is an application to matrices and tells us that the rank of the sum of two matrices cannot exceed the sum of their ranks.

Corollary 6.1 *The following two statements are true:*

(i) Let $\mathcal{S}_1$ and $\mathcal{S}_2$ be two subspaces in $\Re^m$. Then $\dim(\mathcal{S}_1 + \mathcal{S}_2) \leq \dim(\mathcal{S}_1) + \dim(\mathcal{S}_2)$.

(ii) Let $\boldsymbol{A}$ and $\boldsymbol{B}$ be two matrices of the same order. Then $\rho(\boldsymbol{A} + \boldsymbol{B}) \leq \rho(\boldsymbol{A}) + \rho(\boldsymbol{B})$.

Proof. **Proof of (i):** This is immediate from Theorem 6.1:

$$\dim(\mathcal{S}_1 + \mathcal{S}_2) = \dim(\mathcal{S}_1) + \dim(\mathcal{S}_2) - \dim(\mathcal{S}_1 \cap \mathcal{S}_2) \leq \dim(\mathcal{S}_1) + \dim(\mathcal{S}_2) .$$

This proves part (i). □

Proof of (ii): Suppose that $\boldsymbol{A}$ and $\boldsymbol{B}$ are $m \times n$ matrices. Note that $\boldsymbol{u} \in \mathcal{C}(\boldsymbol{A} + \boldsymbol{B})$ implies that $\boldsymbol{u} = (\boldsymbol{A} + \boldsymbol{B})\boldsymbol{x} = \boldsymbol{A}\boldsymbol{x} + \boldsymbol{B}\boldsymbol{x}$ for some $\boldsymbol{x} \in \Re^n$ and, hence, $\boldsymbol{u} \in \mathcal{C}(\boldsymbol{A}) + \mathcal{C}(\boldsymbol{B})$. Therefore, $\mathcal{C}(\boldsymbol{A} + \boldsymbol{B}) \subseteq \mathcal{C}(\boldsymbol{A}) + \mathcal{C}(\boldsymbol{B})$. It now follows that

$$\begin{aligned}\rho(\boldsymbol{A} + \boldsymbol{B}) &= \dim(\mathcal{C}(\boldsymbol{A} + \boldsymbol{B})) \leq \dim(\mathcal{C}(\boldsymbol{A}) + \mathcal{C}(\boldsymbol{B})) \\ &= \dim(\mathcal{C}(\boldsymbol{A})) + \dim(\mathcal{C}(\boldsymbol{B})) - \dim(\mathcal{C}(\boldsymbol{A}) \cap \mathcal{C}(\boldsymbol{B})) \\ &\leq \dim(\mathcal{C}(\boldsymbol{A})) + \dim(\mathcal{C}(\boldsymbol{B})) = \rho(\boldsymbol{A}) + \rho(\boldsymbol{B}) ,\end{aligned}$$

where the second "=" follows from Theorem 6.1. This proves part (ii). □

Given bases of two subspaces $\mathcal{S}_1$ and $\mathcal{S}_2$ of $\Re^n$, how do we obtain a basis for the intersection $\mathcal{S}_1 \cap \mathcal{S}_2$? Suppose $\dim(\mathcal{S}_1) = p$ and $\dim(\mathcal{S}_2) = m$. Let $\boldsymbol{A}$ and $\boldsymbol{B}$ be $n \times p$ and $n \times m$, respectively, such that the columns of $\boldsymbol{A}$ are a basis for $\mathcal{S}_1$ and the columns of $\boldsymbol{B}$ are a basis for $\mathcal{S}_2$. Thus, $\mathcal{C}(\boldsymbol{A}) = \mathcal{S}_1$ and $\mathcal{C}(\boldsymbol{B}) = \mathcal{S}_2$. Clearly, $\mathcal{C}([\boldsymbol{A} : \boldsymbol{B}]) = \mathcal{C}(\boldsymbol{A}) + \mathcal{C}(\boldsymbol{B}) = m + p$. Then,

$$\dim(\mathcal{S}_1 + \mathcal{S}_2) = \dim\left(\mathcal{C}(\boldsymbol{A}) + \mathcal{C}(\boldsymbol{B})\right) = \dim \mathcal{C}([\boldsymbol{A} : \boldsymbol{B}]) \,. \tag{6.1}$$

Observe that $\mathcal{N}(\boldsymbol{A}) = \{\boldsymbol{0}_{p\times 1}\}$ and $\mathcal{N}(\boldsymbol{B}) = \{\boldsymbol{0}_{m\times 1}\}$ because $\boldsymbol{A}$ and $\boldsymbol{B}$ both have linearly independent columns. Consider the subspace $\mathcal{C}(\boldsymbol{A}) \cap \mathcal{C}(\boldsymbol{B})$. Since every $\boldsymbol{x} \in \mathcal{C}(\boldsymbol{A}) \cap \mathcal{C}(\boldsymbol{B})$ is spanned by the columns of $\boldsymbol{A}$ as well as the columns of $\boldsymbol{B}$, there exists some $\boldsymbol{u} \in \Re^p$ and some $\boldsymbol{v} \in \Re^m$ such that $\boldsymbol{x} = \boldsymbol{A}\boldsymbol{u} = \boldsymbol{B}\boldsymbol{v}$. Thus,

$$[\boldsymbol{A} : -\boldsymbol{B}] \begin{bmatrix} \boldsymbol{u} \\ \boldsymbol{v} \end{bmatrix} = \boldsymbol{0} \Longrightarrow \begin{bmatrix} \boldsymbol{u} \\ \boldsymbol{v} \end{bmatrix} \in \mathcal{N}([\boldsymbol{A} : -\boldsymbol{B}]) \,.$$

Let $\boldsymbol{z}_1, \boldsymbol{z}_2, \ldots, \boldsymbol{z}_k$ be a basis for $\mathcal{N}([\boldsymbol{A} : -\boldsymbol{B}])$. Each $\boldsymbol{z}_i$ is $(p+m) \times 1$. Partition $\boldsymbol{z}_i' = [\boldsymbol{u}_i' : \boldsymbol{v}_i']$, where $\boldsymbol{u}_i$ is $p \times 1$ and $\boldsymbol{v}_i$ is $m \times 1$. This means that $\boldsymbol{A}\boldsymbol{u}_i = \boldsymbol{B}\boldsymbol{v}_i$ for each $i = 1, 2, \ldots, k$. Let $\boldsymbol{x}_i = \boldsymbol{A}\boldsymbol{u}_i = \boldsymbol{B}\boldsymbol{v}_i$. Observe that $\boldsymbol{x}_i \neq \boldsymbol{0}$. Otherwise, $\boldsymbol{A}\boldsymbol{u}_i = \boldsymbol{B}\boldsymbol{v}_i = \boldsymbol{0}$, which would imply $\boldsymbol{u}_i = \boldsymbol{0}$ and $\boldsymbol{v}_i = \boldsymbol{0}$ because $\boldsymbol{A}$ and $\boldsymbol{B}$ both have linearly independent columns, and so $\boldsymbol{z}_i = \boldsymbol{0}$. This is impossible because the $\boldsymbol{z}_i$'s are linearly independent.

We claim that $\{\boldsymbol{x}_1, \boldsymbol{x}_2, \ldots, \boldsymbol{x}_k\}$ is a basis for $\mathcal{C}(\boldsymbol{A}) \cap \mathcal{C}(\boldsymbol{B})$. First, we establish linear independence. Let $\sum_{i=1}^k \alpha_i \boldsymbol{x}_i = \boldsymbol{0}$. Since each $\boldsymbol{x}_i = \boldsymbol{A}\boldsymbol{u}_i = \boldsymbol{B}\boldsymbol{v}_i$, we conclude

$$\begin{aligned} \sum_{i=1}^k \alpha_i \boldsymbol{A}\boldsymbol{u}_i = \sum_{i=1}^k \alpha_i \boldsymbol{B}\boldsymbol{v}_i = \boldsymbol{0} &\Longrightarrow \boldsymbol{A}\left(\sum_{i=1}^k \alpha_i \boldsymbol{u}_i\right) = \boldsymbol{B}\left(\sum_{i=1}^k \alpha_i \boldsymbol{v}_i\right) = \boldsymbol{0} \\ &\Longrightarrow \sum_{i=1}^k \alpha_i \boldsymbol{u}_i \in \mathcal{N}(\boldsymbol{A}) \text{ and } \sum_{i=1}^k \alpha_i \boldsymbol{v}_i \in \mathcal{N}(\boldsymbol{B}) \\ &\Longrightarrow \sum_{i=1}^k \alpha_i \begin{bmatrix} \boldsymbol{u}_i \\ \boldsymbol{v}_i \end{bmatrix} = \boldsymbol{0} \Longrightarrow \sum_{i=1}^k \alpha_i \boldsymbol{z}_i = \boldsymbol{0} \Longrightarrow \alpha_1 = \alpha_2 = \cdots = \alpha_k = 0 \,, \end{aligned}$$

where the last step follows from the linear independence of the $\boldsymbol{z}_i$'s. This proves that $\{\boldsymbol{x}_1, \boldsymbol{x}_2, \ldots, \boldsymbol{x}_k\}$ is linearly independent. Next, we show that $\{\boldsymbol{x}_1, \boldsymbol{x}_2, \ldots, \boldsymbol{x}_k\}$ spans $\mathcal{C}(\boldsymbol{A}) \cap \mathcal{C}(\boldsymbol{B})$. For any $\boldsymbol{x} \in \mathcal{C}(\boldsymbol{A}) \cap \mathcal{C}(\boldsymbol{B})$, assume that $\boldsymbol{u} \in \Re^p$ and $\boldsymbol{v} \in \Re^m$ are such that $\boldsymbol{x} = \boldsymbol{A}\boldsymbol{u} = \boldsymbol{B}\boldsymbol{v}$. This implies

$$[\boldsymbol{A} : -\boldsymbol{B}] \begin{bmatrix} \boldsymbol{u} \\ \boldsymbol{v} \end{bmatrix} = \boldsymbol{0} \Longrightarrow \boldsymbol{z} = \begin{bmatrix} \boldsymbol{u} \\ \boldsymbol{v} \end{bmatrix} \in \mathcal{N}([\boldsymbol{A} : -\boldsymbol{B}]) \,.$$

Therefore, $\boldsymbol{z}$ is spanned by $\boldsymbol{z}_1, \boldsymbol{z}_2, \ldots, \boldsymbol{z}_k$ and we can conclude that

$$\begin{aligned} \boldsymbol{z} = \sum_{i=1}^k \theta_i \boldsymbol{z}_i \Longrightarrow \begin{bmatrix} \boldsymbol{u} \\ \boldsymbol{v} \end{bmatrix} = \sum_{i=1}^k \theta_i \begin{bmatrix} \boldsymbol{u}_i \\ \boldsymbol{v}_i \end{bmatrix} &\Longrightarrow \boldsymbol{u} = \sum_{i=1}^k \theta_i \boldsymbol{u}_i \\ &\Longrightarrow \boldsymbol{x} = \boldsymbol{A}\boldsymbol{u} = \sum_{i=1}^k \theta_i \boldsymbol{A}\boldsymbol{u}_i = \sum_{i=1}^k \theta_i \boldsymbol{x}_i \,. \end{aligned}$$

Therefore, $\{\boldsymbol{x}_1, \boldsymbol{x}_2, \ldots, \boldsymbol{x}_k\}$ is a basis for $\mathcal{C}(\boldsymbol{A}) \cap \mathcal{C}(\boldsymbol{B})$ and so

$$\dim(\mathcal{C}(\boldsymbol{A}) \cap \mathcal{C}(\boldsymbol{B})) = \dim(\mathcal{N}([\boldsymbol{A} : -\boldsymbol{B}])) . \tag{6.2}$$

The Rank-Nullity Theorem and (6.2) immediately results in Theorem 6.1:

$$\begin{aligned}\dim(\mathcal{C}(\boldsymbol{A}) + \mathcal{C}(\boldsymbol{B})) &= \dim(\mathcal{C}([\boldsymbol{A} : \boldsymbol{B}])) = \dim(\mathcal{C}([\boldsymbol{A} : -\boldsymbol{B}])) \\ &= m + p - \dim(\mathcal{N}([\boldsymbol{A} : -\boldsymbol{B}])) \\ &= m + p - \dim(\mathcal{C}(\boldsymbol{A}) \cap \mathcal{C}(\boldsymbol{B})) \\ &= \dim(\mathcal{C}(\boldsymbol{A})) + \dim(\mathcal{C}(\boldsymbol{B})) - \dim(\mathcal{C}(\boldsymbol{A}) \cap \mathcal{C}(\boldsymbol{B})) . \end{aligned} \tag{6.3}$$

6.3 Direct sums and complements

Definition 6.1 *Two subspaces $\mathcal{S}_1$ and $\mathcal{S}_2$ in a vector space $\mathcal{V}$ are said to be* **complementary** *whenever*

$$\mathcal{V} = \mathcal{S}_1 + \mathcal{S}_2 \text{ and } \mathcal{S}_1 \cap \mathcal{S}_2 = \{\boldsymbol{0}\}. \tag{6.4}$$

In such cases, $\mathcal{V}$ is said to be the **direct sum** *of $\mathcal{S}_1$ and $\mathcal{S}_2$. We denote a direct sum as $\mathcal{V} = \mathcal{S}_1 \oplus \mathcal{S}_2$.*

The following theorem provides several useful characterizations of direct sums.

Theorem 6.2 *Let $\mathcal{V}$ be a vector space in $\Re^m$ and let $\mathcal{S}_1$ and $\mathcal{S}_2$ be two subspaces of $\mathcal{V}$. The following statements are equivalent:*

(i) $\mathcal{V} = \mathcal{S}_1 \oplus \mathcal{S}_2$.

(ii) $\dim(\mathcal{S}_1 + \mathcal{S}_2) = \dim(\mathcal{S}_1) + \dim(\mathcal{S}_2)$.

(iii) Any vector $\boldsymbol{x} \in \mathcal{S}_1 + \mathcal{S}_2$ can be uniquely *represented as*

$$\boldsymbol{x} = \boldsymbol{x}_1 + \boldsymbol{x}_2 , \quad \text{where} \quad \boldsymbol{x}_1 \in \mathcal{S}_1 \quad \text{and} \quad \boldsymbol{x}_2 \in \mathcal{S}_2 . \tag{6.5}$$

We will refer to this as the **unique representation** *or* **unique decomposition** *property of direct sums.*

(iv) If $\mathcal{B}_1$ and $\mathcal{B}_2$ are any basis of $\mathcal{S}_1$ and $\mathcal{S}_2$, respectively, then $\mathcal{B}_1 \cap \mathcal{B}_2$ is empty (i.e., they are disjoint) and $\mathcal{B}_1 \cup \mathcal{B}_2$ is a basis for $\mathcal{S}_1 + \mathcal{S}_2$.

Proof. We will prove that (i) $\Rightarrow$ (ii) $\Rightarrow$ (iii) $\Rightarrow$ (iv) $\Rightarrow$ (i).

Proof of (i) $\Rightarrow$ (ii): From Definition 6.1 statement (i) implies that $\mathcal{S}_1 \cap \mathcal{S}_2 = \{\boldsymbol{0}\}$. Therefore $\dim(\mathcal{S}_1 \cap \mathcal{S}_2) = 0$ and Theorem 6.1 tells us that $\dim(\mathcal{S}_1 + \mathcal{S}_2) = \dim(\mathcal{S}_1) + \dim(\mathcal{S}_2)$. □

Proof of (ii) $\Rightarrow$ (iii): Statement (ii) implies that $\dim(\mathcal{S}_1 \cap \mathcal{S}_2) = 0$ (from Theorem 6.1) and, therefore, $\mathcal{S}_1 \cap \mathcal{S}_2 = \{\boldsymbol{0}\}$. Now argue as follows.

Let $\boldsymbol{x} \in \mathcal{S}_1 + \mathcal{S}_2$ and suppose we have $\boldsymbol{x} = \boldsymbol{u}_1 + \boldsymbol{u}_2 = \boldsymbol{v}_1 + \boldsymbol{v}_2$ where $\boldsymbol{u}_1, \boldsymbol{v}_1 \in \mathcal{S}_1$

and u_2, $v_2 \in \mathcal{S}_2$. Then, $\mathbf{0} = (u_1 - v_1) + (u_2 - v_2)$. But this means that $(u_1 - v_1)$ is a scalar multiple of $(u_2 - v_2)$ and hence must belong to $\mathcal{S}_2$. Therefore, $(u_1 - v_1) \in \mathcal{S}_1 \cap \mathcal{S}_2$ so $u_1 = v_1$. Analogously we have $(u_2 - v_2) \in \mathcal{S}_1 \cap \mathcal{S}_2$ and so $u_2 = v_2$. Therefore, we have the unique representation. □

Proof of (iii) ⇒ (iv): Statement (iii) insures that $\mathcal{V} = \mathcal{S}_1 + \mathcal{S}_2$. Let $v \in \mathcal{V}$. Then $v = x + y$, where $x \in \mathcal{S}_1$ and $y \in \mathcal{S}_2$. First we prove that $\mathcal{B}_1$ and $\mathcal{B}_2$ are disjoint. Suppose there exists some vector $x \in \mathcal{B}_1 \cap \mathcal{B}_2$. Since $\mathcal{B}_1$ and $\mathcal{B}_2$ are linearly independent, we are insured that $x \neq \mathbf{0}$. But then $\mathbf{0} = x + (-x)$ and $\mathbf{0} = \mathbf{0} + \mathbf{0}$ are two different ways of expressing $\mathbf{0}$ as a sum of a vector from $\mathcal{S}_1$ and $\mathcal{S}_2$. This contradicts (iii). Therefore $\mathcal{B}_1 \cap \mathcal{B}_2$ must be empty.

Next we prove that $\mathcal{B}_1 \cup \mathcal{B}_2$ is a basis for $\mathcal{V}$.

Since x can be written as a linear combination of the vectors in $\mathcal{B}_1$ and y can be expressed as a linear combination of those in $\mathcal{B}_2$, it follows that $\mathcal{B}_1 \cup \mathcal{B}_2$ spans $\mathcal{V}$. It remains to prove that $\mathcal{B}_1 \cup \mathcal{B}_2$ is linearly independent. We argue as below.

Let $\mathcal{B}_1 = \{x_1, x_2, \ldots, x_m\}$ and $\mathcal{B}_2 = \{y_1, y_2, \ldots, y_n\}$. Consider the homogeneous linear combination

$$\mathbf{0} = \sum_{i=1}^{m} \alpha_i x_i + \sum_{j=1}^{n} \beta_j y_j. \tag{6.6}$$

Since $\sum_{i=1}^{m} \alpha_i x_i \in \mathcal{S}_1$ and $\sum_{j=1}^{n} \beta_j y_j \in \mathcal{S}_2$, (6.6) is one way to express $\mathbf{0}$ as a sum of a vector from $\mathcal{S}_1$ and from $\mathcal{S}_2$. Of course $\mathbf{0} = \mathbf{0} + \mathbf{0}$ is another way. Consequently, statement (ii) implies that

$$\sum_{i=1}^{r} \alpha_i x_i = \mathbf{0} \text{ and } \sum_{j=1}^{s} \beta_j y_j = \mathbf{0} .$$

This reveals that $\alpha_1 = \alpha_2 = \cdots = \alpha_m = 0$ and $\beta_1 = \beta_2 = \cdots = \beta_n = 0$ because $\mathcal{B}_1$ and $\mathcal{B}_2$ are both linearly independent. We have therefore proved that the only solution to (6.6) is the trivial solution. Therefore, $\mathcal{B}_1 \cup \mathcal{B}_2$ is linearly independent, and hence it is a basis for $\mathcal{V}$.

Proof of (iv) ⇒ (i): If $\mathcal{B}_1 \cup \mathcal{B}_2$ is a basis for $\mathcal{V}$, then $\mathcal{B}_1 \cup \mathcal{B}_2$ is a linearly independent set. Also, $\mathcal{B}_1 \cup \mathcal{B}_2$ spans $\mathcal{S}_1 + \mathcal{S}_2$. This means that $\mathcal{B}_1 \cup \mathcal{B}_2$ is a basis for $\mathcal{S}_1 + \mathcal{S}_2$ as well as for $\mathcal{V}$. Consequently, $\mathcal{V} = \mathcal{S}_1 + \mathcal{S}_2$, and hence

$$\dim(\mathcal{S}_1) + \dim(\mathcal{S}_2) = \dim(\mathcal{V}) = dim(\mathcal{S}_1 + \mathcal{S}_2) = \dim(\mathcal{S}_1) + \dim(\mathcal{S}_2) - \dim(\mathcal{S}_1 \cap \mathcal{S}_2),$$

so $\dim(\mathcal{S}_1 \cap \mathcal{S}_2) = 0$ or, equivalently, $\mathcal{S}_1 \cap \mathcal{S}_2 = \mathbf{0}$. □

The following corollary follows immediately from the equivalence of (i) and (iii) in Theorem 6.2

Corollary 6.2 *Let $\mathcal{S}_1$ and $\mathcal{S}_2$ be two subspaces of a vector space $\mathcal{V}$ in $\Re^m$. Then, the following statements are equivalent:*

(i) $\mathcal{V} = \mathcal{S}_1 \oplus \mathcal{S}_2$;

(ii) $\mathbf{0} = \boldsymbol{x}_1 + \boldsymbol{x}_2$, $\boldsymbol{x}_1 \in \mathcal{S}_1$, $\boldsymbol{x}_2 \in \mathcal{S}_2 \Rightarrow \boldsymbol{x}_1 = \mathbf{0}$ *and* $\boldsymbol{x}_2 = \mathbf{0}$.

Proof. From Theorem 6.2 we know that $\mathcal{V} = \mathcal{S}_1 \oplus \mathcal{S}_2$ if and only if every vector in $\mathcal{V}$ satisfies the unique representation property. Since $\mathbf{0} = \mathbf{0} + \mathbf{0}$, it follows $\mathbf{0} = \boldsymbol{x}_1 + \boldsymbol{x}_2$, where $\boldsymbol{x}_1 \in \mathcal{S}_1$ and $\boldsymbol{x}_2 \in \mathcal{S}_2$, must yield $\boldsymbol{x}_1 = \boldsymbol{x}_2 = \mathbf{0}$. □

The following lemma ensures that every subspace of a vector space has a complement.

Lemma 6.1 *Let $\mathcal{S}$ be a subspace of a vector space $\mathcal{V}$. Then there exists a subspace $\mathcal{T}$ such that $\mathcal{V} = \mathcal{S} \oplus \mathcal{T}$.*

Proof. Let $\dim(\mathcal{S}) = r < n = \dim(\mathcal{V})$ and suppose $\{\boldsymbol{x}_1, \boldsymbol{x}_2, \ldots, \boldsymbol{x}_r\}$ is a basis for $\mathcal{S}$. Since $\mathcal{S}$ is a subspace of $\mathcal{V}$, Lemma 4.21 ensures that there are vectors $\boldsymbol{x}_{r+1}, \boldsymbol{x}_{r+2}, \ldots, \boldsymbol{x}_n$ so that $\{\boldsymbol{x}_1, \boldsymbol{x}_2, \ldots, \boldsymbol{x}_r, \boldsymbol{x}_{r+1}, \ldots, \boldsymbol{x}_n\}$ is a basis of $\mathcal{V}$.

Let $\mathcal{T} = Sp\{\boldsymbol{x}_{r+1}, \ldots, \boldsymbol{x}_n\}$. We claim that $\mathcal{V} = \mathcal{S} \oplus \mathcal{T}$. Consider any $\boldsymbol{y} \in \mathcal{V}$. Then,

$$\boldsymbol{y} = \alpha_1 \boldsymbol{x}_1 + \alpha_2 \boldsymbol{x}_2 + \cdots \alpha_r \boldsymbol{x}_r + \alpha_{r+1} \boldsymbol{x}_{r+1} + \cdots \alpha_n \boldsymbol{x}_n = \boldsymbol{u} + \boldsymbol{v} ,$$

where $\boldsymbol{u} = \sum_{i=1}^{r} \alpha_i \boldsymbol{x}_i$ is in $\mathcal{S}$ and $\boldsymbol{v} = \sum_{i=r+1}^{n} \alpha_i \boldsymbol{x}_i$ resides in $\mathcal{T}$. This shows that $\mathcal{V} = \mathcal{S} + \mathcal{T}$.

It remains to show that $\mathcal{S} \cap \mathcal{T} = \{\mathbf{0}\}$. Suppose there exists some vector $\boldsymbol{z}$ such that $\boldsymbol{z} \in \mathcal{S} \cap \mathcal{T}$. Then $\boldsymbol{z}$ can be spanned by $\{\boldsymbol{x}_1, \boldsymbol{x}_2, \ldots, \boldsymbol{x}_r\}$ and $\{\boldsymbol{x}_{r+1}, \boldsymbol{x}_{r+2}, \ldots, \boldsymbol{x}_n\}$. Therefore, we can write

$$\boldsymbol{z} = \theta_1 \boldsymbol{x}_1 + \theta_2 \boldsymbol{x}_2 + \cdots + \theta_r \boldsymbol{x}_r = \theta_{r+1} \boldsymbol{x}_{r+1} + \theta_{r+2} \boldsymbol{x}_{r+2} + \cdots + \theta_n \boldsymbol{x}_n$$
$$\Rightarrow \mathbf{0} = \sum_{i=1}^{r} \theta_i \boldsymbol{x}_i + \sum_{i=r+1}^{n} (-\theta_i) \boldsymbol{x}_i \Rightarrow \theta_1 = \theta_2 = \ldots = \theta_n = 0$$

because $\{\boldsymbol{x}_1, \boldsymbol{x}_2, \ldots, \boldsymbol{x}_r, \boldsymbol{x}_{r+1}, \ldots, \boldsymbol{x}_n\}$ is linearly independent. Therefore $\boldsymbol{z} = \mathbf{0}$. This proves that $\mathcal{S} \cap \mathcal{T} = \{\mathbf{0}\}$ and, hence, $\mathcal{V} = \mathcal{S} \oplus \mathcal{T}$. □

Theorem 6.2 shows us that a sum $\mathcal{S}_1 + \mathcal{S}_2$ is direct if and only if every vector in $\mathcal{S}_1 + \mathcal{S}_2$ can be expressed in a unique way as $\boldsymbol{x}_1 + \boldsymbol{x}_2$ with $\boldsymbol{x}_1 \in \mathcal{S}_1$ and $\boldsymbol{x}_2 \in \mathcal{S}_2$. This property is often taken as the definition of a direct sum. In fact, it comes in quite handy if we want to extend the concept of a direct sum to more than two subspaces as we see in the following definition.

Definition 6.2 *Let $\mathcal{S}_1, \mathcal{S}_2, \ldots, \mathcal{S}_k$ be subspaces of a vector space. The sum $\mathcal{S}_1 + \mathcal{S}_2 + \cdots + \mathcal{S}_k$ is said to be a* **direct sum** *if any vector in $\mathcal{S}_1 + \mathcal{S}_2 + \cdots + \mathcal{S}_k$ can be expressed uniquely as $\boldsymbol{x}_1 + \boldsymbol{x}_2 + \cdots + \boldsymbol{x}_k$ with $\boldsymbol{x}_i \in \mathcal{S}_i$ for each $i = 1, 2, \ldots, k$. In that case, we write the sum as $\mathcal{S}_1 \oplus \mathcal{S}_2 \oplus \cdots \oplus \mathcal{S}_k$.*

The following theorem is the analogue of Theorem 6.2 for general direct sums.

Theorem 6.3 *Let $\mathcal{S}_1, \mathcal{S}_2, \ldots, \mathcal{S}_k$ be subspaces of a vector space. Then the following are equivalent:*

(i) $\mathcal{S}_1 + \mathcal{S}_2 + \cdots + \mathcal{S}_k$ *is a direct sum;*

(ii) $(\mathcal{S}_1 + \mathcal{S}_2 + \cdots + \mathcal{S}_i) \cap \mathcal{S}_{i+1} = \{\mathbf{0}\}$ *for* $i = 1, 2, \ldots, k-1$*;*

(iii) $\mathbf{0} = \boldsymbol{x}_1 + \boldsymbol{x}_2 + \cdots + \boldsymbol{x}_k$, $\boldsymbol{x}_i \in \mathcal{S}_i$ *for each* $i = 1, 2, \ldots, k \Rightarrow \boldsymbol{x}_1 = \boldsymbol{x}_2 = \cdots = \boldsymbol{x}_k = \mathbf{0}$.

Proof. We will prove (i) $\Rightarrow$ (ii) $\Rightarrow$ (iii) $\Rightarrow$ (i). **Proof of (i) $\Rightarrow$ (ii):** Consider any $i = 1, 2, \ldots, k-1$ and suppose $\boldsymbol{x} \in (\mathcal{S}_1 + \mathcal{S}_2 + \mathcal{S}_i) \cap \mathcal{S}_{i+1}$. Since $\boldsymbol{x} \in (\mathcal{S}_1 + \mathcal{S}_2 + \mathcal{S}_i)$ we can write

$$\boldsymbol{x} = \boldsymbol{x}_1 + \cdots + \boldsymbol{x}_i = \boldsymbol{x}_1 + \cdots + \boldsymbol{x}_i + \mathbf{0} + \mathbf{0} + \cdots + \mathbf{0} \,,$$

where there are $k - i$ zeros added at the end to express $\boldsymbol{x}$ as a sum of k terms, one each from $\mathcal{S}_j$ for $j = 1, 2, \ldots, k$. Also, we because $\boldsymbol{x} \in \mathcal{S}_{i+1}$, we can write

$$\boldsymbol{x} = \mathbf{0} + \cdots + \mathbf{0} + \boldsymbol{x} + \mathbf{0} + \mathbf{0} + \cdots + \mathbf{0} \,,$$

where the $\boldsymbol{x}$ on the left-hand side is again expressed as a sum of k terms with $\boldsymbol{x}$ appearing as the $(i+1)$-th term in the sum on the right-hand side. We equate the above two representations for $\boldsymbol{x}$ to obtain

$$\begin{aligned} \boldsymbol{x} &= \boldsymbol{x}_1 + \cdots + \boldsymbol{x}_i + \mathbf{0} + \mathbf{0} + \cdots + \mathbf{0} \\ &= \mathbf{0} + \cdots + \mathbf{0} + \boldsymbol{x} + \mathbf{0} + \mathbf{0} + \cdots + \mathbf{0} \,. \end{aligned}$$

Since $\mathcal{S}_1 + \mathcal{S}_2 + \cdots + \mathcal{S}_k$ is direct, we can use the unique representation property in Definition 6.2 to conclude that $\boldsymbol{x} = \mathbf{0}$ by comparing the $(i+1)$-th terms. This proves (i) $\Rightarrow$ (ii).

Proof of (ii) $\Rightarrow$ (iii): Let $\mathbf{0} = \boldsymbol{x}_1 + \boldsymbol{x}_2 + \cdots + \boldsymbol{x}_k$, where $\boldsymbol{x}_i \in \mathcal{S}_i$ for each $i = 1, 2, \ldots, k$. Therefore, $\boldsymbol{x}_k = -(\boldsymbol{x}_1 + \boldsymbol{x}_2 + \cdots + \boldsymbol{x}_{k-1}) \in (\mathcal{S}_1 + \cdots + \mathcal{S}_{k-1}) \cap \mathcal{S}_k$, which means that $\boldsymbol{x}_k = \mathbf{0}$. This means that $\mathbf{0} = \boldsymbol{x}_1 + \boldsymbol{x}_2 + \cdots + \boldsymbol{x}_{k-1}$ and repeating the preceding argument yields $\boldsymbol{x}_{k-1} = \mathbf{0}$. Proceeding in this manner, we eventually obtain $\boldsymbol{x}_i = \mathbf{0}$ for $i = 1, 2, \ldots, k$. This proves (ii) $\Rightarrow$ (iii).

Proof of (iii) $\Rightarrow$ (i): The proof of this is analogous to Corollary 6.2 and is left to the reader.

This proves the equivalence of statements (i), (ii) and (iii). □

Corollary 6.3 *Let* $\mathcal{S}_1, \mathcal{S}_2, \ldots, \mathcal{S}_k$ *be subspaces of a vector space. Then the following are equivalent:*

(i) $\mathcal{S}_1 + \mathcal{S}_2 + \cdots + \mathcal{S}_k$ *is a direct sum;*

(ii) $\dim(\mathcal{S}_1 + \mathcal{S}_2 + \cdots + \mathcal{S}_k) = \dim(\mathcal{S}_1) + \dim(\mathcal{S}_2) + \cdots + \dim(\mathcal{S}_k)$.

Proof. We prove (i) $\Longleftrightarrow$ (ii).

Proof of (i) $\Rightarrow$ (ii): From Theorem 6.3, we know that when $\mathcal{S}_1 + \mathcal{S}_2 + \cdots + \mathcal{S}_k$ is direct, we have $(\mathcal{S}_1 + \mathcal{S}_2 + \cdots + \mathcal{S}_i) \cap \mathcal{S}_{i+1} = \{\mathbf{0}\}$ for $i = 1, 2, \ldots, k-1$.

Therefore, using the dimension of the sum of subspaces (Theorem 6.1) successively for $i = k-1$, $i = k-2$ and so on yields

$$\begin{aligned}\dim(\mathcal{S}_1 + \mathcal{S}_2 + \cdots + \mathcal{S}_k) &= \dim(\mathcal{S}_1 + \mathcal{S}_2 + \cdots + \mathcal{S}_{k-1}) + \dim(\mathcal{S}_k) \\ &= \dim(\mathcal{S}_1 + \mathcal{S}_2 + \cdots + \mathcal{S}_{k-2}) + \dim(\mathcal{S}_{k-1}) + \dim(\mathcal{S}_k) \\ &\cdots\cdots \\ &= \dim(\mathcal{S}_1) + \dim(\mathcal{S}_2) + \cdots + \dim(\mathcal{S}_k) \,.\end{aligned}$$

This proves (i) $\Rightarrow$ (ii).

Proof of (ii) $\Rightarrow$ (i): Successive application of Part (i) of Corollary 6.1 yields

$$\begin{aligned}\dim(\mathcal{S}_1 + \mathcal{S}_2 + \cdots + \mathcal{S}_k) &\leq \dim(\mathcal{S}_1 + \mathcal{S}_2 + \cdots + \mathcal{S}_{k-1}) + \dim(\mathcal{S}_k) \\ &\leq \dim(\mathcal{S}_1 + \mathcal{S}_2 + \cdots + \mathcal{S}_{k-2}) + \dim(\mathcal{S}_{k-1}) + \dim(\mathcal{S}_k) \\ &\cdots\cdots \\ &\leq \dim(\mathcal{S}_1) + \dim(\mathcal{S}_2) + \cdots + \dim(\mathcal{S}_k) \,.\end{aligned}$$

Now, (ii) implies that the first and last terms in the above chain of inequalities are equal. Therefore, equality holds throughout. This means, by virtue of Theorem 6.1, that $\dim((\mathcal{S}_1+\mathcal{S}_2+\cdots+\mathcal{S}_i)\cap\mathcal{S}_{i+1}) = 0$ and, so, $(\mathcal{S}_1+\mathcal{S}_2+\cdots+\mathcal{S}_i)\cap\mathcal{S}_{i+1} = \{\mathbf{0}\}$ for $i = 1, 2, \ldots, k-1$. This implies that $\mathcal{S}_1+\mathcal{S}_2+\cdots+\mathcal{S}_k$ is direct (from Theorem 6.3), which proves (ii) $\Rightarrow$ (i). □

6.4 Projectors

Let $\mathcal{S}$ be a subspace of $\Re^n$ and let $\mathcal{T}$ be its complement so that $\Re^n = \mathcal{S} \oplus \mathcal{T}$. Part (iii) of Theorem 6.2 ensures that for every vector $\boldsymbol{y}$ in $\Re^n$, there are unique vectors $\boldsymbol{x}$ in $\mathcal{S}$ and $\boldsymbol{z}$ in $\mathcal{T}$ such that $\boldsymbol{y} = \boldsymbol{x} + \boldsymbol{z}$. This leads to the following definition.

Definition 6.3 *Let $\Re^n = \mathcal{S} \oplus \mathcal{T}$ and let $\boldsymbol{y} = \boldsymbol{x} + \boldsymbol{z}$ be the unique representation of $\boldsymbol{y} \in \Re^n$ with $\boldsymbol{x} \in \mathcal{S}$ and $\boldsymbol{z} \in \mathcal{T}$. Then:*

- *the vector $\boldsymbol{x}$ is called the* **projection** *of $\boldsymbol{y}$ into $\mathcal{S}$ along $\mathcal{T}$;*
- *the vector $\boldsymbol{z}$ is called the* **projection** *of $\boldsymbol{y}$ into $\mathcal{T}$ along $\mathcal{S}$.*

Given a vector $\boldsymbol{y} \in \Re^n$ and a pair of complementary subspaces $\mathcal{S}$ and $\mathcal{T}$ of $\mathcal{V}$, how do we compute the projection of $\boldsymbol{y}$ into $\mathcal{S}$? It will be especially convenient if we can find a matrix $\boldsymbol{P}$ such that $\boldsymbol{P}\boldsymbol{y}$ will be the projection of $\boldsymbol{y}$ into $\mathcal{S}$ along $\mathcal{T}$.

A matrix $\boldsymbol{P}$ that will project every vector $\boldsymbol{y} \in \Re^n$ into $\mathcal{S}$ so that $\boldsymbol{P}\boldsymbol{y} \in \mathcal{S}$ *must* be a square matrix. Why? Because $\boldsymbol{P}\boldsymbol{y}$ is itself a vector that belongs to $\Re^n$ (remember $\mathcal{S} \subseteq \Re^n$) and we should be able to obtain a projection of $\boldsymbol{P}\boldsymbol{y}$ by pre-multiplying it with $\boldsymbol{P}$. In other words, $\boldsymbol{P}(\boldsymbol{P}\boldsymbol{y}) = \boldsymbol{P}^2\boldsymbol{y}$ must make sense. This means that $\boldsymbol{P}$ must be conformable to multiplication with itself and, hence, $\boldsymbol{P}$ must be a square matrix.

We define a *projector* for complementary subspaces in $\Re^n$ as follows.

Definition 6.4 *Let $\Re^n = \mathcal{S} \oplus \mathcal{T}$. An $n \times n$ matrix $\boldsymbol{P}$ is said to be the* **projector** *into $\mathcal{S}$ along $\mathcal{T}$ if, for every $\boldsymbol{y} \in \Re^n$, $\boldsymbol{P}\boldsymbol{y}$ is the projection of $\boldsymbol{y}$ onto $\mathcal{S}$ along $\mathcal{T}$. We say that a matrix $\boldsymbol{P}$ is* **a projector** *if it is a projector into some subspace $\mathcal{S}$ along its complement $\mathcal{T}$.*

What properties should such a projector have? The next theorem summarizes some useful properties of a projector. Before that, it will be useful to present the following definition.

Definition 6.5 *A square matrix $\boldsymbol{P}$ is said to be* **idempotent** *if $\boldsymbol{P}^2 = \boldsymbol{P}$.*

Theorem 6.4 *The following statements about an $n \times n$ matrix $\boldsymbol{P}$ are equivalent:*

(i) $\boldsymbol{P}$ *is a projector.*
(ii) $\boldsymbol{P}$ *and* $\boldsymbol{I} - \boldsymbol{P}$ *are idempotent, i.e.,* $\boldsymbol{P}^2 = \boldsymbol{P}$ *and* $(\boldsymbol{I} - \boldsymbol{P})^2 = \boldsymbol{I} - \boldsymbol{P}$.
(iii) $\mathcal{N}(\boldsymbol{P}) = \mathcal{C}(\boldsymbol{I} - \boldsymbol{P})$.
(iv) $\rho(\boldsymbol{P}) + \rho(\boldsymbol{I} - \boldsymbol{P}) = n$.
(v) $\Re^n = \mathcal{C}(\boldsymbol{P}) \oplus \mathcal{C}(\boldsymbol{I} - \boldsymbol{P})$.

Proof. We will show that (i) ⇒ (ii) ⇒ (iii) ⇒ (iv) ⇒ (v) ⇒ (i).

Proof of (i) ⇒ (ii): Let $\mathcal{S}$ and $\mathcal{T}$ be complementary subspaces of $\Re^n$ and suppose that $\boldsymbol{P}$ is a projector into $\mathcal{S}$ along $\mathcal{T}$. Then for every $\boldsymbol{y} \in \Re^n$, $\boldsymbol{P}\boldsymbol{y} \in \mathcal{S}$. Therefore, $\boldsymbol{P}^2\boldsymbol{y} = \boldsymbol{P}(\boldsymbol{P}\boldsymbol{y}) = \boldsymbol{P}\boldsymbol{y}$ for every $\boldsymbol{y} \in \Re^n$. In particular, we obtain

$$\boldsymbol{P}^2_{*j} = \boldsymbol{P}^2\boldsymbol{e}_{*j} = \boldsymbol{P}\boldsymbol{e}_{*j} = \boldsymbol{P}_{*j} \text{ for } j = 1, 2, \ldots, n \,.$$

This means that the j-th column $\boldsymbol{P}^2$ equals the j-th column of $\boldsymbol{P}$ for $j = 1, 2, \ldots, n$. Therefore $\boldsymbol{P}^2 = \boldsymbol{P}$, which shows that $\boldsymbol{P}$ is idempotent. This immediately yields

$$(\boldsymbol{I} - \boldsymbol{P})^2 = \boldsymbol{I} - 2\boldsymbol{P} + \boldsymbol{P}^2 = \boldsymbol{I} - 2\boldsymbol{P} + \boldsymbol{P} = \boldsymbol{I} - \boldsymbol{P} \,,$$

which proves that $\boldsymbol{I} - \boldsymbol{P}$ is also idempotent. Therefore, (i) ⇒ (ii).

Proof of (ii) ⇒ (iii): From (ii), we obtain $\boldsymbol{P}(\boldsymbol{I} - \boldsymbol{P}) = \boldsymbol{P} - \boldsymbol{P}^2 = \boldsymbol{O}$. Now,

$$\begin{aligned} \boldsymbol{x} \in \mathcal{C}(\boldsymbol{I} - \boldsymbol{P}) &\Longrightarrow \boldsymbol{x} = (\boldsymbol{I} - \boldsymbol{P})\boldsymbol{v} \quad \text{for some} \quad \boldsymbol{v} \in \Re^n \\ &\Longrightarrow \boldsymbol{P}\boldsymbol{x} = \boldsymbol{P}(\boldsymbol{I} - \boldsymbol{P})\boldsymbol{v} = \boldsymbol{O}\boldsymbol{v} = \boldsymbol{0} \Longrightarrow \mathcal{C}(\boldsymbol{I} - \boldsymbol{P}) \subseteq \mathcal{N}(\boldsymbol{P}) \,. \end{aligned}$$

For the reverse inclusion, note that

$$\boldsymbol{x} \in \mathcal{N}(\boldsymbol{P}) \Longrightarrow \boldsymbol{P}\boldsymbol{x} = \boldsymbol{0} \Longrightarrow \boldsymbol{x} = \boldsymbol{x} - \boldsymbol{P}\boldsymbol{x} = (\boldsymbol{I} - \boldsymbol{P})\boldsymbol{x} \in \mathcal{C}(\boldsymbol{I} - \boldsymbol{P}) \,.$$

Therefore, $\mathcal{N}(\boldsymbol{P}) = \mathcal{C}(\boldsymbol{I} - \boldsymbol{P})$ and we have proved (ii) ⇒ (iii).

Proof of (iii) ⇒ (iv): Equating the dimensions of the two sides in (iii), we obtain

$$\rho(\boldsymbol{I} - \boldsymbol{P}) = \dim(\mathcal{C}(\boldsymbol{I} - \boldsymbol{P})) = \dim(\mathcal{N}(\boldsymbol{P})) = \nu(\boldsymbol{P}) = n - \rho(\boldsymbol{P}) \,,$$

which implies (iv).

Proof of (iv) $\Rightarrow$ (v): Any $\boldsymbol{x} \in \Re^n$ can be written as $\boldsymbol{x} = \boldsymbol{P}\boldsymbol{x} + (\boldsymbol{I} - \boldsymbol{P})\boldsymbol{x}$, which is a sum of a vector from $\mathcal{C}(\boldsymbol{P})$ and a vector from $\mathcal{C}(\boldsymbol{I} - \boldsymbol{P})$. This shows that $\Re^n = \mathcal{C}(\boldsymbol{P}) + \mathcal{C}(\boldsymbol{I} - \boldsymbol{P})$. (Note: This is true for any $n \times n$ matrix $\boldsymbol{P}$, not just idempotent matrices.) Part (iv) shows that the dimensions are additive:

$$\dim(\mathcal{C}(\boldsymbol{P})) + \dim(\mathcal{C}(\boldsymbol{I} - \boldsymbol{P})) = n = \dim(\Re^n) = \dim(\mathcal{C}(\boldsymbol{P}) + \mathcal{C}(\boldsymbol{I} - \boldsymbol{P})) .$$

Part (ii) of Theorem 6.2, therefore, says that the sum is direct. In other words, $\Re^n = \mathcal{C}(\boldsymbol{P}) \oplus \mathcal{C}(\boldsymbol{I} - \boldsymbol{P})$.

Proof of (v) $\Rightarrow$ (i) The fact that any vector $\boldsymbol{x}$ in $\Re^n$ can be written as $\boldsymbol{x} = \boldsymbol{P}\boldsymbol{x} + (\boldsymbol{I} - \boldsymbol{P})\boldsymbol{x}$, together with (v) implies that $\mathcal{C}(\boldsymbol{I} - \boldsymbol{P})$ is the complement of $\mathcal{C}(\boldsymbol{P})$ and that $\boldsymbol{P}\boldsymbol{x}$ is the projection of $\boldsymbol{x}$ into $\mathcal{C}(\boldsymbol{P})$ along $\mathcal{C}(\boldsymbol{I} - \boldsymbol{P})$. This proves (v). □

The equivalence of (i) and (ii) in Theorem 6.4 leads to the important characterization that every projector is an idempotent matrix and every idempotent matrix is a projector. In fact, some authors often define a projector to be simply an idempotent matrix. The following corollary follows from the equivalence of parts (ii) and (v) in Theorem 6.4, but a direct proof can be constructed easily.

Corollary 6.4 *Let $\boldsymbol{P}$ be square. Then $\mathcal{C}(\boldsymbol{P}) \cap \mathcal{C}(\boldsymbol{I} - \boldsymbol{P}) = \{\boldsymbol{0}\}$ if and only if $\boldsymbol{P}^2 = \boldsymbol{P}$.*

Proof. Let $\boldsymbol{P}$ be $n \times n$. Suppose $\boldsymbol{x} \in \mathcal{C}(\boldsymbol{P}) \cap \mathcal{C}(\boldsymbol{I} - \boldsymbol{P})$. Therefore, there exists $\boldsymbol{u}$ and $\boldsymbol{v}$ in $\Re^n$ such that $\boldsymbol{x} = \boldsymbol{P}\boldsymbol{u} = (\boldsymbol{I} - \boldsymbol{P})\boldsymbol{v}$ and we have

$$\boldsymbol{x} = \boldsymbol{P}\boldsymbol{u} = \boldsymbol{P}^2\boldsymbol{u} = \boldsymbol{P}\boldsymbol{P}\boldsymbol{u} = \boldsymbol{P}(\boldsymbol{I} - \boldsymbol{P})\boldsymbol{v} = (\boldsymbol{P} - \boldsymbol{P}^2)\boldsymbol{v} = \boldsymbol{O}\boldsymbol{v} = \boldsymbol{0} .$$

This proves that $\mathcal{C}(\boldsymbol{P}) \cap \mathcal{C}(\boldsymbol{I} - \boldsymbol{P}) = \{\boldsymbol{0}\}$.

For the reverse implication, assume that $\mathcal{C}(\boldsymbol{P}) \cap \mathcal{C}(\boldsymbol{I} - \boldsymbol{P}) = \{\boldsymbol{0}\}$. Let $\boldsymbol{u}$ be any vector in $\Re^n$ and let $\boldsymbol{v} = (\boldsymbol{I} - \boldsymbol{P})\boldsymbol{P}\boldsymbol{u}$. Clearly, $\boldsymbol{v} \in \mathcal{C}(\boldsymbol{I} - \boldsymbol{P})$ Also,

$$\boldsymbol{v} = (\boldsymbol{I} - \boldsymbol{P})\boldsymbol{P}\boldsymbol{u} = (\boldsymbol{P} - \boldsymbol{P}^2)\boldsymbol{u} = \boldsymbol{P}(\boldsymbol{I} - \boldsymbol{P})\boldsymbol{u} \in \mathcal{C}(\boldsymbol{P}) .$$

Therefore, $\boldsymbol{v} \in \mathcal{C}(\boldsymbol{P}) \cap \mathcal{C}(\boldsymbol{I} - \boldsymbol{P})$ and so $\boldsymbol{v} = \boldsymbol{0}$. This proves that $(\boldsymbol{I} - \boldsymbol{P})\boldsymbol{P}\boldsymbol{u} = \boldsymbol{0}$ for all $\boldsymbol{u} \in \Re^n$ and so $(\boldsymbol{I} - \boldsymbol{P})\boldsymbol{P} = \boldsymbol{O}$. Therefore, $\boldsymbol{P} = \boldsymbol{P}^2$. □

The following corollary lists a few other useful properties of a projector that follow immediately from Definition 6.4 and Theorem 6.4.

Corollary 6.5 *Let $\Re^n = \mathcal{S} \oplus \mathcal{T}$ and let $\boldsymbol{P}$ be a projector into $\mathcal{S}$ along $\mathcal{T}$.*

(i) $\boldsymbol{P}\boldsymbol{x} = \boldsymbol{x}$ for all $\boldsymbol{x} \in \mathcal{S}$ and $\boldsymbol{P}\boldsymbol{z} = \boldsymbol{0}$ for all $\boldsymbol{z} \in \mathcal{T}$.

(ii) $\mathcal{C}(\boldsymbol{P}) = \mathcal{S}$ and $\mathcal{N}(\boldsymbol{P}) = \mathcal{C}(\boldsymbol{I} - \boldsymbol{P}) = \mathcal{T}$.

(iii) $(\boldsymbol{I} - \boldsymbol{P})\boldsymbol{y}$ is the projection of $\boldsymbol{y}$ into $\mathcal{T}$ along $\mathcal{S}$.

*(iv) $\boldsymbol{P}$ is the **unique** projector.*

Proof. **Proof of (i):** Consider any $\boldsymbol{x} \in \mathcal{S}$. We can always write $\boldsymbol{x} = \boldsymbol{x} + \boldsymbol{0}$ and Part (iii) of Theorem 6.2 tells us that this must be the unique representation of $\boldsymbol{x}$ in terms of a vector in $\mathcal{S}$ ($\boldsymbol{x}$ itself) and a vector ($\boldsymbol{0}$) in $\mathcal{T}$. So, in this case, the projection of $\boldsymbol{x}$ is $\boldsymbol{x}$ itself. Therefore, $\boldsymbol{P}\boldsymbol{x} = \boldsymbol{x}$.

Next suppose $\boldsymbol{z} \in \mathcal{T}$. We can express $\boldsymbol{z}$ in two ways:

$$\boldsymbol{z} = \boldsymbol{0} + \boldsymbol{z} \quad \text{and} \quad \boldsymbol{z} = \boldsymbol{P}\boldsymbol{z} + (\boldsymbol{I} - \boldsymbol{P})\boldsymbol{z} \, .$$

Since $\mathcal{S}$ and $\mathcal{T}$ are complementary subspaces, $\boldsymbol{z}$ must have a unique decomposition in terms of vectors from $\mathcal{S}$ and $\mathcal{T}$. This implies that $\boldsymbol{P}\boldsymbol{z} = \boldsymbol{0}$.

Proof of (ii): Consider any vector $\boldsymbol{x} \in \mathcal{S}$. From (i), we have that $\boldsymbol{x} = \boldsymbol{P}\boldsymbol{x} \in \mathcal{C}(\boldsymbol{P})$. Therefore, $\mathcal{S} \subseteq \mathcal{C}(\boldsymbol{P})$. Also, since $\boldsymbol{P}$ is a projector into $\mathcal{S}$, by definition we have that $\boldsymbol{P}\boldsymbol{y} \in \mathcal{S}$ for every $\boldsymbol{y} \in \Re^n$. This implies $\mathcal{C}(\boldsymbol{P}) \subseteq \mathcal{S}$. Therefore, $\mathcal{C}(\boldsymbol{P}) = \mathcal{S}$.

We already saw in Theorem 6.4 that $\mathcal{N}(\boldsymbol{P}) = \mathcal{C}(\boldsymbol{I} - \boldsymbol{P})$. Now consider any vector $\boldsymbol{x} \in \mathcal{T}$. From part (i), we have that $\boldsymbol{P}\boldsymbol{x} = \boldsymbol{0}$. This means that we can write $\boldsymbol{x} = \boldsymbol{x} - \boldsymbol{P}\boldsymbol{x} = (\boldsymbol{I} - \boldsymbol{P})\boldsymbol{x} \in \mathcal{C}(\boldsymbol{I} - \boldsymbol{P})$. Therefore, $\mathcal{T} \subseteq \mathcal{C}(\boldsymbol{I} - \boldsymbol{P})$. Also, since $\boldsymbol{y} = \boldsymbol{P}\boldsymbol{y} + (\boldsymbol{I} - \boldsymbol{P})\boldsymbol{y}$ for any $\boldsymbol{y} \in \Re^n$, and $\boldsymbol{P}\boldsymbol{y} \in \mathcal{S}$, it follows that $(\boldsymbol{I} - \boldsymbol{P})\boldsymbol{y} \in \mathcal{T}$ (because $\mathcal{S}$ and $\mathcal{T}$ are complementary). This proves the reverse inclusion $\mathcal{C}(\boldsymbol{I} - \boldsymbol{P}) \subseteq \mathcal{T}$ and, hence, we have $\mathcal{C}(\boldsymbol{I} - \boldsymbol{P}) = \mathcal{T}$.

Proof of (iii): Since $\boldsymbol{y} = \boldsymbol{P}\boldsymbol{y} + (\boldsymbol{I} - \boldsymbol{P})\boldsymbol{y}$ and $\boldsymbol{P}\boldsymbol{y} \in \mathcal{S}$ it follows from the definition of the projector that $(\boldsymbol{I} - \boldsymbol{P})\boldsymbol{y}$ is the projection of $\boldsymbol{y}$ into $\mathcal{T}$ along $\mathcal{S}$.

Proof of (iv): It remains to show that $\boldsymbol{P}$ is unique. This is evident from the criteria of uniqueness of the projection (Part (iii) of Theorem 6.2) because if $\boldsymbol{P}_1$ and $\boldsymbol{P}_2$ are both projectors, then $\boldsymbol{P}_1\boldsymbol{x} = \boldsymbol{P}_2\boldsymbol{x}$ for all $\boldsymbol{x} \in \Re^n$, which implies that $\boldsymbol{P}_1 = \boldsymbol{P}_2$. □

Corollary 6.6 *If $\boldsymbol{P}$ is an idempotent matrix it is a projector into $\mathcal{C}(\boldsymbol{P})$ along $\mathcal{N}(\boldsymbol{P})$. Also, $\boldsymbol{I} - \boldsymbol{P}$ is the projector into $\mathcal{N}(\boldsymbol{P})$ along $\mathcal{C}(\boldsymbol{P})$.*

Proof. From Lemma 6.4, we know that if $\boldsymbol{P}$ is an idempotent matrix, it is a projector. Parts (ii) and (iii) of Corollary 6.5 show that $\boldsymbol{P}$ is a projector into $\mathcal{C}(\boldsymbol{P})$ along $\mathcal{N}(\boldsymbol{P})$ and that $\boldsymbol{I} - \boldsymbol{P}$ is a projector $\boldsymbol{I} - \boldsymbol{P}$ is the projector into $\mathcal{N}(\boldsymbol{P})$ along $\mathcal{C}(\boldsymbol{P})$. □

It is often useful to characterize projectors in terms of rank factorizations. The following theorem does that.

Theorem 6.5 *Let $\boldsymbol{P}$ be an $n \times n$ matrix of rank r and suppose that $\boldsymbol{P} = \boldsymbol{S}\boldsymbol{U}'$ is a rank-factorization of $\boldsymbol{P}$. Then $\boldsymbol{P}$ is a projector if and only if $\boldsymbol{U}'\boldsymbol{S} = \boldsymbol{I}_r$.*

Proof. Since $\boldsymbol{P} = \boldsymbol{S}\boldsymbol{U}'$ is a rank factorization, it follows that $\boldsymbol{S}$ and $\boldsymbol{U}'$ are $n \times r$ and $r \times n$ matrices, each of rank r.

First suppose that $\boldsymbol{U}'\boldsymbol{S} = \boldsymbol{I}_r$. Then,

$$\boldsymbol{P}^2 = \boldsymbol{P}\boldsymbol{P} = (\boldsymbol{S}\boldsymbol{U}')(\boldsymbol{S}\boldsymbol{U}') = \boldsymbol{S}(\boldsymbol{U}'\boldsymbol{S})\boldsymbol{U}' = \boldsymbol{S}\boldsymbol{I}_r\boldsymbol{U}' = \boldsymbol{S}\boldsymbol{U}' = \boldsymbol{P} \, .$$

This proves that $P^2 = P$ and, hence, that P is a projector.

Now we prove the converse. Since S is of full column rank and U' is of full row rank, there exists a left-inverse, say S_L, for S and a right inverse, say U'_R for U'. This means that S can be canceled from the left and U' can be canceled from the right. Therefore,

$$P^2 = P \Rightarrow SU'SU' = SU' \Rightarrow S_LSU'SU'U'_R = S_LSU'U'_R \Rightarrow U'S = I_r \, .$$

This completes the proof. □

The following theorem says that the rank and trace of a projector is the same. This is an extremely useful property of projectors that is used widely in statistics.

Theorem 6.6 *Let P be a projector (or, equivalently, an idempotent matrix). Then $tr(P) = \rho(P)$.*

Proof. Let $P = SU'$ be a rank factorization of P, where $\rho(P) = r$, S is $n \times r$ and U' is $r \times n$. From Theorem 6.5, we know that $U'S = I_r$. Therefore, we can write

$$\text{tr}(P) = \text{tr}(SU') = \text{tr}(U'S) = \text{tr}(I_r) = r = \rho(P) \, .$$

This completes the proof. □

We have an important task at hand: Given a pair of complementary subspaces in $\Re^n$, how do we construct a projector? The following lemma uses Theorem 6.5 to provide us with an explicit construction.

Theorem 6.7 *Let $\Re^n = \mathcal{S} \oplus \mathcal{T}$ such that $\dim(\mathcal{S}) = r < n$ and $\dim(\mathcal{T}) = n - r$. Let S be an $n \times r$ matrix and let T be a an $n \times (n-r)$ matrix whose column vectors constitute bases for $\mathcal{S}$ and $\mathcal{T}$, respectively. Form the matrices*

$$A = [S : T] \quad \textit{and} \quad A^{-1} = \begin{bmatrix} U' \\ V' \end{bmatrix} \, , \tag{6.7}$$

where U' is $r \times n$ so that it is conformable to multiplication with S. Then $P = SU'$ is the projector onto $\mathcal{S}$ along $\mathcal{T}$.

Proof. By virtue of Theorem 6.5, it suffices to show that (i) $P = SU'$ is a rank factorization, and (ii) $U'S = I_r$.

By construction, S has full column rank—its columns form a basis for $\mathcal{S}$ and, hence, must be linearly independent. Also, U' has full row rank. This is because the rows of U' are a subset of the rows of a nonsingular matrix (the inverse of A) and must be linearly independent. This means that $U'U$ is invertible and $U'G = I_r$, where $G = U(U'U)^{-1}$. Therefore,

$$\mathcal{C}(P) = \mathcal{C}(SU') \subseteq \mathcal{C}(S) = \mathcal{C}(SU'G) \subseteq \mathcal{C}(SU') = \mathcal{C}(P) \, .$$

This proves $\mathcal{C}(P) = \mathcal{C}(S) = \mathcal{S}$. Therefore, $\rho(P) = \rho(S) = r$ and $P = SU'$ is indeed a rank factorization.

It remains to show that $\boldsymbol{U}'\boldsymbol{S} = \boldsymbol{I}_r$. Since $\boldsymbol{A}^{-1}\boldsymbol{A} = \boldsymbol{I}_n$, the construction in (6.7) immediately yields

$$\begin{bmatrix} \boldsymbol{I}_r & \boldsymbol{O}_{r\times(n-r)} \\ \boldsymbol{O}_{(n-r)\times r} & \boldsymbol{I}_{n-r} \end{bmatrix} = \boldsymbol{A}^{-1}\boldsymbol{A} = \begin{bmatrix} \boldsymbol{U}' \\ \boldsymbol{V}' \end{bmatrix} [\boldsymbol{S} : \boldsymbol{T}] = \begin{bmatrix} \boldsymbol{U}'\boldsymbol{S} & \boldsymbol{U}'\boldsymbol{T} \\ \boldsymbol{V}'\boldsymbol{S} & \boldsymbol{V}'\boldsymbol{T} \end{bmatrix} .$$

This shows that $\boldsymbol{U}'\boldsymbol{S} = \boldsymbol{I}_r$ and the proof is complete. □

An alternative expression for the projector can be derived as follows. Using the same notations as in Theorem 6.7, we find that any projector $\boldsymbol{P}$ must satisfy

$$\boldsymbol{P}\boldsymbol{A} = \boldsymbol{P}[\boldsymbol{S} : \boldsymbol{T}] = [\boldsymbol{P}\boldsymbol{S} : \boldsymbol{P}\boldsymbol{T}] = [\boldsymbol{S} : \boldsymbol{O}] \Longrightarrow \boldsymbol{P} = [\boldsymbol{S} : \boldsymbol{O}]\boldsymbol{A}^{-1}$$

$$\Longrightarrow \boldsymbol{P} = [\boldsymbol{S} : \boldsymbol{T}] \begin{bmatrix} \boldsymbol{I}_{r\times r} & \boldsymbol{O}_{r\times(n-r)} \\ \boldsymbol{O}_{(n-r)\times r} & \boldsymbol{O}_{(n-r)\times(n-r)} \end{bmatrix} \boldsymbol{A}^{-1} = \boldsymbol{A}\boldsymbol{N}_r\boldsymbol{A}^{-1} , \tag{6.8}$$

where $\boldsymbol{N}_r$ is a matrix in rank-normal form (recall Definition 5.4).

Yet another expression for the projector onto $\mathcal{S}$ along $\mathcal{T}$ can be derived by considering a matrix whose null space coincides with the column space of $\boldsymbol{T}$ in the setting of Theorem 6.7. The next theorem reveals this.

Theorem 6.8 *Let $\Re^n = \mathcal{S} \oplus \mathcal{T}$ such that $\dim(\mathcal{S}) = r < n$ and $\dim(\mathcal{T}) = n - r$. Let $\boldsymbol{S}$ be an $n \times r$ matrix and let $\boldsymbol{T}$ be a an $n \times (n-r)$ matrix whose column vectors constitute bases for $\mathcal{S}$ and $\mathcal{T}$, respectively. Let $\boldsymbol{V}$ be an $n \times r$ matrix such that $\mathcal{N}(\boldsymbol{V}') = \mathcal{C}(\boldsymbol{T})$. The projector onto $\mathcal{S}$ along $\mathcal{T}$ is*

$$\boldsymbol{P} = \boldsymbol{S}(\boldsymbol{V}'\boldsymbol{S})^{-1}\boldsymbol{V}' .$$

Proof. Observe that $\mathcal{N}(\boldsymbol{V}') = \mathcal{C}(\boldsymbol{T}) \subseteq \Re^n$. Since

$$n - r = \dim(\mathcal{C}(\boldsymbol{T})) = \dim(\mathcal{N}(\boldsymbol{V}')) = n - \rho(\boldsymbol{V}') = n - \rho(\boldsymbol{V}) ,$$

it follows that the rank of $\boldsymbol{V}$ is r. Therefore, $\boldsymbol{V}$ is $n \times r$ with r linearly independent columns. Suppose $\boldsymbol{P} = \boldsymbol{S}\boldsymbol{U}'$ is the projector onto $\mathcal{S}$ along $\mathcal{T}$ as in Theorem 6.7. Since $\mathcal{C}(\boldsymbol{T}) = \mathcal{N}(\boldsymbol{P}) = \mathcal{C}(\boldsymbol{I} - \boldsymbol{P})$, we have $\boldsymbol{V}'(\boldsymbol{I} - \boldsymbol{P}) = \boldsymbol{O}$ or

$$\boldsymbol{V}' = \boldsymbol{V}'\boldsymbol{P} = \boldsymbol{V}'\boldsymbol{S}\boldsymbol{U}' . \tag{6.9}$$

We claim that that $\boldsymbol{V}'\boldsymbol{S}$ is nonsingular. To prove this, we need to show that the rank of $\boldsymbol{V}'\boldsymbol{S}$ is equal to r. Using (6.9), we obtain

$$\mathcal{C}(\boldsymbol{V}'\boldsymbol{S}) \subseteq \mathcal{C}(\boldsymbol{V}') = \mathcal{C}(\boldsymbol{V}'\boldsymbol{S}\boldsymbol{U}') \subseteq \mathcal{C}(\boldsymbol{V}'\boldsymbol{S}) \Longrightarrow \mathcal{C}(\boldsymbol{V}'\boldsymbol{S}) = \mathcal{C}(\boldsymbol{V}') . \tag{6.10}$$

Hence, $\rho(\boldsymbol{V}'\boldsymbol{S}) = \rho(\boldsymbol{V}') = \rho(\boldsymbol{V}) = r$ and $\boldsymbol{V}'\boldsymbol{S}$ is nonsingular. This allows us to write $\boldsymbol{U}' = (\boldsymbol{V}'\boldsymbol{S})^{-1}\boldsymbol{V}'$ (from (6.9)), which yields $\boldsymbol{P} = \boldsymbol{S}\boldsymbol{U}' = \boldsymbol{S}(\boldsymbol{V}'\boldsymbol{S})^{-1}\boldsymbol{V}'$. □

In the above theorem, $\mathcal{N}(\boldsymbol{V}') = \mathcal{C}(\boldsymbol{T}) = \mathcal{N}(\boldsymbol{P})$ and $\mathcal{C}(\boldsymbol{S}) = \mathcal{C}(\boldsymbol{P})$. Theorem 6.8 reveals that the projector is defined by the column space and null space of $\boldsymbol{P}$.

We next turn to an important characterization known as *Cochran's Theorem.*

Theorem 6.9 Cochran's theorem. *Let $\boldsymbol{P}_1, \boldsymbol{P}_2, \ldots, \boldsymbol{P}_k$ be $n \times n$ matrices with $\sum_{i=1}^{k} \boldsymbol{P}_i = \boldsymbol{I}_n$. Then the following statements are equivalent:*

(i) $\sum_{i=1}^{k} \rho(\boldsymbol{P}_i) = n.$

(ii) $\boldsymbol{P}_i\boldsymbol{P}_j = \boldsymbol{O}$ *for* $i \neq j$.

(iii) $\boldsymbol{P}_i^2 = \boldsymbol{P}_i$ *for* $i = 1, 2, \ldots, k$, *i.e., each* $\boldsymbol{P}_i$ *is a projector.*

Proof. We will prove that (i) ⇒ (ii) ⇒ (iii) ⇒ (i).

Proof of (i) ⇒ (ii): Let $\rho(\boldsymbol{P}_i) = r_i$ and suppose that $\boldsymbol{P}_i = \boldsymbol{S}_i\boldsymbol{U}_i'$ is a rank factorization for $i = 1, 2, \ldots, k$, where each $\boldsymbol{S}_i$ is $n \times r_i$ and $\boldsymbol{U}_i'$ is $r_i \times n$ with $\rho(\boldsymbol{S}_i) = \rho(\boldsymbol{U}_i') = r_i$.

We can now write

$$\sum_{i=1}^{k} \boldsymbol{P}_i = \boldsymbol{I}_k \Longrightarrow \boldsymbol{S}_1\boldsymbol{U}_1' + \boldsymbol{S}_2\boldsymbol{U}_2' + \cdots \boldsymbol{S}_k\boldsymbol{U}_k' = \boldsymbol{I}_n$$

$$\Longrightarrow \left[\boldsymbol{S}_1 : \boldsymbol{S}_2 : \ldots : \boldsymbol{S}_k\right] \begin{bmatrix} \boldsymbol{U}_1' \\ \boldsymbol{U}_2' \\ \vdots \\ \boldsymbol{U}_k' \end{bmatrix} = \boldsymbol{I}_k \Longrightarrow \boldsymbol{S}\boldsymbol{U}' = \boldsymbol{I}_n \,,$$

where $\boldsymbol{S} = [\boldsymbol{S}_1 : \boldsymbol{S}_2 : \ldots : \boldsymbol{S}_k]$ and $\boldsymbol{U}' = \begin{bmatrix} \boldsymbol{U}_1' \\ \boldsymbol{U}_2' \\ \vdots \\ \boldsymbol{U}_k' \end{bmatrix}$.

Since $\sum_{i=1}^{k} r_i = \sum_{i=1}^{n} \rho(\boldsymbol{P}_i) = n$, we have that $\boldsymbol{S}$ and $\boldsymbol{U}$ are both $n \times n$ square matrices that satisfy $\boldsymbol{S}\boldsymbol{U}' = \boldsymbol{I}_n$. Therefore, we have $\boldsymbol{U}'\boldsymbol{S} = \boldsymbol{I}_n$ and

$$\begin{bmatrix} \boldsymbol{U}_1' \\ \boldsymbol{U}_2' \\ \vdots \\ \boldsymbol{U}_k' \end{bmatrix} \left[\boldsymbol{S}_1 : \boldsymbol{S}_2 : \ldots : \boldsymbol{S}_k\right] = \begin{bmatrix} \boldsymbol{I}_{r_1} & \boldsymbol{O} & \ldots & \boldsymbol{O} \\ \boldsymbol{O} & \boldsymbol{I}_{r_2} & \ldots & \boldsymbol{O} \\ \vdots & \vdots & \ddots & \vdots \\ \boldsymbol{O} & \boldsymbol{O} & \ldots & \boldsymbol{I}_{r_k} \end{bmatrix} .$$

This means that $\boldsymbol{U}_i'\boldsymbol{S}_j = \boldsymbol{O}$ whenever $i \neq j$, which implies

$$\boldsymbol{P}_i\boldsymbol{P}_j = \boldsymbol{S}_i\boldsymbol{U}_i'\boldsymbol{S}_j\boldsymbol{U}_j' = \boldsymbol{S}_i(\boldsymbol{U}_i'\boldsymbol{S}_j)\boldsymbol{U}_j' = \boldsymbol{S}_i\boldsymbol{O}\boldsymbol{U}_j' = \boldsymbol{O} \text{ whenever } i \neq j \,.$$

This proves (i) ⇒ (ii).

Proof of (ii) ⇒ (iii): Because $\sum_{i=1}^{k} \boldsymbol{P}_i = \boldsymbol{I}$, it follows from (ii) that

$$\boldsymbol{P}_j = \boldsymbol{P}_j \left(\sum_{i=1}^{k} \boldsymbol{P}_i\right) = \boldsymbol{P}_j^2 + \sum_{i=1,\, i\neq j}^{k} \boldsymbol{P}_i\boldsymbol{P}_j = \boldsymbol{P}_j^2 \,.$$

This proves (ii) ⇒ (iii).

Proof of (iii) $\Rightarrow$ (i): Since P_i is idempotent, $\rho(P_i) = \text{tr}(P_i)$. We now obtain

$$\sum_{i=1}^{k} \rho(P_i) = \sum_{i=1}^{k} \text{tr}(P_i) = \text{tr}\left(\sum_{i=1}^{k} P_i\right) = \text{tr}(I_n) = n .$$

This proves (iii) $\Rightarrow$ (i) and the theorem is proved. □

6.5 The column space-null space decomposition

If $\mathcal{S}$ and $\mathcal{T}$ are two complementary subspaces in $\Re^n$, then we have seen that $\dim(\mathcal{S}) + \dim(\mathcal{T}) = n$. The Rank-Nullity Theorem says that $\dim(\mathcal{C}(A)) + \dim(\mathcal{N}(A)) = n$ for every $m \times n$ matrix A. Suppose $m = n$, i.e., A is a square matrix. A reasonable question to ask is whether the column space and null space of A are complementary subspaces of $\Re^n$. If A is nonsingular, then this is trivially true because $\mathcal{N}(A) = \{0\}$ so $\mathcal{C}(A) \cap \mathcal{N}(A) = \{0\}$. We have also seen that $\mathcal{C}(A)$ and $\mathcal{N}(A)$ are complementary subspaces when A is a projector. However, when A is singular in general $\mathcal{C}(A)$ and $\mathcal{N}(A)$ need not be complementary because their intersection can include nonzero vectors. The following is a classic example.

Example 6.1 Consider the matrix

$$A = \begin{bmatrix} 0 & 1 \\ 0 & 0 \end{bmatrix} .$$

It is easy to verify that the vector $\begin{bmatrix} 1 \\ 0 \end{bmatrix} \in \mathcal{C}(A) \cap \mathcal{N}(A)$. In fact, this vector is a basis for both $\mathcal{C}(A)$ and $\mathcal{N}(A)$. ■

It turns out that there exists some positive integer k such that $\mathcal{C}(A^k)$ and $\mathcal{N}(A^k)$ are complementary subspaces. What is this integer k? We will explore this now in a series of results that will lead us to the main theorem that $\Re^n = \mathcal{C}(A^k) \oplus \mathcal{N}(A^k)$ for some positive integer k.

Lemma 6.2 *Let A be an $n \times n$ singular matrix. Then,*

$$\mathcal{N}(A^0) \subseteq \mathcal{N}(A) \subseteq \mathcal{N}(A^2) \subseteq \cdots \subseteq \mathcal{N}(A^p) \subseteq \mathcal{N}(A^{p+1}) \subseteq \cdots$$
$$\mathcal{C}(A^0) \supseteq \mathcal{C}(A) \supseteq \mathcal{C}(A^2) \supseteq \cdots \supseteq \mathcal{C}(A^p) \supseteq \mathcal{C}(A^{p+1}) \supseteq \cdots .$$

Proof. If $x \in \mathcal{N}(A^p)$, then $A^p x = 0$ and $A^{p+1} x = A(A^p x) = A0 = 0$. So, $x \in \mathcal{N}(A^{p+1})$. This shows that $\mathcal{N}(A^p) \subseteq \mathcal{N}(A^{p+1})$ for any positive integer p and establishes the inclusion of the null spaces in the lemma.

Now consider the inclusion of the column spaces. If $x \in \mathcal{C}(A^{p+1})$, then there exists a vector u such that $A^{p+1} u = x$. Therefore, $x = A^{p+1} u = A^p(Au) = A^p v$, where $v = A^p u$, so $x \in \mathcal{C}(A^p)$. This proves that $\mathcal{C}(A^{p+1}) \subseteq \mathcal{C}(A^p)$. □

Lemma 6.3 *There exists positive integers k and m such that $\mathcal{N}(\boldsymbol{A}^k) = \mathcal{N}(\boldsymbol{A}^{k+1})$ and $\mathcal{C}(\boldsymbol{A}^m) = \mathcal{C}(\boldsymbol{A}^{m+1})$.*

Proof. Taking dimensions in the chains of inclusions in Lemma 6.2, we obtain

$$0 \leq \nu(\boldsymbol{A}) \leq \nu(\boldsymbol{A}^2) \leq \cdots \leq \nu(\boldsymbol{A}^p) \leq \nu(\boldsymbol{A}^{p+1}) \leq \cdots \leq n$$
$$n \geq \rho(\boldsymbol{A}) \geq \rho(\boldsymbol{A}^2) \geq \cdots \geq \rho(\boldsymbol{A}^p) \geq \rho(\boldsymbol{A}^{p+1}) \geq \cdots \geq 0 \,.$$

Since these inequalities hold for all $p > 0$, they cannot always be strict. Otherwise, $\nu(\boldsymbol{A}^p)$ would eventually exceed n and $\rho(\boldsymbol{A}^p)$ would eventually become negative, which are impossible. Therefore, there must exist positive integers k and m such that $\nu(\boldsymbol{A}^k) = \nu(\boldsymbol{A}^{k+1})$ and $\rho(\boldsymbol{A}^m) = \rho(\boldsymbol{A}^{m+1})$. The equality of these dimensions together with the facts that $\mathcal{N}(\boldsymbol{A}^k) \subseteq \mathcal{N}(\boldsymbol{A}^{k+1})$ and $\mathcal{C}(\boldsymbol{A}^{m+1}) \subseteq \mathcal{C}(\boldsymbol{A}^m)$ implies that $\mathcal{N}(\boldsymbol{A}^k) = \mathcal{N}(\boldsymbol{A}^{k+1})$ and $\mathcal{C}(\boldsymbol{A}^m) = \mathcal{C}(\boldsymbol{A}^{m+1})$. $\square$

We write $\boldsymbol{A}\mathcal{C}(\boldsymbol{B})$ to mean the set $\{\boldsymbol{A}\boldsymbol{x} : \boldsymbol{x} \in \mathcal{C}(\boldsymbol{B})\}$. Using this notation we establish the following lemma.

Lemma 6.4 *Let $\boldsymbol{A}$ be an $n \times n$ matrix. Then $\boldsymbol{A}^i\mathcal{C}(\boldsymbol{A}^k) = \mathcal{C}(\boldsymbol{A}^{k+i})$ for all positive integers i and k.*

Proof. We first prove this result when $i = 1$ and k is any positive integer. Suppose $\boldsymbol{y} \in \mathcal{C}(\boldsymbol{A}^{k+1})$. Then $\boldsymbol{y} = \boldsymbol{A}^{k+1}\boldsymbol{u} = \boldsymbol{A}(\boldsymbol{A}^k\boldsymbol{u})$ for some $\boldsymbol{u} \in \Re^n$, which implies that $\boldsymbol{y} \in \boldsymbol{A}\mathcal{C}(\boldsymbol{A}^k\boldsymbol{u})$. Therefore, $\mathcal{C}(\boldsymbol{A}^{k+1}) \subseteq \boldsymbol{A}\mathcal{C}(\boldsymbol{A}^k)$.

To prove the reverse inclusion, suppose $\boldsymbol{y} \in \boldsymbol{A}\mathcal{C}(\boldsymbol{A}^k)$. Therefore, $\boldsymbol{y} = \boldsymbol{A}\boldsymbol{x}$ for some $\boldsymbol{x} \in \mathcal{C}(\boldsymbol{A}^k)$, which means that $\boldsymbol{x} = \boldsymbol{A}^k\boldsymbol{v}$ for some $\boldsymbol{v} \in \Re^n$. Therefore, $\boldsymbol{y} = \boldsymbol{A}^{k+1}\boldsymbol{v} \in \mathcal{C}(\boldsymbol{A}^{k+1})$ and so $\boldsymbol{A}\mathcal{C}(\boldsymbol{A}^k) \subseteq \mathcal{C}(\boldsymbol{A}^{k+1})$.

This proves that $\boldsymbol{A}\mathcal{C}(\boldsymbol{A}^k) = \mathcal{C}(\boldsymbol{A}^{k+1})$. When i is any positive integer then we proceed inductively and establish

$$\boldsymbol{A}^i\mathcal{C}(\boldsymbol{A}^k) = \boldsymbol{A}^{i-1}(\boldsymbol{A}\mathcal{C}(\boldsymbol{A}^k)) = \boldsymbol{A}^{i-1}\mathcal{C}(\boldsymbol{A}^{k+1}) = \cdots = \boldsymbol{A}\mathcal{C}(\boldsymbol{A}^{k+i-1}) = \mathcal{C}(\boldsymbol{A}^{k+i}) \,.$$

This completes the proof. $\square$

The above result has an important implication. If $\mathcal{C}(\boldsymbol{A}^k) = \mathcal{C}(\boldsymbol{A}^{k+1})$, then Lemma 6.4 ensures that $\mathcal{C}(\boldsymbol{A}^{k+i}) = \mathcal{C}(\boldsymbol{A}^k)$ for any positive integer i. So, once equality is attained in the chain of inclusions in the column spaces in Lemma 6.2, the equality persists throughout the remainder of the chain. Quite remarkably, the null spaces in Lemma 6.2 also stop growing exactly at the same place as seen below.

Lemma 6.5 *Let k denote the smallest positive integer such that $\mathcal{C}(\boldsymbol{A}^k) = \mathcal{C}(\boldsymbol{A}^{k+1})$. Then, $\mathcal{C}(\boldsymbol{A}^k) = \mathcal{C}(\boldsymbol{A}^{k+i})$ and $\mathcal{N}(\boldsymbol{A}^k) = \mathcal{N}(\boldsymbol{A}^{k+i})$ for any positive integer i.*

Proof. Lemma 6.4 tells us that

$$\mathcal{C}(\boldsymbol{A}^{k+i}) = \mathcal{C}(\boldsymbol{A}^i\boldsymbol{A}^k) = \boldsymbol{A}^i\mathcal{C}(\boldsymbol{A}^k) = \boldsymbol{A}^i\mathcal{C}(\boldsymbol{A}^{k+1}) = \mathcal{C}(\boldsymbol{A}^{k+i+1}) \text{ for } i = 1, 2, \ldots \,.$$

Therefore, $\mathcal{C}(\boldsymbol{A}^k) = \mathcal{C}(\boldsymbol{A}^{k+i})$ for any positive integer i. The Rank-Nullity Theorem tells us

$$\dim(\mathcal{N}(\boldsymbol{A}^{k+i})) = n - \dim(\mathcal{C}(\boldsymbol{A}^{k+i})) = n - \dim(\mathcal{C}(\boldsymbol{A}^k)) = \dim(\mathcal{N}(\boldsymbol{A}^k)) .$$

So $\mathcal{N}(\boldsymbol{A}^k) \subseteq \mathcal{N}(\boldsymbol{A}^{k+i})$ and these two subspaces have the same dimension. Therefore, $\mathcal{N}(\boldsymbol{A}^k) = \mathcal{N}(\boldsymbol{A}^{k+i})$ for any positive integer i. □

Based upon what we have established above, the inclusions in Lemma 6.2 can be rewritten as

$$\begin{gathered}\mathcal{N}(\boldsymbol{A}^0) \subseteq \mathcal{N}(\boldsymbol{A}) \subseteq \mathcal{N}(\boldsymbol{A}^2) \subseteq \cdots \subseteq \mathcal{N}(\boldsymbol{A}^k) = \mathcal{N}(\boldsymbol{A}^{k+1}) \subseteq \cdots \\ \mathcal{C}(\boldsymbol{A}^0) \supseteq \mathcal{C}(\boldsymbol{A}) \supseteq \mathcal{C}(\boldsymbol{A}^2) \supseteq \cdots \supseteq \mathcal{C}(\boldsymbol{A}^k) = \mathcal{C}(\boldsymbol{A}^{k+1}) \supseteq \cdots ,\end{gathered}$$

where k is the smallest integer for which $\mathcal{C}(\boldsymbol{A}^k) = \mathcal{C}(\boldsymbol{A}^{k+1})$. By virtue of Lemma 6.5, k is also the smallest integer for which $\mathcal{N}(\boldsymbol{A}^k) = \mathcal{N}(\boldsymbol{A}^{k+1})$. We call this integer k the ***index*** of $\boldsymbol{A}$. Put differently, the index of $\boldsymbol{A}$ is the smallest integer for which the column spaces of powers of $\boldsymbol{A}$ stops shrinking and the null spaces of powers of $\boldsymbol{A}$ stop growing. For nonsingular matrices we define their idex to be 0.

We now prove the main theorem of this section.

Theorem 6.10 *For every $n \times n$ singular matrix $\boldsymbol{A}$, there exists a positive integer k such that*

$$\Re^n = \mathcal{C}(\boldsymbol{A}^k) \oplus \mathcal{N}(\boldsymbol{A}^k) .$$

The smallest positive integer k for which the above holds is called the ***index*** *of $\boldsymbol{A}$.*

Proof. Let k be the smallest integer for which $\mathcal{N}(\boldsymbol{A}^k) = \mathcal{N}(\boldsymbol{A}^{k+i})$. To prove that $\mathcal{C}(\boldsymbol{A}^k)$ and $\mathcal{N}(\boldsymbol{A}^k)$ are complementary subspaces of $\Re^n$, we will show that they are essentially disjoint and their sum is $\Re^n$.

Suppose $\boldsymbol{x} \in \mathcal{C}(\boldsymbol{A}^k) \cap \mathcal{N}(\boldsymbol{A}^k)$. Then $\boldsymbol{x} = \boldsymbol{A}^k\boldsymbol{u}$ for some vector $\boldsymbol{u} \in \Re^n$. Also, $\boldsymbol{A}^k\boldsymbol{x} = \boldsymbol{0}$. Therefore, $\boldsymbol{A}^{2k}\boldsymbol{u} = \boldsymbol{A}^k\boldsymbol{x} = \boldsymbol{0}$ so $\boldsymbol{u} \in \mathcal{N}(\boldsymbol{A}^{2k}) = \mathcal{N}(\boldsymbol{A}^k)$. Hence, $\boldsymbol{x} = \boldsymbol{A}^k\boldsymbol{u} = \boldsymbol{0}$. This proves that $\mathcal{C}(\boldsymbol{A}^k) \cap \mathcal{N}(\boldsymbol{A}^k) = \{\boldsymbol{0}\}$.

We next show that $\mathcal{C}(\boldsymbol{A}^k) + \mathcal{N}(\boldsymbol{A}^k) = \Re^n$. Since $\mathcal{C}(\boldsymbol{A}^k) + \mathcal{N}(\boldsymbol{A}^k) \subseteq \Re^n$, all we need to show is that the dimension of $\mathcal{C}(\boldsymbol{A}^k) + \mathcal{N}(\boldsymbol{A}^k)$ is equal to n. This is true because

$$\begin{aligned}\dim\left[\mathcal{C}(\boldsymbol{A}^k) + \mathcal{N}(\boldsymbol{A}^k)\right] &= \dim\left[\mathcal{C}(\boldsymbol{A}^k)\right] + \dim\left[\mathcal{N}(\boldsymbol{A}^k)\right] - \dim\left[\mathcal{C}(\boldsymbol{A}^k) \cap \mathcal{N}(\boldsymbol{A}^k)\right] \\ &= \dim\left[\mathcal{C}(\boldsymbol{A}^k)\right] + \dim\left[\mathcal{N}(\boldsymbol{A}^k)\right] = n ,\end{aligned}$$

where the second equality follows because $\mathcal{C}(\boldsymbol{A}^k) \cap \mathcal{N}(\boldsymbol{A}^k) = \{\boldsymbol{0}\}$ and the last equality follows from the Rank-Nullity Theorem. This completes the proof. □

6.6 Invariant subspaces and the Core-Nilpotent decomposition

In Section 4.9 we had briefly seen how invariant subspaces under a matrix $\boldsymbol{A}$ can produce simpler representations. The column space-null space decomposition in the preceding section makes this transparent for singular matrices and we can arrive at a block diagonal matrix decomposition for singular matrices. The key observation is the following lemma.

Lemma 6.6 *If $\boldsymbol{A}$ is an $n \times n$ matrix $\boldsymbol{A}$, then $\mathcal{C}(\boldsymbol{A}^k)$ and $\mathcal{N}(\boldsymbol{A}^k)$ are invariant subspaces for $\boldsymbol{A}$ for any positive integer k.*

Proof. If $\boldsymbol{x} \in \mathcal{C}(\boldsymbol{A}^k)$, then $\boldsymbol{x} = \boldsymbol{A}^k \boldsymbol{u}$ for some vector $\boldsymbol{u} \in \Re^n$. Therefore,

$$\boldsymbol{A}\boldsymbol{x} = \boldsymbol{A}(\boldsymbol{A}^k \boldsymbol{u}) = \boldsymbol{A}^{k+1}\boldsymbol{u} = \boldsymbol{A}^k(\boldsymbol{A}\boldsymbol{u}) \in \mathcal{C}(\boldsymbol{A}^k) .$$

This proves that $\mathcal{C}(\boldsymbol{A}^k)$ is an invariant subspace under $\boldsymbol{A}$.

That $\mathcal{N}(\boldsymbol{A}^k)$ is invariant under $\boldsymbol{A}$ can be seen from

$$\begin{aligned} \boldsymbol{x} \in \mathcal{N}(\boldsymbol{A}^k) &\Longrightarrow \boldsymbol{A}^k \boldsymbol{x} = \boldsymbol{0} \Longrightarrow \boldsymbol{A}^{k+1}\boldsymbol{x} = \boldsymbol{0} \\ &\Longrightarrow \boldsymbol{A}^k(\boldsymbol{A}\boldsymbol{x}) = \boldsymbol{0} \Longrightarrow \boldsymbol{A}\boldsymbol{x} \in \mathcal{N}(\boldsymbol{A}^k) . \end{aligned}$$

This completes the proof. □

The above result, with $k = 1$, implies that $\mathcal{C}(\boldsymbol{A})$ and $\mathcal{N}(\boldsymbol{A})$ are both invariant subspaces under $\boldsymbol{A}$. What is a bit more interesting is the case when k is the index of $\boldsymbol{A}$, which leads to a particular matrix decomposition under similarity transformations. Before we derive this decomposition, we define a special type of matrix and show its relationship with its index.

Definition 6.6 Nilpotent matrix. *A square matrix $\boldsymbol{N}$ is said to be* **nilpotent** *if $\boldsymbol{N}^k = \boldsymbol{O}$ for some positive integer k.*

Example 6.2 Triangular matrices with zeros along the diagonal are nilpotent. Recall Lemma 1.2. Consider the following two matrices:

$$\boldsymbol{A} = \begin{bmatrix} 0 & 1 & 0 \\ 0 & 0 & 1 \\ 0 & 0 & 0 \end{bmatrix} \text{ and } \boldsymbol{B} = \begin{bmatrix} 0 & 1 & 0 \\ 0 & 0 & 0 \\ 0 & 0 & 0 \end{bmatrix} .$$

It is easily verified that $\boldsymbol{A}^3 = \boldsymbol{O}$ (but $\boldsymbol{A}^2 \neq \boldsymbol{O}$) and $\boldsymbol{B}^2 = \boldsymbol{O}$. Therefore, $\boldsymbol{A}$ and $\boldsymbol{B}$ are both nilpotent matrices. Nilpotent matrices do not have to be triangular with zeros along the diagonal. The matrix

$$\boldsymbol{C} = \begin{bmatrix} 6 & -9 \\ 4 & -6 \end{bmatrix}$$

is nilpotent because $\boldsymbol{C}^2 = \boldsymbol{O}$. ■

The following lemma gives a characterization for the index of a nilpotent matrix.

Lemma 6.7 *The index k of a nilpotent matrix $\boldsymbol{N}$ is the smallest integer for which $\boldsymbol{N}^k = \boldsymbol{O}$.*

Proof. Let p be a positive integer such that $\boldsymbol{N}^p = \boldsymbol{O}$ and $\boldsymbol{N}^{p-1} \neq \boldsymbol{O}$. From the results in the preceding section, in particular Lemma 6.5, we know that

$$\mathcal{C}(\boldsymbol{N}^0) \supseteq \mathcal{C}(\boldsymbol{N}) \supseteq \mathcal{C}(\boldsymbol{N}^2) \supseteq \cdots \supseteq \mathcal{C}(\boldsymbol{N}^k) = \mathcal{C}(\boldsymbol{N}^{k+1}) = \cdots = \cdots .$$

The cases $p < k$ and $p > k$ will both contradict the fact that k is the index. Therefore, $p = k$ is the only choice. □

The above characterization is especially helpful to find the index of a nilpotent matrix. It is easier to check the smallest integer for which $\boldsymbol{N}^k = \boldsymbol{O}$ rather than check when the null spaces stop growing or the column spaces stop shrinking. In Example 6.2, $\boldsymbol{A}$ has index 3, while $\boldsymbol{B}$ and $\boldsymbol{C}$ have index 2.

Since $\mathcal{C}(\boldsymbol{A}^k)$ is an invariant subspace under $\boldsymbol{A}$, any vector in $\mathcal{C}(\boldsymbol{A}^k)$ remains in it after being transformed by $\boldsymbol{A}$. In particular, if $\boldsymbol{X}$ is a matrix whose columns are a basis for $\mathcal{C}(\boldsymbol{A}^k)$, then the image of each column after being multiplied by $\boldsymbol{A}$ can be expressed as a linear combination of the columns in $\boldsymbol{X}$. In other words, there exists a matrix $\boldsymbol{C}$ such that $\boldsymbol{AX} = \boldsymbol{XC}$. Each column of $\boldsymbol{C}$ holds the coordinates of the image of a basis vector when multiplied by $\boldsymbol{A}$. The following lemma formalizes this and also shows that the matrix of coordinates $\boldsymbol{C}$ is nonsingular.

Lemma 6.8 *Let $\boldsymbol{A}$ be an $n \times n$ singular matrix with index k and $\rho(\boldsymbol{A}^k) = r < n$. Suppose $\boldsymbol{X}$ is an $n \times r$ matrix whose columns form a basis for $\mathcal{C}(\boldsymbol{A}^k)$. Then, there exists an $r \times r$ nonsingular matrix $\boldsymbol{C}$ such that $\boldsymbol{AX} = \boldsymbol{XC}$.*

Proof. Since $\mathcal{C}(\boldsymbol{A}^k)$ is invariant under $\boldsymbol{A}$ and $\mathcal{C}(\boldsymbol{X}) = \mathcal{C}(\boldsymbol{A}^k)$, we obtain

$$\mathcal{C}(\boldsymbol{AX}) \subseteq \mathcal{C}(\boldsymbol{A}^k) = \mathcal{C}(\boldsymbol{X}) \Longrightarrow \boldsymbol{AX} = \boldsymbol{XC}$$

for some $r \times r$ matrix $\boldsymbol{C}$ (recall Theorem 4.6). It remains to prove that $\boldsymbol{C}$ is nonsingular. Since $\mathcal{C}(\boldsymbol{X}) = \mathcal{C}(\boldsymbol{A}^k)$, it follows that there exists an $n \times r$ matrix $\boldsymbol{B}$ such that $\boldsymbol{X} = \boldsymbol{A}^k\boldsymbol{B}$. Using this fact, we can establish the nonsingularity of $\boldsymbol{C}$ as follows:

$$\begin{aligned} \boldsymbol{Cw} = \boldsymbol{0} &\Longrightarrow \boldsymbol{XCw} = \boldsymbol{0} \Longrightarrow \boldsymbol{AXw} = \boldsymbol{0} \Longrightarrow \boldsymbol{A}^{k+1}\boldsymbol{Bw} = \boldsymbol{0} \\ &\Longrightarrow \boldsymbol{Bw} \in \mathcal{N}(\boldsymbol{A}^{k+1}) = \mathcal{N}(\boldsymbol{A}^k) \Longrightarrow \boldsymbol{A}^k\boldsymbol{Bw} = \boldsymbol{0} \\ &\Longrightarrow \boldsymbol{Xw} = \boldsymbol{0} \Longrightarrow \boldsymbol{w} = \boldsymbol{0}\,, \end{aligned}$$

where we used $\mathcal{N}(\boldsymbol{A}^{k+1}) = \mathcal{N}(\boldsymbol{A}^k)$ since k is the index of $\boldsymbol{A}$. □

The next lemma presents an analogous result for null spaces. Interestingly, the coordinate matrix is now nilpotent and has the same index as $\boldsymbol{A}$.

Lemma 6.9 *Let $\boldsymbol{A}$ be an $n \times n$ singular matrix with index k and $\rho(\boldsymbol{A}^k) = r < n$. If $\boldsymbol{Y}$ is an $n \times (n-r)$ matrix whose columns are a basis for $\mathcal{N}(\boldsymbol{A}^k)$, then there exists a nilpotent matrix $\boldsymbol{N}$ with index k such that $\boldsymbol{AY} = \boldsymbol{YN}$.*

Proof. Since $\mathcal{N}(\boldsymbol{A}^k)$ is invariant under $\boldsymbol{A}$ and $\mathcal{C}(\boldsymbol{Y}) = \mathcal{N}(\boldsymbol{A}^k)$, we obtain

$$\mathcal{C}(\boldsymbol{AY}) \subseteq \mathcal{N}(\boldsymbol{A}^k) = \mathcal{C}(\boldsymbol{Y}) \Longrightarrow \boldsymbol{AY} = \boldsymbol{YN}$$

for some $(n-r) \times (n-r)$ matrix $\boldsymbol{N}$ (recall Theorem 4.6). It remains to prove that $\boldsymbol{N}$ is nilpotent with index k. Since the columns of $\boldsymbol{Y}$ are a basis for the null space of $\boldsymbol{A}^k$, we find that

$$\begin{aligned}\boldsymbol{O} = \boldsymbol{A}^k\boldsymbol{Y} &= \boldsymbol{A}^{k-1}\boldsymbol{A}\boldsymbol{Y} = \boldsymbol{A}^{k-1}\boldsymbol{Y}\boldsymbol{N} = \boldsymbol{A}^{k-2}(\boldsymbol{A}\boldsymbol{Y})\boldsymbol{N} = \boldsymbol{A}^{k-2}\boldsymbol{Y}\boldsymbol{N}^2 \\ &= \cdots = \boldsymbol{A}\boldsymbol{Y}\boldsymbol{N}^{k-1} = (\boldsymbol{A}\boldsymbol{Y})\boldsymbol{N}^{k-1} = \boldsymbol{Y}\boldsymbol{N}^k \,.\end{aligned}$$

Therefore, $\boldsymbol{Y}\boldsymbol{N}^k = \boldsymbol{O}$. The columns of $\boldsymbol{Y}$ are linearly independent, so each column of $\boldsymbol{N}^k$ must be $\boldsymbol{0}$. This proves that $\boldsymbol{N}^k = \boldsymbol{O}$, so $\boldsymbol{N}$ is nilpotent.

To complete the proof we need to show that $\boldsymbol{N}^{k-1} \neq \boldsymbol{O}$. Suppose, if possible, the contrary, i.e., $\boldsymbol{N}^{k-1} = \boldsymbol{O}$. Then, $\boldsymbol{A}^{k-1}\boldsymbol{Y} = \boldsymbol{Y}\boldsymbol{N}^{k-1} = \boldsymbol{O}$, which would imply that

$$\mathcal{C}(\boldsymbol{Y}) \subseteq \mathcal{N}(\boldsymbol{A}^{k-1}) \subseteq \mathcal{N}(\boldsymbol{A}^k) = \mathcal{C}(\boldsymbol{Y})\,.$$

But the above inclusion would mean that $\mathcal{N}(\boldsymbol{A}^{k-1}) = \mathcal{N}(\boldsymbol{A}^k)$. This contradicts k being the index of $\boldsymbol{A}$. Therefore, $\boldsymbol{N}^{k-1} \neq \boldsymbol{O}$ and $\boldsymbol{N}$ is nilpotent with index k. □

We now come to the promised decomposition, known as the ***core-nilpotent decomposition***, which is the main theorem of this section. It follows easily from the above lemmas.

Theorem 6.11 *Let $\boldsymbol{A}$ be an $n \times n$ matrix with index k and $\rho(\boldsymbol{A}^k) = r$. There exists an $n \times n$ nonsingular matrix $\boldsymbol{Q}$ such that*

$$\boldsymbol{Q}^{-1}\boldsymbol{A}\boldsymbol{Q} = \begin{bmatrix} \boldsymbol{C} & \boldsymbol{O} \\ \boldsymbol{O} & \boldsymbol{N} \end{bmatrix},$$

where $\boldsymbol{C}$ is an $r \times r$ nonsingular matrix and $\boldsymbol{N}$ is a nilpotent matrix with index k.

Proof. Let $\boldsymbol{X}$ be an $n \times r$ matrix whose columns form a basis for $\mathcal{C}(\boldsymbol{A}^k)$ and let $\boldsymbol{Y}$ be an $n \times (n-r)$ matrix whose columns are a basis for $\mathcal{N}(\boldsymbol{A}^k)$. Lemma 6.8 ensures that there exists an $r \times r$ nonsingular matrix $\boldsymbol{C}$ such that $\boldsymbol{A}\boldsymbol{X} = \boldsymbol{X}\boldsymbol{C}$. Lemma 6.9 ensures that there exists an $(n-r) \times (n-r)$ nilpotent matrix $\boldsymbol{N}$ of index k such that $\boldsymbol{A}\boldsymbol{Y} = \boldsymbol{Y}\boldsymbol{N}$.

Let $\boldsymbol{Q} = [\boldsymbol{X} : \boldsymbol{Y}]$ be an $n \times n$ matrix. Since $\mathcal{C}(\boldsymbol{A}^k)$ and $\mathcal{N}(\boldsymbol{A}^k)$ are complementary subspaces of $\Re^n$, $\boldsymbol{Q}$ is nonsingular. We now see that

$$\boldsymbol{A}\boldsymbol{Q} = [\boldsymbol{A}\boldsymbol{X} : \boldsymbol{A}\boldsymbol{Y}] = [\boldsymbol{X}\boldsymbol{C} : \boldsymbol{Y}\boldsymbol{N}] = [\boldsymbol{X} : \boldsymbol{Y}]\begin{bmatrix} \boldsymbol{C} & \boldsymbol{O} \\ \boldsymbol{O} & \boldsymbol{N} \end{bmatrix} = \boldsymbol{Q}\begin{bmatrix} \boldsymbol{C} & \boldsymbol{O} \\ \boldsymbol{O} & \boldsymbol{N} \end{bmatrix},$$

which implies that

$$\boldsymbol{Q}^{-1}\boldsymbol{A}\boldsymbol{Q} = \begin{bmatrix} \boldsymbol{C} & \boldsymbol{O} \\ \boldsymbol{O} & \boldsymbol{N} \end{bmatrix}$$

and concludes the proof. □

The core-nilpotent decomposition is so named because it decomposes a singular matrix in terms of a nonsingular "core" component $\boldsymbol{C}$ and a nilpotent component $\boldsymbol{N}$. It is worth pointing out that when $\boldsymbol{A}$ is nonsingular the core-nilpotent decomposition does not offer anything useful. In that case, the nilpotent component disappears and the "core" becomes $C = A$ with $Q = Q^{-1}$.

The core-nilpotent decomposition does contain useful information on $\mathcal{C}(A^k)$ and $\mathcal{N}(A^k)$. For example, it provides us with projectors onto $\mathcal{C}(A^k)$ and $\mathcal{N}(A^k)$. Suppose Q in Theorem 6.11 is partitioned conformably as

$$Q = [X : Y] \text{ and } Q^{-1} = \begin{bmatrix} U' \\ V' \end{bmatrix},$$

where U' is $r \times n$ and V' is $(n-r) \times n$. Since $\mathcal{C}(A^k)$ and $\mathcal{N}(A^k)$ are complementary subspaces for $\Re^n$, Theorem 6.7 tells us that XU' is the projector onto $\mathcal{C}(A^k)$ along $\mathcal{N}(A^k)$ and YV' is the complementary projector onto $\mathcal{N}(A^k)$ along $\mathcal{C}(A^k)$.

6.7 Exercises

1. Find the dimension of $\mathcal{C}(A) + \mathcal{N}(A)$ and of $\mathcal{C}(A) \cap \mathcal{N}(A)$, where

$$A = \begin{bmatrix} 2 & -2 & -5 & -3 & -1 & 2 \\ 2 & -1 & -3 & 2 & 3 & 2 \\ 4 & -1 & -4 & 10 & 11 & 4 \\ 0 & 1 & 2 & 5 & 4 & 0 \end{bmatrix}.$$

   Also find the dimension of $\mathcal{R}(A) + \mathcal{N}(A')$ and of $\mathcal{R}(A) \cap \mathcal{N}(A')$.
2. Prove that $\rho([A : B]) = \rho(A) + \rho(B) - \dim(\mathcal{C}(A) \cap \mathcal{C}(B))$.
3. Prove that $\nu([A : B]) = \nu(A) + \nu(B) + \dim(\mathcal{C}(A) \cap \mathcal{C}(B))$.
4. Let $\mathcal{S} = \mathcal{N}(A)$ and $\mathcal{T} = \mathcal{N}(B)$, where

$$A = \begin{bmatrix} 1 & 1 & 0 & 0 \\ 0 & 0 & 1 & 1 \end{bmatrix} \text{ and } B = \begin{bmatrix} 1 & 0 & 1 & 0 \\ 0 & 1 & 0 & 1 \end{bmatrix}.$$

   Find a basis for each of $\mathcal{S} + \mathcal{T}$ and $\mathcal{S} \cap \mathcal{T}$.
5. Let A and B be $m \times n$ and $n \times p$, respectively, and let $\rho(B) = r$. If X is an $n \times r$ matrix whose columns form a basis for $\mathcal{C}(B)$, and Y is an $r \times s$ matrix whose columns are a basis for $\mathcal{N}(AX)$, then prove that the columns of XY form a basis for $\mathcal{N}(A) \cap \mathcal{C}(B)$.
6. Use the result of the preceding exercise to find a basis for $\mathcal{C}(A) \cap \mathcal{N}(A)$, where

$$A = \begin{bmatrix} 1 & 2 & 3 \\ 4 & 5 & 6 \\ 7 & 8 & 9 \end{bmatrix}.$$

   Find the dimension and a basis for $\mathcal{C}(A) \cap \mathcal{R}(A)$.
7. Theorem 6.1 resembles the familiar formula to count the number of elements in the union of two finite sets. However, this analogy does not carry through for three or more sets. Let $\mathcal{S}_1$, $\mathcal{S}_2$ and $\mathcal{S}_3$ be subspaces in $\Re^n$. Construct an example to show

that the following "extension" of Theorem 6.1 is *not* necessarily true:

$$\begin{aligned}\dim(\mathcal{S}_1+\mathcal{S}_2+\mathcal{S}_3) = {} & \dim(\mathcal{S}_1)+\dim(\mathcal{S}_2)+\dim(\mathcal{S}_3)\\ & -\dim(\mathcal{S}_1\cap\mathcal{S}_2)-\dim(\mathcal{S}_1\cap\mathcal{S}_3)-\dim(\mathcal{S}_2\cap\mathcal{S}_3)\\ & +\dim(\mathcal{S}_1\cap\mathcal{S}_2\cap\mathcal{S}_3)\ .\end{aligned}$$

8. Prove that $|\rho(\boldsymbol{A})-\rho(\boldsymbol{B})| \leq \rho(\boldsymbol{A}-\boldsymbol{B})$.
9. If $\boldsymbol{A}$ and $\boldsymbol{B}$ are $m\times n$ matrices such that $\rho(\boldsymbol{A}) = s$ and $\rho(\boldsymbol{B}) = r \leq s$, then show that $s-r \leq \rho(\boldsymbol{A}+\boldsymbol{B}) \leq s+r$.
10. Show that $\rho(\boldsymbol{A}+\boldsymbol{B}) = \rho(\boldsymbol{A})+\rho(\boldsymbol{B})$ if and only if $\mathcal{C}(\boldsymbol{A})\cap\mathcal{C}(\boldsymbol{B}) = \{\boldsymbol{0}\}$ and $\mathcal{R}(\boldsymbol{A})\cap\mathcal{R}(\boldsymbol{B}) = \{\boldsymbol{0}\}$, where $\boldsymbol{A}$ and $\boldsymbol{B}$ are matrices of the same size.
11. If $\boldsymbol{A}$ is a square matrix, then prove that the following statements are equivalent: (i) $\mathcal{N}(\boldsymbol{A}^2)=\mathcal{N}(\boldsymbol{A})$, (ii) $\mathcal{C}(\boldsymbol{A}^2)=\mathcal{C}(\boldsymbol{A})$, and (iii) $\mathcal{C}(\boldsymbol{A})\cap\mathcal{N}(\boldsymbol{A})=\{\boldsymbol{0}\}$.
12. Find two different complements of the subspace $\mathcal{S}$ in $\Re^4$, where

$$\mathcal{S} = \left\{\begin{bmatrix}x_1\\x_2\\x_3\\x_4\end{bmatrix} : x_3 - x_4 = 0\right\}\ .$$

13. Find a complement of $\mathcal{N}(\boldsymbol{A})$ in $\Re^5$, where

$$\boldsymbol{A} = \begin{bmatrix}1&0&0&1&0\\2&0&1&0&1\end{bmatrix}\ .$$

14. True or false: If $\mathcal{S}$, $\mathcal{T}$ and $\mathcal{W}$ are subspaces in $\Re^n$ such that $\mathcal{S}\oplus\mathcal{W} = \Re^n$ and $\mathcal{T}\oplus\mathcal{W}=\Re^n$, then $\mathcal{S}=\mathcal{T}$.
15. Construct an example to show that $\mathcal{S}_i\cap\mathcal{S}_j = \{\boldsymbol{0}\}$ for $i,j = 1,2,3$ does *not* necessarily imply that $\mathcal{S}_1+\mathcal{S}_2+\mathcal{S}_3$ is a direct sum.
16. Let $\Re^n = \mathcal{S}+\mathcal{T}$ but $\mathcal{S}\cap\mathcal{T}\neq\{\boldsymbol{0}\}$. Construct an example to show that a vector $\boldsymbol{x}\in\Re^n$ can be expressed in more than one way as $\boldsymbol{x}=\boldsymbol{u}+\boldsymbol{v}$, where $\boldsymbol{u}\in\mathcal{S}$ and $\boldsymbol{v}\in\mathcal{T}$.
17. Let $\mathcal{S}$ and $\mathcal{T}$ be two subspaces in $\Re^n$ such that $\mathcal{S}\subseteq\mathcal{T}$. Let $\mathcal{W}$ be a subspace such that $\mathcal{S}\oplus\mathcal{W}=\Re^n$. Prove that $\mathcal{S}\oplus(\mathcal{W}\cap\mathcal{T})=\mathcal{T}$.
18. True or false: If $\boldsymbol{P}$ is an $n\times n$ projector, then $\mathcal{C}(\boldsymbol{P}) = \{\boldsymbol{x}\in\Re^n : \boldsymbol{P}\boldsymbol{x}=\boldsymbol{x}\}$.
19. Show that $\begin{bmatrix}\boldsymbol{I}&\boldsymbol{B}\\\boldsymbol{O}&\boldsymbol{O}\end{bmatrix}$ is idempotent (hence a projector) for any matrix $\boldsymbol{B}$.
20. Let $\boldsymbol{A}$ be a projector. Prove that (i) $\mathcal{C}(\boldsymbol{A}\boldsymbol{B})\subseteq\mathcal{C}(\boldsymbol{B})$ if and only if $\boldsymbol{A}\boldsymbol{B}=\boldsymbol{B}$, and (ii) $\mathcal{R}(\boldsymbol{B}\boldsymbol{A})\subseteq\mathcal{R}(\boldsymbol{B})$ if and only if $\boldsymbol{B}\boldsymbol{A}=\boldsymbol{B}$.
21. If $\boldsymbol{A}$ is idempotent and triangular (upper or lower), then prove that each diagonal entry of $\boldsymbol{A}$ is either 0 or 1.
22. Consider the following two sets of vectors:

$$\mathcal{X} = \left\{\begin{bmatrix}0\\-1\\2\end{bmatrix}, \begin{bmatrix}-1\\1\\1\end{bmatrix}\right\} \text{ and } \mathcal{Y} = \left\{\begin{bmatrix}1\\-1\\0\end{bmatrix}\right\}\ .$$

Prove that $Sp(\mathcal{X})$ and $Sp(\{\mathcal{Y}\})$ are complementary subspaces. Derive an expression for the projector onto $Sp(\mathcal{X})$ along $Sp(\{\mathcal{Y}\})$.

23. Find a complement $\mathcal{T}$ of $Sp(\mathcal{X})$ in $\Re^4$, where

$$\mathcal{X} = \left\{ \begin{bmatrix} 2 \\ 0 \\ 1 \\ 3 \end{bmatrix}, \begin{bmatrix} 0 \\ 3 \\ 1 \\ 1 \end{bmatrix}, \begin{bmatrix} 2 \\ -6 \\ -1 \\ -1 \end{bmatrix} \right\}.$$

Find an expression for the projector onto $Sp(\mathcal{X})$ along $\mathcal{T}$.

24. Let $\boldsymbol{A}$ and $\boldsymbol{B}$ be two $n \times n$ projectors such that $\boldsymbol{AB} = \boldsymbol{BA} = \boldsymbol{O}$. Prove that

$$\rho(\boldsymbol{A} + \boldsymbol{B}) = \rho(\boldsymbol{A}) + \rho(\boldsymbol{B}) .$$

25. The above result holds true under milder conditions: Show that if $\rho(\boldsymbol{A}) = \rho(\boldsymbol{A}^2)$ and $\boldsymbol{AB} = \boldsymbol{BA} = \boldsymbol{O}$, then $\rho(\boldsymbol{A} + \boldsymbol{B}) = \rho(\boldsymbol{A}) + \rho(\boldsymbol{B})$.

26. Let $\boldsymbol{A}$ and $\boldsymbol{B}$ be $n \times n$ projectors. Show that $\boldsymbol{A} + \boldsymbol{B}$ is a projector if and only if $\mathcal{C}(\boldsymbol{A}) \subseteq \mathcal{N}(\boldsymbol{B})$ and $\mathcal{C}(\boldsymbol{B}) \subseteq \mathcal{N}(\boldsymbol{A})$.

27. Let $\boldsymbol{A}$ and $\boldsymbol{B}$ be two $n \times n$ projectors. Show that $\boldsymbol{A} + \boldsymbol{B}$ is a projector if and only if $\boldsymbol{AB} = \boldsymbol{BA} = \boldsymbol{O}$. Under this condition, show that $\boldsymbol{A} + \boldsymbol{B}$ is the projector onto $\mathcal{C}(\boldsymbol{A} + \boldsymbol{B}) = \mathcal{C}(\boldsymbol{A}) \oplus \mathcal{C}(\boldsymbol{B})$ along $\mathcal{N}(\boldsymbol{A} + \boldsymbol{B}) = \mathcal{N}(\boldsymbol{A}) \cap \mathcal{N}(\boldsymbol{B})$.

28. Prove that $\boldsymbol{A} - \boldsymbol{B}$ is a projector if and only if $\mathcal{C}(\boldsymbol{B}) \subseteq \mathcal{C}(\boldsymbol{A})$ and $\mathcal{R}(\boldsymbol{B}) \subseteq \mathcal{R}(\boldsymbol{A})$.

29. Let $\boldsymbol{A}$ and $\boldsymbol{B}$ be two $n \times n$ projectors. Show that $\boldsymbol{A} - \boldsymbol{B}$ is a projector if and only if $\boldsymbol{AB} = \boldsymbol{BA} = \boldsymbol{B}$. Under this condition, show that $\boldsymbol{A} - \boldsymbol{B}$ is the projector onto $\mathcal{C}(\boldsymbol{A}) \cap \mathcal{N}(\boldsymbol{B})$ along $\mathcal{N}(\boldsymbol{A}) \oplus \mathcal{C}(\boldsymbol{B})$.

30. Let $\boldsymbol{A}$ and $\boldsymbol{B}$ be two $n \times n$ projectors such that $\boldsymbol{AB} = \boldsymbol{BA} = \boldsymbol{P}$. Prove that $\boldsymbol{P}$ is the projector onto $\mathcal{C}(\boldsymbol{A}) \cap \mathcal{C}(\boldsymbol{B})$ along $\mathcal{N}(\boldsymbol{A}) + \mathcal{N}(\boldsymbol{B})$.

31. Find the index of the following matrices:

$$\begin{bmatrix} 1 & 0 & 0 \\ 0 & 1 & 1 \\ 0 & -1 & -1 \end{bmatrix} \text{ and } \begin{bmatrix} 0 & 1 & 0 \\ 0 & 0 & 1 \\ 0 & 0 & 0 \end{bmatrix}.$$

32. Let $\boldsymbol{N}$ be a nilpotent matrix of index k and let $\boldsymbol{x}$ be vector such that $\boldsymbol{N}^{k-1}\boldsymbol{x} \neq \boldsymbol{0}$. Prove that $\{\boldsymbol{x}, \boldsymbol{Nx}, \boldsymbol{N}^2\boldsymbol{x}, \ldots, \boldsymbol{N}^{k-1}\boldsymbol{x}\}$ is a linearly independent set. Such sets are called ***Krylov sequences***.

33. What is the index of an identity matrix?

34. Let $\boldsymbol{P} \neq \boldsymbol{I}$ be a projector? What is the index of $\boldsymbol{P}$?

35. Find a core-nilpotent decomposition for any projector $\boldsymbol{P}$.

CHAPTER 7

Orthogonality, Orthogonal Subspaces and Projections

7.1 Inner product, norms and orthogonality

We will now study in greater detail the concepts of inner product and orthogonality. We had very briefly introduced the (standard) inner product between two vectors and the norm of a vector way back in Definition 1.9 (see (1.5)). The inner product is the analogue of the *dot product* of two vectors in analytical geometry when $m = 2$ or $m = 3$. The following result outlines some basic properties of the inner product.

Lemma 7.1 *Let $\boldsymbol{x}$, $\boldsymbol{y}$ and $\boldsymbol{z}$ be vectors in $\Re^m$.*

(i) The inner product is symmetric, i.e., $\langle \boldsymbol{x}, \boldsymbol{y} \rangle = \langle \boldsymbol{y}, \boldsymbol{x} \rangle$.

(ii) The inner product is linear in both its arguments (bilinear), i.e.,

$$\langle a\boldsymbol{x}, b\boldsymbol{y}+c\boldsymbol{z} \rangle = ab\langle \boldsymbol{x}, \boldsymbol{y} \rangle + ac\langle \boldsymbol{x}, \boldsymbol{z} \rangle \text{ and } \langle a\boldsymbol{x}+b\boldsymbol{y}, c\boldsymbol{z} \rangle = ac\langle \boldsymbol{x}, \boldsymbol{z} \rangle + bc\langle \boldsymbol{y}, \boldsymbol{z} \rangle \,.$$

(iii) The inner product of a vector with itself is always non-negative, i.e., $\langle \boldsymbol{x}, \boldsymbol{x} \rangle \geq 0$, with $\langle \boldsymbol{x}, \boldsymbol{x} \rangle = 0$ if and only if $\boldsymbol{x} = \boldsymbol{0}$.

Proof. The proof for (i) follows immediately from the definition in (1.5):

$$\langle \boldsymbol{x}, \boldsymbol{y} \rangle = \boldsymbol{y}'\boldsymbol{x} = \sum_{i=1}^{m} y_i x_i = \sum_{i=1}^{m} x_i y_i = \boldsymbol{x}'\boldsymbol{y} = \langle \boldsymbol{y}, \boldsymbol{x} \rangle.$$

For (ii), the linearity in the first argument is obtained as

$$\langle a\boldsymbol{x}, b\boldsymbol{y}+c\boldsymbol{z} \rangle = \sum_{i=1}^{m} a x_i (b y_i + c z_i) = ab \sum_{i=1}^{m} x_i y_i + ac \sum_{i=1}^{m} x_i z_i = ab\langle \boldsymbol{x}, \boldsymbol{y} \rangle + ac\langle \boldsymbol{x}, \boldsymbol{z} \rangle \,,$$

while the second follows from the symmetric property in (i) and the linearity in the first argument: $\langle a\boldsymbol{x} + b\boldsymbol{y}, c\boldsymbol{z} \rangle = \langle c\boldsymbol{z}, a\boldsymbol{x} + b\boldsymbol{y} \rangle = ac\langle \boldsymbol{x}, \boldsymbol{z} \rangle + bc\langle \boldsymbol{y}, \boldsymbol{z} \rangle$.

For (iii), we first note that $\langle \boldsymbol{x}, \boldsymbol{x} \rangle = \sum_{i=1}^{m} x_i^2 \geq 0$. Also, $\langle \boldsymbol{x}, \boldsymbol{x} \rangle = 0$ if and only if $\sum_{i=1}^{m} x_i^2 = 0$, which occurs if and only if each $x_i = 0$ or, in other words, $\boldsymbol{x} = \boldsymbol{0}$. □

The following property is an immediate consequence of Definition 1.9.

Lemma 7.2 *Let $\boldsymbol{x}$ be a vector in $\Re^m$. Then $\|a\boldsymbol{x}\| = |a|\|\boldsymbol{x}\|$ for any real number a.*

Proof. When $a = 0$, $\|a\boldsymbol{x}\| = \|\boldsymbol{0}\| = 0 = |a|\|\boldsymbol{x}\|$. If $a \neq 0$, we have $\|a\boldsymbol{x}\| = \sqrt{\langle a\boldsymbol{x}, a\boldsymbol{x}\rangle} = \sqrt{a^2\langle \boldsymbol{x}, \boldsymbol{x}\rangle} = |a|\|\boldsymbol{x}\|$. □

We now derive an inequality of exceptional importance.

Theorem 7.1 The Cauchy-Schwarz inequality. *Let $\boldsymbol{x}$ and $\boldsymbol{y}$ be two real vectors of order m. Then,*

$$\langle \boldsymbol{x}, \boldsymbol{y}\rangle^2 \leq \|\boldsymbol{x}\|^2\|\boldsymbol{y}\|^2 \tag{7.1}$$

with equality holding if and only if $\boldsymbol{x} = a\boldsymbol{y}$ for some real number a.

Proof. For any real number a, we have

$$\begin{aligned} 0 \leq \|\boldsymbol{x} - a\boldsymbol{y}\|^2 &= \langle \boldsymbol{x} - a\boldsymbol{y}, \boldsymbol{x} - a\boldsymbol{y}\rangle = \langle \boldsymbol{x}, \boldsymbol{x}\rangle - 2a\langle \boldsymbol{x}, \boldsymbol{y}\rangle + a^2\langle \boldsymbol{y}, \boldsymbol{y}\rangle \\ &= \|\boldsymbol{x}\|^2 - 2a\langle \boldsymbol{x}, \boldsymbol{y}\rangle + a^2\|\boldsymbol{y}\|^2. \end{aligned}$$

If $\boldsymbol{y} = \boldsymbol{0}$, then we have equality in (7.1). Let $\boldsymbol{y} \neq \boldsymbol{0}$ and set $a = \langle \boldsymbol{x}, \boldsymbol{y}\rangle / \|\boldsymbol{y}\|^2$ as a particular choice. Then the above expression becomes

$$0 \leq \|\boldsymbol{x}\|^2 - 2\frac{\langle \boldsymbol{x}, \boldsymbol{y}\rangle^2}{\|\boldsymbol{y}\|^2} + \frac{\langle \boldsymbol{x}, \boldsymbol{y}\rangle^2}{\|\boldsymbol{y}\|^2} = \|\boldsymbol{x}\|^2 - \frac{\langle \boldsymbol{x}, \boldsymbol{y}\rangle^2}{\|\boldsymbol{y}\|^2},$$

which proves (7.1). When $\boldsymbol{y} = \boldsymbol{0}$ or $\boldsymbol{x} = \boldsymbol{0}$ we have equality. Otherwise, equality follows if and only if $\langle \boldsymbol{x} - a\boldsymbol{y}, \boldsymbol{x} - a\boldsymbol{y}\rangle = 0$, which implies $\boldsymbol{x} = a\boldsymbol{y}$. □

A direct consequence of the Cauchy-Schwarz inequality is the *triangle inequality* for vector norms. The triangle inequality, when applied to $\Re^2$, says that the sum of two sides of a triangle will be greater than the third.

Corollary 7.1 The triangle inequality. *Let $\boldsymbol{x}$ and $\boldsymbol{y}$ be two non-null vectors in $\Re^m$. Then,*

$$\|\boldsymbol{x} + \boldsymbol{y}\| \leq \|\boldsymbol{x}\| + \|\boldsymbol{y}\|, \tag{7.2}$$

with equality if and only if $\boldsymbol{x} = a\boldsymbol{y}$ for some non-negative real number a.

Proof. The Cauchy-Schwarz inequality gives $\langle \boldsymbol{x}, \boldsymbol{y}\rangle \leq \|\boldsymbol{x}\|\|\boldsymbol{y}\|$. Therefore,

$$\begin{aligned} \|\boldsymbol{x} + \boldsymbol{y}\|^2 &= \langle \boldsymbol{x} + \boldsymbol{y}, \boldsymbol{x} + \boldsymbol{y}\rangle = \|\boldsymbol{x}\|^2 + \|\boldsymbol{y}\|^2 + 2\langle \boldsymbol{x}, \boldsymbol{y}\rangle \\ &\leq \|\boldsymbol{x}\|^2 + \|\boldsymbol{y}\|^2 + 2\|\boldsymbol{x}\|\|\boldsymbol{y}\| = (\|\boldsymbol{x}\| + \|\boldsymbol{y}\|)^2. \end{aligned}$$

This proves the triangle inequality. Note that equality occurs if and only if we have equality in the Cauchy-Schwarz inequality, i.e., when $\boldsymbol{x} = a\boldsymbol{y}$ for some non-negative real number a. □

The Cauchy-Schwarz inequality helps us extend the concept of *angle* to real vectors of arbitrary order. Using the linearity of inner products (see part (ii) of Lemma 7.1) and the definition of the norm, we obtain

$$\|\boldsymbol{x} - \boldsymbol{y}\|^2 = \langle \boldsymbol{x} - \boldsymbol{y}, \boldsymbol{x} - \boldsymbol{y} \rangle = \|\boldsymbol{x}\|^2 + \|\boldsymbol{y}\|^2 - 2\langle \boldsymbol{x}, \boldsymbol{y} \rangle \tag{7.3}$$

for any two vectors $\boldsymbol{x}$ and $\boldsymbol{y}$ in $\Re^m$. On the plane, i.e., when $m = 2$, the cosine law of plane trigonometry yields

$$\|\boldsymbol{x} - \boldsymbol{y}\|^2 = \|\boldsymbol{x}\|^2 + \|\boldsymbol{y}\|^2 - 2\|\boldsymbol{x}\|\|\boldsymbol{y}\| \cos(\theta), \tag{7.4}$$

where θ is the angle between the vectors $\boldsymbol{x}$ and $\boldsymbol{y}$. Equating (7.3) and (7.4) motivates the following definition of the *angle* between two vectors in $\Re^m$.

Definition 7.1 Angle between vectors. *In the Euclidean real coordinate spaces, say $\Re^m$, the radian measure of the angle between nonzero vectors $\boldsymbol{x}$ and $\boldsymbol{y}$ is defined to be the number $\theta \in [0, \pi]$ such that*

$$\cos(\theta) = \frac{\langle \boldsymbol{x}, \boldsymbol{y} \rangle}{\|\boldsymbol{x}\|\|\boldsymbol{y}\|} = \langle \boldsymbol{u}_x, \boldsymbol{u}_y \rangle \ \ (0 \leq \theta \leq \pi) \ , \tag{7.5}$$

where $\boldsymbol{u}_x = \dfrac{\boldsymbol{x}}{\|\boldsymbol{x}\|}$ *and* $\boldsymbol{u}_y = \dfrac{\boldsymbol{y}}{\|\boldsymbol{y}\|}$.

The Cauchy-Schwarz inequality affirms that the right-hand side of (7.5) lies between -1 and 1, so there is a unique value $\theta \in [0, \pi]$ for which (7.5) will hold.

From Lemma 7.2, it is immediate that $\|\boldsymbol{u}_x\| = \|\boldsymbol{u}_y\| = 1$. We say that we have *normalized* $\boldsymbol{x}$ and $\boldsymbol{y}$ to $\boldsymbol{u}_x$ and $\boldsymbol{u}_y$. In other words, the angle between $\boldsymbol{x}$ and $\boldsymbol{y}$ does not depend upon the lengths of $\boldsymbol{x}$ and $\boldsymbol{y}$ but only on their *directions*. The vectors $\boldsymbol{u}_x$ and $\boldsymbol{u}_y$ have the same directions as $\boldsymbol{x}$ and $\boldsymbol{y}$, so the angle between $\boldsymbol{x}$ and $\boldsymbol{y}$ is the same as the angle between $\boldsymbol{u}_x$ and $\boldsymbol{u}_y$. The following definition is useful.

Definition 7.2 *A* **unit vector** *is any vector with a length of one. Unit vectors are often used to indicate direction and are also called* **direction vectors**. *A vector of arbitrary length can be divided by its length to create a unit vector. This is known as* **normalizing** *a vector.*

In plane geometry, when two vectors are orthogonal, they are perpendicular to each other and the angle between them is equal to 90 degrees or $\pi/2$ radians. In that case, $\cos(\theta) = 0$. The definition of the "angle" between two vectors in $\Re^m$, as given in Definition 7.1, suggests a natural definition for *orthogonal* vectors.

Definition 7.3 Orthogonal vectors. *Two vectors, $\boldsymbol{x}$ and $\boldsymbol{y}$, are said to be* **orthogonal** *if the angle between them, θ, is such that $\cos(\theta) = 0$ or, in terms of the inner product, $\langle \boldsymbol{x}, \boldsymbol{y} \rangle = 0$. We denote two orthogonal vectors as $\boldsymbol{x} \perp \boldsymbol{y}$.*

Note that when $\theta = 0$, the cosine is equal to 1. In this case, $\boldsymbol{x}$ and $\boldsymbol{y}$ satisfy $\langle \boldsymbol{x}, \boldsymbol{y} \rangle = \|\boldsymbol{x}\|\|\boldsymbol{y}\|$ and we say that the vectors $\boldsymbol{x}$ and $\boldsymbol{y}$ are ***parallel***.

Given two vectors in $\Re^3$, we can find a vector orthogonal to both those vectors using the *cross product* usually introduced in analytical geometry.

Definition 7.4 *Let $\boldsymbol{u}$ and $\boldsymbol{v}$ be two vectors in $\Re^3$. Then the* **cross product** *or* **vector product** *between $\boldsymbol{u}$ and $\boldsymbol{v}$ is defined as the 3×1 vector*

$$\boldsymbol{u} \times \boldsymbol{v} = \begin{bmatrix} u_2 v_3 - u_3 v_2 \\ u_3 v_1 - u_1 v_3 \\ u_1 v_2 - u_2 v_1 \end{bmatrix} .$$

The vector $\boldsymbol{u} \times \boldsymbol{v}$ is normal to both $\boldsymbol{u}$ and $\boldsymbol{v}$. This is easily verified as follows:

$$\begin{aligned} \langle \boldsymbol{u} \times \boldsymbol{v}, \boldsymbol{u} \rangle &= \boldsymbol{u}'(\boldsymbol{u} \times \boldsymbol{v}) \\ &= u_1(u_2v_3 - u_3v_2) + u_2(u_3v_1 - u_1v_3) + u_3(u_1v_2 - u_2v_1) \\ &= u_1u_2v_3 - u_1u_3v_2 + u_2u_3v_1 - u_2u_1v_3 + u_3u_1v_2 - u_3u_2v_1 \\ &= u_1u_2(v_3 - v_3) + u_2u_3(v_1 - v_1) + u_3u_1(v_2 - v_2) = 0 . \end{aligned}$$

Similarly, $\langle \boldsymbol{u} \times \boldsymbol{v}, \boldsymbol{v} \rangle = 0$.

Until now we have discussed orthogonality of two vectors. We can also talk about an *orthogonal set* of vectors when any two members of the set are orthogonal. A formal definition is provided below.

Definition 7.5 *A set $\mathcal{X} = \{\boldsymbol{x}_1, \boldsymbol{x}_2, \ldots\}$ of non-null vectors is said to be an* **orthogonal set** *if any two vectors in it are orthogonal, i.e., $\langle \boldsymbol{x}_i, \boldsymbol{x}_j \rangle = 0$ for any two vectors $\boldsymbol{x}_i$ and $\boldsymbol{x}_j$ in $\mathcal{X}$. The set $\mathcal{X}$ is said to be* **orthonormal** *if, in addition to any two vectors in it being orthogonal, each vector has unit length or norm, i.e., $\|\boldsymbol{x}_i\| = 1$ for every $\boldsymbol{x}_i$. Thus, for any two vectors in an orthonormal set $\langle \boldsymbol{x}_i, \boldsymbol{x}_j \rangle = \delta_{ij}$, where $\delta_{ij} = 1$ if $i = j$ and 0 otherwise (it is called the Kronecker delta). We call an orthonormal set* **complete** *if it is not contained in any larger orthonormal set.*

Example 7.1 Consider the standard basis $\{\boldsymbol{e}_1, \boldsymbol{e}_2, \ldots, \boldsymbol{e}_m\}$ in $\Re^m$, where $\boldsymbol{e}_i$ is the $n \times 1$ vector with a 1 as its i-th element and 0 everywhere else. It is easy to verify that $\langle \boldsymbol{e}_i, \boldsymbol{e}_j \rangle = 0$ for all $i \neq j$ and $\langle \boldsymbol{e}_i, \boldsymbol{e}_i \rangle = \|\boldsymbol{e}_1\|^2 = 1$ for $i = 1, 2, \ldots, m$. More succinctly, we write $\langle \boldsymbol{e}_i, \boldsymbol{j}_j \rangle = \delta_{ij}$ and, from Definition 7.5, it follows that the standard basis in $\Re^m$ is an orthonormal set.

Orthogonal vectors in $\Re^m$ satisfy the *Pythagorean identity.*

Corollary 7.2 The Pythagorean identity. *Let $\boldsymbol{x}$ and $\boldsymbol{y}$ be two orthogonal vectors in $\Re^m$. Then,*

$$\|\boldsymbol{x} + \boldsymbol{y}\|^2 = \|\boldsymbol{x}\|^2 + \|\boldsymbol{y}\|^2. \tag{7.6}$$

More generally, $\|\sum_{i=1}^{k} \boldsymbol{x}_i\|^2 = \sum_{i=1}^{k} \|\boldsymbol{x}_i\|^2$ if $\{\boldsymbol{x}_1, \boldsymbol{x}_2, \ldots, \boldsymbol{x}_k\}$ is an orthogonal set.

Proof. The proof follows from the linearity of the inner product and the definition of orthogonality:

$$\|\boldsymbol{x}+\boldsymbol{y}\|^2 = \langle \boldsymbol{x}+\boldsymbol{y}, \boldsymbol{x}+\boldsymbol{y}\rangle = \|\boldsymbol{x}\|^2 + \|\boldsymbol{y}\|^2 + 2\langle \boldsymbol{x}, \boldsymbol{y}\rangle = \|\boldsymbol{x}\|^2 + \|\boldsymbol{y}\|^2.$$

The more general case follows easily by induction and is left to the reader. □

Example 7.1 demonstrated a set of vectors that are orthogonal and linearly independent. It is not necessary that every linearly independent set of vectors is orthogonal. Consider for example, the set $\{(1,0),(2,3)\}$ of two vectors in $\Re^2$. These two vectors are clearly linearly independent, but they are not orthogonal (their inner product equals 2). What about the reverse: Do orthogonal vectors always form a linearly independent set? Well, notice that the null vector, $\boldsymbol{0}$, is orthogonal to every other vector, yet it cannot be a part of a linearly independent set. But we have been careful enough to *exclude* the null vector from the definition of an orthogonal set in Definition 7.5.

Lemma 7.3 *A set of non-null orthogonal vectors is linearly independent. In other words, an orthogonal set is linearly independent.*

Proof. Let $\mathcal{X} = \{\boldsymbol{x}_1, \boldsymbol{x}_2, \ldots, \boldsymbol{x}_n\}$ be an orthogonal set such that none of the $\boldsymbol{x}_i$'s is the $\boldsymbol{0}$ vector. Consider the homogenous relation $\sum_{i=1}^{n} \alpha_i \boldsymbol{x}_i = \boldsymbol{0}$. Consider any $\boldsymbol{x}_j$ in $\mathcal{X}$ and take the inner product of $\sum_{i=1}^{n} \alpha_i \boldsymbol{x}_i$ and $\boldsymbol{x}_j$. We note the following:

$$\sum_{i=1}^{n} \alpha_i \boldsymbol{x}_i = \boldsymbol{0} \Rightarrow \langle \sum_{i=1}^{n} \alpha_i \boldsymbol{x}_i, \boldsymbol{x}_j \rangle = \alpha_j \langle \boldsymbol{x}_j, \boldsymbol{x}_j \rangle = \boldsymbol{0} \Rightarrow \alpha_j = 0\,.$$

Repeating this for each $\boldsymbol{x}_j$ in $\mathcal{X}$ yields $\alpha_j = 0$ for $j = 1, 2, \ldots, n$, which implies that $\mathcal{X}$ is a linearly independent set. □

Usually, when we mention sets of orthogonal vectors we will exclude $\boldsymbol{0}$. So we simply say that any set of orthogonal vectors is linearly independent.

In Example 4.3 we saw that a plane in $\Re^3$ was determined by two non-parallel vectors $\boldsymbol{u}$ and $\boldsymbol{v}$ and the position vector $\boldsymbol{r}_0$ for a known point on the plane. The ***normal*** to this plane is the vector that is perpendicular to the plane, which means that it is orthogonal to every vector lying on the plane. This normal must be perpendicular to both $\boldsymbol{u}$ and $\boldsymbol{v}$ and can be computed easily as $\boldsymbol{n} = \boldsymbol{u} \times \boldsymbol{v}$. When the vector $\boldsymbol{n}$ is normalized to be of unit length we set $\boldsymbol{n} = \dfrac{\boldsymbol{u} \times \boldsymbol{v}}{\|\boldsymbol{u} \times \boldsymbol{v}\|}$.

Lemma 7.3 ensures that this normal is unique in the sense that any other vector perpendicular to the plane must be a scalar multiple of $\boldsymbol{n}$ or, in the geometric sense, must be parallel to $\boldsymbol{n}$. Otherwise, if $\boldsymbol{n}_1$ and $\boldsymbol{n}_2$ were both normals to the plane that were not scalar multiples of each other, then the set $\{\boldsymbol{u}, \boldsymbol{v}, \boldsymbol{n}_1, \boldsymbol{n}_2\}$ would be a linearly independent set of four vectors in $\Re^3$, which is impossible. It makes sense, therefore, to characterize a plane in terms of its normal direction. We do this in the example below, which should be familiar from analytical geometry.

Example 7.2 The equation of a plane. In three-dimensional space, another important way of defining a plane is by specifying a point and a normal vector to the plane.

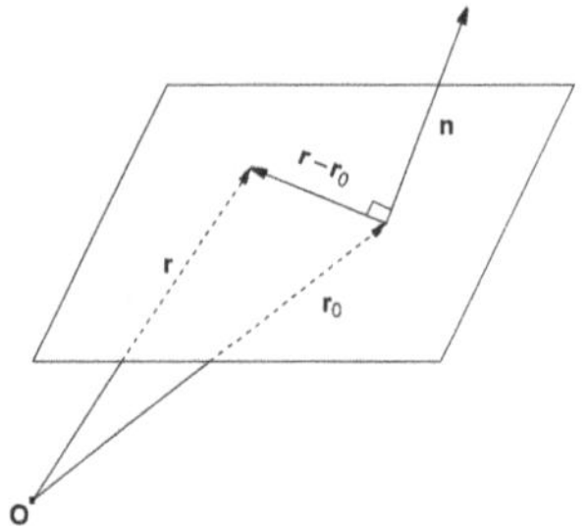

Figure 7.1 *The vector equation of a plane using the normal vector. The vector $\boldsymbol{r}_0$ is the position vector for a known point on the plane, $\boldsymbol{r}$ is the position vector for any arbitrary point on the plane and $\boldsymbol{n}$ is the normal to the plane.*

Let $\boldsymbol{r}_0$ be the position vector of a known point on the plane, and let $\boldsymbol{n}$ be a nonzero vector normal to the plane. Suppose that $\boldsymbol{r}$ is the position vector of any arbitrary point on the plane. This means that the vector $\boldsymbol{r} - \boldsymbol{r}_0$ lies on the plane and must be orthogonal to $\boldsymbol{n}$ (see Figure 7.1). Therefore, the vector equation of the plane in terms of a point and the normal to the plane is given by

$$\boldsymbol{0} = \langle \boldsymbol{r} - \boldsymbol{r}_0, \boldsymbol{n} \rangle = \boldsymbol{n}'(\boldsymbol{r} - \boldsymbol{r}_0) \Longleftrightarrow \boldsymbol{n}'\boldsymbol{r} = \boldsymbol{n}'\boldsymbol{r}_0 = d \;, \tag{7.7}$$

where $\boldsymbol{n}'\boldsymbol{r}_0 = d$. In other words, any plane in $\Re^3$ can be defined as the set $\{\boldsymbol{r} \in \Re^3 : \boldsymbol{n}'\boldsymbol{r} = d\}$ for some fixed number d.

To obtain a scalar equation for the plane, let $\boldsymbol{n} = (a, b, c)'$, $\boldsymbol{r} = (x, y, z)'$ and $\boldsymbol{r}_0 = (x_0, y_0, z_0)'$. Then the vector equation in (7.7) becomes

$$a(x - x_0) + b(y - y_0) + c(z - z_0) = 0 \Longleftrightarrow ax + by + cz - d = 0 \;,$$

where $d = \boldsymbol{n}'\boldsymbol{r}_0 = ax_0 + by_0 + cz_0$. ■

7.2 Row rank = column rank: A proof using orthogonality

We have already seen a few different proofs of the fact that the row rank and column rank of a matrix are equal or, equivalently, that $\rho(\boldsymbol{A}) = \rho(\boldsymbol{A}')$. Here, we present yet another elegant exposition using orthogonality based upon Mackiw (1995).

Let $\boldsymbol{A}$ be an $m \times n$ matrix whose row rank is r. Therefore, the dimension of the row space of $\boldsymbol{A}$ is r and suppose that $\boldsymbol{x}_1, \boldsymbol{x}_2, \ldots, \boldsymbol{x}_r$ is a basis for the row space of $\boldsymbol{A}$.

We claim that the vectors $\boldsymbol{A}\boldsymbol{x}_1, \boldsymbol{A}\boldsymbol{x}_2, \ldots, \boldsymbol{A}\boldsymbol{x}_r$ are linearly independent.

To see why, consider the linear homogeneous relation involving these vectors with scalar coefficients $c_1, c_2, \ldots, c_r$:

$$c_1\boldsymbol{A}\boldsymbol{x}_1 + c_2\boldsymbol{A}\boldsymbol{x}_2 + \cdots c_r\boldsymbol{A}\boldsymbol{x}_r = \boldsymbol{A}(c_1\boldsymbol{x}_1 + c_2\boldsymbol{x}_2 + \cdots + c_r\boldsymbol{x}_r) = \boldsymbol{A}\boldsymbol{v} = \boldsymbol{0},$$

where $\boldsymbol{v} = c_1\boldsymbol{x}_1 + c_2\boldsymbol{x}_2 + \ldots, c_r\boldsymbol{x}_r$. We want to show that each of these coefficients is zero. We make two easy observations below.

(a) The vector $\boldsymbol{v}$ is a linear combination of vectors in the row space of $\boldsymbol{A}$, so $\boldsymbol{v}$ belongs to the row space of $\boldsymbol{A}$.

(b) Let $\boldsymbol{a}_i'$ be the i-th row of $\boldsymbol{A}$. Since $\boldsymbol{A}\boldsymbol{v} = \boldsymbol{0}$, we obtain $\boldsymbol{a}_i'\boldsymbol{v} = 0$ for $i = 1, 2, \ldots, m$. Therefore, $\boldsymbol{v}$ is orthogonal to every row vector of $\boldsymbol{A}$ and, hence, is orthogonal to every vector in the row space of $\boldsymbol{A}$.

Now comes a crucial observation: the facts (a) and (b) together imply that $\boldsymbol{v}$ is orthogonal to itself. So $\boldsymbol{v}$ must be equal to $\boldsymbol{0}$ and, from the definition of $\boldsymbol{v}$, we see

$$\boldsymbol{0} = \boldsymbol{v} = c_1\boldsymbol{x}_1 + c_2\boldsymbol{x}_2 + \ldots + c_r\boldsymbol{x}_r \; .$$

But recall that the $\boldsymbol{x}_i$'s are linearly independent because they are a basis for the row space of $\boldsymbol{A}$, which implies that $c_1 = c_2 = \cdots = c_r = 0$. This proves our claim that $\boldsymbol{A}\boldsymbol{x}_1, \boldsymbol{A}\boldsymbol{x}_2, \ldots, \boldsymbol{A}\boldsymbol{x}_r$ are linearly independent.

Now, each $\boldsymbol{A}\boldsymbol{x}_i$ is obviously a vector in the column space of $\boldsymbol{A}$. So, $\boldsymbol{A}\boldsymbol{x}_1, \boldsymbol{A}\boldsymbol{x}_2, \ldots, \boldsymbol{A}\boldsymbol{x}_r$ is a set of r linearly independent vectors in the column space of $\boldsymbol{A}$. Therefore, the dimension of the column space of $\boldsymbol{A}$ (i.e., the column rank of $\boldsymbol{A}$) must be at least as big as r. This proves that

$$\text{“row rank”} = \dim(\mathcal{R}(\boldsymbol{A})) \leq \dim(\mathcal{C}(\boldsymbol{A})) = \text{“column rank.”}$$

Now use a familiar trick: apply this result to $\boldsymbol{A}'$ to get the reverse inequality:

$$\text{“column rank”} = \dim(\mathcal{C}(\boldsymbol{A})) = \dim(\mathcal{R}(\boldsymbol{A}')) \leq \dim(\mathcal{C}(\boldsymbol{A}')) = \text{“row rank.”}$$

The two preceding inequalities together imply that the column rank and row rank of $\boldsymbol{A}$ are the same or, equivalently, $\rho(\boldsymbol{A}) = \rho(\boldsymbol{A}')$. We have just seen that if $\{\boldsymbol{x}_1, \boldsymbol{x}_2, \ldots, \boldsymbol{x}_r\}$ is a basis for the row space of $\boldsymbol{A}$, then $\{\boldsymbol{A}\boldsymbol{x}_1, \boldsymbol{A}\boldsymbol{x}_2, \ldots, \boldsymbol{A}\boldsymbol{x}_r\}$ is a linearly independent set of r vectors in the column space of $\boldsymbol{A}$, where r is the dimension of the column space of $\boldsymbol{A}$. Therefore, $\{\boldsymbol{A}\boldsymbol{x}_1, \boldsymbol{A}\boldsymbol{x}_2, \ldots, \boldsymbol{A}\boldsymbol{x}_r\}$ is a basis for the column space of $\boldsymbol{A}$. This is a simple way to produce a basis for $\mathcal{C}(\boldsymbol{A})$ when we are given a basis for $\mathcal{R}(\boldsymbol{A})$.

7.3 Orthogonal projections

Suppose $\mathcal{X} = \{\boldsymbol{x}_1, \boldsymbol{x}_2, \ldots, \boldsymbol{x}_k\}$ is an orthogonal set and let $\boldsymbol{x}$ be a vector that belongs to the span of $\mathcal{X}$. Orthogonality of the vectors in $\mathcal{X}$ makes it really easy to find the coefficients when $\boldsymbol{x}$ is expressed as a linear combination of the vectors in $\mathcal{X}$. To be precise, suppose that

$$\boldsymbol{x} = \alpha_1\boldsymbol{x}_1 + \alpha_2\boldsymbol{x}_2 + \cdots + \alpha_k\boldsymbol{x}_k = \sum_{i=1}^{k} \alpha_i\boldsymbol{x}_i \; . \tag{7.8}$$

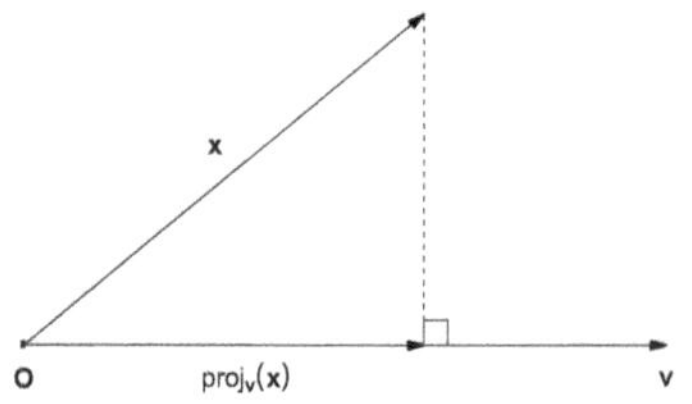

Figure 7.2 *The vector* $\text{proj}_{\boldsymbol{v}}(\boldsymbol{x})$ *is the projection of* $\boldsymbol{x}$ *onto the vector* $\boldsymbol{v}$.

Taking the inner product of both sides with any $\boldsymbol{x}_j \in \mathcal{X}$ yields

$$\langle \boldsymbol{x}, \boldsymbol{x}_j \rangle = \left\langle \sum_{i=1}^{k} \alpha_i \boldsymbol{x}_i, \boldsymbol{x}_j \right\rangle = \sum_{i=1}^{k} \alpha_i \langle \boldsymbol{x}_i, \boldsymbol{x}_j \rangle = \alpha_j \langle \boldsymbol{x}_j, \boldsymbol{x}_j \rangle$$

because $\langle \boldsymbol{x}_i, \boldsymbol{x}_j \rangle = 0$ for all $i \neq j$. This shows that $\alpha_j = \dfrac{\langle \boldsymbol{x}, \boldsymbol{x}_j \rangle}{\langle \boldsymbol{x}_j, \boldsymbol{x}_j \rangle}$ for every $j = 1, 2, \ldots, k$. Substituting these values for the α_j's in (7.8) yields

$$\boldsymbol{x} = \frac{\langle \boldsymbol{x}, \boldsymbol{x}_1 \rangle}{\langle \boldsymbol{x}_1, \boldsymbol{x}_1 \rangle} \boldsymbol{x}_1 + \frac{\langle \boldsymbol{x}, \boldsymbol{x}_2 \rangle}{\langle \boldsymbol{x}_2, \boldsymbol{x}_2 \rangle} \boldsymbol{x}_2 + \cdots + \frac{\langle \boldsymbol{x}, \boldsymbol{x}_k \rangle}{\langle \boldsymbol{x}_k, \boldsymbol{x}_k \rangle} \boldsymbol{x}_k = \sum_{i=1}^{k} \frac{\langle \boldsymbol{x}, \boldsymbol{x}_i \rangle}{\langle \boldsymbol{x}_i, \boldsymbol{x}_i \rangle} \boldsymbol{x}_i \,. \tag{7.9}$$

Matters are even simpler when $\mathcal{X} = \{\boldsymbol{x}_1, \boldsymbol{x}_2, \ldots, \boldsymbol{x}_k\}$ is an orthonormal set. Then we have that each $\alpha_j = \langle \boldsymbol{x}, \boldsymbol{x}_j \rangle$ and $\boldsymbol{x} = \sum_{i=1}^{k} \langle \boldsymbol{x}, \boldsymbol{x}_i \rangle \boldsymbol{x}_i$.

Orthogonal expansions such as in (7.9) can be developed using some geometric insight and, in particular, using the *projection function*, which we define below.

Definition 7.6 The projection function. *Let* $\boldsymbol{v}$ *and* $\boldsymbol{x}$ *be two vectors in* $\Re^n$ *such that* $\boldsymbol{v} \neq \boldsymbol{0}$. *Then, a* **projection** *of* $\boldsymbol{x}$ *onto the vector* $\boldsymbol{v}$ *is given by the vector function*

$$\text{proj}_{\boldsymbol{v}}(\boldsymbol{x}) = \frac{\langle \boldsymbol{x}, \boldsymbol{v} \rangle}{\langle \boldsymbol{v}, \boldsymbol{v} \rangle} \boldsymbol{v} \,, \quad \textit{whenever} \;\; \boldsymbol{v} \neq \boldsymbol{0} \,. \tag{7.10}$$

Geometrically, this function gives the ***vector projection*** in analytical geometry (also known as the ***vector resolute***, or ***vector component***) of a vector $\boldsymbol{x}$ in the direction of a vector $\boldsymbol{v}$. Alternatively, we say that $\text{proj}_{\boldsymbol{v}}(\boldsymbol{x})$ is the ***orthogonal projection*** of $\boldsymbol{x}$ on (or onto) $\boldsymbol{v}$. Clearly, $\text{proj}_{\boldsymbol{v}}(\boldsymbol{x})$ is well-defined whenever $\boldsymbol{v}$ is not the null vector. It does not make sense to project onto the null vector. When $\boldsymbol{v}$ is normalized to be unit length, we obtain $\text{proj}_{\boldsymbol{v}}(\boldsymbol{x}) = \langle \boldsymbol{x}, \boldsymbol{v} \rangle \boldsymbol{v} = (\boldsymbol{v}'\boldsymbol{x})\boldsymbol{v}$.

In analytical geometry the quantity $\dfrac{\boldsymbol{v}'\boldsymbol{x}}{\|\boldsymbol{v}\|}$ is often referred to as the ***component*** of $\boldsymbol{x}$ in the direction of $\boldsymbol{v}$ and is denoted by $\text{comp}_{\boldsymbol{v}}(\boldsymbol{x})$. The length of the projection of $\boldsymbol{x}$

onto $\boldsymbol{v}$ is equal to the absolute value of the component of $\boldsymbol{x}$ along $\boldsymbol{v}$:

$$\|\text{proj}_{\boldsymbol{v}}(\boldsymbol{x})\| = \left(\frac{|\boldsymbol{v}'\boldsymbol{x}|}{\boldsymbol{v}'\boldsymbol{v}}\right)\|\boldsymbol{v}\| = \frac{|\boldsymbol{v}'\boldsymbol{x}|}{\|\boldsymbol{v}\|} = |\text{comp}_{\boldsymbol{v}}(\boldsymbol{x})| \ .$$

In two-dimensions, this is clear from basic trigonometry. The length of the projection of $\boldsymbol{x}$ onto $\boldsymbol{v}$ is given by $\|\boldsymbol{x}\|\cos\theta$, where θ is the acute angle between $\boldsymbol{x}$ and $\boldsymbol{v}$. From Definition 7.1, we see that $\|\boldsymbol{x}\|\cos\theta = \dfrac{|\boldsymbol{v}'\boldsymbol{x}|}{\|\boldsymbol{v}\|} = |\text{comp}_{\boldsymbol{v}}(\boldsymbol{x})|$.

A familiar application of the component in $\Re^3$ is to find the distance of a point from the plane.

Example 7.3 Consider a plane in $\Re^3$. Let $\boldsymbol{n}$ be a normal vector to the plane and $\boldsymbol{r}_0$ be the position vector of a known point $P_0 = (x_0, y_0, z_0)$ on the plane. Let $\boldsymbol{x}$ be the position vector of a point $P = (x, y, z)$ in $\Re^3$. The distance D of P from the plane is the length of the perpendicular dropped from P onto the plane, which is precisely the absolute value of the component of $\boldsymbol{b} = \boldsymbol{x} - \boldsymbol{r}_0$ along the direction of $\boldsymbol{n}$. Therefore,

$$D = |\text{comp}_{\boldsymbol{n}}(\boldsymbol{b})| = \frac{|\boldsymbol{n}'\boldsymbol{b}|}{\|\boldsymbol{n}\|} = \frac{|\boldsymbol{n}'(\boldsymbol{x} - \boldsymbol{r}_0)|}{\|\boldsymbol{n}\|} = \frac{a(x - x_0) + b(y - y_0) + c(z - z_0)}{\sqrt{a^2 + b^2 + c^2}}$$
$$= \frac{ax + by + cz - d}{\sqrt{a^2 + b^2 + c^2}} \ ,$$

where $\boldsymbol{n} = (a, b, c)'$ and $d = \boldsymbol{n}'\boldsymbol{r}_0 = ax_0 + by_0 + cz_0$. ■

Recall that we have already encountered "projections" in Section 6.4. How, then, is $\text{proj}_{\boldsymbol{v}}(\boldsymbol{x})$ related to the definition of the projection given in Definition 6.3? This is best understood by considering $\mathcal{X} = \{\boldsymbol{x}_1, \boldsymbol{x}_2\}$, where $\boldsymbol{x}_1$ and $\boldsymbol{x}_2$ are orthogonal to each other. Let $\boldsymbol{x} \in \mathcal{S} = Sp(\mathcal{X})$. From (7.9) and the definition of the proj function in (7.10), we find that

$$\boldsymbol{x} = \frac{\langle \boldsymbol{x}, \boldsymbol{x}_1 \rangle}{\langle \boldsymbol{x}_1, \boldsymbol{x}_1 \rangle}\boldsymbol{x}_1 + \frac{\langle \boldsymbol{x}, \boldsymbol{x}_2 \rangle}{\langle \boldsymbol{x}_2, \boldsymbol{x}_2 \rangle}\boldsymbol{x}_2 = \text{proj}_{\boldsymbol{x}_1}(\boldsymbol{x}) + \text{proj}_{\boldsymbol{x}_2}(\boldsymbol{x}) \ . \tag{7.11}$$

Therefore, $\boldsymbol{x}$ is the sum of two vectors $\text{proj}_{\boldsymbol{x}_1}(\boldsymbol{x})$ and $\text{proj}_{\boldsymbol{x}_2}(\boldsymbol{x})$, where the first vector belongs to $\mathcal{S}_1 = Sp(\{\boldsymbol{x}_1\})$ and the second belongs to $\mathcal{S}_2 = Sp(\{\boldsymbol{x}_2\})$. Equation 7.11 reveals that $\mathcal{S} = Sp(\{\mathcal{X}\})$ is a sum of the subspaces $\mathcal{S}_1$ and $\mathcal{S}_2$, i.e., $\mathcal{S} = \mathcal{S}_1 + \mathcal{S}_2$. We now claim that this sum is a direct sum. A simple way to verify this is to consider any $\boldsymbol{v} \in \mathcal{S}_1 \cap \mathcal{S}_2$. Since $\boldsymbol{v} \in \mathcal{S}_1 = Sp(\{\boldsymbol{x}_1\})$ and $\boldsymbol{v} \in \mathcal{S}_2 = Sp(\{\boldsymbol{x}_2\})$, it follows that $\boldsymbol{v} = \theta_1\boldsymbol{x}_1$ and $\boldsymbol{v} = \theta_2\boldsymbol{x}_2$ for scalars θ_1 and θ_2. So $\boldsymbol{0} = \boldsymbol{v} - \boldsymbol{v} = \theta_1\boldsymbol{x}_1 - \theta_2\boldsymbol{x}_2$, which implies $\theta_1 = \theta_2 = 0$ because $\{\boldsymbol{x}_1, \boldsymbol{x}_2\}$ is linearly independent ($\boldsymbol{x}_1$ is orthogonal to $\boldsymbol{x}_2$). This shows that $\boldsymbol{v} = \boldsymbol{0}$. Therefore, $\mathcal{S}_1 \cap \mathcal{S}_2 = \{\boldsymbol{0}\}$ and so $\mathcal{S} = \mathcal{S}_1 \oplus \mathcal{S}_2$. Using the language of Section 6.4, we can now say that $\text{proj}_{\boldsymbol{x}_1}(\boldsymbol{x})$ is the projection of $\boldsymbol{x}$ onto $\boldsymbol{x}_1$ along the span of $\{\boldsymbol{x}_2\}$.

For the general case, let $\mathcal{S}_i = Sp(\{\boldsymbol{x}_i\})$ for $i = 1, 2, \ldots, k$. First of all, we note that $\mathcal{S}_1 + \mathcal{S}_2 + \cdots + \mathcal{S}_k$ is direct. To see why this is true, consider $\boldsymbol{0} = \boldsymbol{v}_1 + \boldsymbol{v}_2 + \cdots + \boldsymbol{v}_k$, where $\boldsymbol{v}_i = \alpha_i\boldsymbol{x}_i \in \mathcal{S}_i$, where α_i is a scalar for each $i = 1, 2, \ldots, k$. This means that

$\mathbf{0} = \alpha_1 \boldsymbol{x}_1 + \alpha_2 \boldsymbol{x}_2 + \cdots + \alpha_k \boldsymbol{x}_k$. Since the $\boldsymbol{x}_i$'s are orthogonal, they are also linearly independent (Lemma 8.1). Therefore, $\alpha_i = 0$ and so $\boldsymbol{v}_i = \mathbf{0}$ for each $i = 1, 2, \ldots, k$. Therefore, $\mathcal{S}_1 + \mathcal{S}_2 + \cdots + \mathcal{S}_k$ is a direct sum. Using the orthogonal projection function defined in (7.10), we can rewrite (7.9) as

$$\boldsymbol{x} = \mathrm{proj}_{\boldsymbol{x}_1}(\boldsymbol{x}) + \mathrm{proj}_{\boldsymbol{x}_2}(\boldsymbol{x}) + \cdots + \mathrm{proj}_{\boldsymbol{x}_k}(\boldsymbol{x}) = \sum_{i=1}^{k} \mathrm{proj}_{\boldsymbol{x}_i}(\boldsymbol{x}) . \tag{7.12}$$

In other words, $\boldsymbol{x}$ is "decomposed" into a sum of its projections onto each of the subspaces $\mathcal{S}_i$ in a direct sum. Here each $\mathcal{S}_i$ is the span of a single vector—a "line"—and $\mathrm{proj}_{\boldsymbol{x}_i}(\boldsymbol{x})$ sometimes called the projection of $\boldsymbol{x}$ on the "line" $\boldsymbol{x}_i$. This is consistent with the usual definitions of projections in analytical geometry of two and three dimensional spaces. Later we will generalize this concept to orthogonal projections onto general subspaces.

Given that $\mathrm{proj}_{\boldsymbol{v}}(\boldsymbol{x})$ is an orthogonal projection of $\boldsymbol{x}$ onto the vector spanned by $\boldsymbol{v}$, we should be able to find a projector matrix, say $\boldsymbol{P}_{\boldsymbol{v}}$, such that $\mathrm{proj}_{\boldsymbol{v}}(\boldsymbol{x}) = \boldsymbol{P}_{\boldsymbol{v}}\boldsymbol{x}$. This helps us elicit some useful properties of orthogonal projections and facilitates further generalizations that we will study later. The following lemma shows a few basic properties of orthogonal projection.

Lemma 7.4 *Let $\boldsymbol{v}$ and $\boldsymbol{x}$ be two $n \times 1$ vectors in $\Re^n$. Assume $\boldsymbol{v} \neq \mathbf{0}$ and define*

$$\boldsymbol{P}_{\boldsymbol{v}} = \frac{\boldsymbol{v}\boldsymbol{v}'}{\boldsymbol{v}'\boldsymbol{v}} = \frac{1}{\|\boldsymbol{v}\|^2}\boldsymbol{v}\boldsymbol{v}' .$$

(i) $\boldsymbol{P}_{\boldsymbol{v}}$ *is an $n \times n$ matrix of rank one that is symmetric and idempotent.*

(ii) $\mathrm{proj}_{\boldsymbol{v}}(\boldsymbol{x}) = \boldsymbol{P}_{\boldsymbol{v}}\boldsymbol{x}$.

(iii) $\mathrm{proj}_{\boldsymbol{v}}(\boldsymbol{x})$ *is linear in* $\boldsymbol{x}$.

(iv) If $\boldsymbol{x}$ and $\boldsymbol{v}$ are orthogonal, then $\mathrm{proj}_{\boldsymbol{v}}(\boldsymbol{x}) = \mathbf{0}$ *and* $\boldsymbol{P}_{\boldsymbol{v}}\boldsymbol{P}_{\boldsymbol{x}} = \boldsymbol{O}$.

(v) $\boldsymbol{P}_{\boldsymbol{v}}\boldsymbol{v} = \boldsymbol{v}$.

(vi) If $\boldsymbol{x} = \alpha\boldsymbol{v}$ for some scalar α, then $\mathrm{proj}_{\boldsymbol{v}}(\boldsymbol{x}) = \boldsymbol{x}$.

(vii) If $\boldsymbol{u} = \alpha\boldsymbol{v}$ for some nonzero scalar α, then $\boldsymbol{P}_{\boldsymbol{u}} = \boldsymbol{P}_{\boldsymbol{v}}$ and $\mathrm{proj}_{\boldsymbol{v}}(\boldsymbol{x}) = \mathrm{proj}_{\boldsymbol{u}}(\boldsymbol{x})$.

Proof. **Proof of (i):** Since $\boldsymbol{v}$ is $n \times 1$, $\boldsymbol{v}\boldsymbol{v}'$ is $n \times n$. Clearly $\rho(\boldsymbol{P}_{\boldsymbol{v}}) = \rho(\boldsymbol{v}) = 1$. Also,

$$\boldsymbol{P}_{\boldsymbol{v}}' = \left(\frac{1}{\|\boldsymbol{v}\|^2}\boldsymbol{v}\boldsymbol{v}'\right)' = \frac{1}{\|\boldsymbol{v}\|^2}(\boldsymbol{v}\boldsymbol{v}')' = \frac{1}{\|\boldsymbol{v}\|^2}(\boldsymbol{v}')'\boldsymbol{v}' = \frac{1}{\|\boldsymbol{v}\|^2}\boldsymbol{v}\boldsymbol{v}' = \boldsymbol{P}_{\boldsymbol{v}}$$

$$\text{and } \boldsymbol{P}_{\boldsymbol{v}}^2 = \frac{1}{\|\boldsymbol{v}\|^4}\boldsymbol{v}\boldsymbol{v}'\boldsymbol{v}\boldsymbol{v}' = \frac{1}{\|\boldsymbol{v}\|^4}\boldsymbol{v}(\boldsymbol{v}'\boldsymbol{v})\boldsymbol{v}' = \frac{(\boldsymbol{v}'\boldsymbol{v})}{\|\boldsymbol{v}\|^4}\boldsymbol{v}\boldsymbol{v}' = \frac{1}{\|\boldsymbol{v}\|^2}\boldsymbol{v}\boldsymbol{v}' = \boldsymbol{P}_{\boldsymbol{v}} .$$

(In proving $\boldsymbol{P}_{\boldsymbol{v}}^2 = \boldsymbol{P}_{\boldsymbol{v}}$, we used the fact that $(\boldsymbol{v}'\boldsymbol{v})$ is a scalar so $\boldsymbol{v}(\boldsymbol{v}'\boldsymbol{v})\boldsymbol{v}' = (\boldsymbol{v}'\boldsymbol{v})\boldsymbol{v}\boldsymbol{v}'$.) The above equations prove $\boldsymbol{P}_{\boldsymbol{v}}$ is symmetric and idempotent.

Proof of (ii): This follows from the definition of $\text{proj}_{\boldsymbol{v}}(\boldsymbol{x})$ in (7.10):

$$\text{proj}_{\boldsymbol{v}}(\boldsymbol{x}) = \frac{\langle \boldsymbol{x}, \boldsymbol{v} \rangle}{\langle \boldsymbol{v}, \boldsymbol{v} \rangle}\boldsymbol{v} = \frac{1}{\|\boldsymbol{v}\|^2}(\boldsymbol{v}'\boldsymbol{x})\boldsymbol{v} = \frac{1}{\|\boldsymbol{v}\|^2}\boldsymbol{v}(\boldsymbol{v}'\boldsymbol{x}) = \frac{1}{\|\boldsymbol{v}\|^2}(\boldsymbol{v}\boldsymbol{v}')\boldsymbol{x} = \boldsymbol{P}_{\boldsymbol{v}}\boldsymbol{x} ,$$

where the third equality uses the fact that $\langle \boldsymbol{x}\boldsymbol{v} \rangle = \boldsymbol{v}'\boldsymbol{x}$ is a scalar and the fourth equality uses the associative law of matrix multiplication. This proves (ii).

Proof of (iii): This follows immediately from (ii) and the linearity of matrix multiplication:

$$\begin{aligned} \text{proj}_{\boldsymbol{v}}(\alpha_1\boldsymbol{x}_1 + \alpha_2\boldsymbol{x}_2) &= \boldsymbol{P}_{\boldsymbol{v}}(\alpha_1\boldsymbol{x}_1 + \alpha_2\boldsymbol{x}_2) = \alpha_1\boldsymbol{P}_{\boldsymbol{v}}\boldsymbol{x}_1 + \alpha_2\boldsymbol{P}_{\boldsymbol{v}}\boldsymbol{x}_2 \\ &= \alpha_1\text{proj}_{\boldsymbol{v}}(\boldsymbol{x}_1) + \alpha_2\text{proj}_{\boldsymbol{v}}(\boldsymbol{x}_2) . \end{aligned}$$

Proof of (iv): If $\boldsymbol{x}$ and $\boldsymbol{v}$ are orthogonal then clearly $\langle \boldsymbol{x}, \boldsymbol{v} \rangle = \boldsymbol{v}'\boldsymbol{x} = 0$ and $\text{proj}_{\boldsymbol{v}}(\boldsymbol{x}) = 0$ from (7.10). Also,

$$\boldsymbol{P}_{\boldsymbol{v}}\boldsymbol{P}_{\boldsymbol{x}} = \frac{1}{\|\boldsymbol{v}\|^2}\boldsymbol{v}\boldsymbol{v}'\frac{1}{\|\boldsymbol{x}\|^2}\boldsymbol{x}\boldsymbol{x}' = \frac{1}{\|\boldsymbol{v}\|^2\|\boldsymbol{x}\|^2}\boldsymbol{v}(\boldsymbol{v}'\boldsymbol{x})\boldsymbol{x}' = \frac{1}{\|\boldsymbol{v}\|^2\|\boldsymbol{x}\|^2}\boldsymbol{v}0\boldsymbol{x}' = \boldsymbol{O} .$$

Proof of (v): $\boldsymbol{P}_{\boldsymbol{v}}\boldsymbol{v} = \frac{1}{\|\boldsymbol{v}\|^2}(\boldsymbol{v}\boldsymbol{v}')\boldsymbol{v} = \frac{1}{\|\boldsymbol{v}\|^2}\boldsymbol{v}\,(\boldsymbol{v}'\boldsymbol{v}) = \frac{1}{\|\boldsymbol{v}\|^2}\boldsymbol{v}(\|\boldsymbol{v}\|^2) = \boldsymbol{v}.$

Proof of (vi): If $\boldsymbol{x} = \alpha\boldsymbol{v}$ for some scalar α, then

$$\text{proj}_{\boldsymbol{v}}(\boldsymbol{x}) = \boldsymbol{P}_{\boldsymbol{v}}\boldsymbol{x} = \alpha\boldsymbol{P}_{\boldsymbol{v}}\boldsymbol{v} = \alpha\boldsymbol{v} = \boldsymbol{x} ,$$

where the third equality follows from part (v).

Proof of (vii): Let $\boldsymbol{u} = \alpha\boldsymbol{v}$ for some $\alpha \neq 0$. Then

$$\boldsymbol{P}_{\boldsymbol{u}} = \boldsymbol{P}_{\alpha\boldsymbol{v}} = \frac{1}{\alpha^2\|\boldsymbol{v}\|^2}(\alpha\boldsymbol{v})(\alpha\boldsymbol{v})' = \frac{1}{\|\boldsymbol{v}\|^2}\boldsymbol{v}\boldsymbol{v}' = \boldsymbol{P}_{\boldsymbol{v}} .$$

Therefore $\text{proj}_{\boldsymbol{u}}(\boldsymbol{x}) = \boldsymbol{P}_{\boldsymbol{u}}\boldsymbol{x} = \boldsymbol{P}_{\boldsymbol{v}}\boldsymbol{x} = \text{proj}_{\boldsymbol{v}}(\boldsymbol{x})$. □

Corollary 7.3 *Let $\boldsymbol{v}$ be a non-null vector in $\Re^n$ and let $\boldsymbol{u} = \frac{\boldsymbol{v}}{\|\boldsymbol{v}\|}$ be the unit vector in the direction of $\boldsymbol{v}$. Then $\boldsymbol{P}_{\boldsymbol{u}} = \boldsymbol{P}_{\boldsymbol{v}}$ and $\text{proj}_{\boldsymbol{u}}(\boldsymbol{x}) = \text{proj}_{\boldsymbol{v}}(\boldsymbol{x})$ for all $\boldsymbol{x} \in \Re^n$.*

Proof. This follows by setting $\alpha = \frac{1}{\|\boldsymbol{v}\|}$ in part (vii) of Lemma 7.4. □

Since $\boldsymbol{P}_{\boldsymbol{v}}$ is idempotent (part (i) of Lemma 7.4), it is a projector. The fact that it is symmetric makes it special—and makes the corresponding projection "orthogonal." Clearly $\boldsymbol{I} - \boldsymbol{P}_{\boldsymbol{v}}$ is also idempotent and symmetric. Corollary 7.3 shows us that we lose nothing by restricting ourselves to vectors of unit length when studying these matrices. When $\boldsymbol{v}$ is a vector of unit length, we call these matrices the two *elementary orthogonal projectors* corresponding to $\boldsymbol{v}$.

Definition 7.7 *Let $\boldsymbol{u} \in \Re^n$ and $\|\boldsymbol{u}\| = 1$. Then the matrices $\boldsymbol{P}_{\boldsymbol{u}} = \boldsymbol{u}\boldsymbol{u}'$ and $\boldsymbol{I} - \boldsymbol{P}_{\boldsymbol{u}}$ are called the two* **elementary orthogonal projectors** *corresponding to $\boldsymbol{u}$.*

To obtain some further geometric insight into the nature of $\boldsymbol{P_u}$ and $\boldsymbol{I} - \boldsymbol{P_u}$, consider a vector $\boldsymbol{u}$ of unit length (i.e. $\|\boldsymbol{u}\| = 1$) in $\Re^3$. Let $\boldsymbol{u}^{\perp} = \{\boldsymbol{w} \in \Re^3 : \langle \boldsymbol{w}, \boldsymbol{u} \rangle = 0\}$ denote the space consisting of all vectors in $\Re^3$ that are perpendicular to $\boldsymbol{u}$. In the language of analytical geometry, $\boldsymbol{u}^{\perp}$ is the plane through the origin that is perpendicular to $\boldsymbol{u}$—the normal direction to the plane is given by $\boldsymbol{u}$. The matrix $\boldsymbol{P_u}$ is the *orthogonal projector* onto $\boldsymbol{u}$ in the sense that $\boldsymbol{P_u x}$ maps any $\boldsymbol{x} \in \Re^3$ onto its orthogonal projection on the line determined by the direction $\boldsymbol{u}$. In other words, $\boldsymbol{P_u x}$ is precisely the point obtained by dropping a perpendicular from $\boldsymbol{x}$ onto the straight line given by the direction vector $\boldsymbol{u}$. The matrix $\boldsymbol{I} - \boldsymbol{P_u} = \boldsymbol{I} - \boldsymbol{u}\boldsymbol{u}'$ is again an idempotent matrix and is the orthogonal projector onto $\boldsymbol{u}^{\perp}$. This means that $(\boldsymbol{I} - \boldsymbol{P_u})\boldsymbol{x}$ is the point obtained by dropping a perpendicular from $\boldsymbol{x}$ onto the plane perpendicular to $\boldsymbol{u}$. The plane $\boldsymbol{u}^{\perp}$ is a subspace (it passes through the origin) and is called the ***orthogonal complement*** or ***orthocomplement*** of $\boldsymbol{u}$. We will study such subspaces in general $\Re^n$ later. The geometric insight obtained from $\Re^3$ is, nevertheless, helpful.

We conclude this section by finding the orthogonal projection onto a plane (not necessarily a subspace). In Example 7.3 we found the distance of a point in $\Re^3$ from a plane without explicitly computing the orthogonal projection. Alternatively, one could first find the orthogonal projection onto the plane and then find its distance from the point. We demonstrate this in the following example.

Example 7.4 Orthogonal projection onto a plane. Let $\boldsymbol{x}$ be a known point in $\Re^3$ and consider the plane defined by the set of points $\mathcal{P} = \{\boldsymbol{r} \in \Re^3 : \boldsymbol{n}'\boldsymbol{r} = \boldsymbol{n}'\boldsymbol{r}_0\}$, where $\boldsymbol{n}$ is a normal to the plane and $\boldsymbol{r}_0$ is a known point on the plane. Suppose we want to find the orthogonal projection, say $\boldsymbol{p}$, of $\boldsymbol{x}$ onto the plane $\mathcal{P}$. Geometrically, $\boldsymbol{p}$ is the position vector of the foot of the perpendicular dropped from $\boldsymbol{x}$ onto $\mathcal{P}$.

We make the following two observations:

(a) $\boldsymbol{p}$ is the position vector of a point on the plane and so it satisfies $\boldsymbol{n}'(\boldsymbol{p} - \boldsymbol{r}_0) = 0$;

(b) $\boldsymbol{x} - \boldsymbol{p}$ is parallel to (i.e., in the same direction as) the normal to the plane and so $\boldsymbol{x} - \boldsymbol{p} = \alpha\boldsymbol{n}$ for some scalar α.

From (b), we can write $\boldsymbol{p} = \boldsymbol{x} - \alpha\boldsymbol{n}$ and obtain α using the observation in (a) thus:

$$\boldsymbol{n}'(\boldsymbol{p} - \boldsymbol{r}_0) = 0 \Longrightarrow \boldsymbol{n}'(\boldsymbol{x} - \boldsymbol{r}_0 - \alpha\boldsymbol{n}) = 0 \Longrightarrow \alpha = \frac{\boldsymbol{n}'(\boldsymbol{x} - \boldsymbol{r}_0)}{\boldsymbol{n}'\boldsymbol{n}} = \frac{1}{\|\boldsymbol{n}\|^2}\boldsymbol{n}'\boldsymbol{b}\,,$$

where $\boldsymbol{b} = \boldsymbol{x} - \boldsymbol{r}_0$. Substituting α in $\boldsymbol{p} = \boldsymbol{x} - \alpha\boldsymbol{n}$, we find

$$\begin{aligned} \boldsymbol{p} &= \boldsymbol{x} - \frac{\boldsymbol{n}'\boldsymbol{b}}{\boldsymbol{n}'\boldsymbol{n}}\boldsymbol{n} = \boldsymbol{x} - \frac{1}{\|\boldsymbol{n}\|^2}\boldsymbol{n}\boldsymbol{n}'\boldsymbol{b} = \boldsymbol{x} - \boldsymbol{P_n}\boldsymbol{b} = \boldsymbol{r}_0 + \boldsymbol{b} - \boldsymbol{P_n}\boldsymbol{b} \\ &= \boldsymbol{r}_0 + (\boldsymbol{I} - \boldsymbol{P_n})\boldsymbol{b}\,. \end{aligned} \tag{7.13}$$

The last expression is intuitive. If the origin is shifted to $\boldsymbol{r}_0$, then under the new coordinate system $\mathcal{P}$ is a plane passing through the origin and, hence, a subspace. In fact, the plane then is the orthocomplementary subspace of the normal vector $\boldsymbol{n}$ and indeed the projector onto this subspace is simply $\boldsymbol{I} - \boldsymbol{P_n}$. Shifting the origin to $\boldsymbol{r}_0$

simply means subtracting the position vector $\boldsymbol{r}_0$ from other position vectors in the original coordinate system and so we obtain $\boldsymbol{p} - \boldsymbol{r}_0 = (\boldsymbol{I} - \boldsymbol{P}_{\boldsymbol{n}})(\boldsymbol{x} - \boldsymbol{r}_0)$.

Note that the distance from $\boldsymbol{x}$ to $\mathcal{P}$ is now simply the distance from $\boldsymbol{x}$ to $\boldsymbol{p}$ and is

$$\|\boldsymbol{x} - \boldsymbol{p}\| = \|\boldsymbol{b} - (\boldsymbol{I} - \boldsymbol{P}_{\boldsymbol{n}})\boldsymbol{b}\| = \|\boldsymbol{P}_{\boldsymbol{n}}\boldsymbol{b}\| = \frac{\boldsymbol{n}'\boldsymbol{b}}{\|\boldsymbol{n}\|} = \text{comp}_{\boldsymbol{n}}(\boldsymbol{b}) ,$$

which is precisely the distance formula derived in Example 7.3. ■

7.4 Gram-Schmidt orthogonalization

Let $\mathcal{X} = \{\boldsymbol{x}_1, \boldsymbol{x}_2, \ldots, \boldsymbol{x}_m\}$ be an orthogonal set. Therefore, it is also a linearly independent set and forms a basis of $Sp(\mathcal{X})$. The preceding discussions show that it is especially easy to find the coefficients of any vector in $Sp(\mathcal{X})$ with respect to the basis $\mathcal{X}$. This shows the utility of having a basis consisting of orthogonal or orthonormal vectors.

Definition 7.8 *A basis of a subspace consisting of* **orthogonal** *vectors is called an* **orthogonal basis**. *A basis of a subspace consisting of* **orthonormal** *vectors is called an* **orthonormal basis**.

The following is an example of such a basis.

Example 7.5 The standard basis $\{\boldsymbol{e}_1, \boldsymbol{e}_2, \ldots, \boldsymbol{e}_m\}$ in $\Re^m$, where $\boldsymbol{e}_i$ is the $n \times 1$ vector with a 1 as its i-th element and 0 everywhere else, is an orthonormal basis for $\Re^m$.

The elements of any vector $\boldsymbol{x} = (x_1, x_2, \ldots, x_m)'$ in $\Re^m$ are simply the coordinates with respect to the standard basis. Indeed $x_i = \langle \boldsymbol{x}, \boldsymbol{e}_i \rangle$ for each $i = 1, 2, \ldots, m$.

We pose two questions. Does every subspace of $\Re^m$ admit an orthogonal or orthonormal basis? Given a regular basis, can we devise an algorithm to construct an orthogonal basis from it? If the answer to the second question is yes, then that would answer the first question as well. Since every subspace of $\Re^m$ has a basis, we could apply our algorithm to any basis and convert it to an orthogonal basis. Normalizing the vectors in an orthogonal basis will produce an orthonormal basis.

To see how one may obtain such a basis, let us start with a simple setting. Let $\mathcal{X} = \{\boldsymbol{x}_1, \boldsymbol{x}_2\}$ be a linearly independent set of two vectors that are not orthogonal. We want to find a set $\mathcal{Y} = \{\boldsymbol{y}_1, \boldsymbol{y}_2\}$ such that $\boldsymbol{y}_1 \perp \boldsymbol{y}_2$ and the span of the $\boldsymbol{y}$'s is the same as the span of the $\boldsymbol{x}$'s. Suppose we decide not to alter the first vector and set $\boldsymbol{y}_1 = \boldsymbol{x}_1$. Now, we want to find $\boldsymbol{y}_2$ that (i) belongs to the span of $\{\boldsymbol{y}_1, \boldsymbol{x}_2\}$, and (ii) is orthogonal to $\boldsymbol{y}_1$. Assume $\boldsymbol{y}_2 = \boldsymbol{x}_2 - \alpha \boldsymbol{y}_1$ and let us see if we can find an α so that $\langle \boldsymbol{y}_1, \boldsymbol{y}_2 \rangle = 0$. This is easy:

$$\langle \boldsymbol{y}_1, \boldsymbol{y}_2 \rangle = 0 \Rightarrow \langle \boldsymbol{y}_1, \boldsymbol{x}_2 - \alpha \boldsymbol{y}_1 \rangle = 0 \Rightarrow \langle \boldsymbol{y}_1, \boldsymbol{x}_2 \rangle - \alpha \langle \boldsymbol{y}_1, \boldsymbol{y}_1 \rangle = 0$$
$$\Rightarrow \alpha = \frac{\langle \boldsymbol{y}_1, \boldsymbol{x}_2 \rangle}{\langle \boldsymbol{y}_1, \boldsymbol{y}_1 \rangle} \Rightarrow \boldsymbol{y}_2 = \boldsymbol{x}_2 - \alpha \boldsymbol{y}_1 = \boldsymbol{x}_2 - \mathrm{proj}_{\boldsymbol{y}_1}(\boldsymbol{x}_2) \,.$$

Because $\mathcal{X}$ is linearly independent, it cannot include the null vector. In particular, this means $\boldsymbol{x}_1 \neq \boldsymbol{0}$ and so $\boldsymbol{y}_1 \neq 0$, which implies that α is well-defined (i.e., there is no division by zero) and so is $\boldsymbol{y}_2$. Thus, $\boldsymbol{y}_1 = \boldsymbol{x}_1$ and $\boldsymbol{y}_2 = \boldsymbol{x}_2 - \mathrm{proj}_{\boldsymbol{y}_1}(\boldsymbol{x}_2)$ forms a set such that $\boldsymbol{y}_1 \perp \boldsymbol{y}_2$ and the span of the $\boldsymbol{y}$'s is the same as the span of the $\boldsymbol{x}$'s. Also, note what happens if the set $\mathcal{X}$ is orthogonal to begin with. In that case, $\boldsymbol{x}_1 \perp \boldsymbol{x}_2$ and so $\boldsymbol{y}_1 \perp \boldsymbol{x}_2$, which means that $\alpha = 0$. Thus, $\boldsymbol{y}_2 = \boldsymbol{x}_2$. In other words, our strategy will not alter $\mathcal{X}$ when the $\boldsymbol{x}$'s are orthogonal to begin with.

Next consider the situation with three linearly independent vectors $\mathcal{X} = \{\boldsymbol{x}_1, \boldsymbol{x}_2, \boldsymbol{x}_3\}$. We define $\boldsymbol{y}_1$ and $\boldsymbol{y}_2$ as above, so that $\{\boldsymbol{y}_1, \boldsymbol{y}_2\}$ is an orthogonal set and try to find a linear combination of $\boldsymbol{x}_3$, $\boldsymbol{y}_1$ and $\boldsymbol{y}_2$ that will be orthogonal to $\boldsymbol{y}_1$ and $\boldsymbol{y}_2$. We write

$$\begin{aligned} \boldsymbol{y}_1 &= \boldsymbol{x}_1 \,; \\ \boldsymbol{y}_2 &= \boldsymbol{x}_2 - \mathrm{proj}_{\boldsymbol{y}_1}(\boldsymbol{x}_2); \\ \boldsymbol{y}_3 &= \boldsymbol{x}_3 - \alpha_1 \boldsymbol{y}_1 - \alpha_2 \boldsymbol{y}_2 \,. \end{aligned}$$

Clearly $\boldsymbol{y}_3$ belongs to the span of $\{\boldsymbol{y}_1, \boldsymbol{y}_2, \boldsymbol{x}_3\}$ and $\{\boldsymbol{y}_1, \boldsymbol{y}_2\}$ belongs to the span of $\{\boldsymbol{x}_1, \boldsymbol{x}_2\}$. Therefore, $\boldsymbol{y}_3$ belongs to the span of $\{\boldsymbol{x}_1, \boldsymbol{x}_2, \boldsymbol{x}_3\}$ and the span of $\{\boldsymbol{y}_1, \boldsymbol{y}_2, \boldsymbol{y}_3\}$ is the same as the span of $\{\boldsymbol{x}_1, \boldsymbol{x}_2, \boldsymbol{x}_3\}$. We want to solve for the α's so that $\boldsymbol{y}_3 \perp \boldsymbol{y}_1$ and $\boldsymbol{y}_3 \perp \boldsymbol{y}_2$ are satisfied. Since $\boldsymbol{y}_1 \perp \boldsymbol{y}_2$, we can easily solve for α_1:

$$0 = \langle \boldsymbol{y}_3, \boldsymbol{y}_1 \rangle = \langle \boldsymbol{x}_3, \boldsymbol{y}_1 \rangle - \alpha_1 \langle \boldsymbol{y}_1, \boldsymbol{y}_1 \rangle \Rightarrow \alpha_1 = \frac{\langle \boldsymbol{x}_3, \boldsymbol{y}_1 \rangle}{\langle \boldsymbol{y}_1, \boldsymbol{y}_1 \rangle} \,.$$

Similarly, we solve for α_2:

$$0 = \langle \boldsymbol{y}_3, \boldsymbol{y}_2 \rangle = \langle \boldsymbol{x}_3, \boldsymbol{y}_2 \rangle - \alpha_2 \langle \boldsymbol{y}_2, \boldsymbol{y}_2 \rangle \Rightarrow \alpha_2 = \frac{\langle \boldsymbol{x}_3, \boldsymbol{y}_2 \rangle}{\langle \boldsymbol{y}_2, \boldsymbol{y}_2 \rangle} \,.$$

Therefore,

$$\boldsymbol{y}_3 = \boldsymbol{x}_3 - \sum_{j=1}^{2} \frac{\langle \boldsymbol{x}_3, \boldsymbol{y}_j \rangle}{\langle \boldsymbol{y}_j, \boldsymbol{y}_j \rangle} \boldsymbol{y}_j = \boldsymbol{x}_3 - \sum_{j=1}^{2} \mathrm{proj}_{\boldsymbol{y}_i}(\boldsymbol{x}_3) \,.$$

It is easy to verify that $\{\boldsymbol{y}_1, \boldsymbol{y}_2, \boldsymbol{y}_3\}$ is the required orthogonal set that spans the same subspace as $\{\boldsymbol{x}_1, \boldsymbol{x}_2, \boldsymbol{x}_3\}$. Note that $\frac{\boldsymbol{y}_i}{\|\boldsymbol{y}_i\|}$, $i = 1, 2, 3$, is an orthonormal basis.

We are now ready to state and prove the following theorem.

Theorem 7.2 The Gram-Schmidt orthogonalization. *Let* $\mathcal{X} = \{\boldsymbol{x}_1, \boldsymbol{x}_1, \ldots, \boldsymbol{x}_n\}$ *be a basis for some subspace* $\mathcal{S}$. *Suppose we construct the following set of vectors* *from those in $\mathcal{X}$:*

$$\boldsymbol{y}_1 = \boldsymbol{x}_1 \; ; \quad \boldsymbol{q}_1 = \frac{\boldsymbol{y}_1}{\|\boldsymbol{y}_1\|} \; ;$$

$$\boldsymbol{y}_2 = \boldsymbol{x}_2 - \text{proj}_{\boldsymbol{y}_1}(\boldsymbol{x}_2) \; ; \quad \boldsymbol{q}_2 = \frac{\boldsymbol{y}_2}{\|\boldsymbol{y}_2\|} \; ;$$

$$\vdots$$

$$\boldsymbol{y}_k = \boldsymbol{x}_k - \text{proj}_{\boldsymbol{y}_1}(\boldsymbol{x}_k) - \text{proj}_{\boldsymbol{y}_2}(\boldsymbol{x}_k) - \cdots - \text{proj}_{\boldsymbol{y}_{k-1}}(\boldsymbol{x}_k) \; ; \quad \boldsymbol{q}_k = \frac{\boldsymbol{y}_k}{\|\boldsymbol{y}_k\|} \; ;$$

$$\vdots$$

$$\boldsymbol{y}_n = \boldsymbol{x}_n - \text{proj}_{\boldsymbol{y}_1}(\boldsymbol{x}_n) - \text{proj}_{\boldsymbol{y}_2}(\boldsymbol{x}_n) - \cdots - \cdots - \text{proj}_{\boldsymbol{y}_{n-1}}(\boldsymbol{x}_n) \; ; \; \boldsymbol{q}_n = \frac{\boldsymbol{y}_n}{\|\boldsymbol{y}_n\|} .$$

Then the set $\mathcal{Y} = \{\boldsymbol{y}_1, \boldsymbol{y}_2, \ldots, \boldsymbol{y}_n\}$ is an orthogonal basis and $\mathcal{Q} = \{\boldsymbol{q}_1, \boldsymbol{q}_2, \ldots, \boldsymbol{q}_n\}$ is an orthonormal basis for $\mathcal{S}$.

Proof. First, we show that $\boldsymbol{y}_i \neq \boldsymbol{0}$ and that $Sp(\{\boldsymbol{y}_1, \ldots, \boldsymbol{y}_i\}) = Sp(\{\boldsymbol{x}_1, \ldots, \boldsymbol{x}_i\})$ for each $i = 1, 2, \ldots, n$. Since $\mathcal{X} = \{\boldsymbol{x}_1, \boldsymbol{x}_2, \ldots, \boldsymbol{x}_n\}$ is a linearly independent set, each $\boldsymbol{x}_i \neq \boldsymbol{0}$. Therefore, $\boldsymbol{y}_1 \neq \boldsymbol{0}$. Suppose, if possible, $\boldsymbol{y}_2 = \boldsymbol{0}$. This means that

$$\boldsymbol{0} = \boldsymbol{x}_2 - \text{proj}_{\boldsymbol{y}_1}(\boldsymbol{x}_2) = \boldsymbol{x}_2 - \alpha \boldsymbol{y}_1 = \boldsymbol{x}_2 - \alpha \boldsymbol{x}_1 \; , \quad \text{where } \alpha = \frac{\langle \boldsymbol{y}_1, \boldsymbol{x}_2 \rangle}{\langle \boldsymbol{y}_1, \boldsymbol{y}_1 \rangle} \neq 0 \; ,$$

which is impossible because $\{\boldsymbol{x}_1, \boldsymbol{x}_2\}$ is a linearly independent set (since $\mathcal{X}$ is linearly independent, no subset of it can be linearly dependent). Therefore, $\boldsymbol{y}_2 \neq \boldsymbol{0}$ and $\boldsymbol{y}_2 \in Sp(\{\boldsymbol{x}_1, \boldsymbol{x}_2\})$, which implies that $Sp(\{\boldsymbol{y}_1, \boldsymbol{y}_2\}) = Sp(\{\boldsymbol{x}_1, \boldsymbol{x}_2\})$.

We now proceed by induction. Suppose $\boldsymbol{y}_1, \boldsymbol{y}_2, \ldots, \boldsymbol{y}_{k-1}$ are all non-null vectors and that $Sp(\{\boldsymbol{y}_1, \boldsymbol{y}_2, \ldots, \boldsymbol{y}_{k-1}\}) = Sp(\{\boldsymbol{x}_1, \boldsymbol{x}_2, \ldots, \boldsymbol{x}_{k-1}\})$. Observe that

$$\begin{aligned} \boldsymbol{y}_k &= \boldsymbol{x}_k - \text{proj}_{\boldsymbol{y}_1}(\boldsymbol{x}_k) - \text{proj}_{\boldsymbol{y}_2}(\boldsymbol{x}_k) - \cdots - \text{proj}_{\boldsymbol{y}_{k-1}}(\boldsymbol{x}_k) \\ &= \boldsymbol{x}_k - \alpha_{1k}\boldsymbol{y}_1 - \alpha_{2k}\boldsymbol{y}_2 - \cdots - \alpha_{k-1,k}\boldsymbol{y}_{k-1} \; , \quad \text{where } \alpha_{ik} = \frac{\langle \boldsymbol{y}_i, \boldsymbol{x}_k \rangle}{\langle \boldsymbol{y}_i, \boldsymbol{y}_i \rangle} \; . \end{aligned}$$

Since the $\boldsymbol{y}_i$'s are non-null, $\alpha_{ik} \neq 0$ for each $i = 1, 2, \ldots, k-1$. This implies that $\boldsymbol{y}_k \in Sp(\{\boldsymbol{y}_1, \boldsymbol{y}_2, \ldots, \boldsymbol{y}_{k-1}, \boldsymbol{x}_k\}) = Sp(\{\boldsymbol{x}_1, \boldsymbol{x}_2, \ldots, \boldsymbol{x}_{k-1}, \boldsymbol{x}_k\})$ and so $\boldsymbol{y}_k$ is a linear combination of $\boldsymbol{x}_1, \boldsymbol{x}_2, \ldots, \boldsymbol{x}_k$. If it were true that $\boldsymbol{y}_k = \boldsymbol{0}$, then $\{\boldsymbol{x}_1, \boldsymbol{x}_2, \ldots, \boldsymbol{x}_k$ would be linearly dependent. Since the $\mathcal{X}$ is a basis, it is a linearly independent set and no subset of it can be linearly dependent. Therefore, $\boldsymbol{y}_k \neq \boldsymbol{0}$ and indeed $\boldsymbol{y}_k \in Sp(\{\boldsymbol{x}_1, \boldsymbol{x}_2, \ldots, \boldsymbol{x}_{k-1}, \boldsymbol{x}_k\})$, which means that $Sp(\{\boldsymbol{y}_1, \boldsymbol{y}_2, \ldots, \boldsymbol{y}_k\}) = Sp(\{\boldsymbol{x}_1, \boldsymbol{x}_2, \ldots, \boldsymbol{x}_k\})$. This completes the induction step and we have proved that $\boldsymbol{y}_i \neq \boldsymbol{0}$ and that $Sp(\{\boldsymbol{y}_1, \ldots, \boldsymbol{y}_i\}) = Sp(\{\boldsymbol{x}_1, \ldots, \boldsymbol{x}_i\})$ for each $i = 1, 2, \ldots, n$.

In particular, $Sp(\{\boldsymbol{y}_1, \ldots, \boldsymbol{y}_n\}) = Sp(\{\boldsymbol{x}_1, \ldots, \boldsymbol{x}_n\})$. Therefore, $Sp(\mathcal{Y}) = Sp(\mathcal{X}) = \mathcal{S}$ and the $\boldsymbol{y}_i$'s form a spanning set for $\mathcal{S}$.

We claim that the $\boldsymbol{y}_i$'s are orthogonal.

Again, we proceed by induction. It is easy to verify that $\{\boldsymbol{y}_1, \boldsymbol{y}_2\}$ is orthogonal. Suppose that $\{\boldsymbol{y}_1, \boldsymbol{y}_2, \ldots, \boldsymbol{y}_{k-1}\}$ is an orthogonal set and consider the inner product of any one of them with $\boldsymbol{y}_k$. We obtain, for each $i = 1, 2, \ldots, k-1$,

$$\begin{aligned}\langle \boldsymbol{y}_i, \boldsymbol{y}_k \rangle &= \langle \boldsymbol{y}_i, \boldsymbol{x}_k - \mathrm{proj}_{\boldsymbol{y}_1}(\boldsymbol{x}_k) - \mathrm{proj}_{\boldsymbol{y}_2}(\boldsymbol{x}_k) - \cdots - \mathrm{proj}_{\boldsymbol{y}_{k-1}}(\boldsymbol{x}_k) \rangle \\ &= \langle \boldsymbol{y}_i, \boldsymbol{x}_k \rangle - \sum_{j=1}^{k-1} \langle \boldsymbol{y}_i, \mathrm{proj}_{\boldsymbol{y}_j}(\boldsymbol{x}_k) \rangle = \langle \boldsymbol{y}_i, \boldsymbol{x}_k \rangle - \langle \boldsymbol{y}_i, \boldsymbol{x}_k \rangle = 0 ,\end{aligned}$$

where we have used the fact that for $i = 1, 2, \ldots, k-1$

$$\sum_{j=1}^{k-1} \langle \boldsymbol{y}_i, \mathrm{proj}_{\boldsymbol{y}_j}(\boldsymbol{x}_k) \rangle = \sum_{j=1}^{k-1} \frac{\langle \boldsymbol{y}_j, \boldsymbol{x}_k \rangle}{\langle \boldsymbol{y}_j, \boldsymbol{y}_j \rangle} \langle \boldsymbol{y}_i, \boldsymbol{y}_j \rangle = \frac{\langle \boldsymbol{y}_i, \boldsymbol{x}_k \rangle}{\langle \boldsymbol{y}_i, \boldsymbol{y}_i \rangle} \langle \boldsymbol{y}_i, \boldsymbol{y}_i \rangle = \langle \boldsymbol{y}_i, \boldsymbol{x}_k \rangle .$$

Therefore, $\{\boldsymbol{y}_1, \boldsymbol{y}_2, \ldots, \boldsymbol{y}_{k+1}\}$ is an orthogonal set. This completes the induction step, thereby proving that $\mathcal{Y}$ is an orthogonal set.

We have already proved that $\mathcal{Y}$ spans $\mathcal{S}$, which proves that it is an orthogonal basis.

Turning to the $\boldsymbol{q}_i$'s, note that it is clear that each $\boldsymbol{y}_i \neq \boldsymbol{0}$, which ensures that each of the $\boldsymbol{q}_i$'s are well-defined. Also, since each $\boldsymbol{q}_i$ is a scalar multiple of the corresponding $\boldsymbol{y}_i$, the subspace spanned by $\mathcal{Q} = \{\boldsymbol{q}_1, \boldsymbol{q}_2, \ldots, \boldsymbol{q}_n\}$ is equal to $Sp(\mathcal{Y})$. Therefore, $Sp(\mathcal{Q}) = \mathcal{S}$. Finally, note that

$$\langle \boldsymbol{q}_i, \boldsymbol{q}_j \rangle = \left\langle \frac{\boldsymbol{y}_i}{\|\boldsymbol{y}_i\|}, \frac{\boldsymbol{y}_j}{\|\boldsymbol{y}_j\|} \right\rangle = \frac{1}{\|\boldsymbol{y}_i\| \|\boldsymbol{y}_j\|} \langle \boldsymbol{y}_i, \boldsymbol{y}_j \rangle = \delta_{ij} ,$$

where $\delta_{ij} = 1$ if $i = j$ and $\delta_{ij} = 0$ whenever $i \neq j$. This proves that $\mathcal{Q}$ is a set of orthonormal vectors and is an orthonormal basis for $\mathcal{S}$. □

The above theorem proves that producing an orthonormal basis from the Gram-Schmidt orthogonalization process involves just one additional step—that of normalizing the orthogonal vectors. We could, therefore, easily call this the *Gram-Schmidt orthonormalization* procedure. Usually we just refer to this as the "Gram-Schmidt procedure."

Example 7.6 Gram-Schmidt orthogonalization. Consider the following three vectors in $\Re^3$:

$$\boldsymbol{x}_1 = \begin{bmatrix} 1 \\ 2 \\ 3 \end{bmatrix} , \quad \boldsymbol{x}_2 = \begin{bmatrix} 3 \\ 7 \\ 10 \end{bmatrix} , \quad \boldsymbol{x}_3 = \begin{bmatrix} 0 \\ 1 \\ 0 \end{bmatrix} .$$

We illustrate an application of Gram-Schmidt orthogonalization to these vectors.

Let $\boldsymbol{y}_1 = \boldsymbol{x}_1$. Therefore, $\langle \boldsymbol{y}_1, \boldsymbol{y}_1 \rangle = \|\boldsymbol{y}_1\|^2 = 1^2 + 2^2 + 3^2 = 14$. We now construct

$$\boldsymbol{y}_1 = \boldsymbol{x}_1 = \begin{bmatrix} 1 \\ 2 \\ 3 \end{bmatrix} \; ; \quad \|\boldsymbol{y}_1\|^2 = 1^2 + 2^2 + 3^2 = 14 \; ;$$

$$\begin{aligned} \boldsymbol{y}_2 &= \boldsymbol{x}_2 - \text{proj}_{\boldsymbol{y}_1}(\boldsymbol{x}_2) = \boldsymbol{x}_2 - \left(\frac{\boldsymbol{x}_2' \boldsymbol{y}_1}{\|\boldsymbol{y}_1\|^2} \right) \boldsymbol{y}_1 = \begin{bmatrix} 3 \\ 7 \\ 10 \end{bmatrix} - \frac{47}{14} \begin{bmatrix} 1 \\ 2 \\ 3 \end{bmatrix} \\ &= \frac{1}{14} \begin{bmatrix} -5 \\ 4 \\ -1 \end{bmatrix} \; ; \quad \|\boldsymbol{y}_2\|^2 = \frac{1}{14^2}(5^2 + 4^2 + 1^2) = \frac{42}{14^2} = \frac{3}{14} \; ; \\ \boldsymbol{y}_3 &= \boldsymbol{x}_3 - \text{proj}_{\boldsymbol{y}_1}(\boldsymbol{x}_3) - \text{proj}_{\boldsymbol{y}_2}(\boldsymbol{x}_3) = \boldsymbol{x}_3 - \left(\frac{\boldsymbol{x}_3' \boldsymbol{y}_1}{\|\boldsymbol{y}_1\|^2} \right) \boldsymbol{y}_1 - \left(\frac{\boldsymbol{x}_3' \boldsymbol{y}_2}{\|\boldsymbol{y}_2\|^2} \right) \boldsymbol{y}_2 \\ &= \begin{bmatrix} 0 \\ 1 \\ 0 \end{bmatrix} - \frac{1}{7} \begin{bmatrix} 1 \\ 2 \\ 3 \end{bmatrix} - \frac{2}{21} \begin{bmatrix} -5 \\ 4 \\ -1 \end{bmatrix} = \begin{bmatrix} 0 \\ 1 \\ 0 \end{bmatrix} - \frac{3}{21} \begin{bmatrix} 1 \\ 2 \\ 3 \end{bmatrix} - \frac{2}{21} \begin{bmatrix} -5 \\ 4 \\ -1 \end{bmatrix} \\ &= \begin{bmatrix} 0 \\ 1 \\ 0 \end{bmatrix} - \frac{1}{21} \begin{bmatrix} -7 \\ 14 \\ 7 \end{bmatrix} = \begin{bmatrix} 0 \\ 1 \\ 0 \end{bmatrix} - \frac{1}{3} \begin{bmatrix} -1 \\ 2 \\ 1 \end{bmatrix} \\ &= \frac{1}{3} \begin{bmatrix} 1 \\ 1 \\ -1 \end{bmatrix} \; ; \quad \|\boldsymbol{y}_3\|^2 = \frac{1}{3^2}(1^2 + 1^2 + (-1)^2) = \frac{3}{9} = \frac{1}{3} \, . \end{aligned}$$

Once the process is complete, we can ignore the scalar multiples of the $\boldsymbol{y}_i$'s as they do not affect the orthogonality. Therefore, $\boldsymbol{z}_1 = \begin{bmatrix} 1 \\ 2 \\ 3 \end{bmatrix}$, $\boldsymbol{z}_2 = \begin{bmatrix} -5 \\ 4 \\ -1 \end{bmatrix}$ and $\boldsymbol{z}_3 = \begin{bmatrix} 1 \\ 1 \\ -1 \end{bmatrix}$ are a set of orthogonal vectors as can be easily verified:

$$\begin{aligned} \boldsymbol{z}_2' \boldsymbol{z}_1 &= (-5) \times 1 + 4 \times 2 + (-1) \times 3 = -5 + 8 - 3 = 0 \; ; \\ \boldsymbol{z}_3' \boldsymbol{z}_1 &= 1 \times 1 + 1 \times 2 + (-1) \times 3 = 1 + 2 - 3 = 0 \; ; \\ \boldsymbol{z}_3' \boldsymbol{z}_2 &= 1 \times (-5) + 1 \times 4 + (-1) \times (-1) = (-5 + 4 + 1) = 0 \, . \end{aligned}$$

Producing the corresponding set of orthonormal vectors is easy—we simply divide the orthogonal vectors by their lengths. Therefore,

$$\boldsymbol{q}_1 = \frac{1}{\sqrt{14}} \begin{bmatrix} 1 \\ 2 \\ 3 \end{bmatrix} , \quad \boldsymbol{q}_2 = \frac{1}{\sqrt{42}} \begin{bmatrix} -5 \\ 4 \\ -1 \end{bmatrix} , \quad \boldsymbol{q}_3 = \frac{1}{\sqrt{3}} \begin{bmatrix} 1 \\ 1 \\ -1 \end{bmatrix}$$

are orthonormal vectors. Since these are three orthonormal vectors in $\Re^3$, they form an orthonormal basis for $\Re^3$. ■

What happens if we apply the Gram-Schmidt procedure to a set that already comprises orthonormal vectors? The following corollary has the answer.

Corollary 7.4 *Let $\mathcal{X} = \{\boldsymbol{x}_1, \boldsymbol{x}_2, \ldots, \boldsymbol{x}_n\}$ be an orthonormal set in some subspace $\mathcal{S}$. The vectors in this set are not altered by Gram-Schmidt orthogonalization.*

Proof. Since $\mathcal{X} = \{\boldsymbol{x}_1, \boldsymbol{x}_2, \ldots, \boldsymbol{x}_n\}$ is an orthonormal set, each $\|\boldsymbol{x}_i\| = 1$. Gram-Schmidt orthogonalization sets $\boldsymbol{y}_1 = \boldsymbol{x}_1$ and $\boldsymbol{y}_2 = \boldsymbol{x}_2 - \text{proj}_{\boldsymbol{y}_1}(\boldsymbol{x}_2)$. Since $\boldsymbol{x}_1 \perp \boldsymbol{x}_2$, it follows that $\boldsymbol{y}_1 \perp \boldsymbol{x}_2$ and, hence, $\text{proj}_{\boldsymbol{y}_1}(\boldsymbol{x}_2) = \boldsymbol{0}$. Therefore, $\boldsymbol{y}_2 = \boldsymbol{x}_2$. Continuing in this fashion, we easily see that $\boldsymbol{y}_i = \boldsymbol{x}_i$ for $i = 1, 2, \ldots, n$. This proves that $\mathcal{X}$ is not altered by Gram-Schmidt orthogonalization. □

The following result is another immediate consequence of Gram-Schmidt orthogonalization. It says that every finite-dimensional linear subspace has an orthonormal basis and that any orthonormal set of vectors in a linear subspace can be extended to an orthonormal basis of that subspace.

Theorem 7.3 *Let $\mathcal{S}$ be a vector space such that* $\dim(\mathcal{S}) = m \geq 1$. *Then the following statements are true.*

(i) $\mathcal{S}$ has an orthonormal basis.

(ii) Let $\{\boldsymbol{q}_1, \boldsymbol{q}_2, \ldots, \boldsymbol{q}_r\}$ be an orthonormal set in $\mathcal{S}$. Then we can find orthonormal vectors $\{\boldsymbol{q}_{r+1}, \boldsymbol{q}_{r+2}, \ldots, \boldsymbol{q}_m\}$ such that $\{\boldsymbol{q}_1, \boldsymbol{q}_2, \ldots, \boldsymbol{q}_r, \boldsymbol{q}_{r+1}, \ldots, \boldsymbol{q}_m\}$ is an orthonormal basis for $\mathcal{S}$.

Proof. **Proof of (i):** Consider any basis $\mathcal{B}$ of $\mathcal{S}$ and apply the Gram-Schmidt process as in Theorem 7.2. This yields an orthonormal basis for $\mathcal{S}$. This proves (i).

Proof of (ii): Since $\mathcal{Q}_1 = \{\boldsymbol{q}_1, \boldsymbol{q}_2, \ldots, \boldsymbol{q}_r\}$ is an orthonormal set in $\mathcal{S}$, it is also a linearly independent set of vectors. This means that $r \leq m$. If $r = m$, then $\mathcal{Q}_1$ is itself an orthonormal basis and there is nothing to prove.

If $r < m$, we can extend such a set to a basis for $\mathcal{S}$. Let $\mathcal{X} = \{\boldsymbol{x}_{r+1}, \boldsymbol{x}_{r+2}, \ldots, \boldsymbol{x}_m\}$ be the extension vectors such that $\mathcal{Q}_1 \cup \mathcal{X}$ is a basis for $\mathcal{S}$.

We now apply the Gram-Schmidt orthogonalization process to $\mathcal{Q}_1 \cup \mathcal{X}$. Since the subset comprising the first r vectors in $\{\boldsymbol{q}_1, \boldsymbol{q}_2, \ldots, \boldsymbol{q}_r, \boldsymbol{x}_{r+1}, \ldots, \boldsymbol{x}_m\}$ is an orthonormal set, Corollary 7.4 tells us that the first r vectors (i.e., the $\boldsymbol{q}_i$'s) are not altered by Gram-Schmidt orthogonalization. Therefore, the process in Theorem 7.2 yields $\{\boldsymbol{q}_1, \boldsymbol{q}_2, \ldots, \boldsymbol{q}_r, \boldsymbol{q}_{r+1}, \ldots, \boldsymbol{q}_m\}$, which is an orthonormal basis for $\mathcal{S}$ obtained by extending the orthonormal set $\{\boldsymbol{q}_1, \boldsymbol{q}_2, \ldots, \boldsymbol{q}_r\}$. □

Lemma 7.5 *Let $\{\boldsymbol{q}_1, \boldsymbol{q}_2, \ldots, \boldsymbol{q}_m\}$ be an orthonormal basis for a vector space $\mathcal{V}$. Then any vector $\boldsymbol{y} \in \mathcal{V}$ can be written as*

$$\boldsymbol{y} = \langle \boldsymbol{y}, \boldsymbol{q}_1 \rangle + \langle \boldsymbol{y}, \boldsymbol{q}_2 \rangle + \cdots + \langle \boldsymbol{y}, \boldsymbol{q}_m \rangle = \sum_{i=1}^{m} \langle \boldsymbol{y}, \boldsymbol{q}_i \rangle \boldsymbol{q}_i \,.$$

Proof. Since $\boldsymbol{q}_1, \boldsymbol{q}_2, \ldots, \boldsymbol{q}_m$ form a basis, any vector $\boldsymbol{y}$ can be expressed as a linear combination of the $\boldsymbol{q}_i$'s. The desired result now immediately follows from (7.9) and by noting that $\langle \boldsymbol{q}_i, \boldsymbol{q}_i \rangle = 1$ for each $i = 1, 2, , \ldots, m$. □

7.5 Orthocomplementary subspaces

The preceding development has focused upon orthogonality between two vectors. It will be useful to extend this concept so that one can talk when two sets are orthogonal. For this, we have the following definition.

Definition 7.9 *Let $\mathcal{X}$ be a set (not necessarily a subspace) of vectors in some linear subspace of $\Re^m$, say $\mathcal{S}$. An* **orthocomplement set** *of $\mathcal{X}$ (with respect to $\mathcal{S}$) is defined to be the set comprising all vectors in $\mathcal{S}$ that are orthogonal to every vector in $\mathcal{X}$. We denote this set as $\mathcal{X}^\perp$ (often pronounced as the "perp" of $\mathcal{X}$ or simply "$\mathcal{X}$ perp") and formally define it as*

$$\mathcal{X}^\perp = \{\boldsymbol{u} \in \mathcal{S} : \langle \boldsymbol{x}, \boldsymbol{u} \rangle = 0 \; \forall \, \boldsymbol{x} \in \mathcal{X}\} \, .$$

The following elementary properties are immediate from the above definition.

Lemma 7.6 *Let $\mathcal{S}$ be a linear subspace of $\Re^m$ and let $\mathcal{X}$ be any subset (not necessarily a subspace) of $\mathcal{S}$. Then the following statements are true.*

(i) If $\boldsymbol{u} \in \mathcal{X}^\perp$ then $\boldsymbol{u} \in \{Sp(X)\}^\perp$.

(ii) $\mathcal{X}^\perp$ is always a subspace, whether or not $\mathcal{X}$ is a subspace.

(iii) $Sp(\mathcal{X}) \subseteq [\mathcal{X}^\perp]^\perp$ is always true, whether or not $\mathcal{X}$ is a subspace.

Proof. **Proof of (i):** Suppose that $\boldsymbol{u} \in \mathcal{X}^\perp$. This means that $\boldsymbol{u}$ is orthogonal to every vector in $\mathcal{X}$ and so $\langle \boldsymbol{u}, \boldsymbol{x}_i \rangle = 0$ for $i = 1, 2, \ldots, k$. Let $\boldsymbol{v} \in Sp(\mathcal{X})$. Then, $\boldsymbol{v}$ is a linear combination of vectors in $\mathcal{X}$, say, $\boldsymbol{v} = \sum_{i=1}^k a_i \boldsymbol{x}_i$. This means that

$$\langle \boldsymbol{u}, \boldsymbol{v} \rangle = \left\langle \boldsymbol{u}, \sum_{i=1}^k a_i \boldsymbol{x}_i \right\rangle = \sum_{i=1}^k a_i \langle \boldsymbol{u}, \boldsymbol{x}_i \rangle = 0 \, .$$

Hence, $\boldsymbol{u}$ is perpendicular to any linear combination of vectors in $\mathcal{X}$ and so $\boldsymbol{u} \in \{Sp(X)\}^\perp$.

Proof of (ii): Since $\mathcal{S}$ is a subspace, $\boldsymbol{0} \in \mathcal{S}$ and of course $\langle \boldsymbol{x}, \boldsymbol{0} \rangle = 0$ for every $\boldsymbol{x} \in \mathcal{X}$. This means that $\boldsymbol{0} \in \mathcal{X}^\perp$. Next suppose that $\boldsymbol{u}$ and $\boldsymbol{v}$ are two vectors in $\mathcal{X}^\perp$. Consider the linear combination $\boldsymbol{u} + \alpha \boldsymbol{v}$ and observe that

$$\langle \boldsymbol{x}, \boldsymbol{u} + \alpha \boldsymbol{v} \rangle = \langle \boldsymbol{x}, \boldsymbol{u} \rangle + \alpha \langle \boldsymbol{x}, \boldsymbol{v} \rangle = 0$$

for any α. Therefore, $\boldsymbol{u} + \alpha \boldsymbol{v} \in \mathcal{X}^\perp$. This proves that $\mathcal{X}^\perp$ is a subspace of $\mathcal{S}$. (Note: We have not assumed that $\mathcal{X}$ is a subspace.)

Proof of (iii): Suppose $\boldsymbol{x}$ is a vector in $Sp(\mathcal{X})$. This means that $\boldsymbol{x}$ is orthogonal to all $\boldsymbol{u} \in \mathcal{X}^\perp$ and, hence, $\boldsymbol{x} \in [\mathcal{X}^\perp]^\perp$. This proves (iii). □

Theorem 7.4 The direct sum theorem for orthocomplementary subspaces. *Let $\mathcal{S}$ be a subspace of a vector space $\mathcal{V}$ with $\dim(\mathcal{V}) = m$.*

(i) Every vector $\boldsymbol{y} \in \mathcal{V}$ can be expressed as $\boldsymbol{y} = \boldsymbol{u} + \boldsymbol{v}$, where $\boldsymbol{u} \in \mathcal{S}$ and $\boldsymbol{v} \in \mathcal{S}^\perp$.

(ii) $\mathcal{S} \cap \mathcal{S}^{\perp} = \{\mathbf{0}\}$.

(iii) $\mathcal{V} = \mathcal{S} \oplus \mathcal{S}^{\perp}$.

(iv) $\dim(\mathcal{S}^{\perp}) = m - \dim(\mathcal{S})$.

Proof. **Proof of (i):** Let $\{\boldsymbol{z}_1, \boldsymbol{z}_2, \ldots, \boldsymbol{z}_r\}$ be an orthonormal basis of $\mathcal{S}$ and suppose we extend it to an orthonormal basis $\mathcal{B} = \{\boldsymbol{z}_1, \boldsymbol{z}_2, \ldots, \boldsymbol{z}_r, \boldsymbol{z}_{r+1}, \ldots, \boldsymbol{z}_m\}$ for $\mathcal{V}$ (see part (ii) of Theorem 7.3). Then, Lemma 7.5 tells us that any $\boldsymbol{y} \in \mathcal{V}$ can be expressed as

$$\begin{aligned} \boldsymbol{y} &= \underbrace{\sum_{i=1}^{r} \langle \boldsymbol{y}, \boldsymbol{z}_i \rangle \boldsymbol{z}_i}_{} + \underbrace{\sum_{i=r+1}^{m} \langle \boldsymbol{y}, \boldsymbol{z}_i \rangle \boldsymbol{z}_i}_{}, \\ &= \boldsymbol{u} + \boldsymbol{v} \end{aligned}$$

where $\boldsymbol{u} \in Sp(\{\boldsymbol{z}_1, \boldsymbol{z}_2, \ldots, \boldsymbol{z}_r\}) = \mathcal{S}$ and $\boldsymbol{v} \in Sp(\{\boldsymbol{z}_{r+1}, \boldsymbol{z}_{r+2}, \ldots, \boldsymbol{z}_m\})$. Clearly, the vectors $\boldsymbol{z}_{r+1}, \boldsymbol{z}_{r+2}, \ldots, \boldsymbol{z}_m$ belong to $\mathcal{S}^{\perp}$. We now show that these vectors span $\mathcal{S}^{\perp}$.

Consider any $\boldsymbol{x} \in \mathcal{S}^{\perp}$. Since $\mathcal{S}^{\perp}$ is a subspace of $\mathcal{V}$ and $\mathcal{B}$ is a basis for $\mathcal{V}$, we can write

$$\boldsymbol{x} = \sum_{i=1}^{m} \langle \boldsymbol{x}, \boldsymbol{z}_i \rangle \boldsymbol{z}_i = \sum_{i=r+1}^{m} \langle \boldsymbol{x}, \boldsymbol{z}_i \rangle \boldsymbol{z}_i \in Sp(\{\boldsymbol{z}_{r+1}, \boldsymbol{z}_{r+2}, \ldots, \boldsymbol{z}_m\}),$$

where the second "=" follows from the fact that $\langle \boldsymbol{x}, \boldsymbol{z}_i \rangle = 0$ for $i = 1, \ldots, r$. This proves that $Sp(\{\boldsymbol{z}_{r+1}, \boldsymbol{z}_{r+2}, \ldots, \boldsymbol{z}_m\}) = \mathcal{S}^{\perp}$ and, therefore, $\boldsymbol{v} \in \mathcal{S}^{\perp}$.

Proof of (ii): Suppose $\boldsymbol{x} \in \mathcal{S} \cap \mathcal{S}^{\perp}$. Then $\boldsymbol{x} \perp \boldsymbol{x}$, which implies that $\boldsymbol{x} = \mathbf{0}$. This proves that $\mathcal{S} \cap \mathcal{S}^{\perp} = \{\mathbf{0}\}$.

Proof of (iii): Part (i) tells us that $\mathcal{V} = \mathcal{S} + \mathcal{S}^{\perp}$ and part (ii) says that $\mathcal{S} \cap \mathcal{S}^{\perp} = \{\mathbf{0}\}$. Therefore, the sum is direct and $\mathcal{V} = \mathcal{S} \oplus \mathcal{S}^{\perp}$.

Proof of (iv): This follows immediately from the properties of direct sums (see Theorem 6.1). Alternatively, it follows from the proof of part (i), where we proved that $\{\boldsymbol{z}_{r+1}, \ldots, \boldsymbol{z}_{r+m}\}$ was a basis for $\mathcal{S}^{\perp}$ and r was the dimension of $\mathcal{S}$. Therefore, $\dim(\mathcal{S})^{\perp} = m - r = m - \dim(\mathcal{S})$. □

Corollary 7.5 *Let $\mathcal{S}$ be a subspace of a vector space $\mathcal{V}$. Then every vector $\boldsymbol{y} \in \mathcal{V}$ can be expressed* **uniquely** *as $\boldsymbol{y} = \boldsymbol{u} + \boldsymbol{v}$, where $\boldsymbol{u} \in \mathcal{S}$ and $\boldsymbol{v} \in \mathcal{S}^{\perp}$.*

Proof. Part (i) of Theorem 7.4 tells us that every vector $\boldsymbol{y} \in \mathcal{V}$ can be expressed as $\boldsymbol{y} = \boldsymbol{u} + \boldsymbol{v}$, where $\boldsymbol{u} \in \mathcal{S}$ and $\boldsymbol{v} \in \mathcal{S}^{\perp}$. The vectors $\boldsymbol{u}$ and $\boldsymbol{v}$ are uniquely determined from $\boldsymbol{y}$ because $\mathcal{S} + \mathcal{S}^{\perp}$ is direct (see part (iii) of Theorem 6.2). □

Below we note a few important properties of orthocomplementary spaces.

Lemma 7.7 Orthocomplement of an orthocomplement. *Let $\mathcal{S}$ be any subspace of $\mathcal{V}$ with $\dim(\mathcal{V}) = m$. Then, $\left(\mathcal{S}^{\perp}\right)^{\perp} = \mathcal{S}$.*

Proof. The easiest way to prove this is to show that (a) $\mathcal{S}$ is a subspace of $\left[\mathcal{S}^{\perp}\right]^{\perp}$ and that (b) $\dim(\mathcal{S}) = \dim\left([\mathcal{S}^{\perp}]^{\perp}\right)$.

The first part follows immediately from part (iii) of Lemma 7.6: since $\mathcal{S}$ is a subspace, $\mathcal{S} = Sp(\mathcal{S}) \subseteq (\mathcal{S}^{\perp})^{\perp}$. The next part follows immediately from part (iv) of Theorem 7.4 and keeping in mind that $\left[\mathcal{S}^{\perp}\right]^{\perp}$ is the orthocomplement of $\mathcal{S}^{\perp}$:

$$\dim\left([\mathcal{S}^{\perp}]^{\perp}\right) = m - \dim\left(\mathcal{S}^{\perp}\right) = m - (m - \dim(\mathcal{S})) = \dim(\mathcal{S})\ .$$

This proves the theorem. □

It may be interesting to construct a direct proof of the inclusion $(\mathcal{S}^{\perp})^{\perp} \subseteq \mathcal{S}$ as a part of the above lemma. Consider the direct sum decomposition of any vector $\boldsymbol{x} \in (\mathcal{S}^{\perp})^{\perp}$. Since that vector is in $\mathcal{V}$, we must have $\boldsymbol{x} = \boldsymbol{u} + \boldsymbol{v}$ with $\boldsymbol{u} \in \mathcal{S}$ and $\boldsymbol{v} \in \mathcal{S}^{\perp}$. Therefore, $\boldsymbol{v}'\boldsymbol{x} = \boldsymbol{v}'\boldsymbol{u} + \boldsymbol{v}'\boldsymbol{v} = \|\boldsymbol{v}\|^2$ (since $\boldsymbol{v}'\boldsymbol{u} = 0$). But $\boldsymbol{x} \perp \boldsymbol{v}$, so $\|\boldsymbol{v}\| = 0$ implying $\boldsymbol{v} = \boldsymbol{0}$. Therefore, $\boldsymbol{x} = \boldsymbol{u} \in \mathcal{S}$, which proves that $(\mathcal{S}^{\perp})^{\perp} \subseteq \mathcal{S}$.

Lemma 7.7 ensures a symmetry in the orthogonality of subspaces: if $\mathcal{S}$ and $\mathcal{T}$ are two subspaces such that $\mathcal{T} = \mathcal{S}^{\perp}$, then $\mathcal{S} = \left(\mathcal{S}^{\perp}\right)^{\perp} = \mathcal{T}^{\perp}$ as well and we can simply write $\mathcal{S} \perp \mathcal{T}$ without any confusion. Below we find conditions for $\mathcal{C}(\boldsymbol{A}) \perp \mathcal{C}(\boldsymbol{B})$.

Lemma 7.8 *Let $\boldsymbol{A}$ and $\boldsymbol{B}$ be two matrices with the same number of rows. Then $\mathcal{C}(\boldsymbol{A}) \perp \mathcal{C}(\boldsymbol{B})$ if and only if $\boldsymbol{A}'\boldsymbol{B} = \boldsymbol{B}'\boldsymbol{A} = \boldsymbol{O}$.*

Proof. Let $\boldsymbol{A}$ and $\boldsymbol{B}$ be $m \times p_1$ and $m \times p_2$, respectively. Any vector in $\mathcal{C}(\boldsymbol{A})$ can be written as $\boldsymbol{A}\boldsymbol{x}$ for some $\boldsymbol{x} \in \Re^{p_1}$ and any vector in $\mathcal{C}(\boldsymbol{B})$ can be written as $\boldsymbol{B}\boldsymbol{y}$ for some $\boldsymbol{y} \in \Re^{p_2}$. We conclude

$$\begin{aligned}\mathcal{C}(\boldsymbol{A}) \perp \mathcal{C}(\boldsymbol{B}) &\Longleftrightarrow \langle \boldsymbol{A}\boldsymbol{x}, \boldsymbol{B}\boldsymbol{y} \rangle = 0 \Longleftrightarrow \boldsymbol{y}'\boldsymbol{B}'\boldsymbol{A}'\boldsymbol{x} = 0 \ \ \forall\ \ \boldsymbol{x} \in \Re^{p_1}, \boldsymbol{y} \in \Re^{p_2} \\ &\Longleftrightarrow \boldsymbol{B}'\boldsymbol{A} = \boldsymbol{O} = (\boldsymbol{A}'\boldsymbol{B})' \Longleftrightarrow \boldsymbol{A}'\boldsymbol{B} = (\boldsymbol{B}'\boldsymbol{A})' = \boldsymbol{O}\ .\end{aligned}$$

The second-to-last equivalence follows by taking $\boldsymbol{y}$ as the i-th column of $\boldsymbol{I}_{p_2}$ and $\boldsymbol{x}$ as the j-th column of $\boldsymbol{I}_{p_1}$, for $i = 1, 2, \ldots, p_2$ and $j = 1, 2, \ldots, p_1$. Then $\boldsymbol{y}'\boldsymbol{B}'\boldsymbol{A}'\boldsymbol{x} = 0$ means that the (i, j)-th element of $\boldsymbol{B}'\boldsymbol{A}$ is zero. □

A few other interesting, and not entirely obvious, properties of orthocomplementary subspaces are provided in the following lemmas.

Lemma 7.9 **Reversion of inclusion.** *Let $\mathcal{S}_1 \subseteq \mathcal{S}_2$ be subsets of $\mathcal{V}$. Then, $\mathcal{S}_1^{\perp} \supseteq \mathcal{S}_2^{\perp}$.*

Proof. Let $\boldsymbol{x}_2 \in \mathcal{S}_2^{\perp}$. Therefore, $\langle \boldsymbol{x}_2, \boldsymbol{u} \rangle = 0$ for all $\boldsymbol{u} \in \mathcal{S}_2$. But as $\mathcal{S}_1$ is included in $\mathcal{S}_2$, we have $\langle \boldsymbol{x}_2, \boldsymbol{u} \rangle = 0$ for all $\boldsymbol{u}$ in $\mathcal{S}_1$. Hence $\boldsymbol{x}_2 \in \mathcal{S}_1^{\perp}$. This proves the reversion of inclusion. □

Lemma 7.10 **Equality of orthocomplements.** *Let $\mathcal{S}_1 = \mathcal{S}_2$ be two subspaces in $\Re^m$. Then $\mathcal{S}_1^{\perp} = \mathcal{S}_2^{\perp}$.*

Proof. This follows easily from the preceding lemma. In particular, if $\mathcal{S}_1 = \mathcal{S}_2$ then both $\mathcal{S}_1 \subseteq \mathcal{S}_2$ *and* $\mathcal{S}_2 \subseteq \mathcal{S}_1$ are true. Applying the reversion of inclusion lemma we have $\mathcal{S}_2^\perp \subseteq \mathcal{S}_1^\perp$ from the first and $\mathcal{S}_1^\perp \subset \mathcal{S}_2^\perp$ from the second. This proves $\mathcal{S}_1^\perp = \mathcal{S}_2^\perp$. □

Finally, we have an interesting "analogue" of De Morgan's law in set theory for sums and intersection of subspaces.

Lemma 7.11 *Let $\mathcal{S}_1$ and $\mathcal{S}_2$ be two subspaces in $\Re^n$. Then:*

(i) $(\mathcal{S}_1 + \mathcal{S}_2)^\perp = \mathcal{S}_1^\perp \cap \mathcal{S}_2^\perp$.

(ii) $(\mathcal{S}_1 \cap \mathcal{S}_2)^\perp = \mathcal{S}_1^\perp + \mathcal{S}_2^\perp$.

Proof. **Proof of (i):** Note that $\mathcal{S}_1 \subseteq \mathcal{S}_1 + \mathcal{S}_2$ and $\mathcal{S}_2 \subseteq \mathcal{S}_1 + \mathcal{S}_2$. Therefore, using Lemma 7.9, we obtain

$$(\mathcal{S}_1 + \mathcal{S}_2)^\perp \subseteq \mathcal{S}_1^\perp \text{ and } (\mathcal{S}_1 + \mathcal{S}_2)^\perp \subseteq \mathcal{S}_2^\perp ,$$

which implies that $(\mathcal{S}_1 + \mathcal{S}_2)^\perp \subseteq \mathcal{S}_1^\perp \cap \mathcal{S}_2^\perp$. To prove the reverse inclusion, consider any $\boldsymbol{x} \in \mathcal{S}_1^\perp \cap \mathcal{S}_2^\perp$. If $\boldsymbol{y}$ is any vector in $\mathcal{S}_1 + \mathcal{S}_2$, then $\boldsymbol{y} = \boldsymbol{u} + \boldsymbol{v}$ for some $\boldsymbol{u} \in \mathcal{S}_1$ and $\boldsymbol{v} \in \mathcal{S}_2$ and we find

$$\langle \boldsymbol{x}, \boldsymbol{y} \rangle = \langle \boldsymbol{x}, \boldsymbol{u} + \boldsymbol{v} \rangle = \langle \boldsymbol{x}, \boldsymbol{u} \rangle + \langle \boldsymbol{x}, \boldsymbol{v} \rangle = 0 + 0 = 0 .$$

This shows that $\boldsymbol{x} \perp \boldsymbol{y}$ for any vector $\boldsymbol{y} \in \mathcal{S}_1 + \mathcal{S}_2$. This proves that $\boldsymbol{x} \in (\mathcal{S}_1 + \mathcal{S}_2)^\perp$ and, therefore, $\mathcal{S}_1^\perp \cap \mathcal{S}_2^\perp \subseteq (\mathcal{S}_1 + \mathcal{S}_2)^\perp$. This proves (i).

Proof of (ii): This can be proved from (i) as follows. Apply part (i) to $\mathcal{S}_1^\perp + \mathcal{S}_2^\perp$ to obtain

$$\left(\mathcal{S}_1^\perp + \mathcal{S}_2^\perp\right)^\perp = \left(\mathcal{S}_1^\perp\right)^\perp \cap \left(\mathcal{S}_2^\perp\right)^\perp = \mathcal{S}_1 \cap \mathcal{S}_2 \Longrightarrow \left(\mathcal{S}_1^\perp + \mathcal{S}_2^\perp\right) = (\mathcal{S}_1 \cap \mathcal{S}_2)^\perp ,$$

where the last equality follows from applications of Lemmas 7.7 and 7.10. □

7.6 The Fundamental Theorem of Linear Algebra

Let us revisit the four fundamental subspaces of a matrix introduced in Section 4.6. There exist some beautiful relationships among the four fundamental subspaces. It relates the null spaces of $\boldsymbol{A}$ and $\boldsymbol{A}'$ (i.e., the null space and left-hand null space of $\boldsymbol{A}$) to the orthocomplements of the column spaces of $\boldsymbol{A}$ and $\boldsymbol{A}'$ (i.e., the column space and row space of $\boldsymbol{A}$).

Theorem 7.5 *Let $\boldsymbol{A}$ be an $m \times n$ matrix. Then:*

(i) $\mathcal{C}(\boldsymbol{A})^\perp = \mathcal{N}(\boldsymbol{A}')$ *and* $\Re^m = \mathcal{C}(\boldsymbol{A}) \oplus \mathcal{N}(\boldsymbol{A}')$;

(ii) $\mathcal{C}(\boldsymbol{A}) = \mathcal{N}(\boldsymbol{A}')^\perp$;

(iii) $\mathcal{N}(\boldsymbol{A})^\perp = \mathcal{C}(\boldsymbol{A}')$ *and* $\Re^n = \mathcal{N}(\boldsymbol{A}) \oplus \mathcal{C}(\boldsymbol{A}')$.

Proof. **Proof of (i):** Let $\boldsymbol{x}$ be any element in $\mathcal{C}(\boldsymbol{A})^{\perp}$. Since $\boldsymbol{A}\boldsymbol{u}$ is in the column space of $\boldsymbol{A}$ for any element $\boldsymbol{u}$, observe that

$$\begin{aligned} \boldsymbol{x} \in \mathcal{C}(\boldsymbol{A})^{\perp} &\iff \langle \boldsymbol{x}, \boldsymbol{A}\boldsymbol{u} \rangle = \boldsymbol{0} \iff \boldsymbol{u}'\boldsymbol{A}'\boldsymbol{x} = \boldsymbol{0} \quad \text{for all} \quad \boldsymbol{u} \in \Re^n \\ &\iff \boldsymbol{A}'\boldsymbol{x} = \boldsymbol{0} \iff \boldsymbol{x} \in \mathcal{N}(\boldsymbol{A}') \ . \end{aligned}$$

This proves that $\mathcal{C}(\boldsymbol{A})^{\perp} = \mathcal{N}(\boldsymbol{A}')$. Part (iii) of Theorem 7.4 says that $\mathcal{C}(\boldsymbol{A})$ and $\mathcal{C}(\boldsymbol{A})^{\perp}$ form a direct sum for $\Re^m$. Therefore,

$$\Re^m = \mathcal{C}(\boldsymbol{A}) \oplus \mathcal{C}(\boldsymbol{A})^{\perp} = \mathcal{C}(\boldsymbol{A}) \oplus \mathcal{N}(\boldsymbol{A}') \ .$$

This proves part (i).

Proof of (ii): Applying Lemma 7.7 to the result in part (1), we obtain

$$\mathcal{C}(\boldsymbol{A}) = [\mathcal{C}(\boldsymbol{A})^{\perp}]^{\perp} = \mathcal{N}(\boldsymbol{A}')^{\perp} \ .$$

This proves (ii).

Proof of (iii): Applying the result in part (ii) to $\boldsymbol{A}'$ yields

$$\begin{aligned} \mathcal{N}(\boldsymbol{A})^{\perp} &= \mathcal{N}((\boldsymbol{A}')')^{\perp} = \mathcal{C}(\boldsymbol{A}') \ \text{ and} \\ \Re^n &= \mathcal{N}(\boldsymbol{A}) \oplus \mathcal{N}(\boldsymbol{A}')^{\perp} = \mathcal{N}(\boldsymbol{A}) \oplus \mathcal{C}(\boldsymbol{A}') \ . \end{aligned}$$

This proves the theorem. □

The following corollary translates the above result in terms of the dimensions of the four fundamental subspaces. It is equally important and reveals yet another proof of why the rank of a matrix is equal to that of its transpose.

Corollary 7.6 *Let $\boldsymbol{A}$ be an $m \times n$ matrix.*

(i) $\dim(\mathcal{C}(\boldsymbol{A}')) + \dim(\mathcal{N}(\boldsymbol{A})) = n$ and $\dim(\mathcal{C}(\boldsymbol{A})) + \dim(\mathcal{N}(\boldsymbol{A}') = m$;

(ii) $\dim(\mathcal{C}(\boldsymbol{A}')) = \dim(\mathcal{C}(\boldsymbol{A}))$. That is, $\rho(\boldsymbol{A}') = \rho(\boldsymbol{A})$.

Proof. **Proof of (i):** This follows from Theorem 7.5 and part (iv) of Theorem 7.4.

Proof of (ii): A combination of the *Rank-Nullity Theorem* (recall Theorem 5.1) and part (i) yields the desired result. To be precise, we have the following.

$$\begin{aligned} \dim(\mathcal{C}(\boldsymbol{A})) + \dim(\mathcal{N}(\boldsymbol{A})) &= n \qquad \text{from the Rank-Nullity Theorem;} \\ \dim(\mathcal{C}(\boldsymbol{A}')) + \dim(\mathcal{N}(\boldsymbol{A})) &= n \qquad \text{from part (i)} \ . \end{aligned}$$

This implies that $\dim(\mathcal{C}(\boldsymbol{A}')) = \dim(\mathcal{C}(\boldsymbol{A}))$, i.e., $\rho(\boldsymbol{A}) = \rho(\boldsymbol{A}')$. □

In the last part of Corollary 7.6, we used the Rank-Nullity Theorem (Theorem 5.1) to prove $\rho(\boldsymbol{A}) = \rho(\boldsymbol{A}')$. Recall that $\rho(\boldsymbol{A}) = \rho(\boldsymbol{A}')$ was already proved by other means on more than one occasion. See Theorem 5.2, Theorem 5.8 or even the argument using orthogonality in Section 7.2. If, therefore, we decide to use the fact that the ranks of $\boldsymbol{A}$ and $\boldsymbol{A}'$ are equal, then we can easily establish the Rank-Nullity Theorem—use part (i) of Corollary 7.6 and the fact that $\rho(\boldsymbol{A}) = \rho(\boldsymbol{A}')$ to obtain

$$\nu(\boldsymbol{A}) = \dim(\mathcal{N}(\boldsymbol{A})) = n - \dim(\mathcal{C}(\boldsymbol{A}')) = n - \rho(\boldsymbol{A}') = n - \rho(\boldsymbol{A}) \ .$$

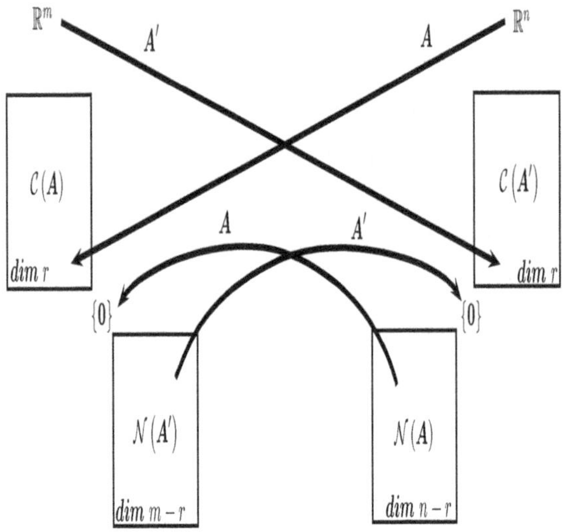

Figure 7.3 *The four fundamental subspaces of an $m \times n$ matrix $\boldsymbol{A}$.*

Theorem 7.5 can help us concretely identify a matrix whose null space coincides with the column space of a given matrix. For example, if we are given an $n \times p$ matrix $\boldsymbol{A}$ with rank r how can we find a matrix $\boldsymbol{B}'$ such that $\mathcal{N}(\boldsymbol{B}') = \mathcal{C}(\boldsymbol{A}) \subseteq \Re^n$? Clearly, such a matrix $\boldsymbol{B}$ must have n rows. Theorem 7.5 ensures that

$$\mathcal{C}(\boldsymbol{A}) = \mathcal{N}(\boldsymbol{B}') = \mathcal{C}(\boldsymbol{B})^{\perp} \Longrightarrow \mathcal{C}(\boldsymbol{B}) = \left[\mathcal{C}(\boldsymbol{B})^{\perp}\right]^{\perp} = \mathcal{C}(\boldsymbol{A})^{\perp} . \tag{7.14}$$

Therefore, we can find any set of $n - r$ basis vectors for $\mathcal{C}(\boldsymbol{A})^{\perp}$ and place them as columns of $\boldsymbol{B}$. Then, $\boldsymbol{B}$ is an $n \times (n - r)$ matrix such that $\mathcal{N}(\boldsymbol{B}') = \mathcal{C}(\boldsymbol{A})$.

Figure 7.3 presents the four fundamental subspaces associated with a matrix. In a beautiful expository article, Strang (1993) presented the essence of the fundamental theorem of linear algebra using diagrams and pictures; also see Strang (2005, 2009). The fundamental theorem explains how an $m \times n$ matrix $\boldsymbol{A}$ transforms a vector in $\Re^n$. If $\boldsymbol{x} \in \Re^n$, then we have the direct sum decomposition $\boldsymbol{x} = \boldsymbol{x}_r + \boldsymbol{x}_n$, where $\boldsymbol{x}_r$ is in the row space of $\boldsymbol{A}$ and $\boldsymbol{x}_n$ is in the null space of $\boldsymbol{A}$ (recall Corollary 7.5). We say that $\boldsymbol{x}_r$ is the orthogonal projection of $\boldsymbol{x}$ into the row space of $\boldsymbol{A}$ along the null space of $\boldsymbol{A}$. Since $\boldsymbol{A}\boldsymbol{x}_n = \boldsymbol{0}$, it follows that $\boldsymbol{A}\boldsymbol{x} = \boldsymbol{A}\boldsymbol{x}_r$. Thus, every vector $\boldsymbol{A}\boldsymbol{x}$ in the $\mathcal{C}(\boldsymbol{A})$ can be obtained by multiplying $\boldsymbol{A}$ with a vector $\boldsymbol{x}_r$ in $\mathcal{R}(\boldsymbol{A})$.

If $\{\boldsymbol{v}_1, \boldsymbol{v}_2, \ldots, \boldsymbol{v}_r\}$ is a basis of the row space of $\boldsymbol{A}$, then $\{\boldsymbol{A}\boldsymbol{v}_1, \boldsymbol{A}\boldsymbol{v}_2, \ldots, \boldsymbol{A}\boldsymbol{v}_r\}$ is a basis for the column space of $\boldsymbol{A}$ (see Section 7.2). This number r is the rank of $\boldsymbol{A}$. In fact, there is even more to this. Later, when we discuss the Singular Value Decomposition (SVD) in Section 12.1, we will find that we can choose an orthonormal basis $\{\boldsymbol{v}_1, \boldsymbol{v}_2, \ldots, \boldsymbol{v}_r\}$ of the row space of $\boldsymbol{A}$ and an orthonormal basis $\{\boldsymbol{u}_1, \boldsymbol{u}_2, \ldots, \boldsymbol{u}_r\}$

of the column space of $\boldsymbol{A}$ such that $\boldsymbol{A}\boldsymbol{v}_i = \sigma_i \boldsymbol{u}_i$ for $i = 1, 2, \ldots, r$, where σ_i is a positive real number. The SVD provides a diagonal representation of $\boldsymbol{A}$ with respect to these orthogonal bases and is a numerically stable method (unlike row-reduced echelon forms) to obtain basis vectors for the four fundamental subspaces.

Furthermore, $\boldsymbol{A}$ defines a one-one map between the row space and the column space of $\boldsymbol{A}$. To see why this is true, let $\boldsymbol{u}$ be a vector in $\mathcal{C}(\boldsymbol{A})$ and suppose, if possible, that there are two vectors $\boldsymbol{v}_1$ and $\boldsymbol{v}_2$ in $\mathcal{R}(\boldsymbol{A})$ such that $\boldsymbol{u} = \boldsymbol{A}\boldsymbol{v}_1 = \boldsymbol{A}\boldsymbol{v}_2$. Then, $\boldsymbol{A}(\boldsymbol{v}_1 - \boldsymbol{v}_2) = \boldsymbol{0}$. This means that $\boldsymbol{v}_1 - \boldsymbol{v}_2$ is in $\mathcal{N}(\boldsymbol{A})$. Since the null space and row space are orthocomplementary subspaces, $\boldsymbol{v}_1 - \boldsymbol{v}_2 = \boldsymbol{0}$ and $\boldsymbol{v}_1 = \boldsymbol{v}_2 = \boldsymbol{v}$. Thus, each vector $\boldsymbol{u}$ in $\mathcal{C}(\boldsymbol{A})$ corresponds to a *unique* vector $\boldsymbol{v}$ in $\mathcal{R}(\boldsymbol{A})$. Therefore, $\boldsymbol{A}$ yields an invertible map from its r-dimensional row space to its r-dimensional column space. What is the inverse of this map? Surely it cannot be the usual matrix inverse of $\boldsymbol{A}$ because $\boldsymbol{A}$ need not be square, let alone nonsingular. The inverse map is given by a matrix that will map the column space to the row space in one-one fashion. This matrix is called a ***pseudo-inverse*** or ***generalized inverse*** of $\boldsymbol{A}$ (see Section 9.4).

We conclude this chapter by revisiting the derivation of the projector in Theorem 6.8. Let $\Re^n = \mathcal{S} \oplus \mathcal{T}$ such that $\dim(\mathcal{S}) = r < n$ and $\dim(\mathcal{T}) = n - r$. Let $\boldsymbol{S}$ be an $n \times r$ matrix whose column vectors constitute basis for $\mathcal{S}$. Note that $\dim(\mathcal{T}^{\perp}) = r$. We find a basis $\boldsymbol{v}_1, \boldsymbol{v}_2, \ldots, \boldsymbol{v}_r$ for $\mathcal{T}^{\perp}$, where each $\boldsymbol{v}_i \in \Re^n$, and place them as columns in the $n \times r$ matrix $\boldsymbol{V}$. Any vector $\boldsymbol{x} \in \Re^n$ can be written as

$$\boldsymbol{x} = \boldsymbol{S}\boldsymbol{\beta} + (\boldsymbol{x} - \boldsymbol{S}\boldsymbol{\beta}) = \boldsymbol{S}\boldsymbol{\beta} + \boldsymbol{w}\ , \quad \text{where } \boldsymbol{w} = \boldsymbol{x} - \boldsymbol{S}\boldsymbol{\beta}\ . \tag{7.15}$$

The projection of $\boldsymbol{x}$ onto $\mathcal{S}$ along $\mathcal{T}$ is $\boldsymbol{S}\boldsymbol{\beta}$ and $\boldsymbol{w} \in \mathcal{T}$. Since the columns of $\boldsymbol{V}$ are orthogonal to any vector in $\mathcal{T}$, we have that $\boldsymbol{v}_i'\boldsymbol{w} = 0$ for each $i = 1, 2, \ldots, r$. Thus, $\boldsymbol{V}'\boldsymbol{w} = \boldsymbol{0}$ or $\boldsymbol{w} \in \mathcal{N}(\boldsymbol{V}')$. Multiplying both sides of (7.15) with $\boldsymbol{V}'$ yields

$$\boldsymbol{V}'\boldsymbol{x} = \boldsymbol{V}'\boldsymbol{S}\boldsymbol{\beta} + \boldsymbol{V}'\boldsymbol{w} = \boldsymbol{V}'\boldsymbol{S}\boldsymbol{\beta}\ , \quad \text{which implies that } \boldsymbol{\beta} = (\boldsymbol{V}'\boldsymbol{S})^{-1}\boldsymbol{V}'\boldsymbol{x}$$

because $\boldsymbol{V}'\boldsymbol{S}$ is nonsingular (recall (6.10)). If $\boldsymbol{P}$ is the projector onto $\mathcal{S}$ along $\mathcal{T}$, then $\boldsymbol{P}\boldsymbol{x} = \boldsymbol{S}\boldsymbol{\beta} = \boldsymbol{S}(\boldsymbol{V}'\boldsymbol{S})^{-1}\boldsymbol{V}'\boldsymbol{x}$ for every $\boldsymbol{x} \in \Re^n$. Therefore,

$$\boldsymbol{P} = \boldsymbol{S}(\boldsymbol{V}'\boldsymbol{S})^{-1}\boldsymbol{V}'\ . \tag{7.16}$$

This is an alternate derivation of the formula in Theorem 6.8.

7.7 Exercises

1. Verify that the following set of vectors are orthogonal to each other:

$$\left\{ \boldsymbol{x}_1 = \begin{bmatrix} 1 \\ 1 \\ 1 \end{bmatrix},\ \boldsymbol{x}_2 = \begin{bmatrix} -1 \\ 1 \\ 0 \end{bmatrix},\ \boldsymbol{x}_3 = \begin{bmatrix} 1 \\ 1 \\ -2 \end{bmatrix} \right\} .$$

   Find $\|\boldsymbol{x}_1\|$, $\|\boldsymbol{x}_2\|$ and $\|\boldsymbol{x}_3\|$.

2. True or false: $\langle \boldsymbol{x}, \boldsymbol{y} \rangle = 0$ for all $\boldsymbol{y}$ if and only if $\boldsymbol{x} = \boldsymbol{0}$.

3. Let $\boldsymbol{A} = \boldsymbol{A}'$ be a real symmetric matrix. Prove that

$$\langle \boldsymbol{A}\boldsymbol{x}, \boldsymbol{y} \rangle = \langle \boldsymbol{x}, \boldsymbol{A}\boldsymbol{y} \rangle \ \text{ for all } \ \boldsymbol{x}, \boldsymbol{y} \in \Re^n\ .$$

4. Recall from (1.18) that $\text{tr}(\boldsymbol{A}'\boldsymbol{A}) \geq 0$. Derive the Cauchy-Schwarz inequality by applying this to the matrix $\boldsymbol{A} = \boldsymbol{x}\boldsymbol{y}' - \boldsymbol{y}\boldsymbol{x}'$.

5. Prove the ***Parallelogram identity***

$$\|\boldsymbol{x}+\boldsymbol{y}\|^2 + \|\boldsymbol{x}-\boldsymbol{y}\|^2 = 2\|\boldsymbol{x}\|^2 + 2\|\boldsymbol{y}\|^2 ,$$

where $\boldsymbol{x}$ and $\boldsymbol{y}$ are vectors in $\Re^n$.

6. Find the equation of the plane that passes through the point $(-1,-1,2)$ and has normal vector $\boldsymbol{n}' = (2,3,4)$.

7. Find the equation of a plane that passes through the points $(2,1,3)$, $(4,2,-1)$ and $(6,2,2)$.

8. Find the angle between the vectors $\begin{bmatrix} 2 \\ -1 \\ 1 \end{bmatrix}$ and $\begin{bmatrix} 1 \\ 1 \\ 2 \end{bmatrix}$.

9. Verify each of the following identities for the cross product (Definition 7.4):
   (a) $\boldsymbol{u} \times \boldsymbol{v} = -(\boldsymbol{v} \times \boldsymbol{u})$.
   (b) $\boldsymbol{u} \times (\boldsymbol{v} + \boldsymbol{w}) = (\boldsymbol{u} \times \boldsymbol{v}) + (\boldsymbol{u} \times \boldsymbol{w})$.
   (c) $(\boldsymbol{u} + \boldsymbol{v}) \times \boldsymbol{w} = (\boldsymbol{u} \times \boldsymbol{w}) + (\boldsymbol{v} \times \boldsymbol{w})$.
   (d) $\alpha(\boldsymbol{u} \times \boldsymbol{v}) = (\alpha\boldsymbol{u}) \times \boldsymbol{v} = \boldsymbol{u} \times (\alpha\boldsymbol{v})$ for any real number α.
   (e) $\boldsymbol{u} \times \boldsymbol{0} = \boldsymbol{0} \times \boldsymbol{u} = \boldsymbol{0}$.
   (f) $\boldsymbol{u} \times \boldsymbol{u} = \boldsymbol{0}$.
   (g) $\|\boldsymbol{u} \times \boldsymbol{v}\| = \|\boldsymbol{u}\|\|\boldsymbol{v}\| \sin(\theta) = \sqrt{\|\boldsymbol{u}\|^2\|\boldsymbol{v}\|^2 - |\langle \boldsymbol{u}, \boldsymbol{v}\rangle|^2}$, where θ is the angle between $\boldsymbol{u}$ and $\boldsymbol{v}$.

10. Find a vector that is orthogonal to each of the vectors $\boldsymbol{u} = \begin{bmatrix} 1 \\ 2 \\ -1 \end{bmatrix}$ and $\boldsymbol{v} = \begin{bmatrix} 0 \\ 2 \\ 3 \end{bmatrix}$.

11. Find the orthogonal projection of the point $(1,1,1)$ onto the plane spanned by the vectors $\begin{bmatrix} 1 \\ 1 \\ 0 \end{bmatrix}$ and $\begin{bmatrix} -2 \\ 2 \\ 1 \end{bmatrix}$.

12. Find the orthogonal projection of the point $(3,2,3)$ onto the plane $x-2y+3z=5$.

13. Find the distance from the point $(3,2,3)$ to the plane $x-2y+3z=5$.

14. Find the rank of an elementary projector $\boldsymbol{P}_{\boldsymbol{u}}$, where $\boldsymbol{u} \neq \boldsymbol{0}$.

15. Prove ***Bessel's inequality***: If $\mathcal{X} = \{\boldsymbol{x}_1, \ldots, \boldsymbol{x}_k\}$ is an orthonormal set of vectors in $\Re^n$, then

$$\sum_{i=1}^{k} |\langle \boldsymbol{x}, \boldsymbol{x}_i\rangle|^2 \leq \|\boldsymbol{x}\|^2 .$$

   Equality holds only when $\boldsymbol{x} \in Sp(\mathcal{X})$.

16. Let $\boldsymbol{U}$ be an $n \times n$ matrix whose columns form an orthonormal basis for $\Re^n$. Prove that $\boldsymbol{U}' = \boldsymbol{U}^{-1}$.

17. Let $\boldsymbol{U}$ be an $n \times p$ matrix with orthogonal columns. Show that $\boldsymbol{U}'\boldsymbol{U} = \boldsymbol{I}_p$. Can $\boldsymbol{U}\boldsymbol{U}' = \boldsymbol{I}_n$ when $p < n$?

18. Consider the following vectors

$$\left\{ \begin{bmatrix} 1 \\ 1 \\ 1 \\ 1 \end{bmatrix}, \begin{bmatrix} 0 \\ 1 \\ 1 \\ 1 \end{bmatrix}, \begin{bmatrix} 0 \\ 0 \\ 1 \\ 1 \end{bmatrix}, \begin{bmatrix} 0 \\ 0 \\ 0 \\ 1 \end{bmatrix} \right\}$$

in $\Re^4$. Apply the Gram-Schmidt procedure to these vectors to produce an orthonormal basis for $\Re^4$.

19. Find an orthonormal basis for the column space of the matrix

$$\begin{bmatrix} 2 & 6 & 4 \\ -1 & 1 & 1 \\ 0 & 4 & 3 \\ 1 & -5 & -4 \end{bmatrix}.$$

20. Find an orthonormal basis for the orthogonal complement of the subspace $\{\boldsymbol{x} \in \Re^3 : \boldsymbol{1}'\boldsymbol{x} = 0\}$.

21. Let $\boldsymbol{a} \in \Re^3$ be a fixed vector. Find an orthogonal basis for the orthogonal complement of the subspace $\{\boldsymbol{x} \in \Re^3 : \langle \boldsymbol{a}, \boldsymbol{x} \rangle = 0\}$.

22. Let $\boldsymbol{u} \in \Re^n$ be a nonzero vector. What is $\dim([Sp\{\boldsymbol{u}\}]^\perp)$?

23. Prove that if $\mathcal{S}$ and $\mathcal{T}$ are complementary subspaces of $\Re^n$, then $\mathcal{S}^\perp$ and $\mathcal{T}^\perp$ are also complementary subspaces.

24. In Example 5.1, we found a basis for each of the four fundamental subspaces of

$$\boldsymbol{A} = \begin{bmatrix} 2 & -2 & -5 & -3 & -1 & 2 \\ 2 & -1 & -3 & 2 & 3 & 2 \\ 4 & -1 & -4 & 10 & 11 & 4 \\ 0 & 1 & 2 & 5 & 4 & 0 \end{bmatrix}.$$

Using the Gram-Schmidt orthogonalization procedure find an orthonormal basis for each of the four fundamental subspaces.

25. Let $\boldsymbol{A}$ be a real symmetric matrix. Show that the system $\boldsymbol{A}\boldsymbol{x} = \boldsymbol{b}$ has a solution if and only if $\boldsymbol{b}$ is orthogonal to $\mathcal{N}(\boldsymbol{A})$.

26. Suppose that the columns of a matrix $\boldsymbol{U}$ span a subspace $\mathcal{S} \subset \Re^n$. Prove that a vector $\boldsymbol{v} \in \Re^n$ is orthogonal to every vector in $\mathcal{S}$ if and only if $\boldsymbol{v} \in \mathcal{N}(\boldsymbol{A}')$.

27. Matrices that satisfy $\boldsymbol{A}'\boldsymbol{A} = \boldsymbol{A}\boldsymbol{A}'$ are said to be *normal*. If $\boldsymbol{A}$ is normal, then prove that every vector in $\mathcal{C}(\boldsymbol{A})$ is orthogonal to every vector in $\mathcal{N}(\boldsymbol{A})$.

28. If $\boldsymbol{x}, \boldsymbol{y} \in \Re^n$, then prove that $(\boldsymbol{x} + \boldsymbol{y}) \perp (\boldsymbol{x} - \boldsymbol{y})$ if and only if $\|\boldsymbol{x}\| = \|\boldsymbol{y}\|$.

29. Suppose $\mathcal{S}$ is a subspace of $\Re^n$. Prove that $\mathcal{S}^\perp = \{\boldsymbol{0}\}$ if and only if $\mathcal{S} = \Re^n$.

30. Find $\mathcal{S} + \mathcal{T}$, $\mathcal{S}^\perp$, $\mathcal{T}^\perp$ and verify that $(\mathcal{S} + \mathcal{T})^\perp = \mathcal{S}^\perp \cap \mathcal{T}^\perp$, where

$$\mathcal{S} = \left\{ \begin{bmatrix} x_1 \\ x_2 \\ x_3 \\ x_4 \end{bmatrix} \in \Re^4 : x_1 = x_2 = x_3 \right\} \text{ and } \mathcal{T} = \left\{ \begin{bmatrix} x_1 \\ x_2 \\ x_3 \\ 0 \end{bmatrix} \in \Re^4 : x_1 = x_2 \right\}.$$

CHAPTER 8

More on Orthogonality

8.1 Orthogonal matrices

In linear algebra, when we look at vectors with special attributes, we also look at properties of matrices that have been formed by placing such vectors as columns or rows. We now do so for orthogonal vectors.

Let us suppose that we have formed an $n \times p$ matrix $\boldsymbol{U} = [\boldsymbol{u}_1 : \boldsymbol{u}_2 : \ldots : \boldsymbol{u}_p]$ such that its column vectors are a set of p non-null orthogonal vectors in $\Re^n$. Note that p cannot exceed n because Lemma 7.3 insures that the p columns of $\boldsymbol{U}$ are linearly independent vectors in $\Re^n$. Because the $\boldsymbol{u}_i$'s are orthogonal, it follows that $\boldsymbol{u}_i'\boldsymbol{u}_j = 0$ whenever $i \neq j$ and that $\boldsymbol{u}_i'\boldsymbol{u}_i = \|\boldsymbol{u}_i\|^2 \neq 0$ because none of the $\boldsymbol{u}_i$'s is the null vector. The matrix $\boldsymbol{U}'\boldsymbol{U}$, therefore, is a $p \times p$ diagonal matrix with i-th diagonal element given by $\|\boldsymbol{u}_i\|^2$:

$$\boldsymbol{U}'\boldsymbol{U} = \begin{bmatrix} \boldsymbol{u}_1' \\ \boldsymbol{u}_2' \\ \vdots \\ \boldsymbol{u}_p' \end{bmatrix} [\boldsymbol{u}_1 : \boldsymbol{u}_2 : \ldots : \boldsymbol{u}_p] = \begin{bmatrix} \|\boldsymbol{u}_1\|^2 & 0 & \ldots & 0 \\ 0 & \|\boldsymbol{u}_2\|^2 & \ldots & 0 \\ \vdots & \vdots & \ddots & \vdots \\ 0 & 0 & \ldots & \|\boldsymbol{u}_n\|^2 \end{bmatrix} . \tag{8.1}$$

When the columns of $\boldsymbol{U}$ are *orthonormal*, then $\|\boldsymbol{u}_i\|^2 = 1$ for $i = 1, 2, \ldots, p$ and (8.1) yields $\boldsymbol{U}'\boldsymbol{U} = \boldsymbol{I}_p$.

Orthogonal matrices in $\Re^1$ are the simplest. These are the 1×1 matrices $[1]$ and $[1]$ which we can interpret, respectively, as the identity matrix and a reflection of the real line across the origin. The following example elucidates what happens in $\Re^2$.

Example 8.1 It is easily verified that that the 2×2 matrices

$$\boldsymbol{Q}_1(\theta) = \begin{bmatrix} \cos\theta & -\sin\theta \\ \sin\theta & \cos\theta \end{bmatrix} \quad \text{and} \quad \boldsymbol{Q}_2(\theta) = \begin{bmatrix} \cos\theta & \sin\theta \\ \sin\theta & -\cos\theta \end{bmatrix}$$

are orthogonal matrices. Observe that $\boldsymbol{Q}_1(\theta)\boldsymbol{u}$ gives the point obtained by rotating $\boldsymbol{u}$ about the origin by an angle θ in the counterclockwise direction, while $\boldsymbol{Q}_2(\theta)\boldsymbol{u}$ is the reflection of $\boldsymbol{u}$ in the line which makes an angle of $\theta/2$ with the x-axis (or the

$e_1 = (1, 0)'$ vector). It is easily verified that

$$\begin{bmatrix} \cos\theta & \sin\theta \\ -\sin\theta & \cos\theta \end{bmatrix} = \boldsymbol{Q}_1(-\theta) = \boldsymbol{Q}_1(\theta)' = \boldsymbol{Q}_1(\theta)' = \boldsymbol{Q}_1(\theta)^{-1} .$$

This matrix represents rotation about the origin by an angle θ in the clockwise direction. It is geometrically obvious that rotating a point counterclockwise by an angle θ and then clockwise by the same angle returns the original point. On the other hand, for the reflector, it is easily verified that $\boldsymbol{Q}_2(\boldsymbol{\theta})^2 = \boldsymbol{I}_2$. The geometric intuition here is also clear: reflecting a point about a line and then reflecting the reflection about the same line will return the original point.

In fact, it can be argued that *any* orthogonal matrix in $\Re^2$ must be either $\boldsymbol{Q}_1(\theta)$ or $\boldsymbol{Q}_2(\theta)$ for some θ. To see why, let $\boldsymbol{A} = [\boldsymbol{a}_1 : \boldsymbol{a}_2]$ be a 2×2 orthogonal matrix with columns $\boldsymbol{a}_1$ and $\boldsymbol{a}_2$. Observe that $\|\boldsymbol{a}_1\| = \|\boldsymbol{a}_2\| = 1$ and let θ be the angle between $\boldsymbol{a}_1$ and $\boldsymbol{e}_1$. To be precise,

$$\arccos\theta = \boldsymbol{a}_1'\boldsymbol{e}_1 \quad \text{and} \quad \boldsymbol{a}_1 = \begin{bmatrix} \cos\theta \\ \sin\theta \end{bmatrix} .$$

Similarly, $\boldsymbol{a}_2$ is a vector of length one and can be represented as $\boldsymbol{a}_2 = (\cos\phi, \sin\phi)'$. Because $\boldsymbol{A}$ is an orthogonal matrix, $\boldsymbol{a}_1$ and $\boldsymbol{a}_2$ are orthogonal vectors, which means that the angle between them is $\pi/2$. This implies that ϕ is either $\theta + \dfrac{\pi}{2}$ or $\phi = \theta - \dfrac{\pi}{2}$, which corresponds to $\boldsymbol{A} = \boldsymbol{Q}_1(\theta)$ or $\boldsymbol{A} = \boldsymbol{Q}_2(\theta)$, respectively. ■

If $\boldsymbol{V} = [\boldsymbol{v}_1 : \boldsymbol{v}_2 : \ldots : \boldsymbol{v}_k]$ is an $n \times k$ matrix such that each $\boldsymbol{v}_i$ is orthogonal to each $\boldsymbol{u}_j$, then $\boldsymbol{v}_i'\boldsymbol{u}_j = \boldsymbol{u}_j'\boldsymbol{v}_i = 0$ for all $i = 1, 2, \ldots, k$ and $j = 1, 2, \ldots, p$. This means that $\boldsymbol{V}'\boldsymbol{U} = \boldsymbol{O}$ and $\boldsymbol{U}'\boldsymbol{V} = \boldsymbol{O}$. (Note both $\boldsymbol{O}$'s are null matrices, but the first is $k \times k$ and the second is $p \times p$.) These properties can be useful in proving certain properties of orthogonal vectors. The following lemma is an example.

Lemma 8.1 *Let $\mathcal{U} = \{\boldsymbol{u}_1, \boldsymbol{u}_2, \ldots, \boldsymbol{u}_p\}$ and $\mathcal{V} = \{\boldsymbol{v}_1, \boldsymbol{v}_2, \ldots, \boldsymbol{v}_k\}$ be linearly independent sets of vectors in $\Re^n$ such that $\boldsymbol{v}_i \perp \boldsymbol{u}_j$ for every $\boldsymbol{v}_i$ in $\mathcal{V}$ and every $\boldsymbol{u}_i$ in $\mathcal{U}$. Then $\mathcal{U} \cup \mathcal{V}$ is a linearly independent set.*

Proof. Let us form the matrices $\boldsymbol{U} = [\boldsymbol{u}_1 : \boldsymbol{u}_2 : \ldots : \boldsymbol{u}_p]$ and $\boldsymbol{V} = [\boldsymbol{v}_1 : \boldsymbol{v}_2 : \ldots : \boldsymbol{v}_k]$. Note that the column vectors of $\boldsymbol{U}$ and those of $\boldsymbol{V}$ are linearly independent. Therefore, $\boldsymbol{U}'\boldsymbol{U}$ and $\boldsymbol{V}'\boldsymbol{V}$ are both nonsingular (recall Theorem 5.7).

We want to show that the linear homogeneous equation $\boldsymbol{U\alpha} + \boldsymbol{V\beta} = \boldsymbol{0}$ implies that the $p \times 1$ vector $\boldsymbol{\alpha} = \boldsymbol{0}$ and the $k \times 1$ vector $\boldsymbol{\beta} = \boldsymbol{0}$. Since each column vector in $\boldsymbol{V}$ is orthogonal to every column vector in $\boldsymbol{U}$, it follows that $\boldsymbol{V}'\boldsymbol{U} = \boldsymbol{O}$ and we obtain

$$\boldsymbol{U\alpha} + \boldsymbol{V\beta} = \boldsymbol{0} \Longrightarrow \boldsymbol{V}'\boldsymbol{U\alpha} + \boldsymbol{V}'\boldsymbol{V\beta} = \boldsymbol{V}'\boldsymbol{0} \Longrightarrow \boldsymbol{V}'\boldsymbol{V\beta} = \boldsymbol{0} \Longrightarrow \boldsymbol{\beta} = \boldsymbol{0} ,$$

where the last implication follows from $\boldsymbol{V}'\boldsymbol{V}$ being nonsingular. Similarly, premultiplying $\boldsymbol{U\alpha} + \boldsymbol{V\beta} = \boldsymbol{0}$ with $\boldsymbol{U}'$ results in $\boldsymbol{U}'\boldsymbol{U\alpha} = \boldsymbol{0}$ because $\boldsymbol{U}'\boldsymbol{V} = \boldsymbol{O}$. The nonsingularity of $\boldsymbol{U}'\boldsymbol{U}$ now implies that $\boldsymbol{\alpha} = \boldsymbol{0}$. This proves that $\boldsymbol{U\alpha} + \boldsymbol{V\beta} = \boldsymbol{0}$ implies that $\boldsymbol{\alpha} = \boldsymbol{0}$ and $\boldsymbol{\beta} = \boldsymbol{0}$. Hence the columns in $[\boldsymbol{U} : \boldsymbol{V}]$ are linearly independent, which means that $\mathcal{U} \cup \mathcal{V}$ is a linearly independent set. □

The above result can be looked upon as a generalization of Lemma 7.3. To be precise, when $\mathcal{U} = \mathcal{V}$ in Lemma 8.1 we obtain the result that a set of orthogonal vectors is linearly independent (Lemma 7.3).

Having orthogonal columns yields identities such as (8.1), which often assist in simplifying computations. It is important to note that a matrix can have orthogonal columns, but its rows may not be orthogonal. For example, the two columns of $\begin{bmatrix} 1 & -1 \\ 2 & \frac{1}{2} \end{bmatrix}$ are orthogonal, but its rows are not. However, if the columns of a square matrix are *orthonormal*, i.e., orthogonal vectors of unit length, then it can be shown that its row vectors are orthonormal as well. We summarize this below.

Theorem 8.1 *Let $\boldsymbol{Q}$ be an $n \times n$ square matrix. The following are equivalent:*

(i) The columns of $\boldsymbol{Q}$ are orthonormal vectors.

(ii) $\boldsymbol{Q}'\boldsymbol{Q} = \boldsymbol{I} = \boldsymbol{Q}\boldsymbol{Q}'$.

(iii) The rows of $\boldsymbol{Q}$ are orthonormal vectors.

Proof. We will prove that (i) $\Rightarrow$ (ii) $\Rightarrow$ (iii) $\Rightarrow$ (i).

Proof of (i) $\Rightarrow$ (ii): Let $\boldsymbol{Q} = [\boldsymbol{q}_{*1} : \boldsymbol{q}_{*2} : \ldots : \boldsymbol{q}_{*n}]$, where the $\boldsymbol{q}_{*j}$'s are the n orthonormal vectors in $\Re^n$. Then, as in (8.1), $\boldsymbol{Q}'\boldsymbol{Q}$ is a diagonal matrix with j-th diagonal element $\|\boldsymbol{q}_{*j}\|^2 = 1$, which proves $\boldsymbol{Q}'\boldsymbol{Q} = \boldsymbol{I}$. Also, because $\boldsymbol{Q}$ is a square matrix with full column rank (recall from Lemma 7.3 that orthogonal vectors are linearly independent), it is nonsingular. This means that $\boldsymbol{Q}^{-1} = \boldsymbol{Q}'$ and so $\boldsymbol{Q}\boldsymbol{Q}' = \boldsymbol{Q}\boldsymbol{Q}^{-1} = \boldsymbol{I}$. This proves (i) $\Rightarrow$ (ii).

Proof of (ii) $\Rightarrow$ (iii): Since $\boldsymbol{Q}\boldsymbol{Q}' = \boldsymbol{I}$, the (i, j)-th element of $\boldsymbol{Q}\boldsymbol{Q}'$ is δ_{ij}, which is also the inner product of the i-th and j-th row vectors of $\boldsymbol{Q}$ (i.e., $\boldsymbol{q}_{i*}'\boldsymbol{q}_{j*}$). Therefore the rows of $\boldsymbol{Q}$ are orthonormal and we have (ii) $\Rightarrow$ (iii).

Proof of (iii) $\Rightarrow$ (i): The rows of $\boldsymbol{Q}$ are the columns of $\boldsymbol{Q}'$. Therefore, the columns of $\boldsymbol{Q}'$ are orthonormal vectors. Note that we have proved (i) $\Rightarrow$ (ii) $\Rightarrow$ (iii), which means that (i) $\Rightarrow$ (iii). Applying this to $\boldsymbol{Q}'$, we conclude that the rows of $\boldsymbol{Q}'$ are orthonormal vectors. Therefore, the columns of $\boldsymbol{Q}$ are orthonormal. This proves (iii) $\Rightarrow$ (i).

This completes the proof of (i) $\Rightarrow$ (ii) $\Rightarrow$ (iii) $\Rightarrow$ (i). $\square$

The above result leads to the following definition.

Definition 8.1 Orthogonal matrix. *An $n \times n$ square matrix is called* **orthogonal** *if its columns (or rows) are orthonormal vectors (i.e., orthogonal vectors of unit length). Equivalently, a square matrix $\boldsymbol{Q}$ that has its transpose as its inverse, i.e., $\boldsymbol{Q}\boldsymbol{Q}' = \boldsymbol{Q}'\boldsymbol{Q} = \boldsymbol{I}$ is said to be an orthogonal matrix.*

One might call such a matrix an "orthonormal" matrix, but the more conventional name is an "orthogonal" matrix. The columns of an *orthogonal* matrix are *orthonormal* vectors. Theorem 8.1 ensures that a square matrix with orthonormal columns *or* rows is an orthogonal matrix. A square matrix is orthogonal if $\boldsymbol{Q}^{-1} = \boldsymbol{Q}'$.

Example 8.2 Rotation matrices. An important class of orthogonal matrices is that of ***rotation matrices***. Let $\boldsymbol{u} \in \Re^3$. The vector $\boldsymbol{P_x u}$ (see Figure 8.1) is the point obtained by rotating $\boldsymbol{u}$ counterclockwise about the x-axis through an angle θ. We use "counterclockwise" assuming that the origin is the center of the clock.

$$\boldsymbol{P_x} = \begin{bmatrix} 1 & 0 & 0 \\ 0 & \cos\theta & -\sin\theta \\ 0 & \sin\theta & \cos\theta \end{bmatrix}$$

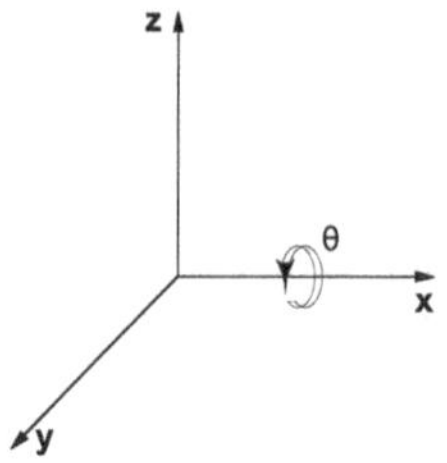

Figure 8.1 *Counterclockwise rotation around the x-axis in $\Re^3$.*

Similarly, $\boldsymbol{P_y u}$ (see Figure 8.2) is the point obtained by rotating $\boldsymbol{u}$ counterclockwise around the y-axis through an angle θ.

$$\boldsymbol{P_y} = \begin{bmatrix} \cos\theta & 0 & \sin\theta \\ 0 & 1 & 0 \\ -\sin\theta & 0 & \cos\theta \end{bmatrix}$$

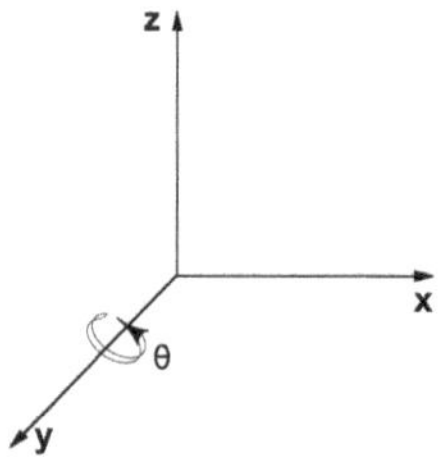

Figure 8.2 *Counterclockwise rotation around the y-axis in $\Re^3$.*

And $\boldsymbol{P_z u}$ (see Figure 8.3) is the point obtained by rotating $\boldsymbol{u}$ counterclockwise around the z-axis through an angle θ.

$$\boldsymbol{P_z} = \begin{bmatrix} \cos\theta & -\sin\theta & 0 \\ \sin\theta & \cos\theta & 0 \\ 0 & 0 & 1 \end{bmatrix}$$

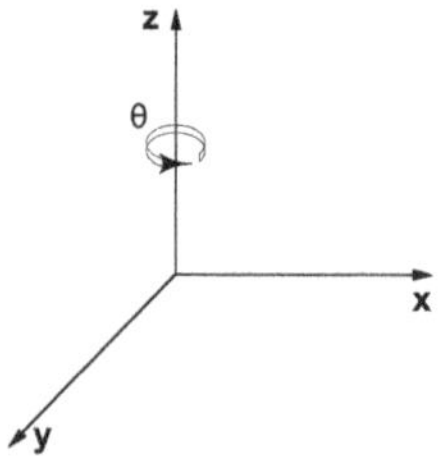

Figure 8.3 *Counterclockwise rotation around the z-axis in $\Re^3$.*

A matrix with orthonormal columns can always be "extended" to an orthogonal matrix in the following sense.

Lemma 8.2 *Let Q_1 be an $n \times p$ matrix with p orthonormal column vectors. Then, there exists an $n \times (n-p)$ matrix Q_2 such that $Q = [Q_1 : Q_2]$ is orthogonal.*

Proof. First of all note that $p \le n$ in the above. Let $Q_1 = [q_1 : q_2 : \ldots : q_p]$, where the q_i's form a set of orthonormal vectors. Part (ii) of Theorem 7.3 tells us that we can find orthonormal vectors $q_{p+1}, q_{p+2}, \ldots, q_n$ such that $\{q_1, q_2, \ldots, q_n\}$ is an orthonormal basis of $\Re^n$. Let $Q_2 = [q_{p+1} : q_{p+2} : \ldots : q_{p+n}]$. Then $Q = [Q_1 : Q_2]$ is $n \times n$ and has n orthonormal columns. Hence Q is an orthogonal matrix. □

This result is the "orthogonal" analogue of Corollary 4.3.

8.2 The QR decomposition

The Gram-Schmidt procedure leads to an extremely important matrix factorization that is of central importance in the study of linear systems (and statistical linear regression models in particular). It says that any matrix with full column rank can be expressed as the product of a matrix with orthogonal columns and an upper-triangular matrix. This factorization is hidden in the Gram-Schmidt procedure. To see how, consider the situation with just two linearly independent vectors x_1 and x_2. The Gram-Schmidt procedure (Theorem 7.2) yields orthogonal vectors y_1 and y_2, where

$$y_1 = x_1 \text{ and } y_2 = x_2 - \text{proj}_{y_1}(x_2) = x_2 - \frac{\langle y_1, x_2\rangle}{\langle y_1, y_1\rangle} y_1 \,. \tag{8.2}$$

Define $q_1 = \dfrac{y_1}{\|y_1\|}$ and $q_2 = \dfrac{y_2}{\|y_2\|}$, which are orthonormal vectors. Now, from (8.2), we can express the x_i's as linear combinations of q_i's as

$$x_1 = r_{11} q_1 \,, \text{ where } r_{11} = \|y_1\| \,,$$
$$x_2 = \frac{\langle y_1, x_2\rangle}{\|y_1\|^2} y_1 + y_2 = \left\langle \frac{y_1}{\|y_1\|}, x_2 \right\rangle \frac{y_1}{\|y_1\|} + y_2 = r_{12} q_1 + r_{22} q_2 \,,$$

where $r_{12} = \left\langle \dfrac{y_1}{\|y_1\|}, x_2 \right\rangle = \langle q_1, x_2\rangle = q_1' x_2$ and $r_{22} = \|y_2\|$. Writing the above in terms of matrices, we obtain

$$[x_1 : x_2] = [q_1 : q_2] \begin{bmatrix} r_{11} & r_{12} \\ 0 & r_{22} \end{bmatrix} .$$

Thus, $X = QR$, where $X = [x_1 : x_2]$ has two linearly independent columns, $Q = [q_1 : q_2]$ is a matrix with orthogonal columns and $R = \begin{bmatrix} r_{11} & r_{12} \\ 0 & r_{22} \end{bmatrix}$ is upper-triangular.

The same strategy yields a QR decomposition for any matrix with full column rank. We present this in the form of the theorem below.

Theorem 8.2 *Let X be an $n \times p$ matrix with $r \le n$ and $\rho(X) = p$ (i.e., X has* *linearly independent columns). Then $\boldsymbol{X}$ can be factored as $\boldsymbol{X} = \boldsymbol{QR}$, where $\boldsymbol{Q}$ is an $n \times p$ matrix whose columns form an orthonormal basis for $\mathcal{C}(\boldsymbol{X})$ and $\boldsymbol{R}$ is an upper-triangular $p \times p$ matrix with positive diagonal entries.*

Proof. Let $\boldsymbol{X} = [\boldsymbol{x}_1 : \boldsymbol{x}_2 : \ldots : \boldsymbol{x}_p]$ be an $n \times p$ matrix whose columns are linearly independent. Suppose we apply the Gram-Schmidt procedure to the columns of $\boldsymbol{X}$ to produce an orthogonal set of vectors $\boldsymbol{y}_1, \boldsymbol{y}_2, \ldots, \boldsymbol{y}_p$ as in Theorem 7.2. Let $\boldsymbol{q}_i = \frac{\boldsymbol{y}_i}{\|\boldsymbol{y}_i\|}$ be the normalized $\boldsymbol{y}_i$'s so that $\boldsymbol{q}_1, \boldsymbol{q}_2, \ldots, \boldsymbol{q}_p$ form an orthonormal basis for $\mathcal{C}(\boldsymbol{X})$. Also, define $r_{ii} = \|\boldsymbol{y}_i\|$ so that $\boldsymbol{y}_i = r_{ii}\boldsymbol{q}_i$ for $i = 1, 2, \ldots, p$.

Consider the k-th step of the Gram-Schmidt procedure in Theorem 7.2. For $k = 1$, we have $\boldsymbol{x}_1 = r_{11}\boldsymbol{q}_1$. Also, since $\text{proj}_{\boldsymbol{y}_i}(\boldsymbol{x}_k) = \text{proj}_{\boldsymbol{q}_i}(\boldsymbol{x}_k)$ (see Corollary 7.3), the k-th step ($2 \le k \le p$) of the Gram-Schmidt procedure in Theorem 7.2 is

$$\begin{aligned} \boldsymbol{x}_k &= \text{proj}_{\boldsymbol{y}_1}(\boldsymbol{x}_k) + \text{proj}_{\boldsymbol{y}_2}(\boldsymbol{x}_k) + \cdots + \text{proj}_{\boldsymbol{y}_{k-1}}(\boldsymbol{x}_k) + r_{kk}\boldsymbol{q}_k \\ &= \text{proj}_{\boldsymbol{q}_1}(\boldsymbol{x}_k) + \text{proj}_{\boldsymbol{q}_2}(\boldsymbol{x}_k) + \cdots + \text{proj}_{\boldsymbol{q}_{k-1}}(\boldsymbol{x}_k) + r_{kk}\boldsymbol{q}_k \\ &= r_{1k}\boldsymbol{q}_1 + r_{2k}\boldsymbol{q}_2 + \cdots + r_{k-1,k}\boldsymbol{q}_{k-1} + r_{kk}\boldsymbol{q}_k \,, \end{aligned} \tag{8.3}$$

where we have defined $r_{ik} = \langle \boldsymbol{q}_i, \boldsymbol{x}_k \rangle = \boldsymbol{q}_i'\boldsymbol{x}_k$ for $i < k$. Writing (8.3) in matrix form by placing the $\boldsymbol{x}_k$'s as columns in a matrix, we obtain

$$[\boldsymbol{x}_1 : \boldsymbol{x}_2 : \ldots : \boldsymbol{x}_p] = [\boldsymbol{q}_1 : \boldsymbol{q}_2 : \ldots : \boldsymbol{q}_p] \begin{bmatrix} r_{11} & r_{12} & r_{13} & \ldots & r_{1p} \\ 0 & r_{22} & r_{23} & \ldots & r_{2p} \\ 0 & 0 & r_{33} & \ldots & r_{3p} \\ \vdots & \vdots & \vdots & \ddots & \vdots \\ 0 & 0 & 0 & \ldots & r_{pp} \end{bmatrix} .$$

Therefore, $\boldsymbol{X} = \boldsymbol{QR}$, where $\boldsymbol{Q} = [\boldsymbol{q}_1 : \boldsymbol{q}_2 : \ldots : \boldsymbol{q}_p]$ is $n \times p$ and $\boldsymbol{R}$ is the $p \times p$ matrix whose (i, j)-th element is given by

$$r_{ij} = \begin{cases} \|\boldsymbol{y}_i\|^2 & \text{if } i < j \,, \\ \boldsymbol{q}_i'\boldsymbol{x}_j & \text{if } i < j \,, \\ 0 & \text{if } i > j \,. \end{cases}$$

The Gram-Schmidt procedure ensures that each $\boldsymbol{y}_i$ is non-null, so each diagonal element of $\boldsymbol{R}$ is nonzero. This shows that $\boldsymbol{R}$ is nonsingular. Therefore, $\mathcal{C}(\boldsymbol{X}) = \mathcal{C}(\boldsymbol{Q})$ and indeed the columns of $\boldsymbol{Q}$ form an orthonormal basis for $\mathcal{C}(\boldsymbol{X})$. $\square$

In practice, the $\boldsymbol{R}$ matrix can be determined only from columns of $\boldsymbol{X}$ and $\boldsymbol{Q}$. As we saw in the above proof, $r_{ij} = \boldsymbol{q}_i'\boldsymbol{x}_j$ if $i < j$ and 0 for $i > j$. The diagonal elements, r_{kk} can also be expressed as $\boldsymbol{q}_k'\boldsymbol{x}_i$. Premultiplying both sides of (8.3) with $\boldsymbol{q}_k'$ yields

$$\boldsymbol{q}_k'\boldsymbol{x}_k = r_{1k}\boldsymbol{q}_k'\boldsymbol{q}_1 + r_{2k}\boldsymbol{q}_k'\boldsymbol{q}_2 + \cdots + r_{k-1,k}\boldsymbol{q}_k'\boldsymbol{q}_{k-1} + r_{kk}\boldsymbol{q}_k'\boldsymbol{q}_k = r_{kk}$$

because the $\boldsymbol{q}_i$'s are orthonormal vectors. This means that the (i, j)-th element of $\boldsymbol{R}$ is given by the inner product between $\boldsymbol{q}_i$ and $\boldsymbol{x}_j$ for elements on or above the diagonal, while they are zero below the diagonal. We write

$$\boldsymbol{X} = \boldsymbol{Q} \begin{bmatrix} \boldsymbol{q}_1'\boldsymbol{x}_1 & \boldsymbol{q}_1'\boldsymbol{x}_2 & \boldsymbol{q}_1'\boldsymbol{x}_3 & \dots & \boldsymbol{q}_1'\boldsymbol{x}_p \\ 0 & \boldsymbol{q}_2'\boldsymbol{x}_2 & \boldsymbol{q}_2'\boldsymbol{x}_3 & \dots & \boldsymbol{q}_2'\boldsymbol{x}_p \\ 0 & 0 & \boldsymbol{q}_3'\boldsymbol{x}_3 & \dots & \boldsymbol{q}_3'\boldsymbol{x}_p \\ \vdots & \vdots & \vdots & \ddots & \vdots \\ 0 & 0 & 0 & \dots & \boldsymbol{q}_p'\boldsymbol{x}_p \end{bmatrix} . \tag{8.4}$$

Example 8.3 Let $\boldsymbol{X} = [\boldsymbol{x}_1 : \boldsymbol{x}_2 : \boldsymbol{x}_3]$ be the matrix formed by placing the three vectors in Example 7.6 as its columns. Let $\boldsymbol{Q} = [\boldsymbol{q}_1 : \boldsymbol{q}_2 : \boldsymbol{q}_3]$ be the orthogonal matrix formed by placing the three orthonormal vectors obtained in Example 7.6 by applying the Gram-Schmidt process to the column vectors of $\boldsymbol{X}$. Therefore,

$$\boldsymbol{X} = \begin{bmatrix} 1 & 3 & 0 \\ 2 & 7 & 1 \\ 3 & 10 & 0 \end{bmatrix} \quad \text{and} \quad \boldsymbol{Q} = \begin{bmatrix} \frac{1}{\sqrt{14}} & -\frac{5}{\sqrt{42}} & \frac{1}{\sqrt{3}} \\ \frac{2}{\sqrt{14}} & \frac{4}{\sqrt{42}} & \frac{1}{\sqrt{3}} \\ \frac{3}{\sqrt{14}} & -\frac{1}{\sqrt{42}} & -\frac{1}{\sqrt{3}} \end{bmatrix} .$$

Suppose we want to find the upper-triangular matrix $\boldsymbol{R}$ such that $\boldsymbol{X} = \boldsymbol{Q}\boldsymbol{R}$. One could simply supply the entries in $\boldsymbol{R}$ according to (8.4). Alternatively, because $\boldsymbol{Q}$ is orthogonal, one could simply compute

$$\boldsymbol{Q}'\boldsymbol{X} = \boldsymbol{Q}'(\boldsymbol{Q}\boldsymbol{R}) = (\boldsymbol{Q}'\boldsymbol{Q})\boldsymbol{R} = \boldsymbol{R} .$$

Thus, in our example

$$\boldsymbol{Q}'\boldsymbol{X} = \begin{bmatrix} \sqrt{14} & \frac{47}{\sqrt{14}} & \frac{2}{\sqrt{14}} \\ 0 & \frac{3}{\sqrt{42}} & \frac{4}{\sqrt{42}} \\ 0 & 0 & \frac{1}{\sqrt{3}} \end{bmatrix} .$$

Note that the entries here are simply computed as $\boldsymbol{q}_i'\boldsymbol{x}_j$'s. The diagonal entries are also equal to the lengths of the vectors (i.e., $\|\boldsymbol{y}_i\|^2$'s for $i = 1, 2, 3$) obtained from the Gram-Schmidt process in Example 7.6. ■

Remark: Since the columns of $\boldsymbol{Q}$ are orthonormal, it is always true that $\boldsymbol{Q}'\boldsymbol{Q} = \boldsymbol{I}$ (see (8.1)). When $p = n$, i.e., $\boldsymbol{X}$ is an $n \times n$ square matrix with full rank, then Theorem 8.2 implies that $\boldsymbol{Q}$ is also a square matrix with orthonormal columns. In that case, we can write $\boldsymbol{Q}^{-1} = \boldsymbol{Q}'$ and in fact $\boldsymbol{Q}$ is an orthogonal matrix satisfying $\boldsymbol{Q}'\boldsymbol{Q} = \boldsymbol{Q}\boldsymbol{Q}' = \boldsymbol{I}$. When $p < n$, $\boldsymbol{X}$ and $\boldsymbol{Q}$ are rectangular matrices (with more rows than columns) and some authors prefer to emphasize the point by calling it the *rectangular* $\boldsymbol{QR}$ decomposition. Remember that $\boldsymbol{R}$ is always a square matrix in a $\boldsymbol{QR}$ decomposition.

The factorization in Theorem 8.2 is unique. The proof is a bit tricky, but once we get the trick, it is short and elegant. The fact that the diagonal elements of the upper-triangular matrix in Theorem 8.2 is positive plays a crucial role in the proof.

Lemma 8.3 *The $\boldsymbol{QR}$ decomposition, as described in Theorem 8.2, is unique.*

Proof. Let $\boldsymbol{X} = \boldsymbol{Q}_1\boldsymbol{R}_1 = \boldsymbol{Q}_2\boldsymbol{R}_2$ be two possibly different $\boldsymbol{QR}$ decompositions of $\boldsymbol{X}$ as described in Theorem 8.2. We will prove that $\boldsymbol{R}_1 = \boldsymbol{R}_2$ which will imply $\boldsymbol{Q}_1 = \boldsymbol{Q}_2$. Since $\boldsymbol{R}_1$ is nonsingular, we can write $\boldsymbol{Q}_1 = \boldsymbol{Q}_2\boldsymbol{T}$, where $\boldsymbol{T} = \boldsymbol{R}_2\boldsymbol{R}_1^{-1}$ is again an upper-triangular matrix with positive diagonal elements. In fact, the i-th diagonal element of $\boldsymbol{T}$ is the product of the i-th diagonal element of $\boldsymbol{R}_2$ divided by the i-th diagonal element of $\boldsymbol{R}_1$ (both of which are positive). Also, it is easily verified that if $\boldsymbol{T}'\boldsymbol{T} = \boldsymbol{I}$ for any upper-triangular matrix $\boldsymbol{T}$ with positive diagonal elements, then $\boldsymbol{T} = \boldsymbol{I}$. Using these facts, we can argue:

$$\begin{aligned}\boldsymbol{Q}_1\boldsymbol{R}_1 = \boldsymbol{Q}_2\boldsymbol{R}_2 &\Longrightarrow \boldsymbol{Q}_1 = \boldsymbol{Q}_2\boldsymbol{T} \Longrightarrow \boldsymbol{T} = \boldsymbol{Q}_2'\boldsymbol{Q}_1 \text{ (since } \boldsymbol{Q}_2'\boldsymbol{Q}_2 = \boldsymbol{I}) \\ &\Longrightarrow \boldsymbol{I} = \boldsymbol{Q}_1'\boldsymbol{Q}_1 = \boldsymbol{T}'\boldsymbol{Q}_2'\boldsymbol{Q}_2\boldsymbol{T} = \boldsymbol{T}'\boldsymbol{T} \Longrightarrow \boldsymbol{T} = \boldsymbol{I}\,.\end{aligned}$$

Since $\boldsymbol{T} = \boldsymbol{R}_2\boldsymbol{R}_1^{-1} = \boldsymbol{I}$, we have $\boldsymbol{R}_2 = \boldsymbol{R}_1$ and $\boldsymbol{Q}_1 = \boldsymbol{Q}_2\boldsymbol{T} = \boldsymbol{Q}_2$. □

The $\boldsymbol{QR}$ decomposition plays as important a role in solving linear systems as does the $\boldsymbol{LU}$ decomposition. Consider a linear system $\boldsymbol{Ax} = \boldsymbol{b}$, where $\boldsymbol{A}$ is an $n \times n$ nonsingular matrix. Recall that once the $\boldsymbol{LU}$ factors (assuming they exist) of the nonsingular matrix $\boldsymbol{A}$ have been found, the solution of $\boldsymbol{Ax} = \boldsymbol{b}$ is easily computed: we first solve $\boldsymbol{Ly} = \boldsymbol{b}$ by forward substitution, and then we solve $\boldsymbol{Ux} = \boldsymbol{y}$ by back substitution. The $\boldsymbol{QR}$ decomposition can be used in a similar manner. If $\boldsymbol{A}$ is nonsingular, then Theorem 8.2 ensures that $\boldsymbol{A} = \boldsymbol{QR}$ with $\boldsymbol{Q}^{-1} = \boldsymbol{Q}'$ and we obtain

$$\boldsymbol{Ax} = \boldsymbol{b} \Longrightarrow \boldsymbol{QRx} = \boldsymbol{b} \Longrightarrow \boldsymbol{Rx} = \boldsymbol{Q}'\boldsymbol{b}\,. \tag{8.5}$$

Therefore, to solve $\boldsymbol{Ax} = \boldsymbol{b}$ using the $\boldsymbol{QR}$ factors of $\boldsymbol{A}$, we first compute $\boldsymbol{y} = \boldsymbol{Q}'\boldsymbol{b}$ and then solve the upper-triangular system $\boldsymbol{Rx} = \boldsymbol{y}$ using back substitution.

While the $\boldsymbol{LU}$ and $\boldsymbol{QR}$ factors can be used in similar fashion to solve nonsingular linear systems, matters become rather different when $\boldsymbol{A}$ is singular or a rectangular matrix with full column rank. Such systems appear frequently in statistical modeling. Unfortunately the $\boldsymbol{LU}$ factors do not exist for singular or rectangular systems. In fact, the system under consideration may even be inconsistent. When $\boldsymbol{A}$ is of full column rank, however, we can still arrive at what is called a ***least squares solution*** for the system. We will take a closer look at least squares elsewhere in the book but, for now, here is the basic idea. Suppose $\boldsymbol{A}$ is $n \times p$ with full column rank p. Then, $\boldsymbol{A}'\boldsymbol{A}$ is $p \times p$ and has rank p, which implies that it is nonsingular. We can now convert the system $\boldsymbol{Ax} = \boldsymbol{b}$ to a nonsingular system by multiplying both sides of the equation with $\boldsymbol{A}'$:

$$\boldsymbol{Ax} = \boldsymbol{b} \Longrightarrow \boldsymbol{A}'\boldsymbol{Ax} = \boldsymbol{A}'\boldsymbol{b} \Longrightarrow \boldsymbol{x} = (\boldsymbol{A}'\boldsymbol{A})^{-1}\boldsymbol{A}'\boldsymbol{b}\,. \tag{8.6}$$

The resulting nonsingular system in (8.6) is often called the ***normal equations*** of linear regression. The $\boldsymbol{QR}$ decomposition is particularly helpful in solving the normal equations: if $\boldsymbol{A} = \boldsymbol{QR}$, then

$$\boldsymbol{A}'\boldsymbol{Ax} = \boldsymbol{A}'\boldsymbol{b} \Longrightarrow \boldsymbol{R}'\boldsymbol{Q}'\boldsymbol{QRx} = \boldsymbol{R}'\boldsymbol{Q}'\boldsymbol{b} \Longrightarrow \boldsymbol{R}'\boldsymbol{Rx} = \boldsymbol{R}'\boldsymbol{Q}'\boldsymbol{b} \Longrightarrow \boldsymbol{Rx} = \boldsymbol{Q}'\boldsymbol{b}\,, \tag{8.7}$$

where the last equality follows because $\boldsymbol{R}'$ is nonsingular. Equation (8.7) again yields the triangular system $Rx = Q'b$, which can be solved using back substitution. Comparing with (8.5), we see that the QR factors of A solve the system $Ax = b$ in the same way as they solve the normal equations.

The LU factors do not exist for singular or rectangular matrices. And while the LU factors will exist for $A'A$ in the normal equations, they are less helpful than the QR factors for solving (8.7). To use the LU factors, we will first need to compute $A'A$ and then apply the LU decomposition to it. This requires computation of $A'A$, which may be less efficient with floating point arithmetic. The QR decomposition avoids this unnecessary computation. The QR factors of an $n \times p$ matrix A will exist whenever A has linearly independent columns, and, as seen in (8.7) provide the least squares solutions from the normal equations in exactly the same way as they provide the solution for a consistent or nonsingular system (as in (8.5)).

8.3 Orthogonal projection and projector

From Theorem 7.4 we know that the sum of a subspace and its orthocomplement is a direct sum. This means that we can meaningfully talk about the projection of a vector into a subspace along its orthocomplement. In fact, part (i) of Theorem 7.4 and Corollary 7.5 ensures that we can have a meaningful definition of such a projection.

Definition 8.2 *If $\mathcal{S}$ is a subspace of some vector space $\mathcal{V}$ and y is any vector in $\mathcal{V}$, then the projection of y into $\mathcal{S}$ along $\mathcal{S}^\perp$ is called the* **orthogonal projection** *of y into $\mathcal{S}$. More explicitly, if $y = u + v$, where $u \in \mathcal{S}$ and $v \in \mathcal{S}^\perp$, then u is the orthogonal projection of y into $\mathcal{S}$.*

The above definition is unambiguous because u is uniquely determined from y. Geometrically, the orthogonal projection u is precisely the foot of the perpendicular dropped from y onto the subspace $\mathcal{S}$. Since $\left(\mathcal{S}^\perp\right)^\perp = \mathcal{S}$, Definition 8.2 applies equally well to the subspace $\mathcal{S}^\perp$ and $y - u = v$ is the orthogonal projection of y into $\mathcal{S}^\perp$.

The geometry of orthogonal projections in $\Re^2$ or $\Re^3$ suggests that u is the point on a line or a plane that is closest to y. This is true for higher-dimensional spaces as well and proved in the following important theorem.

Theorem 8.3 The closest point theorem. *Let $\mathcal{S}$ be a subspace of $\Re^n$ and let y be a vector in $\Re^n$. The orthogonal projection of y into $\mathcal{S}$ is the* **unique** *point in $\mathcal{S}$ that is closest to y. In other words, if u is the orthogonal projection of y into $\mathcal{S}$, then*

$$\|y - u\|^2 \le \|y - w\|^2 \text{ for all } w \in \mathcal{S} ,$$

with equality holding only when $w = u$.

Proof. Let u be the orthogonal projection of y into $\mathcal{S}$ and let w be any vector in $\mathcal{S}$. Note that $v = y - u$ is a vector in $\mathcal{S}^\perp$ and $z = u - w$ is a vector in $\mathcal{S}$. Therefore, $z \perp$

$\boldsymbol{v}$ and the Pythagorean identity (Corollary 7.2) tells us that $\|\boldsymbol{v}+\boldsymbol{z}\|^2 = \|\boldsymbol{v}\|^2+\|\boldsymbol{z}\|^2$. This reveals

$$\begin{aligned}\|\boldsymbol{y}-\boldsymbol{w}\|^2 &= \|(\boldsymbol{y}-\boldsymbol{u})+(\boldsymbol{u}-\boldsymbol{w})\|^2 = \|\boldsymbol{v}+\boldsymbol{z}\|^2 = \|\boldsymbol{v}\|^2+\|\boldsymbol{z}\|^2 \\ &= \|\boldsymbol{y}-\boldsymbol{u}\|^2+\|\boldsymbol{u}-\boldsymbol{w}\|^2 \geq \|\boldsymbol{y}-\boldsymbol{u}\|^2 .\end{aligned}$$

Clearly, equality will hold only when $\boldsymbol{z} = \boldsymbol{u}-\boldsymbol{w} = \boldsymbol{0}$, i.e., when $\boldsymbol{w} = \boldsymbol{u}$. Furthermore, the orthogonal projection is unique, so it is the unique point in $\mathcal{S}$ that is closest to $\boldsymbol{y}$. □

Recall from Section 6.4 that every pair of complementary subspaces defines a projector matrix. When the complementary subspaces happen to be orthogonal complements, the resulting projector is referred to as an *orthogonal projector* and has some special and rather attractive properties.

Definition 8.3 *Let $\Re^n = \mathcal{S}\oplus\mathcal{S}^{\perp}$. An $n\times n$ matrix $\boldsymbol{P}_{\mathcal{S}}$ is said to be the* **orthogonal projector** *into $\mathcal{S}$ if, for every $\boldsymbol{y}\in\Re^n$, $\boldsymbol{P}_{\mathcal{S}}\boldsymbol{y}$ is the projection of $\boldsymbol{y}$ onto $\mathcal{S}$ along $\mathcal{S}^{\perp}$. If $\boldsymbol{X}$ is a matrix whose columns are a basis for $\mathcal{S}$, i.e., $\mathcal{C}(\boldsymbol{X}) = \mathcal{S}$, then we write the orthogonal projector onto $\mathcal{C}(\boldsymbol{X})$ as $\boldsymbol{P}_{\boldsymbol{X}}$.*

A few remarks about the above definition are in order. If we strictly follow the language for general projectors, as in Section 6.4, we should say that $\boldsymbol{P}_{\mathcal{S}}$ is the projector into $\mathcal{S}$ along $\mathcal{S}^{\perp}$. However, if we say that $\boldsymbol{P}_{\mathcal{S}}$ is an *orthogonal projector*, then it is implicit that this projection is "along $\mathcal{S}^{\perp}$." Therefore, we simply write $\boldsymbol{P}_{\mathcal{S}}$ is an orthogonal projector into $\mathcal{S}$, or even just that $\boldsymbol{P}_{\mathcal{S}}$ is an orthogonal projector; the suffix denotes that it projects into $\mathcal{S}$. For most of our subsequent development we will deal with orthogonal projectors into column spaces of matrices with full column rank. Here, again, strictly notation suggests that we write $\boldsymbol{P}_{\mathcal{C}(\boldsymbol{X})}$, but we will abbreviate this as $\boldsymbol{P}_{\boldsymbol{X}}$. And, yes, this projector projects into $\mathcal{C}(\boldsymbol{X})$ along $\mathcal{C}(\boldsymbol{X})^{\perp}$.

It is worth pointing out that when we talk about the projector into the column space of a matrix, we can easily restrict attention to the case when the matrix has linearly independent columns, i.e., full column rank. This is because when the matrix has dependent columns, we can always eliminate them and retain only the linearly independent (basic) columns of the matrix. These columns constitute a basis for the column space and the projector can be computed from them.

Theorem 8.4 *Let $\boldsymbol{y}$ be a vector in $\Re^n$ and let $\boldsymbol{X}$ be an $n\times p$ matrix.*

(i) The orthogonal projection of $\boldsymbol{y}$ into the column space of $\boldsymbol{X}$ is given by the $n\times 1$ vector $\boldsymbol{u} = \boldsymbol{X}\boldsymbol{\beta}$, where $\boldsymbol{\beta}\in\Re^p$ satisfies the **normal equations***:*

$$\boldsymbol{X}'\boldsymbol{X}\boldsymbol{\beta} = \boldsymbol{X}'\boldsymbol{y} . \tag{8.8}$$

(ii) If $\boldsymbol{X}$ has full column rank, i.e., the columns of $\boldsymbol{X}$ form a basis for $\mathcal{C}(\boldsymbol{X})$, then the orthogonal projector into the column space of $\boldsymbol{X}$ is given by

$$\boldsymbol{P}_{\boldsymbol{X}} = \boldsymbol{X}\left(\boldsymbol{X}'\boldsymbol{X}\right)^{-1}\boldsymbol{X}' . \tag{8.9}$$

Proof. **Proof of (i):** Let $y = u + v$, where $u \in \mathcal{C}(X)$ and $v \in \mathcal{C}(X)^\perp$. Since u is the orthogonal projection of y into the subspace spanned by the columns of X, it follows that there exists an $\beta \in \Re^p$ such that $u = X\beta$. Furthermore, $v = y - X\beta$ belongs to $\mathcal{C}(X)^\perp$, which means that it must be orthogonal to every vector in $\mathcal{C}(X)$. In particular, $v \perp x_{*j}$ for $j = 1, 2, \ldots, p$, where x_{*j} is the j-th column of X. Thus, $\langle v, x_{*j} \rangle = x'_{*j} v = x'_{*j}(y - X\beta) = 0$ for $j = 1, 2, \ldots, p$. Writing this system in matrix format yields

$$\begin{bmatrix} x'_{*1} \\ x'_{*2} \\ \vdots \\ x'_{*p} \end{bmatrix} (y - X\beta) = 0 \Longrightarrow X'(y - X\beta) = 0 \Longrightarrow X'X\beta = X'y \; .$$

This proves (i).

Proof of (ii): If X has full column rank, then $p = \rho(X) = \rho(X'X)$. This means that $X'X$ is a $p \times p$ matrix with rank p and, hence, is invertible. Now the normal equations in (8.8) can be uniquely solved to yield

$$\beta = (X'X)^{-1} X'y \Longrightarrow u = X\beta = X(X'X)^{-1} X'y = P_X y \; ,$$

where $P_X = X(X'X)^{-1} X'$. This proves (ii). □

Let $\beta \in \Re^p$ be a solution of the normal equations. It follows from the closest point theorem (Theorem 8.3) that $X\beta = P_X y$ yields the point in the column space of X that is closest to y. In fact, the normal equations emerge quite naturally when we try to minimize the distance of y from points in $\mathcal{C}(X)$. To see what we mean, let $X\theta$ be any point in $\mathcal{C}(X)$ and let $\gamma = \beta - \theta$. We can now write

$$\begin{aligned} \|y - X\theta\|^2 &= \|y - X\beta + X\beta - X\theta\|^2 = (y - X\beta + X\gamma)'(y - X\beta + X\gamma) \\ &= (y - X\beta)'(y - X\beta) + 2\gamma'X'(y - X\beta) + \gamma'X'X\gamma \\ &= \|y - X\beta\|^2 + 2\gamma'X'(y - X\beta) + \|X\gamma\|^2 \; . \end{aligned} \tag{8.10}$$

Now, no matter what the vector θ is, because β satisfies the normal equations we obtain $X'(y - X\beta) = 0$. Therefore, (8.10) yields

$$\|y - X\theta\|^2 = \|y - X\beta\|^2 + \|X\gamma\|^2 \geq \|y - X\beta\|^2 \; ,$$

where equality occurs only when $\gamma = 0$, i.e., $\beta = \theta$. This, in fact, is an alternative derivation of the closest point theorem (Theorem 8.3) that explicitly shows the role played by the normal equations in finding the closest point to y in the column space of X. In practical computations, we first obtain β from the normal equations and then compute $P_X = X\beta$. In Section 8.2 we have seen how the QR decomposition of X can be used efficiently to solve the normal equations in (8.8) when X has full column rank.

The orthogonal projector should not depend upon the basis we use to compute it. The next lemma shows that (8.9) is indeed invariant to such choices.

Lemma 8.4 *Let $\boldsymbol{A}$ and $\boldsymbol{B}$ be two $n \times p$ matrices, both with full column rank and such that $\mathcal{C}(\boldsymbol{A}) = \mathcal{C}(\boldsymbol{B})$. Then $\boldsymbol{P_A} = \boldsymbol{P_B}$.*

Proof. Since $\mathcal{C}(\boldsymbol{A}) = \mathcal{C}(\boldsymbol{B})$, there exists a $p \times p$ nonsingular matrix $\boldsymbol{C}$ such that $\boldsymbol{A} = \boldsymbol{BC}$. Also, since $\boldsymbol{A}$ and $\boldsymbol{B}$ have full column rank, $\boldsymbol{A}'\boldsymbol{A}$ and $\boldsymbol{B}'\boldsymbol{B}$ are nonsingular. Now we can write

$$\begin{aligned}\boldsymbol{P_A} &= \boldsymbol{A}\left(\boldsymbol{A}'\boldsymbol{A}\right)^{-1}\boldsymbol{A}' = \boldsymbol{BC}\left(\boldsymbol{C}'\boldsymbol{B}'\boldsymbol{BC}\right)^{-1}\boldsymbol{C}'\boldsymbol{B}' \\ &= \boldsymbol{BCC}^{-1}\left(\boldsymbol{B}'\boldsymbol{B}\right)^{-1}\boldsymbol{C}'^{-1}\boldsymbol{C}'\boldsymbol{B}' = \boldsymbol{B}\left(\boldsymbol{B}'\boldsymbol{B}\right)^{-1}\boldsymbol{B}' = \boldsymbol{P_B}\,.\end{aligned}$$

This proves that the orthogonal projector is invariant to the choice of basis used for computing it. □

It should be clear from the above developments that $\boldsymbol{P_X}$ is a projector corresponding to a projection and, hence, must satisfy all the properties of general projectors that we saw in Section 6.4. Several properties of the orthogonal projector can be derived directly from the formula in (8.9). We do so in the following theorem.

Theorem 8.5 *Let $\boldsymbol{P_X}$ be the orthogonal projector into $\mathcal{C}(\boldsymbol{X})$, where $\boldsymbol{X}$ is an $n \times p$ matrix with full column rank.*

(i) $\boldsymbol{P_X}$ and $\boldsymbol{I} - \boldsymbol{P_X}$ are both symmetric and idempotent.

(ii) $\boldsymbol{P_X X} = \boldsymbol{X}$ and $(\boldsymbol{I} - \boldsymbol{P_X})\boldsymbol{X} = \boldsymbol{O}$.

(iii) $\mathcal{C}(\boldsymbol{X}) = \mathcal{C}(\boldsymbol{P_X})$ and $\rho(\boldsymbol{P_X}) = \rho(\boldsymbol{X}) = p$.

(iv) $\mathcal{C}(\boldsymbol{I} - \boldsymbol{P_X}) = \mathcal{N}(\boldsymbol{X}') = \mathcal{C}(\boldsymbol{X})^{\perp}$.

(v) $\boldsymbol{I} - \boldsymbol{P_X}$ is the orthogonal projector into $\mathcal{N}(\boldsymbol{X}')$ (or $\mathcal{C}(\boldsymbol{X})^{\perp}$).

Proof. **Proof of (i):** These are direct verifications. For idempotence,

$$\begin{aligned}\boldsymbol{P}_{\boldsymbol{X}}^2 &= \boldsymbol{P_X P_X} = \boldsymbol{X}\left(\boldsymbol{X}'\boldsymbol{X}\right)^{-1}\boldsymbol{X}'\boldsymbol{X}\left(\boldsymbol{X}'\boldsymbol{X}\right)^{-1}\boldsymbol{X}' \\ &= \boldsymbol{X}\left(\boldsymbol{X}'\boldsymbol{X}\right)^{-1}\left(\boldsymbol{X}'\boldsymbol{X}\right)\left(\boldsymbol{X}'\boldsymbol{X}\right)^{-1}\boldsymbol{X}' = \boldsymbol{X}\left(\boldsymbol{X}'\boldsymbol{X}\right)^{-1}\boldsymbol{X}' = \boldsymbol{P_X}\,.\end{aligned}$$

For symmetry, note that $\boldsymbol{X}'\boldsymbol{X}$ is symmetric and so is its inverse. Therefore,

$$\boldsymbol{P}'_{\boldsymbol{X}} = \left[\boldsymbol{X}\left(\boldsymbol{X}'\boldsymbol{X}\right)^{-1}\boldsymbol{X}'\right]' = [\boldsymbol{X}']'\left[\left(\boldsymbol{X}'\boldsymbol{X}\right)^{-1}\right]'[\boldsymbol{X}]' = \boldsymbol{X}\left(\boldsymbol{X}'\boldsymbol{X}\right)^{-1}\boldsymbol{X}' = \boldsymbol{P_X}\,.$$

For $\boldsymbol{I} - \boldsymbol{P_X}$, we use the idempotence of $\boldsymbol{P_X}$ to obtain

$$(\boldsymbol{I} - \boldsymbol{P_X})^2 = \boldsymbol{I} - 2\boldsymbol{P_X} + \boldsymbol{P}_{\boldsymbol{X}}^2 = \boldsymbol{I} - 2\boldsymbol{P_X} + \boldsymbol{P_X} = \boldsymbol{I} - \boldsymbol{P_X}$$

and the symmetry of $\boldsymbol{P_X}$ to obtain

$$(\boldsymbol{I} - \boldsymbol{P_X})' = \boldsymbol{I} - \boldsymbol{P}'_{\boldsymbol{X}} = \boldsymbol{I} - \boldsymbol{P_X}\,.$$

This proves (i).

Proof of (ii): These, again, are easy verifications:

$$\boldsymbol{P_X X} = \boldsymbol{X}\left(\boldsymbol{X}'\boldsymbol{X}\right)^{-1}\boldsymbol{X}'\boldsymbol{X} = \boldsymbol{X}\left(\boldsymbol{X}'\boldsymbol{X}\right)^{-1}\left(\boldsymbol{X}'\boldsymbol{X}\right) = \boldsymbol{X}$$

and use this to obtain

$$(\boldsymbol{I}-\boldsymbol{P_X})\boldsymbol{X}=\boldsymbol{X}-\boldsymbol{P_X}\boldsymbol{X}=\boldsymbol{X}-\boldsymbol{X}=\boldsymbol{O}\,.$$

Proof of (iii): This follows from

$$\mathcal{C}(\boldsymbol{P_X})=\mathcal{C}(\boldsymbol{XB})\subseteq\mathcal{C}(\boldsymbol{X})=\mathcal{C}(\boldsymbol{P_X}\boldsymbol{X})\subseteq\mathcal{C}(\boldsymbol{P_X})\,,$$

where $\boldsymbol{B}=(\boldsymbol{X}'\boldsymbol{X})^{-1}\boldsymbol{X}'$. And $\rho(\boldsymbol{P_X})=\dim(\mathcal{C}(\boldsymbol{P_X}))=\dim(\mathcal{C}(\boldsymbol{X}))=\rho(\boldsymbol{X})=p$.

Proof of (iv): Let $\boldsymbol{u}\in\mathcal{C}(\boldsymbol{I}-\boldsymbol{P_X})$. Then $\boldsymbol{u}=(\boldsymbol{I}-\boldsymbol{P_X})\boldsymbol{\alpha}$ for some vector $\boldsymbol{\alpha}\in\Re^n$ and we have

$$\boldsymbol{X}'\boldsymbol{u}=\boldsymbol{X}'(\boldsymbol{I}-\boldsymbol{P_X})\boldsymbol{\alpha}=[(\boldsymbol{I}-\boldsymbol{P_X})\boldsymbol{X}]'\,\boldsymbol{\alpha}=\boldsymbol{O}\boldsymbol{\alpha}=\boldsymbol{0}\,.$$

Therefore, $\boldsymbol{u}\in\mathcal{N}(\boldsymbol{X}')$ and we have $\mathcal{C}(\boldsymbol{I}-\boldsymbol{P_X})\subset\mathcal{N}(\boldsymbol{X}')$.

Next, suppose $\boldsymbol{u}\in\mathcal{N}(\boldsymbol{X}')$ so $\boldsymbol{X}'\boldsymbol{u}=\boldsymbol{0}$ and so $\boldsymbol{P_X}\boldsymbol{u}=\boldsymbol{X}(\boldsymbol{X}'\boldsymbol{X})^{-1}\boldsymbol{X}'\boldsymbol{u}=\boldsymbol{0}$. Therefore, we can write

$$\boldsymbol{u}=\boldsymbol{u}-\boldsymbol{P_X}\boldsymbol{u}=(\boldsymbol{I}-\boldsymbol{P_X})\boldsymbol{u}\in\mathcal{C}(\boldsymbol{I}-\boldsymbol{P_X})\,.$$

This proves that $\mathcal{N}(\boldsymbol{X}')\subseteq\mathcal{C}(\boldsymbol{I}-\boldsymbol{P_X})$. We already showed the reverse inclusion was true and so $\mathcal{C}(\boldsymbol{I}-\boldsymbol{P_X})=\mathcal{N}(\boldsymbol{X}')$. Also,

$$\boldsymbol{u}\in\mathcal{N}(\boldsymbol{X}')\iff\boldsymbol{X}'\boldsymbol{u}=\boldsymbol{0}\iff\boldsymbol{u}'\boldsymbol{X}=\boldsymbol{0}'\iff\boldsymbol{u}\in\mathcal{C}(\boldsymbol{X})^{\perp}\,,$$

which shows $\mathcal{N}(\boldsymbol{X}')=\mathcal{C}(\boldsymbol{X})^{\perp}$.

Proof of (v): Clearly we can write $\boldsymbol{y}=(\boldsymbol{I}-\boldsymbol{P_X})\boldsymbol{y}+\boldsymbol{P_X}\boldsymbol{y}$. Clearly $\boldsymbol{P_X}\boldsymbol{y}$ and $(\boldsymbol{I}-\boldsymbol{P_X})\boldsymbol{y}$ are orthogonal because

$$\langle\boldsymbol{P_X}\boldsymbol{y},(\boldsymbol{I}-\boldsymbol{P_X})\boldsymbol{y}\rangle=\boldsymbol{y}'(\boldsymbol{I}-\boldsymbol{P_X})\boldsymbol{P_X}\boldsymbol{y}=\boldsymbol{y}'\boldsymbol{O}\boldsymbol{y}=0.$$

Therefore, $(\boldsymbol{I}-\boldsymbol{P_X})$ is the orthogonal projector into $\mathcal{C}(\boldsymbol{I}-\boldsymbol{P_X})$. And from part (iv) we know that $\mathcal{C}(\boldsymbol{I}-\boldsymbol{P_X})=\mathcal{N}(\boldsymbol{X}')=\mathcal{C}(\boldsymbol{X})^{\perp}$. □

Clearly, an orthogonal projector satisfies the definition of a projector (i.e., it is idempotent) and, therefore, enjoys all its properties. Orthogonality, in addition, ensures that the projector is symmetric. In fact, any symmetric and idempotent matrix must be an orthogonal projector.

Theorem 8.6 *The following statements about an $n\times n$ matrix $\boldsymbol{P}$ are equivalent:*

(i) $\boldsymbol{P}$ is an orthogonal projector.

(ii) $\boldsymbol{P}'=\boldsymbol{P}'\boldsymbol{P}$.

(iii) $\boldsymbol{P}$ is symmetric and idempotent.

Proof. **Proof of (i) ⇒ (ii):** Since $\boldsymbol{P}$ is an orthogonal projector, we have $\mathcal{C}(\boldsymbol{P})\perp\mathcal{C}(\boldsymbol{I}-\boldsymbol{P})$. Therefore, $\langle\boldsymbol{Px},(\boldsymbol{I}-\boldsymbol{P})\boldsymbol{y}\rangle=0$ for all vectors $\boldsymbol{x}$ and $\boldsymbol{y}$ in $\Re^n$ and we conclude that

$$\boldsymbol{x}'\boldsymbol{P}'(\boldsymbol{I}-\boldsymbol{P})\boldsymbol{y}=0\quad\forall\quad\boldsymbol{x},\boldsymbol{y}\in\Re^n\Longrightarrow\boldsymbol{P}'(\boldsymbol{I}-\boldsymbol{P})=\boldsymbol{O}\Longrightarrow\boldsymbol{P}'=\boldsymbol{P}'\boldsymbol{P}\,.$$

Proof of (ii) $\Rightarrow$ (iii): Since $\boldsymbol{P}'\boldsymbol{P}$ is a symmetric matrix, it follows from (ii) that

$$\boldsymbol{P}' = \boldsymbol{P}'\boldsymbol{P} = (\boldsymbol{P}'\boldsymbol{P})' = (\boldsymbol{P}')' = \boldsymbol{P} .$$

This proves that $\boldsymbol{P}$ is symmetric. To see why $\boldsymbol{P}$ is idempotent, use the symmetry of $\boldsymbol{P}$ and (ii) to argue that

$$\boldsymbol{P} = \boldsymbol{P}' = \boldsymbol{P}'\boldsymbol{P} = \boldsymbol{P}\boldsymbol{P} = \boldsymbol{P}^2 .$$

Proof of (iii) $\Rightarrow$ (i): Any vector $\boldsymbol{y} \in \Re^n$ can be expressed as $\boldsymbol{y} = \boldsymbol{u} + \boldsymbol{v}$, where $\boldsymbol{u} = \boldsymbol{P}\boldsymbol{y}$ and $\boldsymbol{v} = (\boldsymbol{I} - \boldsymbol{P})\boldsymbol{y}$. Note that if $\boldsymbol{P}$ is symmetric and idempotent, then

$$\boldsymbol{P}'(\boldsymbol{I} - \boldsymbol{P}) = \boldsymbol{P}(\boldsymbol{I} - \boldsymbol{P}) = \boldsymbol{P} - \boldsymbol{P}^2 = \boldsymbol{O} .$$

This means that $\boldsymbol{u}'\boldsymbol{v} = \boldsymbol{y}'\boldsymbol{P}'(\boldsymbol{I} - \boldsymbol{P})\boldsymbol{y} = 0$ and, hence, $\boldsymbol{u} \perp \boldsymbol{v}$. It follows that $\boldsymbol{u}$ is the orthogonal projection of $\boldsymbol{y}$ into $\mathcal{C}(\boldsymbol{P})$ and that $\boldsymbol{P}$ is the corresponding orthogonal projector. □

8.4 Orthogonal projector: Alternative derivations

The formula in (8.9) is consistent with that of the elementary orthogonal projector in Definition 7.7. We obtain the elementary orthogonal projector when $p = 1$ in Theorem 8.4. If $\boldsymbol{q}$ is a vector of unit length, then the orthogonal projector formula in (8.9) produces the elementary orthogonal projector into $\boldsymbol{q}$:

$$\boldsymbol{P}_{\boldsymbol{q}} = \boldsymbol{q}(\boldsymbol{q}'\boldsymbol{q})^{-1}\boldsymbol{q}' = \frac{1}{\|\boldsymbol{q}\|^2}\boldsymbol{q}\boldsymbol{q}' = \boldsymbol{q}\boldsymbol{q}' .$$

The expression for the orthogonal projector in (8.9) becomes especially simple when we consider projecting onto spans of orthonormal vectors. In particular, if $\boldsymbol{Q}$ is an $n \times p$ matrix with orthonormal columns, then $\boldsymbol{Q}'\boldsymbol{Q} = \boldsymbol{I}_p$ and the orthogonal projector into the space spanned by the columns of $\boldsymbol{Q}$ becomes

$$\boldsymbol{P}_{\boldsymbol{Q}} = \boldsymbol{Q}\left(\boldsymbol{Q}'\boldsymbol{Q}\right)^{-1}\boldsymbol{Q}' = \boldsymbol{Q}\boldsymbol{Q}' . \tag{8.11}$$

The expression in (8.11) can be derived independently of (8.9). In fact, we can go in the reverse direction: we can first derive (8.11) directly from an orthogonal basis representation and then derive the more general expression in (8.9) using the $\boldsymbol{QR}$ decomposition. We now describe this route.

Recall the proof of part (i) of Theorem 7.4. Suppose $\dim(\mathcal{V}) = n$ and $\boldsymbol{q}_1, \boldsymbol{q}_2, \ldots, \boldsymbol{q}_n$ is an orthonormal basis for $\mathcal{V}$ such that $\boldsymbol{q}_1, \boldsymbol{q}_2, \ldots, \boldsymbol{q}_p$, where $p \leq n$, is an orthonormal basis for $\mathcal{S}$, then

$$\begin{aligned} \boldsymbol{y} &= \underbrace{\sum_{i=1}^{p}\langle \boldsymbol{y}, \boldsymbol{q}_i\rangle \boldsymbol{q}_i}_{} + \underbrace{\sum_{i=p+1}^{n}\langle \boldsymbol{y}, \boldsymbol{q}_i\rangle \boldsymbol{q}_i}_{} \\ &= \qquad \boldsymbol{u} \qquad + \qquad \boldsymbol{v} , \end{aligned}$$

where $\boldsymbol{u}$ is the orthogonal projection of $\boldsymbol{y}$ into $\mathcal{S}$. This can also be expressed in terms of the proj function:

$$u = \sum_{i=1}^{p} \langle y, q_i \rangle q_i = \sum_{i=1}^{p} \text{proj}_{q_i}(y) = \sum_{i=1}^{p} P_{q_i} y \,, \tag{8.12}$$

where P_{q_i} is the elementary orthogonal projector into q_i. Equation (7.12) shows that the orthogonal projection of y into the span of a set of orthonormal vectors is simply the sum of the individual vector projections of y into each of the orthonormal vectors. Now (8.12) yields

$$\begin{aligned} u &= \sum_{i=1}^{p} \text{proj}_{q_i}(y) = \sum_{i=1}^{p} (q_i' y) q_i = \sum_{i=1}^{p} q_i (q_i' y) = \sum_{i=1}^{p} (q_i q_i') y \\ &= [q_1 : q_2 : \ldots : q_p] \begin{bmatrix} q_1' \\ q_2' \\ \vdots \\ q_p' \end{bmatrix} y = QQ'y = P_Q y \,, \end{aligned}$$

where $Q = [q_1 : q_2 : \ldots : q_p]$. This shows that $P_Q = QQ'$ (as we saw in (8.11)). One other point emerges from these simple but elegant manipulations: the orthogonal projector into a span of orthonormal vectors is the sum of the elementary orthogonal projectors onto the individual vectors. In other words,

$$P_Q = QQ' = \sum_{i=1}^{p} q_i q_i' = \sum_{i=1}^{p} P_{q_i} \,. \tag{8.13}$$

We can now construct an orthogonal projector into the span of linearly independent, but not necessarily orthogonal, vectors by applying a Gram-Schmidt process to the linearly independent vectors. More precisely, let X be an $n \times p$ matrix with p linearly independent columns and let us compute the QR decomposition of X, which results from applying the Gram-Schmidt process to the columns of X (see Section 8.2). Thus we obtain $X = QR$, where Q is an $n \times p$ matrix with orthonormal columns (hence $Q'Q = I_p$) and R is a $p \times p$ upper-triangular matrix with rank p. Since R is nonsingular, we have $\mathcal{C}(X) = \mathcal{C}(Q)$; indeed the columns of Q form an orthonormal basis for $\mathcal{C}(X)$. Also note that $Q = XR^{-1}$ and $X'X = R'Q'QR = R'R$. Therefore, we can construct P_X as

$$P_X = P_Q = QQ' = XR^{-1}R'^{-1}X' = X\left(R'R\right)^{-1}X' = X\left(X'X\right)^{-1}X' \,,$$

which is exactly the same expression as in (8.9).

Let us turn to another derivation. In Theorem 8.6 we saw that every symmetric and idempotent matrix P is an orthogonal projector. Without assuming any explicit form for P, we can argue that every symmetric and idempotent matrix P must be of the form in (8.9). To make matters more precise, let P be an $n \times n$ matrix that is symmetric and idempotent and has rank p. Let X be the $n \times p$ matrix whose columns form a basis for $\mathcal{C}(P)$. Then, $P = XB$ for some $p \times n$ matrix B that is of full row rank. Since P is symmetric and idempotent, it follows that $P = P'P$ and we have

$XB = B'X'XB$. Now, since B is of full row rank, it has a right inverse, which means that we can cancel B on the right and obtain $X = B'X'X$. Since X is of full column rank, $X'X$ is nonsingular and so $X(X'X)^{-1} = B'$. Therefore,

$$P = P' = B'X' = X(X'X)^{-1}X' .$$

Finally, we turn to yet another derivation of (8.9). Our discussion of orthogonal projectors so far has, for the most part, been independent of the form of general (oblique) projectors that we derived in Section 6.4. Recall the setting (and notations) in Lemma 6.7. Let us assume now that $\mathcal{S}$ and $\mathcal{T}$ are not just complements, but orthocomplements, i.e., $\mathcal{S} \perp \mathcal{T}$. This means that $S'T = O$ (see Lemma 7.8). With $A = [S : T]$ and $A^{-1} = \begin{bmatrix} U' \\ V' \end{bmatrix}$, Lemma 6.7 tells us that the projector is $P = SU'$. We now argue that

$$SU' + TV' = AA^{-1} = I \Longrightarrow S'SU' + S'TV' = S' \Longrightarrow S'SU' = S'$$
$$\Longrightarrow U' = (S'S)^{-1} S' \Longrightarrow P = SU' = S(S'S)^{-1} S' .$$

Thus, we obtain the expression in (8.9). Note that $S'S$ is invertible because S has linearly independent columns by construction (see Lemma 6.7). In fact, the formula for the general projector in Theorem 6.8 (see also (7.16)) immediately yields the formula for the orthogonal projector because when $\mathcal{S} \perp \mathcal{T}$, the columns of S are a basis for the orthocomplement space of $\mathcal{T}$ and, hence the matrix V in Theorem 6.8, or in (7.16), can be chosen to be equal to S.

8.5 Sum of orthogonal projectors

This section explores some important relationships concerning the sums of orthogonal projectors. This, as we see, is also related to deriving projectors for partitioned matrices. Let us start with a fairly simple and intuitive lemma.

Lemma 8.5 *Let $A = [A_1 : A_2]$ be a matrix with full column rank. The following statements are equivalent.*

(i) $\mathcal{C}(A_1) \perp \mathcal{C}(A_2)$.

(ii) $P_A = P_{A_1} + P_{A_2}$.

(iii) $P_{A_1}P_{A_2} = P_{A_2}P_{A_1} = O$.

Proof. We will prove that (i) $\Longrightarrow$ (ii) $\Longrightarrow$ (iii) $\Longrightarrow$ (i).

Proof of (i) $\Longrightarrow$ (ii): Since $\mathcal{C}(A_1) \perp \mathcal{C}(A_2)$, it follows from Lemma 7.8 that $A_1'A_2 = A_2'A_1 = O$. Also, because A has full column rank so does A_1 and A_2, which means $A_1'A_1$ and $A_2'A_2$ are both nonsingular. Therefore,

$$\begin{aligned} P_A &= [A_1 : A_2] \begin{bmatrix} A_1'A_1 & A_1'A_2 \\ A_2'A_1 & A_2'A_2 \end{bmatrix}^{-1} \begin{bmatrix} A_1' \\ A_2' \end{bmatrix} \\ &= [A_1 : A_2] \begin{bmatrix} (A_1'A_1)^{-1} & O \\ O & (A_2'A_2)^{-1} \end{bmatrix} \begin{bmatrix} A_1' \\ A_2' \end{bmatrix} \\ &= A_1(A_1'A_1)^{-1}A_1' + A_2(A_2'A_2)^{-1}A_2' = P_{A_1} + P_{A_2} . \end{aligned}$$

Proof of (ii) $\Longrightarrow$ (iii): This follows from Cochran's Theorem (Theorem 6.9) for any projector (not just), but here is another proof. If $P_{A_1}+P_{A_2} = P_A$, then $P_{A_1}+P_{A_2}$ is an symmetric and idempotent. We use only the idempotence below, so this part is true for any projector (not just orthogonal).

Since P_{A_1}, P_{A_2} and $P_{A_1} + P_{A_2}$ are all idempotent, we can write

$$\begin{aligned} &P_{A_1} + P_{A_2} = (P_{A_1} + P_{A_2})^2 = P_{A_1} + P_{A_2} + P_{A_1}P_{A_2} + P_{A_2}P_{A_1} \\ &\Longrightarrow P_{A_1}P_{A_2} + P_{A_2}P_{A_1} = O . \end{aligned} \tag{8.14}$$

Now, multiply (8.14) by P_{A_1} *both* on the left and the right to obtain

$$P_{A_1}(P_{A_1}P_{A_2} + P_{A_2}P_{A_1}) = O \text{ and } (P_{A_1}P_{A_2} + P_{A_2}P_{A_1})P_{A_1} = O .$$

The above two equations yield $P_{A_1}P_{A_2} = -P_{A_1}P_{A_2}P_{A_1} = P_{A_2}P_{A_1}$. It now follows from (8.14) that $P_{A_1}P_{A_2} = P_{A_2}P_{A_1} = O$.

Proof of (iii) $\Longrightarrow$ (i): Using basic properties of orthogonal projectors,

$$P_{A_1}P_{A_2} = O \Longrightarrow A_1'P_{A_1}P_{A_2}A_2 = O \Longrightarrow A_1'A_2 = O \Longrightarrow \mathcal{C}(A_1) \perp \mathcal{C}(A_2) ,$$

where the last implication is true by virtue of Lemma 7.8. □

The following theorem generalizes the preceding lemma to settings where the column spaces of the two submatrices are not necessarily orthocomplementary.

Theorem 8.7 *Let $X = [X_1 : X_2]$ be a matrix of full column rank and let $Z = (I - P_{X_1})X_2$. Then:*

(i) $\mathcal{C}(X_1) \perp \mathcal{C}(Z)$.

(ii) $\mathcal{C}(X) = \mathcal{C}([X_1 : Z]) = \mathcal{C}(X_1) \oplus \mathcal{C}(Z)$.

(iii) $P_X = P_{X_1} + P_Z$.

Proof. Let X be an $n \times p$ matrix, where X_1 is $n \times p_1$, X_2 is $n \times p_2$ and $p_1 + p_2 = p$. Note that Z is $n \times p_2$.

Proof of (i): By virtue of Lemma 7.8, it is enough to verify that $X_1'Z = O$. Since $X_1'(I - P_{X_1}) = O$, this is immediate:

$$X_1'Z = X_1'(I - P_{X_1})X_2 = OX_2 = O .$$

Proof of (ii): We will first prove that $\mathcal{C}(\boldsymbol{X}) = \mathcal{C}(\boldsymbol{X}_1) + \mathcal{C}(\boldsymbol{Z})$. Let $\boldsymbol{u} \in \mathcal{C}(\boldsymbol{X}_1) + \mathcal{C}(\boldsymbol{Z})$. This means that we can find vectors $\boldsymbol{\alpha} \in \Re^{p_1}$ and $\boldsymbol{\beta} \in \Re^{p_2}$ such that

$$\begin{aligned}\boldsymbol{u} &= \boldsymbol{X}_1\boldsymbol{\alpha} + \boldsymbol{Z}\boldsymbol{\beta} = \boldsymbol{X}_1\boldsymbol{\alpha} + (\boldsymbol{I} - \boldsymbol{P}_{\boldsymbol{X}_1})\boldsymbol{X}_2\boldsymbol{\beta} = \boldsymbol{X}_1\boldsymbol{\alpha} - \boldsymbol{P}_{\boldsymbol{X}_1}\boldsymbol{X}_2\boldsymbol{\beta} + \boldsymbol{X}_2\boldsymbol{\beta} \\ &= \boldsymbol{X}_1\left(\boldsymbol{\alpha} - (\boldsymbol{X}_1'\boldsymbol{X}_1)^{-1}\boldsymbol{X}_1'\boldsymbol{X}_2'\boldsymbol{\beta}\right) + \boldsymbol{X}_2\boldsymbol{\beta} = \boldsymbol{X}_1\boldsymbol{\gamma} + \boldsymbol{X}_2\boldsymbol{\beta} \in \mathcal{C}(\boldsymbol{X}),\end{aligned}$$

where $\boldsymbol{\gamma} = \boldsymbol{\alpha} - (\boldsymbol{X}_1'\boldsymbol{X}_1)^{-1}\boldsymbol{X}_1'\boldsymbol{X}_2'\boldsymbol{\beta}$ is a $p_1 \times 1$ vector in the above. This proves that $\mathcal{C}(\boldsymbol{X}_1) + \mathcal{C}(\boldsymbol{Z}) \subseteq \mathcal{C}(\boldsymbol{X})$.

To prove the reverse inclusion, suppose $\boldsymbol{u} \in \mathcal{C}(\boldsymbol{X})$. Then, we can find vectors $\boldsymbol{\alpha} \in \Re^{p_1}$ and $\boldsymbol{\beta} \in \Re^{p_2}$ such that $\boldsymbol{u} = \boldsymbol{X}_1\boldsymbol{\alpha} + \boldsymbol{X}_2\boldsymbol{\beta}$. Writing $\boldsymbol{X}_2 = \boldsymbol{P}_{\boldsymbol{X}_1}\boldsymbol{X}_2 + (\boldsymbol{I} - \boldsymbol{P}_{\boldsymbol{X}_1}\boldsymbol{X}_2)$, we obtain

$$\begin{aligned}\boldsymbol{u} &= \boldsymbol{X}_1\boldsymbol{\alpha} + \boldsymbol{X}_2\boldsymbol{\beta} = \boldsymbol{X}_1\boldsymbol{\alpha} + \boldsymbol{P}_{\boldsymbol{X}_1}\boldsymbol{X}_2\boldsymbol{\beta} + (\boldsymbol{I} - \boldsymbol{P}_{\boldsymbol{X}_1}\boldsymbol{X}_2)\boldsymbol{\beta} \\ &= \boldsymbol{X}_1\left(\boldsymbol{\alpha} + (\boldsymbol{X}_1'\boldsymbol{X}_1)^{-1}\boldsymbol{X}_1'\boldsymbol{X}_2'\boldsymbol{\beta}\right) + \boldsymbol{Z}\boldsymbol{\beta} = \boldsymbol{X}_1\boldsymbol{\theta} + \boldsymbol{Z}\boldsymbol{\beta} \in \mathcal{C}(\boldsymbol{X}_1) + \mathcal{C}(\boldsymbol{Z}),\end{aligned}$$

where $\boldsymbol{\theta} = \boldsymbol{\alpha} + (\boldsymbol{X}_1'\boldsymbol{X}_1)^{-1}\boldsymbol{X}_1'\boldsymbol{X}_2'\boldsymbol{\beta}$ in the above. This proves $\mathcal{C}(\boldsymbol{X}) \subseteq \mathcal{C}(\boldsymbol{X}_1) + \mathcal{C}(\boldsymbol{Z})$ and, hence, that $\mathcal{C}(\boldsymbol{X}) = \mathcal{C}(\boldsymbol{X}_1) + \mathcal{C}(\boldsymbol{Z})$.

From Theorem 4.9, we know that $\mathcal{C}([\boldsymbol{X}_1 : \boldsymbol{Z}]) = \mathcal{C}(\boldsymbol{X}_1) + \mathcal{C}(\boldsymbol{Z})$. Therefore, $\mathcal{C}(\boldsymbol{X}) = \mathcal{C}([\boldsymbol{X}_1 : \boldsymbol{Z}])$.

Finally, the sum $\mathcal{C}(\boldsymbol{X}_1) + \mathcal{C}(\boldsymbol{Z})$ is direct because $\mathcal{C}(\boldsymbol{X}_1) \perp \mathcal{C}(\boldsymbol{Z})$.

Proof of (iii): We can now apply Lemma 8.5 to $[\boldsymbol{X}_1 : \boldsymbol{Z}]$ and obtain $\boldsymbol{P}_{\boldsymbol{X}} = \boldsymbol{P}_{[\boldsymbol{X}_1:\boldsymbol{Z}]} = \boldsymbol{P}_{\boldsymbol{X}_1} + \boldsymbol{P}_{\boldsymbol{Z}}$. □

Part (iii) of Theorem 8.7 also follows easily from the $\boldsymbol{QR}$ decomposition of $\boldsymbol{X}$. Consider the partitioned $n \times p$ matrix $\boldsymbol{X} = [\boldsymbol{X}_1 : \boldsymbol{X}_2]$, where $\boldsymbol{X}_1$ is $n \times p_1$, $\boldsymbol{X}_2$ is $n \times p_2$ and $p_1 + p_2 = p$. Suppose $\boldsymbol{X} = \boldsymbol{QR}$ is the $\boldsymbol{QR}$ decomposition of this matrix, so $\boldsymbol{P}_{\boldsymbol{X}} = \boldsymbol{QQ}'$. We partition $\boldsymbol{X}$ and its $\boldsymbol{QR}$ decomposition in the following conformable manner and deduce expressions for $\boldsymbol{X}_1$ and $\boldsymbol{X}_2$:

$$\begin{aligned}&\boldsymbol{X} = [\boldsymbol{X}_1 : \boldsymbol{X}_2] = [\boldsymbol{Q}_1 : \boldsymbol{Q}_2]\begin{bmatrix}\boldsymbol{R}_{11} & \boldsymbol{R}_{12} \\ \boldsymbol{O} & \boldsymbol{R}_{22}\end{bmatrix} \\ &\Longrightarrow \boldsymbol{X}_1 = \boldsymbol{Q}_1\boldsymbol{R}_{11} \text{ and } \boldsymbol{X}_2 = \boldsymbol{Q}_1\boldsymbol{R}_{12} + \boldsymbol{Q}_2\boldsymbol{R}_{22}. \qquad (8.15)\end{aligned}$$

Here $\boldsymbol{Q}_1$ and $\boldsymbol{Q}_2$ have the same dimensions as $\boldsymbol{X}_1$ and $\boldsymbol{X}_2$, respectively, $\boldsymbol{R}_{11}$ is $p_1 \times p_1$ upper-triangular, $\boldsymbol{R}_{22}$ is $p_2 \times p_2$ upper-triangular and $\boldsymbol{R}_{12}$ is $p_1 \times p_2$ (and not necessarily upper-triangular). Also note that since $\boldsymbol{Q}$ has orthonormal columns, $\boldsymbol{Q}_1'\boldsymbol{Q}_2 = \boldsymbol{Q}_2'\boldsymbol{Q}_1 = \boldsymbol{O}$. The expression for $\boldsymbol{X}_1$ in 8.15 reveals that the columns of $\boldsymbol{Q}_1$ form an orthonormal basis for $\mathcal{C}(\boldsymbol{X}_1)$ and, hence, $\boldsymbol{P}_{\boldsymbol{X}_1} = \boldsymbol{Q}_1\boldsymbol{Q}_1'$. Also, from the expression for $\boldsymbol{X}_2$ in (8.15) we obtain

$$\boldsymbol{Q}_1'\boldsymbol{X}_2 = \boldsymbol{Q}_1'\boldsymbol{Q}_1\boldsymbol{R}_{12} + \boldsymbol{Q}_1'\boldsymbol{Q}_2\boldsymbol{R}_{22} = \boldsymbol{R}_{12},$$

which means that $\boldsymbol{Q}_1\boldsymbol{R}_{12} = \boldsymbol{Q}_1\boldsymbol{Q}_1'\boldsymbol{X}_2 = \boldsymbol{P}_{\boldsymbol{X}_1}\boldsymbol{X}_2$. Substituting this back into the expression for $\boldsymbol{X}_2$ from (8.15) yields

$$\boldsymbol{X}_2 = \boldsymbol{P}_{\boldsymbol{X}_1}\boldsymbol{X}_2 + \boldsymbol{Q}_2\boldsymbol{R}_{22} \Longrightarrow (I - \boldsymbol{P}_{\boldsymbol{X}_1})\boldsymbol{X}_2 = \boldsymbol{Q}_2\boldsymbol{R}_{22}.$$

Therefore, $\boldsymbol{Q}_2\boldsymbol{R}_{22}$ is the $\boldsymbol{QR}$ decomposition of $(I - \boldsymbol{P}_{\boldsymbol{X}_1})\boldsymbol{X}_2$ and so $\boldsymbol{Q}_2\boldsymbol{Q}_2' = \boldsymbol{P}_{(I-\boldsymbol{P}_{\boldsymbol{X}_1})\boldsymbol{X}_2}$. From this we conclude

$$\boldsymbol{P}_{\boldsymbol{X}} = \boldsymbol{Q}\boldsymbol{Q}' = \boldsymbol{Q}_1\boldsymbol{Q}_1' + \boldsymbol{Q}_2\boldsymbol{Q}_2' = \boldsymbol{P}_{\boldsymbol{X}_1} + \boldsymbol{P}_{(I-\boldsymbol{P}_{\boldsymbol{X}_1})\boldsymbol{X}_2} \, .$$

There is yet another way to derive the above equation. It emerges naturally from solving normal equations using block matrices:

$$\boldsymbol{X}'\boldsymbol{X}\boldsymbol{\beta} = \boldsymbol{X}'\boldsymbol{y} \Longrightarrow \begin{bmatrix} \boldsymbol{X}_1'\boldsymbol{X}_1 & \boldsymbol{X}_1'\boldsymbol{X}_2 \\ \boldsymbol{X}_2'\boldsymbol{X}_1 & \boldsymbol{X}_2'\boldsymbol{X}_2 \end{bmatrix} \begin{bmatrix} \boldsymbol{\beta}_1 \\ \boldsymbol{\beta}_2 \end{bmatrix} = \begin{bmatrix} \boldsymbol{X}_1' \\ \boldsymbol{X}_2' \end{bmatrix} \boldsymbol{y} \, .$$

Subtracting $\boldsymbol{X}_2'\boldsymbol{X}_1(\boldsymbol{X}_1'\boldsymbol{X}_1)^{-1}$ times the first row from the second row in the partitioned $\boldsymbol{X}'\boldsymbol{X}$ reduces the second equation to

$$\boldsymbol{X}_2'\left(\boldsymbol{I} - \boldsymbol{P}_{\boldsymbol{X}_1}\right)\boldsymbol{X}_2\boldsymbol{\beta}_2 = \boldsymbol{X}_2'\left(\boldsymbol{I} - \boldsymbol{P}_{\boldsymbol{X}_1}\right)\boldsymbol{y} \, .$$

This can be rewritten as normal equations $\boldsymbol{Z}'\boldsymbol{Z}\boldsymbol{\beta}_2 = \boldsymbol{Z}'\boldsymbol{y}$, where $\boldsymbol{Z}$ is as in Theorem 8.7. The above row operation also reveals that $\rho(\boldsymbol{Z}) = p_2$, which is equal to the rank of $\boldsymbol{X}_2$ (this also follows from Theorem 8.7). Thus, $\boldsymbol{Z}$ has full column rank so $\boldsymbol{Z}'\boldsymbol{Z}$ is nonsingular and $\boldsymbol{\beta}_2 = (\boldsymbol{Z}'\boldsymbol{Z})^{-1}\boldsymbol{Z}'\boldsymbol{y}$. Solving for $\boldsymbol{\beta}_1$ from the first block of equations, we obtain $\boldsymbol{\beta}_1 = \left(\boldsymbol{X}_1'\boldsymbol{X}_1\right)^{-1}\left(\boldsymbol{X}_1'\boldsymbol{y} - \boldsymbol{X}_1'\boldsymbol{X}_2\boldsymbol{\beta}_2\right)$. Substituting the solutions for $\boldsymbol{\beta}_1$ and $\boldsymbol{\beta}_2$ in $\boldsymbol{P}_{\boldsymbol{X}}\boldsymbol{y} = \boldsymbol{X}_1\boldsymbol{\beta}_1 + \boldsymbol{X}_2\boldsymbol{\beta}_2$ yields

$$\begin{aligned}
\boldsymbol{P}_{\boldsymbol{X}}\boldsymbol{y} &= \boldsymbol{X}_1\left(\boldsymbol{X}_1'\boldsymbol{X}_1\right)^{-1}\left(\boldsymbol{X}_1'\boldsymbol{y} - \boldsymbol{X}_1'\boldsymbol{X}_2\boldsymbol{\beta}_2\right) + \boldsymbol{X}_2\boldsymbol{\beta}_2 \\
&= \boldsymbol{P}_{\boldsymbol{X}_1}\boldsymbol{y} - \boldsymbol{P}_{\boldsymbol{X}_1}\boldsymbol{X}_2\boldsymbol{\beta}_2 + \boldsymbol{X}_2\boldsymbol{\beta}_2 = \boldsymbol{P}_{\boldsymbol{X}_1}\boldsymbol{y} + (\boldsymbol{I} - \boldsymbol{P}_{\boldsymbol{X}_1})\boldsymbol{X}_2\boldsymbol{\beta}_2 \\
&= \boldsymbol{P}_{\boldsymbol{X}_1}\boldsymbol{y} + \boldsymbol{Z}\boldsymbol{\beta}_2 = \boldsymbol{P}_{\boldsymbol{X}_1}\boldsymbol{y} + \boldsymbol{Z}(\boldsymbol{Z}'\boldsymbol{Z})^{-1}\boldsymbol{Z}'\boldsymbol{y} = (\boldsymbol{P}_{\boldsymbol{X}_1} + \boldsymbol{P}_{\boldsymbol{Z}})\boldsymbol{y} \, .
\end{aligned}$$

Since this is true for any vector $\boldsymbol{y} \in \Re^n$, we obtain part (iii) of Theorem 8.7.

8.6 Orthogonal triangularization

Gaussian elimination reveals one way of reducing a matrix $\boldsymbol{A}$ to row echelon form by elementary row operations. This is achieved by premultiplying $\boldsymbol{A}$ with elementary lower-triangular matrices. For square matrices, as we have seen earlier, the original matrix is reduced to a triangular matrix yielding the $\boldsymbol{LU}$ decomposition. It is natural to ask if we can use orthogonal transformations, instead of elementary row operations to reduce a square matrix to a triangular form yielding the $\boldsymbol{QR}$ decomposition. The answer is simple: if $\boldsymbol{A} = \boldsymbol{QR}$ then, using the orthogonality of $\boldsymbol{Q}$, we see $\boldsymbol{Q}'\boldsymbol{A} = \boldsymbol{R}$. Thus, $\boldsymbol{Q}'$ represents the orthogonal transformations on the rows of $\boldsymbol{A}$ that produce an upper-triangular matrix.

The Gram-Schmidt process is one way of determining $\boldsymbol{Q}$. However, numerical computations with floating point arithmetic must account for rounding errors and the Gram-Schmidt process is notoriously bad with rounding errors. Here we discuss three methods that are deemed much more efficient for numerical computations.

8.6.1 The modified Gram-Schmidt process

The Gram-Schmidt process described in Section 7.4 is often called the ***classical Gram-Schmidt*** process. When implemented on a computer, however, it produces $\boldsymbol{y}_k$'s that are often not orthogonal, not even approximately, due to rounding errors in floating-point arithmetic. In some cases this loss of orthogonality can be particularly bad. Hence, the classical Gram-Schmidt process is termed *numerically unstable*.

A small modification, known as the ***modified Gram-Schmidt*** process, can stabilize the classical algorithm. The modified approach produces the same result as the classical one in exact arithmetic but introduces much smaller errors in finite-precision or floating point arithmetic. The idea is really simple and we outline it below.

It is helpful to describe this using orthogonal projectors. Note that the classical Gram-Schmidt process can be computed

$$\begin{aligned}
\boldsymbol{y}_1 &= \boldsymbol{x}_1 \, ; \quad \boldsymbol{q}_1 = \frac{\boldsymbol{y}_1}{\|\boldsymbol{y}_1\|} \, ; \\
\boldsymbol{y}_2 &= \boldsymbol{x}_2 - \mathrm{proj}_{\boldsymbol{q}_1}(\boldsymbol{x}_2) = \left(\boldsymbol{I} - \boldsymbol{P}_{\boldsymbol{q}_1}\right)\boldsymbol{x}_2 \, ; \quad \boldsymbol{q}_2 = \frac{\boldsymbol{y}_2}{\|\boldsymbol{y}_2\|} \, ; \\
\boldsymbol{y}_3 &= \boldsymbol{x}_2 - \sum_{i=1}^{2} \mathrm{proj}_{\boldsymbol{q}_i}(\boldsymbol{x}_2) = \left(\boldsymbol{I} - \boldsymbol{P}_{\boldsymbol{Q}_2}\right)\boldsymbol{x}_2 \, ; \quad \boldsymbol{q}_3 = \frac{\boldsymbol{y}_3}{\|\boldsymbol{y}_3\|} \, ,
\end{aligned}$$

where $\boldsymbol{Q}_2 = [\boldsymbol{q}_1 : \boldsymbol{q}_2]$. In general, the k-th step is computed as

$$\boldsymbol{y}_k = \boldsymbol{x}_k - \sum_{i=1}^{k-1} \mathrm{proj}_{\boldsymbol{q}_i}(\boldsymbol{x}_k) = \boldsymbol{x}_k - \sum_{i=1}^{k-1} \boldsymbol{q}_i \boldsymbol{q}_i' \boldsymbol{x}_k = \left(\boldsymbol{I} - \boldsymbol{P}_{\boldsymbol{Q}_{k-1}}\right)\boldsymbol{x}_k \, ; \ \boldsymbol{q}_k = \frac{\boldsymbol{y}_k}{\|\boldsymbol{y}_k\|} \, ,$$

where $\boldsymbol{Q}_{k-1} = [\boldsymbol{q}_1 : \boldsymbol{q}_2 : \ldots : \boldsymbol{q}_{k-1}]$.

The key observation underlying the modified Gram-Schmidt process is that the orthogonal projector $\boldsymbol{I} - \boldsymbol{P}_{\boldsymbol{Q}_{k-1}}$ can be written as the product of $\boldsymbol{I} - \boldsymbol{P}_{\boldsymbol{q}_j}$'s:

$$\begin{aligned}
\boldsymbol{I} - \boldsymbol{P}_{\boldsymbol{Q}_{k-1}} &= \boldsymbol{I} - \boldsymbol{Q}_{k-1}\boldsymbol{Q}_{k-1}' = \boldsymbol{I} - \sum_{j=1}^{k-1} \boldsymbol{q}_j \boldsymbol{q}_j' = \boldsymbol{I} - \sum_{j=1}^{k-1} \boldsymbol{P}_{\boldsymbol{q}_j} \\
&= \left(\boldsymbol{I} - \boldsymbol{P}_{\boldsymbol{q}_{k-1}}\right)\left(\boldsymbol{I} - \boldsymbol{P}_{\boldsymbol{q}_{k-2}}\right)\cdots\left(\boldsymbol{I} - \boldsymbol{P}_{\boldsymbol{q}_2}\right)\left(\boldsymbol{I} - \boldsymbol{P}_{\boldsymbol{q}_1}\right) \, , \qquad (8.16)
\end{aligned}$$

because $\boldsymbol{P}_{\boldsymbol{q}_i}\boldsymbol{P}_{\boldsymbol{q}_j} = \boldsymbol{O}$ whenever $i \neq j$.

Suppose we have completed $k - 1$ steps and obtained orthogonal vectors $\boldsymbol{y}_1, \boldsymbol{y}_2, \ldots, \boldsymbol{y}_{k-1}$. In the classical process, we would compute $\boldsymbol{y}_k = \left(\boldsymbol{I} - \boldsymbol{P}_{\boldsymbol{Q}_{k-1}}\right)\boldsymbol{x}_k$. Instead of this, the modified Gram-Schmidt process suggests using the result in (8.16) to compute $\boldsymbol{y}_k$ in a sequence of $k - 1$ steps. We construct a sequence $\boldsymbol{y}_k^{(1)}, \boldsymbol{y}_k^{(2)}, \ldots, \boldsymbol{y}_k^{(k-1)}$ in the following manner:

$$\begin{aligned}
&\boldsymbol{y}_k^{(1)} = \left(\boldsymbol{I} - \boldsymbol{P}_{\boldsymbol{q}_1}\right)\boldsymbol{x}_k \, ; \quad \boldsymbol{y}_k^{(2)} = \left(\boldsymbol{I} - \boldsymbol{P}_{\boldsymbol{q}_2}\right)\boldsymbol{y}_k^{(1)} \, ; \quad \boldsymbol{y}_k^{(3)} = \left(\boldsymbol{I} - \boldsymbol{P}_{\boldsymbol{q}_3}\right)\boldsymbol{y}_k^{(2)} \, ; \ \ldots \\
&\ldots ; \quad \boldsymbol{y}_k^{(k-2)} = \left(\boldsymbol{I} - \boldsymbol{P}_{\boldsymbol{q}_{k-2}}\right)\boldsymbol{y}_k^{(k-3)} \, ; \quad \boldsymbol{y}_k^{(k-1)} = \left(\boldsymbol{I} - \boldsymbol{P}_{\boldsymbol{q}_{k-1}}\right)\boldsymbol{y}_k^{(k-2)} \, .
\end{aligned}$$

From (8.16) it is immediate that $\boldsymbol{y}_k^{(k-1)} = \boldsymbol{y}_k$.

8.6.2 Reflectors

A different approach to computing the $\boldsymbol{QR}$ decomposition of a matrix relies upon *reflecting* a point about a plane (in $\Re^3$) or a subspace. In $\Re^3$, the mid-point of the original point and its reflection about a plane is the orthogonal projection of the original point on the plane.

In higher dimensions, we reflect a point about a hyperplane containing the origin that is orthogonal to a given vector (i.e., the orthocomplement of the span of a given vector). In Figure 8.6.2, let $\boldsymbol{x}$ be a point in $\Re^n$ and let $\boldsymbol{v}$ be a non-null vector in $\Re^n$. The *reflection* of $\boldsymbol{x}$ about the orthocomplementary subspace of $\boldsymbol{v}$ is the vector $\boldsymbol{y}$.

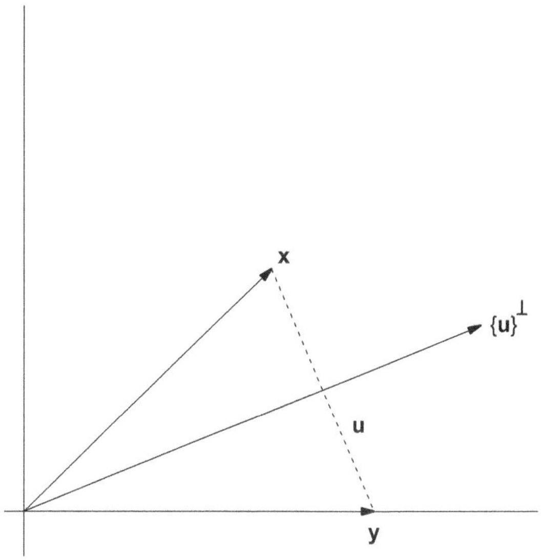

Assume that $\boldsymbol{u} = \frac{\boldsymbol{v}}{\|\boldsymbol{v}\|}$ is normalized to be of unit length. The orthogonal projection of $\boldsymbol{x}$ onto the hyperplane containing the origin and orthogonal to $\boldsymbol{u}$ (i.e., orthocomplementary subspace of $\boldsymbol{u}$) is given by $(\boldsymbol{I} - \boldsymbol{P_u})\boldsymbol{x} = (\boldsymbol{I} - \boldsymbol{uu}')\boldsymbol{x}$. Let $\boldsymbol{y}$ be the reflection of $\boldsymbol{x}$ about the orthocomplementary subspace of $\boldsymbol{u}$. Then,

$$\frac{\boldsymbol{x}+\boldsymbol{y}}{2} = (\boldsymbol{I} - \boldsymbol{uu}')\boldsymbol{x} \Longrightarrow \boldsymbol{y} = (\boldsymbol{I} - 2\boldsymbol{uu}')\boldsymbol{x} \Longrightarrow \boldsymbol{y} = \boldsymbol{H_u}\boldsymbol{x} \; ,$$

where $\boldsymbol{H_u} = \boldsymbol{I} - 2\boldsymbol{uu}' = \boldsymbol{I} - 2\frac{\boldsymbol{vv}'}{\boldsymbol{v}'\boldsymbol{v}}$ reflects $\boldsymbol{x}$ about the orthocomplementary subspace spanned by $\boldsymbol{u}$.

Here is another way to derive $\boldsymbol{H_u}$. Projection and reflection of $\boldsymbol{x}$ about a hyperplane are both obtained by traveling a certain distance from $\boldsymbol{x}$ along the direction $\boldsymbol{u}$, where $\boldsymbol{u}$ is a vector of unit length. Let α be the distance we need to travel from $\boldsymbol{x}$ in the direction of $\boldsymbol{u}$ to arrive at the projection of $\boldsymbol{x}$ onto the orthocomplementary subspace of $\boldsymbol{u}$. Thus, $\boldsymbol{x} + \alpha\boldsymbol{u} = (\boldsymbol{I} - \boldsymbol{u}\boldsymbol{u}')\boldsymbol{x}$ and, hence, $\alpha\boldsymbol{u} = -(\boldsymbol{u}'\boldsymbol{x})\boldsymbol{u}$. This means that $(\alpha + \boldsymbol{u}'\boldsymbol{x})\boldsymbol{u} = \boldsymbol{0}$ and, since $\boldsymbol{u} \neq \boldsymbol{0}$, this implies that $\alpha = -\boldsymbol{u}'\boldsymbol{x}$. Now, to arrive at the reflection of $\boldsymbol{x}$ about the same subspace we need to travel a further α units from the projection along the direction $\boldsymbol{u}$. In other words, the reflection is given by $\boldsymbol{x} + 2\alpha\boldsymbol{u}$, which simplifies to

$$\boldsymbol{x} + 2\alpha\boldsymbol{u} = \boldsymbol{x} - 2(\boldsymbol{u}'\boldsymbol{x})\boldsymbol{u} = \boldsymbol{x} - 2\boldsymbol{u}(\boldsymbol{u}'\boldsymbol{x}) = \boldsymbol{x} - 2\boldsymbol{u}\boldsymbol{u}'\boldsymbol{x} = (\boldsymbol{I} - 2\boldsymbol{u}\boldsymbol{u}')\boldsymbol{x} = \tilde{\boldsymbol{Q}}_{\boldsymbol{u}}\boldsymbol{x} .$$

Reflection is a linear transformation and is often referred to as a Householder transformation (also known as a ***Householder reflection*** or an ***elementary reflector***). The following definition is motivated from the above properties of reflectors.

Definition 8.4 *Let $\boldsymbol{v} \neq \boldsymbol{0}$ be an $n \times 1$ vector. The* **elementary reflector** *about $\{\boldsymbol{v}\}^{\perp}$ is defined as*

$$\boldsymbol{H}_{\boldsymbol{v}} = \boldsymbol{I} - 2\frac{\boldsymbol{v}\boldsymbol{v}'}{\boldsymbol{v}'\boldsymbol{v}} . \tag{8.17}$$

For normalized vectors, say $\boldsymbol{u} = \dfrac{\boldsymbol{v}}{\|\boldsymbol{v}\|}$, the elementary reflector about $\{\boldsymbol{u}\}^{\perp}$ is

$$\boldsymbol{H}_{\boldsymbol{u}} = \boldsymbol{I} - 2\boldsymbol{u}\boldsymbol{u}' .$$

Henceforth we will assume that $\boldsymbol{u}$ is a unit vector in the definition of $\boldsymbol{H}_{\boldsymbol{u}}$.

From the above definition it is clear that $\boldsymbol{H}_{\boldsymbol{u}}$ is a symmetric matrix. Geometric intuition suggests that if we reflect $\boldsymbol{x}$ about $\{\boldsymbol{u}\}^{\perp}$ and then reflect the reflection (i.e., $\boldsymbol{H}_{\boldsymbol{u}}\boldsymbol{x}$) again about $\{\boldsymbol{u}\}^{\perp}$, then we should obtain $\boldsymbol{x}$. Thus, $\boldsymbol{H}_{\boldsymbol{u}}^2\boldsymbol{x} = \boldsymbol{H}_{\boldsymbol{u}}(\boldsymbol{H}_{\boldsymbol{u}}\boldsymbol{x}) = \boldsymbol{x}$ for every $\boldsymbol{x}$ and so $\boldsymbol{H}_{\boldsymbol{u}}^2 = \boldsymbol{I}$. An algebraic confirmation is immediate:

$$\boldsymbol{H}_{\boldsymbol{u}}^2 = (\boldsymbol{I} - 2\boldsymbol{u}\boldsymbol{u}')(\boldsymbol{I} - 2\boldsymbol{u}\boldsymbol{u}') = \boldsymbol{I} - 4\boldsymbol{u}\boldsymbol{u}' + 4\boldsymbol{u}\boldsymbol{u}'\boldsymbol{u}\boldsymbol{u}' = \boldsymbol{I} . \tag{8.18}$$

Also, $\boldsymbol{H}_{\boldsymbol{u}}$ is symmetric and, from (8.18), we find $\boldsymbol{H}_{\boldsymbol{u}}$ is orthogonal:

$$\boldsymbol{H}_{\boldsymbol{u}}'\boldsymbol{H}_{\boldsymbol{u}} = \boldsymbol{H}_{\boldsymbol{u}}\boldsymbol{H}_{\boldsymbol{u}}' = \boldsymbol{H}_{\boldsymbol{u}}^2 = \boldsymbol{I} . \tag{8.19}$$

Since every orthogonal matrix in $\Re^2$ or $\Re^3$ is a rotation by some angle θ (recall Example 8.1), it follows that every reflection is also a rotation. This is obvious from Figure 8.6.2, where the reflection of $\boldsymbol{x}$ can be achieved by a clockwise rotation of $\boldsymbol{x}$ by the angle formed by $\boldsymbol{x}$ and $\boldsymbol{y}$.

Suppose we are given two vectors, $\boldsymbol{x}$ and $\boldsymbol{y}$, that have the same length, i.e., $\|\boldsymbol{x}\| = \|\boldsymbol{y}\|$. Can we find a reflector that will map one to the other? For this, we need to find a normalized vector $\boldsymbol{u}$ that satisfies

$$\boldsymbol{H}_{\boldsymbol{u}}\boldsymbol{x} = (\boldsymbol{I} - 2\boldsymbol{u}\boldsymbol{u}')\boldsymbol{x} = \boldsymbol{x} - 2(\boldsymbol{u}'\boldsymbol{x})\boldsymbol{u} = \boldsymbol{y} . \tag{8.20}$$

But will such a $\boldsymbol{u}$ always exist? The answer is yes. Some geometric intuition is useful. If $\boldsymbol{y}$ is indeed a reflection of $\boldsymbol{x}$ about $\{\boldsymbol{u}\}^{\perp}$, then $\boldsymbol{x} - \boldsymbol{y}$ is in the same direction as $\boldsymbol{u}$. This prompts us to take $\boldsymbol{u}$ to be the normalized vector in the direction of $\boldsymbol{x} - \boldsymbol{y}$ and check if such a $\boldsymbol{u}$ can satisfy (8.20). To be precise, define

$$\boldsymbol{v} = \boldsymbol{x} - \boldsymbol{y} \text{ and set } \boldsymbol{u} = \frac{\boldsymbol{v}}{\|\boldsymbol{v}\|} = \frac{\boldsymbol{x} - \boldsymbol{y}}{\|\boldsymbol{x} - \boldsymbol{y}\|} . \tag{8.21}$$

We make the following observation:

$$v'v = x'x + y'y - 2x'y = 2x'x - 2x'y = 2x'(x - y) = 2v'x \,, \qquad (8.22)$$

where we have used the fact that $x'x = \|x\|^2 = \|y\|^2 = y'y$. Therefore,

$$H_u x = x - 2(u'x)u = x - \frac{2(v'x)}{v'v}v = x - v = x - (x - y) = y \,,$$

which confirms that u satisfies (8.20). Some geometric reasoning behind (8.22) is also helpful. The vectors x and y form two sides of an isosceles triangle and the vector v is the base of the triangle. It follows from elementary trigonometry that the cosine of the angle between x and v is given by $\frac{\|v\|}{2\|x\|}$. Therefore, $\frac{v'x}{\|v\|\|x\|} = \frac{\|v\|}{2\|x\|}$, which yields $2v'x = v'v$.

The above discussion suggests the following application for elementary reflectors. Like elementary lower-triangular matrices, elementary reflectors can also be used to zero all entries below a given element (e.g. a pivot) in a vector. Suppose we want to zero all entries below the first element in a vector x. We can achieve this by setting $y = \pm\|x\|e_1$ and find the direction u such that the reflection of x about the normal perpendicular (or normal) is y. That direction is given by (8.21). To be precise, define

$$v = x \pm \|x\|e_1 \text{ and set } u = \frac{v}{\|v\|} = \frac{x \pm \|x\|e_1}{\|x \pm \|x\|e_1\|} \,. \qquad (8.23)$$

The elementary reflector H_u is then called the ***Householder reflector*** and

$$H_u x = (I - 2uu')x = x - \frac{2(v'x)}{v'v}v = x - v = \pm\|x\|e_1 = \begin{bmatrix} \pm\|x\| \\ 0 \\ \vdots \\ 0 \end{bmatrix} . \qquad (8.24)$$

Therefore, with proper choice of the hyperplane about which to reflect, an elementary reflector can sweep out the elements of a vector below a pivot. In theory, the zeroing process will work well whether we choose $v = x + \|x\|e_1$ or $v = x - \|x\|e_1$ to construct H_u. In practice, to avoid cancellation with floating point arithmetic for real matrices, we choose $v = x + \text{sign}(x_1)\|x\|e_1$, where $\text{sign}(x_1)$ is the sign of the first element in x.

Householder reduction

We now show how (8.24) can be used to reduce a square matrix to upper-triangular form. Let $A = [a_{*1} : a_{*2} : \cdots : a_{*n}]$ be an $n \times n$ matrix. We construct the Householder elementary reflector using u in (8.23) with x as the first column of A.

We denote this by $\boldsymbol{H}_1 = \boldsymbol{I}_n - 2\boldsymbol{u}\boldsymbol{u}'$, so that

$$\boldsymbol{H}_1 \boldsymbol{a}_{*1} = \boldsymbol{H}_1 \boldsymbol{a}_{*1} = \begin{bmatrix} r_{11} \\ 0 \\ \vdots \\ 0 \end{bmatrix}, \tag{8.25}$$

where $r_{11} = \|\boldsymbol{a}_{*1}\|$. If $\boldsymbol{A}_2 = \boldsymbol{H}_1\boldsymbol{A}$, then

$$\boldsymbol{A}_2 = \boldsymbol{H}_1\boldsymbol{A} = [\boldsymbol{H}_1\boldsymbol{a}_{*1} : \boldsymbol{H}_1\boldsymbol{a}_{*2} : \ldots : \boldsymbol{H}_1\boldsymbol{a}_{*n}] = \begin{bmatrix} r_{11} & \boldsymbol{r}_1' \\ \boldsymbol{0} & \tilde{\boldsymbol{A}}_2 \end{bmatrix},$$

where $\tilde{\boldsymbol{A}}_2$ is $(n-1) \times (n-1)$. Therefore, all entries below the $(1,1)$-element of $\boldsymbol{A}$ have been swept out by $\boldsymbol{H}_1$. It is instructive to provide a schematic presentation that keeps track of what elements change in $\boldsymbol{A}$. With $n = 4$, the first step yields

$$\boldsymbol{A}_2 = \boldsymbol{H}_1 \begin{bmatrix} * & * & * & * \\ * & * & * & * \\ * & * & * & * \\ * & * & * & * \end{bmatrix} = \begin{bmatrix} + & + & + & + \\ 0 & + & + & + \\ 0 & + & + & + \\ 0 & + & + & + \end{bmatrix},$$

where, on the right-hand side, the $+$'s indicate nonzero elements that have been altered by the preceding operation (i.e., multiplication from the left by $\boldsymbol{H}_1$), while the $*$'s denote entries that are not necessarily 0 and are unaltered.

Next, we apply the same procedure to the $(n-1)$-dimensional square matrix $\tilde{\boldsymbol{A}}_2$. Again we construct an elementary reflector from $\boldsymbol{u}$ in (8.23) but now with $\boldsymbol{x}$ as the first column of $\tilde{\boldsymbol{A}}_2$. (Note: $\boldsymbol{u}$ is now an $(n-1) \times 1$ vector). Call this reflector $\tilde{\boldsymbol{H}}_2$ and note that it annihilates all entries below the $(1,1)$-position in $\tilde{\boldsymbol{A}}_2$.

Set $\boldsymbol{H}_2 = \begin{bmatrix} 1 & \boldsymbol{0}' \\ \boldsymbol{0} & \tilde{\boldsymbol{H}}_2 \end{bmatrix}$. Since $\tilde{\boldsymbol{H}}_2 = \boldsymbol{I}_{n-1} - 2\boldsymbol{u}\boldsymbol{u}'$, we can write

$$\boldsymbol{H}_2 = \begin{bmatrix} 1 & \boldsymbol{0} \\ \boldsymbol{0} & \boldsymbol{I}_{n-1} - 2\boldsymbol{u}\boldsymbol{u}' \end{bmatrix} = \boldsymbol{I}_n - 2\begin{bmatrix} 0 \\ \boldsymbol{u} \end{bmatrix}[0 : \boldsymbol{u}'] = \boldsymbol{I}_n - 2\tilde{\boldsymbol{u}}\tilde{\boldsymbol{u}}',$$

where $\tilde{\boldsymbol{u}} = \begin{bmatrix} 0 \\ \boldsymbol{u} \end{bmatrix}$ is now an $(n-1) \times 1$ vector such that $\|\boldsymbol{u}\| = 1$. This shows that $\boldsymbol{H}_2$ is itself an elementary reflector and, hence, an orthogonal matrix. Therefore, $\boldsymbol{H}_2\boldsymbol{H}_1$ is also an orthogonal matrix (being the product of orthogonal matrices) and we obtain

$$\boldsymbol{A}_3 = \boldsymbol{H}_2\boldsymbol{A}_2 = \begin{bmatrix} r_{11} & \boldsymbol{r}_1' \\ \boldsymbol{0} & \tilde{\boldsymbol{A}}_3 \end{bmatrix}, \quad \text{where } \tilde{\boldsymbol{A}}_3 = \tilde{\boldsymbol{H}}_2\tilde{\boldsymbol{A}}_2 .$$

With $n = 4$, this step gives us

$$\boldsymbol{A}_3 = \boldsymbol{H}_2 \begin{bmatrix} * & * & * & * \\ 0 & * & * & * \\ 0 & * & * & * \\ 0 & * & * & * \end{bmatrix} = \begin{bmatrix} * & * & * & * \\ 0 & + & + & + \\ 0 & 0 & + & + \\ 0 & 0 & + & + \end{bmatrix}.$$

Because of its structure, $\boldsymbol{H}_2$ does not alter the first row or the first column of $\boldsymbol{A}_2$. So, the zeroes in the first column of $\boldsymbol{A}_2$ remain intact.

A generic step is easy to describe. After $k-1$ steps, we have $\boldsymbol{A}_k = \boldsymbol{H}_{k-1}\boldsymbol{A}_{k-1} = \begin{bmatrix} \boldsymbol{R}_{k-1} & \tilde{\boldsymbol{R}}_{k-1} \\ \boldsymbol{0} & \tilde{\boldsymbol{A}}_k \end{bmatrix}$. At step k an elementary reflector $\tilde{\boldsymbol{H}}_k$ is constructed to sweep out all entries below the $(1,1)$-th entry in $\tilde{\boldsymbol{A}}_k$. We then define $\boldsymbol{H}_k = \begin{bmatrix} \boldsymbol{I}_{k-1} & \boldsymbol{0} \\ \boldsymbol{0} & \tilde{\boldsymbol{H}}_k \end{bmatrix}$, which is another elementary reflector. Observe that $\boldsymbol{H}_k$ will not affect the first $k-1$ rows or columns of a matrix it premultiplies. With $n = 4$ and $k = 3$, we obtain

$$\boldsymbol{A}_4 = \boldsymbol{H}_3 \begin{bmatrix} * & * & * & * \\ 0 & * & * & * \\ 0 & 0 & * & * \\ 0 & 0 & * & * \end{bmatrix} = \begin{bmatrix} * & * & * & * \\ 0 & * & * & * \\ 0 & 0 & + & + \\ 0 & 0 & 0 & + \end{bmatrix}.$$

From the $+$'s we see that $\boldsymbol{H}_3$ zeroes the last element in the third column without affecting the first two rows and columns from the previous step. $\boldsymbol{A}_4$ is upper-triangular.

In general, after completing $n-1$ reflections we obtain an upper-triangular matrix $\boldsymbol{A}_n = \boldsymbol{H}_{n-1}\cdots\boldsymbol{H}_2\boldsymbol{H}_1\boldsymbol{A}$. However, one final reflection, the n-th, may be required to ensure that all the diagonal elements in the upper-triangular matrix are positive (see Example 8.4 below). The product of elementary reflectors is not an elementary reflector but the product of orthogonal matrices is another orthogonal matrix. Therefore, $\boldsymbol{H} = \boldsymbol{H}_n\boldsymbol{H}_{n-1}\cdots\boldsymbol{H}_2\boldsymbol{H}_1$ is an orthogonal matrix (though not necessarily a reflector) and $\boldsymbol{H}\boldsymbol{A} = \boldsymbol{R}$, where $\boldsymbol{R}$ is upper-triangular with positive diagonal elements. This reveals the $\boldsymbol{QR}$ factorization: $\boldsymbol{A} = \boldsymbol{QR}$, where $\boldsymbol{Q} = \boldsymbol{H}'$.

Example 8.4 Consider the matrix $\boldsymbol{X}$ in Example 8.3, where we obtained the $\boldsymbol{QR}$ factorization of $\boldsymbol{X}$ using exact arithmetic. We will now use Householder reflections in floating point arithmetic, rounding up to four decimal places, to reduce $\boldsymbol{X}$ to an upper-triangular form.

We find the vector $\boldsymbol{u}$ in (8.23) with $\boldsymbol{x}$ as the first column vector of $\boldsymbol{X}$ and form $\boldsymbol{H}_1$. These are obtained as

$$\boldsymbol{u} = \begin{bmatrix} -0.6053 \\ 0.4415 \\ 0.6623 \end{bmatrix} \text{ and } \boldsymbol{H}_1 = \begin{bmatrix} 0.2673 & 0.5345 & 0.8018 \\ 0.5345 & 0.6101 & -0.5849 \\ 0.8018 & -0.5849 & 0.1227 \end{bmatrix}.$$

This yields

$$\boldsymbol{H}_1\boldsymbol{X} = \begin{bmatrix} 3.7417 & 12.5613 & 0.5345 \\ 0.0000 & 0.0252 & 0.6101 \\ 0.0000 & -0.4622 & -0.5849 \end{bmatrix}.$$

Next, we consider the 2×2 submatrix obtained by deleting the first row and first column from $\boldsymbol{H}_1\boldsymbol{X}$. This is the matrix $\boldsymbol{X}_2 = \begin{bmatrix} 0.0252 & 0.6101 \\ -0.4622 & -0.5849 \end{bmatrix}$. We now form $\boldsymbol{u}$ setting $\boldsymbol{x}$ as the first column of $\boldsymbol{X}_2$ and obtain

$$\boldsymbol{u} = \begin{bmatrix} -0.6876 \\ -0.7261 \end{bmatrix} \text{ and } \tilde{\boldsymbol{H}}_2 = \begin{bmatrix} 0.0544 & -0.9985 \\ -0.9985 & -0.0544 \end{bmatrix}.$$

We form $\boldsymbol{H}_2 = \begin{bmatrix} 1 & \boldsymbol{0}' \\ \boldsymbol{0} & \tilde{\boldsymbol{H}}_2 \end{bmatrix} = \begin{bmatrix} 1 & 0 & 0 \\ 0 & 0.0544 & -0.9985 \\ 0 & -0.9985 & -0.0544 \end{bmatrix}$ and obtain

$$\boldsymbol{H}_2\boldsymbol{H}_1\boldsymbol{X} = \begin{bmatrix} 3.7417 & 12.5613 & 0.5345 \\ 0.0000 & 0.4629 & 0.6172 \\ 0.0000 & 0.0000 & -0.5774 \end{bmatrix}.$$

Note that the above matrix is already upper-triangular. However, the conventional $\boldsymbol{QR}$ decomposition restricts the diagonal elements of $\boldsymbol{R}$ to be positive. This makes $\boldsymbol{Q}$ and $\boldsymbol{R}$ unique (recall Lemma 8.3). So we perform one last reflection using

$$\boldsymbol{u} = \begin{bmatrix} 0 \\ 0 \\ 1 \end{bmatrix} \text{ and } \boldsymbol{H}_3 = \boldsymbol{I} - 2\boldsymbol{u}\boldsymbol{u}' = \begin{bmatrix} 1 & 0 & 0 \\ 0 & 1 & 0 \\ 0 & 0 & -1 \end{bmatrix}.$$

This produces

$$\boldsymbol{R} = \boldsymbol{H}_3\boldsymbol{H}_2\boldsymbol{H}_1\boldsymbol{X} = \begin{bmatrix} 3.7417 & 12.5613 & 0.5345 \\ 0.0000 & 0.4629 & 0.6172 \\ 0.0000 & 0.0000 & 0.5774 \end{bmatrix} \text{ and}$$

$$\boldsymbol{Q}' = \boldsymbol{H}_3\boldsymbol{H}_2\boldsymbol{H}_1 = \begin{bmatrix} 0.2673 & 0.5345 & 0.8018 \\ -0.7715 & 0.6172 & -0.1543 \\ 0.5774 & 0.5774 & -0.5774 \end{bmatrix}.$$

Note that the $\boldsymbol{Q}$ and $\boldsymbol{R}$ obtained from the above using floating point arithmetic are the same (up to negligible round-off errors) as those obtained in Example 8.3 using exact arithmetic. ■

8.6.3 *Rotations*

In Example 8.1, we saw that every orthogonal matrix was either a rotation or a reflection. A 2×2 rotation matrix, often called a ***plane rotation matrix***, has the form

$$\boldsymbol{G} = \begin{bmatrix} c & s \\ -s & c \end{bmatrix}, \quad \text{where } c^2 + s^2 = 1. \tag{8.26}$$

Clearly, $\boldsymbol{G}'\boldsymbol{G} = \boldsymbol{G}\boldsymbol{G}' = \boldsymbol{I}_2$, which shows that $\boldsymbol{G}$ is orthogonal.Since $c^2 + s^2 = 1$, both c and s are real numbers between -1 and 1. This means that there exists an angle of rotation θ such that $c = \cos\theta$ and $s = \sin\theta$. This angle is often referred to as the ***Givens angle***. The vector $\boldsymbol{y} = \boldsymbol{G}\boldsymbol{x}$ is the point obtained by rotating $\boldsymbol{x}$ in a clockwise direction by an angle θ, where $c = \cos\theta$ and $s = \sin\theta$.

A 2×2 plane rotation can be used to annihilate the second element of a nonzero vector $\boldsymbol{x} = (x_1, x_2)' \in \Re^2$ by rotating $\boldsymbol{x}$ so that it aligns with (i.e., becomes parallel to) the $(1, 0)$ vector. To be precise, if we choose $c = x_1/\sqrt{x_1^2 + x_2^2}$ and $s = x_2/\sqrt{x_1^2 + x_2^2}$, then

$$\boldsymbol{G}\boldsymbol{x} = \begin{bmatrix} c & s \\ -s & c \end{bmatrix}\begin{bmatrix} x_1 \\ x_2 \end{bmatrix} = \frac{1}{\sqrt{x_1^2 + x_2^2}}\begin{bmatrix} x_1 & x_2 \\ -x_2 & x_1 \end{bmatrix}\begin{bmatrix} x_1 \\ x_2 \end{bmatrix} = \begin{bmatrix} \sqrt{x_1^2 + x_2^2} \\ 0 \end{bmatrix}.$$

This simple result can be used in higher dimensions to annihilate specific elements in a vector. Before, we consider the general case, let us consider what happens in $\Re^3$ (see Example 8.2). Here, rotating a point about a given axis is equivalent to rotation in the orthocomplementary space of that axis. The corresponding 3×3 rotation matrix is constructed by embedding a 2×2 rotation in the appropriate position within the 3×3 identity matrix. For example, rotation around the y-axis is rotation in the xz-plane, and the corresponding (clockwise) rotator is produced by embedding the 2×2 rotator in the "xz-position" of $\boldsymbol{I}_{3\times3}$. We write this rotator as

$$\boldsymbol{G}_{13} = \begin{bmatrix} c & 0 & s \\ 0 & 1 & 0 \\ -s & 0 & c \end{bmatrix}, \quad \text{where } c^2 + s^2 = 1.$$

A quick word about the notation $\boldsymbol{G}_{13}$. We should, strictly speaking, have written this as $\boldsymbol{G}_{xz}$ to indicate that the rotation takes place in the xz plane. However, letters becomes clumsy in higher dimensions, so we replace letters by numbers and refer to the xyz-plane as the 123-plane. We say that $\boldsymbol{G}_{13}$ rotates points in the $(1,3)$ plane. The real numbers c, s, $-s$ and c occupy the four positions in $\boldsymbol{G}_{13}$ where the first and third rows intersect with the first and third columns. The remaining diagonal elements in $\boldsymbol{G}_{13}$ are equal to 1 and the remaining off-diagonal elements are all equal to 0. Since the $(1,3)$ plane and the $(3,1)$ plane are one and the same, note that $\boldsymbol{G}_{13} = \boldsymbol{G}_{31}$. This is also clear from the structure of $\boldsymbol{G}_{13}$.

To elucidate further, consider the effect of $\boldsymbol{G}_{13}$ on a 3×1 vector $\boldsymbol{x}$:

$$\boldsymbol{G}_{13}\boldsymbol{x} = \begin{bmatrix} c & 0 & s \\ 0 & 1 & 0 \\ -s & 0 & c \end{bmatrix} \begin{bmatrix} x_1 \\ x_2 \\ x_3 \end{bmatrix} = \begin{bmatrix} cx_1 + sx_3 \\ x_2 \\ -sx_1 + cx_3 \end{bmatrix}.$$

Note that $\boldsymbol{G}_{13}$ alters only the first and third elements in $\boldsymbol{x}$ and leaves the second element unchanged. The embedded 2×2 rotator rotates the point (x_1, x_3) in the $(1,3)$-plane by an angle θ, where $c = \cos\theta$ and $s = \sin\theta$. Choosing $c = x_1/\sqrt{x_1^2 + x_3^2}$ and $s = x_3/\sqrt{x_1^2 + x_3^2}$ rotates $(x_1, x_3)'$ to the point $(\sqrt{x_1^2 + x_3^2}, 0)'$. So, the effect of $\boldsymbol{G}_{13}$ on $\boldsymbol{x} = (x_1, x_2, x_3)'$ is to zero the third element:

$$\boldsymbol{G}_{13}\boldsymbol{x} = \begin{bmatrix} \frac{x_1}{\sqrt{x_1^2+x_3^2}} & 0 & \frac{x_3}{\sqrt{x_1^2+x_3^2}} \\ 0 & 1 & 0 \\ -\frac{x_3}{\sqrt{x_1^2+x_3^2}} & 0 & \frac{x_1}{\sqrt{x_1^2+x_3^2}} \end{bmatrix} \begin{bmatrix} x_1 \\ x_2 \\ x_3 \end{bmatrix} = \begin{bmatrix} \sqrt{x_1^2 + x_3^2} \\ x_2 \\ 0 \end{bmatrix}.$$

Thus, 2×2 and 3×3 plane rotators can be used to zero certain elements in a vector, while not affecting other elements.

For $\Re^2$ and $\Re^3$, representing c and s in terms of the angle of rotation is helpful because of the geometric interpretation and our ability to visualize. However, for higher dimensions these angles are more of a distraction and are often omitted from the notation. Let us present another schematic example with a 4×4 rotator, where all elements below the first in a 4×1 vector are zeroed.

Example 8.5 Consider a vector $\boldsymbol{x}$ in $\Re^4$. We can use a plane-rotation in the $(1,2)$-

plane to zero the second element by choosing c and s appropriately.

$$G_{12}x = \begin{bmatrix} c & s & 0 & 0 \\ -s & c & 0 & 0 \\ 0 & 0 & 1 & 0 \\ 0 & 0 & 0 & 1 \end{bmatrix} \begin{bmatrix} * \\ * \\ * \\ * \end{bmatrix} = \begin{bmatrix} + \\ 0 \\ * \\ * \end{bmatrix},$$

where, on the right-hand side, the $+$'s indicate possibly nonzero elements that have been changed by the operation on the left and the $*$'s indicate elements that have remained unaltered. Since G_{12} operates on the $(1,2)$ plane, it affects only the first and second elements in x. It zeroes the second element, possibly alters the first element and leaves the last two elements intact.

In the next step, we choose a rotator G_{13} in the $(1,3)$ plane to zero the third element in the vector obtained from the first step.

$$G_{13}(G_{12}x) = \begin{bmatrix} c & 0 & s & 0 \\ 0 & 1 & 0 & 0 \\ -s & 0 & c & 0 \\ 0 & 0 & 0 & 1 \end{bmatrix} \begin{bmatrix} * \\ 0 \\ * \\ * \end{bmatrix} = \begin{bmatrix} + \\ 0 \\ 0 \\ * \end{bmatrix}.$$

Since G_{13} operates on the $(1,3)$ plane, it affects only the first and third elements of the vector. It zeroes the third element, alters the first, and leaves the second and fourth unchanged. Importantly, the 0 introduced in the second position remains unaltered. This is critical. Otherwise, we would have undone our work in the first step.

The third and final rotation zeroes the fourth element in the vector obtained from the previous step. We achieve this by using an appropriate rotator G_{14} that operates on the $(1,4)$ plane.

$$G_{14}(G_{13}G_{12}x) = \begin{bmatrix} c & 0 & 0 & s \\ 0 & 1 & 0 & 0 \\ 0 & 0 & 1 & 0 \\ -s & 0 & 0 & c \end{bmatrix} \begin{bmatrix} * \\ 0 \\ 0 \\ * \end{bmatrix} = \begin{bmatrix} + \\ 0 \\ 0 \\ 0 \end{bmatrix}.$$

The rotator G_{14} alters only the first and fourth elements in the vector and zeroes the third element. This way, a sequence of rotations $G_{14}G_{13}G_{12}$ can be constructed to zero all elements below the first in a 4×1 vector. ■

The preceding example illustrates how we can choose a sequence of rotators to annihilate specific entries in a 4×1 vector. It is worth pointing out that there are alternative sequences of rotators that can also do this job. For instance, in Example 8.5, we successively annihilated the second, third and fourth elements. This is a "top-down" approach in the sense that we start at the top and proceed downward in the zeroing process. We could have easily gone in the reverse direction by first zeroing the fourth element, then the third and finally the second. This is the "bottom-up" approach as we start at the bottom of the vector and proceed upward. The bottom-up approach for the setting in Example 8.5 could have used $G_{12}G_{13}G_{14}x$ to zero all the elements below the first in x by choosing the appropriate rotators. Another sequence for the bottom-up approach could have been $G_{12}G_{23}G_{34}x$.

Givens reduction

To generalize the above concepts to $\Re^n$ and see how we can use rotations to orthogonally reduce a matrix to a triangular form, we will generalize the notion of rotators in higher dimensions.

Definition 8.5 Givens rotations. *An $n \times n$ square matrix is said to be a* **Givens rotation matrix** *or* **Givens rotator**, *denoted G_{ij} if the (i,i)-th, (i,j)-th, (j,i)-th and (j,j)-th elements (i.e., the intersection of the i-th and j-th rows with the i-th and j-th columns) are c, s, $-s$ and c, where c and s are real numbers satisfying $c^2 + s^2 = 1$, all other diagonal elements are 1, and all other off-diagonal elements are 0. These are called* **plane rotation matrices** *because they perform a clockwise rotation in the (i,j)-plane of $\Re^n$.*

A Givens rotation matrix can be explicitly written down as

$$G_{ij} = \begin{bmatrix} 1 & \cdots & 0 & \cdots & 0 & \cdots & 0 \\ \vdots & \ddots & \vdots & & \vdots & & \vdots \\ 0 & \cdots & c & \cdots & s & \cdots & 0 \\ \vdots & & \vdots & \ddots & \vdots & & \vdots \\ 0 & \cdots & -s & \cdots & c & \cdots & 0 \\ \vdots & & \vdots & & \vdots & \ddots & \vdots \\ 0 & \cdots & 0 & \cdots & 0 & \cdots & 1 \end{bmatrix},$$

where the c's and s's appear at the intersections i-th and j-th rows and columns. From the structure of G_{ij}, it is clear that $G_{ij} = G_{ji}$. We prefer to write rotators as G_{ij} with $i < j$. Direct matrix multiplication reveals that $G_{ij}G'_{ij} = G'_{ij}G_{ij} = I_n$, which shows that Givens rotators in higher dimensions are orthogonal matrices.

Applying G_{ij} to a nonzero vector x affects only the i-th and j-th elements of x and leaves all other elements unchanged. To be precise, if $G_{ij}x = y$, then the k-th element of y is given by

$$y_k = \begin{cases} cx_i + sx_j & \text{if } k = i \\ -sx_i + cx_j & \text{if } k = j \\ x_k & \text{if } k \neq i, j \end{cases} \quad \text{or} \quad G_{ji}x = \begin{pmatrix} x_1 \\ \vdots \\ cx_i + sx_j \\ \vdots \\ -sx_i + cx_j \\ \vdots \\ x_n \end{pmatrix} \begin{matrix} \\ \\ \longleftarrow \; i \\ \\ \longleftarrow \; j \\ \\ \\ \end{matrix}.$$

Premultiplication by G_{ij} amounts to a clockwise rotation by an angle of θ radians about the (i,j) coordinate plane. This suggests that the j-th element of x can be made zero by rotating x so that it aligns with the i-axis. Explicit calculation of the angle of rotation is rarely necessary or even desirable from a numerical stability standpoint. Instead, we achieve the desired rotation by directly seeking values of c and s in terms of the elements of $\boldsymbol{x}$. Suppose x_i and x_j are not both zero and let $\boldsymbol{G}_{ij}$ be a Givens rotation matrix with

$$c = \frac{x_i}{\sqrt{x_i^2 + x_j^2}} \text{ and } s = \frac{x_j}{\sqrt{x_i^2 + x_j^2}} . \tag{8.27}$$

If $\boldsymbol{y} = \boldsymbol{G}_{ij}\boldsymbol{x}$, then the k-th element of $\boldsymbol{y}$ is given by

$$y_k = \begin{cases} \sqrt{x_i^2 + x_j^2} & \text{if } k = i \\ 0 & \text{if } k = j \\ x_k & \text{if } k \neq i, j \end{cases} \quad \text{or, equivalently,} \quad \boldsymbol{y} = \begin{pmatrix} x_1 \\ \vdots \\ \sqrt{x_i^2 + x_j^2} \\ \vdots \\ 0 \\ \vdots \\ x_n \end{pmatrix} \begin{matrix} \\ \\ \longleftarrow \quad i \\ \\ \longleftarrow \quad j \\ \\ \\ \end{matrix} .$$

This suggests that we can selectively sweep out any component j in $\boldsymbol{x}$ without altering any other entry except x_i and x_j. Consequently, plane rotations can be applied to sweep out all elements below any particular "pivot." For example, to annihilate *all* entries below the first position in $\boldsymbol{x}$, we can apply a sequence of plane rotations analogous to Example 8.5:

$$\boldsymbol{G}_{12}\boldsymbol{x} = \begin{bmatrix} \sqrt{x_1^2 + x_2^2} \\ 0 \\ x_3 \\ x_4 \\ \vdots \\ x_n \end{bmatrix}, \quad \boldsymbol{G}_{13}\boldsymbol{G}_{12}\boldsymbol{x} = \begin{bmatrix} \sqrt{x_1^2 + x_2^2 + x_3^2} \\ 0 \\ 0 \\ x_4 \\ \vdots \\ x_n \end{bmatrix}, \ldots,$$

$$\boldsymbol{G}_{1n} \cdots \boldsymbol{G}_{13}\boldsymbol{G}_{12}\boldsymbol{x} = \begin{bmatrix} \|\boldsymbol{x}\| \\ 0 \\ 0 \\ 0 \\ \vdots \\ 0 \end{bmatrix} .$$

Let us now turn to the effect of Givens rotations on matrices. Consider an $n \times n$ square matrix $\boldsymbol{A}$. Then $\boldsymbol{G}_{ij}\boldsymbol{A}$ has altered only rows i and j of $\boldsymbol{A}$. To sweep out all the elements below the $(1,1)$-th element, we use $\boldsymbol{Z}_1 = \boldsymbol{G}_{1n}\boldsymbol{G}_{1,n-1} \cdots \boldsymbol{G}_{12}$, with appropriately chosen $\boldsymbol{G}_{ij}$'s, to form the matrix $\boldsymbol{A}_2 = \boldsymbol{Z}_1\boldsymbol{A}$. Schematically, for

$n = 4$, we obtain

$$A_2 = Z_1 \begin{bmatrix} * & * & * & * \\ * & * & * & * \\ * & * & * & * \\ * & * & * & * \end{bmatrix} = \begin{bmatrix} + & + & + & + \\ 0 & + & + & + \\ 0 & + & + & + \\ 0 & + & + & + \end{bmatrix},$$

where, as usual, the $+$'s on the right-hand side indicate nonzero elements that have been altered from the $*$'s on the left-hand side. Similarly, to annihilate all entries below the $(2, 2)$-th element in the second column of A_2, we first form the Givens rotation matrices $G_{2,j}$ to annihilate the $(j, 2)$-th element in A_2, for $j = 3, 4, \ldots, n$. We then apply $Z_2 = G_{2n}G_{2,n-1} \cdots G_{2,3}$ and form $A_3 = Z_2A_2 = Z_2Z_1A$. Symbolically, for $n = 4$, this step produces

$$A_3 = Z_2 \begin{bmatrix} * & * & * & * \\ 0 & * & * & * \\ 0 & * & * & * \\ 0 & * & * & * \end{bmatrix} = \begin{bmatrix} * & * & * & * \\ 0 & + & + & + \\ 0 & 0 & + & + \\ 0 & 0 & + & + \end{bmatrix}.$$

Z_2 does not affect the first row of A_2 because it rotates elements that are below the $(2, 2)$-th element. It also does not affect the first column of A_2 because the j-th element in the first column is already zero for $j = 2, 3, \ldots, n$. At the $k - 1$-th step we have obtained A_k, which has zeroes in all positions below the diagonal elements for the first until the $k - 1$-th column. In the k-th step, we define $Z_k = G_{kn}G_{k,n-1} \cdots G_{k,k+1}$ and compute $A_{k+1} = Z_kA_k$, where A_{k+1} has zeroes as its entries below the diagonal elements for the first k columns. With $n = 4$ and $k = 3$, we find

$$A_4 = Z_3 \begin{bmatrix} * & * & * & * \\ 0 & * & * & * \\ 0 & 0 & * & * \\ 0 & 0 & * & * \end{bmatrix} = \begin{bmatrix} * & * & * & * \\ 0 & * & * & * \\ 0 & 0 & + & + \\ 0 & 0 & 0 & + \end{bmatrix}.$$

The matrix on the right is upper-triangular.

In general, after completing $n - 1$ premultiplications by the Z_k's, we arrive at an upper-triangular matrix $A_n = Z_{n-1} \cdots Z_2Z_1A$. As with Householder reflectors, a final reflection may be required to ensure that all diagonal elements in the upper-triangular matrix are positive (see Example 8.6 below). Therefore, $ZA = R$ is upper-triangular with positive diagonal elements, where $Z = H_nZ_{n-1} \cdots Z_2Z_1$. Each Z_k is an orthogonal matrix because it is a product of Givens rotators and H_n is orthogonal as well. So, Z is orthogonal as well and we arrive at the QR factorization: $A = QR$, where $Q = Z'$. This is called a ***Givens reduction***. We present a numerical example below.

Example 8.6 Let X be as in Example 8.3. We will now use Givens rotations in floating point arithmetic, rounding up to four decimal places, to reduce X to an upper-triangular form.

We start by making the $(2, 1)$-th element in $\boldsymbol{X}$ equal to zero. For this we need to calculate the Givens rotations $\boldsymbol{G}_{12}$. Using the procedure described above, we obtain

$$c = 0.4472\ , \quad s = 0.8944 \text{ and } \boldsymbol{G}_{12} = \begin{bmatrix} 0.4472 & 0.8944 & 0 \\ -0.8944 & 0.4472 & 0 \\ 0.0000 & 0.0000 & 1 \end{bmatrix} .$$

We compute $\boldsymbol{X}_2 = \boldsymbol{G}_{12}\boldsymbol{X} = \begin{bmatrix} 2.2361 & 7.6026 & 0.8944 \\ 0.0000 & 0.4472 & 0.4472 \\ 3.0000 & 10.0000 & 0.0000 \end{bmatrix}$.

Next, we form the Givens rotation to eliminate the $(1, 3)$-th element in $\boldsymbol{X}_2$:

$$c = 0.5976\ , \quad s = 0.8018 \text{ and } \boldsymbol{G}_{13} = \begin{bmatrix} 0.5976 & 0 & 0.8018 \\ 0.0000 & 1 & 0.0000 \\ -0.8018 & 0 & 0.5976 \end{bmatrix} .$$

The resulting matrix is $\boldsymbol{X}_3 = \boldsymbol{G}_{13}\boldsymbol{G}_{12}\boldsymbol{X} = \begin{bmatrix} 3.7417 & 12.5613 & 0.5345 \\ 0.0000 & 0.4472 & 0.4472 \\ 0.0000 & -0.1195 & -0.7171 \end{bmatrix}$. This matrix has its first column swept out below the pivot.

Turning to the second column, we need to annihilate $(3, 2)$-th element in $\boldsymbol{X}_3$. The Givens rotation for this is obtained as

$$c = 0.5976\ , \quad s = 0.8018 \text{ and } \boldsymbol{G}_{23} = \begin{bmatrix} 1 & 0 & 0 \\ 0 & 0.9661 & -0.2582 \\ 0 & 0.2582 & 0.9661 \end{bmatrix} .$$

The resulting matrix is upper-triangular:

$$\boldsymbol{G}_{23}\boldsymbol{G}_{13}\boldsymbol{G}_{12}\boldsymbol{X} = \begin{bmatrix} 3.7417 & 12.5613 & 0.5345 \\ 0.0000 & 0.4629 & 0.6172 \\ 0.0000 & 0.0000 & -0.5774 \end{bmatrix} .$$

As in Example 8.4, here also the upper-triangular matrix has a negative element in its $(3, 3)$-th position. A premultiplication by $\boldsymbol{H}_3 = \begin{bmatrix} 1 & 0 & 0 \\ 0 & 1 & 0 \\ 0 & 0 & -1 \end{bmatrix}$ makes that element positive. This, of course, is not necessary but we do so to reveal how the $\boldsymbol{R}$ and $\boldsymbol{Q}$ matrices obtained here are the same as those obtained from Householder reflectors (ensured by Lemma 8.3). Thus,

$$\boldsymbol{R} = \boldsymbol{H}_3\boldsymbol{G}_{23}\boldsymbol{G}_{13}\boldsymbol{G}_{12}\boldsymbol{X} = \begin{bmatrix} 3.7417 & 12.5613 & 0.5345 \\ 0.0000 & 0.4629 & 0.6172 \\ 0.0000 & 0.0000 & 0.5774 \end{bmatrix} \text{ and}$$

$$\boldsymbol{Q}' = \boldsymbol{H}_3\boldsymbol{G}_{23}\boldsymbol{G}_{13}\boldsymbol{G}_{21} = \begin{bmatrix} 0.2673 & 0.5345 & 0.8018 \\ -0.7715 & 0.6172 & -0.1543 \\ 0.5774 & 0.5774 & -0.5774 \end{bmatrix} ,$$

which are the same (up to four decimal places) as in Examples 8.3 and 8.4. ■

8.6.4 *The rectangular QR decomposition*

This procedure also works when A is an $m \times n$ (i.e., rectangular) matrix. In that case, the procedure continues until all of the rows or all of the columns (whichever is smaller) will be exhausted. The final result is one of the two following **upper-trapezoidal** forms:

$$
Q'A_{m \times n} =
\left.
\begin{bmatrix}
* & * & \cdots & * \\
0 & * & \cdots & * \\
\vdots & & \ddots & \vdots \\
0 & 0 & \cdots & * \\
0 & 0 & \cdots & 0 \\
\vdots & \vdots & & \vdots \\
0 & 0 & \cdots & 0
\end{bmatrix}
\right\} n \times n
\qquad \text{when } m > n,
$$

$$
Q'A_{m \times n} =
\underbrace{
\left(
\begin{array}{cccc|ccc}
* & * & \cdots & * & * & \cdots & * \\
0 & * & \cdots & * & * & \cdots & * \\
\vdots & & \ddots & \vdots & \vdots & & \vdots \\
0 & 0 & \cdots & * & * & \cdots & *
\end{array}
\right)
}_{m \times m}
\qquad \text{when } m < n.
$$

As for the $m = n$ case, the matrix Q' is formed either by a composition of Householder reflections or Givens rotations.

Let us provide a symbolic presentation of the dynamics. Consider a 5×3 matrix A:

$$
A =
\begin{bmatrix}
* & * & * \\
* & * & * \\
* & * & * \\
* & * & * \\
* & * & *
\end{bmatrix}.
$$

Let H_1 be a Householder reflector that zeroes all the entries below the $(1,1)$-th in A. In the first step, we compute

$$
A_2 = H_1 A = H_1
\begin{bmatrix}
* & * & * \\
* & * & * \\
* & * & * \\
* & * & * \\
* & * & *
\end{bmatrix}
=
\begin{bmatrix}
+ & + & + \\
0 & + & + \\
0 & + & + \\
0 & + & + \\
0 & + & +
\end{bmatrix},
$$

where $+$'s denote entries that have changed in the transformation.

In the second step, we compute $A_3 = H_2 A_2 = H_2 H_1 A$, where H_2 is a Householder reflector that zeroes all entries below the diagonal in the second column. Thus,

$$A_3 = H_2 A_2 = H_2 \begin{bmatrix} * & * & * \\ 0 & * & * \\ 0 & * & * \\ 0 & * & * \\ 0 & * & * \end{bmatrix} = \begin{bmatrix} * & * & * \\ 0 & + & + \\ 0 & 0 & + \\ 0 & 0 & + \\ 0 & 0 & + \end{bmatrix} .$$

Again, $+$'s denote entries that have changed in the last transformation, while $*$'s indicate entries that have not been changed.

In the third step, we use a Householder reflector H_3 to zero entries below $(3,3)$ in A_3:

$$A_4 = H_3 A_3 = H_3 \begin{bmatrix} * & * & * \\ 0 & * & * \\ 0 & 0 & * \\ 0 & 0 & * \\ 0 & 0 & * \end{bmatrix} = \begin{bmatrix} * & * & * \\ 0 & * & * \\ 0 & 0 & + \\ 0 & 0 & 0 \\ 0 & 0 & 0 \end{bmatrix} = \begin{bmatrix} R \\ O \end{bmatrix} ,$$

which is upper-trapezoidal. Thus, we have $Q'A = A_4$, where $Q' = H_3 H_2 H_1$.

In general, the above procedure, when applied to an $m \times n$ matrix A with $m \geq n$ will eventually produce the rectangular QR decomposition:

$$A = Q \begin{bmatrix} R \\ O \end{bmatrix} ,$$

where Q is $m \times m$ orthogonal, R is $n \times n$ upper-triangular and Q is $(m-n) \times n$. If A has linearly independent columns, then the matrix R is nonsingular. Since R is upper-triangular, it is nonsingular if and only if each of its diagonal entries is nonzero. So, A has linearly independent columns if and only if all the diagonal elements of R are nonzero.

A *thin* QR decomposition can be derived by partitioning $Q = [Q_1 : Q_2]$, where Q_1 is $m \times n$ and Q_2 is $m \times (m-n)$. Therefore,

$$A = Q \begin{bmatrix} R \\ O \end{bmatrix} = [Q_1 : Q_2] \begin{bmatrix} R \\ O \end{bmatrix} = Q_1 R .$$

The columns of Q_1 constitute an orthonormal basis for the column space of A. The j-th column of A can be written as a linear combination of the columns of Q_1 with coordinates (or coefficients) given by the entries in the j-th column of R.

8.6.5 Computational effort

Both Householder reflectors and Givens rotations can be used to reduce a matrix into triangular form. It is, therefore, natural to compare the number of operations (flops) they entail. Consider Givens rotations first and note that the operation

$$\begin{bmatrix} c & s \\ -s & c \end{bmatrix} \begin{bmatrix} x_i \\ x_j \end{bmatrix} = \begin{bmatrix} cx_i + sx_j \\ -sx_i + cx_j \end{bmatrix}$$

requires four multiplications and two additions. Considering each basic arithmetic operation as one flop, this operation involves 6 flops. Hence, a Givens rotation requires 6 flops to zero any one entry in a column vector. So, if we want to zero all the $m-1$ entries below the pivot in an $m \times 1$ vector, we will require $6(m-1) \approx 6m$ flops.

Now consider the Householder reflector to zero $m-1$ entries below the pivot in an $m \times 1$ vector:

$$\boldsymbol{H}_{\boldsymbol{u}}\boldsymbol{x} = (\boldsymbol{I} - 2\boldsymbol{u}\boldsymbol{u}')\boldsymbol{x} = \boldsymbol{x} - 2(\boldsymbol{u}'\boldsymbol{x})\boldsymbol{u} \ , \ \text{ where } \boldsymbol{u} = \frac{\boldsymbol{x} - \|\boldsymbol{x}\|\boldsymbol{e}_1}{\|\boldsymbol{x} - \|\boldsymbol{x}\|\boldsymbol{e}_1\|} \ .$$

It will be wasteful to construct the matrix $\boldsymbol{H}_{\boldsymbol{u}}$ first and then compute $\boldsymbol{H}_{\boldsymbol{u}}\boldsymbol{x}$. The dot-product between a row of $\boldsymbol{H}_{\boldsymbol{u}}$ and $\boldsymbol{x}$ will entail m multiplications plus $m-1$ additions, which add up to $2m-1$ flops. Doing this for all the m rows of $\boldsymbol{H}_{\boldsymbol{u}}$ to compute $\boldsymbol{H}_{\boldsymbol{u}}\boldsymbol{x}$ will cost $m(2m-1) \approx 2m^2$ flops. Instead, we should simply compute $\boldsymbol{x} - 2(\boldsymbol{u}'\boldsymbol{x})\boldsymbol{u}$. This will involve m subtractions, m scalar multiplications and another $2m-1$ operations for computing the dot product $\boldsymbol{u}'\boldsymbol{x}$. These add up to $4m-1 \approx 4m$ flops.

Thus, the Householder reflector is cheaper than the Givens rotations by approximately $2m$ flops when operating on an $m \times 1$ vector. Suppose now that the Householder reflector operates on an $m \times n$ matrix to zero the entries below the first element in the first column. Then, we are operating on all n columns, so the cost is $(4m-1)n \approx 4mn$ flops. The Givens, on the other hand, will need $6(m-1)n \approx 6mn$ flops, which is $3/2$ times more expensive.

Earlier, in describing these methods, we used partitioned matrices with elementary reflectors and 2×2 plane rotators embedded in larger matrices. In practice, such embedding is a waste of resources—both in terms of storage as well as the number of flops—because it does not utilize the fact that only a few rows of the matrix are changed in each iteration. For example, consider using Householder reflectors to reduce an $m \times n$ matrix $\boldsymbol{A}$ to an upper-trapezoidal matrix. After the entries below the diagonal in the first column have been zeroed, the reflectors do not affect the first row and column in any of the subsequent steps. The second step applies a reflector of dimension one less than the preceding one to the $(m-1) \times (n-1)$ submatrix formed by rows $2, 3, \ldots, m$ and columns $2, 3, \ldots, n$. So, the number of flops in the second stage is $4(m-1)(n-1)$. Continuing in this manner, we see that the cost at the k-th stage is $4(m-k+1)(n-k+1)$ flops. Hence, the total cost of Householder reduction for an $m \times n$ matrix with $m > n$ is

$$4\sum_{k=1}^{n}(m-k+1)(n-k+1) \approx 2mn^2 - \frac{2}{3}n^3 \text{ flops.}$$

Givens reduction costs approximately $3mn^2 - n^3$ flops, which is about 50% more than Householder reduction. With the above operations, the $\boldsymbol{Q}$ matrix is obtained in factorized form. Recovering the full $\boldsymbol{Q}$ matrix entails an additional cost of approximately $4m^2n - 2mn^2$ operations.

When $m = n$, i.e., for square matrices, the number of flops for Householder reductions is approximately $4n^3/3$, while that for Givens is about $2n^3$. Gram-Schmidt orthogonalization, both classical and modified, needs approximately $2n^3$ flops to reduce a general square matrix to triangular form, while Gaussian elimination (with partial pivoting), or the $\boldsymbol{LU}$ decomposition, requires about $2n^3/3$ flops. Therefore, both Gram-Schmidt and Gaussian elimination can arrive at triangular forms faster than Householder or Givens reductions. The additional cost for the latter two come with the benefit of their being unconditionally stable algorithms.

We remark that no one triangularization strategy can work for all settings and there usually is a cost-stability trade-off that comes into consideration. For example, Gaussian elimination with partial pivoting is perhaps among the most widely used methods for solving linear systems without any special structure because its relatively minor computational pitfalls do not justify complete pivoting or slightly more stable but more expensive algorithms. On the other hand, Householder reduction or modified Gram-Schmidt are frequently for least square problems to protect against sensitivities such problems often entail.

We have seen that the modified Gram-Schmidt, Householder and Givens reductions can all be used to obtain a $\boldsymbol{QR}$ decomposition or find an orthonormal basis for the column space of $\boldsymbol{A}$. When $\boldsymbol{A}$ is dense and unstructured, usually Householder reduction is the preferred choice. The Givens is approximately 50% more costly and the Gram-Schmidt procedures are unstable. However, when $\boldsymbol{A}$ has a pattern, perhaps with a number of zeroes, Givens reductions can make use of the structure and be more efficient. Below is an example with a very important class of sparse matrices.

Example 8.7 $\boldsymbol{QR}$ decomposition of a Hessenberg matrix. An *upper-Hessenberg* matrix is almost an upper-triangular matrix except that the entries along the sub-diagonal are also nonzero. A 5×5 Hessenberg matrix has the following structure

$$\boldsymbol{A} = \begin{bmatrix} * & * & * & * & * \\ * & * & * & * & * \\ 0 & * & * & * & * \\ 0 & 0 & * & * & * \\ 0 & 0 & 0 & * & * \end{bmatrix}.$$

Suppose we want to compute a $\boldsymbol{QR}$ decomposition of $\boldsymbol{A}$. Usually, because of fewer number of operations, Householder reflections are preferred to Givens rotations for bringing $\boldsymbol{A}$ to its triangular form. Note that the 0's in an upper-Hessenberg matrix remain as 0's in the final triangular form. Ideally, therefore, we would want the 0's in $\boldsymbol{A}$ to remain unharmed.

Unfortunately, the very first step of Householder reduction distorts most of the zeros in $\boldsymbol{A}$. They will eventually reappear in the final upper-triangular form, but this losing the 0's before recovering them seems wasteful. On the other hand, plane rotations can exploit the structure in $\boldsymbol{A}$ and arrive at the upper-triangular form without harming the existing 0's. We make use of the property that a Given's rotation $\boldsymbol{G}_{ij}$ only affects rows i and j of the matrix it multiplies from the left. We provide a schematic demonstration of the sequence of rotations that do the job on a 4×4 upper-Hessenberg matrix.

In the first step, we use a plane rotation in the $(1, 2)$ plane to zero the possibly nonzero $(2, 1)$-th element the first column of $\boldsymbol{A}$. Therefore,

$$\boldsymbol{A}_2 = \boldsymbol{G}_{12}\boldsymbol{A} = \boldsymbol{G}_{12} \begin{bmatrix} * & * & * & * \\ * & * & * & * \\ 0 & * & * & * \\ 0 & 0 & * & * \end{bmatrix} = \begin{bmatrix} + & + & + & + \\ 0 & + & + & + \\ 0 & * & * & * \\ 0 & 0 & * & * \end{bmatrix},$$

where $\boldsymbol{G}_{12}$ is an appropriately chosen 4×4 Givens rotator. Keep in mind that $\boldsymbol{G}_{12}$ affects only the first and second rows of $\boldsymbol{A}$, which results in the structure on the right. Again, $+$'s on the right-hand side indicate entries that are not necessarily 0 and are altered from what they were in $\boldsymbol{A}$, while the $*$'s represent entries that are not necessarily 0 and remain unaltered.

In the second step, we zero the $(3, 2)$-th in $\boldsymbol{A}_1$:

$$\boldsymbol{A}_3 = \boldsymbol{G}_{23}\boldsymbol{A}_2 = \boldsymbol{G}_{23} \begin{bmatrix} * & * & * & * \\ 0 & * & * & * \\ 0 & * & * & * \\ 0 & 0 & * & * \end{bmatrix} = \begin{bmatrix} * & * & * & * \\ 0 & + & + & + \\ 0 & 0 & + & + \\ 0 & 0 & * & * \end{bmatrix},$$

where $\boldsymbol{G}_{23}$ is a Givens rotator that zeroes the $(3, 2)$-th entry while altering only the second and third rows of $\boldsymbol{A}_2$.

The third, and final, step zeroes the $(4, 3)$ entry in $\boldsymbol{A}_3$:

$$\boldsymbol{A}_4 = \boldsymbol{G}_{34}\boldsymbol{A}_3 = \begin{bmatrix} * & * & * & * \\ 0 & * & * & * \\ 0 & 0 & * & * \\ 0 & 0 & * & * \end{bmatrix} = \begin{bmatrix} * & * & * & * \\ 0 & * & * & * \\ 0 & 0 & + & + \\ 0 & 0 & 0 & + \end{bmatrix},$$

where $\boldsymbol{G}_{34}$ zeroes the $(4, 3)$ entry in $\boldsymbol{A}_3$, while affecting only the third and fourth rows of $\boldsymbol{A}_3$. $\boldsymbol{A}_4 = \boldsymbol{R}$ and $\boldsymbol{Q} = \boldsymbol{G}_{34}'\boldsymbol{G}_{23}'\boldsymbol{G}_{12}'$ are the factors in the $\boldsymbol{QR}$ decomposition of $\boldsymbol{A}$. ■

Practical implementations of orthogonal triangularizations and the $\boldsymbol{QR}$ decompositions sometimes proceeds in two steps. First, a dense unstructured matrix is reduced to an upper-Hessenberg form using Householder transformations. Then, the $\boldsymbol{QR}$ decomposition is computed for this upper-Hessenberg matrix using Givens rotations. We just saw how to accomplish the second task in Example 8.7. In the next section, we demonstrate the first task: how a regular dense matrix can be brought to an upper-Hessenberg matrix using orthogonal similarity transformations.

8.7 Orthogonal similarity reduction to Hessenberg forms

We have seen how Householder reflections and Givens rotations can be used to reduce a matrix to triangular form, say $\boldsymbol{Q}'\boldsymbol{A} = \boldsymbol{R}$, where $\boldsymbol{Q}$ is orthogonal and

$\boldsymbol{R}$ is upper-triangular. Notice, however, that the orthogonal transformation operates only from the left. As briefly mentioned in Section 4.9, we often want to use similarity transformations (recall Definition 4.15) to reduce a matrix $\boldsymbol{A}$ to a simpler form. This raises the question: Can we find an orthogonal matrix $\boldsymbol{Q}$ such that $\boldsymbol{Q}'\boldsymbol{A}\boldsymbol{Q} = \boldsymbol{T}$ is upper-triangular? Theoretical issues, which we discuss later in Sections 11.5 and 11.8, establish the fact that this may not be possible using algorithms with a finite number of steps. An infinite number of iterations will be needed to *converge* to the desired triangularization.

If we restrict ourselves to procedures with a finite number of steps, what is the next best thing we can do? It turns out that we can construct an orthogonal matrix $\boldsymbol{Q}$, using Householder reflections or Givens rotations, such that $\boldsymbol{Q}'\boldsymbol{A}\boldsymbol{Q}$ is *almost* upper-triangular in which all entries below the first subdiagonal are zero. Such a matrix is called an ***upper-Hessenberg*** matrix. Below is the structure for a 5×5 upper-Hessenberg matrix:

$$\boldsymbol{H} = \begin{bmatrix} * & * & * & * & * \\ * & * & * & * & * \\ 0 & * & * & * & * \\ 0 & 0 & * & * & * \\ 0 & 0 & 0 & * & * \end{bmatrix}.$$

Remark: In this section, we will use $\boldsymbol{H}$ to denote an upper-Hessenberg matrix. Earlier, we had used $\boldsymbol{H}$ for Householder reflections but we will not do so here.

We demonstrate how to obtain $\boldsymbol{Q}'\boldsymbol{A}\boldsymbol{Q} = \boldsymbol{H}$ using 5×5 matrices as an example. We will first present a schematic overview of the process, which may suffice for readers who may not seek too many details. We will supply the details after the overview.

Schematic overview

The idea is to zero entries below the subdiagonal. We begin by choosing an orthogonal matrix $\boldsymbol{Q}_1$ that leaves the first row unchanged and zeroes everything *below* the $(2,1)$-th element. $\boldsymbol{Q}_1$ can be constructed using either a Householder reflector or a sequence of Givens rotations. Thus,

$$\boldsymbol{Q}_1'\boldsymbol{A} = \boldsymbol{Q}_1' \begin{bmatrix} * & * & * & * & * \\ * & * & * & * & * \\ * & * & * & * & * \\ * & * & * & * & * \\ * & * & * & * & * \end{bmatrix} = \begin{bmatrix} * & * & * & * & * \\ + & + & + & + & + \\ 0 & + & + & + & + \\ 0 & + & + & + & + \\ 0 & + & + & + & + \end{bmatrix},$$

where the $+$'s on the right-hand side indicate possibly nonzero entries that have changed from what they were in $\boldsymbol{A}$. The $*$'s on the right-hand side indicate entries that have not changed from $\boldsymbol{A}$.

The next observation is crucial. The first column does not change between $\boldsymbol{Q}_1'\boldsymbol{A}$ and $\boldsymbol{Q}_1'\boldsymbol{A}\boldsymbol{Q}_1$ because of the following: $\boldsymbol{Q}_1'\boldsymbol{A}$ has the same first row as $\boldsymbol{A}$, while each of its other rows is a linear combination of the second through fifth rows of $\boldsymbol{A}$. Post-multiplication by $\boldsymbol{Q}_1$ has the same effect on the columns of $\boldsymbol{Q}_1'\boldsymbol{A}$. Therefore, just as the first row of $\boldsymbol{A}$ is not altered in $\boldsymbol{Q}_1'\boldsymbol{A}$, so the first column of $\boldsymbol{Q}_1'\boldsymbol{A}$ is not altered in $\boldsymbol{Q}_1'\boldsymbol{A}\boldsymbol{Q}_1$. This is important to the success of this procedure because the zeroes in the first column of $\boldsymbol{Q}_1'\boldsymbol{A}$ are not distorted in $\boldsymbol{Q}_1'\boldsymbol{A}\boldsymbol{Q}_1$.

Each of the other columns of $\boldsymbol{Q}_1'\boldsymbol{A}\boldsymbol{Q}_1$ is a linear combination of the second through fifth columns of $\boldsymbol{Q}_1'\boldsymbol{A}$. This means that all the rows can change when moving from $\boldsymbol{Q}_1'\boldsymbol{A}$ to $\boldsymbol{Q}_1'\boldsymbol{A}\boldsymbol{Q}_1$. Schematically, we denote the dynamic effect of $\boldsymbol{Q}_1$ as follows:

$$\boldsymbol{Q}_1'\boldsymbol{A}\boldsymbol{Q}_1 = \begin{bmatrix} * & * & * & * & * \\ + & + & + & + & + \\ 0 & + & + & + & + \\ 0 & + & + & + & + \\ 0 & + & + & + & + \end{bmatrix} \boldsymbol{Q}_1 = \begin{bmatrix} * & + & + & + & + \\ * & + & + & + & + \\ 0 & + & + & + & + \\ 0 & + & + & + & + \\ 0 & + & + & + & + \end{bmatrix},$$

where $+$'s indicate elements that can get altered from the immediately preceding operation. For instance, the first column does not change between $\boldsymbol{Q}_1'\boldsymbol{A}$ and $\boldsymbol{Q}_1'\boldsymbol{A}\boldsymbol{Q}_1$, but all the rows possibly change.

The rest of the procedure now follows a pattern. Let $\boldsymbol{A}_2 = \boldsymbol{Q}_1'\boldsymbol{A}\boldsymbol{Q}_1$. In the second step, we construct an orthogonal matrix $\boldsymbol{Q}_2$ that zeroes entries below the $(3,2)$-th element in $\boldsymbol{A}_2$, while not affecting the first *two* rows of $\boldsymbol{A}_2$. Following steps analogous to the above, we obtain the structure of $\boldsymbol{Q}_2'\boldsymbol{A}_2\boldsymbol{Q}_2$ as

$$\boldsymbol{Q}_2' \begin{bmatrix} * & * & * & * & * \\ * & * & * & * & * \\ 0 & * & * & * & * \\ 0 & * & * & * & * \\ 0 & * & * & * & * \end{bmatrix} \boldsymbol{Q}_2 = \begin{bmatrix} * & * & * & * & * \\ * & * & * & * & * \\ 0 & + & + & + & + \\ 0 & 0 & + & + & + \\ 0 & 0 & + & + & + \end{bmatrix} \boldsymbol{Q}_2 = \begin{bmatrix} * & * & + & + & + \\ * & * & + & + & + \\ 0 & * & + & + & + \\ 0 & 0 & + & + & + \\ 0 & 0 & + & + & + \end{bmatrix}.$$

The first two columns do not change when moving from $\boldsymbol{Q}_2'\boldsymbol{A}$ to $\boldsymbol{Q}_2'\boldsymbol{A}\boldsymbol{Q}_2$, but all the rows possibly do.

Finally, letting $\boldsymbol{A}_3 = \boldsymbol{Q}_2'\boldsymbol{A}_2\boldsymbol{Q}_2 = \boldsymbol{Q}_2'\boldsymbol{Q}_1'\boldsymbol{A}\boldsymbol{Q}_1\boldsymbol{Q}_2$, we construct an orthogonal matrix $\boldsymbol{Q}_3$ that zeroes entries below the $(4,3)$-th element in $\boldsymbol{A}_3$, while not affecting the first *three* rows of $\boldsymbol{A}_3$. Then, we obtain $\boldsymbol{A}_4 = \boldsymbol{Q}_3'\boldsymbol{A}_3\boldsymbol{Q}_3$ as

$$\boldsymbol{Q}_3' \begin{bmatrix} * & * & * & * & * \\ * & * & * & * & * \\ 0 & * & * & * & * \\ 0 & 0 & * & * & * \\ 0 & 0 & * & * & * \end{bmatrix} \boldsymbol{Q}_3 = \begin{bmatrix} * & * & * & * & * \\ * & * & * & * & * \\ 0 & * & * & * & * \\ 0 & 0 & + & + & + \\ 0 & 0 & 0 & + & + \end{bmatrix} \boldsymbol{Q}_3 = \begin{bmatrix} * & * & * & + & + \\ * & * & * & + & + \\ 0 & * & * & + & + \\ 0 & 0 & * & + & + \\ 0 & 0 & 0 & + & + \end{bmatrix},$$

which is in upper-Hessenberg form.

The $n \times n$ case

The procedure for reducing $n \times n$ matrices to Hessenberg form proceeds by repeating the above steps $n-2$ times. Letting $\boldsymbol{Q} = \boldsymbol{Q}_1\boldsymbol{Q}_2 \cdots \boldsymbol{Q}_{n-2}$, we will obtain

$$\boldsymbol{Q}'\boldsymbol{A}\boldsymbol{Q} = \boldsymbol{Q}'_{n-2} \cdots \boldsymbol{Q}'_2\boldsymbol{Q}'_1\boldsymbol{A}\boldsymbol{Q}_1\boldsymbol{Q}_2 \cdots \boldsymbol{Q}_{n-2} = \boldsymbol{H}$$

as an $n \times n$ upper-Hessenberg matrix. This procedure establishes the following result: *Every square matrix is orthogonally similar to an upper-Hessenberg matrix.* Householder reductions can be used to arrive at this form in a finite number of iterations.

Details of a 5×5 Hessenberg reduction

Here we offer a bit more detail on how the orthogonal $\boldsymbol{Q}_i$'s are constructed. Using suitably partitioned matrices can be helpful.

Our first step begins with the following partition of $\boldsymbol{A}$:

$$\boldsymbol{A} = \begin{bmatrix} * & * & * & * & * \\ * & * & * & * & * \\ * & * & * & * & * \\ * & * & * & * & * \\ * & * & * & * & * \end{bmatrix} = \begin{bmatrix} a_{11} & \boldsymbol{a}'_{12} \\ \boldsymbol{a}_{21} & \boldsymbol{A}_{22} \end{bmatrix},$$

where a_{11} is a scalar, $\boldsymbol{a}'_{12}$ is 1×2, $\boldsymbol{a}_{21}$ is 2×1 and $\boldsymbol{A}_{22}$ is 4×4. Let $\boldsymbol{P}_1$ be an orthogonal matrix that zeroes all elements below the first term of $\boldsymbol{a}_{21}$. Thus,

$$\boldsymbol{P}_1\boldsymbol{a}_{21} = \boldsymbol{P}_1 \begin{bmatrix} * \\ * \\ * \\ * \end{bmatrix} = \begin{bmatrix} + \\ 0 \\ 0 \\ 0 \end{bmatrix}.$$

We can choose $\boldsymbol{P}_1$ either to be an appropriate 4×4 Householder reflector or an appropriate sequence of Givens rotations. Below, we choose the Householder reflector, so $\boldsymbol{P}_1 = \boldsymbol{P}'_1 = \boldsymbol{P}_1^{-1}$. Choosing a Householder reflector has the advantage that we do not need to deal with transposes.

Let us construct $\boldsymbol{Q}_1 = \begin{bmatrix} 1 & \boldsymbol{0}' \\ \boldsymbol{0} & \boldsymbol{P}_1 \end{bmatrix}$, which means $\boldsymbol{Q}_1 = \boldsymbol{Q}'_1$ because $\boldsymbol{P}_1 = \boldsymbol{P}'_1$. Then

$$\boldsymbol{A}_2 = \boldsymbol{Q}_1\boldsymbol{A}\boldsymbol{Q}_1 = \begin{bmatrix} a_{11} & \boldsymbol{a}'_{12}\boldsymbol{P}_1 \\ \boldsymbol{P}_1\boldsymbol{a}_{21} & \boldsymbol{P}_1\boldsymbol{A}_{22}\boldsymbol{P}_1 \end{bmatrix} = \begin{bmatrix} * & + & + & + & + \\ + & + & + & + & + \\ 0 & + & + & + & + \\ 0 & + & + & + & + \\ 0 & + & + & + & + \end{bmatrix},$$

where $+$'s stand for entries that are not necessarily 0 and have possibly changed from those in $\boldsymbol{A}$, while $*$ denotes entries that are not necessarily 0 and remain unchanged.

In the second step, consider the following partition of $\boldsymbol{A}_2 = \boldsymbol{Q}_1'\boldsymbol{A}\boldsymbol{Q}_1$:

$$\boldsymbol{A}_2 = \begin{bmatrix} * & * & * & * & * \\ * & * & * & * & * \\ 0 & * & * & * & * \\ 0 & * & * & * & * \\ 0 & * & * & * & * \end{bmatrix} = \begin{bmatrix} \boldsymbol{A}_{2(1,1)} & \boldsymbol{A}_{2(1,2)} \\ \boldsymbol{A}_{2(2,1)} & \boldsymbol{A}_{2(2,2)} \end{bmatrix},$$

where $\boldsymbol{A}_{2(1,1)}$ is 2×2, $\boldsymbol{A}_{2(1,2)}$ is 2×3, $\boldsymbol{A}_{2(2,1)}$ is 3×2 and $\boldsymbol{A}_{2(2,,2)}$ is 3×3. Consider a further partition of the submatrix

$$\boldsymbol{A}_{2(2,1)} = \begin{bmatrix} 0 & * \\ 0 & * \\ 0 & * \end{bmatrix} = \begin{bmatrix} \boldsymbol{0} : \boldsymbol{b} \end{bmatrix},$$

where $\boldsymbol{b}$ is a 3×1 vector. Let $\boldsymbol{P}_2$ be a 3×3 Householder reflector such that

$$\boldsymbol{P}_2\boldsymbol{b} = \boldsymbol{P}_2 \begin{bmatrix} * \\ * \\ * \end{bmatrix} = \begin{bmatrix} + \\ 0 \\ 0 \end{bmatrix}, \text{ hence } \boldsymbol{P}_2\boldsymbol{A}_{2(2,1)} = \begin{bmatrix} \boldsymbol{P}_2\boldsymbol{0} : \boldsymbol{P}_2\boldsymbol{b} \end{bmatrix} = \begin{bmatrix} 0 & + \\ 0 & 0 \\ 0 & 0 \end{bmatrix}.$$

If $\boldsymbol{Q}_2 = \begin{bmatrix} \boldsymbol{I}_2 & \boldsymbol{O} \\ \boldsymbol{O} & \boldsymbol{P}_2 \end{bmatrix} = \boldsymbol{Q}_2'$, then

$$\boldsymbol{A}_3 = \boldsymbol{Q}_2\boldsymbol{A}_2\boldsymbol{Q}_2 = \begin{bmatrix} \boldsymbol{A}_{2(1,1)} & \boldsymbol{A}_{2(1,2)}\boldsymbol{P}_2 \\ \boldsymbol{P}_2\boldsymbol{A}_{2(2,1)} & \boldsymbol{P}_2\boldsymbol{A}_{2(2,2)}\boldsymbol{P}_2 \end{bmatrix} = \begin{bmatrix} * & * & + & + & + \\ * & * & + & + & + \\ 0 & + & + & + & + \\ 0 & 0 & + & + & + \\ 0 & 0 & + & + & + \end{bmatrix},$$

where $+$'s denote the entries in $\boldsymbol{A}_3$ that have possibly changed from $\boldsymbol{A}_2$ and the $*$'s are entries that are not necessarily zero and remain unchanged from $\boldsymbol{A}_2$. This completes the second step.

The third step begins with the following partition of $\boldsymbol{A}_3$:

$$\boldsymbol{A}_3 = \begin{bmatrix} * & * & * & * & * \\ * & * & * & * & * \\ 0 & * & * & * & * \\ 0 & 0 & * & * & * \\ 0 & 0 & * & * & * \end{bmatrix} = \begin{bmatrix} \boldsymbol{A}_{3(1,1)} & \boldsymbol{A}_{3(1,2)} \\ \boldsymbol{A}_{3(2,1)} & \boldsymbol{A}_{3(2,2)} \end{bmatrix},$$

where $\boldsymbol{A}_{3(1,1)}$ is 3×3, $\boldsymbol{A}_{3(1,2)}$ is 3×2, $\boldsymbol{A}_{3(2,1)}$ is 2×3 and $\boldsymbol{A}_{3(2,,2)}$ is 2×2. Consider a further partition of the submatrix

$$\boldsymbol{A}_{3(2,1)} = \begin{bmatrix} 0 & 0 & * \\ 0 & 0 & * \end{bmatrix} = \begin{bmatrix} \boldsymbol{O} : \boldsymbol{d} \end{bmatrix},$$

where $\boldsymbol{O}$ is a 2×2 matrix of zeroes and $\boldsymbol{d}$ is a 2×1 vector. Let $\boldsymbol{P}_3$ be a 2×2 Householder reflector such that

$$\boldsymbol{P}_3\boldsymbol{d} = \boldsymbol{P}_3 \begin{bmatrix} * \\ * \end{bmatrix} = \begin{bmatrix} + \\ 0 \end{bmatrix}, \text{ hence } \boldsymbol{P}_3\boldsymbol{A}_{3(2,1)} = \begin{bmatrix} \boldsymbol{P}_3\boldsymbol{O} : \boldsymbol{P}_3\boldsymbol{d} \end{bmatrix} = \begin{bmatrix} 0 & 0 & + \\ 0 & 0 & 0 \end{bmatrix}.$$

If $\boldsymbol{Q}_3 = \begin{bmatrix} \boldsymbol{I}_3 & \boldsymbol{O} \\ \boldsymbol{O} & \boldsymbol{P}_3 \end{bmatrix} = \boldsymbol{Q}_3'$, then

$$\boldsymbol{A}_4 = \boldsymbol{Q}_3\boldsymbol{A}_3\boldsymbol{Q}_3 = \begin{bmatrix} \boldsymbol{A}_{3(1,1)} & \boldsymbol{A}_{3(1,2)}\boldsymbol{P}_3 \\ \boldsymbol{P}_3\boldsymbol{A}_{3(2,1)} & \boldsymbol{P}_3\boldsymbol{A}_{3(2,2)}\boldsymbol{P}_3 \end{bmatrix} = \begin{bmatrix} * & * & * & + & + \\ * & * & * & + & + \\ 0 & * & * & + & + \\ 0 & 0 & + & + & + \\ 0 & 0 & 0 & + & + \end{bmatrix},$$

where the $+$'s indicate entries that are not necessarily 0 and have possibly changed from $\boldsymbol{A}_3$, while the $*$'s stand for entries that are not necessarily zero and are the same as in $\boldsymbol{A}_3$. $\boldsymbol{A}_4$ is upper-Hessenberg and the process is complete.

The symmetric case: Tridiagonal matrices

What happens if we apply the above procedure to a symmetric matrix? If $\boldsymbol{A}$ is symmetric and $\boldsymbol{Q}'\boldsymbol{A}\boldsymbol{Q} = \boldsymbol{H}$ is upper-Hessenberg, then

$$\boldsymbol{H} = \boldsymbol{Q}'\boldsymbol{A}\boldsymbol{Q} = \boldsymbol{Q}'\boldsymbol{A}'\boldsymbol{Q} = (\boldsymbol{Q}'\boldsymbol{A}\boldsymbol{Q})' = \boldsymbol{H}',$$

which means that $\boldsymbol{H}$ is symmetric. How would a symmetric upper-Hessenberg matrix look? Symmetry would force entries *above* the subdiagonal to be zero because of the zeroes appearing below the subdiagonal in $\boldsymbol{H}$. So, for the 5×5 case,

$$\boldsymbol{H} = \boldsymbol{H}' = \begin{bmatrix} * & * & 0 & 0 & 0 \\ * & * & * & 0 & 0 \\ 0 & * & * & * & 0 \\ 0 & 0 & * & * & * \\ 0 & 0 & 0 & * & * \end{bmatrix},$$

which referred to as a ***tridiagonal*** matrix. Our algorithm establishes the following fact: *Every symmetric matrix is orthogonally similar to a tridiagonal matrix.* Householder reductions can be used to arrive at this form in a finite number of iterations.

8.8 Orthogonal reduction to bidiagonal forms

The previous section showed how similarity transformations could reduce a square matrix $\boldsymbol{A}$ to an upper-Hessenberg matrix. If we relax the use of similarity transformations, i.e., we no longer require the *same* orthogonal transformations be used on the columns as on the rows, then we can go further and reduce $\boldsymbol{A}$ to an upper-bidiagonal matrix. An upper-bidiagonal matrix has nonzero entries only along the diagonal and the super-diagonal. For example, here is how a 5×5 upper-bidiagonal matrix looks:

$$\boldsymbol{B} = \begin{bmatrix} * & * & 0 & 0 & 0 \\ 0 & * & * & 0 & 0 \\ 0 & 0 & * & * & 0 \\ 0 & 0 & 0 & * & * \\ 0 & 0 & 0 & 0 & * \end{bmatrix}.$$

Also, because we are now free to choose different orthogonal transformations for the rows and columns, $\boldsymbol{A}$ can be rectangular and the matrices multiplying $\boldsymbol{A}$ from the left need not have the same dimension as those multiplying $\boldsymbol{A}$ on the right. We illustrate the procedure with a 5×4 matrix $\boldsymbol{A}$.

Our first step begins with the following partition of $\boldsymbol{A}$:

$$\boldsymbol{A} = \begin{bmatrix} \boldsymbol{a}_1 : \boldsymbol{A}_{12} \end{bmatrix} = \begin{bmatrix} * & * & * & * \\ * & * & * & * \\ * & * & * & * \\ * & * & * & * \\ * & * & * & * \end{bmatrix},$$

where $\boldsymbol{a}_1$ is 5×1 and $\boldsymbol{A}_{12}$ is 5×3. Let $\boldsymbol{Q}_1'$ be a 5×5 orthogonal matrix (Householder or Givens) that zeroes all elements below the first term of $\boldsymbol{a}_1$. Thus,

$$\boldsymbol{Q}_1'\boldsymbol{a}_1 = \boldsymbol{Q}_1' \begin{bmatrix} * \\ * \\ * \\ * \\ * \end{bmatrix} = \begin{bmatrix} + \\ 0 \\ 0 \\ 0 \\ 0 \end{bmatrix},$$

which means that

$$\boldsymbol{Q}_1'\boldsymbol{A} = \begin{bmatrix} \boldsymbol{Q}_1'\boldsymbol{a}_1 : \boldsymbol{Q}_1'\boldsymbol{A}_{12} \end{bmatrix} = \begin{bmatrix} t_{11} & \tilde{\boldsymbol{a}}_1' \\ \boldsymbol{0} & \tilde{\boldsymbol{A}}_1 \end{bmatrix} = \begin{bmatrix} + & + & + & + \\ 0 & + & + & + \\ 0 & + & + & + \\ 0 & + & + & + \\ 0 & + & + & + \end{bmatrix}.$$

The $+$'s indicate entries not necessarily zero that have been altered by $\boldsymbol{Q}_1'$. Also, $\boldsymbol{Q}_1'\boldsymbol{A}$ has been partitioned so that t_{11} is a scalar, $\tilde{\boldsymbol{a}}_1'$ is 1×3 and $\tilde{\boldsymbol{A}}_1$ is 4×3. Let $\boldsymbol{W}_1$ be a 3×3 orthogonal matrix such that

$$\boldsymbol{W}_1'\tilde{\boldsymbol{a}}_1 = \boldsymbol{W}_1' \begin{bmatrix} * \\ * \\ * \end{bmatrix} = \begin{bmatrix} + \\ 0 \\ 0 \end{bmatrix}.$$

Embed $\boldsymbol{W}_1$ in $\boldsymbol{P}_1 = \begin{bmatrix} 1 & \boldsymbol{0}' \\ \boldsymbol{0} & \boldsymbol{W}_1 \end{bmatrix}$. Therefore, $\boldsymbol{P}_1$ is orthogonal. Specifically, if $\boldsymbol{W}_1$ is chosen to be a Householder reflector, then $\boldsymbol{P}_1$ is also a Householder reflector and

$$\boldsymbol{A}_2 = \boldsymbol{Q}_1'\boldsymbol{A}\boldsymbol{P}_1 = \begin{bmatrix} t_{11} & \tilde{\boldsymbol{a}}_1'\boldsymbol{W}_1 \\ \boldsymbol{0} & \tilde{\boldsymbol{A}}_1\boldsymbol{W}_1 \end{bmatrix} = \begin{bmatrix} * & + & 0 & 0 \\ 0 & + & + & + \\ 0 & + & + & + \\ 0 & + & + & + \\ 0 & + & + & + \end{bmatrix},$$

where $+$'s stand for entries that are not necessarily 0 and have possibly been altered by $\boldsymbol{P}_1$, while $*$ denotes the lone entry that is not necessarily 0 and remains unchanged. Hence, the first column does not change between $\boldsymbol{Q}_1'\boldsymbol{A}$ and $\boldsymbol{Q}_1'\boldsymbol{A}\boldsymbol{P}_1$, which is crucial because the 0's in the first column of $Q_1'A$ remain intact. This completes the first step toward bidiagonalization. The first row of A_2 has nonzero entries only on the diagonal and super-diagonal.

In the second step, consider the following partition of $A_2 = Q_1'AP_1$:

$$A_2 = \begin{bmatrix} t_{11} & t_{12} & \mathbf{0}' \\ \mathbf{0} & a_{2(2,2)} & A_{2(2,3)} \end{bmatrix} = \begin{bmatrix} * & * & 0 & 0 \\ 0 & * & * & * \\ 0 & * & * & * \\ 0 & * & * & * \\ 0 & * & * & * \end{bmatrix},$$

where t_{11} and t_{12} are scalars, $a_{2(2,2)}$ is 4×1 and $A_{2(2,3)}$ is 4×2. Let V_2' be a 4×4 orthogonal matrix such that

$$V_2'a_{2(2,2)} = V_2' \begin{bmatrix} * \\ * \\ * \\ * \end{bmatrix} = \begin{bmatrix} + \\ 0 \\ 0 \\ 0 \end{bmatrix}.$$

If $Q_2' = \begin{bmatrix} 1 & \mathbf{0}' \\ \mathbf{0} & V_2' \end{bmatrix}$, then

$$Q_2'A_2 = \begin{bmatrix} t_{11} & t_{12} & \mathbf{0}' \\ \mathbf{0} & V_2'a_{2(2,2)} & V_2'A_{2(2,3)} \end{bmatrix} = \begin{bmatrix} * & * & 0 & 0 \\ 0 & + & + & + \\ 0 & 0 & + & + \\ 0 & 0 & + & + \\ 0 & 0 & + & + \end{bmatrix},$$

where $+$'s indicate entries that are not necessarily zero and may have been altered from A_2. Partition $Q_2'A_2$ as

$$Q_2'A_2 = \begin{bmatrix} t_{11} & t_{12} & \mathbf{0}' \\ 0 & t_{22} & \tilde{a}_2' \\ \mathbf{0} & \mathbf{0} & \tilde{A}_2 \end{bmatrix},$$

where $\tilde{a}_2'$ is 1×2 and $\tilde{A}_2$ is 3×2. We now wish to zero all the elements except the first in $\tilde{a}_2'$. Let W_2' be a 2×2 orthogonal matrix such that

$$W_2'\tilde{a}_2 = W_2' \begin{bmatrix} * \\ * \end{bmatrix} = \begin{bmatrix} + \\ 0 \end{bmatrix}.$$

The matrix $P_2 = \begin{bmatrix} I_2 & O \\ O & W_2 \end{bmatrix}$ is orthogonal and

$$A_3 = Q_2'A_2P_2 = \begin{bmatrix} t_{11} & t_{12} & \mathbf{0}' \\ 0 & t_{22} & \tilde{a}_2'W_2 \\ \mathbf{0} & \mathbf{0} & \tilde{A}_2W_2 \end{bmatrix} = \begin{bmatrix} * & * & 0 & 0 \\ 0 & * & + & 0 \\ 0 & 0 & + & + \\ 0 & 0 & + & + \\ 0 & 0 & + & + \end{bmatrix},$$

where $+$'s denote the entries in A_3 that are not necessarily zero and have possibly changed from $Q_2'A_2$ and $*$'s are entries that are not necessarily zero and remain unchanged from $Q_2'A_2$. Note that the first two columns of $Q_2'A_2$ are not altered by P_2, which means that zeros introduced in the first two columns in previous steps are retained. This completes the second step. The first two rows of A_3 have nonzero entries only along the diagonal and super-diagonal.

The third step begins with the following partition of $A_3 = Q_2'A_2P_2$:

$$A_3 = \begin{bmatrix} t_{11} & t_{12} & 0 & 0 \\ 0 & t_{22} & t_{23} & 0 \\ \mathbf{0} & \mathbf{0} & a_{3(3,3)} & a_{3(3,4)} \end{bmatrix} = \begin{bmatrix} * & * & 0 & 0 \\ 0 & * & * & 0 \\ 0 & 0 & * & * \\ 0 & 0 & * & * \\ 0 & 0 & * & * \end{bmatrix},$$

where $a_{3(3,3)}$ and $a_{3(3,4)}$ are 3×1. We now find an orthogonal matrix Q_3' that will zero the entries below the $(3,3)$-th element in A_3, which means all entries below the first in $a_{3(3,3)}$. Let V_3' be a 3×3 orthogonal matrix such that

$$V_3'a_{3(3,3)} = V_3' \begin{bmatrix} * \\ * \\ * \end{bmatrix} = \begin{bmatrix} + \\ 0 \\ 0 \end{bmatrix}.$$

Construct the orthogonal matrix $Q_3' = \begin{bmatrix} I_2 & O \\ O & V_3' \end{bmatrix}$. Then,

$$Q_3'A_3 = \begin{bmatrix} t_{11} & t_{12} & 0 & 0 \\ 0 & t_{22} & t_{23} & 0 \\ \mathbf{0} & \mathbf{0} & V_3'a_{3(3,3)} & V_3'a_{3(3,4)} \end{bmatrix} = \begin{bmatrix} * & * & 0 & 0 \\ 0 & * & * & 0 \\ 0 & 0 & + & + \\ 0 & 0 & 0 & + \\ 0 & 0 & 0 & + \end{bmatrix}.$$

The $+$'s indicate entries that have possibly been altered by Q_3' and are not necessarily zero, while $*$'s indicate entries not necessarily zero that have not been changed by Q_3'. Note that no more column operations on $Q_3'A_3$ are needed. Partition $Q_3'A_3$ as

$$Q_3'A_3 = \begin{bmatrix} t_{11} & t_{12} & 0 & 0 \\ 0 & t_{22} & t_{23} & 0 \\ 0 & 0 & t_{33} & t_{34} \\ \mathbf{0} & \mathbf{0} & \mathbf{0} & \tilde{a}_3 \end{bmatrix}$$

where $\tilde{a}_3$ is 2×1. In the final step, we find an orthogonal matrix Q_4' that will zero all the entries below the $(4,4)$-th entry in $Q_3'A_3$, or below the first element $\tilde{a}_3$. We construct a 2×2 orthogonal matrix V_4', such that

$$V_4'\tilde{a}_3 = V_4' \begin{bmatrix} * \\ * \end{bmatrix} = \begin{bmatrix} + \\ 0 \end{bmatrix},$$

and let $Q_4' = \begin{bmatrix} I_3 & O \\ O & V_4' \end{bmatrix}$. Then,

$$A_4 = Q_4'Q_3'A_3 = \begin{bmatrix} t_{11} & t_{12} & 0 & 0 \\ 0 & t_{22} & t_{23} & 0 \\ 0 & 0 & t_{33} & t_{34} \\ \mathbf{0} & \mathbf{0} & \mathbf{0} & V_4'\tilde{a}_3 \end{bmatrix} = \begin{bmatrix} * & * & 0 & 0 \\ 0 & * & * & 0 \\ 0 & 0 & * & * \\ 0 & 0 & 0 & + \\ 0 & 0 & 0 & 0 \end{bmatrix},$$

where $+$'s are entries that have possibly been changed by Q_4', while $*$'s are entries that have not.

In summary, the above sequence of steps have resulted in

$$Q'AP = Q_4'Q_3'Q_2'Q_1'AP_1P_2 = \begin{bmatrix} + & + & 0 & 0 \\ 0 & + & + & 0 \\ 0 & 0 & + & + \\ 0 & 0 & 0 & + \\ 0 & 0 & 0 & 0 \end{bmatrix},$$

where $Q = Q_1Q_2Q_3Q_4$ and $P = P_1P_2$ are orthogonal matrices. Here, the $+$'s indicate entries that can be nonzero and have changed from A.

In general, if A is an $m \times n$ matrix with $m \geq n$, then the above procedure results in orthogonal matrices Q and P, of order $m \times m$ and $n \times n$, respectively, such that

$$Q'AP = \begin{bmatrix} B \\ O \end{bmatrix}, \quad \text{where } B = \begin{bmatrix} t_{11} & t_{12} & 0 & \cdots & 0 & 0 \\ 0 & t_{22} & t_{23} & \cdots & 0 & 0 \\ 0 & 0 & t_{33} & \cdots & 0 & 0 \\ \vdots & \vdots & \vdots & \ddots & \vdots & \vdots \\ 0 & 0 & 0 & \cdots & t_{n-1,n-1} & t_{n-1,n} \\ 0 & 0 & 0 & \cdots & 0 & t_{nn} \end{bmatrix}$$

is upper-bidiagonal and O is the $(m-n) \times n$ matrix of zeros.

8.9 Some further reading on statistical linear models

We conclude this chapter with a few brief remarks. Orthogonality, orthogonal projections and projectors, as discussed in this chapter and Chapter 7, play a central role in the theory of ***statistical linear regression models***. Such models are usually represented as $y = X\beta + \eta$, where y is an $n \times 1$ vector containing the observations on the outcome (also called the response or the dependent variable), X is an $n \times p$ matrix whose columns comprise observations on regressors (also called predictors or covariates or independent variables) and η is an $n \times 1$ vector of random errors that follow a specified probability law—most commonly the Normal or Gaussian distribution.

While it is tempting to discuss the beautiful theory of statistical linear models, which brings together probability theory and linear algebra, we opt not to pursue that route in order to maintain our focus on linear algebra and matrix analysis. In fact, there are several excellent texts on the subject. The texts by Schott (2005) and Gentle (2010) both include material on linear models, while the somewhat concise text by Bapat (2012) delightfully melds linear algebra with statistical linear models. Rao (1973) is a classic on linear statistical inference that makes heavy use of matrix algebra. Other, more recent, texts that teach linear models using plenty of linear algebra include, but certainly are not limited to, Stapleton (1995), Seber and Lee (2003), Faraway (2005), Monahan (2008) and Christensen (2011). A collection of useful matrix algebra tricks for linear models, with heavy use of orthogonal projectors, have been collected by Putanen, Styan and Isotallo (2011). The reader is encouraged to explore these wonderful resources to see the interplay between linear algebra and statistical linear models.

8.10 Exercises

1. For what values of real numbers a and b will $\begin{bmatrix} a+b & a-b \\ b-a & a+b \end{bmatrix}$ be an orthogonal matrix?
2. Prove that any permutation matrix is an orthogonal matrix.
3. If $\boldsymbol{A}$ and $\boldsymbol{B}$ are two $n \times n$ orthogonal matrices, then show that $\boldsymbol{AB}$ is an orthogonal matrix.
4. Let $\boldsymbol{A}$ and $\boldsymbol{B}$ be $n \times n$ and $p \times p$, respectively. If $\boldsymbol{A}$ and $\boldsymbol{B}$ are orthogonal matrices, prove that $\begin{bmatrix} \boldsymbol{A} & \boldsymbol{O} \\ \boldsymbol{O} & \boldsymbol{B} \end{bmatrix}$ is an orthogonal matrix.
5. Let $\boldsymbol{Q}_1(\theta)$ and $\boldsymbol{Q}_2(\theta)$ be defined as in Example 8.1. Find $\boldsymbol{Q}_1(\theta)\boldsymbol{Q}_1(\eta)$, $\boldsymbol{Q}_1(\theta)\boldsymbol{Q}_2(\eta)$, $\boldsymbol{Q}_2(\eta)\boldsymbol{Q}_1(\theta)$ and $\boldsymbol{Q}_2(\theta)\boldsymbol{Q}_2(\eta)$. Verify that each of these are orthogonal matrices.
6. Let $\boldsymbol{x}_0 = (1, 1, 1)'$. Rotate this point counterclockwise by 45 degrees about the x-axis and call the resulting point $\boldsymbol{x}_1$. Rotate $\boldsymbol{x}_1$ clockwise by 90 degrees about the y-axis and call this point $\boldsymbol{x}_2$. Finally, rotate $\boldsymbol{x}_2$ counterclockwise by 60 degrees to produce $\boldsymbol{x}_3$. Find each of the points $\boldsymbol{x}_1$, $\boldsymbol{x}_2$ and $\boldsymbol{x}_3$. Find the orthogonal matrix $\boldsymbol{Q}$ such that $\boldsymbol{Q}\boldsymbol{x}_0 = \boldsymbol{x}_3$.
7. Let $\boldsymbol{A}$ and $\boldsymbol{B}$ be two $n \times n$ orthogonal matrices. Construct an example to show that $\boldsymbol{A} + \boldsymbol{B}$ need not be orthogonal.
8. Prove that the following statements are equivalent:
   (a) $\boldsymbol{A}$ is an $n \times n$ orthogonal matrix;
   (b) $\langle \boldsymbol{Ax}, \boldsymbol{Ay} \rangle = \langle \boldsymbol{x}, \boldsymbol{y} \rangle$ for every $\boldsymbol{x}, \boldsymbol{y} \in \Re^n$;
   (c) $\|\boldsymbol{Ax}\| = \|\boldsymbol{x}\|$ for every $\boldsymbol{x} \in \Re^n$.
9. True or false: $\boldsymbol{A}$ is an orthogonal matrix if and only if $\|\boldsymbol{Ax} - \boldsymbol{Ay}\| = \|\boldsymbol{x} - \boldsymbol{y}\|$ for every $\boldsymbol{x}, \boldsymbol{y} \in \Re^n$.
10. Let $\boldsymbol{A}$ be an $n \times n$ orthogonal matrix and let $\{\boldsymbol{x}_1, \boldsymbol{x}_2, \ldots, \boldsymbol{x}_n\}$ be an orthonormal basis for $\Re^n$. Show that $\{\boldsymbol{Ax}_1, \boldsymbol{Ax}_2, \ldots, \boldsymbol{Ax}_n\}$ is an orthonormal basis for $\Re^n$.

11. Let $\boldsymbol{A} = [\boldsymbol{A}_1 : \boldsymbol{A}_2]$ be a partitioned orthogonal matrix. Prove that $\mathcal{C}(\boldsymbol{A}_1)$ and $\mathcal{C}(\boldsymbol{A}_2)$ are orthogonal complements of each other.
12. If $\boldsymbol{A}$ is orthogonal, prove that $\mathcal{C}(\boldsymbol{I} - \boldsymbol{A})$ and $\mathcal{N}(\boldsymbol{A})$ are orthogonal complements of each other.
13. Let $\boldsymbol{u} \in \Re^n$ such that $\|\boldsymbol{u}\| = 1$ and let $\boldsymbol{u} = \begin{bmatrix} u_1 \\ \boldsymbol{u}_2 \end{bmatrix}$, where $u_1 \neq 1$ is a scalar and $\boldsymbol{u}_2 \neq \boldsymbol{0}$ is $(n-1) \times 1$. Find a value of α for which
$$\begin{bmatrix} u_1 & \boldsymbol{u}_2' \\ \boldsymbol{u}_2 & \boldsymbol{I} - \alpha \boldsymbol{u}_2 \boldsymbol{u}_2' \end{bmatrix}$$
is an orthogonal matrix.
14. In Example 2.1 we solved the linear system $\boldsymbol{A}\boldsymbol{x} = \boldsymbol{b}$, where
$$\boldsymbol{A} = \begin{bmatrix} 2 & 3 & 0 & 0 \\ 4 & 7 & 2 & 0 \\ -6 & -10 & 0 & 1 \\ 4 & 6 & 4 & 5 \end{bmatrix} \text{ and } \boldsymbol{b} = \begin{bmatrix} 1 \\ 2 \\ 1 \\ 0 \end{bmatrix}.$$
Using the Gram-Schmidt procedure, as described in Section 8.2, find a $\boldsymbol{QR}$ decomposition of $\boldsymbol{A}$ in exact arithmetic. Solve $\boldsymbol{R}\boldsymbol{x} = \boldsymbol{Q}'\boldsymbol{b}$ and verify that this is the same solution obtained in Example 2.1.
15. Consider the 4×3 matrix
$$\boldsymbol{A} = \begin{bmatrix} 2 & 6 & 4 \\ -1 & 1 & 1 \\ 0 & 4 & 3 \\ 1 & -5 & -4 \end{bmatrix}.$$
Find the rank of $\boldsymbol{A}$ and call it p. Find a $4 \times p$ matrix $\boldsymbol{Q}$ whose columns form an orthonormal basis for $\mathcal{C}(\boldsymbol{A})$ and a $p \times p$ upper-triangular matrix $\boldsymbol{R}$ such that $\boldsymbol{A} = \boldsymbol{QR}$.
16. Explore obtaining a $\boldsymbol{QR}$ decomposition for the singular matrix
$$\begin{bmatrix} 1 & 2 & 3 \\ 4 & 5 & 6 \\ 7 & 8 & 9 \end{bmatrix}.$$
17. Find the orthogonal projector onto the column space of $\boldsymbol{X}$, where
$$\begin{bmatrix} 1 & 1 \\ 1 & 2 \\ 1 & 3 \\ 1 & 4 \end{bmatrix}.$$
18. What is the orthogonal projector onto the subspace spanned by $\boldsymbol{1}_{n \times 1}$?
19. True or false: Orthogonal projectors are orthogonal matrices.
20. Let $\boldsymbol{P}$ and $\boldsymbol{Q}$ be $n \times n$ orthogonal projectors. Show that $\boldsymbol{P} + \boldsymbol{Q}$ is an orthogonal projector if and only if $\mathcal{C}(\boldsymbol{P}) \perp \mathcal{C}(\boldsymbol{Q})$. Under this condition, show that $\boldsymbol{P} + \boldsymbol{Q}$ is the orthogonal projector onto $\mathcal{C}(\boldsymbol{P}) + \mathcal{C}(\boldsymbol{Q})$.

21. Let $\boldsymbol{P} = \{p_{ij}\}$ be an $n \times n$ orthogonal projector. Show that (i) $0 \leq p_{ii} \leq 1$ for each $i = 1, 2, \ldots, n$, and (ii) $-1/2 \leq p_{ij} \leq 1/2$ whenever $i \neq j$.

22. If $\mathcal{C}(\boldsymbol{A}) \subset \mathcal{C}(\boldsymbol{B})$, then prove that $\boldsymbol{P_B} - \boldsymbol{P_A}$ is the orthogonal projector onto $\mathcal{C}((\boldsymbol{I} - \boldsymbol{P_A})\boldsymbol{B})$.

23. Let $\mathcal{S}$ and $\mathcal{T}$ be subspaces in $\Re^n$ such that $\mathcal{S} \subset \mathcal{T}$. Let $\boldsymbol{P}_{\mathcal{S}}$ and $\boldsymbol{P}_{\mathcal{T}}$ be the orthogonal projectors onto $\mathcal{S}$ and $\mathcal{T}$, respectively. Prove the following:
   (a) $\boldsymbol{P}_{\mathcal{S}}\boldsymbol{P}_{\mathcal{T}} = \boldsymbol{P}_{\mathcal{T}}\boldsymbol{P}_{\mathcal{S}} = \boldsymbol{P}_{\mathcal{S}}$.
   (b) $\boldsymbol{P}_{\mathcal{T}} = \boldsymbol{P}_{\mathcal{S}} + \boldsymbol{P}_{\mathcal{S}^{\perp} \cap \mathcal{T}}$.
   (c) If $\mathcal{S} = \mathcal{C}(\boldsymbol{A}) \cap \mathcal{T}$, then $\mathcal{S}^{\perp} \cap \mathcal{T} = \mathcal{C}(\boldsymbol{P_T}\boldsymbol{A}')$.

24. If $\boldsymbol{P} = \boldsymbol{P}^2$ and $\mathcal{N}(\boldsymbol{P}) \perp \mathcal{C}(\boldsymbol{P})$, then show that $\boldsymbol{P}$ is an orthogonal projector.

25. If $\boldsymbol{P} = \boldsymbol{P}^2$ is $n \times n$ and $\|\boldsymbol{P}\boldsymbol{x}\| \leq \|\boldsymbol{x}\|$ for every $\boldsymbol{x} \in \Re^n$, then show that $\boldsymbol{P}$ is an orthogonal projector.

26. Use the modified Gram-Schmidt procedure in Section 8.6.1 to obtain the $\boldsymbol{QR}$ decomposition of the matrix $\boldsymbol{A}$ in Exercise 14.

27. Using the modified Gram-Schmidt procedure find an orthonormal basis for each of the four fundamental subspaces of

$$\boldsymbol{A} = \begin{bmatrix} 2 & -2 & -5 & -3 & -1 & 2 \\ 2 & -1 & -3 & 2 & 3 & 2 \\ 4 & -1 & -4 & 10 & 11 & 4 \\ 0 & 1 & 2 & 5 & 4 & 0 \end{bmatrix}.$$

28. Let $\boldsymbol{H} = (\boldsymbol{I} - 2\boldsymbol{u}\boldsymbol{u}')$, where $\|\boldsymbol{u}\| = 1$. If $\boldsymbol{x}$ is a *fixed point* of $\boldsymbol{H}$ in the sense that $\boldsymbol{H}\boldsymbol{x} = \boldsymbol{x}$, then prove that $\boldsymbol{x}$ must be orthogonal to $\boldsymbol{u}$.

29. Let $\boldsymbol{x}$ and $\boldsymbol{y}$ be vectors in $\Re^n$ such that $\|\boldsymbol{x}\| = \|\boldsymbol{y}\|$ but $\boldsymbol{x} \neq \boldsymbol{y}$. Can you find an elementary reflector $\boldsymbol{H_u}$ such that $\boldsymbol{H_u}\boldsymbol{x} = \boldsymbol{y}$?

30. Use Householder reduction to find an orthogonal matrix $\boldsymbol{P}$ such that $\boldsymbol{PA} = \boldsymbol{T}$ is upper-triangular with positive diagonal entries, where $\boldsymbol{A}$ is the matrix in Exercise 14. Repeat the exercise using Givens rotations.

31. Use either Householder or Givens reduction to find the rectangular $\boldsymbol{QR}$ decomposition (as described in Section 8.6.4) of the matrix $\boldsymbol{A}$ in Exercise 15.

32. Use Givens reduction to find the $\boldsymbol{QR}$ decomposition $\boldsymbol{A} = \boldsymbol{QR}$, where

$$\boldsymbol{A} = \begin{bmatrix} 1 & 4 & 2 & 3 \\ 3 & 4 & 1 & 7 \\ 0 & 2 & 3 & 4 \\ 0 & 0 & 1 & 3 \end{bmatrix} \quad \text{is upper-Hessenberg .}$$

CHAPTER 9

Revisiting Linear Equations

9.1 Introduction

In Chapter 2, we discussed systems of linear equations and described the mechanics of Gaussian elimination to solve such systems. In this chapter, we revisit linear systems from a more theoretical perspective and explore how we can understand them better with the help of subspaces and their dimensions.

9.2 Null spaces and the general solution of linear systems

The null space of a matrix $\boldsymbol{A}$ also plays a role in describing the solutions for a consistent non-homogeneous system $\boldsymbol{Ax} = \boldsymbol{b}$. The following result says that any solution for $\boldsymbol{Ax} = \boldsymbol{b}$ must be of the form $\boldsymbol{x}_p + \boldsymbol{w}$, where $\boldsymbol{x}_p$ is any particular solution for $\boldsymbol{Ax} = \boldsymbol{b}$ and $\boldsymbol{w}$ is some vector in the null space of $\boldsymbol{A}$.

Theorem 9.1 *If $\boldsymbol{Ax} = \boldsymbol{b}$ be a consistent linear system, where $\boldsymbol{A}$ is an $m \times n$ matrix, and $S_{\boldsymbol{A},\boldsymbol{b}} = \{\boldsymbol{x} \in \Re^n : \boldsymbol{Ax} = \boldsymbol{b}\}$ is the set of solutions for the system, then*

$$S_{\boldsymbol{A},\boldsymbol{b}} = \{\boldsymbol{x}_p\} + \mathcal{N}(\boldsymbol{A}) \, ,$$

where $\boldsymbol{x}_p \in S$ is any particular solution for $\boldsymbol{Ax} = \boldsymbol{b}$ and $\{\boldsymbol{x}_p\} + \mathcal{N}(\boldsymbol{A})$ denotes the set $\{\boldsymbol{u} : \boldsymbol{u} = \boldsymbol{x}_p + \boldsymbol{w}, \quad \boldsymbol{w} \in \mathcal{N}(\boldsymbol{A})\}$.

Proof. Let $\boldsymbol{u}$ be any member in $\{\boldsymbol{x}_p\} + \mathcal{N}(\boldsymbol{A})$. Then, $\boldsymbol{u} = \boldsymbol{x}_p + \boldsymbol{w}$ for some vector $\boldsymbol{w} \in \mathcal{N}(\boldsymbol{A})$. Clearly $\boldsymbol{u}$ belongs to $S_{\boldsymbol{A},\boldsymbol{b}}$ because

$$\boldsymbol{Au} = \boldsymbol{A}(\boldsymbol{x}_p + \boldsymbol{w}) = \boldsymbol{Ax}_p + \boldsymbol{Aw} = \boldsymbol{b} + \boldsymbol{0} = \boldsymbol{b} \, .$$

This proves that $\{\boldsymbol{x}_p\} + \mathcal{N}(\boldsymbol{A}) \subseteq S$.

To prove the other direction, argue as follows:

$$\begin{aligned} \boldsymbol{x} \in S_{\boldsymbol{A},\boldsymbol{b}} &\Longrightarrow \boldsymbol{Ax} = \boldsymbol{b} = \boldsymbol{Ax}_p \Longrightarrow \boldsymbol{A}(\boldsymbol{x} - \boldsymbol{x}_p) = \boldsymbol{0} \Longrightarrow \boldsymbol{x} - \boldsymbol{x}_p \in \mathcal{N}(\boldsymbol{A}) \\ &\Longrightarrow \boldsymbol{x} = \boldsymbol{x}_p + (\boldsymbol{x} - \boldsymbol{x}_p) = \boldsymbol{x} + \boldsymbol{w} \, , \quad \text{where } \boldsymbol{w} = \boldsymbol{x} - \boldsymbol{x}_p \in \mathcal{N}(\boldsymbol{A}) \\ &\Longrightarrow \boldsymbol{x} \in \{\boldsymbol{x}_p\} + \mathcal{N}(\boldsymbol{A}) \, . \end{aligned}$$

This completes the proof. □

Note that the set $S_{\boldsymbol{A},\boldsymbol{b}}$ defined in Theorem 9.1 is not a subspace unless $\boldsymbol{b} = \boldsymbol{0}$. If $\boldsymbol{b} \neq \boldsymbol{0}$, $S_{\boldsymbol{A},\boldsymbol{b}}$ does not contain the vector $\boldsymbol{0}$ and is not closed under addition or scalar multiplication. This is analogous to why a plane not passing through the origin is also not a subspace of $\Re^3$. Each equation in $\boldsymbol{A}\boldsymbol{x} = \boldsymbol{b}$ describes a hyperplane, so $S_{\boldsymbol{A},\boldsymbol{b}}$ describes the set of points lying on the collection of hyperplanes determined by the rows of $\boldsymbol{A}$. The set of solutions will be a subspace if and only if all these hyperplanes pass through the origin.

The set $S_{\boldsymbol{A},\boldsymbol{b}}$ is often described as a *flat*. Geometrically, flats are sets formed by *translating* (which means shifting without changing the orientation) subspaces by some vector. For example, $S_{\boldsymbol{A},\boldsymbol{b}}$ is formed by translating the subspace $\mathcal{N}(\boldsymbol{A})$ by any vector that is a particular solution for $\boldsymbol{A}\boldsymbol{x} = \boldsymbol{b}$. Lines and planes not passing through the origin are examples of flats. If we shift back a flat by any member in it, we obtain a subspace. Thus, $S_{\boldsymbol{A},\boldsymbol{b}} - \boldsymbol{x}_p$ will be a subspace for any $\boldsymbol{x}_p \in S_{\boldsymbol{A},\boldsymbol{b}}$.

The left null space of $\boldsymbol{A}$, i.e., $\mathcal{N}(\boldsymbol{A}')$ can also be used to characterize consistency of linear systems.

Theorem 9.2 *The system* $\boldsymbol{A}\boldsymbol{x} = \boldsymbol{b}$ *is consistent if and only if*

$$\boldsymbol{A}'\boldsymbol{u} = \boldsymbol{0} \Longrightarrow \boldsymbol{b}'\boldsymbol{u} = 0 \, .$$

In other words, $\boldsymbol{A}\boldsymbol{x} = \boldsymbol{b}$ *is consistent if and only if* $\mathcal{N}(\boldsymbol{A}') \subseteq \mathcal{N}(\boldsymbol{b}')$.

Proof. We provide two proofs of this result. The first uses basic properties of the null space and the rank-plus-nullity theorem. The second uses orthogonality.

First proof: Suppose the system is consistent. Then $\boldsymbol{b} = \boldsymbol{A}\boldsymbol{x}_0$ for some $\boldsymbol{x}_0$ and

$$\boldsymbol{A}'\boldsymbol{u} = \boldsymbol{0} \Longrightarrow \boldsymbol{u}'\boldsymbol{A} = \boldsymbol{0}' \Longrightarrow \boldsymbol{u}'\boldsymbol{A}\boldsymbol{x}_0 = \boldsymbol{0}' \Longrightarrow \boldsymbol{u}'\boldsymbol{b} = 0 \, .$$

This proves the "only if" part.

Now suppose that $\boldsymbol{A}'\boldsymbol{u} = \boldsymbol{0} \Longrightarrow \boldsymbol{b}'\boldsymbol{u} = 0$. This means that $\mathcal{N}(\boldsymbol{A}') \subseteq \mathcal{N}(\boldsymbol{b}')$. So,

$$\mathcal{N}(\boldsymbol{A}') \subseteq \mathcal{N}\left(\begin{bmatrix} \boldsymbol{A}' \\ \boldsymbol{b}' \end{bmatrix}\right) .$$

We now argue that

$$\mathcal{N}\left(\begin{bmatrix} \boldsymbol{A}' \\ \boldsymbol{b}' \end{bmatrix}\right) = \mathcal{N}(\boldsymbol{A}') \cap \mathcal{N}(\boldsymbol{b}') \subseteq \mathcal{N}(\boldsymbol{A}') \subseteq \mathcal{N}\left(\begin{bmatrix} \boldsymbol{A}' \\ \boldsymbol{b}' \end{bmatrix}\right) ,$$

where the first equality follows from Theorem 4.9 Therefore, we have equality:

$$\mathcal{N}(\boldsymbol{A}') = \mathcal{N}\left(\begin{bmatrix} \boldsymbol{A}' \\ \boldsymbol{b}' \end{bmatrix}\right) .$$

Taking dimensions, we have $\nu\left(\begin{bmatrix} \boldsymbol{A}' \\ \boldsymbol{b}' \end{bmatrix}\right) = \nu(\boldsymbol{A}')$ and, because they have the same number of columns, the Rank-Plus-Nullity theorem says that the rank of $\begin{bmatrix} \boldsymbol{A}' \\ \boldsymbol{b}' \end{bmatrix}$ is equal to the rank of A'. Therefore,

$$\dim\left[\mathcal{C}([A : b])\right] = \rho([A : b]) = \rho\left(\begin{bmatrix} A' \\ b' \end{bmatrix}\right) = \rho(A') = \rho(A) = \dim\left[\mathcal{C}(A)\right] .$$

Since $\mathcal{C}(A) \subseteq \mathcal{C}([A : b])$, the equality of their dimensions implies that $\mathcal{C}(A) = \mathcal{C}([A : b])$, which means that $b \in \mathcal{C}(A)$. Hence, the system is consistent.

Second proof (using orthogonality): Suppose the system is consistent. Then, $b \in \mathcal{C}(A)$. Let u be a vector in $\mathcal{N}(A')$ so $A'u = 0$. Then, u is orthogonal to each row of A', which means that u is orthogonal to each column of A. Therefore, $u \in \mathcal{C}(A)^{\perp}$ and, so $u \perp b$, implying that $b'u = 0$.

Now suppose that $b'u = 0$ whenever $A'u = 0$. This means that if a vector is orthogonal to all the rows of A', i.e., all the columns of A, then it must be orthogonal to b. In other words, b is orthogonal to every vector in $\mathcal{N}(A')$ so $b \in \mathcal{N}(A')^{\perp}$. From the Fundamental Theorem of Linear Algebra (Theorem 7.5), we know that $\mathcal{N}(A') = \mathcal{C}(A)^{\perp}$. Therefore, $b \in \mathcal{N}(A')^{\perp} = [\mathcal{C}(A)^{\perp}]^{\perp} = \mathcal{C}(A)$ (recall Theorem 7.7). Therefore, the system is consistent. □

The above theorem says that $Ax = b$ is *inconsistent*, i.e., does not have a solution, if and only if we can find a vector u such that $A'u = 0$ but $b'u \neq 0$.

9.3 Rank and linear systems

In Chapter 2, we studied in detail the mechanics behind solving a system of linear equations. A very important quantity we encountered there was the number of pivots in a matrix. This number was equal to the number of nonzero rows in any row echelon form of the matrix. It was also equal to the number of basic columns in the matrix. For a system of linear equations, the number of pivots also gave the number of basic variables, while all remaining variables were the free variables.

In Section 5.2, we observed that the number of pivots in a matrix is the dimension of the row space and the column space. Therefore, the rank of a matrix is the number of pivots. The advantage of looking upon rank as a dimension of the column space or row space is that it helps elicit less apparent facts about the nature of linear equations. For example, it may not be clear from the mechanics of Gaussian elimination why the number of pivots in a matrix A should be the same as that in A'. Treating rank as the dimension of the column space or row space helped us explain why the rank does not change with transposition. This makes the number of pivots invariant to transposition as well.

Gaussian elimination also showed that $Ax = b$ is consistent, i.e., it has at least one solution, if and only if the number of pivots in A is the same as that in the augmented matrix $[A : b]$. Since the number of pivots equals the rank of A, one could say that $Ax = b$ is consistent if and only if $\rho([A : b]) = \rho(A)$. It is, nevertheless, important to form the habit of deriving these results not using properties of pivots but, instead, using the definition of rank as a dimension. We will provide some characterizations of linear systems in terms of the rank of the coefficient matrix.

Theorem 9.3 *The system $\boldsymbol{Ax} = \boldsymbol{b}$ is consistent if and only if $\rho([\boldsymbol{A} : \boldsymbol{b}]) = \rho(\boldsymbol{A})$.*

Proof. If $\boldsymbol{Ax} = \boldsymbol{b}$ is consistent, then, $\boldsymbol{b} = \boldsymbol{Ax}_0$ for some solution vector $\boldsymbol{x}_0$. Therefore, $\boldsymbol{b} \in \mathcal{C}(\boldsymbol{A})$, which means that the dimension of $\mathcal{C}(\boldsymbol{A})$ will not be increased by augmenting the set of columns of $\boldsymbol{A}$ with $\boldsymbol{b}$. Therefore, $\rho([\boldsymbol{A} : \boldsymbol{b}]) = \rho(\boldsymbol{A})$. This proves the "only if" part.

To prove the "if" part, note that $\mathcal{C}(\boldsymbol{A}) \subseteq \mathcal{C}([\boldsymbol{A} : \boldsymbol{b}])$. If $\rho([\boldsymbol{A} : \boldsymbol{b}]) = \rho(\boldsymbol{A})$, then $\mathcal{C}(\boldsymbol{A}) \subseteq \mathcal{C}([\boldsymbol{A} : \boldsymbol{b}])$ (recall Theorem 5.3). Therefore, $\boldsymbol{b} \in \mathcal{C}(\boldsymbol{A})$ and there exists some vector $\boldsymbol{x}_0$ such that $\boldsymbol{Ax}_0 = \boldsymbol{b}$. Therefore, the system is consistent. □

Remark: The above proof subsumes the case when $\boldsymbol{b} = \boldsymbol{0}$. The homogeneous system $\boldsymbol{Ax} = \boldsymbol{0}$ is always consistent and obviously $\rho([\boldsymbol{A} : \boldsymbol{0}]) = \rho(\boldsymbol{A})$.

Sometimes we want to know if a consistent system $\boldsymbol{Ax} = \boldsymbol{b}$ has a *unique* solution. The mechanics of Gaussian elimination tells us that this will happen if and only if the number of pivots of $\boldsymbol{A}$ is equal to the number of number of variables. Then there will be no free variables and we can use back substitution to solve for $\boldsymbol{x}$. Since the number of variables is the same as the number of columns of $\boldsymbol{A}$, the number of pivots equals the number of columns in this case. And the number of pivots is the rank, so the matrix must have linearly independent columns. It is, nevertheless, instructive to prove this claim using vector spaces. Below we present such a proof.

Theorem 9.4 *A consistent system $\boldsymbol{Ax} = \boldsymbol{b}$ has a unique solution if and only the columns of $\boldsymbol{A}$ are linearly independent.*

Proof. Let $\boldsymbol{A}$ be a matrix with n columns. Assume that $\boldsymbol{A}$ has linearly independent columns. Therefore, $\rho(\boldsymbol{A}) = n$ so $\nu(\boldsymbol{A}) = n - n = 0$ and $\mathcal{N}(\boldsymbol{A}) = \{\boldsymbol{0}\}$. Suppose, if possible, $\boldsymbol{x}_1$ and $\boldsymbol{x}_2$ are two solutions for $\boldsymbol{Ax} = \boldsymbol{b}$. Then,

$$\begin{aligned} \boldsymbol{Ax}_1 = \boldsymbol{b} = \boldsymbol{Ax}_2 &\Longrightarrow \boldsymbol{A}(\boldsymbol{x}_2 - \boldsymbol{x}_1) = \boldsymbol{0} \Longrightarrow \boldsymbol{x}_2 - \boldsymbol{x}_1 \in \mathcal{N}(\boldsymbol{A}) = \{\boldsymbol{0}\} \\ &\Longrightarrow \boldsymbol{x}_2 - \boldsymbol{x}_1 = \boldsymbol{0} \Longrightarrow \boldsymbol{x}_1 = \boldsymbol{x}_2 , \end{aligned}$$

which proves that the solution for $\boldsymbol{Ax} = \boldsymbol{b}$ is unique.

Now suppose that $\boldsymbol{Ax} = \boldsymbol{b}$ has a unique solution, say $\boldsymbol{x}_0$, and let $\boldsymbol{w} \in \mathcal{N}(\boldsymbol{A})$. This means that $\boldsymbol{x}_0 + \boldsymbol{w}$ is also a solution for $\boldsymbol{Ax} = \boldsymbol{b}$ (it may help to recall Theorem 9.1). Since the solution is unique $\boldsymbol{x}_0$ must be equal to $\boldsymbol{x}_0 + \boldsymbol{w}$, which implies that $\boldsymbol{w} = \boldsymbol{0}$. Therefore, $\mathcal{N}(\boldsymbol{A}) = \{\boldsymbol{0}\}$, which means that the columns of $\boldsymbol{A}$ are linearly independent. □

The following theorem exhausts the possibilities for $\boldsymbol{Ax} = \boldsymbol{b}$ using conditions on the rank of $\boldsymbol{A}$ and $[\boldsymbol{A} : \boldsymbol{b}]$.

Theorem 9.5 *Let $\boldsymbol{A}$ be a matrix with n columns. Then $\boldsymbol{Ax} = \boldsymbol{b}$ has:*

(i) no solution if $\rho(\boldsymbol{A}) < \rho([\boldsymbol{A} : \boldsymbol{b}])$,

(ii) a unique solution if and only if $\rho(\boldsymbol{A}) = \rho([\boldsymbol{A} : \boldsymbol{b}]) = n$,

(iii) an infinite number of solutions if $\rho(\boldsymbol{A}) = \rho([\boldsymbol{A} : \boldsymbol{b}]) < n$.

Proof. Part (i) follows immediately from Theorem 9.3, while part (ii) is simply a restatement of Theorem 9.4 in terms of the rank of $\boldsymbol{A}$.

Part (iii) considers the only remaining condition on the ranks. Therefore, with this condition the system must have more than one solution. But if $\boldsymbol{x}_1$ and $\boldsymbol{x}_2$ are two solutions for $\boldsymbol{Ax} = \boldsymbol{b}$, then $\boldsymbol{Ax}_1 = \boldsymbol{b} = \boldsymbol{Ax}_2$ implies that

$$\boldsymbol{A}(\beta\boldsymbol{x}_1 + (1-\beta)\boldsymbol{x}_2) = \beta\boldsymbol{Ax}_1 + (1-\beta)\boldsymbol{Ax}_2 = \beta\boldsymbol{b} + (1-\beta)\boldsymbol{b} = \boldsymbol{b}\,,$$

where β_1 is any real number between 0 and 1. Clearly this means that there are an uncountably infinite number of solutions. □

Consider the system $\boldsymbol{Ax} = \boldsymbol{b}$ with $\rho(\boldsymbol{A}) = r$. Using elementary row operations we can reduce $\boldsymbol{A}$ to an echelon matrix with only r nonzero rows. This produces $\boldsymbol{GA} = \begin{bmatrix} \boldsymbol{U} \\ \boldsymbol{O} \end{bmatrix}$, where $\boldsymbol{G}$ is nonsingular and $\boldsymbol{U}$ is an $r \times n$ echelon matrix with nonzero rows. Partitioning $\boldsymbol{G} = \begin{bmatrix} \boldsymbol{G}_1 \\ \boldsymbol{G}_2 \end{bmatrix}$ conformably, so that $\boldsymbol{G}_1$ has r rows, we see that

$$\boldsymbol{Ax} = \boldsymbol{b} \iff \begin{bmatrix} \boldsymbol{G}_1 \\ \boldsymbol{G}_2 \end{bmatrix} \boldsymbol{Ax} = \begin{bmatrix} \boldsymbol{G}_1 \\ \boldsymbol{G}_2 \end{bmatrix} \boldsymbol{b} \iff \begin{bmatrix} \boldsymbol{U} \\ \boldsymbol{O} \end{bmatrix} \boldsymbol{x} = \begin{bmatrix} \boldsymbol{G}_1 \\ \boldsymbol{G}_2 \end{bmatrix} \boldsymbol{b} \iff \begin{bmatrix} \boldsymbol{Ux} \\ \boldsymbol{0} \end{bmatrix} = \begin{bmatrix} \boldsymbol{G}_1\boldsymbol{b} \\ \boldsymbol{G}_2\boldsymbol{b} \end{bmatrix}.$$

This reveals that the system $\boldsymbol{Ax} = \boldsymbol{b}$ is consistent if and only if $\boldsymbol{b} \in \mathcal{N}(\boldsymbol{G}_2)$. If this is satisfied, then the solution is obtained by simply solving only the r equations given by $\boldsymbol{Ux} = \boldsymbol{G}_1\boldsymbol{b}$. The following theorem shows that we can simply use the reduced system obtained from the r linearly independent rows of $\boldsymbol{A}$. For better clarity, we assume that the first r rows of the matrix are linearly independent.

Theorem 9.6 *Let $\boldsymbol{A}$ be an $m \times n$ matrix with rank r and suppose that $\boldsymbol{Ax} = \boldsymbol{b}$ is consistent. Assume that the first r rows are linearly independent. Then the last $m - r$ equations in $\boldsymbol{Ax} = \boldsymbol{b}$ are redundant and the solution set remains unaltered even if these are dropped.*

Proof. Because the system is consistent, we know that $\rho(\boldsymbol{A}) = \rho([\boldsymbol{A} : \boldsymbol{b}])$. Let $\boldsymbol{A} = \begin{bmatrix} \boldsymbol{A}_1 \\ \boldsymbol{A}_2 \end{bmatrix}$, where $\boldsymbol{A}_1$ is the $r \times n$ submatrix formed from the first r rows of $\boldsymbol{A}$. Conformably partition $\boldsymbol{b} = \begin{bmatrix} \boldsymbol{b}_1 \\ \boldsymbol{b}_2 \end{bmatrix}$ so that $\boldsymbol{b}_1$ is an $r \times 1$ vector. Note that the first r rows of $\boldsymbol{A}$ are linearly independent. By virtue of Lemma 4.4, this means that the first r rows of the augmented matrix $[\boldsymbol{A} : \boldsymbol{b}]$ are also linearly independent. Therefore, $\mathcal{R}([\boldsymbol{A}_2 : \boldsymbol{b}_2]) \subseteq \mathcal{R}([\boldsymbol{A}_1 : \boldsymbol{b}_1])$ so $[\boldsymbol{A}_2 : \boldsymbol{b}_2] = \boldsymbol{D}[\boldsymbol{A}_1 : \boldsymbol{b}_1]$ for some matrix $\boldsymbol{D}$. This means that $\boldsymbol{A}_2 = \boldsymbol{DA}_1$ and $\boldsymbol{b}_1 = \boldsymbol{Db}_1$ and we can argue that

$$\boldsymbol{Ax} = \boldsymbol{b} \iff \begin{bmatrix} \boldsymbol{A}_1 \\ \boldsymbol{DA}_1 \end{bmatrix} \boldsymbol{x} = \begin{bmatrix} \boldsymbol{b}_1 \\ \boldsymbol{Db}_1 \end{bmatrix}.$$

Therefore, $\boldsymbol{Ax} = \boldsymbol{b}$ if and only if $\boldsymbol{A}_1\boldsymbol{x} = \boldsymbol{b}_1$, so the last $m - r$ equations have no impact on the solutions. □

There really is no loss of generality in assuming that the first r rows of $\boldsymbol{A}$ are linearly independent. Since $\rho(\boldsymbol{A}) = r$, we can find r linearly independent rows in $\boldsymbol{A}$. Permuting them so that these r rows become the first r rows of $\boldsymbol{A}$ does not change the solution space for $\boldsymbol{Ax} = \boldsymbol{b}$—we still have the same set of equations but written in a different order.

Take a closer look at the reduced system $\boldsymbol{A}_1\boldsymbol{x} = \boldsymbol{b}_1$, where $\boldsymbol{A}_1$ is an $r \times n$ matrix with $\rho(\boldsymbol{A}_1) = r$. We can, therefore, find r linearly independent columns in this matrix. Permuting these columns so that they occupy the first r columns simply permutes the variables (i.e., the x_i's) accordingly. Therefore, we can assume, again without any loss of generality, that the first r columns of $\boldsymbol{A}_1$ are linearly independent. We can write $\boldsymbol{A}_1 = [\boldsymbol{A}_{11} : \boldsymbol{A}_{12}]$, where $\boldsymbol{A}_{11}$ is an $r \times r$ nonsingular matrix and $\mathcal{C}(\boldsymbol{A}_{12}) \subseteq \mathcal{C}(\boldsymbol{A}_{11})$. Therefore, $\boldsymbol{A}_{12} = \boldsymbol{A}_{11}\boldsymbol{C}$ for some matrix $\boldsymbol{C}$. We can then write

$$\boldsymbol{A}_1\boldsymbol{x} = \boldsymbol{b}_1 \Longrightarrow \boldsymbol{A}_{11}\left[\boldsymbol{I} : \boldsymbol{C}\right]\begin{bmatrix}\boldsymbol{x}_1\\ \boldsymbol{x}_2\end{bmatrix} = \boldsymbol{b}_1 .$$

Therefore, all solutions for $\boldsymbol{A}_1\boldsymbol{x}_1 = \boldsymbol{b}_1$ can be obtained by fixing $\boldsymbol{x}_2$ arbitrarily and then solving the nonsingular system $\boldsymbol{A}_{11}\boldsymbol{x}_1 = \boldsymbol{b}_1 - \boldsymbol{A}_{11}\boldsymbol{C}\boldsymbol{x}_2$ for $\boldsymbol{x}_1$.

The fundamental theorem of ranks also plays an important role in analyzing linear systems. To see why, consider a system $\boldsymbol{Ax} = \boldsymbol{b}$ that may or may not be consistent. Multiplying both sides of the system by $\boldsymbol{A}'$ yields $\boldsymbol{A}'\boldsymbol{Ax} = \boldsymbol{A}'\boldsymbol{b}$. The system $\boldsymbol{A}'\boldsymbol{Ax} = \boldsymbol{A}'\boldsymbol{b}$ plays a central role in the theory of least squares and statistical regression modeling and is important enough to merit its own definition.

Definition 9.1 *Let* $\boldsymbol{A}$ *be an* $m \times n$ *matrix. The system*

$$\boldsymbol{A}'\boldsymbol{Ax} = \boldsymbol{A}'\boldsymbol{b}$$

is known as the **normal equations** *associated with* $\boldsymbol{Ax} = \boldsymbol{b}$.

Interestingly, this system of normal equations is *always* consistent. This is our next theorem.

Theorem 9.7 *Let* $\boldsymbol{A}$ *be an* $m \times n$ *matrix. The system* $\boldsymbol{A}'\boldsymbol{Ax} = \boldsymbol{A}'\boldsymbol{b}$ *is always consistent.*

Proof. Note that $\mathcal{C}(\boldsymbol{A}'\boldsymbol{A}) \subseteq \mathcal{C}(\boldsymbol{A}')$. Theorem 5.2 tells us that $\rho(\boldsymbol{A}'\boldsymbol{A}) = \rho(\boldsymbol{A}')$. Therefore, Theorem 5.3 ensures that $\mathcal{C}(\boldsymbol{A}'\boldsymbol{A}) = \mathcal{C}(\boldsymbol{A}')$. In particular, $\boldsymbol{A}'\boldsymbol{b} \in \mathcal{C}(\boldsymbol{A}'\boldsymbol{A})$ and so there exists a vector $\boldsymbol{x}_0$ such that $\boldsymbol{A}'\boldsymbol{b} = \boldsymbol{A}'\boldsymbol{A}\boldsymbol{x}_0$. □

What is attractive about the normal equations is that if the system $\boldsymbol{Ax} = \boldsymbol{b}$ happens to be consistent, then $\boldsymbol{Ax} = \boldsymbol{b}$ and $\boldsymbol{A}'\boldsymbol{Ax} = \boldsymbol{A}'\boldsymbol{b}$ have exactly the same solutions.

To see why this is true, note that any particular solution, say $\boldsymbol{x}_p$, of $\boldsymbol{A}\boldsymbol{x} = \boldsymbol{b}$ is also a particular solution of the associated normal equations:

$$\boldsymbol{A}\boldsymbol{x}_p = \boldsymbol{b} \Longrightarrow \boldsymbol{A}'\boldsymbol{A}\boldsymbol{x}_p = \boldsymbol{A}'\boldsymbol{b} .$$

Now, the general solution of $\boldsymbol{A}\boldsymbol{x} = \boldsymbol{b}$ is $\boldsymbol{x}_p + \mathcal{N}(\boldsymbol{A})$, while that of the normal equations is $\boldsymbol{x}_p + \mathcal{N}(\boldsymbol{A}'\boldsymbol{A})$. These two general solutions are the same because $\mathcal{N}(\boldsymbol{A}'\boldsymbol{A}) = \mathcal{N}(\boldsymbol{A})$.

The case when $\boldsymbol{A}\boldsymbol{x} = \boldsymbol{b}$ is consistent and $\mathcal{N}(\boldsymbol{A}) = \{\boldsymbol{0}\}$ is particularly interesting. Since $\mathcal{N}(\boldsymbol{A}) = \{\boldsymbol{0}\}$ means that $\boldsymbol{A}$ has full column rank, Theorem 9.4 ensures that $\boldsymbol{A}\boldsymbol{x} = \boldsymbol{b}$ has a unique solution. And so does the associated normal equations. This is because $\mathcal{N}(\boldsymbol{A}'\boldsymbol{A}) = \mathcal{N}(\boldsymbol{A}) = \{\boldsymbol{0}\}$, so $\boldsymbol{A}'\boldsymbol{A}$ is nonsingular. In this case, therefore, we can obtain the unique solution for $\boldsymbol{A}'\boldsymbol{A}\boldsymbol{x} = \boldsymbol{b}$ and verify directly that it also is a solution for $\boldsymbol{A}\boldsymbol{x} = \boldsymbol{b}$. To be precise, if $\boldsymbol{x}_0$ is a solution for the normal equations, then

$$\boldsymbol{A}'\boldsymbol{A}\boldsymbol{x}_0 = \boldsymbol{b} \Longrightarrow \boldsymbol{x}_0 = (\boldsymbol{A}'\boldsymbol{A})^{-1}\boldsymbol{A}'\boldsymbol{b} \Longrightarrow \boldsymbol{A}\boldsymbol{x}_0 = \boldsymbol{A}(\boldsymbol{A}'\boldsymbol{A})^{-1}\boldsymbol{A}'\boldsymbol{b} = \boldsymbol{P}_{\boldsymbol{A}}\boldsymbol{b} . \tag{9.1}$$

Because $\boldsymbol{A}\boldsymbol{x} = \boldsymbol{b}$ is consistent, $\boldsymbol{b}$ belongs to $\mathcal{C}(\boldsymbol{A})$. Therefore, the orthogonal projection of $\boldsymbol{b}$ onto the $\mathcal{C}(\boldsymbol{A})$ is itself. So, $\boldsymbol{P}_{\boldsymbol{A}}\boldsymbol{b} = \boldsymbol{b}$ and (9.1) implies that $\boldsymbol{A}\boldsymbol{x}_0 = \boldsymbol{b}$.

9.4 Generalized inverse of a matrix

Let us consider a consistent system $\boldsymbol{A}\boldsymbol{x} = \boldsymbol{b}$, where $\boldsymbol{A}$ is $m \times n$ and $\rho(\boldsymbol{A}) = r$. If $m = n = r$, then $\boldsymbol{A}$ is nonsingular and $\boldsymbol{x} = \boldsymbol{A}^{-1}\boldsymbol{b}$ is the unique solution for the system. Note that $\boldsymbol{A}^{-1}$ does not depend upon $\boldsymbol{b}$. But what if $\boldsymbol{A}$ is not a square matrix or, even if it square, it is singular and $\boldsymbol{A}^{-1}$ does not exist? While Theorem 9.5 characterizes when a solution exists for a linear system in terms of the rank of $\boldsymbol{A}$, it does not tell us how we can find a solution when $\boldsymbol{A}$ is singular. The case when $\boldsymbol{A}$ has full column rank can be tackled using the normal equations and we obtain the solution (9.1). But what happens if $\boldsymbol{A}$ is not of full column rank?

In general, what we seek is a matrix $\boldsymbol{G}$ such that $\boldsymbol{G}\boldsymbol{b}$ is a solution for $\boldsymbol{A}\boldsymbol{x} = \boldsymbol{b}$. Then the matrix $\boldsymbol{G}$ is called a *generalized inverse* and can provide greater clarity to the study of general linear systems. Below is a formal definition.

Definition 9.2 *An $n \times m$ matrix $\boldsymbol{G}$ is said to be the* **generalized inverse** *(also called a* **g-inverse** *or a* **pseudo-inverse***) of an $m \times n$ matrix $\boldsymbol{A}$ if $\boldsymbol{G}\boldsymbol{b}$ is a solution to $\boldsymbol{A}\boldsymbol{x} = \boldsymbol{b}$ for every vector $\boldsymbol{b} \in \mathcal{C}(\boldsymbol{A})$.*

We know that the system $\boldsymbol{A}\boldsymbol{x} = \boldsymbol{b}$ is consistent for every vector $\boldsymbol{b} \in \mathcal{C}(\boldsymbol{A})$, so the generalized inverse yields a solution for every consistent system. But does a generalized inverse always exist? This brings us to our first theorem on generalized inverses.

Theorem 9.8 *Every matrix has a generalized inverse.*

Proof. Let $\boldsymbol{A}$ be an $m \times n$ matrix. First consider the case $\boldsymbol{A} = \boldsymbol{O}$. Then, the system $\boldsymbol{Ax} = \boldsymbol{b}$ is consistent if and only if $\boldsymbol{b} = \boldsymbol{0}$. This means that every $n \times m$ matrix $\boldsymbol{G}$ is a generalized inverse of $\boldsymbol{A}$ because $\boldsymbol{Gb} = \boldsymbol{G0} = \boldsymbol{0}$ will be a solution.

Now suppose $\boldsymbol{A}$ is not the null matrix and let $\rho(\boldsymbol{A}) = r > 0$. Let $\boldsymbol{A} = \boldsymbol{CR}$ be a rank factorization for $\boldsymbol{A}$. Then $\boldsymbol{C}$ is $m \times r$ and has full column rank, while $\boldsymbol{R}$ is $r \times n$ and has full row rank. Then, $\boldsymbol{C}$ has a left-inverse and $\boldsymbol{R}$ has a right inverse (recall Theorem 5.5). Let $\boldsymbol{B}$ be a left inverse of $\boldsymbol{C}$ and $\boldsymbol{D}$ be a right inverse of $\boldsymbol{R}$.

Note that $\boldsymbol{B}$ is $r \times m$ and $\boldsymbol{D}$ is $n \times r$. Form the $n \times m$ matrix $\boldsymbol{G} = \boldsymbol{DB}$ and note

$$\boldsymbol{AGb} = \boldsymbol{CRDBb} = \boldsymbol{C}(\boldsymbol{RD})\boldsymbol{Bb} = \boldsymbol{CI}_r\boldsymbol{Bb} = \boldsymbol{CBb} = \boldsymbol{I}_m\boldsymbol{b} = \boldsymbol{b} \,.$$

Therefore, $\boldsymbol{G}$ is a generalized inverse of $\boldsymbol{A}$. □

In the above proof we showed how to construct a generalized inverse from a rank factorization. Since every matrix with nonzero rank has a rank factorization, every matrix has a generalized inverse. Furthermore, $\boldsymbol{G}$ was constructed only from a rank factorization of $\boldsymbol{A}$ and, hence, it does not depend upon $\boldsymbol{b}$.

Clearly, a generalized inverse is not unique. Every matrix is a generalized inverse of the null matrix. Furthermore, a matrix can have several full rank factorizations—if $\boldsymbol{A} = \boldsymbol{CR}$ is a rank-factorization, then $\boldsymbol{A} = (\boldsymbol{CD})(\boldsymbol{D}^{-1}\boldsymbol{R})$ is a rank factorization for any $r \times r$ nonsingular matrix $\boldsymbol{D}$. Each rank factorization can be used to construct a generalized inverse as shown in the proof of Theorem 9.8. So a matrix can have several generalized inverses. The key point, though, is that whatever generalized inverse $\boldsymbol{G}$ of $\boldsymbol{A}$ we choose, $\boldsymbol{Gb}$ will be a solution for any consistent system $\boldsymbol{Ax} = \boldsymbol{b}$.

A generalized inverse can be used not only to solve a consistent system but also to check the consistency of a system. To be precise, suppose that $\boldsymbol{G}$ is a generalized inverse of $\boldsymbol{A}$. Then, clearly the system $\boldsymbol{Ax} = \boldsymbol{b}$ is consistent if the vector $\boldsymbol{b}$ satisfies $\boldsymbol{AGb} = \boldsymbol{b}$. On the other hand, if the system is consistent then the definition of the generalized inverse says that $\boldsymbol{Gb}$ is a solution for $\boldsymbol{Ax} = \boldsymbol{b}$. In other words, a system $\boldsymbol{Ax} = \boldsymbol{b}$ is consistent if and only if $\boldsymbol{AGb} = \boldsymbol{b}$.

Lemma 9.1 *Let $\boldsymbol{G}$ be a generalized inverse of $\boldsymbol{A}$. Then $\mathcal{C}(\boldsymbol{C}) \subseteq \mathcal{C}(\boldsymbol{A})$ if and only if $\boldsymbol{C} = \boldsymbol{AGC}$.*

Proof. Note that for $\mathcal{C}(\boldsymbol{C}) \subseteq \mathcal{C}(\boldsymbol{A})$ to make sense, the column vectors of $\boldsymbol{C}$ and $\boldsymbol{A}$ must reside in the same subspace of the Euclidean space. This means that $\boldsymbol{C}$ and $\boldsymbol{B}$ must have the same number of rows. Let $\boldsymbol{A}$ be $m \times n$ and $\boldsymbol{C}$ be an $m \times p$ matrix.

Clearly, if $\boldsymbol{C} = \boldsymbol{AGC}$, then $\mathcal{C}(\boldsymbol{C})\mathcal{C}(\boldsymbol{A})$ (Theorem 4.6).

Now suppose that $\mathcal{C}(\boldsymbol{C}) \subseteq \mathcal{C}(\boldsymbol{A})$. This means that the system $\boldsymbol{Ax} = \boldsymbol{c}_{*j}$ is consistent for every column $\boldsymbol{c}_{*j}$ in $\boldsymbol{C}$. So, $\boldsymbol{Gc}_{*j}$ will be a solution for $\boldsymbol{Ax} = \boldsymbol{c}_{*j}$ and we obtain

$$\begin{aligned}\boldsymbol{C} &= [\boldsymbol{c}_{*1} : \boldsymbol{c}_{*2} : \ldots : \boldsymbol{c}_{*p}] = \boldsymbol{A}[\boldsymbol{Gc}_{*1} : \boldsymbol{Gc}_{*2} : \ldots : \boldsymbol{Gc}_{*p}] \\ &= \boldsymbol{AG}[\boldsymbol{c}_{*1} : \boldsymbol{c}_{*2} : \ldots : \boldsymbol{c}_{*p}] = \boldsymbol{AGC} \,.\end{aligned}$$

This completes the proof. □

It is instructive to compare the above result with Theorem 4.6. There, we showed that if $\mathcal{C}(\boldsymbol{C}) \subseteq \mathcal{C}(\boldsymbol{A})$, then there exists a matrix $\boldsymbol{B}$ for which $\boldsymbol{C} = \boldsymbol{AB}$. Lemma 9.1 shows how this matrix $\boldsymbol{B}$ can be explicitly obtained from $\boldsymbol{C}$ and any generalized inverse of $\boldsymbol{A}$.

The following is a very useful necessary and sufficient condition for a matrix to be a generalized inverse. In fact, this is often taken to be the definition of generalized inverses.

Theorem 9.9 *$\boldsymbol{G}$ is a generalized inverse of $\boldsymbol{A}$ if and only if $\boldsymbol{AGA} = \boldsymbol{A}$.*

Proof. Suppose $\boldsymbol{G}$ is a generalized inverse of the $m \times n$ matrix $\boldsymbol{A}$. Then, $\boldsymbol{AGb} = \boldsymbol{b}$ for all $\boldsymbol{b} \in \mathcal{C}(\boldsymbol{A})$. In particular, this holds when $\boldsymbol{b}$ is any column vector of $\boldsymbol{A}$. Thus, $\boldsymbol{AGa}_{*j} = \boldsymbol{a}_{*j}$ for $j = 1, 2, \ldots, n$ and so

$$\boldsymbol{AGA} = \boldsymbol{AG}[\boldsymbol{a}_{*1} : \boldsymbol{a}_{*2} : \ldots : \boldsymbol{a}_{*n}] = [\boldsymbol{a}_{*1} : \boldsymbol{a}_{*2} : \ldots : \boldsymbol{a}_{*n}] = \boldsymbol{A}\,.$$

Now suppose $\boldsymbol{AGA} = \boldsymbol{A}$. Then,

$$\boldsymbol{b} = \boldsymbol{Ax} \Longrightarrow \boldsymbol{AGb} = \boldsymbol{AGAx} \Longrightarrow \boldsymbol{A}(\boldsymbol{Gb}) = \boldsymbol{Ax} = \boldsymbol{b}\,,$$

which implies that $\boldsymbol{Gb}$ is a solution for $\boldsymbol{Ax} = \boldsymbol{b}$ and so $\boldsymbol{G}$ is a generalized inverse of $\boldsymbol{A}$. □

The characterization in Theorem 9.9 is sometimes helpful in deriving other properties of generalized inverses. One immediately gets a generalized inverse for $\boldsymbol{A}'$.

Corollary 9.1 *If $\boldsymbol{G}$ is a generalized inverse of $\boldsymbol{A}$, then $\boldsymbol{G}'$ is one of $\boldsymbol{A}'$.*

Proof. Taking transposes of $\boldsymbol{AGA} = \boldsymbol{A}$ yields $\boldsymbol{A}'\boldsymbol{G}'\boldsymbol{A}' = \boldsymbol{A}'$. □

Another result that easily follows from Theorem 9.9 is that if $\boldsymbol{A}$ is square and nonsingular, then the generalized inverse of $\boldsymbol{A}$ coincides with $\boldsymbol{A}^{-1}$.

Theorem 9.10 *Let $\boldsymbol{A}$ be an $n \times n$ nonsingular matrix. Then $\boldsymbol{G} = \boldsymbol{A}^{-1}$ is the only generalized inverse of $\boldsymbol{A}$.*

Proof. First of all, observe that $\boldsymbol{A}^{-1}$ is a generalized inverse of $\boldsymbol{A}$ because $\boldsymbol{AA}^{-1}\boldsymbol{A} = \boldsymbol{A}$. Now suppose $\boldsymbol{G}$ is a generalized inverse of $\boldsymbol{A}$. Then, $\boldsymbol{AGA} = \boldsymbol{A}$ and, multiplying both sides with $\boldsymbol{A}^{-1}$, both from the left and right, yields

$$\boldsymbol{A}^{-1}\boldsymbol{AGAA}^{-1} = \boldsymbol{A}^{-1}\boldsymbol{AA}^{-1} = \boldsymbol{A}^{-1}\,.$$

This proves that if $\boldsymbol{G}$ is a generalized inverse, it must be equal to $\boldsymbol{A}^{-1}$. Because $\boldsymbol{A}^{-1}$ is unique, the generalized inverse in this case is also unique. □

Recall that if $\boldsymbol{A}$ is nonsingular and symmetric, then $\boldsymbol{A}^{-1}$ is symmetric as well. The following lemma provides an analogy for generalized inverses.

Theorem 9.11 *Every symmetric matrix has a symmetric generalized inverse.*

Proof. Let G be a generalized inverse of A. Since A is symmetric, taking transposes of both sides of $AGA = A$ yields $AG'A = A$. Therefore,

$$A\left(\frac{G+G'}{2}\right)A = \frac{A+A}{2} = A\,,$$

which implies that $\left(\frac{G+G'}{2}\right)$ is a symmetric generalized inverse of A. □

Suppose that A is a matrix of full column rank. In Theorem 5.7 we showed that $A'A$ is nonsingular and that $(A'A)^{-1}A'$ is a left inverse of A. Note that

$$A\left[(A'A)^{-1}A'\right]A = A(A'A)^{-1}A'A = AI = A\,,$$

which means that $(A'A)^{-1}A'$ is a generalized inverse of A. Theorem 5.7 also tells us that if A is a matrix of full row rank, then AA' is nonsingular and $A'(AA')^{-1}$ is a right inverse. Then,

$$A\left[A'(AA')^{-1}\right]A = AA'(A'A)^{-1}A = IA\,,$$

which implies that $A'(AA')^{-1}$ is a generalized inverse of A.

It is easy to see that every left inverse is a generalized inverse: if $GA = I$, then $AGA = A$. Similarly, every right inverse is a generalized inverse: $AG = I$ implies $AGA = A$. The converse is true for matrices with full column or row rank.

Theorem 9.12 *Suppose A has full column (row) rank. Then, G is a generalized inverse of A if and only if it is a left (right) inverse.*

Proof. We saw above that every left or right inverse of A is a generalized inverse of A. This is true for every matrix, irrespective of whether it has full column or row rank.

Now suppose G is a generalized inverse of A. Since A has full column rank, it has a left inverse B (recall Theorem 5.5). We make use of $AGA = A$ and $BA = I$ to argue that

$$AGA = A \Longrightarrow BAGA = BA \Longrightarrow (BA)GA = (BA) \Longrightarrow GA = I\,,$$

implying that G is a left inverse of A. If A has full row rank, then we can find a right inverse C such that $AC = I$. The remainder of the argument is similar:

$$AGA = A \Longrightarrow AGAC = AC \Longrightarrow AG(AC) = (AC) \Longrightarrow AG = I\,.$$

Therefore, G is a right inverse. This completes the proof. □

The following theorem also provides an important characterization for generalized inverses.

Theorem 9.13 *The following statements are equivalent for any A and G:*

(i) $AGA = A$.

(ii) $\boldsymbol{AG}$ *is idempotent and* $\rho(\boldsymbol{AG}) = \rho(\boldsymbol{A})$.

Proof. **Proof of (i) ⇒ (ii):** If $\boldsymbol{AGA} = \boldsymbol{A}$, then $\boldsymbol{AGAG} = \boldsymbol{AG}$ so $\boldsymbol{AG}$ is idempotent. Also,

$$\rho(\boldsymbol{A}) = \rho(\boldsymbol{AGA}) \leq \rho(\boldsymbol{AG}) \leq \rho(\boldsymbol{A}) ,$$

which implies that $\rho(\boldsymbol{AG}) = \rho(\boldsymbol{A})$.

Proof of (ii) ⇒ (i): Suppose that $\boldsymbol{AG}$ is idempotent and $\rho(\boldsymbol{AG}) = \rho(\boldsymbol{A})$. Because $\boldsymbol{AG}$ is idempotent, we can write $\boldsymbol{AG} = \boldsymbol{AGAG}$. We can now invoke the rank cancellation laws (Lemma 5.4) to *cancel* the $\boldsymbol{G}$ from the right. Thus, we obtain $\boldsymbol{A} = \boldsymbol{AGA}$. □

The following corollary connects the generalized inverse with a projector.

Corollary 9.2 *If* $\boldsymbol{G}$ *is a generalized inverse of* $\boldsymbol{A}$*, then* $\boldsymbol{AG}$ *is the projector onto* $\mathcal{C}(\boldsymbol{A})$ *along* $\mathcal{N}(\boldsymbol{AG})$.

Proof. If $\boldsymbol{G}$ is a generalized inverse of $\boldsymbol{A}$, then, from Theorem 9.13, we find that $\boldsymbol{AG}$ is idempotent. Therefore, $\boldsymbol{AG}$ is a projector onto $\mathcal{C}(\boldsymbol{AG})$ along $\mathcal{N}(\boldsymbol{AG})$. But $\mathcal{C}(\boldsymbol{AG}) = \mathcal{C}(\boldsymbol{A})$ because $\rho(\boldsymbol{AG}) = \rho(\boldsymbol{A})$, so $\boldsymbol{AG}$ is the projector onto $\mathcal{C}(\boldsymbol{A})$ along $\mathcal{N}(\boldsymbol{AG})$. □

If $\boldsymbol{G}$ is any generalized inverse of $\boldsymbol{A}$, Theorem 9.9 tells us that $\boldsymbol{AG}$ is a projector (idempotent matrix). Therefore, its rank and trace are the same. Theorem 9.9 also tells us that the rank of $\boldsymbol{AG}$ is the same as the rank of $\boldsymbol{A}$. Combining these results, we can conclude that

$$\rho(\boldsymbol{A}) = \rho(\boldsymbol{AG}) = \text{tr}(\boldsymbol{AG}) = \text{tr}(\boldsymbol{GA}) \tag{9.2}$$

for any generalized inverse $\boldsymbol{G}$ of $\boldsymbol{A}$.

We next show that the problem of finding the generalized inverse of an $m \times n$ matrix can be reduced to one of finding a generalized inverse of a $k \times k$ matrix where k is the smaller of m and n.

Lemma 9.2 *Let* $\boldsymbol{A}$ *be an* $m \times n$ *matrix. Let* $(\boldsymbol{A}'\boldsymbol{A})^g$ *and* $(\boldsymbol{A}\boldsymbol{A}')^g$ *be generalized inverses of* $\boldsymbol{A}'\boldsymbol{A}$ *and* $\boldsymbol{A}\boldsymbol{A}'$*, respectively. Then,* $(\boldsymbol{A}'\boldsymbol{A})^g\boldsymbol{A}'$ *and* $\boldsymbol{A}'(\boldsymbol{A}\boldsymbol{A}')^g$ *are generalized inverses of* $\boldsymbol{A}$.

Proof. Clearly $\boldsymbol{A}'\boldsymbol{A}(\boldsymbol{A}'\boldsymbol{A})^g\boldsymbol{A}'\boldsymbol{A} = \boldsymbol{A}'\boldsymbol{A}$. Since $\rho(\boldsymbol{A}'\boldsymbol{A}) = \rho(\boldsymbol{A})$, we can use Lemma 5.4 to cancel $\boldsymbol{A}'$ from the left to obtain

$$\boldsymbol{A}\left[(\boldsymbol{A}'\boldsymbol{A})^g\boldsymbol{A}'\right]\boldsymbol{A} = \boldsymbol{A} ,$$

which proves that $(\boldsymbol{A}'\boldsymbol{A})^g\boldsymbol{A}'$ is a generalized inverse of $\boldsymbol{A}$. An analogous argument will prove that $\boldsymbol{A}'(\boldsymbol{A}\boldsymbol{A}')^g$ is also a generalized inverse of $\boldsymbol{A}$. □

The orthogonal projector onto the column space of a matrix $\boldsymbol{A}$ is given by (8.9) when the matrix $\boldsymbol{A}$ has full column rank. Based upon Theorem 9.2 and Theorem 9.2, we can define the orthogonal projector onto $\mathcal{C}(\boldsymbol{A})$ more generally as

$$\boldsymbol{P_A} = \boldsymbol{A}(\boldsymbol{A}'\boldsymbol{A})^g\boldsymbol{A}' , \tag{9.3}$$

where $(\boldsymbol{A}'\boldsymbol{A})^g$ is a symmetric generalized inverse of $\boldsymbol{A}$. Clearly $\boldsymbol{P_A}$ is symmetric. Since $(\boldsymbol{A}'\boldsymbol{A})^g\boldsymbol{A}'$ is a generalized inverse of $\boldsymbol{A}$ (Theorem 9.2), $\boldsymbol{P_A}$ is idempotent. Therefore, $\boldsymbol{P_A}$ is an orthogonal projector.

One outstanding issue remains—is $\boldsymbol{P_A}$ unique? After all, $\boldsymbol{P_A}$ depends upon a generalized inverse and we know generalized inverses are not unique. The following theorem and its corollary resolves this issue.

Theorem 9.14 *Let $\boldsymbol{A}$, $\boldsymbol{B}$ and $\boldsymbol{C}$ be matrices such that $\mathcal{R}(\boldsymbol{A}) \subseteq \mathcal{R}(\boldsymbol{B})$ and $\mathcal{C}(\boldsymbol{C}) \subseteq \mathcal{C}(\boldsymbol{B})$. Then $\boldsymbol{AB}^g\boldsymbol{C}$ is invariant to different choices of $\boldsymbol{B}^g$, where $\boldsymbol{B}^g$ is a generalized inverse of $\boldsymbol{B}$.*

Proof. Because $\mathcal{R}(\boldsymbol{A}) \subseteq \mathcal{R}(\boldsymbol{B})$ and $\mathcal{C}(\boldsymbol{C}) \subseteq \mathcal{C}(\boldsymbol{B})$, it follows that $\boldsymbol{A} = \boldsymbol{MB}$ and $\boldsymbol{C} = \boldsymbol{BN}$ for some matrices $\boldsymbol{M}$ and $\boldsymbol{N}$. Then,

$$\boldsymbol{AB}^g\boldsymbol{C} = \boldsymbol{MBB}^g\boldsymbol{BN} = \boldsymbol{M}(\boldsymbol{BB}^g\boldsymbol{B})\boldsymbol{N} = \boldsymbol{MBN} ,$$

which does not depend upon $\boldsymbol{B}^g$. Therefore, $\boldsymbol{AB}^g\boldsymbol{C}$ is invariant to the choice of $\boldsymbol{B}^g$. □

Corollary 9.3 *The matrix $\boldsymbol{P_A} = \boldsymbol{A}(\boldsymbol{A}'\boldsymbol{A})^g\boldsymbol{A}'$ is invariant to the choice of $(\boldsymbol{A}'\boldsymbol{A})^g$.*

Proof. This follows from Theorem 9.14 because $\mathcal{R}(\boldsymbol{A}'\boldsymbol{A}) = \mathcal{R}(\boldsymbol{A})$ and $\mathcal{C}(\boldsymbol{A}'\boldsymbol{A}) = \mathcal{C}(\boldsymbol{A}')$. □

9.5 Generalized inverses and linear systems

Generalized inverses can also be used to characterize the solutions of homogeneous and non-homogeneous systems. This follows from the following representation for the null space of a matrix.

Theorem 9.15 *Let $\boldsymbol{A}$ be any matrix and let $\boldsymbol{G}$ be a generalized inverse of $\boldsymbol{A}$. Then*

$$\mathcal{N}(\boldsymbol{A}) = \mathcal{C}(\boldsymbol{I} - \boldsymbol{GA}) \text{ and } \mathcal{N}(\boldsymbol{A}') = \mathcal{C}(\boldsymbol{I} - \boldsymbol{G}'\boldsymbol{A}').$$

Proof. If $\boldsymbol{x} \in \mathcal{C}(\boldsymbol{I} - \boldsymbol{GA})$, then $\boldsymbol{x} = (\boldsymbol{I} - \boldsymbol{GA})\boldsymbol{u}$ for some vector $\boldsymbol{u}$ and

$$\boldsymbol{Ax} = \boldsymbol{A}(\boldsymbol{I} - \boldsymbol{GA})\boldsymbol{u} = (\boldsymbol{A} - \boldsymbol{AGA})\boldsymbol{u} = \boldsymbol{0} \Longrightarrow \boldsymbol{x} \in \mathcal{N}(\boldsymbol{A}) .$$

Therefore, $\mathcal{C}(\boldsymbol{I} - \boldsymbol{GA}) \subseteq \mathcal{N}(\boldsymbol{A})$.

To prove the other direction, suppose $\boldsymbol{x} \in \mathcal{N}(\boldsymbol{A})$ so $\boldsymbol{Ax} = \boldsymbol{0}$. Therefore, we can write

$$\boldsymbol{x} = \boldsymbol{x} - \boldsymbol{GAx} = (\boldsymbol{I} - \boldsymbol{GA})\boldsymbol{x} \in \mathcal{C}(\boldsymbol{I} - \boldsymbol{GA})$$

and so $\mathcal{N}(\boldsymbol{A}) \subseteq \mathcal{C}(\boldsymbol{I}-\boldsymbol{G}\boldsymbol{A})$. This completes the proof for $\mathcal{N}(\boldsymbol{A}) = \mathcal{C}(\boldsymbol{I}-\boldsymbol{G}\boldsymbol{A})$.

Applying the above result to $\boldsymbol{A}'$, and noting that $\boldsymbol{G}'$ is a generalized inverse of $\boldsymbol{A}'$, we obtain $\mathcal{N}(\boldsymbol{A}') = \mathcal{C}(\boldsymbol{I}-\boldsymbol{G}'\boldsymbol{A}')$. $\square$

Theorem 9.15 also yields an alternative proof of the Rank-Plus-Nullity Theorem.

Theorem 9.16 *For any $m \times n$ matrix $\boldsymbol{A}$, $\nu(\boldsymbol{A}) = n - \rho(\boldsymbol{A})$.*

Proof. Since $\boldsymbol{G}$ is a generalized inverse of $\boldsymbol{A}$, we have that $\boldsymbol{AGA} = \boldsymbol{A}$. Multiplying both sides from the left by $\boldsymbol{G}$, we see that $\boldsymbol{GAGA} = \boldsymbol{GA}$, which proves that $\boldsymbol{GA}$ is idempotent. This means that $\boldsymbol{I}-\boldsymbol{GA}$ is also idempotent. Using the fact that the rank and trace of an idempotent matrix are equal, we obtain

$$\nu(\boldsymbol{A}) = \dim[\mathcal{N}(\boldsymbol{A})] = \dim[\mathcal{C}(\boldsymbol{I}-\boldsymbol{GA})] = \rho(\boldsymbol{I}-\boldsymbol{GA}) = \operatorname{tr}(\boldsymbol{I}-\boldsymbol{GA})$$
$$= n - \operatorname{tr}(\boldsymbol{GA}) = n - \rho(\boldsymbol{A})\,.$$

$\square$

Theorem 9.15 characterizes solutions for linear systems in terms of generalized inverses.

Theorem 9.17 *Let $\boldsymbol{A}$ be any matrix and let $\boldsymbol{G}$ be a generalized inverse of $\boldsymbol{A}$.*

(i) If $\boldsymbol{x}_0$ is a solution for $\boldsymbol{Ax} = \boldsymbol{0}$, then $\boldsymbol{x}_0 = (\boldsymbol{I}-\boldsymbol{GA})\boldsymbol{u}$ for some vector $\boldsymbol{u}$.

(ii) If $\boldsymbol{x}_0$ is a solution for a consistent system $\boldsymbol{Ax} = \boldsymbol{b}$, then $\boldsymbol{x}_0 = \boldsymbol{Gb}+(\boldsymbol{I}-\boldsymbol{GA})\boldsymbol{u}$ for some vector $\boldsymbol{u}$.

Proof. **Proof of (i):** This follows from Theorem 9.15. If $\boldsymbol{x}_0$ is a solution for $\boldsymbol{Ax} = \boldsymbol{0}$, then $\boldsymbol{x}_0 \in \mathcal{N}(\boldsymbol{A}) = \mathcal{C}(\boldsymbol{I}-\boldsymbol{GA})$, which means that there must exist some vector $\boldsymbol{u}$ for which $\boldsymbol{x}_0 = (\boldsymbol{I}-\boldsymbol{GA})\boldsymbol{u}$.

Proof of (ii): If $\boldsymbol{x}_0$ is a solution for a consistent system $\boldsymbol{Ax} = \boldsymbol{b}$, then $\boldsymbol{x}_0 = \boldsymbol{x}_p + \boldsymbol{w}$, where $\boldsymbol{x}_p$ is a particular solution for $\boldsymbol{Ax} = \boldsymbol{b}$ and $\boldsymbol{w} \in \mathcal{N}(\boldsymbol{A})$ (Theorem 9.1). Since $\boldsymbol{Gb}$ is a particular solution, we can set $\boldsymbol{x}_p = \boldsymbol{Gb}$ and since $\mathcal{N}(\boldsymbol{A}) = \mathcal{C}(\boldsymbol{I}-\boldsymbol{GA})$ we know that $\boldsymbol{w} = (\boldsymbol{I}-\boldsymbol{GA})\boldsymbol{u}$ for some vector $\boldsymbol{u}$. This proves (ii). $\square$

If $\boldsymbol{Ax} = \boldsymbol{b}$ is a consistent system and $\boldsymbol{G}$ is a generalized inverse of $\boldsymbol{A}$, then $\boldsymbol{Gb}$ is a solution for $\boldsymbol{Ax} = \boldsymbol{b}$. But what about the converse: if $\boldsymbol{x}_0$ is a solution of $\boldsymbol{Ax} = \boldsymbol{b}$, then is it necessary that $\boldsymbol{x}_0 = \boldsymbol{Gb}$ for some generalized inverse of $\boldsymbol{A}$? The answer is clearly no in general. Part (ii) of Theorem 9.17 says that $\boldsymbol{x}_0 = \boldsymbol{Gb} + (\boldsymbol{I}-\boldsymbol{GA})\boldsymbol{u}$ for some vector $\boldsymbol{u}$, which need not belong to $\mathcal{C}(\boldsymbol{G})$. In fact, if $\boldsymbol{b} = \boldsymbol{0}$ (i.e., a homogeneous system) and $\rho(\boldsymbol{A}) < n$, then the system has an infinite number of solutions (Theorem 9.5) but the only solution that belongs to $\mathcal{C}(\boldsymbol{G})$ is the zero vector.

The following lemma describes when every solution of $\boldsymbol{Ax} = \boldsymbol{b}$ can be expressed as $\boldsymbol{Gb}$ for some generalized inverse $\boldsymbol{G}$ of $\boldsymbol{A}$.

Lemma 9.3 *Let $\boldsymbol{Ax} = \boldsymbol{b}$ be a consistent linear system, where $\boldsymbol{A}$ is $m \times n$ and at least one of the following two conditions hold: (a) $\rho(\boldsymbol{A}) = n$ (i.e., $\boldsymbol{A}$ is of full column rank), or (b) $\boldsymbol{b} \neq \boldsymbol{0}$. Then, $\boldsymbol{x}_0$ is a solution to $\boldsymbol{Ax} = \boldsymbol{b}$ if and only if $\boldsymbol{x}_0 = \boldsymbol{Gb}$ for some generalized inverse $\boldsymbol{G}$ of $\boldsymbol{A}$.*

Proof. If $\boldsymbol{x}_0 = \boldsymbol{Gb}$ for some generalized inverse of $\boldsymbol{A}$, then Definition 9.2 itself ensures that $\boldsymbol{x}_0$ is a solution.

Suppose (a) holds. Then $\boldsymbol{A}$ has full column rank, which means that $\boldsymbol{Ax} = \boldsymbol{b}$ has a unique solution (Theorem 9.4). This solution must be of the form $\boldsymbol{Gb}$, where $\boldsymbol{G}$ is a generalized inverse of $\boldsymbol{A}$, because $\boldsymbol{Gb}$ is a solution.

Now suppose that (b) holds. That is, $\boldsymbol{b} \neq \boldsymbol{0}$. If $\boldsymbol{x}_0$ is a solution for $\boldsymbol{Ax} = \boldsymbol{b}$, then $\boldsymbol{x}_0 = \boldsymbol{Gb} + (\boldsymbol{I}_m - \boldsymbol{GA})\boldsymbol{u}$ for some vector $\boldsymbol{u} \in \Re^n$. Since $\boldsymbol{b} = \boldsymbol{Ax}_0$, we have

$$\boldsymbol{x}_0 = \boldsymbol{GAx}_0 + (\boldsymbol{I}_m - \boldsymbol{GA})\boldsymbol{u} = \boldsymbol{u} + \boldsymbol{GA}(\boldsymbol{x}_0 - \boldsymbol{u}) .$$

□

9.6 The Moore-Penrose inverse

The generalized inverse, as defined in Definition 9.2, is not unique. It is, however, possible to construct a *unique* generalized inverse if we impose some additional conditions. We start with the following definition.

Definition 9.3 *Let $\boldsymbol{A}$ be an $m \times n$ matrix. An $n \times m$ matrix $\boldsymbol{G}$ is called the* **Moore-Penrose inverse** *of $\boldsymbol{A}$ if it satisfies the following four conditions:*

(i) $\boldsymbol{AGA} = \boldsymbol{A}$; (ii) $\boldsymbol{GAG} = \boldsymbol{G}$; (iii) $(\boldsymbol{AG})' = \boldsymbol{AG}$ and (iv) $(\boldsymbol{GA})' = \boldsymbol{GA}$.

Our first order of business is to demonstrate that a Moore-Penrose inverse not only exists but it is also unique. We do so in the following two theorems.

Theorem 9.18 *The Moore-Penrose inverse of any matrix $\boldsymbol{A}$ is unique.*

Proof. Suppose, if possible, $\boldsymbol{G}_1$ and $\boldsymbol{G}_2$ are two Moore-Penrose inverses of $\boldsymbol{A}$. We will show that $\boldsymbol{G}_1 = \boldsymbol{G}_2$. We make abundant use of the fact that $\boldsymbol{G}_1$ and $\boldsymbol{G}_2$ *both* satisfy the four conditions in Definition 9.3 in what follows:

$$\begin{aligned}\boldsymbol{G}_1 &= \boldsymbol{G}_1\boldsymbol{AG}_1 = \boldsymbol{G}_1(\boldsymbol{AG}_1) = \boldsymbol{G}_1(\boldsymbol{AG}_1)' = \boldsymbol{G}_1\boldsymbol{G}_1'\boldsymbol{A}' \\ &= \boldsymbol{G}_1\boldsymbol{G}_1'\boldsymbol{A}'\boldsymbol{G}_2'\boldsymbol{A}' = \boldsymbol{G}_1\boldsymbol{G}_1'\boldsymbol{A}'(\boldsymbol{AG}_2)' = \boldsymbol{G}_1\boldsymbol{G}_1'\boldsymbol{A}'\boldsymbol{AG}_2 \\ &= \boldsymbol{G}_1(\boldsymbol{AG}_1)'\boldsymbol{AG}_2 = \boldsymbol{G}_1\boldsymbol{AG}_1\boldsymbol{AG}_2 = \boldsymbol{G}_1(\boldsymbol{AG}_1\boldsymbol{A})\boldsymbol{G}_2 \\ &= \boldsymbol{G}_1\boldsymbol{AG}_2 .\end{aligned}$$

Pause for a while, take a deep breath and continue:

$$\begin{aligned} \boldsymbol{G}_1 &= \boldsymbol{G}_1\boldsymbol{A}\boldsymbol{G}_2 = (\boldsymbol{G}_1\boldsymbol{A})\boldsymbol{G}_2 = (\boldsymbol{G}_1\boldsymbol{A})'\boldsymbol{G}_2\boldsymbol{A}\boldsymbol{G}_2 = \boldsymbol{A}'\boldsymbol{G}_1'(\boldsymbol{G}_2\boldsymbol{A})\boldsymbol{G}_2 \\ &= \boldsymbol{A}'\boldsymbol{G}_1'(\boldsymbol{G}_2\boldsymbol{A})'\boldsymbol{G}_2 = \boldsymbol{A}'\boldsymbol{G}_1'\boldsymbol{A}'\boldsymbol{G}_2'\boldsymbol{G}_2 = (\boldsymbol{A}'\boldsymbol{G}_1'\boldsymbol{A}')\boldsymbol{G}_2'\boldsymbol{G}_2 \\ &= \boldsymbol{A}'\boldsymbol{G}_2'\boldsymbol{G}_2 = (\boldsymbol{A}'\boldsymbol{G}_2')\boldsymbol{G}_2 = (\boldsymbol{G}_2\boldsymbol{A})'\boldsymbol{G}_2 = \boldsymbol{G}_2\boldsymbol{A}\boldsymbol{G}_2 \\ &= \boldsymbol{G}_2 \,. \end{aligned}$$

This proves the uniqueness of the Moore-Penrose inverse. □

We denote the unique Moore-Penrose inverse of $\boldsymbol{A}$ by $\boldsymbol{A}^+$. Several elementary properties of $\boldsymbol{A}^+$ can be derived from Definition 9.3. The following two are examples.

Lemma 9.4 *Let $\boldsymbol{A}^+$ be a Moore-Penrose inverse of $\boldsymbol{A}$. Then:*

(i) $(\boldsymbol{A}^+)' = (\boldsymbol{A}')^+$;

(ii) $(\boldsymbol{A}^+)^+ = \boldsymbol{A}$.

Proof. **Proof of (i):** Let $\boldsymbol{G} = (\boldsymbol{A}^+)'$. We need to prove that $\boldsymbol{G}$ satisfies the four conditions in Definition 9.3 with respect to $\boldsymbol{A}'$.

(i): $\boldsymbol{A}'\boldsymbol{G}\boldsymbol{A}' = (\boldsymbol{A}\boldsymbol{G}'\boldsymbol{A})' = (\boldsymbol{A}\boldsymbol{A}^+\boldsymbol{A})' = \boldsymbol{A}'$.

(ii): $\boldsymbol{G}\boldsymbol{A}'\boldsymbol{G} = (\boldsymbol{G}'\boldsymbol{A}\boldsymbol{G}')' = (\boldsymbol{A}^+\boldsymbol{A}\boldsymbol{A}^+)' = (\boldsymbol{A}^+)' = \boldsymbol{G}$.

(iii): $(\boldsymbol{A}'\boldsymbol{G})' = \boldsymbol{G}'\boldsymbol{A} = \boldsymbol{A}^+\boldsymbol{A} = (\boldsymbol{A}^+\boldsymbol{A})' = \boldsymbol{A}'(\boldsymbol{A}^+)' = \boldsymbol{A}'\boldsymbol{G}$.

(iv): $(\boldsymbol{G}\boldsymbol{A}')' = \boldsymbol{A}\boldsymbol{G}' = \boldsymbol{A}\boldsymbol{A}^+ = (\boldsymbol{A}\boldsymbol{A}^+)' = (\boldsymbol{A}^+)'\boldsymbol{A}' = \boldsymbol{G}\boldsymbol{A}'$.

Therefore $\boldsymbol{G} = (\boldsymbol{A}^+)'$ is a Moore-Penrose inverse of $\boldsymbol{A}'$.

Proof of (ii): The conditions in Definition 9.3 have a symmetry about them that implies that if $\boldsymbol{G}$ is a Moore-Penrose inverse of $\boldsymbol{A}$, then $\boldsymbol{A}$ is a Moore-Penrose of $\boldsymbol{G}$. □

Theorem 9.19 *Let $\boldsymbol{A}$ be an $m \times n$ matrix and suppose that $\boldsymbol{A} = \boldsymbol{C}\boldsymbol{R}$ is a rank factorization for $\boldsymbol{A}$. Then:*

(i) $\boldsymbol{C}^+ = (\boldsymbol{C}'\boldsymbol{C})^{-1}\boldsymbol{C}'$;

(ii) $\boldsymbol{R}^+ = \boldsymbol{R}'(\boldsymbol{R}\boldsymbol{R}')^{-1}$;

(iii) $\boldsymbol{A}^+ = \boldsymbol{R}^+\boldsymbol{C}^+$.

Proof. $\boldsymbol{C}$ is $m \times r$, $\boldsymbol{R}$ is $r \times n$ and $\rho(\boldsymbol{A}) = \rho(\boldsymbol{C}) = \rho(\boldsymbol{R}) = r$. Since $\boldsymbol{C}$ has full column rank and $\boldsymbol{R}$ has full row rank, $\boldsymbol{C}'\boldsymbol{C}$ and $\boldsymbol{R}'\boldsymbol{R}$ are nonsingular. This ensures the existence of $\boldsymbol{C}^+$ and $\boldsymbol{R}^+$ as defined in (i) and (ii) and one simply needs to verify that these matrices are indeed the Moore-Penrose inverses of $\boldsymbol{C}$ and $\boldsymbol{R}$, respectively.

This is straightforward algebra. We show the flavor with C^+:

$$\begin{aligned}
CC^+C &= C(C'C)^{-1}C'C = C(C'C)^{-1}(C'C) = CI = C\,;\\
C^+CC^+ &= (C'C)^{-1}C'C(C'C)^{-1}C' = \left[(C'C)^{-1}C'C\right](C'C)^{-1}C'\\
&= I(C'C)^{-1}C' = IC^+ = C^+\,;\\
(CC^+)' &= (C^+)'C' = C(C'C)^{-1}C' = CC^+\\
(C^+C)' &= C'(C^+)' = C'C(C'C)^{-1} = I = (C'C)^{-1}C'C = C^+C\,.
\end{aligned}$$

Therefore, C^+ is indeed the Moore-Penrose inverse of C. the verification for R^+ is analogous and equally straightforward. Note that C^+ is a left inverse of C and R^+ is a right inverse of R so $C^+C = I_r = RR^+$. We now verify that A is a Moore-Penrose inverse of A:

$$\begin{aligned}
AA^+A &= (CR)R^+C^+(CR) = CRR'(RR')^{-1}(C'C)^{-1}C'CR\\
&= CI_rR = CR = A\,;\\
A^+AA^+ &= R^+C^+(CR)R^+C^+\\
&= R'(RR')^{-1}(C'C)^{-1}C'CRR'(RR')^{-1}(C'C)^{-1}C'\,;\\
&= R'(RR')^{-1}I_rI_r(C'C)^{-1}C' = R^+C^+ = A^+\,;\\
(AA^+)' &= (A^+)'A' = (R^+C^+)'R'C' = (C^+)'(R^+)'R'C'\\
&= (C^+)'(RR^+)'C' = (C^+)'C' = (CC^+)' = CC^+\\
&= CRR^+C^+ = AA^+\,;\\
(A^+A)' &= A'(A^+)' = R'C'(R^+C^+)' = R'C'(C^+)'(R^+)'\\
&= R'(C^+C)'(R^+)' = R'(R^+)' = (R^+R)' = R^+R\\
&= R^+C^+CR = A^+A\,.
\end{aligned}$$

These identities prove that A^+ is a Moore-Penrose inverse of A. □

Theorem 9.13 and its corollary tells us that AG is a projector onto $\mathcal{C}(A)$ along $\mathcal{N}(AG)$ for any generalized inverse G. If, in addition, AG is symmetric (one of the conditions for the Moore-Penrose inverse) then this projector must be an orthogonal projector. This suggests that $x = Gb$ minimizes the distance $\|Ax - b\|$, whether $Ax = b$ is consistent or not.

Theorem 9.20 *Let A be an $m \times n$ matrix. Then,*

$$\|AA^+b - b\| \leq \|Ax - b\| \;\; \textit{for all } x \textit{ and } b\,.$$

Proof. Because AA^+ is symmetric, $AA^+ = (A^+)'A'$ so

$$A'(AA^+ - I) = A'((A^+)'A' - I) = A'(A^+)'A' - A' = (AA^+A - A)' = O\,.$$

We now use this to conclude that

$$\begin{aligned}\|\boldsymbol{Ax}-\boldsymbol{b}\|^2 &= \|(\boldsymbol{Ax}-\boldsymbol{AA}^+\boldsymbol{b})+(\boldsymbol{AA}^+\boldsymbol{b}-\boldsymbol{b})\|^2\\ &= \|\boldsymbol{A}(\boldsymbol{x}-\boldsymbol{A}^+\boldsymbol{b})\|^2+\|(\boldsymbol{AA}^+-\boldsymbol{I})\boldsymbol{b}\|^2\\ &\qquad +2(\boldsymbol{x}-\boldsymbol{A}^+\boldsymbol{b})'\boldsymbol{A}'(\boldsymbol{AA}^+-\boldsymbol{I})\boldsymbol{b}\\ &= \|\boldsymbol{A}(\boldsymbol{x}-\boldsymbol{A}^+\boldsymbol{b})\|^2+\|(\boldsymbol{AA}^+-\boldsymbol{I})\boldsymbol{b}\|^2\\ &\geq \|(\boldsymbol{AA}^+-\boldsymbol{I})\boldsymbol{b}\|^2=\|\boldsymbol{AA}^+\boldsymbol{b}-\boldsymbol{b}\|^2\,.\end{aligned}$$

□

Note that we have used only two properties of the Moore-Penrose inverse, $\boldsymbol{AA}^+\boldsymbol{A}=\boldsymbol{A}$ and $(\boldsymbol{AA}^+)'=\boldsymbol{AA}^+$—to prove Theorem 9.20. In fact, the additional properties of the Moore-Penrose inverse allows us to characterize orthogonal projections onto the four fundamental subspaces in terms of the Moore-Penrose inverse. Since the Moore-Penrose inverse is a generalized inverse, Corollary 9.2 tells us that $\boldsymbol{AA}^+$ is a projector onto $\mathcal{C}(\boldsymbol{A})$ along $\mathcal{N}(\boldsymbol{AA}^+)$. Because $\boldsymbol{AA}^+$ is symmetric, $\boldsymbol{AA}^+$ is also an *orthogonal projector* onto the column space of $\boldsymbol{A}$. Similarly, $\boldsymbol{A}^+\boldsymbol{A}$ is symmetric and idempotent. The symmetry implies $\mathcal{C}(\boldsymbol{A}^+\boldsymbol{A})=\mathcal{C}((\boldsymbol{A}^+\boldsymbol{A})')=\mathcal{R}(\boldsymbol{A}^+\boldsymbol{A})\subseteq\mathcal{R}(\boldsymbol{A})$. Also, $\rho(\boldsymbol{A}^+\boldsymbol{A})\leq\rho(\boldsymbol{A})=\rho(\boldsymbol{AA}^+\boldsymbol{A})\leq\rho(\boldsymbol{A}^+\boldsymbol{A})$, which means that $\rho(\boldsymbol{A}^+\boldsymbol{A})=\rho(\boldsymbol{A})$; this is true for any generalized inverse of $\boldsymbol{A}$, not just $\boldsymbol{A}^+$. Therefore, the dimensions of $\mathcal{C}(\boldsymbol{A}^+\boldsymbol{A})$ and $\mathcal{R}(\boldsymbol{A})$ are the same, which allows us to conclude that $\mathcal{C}(\boldsymbol{A}^+\boldsymbol{A})=\mathcal{R}(\boldsymbol{A})$). This means that $\boldsymbol{A}^+\boldsymbol{A}$ is an orthogonal projector onto $\mathcal{R}(\boldsymbol{A})$.

It is easily verified that $\boldsymbol{I}-\boldsymbol{AA}^+$ and $\boldsymbol{I}-\boldsymbol{A}^+\boldsymbol{A}$ are symmetric and idempotent. So they are orthogonal projectors onto $\mathcal{C}(\boldsymbol{I}-\boldsymbol{AA}^+)$ and $\mathcal{C}(\boldsymbol{I}-\boldsymbol{A}^+\boldsymbol{A})$, respectively. Theorem 9.15 tells us that $\mathcal{N}(\boldsymbol{A})=\mathcal{C}(\boldsymbol{I}-\boldsymbol{A}^+\boldsymbol{A})=\mathcal{N}(\boldsymbol{A})$ and, using the symmetry of $\boldsymbol{AA}^+$, that $\mathcal{N}(\boldsymbol{A}')=\mathcal{C}(\boldsymbol{I}-(\boldsymbol{A}^+)'\boldsymbol{A}')=\mathcal{C}(\boldsymbol{I}-\boldsymbol{AA}^+)$.

The above discussions show how the Moore-Penrose inverse yields orthogonal projectors for each of the four fundamental subspaces. We call these the ***four fundamental projectors*** and summarize below:

(i) $\boldsymbol{AA}^+$ is the orthogonal projector onto the column space of $\boldsymbol{A}$, $\mathcal{C}(\boldsymbol{A})$;

(ii) $\boldsymbol{A}^+\boldsymbol{A}$ is the orthogonal projector onto the row space of $\boldsymbol{A}$, $\mathcal{R}(\boldsymbol{A})=\mathcal{C}(\boldsymbol{A}')$;

(iii) $\boldsymbol{I}-\boldsymbol{A}^+\boldsymbol{A}$ is the orthogonal projector onto the null space of $\boldsymbol{A}$, $\mathcal{N}(\boldsymbol{A})$;

(iv) $\boldsymbol{I}-\boldsymbol{AA}^+$ is the orthogonal projector onto the left null space of $\boldsymbol{A}$, $\mathcal{N}(\boldsymbol{A}')$.

What we have presented in this chapter can be regarded as a rather brief overview of generalized inverses. Dedicated texts on this topic include classics such as Rao and Mitra (1972) and Ben-Israel and Greville (2003). Most texts written for a statistical audience, such as Rao (1973), Searle (1982), Harville (1997), Rao and Bhimasankaran, Healy (2000), Graybill (2001), Schott (2005) and Gentle (2010), discuss generalized inverses with varying levels of detail. Applications abound in statistical linear models and can be found in the texts mentioned in Section 8.9.

9.7 Exercises

1. Let $\boldsymbol{A}$ be an $m \times n$ matrix with $\rho(\boldsymbol{A}) = r$. Assume a consistent linear system $\boldsymbol{Ax} = \boldsymbol{b}$, where $\boldsymbol{b} \neq \boldsymbol{0}$, and let $\boldsymbol{x}_p$ be a particular solution. If $\{\boldsymbol{u}_1, \boldsymbol{u}_2, \ldots, \boldsymbol{u}_{n-r}\}$ is a basis for $\mathcal{N}(\boldsymbol{A})$, then prove that

$$\{\boldsymbol{x}_p, \boldsymbol{x}_p + \boldsymbol{u}_1, \boldsymbol{x}_p + \boldsymbol{u}_2, \ldots, \boldsymbol{x}_p + \boldsymbol{u}_{n-r}\}$$

is a linearly independent set of $n - r + 1$ solutions. Show that no set containing more than $n - r + 1$ solutions can be linearly independent.

2. Prove that $\boldsymbol{Ax} = \boldsymbol{b}$ is consistent for all $\boldsymbol{b}$ if and only if $\boldsymbol{A}$ has full row rank.

3. Prove that $\boldsymbol{Ax} = \boldsymbol{b}$ has a solution in $\mathcal{C}(\boldsymbol{B})$ if and only if the system $\boldsymbol{ABu} = \boldsymbol{b}$ is consistent.

4. True or false: If $\boldsymbol{A}$ is $m \times n$, then $\boldsymbol{Ax} = \boldsymbol{b}$ is a consistent system for every $\boldsymbol{b} \in \Re^m$ if $m < n$.

5. True or false: If $\boldsymbol{b} \in \mathcal{C}(\boldsymbol{A})$, then $\boldsymbol{Ax} = \boldsymbol{b}$ has a unique solution belonging to $\mathcal{R}(\boldsymbol{A})$.

6. For every $\boldsymbol{b} \in \mathcal{C}(\boldsymbol{A})$, where $\boldsymbol{A}$ is square, prove that $\boldsymbol{Ax} = \boldsymbol{b}$ has a solution belonging to $\mathcal{C}(\boldsymbol{A})$ if and only if $\rho(\boldsymbol{A}) = \rho(\boldsymbol{A}^2)$.

7. Let $\boldsymbol{Ax} = \boldsymbol{b}$ be consistent. Show that $(\boldsymbol{A} + \boldsymbol{uv}')\boldsymbol{x} = \boldsymbol{b}$ is consistent whenever $\boldsymbol{v} \notin \mathcal{R}(\boldsymbol{A})$.

8. Let $\boldsymbol{Ax} = \boldsymbol{b}$ be consistent and suppose we add one more equation $\boldsymbol{u}'\boldsymbol{x} = \alpha$ to the system, where $\boldsymbol{u} \notin \mathcal{R}(\boldsymbol{A})$. Show that the new system is still consistent although the solution set for the new system is different from the old system.

9. Let $\boldsymbol{Ax} = \boldsymbol{b}$ be consistent and suppose we add one more equation $\boldsymbol{u}'\boldsymbol{x} = \alpha$ to the system, where $\boldsymbol{u} \in \mathcal{R}(\boldsymbol{A})$. Show that the new system is consistent if and only if the solution set for the old systems is the same as that for the new system.

10. Let $\boldsymbol{A}$ be $m \times n$. If $\boldsymbol{AX} = \boldsymbol{O}_{m \times p}$, prove that $\mathcal{C}(\boldsymbol{X}) \subseteq \mathcal{N}(\boldsymbol{A})$. Prove that if $\rho(\boldsymbol{X}) = n - \rho(\boldsymbol{A})$, then $\mathcal{C}(\boldsymbol{X}) = \mathcal{N}(\boldsymbol{A})$.

11. Let $\boldsymbol{A}$ be $m \times n$ and $\boldsymbol{G}$ be a generalized inverse of $\boldsymbol{A}$. Prove that the $n \times p$ matrix $\boldsymbol{X}_0$ is a solution of the linear system $\boldsymbol{AX} = \boldsymbol{O}$, where $\boldsymbol{O}$ is the $m \times p$ matrix of zeroes if and only if $\boldsymbol{X}_0 = (\boldsymbol{I} - \boldsymbol{GA})\boldsymbol{Y}$ for some matrix $\boldsymbol{Y}$.

12. Let $\boldsymbol{A}$ be $m \times n$ and $\boldsymbol{G}$ be a generalized inverse of $\boldsymbol{A}$. Prove that the $n \times p$ matrix $\boldsymbol{X}_0$ is a solution of the linear system $\boldsymbol{AX} = \boldsymbol{B}$, where $\rho(\boldsymbol{A}) = n$ or $\rho(\boldsymbol{B}) = p$, if and only if $\boldsymbol{X}_0 = \boldsymbol{GB}$.

13. If $\boldsymbol{G}_1$ and $\boldsymbol{G}_2$ are generalized inverses of $\boldsymbol{A}$, then prove that $\alpha\boldsymbol{G}_1 + (1 - \alpha)\boldsymbol{G}_2$ is a generalized inverse of $\boldsymbol{A}$.

14. Let $\boldsymbol{M} = \begin{bmatrix} \boldsymbol{A} & \boldsymbol{O} \\ \boldsymbol{O} & \boldsymbol{O} \end{bmatrix}$, where $\boldsymbol{A}$ is nonsingular. Then prove that

$$\boldsymbol{G} = \begin{bmatrix} \boldsymbol{A}^{-1} & \boldsymbol{B} \\ \boldsymbol{C} & \boldsymbol{D} \end{bmatrix}, \quad \text{where } \boldsymbol{B}, \boldsymbol{C} \text{ and } \boldsymbol{D} \text{ are arbitrary,}$$

is a generalized inverse of $\boldsymbol{M}$. Is it true that every generalized inverse of $\boldsymbol{M}$ will be of the above form for some choice of $\boldsymbol{B}$, $\boldsymbol{C}$ and $\boldsymbol{D}$?

15. True or false: If $\boldsymbol{G}$ is a generalized inverse of a square matrix $\boldsymbol{A}$, then $\boldsymbol{G}^2$ is a generalized inverse of $\boldsymbol{A}^2$.
16. If $\boldsymbol{G}$ is a generalized inverse of $\boldsymbol{A}$, then prove that $\mathcal{N}(\boldsymbol{A}') = \mathcal{R}(\boldsymbol{I} - \boldsymbol{GA})$.
17. If $\boldsymbol{G}$ is a generalized inverse of $\boldsymbol{A}$, then prove that $\mathcal{R}(\boldsymbol{B}) \subseteq \mathcal{R}(\boldsymbol{A})$ if and only if $\boldsymbol{BGA} = \boldsymbol{B}$.
18. Let $\rho(\boldsymbol{A} + \boldsymbol{B}) = \rho(\boldsymbol{A}) + \rho(\boldsymbol{B})$ and let $\boldsymbol{G}$ be any generalized inverse of $\boldsymbol{A} + \boldsymbol{B}$. Show that $\boldsymbol{AGA} = \boldsymbol{A}$ and $\boldsymbol{AGB} = \boldsymbol{O}$.
19. Let $\boldsymbol{x}$ be a nonzero vector. Find a generalized inverse of $\boldsymbol{x}$.
20. Find a generalized inverse of the following matrices:

$$\begin{bmatrix} 1 & 0 \\ 0 & 0 \end{bmatrix}, \quad \begin{bmatrix} 1 & 1 \\ 1 & 1 \end{bmatrix}, \quad \text{and} \quad \begin{bmatrix} 1 & 1 \\ 1 & 2 \\ 1 & 3 \\ 1 & 4 \end{bmatrix}.$$

21. Let $\boldsymbol{A} = \begin{bmatrix} \boldsymbol{A}_{11} & \boldsymbol{A}_{12} \\ \boldsymbol{A}_{21} & \boldsymbol{A}_{22} \end{bmatrix}$, where $\boldsymbol{A}_{11}$ and $\boldsymbol{A}_{22}$ are square with $\mathcal{C}(\boldsymbol{A}_{12}) \subseteq \mathcal{C}(\boldsymbol{A}_{11})$ and $\mathcal{R}(\boldsymbol{A}_{21}) \subseteq \mathcal{R}(\boldsymbol{A}_{22})$. Prove that

$$\rho(\boldsymbol{A}) = \rho(\boldsymbol{A}_{11}) + \rho(\boldsymbol{A}_{22} - \boldsymbol{A}_{21}\boldsymbol{G}\boldsymbol{A}_{12}) ,$$

where $\boldsymbol{G}$ is some generalized inverse of $\boldsymbol{A}_{11}$.

22. Let $\boldsymbol{A}$ be $m \times n$ and $\boldsymbol{G}$ be a generalized inverse of $\boldsymbol{A}$. Let $\boldsymbol{u} \in \mathcal{C}(\boldsymbol{A})$ and $\boldsymbol{v} \in \mathcal{C}(\boldsymbol{A}')$ and define $\delta = 1 + \boldsymbol{v}'\boldsymbol{G}\boldsymbol{u}$. Prove the following:
    (a) If $\delta = 0$, then $\rho(\boldsymbol{A} + \boldsymbol{uv}') = \rho(\boldsymbol{A}) - 1$ and $\boldsymbol{G}$ is a generalized inverse of $\boldsymbol{A} + \boldsymbol{uv}'$.
    (b) If $\delta \neq 0$, then $\rho(\boldsymbol{A} + \boldsymbol{uv}') = \rho(\boldsymbol{A})$ and

$$(\boldsymbol{A} + \boldsymbol{uv}')^g = \boldsymbol{G} - \frac{1}{\delta}\boldsymbol{Guv}'\boldsymbol{G}$$

    is a generalized inverse of $\boldsymbol{A} + \boldsymbol{uv}'$.

23. Provide an example to show that $(\boldsymbol{AB})^+ \neq \boldsymbol{B}^+\boldsymbol{A}^+$ in general.
24. Find the Moore-Penrose inverse of $\boldsymbol{xy}'$, where $\boldsymbol{x}, \boldsymbol{y} \in \Re^n$.
25. Let $\boldsymbol{A}$ and $\boldsymbol{B}$ be $m \times n$ and $m \times p$, respectively. Prove that $\mathcal{C}(\boldsymbol{B}) \subseteq \mathcal{C}(\boldsymbol{A})$ if and only if $\boldsymbol{AA}^+\boldsymbol{B} = \boldsymbol{B}$.
26. True or false: $\mathcal{C}(\boldsymbol{A}^+) = \mathcal{R}(\boldsymbol{A})$.
27. Let $\boldsymbol{A}$ and $\boldsymbol{B}$ be $m \times n$ and $p \times n$, respectively. Prove that $\mathcal{N}(\boldsymbol{A}) \subseteq \mathcal{N}(\boldsymbol{B})$ if and only if $\boldsymbol{BA}^+\boldsymbol{A} = \boldsymbol{B}$.

CHAPTER 10

Determinants

10.1 Introduction

Sometimes a single number associated with a square matrix can provide insight about the properties of the matrix. One such number that we have already encountered is the trace of a matrix. The "determinant" is another example of a single number that conveys an amazing amount of information about the matrix. For example, we will soon see that a square matrix will have no inverse if and only if the determinant is zero. Solutions to linear systems can also be expressed in terms of determinants. In multivariable calculus, probability and statistics, determinants arise as "Jacobians" when transforming variables by differentiable functions.

A formal definition of determinants uses *permutations.* We have already encountered permutations in the definition of permutation matrices. More formally, we can define the permutation as follows.

Definition 10.1 *A* **permutation** *is a one-one map from a finite non-empty set S onto itself. Let $S = \{s_1, s_2 \ldots, s_n\}$. A permutation π is often represented as a $2 \times n$ array*

$$\pi = \begin{pmatrix} s_1 & s_2 & \cdots & s_n \\ \pi(s_1) & \pi(s_2) & \cdots & \pi(s_n) \end{pmatrix}. \tag{10.1}$$

For notational convenience, we will often write $\pi(i)$ as π_i. The finite set S of n elements is often taken as $(1, 2, \ldots, n)$, in which case we write π as $\pi = (\pi_1, \pi_2, ..., \pi_n)$. Simply put, a permutation is any rearrangement of $(1, 2, \ldots, n)$. For example, the complete set of permutations for $\{1, 2, 3\}$ is given by

$$\{(1,2,3), (1,3,2), (2,1,3), (2,3,1), (3,1,2), (3,2,1)\}.$$

In general, the set $(1, 2, \ldots, n)$ has $n! = n \cdot (n-1) \cdots 2 \cdot 1$ different permutations.

Example 10.1 Consider the following two permutations on $S = (1, 2, 3, 4, 5)$: $\pi = (1, 3, 4, 2, 5)$ and $\theta = (1, 2, 5, 3, 4)$. Suppose we apply θ followed by π to $(1, 2, \ldots, 5)$. Then the resultant permutation is is given by

$$\pi\theta = (\pi_{\theta_1}, \pi_{\theta_2}, \ldots, \pi_{\theta_5}) = (\pi_1, \pi_2, \pi_5, \pi_3, \pi_4) = (1, 3, 5, 4, 2).$$

Note that $\pi\theta = \pi \circ \theta$ is formed from the rules of function composition. We call $\pi\theta$ the *composition* or the *product* of π and θ.

The product of two permutations is not necessarily commutative. In other words, $\pi\theta$ need not be equal to $\theta\pi$. For instance, in the above example we find that

$$\theta\pi = (\theta_{\pi_1}, \theta_{\pi_2}, \ldots, \theta_{\pi_5}) = (\theta_1, \theta_3, \theta_4, \theta_2, \theta_5) = (1, 5, 3, 2, 4),$$

which is different from $\pi\theta$ evaluated earlier. ■

Note that the original set $(1, 2, \ldots, n)$ itself is a specific permutation—it is said to be in its ***natural order***. The corresponding permutation, say π_0, is given by $\pi_{0i} = i,\ i = 1, \ldots, n$ and is known as the ***identity permutation***. Also, since any permutation π is a one-one onto mapping, there exists a unique inverse, say π^{-1}, such that $\pi\pi^{-1} = \pi^{-1}\pi = \pi_0$.

Example 10.2 Consider S, π and θ as in Example 10.1. We construct the inverse of π by reversing the action on each element. Since $\pi_1 = 1$, we will have $\pi_1^{-1} = 1$ as well. Next, we see that π maps the second element of S to the third, i.e., $\pi_2 = 3$. Therefore, we have $\pi_3^{-1} = 2$. Similarly, since $\pi_3 = 4$, $\pi_4 = 2$ and $\pi_5 = 5$, we have $\pi_4^{-1} = 3$, $\pi_2^{-1} = 4$ and $\pi_5^{-1} = 5$. Therefore, $\pi^{-1} = (1, 4, 2, 3, 5)$ is the inverse of π.

The inverse of a permutation can be easily found through the following steps. We first represent the permutation in an array, as in (10.1). We then rearrange the columns so that the lower row is in natural order. Finally we interchange the two rows. Here is how we find π^{-1}:

$$\begin{aligned} \pi &= \begin{pmatrix} 1 & 2 & 3 & 4 & 5 \\ 1 & 3 & 4 & 2 & 5 \end{pmatrix} \longrightarrow \begin{pmatrix} 1 & 4 & 2 & 3 & 5 \\ 1 & 2 & 3 & 4 & 5 \end{pmatrix} \\ &\longrightarrow \begin{pmatrix} 1 & 2 & 3 & 4 & 5 \\ 1 & 4 & 2 & 3 & 5 \end{pmatrix} = \pi^{-1}. \end{aligned} \tag{10.2}$$

We leave it to the reader to verify that the inverse permutation of θ is given by $\theta^{-1} = (1, 2, 4, 5, 3)$. ■

Equation (10.2) easily demonstrates how we can move from π to π^{-1}:

$$\begin{aligned} \pi = \begin{pmatrix} 1 & 2 & \cdots & n \\ \pi_1 & \pi_2 & \cdots & \pi_n \end{pmatrix} &\xrightarrow{\text{Bring lower row to natural order}} \begin{pmatrix} \pi_1^{-1} & \pi_2^{-1} & \cdots & \pi_n^{-1} \\ 1 & 2 & \cdots & n \end{pmatrix} \\ &\xrightarrow{\text{Interchange the rows}} \begin{pmatrix} 1 & 2 & \cdots & n \\ \pi_1^{-1} & \pi_2^{-1} & \cdots & \pi_n^{-1} \end{pmatrix} = \pi^{-1}. \end{aligned} \tag{10.3}$$

It is obvious that we can move from π^{-1} to π by simply retracing the arrows in the opposite direction. Put another way, if $\theta = \pi^{-1}$, then the same permutation that changes $(\pi_1, \pi_2, \ldots, \pi_n)$ to $(1, 2, \ldots, n)$ maps $(1, 2, \ldots, n)$ to $(\theta_1, \ldots, \theta_n)$.

It is also easy to see that for any two permutations π and θ, the inverse of their product will be the product of their inverses in reverse order. In other words,

$(\pi\theta)^{-1} = \theta^{-1}\pi^{-1}$. We leave this as an exercise for the reader. Some of these concepts will be useful in deducing properties of determinants.

The number of swaps or interchanges required to restore a permutation π to its natural order is an important concept that arises in the definition of a determinant. Consider the permutation $\pi = (1, 3, 2)$. Only one interchange—swapping 3 and 2—will restore π to its natural order, but one could also use more devious routes, such as $(1, 3, 2) \to (2, 3, 1) \to (2, 1, 3) \to (1, 2, 3)$. The important thing to note is that both these routes involve an odd number of swaps to restore natural order. In fact, it is impossible to restore natural order with an even number of swaps in this case. We say that π is an ***odd permutation***.

Take another example. Let $\pi = (3, 1, 2)$. Now we can restore natural order in two swaps: $(3, 1, 2) \to (1, 3, 2) \to (1, 2, 3)$. Any other route of restoring natural order *must* involve an even number of swaps. We say that π is an ***even permutation***. Note that the natural order $(1, 2, 3)$ requires zero swaps, so is an even permutation.

The above facts hold true in general permutations. Indeed, if we can find a particular sequence with (odd) even number of interchanges to restore a permutation to its natural order, then *all* sequences that restore natural order *must* involve (odd) even number of interchanges. This can be proved using algebraic principles, but providing a proof here will become a digression from our main theme. Hence we omit it here.

We can associate with each permutation π a function called the ***sign of the permutation***. This function, which we will denote as $\sigma(\pi)$, will take on only two values, $+1$ or -1, depending upon the number of interchanges it takes to restore a permutation π to its natural order. If the number of interchanges required is an even number, then $\sigma(\pi) = +1$, while if the number of interchanges required is an odd number, then $\sigma(\pi) = -1$. In other words,

$$\sigma(\pi) = \begin{cases} +1 & \text{if } \pi \text{ is an even permutation} \\ -1 & \text{if } \pi \text{ is an odd permutation.} \end{cases} \tag{10.4}$$

For example, if $\pi = (1, 4, 3, 2)$, then $\sigma(\pi) = 1$, and if $\pi = (4, 3, 2, 1)$, then $\sigma(\pi) = +1$. If $\pi = (1, 2, 3, 4)$ is the natural order then $\sigma(\pi) = +1$. We can now provide the classical definition of the determinant.

Definition 10.2 *The* **determinant** *of an $n \times n$ matrix $\boldsymbol{A} = a_{ij}$, to be denoted by $|\boldsymbol{A}|$ or by $\det(\boldsymbol{A})$, is defined by*

$$|\boldsymbol{A}| = \sum_{\pi} \sigma(\pi) a_{1\pi_1} \cdots a_{n\pi_n}, \tag{10.5}$$

where $(\pi_1, \ldots, \pi_n)$ is a permutation of the first n positive integers and the summation is over all such permutations. Alternatively, rather than the sign function, we can define $\phi(\pi)$ to denote the minimum number of swaps required to restore natural order and define

$$|\boldsymbol{A}| = \sum_{\pi} (-1)^{\phi(\pi)} a_{1\pi_1} \cdots a_{n\pi_n}. \tag{10.6}$$

π	$\sigma(\pi)$	$a_{1\pi_1}a_{2\pi_2}a_{3\pi_3}$
$(1,2,3)$	$+$	$a_{11}a_{22}a_{33}$
$(1,3,2)$	$-$	$a_{11}a_{23}a_{32}$
$(2,1,3)$	$-$	$a_{12}a_{21}a_{33}$
$(2,3,1)$	$+$	$a_{12}a_{23}a_{31}$
$(3,1,2)$	$+$	$a_{13}a_{21}a_{32}$
$(3,2,1)$	$-$	$a_{13}a_{22}a_{31}$

Table 10.1 *List of permutations, their signs and the products appearing in a* 3×3 *determinant.*

Example 10.3 Consider the determinant for 2×2 matrices. Let

$$\boldsymbol{A} = \begin{pmatrix} a_{11} & a_{12} \\ a_{21} & a_{22} \end{pmatrix}.$$

There are $2! = 2$ permutations of $(1,2)$, given by $(1,2)(2,1)$. Note that $\sigma(1,2) = +1$ and $\sigma(2,1) = -1$. Using (10.5), $|\boldsymbol{A}|$ is given by

$$|\boldsymbol{A}| = \sigma(1,2)a_{11}a_{22} + \sigma(2,1)a_{12}a_{21} = a_{11}a_{22} - a_{12}a_{21}.$$

Let us turn to a 3×3 matrix,

$$\boldsymbol{A} = \begin{pmatrix} a_{11} & a_{12} & a_{13} \\ a_{21} & a_{22} & a_{23} \\ a_{31} & a_{32} & a_{33} \end{pmatrix}.$$

Now there are $3! = 3 \times 2 \times 1 = 6$ permutations of $(1,2,3)$. These permutations, together with their associated signs, are presented in Table 10.1. Using the entries in the table, we find

$$|\boldsymbol{A}| = a_{11}a_{22}a_{33} - a_{11}a_{23}a_{32} - a_{12}a_{21}a_{33} + a_{12}a_{23}a_{31} + a_{13}a_{21}a_{32} - a_{13}a_{22}a_{31}. \tag{10.7}$$

■

10.2 Some basic properties of determinants

We will now prove some important properties of determinants.

Theorem 10.1 *The* **determinant of a triangular matrix** *is the product of its diagonal entries.*

Proof. Let us consider an upper-triangular matrix $\boldsymbol{U}$. We want to prove that

$$\boldsymbol{U} = \begin{vmatrix} u_{11} & u_{12} & \dots & u_{1n} \\ 0 & u_{22} & \dots & u_{2n} \\ \vdots & \vdots & \ddots & \vdots \\ 0 & 0 & \dots & u_{nn} \end{vmatrix} = u_{11}u_{22}\dots u_{nn}.$$

Each term $u_{1\pi_1}\cdots u_{n\pi_n}$ in (10.5) contains exactly one entry from each row and each column of $\boldsymbol{U}$. This means that there is only one term in the expansion of the determinant that does not contain an entry above the diagonal, and that term is $u_{11}u_{22}\cdots u_{nn}$. All other terms will contain at least one entry above the diagonal (which is 0) and will vanish from the expansion. This proves the theorem for upper-triangular matrices.

An exactly analogous argument holds for lower-triangular matrices. □

As an obvious special case of the above theorem, we have the following corollary.

Corollary 10.1 *The determinant of a diagonal matrix equals the product of its diagonal elements.*

This immediately yields $|\boldsymbol{0}| = 0$ and $|\boldsymbol{I}| = 1$.

The following theorem gives a simple expression for the determinant of a permutation matrix.

Theorem 10.2 Determinant of a permutation matrix. *Let $\boldsymbol{P} = \{p_{ij}\}$ be an $n \times n$ permutation matrix whose rows are obtained by permuting the rows of the $n \times n$ identity matrix according to the permutation $\theta = (\theta_1, \theta_2, \dots, \theta_n)$. Then $|\boldsymbol{P}| = \sigma(\theta)$.*

Proof. From the definition of $\boldsymbol{P}$, we find that $p_{i\theta_i} = 1$ and $p_{ij} = 0$ whenever $j \neq \theta_i$ for $i = 1, \dots, n$. Therefore, $p_{1\theta_1}p_{2\theta_2}\cdots p_{n\theta_n} = 1$ and all other terms in the expansion of $\boldsymbol{P}$ will be zero so that $|\boldsymbol{P}|$ is given by

$$\begin{aligned}\sum_{\pi} \sigma(\pi)p_{1\pi_1}p_{2\pi_2}\cdots p_{n\pi_n} &= \sigma(\theta)p_{1\theta_1}p_{2\theta_2}\cdots p_{n\theta_n} + \sum_{\pi\neq\theta} \sigma(\pi)p_{1\pi_1}p_{2\pi_2}\cdots p_{n\pi_n} \\ &= \sigma(\theta)p_{1\theta_1}p_{2\theta_2}\cdots p_{n\theta_n} + 0 = \sigma(\theta).\end{aligned}$$

□

From Theorem 10.1, it is easily seen that the determinant of a triangular matrix is equal to that of its transpose. This is because the determinant of a triangular matrix is simply the product of its diagonal elements, which are left unaltered by transposition. Also, from Theorem 10.1, we find that the determinant of a permutation matrix is simply the sign of the corresponding permutation on the rows of $\boldsymbol{I}$. It is easily verified that the inverse of this permutation produces the transposed matrix. Since a permutation and its inverse have the same sign (this is easily verified), it follows that the determinant of a permutation matrix is the same as that of its transpose.

Our next theorem will prove that the determinant of *any* matrix is equal to the determinant of its transpose. Before we prove this, let us make a few observations. Let $\boldsymbol{A}$ be an $n \times n$ matrix and consider $|\boldsymbol{A}|$. Since each term $a_{1\pi_1} \cdots a_{n\pi_n}$ in (10.5) contains exactly one entry from each row and each column of $\boldsymbol{A}$, every product, $a_{1\pi_1} \cdots a_{n\pi_n}$, appearing in (10.5) can be rearranged so that they are ordered by column number, yielding $a_{\theta_1 1} \cdots a_{\theta_n n}$, where $(\theta_1, \ldots, \theta_n)$ is again a permutation of the first n positive integers.

What is the relation between θ and π? To understand this relationship better, consider the setting with $n = 4$ and the term corresponding to $\pi = (4, 1, 3, 2)$. Here we have $a_{14}a_{21}a_{33}a_{42} = a_{21}a_{42}a_{33}a_{14}$. Therefore $\theta = (2, 4, 3, 1)$. Note that π can be looked upon as a map from the row numbers to the column numbers. And, θ is precisely the permutation of $(1, 2, 3, 4)$ when π is rearranged in natural order. In other words, we obtain θ from π

$$\pi = \begin{pmatrix} 1 & 2 & 3 & 4 \\ 4 & 1 & 3 & 2 \end{pmatrix} \xrightarrow{\text{Rearrange columns}} \begin{pmatrix} 2 & 4 & 3 & 1 \\ 1 & 2 & 3 & 4 \end{pmatrix} \xrightarrow{\text{Swap rows}} \begin{pmatrix} 1 & 2 & 3 & 4 \\ 2 & 4 & 3 & 1 \end{pmatrix} = \theta.$$

The above operation is exactly analogous to what we described in (10.2) and, more generally, in (10.3). In fact, we immediately see that θ is the inverse of π. Put another way, the same permutation that changes $(\pi_1, \pi_2, \ldots, \pi_n)$ to $(1, 2, \ldots, n)$ maps $(1, 2, \ldots, n)$ to $(\theta_1, \ldots, \theta_n)$. Therefore, to rearrange the term $a_{1\pi_1}a_{2\pi_2} \cdots a_{n\pi_n}$ so that the column numbers are in natural order, we take the inverse permutation of π, say $\theta = \pi^{-1}$, and form $a_{\theta_1 1} \cdots a_{\theta_n n}$.

We use the above observations to prove the following important theorem.

Theorem 10.3 *For any square matrix* $\boldsymbol{A}$, $|\boldsymbol{A}'| = |\boldsymbol{A}|$.

Proof. Let $\boldsymbol{A}$ be an $n \times n$ matrix. Then the definition in (10.5) says

$$|\boldsymbol{A}| = \sum_{\pi} \sigma(\pi) a_{1\pi_1} a_{2\pi_2} \cdots a_{n\pi_n},$$

where π runs over all permutations of $(1, \ldots, n)$. Let $\boldsymbol{B} = \boldsymbol{A}'$. Since $b_{ij} = a_{ji}$, we can write the determinant of $\boldsymbol{A}'$ as

$$|\boldsymbol{B}| = \sum_{\pi} \sigma(\pi) b_{1\pi_1} b_{2\pi_2} \cdots b_{n\pi_n} = \sum_{\pi} \sigma(\pi) a_{\pi_1 1} a_{\pi_2 2} \cdots a_{\pi_n n}.$$

The theorem will be proved if we can show that the determinant of $\boldsymbol{A}$ can also be expressed as in the right hand side of the above expression. Put another way, we need to show that we can find a permutation θ such that the set $\{\sigma(\pi) a_{1\pi_1} a_{2\pi_2} \cdots a_{n\pi_n}\}$ will equal the set $\{\sigma(\theta) a_{\theta_1 1} a_{\theta_2 2} \cdots a_{\theta_n n}\}$ as π runs through all permutations of the column numbers and θ runs through all permutations of the row numbers.

Since every π gives rise to a unique π^{-1} (and also $\left(\pi^{-1}\right)^{-1} = \pi$), it follows that as π runs over all permutations of $(1, 2, \ldots, n)$, so does π^{-1}. Moreover, we have $\sigma(\pi) = \sigma(\pi^{-1})$ because π^{-1} simply reverses the action that produced π and, therefore, requires the same number of interchanges to restore natural order as were required to produce π. Writing θ for π^{-1}, we see (also see the discussion preceding the theorem) that

$$\sigma(\pi)a_{1\pi_1}a_{2\pi_2}\cdots a_{n\pi_n} = \sigma(\theta)a_{\theta_1 1}\cdots a_{\theta_n n}.$$

Now, letting both θ and π run over all permutations of $(1, 2, \ldots, n)$ yields

$$|\boldsymbol{A}| = \sum_{\pi} \sigma(\pi)a_{1\pi_1}a_{2\pi_2}\cdots a_{n\pi_n} = \sum_{\theta} \sigma(\theta)a_{\theta_1 1}\cdots a_{\theta_n n} = |\boldsymbol{A}'| .$$

□

The above theorem tells us a very important fact: it is not necessary to distinguish between rows and columns when discussing properties of determinants. In fact, an alternative definition of the determinant is given by letting the permutations run over the row indices:

$$|\boldsymbol{A}| = \sum_{\pi} \sigma(\pi)a_{\pi_1 1}\cdots a_{\pi_n n}. \tag{10.8}$$

Therefore, theorems that involve row manipulations will hold true for the corresponding column manipulations.

For example, we can prove the following important relationship between the determinant of a matrix and any of its columns or rows in the same breath.

Theorem 10.4 *Let $\boldsymbol{A}$ be an $n \times n$ matrix. The determinant $|\boldsymbol{A}|$ is a linear function of any row (column), given that all the other rows (columns) are held fixed. Specifically, suppose the i-th row (column) of $\boldsymbol{A}$ is a linear combination of two row (column) vectors, $\boldsymbol{a}'_{i*} = \alpha_1\boldsymbol{b}'_{i*} + \alpha_2\boldsymbol{c}'_{i*}$ ($\boldsymbol{a}_{*i} = \alpha_1\boldsymbol{b}_{*i} + \alpha_2\boldsymbol{c}_{*i}$). Let $\boldsymbol{B}$ and $\boldsymbol{C}$ be $n \times n$ matrices that agree with $\boldsymbol{A}$ everywhere, except that their i-th rows (columns) are given by $\boldsymbol{b}'_{i*}$ ($\boldsymbol{b}_{*i}$) and $\boldsymbol{c}'_{i*}$ ($\boldsymbol{c}_{*i}$), respectively. Then,*

$$|\boldsymbol{A}| = \alpha_1|\boldsymbol{B}| + \alpha_2|\boldsymbol{C}|.$$

Proof. Consider the proof for the rows. This will follow directly from the definition of the determinant in (10.5):

$$\begin{aligned}
|\boldsymbol{A}| &= \sum_{\pi} \sigma(\pi)a_{1\pi_1}a_{2\pi_2}\cdots a_{i\pi_i}\cdots a_{n\pi_n} \\
&= \sum_{\pi} \sigma(\pi)a_{1\pi_1}a_{2\pi_2}\cdots(\alpha_1 b_{i\pi_i} + \alpha_2 c_{i\pi_i})\cdots a_{n\pi_n} \\
&= \sum_{\pi} \sigma(\pi)a_{1\pi_1}a_{2\pi_2}\cdots(\alpha_1 b_{i\pi_i})\cdots a_{n\pi_n} \\
&\quad + \sum_{\pi} \sigma(\pi)a_{1\pi_1}a_{2\pi_2}\cdots(\alpha_2 c_{i\pi_i})\cdots a_{n\pi_n} \\
&= \alpha_1\sum_{\pi} \sigma(\pi)a_{1\pi_1}a_{2\pi_2}\cdots b_{i\pi_i}\cdots a_{n\pi_n} \\
&\quad + \alpha_2\sum_{\pi} \sigma(\pi)a_{1\pi_1}a_{2\pi_2}\cdots c_{i\pi_i}\cdots a_{n\pi_n} \\
&= \alpha_1|\boldsymbol{B}| + \alpha_2|\boldsymbol{C}| .
\end{aligned}$$

The proof for the columns will follow analogously from the expression of the determinant in (10.8), but here is how we can argue using Theorem 10.3.

Suppose now that $\boldsymbol{A}$ is a matrix whose i-th *column* is given by $\boldsymbol{a}_{*i} = \alpha_1 \boldsymbol{b}_{*i} + \alpha_2 \boldsymbol{c}_{*i}$. Then, $\boldsymbol{A}'$'s i-th row can be written as $\boldsymbol{a}'_{*i} = \alpha_1 \boldsymbol{b}'_{*i} + \alpha_2 \boldsymbol{c}'_{*i}$. Note that $\boldsymbol{b}'_{*i}$ and $\boldsymbol{c}'_{*i}$ are the i-th rows of $\boldsymbol{B}'$ and $\boldsymbol{C}'$, respectively. By virtue of the result we proved above for the rows, we have that $|\boldsymbol{A}'| = \alpha_1 |\boldsymbol{B}'| + \alpha_2 |\boldsymbol{C}'|$ which, from Theorem 10.3, immediately yields $|\boldsymbol{A}| = \alpha_1 |\boldsymbol{B}| + \alpha_2 |\boldsymbol{C}|$. □

The above result can be generalized to any linear combination. For example, consider a specific column being given by $\boldsymbol{a}_{*i} = \sum_{i=1}^{k} \alpha_i \boldsymbol{x}_i$. Then,

$$\left| \left[\boldsymbol{a}_{*1} : \boldsymbol{a}_{*2} : \ldots : \sum_{i=1}^{k} \alpha_i \boldsymbol{x}_i : \ldots : \boldsymbol{a}_{*n} \right] \right| = \sum_{i=1}^{k} \alpha_i \left| [\boldsymbol{a}_{*1} : \boldsymbol{a}_{*2} : \ldots : \boldsymbol{x}_i : \ldots : \boldsymbol{a}_{*n}] \right| . \tag{10.9}$$

As another important consequence of Theorem 10.3, to know how elementary row and column operations alter the determinant of a matrix, it suffices to limit the discussion to elementary row operations. We now state and prove an important theorem concerning elementary row (column) operations.

Theorem 10.5 *Let $\boldsymbol{B}$ be the matrix obtained from an $n \times n$ matrix $\boldsymbol{A}$ by applying one of the three elementary row (column) operations:*

Type I: Interchange two rows (columns) of $\boldsymbol{A}$.

Type II: Multiply a row (column) of $\boldsymbol{A}$ by a scalar α.

Type III: Add α times a given row (column) to another row (column).

Then, the determinant of $\boldsymbol{B}$ is given as follows:

$|\boldsymbol{B}| = -|\boldsymbol{A}|$ *for Type I operations;*

$|\boldsymbol{B}| = \alpha|\boldsymbol{A}|$ *for Type II operations;*

$|\boldsymbol{B}| = |\boldsymbol{A}|$ *for Type III operations.*

Proof. We will prove these results only for elementary row operations. The results for the corresponding column operations will follow immediately by an application of Theorem 10.3.

Let us first consider Type I operations. Let $\boldsymbol{B}$ be obtained by interchanging the i-th and j-th rows (say, $i < j$) of $\boldsymbol{A}$. Therefore $\boldsymbol{B}$ agrees with $\boldsymbol{A}$ except that $\boldsymbol{b}'_{i*} = \boldsymbol{a}'_{j*}$ and $\boldsymbol{b}'_{j*} = \boldsymbol{a}'_{i*}$. Then, for each permutation $\pi = (\pi_1, \pi_2, \ldots, \pi_n)$ of $(1, 2, \ldots, n)$,

$$\begin{aligned} b_{1\pi_1} b_{2\pi_2} \cdots b_{i\pi_i} \cdots b_{j\pi_j} \cdots b_{n\pi_n} &= a_{1\pi_1} a_{2\pi_2} \cdots a_{j\pi_i} \cdots a_{i\pi_j} \cdots a_{n\pi_n} \\ &= a_{1\pi_1} a_{2\pi_2} \cdots a_{i\pi_j} \cdots a_{j\pi_i} \cdots a_{n\pi_n} . \end{aligned}$$

Furthermore, $\sigma(\pi_1, \ldots, \pi, \ldots, \pi_j, \ldots, \pi_n) = \sigma(\pi_1, \ldots, \pi_j, \ldots, \pi_i, \ldots, \pi_n)$ because the two permutations differ only by one interchange. Therefore $|\boldsymbol{B}| = -|\boldsymbol{A}|$.

Put another way, when rows i and j are interchanged, we interchange two first subscripts in each term of the sum

$$\sum_{\pi} \sigma(\pi) a_{1\pi_1} a_{2\pi_2} \cdots a_{n\pi_n}.$$

Therefore, each term in the above sum changes sign and so does the determinant.

Next, we turn to Type II operations. Suppose $\boldsymbol{B}$ is obtained from $\boldsymbol{A}$ by multiplying the i-th row of $\boldsymbol{A}$ with $\alpha \neq 0$. Then $\boldsymbol{B}$ agrees with $\boldsymbol{A}$ except that $\boldsymbol{b}'_{i*} = \alpha \boldsymbol{a}'_{i*}$. Therefore, we have

$$\begin{aligned} |\boldsymbol{B}| &= \sum_{\pi} \sigma(\pi) b_{1\pi_1} b_{2\pi_2} \cdots b_{i\pi_i} \cdots b_{n\pi_n} \\ &= \sum_{\pi} \sigma(\pi) a_{1\pi_1} a_{2\pi_2} \cdots (\alpha a_{i\pi_i}) \cdots a_{n\pi_n} \\ &= \alpha \sum_{\pi} \sigma(\pi) a_{1\pi_1} a_{2\pi_2} \cdots a_{i\pi_i} \cdots a_{n\pi_n} = \alpha |\boldsymbol{A}|. \end{aligned}$$

(This also follows from Theorem 10.4.)

Finally, we consider Type III operations, where $\boldsymbol{B}$ is obtained by adding a scalar multiple of the i-th row of $\boldsymbol{A}$ to the j-th row of $\boldsymbol{A}$. Therefore, $\boldsymbol{B}$ agrees with $\boldsymbol{A}$ except that $\boldsymbol{b}'_{j*} = \boldsymbol{a}'_{j*} + \alpha \boldsymbol{a}'_{i*}$. Without loss of generality, assume that $i < j$. Now for each permutation $\pi = (\pi_1, \pi_2, \ldots, \pi_n)$ we have

$$\begin{aligned} b_{1\pi_1} b_{2\pi_2} \cdots b_{i\pi_i} \cdots b_{j\pi_j} \cdots b_{n\pi_n} &= a_{1\pi_1} a_{2\pi_2} \cdots a_{i\pi_i} \cdots (a_{j\pi_j} + \alpha a_{i\pi_i}) \cdots a_{n\pi_n} \\ &= a_{1\pi_1} a_{2\pi_2} \cdots a_{i\pi_i} \cdots a_{j\pi_j} \cdots a_{n\pi_n} + \alpha (a_{1\pi_1} a_{2\pi_2} \cdots a_{i\pi_i} \cdots a_{i\pi_i} \cdots a_{n\pi_n}). \end{aligned}$$

The above, when substituted in the expansion of the determinant, yields

$$\begin{aligned} |\boldsymbol{B}| = \sum_{\pi} \sigma(\pi) & a_{1\pi_1} a_{2\pi_2} \cdots a_{i\pi_i} \cdots a_{j\pi_j} \cdots a_{n\pi_n} \\ &+ \alpha \sum_{\pi} \sigma(\pi) a_{1\pi_1} a_{2\pi_2} \cdots a_{i\pi_i} \cdots a_{i\pi_i} \cdots a_{n\pi_n} = |\boldsymbol{A}| + \alpha |\tilde{\boldsymbol{A}}|, \end{aligned}$$

where $\tilde{\boldsymbol{A}}$ is obtained from $\boldsymbol{A}$ by replacing its j-th row by the i-th row. Thus, the i-th and j-th rows of $\tilde{\boldsymbol{A}}$ are identical. What is the value of $|\tilde{\boldsymbol{A}}|$? To see this, note that interchanging the two rows of $\tilde{\boldsymbol{A}}$ does not alter $\tilde{\boldsymbol{A}}$. Yet, by virtue of what we have already proved for Type I operations, the determinant changes sign. But this means that we must have $|\tilde{\boldsymbol{A}}| = -|\tilde{\boldsymbol{A}}|$, and so $|\tilde{\boldsymbol{A}}| = 0$. Therefore, we obtain $|\boldsymbol{B}| = |\boldsymbol{A}|$.
□

The above theorem also holds when $\alpha = 0$. For Type III operations, $\alpha = 0$ implies that $\boldsymbol{B} = \boldsymbol{A}$, so obviously $|\boldsymbol{B}| = |\boldsymbol{A}|$. For Type II operations, $\alpha = 0$ implies the following corollary.

Corollary 10.2 *If one or more rows (columns) of an $n \times n$ matrix $\boldsymbol{A}$ are null, then* $|\boldsymbol{A}| = 0$.

Proof. Take $\alpha = 0$ in the Type II elementary operation in Theorem 10.5. □

While proving Theorem 10.5 for Type III operations, we have also derived the following important result. We state and prove this as a separate corollary.

Corollary 10.3 *If two rows (columns) of an $n \times n$ matrix $\boldsymbol{A}$ are identical, then* $|\boldsymbol{A}| = 0$.

Proof. Suppose that the i-th and j-th rows (columns) of an $n \times n$ matrix $\boldsymbol{A}$ are identical. Let $\boldsymbol{B}$ represent the matrix formed from $\boldsymbol{A}$ by interchanging its ith and jth rows (columns). Clearly, $\boldsymbol{B} = \boldsymbol{A}$ and hence $|\boldsymbol{B}| = |\boldsymbol{A}|$. Yet, interchanging two rows (columns) is a Type I elementary operation and according to Theorem 10.5, $|\boldsymbol{B}| = -|\boldsymbol{A}|$. Thus, $|\boldsymbol{A}| = |\boldsymbol{B}| = -|\boldsymbol{A}|$, implying that $|\boldsymbol{A}| = 0$. □

The result on Type II operations in Theorem 10.5 tells us the value of the determinant when a scalar is multiplied to a single row (column) of a matrix. The following useful result gives us the determinant of $|\alpha\boldsymbol{A}|$.

Corollary 10.4 *For any $n \times n$ matrix $\boldsymbol{A}$ and any scalar α, $|\alpha\boldsymbol{A}| = \alpha^n|\boldsymbol{A}|$.*

Proof. This result follows from Theorem 10.5 (Type II operation) upon observing that $\alpha\boldsymbol{A}$ can be formed from $\boldsymbol{A}$ by successively multiplying each of the n rows of $\boldsymbol{A}$ by α. A direct verification from Definition 10.2 is also easy. □

Theorem 10.5 immediately yields the determinant of an elementary matrix associated with any of the three types of elementary operations.

Corollary 10.5 *Let $\boldsymbol{E}_{ij}$, $\boldsymbol{E}_i(\alpha)$, and $\boldsymbol{E}_{ij}(\alpha)$ be elementary matrices of Types I, II, and III, respectively, as defined in Section 2.4. Then:*

$|\boldsymbol{E}_{ij}| = -|\boldsymbol{I}| = -1$ *for Type I operations;*
$|\boldsymbol{E}_i(\alpha)| = \alpha|\boldsymbol{I}| = \alpha$ *for Type II operations;*
$|\boldsymbol{E}_{ij}(\alpha)| = |\boldsymbol{I}| = 1$ *for Type III operations.*

Proof. Recall that each of these elementary matrices can be obtained by performing the associated row (column) operation to an identity matrix of appropriate size. The results now follow immediately by taking $\boldsymbol{A} = \boldsymbol{I}$ in Theorem 10.5. □

We have already seen in Section 2.4 that if $\boldsymbol{G}$ is an elementary matrix of Type I, II, or III, and if $\boldsymbol{A}$ is any other matrix, then the product $\boldsymbol{GA}$ is the matrix obtained by performing the elementary operation associated with $\boldsymbol{G}$ to the rows of $\boldsymbol{A}$. This, together with Theorem 10.5 and Corollary 10.5, leads to the conclusion that for every square matrix $\boldsymbol{A}$,

$$\begin{aligned} |\boldsymbol{E}_{ij}\boldsymbol{A}| &= -|\boldsymbol{A}| &= |\boldsymbol{E}_{ij}||\boldsymbol{A}|; \\ |\boldsymbol{E}_i(\alpha)\boldsymbol{A}| &= \alpha|\boldsymbol{A}| &= |\boldsymbol{E}_i(\alpha)||\boldsymbol{A}|; \\ |\boldsymbol{E}_{ij}(\alpha)\boldsymbol{A}| &= |\boldsymbol{A}| &= |\boldsymbol{E}_{ij}(\alpha)||\boldsymbol{A}|. \end{aligned}$$

In other words, $|\boldsymbol{GA}| = |\boldsymbol{G}||\boldsymbol{A}|$ whenever $\boldsymbol{G}$ is an elementary matrix of Type I, II, or III, and $\boldsymbol{A}$ is any square matrix. This is easily generalized to any number of elementary matrices, $\boldsymbol{G}_1, \boldsymbol{G}_2,\ldots,\boldsymbol{G}_k$, by writing

$$\begin{aligned} |\boldsymbol{G}_1\boldsymbol{G}_2\cdots\boldsymbol{G}_k\boldsymbol{A}| &= |\boldsymbol{G}_1||\boldsymbol{G}_2\cdots\boldsymbol{G}_k\boldsymbol{A}| \\ &= |\boldsymbol{G}_1||\boldsymbol{G}_2||\boldsymbol{G}_3\cdots\boldsymbol{G}_k\boldsymbol{A}| \\ &= \vdots \\ &= |\boldsymbol{G}_1||\boldsymbol{G}_2||\boldsymbol{G}_3|\cdots|\boldsymbol{G}_k||\boldsymbol{A}|. \end{aligned} \tag{10.10}$$

In particular, taking $\boldsymbol{A}$ to be the identity matrix we have

$$|\boldsymbol{G}_1\boldsymbol{G}_2\cdots\boldsymbol{G}_k| = |\boldsymbol{G}_1||\boldsymbol{G}_2|\cdots|\boldsymbol{G}_k|. \tag{10.11}$$

In other words, the determinant of the product of elementary matrices is the product of their determinants.

Note that the above results also hold when $\boldsymbol{A}$ is post-multiplied by a sequence of elementary matrices. This can be proved directly or by using Theorem 10.3 and Theorem 2.3 (if $\boldsymbol{G}$ is an elementary matrix, so is $\boldsymbol{G}'$). Then we can write

$$|\boldsymbol{AG}| = |(\boldsymbol{AG})'| = |\boldsymbol{G}'\boldsymbol{A}'| = |\boldsymbol{G}'||\boldsymbol{A}'| = |\boldsymbol{G}||\boldsymbol{A}| = |\boldsymbol{A}||\boldsymbol{G}|.$$

Extending this to post-multiplication with any number of elementary matrices is analogous to how we obtained (10.10). The results proved above provide an efficient method for computing $|\boldsymbol{A}|$. We reduce $\boldsymbol{A}$ to an upper-triangular matrix $\boldsymbol{U}$ using elementary row operations, say $\boldsymbol{G}_1\boldsymbol{G}_2\cdots\boldsymbol{G}_k\boldsymbol{A} = \boldsymbol{U}$. Note that we can achieve this using only Type-I and Type-III operations as long as we do not force $\boldsymbol{U}$ to be unit upper-triangular. Suppose q interchanges of rows (i.e., Type I operations) are used. Then, Theorem 10.5 and Theorem 10.1 yield

$$\begin{aligned} |\boldsymbol{G}_1||\boldsymbol{G}_2|\cdots|\boldsymbol{G}_k||\boldsymbol{A}| = |\boldsymbol{U}| &\Rightarrow (-1)^q|\boldsymbol{A}| = |\boldsymbol{U}| \\ &\Rightarrow |\boldsymbol{A}| = (-1)^q u_{11}u_{22}\cdots u_{nn}, \end{aligned} \tag{10.12}$$

where $u_{11}, u_{22}, \ldots, u_{nn}$ are the diagonal elements of $\boldsymbol{U}$.

The following theorem characterizes determinants of nonsingular matrices and is very useful.

Theorem 10.6 *Any square matrix* $\boldsymbol{A}$ *is non-singular if and only if* $|\boldsymbol{A}| \neq 0$.

Proof. Let $\boldsymbol{A}$ be a square matrix. Then, using elementary row operations we can reduce $\boldsymbol{A}$ to an upper-triangular matrix $\boldsymbol{U}$. If $\boldsymbol{A}$ is non-singular, then each of the diagonal elements of $\boldsymbol{U}$ are nonzero and it follows from (10.12) that $|\boldsymbol{A}| \neq 0$.

On the other hand, if $\boldsymbol{A}$ is singular, then at least one of the diagonal elements in $\boldsymbol{U}$ will be zero so (10.12) implies that $|\boldsymbol{A}| = 0$. □

The following corollary follows immediately.

Corollary 10.6 *Let* $\boldsymbol{A}$ *be an* $n \times n$ *square matrix. Then* $|\boldsymbol{A}| = 0$ *if and only if at least one of its rows (columns) is a linear combination of the other rows (columns).*

Proof. $\boldsymbol{A}$ is singular if and only if its rank is smaller than n, which means that the maximum number of linearly independent rows (columns) must be smaller than n. Therefore, $\boldsymbol{A}$ is singular if and only if some row (column) is a linear combination of the some other rows (columns). The result now follows from Theorem 10.6. □

Note that if a row (column) is a linear combination of other rows (columns) in a matrix, then Type-III operations can be used to sweep out that row (column) and make it zero. The determinant of this transformed matrix is obviously zero (Corollary 10.3). But Type-III operations do not change the value of the determinant, so the original determinant is zero as well.

In the following section, we take up the matter of investigating the determinant of the product of two general matrices.

10.3 Determinant of products

The key result we will prove here is that if $\boldsymbol{A}$ and $\boldsymbol{B}$ are two $n \times n$ matrices, then $|\boldsymbol{AB}| = |\boldsymbol{A}||\boldsymbol{B}|$. We have already seen the result is true if either $\boldsymbol{A}$ or $\boldsymbol{B}$ are elementary matrices. We now extend this to general square matrices.

Theorem 10.7 *Let $\boldsymbol{A}$ and $\boldsymbol{B}$ be two $n \times n$ matrices. Then $|\boldsymbol{AB}| = |\boldsymbol{A}||\boldsymbol{B}|$.*

Proof. First consider the case where at least one of $\boldsymbol{A}$ or $\boldsymbol{B}$ is singular, say $\boldsymbol{A}$. Since the rank of $\boldsymbol{AB}$ cannot exceed the rank of $\boldsymbol{A}$, it follows that $\boldsymbol{AB}$ must also be singular. Consequently, from Theorem 10.6, we have

$$|\boldsymbol{AB}| = 0 = |\boldsymbol{A}||\boldsymbol{B}|.$$

Now suppose $\boldsymbol{A}$ and $\boldsymbol{B}$ are both non-singular. From Theorem 2.6, we know that $\boldsymbol{A}$ can be written as a product of elementary matrices so that $\boldsymbol{A} = \boldsymbol{G}_1\boldsymbol{G}_2\cdots\boldsymbol{G}_k$, where each $\boldsymbol{G}_i$ is an elementary matrix of Type I, II or III. Now we can apply (10.10) and (10.11) to produce

$$\begin{aligned}|\boldsymbol{AB}| &= |\boldsymbol{G}_1\boldsymbol{G}_2\cdots\boldsymbol{G}_k\boldsymbol{B}| = |\boldsymbol{G}_1||\boldsymbol{G}_2|\cdots|\boldsymbol{G}_k||\boldsymbol{B}| \\ &= |\boldsymbol{G}_1\boldsymbol{G}_2\cdots\boldsymbol{G}_k||\boldsymbol{B}| = |\boldsymbol{A}||\boldsymbol{B}|.\end{aligned}$$

□

A different proof of Theorem 10.7 uses the linearity in a given column for determinants.

Proof. Writing $\boldsymbol{AB} = [\boldsymbol{Ab}_{*1} : \boldsymbol{Ab}_{*2} : \ldots : \boldsymbol{Ab}_{*n}]$ and noting that $\boldsymbol{Ab}_{*1} = \boldsymbol{A}\left(\sum_{i=1}^n b_{i1}\boldsymbol{e}_i\right) = \sum_{i=1}^n b_{i1}\boldsymbol{Ae}_i$, using (10.9) we obtain

$$|\boldsymbol{AB}| = \left|\left[\sum_{i=1}^n b_{i1}\boldsymbol{Ae}_i : \boldsymbol{Ab}_{*2} : \ldots : \boldsymbol{Ab}_{*n}\right]\right| = \sum_{i=1}^n b_{i1}|\boldsymbol{Ae}_i : \boldsymbol{Ab}_{*2} : \ldots : \boldsymbol{Ab}_{*n}|.$$

By repeated application of the above to the columns $b_{*2}, \ldots, b_{*n}$, we obtain

$$\begin{aligned} |AB| &= \sum_{i_1}\sum_{i_2}\cdots\sum_{i_n} b_{i_1 1} b_{i_2 2}\cdots b_{i_n n}\,|[Ae_{i_1} : Ae_{i_2} : \ldots : Ae_{i_n}]| \\ &= \sum_{i_1}\sum_{i_2}\cdots\sum_{i_n} b_{i_1 1} b_{i_2 2}\cdots b_{i_n n}\,|[a_{*i_1} : a_{*i_2} : \ldots : a_{*i_n}]| . \end{aligned} \tag{10.13}$$

Note that $i_j = i_k$ for any pair (j,k) will result in two identical columns in $[a_{*i_1} : a_{i_2} : \ldots : a_{*i_n}]$ and, hence, by virtue of Corollary 10.3, its determinant will be zero. Therefore, the only nonzero terms in the above sum arise when $(i_1, i_2, \ldots, i_n)$ is a permutation of $(1, 2, \ldots, n)$. Also, it can be easily verified that

$$|[a_{*i_1} : a_{*i_2} : \ldots : a_{*i_n}]| = \sigma(i_1, i_2, \ldots, i_n)\,|A| .$$

Substituting in (10.13), we obtain

$$\begin{aligned} |AB| &= \sum_{(i_1,i_2,\ldots,i_n)} \sigma(i_1, i_2, \ldots, i_n) b_{i_1 1} b_{i_2 2}\cdots b_{i_n n}|A| \\ &= |A| \left(\sum_{(i_1,i_2,\ldots,i_n)} \sigma(i_1, i_2, \ldots, i_n) b_{i_1 1} b_{i_2 2}\cdots b_{i_n n} \right) = |A||B| . \end{aligned}$$

□

The repeated application of Theorem 10.7 leads to the following formula for the product of any finite number of $n \times n$ matrices $A_1, A_2, \ldots, A_k$:

$$|A_1, A_2, \ldots, A_k| = |A_1||A_2|\cdots|A_k| . \tag{10.14}$$

As a special case of (10.14), we obtain the following formula for the determinant of the k-th power of an $n \times n$ matrix A:

$$|A^k| = |A|^k . \tag{10.15}$$

Applying Theorems 10.7 and 10.3 to $|A'A|$, where A is an $n \times n$ square matrix, yields

$$|A'A| = |A'||A| = |A||A| = |A|^2 . \tag{10.16}$$

Corollary 10.7 *The determinant of any orthogonal matrix Q is given by*

$$|Q| = \pm 1 .$$

Proof. Using (10.16), together with the definition of an orthogonal matrix, we find

$$|Q|^2 = |Q'Q| = |I| = 1,$$

which implies that $|Q| = \pm 1$. □

Theorem 10.6 tells us that a square matrix A is nonsingular if and only if $|A|$ is nonzero. The following corollary of the product rule tells us what the determinant of the inverse is.

Corollary 10.8 *For a nonsingular matrix A, $|A^{-1}| = 1/|A|$.*

Proof. If $\boldsymbol{A}$ is nonsingular, then $\boldsymbol{A}^{-1}$ exists and using the product rule we have that

$$|\boldsymbol{A}||\boldsymbol{A}^{-1}| = |\boldsymbol{A}\boldsymbol{A}^{-1}| = |\boldsymbol{I}| = 1 \Longrightarrow |\boldsymbol{A}^{-1}| = \frac{1}{|\boldsymbol{A}|} .$$

□

Two square matrices $\boldsymbol{A}$ and $\boldsymbol{B}$ are said to be *similar* if there exists a nonsingular matrix $\boldsymbol{P}$ such that $\boldsymbol{B} = \boldsymbol{P}^{-1}\boldsymbol{A}\boldsymbol{P}$. The following corollary says that similar matrices have the same determinant.

Corollary 10.9 *Similar matrices have the same determinant.*

Proof. This is because $|\boldsymbol{P}^{-1}\boldsymbol{A}\boldsymbol{P}| = |\boldsymbol{P}^{-1}||\boldsymbol{A}||\boldsymbol{P}| = |\boldsymbol{A}|$. □

10.4 Computing determinants

Practical algorithms for computing determinants rely upon elementary operations. In fact, (10.12) gives us an explicit formula to compute determinants. The $\boldsymbol{LU}$ decomposition with row interchanges is another way to arrive at the formula in (10.12).

Recall from Section 3.3 that every nonsingular matrix $\boldsymbol{A}$ admits the factorization $\boldsymbol{PA} = \boldsymbol{LU}$, where $\boldsymbol{P}$ is a permutation matrix (which is a product of elementary interchange matrices), $\boldsymbol{L}$ is a unit lower-triangular matrix with 1's on its diagonal, and $\boldsymbol{U}$ is upper-triangular with the pivots on its diagonal. Furthermore, Theorem 10.2 tells us that

$$|\boldsymbol{P}| = |\boldsymbol{P}'| = \begin{cases} +1 & \text{if } \boldsymbol{P} \text{ is a product of an } \textit{even} \text{ number of interchanges} \\ -1 & \text{if } \boldsymbol{P} \text{ is a product of an } \textit{odd} \text{ number of interchanges.} \end{cases}$$

Putting these observations together yields

$$|\boldsymbol{A}| = |\boldsymbol{P}'\boldsymbol{L}\boldsymbol{U}| = |\boldsymbol{P}'||\boldsymbol{L}||\boldsymbol{U}| = |\boldsymbol{P}||\boldsymbol{U}| = \pm u_{11}u_{22}\cdots u_{nn},$$

where the sign is plus (+) if the number of row interchanges is even and is minus (−) if the number of row interchanges is odd. This is the strategy adopted by most computing packages to evaluate the determinant.

Example 10.4 Let us find the determinant of the matrix

$$\boldsymbol{A} = \begin{bmatrix} 2 & 3 & 0 & 0 \\ 4 & 7 & 2 & 0 \\ -6 & -10 & 0 & 1 \\ 4 & 6 & 4 & 5 \end{bmatrix}.$$

Recall from Example 3.1 that $\boldsymbol{LU}$ decomposition of the matrix $\boldsymbol{A}$ was given by

$$\boldsymbol{L} = \begin{bmatrix} 1 & 0 & 0 & 0 \\ 2 & 1 & 0 & 0 \\ -3 & -1 & 1 & 0 \\ 2 & 0 & 2 & 1 \end{bmatrix}; \quad \boldsymbol{U} = \begin{bmatrix} 2 & 3 & 0 & 0 \\ 0 & 1 & 2 & 0 \\ 0 & 0 & 2 & 1 \\ 0 & 0 & 0 & 3 \end{bmatrix}.$$

No row interchanges were required. By the preceding discussion,

$$|\boldsymbol{A}| = u_{11}u_{22}u_{33}u_{44} = 2 \times 1 \times 2 \times 3 = 12.$$

In this example, row interchanges were not necessary to obtain the $\boldsymbol{LU}$ decomposition for $\boldsymbol{A}$ (i.e., no zero pivots are encountered). Nevertheless, a partial pivoting algorithm using row interchanges has numerical benefits. Keeping track of the number of row interchanges is not computationally expensive. As an illustration, recall the partial pivoting algorithm in Example 3.2 that arrived at $\boldsymbol{PA} = \boldsymbol{LU}$, where

$$\boldsymbol{L} = \begin{bmatrix} 1 & 0 & 0 & 0 \\ -2/3 & 1 & 0 & 0 \\ -2/3 & -1/2 & 1 & 0 \\ -1/3 & 1/2 & -1/2 & 1 \end{bmatrix}; \quad \boldsymbol{U} = \begin{bmatrix} -6 & -10 & 0 & 1 \\ 0 & -2/3 & 4 & 17/3 \\ 0 & 0 & 4 & 7/2 \\ 0 & 0 & 0 & -3/4 \end{bmatrix};$$

$$\boldsymbol{P} = \begin{bmatrix} 0 & 0 & 1 & 0 \\ 0 & 0 & 0 & 1 \\ 0 & 1 & 0 & 0 \\ 1 & 0 & 0 & 0 \end{bmatrix}.$$

Note that the permutation matrix $\boldsymbol{P}$ corresponds to $\pi = (3, 4, 2, 1)$ and is easily seen to have negative sign. Therefore, from Theorem 10.2, we find $|\boldsymbol{P}| = |\boldsymbol{P}'| = -1$. Also, $\boldsymbol{L}$ is unit lower-triangular, so $|\boldsymbol{L}| = 1$, and $|\boldsymbol{U}| = u_{11}u_{22}u_{33}u_{44} = -12$. Putting these observations together, we obtain.

$$|\boldsymbol{A}| = |\boldsymbol{P}'\boldsymbol{LU}| = |\boldsymbol{P}'||\boldsymbol{L}||\boldsymbol{U}| = |\boldsymbol{P}'||\boldsymbol{U}| = |\boldsymbol{P}||\boldsymbol{U}| = (-1) \times (-12) = 12. \quad \blacksquare$$

The determinant can also be computed from the $\boldsymbol{QR}$ decomposition. If $\boldsymbol{A} = \boldsymbol{QR}$ is the $\boldsymbol{QR}$ decomposition of $\boldsymbol{A}$, then

$$|\boldsymbol{A}| = |\boldsymbol{Q}||\boldsymbol{R}| = \pm|\boldsymbol{R}| = \pm r_{11}r_{22}\cdots r_{nn}\,. \tag{10.17}$$

10.5 The determinant of the transpose of a matrix—revisited

Apart from being a numerically stable and convenient way to compute determinants, the $\boldsymbol{LU}$ decomposition can also be used to provide an alternative, perhaps more transparent, view about why the determinant of a matrix equals the determinant of its transpose. First of all, we make the following elementary observations:

(a) It is easily verified that $|\boldsymbol{T}| = |\boldsymbol{T}'|$ when $\boldsymbol{T}$ is a triangular matrix. This is because the diagonal elements of a matrix are not altered by transposition and the determinant of a triangular matrix is simply the product of its diagonal elements (Theorem 10.1). Therefore, $|\boldsymbol{L}| = |\boldsymbol{L}'| = 1$ and $|\boldsymbol{U}| = |\boldsymbol{U}'|$ in the $\boldsymbol{LU}$ decomposition.

(b) Note that the proof of Theorem 10.7 that used elementary row operations did not use $|\boldsymbol{A}| = |\boldsymbol{A}'|$ anywhere. So, we can use the fact that $|\boldsymbol{AB}| = |\boldsymbol{A}||\boldsymbol{B}|$.

(c) If $\boldsymbol{P}$ is a permutation matrix, then from Theorem 10.2, we know that $|\boldsymbol{P}|$ is equal to the sign of the permutation on the rows of the identity matrix that produced $\boldsymbol{P}$. It is easily verified that the sign of the permutations producing $\boldsymbol{P}$ and $\boldsymbol{P}'$ are the same. Therefore, $|\boldsymbol{P}| = |\boldsymbol{P}'|$.

Assume that $\boldsymbol{A}$ is nonsingular and let $\boldsymbol{A} = \boldsymbol{P}'\boldsymbol{L}\boldsymbol{U}$ be the $\boldsymbol{L}\boldsymbol{U}$ decomposition of $\boldsymbol{A}$. Using the above facts we can conclude that

$$|\boldsymbol{A}| = |\boldsymbol{P}'\boldsymbol{L}\boldsymbol{U}| = |\boldsymbol{P}'||\boldsymbol{L}||\boldsymbol{U}| = |\boldsymbol{P}||\boldsymbol{L}'||\boldsymbol{U}'| = |\boldsymbol{U}'||\boldsymbol{L}'||\boldsymbol{P}| = |\boldsymbol{U}'\boldsymbol{L}'\boldsymbol{P}| = |\boldsymbol{A}'| . \tag{10.18}$$

This proves that $|\boldsymbol{A}| = |\boldsymbol{A}'|$ for nonsingular $\boldsymbol{A}$. When $\boldsymbol{A}$ is singular, so is $\boldsymbol{A}'$ and we obtain $|\boldsymbol{A}| = |\boldsymbol{A}'| = 0$.

10.6 Determinants of partitioned matrices

This section will state and prove some important results for partitioned matrices. Our first result is a simple consequence of counting Type-I operations (i.e., row interchanges) to arrive at an $(m+n) \times (m+n)$ identity matrix.

Theorem 10.8

$$\begin{vmatrix} \boldsymbol{O} & \boldsymbol{I}_m \\ \boldsymbol{I}_n & \boldsymbol{O} \end{vmatrix} = (-1)^{mn}$$

Proof. This is proved by simply counting the number of row interchanges to produce the identity matrix. Suppose, without loss of generality, that $m < n$. To move the $(m+1)$-th row to the top will require m transpositions. Similarly bringing the $(m+2)$-th row to the second row will require another m transpositions. In general, for $j = 1, 2, \ldots, n$ we can move row $m+j$ to the j-th position using m transpositions. This procedure will end up with the lower n rows occupying the first n rows while pushing the top m rows to the bottom. In other words, we have arrived at the $(m+n)\times(m+n)$ identity matrix using a total of mn transpositions. For each of the transpositions, the determinant changes sign (these are Type-II operations) and we see the value of the determinant to be $(-1)^{mn}$.

□

The following is an especially useful result that yields several special cases.

Theorem 10.9 *Let $\boldsymbol{A}$ be an $m \times m$ matrix, $\boldsymbol{B}$ an $m \times n$ matrix and $\boldsymbol{D}$ an $n \times n$ matrix. Then,*

$$\begin{vmatrix} \boldsymbol{A} & \boldsymbol{B} \\ \boldsymbol{O} & \boldsymbol{D} \end{vmatrix} = |\boldsymbol{A}||\boldsymbol{D}| .$$

Proof. Let us write $\boldsymbol{P} = \begin{bmatrix} \boldsymbol{A} & \boldsymbol{B} \\ \boldsymbol{O} & \boldsymbol{D} \end{bmatrix}$. Then, the determinant of $\boldsymbol{P} = \{p_{ij}\}$ is

$$|\boldsymbol{P}| = \sum_{\pi} \sigma(\pi) p_{1\pi_1} p_{2\pi_2} \cdots p_{m\pi_m} p_{m+1,\pi_{m+1}} \cdots p_{m+n,\pi_{m+n}} . \tag{10.19}$$

Note that $p_{m+1,\pi_{m+1}} \cdots p_{m+n,\pi_{m+n}} = 0$ when at least one of $\pi_{m+1}, \pi_{m+2}, \ldots \pi_{m+n}$ equals any integer between 1 and m. This implies that the only possibly nonzero terms in the sum in (10.19) are those where $(\pi_{m+1}, \pi_{m+2}, \ldots, \pi_{m+n})$ is restricted to be a permutation over $(m+1, m+2, \ldots, m+n)$. But this implies that $(\pi_1, \pi_2, \ldots, \pi_m)$ must be restricted to be a permutation over $(1, 2, \ldots, m)$. For any such permutation, we can write any term in (10.19) as

$$\begin{aligned} p_{1\pi_1} p_{2\pi_2} & \cdots p_{m\pi_m} p_{m+1,\pi_{m+1}} \cdots p_{m+n,\pi_{m+n}} \\ &= a_{1\pi_1} a_{2\pi_2} \cdots a_{m\pi_m} d_{1\pi_{m+1}-m} d_{2\pi_{m+2}-m} \cdots d_{n\pi_{m+n}-m} \\ &= a_{1\theta_1} a_{2\theta_2} \cdots a_{m\theta_m} d_{1\gamma_1} d_{2\gamma_2} \cdots d_{n\gamma_n} , \end{aligned}$$

where $\theta_1 = \pi_1, \theta_2 = \pi_2, \ldots, \theta_m = \pi_m$ and $\gamma_1 = \pi_{m+1} - m, \gamma_2 = \pi_{m+2} - m, \ldots, \gamma_n = \pi_{m+n} - m$. From this, we find that θ and γ are permutations over $(1, 2, \ldots, m)$ and $(1, 2, \ldots, n)$, respectively. In other words,

$$\theta = \begin{pmatrix} 1 & 2 & \ldots & m \\ \theta_1 & \theta_2 & \ldots & \theta_m \end{pmatrix} \text{ and } \gamma = \begin{pmatrix} 1 & 2 & \ldots & n \\ \gamma_1 & \gamma_2 & \ldots & \gamma_n \end{pmatrix} .$$

Furthermore, since θ and γ are obtained by essentially "splitting up" π in a disjoint manner, it can be easily verified that $\sigma(\pi) = \sigma(\theta)\sigma(\gamma)$.

Therefore, we can rewrite the sum in (10.19) as

$$\begin{aligned} |\boldsymbol{P}| &= \sum_{\pi} \sigma(\pi) p_{1\pi_1} p_{2\pi_2} \cdots p_{m\pi_m} p_{m+1,\pi_{m+1}} \cdots p_{m+n,\pi_{m+n}} \\ &= \sum_{\theta,\gamma} \sigma(\theta)\sigma(\gamma) a_{1\theta_1} a_{2\theta_2} \cdots a_{m\theta_m} d_{1,\gamma_1} \cdots d_{n,\gamma_n} \\ &= \sum_{\theta} \sum_{\gamma} \sigma(\theta)\sigma(\gamma) a_{1\theta_1} a_{2\theta_2} \cdots a_{m\theta_m} d_{1,\gamma_1} \cdots d_{n,\gamma_n} \\ &= \left(\sum_{\theta} \sigma(\theta) a_{1\theta_1} a_{2\theta_2} \cdots a_{m\theta_m} \right) \left(\sum_{\gamma} \sigma(\gamma) d_{1,\gamma_1} \cdots d_{n,\gamma_n} \right) \\ &= |\boldsymbol{A}||\boldsymbol{D}| . \end{aligned}$$

□

A similar result can be proved for upper block-triangular matrices.

Corollary 10.10 *Let $\boldsymbol{A}$ be an $m \times m$ matrix, $\boldsymbol{C}$ an $n \times m$ matrix and $\boldsymbol{D}$ an $n \times n$ matrix. Then,*

$$\begin{vmatrix} \boldsymbol{A} & \boldsymbol{O} \\ \boldsymbol{C} & \boldsymbol{D} \end{vmatrix} = |\boldsymbol{A}||\boldsymbol{D}| .$$

Proof. This can be proved using an argument analogous to Theorem 10.9. Alternatively, we can use Theorem 10.3 and consider the transpose of the partitioned matrix. To be precise,

$$\begin{vmatrix} \boldsymbol{A} & \boldsymbol{O} \\ \boldsymbol{C} & \boldsymbol{D} \end{vmatrix} = \begin{vmatrix} \boldsymbol{A}' & \boldsymbol{C}' \\ \boldsymbol{O} & \boldsymbol{D}' \end{vmatrix} = |\boldsymbol{A}'||\boldsymbol{D}'| = |\boldsymbol{A}||\boldsymbol{D}| .$$

□

Theorem 10.9 also immediately implies the following corollary.

Corollary 10.11 *Let $\boldsymbol{A}$ be an $m \times m$ matrix and $\boldsymbol{D}$ an $n \times n$ matrix. Then,*

$$\begin{vmatrix} \boldsymbol{A} & \boldsymbol{O} \\ \boldsymbol{O} & \boldsymbol{D} \end{vmatrix} = |\boldsymbol{A}||\boldsymbol{D}| \, .$$

Proof. This follows by taking $\boldsymbol{B} = \boldsymbol{O}$ in Theorem 10.9. □

The repeated application of Theorem 10.9 gives the following general result for upper block-triangular matrices,

$$\begin{vmatrix} \boldsymbol{A}_{11} & \boldsymbol{A}_{12} & \cdots & \boldsymbol{A}_{1n} \\ \boldsymbol{0} & \boldsymbol{A}_{22} & \cdots & \boldsymbol{A}_{2n} \\ \vdots & & \ddots & \vdots \\ \boldsymbol{0} & \boldsymbol{0} & & \boldsymbol{A}_{nn} \end{vmatrix} = |\boldsymbol{A}_{11}||\boldsymbol{A}_{22}|\cdots|\boldsymbol{A}_{nn}| \, , \tag{10.20}$$

while repeated application of Corollary 10.10 yields the analogue for lower block-triangular matrices:

$$\begin{vmatrix} \boldsymbol{A}_{11} & \boldsymbol{0} & \cdots & \boldsymbol{0} \\ \boldsymbol{A}_{21} & \boldsymbol{A}_{22} & & \boldsymbol{0} \\ \vdots & \vdots & \ddots & \\ \boldsymbol{A}_{n1} & \boldsymbol{A}_{n2} & \cdots & \boldsymbol{A}_{nn} \end{vmatrix} = |\boldsymbol{A}_{11}||\boldsymbol{A}_{22}|\cdots|\boldsymbol{A}_{nn}| \, . \tag{10.21}$$

The determinant of a general block-diagonal matrix with the diagonal blocks $\boldsymbol{A}_{11}, \boldsymbol{A}_{22}, \ldots, \boldsymbol{A}_{nn}$ is a special case of (10.20) or of (10.21) and is given by

$$|\boldsymbol{A}_{11}||\boldsymbol{A}_{22}|\cdots|\boldsymbol{A}_{nn}| \, .$$

By making use of Theorem 10.8, we obtain the following corollary of Theorem 10.9.

Corollary 10.12 *Let $\boldsymbol{B}$ represent an $m \times m$ matrix, $\boldsymbol{C}$ an $n \times m$ matrix and $\boldsymbol{D}$ an $n \times n$ matrix. Then,*

$$\begin{vmatrix} \boldsymbol{0} & \boldsymbol{B} \\ \boldsymbol{C} & \boldsymbol{D} \end{vmatrix} = \begin{vmatrix} \boldsymbol{D} & \boldsymbol{C} \\ \boldsymbol{B} & \boldsymbol{0} \end{vmatrix} = (-1)^{mn}|\boldsymbol{B}||\boldsymbol{C}| \, .$$

Corollary 10.13 *For $n \times n$ matrices $\boldsymbol{C}$ and $\boldsymbol{D}$,*

$$\begin{vmatrix} \boldsymbol{0} & -\boldsymbol{I}_n \\ \boldsymbol{C} & \boldsymbol{D} \end{vmatrix} = \begin{vmatrix} \boldsymbol{D} & \boldsymbol{C} \\ -\boldsymbol{I}_n & \boldsymbol{0} \end{vmatrix} = |\boldsymbol{C}| \, .$$

Proof. This follows as a special case of Corollary 10.12 where $m = n$ and $\boldsymbol{A} = -\boldsymbol{I}_n$. We have

$$(-1)^{nn}|-\boldsymbol{I}_n||\boldsymbol{C}| = (-1)^{nn}(-1)^n|\boldsymbol{C}| = (-1)^{n(n+1)}|\boldsymbol{C}| = |\boldsymbol{C}|$$

since $n(n+1)$ is always an even number. □

Now we discuss one of the most important and general results on the determinant of partitioned matrices.

Theorem 10.10 Determinant of a partitioned matrix. *Let $\boldsymbol{A}$ represent an $m \times m$ matrix, $\boldsymbol{B}$ an $m \times n$ matrix, $\boldsymbol{C}$ an $n \times m$ matrix, and $\boldsymbol{D}$ an $n \times n$ matrix. Then*

$$\begin{vmatrix} \boldsymbol{A} & \boldsymbol{B} \\ \boldsymbol{C} & \boldsymbol{D} \end{vmatrix} = \begin{cases} |\boldsymbol{A}|\,|\boldsymbol{D} - \boldsymbol{C}\boldsymbol{A}^{-1}\boldsymbol{B}| & \text{when } \boldsymbol{A}^{-1} \text{ exists,} \\ |\boldsymbol{D}|\,|\boldsymbol{A} - \boldsymbol{B}\boldsymbol{D}^{-1}\boldsymbol{C}| & \text{when } \boldsymbol{D}^{-1} \text{ exists.} \end{cases} \tag{10.22}$$

Proof. Suppose that $\boldsymbol{A}$ is nonsingular. Then we can write

$$\begin{bmatrix} \boldsymbol{A} & \boldsymbol{B} \\ \boldsymbol{C} & \boldsymbol{D} \end{bmatrix} = \begin{bmatrix} \boldsymbol{I} & \boldsymbol{0} \\ \boldsymbol{C}\boldsymbol{A}^{-1} & \boldsymbol{I} \end{bmatrix} \begin{bmatrix} \boldsymbol{A} & \boldsymbol{B} \\ \boldsymbol{0} & \boldsymbol{D} - \boldsymbol{C}\boldsymbol{A}^{-1}\boldsymbol{B} \end{bmatrix} .$$

Applying the product rule (Theorem 10.7) followed by Theorem 10.9, we obtain

$$\begin{vmatrix} \boldsymbol{A} & \boldsymbol{B} \\ \boldsymbol{C} & \boldsymbol{D} \end{vmatrix} = \begin{vmatrix} \boldsymbol{I} & \boldsymbol{0} \\ \boldsymbol{C}\boldsymbol{A}^{-1} & \boldsymbol{I} \end{vmatrix} \begin{vmatrix} \boldsymbol{A} & \boldsymbol{B} \\ \boldsymbol{0} & \boldsymbol{D} - \boldsymbol{C}\boldsymbol{A}^{-1}\boldsymbol{B} \end{vmatrix} = |\boldsymbol{A}|\,|\boldsymbol{D} - \boldsymbol{C}\boldsymbol{A}^{-1}\boldsymbol{B}| .$$

The second formula (when $\boldsymbol{D}^{-1}$ exists) follows using an analogous argument. □

The above identities were presented by Schur (1917) and are sometimes referred to as ***Schur's formulas***. A special case of Theorem 10.10 provides a neat formula for the determinant of so-called *bordered* matrices:

$$\begin{vmatrix} \boldsymbol{A} & \boldsymbol{u} \\ \boldsymbol{v}' & \alpha \end{vmatrix} = |\boldsymbol{A}|(\alpha - \boldsymbol{v}'\boldsymbol{A}^{-1}\boldsymbol{u}) , \tag{10.23}$$

where $\boldsymbol{A}$ is assumed to be nonsingular.

Recall the *Sherman-Woodbury-Morrison* formula we derived in Section 3.7. The following theorem is an immediate consequence of Theorem 10.10 and is often called the *Sherman-Woodbury-Morrison formula for determinants.*

Theorem 10.11 *Let $\boldsymbol{A}$ be an $m \times m$ matrix, $\boldsymbol{B}$ an $m \times n$ matrix, $\boldsymbol{C}$ an $n \times m$ matrix, and $\boldsymbol{D}$ an $n \times n$ matrix. Suppose $\boldsymbol{A}$ and $\boldsymbol{D}$ are both non-singular. Then,*

$$|\boldsymbol{D} - \boldsymbol{C}\boldsymbol{A}^{-1}\boldsymbol{B}| = \frac{|\boldsymbol{D}|}{|\boldsymbol{A}|} \times |\boldsymbol{A} - \boldsymbol{B}\boldsymbol{D}^{-1}\boldsymbol{C}| .$$

Proof. Since $\boldsymbol{A}^{-1}$ and $\boldsymbol{D}^{-1}$ both exist, Theorem 10.10 tells us that we can equate the two expressions in (10.22):

$$|\boldsymbol{A}|\,|\boldsymbol{D} - \boldsymbol{C}\boldsymbol{A}^{-1}\boldsymbol{B}| = \begin{vmatrix} \boldsymbol{A} & \boldsymbol{B} \\ \boldsymbol{C} & \boldsymbol{D} \end{vmatrix} = |\boldsymbol{D}|\,|\boldsymbol{A} - \boldsymbol{B}\boldsymbol{D}^{-1}\boldsymbol{C}| .$$

This yields

$$|\boldsymbol{A}|\,|\boldsymbol{D} - \boldsymbol{C}\boldsymbol{A}^{-1}\boldsymbol{B}| = |\boldsymbol{D}|\,|\boldsymbol{A} - \boldsymbol{B}\boldsymbol{D}^{-1}\boldsymbol{C}|$$

and we complete the proof by dividing throughout by $|\boldsymbol{A}|$. □

The above formula is especially useful when the dimension of $\boldsymbol{A}$ is much smaller than the dimension of $\boldsymbol{D}$ (i.e., $m << n$) and $|\boldsymbol{D}|$ is easily available or easy to compute (e.g., $\boldsymbol{D}$ may be diagonal or triangular). Then, the determinant of $n \times n$ matrices of the form $\boldsymbol{D} - \boldsymbol{C}\boldsymbol{A}^{-1}\boldsymbol{B}$ can be evaluated by evaluating $|\boldsymbol{A}|$ and $|\boldsymbol{A} - \boldsymbol{B}\boldsymbol{D}^{-1}\boldsymbol{C}|$, both of which are $m \times m$. See, e.g., Henderson and Searle (1981) for an extensive review of similar identities.

The above result is often presented in the following form:

$$|\boldsymbol{A} + \boldsymbol{B}\boldsymbol{D}\boldsymbol{C}| = |\boldsymbol{A}||\boldsymbol{D}|\left|\boldsymbol{D}^{-1} + \boldsymbol{C}\boldsymbol{A}^{-1}\boldsymbol{B}\right|.$$

This is obtained by replacing $\boldsymbol{D}$ with $-\boldsymbol{D}^{-1}$ in Theorem 10.10.

The following corollary to Theorem 10.10 reveals the determinant for rank-one updates of a square matrix.

Corollary 10.14 Rank-one updates. *Let $\boldsymbol{A}$ be an $n \times n$ non-singular matrix, and let $\boldsymbol{u}$ and $\boldsymbol{v}$ be $n \times 1$ column vectors. Then*

$$|\boldsymbol{A} + \boldsymbol{u}\boldsymbol{v}'| = |\boldsymbol{A}|\left(1 + \boldsymbol{v}'\boldsymbol{A}^{-1}\boldsymbol{u}\right). \tag{10.24}$$

Proof. The proof is an immediate consequence of Theorem 10.10 with $\boldsymbol{B} = \boldsymbol{u}$, $\boldsymbol{C} = -\boldsymbol{v}'$ and $\boldsymbol{D} = 1$ (a 1×1 nonzero scalar, hence non-singular). To be precise, applying Theorem 10.11 to the matrix

$$\begin{pmatrix} \boldsymbol{A} & \boldsymbol{u} \\ -\boldsymbol{v}' & 1 \end{pmatrix}$$

yields $|\boldsymbol{A}|\left(1 + \boldsymbol{v}'\boldsymbol{A}^{-1}\boldsymbol{u}\right) = |\boldsymbol{A} + \boldsymbol{u}\boldsymbol{v}'|$. □

Taking $\boldsymbol{A} = \boldsymbol{I}$ yields an important special case: $|\boldsymbol{I} + \boldsymbol{u}\boldsymbol{v}'| = 1 + \boldsymbol{v}'\boldsymbol{u}$.

Readers may already be familiar with a classical result, known as Cramer's rule, that expresses the solution of a non-singular linear system of equations in terms of determinants. We present this as a consequence of the preceding result on rank-one updates.

Corollary 10.15 Cramer's rule. *Let $\boldsymbol{A}$ be an $n \times n$ non-singular matrix and let $\boldsymbol{b}$ be any $n \times 1$ vector. Consider the linear system $\boldsymbol{A}\boldsymbol{x} = \boldsymbol{b}$. The i-th unknown in the solution vector $\boldsymbol{x} = \boldsymbol{A}^{-1}\boldsymbol{b}$ is given by*

$$x_i = \frac{|\boldsymbol{A}_i|}{|\boldsymbol{A}|},$$

where $\boldsymbol{A}_i = [\boldsymbol{a}_{*1} : \boldsymbol{a}_{*2} : \ldots : \boldsymbol{a}_{*(i-1)} : \boldsymbol{b} : \boldsymbol{a}_{*(i+1)} : \ldots : \boldsymbol{a}_{*n}]$ is the $n \times n$ matrix that is identical to $\boldsymbol{A}$ except that the i-th column, $\boldsymbol{a}_{*i}$, has been replaced by $\boldsymbol{b}$.

Proof. We can write the matrix $\boldsymbol{A}_i$ as a rank-one update of the matrix $\boldsymbol{A}$ as follows:

$$\boldsymbol{A}_i = \boldsymbol{A} + (\boldsymbol{b} - \boldsymbol{a}_{*i})\boldsymbol{e}_i',$$

where e_i' is the i-th unit vector. Applying Corollary 10.14 and using the facts that $\boldsymbol{x} = \boldsymbol{A}^{-1}\boldsymbol{b}$ and $\boldsymbol{A}^{-1}\boldsymbol{a}_{*i} = \boldsymbol{e}_i$ yields

$$\begin{aligned} |\boldsymbol{A}_i| &= |\boldsymbol{A} + (\boldsymbol{b} - \boldsymbol{a}_{*i})\boldsymbol{e}_i'| = |\boldsymbol{A}|\left(1 + \boldsymbol{e}_i'\boldsymbol{A}^{-1}(\boldsymbol{b} - \boldsymbol{a}_{*i})\right) \\ &= |\boldsymbol{A}|\left(1 + \boldsymbol{e}_i'(\boldsymbol{x} - \boldsymbol{A}^{-1}\boldsymbol{a}_{*i})\right) = |\boldsymbol{A}|(1 + x_i - 1) = |\boldsymbol{A}|x_i. \end{aligned} \tag{10.25}$$

Since $\boldsymbol{A}$ is non-singular, we have $|\boldsymbol{A}| \neq 0$ so that (10.25) implies $x_i = |\boldsymbol{A}_i|/|\boldsymbol{A}|$. □

Cramer's rule is rarely, if ever, used in software packages on a computer as it is computationally more expensive than solving linear systems using the $\boldsymbol{LU}$ decomposition. After all, to compute determinants one needs to obtain the $\boldsymbol{LU}$ decomposition anyway. It can, however, be a handy result to keep in mind when one needs to compute one element of the solution vector in a relatively small linear system (say, manageable with hand calculations).

10.7 Cofactors and expansion theorems

Definition 10.3 *Let $\boldsymbol{A}$ by an $n \times n$ matrix with $n \geq 2$. Then the* **cofactor** *of the element a_{ij} in $\boldsymbol{A}$ is written as A_{ij} and is given by*

$$A_{ij} = (-1)^{i+j}|\boldsymbol{M}_{ij}|, \tag{10.26}$$

where $\boldsymbol{M}_{ij}$ is the $(n-1) \times (n-1)$ matrix obtained from $\boldsymbol{A}$ by deleting the ith row and jth column.

The cofactors of a square matrix $\boldsymbol{A}$ appear naturally in the expansion (10.5) of $|\boldsymbol{A}|$. Consider, for example, the expression for the 3×3 determinant in (10.7). Rearranging the terms, we obtain

$$\begin{aligned} |\boldsymbol{A}| &= a_{11}a_{22}a_{33} - a_{11}a_{23}a_{32} - a_{12}a_{21}a_{33} + a_{12}a_{23}a_{31} + a_{13}a_{21}a_{32} - a_{13}a_{22}a_{31} \\ &= a_{11}(a_{22}a_{33} - a_{23}a_{32}) + a_{12}(a_{23}a_{31} - a_{21}a_{33}) + a_{13}(a_{21}a_{32} - a_{22}a_{31}) \\ &= a_{11}A_{11} + a_{12}A_{12} + a_{13}A_{13}. \end{aligned}$$

Because this expansion is in terms of the entries of the first row and the corresponding cofactors, the above term is called the ***cofactor expansion*** of $|\boldsymbol{A}|$ in terms of the first row. It should be clear that there is nothing special about the first row of $\boldsymbol{A}$. That is, it is just as easy to write an analogous expression in which entries from any other row or column appear. For example, a different rearrangement of the terms in (10.7) produces

$$\begin{aligned} |\boldsymbol{A}| &= a_{12}(a_{23}a_{31} - a_{21}a_{33}) + a_{22}(a_{11}a_{33} - a_{13}a_{31}) + a_{32}(a_{13}a_{21} - a_{11}a_{23}) \\ &= a_{12}A_{21} + a_{22}A_{22} + a_{32}A_{32}\ . \end{aligned}$$

This is called the cofactor expansion for $|\boldsymbol{A}|$ in terms of the second column. The 3×3 case is symptomatic of the reasoning for a more general $n \times n$ matrix which yields the general cofactor expansion about any given row,

$$|\boldsymbol{A}| = a_{i1}A_{i1} + a_{i2}A_{i2} + \cdots + a_{in}A_{in}, \tag{10.27}$$

where i is any fixed row number. Analogously, if j denotes any fixed column number, we have the cofactor expansion about the j-th column as

$$|\boldsymbol{A}| = a_{1j}A_{1j} + a_{2j}A_{2j} + \cdots + a_{nj}A_{nj}. \tag{10.28}$$

The formulas in (10.27) or (10.28) can be derived directly from the definition of the determinant in (10.5). We leave this as an exercise for the reader, but provide a different proof below in Theorem 10.12 using results we have already derived. We prove the result for (10.27). The formula in (10.28) can be proved analogously or by transposing the matrix.

First, we prove the following lemma.

Lemma 10.1 *Let $\boldsymbol{A} = \{a_{ij}\}$ be an $n \times n$ matrix. For any given row and column number, say i and j, respectively, if $a_{ik} = 0$ for all $k \neq j$, then $|\boldsymbol{A}| = a_{ij}A_{ij}$, where A_{ij} is the cofactor of a_{ij}.*

Proof. Using $n - i$ interchanges of rows and $n - j$ interchanges of columns, we can take the i-th row and the j-th column of $\boldsymbol{A}$ to the last positions, without disturbing the others. This produces the matrix

$$\boldsymbol{B} = \begin{bmatrix} \boldsymbol{M}_{ij} & * \\ \boldsymbol{0}' & a_{ij} \end{bmatrix},$$

where $\boldsymbol{M}_{ij}$ is the matrix obtained from $\boldsymbol{A}$ by deleting the i-th row and j-th column. By Theorem 10.9, it follows that $|\boldsymbol{B}| = a_{ij}|\boldsymbol{M}_{ij}|$ which yields

$$\begin{aligned} a_{ij}|\boldsymbol{M}_{ij}| = |\boldsymbol{B}| = (-1)^{n-i+n-j}|\boldsymbol{A}| = (-1)^{2n-(i+j)}|\boldsymbol{A}| = (-1)^{2n}(-1)^{-(i+j)}|\boldsymbol{A}| \\ = (-1)^{-(i+j)}|\boldsymbol{A}| . \end{aligned}$$

This implies that $|\boldsymbol{A}| = a_{ij}(-1)^{i+j}|\boldsymbol{M}_{ij}| = a_{ij}A_{ij}$, where A_{ij} is the cofactor of a_{ij} as defined in (10.26). □

The following theorem proves (10.27).

Theorem 10.12 *Let $\boldsymbol{A}$ be an $n \times n$ square matrix and let i be any fixed row number (an integer such that $1 \leq i \leq n$). Then the determinant of $\boldsymbol{A}$ can expanded in terms of its cofactors as*

$$|\boldsymbol{A}| = \sum_{j=1}^{n} a_{ij}A_{ij}.$$

Proof. The i-th row vector of $\boldsymbol{A}$ can be written as the sum of row vectors:

$$\boldsymbol{a}'_{i*} = \boldsymbol{x}'_1 + \boldsymbol{x}'_2 + \cdots + \boldsymbol{x}'_n,$$

where $\boldsymbol{x}'_j = (0, \ldots, 0, a_{ij}, 0, \ldots, 0)$ with a_{ij} occurring in the j-th position and every other entry is 0. From Theorem 10.4 we find that $|\boldsymbol{A}| = \sum_{j=1}^{n} |\boldsymbol{B}_j|$, where the matrix $\boldsymbol{B}_j$ is obtained from $\boldsymbol{A}$ by replacing its j-th row with $\boldsymbol{x}'_j$.

Also, the (i, j)-th element and its cofactor in $\boldsymbol{B}_j$ are the same as those in $\boldsymbol{A}$. Therefore, Lemma 10.1 tells us that $|\boldsymbol{B}_j| = a_{ij}A_{ij}$ and the theorem follows. □

The following corollary to the above theorem is also useful.

Corollary 10.16 *Let $\boldsymbol{A}$ be an $n \times n$ matrix. Then*

$$\sum_{k=1}^{n} a_{ik}A_{jk} = \begin{cases} |\boldsymbol{A}| & \textit{if } \; i = j \\ 0 & \textit{otherwise.} \end{cases} \tag{10.29}$$

Proof. When $i = j$, (10.29) is a restatement of Theorem 10.12. So let $i \neq j$. Let $\boldsymbol{B}$ be the matrix obtained from $\boldsymbol{A}$ by replacing the j-th row by the i-th row. Expanding $|\boldsymbol{B}|$ by the j-th row, $|\boldsymbol{B}| = \sum_{k=1}^{n} a_{ik}A_{jk}$. But $|\boldsymbol{B}| = 0$ since $\boldsymbol{B}$ has two equal rows and the result follows. □

The algorithms presented in Section 10.4 are much more efficient than (10.27) for evaluating determinants of larger sizes. Cofactor expansions can be handy when a row or a column contains several zeroes or for computing small determinants (say, of dimension not exceeding four) as they reduce the problem in terms of determinants of smaller sizes. Here is an example.

Example 10.5 Computing determinants using cofactors. Let us evaluate the determinant of the same matrix $\boldsymbol{A}$ as in Example 10.4, but using (10.27). An important property of the cofactor expansion formula is that we can choose any row (column) to expand about. Clearly, choosing a row with as many zeroes as possible should simplify matters. Expanding $|\boldsymbol{A}|$ about its first row (has two zeroes) yields:

$$\begin{vmatrix} 2 & 3 & 0 & 0 \\ 4 & 7 & 2 & 0 \\ -6 & -10 & 0 & 1 \\ 4 & 6 & 4 & 5 \end{vmatrix} = 2 \times (-1)^{1+1} \begin{vmatrix} 7 & 2 & 0 \\ -10 & 0 & 1 \\ 6 & 4 & 5 \end{vmatrix} + 3 \times (-1)^{1+2} \begin{vmatrix} 4 & 2 & 0 \\ -6 & 0 & 1 \\ 4 & 4 & 5 \end{vmatrix}. \tag{10.30}$$

Next we expand the two 3×3 determinants on the right hand side. Each of them contain a zero in their first and second rows. We will, therefore, benefit by expanding them about either their first or second rows. We choose the first row. Expanding the first determinant about its first row produces

$$\begin{aligned} \begin{vmatrix} 7 & 2 & 0 \\ -10 & 0 & 1 \\ 6 & 4 & 5 \end{vmatrix} &= 7 \times (-1)^{1+1} \begin{vmatrix} 0 & 1 \\ 4 & 5 \end{vmatrix} + 2 \times (-1)^{1+2} \begin{vmatrix} -10 & 1 \\ 6 & 5 \end{vmatrix} \\ &= 7 \times (-1)^{2}(-4) + 2 \times (-1)^{3}(-44) = -28 + 88 = 60, \end{aligned}$$

while expanding the second determinant about its first row yields

$$\begin{vmatrix} 4 & 2 & 0 \\ -6 & 0 & 1 \\ 4 & 4 & 5 \end{vmatrix} = 4 \times (-1)^{1+1} \begin{vmatrix} 0 & 1 \\ 4 & 5 \end{vmatrix} + 2 \times (-1)^{1+2} \begin{vmatrix} -6 & 1 \\ 4 & 5 \end{vmatrix}$$
$$= 4 \times (-1)^2(-4) + 2 \times (-1)^3(-26) = -16 + 52 = 36.$$

Substituting these in (10.30), we obtain

$$|\boldsymbol{A}| = 2 \times (-1)^2(60) + 3 \times (-1)^3(36) = 120 - 108 = 12.$$

Indeed, Example 10.4 revealed the same value of $|\boldsymbol{A}|$. ■

Although not numerically efficient for evaluating general determinants, cofactors offer interesting theoretical characterizations. One such characterization involves the following matrix, whose (i, j)-th element is given by the cofactor A_{ij} of $\boldsymbol{A}$. We call such a matrix the *adjugate* of $\boldsymbol{A}$.

Definition 10.4 *Let $\boldsymbol{A}$ be an $n \times n$ matrix. The* **adjugate** *of $\boldsymbol{A}$ is the $n \times n$ matrix formed by letting its (i, j)-th element be the cofactor of the (j, i)-th element of $\boldsymbol{A}$, i.e., A_{ji}. We denote this matrix by by $\boldsymbol{A}^{\circledast}$. More succinctly, $\boldsymbol{A}^{\circledast} = [A_{ij}]'$.*

The following is a useful characterization that is often used to obtain algebraically explicit formulas for $\boldsymbol{A}^{-1}$ when $\boldsymbol{A}$ is of small dimension.

Theorem 10.13 *If $\boldsymbol{A}$ is non-singular, then $\boldsymbol{A}^{-1}$ can be written in terms of $\boldsymbol{A}^{\circledast}$ as*

$$\boldsymbol{A}^{-1} = \frac{1}{|\boldsymbol{A}|}\boldsymbol{A}^{\circledast}.$$

Proof. Consider the (i, j)th element of $\boldsymbol{A}\boldsymbol{A}^{\circledast}$. It is given by

$$\boldsymbol{a}'_{i*}\boldsymbol{a}^{\circledast}_{*j} = \sum_{k=1}^{n} a_{ik}A_{jk},$$

where $\boldsymbol{a}^{\circledast}_{*j}$ denotes the j-th column of $\boldsymbol{A}^{\circledast}$. Then, (10.29) tells us that $\boldsymbol{a}'_{i*}\boldsymbol{a}^{\circledast}_{*j} = |\boldsymbol{A}|\delta_{ij}$, where $\delta_{ij} = 1$ if $i = j$ and $\delta_{ij} = 0$ otherwise. This implies

$$\boldsymbol{A}\boldsymbol{A}^{\circledast} = |\boldsymbol{A}|\boldsymbol{I}. \tag{10.31}$$

Using the cofactor expansion about a column, it follows similarly that $\boldsymbol{A}^{\circledast}\boldsymbol{A} = |\boldsymbol{A}|\boldsymbol{I}$ and the result follows. □

10.8 The minor and the rank of a matrix

We have now seen some relationships that exist between the determinant of a matrix and certain special submatrices. Theorems 10.12 and 10.13 are such examples. We now introduce another important concept with regard to submatrices—the ***minor***.

Definition 10.5 *Let $A = \{a_{ij}\}$ be an $m \times n$ matrix. Let $k \leq \min\{m, n\}$ and let $I = \{i_1, i_2, \ldots, i_k\}$ be a subset of the row numbers such that $i_1 < i_2 < \cdots < i_k$ and let $J = \{j_1, j_2, \ldots, j_k\}$ be a subset of column numbers such that $j_1 < j_2 < \cdots < j_k$. Then the* **submatrix** *formed from the* intersection *of I and J is given by*

$$A_{IJ} = \begin{bmatrix} a_{i_1 j_1} & a_{i_1 j_2} & \cdots & a_{i_1 j_k} \\ a_{i_2 j_1} & a_{i_2 j_2} & \cdots & a_{i_2 j_k} \\ \vdots & \vdots & \ddots & \vdots \\ a_{i_k j_1} & a_{i_k j_2} & \cdots & a_{i_k j_k} \end{bmatrix}. \tag{10.32}$$

The determinant $|A_{IJ}|$ is called a **minor** *of A or a* **minor of order** *k of A. Note that these terms are used even when A is not square.*

The definition of the "minor" helps us characterize the rank of a matrix in a different manner. Non-zero minors are often referred to as ***non-vanishing minors*** and they play a useful role in characterizing familiar quantities such as the rank of a matrix. We see this in the following theorem.

Theorem 10.14 Determinants and rank. *The rank of a non-null $m \times n$ matrix A is the largest integer r for which A has a non-vanishing minor of order r.*

Proof. Let A be an $m \times n$ matrix with rank $\rho(A) = r$. We will show that this means that there is at least one $r \times r$ nonsingular submatrix in A, and there are no nonsingular submatrices of larger order.

The theorem will then follow by virtue of Theorem 10.6.

Because $\rho(A) = r$, we can find a linearly independent set of r row vectors as well as a linear independent set of r column vectors in A. Suppose $I = \{i_1, i_2, \ldots, i_r\}$ is the index of the linearly independent row vectors and $J = \{j_1, j_2, \ldots, j_r\}$ is the index of the linearly independent column vectors.

We will prove that the $r \times r$ submatrix B lying on the intersection of these r rows and r columns is nonsingular, i.e., $B = A_{IJ}$ is non-singular.

Elementary row operations can bring the linearly independent row vectors to the first r positions and sweep out the remaining rows. The product of the corresponding elementary matrices is the non-singular matrix, say G. Therefore,

$$GA = \begin{bmatrix} U_{r \times n} \\ O_{(m-r) \times n} \end{bmatrix}.$$

Elementary column operations can bring the r linearly independent columns of U to the first r positions and then annihilate the remaining $n - r$ columns. This amounts to post-multiplying by a non-singular matrix H to get

$$GAH = \begin{bmatrix} B_{r \times r} & O_{r \times (n-r)} \\ O_{(m-r) \times r} & O_{(m-r) \times (n-r)} \end{bmatrix}.$$

Since rank does not change by pre- or post-multiplication with non-singular matrices, we have that $r = \rho(\boldsymbol{A}) = \rho(\boldsymbol{GAH}) = \rho(\boldsymbol{B})$. This proves that $\boldsymbol{B}$ is a non-singular submatrix of $\boldsymbol{A}$. Theorem 10.6 now implies that $|\boldsymbol{B}| \neq 0$, so $|\boldsymbol{B}| = |\boldsymbol{A}_{IJ}|$ serves as the non-vanishing minor of order r in the statement of the theorem.

To see why there can be no non-vanishing minor of order larger than r, note that if $\boldsymbol{C}$ is any $k \times k$ submatrix of $\boldsymbol{A}$ and $|\boldsymbol{C}| \neq 0$, then Theorem 10.6 tells us that $\boldsymbol{C}$ is non-singular. But this implies $k = \rho(\boldsymbol{C}) \leq \rho(\boldsymbol{A}) = r$. □

In Theorem 10.14, we used the definition of rank to be the maximum number of linearly independent columns or rows in a matrix and then showed that a matrix must have a non-vanishing minor of that size. It is possible to go the other way round. One could first define the rank of a matrix to be the largest integer r for which there exists an $r \times r$ non-vanishing minor. The existence of such an r is never in doubt for matrices with a finite number of rows or columns—indeed, there must exist such a non-negative integer bounded above by $\min\{m, n\}$. Note that $r = 0$ if and only if the matrix is null. One can then argue that r is, in fact, the maximum number of linearly independent columns (or rows) of a matrix.

To be precise, let $\boldsymbol{A}$ be an $m \times n$ matrix and suppose that $I = \{i_1 < i_2 < \cdots < i_r\}$ indexes the rows and $J = \{j_1 < j_2 < \cdots < j_r\}$ indexes the columns for which the minor $|\boldsymbol{A}_{IJ}| \neq 0$, where $r \leq \min\{m, n\}$. This means that $\boldsymbol{A}_{IJ}$ is nonsingular (recall Theorem 10.6) and so the columns of $\boldsymbol{A}_{IJ}$ are linearly independent.

But this also means that the corresponding r columns of $\boldsymbol{A}$ *are linearly independent.*

To see why this is true, observe that $\boldsymbol{A}_{IJ}\boldsymbol{x} = \boldsymbol{0} \Longrightarrow \boldsymbol{x} = \boldsymbol{0}$ and this solution remains the same no matter how many equations we add to the system. In particular, it does not change when we add the $(n-r)$ equations corresponding to the remaining $(m-r)$ rows of $\boldsymbol{A}$. Here is an argument using more formal notations: let $\boldsymbol{A}_{*J}$ denote the $m \times r$ matrix formed by selecting the r columns in J and all the m rows of $\boldsymbol{A}$. Since we will consider the homogeneous system $\boldsymbol{Ax} = \boldsymbol{0}$, the solution space does not depend how we arrange the individual equations. So,

$$\boldsymbol{A}_{*J}\boldsymbol{x} = \boldsymbol{0} \Longrightarrow \begin{bmatrix} \boldsymbol{A}_{IJ} \\ \boldsymbol{A}_{\bar{I}J} \end{bmatrix} \boldsymbol{x} = \boldsymbol{0} \Longrightarrow \boldsymbol{A}_{IJ}\boldsymbol{x} = \boldsymbol{0} \Longrightarrow \boldsymbol{x} = \boldsymbol{0} \; ,$$

where $\bar{I}$ is the set of integers in $\{1, 2, \ldots, m\}$ that are not in I. Therefore, if $\boldsymbol{A}$ has an $r \times r$ non-vanishing minor, then it has r linearly independent columns.

Applying the above argument to the transpose of $\boldsymbol{A}$ and keeping in mind that the determinant of $\boldsymbol{A}_{IJ}$ will be the same as that of its transpose, we can conclude that the r rows of $\boldsymbol{A}$ indexed by I are linearly independent. More precisely, let $\boldsymbol{B} = \boldsymbol{A}'$. Then, the transpose of $\boldsymbol{A}_{IJ}$ is $\boldsymbol{B}_{JI}$ and so $|\boldsymbol{B}_{JI}| = |\boldsymbol{A}_{IJ}| \neq 0$. Therefore, by the preceding arguments, the columns of $\boldsymbol{B}$ indexed by I are linearly independent. But these are precisely the r rows of $\boldsymbol{A}$.

Also, *if* all *minors of order greater than r are zero, then* $\boldsymbol{A}$ *cannot have more than r linearly independent columns.*

This is because if $\boldsymbol{A}$ has $r+1$ linearly independent columns, then we would be able to find a non-vanishing minor of order $r+1$. We form the submatrix consisting of these

$r+1$ columns. Since this sub-matrix has linearly independent columns, it will have $r+1$ pivots. We can use elementary row operations to reduce it to an echelon form with $r+1$ nonzero rows. Keeping track of the row permutations in those elementary operations can help us identify a set of $r+1$ rows in $\boldsymbol{A}$ that are linearly independent. The minor corresponding to the intersection of these linearly independent rows and columns must be nonzero (see the proof of Theorem 10.14).

The above arguments prove that defining the rank of a matrix in terms of non-vanishing minors is equivalent to the usual definition of rank as the maximum number of linearly independent columns or rows in a matrix. Observe that in the process of clarifying this equivalence, we constructed yet another argument why the rank of a matrix must be equal to its transpose.

10.9 The Cauchy-Binet formula

We consider a generalization of the determinant of the product of two matrices. Recall that in Theorem 10.7, we proved that $|\boldsymbol{AB}| = |\boldsymbol{A}||\boldsymbol{B}|$ for any two square matrices $\boldsymbol{A}$ and $\boldsymbol{B}$. What happens if $\boldsymbol{A}$ and $\boldsymbol{B}$ are not themselves square but their product is a square matrix so that $|\boldsymbol{AB}|$ is still defined? For example, $\boldsymbol{A}$ could be $n \times p$ and $\boldsymbol{B}$ could be $p \times n$ so that $\boldsymbol{AB}$ is $n \times n$.

The answer is easy if $n > p$. Recall that the rank of $\boldsymbol{AB}$ cannot exceed the rank of $\boldsymbol{A}$ or $\boldsymbol{B}$ and so must be smaller than p. Therefore, $\boldsymbol{AB}$ is an $n \times n$ matrix whose rank is $p < n$, which implies that $\boldsymbol{AB}$ is singular and so $|\boldsymbol{AB}| = 0$ (Theorem 10.6).

Matters get a bit more complex when $n \leq p$. In that case, one can expand the determinant using the determinant of some submatrices. The resulting formula is known as the *Cauchy-Binet Product Formula*. We will present it as a theorem. The Cauchy-Binet formula essentially uses Definition 10.2 and manipulates the permutation functions to arrive at the final result. The algebra may seem somewhat daunting at first but is really not that difficult. Usually a simpler example with small values of n and p help clarify what is really going on in the proof.

Before we present its proof, it will be helpful to see how we can derive $|\boldsymbol{AB}| = |\boldsymbol{A}||\boldsymbol{B}|$ directly from Definition 10.2. This provides some practice into the type of manipulations that will go into the Cauchy-Binet formula. If the argument below also seems daunting, try working it out with $n = 2$ (which is too easy!) and then $n = 3$.

First suppose that $n = p$ so that both $\boldsymbol{A}$ and $\boldsymbol{B}$ are $n \times n$ matrices and $\boldsymbol{C} = \boldsymbol{AB}$.

The (i, j)-th element of $\boldsymbol{C}$ is $c_{ij} = \sum_{k=1}^{n} a_{ik}b_{kj}$. From Definition 10.2,

$$
\begin{aligned}
|\boldsymbol{C}| &= \sum_{\pi} \sigma(\pi) c_{1\pi_1} c_{2\pi_2} \cdots c_{n\pi_n} \\
&= \sum_{\pi} \sigma(\pi) \left(\sum_{k=1}^{n} a_{1k} b_{k\pi_1} \right) \left(\sum_{k=1}^{n} a_{2k} b_{k\pi_2} \right) \cdots \left(\sum_{k=1}^{n} a_{nk} b_{k\pi_n} \right) \\
&= \sum_{\pi} \sigma(\pi) \sum_{k_1=1}^{n} \sum_{k_2=1}^{n} \cdots \sum_{k_n=1}^{n} a_{1k_1} b_{k_1\pi_1} a_{2k_2} b_{k_2\pi_2} \cdots a_{nk_n} b_{k_n\pi_n} \,.
\end{aligned}
$$

The last step follows because $(\sum_{k=1}^{n} a_{2k}b_{k\pi_2}) \cdots (\sum_{k=1}^{n} a_{nk}b_{k\pi_n})$ is a sum of n^n terms, each term arising as a product of exactly one term from each of the n sums. This yields the following expressions for $|\boldsymbol{C}|$:

$$
\begin{aligned}
|\boldsymbol{C}| &= \sum_{\pi} \sigma(\pi) \sum_{k_1=1}^{n} \sum_{k_2=1}^{n} \cdots \sum_{k_n=1}^{n} \left(a_{1k_1} a_{2k_2} \cdots a_{nk_n} \right) \left(b_{k_1\pi_1} b_{k_2\pi_2} \cdots b_{k_n\pi_n} \right) \\
&= \sum_{k_1=1}^{n} \sum_{k_2=1}^{n} \cdots \sum_{k_n=1}^{n} \left(a_{1k_1} a_{2k_2} a_{nk_n} \right) \left\{ \sum_{\pi} \sigma(\pi) \left(b_{k_1\pi_1} b_{k_2\pi_2} \cdots b_{k_n\pi_n} \right) \right\} ,
\end{aligned}
\tag{10.33}
$$

where the last step follows from interchanging the sums. Let $K = \{k_1, k_2, \ldots, k_n\}$ and make the following observations:

(a) For any fixed choice of K, the expression $\sum_{\pi} \sigma(\pi) \left(b_{k_1\pi_1} b_{k_2\pi_2} \cdots b_{k_n\pi_n} \right)$ is the determinant of an $n \times n$ matrix whose i-th row is the k_i-th row of $\boldsymbol{B}$.

(b) If any two elements in K are equal, then $\sum_{\pi} \sigma(\pi) \left(b_{k_1\pi_1} b_{k_2\pi_2} \cdots b_{k_n\pi_n} \right) = 0$ because it is the determinant of a matrix two of whose rows are identical.

(c) When no two elements in K are equal, K is a permutation of $\{1, 2, \ldots, n\}$ and

$$
\sum_{\pi} \sigma(\pi) \left(b_{k_1\pi_1} b_{k_2\pi_2} \cdots b_{k_n\pi_n} \right) = \sigma(K) |\boldsymbol{B}| \,.
$$

By virtue of the above observations, the sum over all the k_i's in (10.33) reduces to a sum over all distinct indices, i.e., over choices of K that are permutations of $\{1, 2, \ldots, n\}$. Therefore, we can write

$$
\begin{aligned}
|\boldsymbol{C}| &= \sum_{\kappa} \left(a_{1\kappa_1} a_{2\kappa_2} \cdots a_{n\kappa_n} \right) \sigma(\kappa) |\boldsymbol{B}| = \left(\sum_{\kappa} \sigma(\kappa) a_{1\kappa_1} a_{2\kappa_2} \cdots a_{n\kappa_n} \right) |\boldsymbol{B}| \\
&= |\boldsymbol{A}||\boldsymbol{B}| \,,
\end{aligned}
$$

where κ runs over all permutations of $\{1, 2, \ldots, n\}$. This proves $|\boldsymbol{AB}| = |\boldsymbol{A}||\boldsymbol{B}|$.

We now present the Cauchy-Binet formula as a theorem. Its proof follows essentially the same idea as above.

Theorem 10.15 Cauchy-Binet product formula. *Let $C = AB$ where A is an $n \times p$ matrix and B is a $p \times n$ matrix. Let $E = \{1, 2, \ldots, n\}$ and $J = \{j_1, j_2, \ldots, j_n\}$. Then,*

$$|C| = \begin{cases} 0 & \text{if } n < p \\ \sum_{1 \le j_1 < \cdots < j_n \le p} |A_{EJ}||B_{JE}| & \text{if } n \ge p\,. \end{cases}$$

Proof. Since $\rho(C) = \rho(AB) \le \rho(A) \le p$, it follows that C is singular when $n > p$ and, hence, $|C| = 0$.

Now suppose $n \le p$. Then,

$$\begin{aligned} |C| &= \sum_{\pi} \sigma(\pi) c_{1\pi_1} \cdots c_{n\pi_n} \\ &= \sum_{\pi} \sigma(\pi) \left(\sum_{k_1=1}^{p} a_{1k_1} b_{k_1\pi_1} \right) \cdots \left(\sum_{k_1=1}^{p} a_{nk_n} b_{k_n\pi_n} \right) \\ &= \sum_{k_1} \cdots \sum_{k_n} a_{1k_1} \cdots a_{nk_n} \sum_{\pi} \sigma(\pi) b_{k_1\pi_1} \cdots b_{k_n\pi_n} \\ &= \sum_{k_1} \cdots \sum_{k_n} a_{1k_1} \cdots a_{nk_n} |B_{KE}|\,, \end{aligned} \tag{10.34}$$

where $K = \{k_1, k_2, \ldots, k_n\}$ and, for any choice of K, B_{KE} is the matrix whose i-th row is the k_i-th row of B.

If any two k_i's are equal, then $|B_{K*}| = 0$. Therefore, it is enough to take the summation above over distinct values of $k_1, \ldots, k_n$. This corresponds to choices of K that are permutations of $\{1, 2, \ldots, n\}$.

One way to enumerate such permutations is to first choose $j_1, \ldots, j_n$ such that $1 \le j_1 < \cdots < j_n \le p$ and then permute them. Letting $J_\theta = \{j_{\theta_1}, j_{\theta_2}, \ldots, j_{\theta_n}\}$, we can now write (10.34) as

$$\begin{aligned} |C| &= \sum_{1 \le j_1 < \cdots < j_n \le p} \sum_{\theta} a_{1j_{\theta_1}} \cdots a_{nj_{\theta_n}} |B_{J_\theta E}| \\ &= \sum_{1 \le j_1 < \cdots < j_n \le p} \sum_{\theta} a_{1j_{\theta_1}} \cdots a_{nj_{\theta_n}} \sigma(\theta) |B_{(JE}| \\ &= \sum_{1 \le j_1 < \cdots < j_n \le p} |A_{EJ}||B_{JE}|\,. \end{aligned}$$

This completes the proof. □

Here is an example that verifies the above identity.

Example 10.6 Consider the matrices

$$A = \begin{bmatrix} 1 & 2 & 0 \\ 3 & 4 & 0 \end{bmatrix} \text{ and } B = \begin{bmatrix} 1 & 2 \\ 2 & 3 \\ 3 & 4 \end{bmatrix}.$$

Note that $|\boldsymbol{AB}| = \begin{vmatrix} 5 & 8 \\ 11 & 18 \end{vmatrix} = 2.$

We now apply Theorem 10.15 with $n = 2$ and $p = 3$. It is easily seen that $|\boldsymbol{A}_{JE}| \neq 0$ only with $J = \{1, 2\}$ and

$$|\boldsymbol{A}_{EJ}||\boldsymbol{B}_{JE}| = \begin{vmatrix} 1 & 2 \\ 3 & 4 \end{vmatrix} \times \begin{vmatrix} 1 & 2 \\ 2 & 3 \end{vmatrix} = (-2) \times (-1) = 2 = |\boldsymbol{AB}| \ . \ \blacksquare$$

10.10 The Laplace expansion

We derive an expansion formula for determinants involving submatrices known as the *Laplace expansion.* We first generalize the definition of cofactors to encompass submatrices.

Definition 10.6 *The* **cofactor of a submatrix** $\boldsymbol{A}_{IJ}$ *of* $\boldsymbol{A}$*, where* I *and* J *are as defined in Definition 10.5, is denoted as* A_{IJ} *and is given by*

$$A_{IJ} = (-1)^{i_1+\cdots+i_k+j_1+\cdots+j_k} |\boldsymbol{A}_{\bar{I}\bar{J}}|,$$

where $\bar{I}$ *and* $\bar{J}$ *are the complements of* I *and* J*, respectively in* $\{1, 2, \ldots, n\}$.

Note that this is consistent with the definition of the cofactor of a single entry given earlier in Definition 10.3 because if $I = \{i\}$ and $J = \{j\}$, then $\boldsymbol{A}_{IJ} = a_{ij}$ and $A_{IJ} = (-1)^{i+j}|\boldsymbol{M}_{ij}| = A_{ij}$, where A_{ij} was defined in (10.26).

Example 10.7 Consider the matrix $\boldsymbol{A}$ in Example 10.4. Let $I = \{1, 3\}$ and $J = \{1, 2\}$. so $\bar{I} = \{2, 4\}$ and $\bar{J} = \{3, 4\}$. The corresponding cofactor is

$$A_{IJ} = (-1)^{(1+3)+(1+2)} |\boldsymbol{A}(\bar{I}|\bar{J})| = (-1)^{4+3} \begin{vmatrix} 2 & 0 \\ 4 & 5 \end{vmatrix} = -10 \ . \ \blacksquare$$

With the above generalization, we can derive the following expansion known as the Laplace expansion.

Theorem 10.16 Laplace expansion. *Let* $\boldsymbol{A}$ *be an* $n \times n$ *matrix and let* $I = \{i_1, i_2, \ldots, i_k\}$ *be a subset of row indices such that* $i_1 < i_2 < \cdots < i_k$*. Then*

$$|\boldsymbol{A}| = \sum_J |\boldsymbol{A}_{IJ}| A_{IJ} = \sum_J (-1)^{i_1+\cdots+i_k+j_1+\cdots+j_k} |\boldsymbol{A}(\bar{I}|\bar{J})|,$$

where J *runs over all subsets of* $\{1, 2, \ldots, n\}$ *that contain* k *elements.*

Proof. Let $\bar{I} = \{i_{k+1}, i_{k+2}, \ldots, i_n\}$ where $i_{k+1} < i_{k+2} < \cdots < i_n$ and let $\mu = (i_1, i_2, \ldots, i_n)$. Now it is easily verified that

$$|\boldsymbol{A}| = \sum_\pi \sigma(\pi) a_{1\pi_1} a_{2\pi_2} \cdots a_{n\pi_n} = \sum_\pi \sigma(\pi) a_{i_1\pi(i_1)} a_{i_2\pi(i_2)} \cdots a_{i_n\pi(i_n)}. \quad (10.35)$$

Let π be a permutation of $(1,2,\ldots,n)$ and let $J = \{j_1, j_2, \ldots, j_k\}$, where $j_1, j_2, \ldots, j_k$ are $\pi(i_1), \pi(i_2), \ldots, \pi(i_k)$ written in increasing order. Also let $\bar{J} = \{j_{k+1}, j_{k+2}, \ldots, j_n\}$ where $j_{k+1} < j_{k+2} < \cdots < j_n$ and let $\gamma = (j_1, j_2, \ldots, j_n)$. Let θ by the permutation of $(1,2,\ldots,k)$ defined by

$$\begin{pmatrix} i_1 & i_2 & \ldots & i_k \\ \pi_{i_1} & \pi_{i_2} & \ldots & \pi_{i_k} \end{pmatrix} = \begin{pmatrix} \theta_1 & \theta_2 & \ldots & \theta_k \\ \gamma_{\theta_1} & \gamma_{\theta_2} & \ldots & \gamma_{\theta_k} \end{pmatrix}.$$

Put another way, $\pi(i_s) = \gamma(\theta(s))$ for $s = 1, 2, \ldots, k$. This means that θ represents that permutation on the first k integers which, when permuted according to γ, produces the permutation $(\pi_{i_1}, \pi_{i_2}, \ldots, \pi_{i_k})$. Let τ be the permutation of $\{k+1, k+2, \ldots, n\}$ defined by

$$\pi(i_t) = \gamma(\tau(t)) \qquad \text{for} \quad t = k+1, k+2, \ldots, n\,.$$

Then it is easy to check that $\pi = \gamma\theta\tau\mu^{-1}$. For example, if $1 \leq s \leq k$, then $(\gamma\theta\tau\mu^{-1})(i_s) = (\gamma\theta\tau)(s) = (\gamma\theta)(s) = \pi(i_s)$. Therefore,

$$\sigma(\pi) = \sigma(\mu)\sigma(\gamma)\sigma(\theta)\sigma(\tau).$$

It is also easy to check that the map $\pi \mapsto (J, \theta, \tau)$ is a $1-1$ correspondence between permutations of $(1,2,\ldots,n)$ and triples (J, θ, τ) where J is as in Definition 10.5, θ is a permutation of $(1,2,\ldots,k)$ and τ is a permutation of $k+1, k+2, \ldots, n$. Hence we can rewrite (10.35) as

$$|\boldsymbol{A}| = \sum_J \sigma(\mu)\sigma(\gamma)\Big(\sum_\theta \sigma(\theta) a_{i_1 j_{\theta(1)}} a_{i_2 j_{\theta(2)}} \cdots a_{i_k j_{\theta(k)}}\Big)$$
$$\times \left(\sum_\tau \sigma(\tau) a_{i_{k+1} j_{\tau(k+1)}} a_{i_{k+2} j_{\tau(k+2)}} \cdots a_{i_n j_{\tau(n)}}\right). \qquad (10.36)$$

The first sum in parentheses here is $|\boldsymbol{A}_{IJ}|$ (this is easily seen by calling $|\boldsymbol{A}_{IJ}|$ as $\boldsymbol{B}$ and using the definition of a determinant). Similarly, the second sum in parentheses is $\boldsymbol{A}_{\bar{I}\bar{J}}$. Also,

$$\sigma(\mu)\sigma(\gamma) = (-1)^{i_1 + \cdots + i_k + j_1 + \cdots + j_k - k(k-1)}\,.$$

Since $k(k+1)$ is always even, it can be dropped and $\sigma(\mu)\sigma(\gamma)|\boldsymbol{A}_{\bar{I}\bar{J}}| = A_{IJ}$. This completes the proof. □

Here is an example.

Example 10.8 Consider again the determinant in Example 10.5. We now evaluate it using the Laplace expansion. We fix $I = \{1, 2\}$ and let $J = \{j_1, j_2\}$ run over all subsets of $\{1, 2, 3, 4\}$ such that $j_1 < j_2$. For each such J, we compute the product $A_{IJ} \times \boldsymbol{A}_{IJ}$ and sum them up. The calculations are presented in Table 10.8.

The determinant is the sum of the numbers in the last column:

$$|\boldsymbol{A}| = 20 - 32 + 24 = 12\,,$$

which is what we had obtained in Examples 10.4 and 10.5. ■

J	A_{IJ}	$\lvert \boldsymbol{A}_{IJ} \rvert$	$A_{IJ} \times \lvert \boldsymbol{A}_{IJ} \rvert$
$\{1,2\}$	-10	-2	20
$\{1,3\}$	35	0	0
$\{1,4\}$	-16	2	-32
$\{2,3\}$	-20	0	0
$\{2,4\}$	8	3	24
$\{3,4\}$	4	0	0

Table 10.2 *Computations in the Laplace expansion for the determinant in Example 10.5.*

Remark: The most numerically stable manner of computing determinants is not using cofactor or Laplace expansions, but by the $\boldsymbol{LU}$ method outlined in Example 10.4.

The Laplace expansion can lead to certain properties almost immediately. For example, if $\boldsymbol{A}$ and $\boldsymbol{D}$ are square matrices, possibly of different dimensions, then

$$\begin{vmatrix} \boldsymbol{A} & \boldsymbol{B} \\ \boldsymbol{O} & \boldsymbol{D} \end{vmatrix} = \begin{vmatrix} \boldsymbol{A} & \boldsymbol{O} \\ \boldsymbol{C} & \boldsymbol{D} \end{vmatrix} = |\boldsymbol{A}||\boldsymbol{D}| \; .$$

Recall that we had proved these results in Theorem 10.9 but they follow immediately from Theorem 10.16.

10.11 Exercises

1. Find the $|\boldsymbol{A}|$ without doing any computations, where

$$\begin{bmatrix} 0 & 0 & 1 \\ 0 & 1 & 0 \\ 1 & 0 & 0 \end{bmatrix} \text{ and } \begin{bmatrix} 0 & 0 & 1 & 0 \\ 0 & 1 & 0 & 0 \\ 1 & 0 & 0 & 0 \\ 0 & 0 & 0 & 1 \end{bmatrix} .$$

2. Let $\boldsymbol{A}$ be a 5×5 matrix such that $|\boldsymbol{A}| = -3$. Find $|\boldsymbol{A}^3|$, $|\boldsymbol{A}^{-1}|$ and $|2\boldsymbol{A}|$.

3. Find $|\boldsymbol{A}|$ by identifying the permutations that make $\boldsymbol{A}$ upper-triangular, where

$$\boldsymbol{A} = \begin{bmatrix} 0 & 0 & 1 \\ 2 & 3 & 4 \\ 0 & 5 & 6 \end{bmatrix} .$$

4. Prove that the area of a triangle formed by the three points (x_1, y_1), (x_2, y_2) and (x_3, y_3) is given by

$$\frac{1}{2} \begin{vmatrix} x_1 & y_1 & 1 \\ x_2 & y_2 & 1 \\ x_3 & y_3 & 1 \end{vmatrix} .$$

5. Construct an example to show that $|\boldsymbol{A} + \boldsymbol{B}| \neq |\boldsymbol{A}| + |\boldsymbol{B}|$.

6. Using the method described in Section 10.4, find the following 3×3 determinants:

$$\begin{vmatrix} 2 & 6 & 1 \\ 3 & 9 & 2 \\ 0 & -1 & 3 \end{vmatrix}, \quad \begin{vmatrix} 0 & 2 & 5 \\ -1 & 7 & -5 \\ -1 & 8 & 3 \end{vmatrix}, \quad \text{and} \quad \begin{vmatrix} 0 & 0 & 1 & 2 \\ 2 & 4 & 6 & 2 \\ 1 & 3 & 5 & 1 \\ 4 & 5 & 9 & 6 \end{vmatrix}.$$

7. Using the method described in Section 10.4, verify that

$$\begin{vmatrix} 1 & 2 & 3 \\ 4 & 5 & 6 \\ 7 & 8 & 9 \end{vmatrix} = 0 .$$

8. Prove that $|\boldsymbol{A}'\boldsymbol{A}| \geq 0$ and $|\boldsymbol{A}'\boldsymbol{A}| > 0$ if and only if $\boldsymbol{A}$ has full column rank.

9. Making use of rank-one updated matrices (Corollary 10.14), find the value of $|\boldsymbol{A}|$, where $\boldsymbol{A} = \{a_{ij}\}$ is $n \times n$ with $a_{ii} = \alpha$ for each $i = 1, 2, \ldots, n$ and $a_{ij} = \beta$ whenever $i \neq j$ and $\alpha \neq \beta$.

10. Prove that

$$\begin{vmatrix} 1+\lambda_1 & \lambda_2 & \cdots & \lambda_n \\ \lambda_1 & 1+\lambda_2 & \cdots & \lambda_n \\ \vdots & \vdots & \ddots & \vdots \\ \lambda_1 & \lambda_2 & \cdots & 1+\lambda_n \end{vmatrix} = 1 + \lambda_1 + \lambda_2 + \cdots + \lambda_n .$$

11. **Vandermonde determinants.** Consider the 2×2 and 3×3 determinants:

$$\begin{vmatrix} 1 & x_1 \\ 1 & x_2 \end{vmatrix} \quad \text{and} \quad \begin{vmatrix} 1 & x_1 & x_1^2 \\ 1 & x_2 & x_2^2 \end{vmatrix} .$$

Prove that these determinants are equal to $x_2 - x_1$ and $(x_3-x_1)(x_3-x_2)(x_2-x_1)$, respectively. Now consider the $n \times n$ Vandermonde matrix (recall Example 2.9). Prove by induction that the determinant of the Vandermonde matrix is

$$|\boldsymbol{V}| = \begin{vmatrix} 1 & x_1 & x_1^2 & \cdots & x_1^{n-1} \\ 1 & x_2 & x_2^2 & \cdots & x_2^{n-1} \\ 1 & x_3 & x_3^2 & \cdots & x_3^{n-1} \\ \vdots & \vdots & \vdots & \ddots & \vdots \\ 1 & x_n & x_n^2 & \cdots & x_n^{n-1} \end{vmatrix} = \prod_{1 \leq i < j \leq n} (x_j - x_i) .$$

12. If $\boldsymbol{A}$, $\boldsymbol{B}$ and $\boldsymbol{C}$ are $n \times n$, $n \times p$ and $p \times n$, respectively, and $\boldsymbol{A}$ is nonsingular, then prove that $|\boldsymbol{A} + \boldsymbol{BC}| = |\boldsymbol{A}||\boldsymbol{I}_k + \boldsymbol{CA}^{-1}\boldsymbol{B}|$.

13. Using square matrices $\boldsymbol{A}$, $\boldsymbol{B}$, $\boldsymbol{C}$ and $\boldsymbol{D}$, construct an example showing

$$\begin{vmatrix} \boldsymbol{A} & \boldsymbol{B} \\ \boldsymbol{C} & \boldsymbol{D} \end{vmatrix} \neq |\boldsymbol{A}||\boldsymbol{D}| - |\boldsymbol{C}||\boldsymbol{B}| .$$

14. If $\boldsymbol{A}$, $\boldsymbol{B}$, $\boldsymbol{C}$ and $\boldsymbol{D}$ are all $n \times n$ and $\boldsymbol{AC} = \boldsymbol{CA}$, then prove that

$$\begin{vmatrix} \boldsymbol{A} & \boldsymbol{B} \\ \boldsymbol{C} & \boldsymbol{D} \end{vmatrix} \neq |\boldsymbol{AD} - \boldsymbol{CB}| .$$

15. **An alternative derivation of the product rule.** Prove that the product rule for determinants can be derived by computing the determinant $\begin{vmatrix} \boldsymbol{A} & \boldsymbol{O} \\ -\boldsymbol{I} & \boldsymbol{B} \end{vmatrix}$ in two different ways: (i) by direct computations (as in Theorem 10.12), and (ii) by reducing $\boldsymbol{B}$ to $\boldsymbol{O}$ using suitable elementary column operations. The first method yields $|\boldsymbol{A}||\boldsymbol{B}|$, while the second will yield $|\boldsymbol{AB}|$.

16. Use Cramer's rule (Theorem 10.15) to solve the system of linear equations in Example 2.1. Note that the coefficient matrix is the same as in Example 10.4.

17. Use Cramer's rule (Theorem 10.15) to solve $\boldsymbol{Ax} = \boldsymbol{b}$ for each $\boldsymbol{A}$ and $\boldsymbol{b}$ in Exercise 2 of Section 2.7.

18. Use the cofactor expansion formula (see, e.g., Example 10.5) to evaluate the determinants in Exercise 6.

19. Use the Laplace expansion (see, e.g., Example 10.8) to evaluate the determinants in Exercise 6.

20. Find the adjugate of $\boldsymbol{A}$ in Example 10.4. Applying Theorem 10.8, find the inverse of $\boldsymbol{A}$.

21. **Axiomatic definition of a determinant.** Let f be a map from $\Re^n \times \Re^n \times \Re^n$ (n times) to $\Re$. That is, $f(\boldsymbol{x}_1, \boldsymbol{x}_2, \ldots, \boldsymbol{x}_n)$ maps a collection of n vectors in $\Re^n$ to the real line. Suppose that f satisfies the following "axioms":
    (a) *Alternating:* $f(\boldsymbol{x}_1, \boldsymbol{x}_2, \ldots, \boldsymbol{x}_n) = 0$ whenever any two of the $\boldsymbol{x}_i$'s are equal;
    (b) *Multilinear:* $f(\boldsymbol{x}_1, \boldsymbol{x}_2, \ldots, \boldsymbol{x}_n)$ is a linear function in each $\boldsymbol{x}_i$ when all the other $\boldsymbol{x}_j$'s are treated as fixed, i.e.,
    $$\begin{aligned} f(\boldsymbol{x}_1, \boldsymbol{x}_2, \ldots, \alpha\boldsymbol{u} + \beta\boldsymbol{v}, \ldots, \boldsymbol{x}_n) &= \alpha f(\boldsymbol{x}_1, \boldsymbol{x}_2, \ldots, \boldsymbol{u}, \ldots, \boldsymbol{x}_n) \\ &\quad + \beta f(\boldsymbol{x}_1, \boldsymbol{x}_2, \ldots, \boldsymbol{v}, \ldots, \boldsymbol{x}_n) ; \end{aligned}$$
    (c) $f(\boldsymbol{e}_1, \boldsymbol{e}_2, \ldots, \boldsymbol{e}_n) = 1$, where $\boldsymbol{e}_1, \boldsymbol{e}_2, \ldots, \boldsymbol{e}_n$ are the n rows of $\boldsymbol{I}_n$.

    Prove that if f satisfies (i) and (ii), then $f(\boldsymbol{x}_1, \boldsymbol{x}_2, \ldots, \boldsymbol{x}_n) = c|\boldsymbol{A}|$ for some real number c, where $\boldsymbol{A}$ is the $n \times n$ matrix with $\boldsymbol{x}_i'$ as its i-th row. In addition, if f satisfies (iii), then prove that $c = 1$ and, hence,
    $$f(\boldsymbol{x}_1, \boldsymbol{x}_2, \ldots, \boldsymbol{x}_n) = |\boldsymbol{A}| .$$

22. **Yet another derivation of the product rule.** Let $\boldsymbol{A}$ and $\boldsymbol{B}$ be $n \times n$ matrices. Show that $|\boldsymbol{AB}|$ is an alternating multilinear function of the rows of $\boldsymbol{A}$ (as in the previous exercise) when $\boldsymbol{B}$ is fixed. The previous exercise now implies that $|\boldsymbol{AB}| = c|\boldsymbol{A}|$ for some scalar c. Show that $c = |\boldsymbol{B}|$ and, hence, $|\boldsymbol{AB}| = |\boldsymbol{A}||\boldsymbol{B}|$.

CHAPTER 11

Eigenvalues and Eigenvectors

When we multiply a vector $\boldsymbol{x}$ with a matrix $\boldsymbol{A}$, we transform the vector $\boldsymbol{x}$ to a new vector $\boldsymbol{Ax}$. Usually, this changes the direction of the vector. Certain *exceptional* vectors, $\boldsymbol{x}$, which have the same direction as $\boldsymbol{Ax}$, play a central role in linear algebra. For such an exceptional vector, the vector $\boldsymbol{Ax}$ will be a scalar λ times the original $\boldsymbol{x}$. For this to make sense, clearly the matrix $\boldsymbol{A}$ must be a square matrix. Otherwise, the number of elements in $\boldsymbol{Ax}$ will be different from $\boldsymbol{x}$ and they will not reside in the same subspace. (In this chapter we will assume all matrices are square.) This yields the following equation for a square matrix $\boldsymbol{A}$:

$$\boldsymbol{Ax} = \lambda\boldsymbol{x} \text{ or } (\boldsymbol{A} - \lambda\boldsymbol{I})\boldsymbol{x} = \boldsymbol{0}\,. \tag{11.1}$$

Obviously $\boldsymbol{x} = \boldsymbol{0}$ always satisfies the above equation for every λ. That is the trivial solution. The more interesting case is when $\boldsymbol{x} \neq \boldsymbol{0}$. A nonzero vector $\boldsymbol{x}$ satisfying (11.1) is called an ***eigenvector*** of $\boldsymbol{A}$ and the scalar λ is called an ***eigenvalue*** of $\boldsymbol{A}$. Clearly, if the eigenvalue is zero, then any vector in the null space of $\boldsymbol{A}$ is an eigenvector. This shows that there can be more than one eigenvector associated with an eigenvalue. Any nonzero $\boldsymbol{x}$ satisfying (11.1) for a particular eigenvector λ is referred to as an eigenvector associated with the eigenvalue λ.

Example 11.1 Let $\boldsymbol{A} = \begin{bmatrix} 6 & 4 \\ -1 & 1 \end{bmatrix}$. Let us find the eigenvalues of this matrix. We want to solve the equation in (11.1), which is

$$\begin{array}{rcrcl} (6-\lambda)x_1 & + & 4x_2 & = & 0, \\ -x_1 & + & (1-\lambda)x_2 & = & 0. \end{array} \tag{11.2}$$

What makes this different from a usual homogeneous system is that apart from the unknown variable $\boldsymbol{x} = (x_1, x_2)'$, the scalar λ in the coefficient matrix is also unknown. We can, nevertheless, proceed as if we were solving a linear system. Let us eliminate x_1. From the second equation, we see that $x_1 = (1-\lambda)x_2$. Substituting this into the first equation produces

$$(6-\lambda)(1-\lambda)x_2 + 4x_2 = 0 \Longrightarrow (\lambda^2 - 7\lambda + 10)x_2 = 0 \Longrightarrow (\lambda-5)(\lambda-2)x_2 = 0\,.$$

If we set $\lambda = 5$ or $\lambda = 2$, any nonzero value of x_2 will result in a non-trivial solution for the equations in (11.2). For example, take $\lambda = 5$ and set $x_2 = 1$. Then, $x_1 = (1-\lambda)x_2 = -4$. This means $\boldsymbol{x} = (-4, 1)'$ is a solution of (11.2) when $\lambda = 5$.

Therefore, $\lambda = 5$ is an eigenvalue of $\boldsymbol{A}$ and $\boldsymbol{x} = (-4, 1)'$ is an eigenvector associated with the eigenvalue 5. Now set $\lambda = 2$. The solution $\boldsymbol{x} = (-1, 1)'$ is an eigenvector associated with the eigenvalue 2.

What is interesting to note is that $\lambda = 5$ and $\lambda = 2$ are the only eigenvalues of this system. This can be easily verified. There are, however, an infinite number of eigenvectors associated with each of these eigenvalues. For example, any scalar multiple of $(-4, 1)'$ is an eigenvector of $\boldsymbol{A}$ associated with eigenvalue 5 and any scalar multiple of $(-1, 1)'$ is an eigenvector of $\boldsymbol{A}$ associated with eigenvalue 2. We will explore these concepts in greater detail. ■

Does every square matrix $\boldsymbol{A}$ have a nonzero eigenvalue and eigenvector associated with it? Consider any rotation matrix that rotates every vector by 90 degrees in $\Re^2$. This means that every vector, except the null vector $\boldsymbol{0}$, will change its direction when multiplied by $\boldsymbol{A}$. Therefore, we cannot find a nonzero vector $\boldsymbol{x}$ in $\Re^2$ that will have the same direction as $\boldsymbol{A}$ above. Hence, we cannot also have a real eigenvalue. Below is a specific example.

Example 11.2 Consider the matrix

$$\boldsymbol{A} = \begin{bmatrix} 0 & 1 \\ -1 & 0 \end{bmatrix} .$$

Suppose we want to find λ and $\boldsymbol{x}$ such that $\boldsymbol{A}\boldsymbol{x} = \lambda\boldsymbol{x}$. Let $\boldsymbol{x} = (x_1, x_2)'$. From the first row we obtain $x_2 = \lambda x_1$ and from the second row we obtain $x_1 = -\lambda x_2$. Note that if either x_1 or x_2 is zero, then $\boldsymbol{x} = \boldsymbol{0}$. If $\boldsymbol{x} \neq \boldsymbol{0}$, then we obtain

$$x_2 = \lambda(-\lambda x_2) \Longrightarrow (\lambda^2 + 1)x_2 = 0 \Longrightarrow \lambda^2 + 1 = 0 ,$$

which implies that $\lambda = \sqrt{-1}$ or $-\sqrt{-1}$. The corresponding eigenvectors are $\begin{bmatrix} 1 \\ \sqrt{-1} \end{bmatrix}$ and $\begin{bmatrix} \sqrt{-1} \\ 1 \end{bmatrix}$. This shows that although every element in $\boldsymbol{A}$ is real, it does not have a real eigenvalue or a real eigenvector. ■

Hence, we will allow eigenvalues to be complex numbers and eigenvectors to be vectors of complex numbers. We the complex number plane by $\mathbb{C}$ and the n-dimensional complex field by $\mathbb{C}^n$. Note that when we write $\lambda \in \mathbb{C}$ we mean that λ can be a real number or a complex number because the complex number plane includes the real line. Similarly, when we write $\boldsymbol{x} \in \mathbb{C}^n$ we include vectors in $\Re^n$ as well. We now provide the following definition.

Definition 11.1 *An* **eigenvalue** *of a square matrix* $\boldsymbol{A}$ *is a real or complex number* $\lambda \in \mathbb{C}$ *that satisfies the equation* $\boldsymbol{A}\boldsymbol{x} = \lambda\boldsymbol{x}$ *for some nonzero vector* $\boldsymbol{x} \in \mathbb{C}^n$. *Any such* $\boldsymbol{x} \neq \boldsymbol{0}$ *is called an* **eigenvector** *of* $\boldsymbol{A}$ *corresponding to the eigenvalue* $\boldsymbol{A}$.

When we say that $\boldsymbol{x}$ is an eigenvector of $\boldsymbol{A}$, we mean that $\boldsymbol{x}$ is an eigenvector corresponding to some eigenvalue of $\boldsymbol{A}$. An eigenvector of $\boldsymbol{A}$ paired with a corresponding eigenvalue is often called an ***eigen-pair*** of $\boldsymbol{A}$. Thus, if $(\lambda, \boldsymbol{x})$ is an eigen-pair of $\boldsymbol{A}$, we know that $\boldsymbol{A}\boldsymbol{x} = \lambda\boldsymbol{x}$.

The importance of eigenvalues and eigenvectors lies in the fact that almost all matrix results associated with square matrices can be derived from representations of square matrices in terms of their eigenvalues and eigenvectors. Eigenvalues and their associated eigenvectors form a fundamentally important set of scalars and vectors that uniquely identify matrices. They hold special importance in numerous applications in fields as diverse as mathematical physics, sociology, economics and statistics. Among the class of square matrices, real symmetric matrices hold a very special place in many applications. Fortunately, the eigen-analysis of real symmetric matrices yield many astonishing simplifications, to the point that matrix analysis of real symmetric matrices can be done with the same ease and elegance of scalar analysis. We will see such properties later.

11.1 The Eigenvalue equation

Equation (11.1) is sometimes referred to as the ***eigenvalue equation***. Definition 11.1 says that λ is an eigenvalue of an $n \times n$ matrix $\boldsymbol{A}$ if and only if the homogenous system $(\boldsymbol{A} - \lambda\boldsymbol{I})\boldsymbol{x} = \boldsymbol{0}$ has a non-trivial solution $\boldsymbol{x} \neq \boldsymbol{0}$. This means that any λ for which $\boldsymbol{A} - \lambda\boldsymbol{I}$ is singular is an eigenvalue of $\boldsymbol{A}$.

Equivalently, λ is an eigenvalue of an $n \times n$ matrix $\boldsymbol{A}$ if and only if $\mathcal{N}(\boldsymbol{A} - \lambda\boldsymbol{I})$ has a nonzero member. In terms of rank and nullity, we can say the following: λ is an eigenvalue of $\boldsymbol{A}$ if and only if $\nu(\boldsymbol{A} - \lambda\boldsymbol{I}) > 0$ or if and only if $\rho(\boldsymbol{A} - \lambda\boldsymbol{I}) < n$.

For some simple matrices the eigenvalues and eigenvectors can be obtained quite easily. Finding eigenvalues and eigenvectors of a triangular matrix is easy.

Example 11.3 Eigenvalues of a triangular matrix. Let $\boldsymbol{U}$ be an upper-triangular matrix

$$\boldsymbol{U} = \begin{bmatrix} u_{11} & u_{12} & \dots & u_{1n} \\ 0 & u_{22} & \dots & u_{2n} \\ \vdots & \vdots & \ddots & \vdots \\ 0 & 0 & \dots & u_{nn} \end{bmatrix}.$$

The matrix $\boldsymbol{U} - \lambda\boldsymbol{I}$ is also upper-triangular with $u_{ii} - \lambda$ as its i-th diagonal element. Recall that a triangular matrix is invertible whenever all its diagonal elements are nonzero—see Theorem 2.7 and the discussion following it. Therefore, $\boldsymbol{U} - \lambda\boldsymbol{I}$ will be singular whenever $\lambda = d_{ii}$ for any $i = 1, 2, \dots, n$. In other words, any diagonal element of $\boldsymbol{U}$ is an eigenvalue.

Suppose $\lambda = u_{ii}$. Let $\boldsymbol{x} = (0, 0, \dots, 0, x_i, 0, \dots, 0)'$ be an $n \times 1$ vector which has a nonzero element x_i as its i-th element and a 0 for every other element. It is easily seen that $\boldsymbol{x}$ is a nontrivial solution for $(\boldsymbol{U} - \lambda\boldsymbol{I})\boldsymbol{x} = \boldsymbol{0}$. Any such $\boldsymbol{x}$ is an eigenvector for the eigenvalue u_{ii}.

In particular, every diagonal element of a diagonal matrix is an eigenvalue. ■

Given that the rank of a matrix is equal to the rank of its transpose, can we deduce something useful about eigenvalues of the transpose of a matrix? The following theorem provides the answer.

Theorem 11.1 *If λ is an eigenvalue of $\boldsymbol{A}$, then it is also an eigenvalue of $\boldsymbol{A}'$.*

Proof. Suppose that λ is an eigenvalue of an $n \times n$ matrix $\boldsymbol{A}$. This means that $\boldsymbol{A} - \lambda \boldsymbol{I}$ must be singular, which means that $\rho(\boldsymbol{A} - \lambda \boldsymbol{I}) < n$. Now use the fact that rank of a matrix is equal to its transpose and note that

$$n > \rho(\boldsymbol{A} - \lambda \boldsymbol{I}) = \rho\left[(\boldsymbol{A} - \lambda \boldsymbol{I})'\right] = \rho(\boldsymbol{A}' - \lambda \boldsymbol{I}) \ .$$

So the matrix $\boldsymbol{A}' - \lambda \boldsymbol{I}$ is also singular, which means that there exists a nonzero vector $\boldsymbol{x}$ that is a solution of the eigenvalue equation $(\boldsymbol{A} - \lambda \boldsymbol{I})\boldsymbol{x} = \boldsymbol{0}$. Therefore, λ is an eigenvalue of $\boldsymbol{A}'$. □

Suppose that λ is an eigenvalue of $\boldsymbol{A}$. Theorem 11.1 says that λ is also an eigenvalue of $\boldsymbol{A}'$. Let $\boldsymbol{x}$ be an eigenvector of $\boldsymbol{A}'$ associated with λ. This means that $\boldsymbol{A}'\boldsymbol{x} = \lambda \boldsymbol{x}$ or, by taking transposes, that $\boldsymbol{x}'\boldsymbol{A} = \lambda \boldsymbol{x}'$. This motivates the following definitions.

Definition 11.2 *Let λ be an eigenvalue of a matrix $\boldsymbol{A}$. A nonzero vector $\boldsymbol{x}$ is said to be a* **left eigenvector** *of $\boldsymbol{A}$ if $\boldsymbol{x}'\boldsymbol{A} = \lambda \boldsymbol{x}'$. In the same spirit, a nonzero vector $\boldsymbol{x}$ satisfying $\boldsymbol{A}\boldsymbol{x} = \lambda \boldsymbol{x}$ (as in Definition 11.1) is sometimes called a* **right eigenvector** *of $\boldsymbol{A}$.*

The following result concerns the orthogonality of right and left eigenvectors.

Theorem 11.2 *Let λ_1 and λ_2 be two distinct (i.e., $\lambda_1 \neq \lambda_2$) eigenvalues of $\boldsymbol{A}$. If $\boldsymbol{x}_1$ is a left eigenvector of $\boldsymbol{A}$ corresponding to eigenvalue λ_1 and $\boldsymbol{x}_2$ is a right eigenvector of $\boldsymbol{A}$ corresponding to the eigenvalue λ_2, then $\boldsymbol{x}_1 \perp \boldsymbol{x}_2$.*

Proof. Note that $\boldsymbol{x}_1'\boldsymbol{A} = \lambda_1 \boldsymbol{x}_1'$ and $\boldsymbol{A}\boldsymbol{x}_2 = \lambda_2 \boldsymbol{x}_2$. Post-multiply the first equation by $\boldsymbol{x}_2$ and make use of the second equation to obtain

$$\begin{aligned} \boldsymbol{x}_1'\boldsymbol{A} = \lambda_1 \boldsymbol{x}_1' &\Longrightarrow \boldsymbol{x}_1'\boldsymbol{A}\boldsymbol{x}_2 = \lambda_1 \boldsymbol{x}_1'\boldsymbol{x}_2 \Longrightarrow \lambda_2 \boldsymbol{x}_1'\boldsymbol{x}_2 = \lambda_1 \boldsymbol{x}_1'\boldsymbol{x}_2 \\ &\Longrightarrow (\lambda_2 - \lambda_1)\boldsymbol{x}_1'\boldsymbol{x}_2 = 0 \ , \end{aligned}$$

which implies that $\boldsymbol{x}_1'\boldsymbol{x}_2 = 0$ because $\lambda_1 \neq \lambda_2$. □

Here is another important result that can be obtained from the basic definitions of eigenvalues and eigenvectors.

Theorem 11.3 *Let $\boldsymbol{A}$ be an $m \times n$ matrix and $\boldsymbol{B}$ an $n \times m$ matrix. If λ is a nonzero eigenvalue of $\boldsymbol{AB}$, then it is also an eigenvalue of $\boldsymbol{BA}$.*

Proof. If λ is an eigenvalue of the matrix $\boldsymbol{AB}$, then there exists an eigenvector $\boldsymbol{x} \neq \boldsymbol{0}$ such that $\boldsymbol{AB}\boldsymbol{x} = \lambda \boldsymbol{x}$. Note that $\boldsymbol{B}\boldsymbol{x} \neq \boldsymbol{0}$ because $\lambda \neq 0$. We now obtain

$$\boldsymbol{AB}\boldsymbol{x} = \lambda \boldsymbol{x} \Longrightarrow \boldsymbol{BAB}\boldsymbol{x} = \lambda \boldsymbol{B}\boldsymbol{x} \Longrightarrow \boldsymbol{BA}(\boldsymbol{B}\boldsymbol{x}) = \lambda(\boldsymbol{B}\boldsymbol{x}) \ .$$

This implies that $\boldsymbol{BAu} = \lambda \boldsymbol{u}$, which implies that λ is an eigenvalue of $\boldsymbol{BA}$ with $\boldsymbol{u} = \boldsymbol{Bx} \neq \boldsymbol{0}$ an eigenvector associated with λ. □

To find eigenvalues and eigenvectors of a square matrix $\boldsymbol{A}$ we need to find scalars λ and nonzero vectors $\boldsymbol{x}$ that satisfy the eigenvalue equation $\boldsymbol{Ax} = \lambda\boldsymbol{x}$. Writing $\boldsymbol{Ax} = \lambda\boldsymbol{x}$ as $(\boldsymbol{A}\lambda\boldsymbol{I})\boldsymbol{x} = \boldsymbol{0}$ reveals that the eigenvectors are precisely the nonzero vectors in $\mathcal{N}(\boldsymbol{A} - \lambda\boldsymbol{I})$. But $\mathcal{N}(\boldsymbol{A} - \lambda\boldsymbol{I})$ contains nonzero vectors if and only if $\boldsymbol{A} - \lambda\boldsymbol{I}$ is singular, which means that the eigenvalues are precisely the values of λ that make $\boldsymbol{A} - \lambda\boldsymbol{I}$ singular.

A matrix is singular if and only if its determinant is zero. Therefore, the eigenvalues are precisely those values of λ for which $|\boldsymbol{A} - \lambda\boldsymbol{I}| = 0$. This gives us another way to recognize eigenvalues. The eigenvectors corresponding to an eigenvalue λ are simply the nonzero vectors in $\mathcal{N}(\boldsymbol{A} - \lambda\boldsymbol{I})$. To find eigenvectors corresponding to λ, one could simply find a basis for $\mathcal{N}(\boldsymbol{A} - \lambda\boldsymbol{I})$. The null space of $\boldsymbol{A} - \lambda\boldsymbol{I}$ is called the ***eigenspace*** of $\boldsymbol{A}$ corresponding to the eigenvalue λ. We illustrate with an example, revisiting Example 11.1 in greater detail.

Example 11.4 Let $\boldsymbol{A} = \begin{bmatrix} 6 & 4 \\ -1 & 1 \end{bmatrix}$. Its eigenvalues are precisely those values of λ for which $|\boldsymbol{A} - \lambda\boldsymbol{I}| = 0$. This determinant can be computed in terms of λ as

$$\begin{vmatrix} 6-\lambda & 4 \\ -1 & 1-\lambda \end{vmatrix} = \lambda^2 - 7\lambda + 10 = (\lambda - 2)(\lambda - 5) \,,$$

which implies that the values of λ for which $|\boldsymbol{A} - \lambda\boldsymbol{I}| = 0$ are $\lambda = 2$ and $\lambda = 5$. These are the eigenvalues of $\boldsymbol{A}$.

We now find the eigenvalues associated with $\lambda = 2$ and $\lambda = 5$, which amounts to finding linearly independent solutions for the two homogeneous systems:

$$(\boldsymbol{A} - 2\boldsymbol{I})\boldsymbol{x} = \boldsymbol{0} \text{ and } (\boldsymbol{A} - 5\boldsymbol{I})\boldsymbol{x} = \boldsymbol{0} \,.$$

Note that

$$\boldsymbol{A} - 2\boldsymbol{I} = \begin{bmatrix} 4 & 4 \\ -1 & -1 \end{bmatrix} \text{ and } \boldsymbol{A} - 5\boldsymbol{I} = \begin{bmatrix} 1 & 4 \\ -1 & -4 \end{bmatrix} \,.$$

From the above it follows that $(\boldsymbol{A} - 2\boldsymbol{I})\boldsymbol{x} = \boldsymbol{0}$ has solutions of the form $\boldsymbol{x} = \begin{pmatrix} -1 \\ 1 \end{pmatrix} x_2$, where x_2 is the free variable. Similarly, $(\boldsymbol{A} - 5\boldsymbol{I})\boldsymbol{x} = \boldsymbol{0}$ has solutions of the form $\boldsymbol{x} = \begin{pmatrix} -4 \\ 1 \end{pmatrix} x_2$, where x_2 is, again, the free variable. Therefore,

$$\mathcal{N}(\boldsymbol{A} - 2\boldsymbol{I}) = \left\{\boldsymbol{x} : \boldsymbol{x} = \begin{pmatrix} -1 \\ 1 \end{pmatrix} \alpha,\ \alpha \in \Re^1 \right\} \text{ and}$$

$$\mathcal{N}(\boldsymbol{A} - 5\boldsymbol{I}) = \left\{\boldsymbol{x} : \boldsymbol{x} = \begin{pmatrix} -4 \\ 1 \end{pmatrix} \alpha,\ \alpha \in \Re^1 \right\}$$

fully describe the eigenspaces of $\boldsymbol{A}$. All nonzero multiples of $\boldsymbol{x} = (-1, 1)'$ are eigenvectors of $\boldsymbol{A}$ corresponding to the eigenvalue $\lambda = 2$, while all nonzero multiples of $\boldsymbol{x} = (-4, 1)'$ are eigenvectors of $\boldsymbol{A}$ corresponding to the eigenvalue $\lambda = 5$. ■

11.2 Characteristic polynomial and its roots

It is useful to characterize eigenvalues in terms of the determinant. Theorem 10.6 tells us that a matrix is singular if and only if its determinant is zero. Expanding $|\boldsymbol{A} - \lambda \boldsymbol{I}| = 0$ in Example 11.4 produces a quadratic equation $\lambda^2 - 7\lambda + 10 = 0$. The eigenvalues of $\boldsymbol{A}$ are the roots of this quadratic equation. The second-degree polynomial $p_{\boldsymbol{A}}(\lambda) = \lambda^2 - 7\lambda + 10$ is called the *characteristic polynomial* of $\boldsymbol{A}$. We offer the following general definition.

Definition 11.3 *The* **characteristic polynomial** *of an $n \times n$ matrix $\boldsymbol{A}$ is the n-th degree polynomial*

$$p_{\boldsymbol{A}}(\lambda) = |\lambda \boldsymbol{I} - \boldsymbol{A}| \tag{11.3}$$

in the indeterminate λ.

The determinant $|\lambda \boldsymbol{I} - \boldsymbol{A}|$ is a polynomial of degree n and the coefficient of λ^n is one. This can be seen from expanding the determinant using Definition 10.2. As we sum over all possible permutations of the column indices, we find that the permutation $\pi_i = i$ generates the term

$$(\lambda - a_{11})(\lambda - a_{22}) \cdots (\lambda - a_{nn}) = \prod_{i=1}^{n} (\lambda - a_{ii}) \tag{11.4}$$

in the summand. This is the only term that involves λ^n and clearly its coefficient is one.

Here is a bit more detail. Each element of the matrix $\lambda \boldsymbol{I} - \boldsymbol{A}$ can be written as $\lambda \delta_{ij} - a_{ij}$, where $\delta_{ij} = 1$ if $i = j$ and $\delta_{ij} = 0$ if not. Then, the determinant is

$$|\lambda \boldsymbol{I} - \boldsymbol{A}| = \sum_{\pi} \sigma(\pi)(\lambda \delta_{1\pi_1} - a_{1\pi_1})(\lambda \delta_{2\pi_2} - a_{2\pi_2}) \cdots (\lambda \delta_{n\pi_n} - a_{n\pi_n}) , \tag{11.5}$$

which is a polynomial in λ. The highest power of λ is produced by the term in (11.4), so the degree of the polynomial is n and the leading coefficient is 1 (such polynomials are called ***monic polynomials***). Some authors define the characteristic polynomial as $|\boldsymbol{A} - \lambda \boldsymbol{I}|$, which has leading coefficient $(-1)^n$.

The roots of $p_{\boldsymbol{A}}(\lambda)$ are the eigenvalues of $\boldsymbol{A}$ and also referred to as the ***characteristic roots*** of $\boldsymbol{A}$. In fact, λ is an eigenvalue of $\boldsymbol{A}$ if and only if λ is a root of the characteristic polynomial $p_{\boldsymbol{A}}(\lambda)$ because

$$\begin{aligned} \boldsymbol{A}\boldsymbol{x} = \lambda \boldsymbol{x} \text{ for some } \boldsymbol{x} \neq \boldsymbol{0} &\iff (\lambda \boldsymbol{I} - \boldsymbol{A})\boldsymbol{x} = \boldsymbol{0} \text{ for some } \boldsymbol{x} \neq \boldsymbol{0} \\ \iff \lambda \boldsymbol{I} - \boldsymbol{A} \text{ is singular } &\iff |\lambda \boldsymbol{I} - \boldsymbol{A}| = 0 . \end{aligned}$$

The fundamental theorem of algebra states that every polynomial of degree n with real or complex coefficients has n roots. Even if the coefficients of the polynomial are real, some roots may be complex numbers but the complex roots must appear in conjugate pairs. In other words if a complex number is a root of a polynomial with real coefficients, then its complex conjugate must also be a root of that polynomial.

Also some roots may be repeated. These facts have the following implications for eigenvalues of matrices:

- Every $n \times n$ matrix has n eigenvalues. Some of the eigenvalues may be complex numbers and some of them may be repeated.
- As is easily seen from (11.5), if all the entries of $\boldsymbol{A}$ are real numbers, then the characteristic polynomial has real coefficients. This implies that complex eigenvalues of real matrices must occur in conjugate pairs.
- If all the entries of an $n \times n$ matrix $\boldsymbol{A}$ are real and if λ is a *real* eigenvalue of $\boldsymbol{A}$, then $(\boldsymbol{A} - \lambda \boldsymbol{I})\boldsymbol{x} = \boldsymbol{0}$ has a non-trivial solution over $\Re^n$, which ensures that their will exist a *real* eigenvector of $\boldsymbol{A}$ associated with λ. Note that this does not rule out the existence of complex eigenvectors for the real eigenvalue λ. For example, if $\boldsymbol{x}$ is an eigenvector, so is $\sqrt{-1}\boldsymbol{x}$.

Earlier we saw that the diagonal elements of a triangular matrix are its eigenvalues. This is easily seen from the characteristic polynomial as well. If $\boldsymbol{A}$ is a triangular matrix, then $|\lambda \boldsymbol{I} - \boldsymbol{A}| = \prod_{i=1}^{n}(\lambda - a_{ii})$. The a_{ii}'s are the characteristic roots and, hence, for a triangular matrix the diagonal entries, a_{ii}, are the characteristic roots or eigenvalues.

Since the determinant of a matrix is equal to the determinant of its transpose, we have the following result.

Theorem 11.4 *The characteristic polynomial of $\boldsymbol{A}$ is the same as that of $\boldsymbol{A}'$, i.e., $p_{\boldsymbol{A}}(\lambda) = p_{\boldsymbol{A}'}(\lambda)$.*

Proof. This follows immediately from the fact that $|\lambda \boldsymbol{I} - \boldsymbol{A}| = |\lambda \boldsymbol{I} - \boldsymbol{A}'|$. □

Since the characteristic polynomial of $\boldsymbol{A}$ and $\boldsymbol{A}'$ are the same, their characteristic roots are the same as well. The characteristic roots are precisely the eigenvalues of $\boldsymbol{A}$ and $\boldsymbol{A}'$. Therefore, the eigenvalues of $\boldsymbol{A}$ are the same as the eigenvalues of $\boldsymbol{A}'$, which is a result we had proved differently in Theorem 11.1.

The following result provides two useful characterization for the eigenvalues of a matrix.

Theorem 11.5 *The product of the eigenvalues of $\boldsymbol{A}$ is equal to $|\boldsymbol{A}|$ and the sum of the eigenvalues of $\boldsymbol{A}$ is equal to tr$(\boldsymbol{A})$.*

Proof. Let $\lambda_1, \lambda_2, \ldots, \lambda_n$ be the eigenvalues of the matrix $\boldsymbol{A}$. Because they are the roots of the characteristic polynomial, we can write the characteristic polynomial as

$$p_{\boldsymbol{A}}(\lambda) = (\lambda - \lambda_1)(\lambda - \lambda_2)\cdots(\lambda - \lambda_n) .$$

We can also express the characteristic polynomial in terms of its coefficients as

$$p_{\boldsymbol{A}}(\lambda) = \lambda^n + c_{n-1}\lambda^{n-1} + c_{n-2}\lambda^{n-2} + \cdots + c_1\lambda + c_0 .$$

Comparing the constant term in the above expressions for $p_{\boldsymbol{A}}(\lambda)$, we see that $c_0 =$

$(-1)^n\lambda_1\lambda_2\cdots\lambda_n$. Therefore, c_0 is the product of the eigenvalues of $\boldsymbol{A}$. But we can obtain c_0 in another way: set $\lambda = 0$ to see that $p_{\boldsymbol{A}}(0) = c_0$. Therefore,

$$c_0 = p_{\boldsymbol{A}}(0) = |0 \cdot \boldsymbol{I} - \boldsymbol{A}| = |-\boldsymbol{A}| = (-1)^n|\boldsymbol{A}| \, .$$

This proves that $(-1)^n \prod_{i=1}^n \lambda_i = c_0 = (-1)^n|\boldsymbol{A}|$, which implies that the product of the eigenvalues is equal to $|\boldsymbol{A}|$. This proves the first statement.

Turning to the sum of the eigenvalues, comparing coefficients of λ^{n-1} shows that

$$c_{n-1} = -(\lambda_1 + \lambda_2 + \cdots + \lambda_n) \, .$$

It is also easily verified from (11.5) that λ^{n-1} appears in $p_{\boldsymbol{A}}(\lambda) = |\lambda\boldsymbol{I} - \boldsymbol{A}|$ only in the term given by (11.4). Therefore, c_{n-1}, which is the coefficient of λ^{n-1} in $p_{\boldsymbol{A}}(\lambda)$, is equal to the coefficient of λ^{n-1} in $\prod_{i=1}^n(\lambda - a_{ii})$. This implies that $c_{n-1} = -\sum_{i=1}^n a_{ii} = -\text{tr}(\boldsymbol{A})$. We have now shown that

$$\lambda_1 + \lambda_2 + \cdots + \lambda_n = -c_{n-1} = \text{tr}(\boldsymbol{A}) \, ,$$

which completes the proof that the sum of the eigenvalues of $\boldsymbol{A}$ is the trace of $\boldsymbol{A}$. □

Remark: Note that in the above proof we noted that λ^{n-1} appears in $p_{\boldsymbol{A}}(\lambda) = |\lambda\boldsymbol{I} - \boldsymbol{A}|$ only through the term in (11.4), which is also the only term through which λ^n appears in $p_{\boldsymbol{A}}(\lambda)$. Observe that λ^{n-1} appears only if the permutation π chooses $n-1$ of the diagonal elements in $\lambda\boldsymbol{I} - \boldsymbol{A}$. But once we choose $n-1$ elements in a permutation of n elements, the n-th element is automatically chosen as the remaining one. Therefore, if $n-1$ of the diagonal elements in $\lambda\boldsymbol{I} - \boldsymbol{A}$ appear, then the n-th diagonal element in $\lambda\boldsymbol{I} - \boldsymbol{A}$ also appears. That is why both λ^{n-1} and λ^n appear through the same term in the expansion of $p_{\boldsymbol{A}}(\lambda)$.

Recall that two square matrices $\boldsymbol{A}$ and $\boldsymbol{B}$ are said to be ***similar*** if there exists a nonsingular matrix $\boldsymbol{P}$ such that $\boldsymbol{B} = \boldsymbol{P}^{-1}\boldsymbol{A}\boldsymbol{P}$. The following result shows that similar matrices have the same characteristic polynomial and, hence, the same set of eigenvalues.

Theorem 11.6 *Similar matrices have identical characteristic polynomials.*

Proof. If $\boldsymbol{A}$ and $\boldsymbol{B}$ are similar, then there exists a nonsingular matrix $\boldsymbol{P}$ such that $\boldsymbol{B} = \boldsymbol{P}\boldsymbol{A}\boldsymbol{P}^{-1}$. Therefore,

$$\lambda\boldsymbol{I} - \boldsymbol{B} = \lambda\boldsymbol{I} - \boldsymbol{P}\boldsymbol{A}\boldsymbol{P}^{-1} = \lambda\boldsymbol{P}\boldsymbol{P}^{-1} - \boldsymbol{P}\boldsymbol{A}\boldsymbol{P}^{-1} = \boldsymbol{P}(\lambda\boldsymbol{I} - \boldsymbol{A})\boldsymbol{P}^{-1} \, ,$$

which proves that $\lambda\boldsymbol{I} - \boldsymbol{A}$ and $\lambda\boldsymbol{I} - \boldsymbol{B}$ are also similar. Also, recall from Theorem 10.9 that similar matrices have the same determinant. Therefore, $|\lambda\boldsymbol{I} - \boldsymbol{B}| = |\lambda\boldsymbol{I} - \boldsymbol{A}|$ so $\boldsymbol{A}$ and $\boldsymbol{B}$ will have the same characteristic polynomials. □

We have seen in Theorem 11.3 that a nonzero eigenvalue of $\boldsymbol{AB}$ is also an eigenvalue for $\boldsymbol{BA}$ whenever the matrix products are well-defined. What can we say about the characteristic polynomials of $\boldsymbol{AB}$ and $\boldsymbol{BA}$? To unravel this, we first look at the following lemma.

Lemma 11.1 *If A is an $m \times n$ matrix and B an $n \times m$ matrix, then*

$$\lambda^m |\lambda I_n - BA| = \lambda^n |\lambda I_m - AB| \text{ for any nonzero scalar } \lambda .$$

Proof. Applying the Sherman-Woodbury-Morrison formula for determinants (recall Theorem 10.11) to the matrix

$$\begin{bmatrix} \lambda I_m & \lambda A \\ B & \lambda I_n \end{bmatrix}$$

yields $|\lambda I_m||\lambda I_n - BA| = |\lambda I_n||\lambda I_m - AB|$. Since $|\lambda I_m| = \lambda^m$ and $|\lambda I_n| = \lambda^n$, the result follows immediately. □

The following result is a generalization of Theorem 11.3.

Theorem 11.7 *Let A and B be $m \times n$ and $n \times m$ matrices, respectively, and let $m \leq n$. Then $p_{BA}(\lambda) = \lambda^{n-m} p_{AB}(\lambda)$.*

Proof. We give two proofs of this result. The first proof uses the previous Lemma, while the second proof makes use of Theorem 11.6.

First proof: This follows immediately from Lemma 11.1.

Second proof: This proof uses the rank normal form of matrices (Section 5.5). Recall from Theorem 5.10 that there exists nonsingular matrices, P and Q, such that

$$PAQ = \begin{bmatrix} I_r & O \\ O & O \end{bmatrix} ,$$

where the order of the identity matrix, r, is equal to rank of A and the zeros are of appropriate orders. Suppose that a conforming partition of $Q^{-1}BP^{-1}$ is

$$Q^{-1}BP^{-1} = \begin{bmatrix} C & D \\ E & F \end{bmatrix} ,$$

where C is $r \times r$. Then,

$$PABP^{-1} = PAQQ^{-1}BP^{-1} = \begin{bmatrix} I_r & O \\ O & O \end{bmatrix} \begin{bmatrix} C & D \\ E & F \end{bmatrix} = \begin{bmatrix} C & D \\ O & O \end{bmatrix} \text{ and}$$

$$Q^{-1}BAQ = Q^{-1}BP^{-1}PAQ = \begin{bmatrix} C & D \\ E & F \end{bmatrix} \begin{bmatrix} I_r & O \\ O & O \end{bmatrix} = \begin{bmatrix} C & O \\ E & O \end{bmatrix} .$$

Theorem 11.6 tells us that the characteristic polynomial of AB is the same as that of $PABP^{-1}$. Therefore,

$$p_{AB}(\lambda) = p_{PABP^{-1}}(\lambda) = \begin{vmatrix} \lambda I_r - C & -D \\ O & \lambda I_{m-r} \end{vmatrix} = |\lambda I_r - C| \lambda^{m-r} .$$

Analogously, we have the characteristic polynomial for BA as

$$p_{BA}(\lambda) = p_{Q^{-1}BAQ}(\lambda) = \begin{vmatrix} \lambda I_r - C & O \\ -E & \lambda I_{n-r} \end{vmatrix} = |\lambda I_r - C| \lambda^{n-r} .$$

Therefore,

$$p_{\boldsymbol{BA}}(\lambda) = \lambda^{n-r}|\lambda \boldsymbol{I}_r - \boldsymbol{C}| = \lambda^{n-r}\frac{p_{\boldsymbol{AB}}(\lambda)}{\lambda^{m-r}} = \lambda^{n-m}p_{\boldsymbol{AB}}(\lambda) ,$$

which completes the proof. □

If $m \neq n$ in the above theorem, i.e., neither $\boldsymbol{A}$ nor $\boldsymbol{B}$ is square, then the characteristic polynomials for $\boldsymbol{AB}$ and $\boldsymbol{BA}$ cannot be identical and the number of zero eigenvalues in $\boldsymbol{BA}$ is different from that in $\boldsymbol{AB}$. However, when $\boldsymbol{A}$ and $\boldsymbol{B}$ are both square matrices, we have the following corollary.

Corollary 11.1 *For any two $n \times n$ matrices $\boldsymbol{A}$ and $\boldsymbol{B}$, the characteristic polynomials of $\boldsymbol{AB}$ and $\boldsymbol{BA}$ are the same.*

Proof. Theorem 11.7 shows that $p_{\boldsymbol{BA}}(\lambda) = p_{\boldsymbol{AB}}(\lambda)$ when $m = n$. □

Theorem 11.7 implies that whenever $\boldsymbol{AB}$ is square, the nonzero eigenvalues of $\boldsymbol{AB}$ are the same as those of $\boldsymbol{BA}$—a result we had proved independently in Theorem 11.3.

11.3 Eigenspaces and multiplicities

For most of this book we have considered subspaces of $\Re^n$. However, when discussing subspaces associated with eigenvectors, it is better to consider subspaces of the n-dimensional complex field $\mathbb{C}^n$. If λ is an eigenvalue of $\boldsymbol{A}$, then $\mathcal{N}(\boldsymbol{A} - \lambda\boldsymbol{I})$ is a subspace of $\mathbb{C}^n$. In fact, $\mathcal{N}(\boldsymbol{A} - \lambda\boldsymbol{I})$ comprises all eigenvectors of $\boldsymbol{A}$ associated with the eigenvalue λ and the null vector $\boldsymbol{0}$.

Definition 11.4 *If λ is an eigenvalue of a square matrix $\boldsymbol{A}$, then $\mathcal{N}(\boldsymbol{A}-\lambda\boldsymbol{I})$ is called the* **eigenspace** *of $\boldsymbol{A}$ corresponding to λ. It is sometimes denoted by ES$(\boldsymbol{A}, \lambda)$.*

If $\boldsymbol{A}$ has a zero eigenvalue, then the eigenspace corresponding to the zero eigenvalue is simply $\mathcal{N}(\boldsymbol{A})$. Also, suppose $\lambda \neq 0$ is a nonzero eigenvalue of $\boldsymbol{A}$. Then,

$$\boldsymbol{x} \in \text{ES}(\boldsymbol{A}, \lambda) \Longrightarrow \boldsymbol{Ax} = \lambda\boldsymbol{x} \Longrightarrow \boldsymbol{x} = \boldsymbol{A}\left(\frac{1}{\lambda}\boldsymbol{x}\right) \Longrightarrow \boldsymbol{x} \in \mathcal{C}(\boldsymbol{A}) .$$

Another important concept associated with eigenvalues is that of *multiplicity*. Multiplicity helps us in understanding the total number of eigenvectors associated with an eigenvalue. There are two kinds of multiplicities that are usually associated with eigenvalues.

Definition 11.5 *Let λ be an eigenvalue of $\boldsymbol{A}$. The* **algebraic multiplicity** *of λ is the number of times it appears as a root of the characteristic polynomial of $\boldsymbol{A}$. We write the algebraic multiplicity of λ with respect to $\boldsymbol{A}$ as $AM_{\boldsymbol{A}}(\lambda)$. The* **geometric multiplicity** *of λ is the dimension of $\mathcal{N}(\boldsymbol{A} - \lambda\boldsymbol{I})$ or, equivalently, is equal to the* *dimension of the eigenspace associated with λ. We write the geometric multiplicity of λ with respect to $\boldsymbol{A}$ as $GM_{\boldsymbol{A}}(\lambda)$.*

Both algebraic and geometric multiplicity of an $n \times n$ matrix are integers between (including) 1 and n. Let us suppose that an $n \times n$ matrix $\boldsymbol{A}$ has r distinct eigenvalues $\{\lambda_1, \lambda_2, \ldots, \lambda_r\}$. Then, $\text{AM}_{\boldsymbol{A}}(\lambda_i) = m_i$, for $i = 1, 2, \ldots, r$ if and only if the characteristic polynomial of $\boldsymbol{A}$ can be factorized as

$$p_{\boldsymbol{A}}(\lambda) = (\lambda - \lambda_1)^{m_1}(\lambda - \lambda_2)^{m_2} \cdots (\lambda - \lambda_r)^{m_r} .$$

The geometric multiplicity of an eigenvalue λ is equal to the nullity of $\boldsymbol{A} - \lambda\boldsymbol{I}$.

The total number of linearly independent eigenvectors for a matrix is given by the sum of the geometric multiplicities. Over a complex vector space, the sum of the algebraic multiplicities will equal the dimension of the vector space. There is no apparent reason why these two multiplicities should be the same. In fact, they need not be equal and satisfy the following important inequality.

Theorem 11.8 *For any eigenvalue λ, the geometric multiplicity is less than or equal to its algebraic multiplicity.*

Proof. Let λ be an eigenvalue of the $n \times n$ matrix $\boldsymbol{A}$ such that $\text{GM}_{\boldsymbol{A}}(\lambda) = k$. This means that the dimension of the eigenspace $\text{ES}(\boldsymbol{A}, \lambda)$ is k. Let $\{\boldsymbol{x}_1, \boldsymbol{x}_2, \ldots, \boldsymbol{x}_k\}$ be a basis for the eigenspace corresponding to λ. Since $\{\boldsymbol{x}_1, \boldsymbol{x}_2, \ldots, \boldsymbol{x}_k\}$ is a linearly independent set in $\mathbb{C}^n$, it can be extended to a basis $\{\boldsymbol{x}_1, \boldsymbol{x}_2, \ldots, \boldsymbol{x}_k, \boldsymbol{x}_{k+1}, \ldots, \boldsymbol{x}_n\}$ for $\mathbb{C}^n$. Construct the $n \times n$ matrix $\boldsymbol{P} = [\boldsymbol{x}_1 : \boldsymbol{x}_2 : \ldots : \boldsymbol{x}_k : \boldsymbol{x}_{k+1} : \ldots : \boldsymbol{x}_n]$, which is clearly nonsingular. Then, pre-multiplying $\boldsymbol{A}$ by $\boldsymbol{P}^{-1}$ and post-multiplying by $\boldsymbol{P}$ yields

$$\begin{aligned}
\boldsymbol{P}^{-1}\boldsymbol{A}\boldsymbol{P} &= \boldsymbol{P}^{-1}[\boldsymbol{A}\boldsymbol{x}_1 : \boldsymbol{A}\boldsymbol{x}_2 : \ldots : \boldsymbol{A}\boldsymbol{x}_k : \boldsymbol{A}\boldsymbol{x}_{k+1} : \ldots : \boldsymbol{A}\boldsymbol{x}_n] \\
&= \boldsymbol{P}^{-1}[\lambda\boldsymbol{x}_1 : \lambda\boldsymbol{x}_2 : \ldots : \lambda\boldsymbol{x}_k : \boldsymbol{A}\boldsymbol{x}_{k+1} : \ldots : \boldsymbol{A}\boldsymbol{x}_n] \\
&= [\lambda\boldsymbol{e}_1 : \lambda\boldsymbol{e}_2 : \ldots : \lambda\boldsymbol{e}_k : \boldsymbol{P}^{-1}\boldsymbol{A}\boldsymbol{x}_{k+1} : \ldots : \boldsymbol{P}^{-1}\boldsymbol{A}\boldsymbol{x}_n] ,
\end{aligned}$$

where the last equality follows from the fact that $\boldsymbol{P}^{-1}\boldsymbol{x}_j = \boldsymbol{e}_j$, where $\boldsymbol{e}_j$ is the jth column of the $n \times n$ identity matrix. Therefore, we can write $\boldsymbol{P}^{-1}\boldsymbol{A}\boldsymbol{P}$ in partitioned form as

$$\boldsymbol{P}^{-1}\boldsymbol{A}\boldsymbol{P} = \begin{bmatrix} \lambda\boldsymbol{I}_k & \boldsymbol{B} \\ \boldsymbol{O} & \boldsymbol{D} \end{bmatrix} ,$$

where $\boldsymbol{B}$ is $k \times (n-k)$ and $\boldsymbol{D}$ is $(n-k) \times (n-k)$. Therefore, by Theorem 11.6,

$$p_{\boldsymbol{A}}(t) = p_{\boldsymbol{P}^{-1}\boldsymbol{A}\boldsymbol{P}}(t) = |t\boldsymbol{I} - \boldsymbol{P}^{-1}\boldsymbol{A}\boldsymbol{P}| = (t-\lambda)^k|t\boldsymbol{I}_{n-k} - \boldsymbol{D}| = (t-\lambda)^k p_{\boldsymbol{D}}(t) ,$$

which reveals that the algebraic multiplicity of λ must be at least k. Since $\text{GM}_{\boldsymbol{A}}(\lambda) = k$, we have proved that $\text{AM}_{\boldsymbol{A}}(\lambda) \geq \text{GM}_{\boldsymbol{A}}(\lambda)$. □

The following example shows that the algebraic multiplicity can be strictly greater than the geometric multiplicity.

Example 11.5 Consider the matrix

$$\boldsymbol{A} = \begin{bmatrix} 2 & 1 & 0 \\ 0 & 2 & 1 \\ 0 & 0 & 2 \end{bmatrix} .$$

The characteristic polynomial for $\boldsymbol{A}$ is

$$p_{\boldsymbol{A}}(t) = |t\boldsymbol{I} - \boldsymbol{A}| = \begin{vmatrix} t-2 & -1 & 0 \\ 0 & t-2 & -1 \\ 0 & 0 & t-2 \end{vmatrix} = (t-2)^3 .$$

Therefore, $\boldsymbol{A}$ has 2 as the only eigenvalue, repeated thrice as a root of the characteristic polynomial. So, $\text{AM}_{\boldsymbol{A}}(2) = 3$.

To find the geometric multiplicity of 2, we find the nullity of $\boldsymbol{A} - 2\boldsymbol{I}$. Note that

$$\boldsymbol{A} - 2\boldsymbol{I} = \begin{bmatrix} 0 & 1 & 0 \\ 0 & 0 & 1 \\ 0 & 0 & 0 \end{bmatrix} ,$$

from which it is clear that there are two linearly independent columns (rows) and so $\rho(\boldsymbol{A} - 2\boldsymbol{I}) = 2$. Therefore, from the Rank-Nullity Theorem we have

$$\text{GM}_{\boldsymbol{A}}(2) = \nu(\boldsymbol{A} - 2\boldsymbol{I}) = 3 - \rho(\boldsymbol{A} - 2\boldsymbol{I}) = 3 - 2 = 1 .$$

This shows that $\text{AM}_{\boldsymbol{A}}(2) > \text{GM}_{\boldsymbol{A}}(2)$. ■

Eigenvalues for which the algebraic and geometric multiplicities are the same are somewhat special and are accorded a special name.

Definition 11.6 *An eigenvalue λ of a matrix $\boldsymbol{A}$ is said to be* **regular** *or* **semisimple** *if $AM_{\boldsymbol{A}}(\lambda) = GM_{\boldsymbol{A}}(\lambda)$. An eigenvalue whose algebraic multiplicity is 1 is called* **simple**.

If λ is an eigenvalue with algebraic multiplicity equal to 1 (i.e., it appears only once as a root of the characteristic polynomial), then Theorem 11.8 tells us that the geometric multiplicity must also be equal to 1. Clearly, simple eigenvalues are regular eigenvalues.

Theoretical tools to help determine the number of linearly independent eigenvectors for a given matrix are sometimes useful. Recall from Theorem 11.2 that left and right eigenvectors corresponding to distinct eigenvalues are orthogonal (hence, linearly independent). What can we say about two (right) eigenvectors corresponding to two distinct eigenvalues? To answer this, let us suppose that λ_1 and λ_2 are two distinct eigenvalues of $\boldsymbol{A}$. Suppose that $\boldsymbol{x}_1$ and $\boldsymbol{x}_2$ are eigenvectors of $\boldsymbol{A}$ corresponding to λ_1 and λ_2, respectively. Consider the homogeneous equation

$$c_1\boldsymbol{x}_1 + c_2\boldsymbol{x}_2 = \boldsymbol{0} . \qquad (11.6)$$

Multiplying both sides of (11.6) by $\boldsymbol{A}$ yields $c_1\lambda_1\boldsymbol{x}_1 + c_2\lambda_2\boldsymbol{x}_2 = \boldsymbol{0}$ and multiplying both sides of (11.6) by λ_2 yields $c_1\lambda_2\boldsymbol{x}_1 + c_2\lambda_2\boldsymbol{x}_2 = \boldsymbol{0}$. Subtracting the latter from the former yields

$$(c_1\lambda_1\boldsymbol{x}_1 + c_2\lambda_2\boldsymbol{x}_2) - (c_1\lambda_2\boldsymbol{x}_1 + c_2\lambda_2\boldsymbol{x}_2) = \mathbf{0} - \mathbf{0}$$
$$\Longrightarrow c_1(\lambda_1 - \lambda_2)\boldsymbol{x}_1 = \mathbf{0} \Longrightarrow c_1 = 0\ ,$$

where the last implication follows from the facts that $\boldsymbol{x}_1 \neq 0$ and $\lambda_1 \neq \lambda_2$. Once we are forced to conclude that $c_1 = 0$, it follows that $c_2 = 0$ because $\boldsymbol{x}_2 \neq \mathbf{0}$. Therefore, $c_1\boldsymbol{x}_1 + c_2\boldsymbol{x}_2 = \mathbf{0}$ implies that $c_1 = c_2 = 0$, so the eigenvectors $\boldsymbol{x}_1$ and $\boldsymbol{x}_2$ must be linearly independent. The following theorem generalizes this observation.

Theorem 11.9 *Let $\{\lambda_1, \lambda_2, \ldots, \lambda_k\}$ be a set of distinct eigenvalues of $\boldsymbol{A}$:*

(i) If $\boldsymbol{x}_i$ is an eigenvector of $\boldsymbol{A}$ associated with λ_i, for $i = 1, 2, \ldots, k$, then $\{\boldsymbol{x}_1, \boldsymbol{x}_2, \ldots, \boldsymbol{x}_k\}$ is a linearly independent set.

(ii) The sum of eigenspaces associated with distinct eigenvalues is direct:

$$\mathcal{N}(\boldsymbol{A} - \lambda_1\boldsymbol{I}) \oplus \mathcal{N}(\boldsymbol{A} - \lambda_2\boldsymbol{I}) \oplus \cdots \oplus \mathcal{N}(\boldsymbol{A} - \lambda_k\boldsymbol{I})\ .$$

Proof. **Proof of (i):** Suppose, if possible, that the result is false and the set $\{\boldsymbol{x}_1, \boldsymbol{x}_2, \ldots, \boldsymbol{x}_k\}$ is linearly dependent. Let $r < k$ be the largest integer for which $\{\boldsymbol{x}_1, \boldsymbol{x}_2, \ldots, \boldsymbol{x}_r\}$ is a linearly independent set. Then, $\boldsymbol{x}_{r+1}$ can be expressed as a linear combination of $\boldsymbol{x}_1, \boldsymbol{x}_2, \ldots, \boldsymbol{x}_r$, which we write as

$$\boldsymbol{x}_{r+1} = \alpha_1\boldsymbol{x}_1 + \alpha_2\boldsymbol{x}_2 + \cdots + \alpha_r\boldsymbol{x}_r = \sum_{i=1}^{r} \alpha_i\boldsymbol{x}_i\ .$$

Multiplying both sides by $\boldsymbol{A} - \lambda_{r+1}\boldsymbol{I}$ from the left produces

$$(\boldsymbol{A} - \lambda_{r+1})\boldsymbol{x}_{r+1} = \sum_{i=1}^{r}(\boldsymbol{A} - \lambda_{r+1}\boldsymbol{I})\boldsymbol{x}_i\ . \tag{11.7}$$

Since $\boldsymbol{x}_{r+1}$ is an eigenvector associated with λ_{r+1}, the left hand side of the above equation is zero, i.e., $(\boldsymbol{A} - \lambda_{r+1}\boldsymbol{I})\boldsymbol{x}_{r+1} = \mathbf{0}$. Also,

$$(\boldsymbol{A} - \lambda_{r+1}\boldsymbol{I})\boldsymbol{x}_i = \boldsymbol{A}\boldsymbol{x}_i - \lambda_{r+1}\boldsymbol{x}_i = (\lambda_i - \lambda_{r+1})\boldsymbol{x}_i \ \text{ for } i = 1, 2, \ldots, r\ .$$

This means that (11.7) can be written as

$$\mathbf{0} = \sum_{i=1}^{r} \alpha_i(\lambda_i - \lambda_{r+1})\boldsymbol{x}_i\ .$$

Since $\{\boldsymbol{x}_1, \boldsymbol{x}_2, \ldots, \boldsymbol{x}_r\}$ is a linearly independent set, the above implies that $\alpha_i(\lambda_i - \lambda_{r+1}) = 0$ for $i = 1, 2, \ldots, r$. Therefore, each $\alpha_i = 0$, for $i = 1, 2, \ldots, r$, because $\{\lambda_1, \lambda_2, \ldots, \lambda_r\}$ is a set of distinct eigenvalues. But this would mean that $\boldsymbol{x}_{r+1} = \mathbf{0}$, which contradicts the fact that the $\boldsymbol{x}_i$'s are all nonzero as they are eigenvectors (recall Definition 11.1). This proves part (i).

Proof of (ii): By virtue of part (ii) of Theorem 6.3, it is enough to show that

$$\mathcal{V}_{i+1} = [\mathrm{ES}(\boldsymbol{A}, \lambda_1) + \mathrm{ES}(\boldsymbol{A}, \lambda_2) + \cdots + \mathrm{ES}(\boldsymbol{A}, \lambda_i)] \cap \mathrm{ES}(\boldsymbol{A}, \lambda_{i+1}) = \{\mathbf{0}\}$$

for $i = 0, 1, \ldots, k-1$. Suppose, if possible, there exists a nonzero vector $\boldsymbol{x} \in \mathcal{V}_{i+1}$. Then, $\boldsymbol{x} = \boldsymbol{u}_1 + \boldsymbol{u}_2 + \cdots + \boldsymbol{u}_i$, where $\boldsymbol{u}_j \in \text{ES}(\boldsymbol{A}, \lambda_j) = \mathcal{N}(\boldsymbol{A} - \lambda_j \boldsymbol{I})$ for $j = 1, 2, \ldots, i$. Also, $\boldsymbol{x}$ is an eigenvector corresponding to λ_{i+1} so $\boldsymbol{A}\boldsymbol{x} = \lambda_{i+1}\boldsymbol{x}$. Therefore,

$$\boldsymbol{0} = \boldsymbol{A}\boldsymbol{x} - \lambda_{i+1}\boldsymbol{x} = \sum_{j=1}^{i} \boldsymbol{A}\boldsymbol{u}_j - \lambda_{i+1} \sum_{j=1}^{i} \boldsymbol{u}_j = \sum_{j=1}^{i} (\lambda_j - \lambda_{i+1})\boldsymbol{u}_j \, .$$

Since each $\boldsymbol{u}_j$ corresponds to a different eigenvalue, the set $\{\boldsymbol{u}_1, \boldsymbol{u}_2, \ldots, \boldsymbol{u}_i\}$ is linearly independent (by part (i)). But this implies that $\lambda_j - \lambda_{i+1} = 0$ for $j = 1, 2, \ldots, i$, which contradicts the fact that the eigenvalues are distinct. Therefore, $\boldsymbol{x} = \boldsymbol{0}$ and indeed $\mathcal{V}_{i+1} = \{\boldsymbol{0}\}$ for each $i = 0, 1, \ldots, k-1$. This proves part (ii). □

Earlier we proved that if $\boldsymbol{AB}$ is a square matrix, then every nonzero eigenvalue of $\boldsymbol{AB}$ is also an eigenvalue of $\boldsymbol{BA}$ (see Theorem 11.3). Theorem 11.7 established an even stronger result: not only is every nonzero eigenvalue of $\boldsymbol{AB}$ also an eigenvalue of $\boldsymbol{BA}$, it also has the same algebraic multiplicity. We now show that its geometric multiplicity also remains the same.

Theorem 11.10 *Let $\boldsymbol{AB}$ and $\boldsymbol{BA}$ be square matrices, where $\boldsymbol{A}$ and $\boldsymbol{B}$ need not be square. If λ is a nonzero eigenvalue of $\boldsymbol{AB}$, then it is also an eigenvalue of $\boldsymbol{BA}$ with the same geometric multiplicity.*

Proof. Let $\{\boldsymbol{x}_1, \boldsymbol{x}_2, \ldots, \boldsymbol{x}_r\}$ be a basis for the eigenspace $\text{ES}(\boldsymbol{AB}, \lambda) = \mathcal{N}(\boldsymbol{AB} - \lambda\boldsymbol{I})$. This means that $\text{GM}_{\boldsymbol{AB}}(\lambda) = r$. Since each $\boldsymbol{x}_i \in \mathcal{N}(\boldsymbol{AB} - \lambda\boldsymbol{I})$, we can conclude that

$$\boldsymbol{AB}\boldsymbol{x}_i = \lambda\boldsymbol{x}_i \Longrightarrow \boldsymbol{BAB}\boldsymbol{x}_i = \lambda\boldsymbol{B}\boldsymbol{x}_i \Longrightarrow \boldsymbol{BA}\boldsymbol{u}_i = \lambda\boldsymbol{u}_i \;\text{ for }\; i = 1, 2, \ldots, r \, ,$$

where $\boldsymbol{u}_i = \boldsymbol{B}\boldsymbol{x}_i$. Note that each $\boldsymbol{u}_i$ is an eigenvector of $\boldsymbol{BA}$ associated with eigenvalue λ. We now show that $\{\boldsymbol{u}_1, \boldsymbol{u}_2, \ldots, \boldsymbol{u}_r\}$ is a linearly independent set.

Consider the homogeneous system

$$\alpha_1\boldsymbol{u}_1 + \alpha_2\boldsymbol{u}_2 + \cdots \alpha_r\boldsymbol{u}_r = \boldsymbol{0}$$

and note that

$$\sum_{i=1}^{r} \alpha_i\boldsymbol{u}_i = \boldsymbol{0} \Longrightarrow \boldsymbol{A}(\sum_{i=1}^{r} \alpha_i\boldsymbol{u}_i) = \boldsymbol{0} \Longrightarrow \sum_{i=1}^{r} \alpha_i\boldsymbol{A}\boldsymbol{u}_i = \boldsymbol{0} \Longrightarrow \sum_{i=1}^{r} \alpha_i\boldsymbol{AB}\boldsymbol{x}_i = \boldsymbol{0}$$

$$\Longrightarrow \lambda\sum_{i=1}^{r} \alpha_i\boldsymbol{x}_i = \boldsymbol{0} \Longrightarrow \sum_{i=1}^{r} \alpha_i\boldsymbol{x}_i = \boldsymbol{0} \Longrightarrow \alpha_i = 0 \;\text{ for }\; i = 1, 2, \ldots, r \, ,$$

where the last two implications follow from the facts that $\lambda \neq 0$ and that the $\boldsymbol{x}_i$'s are linearly independent. This proves that the $\boldsymbol{u}_i$'s are linearly independent. Therefore, there are at least r linearly independent eigenvectors of $\boldsymbol{BA}$ associated with λ, where $r = \text{GM}_{\boldsymbol{AB}}(\lambda)$. This proves that $\text{GM}_{\boldsymbol{BA}}(\lambda) \geq \text{GM}_{\boldsymbol{AB}}(\lambda)$. The reverse inequality follows by symmetry and we conclude that $\text{GM}_{\boldsymbol{BA}}(\lambda) = \text{GM}_{\boldsymbol{AB}}(\lambda)$. □

11.4 Diagonalizable matrices

Recall from Section 4.9 that two $n \times n$ matrices $\boldsymbol{A}$ and $\boldsymbol{B}$ are said to be similar whenever there exists a nonsingular matrix $\boldsymbol{P}$ such that $\boldsymbol{B} = \boldsymbol{P}^{-1}\boldsymbol{A}\boldsymbol{P}$. As mentioned in Section 4.9, similar matrices can be useful because they lead to simpler structures. In fact, a fundamental objective of linear algebra is to reduce a square matrix to the simplest possible form by means of a similarity transformation.

Since diagonal matrices are the simplest, it is natural to explore if a given square matrix is similar to a diagonal matrix. Why would this be useful? Consider an $n \times n$ matrix $\boldsymbol{A}$ and suppose, if possible, that $\boldsymbol{A}$ is similar to a diagonal matrix. Then, we would be able to find a nonsingular matrix $\boldsymbol{P}$ such that $\boldsymbol{P}^{-1}\boldsymbol{A}\boldsymbol{P} = \boldsymbol{D}$, where $\boldsymbol{D}$ is a diagonal matrix. A primary advantage of having such a relationship is that it allows us to easily find powers of $\boldsymbol{A}$, such as $\boldsymbol{A}^k$. This is because

$$\begin{aligned}\boldsymbol{P}^{-1}\boldsymbol{A}\boldsymbol{P} = \boldsymbol{D} &\Longrightarrow (\boldsymbol{P}^{-1}\boldsymbol{A}\boldsymbol{P})(\boldsymbol{P}^{-1}\boldsymbol{A}\boldsymbol{P}) = \boldsymbol{D}^2 \Longrightarrow \boldsymbol{P}^{-1}\boldsymbol{A}^2\boldsymbol{P} = \boldsymbol{D}^2\\ &\Longrightarrow (\boldsymbol{P}^{-1}\boldsymbol{A}\boldsymbol{P})(\boldsymbol{P}^{-1}\boldsymbol{A}^2\boldsymbol{P}) = (\boldsymbol{D})(\boldsymbol{D}^2) \Longrightarrow \boldsymbol{P}^{-1}\boldsymbol{A}^3\boldsymbol{P} = \boldsymbol{D}^3\\ \cdots \quad &\cdots \Longrightarrow \boldsymbol{P}^{-1}\boldsymbol{A}^k\boldsymbol{P} = \boldsymbol{D}^k \Longrightarrow \boldsymbol{A}^k = \boldsymbol{P}\boldsymbol{D}^k\boldsymbol{P}^{-1}\,.\end{aligned}$$

$\boldsymbol{D}^k$ is easy to compute—it is simply the diagonal matrix with k-th powers of the diagonal elements of $\boldsymbol{D}$—so $\boldsymbol{A}^k$ is also easy to compute. This is useful in matrix analysis, where one is often interested in finding limits of powers of matrices (e.g., in studying iterative linear systems).

Definition 11.7 *A square matrix $\boldsymbol{A}$ is called* **diagonalizable** *or* **semisimple** *if it is similar to a diagonal matrix, i.e., if there exists a nonsingular matrix $\boldsymbol{P}$ such that $\boldsymbol{P}^{-1}\boldsymbol{A}\boldsymbol{P}$ is a diagonal matrix.*

Unfortunately, not every matrix is diagonalizable as seen in the example below.

Example 11.6 A counter example can be constructed from the preceding observation regarding powers of a matrix. Suppose there is an $n \times n$ matrix $\boldsymbol{A} \neq \boldsymbol{O}$ such that $\boldsymbol{A}^n = \boldsymbol{O}$. If such a matrix were similar to a diagonal matrix, there would exist a nonsingular $\boldsymbol{P}$ such that $\boldsymbol{P}^{-1}\boldsymbol{A}^n\boldsymbol{P} = \boldsymbol{D}^n$ and we could conclude that

$$\begin{aligned}\boldsymbol{A}^n = \boldsymbol{O} &\Longrightarrow \boldsymbol{P}\boldsymbol{D}^n\boldsymbol{P}^{-1} = \boldsymbol{O} \Longrightarrow \boldsymbol{D}^n = \boldsymbol{O} \Longrightarrow \boldsymbol{D} = \boldsymbol{O}\\ &\Longrightarrow \boldsymbol{P}^{-1}\boldsymbol{A}\boldsymbol{P} = \boldsymbol{O} \Longrightarrow \boldsymbol{A} = \boldsymbol{O}\,.\end{aligned}$$

Here is such a matrix:

$$\boldsymbol{A} = \begin{bmatrix} 0 & 1 \\ 0 & 0 \end{bmatrix}.$$

It is easy to see that $\boldsymbol{A}^2 = \boldsymbol{O}$. So, the above argument shows that $\boldsymbol{A}$ cannot be similar to a diagonal matrix. ■

So, what does it take for a matrix to be diagonalizable? Suppose that $\boldsymbol{A}$ is an $n \times n$

diagonalizable matrix and

$$P^{-1}AP = \Lambda = \begin{bmatrix} \lambda_1 & 0 & \dots & 0 \\ 0 & \lambda_2 & \dots & 0 \\ \vdots & \vdots & \ddots & \vdots \\ 0 & 0 & \dots & \lambda_n \end{bmatrix} .$$

Then $AP = P\Lambda$, so $Ap_{*j} = \lambda_j p_{*j}$ for $j = 1, 2, \dots, n$. Therefore, each p_{*j} is an eigenvector of A corresponding to the eigenvalue λ_j. Since P is nonsingular, its columns are linearly independent, which implies that P is a matrix whose columns constitute a set of n linearly independent eigenvectors and Λ is a diagonal matrix whose diagonal entries are the corresponding eigenvalues. Conversely, if A has n linearly independent eigenvectors, then we can construct an $n \times n$ nonsingular matrix P by placing the n linearly independent eigenvectors as its columns. This matrix P must satisfy $AP = \Lambda P$, where Λ is a diagonal matrix whose diagonal entries are the corresponding eigenvalues. Therefore, $P^{-1}AP = \Lambda$ and A is diagonalizable.

Let A be diagonalizable and $P^{-1}AP = \Lambda$, where Λ is diagonal with λ_i as the i-th diagonal element. If $p_1, p_2, \dots, p_n$ are the columns of P and $q_1', q_2', \dots, q_n'$ are the rows of P^{-1}, then $A = P\Lambda P^{-1} = \lambda_1 p_1 q_1' + \lambda_2 p_2 q_2' + \dots + \lambda_n p_n q_n'$ is an outer product expansion of A. Recall the definition of left and right eigenvectors in Definition 11.2. Since $AP = P\Lambda$, each column of P is a right eigenvector of A. Since $P^{-1}A = \Lambda P^{-1}$, $q_i'A = \lambda_i q_i'$ for $i = 1, 2, \dots, n$. Therefore, each row of P^{-1} is a left eigenvector of A. Here, the λ_i's need not be distinct (i.e., we count multiplicities) nor be nonzero (i.e., some of the eigenvalues can be zero).

We now turn to some other characterizations of diagonalizable matrices.

Theorem 11.11 *The following statements about an $n \times n$ matrix A are equivalent:*

(i) A is diagonalizable.

(ii) $AM_A(\lambda) = GM_A(\lambda)$ for every eigenvalue λ of A. In other words, every eigenvalue of A is regular.

(iii) If $\{\lambda_1, \lambda_2, \dots, \lambda_k\}$ is the complete set of distinct eigenvalues of A, then

$$\mathbb{C}^n = \mathcal{N}(A - \lambda_1 I) \oplus \mathcal{N}(A - \lambda_2 I) \oplus \cdots \oplus \mathcal{N}(A - \lambda_k I) .$$

(iv) A has n linearly independent eigenvectors.

Proof. We will prove that (i) $\Rightarrow$ (ii) $\Rightarrow$ (iii) $\Rightarrow$ (iv) $\Rightarrow$ (i). **Proof of (i) $\Rightarrow$ (ii):** Let λ be an eigenvalue for A with $\text{AM}_A(\lambda) = a$. If A is diagonalizable, there is a nonsingular matrix P such that

$$P^{-1}AP = \Lambda = \begin{bmatrix} \lambda I_a & O \\ O & \Lambda_{22} \end{bmatrix} ,$$

where Λ is diagonal and λ is *not* an eigenvalue of Λ_{22}. Note that

$$P^{-1}(A - \lambda I)P = P^{-1}AP - \lambda I = \begin{bmatrix} O & O \\ O & \Lambda_{22} - \lambda I_{n-a} \end{bmatrix} ,$$

where $\boldsymbol{\Lambda}_{22} - \lambda \boldsymbol{I}_{n-a}$ is an $(n-a) \times (n-a)$ diagonal matrix with all its diagonal elements nonzero. The geometric multiplicity of λ is the dimension of the null space of $\boldsymbol{A} - \lambda \boldsymbol{I}$, which is seen to be

$$\begin{aligned} \text{GM}_{\boldsymbol{A}}(\lambda) &= \dim(\mathcal{N}(\boldsymbol{A} - \lambda \boldsymbol{I})) = n - \rho(\boldsymbol{A} - \lambda \boldsymbol{I}) = n - \rho(\boldsymbol{\Lambda}_{22} - \lambda \boldsymbol{I}_{n-a}) \\ &= n - (n-a) = a \ , \end{aligned}$$

where we have used the fact that the rank of $\boldsymbol{A} - \lambda \boldsymbol{I}$ is equal to the rank of $\boldsymbol{\Lambda}_{22} - \lambda \boldsymbol{I}_{n-a}$, which is $n - a$. Therefore, $\text{GM}_{\boldsymbol{A}}(\lambda) = a = \text{AM}_{\boldsymbol{A}}(\lambda)$ for any eigenvalue λ of $\boldsymbol{A}$, which establishes (i) $\Rightarrow$ (ii).

Proof of (ii) $\Rightarrow$ (iii): Let $\text{AM}_{\boldsymbol{A}}(\lambda_i) = a_i$ for $i = 1, 2, \ldots, k$. The algebraic multiplicities of all distinct eigenvalues always add up to n, so $\sum_{i=1}^{k} a_i = n$. By the hypothesis of (ii), the arithmetic and geometric multiplicity for each eigenvalue is the same, so $\dim(\mathcal{N}(\boldsymbol{A} - \lambda_i \boldsymbol{I})) = \text{GM}_{\boldsymbol{A}}(\lambda_i) = a_i$ and

$$n = \dim(\mathcal{N}(\boldsymbol{A} - \lambda_1 \boldsymbol{I})) + \dim(\mathcal{N}(\boldsymbol{A} - \lambda_2 \boldsymbol{I})) + \cdots + \dim(\mathcal{N}(\boldsymbol{A} - \lambda_k \boldsymbol{I})) \ .$$

Since the sum of the null spaces $\mathcal{N}(\boldsymbol{A} - \lambda_j \boldsymbol{I})$ for distinct eigenvalues is direct (part (ii) of Theorem 11.9), the result follows.

Proof of (iii) $\Rightarrow$ (iv): This follows immediately. If $\mathcal{B}_i$ is a basis for $\mathcal{N}(\boldsymbol{A} - \lambda_i \boldsymbol{I})$, then $\mathcal{B}_1 \cup \mathcal{B}_2 \cup \cdots \cup \mathcal{B}_k$ is a basis for $\mathbb{C}^n$ comprising entirely of eigenvectors of $\boldsymbol{A}$. Therefore, $\boldsymbol{A}$ must have n linearly independent eigenvectors.

Proof of (iv) $\Rightarrow$ (i): Suppose that $\{\boldsymbol{p}_1, \boldsymbol{p}_2, \ldots, \boldsymbol{p}_n\}$ be a linearly independent set of n eigenvectors of $\boldsymbol{A}$. Therefore, $\boldsymbol{A}\boldsymbol{p}_i = \lambda_i \boldsymbol{p}_i$ for $i = 1, 2, \ldots, n$, where λ_i's are the eigenvalues (not necessarily distinct) of $\boldsymbol{A}$. If we construct an $n \times n$ matrix $\boldsymbol{P}$ by placing $\boldsymbol{p}_i$ as its i-th column, $\boldsymbol{P}$ is clearly nonsingular and it follows that

$$\begin{aligned} \boldsymbol{A}\boldsymbol{P} &= [\boldsymbol{A}\boldsymbol{p}_1 : \boldsymbol{A}\boldsymbol{p}_2 : \ldots : \boldsymbol{A}\boldsymbol{p}_n] = [\lambda_1 \boldsymbol{p}_1 : \lambda_2 \boldsymbol{p}_2 : \ldots : \lambda_n \boldsymbol{p}_n] \\ &= [\boldsymbol{p}_1 : \boldsymbol{p}_2 : \ldots : \boldsymbol{p}_n] \begin{bmatrix} \lambda_1 & 0 & \ldots & 0 \\ 0 & \lambda_2 & \ldots & 0 \\ \vdots & \vdots & \ddots & \vdots \\ 0 & 0 & \ldots & \lambda_n \end{bmatrix} = \boldsymbol{P}\boldsymbol{\Lambda} \Longrightarrow \boldsymbol{P}^{-1}\boldsymbol{A}\boldsymbol{P} = \boldsymbol{\Lambda} \ , \end{aligned}$$

which shows that $\boldsymbol{A}$ is diagonalizable. □

The following corollary is useful.

Corollary 11.2 *An $n \times n$ matrix with n distinct eigenvalues is diagonalizable.*

Proof. The n eigenvectors corresponding to the n distinct eigenvalues are linearly independent. □

Note that if none of the eigenvalues of $\boldsymbol{A}$ is repeated, then the algebraic multiplicity of each eigenvalue is 1. That is, each eigenvalue is simple and, therefore, regular. This provides another reason why $\boldsymbol{A}$ will be diagonalizable.

The above characterizations can sometimes be useful to detect if a matrix is diagonalizable or not. For example, consider the matrix in Example 11.5. Since the algebraic multiplicity of that matrix is strictly less than its geometric multiplicity, that matrix is not diagonalizable.

The following theorem provides some alternative representations for diagonalizable matrices.

Theorem 11.12 *Let $\boldsymbol{A}$ be an $n \times n$ diagonalizable matrix of rank r.*

(i) There exists an $n \times n$ nonsingular matrix $\boldsymbol{V}$ such that

$$\boldsymbol{A} = \boldsymbol{V} \begin{bmatrix} \boldsymbol{\Lambda}_1 & \boldsymbol{O} \\ \boldsymbol{O} & \boldsymbol{O} \end{bmatrix} \boldsymbol{V}^{-1} ,$$

where $\boldsymbol{\Lambda}_1$ is an $r \times r$ diagonal matrix with r nonzero diagonal elements.

(ii) There exist r nonzero scalars $\lambda_1, \lambda_2, \ldots, \lambda_r$ such that

$$\boldsymbol{A} = \lambda_1 \boldsymbol{v}_1 \boldsymbol{w}_1' + \lambda_2 \boldsymbol{v}_2 \boldsymbol{w}_2' + \cdots + \lambda_r \boldsymbol{v}_r \boldsymbol{w}_r' ,$$

where $\boldsymbol{v}_1, \boldsymbol{v}_2, \ldots, \boldsymbol{v}_r$ and $\boldsymbol{w}_1, \boldsymbol{w}_2, \ldots, \boldsymbol{w}_r$ are vectors in $\mathbb{C}^n$ such that $\boldsymbol{w}_i'\boldsymbol{v}_i = 1$ and $\boldsymbol{w}_i'\boldsymbol{v}_j = 0$ whenever $i \neq j$ for $i, j = 1, 2, \ldots, r$.

(iii) There exists an $n \times r$ matrix $\boldsymbol{S}$, an $r \times n$ matrix $\boldsymbol{W}'$ and an $r \times r$ diagonal matrix $\boldsymbol{\Lambda}_1$ such that

$$\boldsymbol{A} = \boldsymbol{S}\boldsymbol{\Lambda}_1\boldsymbol{W}' , \quad \text{where} \quad \boldsymbol{W}'\boldsymbol{S} = \boldsymbol{I}_r .$$

Proof. **Proof of (i):** Since $\boldsymbol{A}$ is diagonalizable, there exists an $n \times n$ nonsingular matrix $\boldsymbol{P}$ such that $\boldsymbol{P}^{-1}\boldsymbol{A}\boldsymbol{P} = \boldsymbol{\Lambda}$, where $\boldsymbol{\Lambda}$ is an $n \times n$ diagonal matrix. Clearly $\rho(\boldsymbol{A}) = \rho(\boldsymbol{\Lambda}) = r$, which means that exactly r of the diagonal elements of $\boldsymbol{\Lambda}$ are nonzero. By applying the same permutation to the rows and columns of $\boldsymbol{\Lambda}$, we can bring the r nonzero elements of $\boldsymbol{\Lambda}$ to the first r positions on the diagonal. This means that there exists a permutation matrix $\boldsymbol{Q}$ such that

$$\boldsymbol{\Lambda} = \boldsymbol{Q} \begin{bmatrix} \boldsymbol{\Lambda}_1 & \boldsymbol{O} \\ \boldsymbol{O} & \boldsymbol{O} \end{bmatrix} \boldsymbol{Q}^{-1} .$$

Let $\boldsymbol{V} = \boldsymbol{P}\boldsymbol{Q}$. Clearly $\boldsymbol{V}$ is nonsingular with $\boldsymbol{V}^{-1} = \boldsymbol{Q}^{-1}\boldsymbol{P}^{-1}$ and

$$\boldsymbol{A} = \boldsymbol{P}\boldsymbol{\Lambda}\boldsymbol{P}^{-1} = \boldsymbol{P}\boldsymbol{Q} \begin{bmatrix} \boldsymbol{\Lambda}_1 & \boldsymbol{O} \\ \boldsymbol{O} & \boldsymbol{O} \end{bmatrix} \boldsymbol{Q}^{-1}\boldsymbol{P}^{-1} = \boldsymbol{V} \begin{bmatrix} \boldsymbol{\Lambda}_1 & \boldsymbol{O} \\ \boldsymbol{O} & \boldsymbol{O} \end{bmatrix} \boldsymbol{V}^{-1} .$$

Proof of (ii): We will use the representation derived above in (i). Let $\lambda_1, \lambda_2, \ldots, \lambda_r$ be the r nonzero diagonal elements in $\boldsymbol{\Lambda}_1$. Let $\boldsymbol{v}_1, \boldsymbol{v}_2, \ldots, \boldsymbol{v}_r$ be the first r columns in $\boldsymbol{V}$ and let $\boldsymbol{w}_1', \boldsymbol{w}_2', \ldots, \boldsymbol{w}_r'$ be the first r rows in $\boldsymbol{V}^{-1}$. Therefore,

$$\boldsymbol{A} = \boldsymbol{V} \begin{bmatrix} \boldsymbol{\Lambda}_1 & \boldsymbol{O} \\ \boldsymbol{O} & \boldsymbol{O} \end{bmatrix} \boldsymbol{V}^{-1} = \lambda_1 \boldsymbol{v}_1 \boldsymbol{w}_1' + \lambda_2 \boldsymbol{v}_2 \boldsymbol{w}_2' + \cdots + \lambda_r \boldsymbol{v}_r \boldsymbol{w}_r' .$$

Also, $\boldsymbol{w}_i'\boldsymbol{v}_j$ is the (i, j)-th element in $\boldsymbol{V}^{-1}\boldsymbol{V} = \boldsymbol{I}_n$, which implies that $\boldsymbol{w}_i'\boldsymbol{v}_i = 1$ and $\boldsymbol{w}_i'\boldsymbol{v}_j = 0$ whenever $i \neq j$ for $i, j = 1, 2, \ldots, r$.

Proof of (iii): Let us use the representation derived above in (ii). Let

$$S = [v_1 : v_2 : \ldots : v_r] \text{ and } W = [w_1 : w_2 : \ldots : w_r] .$$

Then, $W'S = I_r$ and $A = \sum_{i=1}^{r} \lambda_i v_i w_i' = S\Lambda_1 W'$. □

The following theorem is known as the ***spectral representation theorem***.

Theorem 11.13 *Let A be an $n \times n$ matrix with k distinct eigenvalues $\{\lambda_1, \lambda_2, \ldots, \lambda_k\}$. Then A is diagonalizable if and only if there exist $n \times n$ nonnull matrices $G_1, G_2, \ldots, G_k$ satisfying the following four properties:*

(i) $A = \lambda_1 G_1 + \lambda_2 G_2 + \cdots + \lambda_k G_k$;

(ii) $G_i G_j = O$ *whenever* $i \neq j$;

(iii) $G_1 + G_2 + \cdots + G_k = I$;

(iv) G_i *is the projector onto* $\mathcal{N}(A - \lambda_i I)$ *along* $\mathcal{C}(A - \lambda_i I)$.

Proof. We will first prove the *"only if"* part. Suppose A is diagonalizable and let r_i be the multiplicity (geometric and algebraic are the same here) of λ_i. Without loss of generality, we may assume that there exists an $n \times n$ nonsingular matrix P such that

$$P^{-1}AP = \Lambda = \begin{bmatrix} \lambda_1 I_{r_1} & O & \ldots & O \\ O & \lambda_2 I_{r_2} & \ldots & O \\ \vdots & \vdots & \ddots & \vdots \\ O & O & \ldots & \lambda_k I_{r_k} \end{bmatrix} .$$

Let us partition P and P^{-1} as

$$P = [P_1 : P_2 : \ldots : P_k] \text{ and } P^{-1} = Q = \begin{bmatrix} Q_1' \\ Q_2' \\ \vdots \\ Q_k' \end{bmatrix} , \tag{11.8}$$

where each Q_i' is $r_i \times n$ and P_i is an $n \times r_i$ matrix whose columns constitute a basis for $\mathcal{N}(A - \lambda_i I)$. From Theorem 11.11 we know that $\sum_{i=1}^{k} r_i = n$. Then, $P = [P_1 : P_2 : \ldots : P_k]$ is an $n \times n$ nonsingular matrix and

$$A = P\Lambda P^{-1} = [P_1 : P_2 : \ldots : P_k] \begin{bmatrix} \lambda_1 I_{r_1} & O & \ldots & O \\ O & \lambda_2 I_{r_2} & \ldots & O \\ \vdots & \vdots & \ddots & \vdots \\ O & O & \ldots & \lambda_k I_{r_k} \end{bmatrix} \begin{bmatrix} Q_1' \\ Q_2' \\ \vdots \\ Q_k' \end{bmatrix}$$
$$= \lambda_1 P_1 Q_1' + \lambda_2 P_2 Q_2' + \cdots + \lambda_k P_k Q_k' = \lambda_1 G_1 + \lambda_2 G_2 + \cdots + \lambda_k G_k ,$$

where each $G_i = P_i Q_i'$ is $n \times n$ for $i = 1, 2, \ldots, k$. This proves (i).

The way we have partitioned P and P^{-1}, it follows that $Q_i' P_j$ is the $r_i \times r_j$ matrix that forms the (i, j)-th block of $\boldsymbol{P}^{-1}\boldsymbol{P} = \boldsymbol{I}_n$. This means that $\boldsymbol{Q}_i'\boldsymbol{P}_j = \boldsymbol{O}$ whenever $i \neq j$ and so

$$\boldsymbol{G}_i\boldsymbol{G}_j = \boldsymbol{P}_i\boldsymbol{Q}_i'\boldsymbol{P}_j\boldsymbol{Q}_j' = \boldsymbol{P}_i(\boldsymbol{O})\boldsymbol{Q}_j' = \boldsymbol{O} \text{ whenever } i \neq j .$$

This establishes property (ii).

Property (iii) follows immediately from

$$\boldsymbol{G}_1 + \boldsymbol{G}_2 + \cdots + \boldsymbol{G}_k = \boldsymbol{P}_1\boldsymbol{Q}_1' + \boldsymbol{P}_2\boldsymbol{Q}_2' + \cdots + \boldsymbol{P}_k\boldsymbol{Q}_k' = \boldsymbol{P}\boldsymbol{P}^{-1} = \boldsymbol{I} .$$

For property (iv), first note that $\boldsymbol{Q}_i'\boldsymbol{P}_i$ is the i-th diagonal block in $\boldsymbol{P}^{-1}\boldsymbol{P} = \boldsymbol{I}_n$ and so $\boldsymbol{Q}_i'\boldsymbol{P}_i = \boldsymbol{I}_{r_i}$. Therefore,

$$\boldsymbol{G}_i^2 = \boldsymbol{P}_i\boldsymbol{Q}_i'\boldsymbol{P}_i\boldsymbol{Q}_i' = \boldsymbol{P}_i(\boldsymbol{I}_{r_i})\boldsymbol{Q}_i' = \boldsymbol{P}_i\boldsymbol{Q}_i' = \boldsymbol{G}_i ,$$

so $\boldsymbol{G}_i$ is idempotent and, hence, a projector onto $\mathcal{C}(\boldsymbol{G}_i)$ along $\mathcal{N}(\boldsymbol{G}_i)$. We now claim

$$\mathcal{C}(\boldsymbol{G}_i) = \mathcal{N}(\boldsymbol{A} - \lambda_i\boldsymbol{I}) \text{ and } \mathcal{N}(\boldsymbol{G}_i) = \mathcal{C}(\boldsymbol{A} - \lambda_i\boldsymbol{I}) .$$

To prove the first claim, observe that

$$\mathcal{C}(\boldsymbol{G}_i) = \mathcal{C}(\boldsymbol{P}_i\boldsymbol{Q}_i') \subseteq \mathcal{C}(\boldsymbol{P}_i) = \mathcal{C}(\boldsymbol{P}_i\boldsymbol{Q}_i'\boldsymbol{P}_i) \subseteq \mathcal{C}(\boldsymbol{G}_i\boldsymbol{P}_i) = \mathcal{C}(\boldsymbol{G}_i) ,$$

which shows that $\mathcal{C}(\boldsymbol{G}_i) = \mathcal{C}(\boldsymbol{P}_i)$. Since the columns of $\boldsymbol{P}_i$ are a basis for $\mathcal{N}(\boldsymbol{A} - \lambda_i\boldsymbol{I})$, we have established our first claim that $\mathcal{C}(\boldsymbol{G}_i) = \mathcal{N}(\boldsymbol{A} - \lambda_i\boldsymbol{I})$. The second claim proceeds from observing that

$$\begin{aligned}\boldsymbol{G}_i(\boldsymbol{A} - \lambda_i\boldsymbol{I}) &= \boldsymbol{G}_i\left(\sum_{j=1}^{k}\lambda_j\boldsymbol{G}_j - \lambda_i\sum_{j=1}^{k}\boldsymbol{G}_j\right) = \boldsymbol{G}_i\sum_{j=1}^{k}(\lambda_j - \lambda_i)\boldsymbol{G}_j \\ &= \sum_{j=1}^{k}(\lambda_j - \lambda_i)\boldsymbol{G}_i\boldsymbol{G}_j = \boldsymbol{O} .\end{aligned}$$

This proves that $\mathcal{C}(\boldsymbol{A} - \lambda_i\boldsymbol{I}) \subseteq \mathcal{N}(\boldsymbol{G}_i)$. To show that these two subspaces are, in fact, the same we show that their dimensions are equal using the just established fact that $\mathcal{C}(\boldsymbol{G}_i) = \mathcal{N}(\boldsymbol{A} - \lambda_i\boldsymbol{I})$:

$$\dim[\mathcal{C}(\boldsymbol{A} - \lambda_i\boldsymbol{I})] = \rho(\boldsymbol{A} - \lambda_i\boldsymbol{I}) = n - \nu(\boldsymbol{A} - \lambda_i\boldsymbol{I}) = n - \rho(\boldsymbol{G}_i) = \dim[\mathcal{N}(\boldsymbol{G}_i)] .$$

This establishes the second claim and property (iv).

We now prove the *"if"* part. Suppose we have matrices $\boldsymbol{G}_1, \boldsymbol{G}_2, \ldots, \boldsymbol{G}_k$ satisfying the four properties. Let $\rho(\boldsymbol{G}_i) = r_i$. Since each $\boldsymbol{G}_i$ is idempotent, it follows that $\rho(\boldsymbol{G}_i) = \text{tr}(\boldsymbol{G}_i)$ and

$$\sum_{i=1}^{k} r_i = \text{tr}(\boldsymbol{G}_1) + \text{tr}(\boldsymbol{G}_2) + \cdots + \text{tr}(\boldsymbol{G}_k) = \text{tr}(\boldsymbol{G}_1 + \boldsymbol{G}_2 + \cdots + \boldsymbol{G}_k) = \text{tr}(\boldsymbol{I}_n) = n .$$

Let $\boldsymbol{G}_i = \boldsymbol{P}_i\boldsymbol{Q}_i'$ be a rank factorization of $\boldsymbol{G}_i$. Therefore, the matrices $\boldsymbol{P}$ and $\boldsymbol{Q}$ formed in (11.8) are $n \times n$ and both are indeed nonsingular. Since

$$\boldsymbol{I} = \boldsymbol{G}_1 + \boldsymbol{G}_2 + \cdots + \boldsymbol{G}_k = \boldsymbol{P}\boldsymbol{Q} ,$$

it follows that $Q = P^{-1}$ and property (i) now says

$$A = \sum_{i=1}^{k} \lambda_i P_i Q_i' = P \begin{bmatrix} \lambda_1 I_{r_1} & O & \dots & O \\ O & \lambda_2 I_{r_2} & \dots & O \\ \vdots & \vdots & \ddots & \vdots \\ O & O & \dots & \lambda_k I_{r_k} \end{bmatrix} P^{-1} ,$$

which proves that A is diagonalizable. $\square$

Remark: The expansion in property (i) is known as the ***spectral representation*** of A and the G_i's are called the ***spectral projectors*** of A.

The spectral representation of the powers of a diagonalizable matrix are also easily obtained as we show in the following corollary.

Corollary 11.3 *Let $A = \lambda_1 G_1 + \lambda_2 G_2 + \cdots + \lambda_k G_k$ be the spectral representation of a diagonalizable matrix A exactly as described in Theorem 11.13. Then the matrix A^m, where m is any positive integer, has the spectral decomposition*

$$A^m = \lambda_1^m G_1 + \lambda_2^m G_2 + \cdots + \lambda_k^m G_k .$$

Proof. Suppose A is diagonalizable and let r_i be the multiplicity of its eigenvalue λ_i. As in Theorem 11.13, there exists an $n \times n$ nonsingular matrix P such that

$$P^{-1}AP = \Lambda = \begin{bmatrix} \lambda_1 I_{r_1} & O & \dots & O \\ O & \lambda_2 I_{r_2} & \dots & O \\ \vdots & \vdots & \ddots & \vdots \\ O & O & \dots & \lambda_k I_{r_k} \end{bmatrix} .$$

Since $(P^{-1}AP)^m = (P^{-1}AP)(P^{-1}AP)\cdots(P^{-1}AP) = P^{-1}A^m P$, we obtain

$$P^{-1}A^m P = \Lambda^m = \begin{bmatrix} \lambda_1^m I_{r_1} & O & \dots & O \\ O & \lambda_2^m I_{r_2} & \dots & O \\ \vdots & \vdots & \ddots & \vdots \\ O & O & \dots & \lambda_k^m I_{r_k} \end{bmatrix} .$$

Partitioning P and P^{-1} as in Theorem 11.13, yields

$$\begin{aligned} A^m &= P\Lambda^m P^{-1} = \lambda_1^m P_1 Q_1' + \lambda_2^m P_2 Q_2' + \cdots + \lambda_k^m P_k Q_k' \\ &= \lambda_1^m G_1 + \lambda_2^m G_2 + \cdots + \lambda_k^m G_k , \end{aligned}$$

where each $G_i = P_i Q_i'$ is the spectral projector of A associated with λ_i. $\square$

The above result shows that the spectral projectors of a diagonalizable matrix A are also the spectral projectors for A^m, where m is any positive integer.

The spectral theorem for diagonalizable matrices allows us to construct well-defined matrix functions, i.e., functions whose arguments are square matrices, say $f(A)$. How should one define such a function? One might consider applying the function to each element of the matrix. Thus, $f(\boldsymbol{A})$ is a matrix whose elements are $f(a_{ij})$. This is how many computer programs evaluate functions when applied to matrices. This is fine when the objective to implement functions efficiently over an array of arguments. However, this is undesirable in modeling where matrix functions need to imitate behavior of their scalar counterparts.

To ensure consistency with their scalar counterparts, matrix functions can be defined using infinite series expansions of functions. For example, expand the function as a power series and replace the scalar argument with the matrix. This is how the *matrix exponential*, arguably one of the most conspicuous matrix functions in applied mathematics, is defined. Recall that the exponential series is given by

$$\exp(x) = 1 + x + \frac{x^2}{2!} + \frac{x^3}{3!} + \cdots + \frac{x^k}{k!} + \cdots = \sum_{k=1}^{\infty} \frac{x^k}{k!} ,$$

which converges for all x (its radius of convergence equal to ∞). The matrix exponential can then be defined as follows.

Definition 11.8 The matrix exponential. *Let $\boldsymbol{A}$ be an $n \times n$ matrix. The* **exponential of $\boldsymbol{A}$**, *denoted by $e^{\boldsymbol{A}}$ or $\exp(\boldsymbol{A})$, is the $n \times n$ matrix given by the power series*

$$\exp(\boldsymbol{A}) = \boldsymbol{I} + \boldsymbol{A} + \frac{\boldsymbol{A}^2}{2!} + \frac{\boldsymbol{A}^3}{3!} + \cdots + \frac{\boldsymbol{A}^k}{k!} + \cdots = \sum_{k=1}^{\infty} \frac{\boldsymbol{A}^k}{k!} ,$$

where we define $\boldsymbol{A}^0 = \boldsymbol{I}$.

It can be proved, using some added machinery, that the above series always converges. Hence, the above definition is well-defined. The proof requires the development of some machinery regarding the limits of the entries in powers of $\boldsymbol{A}$, which will be a bit of a digression for us. When $\boldsymbol{A}$ is diagonalizable we can avoid this machinery and use the spectral theorem to construct well-defined matrix functions. We discuss this here.

First, consider an $n \times n$ diagonal matrix $\boldsymbol{\Lambda} = \mathrm{diag}(\lambda_1, \lambda_2, \ldots, \lambda_n)$. The matrix exponential $\exp(\boldsymbol{\Lambda})$ is well-defined because $\sum_{k=0}^{\infty} \lambda^k/k! = e^{\lambda}$, which implies

$$\exp(\boldsymbol{\Lambda}) = \sum_{k=1}^{\infty} \frac{\boldsymbol{\Lambda}^k}{k!} = \begin{bmatrix} e^{\lambda_1} & 0 & \ldots & 0 \\ 0 & e^{\lambda_2} & \ldots & 0 \\ \vdots & \vdots & \ddots & \vdots \\ 0 & 0 & \ldots & e^{\lambda_n} \end{bmatrix} . \tag{11.9}$$

Suppose $\boldsymbol{A}$ is an $n \times n$ diagonalizable matrix. Then, $\boldsymbol{A} = \boldsymbol{P}\boldsymbol{\Lambda}\boldsymbol{P}^{-1}$, which implies that $\boldsymbol{A}^k = \boldsymbol{P}\boldsymbol{\Lambda}^k\boldsymbol{P}^{-1}$. So, we can write the matrix exponential as

$$\exp(\boldsymbol{A}) = \sum_{k=1}^{\infty} \frac{\boldsymbol{A}^k}{k!} = \sum_{k=1}^{\infty} \frac{\boldsymbol{P}\boldsymbol{\Lambda}^k\boldsymbol{P}^{-1}}{k!} = \boldsymbol{P}\left(\sum_{k=1}^{\infty} \frac{\boldsymbol{\Lambda}^k}{k!}\right)\boldsymbol{P}^{-1} = \boldsymbol{P}\exp(\boldsymbol{\Lambda})\boldsymbol{P}^{-1} , \tag{11.10}$$

where $\exp(\boldsymbol{\Lambda})$ is defined as in (11.9). Equation (11.10) shows how to construct the exponential of a diagonalizable matrix obviating convergence issues. However, there is a subtle issue of uniqueness here. We know that the basis of eigenvectors used to diagonalize $\boldsymbol{A}$, i.e., the columns of $\boldsymbol{P}$ in (11.10), is not unique. How, then, can we ensure that $\exp(\boldsymbol{A})$ is unique? The spectral theorem comes to our rescue. Assume that $\boldsymbol{A}$ has k distinct eigenvalues with λ_1 repeated r_1 times, λ_2 repeated r_2 times and so on. Of course, $\sum_{i=1}^{k} r_i = n$. Suppose $\boldsymbol{P}$ has been constructed to group these eigenvalues together. That is,

$$\boldsymbol{A} = \boldsymbol{P}\boldsymbol{\Lambda}\boldsymbol{P}^{-1} = \boldsymbol{P}\begin{bmatrix} \lambda_1\boldsymbol{I}_{r_1} & \boldsymbol{O} & \dots & \boldsymbol{O} \\ \boldsymbol{O} & \lambda_2\boldsymbol{I}_{r_2} & \dots & \boldsymbol{O} \\ \vdots & \vdots & \ddots & \vdots \\ \boldsymbol{O} & \boldsymbol{O} & \dots & \lambda_k\boldsymbol{I}_{r_k} \end{bmatrix}\boldsymbol{P}^{-1} .$$

Using the spectral theorem, we can rewrite (11.10) as

$$\exp(\boldsymbol{A}) = e^{\lambda_1}\boldsymbol{G}_1 + e^{\lambda_2}\boldsymbol{G}_2 + \cdots + e^{\lambda_k}\boldsymbol{G}_k ,$$

where $\boldsymbol{G}_i$ is the spectral projector onto $\mathcal{N}(\boldsymbol{A} - \lambda_i\boldsymbol{I})$ along $\mathcal{C}(\boldsymbol{A} - \lambda_i\boldsymbol{I})$. These spectral projectors are uniquely determined by $\boldsymbol{A}$ and invariant to the choice of $\boldsymbol{P}$. Hence, $\exp(\boldsymbol{A})$ is well-defined.

The above strategy can be used to construct matrix functions $f(\boldsymbol{A})$ for any function $f(z)$ that is defined for every eigenvalue of a diagonalizable $\boldsymbol{A}$. Note that z can be a complex number to accommodate complex eigenvalues of $\boldsymbol{A}$. We first define $f(\boldsymbol{\Lambda})$ to be the diagonal matrix with $f(\lambda_i)$'s along its diagonal, with same eigenvalues grouped together, and then define $f(\boldsymbol{A}) = \boldsymbol{P}f(\boldsymbol{\Lambda})\boldsymbol{P}^{-1}$. That is,

$$f(\boldsymbol{A}) = \boldsymbol{P}\begin{bmatrix} f(\lambda_1)\boldsymbol{I}_{r_1} & \boldsymbol{O} & \dots & \boldsymbol{O} \\ \boldsymbol{O} & f(\lambda_2)\boldsymbol{I}_{r_2} & \dots & \boldsymbol{O} \\ \vdots & \vdots & \ddots & \vdots \\ \boldsymbol{O} & \boldsymbol{O} & \dots & f(\lambda_k)\boldsymbol{I}_{r_k} \end{bmatrix}\boldsymbol{P}^{-1} = \sum_{i=1}^{k} f(\lambda_i)\boldsymbol{G}_i .$$

Matrix exponentials and solution of linear system of ODE's

Consider the linear system of ordinary differential equations

$$\frac{d}{dt}\boldsymbol{y}(t) = \boldsymbol{W}\boldsymbol{y}(t) + \boldsymbol{g} , \tag{11.11}$$

where $\boldsymbol{y}(t)$ is an $m \times 1$ vector function of t, $\boldsymbol{W}$ is an $m \times m$ matrix (which may depend upon known and unknown inputs but we suppress them in the notation here) that has m real and distinct eigenvalues and $\boldsymbol{g}$ is a vector of length m that does not depend upon t. The $m \times 1$ vector $\frac{d}{dt}\boldsymbol{y}(t)$ has the derivative of each element of $\boldsymbol{y}(t)$ as its entries. Such systems are pervasive in scientific applications and can be solved effectively using matrix exponentials. We will only consider the case when $\boldsymbol{W}$ is diagonalizable. For more general cases, see, e.g., Laub (2005) and Ortega (1987).

When $\boldsymbol{W}$ has m distinct eigenvalues, we can find a non-singular matrix $\boldsymbol{P}$ such that $\boldsymbol{W} = \boldsymbol{P}\boldsymbol{\Lambda}\boldsymbol{P}^{-1}$, where the columns of $\boldsymbol{P}$ are linearly independent eigenvectors. Therefore, $\exp(\boldsymbol{W}) = \boldsymbol{P}\exp(\boldsymbol{\Lambda})\boldsymbol{P}^{-1}$, where $\boldsymbol{\Lambda}$ is a diagonal matrix with λ_i as the i-th diagonal element, $i = \{1, 2, \cdots, m\}$. Now, let $\boldsymbol{G}_i = \boldsymbol{u}_i\boldsymbol{v}_i'$, where $\boldsymbol{u}_i$ is the i-th column of $\boldsymbol{P}$ and $\boldsymbol{v}_i'$ is the i-th row of $\boldsymbol{P}^{-1}$. (These are the *right* and *left* eigenvectors, respectively.) It is straightforward to see that (i) $\boldsymbol{G}_i^2 = \boldsymbol{G}_i$, (ii) $\boldsymbol{G}_i\boldsymbol{G}_j = 0 \ \forall \ i \neq j$ and (iii) $\sum_{i=1}^{m} \boldsymbol{G}_i = \boldsymbol{I}_m$. Each $\boldsymbol{G}_i$ is the spectral projector onto the null space of $\boldsymbol{W} - \lambda_i\boldsymbol{I}_m$ along the column space of $\boldsymbol{W} - \lambda_i\boldsymbol{I}_m$. It is also easily verified that

$$\exp(t\boldsymbol{W})\exp(-t\boldsymbol{W}) = \boldsymbol{I}_m \text{ and } \exp(t\boldsymbol{W})\boldsymbol{W} = \boldsymbol{W}\exp(t\boldsymbol{W}) . \tag{11.12}$$

The above properties of $\boldsymbol{G}_i$ imply that $\exp(t\boldsymbol{W}) = \sum_{i=1}^{m} \exp(\lambda_i t)\boldsymbol{G}_i$. Consequently,

$$\begin{aligned} \frac{d}{dt}\exp(t\boldsymbol{W}) &= \sum_{i=1}^{m} \lambda_i \exp(\lambda_i t)\boldsymbol{G}_i = \sum_{i=1}^{m} \lambda_i \exp(\lambda_i t)\boldsymbol{u}_i\boldsymbol{v}_i' = \boldsymbol{P}\boldsymbol{\Lambda}\exp(t\boldsymbol{\Lambda})\boldsymbol{P}^{-1} \\ &= \boldsymbol{P}\boldsymbol{\Lambda}\boldsymbol{P}^{-1}\boldsymbol{P}\exp(t\boldsymbol{\Lambda})\boldsymbol{P}^{-1} = \boldsymbol{W}\exp(t\boldsymbol{W}) = \exp(t\boldsymbol{W})\boldsymbol{W} . \end{aligned}$$

$$\begin{aligned} \text{Also,} \quad \int \exp(t\boldsymbol{W})dt &= \sum_{i=1}^{m} \frac{1}{\lambda_i}\exp(\lambda_i t)\boldsymbol{G}_i = \boldsymbol{P}\boldsymbol{\Lambda}^{-1}\exp(t\boldsymbol{\Lambda})\boldsymbol{P}^{-1} \\ &= \boldsymbol{P}\boldsymbol{\Lambda}^{-1}\boldsymbol{P}^{-1}\boldsymbol{P}\exp(t\boldsymbol{\Lambda})\boldsymbol{P}^{-1} = \boldsymbol{W}^{-1}\exp(t\boldsymbol{W}) . \end{aligned}$$

Multiplying both sides of (11.11) by $\exp(-t\boldsymbol{W})$ from the left yields

$$\begin{aligned} &\exp(-t\boldsymbol{W})\left[\frac{d}{dt}\boldsymbol{y}(t) - \boldsymbol{W}\boldsymbol{y}(t)\right] = \exp(-t\boldsymbol{W})\boldsymbol{g} \\ &\Longrightarrow \frac{d}{dt}\left[\exp(-t\boldsymbol{W})\boldsymbol{y}(t)\right] = \exp(-t\boldsymbol{W})\boldsymbol{g} . \end{aligned} \tag{11.13}$$

Integrating out both sides of (11.13), we obtain

$$\exp(-t\boldsymbol{W})\boldsymbol{y}(t) = -\boldsymbol{W}^{-1}\exp(-t\boldsymbol{W})\boldsymbol{g} + \boldsymbol{k} ,$$

where $\boldsymbol{k}$ is a constant vector. The initial condition at $t = 0$ yields $\boldsymbol{y}(0) = -\boldsymbol{W}^{-1}\boldsymbol{g} + \boldsymbol{k}$, so $\boldsymbol{k} = \boldsymbol{y}(0) + \boldsymbol{W}^{-1}\boldsymbol{g}$. Consequently,

$$\boldsymbol{y}(t) = \exp(t\boldsymbol{W})\boldsymbol{y}(0) + \boldsymbol{W}^{-1}\left[\exp(t\boldsymbol{W}) - \boldsymbol{I}_m\right]\boldsymbol{g} \tag{11.14}$$

is the solution to (11.11).

11.5 Similarity with triangular matrices

Corollary 11.2 tells us that every matrix with distinct eigenvalues is diagonalizable. However, not every diagonalizable matrix needs to have distinct eigenvalues. So, the class of diagonalizable matrices is bigger than the class of square matrices with distinct eigenvalues. This is still restrictive and a large number of matrices appearing in the applied sciences are not diagonalizable. One important goal of linear algebra is to find the simplest forms of matrices to which any square matrix can be reduced via similarity transformations. This is given by the ***Jordan Canonical Form***, which are special triangular matrices that have zeroes everywhere except on the diagonal and the diagonal immediately above the main diagonal. We will consider this in Chapter 12.

If we relax our requirement from Jordan forms to triangular forms, then we have a remarkable result: *every square matrix is similar to a triangular matrix* with the understanding that we allow complex entries for all matrices we consider here. This result, which is fundamentally important in theory and applications of linear algebra, is named after the mathematician Issai Schur and is presented below.

Theorem 11.14 *Every $n \times n$ matrix $\boldsymbol{A}$ is similar to an upper-triangular matrix whose diagonal entries are the eigenvalues of $\boldsymbol{A}$.*

Proof. We will use induction on n to prove the theorem. If $n = 1$, the result is trivial. Consider $n > 1$ and assume that all $(n-1) \times (n-1)$ matrices are similar to an upper-triangular matrix. Let $\boldsymbol{A}$ be an $n \times n$ matrix and let λ be an eigenvalue of $\boldsymbol{A}$ with associated eigenvector $\boldsymbol{x}$. Therefore, $\boldsymbol{A}\boldsymbol{x} = \lambda\boldsymbol{x}$, where λ and $\boldsymbol{x}$ may be complex.

We construct a nonsingular matrix $\boldsymbol{P}$ with its first column as $\boldsymbol{x}$, say $\boldsymbol{P} = \left[\boldsymbol{x} : \boldsymbol{P}_2\right]$. This is always possible by extending the set $\{\boldsymbol{x}\}$ to a basis for $\mathbb{C}^n$ and placing the extension vectors as the columns of $\boldsymbol{P}_2$ (see Corollary 4.3). Observe that

$$\begin{aligned}\boldsymbol{AP} &= \boldsymbol{A}\left[\boldsymbol{x} : \boldsymbol{P}_2\right] = \left[\boldsymbol{Ax} : \boldsymbol{AP}_2\right] = \left[\lambda\boldsymbol{x} : \boldsymbol{AP}_2\right] \\ &= \left[\boldsymbol{x} : \boldsymbol{P}_2\right]\begin{bmatrix}\lambda & \boldsymbol{v}' \\ \boldsymbol{0} & \boldsymbol{B}\end{bmatrix} = \boldsymbol{P}\left[\boldsymbol{x} : \boldsymbol{P}_2\right]\begin{bmatrix}\lambda & \boldsymbol{v}' \\ \boldsymbol{0} & \boldsymbol{B}\end{bmatrix},\end{aligned}$$

where $\boldsymbol{v}'$ is some $1 \times (n-1)$ vector and $\boldsymbol{B}$ is some $(n-1) \times (n-1)$ matrix. Therefore,

$$\boldsymbol{P}^{-1}\boldsymbol{AP} = \begin{bmatrix}\lambda & \boldsymbol{v}' \\ \boldsymbol{0} & \boldsymbol{B}\end{bmatrix}.$$

By the induction hypothesis, there exists an $(n-1) \times (n-1)$ nonsingular matrix $\boldsymbol{W}$ such that $\boldsymbol{W}^{-1}\boldsymbol{BW} = \boldsymbol{U}$ where $\boldsymbol{U}$ is upper-triangular. Construct the $n \times n$ matrix

$$\boldsymbol{S} = \begin{bmatrix}1 & \boldsymbol{0}' \\ \boldsymbol{0} & \boldsymbol{W}\end{bmatrix},$$

which is clearly nonsingular. Define $\boldsymbol{Q} = \boldsymbol{PS}$ and note that

$$\boldsymbol{Q}^{-1}\boldsymbol{AQ} = \boldsymbol{S}^{-1}\boldsymbol{P}^{-1}\boldsymbol{APS} = \begin{bmatrix}1 & \boldsymbol{0}' \\ \boldsymbol{0} & \boldsymbol{W}^{-1}\end{bmatrix}\begin{bmatrix}\lambda & \boldsymbol{v}' \\ \boldsymbol{0} & \boldsymbol{B}\end{bmatrix}\begin{bmatrix}1 & \boldsymbol{0}' \\ \boldsymbol{0} & \boldsymbol{W}\end{bmatrix} = \begin{bmatrix}\lambda & \boldsymbol{z}' \\ \boldsymbol{0} & \boldsymbol{U}\end{bmatrix} = \boldsymbol{T},$$

where $\boldsymbol{z}' = \boldsymbol{v}'\boldsymbol{W}$ and $\boldsymbol{T}$ is upper-triangular. Since similar matrices have the same eigenvalues, and since the eigenvalues of a triangular matrix are its diagonal entries (Example 11.3), the diagonal entries of $\boldsymbol{T}$ must be the eigenvalues of $\boldsymbol{A}$. □

The above theorem ensures that for any $n \times n$ matrix $\boldsymbol{A}$, we can find a nonsingular matrix $\boldsymbol{P}$ such that $\boldsymbol{P}^{-1}\boldsymbol{A}\boldsymbol{P} = \boldsymbol{T}$ is triangular, where the diagonal entries of $\boldsymbol{T}$ being the eigenvalues of $\boldsymbol{A}$. Not only that, since permutation matrices are nonsingular, we can choose $\boldsymbol{P}$ so that the eigenvalues can appear in any order we like along the diagonal of $\boldsymbol{T}$.

Notably, Schur realized that the similarity transformation to triangularize the matrix can be made ***unitary***. Unitary matrices are analogues of orthogonal matrices in that their columns form an orthonormal set but the columns may now have complex entries. Earlier, when we discussed orthogonality, we considered only vectors and matrices in $\Re^n$. There, two vectors were orthogonal if $\langle \boldsymbol{x}, \boldsymbol{y} \rangle = \boldsymbol{y}'\boldsymbol{x} = 0$. For vectors in $\mathbb{C}^n$ involving complex entries, this definition of orthogonality is ill-conceived. Here, we will make a brief digression and discuss this.

For any vector $\boldsymbol{x} \in \mathbb{C}^n$, we define its *adjoint* to be the $1 \times n$ vector, denoted $\boldsymbol{x}^*$, with each element of $\boldsymbol{x}$ replaced by its complex conjugate. The standard inner product for $\mathbb{C}^n$ is

$$\langle \boldsymbol{y}, \boldsymbol{x} \rangle = \boldsymbol{x}^*\boldsymbol{y} = \overline{\boldsymbol{y}^*\boldsymbol{x}} = \overline{\langle \boldsymbol{x}, \boldsymbol{y} \rangle} \,, \tag{11.15}$$

where $\overline{a}$ denotes the complex conjugate of a scalar a. Orthogonality for complex vectors is defined with respect to this inner product as $\boldsymbol{x}^*\boldsymbol{y} = 0$. The norm of a complex vector is $\|\boldsymbol{x}\| = \sqrt{\langle \boldsymbol{x}, \boldsymbol{x} \rangle} = \sqrt{\boldsymbol{x}^*\boldsymbol{x}}$, which is well-defined because $\boldsymbol{x}^*\boldsymbol{x}$ is always real (note: $\boldsymbol{x}'\boldsymbol{x}$ may not be real).

A square matrix with possibly complex entries is said to be unitary if $\boldsymbol{Q}^*\boldsymbol{Q} = \boldsymbol{Q}\boldsymbol{Q}^* = \boldsymbol{I}$, where $\boldsymbol{Q}^*$ is the conjugate transpose of $\boldsymbol{Q}$ obtained by transposing $\boldsymbol{Q}$ and then replacing each entry by its complex conjugate. This means that the columns of $\boldsymbol{Q}$ constitute an orthonormal set, where orthogonality and norm are defined with respect to the inner product in (11.15). The matrix $\boldsymbol{Q}^*$ is often called the ***adjoint*** of $\boldsymbol{Q}$. In other words, $\boldsymbol{Q}^*$ is obtained from $\boldsymbol{Q}$. If $\boldsymbol{Q}$ is unitary, then $\boldsymbol{Q}^* = \boldsymbol{Q}^{-1}$. Unitary matrices, like orthogonal matrices, are nonsingular and their inverses are obtained very cheaply.

Schur's triangularization theorem, which we present below, says that for any square matrix $\boldsymbol{A}$ there exists a matrix $\boldsymbol{Q}$, which is not just nonsingular but also unitary, such that $\boldsymbol{Q}^*\boldsymbol{A}\boldsymbol{Q}$ is triangular. We say that $\boldsymbol{A}$ is ***unitarily similar*** to an upper-triangular matrix. This can be proved analogous to Theorem 11.14. The key step is that matrix $\boldsymbol{P}$ in Theorem 11.14 can now be constructed as an orthogonal matrix with the eigenvector $\boldsymbol{x}$ as its first column, where $\boldsymbol{x}$ is now *normalized* so that $\|\boldsymbol{x}\| = \boldsymbol{x}^*\boldsymbol{x} = 1$. The induction steps follow with nonsingular matrices being replaced by unitary matrices. We encourage the reader to construct such a proof. We provide a slightly different proof for Schur's triangularization theorem that uses Theorem 11.14 in conjunction with the $\boldsymbol{QR}$ decomposition for possibly complex matrices. Let $\boldsymbol{P}$ be a nonsingular matrix with possibly complex entries. If we apply the Gram-Schmidt procedure to its columns using the inner product in (11.15) to define the proj functions, then we arrive at the decomposition $\boldsymbol{P} = \boldsymbol{QR}$, where $\boldsymbol{R}$ is upper-triangular and $\boldsymbol{Q}$ is unitary. One can force the diagonal elements of $\boldsymbol{R}$ to be real and positive, in which case the $\boldsymbol{QR}$ decomposition is unique. We use this in the proof below.

Theorem 11.15 Schur's triangularization theorem. *If $\boldsymbol{A}$ is any $n \times n$ matrix (with* *real or complex entries), then there exists a unitary matrix Q (not unique and with possibly complex entries) such that $Q^*AQ = T$, where T is an upper-triangular matrix (not unique and with possibly complex entries) whose diagonal entries are the eigenvalues of A.*

Proof. Theorem 11.14 says that every matrix is similar to an upper-triangular matrix. Therefore, for an $n \times n$ matrix A we can find a nonsingular matrix P such that $P^{-1}AP = U$, where U is upper-triangular. Let $P = QR$ be the unique QR decomposition for P, where Q is unitary and R is upper-triangular with real and positive diagonal elements. So, R is nonsingular and

$$U = P^{-1}AP = R^{-1}Q^{-1}AQR = R^{-1}Q^*AQR \Longrightarrow Q^*AQ = RUR^{-1} .$$

Inverses of upper-triangular matrices are upper-triangular. So R^{-1} is upper-triangular (Theorem 2.7). The product of upper-triangular matrices results in another upper-triangular matrix. Therefore, $Q^*AQ = RUR^{-1}$ is upper-triangular, which proves that A is unitarily similar to the upper-triangular matrix $T = RUR^{-1}$. Since similar matrices have the same eigenvalues, and since the eigenvalues of a triangular matrix are its diagonal entries, the diagonal entries of T must be the eigenvalues of A. □

A closer look at the above proof reveals that if all eigenvalues of A are real, then Q can be chosen to be a real orthogonal matrix. However, even when A is real, Q and T may have to be complex if A has complex eigenvalues. Nevertheless, one can devise a ***real Schur form*** for any real matrix A. Indeed, if A is an $n \times n$ matrix with real entries, then there exists an $n \times n$ real orthogonal matrix Q such that

$$Q'AQ = \begin{bmatrix} T_{11} & T_{12} & \dots & T_{1n} \\ O & T_{22} & \dots & T_{2n} \\ \vdots & \vdots & \ddots & \vdots \\ O & O & \dots & T_{nn} \end{bmatrix} , \tag{11.16}$$

where the matrix on the right is block-triangular with each T_{ii} being either 1×1, in which case it is an eigenvalue of A, or 2×2, whereupon its eigenvalues correspond to a conjugate pair of complex eigenvalues of A (recall that the complex eigenvalues of A appear in conjugate pairs). The proof is by induction and is left to the reader.

Theorems 11.14 and 11.15 can be used to derive several properties of eigenvalues. For example, in Theorem 11.5 we proved that the sum and product of the eigenvalues of a matrix were equal to the trace and the determinant, respectively, by comparing the coefficients in the characteristic polynomial. For an $n \times n$ matrix A, the triangularization theorems reveal that

$$\text{tr}(A) = \text{tr}(Q^{-1}TQ) = \text{tr}(TQQ^{-1}) = \text{tr}(T) = \sum_{i=1}^{n} t_{ii} \text{ and}$$

$$|A| = |Q^{-1}TQ| = |Q^{-1}||T||Q| = |T| = \prod_{i=1}^{n} t_{ii} ,$$

where t_{ii}'s are the diagonal elements of the upper-triangular $\boldsymbol{T}$. The diagonal elements of $\boldsymbol{T}$ are precisely the eigenvalues of $\boldsymbol{A}$. Therefore, $\text{tr}(\boldsymbol{A})$ is the sum of the eigenvalues of $\boldsymbol{A}$ and $|\boldsymbol{A}|$ is the product of the eigenvalues.

Another application of these results is the Caley-Hamilton theorem, which we consider in the next section.

11.6 Matrix polynomials and the Caley-Hamilton Theorem

If $\boldsymbol{A}$ is a square matrix, then it commutes with itself and the powers $\boldsymbol{A}^k$ are all well-defined for $k = 1, 2, \ldots,$. For $k = 0$, we define $\boldsymbol{A}^0 = \boldsymbol{I}$ the identity matrix. This leads to the definition of a matrix polynomial.

Definition 11.9 *Let $\boldsymbol{A}$ be an $n \times n$ matrix and let*

$$f(t) = \alpha_0 + \alpha_1 t + \alpha_2 t^2 + \cdots + \alpha_m t^m$$

be a polynomial of degree m, where the coefficients (i.e., the α_i's) are complex numbers. We then define the **matrix polynomial**, *or the polynomial in $\boldsymbol{A}$, as the $n \times n$ matrix*

$$f(\boldsymbol{A}) = \alpha_0 \boldsymbol{I} + \alpha_1 \boldsymbol{A} + \alpha_2 \boldsymbol{A}^2 + \cdots + \alpha_m \boldsymbol{A}^m ,$$

where $\boldsymbol{I}$ is the $n \times n$ identity matrix.

The following theorem lists some basic properties of matrix polynomials.

Theorem 11.16 *Let $f(t)$ and $g(t)$ be two polynomials and let $\boldsymbol{A}$ be a square matrix.*

(i) If $h(t) = \theta f(t)$ for some scalar θ, then $h(\boldsymbol{A}) = \theta f(\boldsymbol{A})$.
(ii) If $h(t) = f(t) + g(t)$ then $h(\boldsymbol{A}) = f(\boldsymbol{A}) + g(\boldsymbol{A})$.
(iii) If $h(t) = f(t)g(t)$, then $h(\boldsymbol{A}) = f(\boldsymbol{A})g(\boldsymbol{A})$.
(iv) $f(\boldsymbol{A})g(\boldsymbol{A}) = g(\boldsymbol{A})f(\boldsymbol{A})$.
(v) $f(\boldsymbol{A}') = f(\boldsymbol{A})'$.
(vi) If $\boldsymbol{A}$ is upper (lower) triangular, then $f(\boldsymbol{A})$ is upper (lower) triangular.

Proof. **Proof of (i):** If $f(t) = \alpha_0 + \alpha_1 t + \alpha_2 t^2 + \cdots + \alpha_l t^l$ and $h(t) = \theta f(t)$, then $h(t) = \theta\alpha_0 + \theta\alpha_1 t + \theta\alpha_2 t^2 + \cdots + \theta\alpha_l t^l$ and

$$h(\boldsymbol{A}) = \theta\alpha_0 \boldsymbol{I} + \theta\alpha_1 \boldsymbol{A} + \theta\alpha_2 \boldsymbol{A}^2 + \cdots + \theta\alpha_l \boldsymbol{A}^l = \theta \left(\sum_{i=0}^{l} \alpha_i \boldsymbol{A}^i \right) = \theta f(\boldsymbol{A}) .$$

Proof of (ii): This is straightforward. If $f(t) = \alpha_0 + \alpha_1 t + \alpha_2 t^2 + \cdots + \alpha_l t^l$ and $g(t) = \beta_0 + \beta_1 t + \beta_2 t^2 + \cdots + \beta_l t^m$ with $l \leq m$, then

$$h(t) = f(t) + g(t) = (\alpha_0 + \beta_0) + (\alpha_1 + \beta_1)t + (\alpha_2 + \beta_2)t^2 + \cdots + (\alpha_m + \beta_m)t^m ,$$

where $\alpha_i = 0$ if $i > l$. Therefore,

$$h(\boldsymbol{A}) = (\alpha_0 + \beta_0) + (\alpha_1 + \beta_1)\boldsymbol{A} + (\alpha_2 + \beta_2)\boldsymbol{A}^2 + \cdots + (\alpha_m + \beta_m)\boldsymbol{A}^m$$
$$= \sum_{i=0}^{l} \alpha_i \boldsymbol{A}^i + \sum_{i=0}^{m} \beta_i \boldsymbol{A}^i = f(\boldsymbol{A}) + g(\boldsymbol{A}) .$$

Proof of (iii): Suppose that

$$f(t) = \alpha_0 + \alpha_1 t + \alpha_2 t^2 + \cdots + \alpha_l t^l \text{ and } g(t) = \beta_0 + \beta_1 t + \beta_2 t^2 + \cdots + \beta_m t^m ,$$

where $l \leq m$. Collecting the coefficients for the different powers of t in $f(t)g(t)$, we find that

$$h(t) = f(t)g(t) = \gamma_0 + \gamma_1 t + \gamma_2 t^2 + \ldots + \gamma_{l+m} t^{l+m} ,$$

is another polynomial of order $l + m$, where

$$\gamma_j = \alpha_0 \beta_j + \alpha_1 \beta_{j-1} + \cdots + \alpha_j \beta_0 , \quad \text{for } j = 0, 1, \ldots, l + m$$

with $\alpha_i = 0$ for all $i > l$ and $\beta_i = 0$ for all $i > m$. Therefore,

$$h(\boldsymbol{A}) = \gamma_0 \boldsymbol{I} + \gamma_1 \boldsymbol{A} + \gamma_2 \boldsymbol{A}^2 + \ldots + \gamma_{l+m} \boldsymbol{A}^{l+m} .$$

Now compute $f(\boldsymbol{A})g(\boldsymbol{A})$ using direct matrix multiplication to obtain

$$f(\boldsymbol{A})g(\boldsymbol{A}) = \left(\sum_{i=0}^{l} \alpha_i \boldsymbol{A}^i\right)\left(\sum_{i=0}^{m} \beta_i \boldsymbol{A}^i\right)$$
$$= \sum_{i=0}^{l}\sum_{j=0}^{m} \alpha_i \beta_j \boldsymbol{A}^{i+j} = \sum_{k=0}^{l+m}\sum_{j=0}^{k} \alpha_j \beta_{k-j} \boldsymbol{A}^k$$
$$= \sum_{k=0}^{l+m}\left(\sum_{j=0}^{k} \alpha_j \beta_{k-j}\right)\boldsymbol{A}^k = \sum_{k=0}^{l+m} \gamma_k \boldsymbol{A}^k , \quad \text{where } \gamma_k = \sum_{j=0}^{k} \alpha_j \beta_{k-j}$$
$$= \gamma_0 \boldsymbol{I} + \gamma_1 \boldsymbol{A} + \gamma_2 \boldsymbol{A}^2 + \ldots + \gamma_{l+m} \boldsymbol{A}^{l+m} = h(\boldsymbol{A}) .$$

Proof of (iv): This can be proved by evaluating $f(\boldsymbol{A})g(\boldsymbol{A})$ and $g(\boldsymbol{A})f(\boldsymbol{A})$ and showing that the two products are equal to $h(\boldsymbol{A})$. Alternatively, we can use the result proved in (iii) and use the fact that polynomials commute, i.e., $h(t) = f(t)g(t) = g(t)f(t)$. Therefore, the matrix polynomial $f(\boldsymbol{A})g(\boldsymbol{A}) = h(\boldsymbol{A})$. But $h(t) = g(t)f(t)$, so $h(\boldsymbol{A}) = g(\boldsymbol{A})f(\boldsymbol{A})$. Thus,

$$f(\boldsymbol{A})g(\boldsymbol{A}) = h(\boldsymbol{A}) = g(\boldsymbol{A})f(\boldsymbol{A}) .$$

Proof of (v): Using the law of the product of transposes, we find that

$$(\boldsymbol{A}^2)' = (\boldsymbol{A}\boldsymbol{A})' = \boldsymbol{A}'\boldsymbol{A}' = (\boldsymbol{A}')^2 .$$

More generally, it is easy to see that $(\boldsymbol{A}^k)' = (\boldsymbol{A}')^k$. Therefore, if $f(t)$ is as defined in (i), then

$$\begin{aligned} f(\boldsymbol{A}') &= \alpha_0 \boldsymbol{I} + \alpha_1 \boldsymbol{A}' + \alpha_2 (\boldsymbol{A}')^2 + \cdots + \alpha_l (\boldsymbol{A}')^l \\ &= \alpha_0 \boldsymbol{I} + \alpha_1 (\boldsymbol{A})' + \alpha_2 (\boldsymbol{A}^2)' + \cdots + \alpha_l (\boldsymbol{A}^l)' \\ &= \left(\alpha_0 \boldsymbol{I} + \alpha_1 \boldsymbol{A} + \alpha_2 (\boldsymbol{A})^2 + \cdots + \alpha_l (\boldsymbol{A})^l\right)' \\ &= f(\boldsymbol{A})' . \end{aligned}$$

Proof of (vi): If $\boldsymbol{A}$ is upper-triangular, then so is $\boldsymbol{A}^k$ for $k = 1, 2, \ldots$ because products of upper-triangular matrices are also upper-triangular. In that case, $f(\boldsymbol{A})$ is a linear combination of upper-triangular matrices (the powers of $\boldsymbol{A}$) and is upper-triangular. This completes the proof when $\boldsymbol{A}$ is upper-triangular.

If $\boldsymbol{A}$ is lower-triangular, then $\boldsymbol{A}'$ is upper-triangular and so is $f(\boldsymbol{A}')$ (by what we have just proved). But (v) tells us that $f(\boldsymbol{A}') = f(\boldsymbol{A})'$, which means that $f(\boldsymbol{A})'$ is also upper-triangular. Therefore, $f(\boldsymbol{A})$ is lower-triangular. □

Matrix polynomials can lead to interesting results concerning eigenvalues and eigenvectors. Let $\boldsymbol{A}$ be an $n \times n$ matrix with an eigenvalue λ and let $\boldsymbol{x}$ be an eigenvector corresponding to the eigenvalue λ. Therefore, $\boldsymbol{Ax} = \lambda \boldsymbol{x}$, which implies that

$$\begin{aligned} \boldsymbol{A}^2\boldsymbol{x} = \boldsymbol{A}(\lambda \boldsymbol{x}) = \lambda \boldsymbol{Ax} = \lambda^2 \boldsymbol{x} &\Longrightarrow \boldsymbol{A}^3 \boldsymbol{x} = \boldsymbol{A}(\lambda^2 \boldsymbol{x}) = \lambda^2 \boldsymbol{Ax} = \lambda^3 \boldsymbol{x} \\ &\Longrightarrow \ldots\ldots \Longrightarrow \boldsymbol{A}^k \boldsymbol{x} = \boldsymbol{A}(\lambda^{k-1}\boldsymbol{x}) = \lambda^{k-1}\boldsymbol{Ax} = \lambda^k \boldsymbol{x} . \end{aligned}$$

Also note that $\boldsymbol{A}^0\boldsymbol{x} = \boldsymbol{Ix} = \lambda^0 \boldsymbol{x}$. So, we can say that $\boldsymbol{A}^k \boldsymbol{x} = \lambda^k \boldsymbol{x}$ for $k = 0, 1, 2, \ldots,$. Let $f(t) = \sum_{i=0}^m \alpha_i t^i$ be a polynomial. Then,

$$f(\boldsymbol{A})\boldsymbol{x} = \left(\sum_{i=0}^m \alpha_i \boldsymbol{A}^i \right) \boldsymbol{x} = \sum_{i=0}^m \alpha_i \boldsymbol{A}^i \boldsymbol{x} = \sum_{i=0}^m \alpha_i \lambda^i \boldsymbol{x} = f(\lambda)\boldsymbol{x} .$$

In other words, $f(\lambda)$ is an eigenvalue of $f(\boldsymbol{A})$. The following theorem provides a more general result.

Theorem 11.17 *Let $\lambda_1, \lambda_2, \ldots, \lambda_n$ be the eigenvalues (counting multiplicities) of an $n \times n$ matrix $\boldsymbol{A}$. If $f(t)$ is a polynomial, then $f(\lambda_1), f(\lambda_2), \ldots, f(\lambda_n)$ are the eigenvalues of $f(\boldsymbol{A})$.*

Proof. By Theorem 11.14 (or Theorem 11.15) we know that there exists a nonsingular matrix $\boldsymbol{P}$ and an upper-triangular matrix $\boldsymbol{T}$ such that $\boldsymbol{P}^{-1}\boldsymbol{AP} = \boldsymbol{T}$, where $\lambda_1, \lambda_2, \ldots, \lambda_n$ appear along the diagonal of $\boldsymbol{T}$. Without loss of generality, we may assume that $t_{ii} = \lambda_i$ for $i = 1, 2, \ldots, n$. Observe that

$$\begin{aligned} \boldsymbol{T}^2 = (\boldsymbol{P}^{-1}\boldsymbol{AP})(\boldsymbol{P}^{-1}\boldsymbol{AP}) = \boldsymbol{P}^{-1}\boldsymbol{A}^2\boldsymbol{P} &\Longrightarrow \boldsymbol{T}^3 = \boldsymbol{P}^{-1}\boldsymbol{A}^3\boldsymbol{P} \\ &\Longrightarrow \ldots\ldots \Longrightarrow \boldsymbol{T}^k = \boldsymbol{P}^{-1}\boldsymbol{A}^k\boldsymbol{P} \text{ for all } k \geq 0. \end{aligned}$$

Suppose $f(t) = \alpha_0 + \alpha_1 t + \alpha_2 t^2 + \cdots + \alpha_m t^m$. Then,

$$\begin{aligned} f(\boldsymbol{T}) &= \alpha_0 \boldsymbol{I} + \alpha_1 \boldsymbol{T} + \alpha_2 \boldsymbol{T}^2 + \cdots + \alpha_m \boldsymbol{T}^m \\ &= \alpha_0 \boldsymbol{P}^{-1}\boldsymbol{A}^0\boldsymbol{P} + \alpha_1 \boldsymbol{P}^{-1}\boldsymbol{A}\boldsymbol{P} + \alpha_2 \boldsymbol{P}^{-1}\boldsymbol{A}^2\boldsymbol{P} + \cdots + \alpha_m \boldsymbol{P}^{-1}\boldsymbol{A}^m\boldsymbol{P} \\ &= \boldsymbol{P}^{-1}\left(\alpha_0 \boldsymbol{I} + \alpha_1 \boldsymbol{A} + \alpha_2 \boldsymbol{A}^2 + \cdots + \alpha_m \boldsymbol{A}^m\right)\boldsymbol{P} \\ &= \boldsymbol{P}^{-1} f(\boldsymbol{A}) \boldsymbol{P} . \end{aligned}$$

From part (vi) of Theorem 11.16, we know that $f(\boldsymbol{T})$ is also upper-triangular. It is easy to verify that the diagonal entries of $f(\boldsymbol{T})$ are $f(t_{ii}) = f(\lambda_i)$ for $i = 1, 2, \ldots, n$, which proves that the eigenvalues of $f(\boldsymbol{A})$ are $f(\lambda_1), f(\lambda_2), \ldots, f(\lambda_n)$. □

A particular consequence of Theorem 11.17 is the following.

Corollary 11.4 *If $f(t)$ is a polynomial such that $f(\boldsymbol{A}) = \boldsymbol{O}$, and if λ is an eigenvalue of $\boldsymbol{A}$, then λ must be root of $f(t)$, i.e., $f(\lambda) = 0$.*

Proof. From Theorem 11.17 we know that $f(\lambda)$ is an eigenvalue of $f(\boldsymbol{A})$. If $f(\boldsymbol{A}) = \boldsymbol{O}$, then $f(\lambda) = 0$ because the only eigenvalue of the null matrix is zero. □

Here is an important special case.

Corollary 11.5 *Each eigenvalue of an idempotent matrix is either* 1 *or* 0.

Proof. Consider the polynomial $f(t) = t^2 - t = t(t-1)$. If λ is an eigenvalue of $\boldsymbol{A}$, then $f(\lambda)$ is an eigenvalue of $f(\boldsymbol{A})$. If $\boldsymbol{A}$ is idempotent, then $f(\boldsymbol{A}) = \boldsymbol{A}^2 - \boldsymbol{A} = \boldsymbol{O}$, which means that $f(\lambda) = 0$ because the null matrix has zero as its only eigenvalue. Thus, $\lambda^2 - \lambda = (\lambda)(\lambda - 1) = 0$, so λ is 0 or 1. □

Yet another consequence of Theorem 11.17 is Corollary 11.6.

Corollary 11.6 *If $\boldsymbol{A}$ is an $n \times n$ singular matrix, then* 0 *is an eigenvalue of $\boldsymbol{A}$ and*

$$AM_{\boldsymbol{A}}(0) = AM_{\boldsymbol{A}^k}(0) \text{ for every positive integer } k .$$

Proof. That zero must be an eigenvalue of a singular matrix follows from the Schur's Triangularization Theorem and the fact that a triangular matrix is singular if and only if one of its diagonal elements is zero. (Alternatively, consider the fact that $|\boldsymbol{A}| = 0$ so $\lambda = 0$ is clearly a root of the characteristic equation.) The remainder of the result follows from the fact that λ^k is an eigenvalue of $\boldsymbol{A}^k$ for any positive integer k if and only if λ is an eigenvalue of $\boldsymbol{A}$ and that $\lambda^k = 0$ if and only if $\lambda = 0$. We leave the details to the reader. □

We have already seen how polynomials for which $f(\boldsymbol{A}) = \boldsymbol{O}$ can be useful to deduce properties of eigenvalues of certain matrices. A polynomial $f(t)$ for which $f(\boldsymbol{A}) = \boldsymbol{O}$ is said to ***annihilate*** the matrix $\boldsymbol{A}$. Annihilating polynomials occupy a conspicuous role in the development of linear algebra, none more so than a remarkable theorem that says that the characteristic polynomial of $\boldsymbol{A}$ annihilates $\boldsymbol{A}$. This result is known as the Caley-Hamilton Theorem. We prove this below, but prove a useful lemma first.

Lemma 11.2 *Let $\boldsymbol{T}$ be an $n \times n$ upper-triangular matrix with t_{ii} as its i-th diagonal element. Construct the following:*

$$\boldsymbol{U}_k = (\boldsymbol{T} - t_{11}\boldsymbol{I})(\boldsymbol{T} - t_{22}\boldsymbol{I}) \cdots (\boldsymbol{T} - t_{kk}\boldsymbol{I}) , \quad \textit{where } 1 \leq k \leq n .$$

Then, the first k columns of $\boldsymbol{U}_k$ are each equal to the $n \times 1$ null vector $\mathbf{0}$.

Proof. We again use induction to prove this. The result holds trivially for $k = 1$. For any integer $k > 1$, assume that

$$\boldsymbol{U}_{k-1} = (\boldsymbol{T} - t_{11}\boldsymbol{I})(\boldsymbol{T} - t_{22}\boldsymbol{I}) \cdots (\boldsymbol{T} - t_{k-1,k-1}\boldsymbol{I})$$

has its first $k - 1$ columns equal to $\mathbf{0}_{n\times 1}$. This means that we can partition $\boldsymbol{U}_{k-1}$ as

$$\boldsymbol{U}_{k-1} = \left[\boldsymbol{O} : \boldsymbol{u} : \boldsymbol{V}\right] ,$$

where $\boldsymbol{O}$ is the $n \times (k-1)$ null matrix, $\boldsymbol{u}$ is some $n \times 1$ column vector (representing the k-th column of $\boldsymbol{U}_{k-1}$) and $\boldsymbol{V}$ is $n \times (n-k)$. Also, $\boldsymbol{T} - t_{kk}\boldsymbol{I}$ is upper-triangular with the k-th diagonal element equal to 0. So, we can partition $\boldsymbol{T} - t_{kk}\boldsymbol{I}$ as

$$\boldsymbol{T} - t_{kk}\boldsymbol{I} = \begin{bmatrix} \boldsymbol{T}_1 & \boldsymbol{w} & \boldsymbol{B} \\ \mathbf{0}' & 0 & \boldsymbol{z}' \\ \boldsymbol{O} & \mathbf{0} & \boldsymbol{T}_2 \end{bmatrix} ,$$

where $\boldsymbol{T}_1$ is $(k-1) \times (k-1)$ upper-triangular, $\boldsymbol{w}$ is some $(k-1) \times 1$ vector, $\boldsymbol{B}$ is some $(k-1)\times(n-k)$ matrix, $\boldsymbol{z}'$ is some $1\times(n-k)$ vector and $\boldsymbol{T}_2$ is $(n-k)\times(n-k)$ upper-triangular. Therefore,

$$\boldsymbol{U}_k = \boldsymbol{U}_{k-1}(\boldsymbol{T} - t_{kk}\boldsymbol{I}) = \left[\boldsymbol{O} : \boldsymbol{u} : \boldsymbol{V}\right] \begin{bmatrix} \boldsymbol{T}_1 & \boldsymbol{w} & \boldsymbol{B} \\ \mathbf{0}' & 0 & \boldsymbol{z}' \\ \boldsymbol{O} & \mathbf{0} & \boldsymbol{T}_2 \end{bmatrix} = \left[\boldsymbol{O} : \mathbf{0} : \boldsymbol{u}\boldsymbol{z}' + \boldsymbol{V}\boldsymbol{T}_2\right] ,$$

which reveals that the k-th column of $\boldsymbol{U}_k$ is $\mathbf{0}$. □

We use the above lemma to prove the Cayley-Hamilton Theorem, which says that every square matrix satisfies its own characteristic equation.

Theorem 11.18 Cayley-Hamilton Theorem. *Let $p_{\boldsymbol{A}}(\lambda)$ be the characteristic polynomial of a square matrix $\boldsymbol{A}$. Then $p_{\boldsymbol{A}}(\boldsymbol{A}) = \boldsymbol{O}$.*

Proof. Let $\lambda_1, \lambda_2, \ldots, \lambda_n$ be the n eigenvalues (counting multiplicities) of an $n \times n$ matrix $\boldsymbol{A}$. By Theorem 11.14, there exists a nonsingular matrix $\boldsymbol{P}$ such that $\boldsymbol{P}^{-1}\boldsymbol{A}\boldsymbol{P} = \boldsymbol{T}$ is upper-triangular with $t_{ii} = \lambda_i$. Let $\boldsymbol{U}_n$ be the value of the characteristic polynomial of $\boldsymbol{A}$ evaluated at $\boldsymbol{T}$. Therefore,

$$\boldsymbol{U}_n = p_{\boldsymbol{A}}(\boldsymbol{T}) = (\boldsymbol{T} - \lambda_1\boldsymbol{I})(\boldsymbol{T} - \lambda_2\boldsymbol{I}) \cdots (\boldsymbol{T} - \lambda_n\boldsymbol{I}) .$$

By Lemma 11.2, $\boldsymbol{U}_n$ will have its first n columns equal to $\mathbf{0}$, which means that

$p_{\boldsymbol{A}}(\boldsymbol{T}) = \boldsymbol{U}_n = \boldsymbol{O}$. Also note that $\boldsymbol{A} = \boldsymbol{P}\boldsymbol{T}\boldsymbol{P}^{-1}$, so

$$\begin{aligned} p_{\boldsymbol{A}}(\boldsymbol{A}) &= (\boldsymbol{PTP}^{-1} - \lambda_1\boldsymbol{I})(\boldsymbol{PTP}^{-1} - \lambda_2\boldsymbol{I})\cdots(\boldsymbol{PTP}^{-1} - \lambda_n\boldsymbol{I}) \\ &= \boldsymbol{P}(\boldsymbol{T} - \lambda_1\boldsymbol{I})\boldsymbol{P}^{-1}\boldsymbol{P}(\boldsymbol{T} - \lambda_2\boldsymbol{I})\boldsymbol{P}^{-1}\cdots\boldsymbol{P}(\boldsymbol{T} - \lambda_n\boldsymbol{I})\boldsymbol{P}^{-1} \\ &= \boldsymbol{P}(\boldsymbol{T} - \lambda_1\boldsymbol{I})(\boldsymbol{T} - \lambda_2\boldsymbol{I})\cdots(\boldsymbol{T} - \lambda_n\boldsymbol{I})\boldsymbol{P}^{-1} = \boldsymbol{P}p_{\boldsymbol{A}}(\boldsymbol{T})\boldsymbol{P}^{-1} = \boldsymbol{O}\,, \end{aligned}$$

which completes the proof. □

Theorem 11.18 ensures the existence of a polynomial that annihilates $\boldsymbol{A}$, i.e., $p(\boldsymbol{A}) = \boldsymbol{O}$. The characteristic polynomial of a matrix is one such example. If $p(x)$ is a polynomial such that $p(\boldsymbol{A}) = \boldsymbol{O}$ and α is any nonzero scalar, then $\alpha p(x)$ is another polynomial such that $\alpha p(\boldsymbol{A}) = \boldsymbol{O}$. This means that there must exist a *monic* polynomial that annihilates $\boldsymbol{A}$. A monic polynomial is a polynomial

$$p(x) = c_n x^n + c_{n-1}x^{n-1} + \cdots + c_2x^2 + c_1x + c_0$$

in which the leading coefficient $c_n = 1$. The characteristic polynomial, as defined in (11.3) is a monic polynomial. Let k be the smallest degree of a nonzero polynomial that annihilates $\boldsymbol{A}$. Can there be two such polynomials? No, because if $p_1(x)$ and $p_2(x)$ are two monic polynomials of degree k that annihilate $\boldsymbol{A}$, then $f(x) = p_1(x) - p_2(x)$ is a polynomial of degree strictly less than k that would annihilate $\boldsymbol{A}$. And $\alpha f(x)$ would be a monic annihilating polynomial of $\boldsymbol{A}$ with degree less than k, where $1/\alpha$ is the leading coefficient of $f(x)$. But this would contradict the definition of k. So $f(x) = p_1(x) - p_2(x) = 0$. This suggests the following definition.

Definition 11.10 *The monic polynomial of smallest degree that annihilates* $\boldsymbol{A}$ *is called the* **minimum** *or* **minimal polynomial** *for* $\boldsymbol{A}$.

Theorem 11.18 guarantees that the degree of the minimum polynomial for an $n \times n$ matrix $\boldsymbol{A}$ cannot exceed n. The following result shows that the minimum polynomial of $\boldsymbol{A}$ divides every nonzero annihilating polynomial of $\boldsymbol{A}$.

Theorem 11.19 *The minimum polynomial of* $\boldsymbol{A}$ *divides every annihilating polynomial of* $\boldsymbol{A}$.

Proof. Let $m_{\boldsymbol{A}}(x)$ be the minimum polynomial of $\boldsymbol{A}$ and let $g(x)$ be any annihilating polynomial of $\boldsymbol{A}$. Since the degree of $m_{\boldsymbol{A}}(x)$ does not exceed that of $g(x)$, the polynomial long division algorithm from basic algebra ensures that there exist polynomials $q(x)$ and $r(x)$ such that

$$g(x) = m_{\boldsymbol{A}}(x)q(x) + r(x)\,,$$

where the degree of $r(x)$ is strictly less than that of $m_{\boldsymbol{A}}(x)$. Note: $q(x)$ is the "quotient" and $r(x)$ is the "remainder." Since $m_{\boldsymbol{A}}(\boldsymbol{A}) = g(\boldsymbol{A}) = \boldsymbol{O}$, we obtain

$$\boldsymbol{O} = g(\boldsymbol{A}) = m_{\boldsymbol{A}}(\boldsymbol{A})q(\boldsymbol{A}) + r(\boldsymbol{A}) = \boldsymbol{O}\cdot q(\boldsymbol{A}) + r(\boldsymbol{A}) = r(\boldsymbol{A})\,,$$

so $r(\boldsymbol{A}) = \boldsymbol{O}$. But this would contradict $m_{\boldsymbol{A}}(x)$ being the minimum polynomial because $r(x)$ has smaller degree than $m(x)$. Therefore, we must have $r(x) \equiv 0$, so $m(x)$ divides $q(x)$. □

Theorem 11.19 implies that the minimum polynomial divides the characteristic polynomial. This sometimes helps in finding the minimum polynomial of a matrix as we need look no further than the factors of the characteristic polynomial, or any annihilating polynomial for that matter. Here is an example.

Example 11.7 The minimum polynomial of an idempotent matrix. Let $\boldsymbol{A}$ be an idempotent matrix. Then, $f(x) = x^2 - x$ is an annihilating polynomial for $\boldsymbol{A}$. Since $f(x) = x(x-1)$, it follows that the minimum polynomial could be one of the following:

$$m(x) = x \quad \text{or} \quad m(x) = x - 1 \quad \text{or} \quad m(x) = x^2 - x \, .$$

But we also know that $m(\boldsymbol{A}) = \boldsymbol{O}$ and only $m(x) = x^2 - x$ satisfies this requirement. So, the minimum polynomial of $\boldsymbol{A}$ is $m_{\boldsymbol{A}}(x) = x^2 - x$. ■

It is easy to see that if λ is a root of the minimum polynomial, it is also a root of the characteristic polynomial. But is a root of the characteristic polynomial also a root of the minimum polynomial? The following theorem lays the matter to rest.

Theorem 11.20 *A complex number λ is a root of the minimum polynomial of $\boldsymbol{A}$ if and only if it is a root of the characteristic polynomial of $\boldsymbol{A}$.*

Proof. Let $m_{\boldsymbol{A}}(x)$ be the minimum polynomial and $p_{\boldsymbol{A}}(x)$ be the characteristic polynomial of $\boldsymbol{A}$. First consider the easy part. Suppose λ is a root of the minimum polynomial. Since the minimum polynomial divides the characteristic polynomial, there is a polynomial $q(x)$ such that $p_{\boldsymbol{A}}(x) = m(x)q(x)$. Therefore,

$$p_{\boldsymbol{A}}(\lambda) = m_{\boldsymbol{A}}(\lambda)q(\lambda) = 0 \cdot q(\lambda) = 0 \, ,$$

which establishes λ as a characteristic root of $\boldsymbol{A}$.

Next suppose that $p_{\boldsymbol{A}}(\lambda) = 0$. Then λ is an eigenvalue of $\boldsymbol{A}$. From Theorem 11.17, we know that $m_{\boldsymbol{A}}(\lambda)$ is an eigenvalue of $m_{\boldsymbol{A}}(\boldsymbol{A})$. But $m(\boldsymbol{A}) = \boldsymbol{O}$. Therefore, every eigenvalue of $m(\boldsymbol{A})$ is zero and so $m_{\boldsymbol{A}}(\lambda) = 0$. □

The above theorem ensures that the distinct roots of the minimum polynomial are the same as those of the characteristic polynomial. Therefore, if an $n \times n$ matrix $\boldsymbol{A}$ has n distinct characteristic roots, then the minimum polynomial of $\boldsymbol{A}$ is the characteristic polynomial of $\boldsymbol{A}$.

Theorem 11.21 *The minimum polynomial of an $n \times n$ diagonal matrix $\boldsymbol{\Lambda}$ is given by*

$$m_{\boldsymbol{\Lambda}}(\lambda) = (\lambda - \lambda_1)(\lambda - \lambda_2) \cdots (\lambda - \lambda_k) \, ,$$

where $\lambda_1, \lambda_2, \ldots, \lambda_k$ are the distinct diagonal entries in $\boldsymbol{\Lambda}$.

Proof. Suppose that m_i is the number of times λ_i appears along the diagonal of $\boldsymbol{\Lambda}$:

$$\boldsymbol{\Lambda} = \begin{bmatrix} \lambda_1 \boldsymbol{I}_{m_1} & \boldsymbol{O} & \dots & \boldsymbol{O} \\ \boldsymbol{O} & \lambda_2 \boldsymbol{I}_{m_2} & \dots & \boldsymbol{O} \\ \vdots & \vdots & \ddots & \vdots \\ \boldsymbol{O} & \boldsymbol{O} & \dots & \lambda_k \boldsymbol{I}_{m_k} \end{bmatrix} .$$

It is easily verified that

$$(\boldsymbol{\Lambda} - \lambda_1 \boldsymbol{I}_{m_1})(\boldsymbol{\Lambda} - \lambda_2 \boldsymbol{I}_{m_2}) \cdots (\boldsymbol{\lambda} - \lambda_k \boldsymbol{I}_{m_k}) = \boldsymbol{O} .$$

This means that the polynomial

$$f(\lambda) = (\lambda - \lambda_1)(\lambda - \lambda_2) \cdots (\lambda - \lambda_k) \tag{11.17}$$

is an annihilating polynomial of $\boldsymbol{A}$. Hence, the minimum polynomial of $\boldsymbol{A}$ divides $f(\lambda)$. Each λ_i is a characteristic root. So, from Theorem 11.20, we know that each λ_i is a root of the minimum polynomial. Because the minimum polynomial is monic, it follows that the minimum polynomial must equal $f(\lambda)$ in (11.17). □

Theorem 11.6 tells us that the characteristic polynomial of similar matrices are the same. The minimum polynomial also enjoys this property.

Theorem 11.22 *Similar matrices have the same minimum polynomial.*

Proof. Let $\boldsymbol{A}$ and $\boldsymbol{B}$ be similar matrices and suppose that $\boldsymbol{P}^{-1}\boldsymbol{A}\boldsymbol{P} = \boldsymbol{B}$. Then, $f(\boldsymbol{B}) = \boldsymbol{P}^{-1} f(\boldsymbol{A})\boldsymbol{P}$ for any polynomial of $f(\lambda)$. Thus, $f(\boldsymbol{B}) = \boldsymbol{O}$ if and only if $f(\boldsymbol{A}) = \boldsymbol{O}$. Hence, $\boldsymbol{A}$ and $\boldsymbol{B}$ have the same minimum polynomial. □

Theorem 11.21 tells us that the minimum polynomial of a diagonal matrix is a product of distinct linear factors of the form $(\lambda - \lambda_i)$. It is not necessary that every minimum polynomial is a product of distinct linear factors. We will show below that the minimum polynomial of a matrix is a product of distinct linear factors if and only if the matrix is diagonalizable. This makes the minimum polynomial a useful tool in detecting when a matrix is diagonalizable.

Theorem 11.23 *An $n \times n$ matrix $\boldsymbol{A}$ is diagonalizable if and only if its minimum polynomial is a product of distinct linear factors.*

Proof. Let $\boldsymbol{A}$ be diagonalizable and $\boldsymbol{P}^{-1}\boldsymbol{A}\boldsymbol{P} = \boldsymbol{\Lambda}$, where $\boldsymbol{\Lambda}$ is diagonal with the eigenvalues (counting multiplicities) of $\boldsymbol{A}$ along its diagonal. Theorem 11.22 tells us that the minimum polynomials of $\boldsymbol{A}$ and $\boldsymbol{\Lambda}$ are the same. Theorem 11.21 tells us that the minimum polynomial of $\boldsymbol{\Lambda}$ is

$$m_{\boldsymbol{A}}(\lambda) = m_{\boldsymbol{\Lambda}}(\boldsymbol{A}) = (\lambda - \lambda_1)(\lambda - \lambda_2) \cdots (\lambda - \lambda_k) ,$$

where λ_i's, $i = 1, 2, \dots, k$ are the distinct eigenvalues of $\boldsymbol{A}$. Therefore, the minimum polynomial of $\boldsymbol{A}$ is a product of distinct linear factors.

To prove the *only if* part, we will show that $\text{AM}_{\boldsymbol{A}}(\lambda) = \text{GM}_{\boldsymbol{A}}(\lambda)$ for every eigenvalue of $\boldsymbol{A}$. Suppose that the minimum polynomial of an $n \times n$ matrix $\boldsymbol{A}$ is $m_{\boldsymbol{A}}(\lambda) = (\lambda - \lambda_1)(\lambda - \lambda_2)\cdots(\lambda - \lambda_k)$. Since the minimum polynomial annihilates $\boldsymbol{A}$, we obtain

$$m_{\boldsymbol{A}}(\boldsymbol{A}) = (\boldsymbol{A} - \lambda_1 \boldsymbol{I})(\boldsymbol{A} - \lambda_2 \boldsymbol{I})\cdots(\boldsymbol{A} - \lambda_k \boldsymbol{I}) = \boldsymbol{O} .$$

Thus, the rank of $m_{\boldsymbol{A}}(\boldsymbol{A})$ is zero, and so the nullity of $m_{\boldsymbol{A}}(\boldsymbol{A})$ is equal to n. Using Sylvester's inequality (recall Equation (5.4)), we obtain that

$$\nu(\boldsymbol{A} - \lambda_1 \boldsymbol{I}) + \nu(\boldsymbol{A} - \lambda_2 \boldsymbol{I}) + \cdots + \nu(\boldsymbol{A} - \lambda_k \boldsymbol{I}) \geq \nu(m_{\boldsymbol{A}}(\boldsymbol{A})) = n . \quad (11.18)$$

Since λ_i is a root of the minimum polynomial if and only if it is an eigenvalue (recall Theorem 11.20), we know that $\lambda_1, \lambda_2, \ldots, \lambda_k$ are the distinct eigenvalues of $\boldsymbol{A}$. Observe that

$$\nu(\boldsymbol{A} - \lambda_i \boldsymbol{I}) = \mathrm{GM}_{\boldsymbol{A}}(\lambda_i) \leq \mathrm{AM}_{\boldsymbol{A}}(\lambda_i) \text{ for } i = 1, 2, \ldots, k .$$

Using (11.18) and the fact that $n = \sum_{i=1}^{k} \mathrm{AM}_{\boldsymbol{A}}(\lambda_i)$, we find that

$$n \leq \sum_{i=1}^{k} \nu(\boldsymbol{A} - \lambda_i \boldsymbol{I}) = \sum_{i=1}^{k} \mathrm{GM}_{\boldsymbol{A}}(\lambda_i) \leq \sum_{i=1}^{k} \mathrm{AM}_{\boldsymbol{A}}(\lambda_i) = n .$$

This proves that $\sum_{i=1}^{k} \mathrm{GM}_{\boldsymbol{A}}(\lambda_i) \leq \sum_{i=1}^{k} \mathrm{AM}_{\boldsymbol{A}}(\lambda_i)$ and, since each term in the first sum cannot exceed that in the second, $\mathrm{GM}_{\boldsymbol{A}}(\lambda_i) = \mathrm{AM}_{\boldsymbol{A}}(\lambda_i)$ for each $i = 1, 2, \ldots, k$. Therefore, $\boldsymbol{A}$ is diagonalizable. □

We conclude this section with a few remarks on the *companion matrix* of a polynomial. We already know that every matrix has a monic polynomial of degree n as its characteristic polynomial. But is the converse true? Is every monic polynomial of degree n the characteristic polynomial of some $n \times n$ matrix? The answer is YES and it lies with the *companion matrix* of a polynomial.

Definition 11.11 *The* **companion matrix** *of the monic polynomial*

$$p(x) = x^n + a_{n-1}x^{n-1} + a_{n-2}x^{n-2} + \cdots + a_1 x + a_0$$

is defined to be the $n \times n$ *matrix*

$$\boldsymbol{A} = \begin{bmatrix} 0 & 0 & \ldots & 0 & -a_0 \\ 1 & 0 & \ldots & 0 & -a_1 \\ 0 & 1 & \ldots & 0 & -a_2 \\ \vdots & \vdots & \ddots & \vdots & \vdots \\ 0 & 0 & \ldots & 1 & -a_{n-1} \end{bmatrix} .$$

Let us consider a 3×3 example.

Example 11.8 The 3×3 matrix

$$\boldsymbol{A} = \begin{bmatrix} 0 & 0 & -a_0 \\ 1 & 0 & -a_1 \\ 0 & 1 & -a_2 \end{bmatrix}$$

is the companion matrix of the polynomial $p(x) = a_0 + a_1 x + a_2 x^2 + x^3$. The characteristic polynomial of $\boldsymbol{A}$ is $|\lambda \boldsymbol{I} - \boldsymbol{A}|$, which is

$$\begin{vmatrix} \lambda & 0 & a_0 \\ -1 & \lambda & a_1 \\ 0 & 1 & \lambda + a_2 \end{vmatrix} = a_0 \begin{vmatrix} -1 & \lambda \\ 0 & -1 \end{vmatrix} - a_1 \begin{vmatrix} \lambda & 0 \\ 0 & -1 \end{vmatrix} + (\lambda + a_2) \begin{vmatrix} \lambda & 0 \\ -1 & \lambda \end{vmatrix}$$
$$= a_0 + a_1 \lambda + (\lambda + a_2)\lambda^2 = a_0 + a_1 \lambda + a_2 \lambda^2 + \lambda^3 \,.$$

This shows that $p(\lambda)$ is the characteristic polynomial of $\boldsymbol{A}$. ■

Example 11.8 shows that every monic polynomial of degree 3 is the characteristic polynomial of its companion matrix. This result can be established for n-th degree monic polynomials. One way to do this is to compute the determinant, as was done in Example 11.8, using a cofactor expansion.

Theorem 11.24 *Every n-th degree monic polynomial $p(x)$ is the characteristic polynomial of its companion matrix.*

Proof. We prove this by induction. When $n = 1$, the result is trivial, when $n = 2$ it is easy, and $n = 3$ has been established in Example 11.8. Suppose the result is true for all monic polynomials of degree $n - 1$. If $p(x) = a_0 + a_1 \lambda + \cdots + a_n \lambda^n$ and $\boldsymbol{A}$ is its companion matrix, then define $\boldsymbol{B} = \lambda \boldsymbol{I} - \boldsymbol{A}$ and note that

$$|\boldsymbol{B}| = |\lambda \boldsymbol{I}_n - \boldsymbol{A}| = \begin{vmatrix} \lambda & 0 & \dots & 0 & a_0 \\ -1 & \lambda & \dots & 0 & a_1 \\ 0 & -1 & \dots & 0 & a_2 \\ \vdots & \vdots & \ddots & \vdots & \vdots \\ 0 & 0 & \dots & -1 & \lambda + a_{n-1} \end{vmatrix} .$$

Consider the cofactors of λ and a_0 in $\boldsymbol{B}$ (recall Section 10.7). The cofactor of λ in $\boldsymbol{B}$ is B_{11}, which is equal to $|\lambda \boldsymbol{I}_{n-1} - \boldsymbol{M}_{11}|$, where $\boldsymbol{M}_{11}$ is the $(n-1) \times (n-1)$ matrix obtained from $\boldsymbol{A}$ by deleting the first row and first column. Note that $\boldsymbol{M}_{11}$ is the companion matrix of the polynomial $x^{n-1} + a_{n-1}x^{n-2} + a_{n-2}x^{n-3} + \cdots + a_2 x + a_1$. Therefore, the induction hypothesis ensures that

$$B_{11} = |\lambda \boldsymbol{I} - \boldsymbol{M}_{11}| = \lambda^{n-1} + a_{n-1}\lambda^{n-2} + a_{n-2}\lambda^{n-3} + \cdots + a_2 \lambda + a_1 \,.$$

The cofactor of a_0 in $\boldsymbol{B}$ is $B_{1n} = (-1)^{n+1}|\boldsymbol{B}_{1n}|$, where $\boldsymbol{B}_{1n}$ is the $(n-1) \times (n-1)$ matrix obtained by deleting the first row and last column from $\boldsymbol{B}$. This is an upper-triangular matrix with -1's along the diagonal. Therefore, $|\boldsymbol{B}_{1n}| = (-1)^{n-1}$.

Evaluating $|\boldsymbol{B}|$ using a cofactor expansion about the first row, we find

$$\begin{aligned}|\boldsymbol{B}| = |\lambda \boldsymbol{I}_n - \boldsymbol{A}| &= \lambda B_{11} + a_0 B_{1n} = \lambda|\lambda \boldsymbol{I} - \boldsymbol{M}_{11}| + a_0(-1)^{n+1}|\boldsymbol{B}_{1n}| \\ &= \lambda\left(\lambda^{n-1} + a_{n-1}\lambda^{n-2} + a_{n-2}\lambda^{n-3} + \cdots + a_2\lambda + a_1\right) + a_0(-1)^{n+1}(-1)^{n-1} \\ &= \lambda^n + a_{n-1}\lambda^{n-1} + a_{n-2}\lambda^{n-2} + \cdots + a_2\lambda^2 + a_1\lambda + a_0 = p(\lambda) ,\end{aligned}$$

which shows that the characteristic polynomial of $\boldsymbol{A}$ is $p(\lambda)$. □

11.7 Spectral decomposition of real symmetric matrices

Symmetric matrices with real entries form an important class of matrices. They are important in statistics because variance-covariance matrices are precisely of this type. They are also special when studying eigenvalues because they admit certain properties that are otherwise untrue. When matrices have entries that are complex numbers, the analogue of a symmetric matrix is a ***Hermitian matrix***. If $\boldsymbol{A}$ is a Hermitian matrix, then it is equal to its adjoint or conjugate transpose (recall Definition 5.5). In other words, $\boldsymbol{A} = \boldsymbol{A}^*$. If $\boldsymbol{A}$ is a real matrix (i.e., all its elements are real numbers), then the adjoint is equal to the transpose.

We establish an especially important property of Hermitian and real symmetric matrices below.

Theorem 11.25 *If λ is an eigenvalue of a Hermitian matrix $\boldsymbol{A}$, then λ is a real number. In particular, all eigenvalues of a real symmetric matrix are real.*

Proof. Let $\boldsymbol{x}$ be an eigenvector corresponding to the eigenvalue λ. Then,

$$\boldsymbol{x}^*\boldsymbol{A}\boldsymbol{x} = \lambda\boldsymbol{x}^*\boldsymbol{x} ,$$

where $\boldsymbol{x}^*$ is the adjoint of $\boldsymbol{x}$. Since $\boldsymbol{A} = \boldsymbol{A}^*$, taking the conjugate transpose yields

$$\bar{\lambda}\boldsymbol{x}^*\boldsymbol{x} = (\boldsymbol{x}^*\boldsymbol{A}\boldsymbol{x})^* = \boldsymbol{x}^*\boldsymbol{A}^*\boldsymbol{x} = \boldsymbol{x}^*\boldsymbol{A}\boldsymbol{x} = \lambda\boldsymbol{x}^*\boldsymbol{x} .$$

Since $\boldsymbol{x} \neq \boldsymbol{0}$, it follows that $\boldsymbol{x}^*\boldsymbol{x} > 0$ and we conclude that $\lambda = \bar{\lambda}$. Therefore, λ is a real number. □

Earlier, in Theorem 11.9, we saw that eigenvectors associated with distinct eigenvalues are linearly independent. If the matrix in question is real and symmetric, then we can go further and say that such eigenvectors will be orthogonal to each other. This is our next result.

Theorem 11.26 *Let $\boldsymbol{x}_1$ and $\boldsymbol{x}_2$ be real eigenvectors corresponding to two distinct eigenvalues, λ_1 and λ_2, of a real symmetric matrix $\boldsymbol{A}$. Then $\boldsymbol{x} \perp \boldsymbol{x}_2$.*

Proof. We have $\boldsymbol{A}\boldsymbol{x}_1 = \lambda_1\boldsymbol{x}_1$ and $\boldsymbol{A}\boldsymbol{x}_2 = \lambda_2\boldsymbol{x}_2$. Hence $\boldsymbol{x}_2'\boldsymbol{A}\boldsymbol{x}_1 = \lambda_1\boldsymbol{x}_2'\boldsymbol{x}_1$ and $\boldsymbol{x}_1'\boldsymbol{A}\boldsymbol{x}_2 = \lambda_2\boldsymbol{x}_1'\boldsymbol{x}_2$. Since $\boldsymbol{A}$ is symmetric, subtracting the second equation from the first we have $(\lambda_1 - \lambda_2)\boldsymbol{x}_1'\boldsymbol{x}_2 = 0$. Since $\lambda_1 \neq \lambda_2$, we conclude that $\boldsymbol{x}_1'\boldsymbol{x}_2 = 0$. □

Recall that a matrix is diagonalizable if and only if it has a linearly independent set of eigenvectors. For such a matrix $\boldsymbol{A}$, we can find a nonsingular $\boldsymbol{P}$ such that $\boldsymbol{P}^{-1}\boldsymbol{A}\boldsymbol{P} = \boldsymbol{D}$. If we insist that $\boldsymbol{P}$ be a unitary matrix, then all we can ensure (by Schur's Triangularization Theorem) is that $\boldsymbol{P}^*\boldsymbol{A}\boldsymbol{P} = \boldsymbol{T}$ is upper-triangular. What happens if $\boldsymbol{A}$ is Hermitian, i.e., if $\boldsymbol{A} = \boldsymbol{A}^*$? Then,

$$\boldsymbol{T} = \boldsymbol{P}^*\boldsymbol{A}\boldsymbol{P} = \boldsymbol{P}^*\boldsymbol{A}^*\boldsymbol{P} = (\boldsymbol{P}^*\boldsymbol{A}\boldsymbol{P})^* = \boldsymbol{T}^* .$$

This means that $\boldsymbol{T}$ is Hermitian. But $\boldsymbol{T}$ is upper-triangular as well. A matrix that is both triangular and Hermitian must be diagonal. This would suggest that Hermitian matrices should be unitarily similar to a diagonal matrix.

In fact, if $\boldsymbol{A}$ is real and symmetric, we can claim that it will be *orthogonally similar* to a *diagonal* matrix. This is known as the *spectral theorem* or *spectral decomposition* for real symmetric matrices. We state and prove this without deriving it from the Schur's Triangularization Theorem.

Theorem 11.27 Spectral theorem for real symmetric matrices. *If $\boldsymbol{A}$ is a real symmetric matrix, then there exists an orthogonal matrix $\boldsymbol{P}$ and a real diagonal matrix $\boldsymbol{\Lambda}$ such that*

$$\boldsymbol{P}'\boldsymbol{A}\boldsymbol{P} = \boldsymbol{\Lambda} .$$

The diagonal entries of $\boldsymbol{\Lambda}$ are the eigenvalues of $\boldsymbol{A}$ and the columns of $\boldsymbol{P}$ are the corresponding eigenvectors.

Proof. We will prove the assertion by induction. Obviously the result is true for $n = 1$. Suppose it is true for all $(n-1) \times (n-1)$ real symmetric matrices. Let $\boldsymbol{A}$ be an $n \times n$ real symmetric matrix with eigenvalues $\lambda_1 \leq \lambda_2 \cdots \leq \lambda_n$. Let $\boldsymbol{x}_1$ be an eigenvector corresponding to λ_1. Since $\boldsymbol{A}$ is real and λ_1 is real we can choose $\boldsymbol{x}_1$ to be real. Normalize $\boldsymbol{x}_1$ so that $\boldsymbol{x}_1'\boldsymbol{x}_1 = 1$.

Extend $\boldsymbol{x}_1$ to an orthonormal basis of $\mathbb{R}^n$ as $\{\boldsymbol{x}_1, \boldsymbol{x}_2, \ldots, \boldsymbol{x}_n\}$. Let

$$\boldsymbol{X} = \left[\boldsymbol{x}_1 : \boldsymbol{x}_2 : \ldots : \boldsymbol{x}_n\right] = [\boldsymbol{x}_1 : \boldsymbol{X}_2] .$$

By construction, $\boldsymbol{X}$ is orthogonal, so $\boldsymbol{X}'\boldsymbol{X} = \boldsymbol{X}\boldsymbol{X}' = \boldsymbol{I}_n$. Therefore, $\boldsymbol{X}_2'\boldsymbol{x}_1 = \boldsymbol{0}$ and

$$\boldsymbol{X}'\boldsymbol{A}\boldsymbol{X} = \begin{bmatrix} \boldsymbol{x}_1' \\ \boldsymbol{X}_2' \end{bmatrix} \boldsymbol{A}[\boldsymbol{x}_1 : \boldsymbol{X}_2] = \begin{bmatrix} \lambda_1 & \lambda_1\boldsymbol{x}_1'\boldsymbol{X}_2 \\ \lambda_1\boldsymbol{X}_2'\boldsymbol{x}_1 & \boldsymbol{X}_2'\boldsymbol{A}\boldsymbol{X}_2 \end{bmatrix} = \begin{bmatrix} \lambda_1 & \boldsymbol{0}' \\ \boldsymbol{0} & \boldsymbol{B} \end{bmatrix} ,$$

where $\boldsymbol{B} = \boldsymbol{X}_2'\boldsymbol{A}\boldsymbol{X}_2$. The characteristic polynomial of $\boldsymbol{A}$ is the same as that of $\boldsymbol{X}'\boldsymbol{A}\boldsymbol{X}$, which, from the above, is

$$p_{\boldsymbol{A}}(t) = p_{\boldsymbol{X}'\boldsymbol{A}\boldsymbol{X}}(t) = (t - \lambda_1)|t\boldsymbol{I} - \boldsymbol{B}| = (t - \lambda_1)p_{\boldsymbol{B}}(t) .$$

Therefore, the eigenvalues of $\boldsymbol{B}$ are $\lambda_2 \leq \cdots \leq \lambda_n$. Since $\boldsymbol{B}$ is an $(n-1) \times (n-1)$ real symmetric matrix, by the induction hypothesis there exists an $(n-1) \times (n-1)$

orthogonal matrix $\boldsymbol{S}$ such that

$$\boldsymbol{S}'\boldsymbol{B}\boldsymbol{S} = \begin{bmatrix} \lambda_2 & 0 & \dots & 0 \\ 0 & \lambda_2 & \dots & 0 \\ \vdots & \vdots & \ddots & \vdots \\ 0 & 0 & \dots & \lambda_n \end{bmatrix} = \boldsymbol{\Lambda}_2 \,.$$

Let $\boldsymbol{U} = \begin{bmatrix} 1 & \mathbf{0}' \\ \mathbf{0} & \boldsymbol{S} \end{bmatrix}$. Since $\boldsymbol{S}$ is an orthogonal matrix, it is easily verified that $\boldsymbol{U}$ is also an orthogonal matrix. And since the product of two orthogonal matrices is another orthogonal matrix, $\boldsymbol{P} = \boldsymbol{U}\boldsymbol{X}$ is also orthogonal. Now observe that

$$\boldsymbol{P}'\boldsymbol{A}\boldsymbol{P} = \boldsymbol{X}'\boldsymbol{U}'\boldsymbol{A}\boldsymbol{U}\boldsymbol{X} = \begin{bmatrix} \lambda_1 & \mathbf{0}' \\ \mathbf{0} & \boldsymbol{\Lambda}_2 \end{bmatrix} = \boldsymbol{\Lambda} \,,$$

which completes the proof. □

Remark: If $\lambda_1, \lambda_2, \dots, \lambda_k$ are the *distinct* eigenvalues of an $n \times n$ real symmetric matrix $\boldsymbol{A}$ with multiplicities $m_1, m_2, \dots, m_k$, respectively, so $\sum_{i=1}^{k} m_i = n$, then we can place the corresponding set of orthogonal eigenvectors in $\boldsymbol{P}$ in such a way that the spectral form can be expressed, without loss of generality, as

$$\boldsymbol{P}'\boldsymbol{A}\boldsymbol{P} = \begin{bmatrix} \lambda_1 \boldsymbol{I}_{m_1} & \boldsymbol{O} & \dots & \boldsymbol{O} \\ \boldsymbol{O} & \lambda_2 \boldsymbol{I}_{m_2} & \dots & \boldsymbol{O} \\ \vdots & \vdots & \ddots & \vdots \\ \boldsymbol{O} & \boldsymbol{O} & \dots & \lambda_k \boldsymbol{I}_{m_k} \end{bmatrix} .$$

The spectral decomposition above is often referred to as the *eigenvalue decomposition* of a matrix. Hermitian matrices (with possibly complex entries) admit a similar spectral decomposition with respect to unitary matrices. But Hermitian matrices are not the only matrices that are unitarily diagonalizable. What would be a simple way to describe matrices that are unitarily similar to a diagonal matrix? We describe this class below.

Definition 11.12 *A complex matrix* $\boldsymbol{A}$ *is a* **normal matrix** *if* $\boldsymbol{A}^*\boldsymbol{A} = \boldsymbol{A}\boldsymbol{A}^*$.

For the above definition to make sense, $\boldsymbol{A}$ must be a square matrix. Therefore, when we say that a matrix is normal, we implicitly mean that it is also square. It is clear that Hermitian and real-symmetric matrices are normal.

Theorem 11.28 *A complex matrix is unitarily similar to a diagonal matrix if and only if it is normal.*

Proof. We first prove the "*only if*" part. Suppose $\boldsymbol{A}$ is an $n \times n$ matrix that is unitarily similar to a diagonal matrix. This means that there exists a unitary matrix $\boldsymbol{P}$ and a diagonal matrix $\boldsymbol{\Lambda}$ such that $\boldsymbol{A} = \boldsymbol{P}\boldsymbol{\Lambda}\boldsymbol{P}^*$. Now,

$$\boldsymbol{A}\boldsymbol{A}^* = \boldsymbol{P}\boldsymbol{\Lambda}\boldsymbol{P}^*\boldsymbol{P}\boldsymbol{\Lambda}^*\boldsymbol{P}^* = \boldsymbol{P}\boldsymbol{\Lambda}\boldsymbol{\Lambda}^*\boldsymbol{P}^* = \boldsymbol{P}\boldsymbol{\Lambda}^*\boldsymbol{\Lambda}\boldsymbol{P}^* = \boldsymbol{P}\boldsymbol{\Lambda}^*\boldsymbol{P}^*\boldsymbol{P}\boldsymbol{\Lambda}\boldsymbol{P}^* = \boldsymbol{A}^*\boldsymbol{A} \,,$$

where we have used the fact that $\boldsymbol{\Lambda}\boldsymbol{\Lambda}^* = \boldsymbol{\Lambda}^*\boldsymbol{\Lambda}$ (i.e., $\boldsymbol{\Lambda}$, being a diagonal matrix, is easily verified to be normal).

To prove the "*if*" part, we assume that $\boldsymbol{A}$ is an $n \times n$ normal matrix. By the Schur's Triangularization Theorem, there exists a unitary matrix $\boldsymbol{P}$ and an upper-triangular matrix $\boldsymbol{T}$ such that $\boldsymbol{A} = \boldsymbol{P}\boldsymbol{T}\boldsymbol{P}^*$. Note that

$$\boldsymbol{A}\boldsymbol{A}^* = \boldsymbol{P}\boldsymbol{T}\boldsymbol{P}^*\boldsymbol{P}\boldsymbol{T}^*\boldsymbol{P}^* = \boldsymbol{P}\boldsymbol{T}\boldsymbol{T}^*\boldsymbol{P}^* \text{ and } \boldsymbol{A}^*\boldsymbol{A} = \boldsymbol{P}\boldsymbol{T}^*\boldsymbol{P}^*\boldsymbol{P}\boldsymbol{T}\boldsymbol{P}^* = \boldsymbol{P}\boldsymbol{T}^*\boldsymbol{T}\boldsymbol{P}^* .$$

Since $\boldsymbol{A}$ is normal, the above two matrices are equal, which implies that $\boldsymbol{T}^*\boldsymbol{T} = \boldsymbol{T}\boldsymbol{T}^*$. Therefore, $\boldsymbol{T}$ is a normal matrix. But it is easy to see that an upper-triangular matrix is normal if and only if it is diagonal. Equating the ith diagonal entries of $\boldsymbol{T}\boldsymbol{T}^*$ and $\boldsymbol{T}^*\boldsymbol{T}$, respectively, we see that

$$|t_{ii}|^2 = \sum_{j=i}^{n} |t_{ij}|^2.$$

This implies, that for each i, $t_{ij} = 0$ whenever $j > i$. Hence, $\boldsymbol{T}$ is diagonal. □

Remark: The above result shows that any $n \times n$ normal matrix will possess a linearly independent set of n eigenvectors. However, not all linearly independent sets of n eigenvectors of a normal matrix are necessarily orthonormal.

Let us suppose that $\boldsymbol{A}$ is an $n \times n$ normal matrix with rank $r \leq n$. Consider its spectral decomposition $\boldsymbol{P}'\boldsymbol{A}\boldsymbol{P} = \boldsymbol{\Lambda}$, where $\boldsymbol{\Lambda}$ is an $n \times n$ diagonal matrix with i-th diagonal element λ_i. Since $\rho(\boldsymbol{A}) = \rho(\boldsymbol{\Lambda})$, there are exactly r nonzero entries along the diagonal of $\boldsymbol{\Lambda}$. Assume, without loss of generality, that $\lambda_1 \geq \lambda_2 \geq \cdots \geq \lambda_r$ are the nonzero eigenvalues and $\lambda_i = 0$ for $i = r+1, r+2, \ldots, n$. Let $\boldsymbol{P} = [\boldsymbol{P}_1 : \boldsymbol{P}_2]$, where $\boldsymbol{P}_1$ is $n \times r$. Then, the spectral decomposition implies that

$$[\boldsymbol{A}\boldsymbol{P}_1 : \boldsymbol{A}\boldsymbol{P}_2] = \boldsymbol{A}\boldsymbol{P} = \boldsymbol{P}\boldsymbol{\Lambda} = [\boldsymbol{P}_1 : \boldsymbol{P}_2]\begin{bmatrix} \boldsymbol{\Lambda}_1 & \boldsymbol{O} \\ \boldsymbol{O} & \boldsymbol{O} \end{bmatrix} = [\boldsymbol{P}_1\boldsymbol{\Lambda}_1 : \boldsymbol{O}] ,$$

where $\boldsymbol{\Lambda}_1$ is an $r \times r$ diagonal matrix with λ_i as its i-th diagonal element for $i = 1, 2, \ldots, r$. The above implies that $\boldsymbol{A}\boldsymbol{P}_2 = \boldsymbol{O}$. This means that the $n - r$ columns of $\boldsymbol{P}_2$ form an orthonormal basis for the null space of $\boldsymbol{A}$. Also, $\boldsymbol{A}\boldsymbol{P}_1 = \boldsymbol{P}_1\boldsymbol{\Lambda}_1$. Since $\boldsymbol{\Lambda}_1$ is nonsingular, it follows that $\boldsymbol{P}_1 = \boldsymbol{A}\boldsymbol{P}_1\boldsymbol{\Lambda}_1^{-1}$ and

$$\mathcal{C}(\boldsymbol{P}_1) = \mathcal{C}(\boldsymbol{A}\boldsymbol{P}_1\boldsymbol{\Lambda}_1^{-1}) \subseteq \mathcal{C}(\boldsymbol{A}) = \mathcal{C}(\boldsymbol{P}\boldsymbol{\Lambda}\boldsymbol{P}') = \mathcal{C}(\boldsymbol{P}_1\boldsymbol{\Lambda}_1) \subseteq \mathcal{C}(\boldsymbol{P}_1) .$$

This proves that $\mathcal{C}(\boldsymbol{P}_1) = \mathcal{C}(\boldsymbol{A})$ and the r columns of $\boldsymbol{P}_1$ form an orthonormal basis for the column space of $\boldsymbol{A}$. Thus, the spectral decomposition also produces orthonormal bases for the column space and null space of $\boldsymbol{A}$. The first r columns of $\boldsymbol{P}$, which are the eigenvectors corresponding to the nonzero eigenvalues of $\boldsymbol{A}$, constitute an orthonormal basis for the column space of $\boldsymbol{A}$. The remaining $n - r$ columns of $\boldsymbol{P}$, which are eigenvectors associated with the zero eigenvalues, constitute a basis for the null space of $\boldsymbol{A}$.

Further (geometric) insight may be obtained by taking a closer look at $f(\boldsymbol{x}) = \boldsymbol{A}\boldsymbol{x}$. For a real-symmetric matrix $\boldsymbol{A}$, the transformation $f(\boldsymbol{x})$ takes a vector $\boldsymbol{x} \in \Re^n$ and maps it to another vector in $\Re^n$. If $\boldsymbol{A} = \boldsymbol{P}\boldsymbol{\Lambda}\boldsymbol{P}'$ is the spectral decomposition of $\boldsymbol{A}$, the columns of $\boldsymbol{P}$ are eigenvectors that constitute a very attractive basis for $\Re^n$. To see why, consider a vector $\boldsymbol{x} \in \Re^n$ and express this in terms of the column vectors in $\boldsymbol{P}$. There exists a vector $\boldsymbol{\alpha} \in \Re^n$ such that $\boldsymbol{x} = \boldsymbol{P}\boldsymbol{\alpha}$. Now, $\boldsymbol{A}\boldsymbol{x}$ is again a vector in $\Re^n$ and, therefore, can be expressed in terms of the columns of $\boldsymbol{P}$. This means that there exists a vector $\boldsymbol{\beta} \in \Re^n$ such that $\boldsymbol{A}\boldsymbol{x} = \boldsymbol{P}\boldsymbol{\beta}$. The vectors $\boldsymbol{\alpha}$ and $\boldsymbol{\beta}$ are the coordinates $\boldsymbol{x}$ and $\boldsymbol{A}\boldsymbol{x}$, respectively, with respect to the basis formed by the columns of $\boldsymbol{P}$. Now,

$$\boldsymbol{x} = \boldsymbol{P}\boldsymbol{\alpha} \Longrightarrow \boldsymbol{A}\boldsymbol{x} = \boldsymbol{A}\boldsymbol{P}\boldsymbol{\alpha} \Longrightarrow \boldsymbol{A}\boldsymbol{P}\boldsymbol{\alpha} = \boldsymbol{P}\boldsymbol{\beta} \Longrightarrow \boldsymbol{\beta} = (\boldsymbol{P}'\boldsymbol{A}\boldsymbol{P})\boldsymbol{\alpha} = \boldsymbol{\Lambda}\boldsymbol{\alpha} \, .$$

This means that with respect to the columns of $\boldsymbol{P}$, the coordinates of $\boldsymbol{A}\boldsymbol{x}$ are simply scalar multiples of the coordinates of $\boldsymbol{x}$. Since orthogonal matrices are generalizations of rotations in higher dimensions, the spectral decomposition essentially offers a rotated axes system with respect to which $f(\boldsymbol{x}) = \boldsymbol{A}\boldsymbol{x}$ may dilate some of the coordinates of $\boldsymbol{x}$ (corresponding to eigenvalues with magnitude greater than one) and may contract some others (corresponding to eigenvalues with magnitude less than one) while reflecting some through the origin (when the eigenvalues are negative).

11.8 Computation of eigenvalues

From a theoretical perspective, eigenvalues can be computed by finding the roots of the characteristic polynomial. Once an eigenvalue λ is obtained, one can find the eigenvectors (and characterize the eigenspace associated with λ) by solving the homogeneous linear system $(\boldsymbol{A} - \lambda\boldsymbol{I})\boldsymbol{x} = \boldsymbol{0}$. See Example 11.4 for a simple example. This would seem to suggest that computing eigenvalues and eigenvectors for general matrices comprises two steps: (a) solve a polynomial equation to obtain the eigenvalues, and (b) solve the corresponding eigenvalue equation (a linear homogeneous system) to obtain the eigenvectors.

Unfortunately, matters are more complicated for general $n \times n$ matrices. Determining solutions to polynomial equations turns out to be a formidable task. A classical nineteenth century result, attributed largely to the works of Abel and Galois, states that for every integer $m \geq 5$, there exists a polynomial of degree m whose roots cannot be solved using a finite number of arithmetic operations involving addition, subtraction, multiplication, division and n-th roots, This means that there does not exist a generalization of the quadratic formula for polynomials of degree five or more, and general polynomial equations cannot be solved by a finite number of arithmetic operations. Therefore, unlike algorithms such as Gaussian elimination, $\boldsymbol{LU}$ and $\boldsymbol{QR}$ decompositions, which solve a linear systems $\boldsymbol{A}\boldsymbol{x} = \boldsymbol{b}$ using a finite number of operations, solving the eigenvalue problem must involve *iterative* methods involving an infinite sequence of steps that will ultimately *converge* to the correct answer.

A formal treatment of such algorithms is the topic of *numerical linear algebra*, which is beyond the scope of this text. Excellent references for a more rigorous and detailed treatment include the books by Trefethen and Bau III (1997), Stewart (2001) and Golub and Van Loan (2013). Here, we will provide a brief overview of some of the popular approaches for eigenvalue computations. We will not provide formal proofs and restrict ourselves to heuristic arguments that should inform the reader why, intuitively, these algorithms work.

Power method

The *Power method* is an iterative technique for computing the largest absolute eigenvalue and an associated eigenvector for diagonalizable matrices. Let $\boldsymbol{A}$ be an $n \times n$ diagonalizable real matrix $\boldsymbol{A}$, with eigenvalues

$$|\lambda_1| > |\lambda_2| \geq |\lambda_2| \geq \cdots \geq |\lambda_n| \geq 0\,.$$

Note the strict inequality $|\lambda_1| > |\lambda_2|$ in the above sequence, which implies that the largest eigenvalue of $\boldsymbol{A}$ is assumed to be real and simple (i.e., it has multiplicity equal to one). Otherwise, the complex conjugate of λ_1 would have the same magnitude and there will be two "largest" eigenvalues.

The basic idea of the power iteration is that the sequence

$$\frac{\boldsymbol{A}\boldsymbol{x}}{\|\boldsymbol{A}\boldsymbol{x}\|}, \frac{\boldsymbol{A}^2\boldsymbol{x}}{\|\boldsymbol{A}\boldsymbol{x}\|}, \frac{\boldsymbol{A}^3\boldsymbol{x}}{\|\boldsymbol{A}^3\boldsymbol{x}\|}, \ldots, \frac{\boldsymbol{A}^k\boldsymbol{x}}{\|\boldsymbol{A}^k\boldsymbol{x}\|}, \ldots$$

converges to an eigenvector associated with the largest eigenvalue of $\boldsymbol{A}$. To make matters precise, suppose $\boldsymbol{x}_0$ is an initial guess for an eigenvector of $\boldsymbol{A}$. Assume that $\|\boldsymbol{x}_0\| = 1$ and construct the sequence:

$$\boldsymbol{x}_k = \frac{\boldsymbol{A}\boldsymbol{x}_{k-1}}{\|\boldsymbol{A}\boldsymbol{x}_{k-1}\|} \text{ and } \nu_k = \|\boldsymbol{A}\boldsymbol{x}_k\| \text{ for } k = 1, 2, 3, \ldots\,.$$

It turns out that the sequence $\boldsymbol{x}_k$ converges to an eigenvector (up to a negative or positive sign) associated with λ_1 and ν_k converges to λ_1. To see why this happens, observe that $\boldsymbol{x}_k = \frac{\boldsymbol{A}\boldsymbol{x}_{k-1}}{\|\boldsymbol{A}\boldsymbol{x}_{k-1}\|} = \frac{\boldsymbol{A}^2\boldsymbol{x}_{k-2}}{\|\boldsymbol{A}^2\boldsymbol{x}_{k-2)}\|} = \cdots = \frac{\boldsymbol{A}^k\boldsymbol{x}_0}{\|\boldsymbol{A}^k\boldsymbol{x}_0\|}$. Since $\boldsymbol{A}$ is diagonalizable, there exists a linearly independent set of n normalized eigenvectors of $\boldsymbol{A}$ that is a basis for $\Re^n$. Suppose $\{\boldsymbol{p}_1, \boldsymbol{p}_2, \ldots, \boldsymbol{p}_n\}$ is such a set, where $\boldsymbol{p}_1$ is the eigenvalue associated with λ_1. Let $\boldsymbol{x}_0 = \alpha_1\boldsymbol{p}_1 + \alpha_2\boldsymbol{p}_2 + \cdots + \alpha_n\boldsymbol{p}_n$. Now,

$$\begin{aligned}
\boldsymbol{A}^k\boldsymbol{x}_0 &= \alpha_1\boldsymbol{A}^k\boldsymbol{p}_1 + \alpha_2\boldsymbol{A}^k\boldsymbol{p}_2 + \cdots + \alpha_n\boldsymbol{A}^k\boldsymbol{p}_n \\
&= \alpha_1\lambda_1^k\boldsymbol{p}_1 + \alpha_2\lambda_2^k\boldsymbol{p}_2 + \cdots + \alpha_n\lambda_n^k\boldsymbol{p}_n \\
&= \lambda_1^k\left(\alpha_1\boldsymbol{p}_1 + \alpha_2\left(\frac{\lambda_2}{\lambda_1}\right)^k\boldsymbol{p}_2 + \cdots + \alpha_n\left(\frac{\lambda_n}{\lambda_1}\right)^k\boldsymbol{p}_n\right),
\end{aligned}$$

from which it follows that

$$\boldsymbol{x}_k = \frac{\boldsymbol{A}^k \boldsymbol{x}_0}{\|\boldsymbol{A}^k \boldsymbol{x}_0\|} = \frac{\lambda_1^k \left(\alpha_1 \boldsymbol{p}_1 + \alpha_2 \left(\frac{\lambda_2}{\lambda_1}\right)^k \boldsymbol{p}_2 + \cdots + \alpha_n \left(\frac{\lambda_n}{\lambda_1}\right)^k \boldsymbol{p}_n \right)}{\left\| \lambda_1^k \left(\alpha_1 \boldsymbol{p}_1 + \alpha_2 \left(\frac{\lambda_2}{\lambda_1}\right)^k \boldsymbol{p}_2 + \cdots + \alpha_n \left(\frac{\lambda_n}{\lambda_1}\right)^k \boldsymbol{p}_n \right) \right\|}$$
$$\longrightarrow \frac{\alpha_1}{|\alpha_1|} \boldsymbol{p}_1 = \pm \boldsymbol{p}_1 \text{ as } k \longrightarrow \infty$$

because $(\lambda_i/\lambda_1)^k \to 0$ as $k \to \infty$. Turning to the sequence of ν_k's now, we obtain

$$\lim_{k\to\infty} \nu_k = \lim_{k\to\infty} \|\boldsymbol{A}\boldsymbol{x}_k\| = \lim_{k\to\infty} \sqrt{\boldsymbol{x}_k' \boldsymbol{A}' \boldsymbol{A} \boldsymbol{x}_k} = \sqrt{\lambda_1 \boldsymbol{p}_1' \boldsymbol{p}_1} = \lambda_1 ,$$

so ν_k converges to the largest eigenvalue of $\boldsymbol{A}$.

To summarize, the sequence $(\nu_k, \boldsymbol{x}_k)$, constructed as above, converges to a *dominant* eigen-pair $(\lambda_1, \boldsymbol{p}_1)$. The power method is computationally quite efficient with each iteration requiring just a single matrix-vector multiplication. It is, however, most convenient when only a dominant eigen-pair is sought. Also, while the number of flops in each iteration is small, convergence can be slow. For instance, the rate of convergence of the power method is determined by the ratio λ_2/λ_1, i.e., the ratio of the largest to the smallest eigenvalue. If this ratio is close to 1, then convergence is slow.

Newer, faster and more general algorithms have replaced the power method as the default for computing eigenvalues in general. However, it has not lost relevance and continues to be used, especially for large and sparse matrices. For instance, Google uses it to calculate page-rank in their search engine. For matrices that are well-conditioned, large and sparse, the power method can be very effective for finding the dominant eigenvector.

Inverse iteration

Inverse iteration is a variant of the power iterations to compute any real eigen-pair of a diagonalizable matrix, given an approximation to any real eigenvalue. It specifically tries to avoid the problem of slow convergence of the power method when the two largest eigenvalues are not separated enough. Inverse iteration plays a central role in numerical linear algebra and it is the standard method for calculating eigenvectors of a matrix when the associated eigenvalues have been found.

The basic idea of the inverse iteration is fairly simple—apply the power iterations to the inverse of the matrix for which the eigenvalues are desired. Suppose $(\lambda, \boldsymbol{x})$ is an eigen-pair of $\boldsymbol{A}$, where $\lambda \neq 0$. Then,

$$\boldsymbol{A}\boldsymbol{x} = \lambda \boldsymbol{x} \Longrightarrow \boldsymbol{x} = \lambda \boldsymbol{A}^{-1} \boldsymbol{x} \Longrightarrow \boldsymbol{A}^{-1} \boldsymbol{x} = \frac{1}{\lambda} \boldsymbol{x} ,$$

which shows that $(1/\lambda, \boldsymbol{x})$ is an eigen-pair of $\boldsymbol{A}^{-1}$. More generally, if $\boldsymbol{A}$ is an invertible matrix with real nonzero eigenvalues $\{\lambda_1, \lambda_2, \ldots, \lambda_n\}$, then

$\{1/\lambda_1, 1/\lambda_2, \ldots, 1/\lambda_n\}$ are the real nonzero eigenvalues of $\boldsymbol{A}^{-1}$. If we now apply the power iteration to $\boldsymbol{A}^{-1}$ through the sequence $\boldsymbol{x}_k = \frac{\boldsymbol{A}^{-1}\boldsymbol{x}_{k-1}}{\|\boldsymbol{A}^{-1}\boldsymbol{x}_{k-1}\|}$, it follows from the discussion in the previous section that $\|\boldsymbol{A}^{-1}\boldsymbol{x}_k\|$ will converge to the largest eigenvalue of $\boldsymbol{A}^{-1}$, which is given by $1/\lambda_{\min}$, where $\lambda_{\min}$ is the eigenvalue of $\boldsymbol{A}$ with the smallest absolute value, and $\boldsymbol{x}_k$ will converge to an eigenvector associated with $1/\lambda_{\min}$.

Matters can be vastly improved if we can obtain an approximation τ to any real eigenvalue of $\boldsymbol{A}$, say λ. We now apply the power iteration to the matrix $(\boldsymbol{A} - \tau\boldsymbol{I})^{-1}$. Suppose $\boldsymbol{x}$ is an eigenvector of $\boldsymbol{A}$ associated with eigenvalue λ and let τ be a real number that is not an eigenvalue of $\boldsymbol{A}$. Then, $\boldsymbol{A} - \tau\boldsymbol{I}$ is invertible and we can write

$$(\boldsymbol{A} - \tau\boldsymbol{I})\boldsymbol{x} = \boldsymbol{A}\boldsymbol{x} - \tau\boldsymbol{x} = (\lambda - \tau)\boldsymbol{x} \Longrightarrow (\boldsymbol{A} - \tau\boldsymbol{I})^{-1}\boldsymbol{x} = \frac{1}{\lambda - \tau}\boldsymbol{x} ,$$

as long as $\tau \neq \lambda$. This shows that $(1/(\lambda - \tau), \boldsymbol{x})$ is an eigen-pair for $(\boldsymbol{A} - \tau\boldsymbol{I})^{-1}$. In fact, the entire set of eigenvalues of $(\boldsymbol{A} - \tau\boldsymbol{I})^{-1}$ are $\{1/(\lambda_1 - \tau), 1/(\lambda_2 - \tau), \ldots, 1/(\lambda - \tau_n)\}$, where λ_i's are the eigenvalues of $\boldsymbol{A}$.

Let λ be a particular eigenvalue of $\boldsymbol{A}$, which we seek to estimate along with an associated eigenvector. If τ is chosen such that $|\lambda - \tau|$ is smaller than $|\lambda_i - \tau|$ for any other eigenvalue λ_i of $\boldsymbol{A}$, then $1/(\lambda - \tau)$ is the dominant eigenvalue of $(\boldsymbol{A} - \tau\boldsymbol{I})^{-1}$. To find any eigenvalue λ of $\boldsymbol{A}$ along with an associated eigenvector, the inverse iteration method proceeds as below:

- Choose and fix a τ such that $|\lambda - \tau|$ is smaller than $|\lambda_i - \tau|$ for any other eigenvalue λ_i of $\boldsymbol{A}$.
- Start with an initial vector $\boldsymbol{x}_0$ with $\|\boldsymbol{x}_0\| = 1$.
- Construct the sequences:

$$\boldsymbol{x}_k = \frac{(\boldsymbol{A} - \tau\boldsymbol{I})^{-1}\boldsymbol{x}_{k-1}}{\|(\boldsymbol{A} - \tau\boldsymbol{I})^{-1}\boldsymbol{x}_{k-1}\|} \quad \text{and} \quad \nu_k = \|(\boldsymbol{A} - \tau\boldsymbol{I})^{-1}\boldsymbol{x}_k\| \quad \text{for } k = 1, 2, 3, \ldots .$$

  In practice, for any iteration k, we do not invert the matrix $(\boldsymbol{A} - \tau\boldsymbol{I})$ to compute $(\boldsymbol{A} - \tau\boldsymbol{I})^{-1}\boldsymbol{x}_{k-1}$, but solve $(\boldsymbol{A} - \tau\boldsymbol{I})\boldsymbol{w} = \boldsymbol{x}_{k-1}$ and compute $\boldsymbol{x}_k = \frac{\boldsymbol{w}}{\|\boldsymbol{w}\|}$. Efficiency is further enhanced by noting that $\boldsymbol{A} - \tau\boldsymbol{I}$ remains the same for every k. Therefore, we can compute the $\boldsymbol{LU}$ decomposition of $\boldsymbol{A} - \tau\boldsymbol{I}$ at the outset. Then, each iteration requires only one forward solve and one backward solve (recall Section 3.1) to determine $\boldsymbol{w}$.

The power method will now ensure that the sequence $(\nu_k, \boldsymbol{x}_k)$, as defined above, will converge to the eigen-pair $(1/(\lambda - \tau), \boldsymbol{x})$. Therefore, we can obtain $\lambda = \tau + \lim_{k\to\infty} \nu_k^{-1}$. Since $\boldsymbol{x}$ is an eigenvector associated with λ, we have obtained the eigen-pair $(\lambda, \boldsymbol{x})$.

The inverse iteration allows the computation of *any* eigenvalue and an associated eigenvector of a matrix. In this regard it is more versatile than the power method, which finds only the dominant eigen-pair. However, in order to compute a particular eigenvalue, the inverse iteration requires an initial approximation to start the iteration. In practice, convergence of inverse iteration can be dramatically quick, even if $\boldsymbol{x}_0$ is not a very close to the eigenvector $\boldsymbol{x}$; sometimes convergence is seen in just one or two iterations.

An apparent problem that does not turn out to be a problem in practice is when τ is very close to λ. Then, the system $(\boldsymbol{A} - \tau\boldsymbol{I})\boldsymbol{w} = \boldsymbol{x}_{k-1}$ could be ill-conditioned because $(\boldsymbol{A} - \tau\boldsymbol{I})$ is nearly singular. And we know that solving ill-conditioned systems is a problem. However, what is really important in the iterations is the direction of the solution, which is usually quite robust in spite of conditioning problems. This matter is subtle and more formal treatments can be found in numerical linear algebra texts. Briefly, here is what happens. Suppose we solve an ill-conditioned system $(\boldsymbol{A} - \tau\boldsymbol{I})\boldsymbol{w} = \boldsymbol{x}_{k-1}$, where $\boldsymbol{x}_{k-1}$ is parallel to an eigenvector of $\boldsymbol{A}$. Then, even if the solution $\tilde{\boldsymbol{w}}$ is not very close to the true solution $\boldsymbol{w}$, the normalized vectors $\tilde{\boldsymbol{x}}_k = \tilde{\boldsymbol{w}}/\|\tilde{\boldsymbol{w}}\|$ and $\boldsymbol{x}_{k-1} = \boldsymbol{w}/\|\boldsymbol{w}\|$ are still very close. This makes the iterations proceed smoothly toward convergence.

What may cause problems is the slow rate of convergence of

$$\left(\frac{(\lambda - \tau)}{\lambda_i - \tau}\right)^k \longrightarrow 0 \text{ as } k \longrightarrow \infty ,$$

which can happen when there is another eigenvalue λ_i close to the desired λ. In such cases, roundoff errors can divert the iterations toward an eigenvector associated with λ_i instead of λ, even if α is close to λ.

Rayleigh Quotient iteration

The standard form of the inverse iteration, as defined above, uses a constant value of τ throughout the iterations. This standard implementation can be improved if we allow the τ to vary from iteration to iteration. The key to doing this lies in a very important function in matrix analysis known as the *Rayleigh Quotient.*

Definition 11.13 *Let* $\boldsymbol{A}$ *be a real* $n \times n$ *matrix. The* **Rayleigh Quotient** *of a vector* $\boldsymbol{x} \in \Re^n$ *is defined as*

$$r_{\boldsymbol{A}}(\boldsymbol{x}) = \frac{\boldsymbol{x}'\boldsymbol{A}\boldsymbol{x}}{\boldsymbol{x}'\boldsymbol{x}} .$$

If $\boldsymbol{x}$ is an eigenvector of $\boldsymbol{A}$ associated with the eigenvalue λ, then the Rayleigh Quotient is

$$r_{\boldsymbol{A}}(\boldsymbol{x}) = \frac{\boldsymbol{x}'\boldsymbol{A}\boldsymbol{x}}{\boldsymbol{x}'\boldsymbol{x}} = \frac{\lambda\boldsymbol{x}'\boldsymbol{x}}{\boldsymbol{x}'\boldsymbol{x}} = \lambda .$$

Therefore, if $\boldsymbol{x}$ is an eigenvector of $\boldsymbol{A}$, then $r_{\boldsymbol{A}}(\boldsymbol{x})$ provides the associated eigenvalue.

The Rayleigh Quotient can help us find a scalar α that minimizes $\|\boldsymbol{A}\boldsymbol{x} - \alpha\boldsymbol{x}\|$ for a given $n \times n$ matrix $\boldsymbol{A}$ and a nonzero vector $\boldsymbol{x} \in \Re^n$. This is best seen by recognizing this problem to be one of least-squares, except that we now want to find the 1×1 "vector" α such that $\boldsymbol{x}\alpha$ is closest to the vector $\boldsymbol{A}\boldsymbol{x}$. So, $\boldsymbol{x}$ is the $n \times 1$ coefficient "matrix" and $\boldsymbol{A}\boldsymbol{x}$ is the vector in the right hand side. The resulting normal equations and the least square solution for α is obtained as follows:

$$\boldsymbol{x}'\boldsymbol{x}\alpha = \boldsymbol{x}'\boldsymbol{A}\boldsymbol{x} \Longrightarrow \alpha = (\boldsymbol{x}'\boldsymbol{x})^{-1}\boldsymbol{x}'\boldsymbol{A}\boldsymbol{x} = \frac{\boldsymbol{x}'\boldsymbol{A}\boldsymbol{x}}{\boldsymbol{x}'\boldsymbol{x}} = r_{\boldsymbol{A}}(\boldsymbol{x}) .$$

This least-squares property of the Rayleigh Quotient has the following implication. If we know that $\boldsymbol{x}$ is close to, but not necessarily, an eigenvector of $\boldsymbol{A}$ associated with eigenvalue λ, then $r_{\boldsymbol{A}}(\boldsymbol{x}) = \alpha$ provides an "estimate" of the eigenvalue because $\alpha\boldsymbol{x}$ is the "closest" point to $\boldsymbol{A}\boldsymbol{x}$.

Given an eigenvector, the Rayleigh Quotient provides us with an estimate of the corresponding eigenvalue. The inverse iteration, on the other hand, provides an estimate for the eigenvector given an estimate of an eigenvalue. Combining these two concepts suggests the following "improved" inverse iteration, also known as the ***Rayleigh iteration***:

- Start with a normalized vector $\boldsymbol{x}_0$, i.e., $\|\boldsymbol{x}_0\| = 1$.
- Compute $\tau_0 = r_{\boldsymbol{A}}(\boldsymbol{x}_0)$ from the Rayleigh coefficient.
- Construct the sequence

$$\boldsymbol{x}_k = \frac{(\boldsymbol{A} - \tau_{k-1}\boldsymbol{I})^{-1}\boldsymbol{x}_{k-1}}{\|(\boldsymbol{A} - \tau_{k-1}\boldsymbol{I})^{-1}\boldsymbol{x}_{k-1}\|} \text{ and } \tau_k = r_{\boldsymbol{A}}(\boldsymbol{x}_k) \text{ for } k = 1, 2, \ldots .$$

  In practice, we compute $(\boldsymbol{A} - \tau_{k-1}\boldsymbol{I})^{-1}\boldsymbol{x}_{k-1}$ not by inversion, but by solving $(\boldsymbol{A} - \tau_{k-1}\boldsymbol{I})\boldsymbol{w} = \boldsymbol{x}_{k-1}$ and then setting $\boldsymbol{x}_k = \boldsymbol{w}/\|\boldsymbol{w}\|$.

This is similar to the sequence in inverse iteration except that we now use an improved estimate of the τ in each iteration using the Rayleigh Quotient. Unlike the inverse iterations, we no longer require an initial eigenvalue estimates. We call this the Rayleigh Quotient iteration, sometimes abbreviated as RQI.

The Rayleigh Quotient iterations have remarkably sound convergence properties, especially for symmetric matrices. It converges to an eigenvector very quickly, since the approximations to both the eigenvalue and the eigenvector are improved during each iteration. For symmetric matrices, the convergence rate is *cubic*, which means that the number of correct digits in the approximation triples during each iteration. This is much superior to the power method and inverse iteration. The downside is that it works less effectively for general nonsymmetric matrices.

The $\boldsymbol{QR}$ iteration algorithm

The $\boldsymbol{QR}$ iteration algorithm lies at the core of most of eigenvalue computations that are implemented numerical matrix computations today. It emanates from a fairly simple idea motivated by Schur's Triangularization Theorem and the $\boldsymbol{QR}$ decomposition. Recall from Theorem 11.15 that any $n \times n$ matrix $\boldsymbol{A}$ has is unitarily similar to an upper-triangular matrix, so $\boldsymbol{Q}^*\boldsymbol{A}\boldsymbol{Q} = \boldsymbol{T}$, where $\boldsymbol{T}$ is upper-triangular. What we want to do is use a sequence of orthogonal transformations on $\boldsymbol{A}$ that will result in an upper-triangular matrix $\boldsymbol{T}$. For reasons alluded to earlier, there is no hope of achieving this in a finite number of steps. Therefore, we want to construct a sequence of matrices:

$$\boldsymbol{A}_0 = \boldsymbol{A}\ \boldsymbol{A}_k = \boldsymbol{Q}_k^*\boldsymbol{A}_{k-1}\boldsymbol{Q}_k \text{ for } k = 1, 2, \ldots,$$

such that $\lim_{k\to\infty} \boldsymbol{A}_k = \boldsymbol{T}$ is upper-triangular. In other words, we find a sequence of unitary transformations on $\boldsymbol{A}$ that will converge to to a Schur's triangularization:

$$\boldsymbol{Q}_k^*\boldsymbol{Q}_{k-1}^* \cdots \boldsymbol{Q}_1^*\boldsymbol{A}\boldsymbol{Q}_1\boldsymbol{Q}_2 \cdots \boldsymbol{Q}_k \longrightarrow \boldsymbol{T} \text{ as } k \longrightarrow \infty .$$

The $\boldsymbol{QR}$ iteration provides a method of finding such a sequence of unitary transformations.

The underlying concept is to alternate between computing the $\boldsymbol{QR}$ factors and then constructing a matrix by reversing the factors. To keep matters simple, let us assume that $\boldsymbol{A}$ is an $n \times n$ real matrix whose eigenvalues are real numbers. So it can be triangularized by orthogonal (as opposed to unitary) transformations. The $\boldsymbol{QR}$ iteration proceeds in the following manner:

- Set $\boldsymbol{A}_1 = \boldsymbol{A}$.
- Construct the sequence of matrices for $k = 1, 2, 3 \ldots$:

  Compute $\boldsymbol{QR}$ decomposition: $\boldsymbol{A}_k = \boldsymbol{Q}_k\boldsymbol{R}_k$ and set: $\boldsymbol{A}_{k+1} = \boldsymbol{R}_k\boldsymbol{Q}_k$.

  In general, $\boldsymbol{A}_{k+1} = \boldsymbol{R}_k\boldsymbol{Q}_k$, where $\boldsymbol{Q}_k$ and $\boldsymbol{R}_k$ are obtained from the $\boldsymbol{QR}$ decomposition $\boldsymbol{A}_k = \boldsymbol{Q}_k\boldsymbol{R}_k$.

To see why this works, first note that

$$\boldsymbol{Q}_k'\boldsymbol{A}_k\boldsymbol{Q}_k = \boldsymbol{Q}_k'\boldsymbol{Q}_k\boldsymbol{R}_k\boldsymbol{Q}_k = \boldsymbol{R}_k\boldsymbol{Q}_k = \boldsymbol{A}_{k+1} .$$

Letting $\boldsymbol{P}_k = \boldsymbol{Q}_1\boldsymbol{Q}_2 \cdots \boldsymbol{Q}_k$, we obtain

$$\begin{aligned}\boldsymbol{P}_k'\boldsymbol{A}\boldsymbol{P}_k &= \boldsymbol{Q}_k' \cdots \boldsymbol{Q}_2'\boldsymbol{Q}_1'\boldsymbol{A}\boldsymbol{Q}_1\boldsymbol{Q}_2 \cdots \boldsymbol{Q}_k \\ &= \boldsymbol{Q}_k' \cdots \boldsymbol{Q}_2'(\boldsymbol{Q}_1'\boldsymbol{A}\boldsymbol{Q}_1)\boldsymbol{Q}_2 \cdots \boldsymbol{Q}_k = \boldsymbol{Q}_k' \cdots \boldsymbol{Q}_2'\boldsymbol{A}_2\boldsymbol{Q}_2 \cdots \boldsymbol{Q}_k \\ &= \boldsymbol{Q}_k' \cdots \boldsymbol{Q}_3'(\boldsymbol{Q}_2'\boldsymbol{A}_2\boldsymbol{Q}_2)\boldsymbol{Q}_3 \cdots \boldsymbol{Q}_k = \boldsymbol{Q}_k' \cdots \boldsymbol{Q}_3'\boldsymbol{A}_3\boldsymbol{Q}_3 \cdots \boldsymbol{Q}_k \\ &= \ldots\ldots = \ldots\ldots \\ &= \boldsymbol{Q}_k'\boldsymbol{A}_k\boldsymbol{Q}_k = \boldsymbol{A}_{k+1} \text{ for } k = 1, 2, \ldots .\end{aligned}$$

Since $\boldsymbol{P}_k$ is orthogonal for each k, the above tells us that the $\boldsymbol{A}_k$'s produced by the $\boldsymbol{QR}$ iterations are all orthogonally similar to $\boldsymbol{A}$. This is comforting because similar matrices have the same eigenvalues. So, the procedure will not alter the eigenvalues of $\boldsymbol{A}$.

What is less clear, but can be formally proved, is that if the process converges, it converges to an upper-triangular matrix whose diagonal elements are the eigenvalues of $\boldsymbol{A}$. To see why this may happen, observe that

$$\boldsymbol{P}_{k+1} = \boldsymbol{P}_k\boldsymbol{Q}_{k+1} \Longrightarrow \boldsymbol{Q}_{k+1} = \boldsymbol{P}_k'\boldsymbol{P}_{k+1} .$$

If $\boldsymbol{P}_k \to \boldsymbol{P}$ as $k \to \infty$, then $\boldsymbol{Q}_{k+1} \to \boldsymbol{P}'\boldsymbol{P} = \boldsymbol{I}$ and

$$\boldsymbol{A}_{k+1} = \boldsymbol{R}_k\boldsymbol{Q}_k \approx \boldsymbol{R}_k \to \boldsymbol{T} \, ,$$

which is necessarily upper-triangular. This reveals convergence to a Schur's triangularization and the diagonal elements of $\boldsymbol{T}$ must be the eigenvalues of $\boldsymbol{A}$.

If the matter ended here, numerical linear algebra would not have been as deep a subject as it is. Unfortunately, matters are more complicated than this. Promising as the preceding discussion may sound, there are obvious problems. For example, in the usual $\boldsymbol{QR}$ decomposition, the upper-triangular $\boldsymbol{R}$ has only positive entries. This would mean that the $\boldsymbol{T}$ we obtain would also have positive eigenvalues and the method would collapse for matrices with negative or complex eigenvalues.

Practical considerations also arise. Recall that Householder reflections or Givens rotations are typically used to compute $\boldsymbol{Q}$ in the $\boldsymbol{QR}$ decomposition. This is an effective way of obtaining orthonormal bases for the column space of $\boldsymbol{A}$. Natural as it may sound, applying these transformations directly to $\boldsymbol{A}$ may not be such a good idea for computing Schur's triangularization.

Practical algorithm for the unsymmetric eigenvalue problem

In practice, applying the $\boldsymbol{QR}$ algorithm directly on a dense and unstructured square matrix $\boldsymbol{A}$ is not very efficient. One reason for this is that the zeroes introduced in $\boldsymbol{Q}_1'\boldsymbol{A}$ when an orthogonal transformation $\boldsymbol{Q}_1'$ is applied to the columns of $\boldsymbol{A}$ are destroyed when we post-multiply by $\boldsymbol{Q}_1$. For example, suppose $\boldsymbol{A}$ is 4×4 and $\boldsymbol{Q}_1'$ is an orthogonal matrix (Householder or Givens) that zeroes all elements below the $(1,1)$-th in the first column of $\boldsymbol{A}$. Symbolically, we see that

$$\boldsymbol{Q}_1'\boldsymbol{A}\boldsymbol{Q}_1 = \boldsymbol{Q}_1'\begin{bmatrix} * & * & * & * \\ * & * & * & * \\ * & * & * & * \\ * & * & * & * \end{bmatrix}\boldsymbol{Q}_1 = \begin{bmatrix} + & + & + & + \\ 0 & + & + & + \\ 0 & + & + & + \\ 0 & + & + & + \end{bmatrix}\boldsymbol{Q}_1 = \begin{bmatrix} + & + & + & + \\ + & + & + & + \\ + & + & + & + \\ + & + & + & + \end{bmatrix} ,$$

where, as earlier, $+$'s indicate entries that have changed in the last operation. Thus, the zeroes that were introduced in the first step are destroyed when the similarity transformation is completed. This destruction of zeroes by the similarity transformation may seem to have brought us back to the start but every iteration of the $\boldsymbol{QR}$ algorithm shrinks the sizes of the entries below the diagonal and the procedure eventually converges to an upper-triangular matrix. However, this can be slow—as slow as the power method.

A more efficient strategy would be to first reduce $\boldsymbol{A}$ to an upper-Hessenberg matrix using orthogonal similarity transformations. This step is *not* iterative and can be accomplished in a finite number of steps—recall Section 8.7. The $\boldsymbol{QR}$ algorithm is then applied to this upper-Hessenberg matrix. This preserves the Hessenberg structure in each iteration of the $\boldsymbol{QR}$ algorithm. Suppose $\boldsymbol{H}_k$ is an upper-Hessenberg matrix and let $\boldsymbol{H}_k = \boldsymbol{Q}_k\boldsymbol{R}_k$ be the $\boldsymbol{QR}$ decomposition of $\boldsymbol{H}_k$. Then, the matrix $\boldsymbol{H}_{k+1} = \boldsymbol{R}_k\boldsymbol{Q}_k$ can be written as

$$\boldsymbol{H}_{k+1} = \boldsymbol{R}_k\boldsymbol{Q}_k = \boldsymbol{R}_k\boldsymbol{H}_k\boldsymbol{R}_k^{-1}$$

and it is easily verified that multiplying an upper-Hessenberg matrix ($\boldsymbol{H}_k$) by upper-triangular matrices, either from the left or right, does not destroy the upper-Hessenberg structure (recall Theorem 1.9). Therefore, $\boldsymbol{H}_{k+1}$ is also upper-Hessenberg because $\boldsymbol{R}_k$ and $\boldsymbol{R}_k^{-1}$ are both upper-triangular.

There is one further piece to this algorithm. We apply the algorithm to a shifted matrix. So, at the k-th step we compute the $\boldsymbol{QR}$ decomposition of $\boldsymbol{H}_k - \tau_k\boldsymbol{I}_k = \boldsymbol{Q}_k\boldsymbol{R}_k$, where τ_k is an approximation of a real eigenvalue of $\boldsymbol{H}_k$. Usually, a good candidate for τ_k comes from the one of the two lowest diagonal elements in $\boldsymbol{H}_k$. Then, at the $(k+1)$-th step, we construct $\boldsymbol{H}_{k+1} = \boldsymbol{R}_k\boldsymbol{Q}_k + \tau_k\boldsymbol{I}$. This strategy, called the *single shift* $\boldsymbol{QR}$, works well when we are interested in finding real eigenvalues and usually experiences rapid convergence to real eigenvalues. For complex eigenvalues, a modification of the above, often referred to as a *double shift* $\boldsymbol{QR}$, is used.

We assemble our preceding discussion into the following steps.

1. Using Householder reflectors, construct an orthogonal similarity transformation to reduce the $n \times n$ matrix $\boldsymbol{A}$ into an upper-Hessenberg matrix $\boldsymbol{H}$. Therefore, $\boldsymbol{P}'\boldsymbol{A}\boldsymbol{P} = \boldsymbol{H}$, where $\boldsymbol{P}$ is the orthogonal matrix obtained as a composition of Householder reflections. We have seen the details of this procedure in Section 8.7. We provide a schematic representation with a 5×5 matrix. We do not keep track of changes (using $+$'s) as that would result in a lot of redundant detail. Instead, we now use $\times$ to indicate any entry that is not necessarily zero (that may or may not have been altered). This step yields

$$\boldsymbol{H} = \boldsymbol{P}'\boldsymbol{A}\boldsymbol{P} = \boldsymbol{P}'\begin{bmatrix} \times & \times & \times & \times & \times \\ \times & \times & \times & \times & \times \\ \times & \times & \times & \times & \times \\ \times & \times & \times & \times & \times \\ \times & \times & \times & \times & \times \end{bmatrix}\boldsymbol{P} = \begin{bmatrix} \times & \times & \times & \times & \times \\ \times & \times & \times & \times & \times \\ 0 & \times & \times & \times & \times \\ 0 & 0 & \times & \times & \times \\ 0 & 0 & 0 & \times & \times \end{bmatrix}.$$

2. **Finding real eigenvalues.** We apply the shifted $\boldsymbol{QR}$ algorithm to $\boldsymbol{H}$. Let $\boldsymbol{H}_1 = \boldsymbol{H}$ and compute the following for $k = 1, 2, 3 \ldots$:

$$\boldsymbol{QR} \text{ decomposition: } \boldsymbol{H}_k - \tau_k\boldsymbol{I} = \boldsymbol{Q}_k\boldsymbol{R}_k \text{ and set: } \boldsymbol{H}_{k+1} = \boldsymbol{R}_k\boldsymbol{Q}_k + \tau_k\boldsymbol{I} ,$$

where τ_k is an approximation of a real eigenvalue of the $\boldsymbol{H}_k$; often it suffices to set τ_k to simply be the (n, n)-th element of $\boldsymbol{H}_k$. The relationship between $\boldsymbol{H}_k$ and $\boldsymbol{H}_{k+1}$ is

$$\boldsymbol{H}_{k+1} = \boldsymbol{R}_k\boldsymbol{Q}_k + \tau_k\boldsymbol{I} = \boldsymbol{R}_k(\boldsymbol{H}_k - \tau_k\boldsymbol{I})\boldsymbol{R}_k^{-1} + \tau_k\boldsymbol{I} = \boldsymbol{R}_k\boldsymbol{H}_k\boldsymbol{R}_k^{-1} .$$

This ensures that $\boldsymbol{H}_{k+1}$ is also upper-Hessenberg (Theorem 1.9). Also, the above being a similarity transformation, the eigenvalues of $\boldsymbol{H}_k$ and $\boldsymbol{H}_{k+1}$ are the same. As described in Example 8.7, Givens reductions are especially useful for finding the $\boldsymbol{QR}$ decomposition $\boldsymbol{H}_k - \tau_k\boldsymbol{I} = \boldsymbol{Q}_k\boldsymbol{R}_k$ because it exploits the structure of the zeroes in the upper-Hessenberg matrix. In general, if $\boldsymbol{H}_k - \tau_k \boldsymbol{I}$ is $n \times n$, we can find a sequence of $n-1$ Givens rotations can upper-triangularize $\boldsymbol{H}_k - \tau_k \boldsymbol{I}$:

$$\boldsymbol{G}'_{n-1,n}\boldsymbol{G}'_{n-2,n-1}\cdots\boldsymbol{G}'_{1,2}(\boldsymbol{H}_k - \tau_k\boldsymbol{I}) = \boldsymbol{R}_k \Longrightarrow \boldsymbol{Q}'_k(\boldsymbol{H}_k - \tau_k\boldsymbol{I}) = \boldsymbol{R}_k \ ,$$

where $\boldsymbol{Q}'_k = \boldsymbol{G}'_{n-1,n}\boldsymbol{G}'_{n-2,n-1}\cdots\boldsymbol{G}'_{1,2}$. Incidentally, this $\boldsymbol{Q}_k$ is also upper-Hesenberg as can be verified by direct computations. Symbolically, we have

$$\boldsymbol{R}_k = \boldsymbol{Q}'_k\boldsymbol{H}_k = \boldsymbol{Q}'_k \begin{bmatrix} \times & \times & \times & \times & \times \\ \times & \times & \times & \times & \times \\ 0 & \times & \times & \times & \times \\ 0 & 0 & \times & \times & \times \\ 0 & 0 & 0 & \times & \times \end{bmatrix} = \begin{bmatrix} \times & \times & \times & \times & \times \\ 0 & \times & \times & \times & \times \\ 0 & 0 & \times & \times & \times \\ 0 & 0 & 0 & \times & \times \\ 0 & 0 & 0 & 0 & \times \end{bmatrix} .$$

Next, we compute $\boldsymbol{R}_k\boldsymbol{Q}_k = \boldsymbol{R}_k\boldsymbol{G}_{1,2}\boldsymbol{G}_{2,3}\cdots\boldsymbol{G}_{n-1,n}$ to complete the similarity transformation, which yields

$$\boldsymbol{R}_k\boldsymbol{Q}_k = \begin{bmatrix} \times & \times & \times & \times & \times \\ 0 & \times & \times & \times & \times \\ 0 & 0 & \times & \times & \times \\ 0 & 0 & 0 & \times & \times \\ 0 & 0 & 0 & 0 & \times \end{bmatrix} \boldsymbol{Q}_k = \begin{bmatrix} \times & \times & \times & \times & \times \\ \times & \times & \times & \times & \times \\ 0 & \times & \times & \times & \times \\ 0 & 0 & \times & \times & \times \\ 0 & 0 & 0 & \times & \times \end{bmatrix} .$$

Thus, $\boldsymbol{R}_k\boldsymbol{Q}_k$ is again upper-Hessenberg and so is $\boldsymbol{H}_{k+1} = \boldsymbol{R}_k\boldsymbol{Q}_k + \tau_k\boldsymbol{I}$ (the shift is added back).

Finding complex eigenvalues. The single shift algorithm above is modified to a double shift algorithm when we need to compute complex eigenvalues. Here, at each step of the $\boldsymbol{QR}$ algorithm we look at the 2×2 block in the bottom right corner of the upper-Hessenberg matrix $\boldsymbol{H}_k$ and compute the eigenvalues (possibly complex) of this 2×2 submatrix. Let α_k and β_k be the two eigenvalues. Note that if one of these, say β_k is complex, then the other is the complex conjugate $\alpha_k = \bar{\beta}_k$. These pairs of eigenvalues are now used alternately as shifts. A generic step in the double shift algorithm computes $\boldsymbol{H}_{k+1}$ and $\boldsymbol{H}_{k+2}$ starting with $\boldsymbol{H}_k$ as below:

$$\boldsymbol{QR}\text{: } \boldsymbol{H}_k - \alpha_k\boldsymbol{I} = \boldsymbol{Q}_k\boldsymbol{R}_k \ \text{ and set: } \ \boldsymbol{H}_{k+1} = \boldsymbol{R}_k\boldsymbol{Q}_k + \alpha_k\boldsymbol{I} \ ;$$
$$\boldsymbol{QR}\text{: } \boldsymbol{H}_{k+1} - \beta_k\boldsymbol{I} = \boldsymbol{Q}_{k+1}\boldsymbol{R}_{k+1} \ \text{ and set: } \ \boldsymbol{H}_{k+2} = \boldsymbol{R}_{k+1}\boldsymbol{Q}_{k+1} + \beta_k\boldsymbol{I} \ .$$

It is easy to verify that $\boldsymbol{H}_{k+1} = \boldsymbol{Q}'_k\boldsymbol{H}_k\boldsymbol{Q}_k$ and $\boldsymbol{H}_{k+2} = \boldsymbol{Q}'_{k+1}\boldsymbol{H}_{k+1}\boldsymbol{Q}_{k+1} = \boldsymbol{Q}'_{k+1}\boldsymbol{Q}'_k\boldsymbol{H}_k\boldsymbol{Q}_k\boldsymbol{Q}_{k+1}$, so every matrix in the sequence is orthogonally similar to one another. An attractive property of the double shift method is that the matrix $\boldsymbol{Q}_k\boldsymbol{Q}_{k+1}$ is real even if α_k is complex. The double shift method converges very rapidly using only real arithmetic.

The number of iterations in step 1, i.e., the reduction to Hessenberg form using Householder reflections requires $10n^3/3$ fllops. The second step is iterative and can, in principle require a countably infinite number of iterations to converge. However, convergence to machine precision is usually achieved in $O(n)$ iterations. Each iteration requires $O(n^2)$ flops, which means that the total number of flops in the second step is also $O(n^3)$. So, the total number of flops is approximately $10n^3/3 + O(n^3) \sim O(n^3)$.

A formal proof of why the above $\boldsymbol{QR}$ algorithm converges to the Schur's decomposition of a general matrix $\boldsymbol{A}$ can be found in more theoretical texts on numerical linear algebra. Here is an idea of what happens in practice. As the $\boldsymbol{QR}$ algorithm proceeds toward convergence, the entries at the bottom of the first subdiagonal tend to approach zero and typically one of the following two forms is revealed:

$$\boldsymbol{H}_k = \begin{bmatrix} \times & \times & \times & \times & \times \\ \times & \times & \times & \times & \times \\ 0 & \times & \times & \times & \times \\ 0 & 0 & \times & \times & \times \\ 0 & 0 & 0 & \epsilon & \tau \end{bmatrix} \text{ or } \begin{bmatrix} \times & \times & \times & \times & \times \\ \times & \times & \times & \times & \times \\ 0 & \times & \times & \times & \times \\ 0 & 0 & \epsilon & a & b \\ 0 & 0 & 0 & c & d \end{bmatrix}.$$

When ϵ is sufficiently small so as to be considered equal to 0, the first form suggests τ is an eigenvalue. We can then "deflate" the problem by applying the $\boldsymbol{QR}$ algorithm to the $(n-1) \times (n-1)$ submatrix obtained by deleting this last row and column in subsequent iterations. The second form produces a 2×2 submatrix $\begin{bmatrix} a & b \\ c & d \end{bmatrix}$ in the bottom right corner, which can reveal either two real eigenvalues or a real and complex eigenvalue and we can deflate the problem by deleting the last two rows and columns and considering the $(n-2) \times (n-2)$ submatrix.

This idea of deflating the problem is important in efficient practical implementations. If the subdiagonal entry $h_{j+1,j}$ of $\boldsymbol{H}$ is equal to zero, then the problem of computing the eigenvalues of $\boldsymbol{H}$ "decouples" into two smaller problems of computing the eigenvalues of $\boldsymbol{H}_{11}$ and $\boldsymbol{H}_{22}$, where

$$\boldsymbol{H} = \begin{bmatrix} \boldsymbol{H}_{11} & \boldsymbol{H}_{12} \\ \boldsymbol{O} & \boldsymbol{H}_{22} \end{bmatrix},$$

with $\boldsymbol{H}_{11}$ being $j \times j$. Therefore, effective practical implementations of the $\boldsymbol{QR}$ iteration on an upper-Hessenberg $\boldsymbol{H}$ focus upon submatrices of $\boldsymbol{H}$ that are ***unreduced***—an unreduced upper-Hessenberg matrix has all of its subdiagonal entries as nonzero. We also need to monitor the subdiagonal entries after each iteration to ascertain if any of them have become nearly zero, which would allow further decoupling. Once no further decoupling is possible, $\boldsymbol{H}$ has been reduced to upper-triangular form and the $\boldsymbol{QR}$ iterations can terminate.

Let us suppose that we wish to extract only real eigenvalues. Let $\boldsymbol{P}'\boldsymbol{A}\boldsymbol{P} = \boldsymbol{H}$ be the orthogonally similar transformation to reduce a general matrix $\boldsymbol{A}$ to an upper-Hessenberg matrix $\boldsymbol{H}$. Suppose we now apply the single shift $\boldsymbol{QR}$ algorithm on $\boldsymbol{H}$ to iteratively reduce it to an upper-triangular matrix $\boldsymbol{T}$. In other words, $\boldsymbol{H}_k \to \boldsymbol{T}$ as $k \to \infty$. Observe that

$$\boldsymbol{H}_{k+1} = \boldsymbol{R}_k\boldsymbol{Q}_k + \tau_k\boldsymbol{I} = \boldsymbol{Q}_k'(\boldsymbol{H}_k - \tau_k\boldsymbol{I})\boldsymbol{Q}_k + \tau_k\boldsymbol{I} = \boldsymbol{Q}_k'\boldsymbol{H}_k\boldsymbol{Q}_k\,,$$

which means that $\boldsymbol{H}_{k+1} = \boldsymbol{U}_k'\boldsymbol{H}\boldsymbol{U}_k$, where $\boldsymbol{U}_k = \boldsymbol{Q}_1\boldsymbol{Q}_2\cdots\boldsymbol{Q}_k$. So as the $\boldsymbol{U}_k$'s converge to $\boldsymbol{U}$ and $\boldsymbol{H}_k$ converges to the upper-triangular $\boldsymbol{T}$, we obtain the Schur's decomposition $\boldsymbol{U}'\boldsymbol{H}\boldsymbol{U} = \boldsymbol{T}$. This implies that $\boldsymbol{Z}'\boldsymbol{A}\boldsymbol{Z} = \boldsymbol{T}$, where $\boldsymbol{Z} = \boldsymbol{P}\boldsymbol{U}$, is the Schur's decomposition of $\boldsymbol{A}$. The diagonal entries of $\boldsymbol{T}$ are the eigenvalues of $\boldsymbol{A}$.

We now turn to finding eigenvectors of $\boldsymbol{A}$. If an eigenvector associated with any eigenvalue λ of $\boldsymbol{A}$ is required, we can solve the homogeneous system $(\boldsymbol{A} - \lambda\boldsymbol{I})\boldsymbol{x} = \boldsymbol{0}$. This, however, will be unnecessarily expensive given that we have already computed two structured matrices, $\boldsymbol{H}$ (upper-Hessenberg) and $\boldsymbol{T}$ (upper-triangular), that are orthogonally similar to $\boldsymbol{A}$. We know that similar matrices have the same eigenvalues but they may have different eigenvectors. Thus, the eigenvectors of $\boldsymbol{H}$ and $\boldsymbol{A}$ need not be the same. However, it is easy to obtain the eigenvectors of $\boldsymbol{A}$ from those of $\boldsymbol{H}$. Suppose that $\boldsymbol{x}$ is an eigenvector of $\boldsymbol{H}$ associated with eigenvalue λ. Then,

$$\boldsymbol{H}\boldsymbol{x} = \lambda\boldsymbol{x} \Longrightarrow \boldsymbol{P}'\boldsymbol{A}\boldsymbol{P}\boldsymbol{x} = \lambda\boldsymbol{x} \Longrightarrow \boldsymbol{A}(\boldsymbol{P}\boldsymbol{x}) = \lambda(\boldsymbol{P}\boldsymbol{x}) ,$$

which shows that $\boldsymbol{P}\boldsymbol{x}$ is an eigenvector of $\boldsymbol{A}$ associated with eigenvalue λ. Similarly, if $(\lambda, \boldsymbol{x})$ is an eigen-pair of $\boldsymbol{T}$, then $(\lambda, \boldsymbol{Z}\boldsymbol{x})$ is an eigen-pair for $\boldsymbol{A}$. Therefore, we first compute an eigenvector of $\boldsymbol{H}$ or $\boldsymbol{T}$, associated with eigenvalue λ. Suppose we do this for the upper-Hessenberg $\boldsymbol{H}$. Instead of using plain Gaussian elimination to solve $(\boldsymbol{H} - \lambda\boldsymbol{I})\boldsymbol{x} = \boldsymbol{0}$, a more numerically efficient way of finding such an eigenvector is the inverse iteration method we discussed earlier. Once $\boldsymbol{x}$ is obtained, we simply compute $\boldsymbol{P}\boldsymbol{x}$ as the eigenvector of $\boldsymbol{A}$. Analogously, if we found an eigenvector of the upper-triangular $\boldsymbol{T}$, then $\boldsymbol{Z}\boldsymbol{x}$ will be the eigenvector for $\boldsymbol{A}$ associated with λ.

The symmetric eigenvalue problem

If $\boldsymbol{A}$ is symmetric, several simplifications result. If $\boldsymbol{A} = \boldsymbol{A}'$, then $\boldsymbol{H} = \boldsymbol{H}'$. A symmetric Hessenberg matrix is tridiagonal. Therefore, the same Householder reflections that reduce a general matrix to Hessenberg form can be used to reduce a symmetric matrix $\boldsymbol{A}$ to a symmetric tridiagonal matrix $\boldsymbol{T}$. However, the symmetry of $\boldsymbol{A}$ can be exploited to reduce the number of operations needed to apply each Householder reflection on the left and right of $\boldsymbol{A}$.

In each iteration of the $\boldsymbol{QR}$ algorithm, $\boldsymbol{T}_k$ is a symmetric tridiagonal matrix and $\boldsymbol{T}_k = \boldsymbol{Q}_k\boldsymbol{R}_k$ is its $\boldsymbol{QR}$ decomposition computed using the Givens rotations as in the preceding section. Each $\boldsymbol{Q}_k$ is upper-Hessenberg, and $\boldsymbol{R}_k$ is upper-bidiagonal (i.e., it is upper-triangular, with upper bandwidth 1, so that all entries below the main diagonal and above the superdiagonal are zero). Furthermore, $\boldsymbol{T}_{k+1} = \boldsymbol{R}_k\boldsymbol{Q}_k$ is again a symmetric and tridiagonal matrix.

The reduction to tridiagonal form, say $\boldsymbol{P}'\boldsymbol{A}\boldsymbol{P} = \boldsymbol{T}$, using Householder reductions requires $4n^3/3$ flops if $\boldsymbol{P}$ need not be computed explicitly. If $\boldsymbol{P}$ is required, this will involve another additional $4n^3/3$ flops. When the $\boldsymbol{QR}$ algorithm is being implemented, the Givens rotations require $O(n)$ operations to compute the $\boldsymbol{T} = \boldsymbol{QR}$ decomposition of a tridiagonal matrix, and to multiply the factors in reverse order. However, to compute the eigenvectors of $\boldsymbol{A}$ as well as the eigenvalues, it is necessary to compute the product of all of the Givens rotations, which will require an additional $O(n^2)$ flops. The complete procedure, including the reduction to tridiagonal form, requires approximately $4/3n^3 + O(n^2)$ flops if only eigenvalues are required. If the entire set of eigenvectors are required, we need to accumulate the orthogonal transformations and this requires another $9n^3$ flops approximately.

Krylov subspace methods for large and sparse matrices—the Arnoldi iteration

The $\boldsymbol{QR}$ algorithm with well-chosen shifts has been the standard algorithm for general eigenvalue computations since the 1960's and continues to be among the most widely used algorithm implemented in popular numerical linear algebra software today. However, in emerging and rapidly evolving fields such as data mining and pattern recognition, where matrix algorithms are becoming increasingly conspicuous, we often require eigenvalues of matrices that are very large and sparse. Application of the $\boldsymbol{QR}$ algorithm to arrive at Schur's triangularization becomes infeasible due to prohibitive increases in storage requirements (the Schur decompositions for sparse matrices are usually dense and almost all elements are nonzero).

The power iteration suggests an alternative that avoids transforming the matrix itself and relies primarily upon matrix vector multiplications of the form $\boldsymbol{y} = \boldsymbol{A}\boldsymbol{x}$. Indeed, the power method is one such method, where we construct a sequence of vectors, $\boldsymbol{A}\boldsymbol{x}_0, \boldsymbol{A}^2\boldsymbol{x}_0, \ldots$ that converges to the eigenvector associated with the dominant eigenvalue. In the power iteration, once we compute $\boldsymbol{y}_k = \boldsymbol{A}\boldsymbol{y}_{k-1} = \boldsymbol{A}^k\boldsymbol{x}_0$, where $\boldsymbol{y}_{k-1} = \boldsymbol{A}^{k-1}\boldsymbol{x}_0$, we completely ignore the $\boldsymbol{y}_{k-1}$ and all the information contained in the earlier iterations. A natural enhancement of the power method would be to use all the information in the sequence, and to do this in a computationally effective manner.

This brings us to *Krylov subspaces*. The sequence of vectors arising in the power iterations are called a *Krylov sequence*, named after the Russian mathematician Aleksei Nikolaevich Krylov, and the span of a *Krylov sequence* is known as a *Krylov subspace*. The following definitions make this precise.

Definition 11.14 *Let $\boldsymbol{A}$ be an $n \times n$ matrix whose elements are real or complex numbers and let $\boldsymbol{x}_0 \neq \boldsymbol{0}$ be an $n \times 1$ vector in $\mathbb{C}^n$. Then,*

1. *the sequence of vectors $\boldsymbol{x}_0, \boldsymbol{A}\boldsymbol{x}_0, \boldsymbol{A}^2\boldsymbol{x}_0, \ldots, \boldsymbol{A}^{m-1}\boldsymbol{x}_0$ is called a* **Krylov sequence***;*
2. *the subspace $K_m(\boldsymbol{A}, \boldsymbol{x}_0) = Sp\{\boldsymbol{x}_0, \boldsymbol{A}\boldsymbol{x}_0, \boldsymbol{A}^2\boldsymbol{x}_0, \ldots, \boldsymbol{A}^{m-1}\boldsymbol{x}_0\}$ is called a* **Krylov subspace***;*
3. *the matrix $\boldsymbol{K} = \left[\boldsymbol{x}_0 : \boldsymbol{A}\boldsymbol{x}_0 : \ldots : \boldsymbol{A}^{m-1}\boldsymbol{x}_0\right]$ is called a* **Krylov matrix**.

We will not need too many technical details on the above. Krylov subspaces are, however, important enough in modern linear algebra for us to be familiar with Definition 11.14.

A suite of eigenvalue algorithms, known as Krylov subspace methods, seek to extract as good an approximation of the eigenvector as possible from a Krylov subspace.

There are several algorithms that fall under this category. We will briefly explain one of them, known as the ***Arnoldi iteration***, named after an American engineer Walter Edwin Arnoldi.

Let $\boldsymbol{A}$ be an $n \times n$ unsymmetric matrix with real elements and we wish to compute the eigenvalues of $\boldsymbol{A}$ by finding its Schur's decomposition $\boldsymbol{P}'\boldsymbol{A}\boldsymbol{P} = \boldsymbol{T}$, where $\boldsymbol{T}$ is upper-triangular with the eigenvalues along the diagonal of $\boldsymbol{T}$. As discussed earlier, the search for the Schur's decomposition begins with an orthogonal similarity reduction of $\boldsymbol{A}$ to upper-Hessenberg form $\boldsymbol{H}$. In other words, we find an orthogonal matrix $\boldsymbol{Q}$ such that $\boldsymbol{Q}'\boldsymbol{A}\boldsymbol{Q} = \boldsymbol{H}$ is upper-Hessenberg. Earlier, we saw how Householder reflections can be used to construct such a $\boldsymbol{Q}$. However, when $\boldsymbol{A}$ is large and sparse, this may not be as efficient in terms of the storage and number of flops.

The Arnoldi iteration is an alternative algorithm to reduce $\boldsymbol{A}$ to upper-Hessenberg form, which may be more efficient when $\boldsymbol{A}$ is large and sparse. The idea is really simple and wishes to find an orthogonal $\boldsymbol{Q}$ and an upper-Hessenberg $\boldsymbol{H}$ such that

$$\boldsymbol{Q}'\boldsymbol{A}\boldsymbol{Q} = \boldsymbol{H} \Longrightarrow \boldsymbol{A}\boldsymbol{Q} = \boldsymbol{Q}\boldsymbol{H} ,$$

where $\boldsymbol{Q} = [\boldsymbol{q}_1 : \boldsymbol{q}_2 : \ldots : \boldsymbol{q}_n]$ and $\boldsymbol{H} = \{h_{ij}\}$. Keep in mind that $\boldsymbol{H}$ is upper-Hessenberg, so $h_{ij} = 0$ whenever $i > j + 1$. We now equate each column of $\boldsymbol{A}\boldsymbol{Q}$ with that of $\boldsymbol{Q}\boldsymbol{H}$.

For example, equating the first column yields the following relationship,

$$\boldsymbol{A}\boldsymbol{q}_1 = h_{11}\boldsymbol{q}_1 + h_{21}\boldsymbol{q}_2 . \tag{11.19}$$

Suppose that $\|\boldsymbol{q}_1\| = 1$, i.e., $\boldsymbol{q}_1$ is a normalized nonnull vector. Note that (11.19) can be rewritten as

$$h_{21}\boldsymbol{q}_2 = \boldsymbol{A}\boldsymbol{q}_1 - h_{11}\boldsymbol{q}_1 . \tag{11.20}$$

Look at (11.20). Clearly $\boldsymbol{q}_2 \in Sp\{\boldsymbol{q}_1, \boldsymbol{A}\boldsymbol{q}_1\}$. Suppose we want $\boldsymbol{q}_2$ to be a normalized vector (i.e., $\|\boldsymbol{q}_2\| = 1$) which is orthogonal to $\boldsymbol{q}_1$. Can we find h_{11} and h_{21} to make that happen? This is exactly what is done in the Gram-Schmidt orthogonalization process on the Krylov sequence $\{\boldsymbol{q}_1, \boldsymbol{A}\boldsymbol{q}_1\}$. Premultiply both sides of (11.20) by $\boldsymbol{q}_1'$ and note that $\boldsymbol{q}_1'\boldsymbol{q}_2 = 0$. Therefore,

$$0 = h_{21}\boldsymbol{q}_1'\boldsymbol{q}_2 = \boldsymbol{q}_1'\boldsymbol{A}\boldsymbol{q}_1 - h_{11}\boldsymbol{q}_1'\boldsymbol{q}_1 \Longrightarrow h_{11} = \boldsymbol{q}_1'\boldsymbol{A}\boldsymbol{q}_1 .$$

Next, we set $\boldsymbol{q}_2$ to be the normalized vector along $\boldsymbol{A}\boldsymbol{q}_1 - h_{11}\boldsymbol{q}_1$ and determine h_{21} from the condition that $\|\boldsymbol{q}_2\| = 1$. To be precise, we set

$$\boldsymbol{v} = \boldsymbol{A}\boldsymbol{q}_1 - h_{11}\boldsymbol{q}_1 , \; h_{21} = \|\boldsymbol{v}\| \text{ and } \boldsymbol{q}_2 = \frac{\boldsymbol{v}}{\|\boldsymbol{v}\|} = \frac{\boldsymbol{A}\boldsymbol{q}_1 - h_{11}\boldsymbol{q}_1}{\|\boldsymbol{A}\boldsymbol{q}_1 - h_{11}\boldsymbol{q}_1\|} .$$

Let us summarize the above steps. We started with a normalized vector $\boldsymbol{q}_1$. We found a suitable value of h_{11} that would make $\boldsymbol{q}_2$ to be orthogonal to $\boldsymbol{q}_1$ and be a vector in the span of $\{\boldsymbol{q}_1.\boldsymbol{A}\boldsymbol{q}_1\}$. Finally, we found h_{21} from the requirement that $\boldsymbol{q}_2$ is normalized. This completes the first Arnoldi iteration and we have found the first two columns of $\boldsymbol{Q}$ and the first column of $\boldsymbol{H}$. We have achieved this by finding an orthonormal basis $\{\boldsymbol{q}_1, \boldsymbol{q}_2\}$ for the Krylov subspace $Sp\{\boldsymbol{q}_1, \boldsymbol{A}\boldsymbol{q}_1\}$.

Now consider a generic step. Suppose we have found an orthonormal set of vectors $\{q_1, q_2, \ldots, q_j\}$ that form an orthonormal basis for the Krylov subspace $K_j(A, q_1) = Sp\{q_1, Aq_1, A^2q_1, \ldots, A^{j-1}q_1\}$. The Arnoldi iteration proceeds exactly as if we were applying a Gram-Schmidt procedure to orthogonalize the vectors in the Krylov sequence $\{q_1, Aq_1, A^2q_1, \ldots, A^{j-1}q_1\}$. We wish to find a q_{j+1} that is orthogonal to each of the preceding q_i's, $i = 1, 2, \ldots, j$ and such that $\{q_1, q_2, \ldots, q_{j+1}\}$ will form an orthonormal basis for the Krylov subspace $K_{j+1}(A, q_0) = Sp\{q_1, Aq_1, A^2q_1, \ldots, A^jq_1\}$. Equating the j-th column of AQ with that of QH, we find

$$Aq_j = \sum_{k=1}^{j} h_{kj}q_k + h_{j+1,j}q_{j+1} \Longrightarrow h_{j+1,j}q_{j+1} = Aq_j - \sum_{k=1}^{j} h_{kj}q_k \ .$$

The requirement that $\{q_1, q_2, \ldots, q_j, q_{j+1}\}$ be orthonormal leads to

$$q_i'Aq_j = \sum_{k=1}^{j} h_{kj}q_i'q_k + h_{j+1,j}q_i'q_{j+1} = h_{ij} \ \text{ for } \ i = 1, 2, \ldots, j \ .$$

We take q_{j+1} as the normalized vector along $Aq_j - \sum_{k=1}^{j} h_{kj}q_k$ and determine $h_{j+1,j}$ from the condition $\|q_{j+1}\| = 1$. Thus, we set

$$v = Aq_j - \sum_{k=1}^{j} h_{kj}q_k \ , \ h_{j+1,j} = \|v\| \ \text{ and } \ q_{j+1} = \frac{v}{\|v\|} \ .$$

Once we repeat the above steps for $j = 1, 2, \ldots, n$, we obtain $Q'AQ = H$.

The above results can be assembled into the following Arnoldi iteration algorithm.

1. Start with any normalized vector q_1 such that $\|q_1\| = 1$.
2. Construct the following sequence for $j = 1, 2, \ldots, n$:
   (a) Form $h_{ij} = q_i'Aq_j$ for $i = 1, 2, \ldots, j$.
   (b) Set $v = Aq_j - \sum_{k=1}^{j} h_{kj}q_k$ and set $h_{j+1,j} = \|v\|$.
   (c) Set $q_{j+1} = \frac{v}{\|v\|}$ and $q_{n+1} = 0$.

The above recursive algorithm will, after n steps, produce the desired orthogonal similarity reduction $Q'AQ = H$, where H is upper-Hessenberg. While this can be applied to a general $n \times n$ matrix A, the serious benefits of Arnoldi iterations are seen when A is large and sparse. In matrix format, we can collect the Aq_j's for $j = 1, 2, \ldots, t$ and write

$$AQ_t = Q_tH_t + h_{t+1,t}q_{t+1}e_t' \ , \tag{11.21}$$

where $Q_t = [q_1 : q_2 : \ldots : q_t]$ is an $n \times t$ matrix with orthogonal columns, e_t is the t-th column of the $t \times t$ identity matrix and

$$H_t = \begin{bmatrix} h_{11} & h_{12} & \cdots & h_{1,t-1} & h_{1t} \\ h_{21} & h_{22} & \cdots & h_{2,t-1} & h_{2t} \\ 0 & h_{32} & \cdots & h_{3,t-1} & h_{3t} \\ \vdots & \vdots & \ddots & \vdots & \vdots \\ 0 & 0 & \cdots & h_{t-1,t} & h_{tt} \end{bmatrix}$$

is $t \times t$ upper-Hessenberg. Decompositions such as (11.21) are called *Arnoldi decompositions*. Note that after t steps of the recursion we have performed t matrix-vector multiplications. Also, by construction $\{q_1, q_2, \ldots, q_t\}$ is an orthonormal basis for the Krylov subspace $K_t(A, q_1) = \{q_1, Aq_1, \ldots, A^{t-1}q_1\}$. Thus, we have retained all the information produced in the first t steps. This is in stark contrast to the power method, where only the information in the $t-1$-th step is used.

The Arnoldi iterations can be used as an eigenvalue algorithm by finding the eigenvalues of H_t for each t. The eigenvalues of H_t are called the *Ritz eigenvalues*. These are the eigenvalues of the orthogonal projection of A onto the Krylov subspace $K_t(A, q_1)$. To see why this true, let λ_t be an eigenvalue of H_t. If x is an associated eigenvector of λ_t, then

$$Q_t' A Q_t x = H_t x = \lambda_t x \Longrightarrow Q_t Q_t' A (Q_t x) = \lambda_t (Q_t x) \; .$$

Therefore, λ_t is an eigenvalue of $Q_t Q_t' A$ with associated eigenvector $Q_t x$. The columns of Q_t are an orthonormal basis for the Krylov subspace $K_t(A, q_1)$. Therefore, the orthogonal projector onto this subspace is given by $Q_t Q_t'$ and $Q_t Q_t' A$ is the orthogonal projection of A onto $K_t(A, q_1)$.

Since H_t is a Hessenberg matrix of modest size ($t < n$), its eigenvalues can be computed efficiently using, for instance, the QR algorithm. In practice, we often find that some of the Ritz eigenvalues converge to eigenvalues of A. Note that H_t, being $t \times t$, has at most t eigenvalues, so not all eigenvalues of A can be approximated. However, practical experience suggests that the Ritz eigenvalues converge to the extreme eigenvalues of A – those near the largest and smallest eigenvalues. The reason this happens is a particular characterization of H_t: It is the matrix whose characteristic polynomial minimizes $p\|(A)q_1\|$ among all monic polynomials of degree t. This suggests that a good way to get $p(A)$ small is to choose a polynomial $p(\lambda)$ that is small whenever λ is an eigenvalue of A. Therefore, the roots of $p(\lambda)$ (and thus the Ritz eigenvalues) will be close to the eigenvalues of A.

The special case of the Arnoldi iteration when A is symmetric is known as the ***Lanczos iteration*** . Its theoretical properties are better understood and more complete than for the unsymmetric case.

11.9 Exercises

1. Find the eigenvalues for each of the following matrices:

$$\text{(a)} \begin{bmatrix} 3 & 2 \\ -4 & -6 \end{bmatrix}, \quad \text{(b)} \begin{bmatrix} 5 & 6 \\ 3 & -2 \end{bmatrix}, \quad \text{(c)} \begin{bmatrix} 5 & 1 \\ -1 & 3 \end{bmatrix}, \quad \text{(d)} \begin{bmatrix} 1 & 2 \\ 3 & 2 \end{bmatrix}.$$

For an associated eigenvector for each of the eigenvalues you obtained.

2. For which of the matrices in the above exercise does there exist a linearly independent set of eigenvectors.

3. If λ is an eigenvalue of $\boldsymbol{A}$, then prove that (i) $c\lambda$ is an eigenvalue of $c\boldsymbol{A}$ for any scalar c, and (ii) $\lambda + d$ is an eigenvalue of $\boldsymbol{A} + d\boldsymbol{I}$ for any scalar d.

4. If $\boldsymbol{A}$ is a 2×2 matrix, then show that the characteristic polynomial of $\boldsymbol{A}$ is

$$p_{\boldsymbol{A}}(\lambda) = \lambda^2 - \text{tr}(\boldsymbol{A})\lambda + |\boldsymbol{A}| .$$

5. Show that the characteristic polynomial of $\begin{bmatrix} 1 & -1 \\ 2 & -1 \end{bmatrix}$ does not have any real roots.

6. Find the eigenvalues of the matrix

$$\boldsymbol{A} = \begin{bmatrix} 4 & 1 & -1 \\ 2 & 5 & -2 \\ 1 & 1 & 2 \end{bmatrix}.$$

Find a basis of $\mathcal{N}(\boldsymbol{A} - \lambda\boldsymbol{I})$ for each eigenvalue λ. Find the algebraic and geometric multiplicities of each of the eigenvalues.

7. If $\boldsymbol{A} = \{a_{ij}\}$ is a 3×3 matrix, then show that the characteristic polynomial of $\boldsymbol{A}$ is

$$p_{\boldsymbol{A}}(\lambda) = \lambda^3 - \text{tr}(\boldsymbol{A})\lambda^2 + \left(\sum_{i=1}^{n} A_{ii}\right)\lambda - |\boldsymbol{A}| ,$$

where each A_{ii} is the cofactor of a_{ii}. Verify that the characteristic polynomial of $\boldsymbol{A}$ in the previous exercise is $p_{\boldsymbol{A}}(\lambda) = \lambda^3 - 11\lambda^2 + 39\lambda - 45$.

8. Suppose the eigenvalues of $\boldsymbol{A}$ are of the form λ, $\lambda + a$ and $\lambda + 2a$, where a is a real number. Suppose you are given $\text{tr}(\boldsymbol{A}) = 15$ and $|\boldsymbol{A}| = 80$. Find the eigenvalues.

9. True or false: The eigenvalues of a permutation matrix must be real numbers.

10. If $\boldsymbol{A}$ is 2×2, then prove that $|\boldsymbol{I} + \boldsymbol{A}| = 1 + |\boldsymbol{A}|$ if and only if $\text{tr}(\boldsymbol{A}) = \boldsymbol{O}$.

11. If $\lambda \neq 0$ is an eigenvalue of a nonsingular matrix $\boldsymbol{A}$, then prove that $1/\lambda$ is an eigenvalue of $\boldsymbol{A}^{-1}$.

12. Let $\boldsymbol{A}$ and $\boldsymbol{B}$ be similar matrices so that $\boldsymbol{B} = \boldsymbol{P}^{-1}\boldsymbol{A}\boldsymbol{P}$. If $(\lambda, \boldsymbol{x})$ is an eigen-pair for $\boldsymbol{B}$, then prove that $(\lambda, \boldsymbol{P}\boldsymbol{x})$ is an eigen-pair for $\boldsymbol{A}$.

13. Let λ be an eigenvalue of $\boldsymbol{A}$ such that $\text{AM}_{\boldsymbol{A}}(\lambda) = m$. If $\alpha \neq 0$, then prove that $\beta + \alpha\lambda$ is an eigenvalue of $\beta\boldsymbol{I} + \alpha\boldsymbol{A}$ and $\text{AM}_{\boldsymbol{A}}(\beta + \alpha\lambda) = m$.

14. Let λ be an eigenvalue of $\boldsymbol{A}$. Then prove that $\text{ES}(\boldsymbol{A}, \lambda) = \text{ES}(\beta\boldsymbol{I} + \alpha\boldsymbol{A}, \beta + \alpha\lambda)$, where $\alpha \neq 0$ is any nonzero scalar.

15. If $\boldsymbol{A}$ is an $n \times n$ singular matrix with k distinct eigenvalues, then prove that $k-1 \leq \rho(\boldsymbol{A}) \leq n-1$.
16. Let $\boldsymbol{A}$ be $n \times n$ with $\rho(\boldsymbol{A}) = r < n$. If 0 is an eigenvalue of $\boldsymbol{A}$ with $\text{AM}_{\boldsymbol{A}}(0) = \text{GM}_{\boldsymbol{A}}(0)$, then prove that $\boldsymbol{A}$ is similar to $\begin{bmatrix} \boldsymbol{O} & \boldsymbol{O} \\ \boldsymbol{O} & \boldsymbol{C} \end{bmatrix}$, where $\boldsymbol{C}$ is an $r \times r$ nonsingular matrix.
17. Let $\boldsymbol{A}$ be $n \times n$ with $\rho(\boldsymbol{A}) = r < n$. Prove that 0 is an eigenvalue of $\boldsymbol{A}$ with $\text{AM}_{\boldsymbol{A}}(0) = \text{GM}_{\boldsymbol{A}}(0)$ if and only if $\rho(\boldsymbol{A}) = \rho(\boldsymbol{A}^2)$.
18. Let $\boldsymbol{A} = \boldsymbol{u}\boldsymbol{u}'$, where $\boldsymbol{u} \in \Re^n$. Find the eigenvalues of $\boldsymbol{A}$ and the identify the eigenspaces associated with the eigenvalues.
19. **Deflation.** Let $(\lambda, \boldsymbol{u})$ be an eigen-pair for an $n \times n$ matrix $\boldsymbol{A}$ and let $\boldsymbol{v} \in \Re^n$. Prove that $(\lambda - \boldsymbol{v}'\boldsymbol{u}, \boldsymbol{v})$ is an eigen-pair for $\boldsymbol{A} - \boldsymbol{u}\boldsymbol{v}'$. If $\mu \neq (\lambda - \boldsymbol{v}'\boldsymbol{u})$ is any other eigenvalue of $\boldsymbol{A}$, then prove that μ is also an eigenvalue of $\boldsymbol{A} - \boldsymbol{u}\boldsymbol{v}'$.
20. Let $\boldsymbol{A}$ be a nonsingular matrix with eigenvalues $\lambda_1, \lambda_2, \ldots, \lambda_n$ such that $\lambda_j \neq \lambda_1$ for $j \geq 2$. Using the previous exercise, explain how we can obtain a deflated matrix $\boldsymbol{B}$ whose eigenvalues are $0, \lambda_2, \lambda_3, \ldots, \lambda_n$.
21. If λ is an eigenvalue of $\boldsymbol{A}$ such that $\text{GM}_{\boldsymbol{A}}(\lambda) = r$, then prove that the dimension of the subspace spanned by the left eigenvectors (recall Definition 11.2) of $\boldsymbol{A}$ corresponding to λ is equal to r.
22. Let $\boldsymbol{P}$ be a nonsingular matrix whose columns are the (right) eigenvectors of $\boldsymbol{A}$. Show that the columns of $\boldsymbol{P}^{-1}$ are the left eigenvectors of $\boldsymbol{A}$.
23. If $\boldsymbol{A} = \boldsymbol{u}\boldsymbol{u}'$, where $\boldsymbol{u} \in \Re^n$, then prove that $\boldsymbol{A}$ is similar to a diagonal matrix.
24. Let $\boldsymbol{P}$ be a projector (i.e., an idempotent matrix). Prove that 1 and 0 are the only eigenvalues of $\boldsymbol{P}$ and that $\mathcal{C}(\boldsymbol{P}) = \text{ES}(\boldsymbol{A}, 1)$ and $\mathcal{C}(\boldsymbol{I} - \boldsymbol{P}) = \text{ES}(\boldsymbol{A}, 0)$. Show that $\boldsymbol{P}$ has n linearly independent eigenvectors.
25. Find a matrix $\boldsymbol{P}$ such that $\boldsymbol{P}^{-1}\boldsymbol{A}\boldsymbol{P}$ is diagonal, where $\boldsymbol{A} = \begin{bmatrix} 7 & 2 & 1 \\ -6 & 0 & 2 \\ 6 & 4 & 6 \end{bmatrix}$.

    Hence, find $\boldsymbol{A}^5$.
26. True or false: If $\boldsymbol{A}$ is diagonalizable, then so is $\boldsymbol{A}^2$.
27. Assume that $\boldsymbol{A}$ is diagonalizable and the matrix exponential (see Definition 11.8) is well-defined. Prove the following:
    (a) $\exp(\boldsymbol{A}') = [\exp(\boldsymbol{A})]'$;
    (b) $\exp((\alpha + \beta)\boldsymbol{A}) = \exp(\alpha\boldsymbol{A})\exp(\beta\boldsymbol{A})$;
    (c) $\exp(\boldsymbol{A})$ is nonsingular and $[\exp(\boldsymbol{A})]^{-1} = \exp(-\boldsymbol{A})$;
    (d) $\boldsymbol{A}\exp(t\boldsymbol{A}) = \exp(t\boldsymbol{A})\boldsymbol{A}$ for any scalar t;
    (e) $\frac{d}{dt}\exp(t\boldsymbol{A}) = \boldsymbol{A}\exp(t\boldsymbol{A}) = \exp(t\boldsymbol{A})\boldsymbol{A}$;
    (f) $|\exp(t\boldsymbol{A})| = \exp(t\text{tr}(\boldsymbol{A}))$ for any scalar t.

    The above properties also hold when $\boldsymbol{A}$ *is not diagonalizable but the proofs are more involved and use the Jordan Canonical Form discussed in Section 12.6.*

28. If $\boldsymbol{A}$ and $\boldsymbol{B}$ are diagonalizable matrices such that $\boldsymbol{AB} = \boldsymbol{BA}$, then prove that
$$\exp(\boldsymbol{A}+\boldsymbol{B}) = \exp(\boldsymbol{A})\exp(\boldsymbol{B}) = \exp(\boldsymbol{B})\exp(\boldsymbol{A}) \; .$$

29. Let $\boldsymbol{A}$ be an $n \times n$ diagonalizable matrix. Consider the system of differential equations:
$$\frac{d}{dt}\boldsymbol{Y}(t) = \boldsymbol{AY}(t) \; , \quad \boldsymbol{Y}(0) = \boldsymbol{I}_n \; ,$$
where $\boldsymbol{Y}(t) = \{y_{ij}(t)\}$ is $n \times n$ with each entry $y_{ij}(t)$ a differentiable function of t and $\frac{d}{dt}\boldsymbol{Y}(t) = \left\{\frac{d}{dt}y_{ij}(t)\right\}$. Explain why there is a unique $n \times n$ matrix $\boldsymbol{Y}(t)$ which is a solution to the above system. If we *define* this solution $\boldsymbol{Y}(t)$ as the *matrix exponential* $\exp(t\boldsymbol{A})$, then prove that this definition coincides with Definition 11.8. Verify that $\boldsymbol{Y}(t) = \exp((t-t_0)\boldsymbol{A})\boldsymbol{Y}_0$ is the solution to the initial value problem: $\frac{d}{dt}\boldsymbol{Y}(t) = \boldsymbol{AY}(t)$, $\boldsymbol{Y}(t_0) = \boldsymbol{Y}_0$.

30. Find the characteristic and minimum polynomials for the following matrices:
$$\text{(a)} \begin{bmatrix} 1 & 2 \\ 0 & 1 \end{bmatrix}, \quad \text{(b)} \begin{bmatrix} 1 & 1 & 2 \\ 0 & 1 & 3 \\ 0 & 0 & 1 \end{bmatrix}, \quad \text{(c)} \begin{bmatrix} 0 & 1 & 0 \\ 0 & 0 & 1 \\ 0 & 0 & 0 \end{bmatrix}.$$

31. Prove that the minimum polynomial of $\begin{bmatrix} \boldsymbol{A} & \boldsymbol{O} \\ \boldsymbol{O} & \boldsymbol{B} \end{bmatrix}$, where $\boldsymbol{A}$ and $\boldsymbol{B}$ are square, is the lowest common multiple of the minimum polynomials of $\boldsymbol{A}$ and $\boldsymbol{B}$.

32. Prove that a matrix is nilpotent if and only if all its eigenvalues are 0.

33. True or false: A nilpotent matrix cannot be similar to a diagonal matrix.

34. True or false: If $\boldsymbol{A}$ is nilpotent, then $\text{tr}(\boldsymbol{A}) = \boldsymbol{O}$.

35. Find the minimum polynomial of $\boldsymbol{A} = \boldsymbol{1}\boldsymbol{1}'$.

36. Let $\boldsymbol{x} \in \Re^n$. Obtain a spectral decomposition of $\boldsymbol{xx}'$.

37. If $\boldsymbol{A}$ is a real symmetric matrix with spectral decomposition $\boldsymbol{P}'\boldsymbol{AP} = \Lambda$, find the spectral decomposition of $\alpha\boldsymbol{I} + \beta\boldsymbol{A}$.

38. Obtain a spectral decomposition of $(\alpha - \beta)\boldsymbol{I} + \beta\boldsymbol{1}\boldsymbol{1}'$.

39. Find a spectral decomposition of $\begin{bmatrix} 0 & \boldsymbol{1}' \\ \boldsymbol{1} & \boldsymbol{O} \end{bmatrix}$.

40. Find a spectral decomposition of $\begin{bmatrix} 2 & 1 & -1 \\ 1 & 2 & 1 \\ -1 & 1 & 2 \end{bmatrix}$.

41. Consider a real $n \times n$ matrix such that $\boldsymbol{A} = -\boldsymbol{A}'$. Such a matrix is said to be ***real skew-symmetric***. Prove that the only possible real eigenvalue of $\boldsymbol{A}$ is $\lambda = 0$. Also prove that 0 is an eigenvalue of $\boldsymbol{A}$ whenever n is odd.

42. Using the $\boldsymbol{QR}$ iteration method and the inverse-power method (Section 11.8), find the eigenvalues and eigenvectors of the matrix in Exercise 25.

CHAPTER 12

Singular Value and Jordan Decompositions

12.1 Singular value decomposition

If $\boldsymbol{A}$ is a normal matrix (which includes real-symmetric or Hermitian), then we have seen that $\boldsymbol{A}$ is unitarily similar to a diagonal matrix. The spectral decomposition does not exist for non-normal matrices, so non-normal matrices are not unitarily or orthogonally similar to diagonal matrices (Theorem 11.28). But to what extent can a non-normal matrix be simplified using orthogonal transformations? In fact, can we say something about general matrices that are not even square? The answers rest with a famous decomposition called the ***singular value decomposition*** or the ***SVD*** in short. This is perhaps the most widely used factorization in modern applications of linear algebra and, because of its widespread use, is sometimes referred to as the "Singularly Valuable Decomposition." See Kalman (1996) for an excellent expository article of the same name.

The SVD considers a rectangular $m \times n$ matrix $\boldsymbol{A}$ with real entries. If the rank of $\boldsymbol{A}$ is r, then the SVD says that there are unique positive constants $\sigma_1 \geq \sigma_2 \geq \cdots \geq \sigma_r > 0$ and orthogonal matrices $\boldsymbol{U}$ and $\boldsymbol{V}$ of order $m \times m$ and $n \times n$, respectively, such that

$$\boldsymbol{A} = \boldsymbol{U} \begin{bmatrix} \boldsymbol{\Sigma} & \boldsymbol{O} \\ \boldsymbol{O} & \boldsymbol{O} \end{bmatrix} \boldsymbol{V}' , \tag{12.1}$$

where $\boldsymbol{\Sigma}$ is an $r \times r$ diagonal matrix whose i-th diagonal element is σ_i. The $\boldsymbol{O}$ matrices have compatible numbers of rows and columns for the above partition to make sense. These σ_i's are called the ***singular values*** of $\boldsymbol{A}$. We will soon see why every rectangular matrix has a singular value decomposition but before that it is worth obtaining some insight into what is needed to construct the orthogonal matrices $\boldsymbol{U}$ and $\boldsymbol{V}$.

While the spectral decomposition uses one orthogonal matrix to diagonalize a real-symmetric matrix, the SVD accommodates the possibility of two orthogonal matrices $\boldsymbol{U}$ and $\boldsymbol{V}$. To understand why these two matrices are needed, consider the SVD for an $m \times n$ matrix $\boldsymbol{A}$. The transformation $f(\boldsymbol{x}) = \boldsymbol{A}\boldsymbol{x}$ takes a vector in $\Re^n$ and maps it to a vector in a different space, $\Re^m$. The columns of $\boldsymbol{V}$ and $\boldsymbol{U}$ provide natural orthogonal bases for the domain and the range of $f(\boldsymbol{x})$. When vectors in the domain and range of $f(\boldsymbol{x})$ are represented in terms of these bases, the nature of the transformation $\boldsymbol{Ax}$ becomes transparent in a manner similar to that of the spectral decomposition: $\boldsymbol{A}$ simply dilates some components of $\boldsymbol{x}$ (when the singular values have magnitude greater than one) and contracts others (when the singular values have magnitude less than one). The transformation possibly discards components or appends zeros as needed to adjust for the different dimensions of the domain and the range. From this perspective, the SVD achieves for a rectangular matrix what the spectral decomposition achieves for a real-symmetric matrix. They both tell us how to choose orthonormal bases with respect to which the transformation $\boldsymbol{A}$ can be represented by a diagonal matrix.

Given a matrix $\boldsymbol{A}$, how do we choose the bases $\{\boldsymbol{u}_1, \boldsymbol{u}_2, \ldots, \boldsymbol{u}_m\}$ for $\Re^m$ and a basis $\{\boldsymbol{v}_1, \boldsymbol{v}_2, \ldots, \boldsymbol{v}_n\}$ for $\Re^n$ so that we obtain the factorization in (12.1)? Theorem 7.5 tells us that row space and the null space of $\boldsymbol{A}$ are orthocomplements of each other. This ensures the existence of an orthonormal basis $\{\boldsymbol{v}_1, \boldsymbol{v}_2, \ldots, \boldsymbol{v}_n\}$ for $\Re^n$ such that the first r vectors are a basis for $\mathcal{C}(\boldsymbol{A}')$, i.e., the row space of $\boldsymbol{A}$, and the remaining $n - r$ vectors are an orthonormal basis for the null space of $\boldsymbol{A}$. Therefore, $\boldsymbol{Av}_i = \boldsymbol{0}$ for $i = r+1, r+2, \ldots, n$. Also, $\boldsymbol{Av}_i \neq \boldsymbol{0}$ for $i = 1, 2, \ldots, r$ because if $\boldsymbol{Av}_i = \boldsymbol{0}$ for some $i = 1, 2, \ldots, r$, then that $\boldsymbol{v}_i$ would belong to the null space of $\boldsymbol{A}$, and hence be spanned by $\{\boldsymbol{v}_{r+1}, \boldsymbol{v}_{r+2}, \ldots, \boldsymbol{v}_n\}$, which would contradict the linear independence of $\{\boldsymbol{v}_1, \boldsymbol{v}_2, \ldots, \boldsymbol{v}_n\}$.

Let us now choose $\boldsymbol{u}_i$ to be the normalized vector parallel to the nonzero vectors $\boldsymbol{Av}_i$ for $i = 1, 2, \ldots, r$. This means that $\boldsymbol{Av}_i = \sigma_i \boldsymbol{u}_i$, where $\sigma_i = \|\boldsymbol{Av}_i\|$. We have seen in Section 7.2 that if $\{\boldsymbol{v}_1, \boldsymbol{v}_2, \ldots, \boldsymbol{v}_r\}$ is a basis for the row space of $\boldsymbol{A}$, then $\{\boldsymbol{Av}_1, \boldsymbol{Av}_2, \ldots, \boldsymbol{Av}_r\}$ is a linearly independent set of r vectors in the column space of $\boldsymbol{A}$. And r is the dimension of the column space of $\boldsymbol{A}$, which implies that $\{\boldsymbol{Av}_1, \boldsymbol{Av}_2, \ldots, \boldsymbol{Av}_r\}$ is a basis for the column space of $\boldsymbol{A}$. This means that $\{\boldsymbol{u}_1, \boldsymbol{u}_2, \ldots, \boldsymbol{u}_r\}$ is a basis for the column space of $\boldsymbol{A}$. Extend this to a basis for $\Re^m$ by appending vectors $\{\boldsymbol{u}_{r+1}, \boldsymbol{u}_{r+2}, \ldots, \boldsymbol{u}_m\}$. Let

$$\boldsymbol{U} = [\boldsymbol{u}_1 : \boldsymbol{u}_2 : \ldots : \boldsymbol{u}_m] \text{ and } \boldsymbol{V} = [\boldsymbol{v}_1 : \boldsymbol{v}_2 : \ldots : \boldsymbol{v}_n] \,. \tag{12.2}$$

Then,

$$\begin{aligned} \boldsymbol{AV} &= [\boldsymbol{Av}_1 : \boldsymbol{Av}_2 : \ldots : \boldsymbol{Av}_r : \boldsymbol{Av}_{r+1} : \ldots : \boldsymbol{Av}_n] \\ &= [\sigma_1 \boldsymbol{u}_1 : \sigma_2 \boldsymbol{u}_2 : \ldots : \sigma_r \boldsymbol{u}_r : \boldsymbol{0} : \ldots : \boldsymbol{0}] = [\boldsymbol{U}_1 : \boldsymbol{U}_2] \begin{bmatrix} \boldsymbol{\Sigma} & \boldsymbol{O} \\ \boldsymbol{O} & \boldsymbol{O} \end{bmatrix} , \end{aligned}$$

where $\boldsymbol{U} = [\boldsymbol{U}_1 : \boldsymbol{U}_2]$ such that $\boldsymbol{U}_1$ is $m \times r$. Since $\boldsymbol{V}$ is an orthogonal matrix, we have a form similar to (12.1).

Using $\boldsymbol{U}$ and $\boldsymbol{V}$ as in (12.2), we have diagonalized a rectangular matrix $\boldsymbol{A}$. But there is one glitch: although the $\boldsymbol{v}_i$'s are orthogonal, the $\boldsymbol{u}_i$'s as constructed above need not be orthogonal. This begs the question: can we choose an orthonormal basis $\{\boldsymbol{v}_1, \boldsymbol{v}_2, \ldots, \boldsymbol{v}_n\}$ so that its orthogonality is preserved under $\boldsymbol{A}$, i.e., the resulting $\boldsymbol{u}_i$'s are also orthogonal?

The answer is yes and the key lies with the spectral decomposition. Consider the

$n \times n$ matrix $\boldsymbol{A}'\boldsymbol{A}$. This matrix is always symmetric and has real entries whenever $\boldsymbol{A}$ is real. By Theorem 11.27, $\boldsymbol{A}'\boldsymbol{A}$ has a spectral decomposition, so there exists an $n \times n$ orthogonal matrix $\boldsymbol{V}$ and an $n \times n$ diagonal matrix $\boldsymbol{\Lambda}$ such that $\boldsymbol{A}'\boldsymbol{A} = \boldsymbol{V}\boldsymbol{\Lambda}\boldsymbol{V}'$. The columns of $\boldsymbol{V}$ are the eigenvectors of $\boldsymbol{A}$ associated with the eigenvalues appearing along the diagonal of $\boldsymbol{\Lambda}$. These eigenvectors constitute an orthonormal basis $\{\boldsymbol{v}_1, \boldsymbol{v}_2, \ldots, \boldsymbol{v}_n\}$ for $\Re^n$. The rank of $\boldsymbol{A}$ is r. From the fundamental theorem of ranks, we know that $\rho(\boldsymbol{A}'\boldsymbol{A}) = \rho(\boldsymbol{A}) = r$. Since $\rho(\boldsymbol{A}'\boldsymbol{A}) = \rho(\boldsymbol{\Lambda})$, there are exactly r nonzero entries along the diagonal of λ.

Without loss of generality, assume that $\lambda_1 \geq \lambda_2 \geq \cdots \geq \lambda_r > 0$ and $\lambda_i = 0$ for $i = r+1, r+2, \ldots, n$. It is easy to verify from the spectral decomposition that $\{\boldsymbol{v}_1, \boldsymbol{v}_2, \ldots, \boldsymbol{v}_r\}$ is a basis for $\mathcal{C}(\boldsymbol{A}'\boldsymbol{A})$, which is the same as the row space of $\boldsymbol{A}$. The spectral decomposition also implies that $\boldsymbol{A}\boldsymbol{v}_i = \boldsymbol{0}$ for $i = r+1, r+2, \ldots, n$, so $\{\boldsymbol{v}_{r+1}, \boldsymbol{v}_{r+2}, \ldots, \boldsymbol{v}_n\}$ is a basis for the null space of $\boldsymbol{A}$. Note that

$$(\boldsymbol{A}\boldsymbol{v}_j)'(\boldsymbol{A}\boldsymbol{v}_i) = \boldsymbol{v}_j'\boldsymbol{A}'\boldsymbol{A}\boldsymbol{v}_i = \boldsymbol{v}_j'(\boldsymbol{A}'\boldsymbol{A}\boldsymbol{v}_i) = \boldsymbol{v}_j'(\lambda_i\boldsymbol{v}_i) = \lambda_i\boldsymbol{v}_j'\boldsymbol{v}_i = \lambda_i\delta_{ij} ,$$

where $\delta_{ij} = 1$ if $i = j$ and 0 if $i \neq j$. This shows that the set $\{\boldsymbol{A}\boldsymbol{v}_1, \boldsymbol{A}\boldsymbol{v}_2, \ldots, \boldsymbol{A}\boldsymbol{v}_r\}$ is an orthonormal set; it is an orthonormal basis for the column space of $\boldsymbol{A}$.

Take $\boldsymbol{u}_i$ to be a vector of unit length that is parallel to $\boldsymbol{A}\boldsymbol{v}_i$. Observe that

$$0 < \|\boldsymbol{A}\boldsymbol{v}_i\|^2 = \boldsymbol{v}_i'\boldsymbol{A}'\boldsymbol{A}\boldsymbol{v}_i = \boldsymbol{v}_i'(\lambda_i\boldsymbol{v}_i) = \lambda_i\boldsymbol{v}_i'\boldsymbol{v}_i = \lambda_i , \tag{12.3}$$

which implies that $\lambda_i > 0$ (i.e., strictly positive) for $i = 1, 2, \ldots, r$. The vector $\boldsymbol{u}_i$ is now explicitly written as

$$\boldsymbol{u}_i = \frac{\boldsymbol{A}\boldsymbol{v}_i}{\|\boldsymbol{A}\boldsymbol{v}_i\|} = \frac{\boldsymbol{A}\boldsymbol{v}_i}{\sqrt{\lambda_i}} \quad \text{for } i = 1, 2, \ldots, r. \tag{12.4}$$

Setting $\sigma_i = \sqrt{\lambda_i}$ yields $\boldsymbol{A}\boldsymbol{v}_i = \sigma_i\boldsymbol{u}_i$ for $i = 1, 2, \ldots, r$ and now $\{\boldsymbol{u}_1, \boldsymbol{u}_2, \ldots, \boldsymbol{u}_r\}$ is an orthonormal set. Extend this to an orthonormal basis for $\Re^m$ by appending vectors $\{\boldsymbol{u}_{r+1}, \boldsymbol{u}_{r+2}, \ldots, \boldsymbol{u}_m\}$. Constructing the matrices $\boldsymbol{U}$ and $\boldsymbol{V}$ as in (12.2) now yields the factorization in (12.1).

We now assemble the above ideas into a formal theorem and proof.

Theorem 12.1 *Let $\boldsymbol{A}$ be an $m \times n$ matrix with real entries and rank equal to r. There exists an $m \times m$ real orthogonal matrix $\boldsymbol{U}$ and an $n \times n$ real orthogonal matrix $\boldsymbol{V}$ such that*

$$\boldsymbol{A} = \boldsymbol{U}\boldsymbol{D}\boldsymbol{V}' , \quad \text{where } \boldsymbol{D} = \begin{bmatrix} \boldsymbol{\Sigma} & \boldsymbol{O} \\ \boldsymbol{O} & \boldsymbol{O} \end{bmatrix} , \quad \text{and } \boldsymbol{\Sigma} = \begin{bmatrix} \sigma_1 & 0 & \ldots & 0 \\ 0 & \sigma_2 & \ldots & 0 \\ \vdots & \vdots & \ddots & \vdots \\ 0 & 0 & \ldots & \sigma_r \end{bmatrix} ,$$

where $\boldsymbol{D}$ is $m \times n$, $\boldsymbol{\Sigma}$ is $r \times r$ and the σ_i's are real numbers such that $\sigma_1 \geq \sigma_2 \geq \cdots \geq \sigma_r > 0$. This decomposition is also expressed using the following partition:

$$\boldsymbol{A} = \boldsymbol{A} = \boldsymbol{U}\boldsymbol{D}\boldsymbol{V}' = [\boldsymbol{U}_1 : \boldsymbol{U}_2] \begin{bmatrix} \boldsymbol{\Sigma} & \boldsymbol{O} \\ \boldsymbol{O} & \boldsymbol{O} \end{bmatrix} \begin{bmatrix} \boldsymbol{V}_1' \\ \boldsymbol{V}_2' \end{bmatrix} = \boldsymbol{U}_1\boldsymbol{\Sigma}\boldsymbol{V}_1' ,$$

where U_1 and V_1 are $m \times r$ and $n \times r$ matrices, respectively, with orthonormal columns and the O submatrices have compatible dimensions for the above partition to be sensible.

Proof. The matrix $A'A$ is an $n \times n$ symmetric matrix. This means that every eigenvalue of $A'A$ is a real number. Also, from (12.3) we know that any nonzero eigenvalue of $A'A$ must be strictly positive. In other words, all eigenvalues of A are nonnegative real numbers.

Let $\{\lambda_1, \lambda_2, \ldots, \lambda_n\}$ denote the entire set of n eigenvalues (counting multiplicities) of $A'A$. The spectral (eigenvalue) decomposition of $A'A$ yields

$$A'A = V\Lambda V' \,, \quad \text{where } \Lambda = \begin{bmatrix} \lambda_1 & 0 & \ldots & 0 \\ 0 & \lambda_2 & \ldots & 0 \\ \vdots & \vdots & \ddots & \vdots \\ 0 & 0 & \ldots & \lambda_n \end{bmatrix} ,$$

where V is an $n \times n$ orthogonal matrix. Without loss of generality we can assume that the eigenvalues are arranged in decreasing order along the diagonal so that $\lambda_1 \geq \lambda_2 \geq \cdots \lambda_n$. Since $\rho(\Lambda) = \rho(A'A) = \rho(A) = r$, it follows that

$$\lambda_1 \geq \lambda_2 \geq \cdots \lambda_r > 0 = \lambda_{r+1} = \lambda_{r+2} = \cdots = \lambda_n \,.$$

In other words, the r largest eigenvalues are strictly positive and the remaining $n - r$ are zero. The n columns of $V = [v_1 : v_2 : \ldots : v_n]$ are the corresponding orthonormal eigenvectors of $A'A$. Partition $V = [V_1 : V_2]$ such that $V_1 = [v_1 : v_2 : \ldots : v_r]$ is $n \times r$ and $V_1'V_1 = I$ because of the orthonormality of the v_i's.

Let Λ_1 be the $r \times r$ diagonal matrix with $\lambda_1 \geq \lambda_2 \geq \cdots \geq \lambda_r > 0$ as its diagonal elements. The spectral decomposition has the following consequence

$$\begin{aligned} A'A = V\Lambda V' &\Longrightarrow A'AV = V\Lambda = [V_1 : V_2]\,\Lambda = [V_1\Lambda_1 : O] \\ &\Longrightarrow A'AV_2 = O \Longrightarrow V_2'A'AV_2 = O \Longrightarrow AV_2 = O \,. \end{aligned} \tag{12.5}$$

This implies that each column of V_2 belongs to $\mathcal{N}(A)$.

Now define Σ to be the $r \times r$ diagonal matrix such that its i-th diagonal element is $\sigma_i = \sqrt{\lambda_i}$ for $i = 1, 2, \ldots, r$. Thus,

$$\Sigma = \Lambda_1^{1/2} = \begin{bmatrix} \sqrt{\lambda_1} & 0 & \ldots & 0 \\ 0 & \sqrt{\lambda_2} & \ldots & 0 \\ \vdots & \vdots & \ddots & \vdots \\ 0 & 0 & \ldots & \sqrt{\lambda_r} \end{bmatrix} = \begin{bmatrix} \sigma_1 & 0 & \ldots & 0 \\ 0 & \sigma_2 & \ldots & 0 \\ \vdots & \vdots & \ddots & \vdots \\ 0 & 0 & \ldots & \sigma_r \end{bmatrix} .$$

Define the $m \times r$ matrix U_1 as

$$U_1 = AV_1\Sigma^{-1} \,. \tag{12.6}$$

Observe that the spectral decomposition also yields the following:

$$\begin{aligned} A'AV &= V\Lambda = [V_1 : V_2]\,\Lambda = [V_1\Lambda_1 : O] \\ &\Longrightarrow A'AV_1 = V_1\Lambda_1 = V_1\Sigma^2 \Longrightarrow V_1'A'AV_1 = V_1'V_1\Sigma^2 = \Sigma^2 \\ &\Longrightarrow \Sigma^{-1}V_1'A'AV_1\Sigma^{-1} = I = U_1'U_1 \,. \end{aligned} \tag{12.7}$$

This reveals that the r columns of U_1 are orthonormal. Also, (12.6) and (12.7) together imply that

$$U_1 = AV_1\Sigma^{-1} \Longrightarrow I = U_1'U_1 = U_1'AV_1\Sigma^{-1} \Longrightarrow \Sigma = U_1'AV_1 \,. \tag{12.8}$$

Extend the set of columns in U_1 to an orthonormal basis for $\Re^m$. In other words, choose any $m \times (m-r)$ matrix U_2 such that $U = [U_1 : U_2]$ is an orthogonal matrix (recall Lemma 8.2). Therefore, each column of U_2 is orthogonal to every column of U_1, which implies that $U_2'U_1 = O$, and so

$$U_2'AV_1 = U_2'U_1\Sigma = O \,.$$

This implies that

$$\begin{aligned} U'AV &= \begin{bmatrix} U_1' \\ U_2' \end{bmatrix} A\,[V_1 : V_2] = \begin{bmatrix} U_1'AV_1 & U_1'AV_2 \\ U_2'AV_1 & U_2'AV_2 \end{bmatrix} \\ &= \begin{bmatrix} U_1'AV_1 & O \\ U_2'AV_1 & O \end{bmatrix} = \begin{bmatrix} U_1'AV_1 & O \\ O & O \end{bmatrix} = \begin{bmatrix} \Sigma & O \\ O & O \end{bmatrix} = D \end{aligned}$$

because $AV_2 = O$, as was seen in (12.5), and $U_1'AV_1 = \Sigma$, as seen in (12.8). From the above, it is immediate that $A = UDV' = U_1\Sigma V_1'$. □

The following definitions, associated with an SVD of a matrix, are widely used.

Definition 12.1 *Let A be an $m \times n$ matrix of rank r and let $A = UDV'$ be its SVD as in Theorem 12.1.*

1. *The set $\{\sigma_1, \sigma_2, \ldots, \sigma_r\}$ is referred to as the* **nonzero singular values** *of A. The nonzero singular values are strictly positive, so they are equivalently called the* **positive singular values**.
2. *The column vectors of U are called the* **left singular vectors** *of A.*
3. *The column vectors of V are called the* **right singular vectors** *of A.*
4. *An alternative expression for the SVD of a matrix is its* **outer product** *form. Let $A = U_1\Sigma V_1'$ be an SVD of A using the notations in Theorem 12.1. Then,*

$$A = [u_1 : u_2 : \ldots u_r] \begin{bmatrix} \sigma_1 & 0 & \ldots & 0 \\ 0 & \sigma_2 & \ldots & 0 \\ \vdots & \vdots & \ddots & \vdots \\ 0 & 0 & \ldots & \sigma_r \end{bmatrix} \begin{bmatrix} v_1' \\ v_2' \\ \vdots \\ v_r' \end{bmatrix} = \sum_{i=1}^{r} \sigma_i u_i v_i' \,. \tag{12.9}$$

This shows that any matrix A of rank r can be expressed as a sum of r rank-one matrices, where each rank one matrix is constructed from the left and right singular vectors.

The singular vectors satisfy the relations

$$\boldsymbol{A}'\boldsymbol{u}_i = \sigma_i \boldsymbol{v}_i \text{ and } \boldsymbol{A}\boldsymbol{v}_i = \sigma_i \boldsymbol{u}_i \text{ for } i = 1, 2, \ldots, r .$$

As we have seen from the proof of Theorem 12.1 (or the motivating discussion preceding it), the r right singular vectors of $\boldsymbol{A}$ are precisely the eigenvectors of $\boldsymbol{A}'\boldsymbol{A}$. Similarly, note that

$$\boldsymbol{A}\boldsymbol{A}' = \boldsymbol{U}\boldsymbol{D}\boldsymbol{V}'\boldsymbol{V}\boldsymbol{D}'\boldsymbol{U}' = \boldsymbol{U}\boldsymbol{D}\boldsymbol{D}'\boldsymbol{U}' = \boldsymbol{U}\boldsymbol{\Lambda}\boldsymbol{U}' \text{ , where } \boldsymbol{\Lambda} = \boldsymbol{D}\boldsymbol{D}' = \begin{bmatrix} \boldsymbol{\Sigma}^2 & \boldsymbol{O} \\ \boldsymbol{O} & \boldsymbol{O} \end{bmatrix} ,$$

is a spectral decomposition of $\boldsymbol{A}\boldsymbol{A}'$. Therefore, the left singular vectors of $\boldsymbol{A}$ are eigenvectors of $\boldsymbol{A}\boldsymbol{A}'$. Equation 12.3 ensures that any nonzero eigenvalue of $\boldsymbol{A}'\boldsymbol{A}$ is positive. The positive singular values are the square roots of the nonzero (positive) eigenvalues of $\boldsymbol{A}'\boldsymbol{A}$, which are also the nonzero (positive) eigenvalues of $\boldsymbol{A}\boldsymbol{A}'$ (recall Theorem 11.3). We can compute the SVD of a matrix in exact arithmetic by first finding a spectral decomposition of $\boldsymbol{A}'\boldsymbol{A}$, which would produce the eigenvalues (hence the singular values) and the right singular vectors of $\boldsymbol{A}$, and then constructing the left singular vectors using (12.4). Below is an example.

Example 12.1 Computing the SVD in exact arithmetic. Suppose we wish to find the SVD of the 3×2 matrix

$$\boldsymbol{A} = \begin{bmatrix} 1 & 1 \\ \sqrt{3} & 0 \\ 0 & \sqrt{3} \end{bmatrix} .$$

We first form the two products:

$$\boldsymbol{A}'\boldsymbol{A} = \begin{bmatrix} 4 & 1 \\ 1 & 4 \end{bmatrix} \text{ and } \boldsymbol{A}\boldsymbol{A}' = \begin{bmatrix} 2 & \sqrt{3} & \sqrt{3} \\ \sqrt{3} & 3 & 0 \\ \sqrt{3} & 0 & 3 \end{bmatrix} .$$

Compute the eigenvalues and eigenvectors of $\boldsymbol{A}'\boldsymbol{A}$, which are the roots of

$$\begin{vmatrix} 4-\lambda & 1 \\ 1 & 4-\lambda \end{vmatrix} = 0 \Longrightarrow \lambda^2 - 8\lambda + 15 = (\lambda - 5)(\lambda - 3) = 0 .$$

Therefore, the eigenvalues of $\boldsymbol{A}'\boldsymbol{A}$ are 5 and 3. The normalized eigenvectors are found by solving the homogeneous systems

$$\begin{bmatrix} 4-\lambda & 1 \\ 1 & 4-\lambda \end{bmatrix} \begin{bmatrix} x_1 \\ x_2 \end{bmatrix} = \begin{bmatrix} 0 \\ 0 \end{bmatrix} \text{ for } \lambda = 5, 3 .$$

The normalized eigenvectors corresponding to $\lambda = 5$ and $\lambda = 3$ are, respectively,

$$\boldsymbol{v}_1 = \frac{1}{\sqrt{2}} \begin{bmatrix} 1 \\ 1 \end{bmatrix} \text{ and } \boldsymbol{v}_2 = \frac{1}{\sqrt{2}} \begin{bmatrix} 1 \\ -1 \end{bmatrix} .$$

Now form the normalized vectors $\boldsymbol{u}_i = \frac{\boldsymbol{A}\boldsymbol{v}_i}{\|\boldsymbol{A}\boldsymbol{v}_i\|}$ for $i = 1, 2$:

$$\boldsymbol{u}_1 = \begin{bmatrix} 1 & 1 \\ \sqrt{3} & 0 \\ 0 & \sqrt{3} \end{bmatrix} \begin{bmatrix} \frac{1}{\sqrt{2}} \\ \frac{1}{\sqrt{2}} \end{bmatrix} = \begin{bmatrix} \frac{2}{\sqrt{10}} \\ \frac{\sqrt{3}}{\sqrt{10}} \\ \frac{\sqrt{3}}{\sqrt{10}} \end{bmatrix} \text{ and } \boldsymbol{u}_2 = \begin{bmatrix} 1 & 1 \\ \sqrt{3} & 0 \\ 0 & \sqrt{3} \end{bmatrix} \begin{bmatrix} \frac{1}{\sqrt{2}} \\ -\frac{1}{\sqrt{2}} \end{bmatrix} = \begin{bmatrix} 0 \\ \frac{1}{\sqrt{2}} \\ -\frac{1}{\sqrt{2}} \end{bmatrix}.$$

From Theorem 11.3 we know that the nonzero eigenvalues of $\boldsymbol{A}\boldsymbol{A}'$ must be the same as those of $\boldsymbol{A}'\boldsymbol{A}$. Thus, $\boldsymbol{A}\boldsymbol{A}'$ has 5 and 3 as its nonzero eigenvalues. The above construction ensures that $\boldsymbol{u}_1$ and $\boldsymbol{u}_2$ are normalized eigenvectors of $\boldsymbol{A}\boldsymbol{A}'$ associated with eigenvalues $\lambda = 5$ and $\lambda = 3$.

Finally, we choose a normalized vector $\boldsymbol{u}_3$ that is orthogonal to $\boldsymbol{u}_1$ and $\boldsymbol{u}_2$. Any such vector will suffice. We can use Gram-Schmidt orthogonalization starting with any vector linearly independent of $\boldsymbol{u}_1$ and $\boldsymbol{u}_2$. Alternatively, we can argue as follows. The third eigenvalue of $\boldsymbol{A}\boldsymbol{A}'$ must be zero. (This is consistent with the fact that $\boldsymbol{A}\boldsymbol{A}'$ is singular and, hence, must have a zero eigenvalue.) We can choose $\boldsymbol{u}_3$ as a normalized eigenvector associated with eigenvalue $\lambda = 0$ of $\boldsymbol{A}\boldsymbol{A}'$. We obtain this by solving the homogeneous system

$$\begin{bmatrix} 2-\lambda & \sqrt{3} & \sqrt{3} \\ \sqrt{3} & 3-\lambda & 0 \\ \sqrt{3} & 0 & 3-\lambda \end{bmatrix} \begin{bmatrix} x_1 \\ x_2 \\ x_3 \end{bmatrix} = \begin{bmatrix} 0 \\ 0 \\ 0 \end{bmatrix} \quad \text{for } \lambda = 0\,.$$

This yields the normalized eigenvector

$$\boldsymbol{u}_3 = \frac{1}{\sqrt{5}} \begin{bmatrix} -\sqrt{3} \\ 1 \\ 1 \end{bmatrix}.$$

We now have all the ingredients to obtain the SVD and it is just a matter of assembling the orthogonal matrices $\boldsymbol{U}$ and $\boldsymbol{V}$ by placing the normalized eigenvectors. We form the orthogonal matrices $\boldsymbol{V}$ and $\boldsymbol{U}$ as below:

$$\boldsymbol{V} = [\boldsymbol{v}_1 : \boldsymbol{v}_2] = \frac{1}{\sqrt{2}} \begin{bmatrix} 1 & 1 \\ -1 & 1 \end{bmatrix} \text{ and } \boldsymbol{U} = [\boldsymbol{u}_1 : \boldsymbol{u}_2 : \boldsymbol{u}_3] = \begin{bmatrix} \frac{2}{\sqrt{10}} & 0 & -\frac{\sqrt{3}}{\sqrt{5}} \\ \frac{\sqrt{3}}{\sqrt{10}} & \frac{1}{\sqrt{2}} & \frac{1}{\sqrt{5}} \\ \frac{\sqrt{3}}{\sqrt{10}} & -\frac{1}{\sqrt{2}} & \frac{1}{\sqrt{5}} \end{bmatrix}.$$

It now follows that the SVD of $\boldsymbol{A}$ is given by

$$\boldsymbol{A} = \boldsymbol{U}\boldsymbol{D}\boldsymbol{V}' = \begin{bmatrix} \frac{2}{\sqrt{10}} & 0 & -\frac{\sqrt{3}}{\sqrt{5}} \\ \frac{\sqrt{3}}{\sqrt{10}} & \frac{1}{\sqrt{2}} & \frac{1}{\sqrt{5}} \\ \frac{\sqrt{3}}{\sqrt{10}} & -\frac{1}{\sqrt{2}} & \frac{1}{\sqrt{5}} \end{bmatrix} \begin{bmatrix} \sqrt{5} & 0 \\ 0 & \sqrt{3} \\ 0 & 0 \end{bmatrix} \begin{bmatrix} \frac{1}{\sqrt{2}} & -\frac{1}{\sqrt{2}} \\ \frac{1}{\sqrt{2}} & \frac{1}{\sqrt{2}} \end{bmatrix}.$$

The above procedure works well in exact arithmetic but not in floating point arithmetic (on computers), where calculating $\boldsymbol{A}'\boldsymbol{A}$ is not advisable for general matrices. Numerically efficient and accurate algorithms for computing the SVD with floating point arithmetic exist and are implemented in linear algebra software. We provide a brief discussion in Section 12.5. ■

Remark: We must throw in a word of caution here. While it is true that the right singular vectors are eigenvectors of $\boldsymbol{A}'\boldsymbol{A}$ and the left singular vectors are eigenvectors of $\boldsymbol{A}\boldsymbol{A}'$, it is *not true* that *any* orthogonal set of eigenvectors of $\boldsymbol{A}'\boldsymbol{A}$ and $\boldsymbol{A}\boldsymbol{A}'$ can be taken as right and left singular vectors of $\boldsymbol{A}$. In other words, if $\boldsymbol{U}$ and $\boldsymbol{V}$ are orthogonal matrices such that $\boldsymbol{U}'\boldsymbol{A}\boldsymbol{A}'\boldsymbol{U}$ and $\boldsymbol{V}'\boldsymbol{A}'\boldsymbol{V}$ produce spectral decompositions for $\boldsymbol{A}\boldsymbol{A}'$ and $\boldsymbol{A}'\boldsymbol{A}$, respectively, it does not imply that $\boldsymbol{A} = \boldsymbol{U}\boldsymbol{D}\boldsymbol{V}'$ for $\boldsymbol{D}$ as defined in (12.1). This should be evident from the derivation of the SVD. We can choose *any* orthonormal set of eigenvectors of $\boldsymbol{A}'\boldsymbol{A}$ as the set of right singular vectors of $\boldsymbol{A}$. So, the columns of $\boldsymbol{V}$ qualify as one such set. But then the corresponding set of left singular vectors, $\boldsymbol{u}_i$'s, must be "special" normalized eigenvectors of $\boldsymbol{A}\boldsymbol{A}'$ that are parallel to $\boldsymbol{A}\boldsymbol{v}_i$, as in (12.4).

Remark: The SVD of $\boldsymbol{A}'$. Using the same notation as in Theorem 12.1, suppose that $\boldsymbol{A} = \boldsymbol{U}\boldsymbol{D}\boldsymbol{V}' = \boldsymbol{U}_1\boldsymbol{\Sigma}\boldsymbol{V}_1'$ is an SVD of an $m \times n$ matrix $\boldsymbol{A}$. Taking transposes, we have $\boldsymbol{A}' = \boldsymbol{V}\boldsymbol{D}'\boldsymbol{U}' = \boldsymbol{V}_1\boldsymbol{\Sigma}\boldsymbol{U}_1'$. Notice that this qualifies as an SVD of $\boldsymbol{A}'$. The matrix $\boldsymbol{D}$ is, in general, rectangular, so $\boldsymbol{D}$ and $\boldsymbol{D}'$ are not the same. But $\boldsymbol{\Sigma}$ is always square and symmetric, so it is equal to its transpose. The nonzero singular values of $\boldsymbol{A}$ and $\boldsymbol{A}'$ are the same.

It is not difficult to imitate the proof of Theorem 12.1 for complex matrices by using adjoints instead of transposes. If $\boldsymbol{A}$ has possibly complex entries, then the SVD will be $\boldsymbol{A} = \boldsymbol{U}\boldsymbol{D}\boldsymbol{V}^*$, where $\boldsymbol{U}$ and $\boldsymbol{V}$ are unitary matrices.

The SVD is not unique. Here is a trivial example.

Example 12.2 Let $\boldsymbol{A} = \boldsymbol{I}$ be an $n \times n$ identity matrix. Then, $\boldsymbol{A} = \boldsymbol{U}\boldsymbol{I}\boldsymbol{U}'$ is an SVD for any arbitrary orthogonal matrix $\boldsymbol{U}$. ■

Here is another example that shows SVD's are not unique.

Example 12.3

$$\boldsymbol{A} = \begin{bmatrix} 1 & 0 \\ 0 & -1 \end{bmatrix} = \begin{bmatrix} \cos\theta & \sin\theta \\ -\sin\theta & \cos\theta \end{bmatrix} \begin{bmatrix} 1 & 0 \\ 0 & 1 \end{bmatrix} \begin{bmatrix} \cos\theta & \sin\theta \\ \sin\theta & -\cos\theta \end{bmatrix}$$

for any angle θ. ■

In fact, from the proof of Theorem 12.1, we find that any matrix $\boldsymbol{U}_2$ such that $\boldsymbol{U} = [\boldsymbol{U}_1 : \boldsymbol{U}_2]$ is an orthogonal matrix will yield an SVD. Also, any orthonormal basis for the null space of $\boldsymbol{A}$ can be used as the columns of $\boldsymbol{V}_2$. The singular values, however, *are unique* and so the matrix $\boldsymbol{\Sigma}$ is unique.

There are several alternative ways of expressing an SVD. Any factorization $\boldsymbol{A} = \boldsymbol{U}\boldsymbol{D}\boldsymbol{V}'$ of an $m \times n$ matrix is an SVD of $\boldsymbol{A}$ as long as $\boldsymbol{U}$ and $\boldsymbol{V}$ are orthogonal matrices and $\boldsymbol{D}$ is an $m \times n$ *diagonal* matrix, where by "diagonal" we mean that

$$\boldsymbol{D} = \begin{bmatrix} \boldsymbol{\Sigma} \\ \boldsymbol{O} \end{bmatrix} \quad \text{or} \quad \boldsymbol{D} = \begin{bmatrix} \boldsymbol{\Sigma} : \boldsymbol{O} \end{bmatrix} \tag{12.10}$$

depending upon whether $m \geq n$ or $m \leq n$, respectively. The matrix $\boldsymbol{\Sigma}$ is a diagonal matrix with nonnegative diagonal elements. It is $n \times n$ in the former case and $m \times m$ in the latter. If $m = n$, the $\boldsymbol{O}$ submatrices collapse and $\boldsymbol{D}$ itself is an $n \times n$ diagonal matrix with the nonnegative singular values along the diagonal.

Notice the difference in the way $\boldsymbol{\Sigma}$ is defined in (12.10) from how it was defined earlier in (12.1) (or in Theorem 12.1). In (12.1), $\boldsymbol{\Sigma}$ was defined to be the diagonal matrix with strictly positive diagonal elements, so the dimension of $\boldsymbol{\Sigma}$ was equal to the rank of $\boldsymbol{A}$. In (12.10) $\boldsymbol{\Sigma}$ is $k \times k$, where $k = \min\{m, n\}$ and is a diagonal matrix that can have $k - r$ zero singular values along the diagonal, where $\rho(\boldsymbol{A}) = r$. Unless stated otherwise, we will use the notations in (12.1) (or in Theorem 12.1).

12.2 The SVD and the four fundamental subspaces

The SVD is intricately related to the four fundamental subspaces associated with a matrix. If $\boldsymbol{A} = \boldsymbol{U}\boldsymbol{D}\boldsymbol{V}'$ is an SVD of $\boldsymbol{A}$, then the column vectors of $\boldsymbol{U}$ and $\boldsymbol{V}$ hide orthonormal bases for the four fundamental subspaces. In the proof of Theorem 12.1, we already saw that $\boldsymbol{A}\boldsymbol{V}_2 = \boldsymbol{O}$. This means that each of the $n - r$ columns of $\boldsymbol{V}_2$ belong to the null space of $\boldsymbol{A}$. Since the dimension of the null space of $\boldsymbol{A}$ is $n - r$, these columns constitute an orthonormal basis for $\mathcal{N}(\boldsymbol{A})$. Theorem 12.1 also showed that $\boldsymbol{A} = \boldsymbol{U}_1\boldsymbol{\Sigma}\boldsymbol{V}_1'$, which implies that each column of $\boldsymbol{A}$ is a linear combination of the r columns of $\boldsymbol{A}$. This means that the column space of $\boldsymbol{A}$ is contained in the column space of $\boldsymbol{U}_1$. Since the dimension of $\mathcal{C}(\boldsymbol{A})$ is r, it follows that the r columns of $\boldsymbol{U}_1$ are an orthonormal basis for $\mathcal{C}(\boldsymbol{A})$.

At this point we have found orthonormal bases for two of the four fundamental subspaces $\mathcal{C}(\boldsymbol{A})$ and $\mathcal{N}(\boldsymbol{A})$. For the remaining two fundamental subspaces, consider the SVD of $\boldsymbol{A}'$: $\boldsymbol{A}' = \boldsymbol{V}\boldsymbol{D}'\boldsymbol{U}$. Writing this out explicitly, immediately reveals that $\mathcal{C}(\boldsymbol{A}')$ is a subset of the column space of $\boldsymbol{V}_1$. Since $\dim(\mathcal{C}(\boldsymbol{A})) = \dim(\mathcal{C}(\boldsymbol{A}')) = r$, the r columns of $\boldsymbol{V}_1$ are an orthonormal basis for $\mathcal{C}(\boldsymbol{A}')$ or, equivalently, the row space of $\boldsymbol{A}$. Finally, we see that $\boldsymbol{A}'\boldsymbol{U}_2 = \boldsymbol{O}$. This means that the $m - r$ columns of $\boldsymbol{U}_2$ lie in $\mathcal{N}(\boldsymbol{A}')$. Since the dimension of the null space of $\boldsymbol{A}'$ is $m - r$, the columns of $\boldsymbol{U}_2$ constitute an orthonormal basis for $\mathcal{N}(\boldsymbol{A}')$. The above argument uses our knowledge of the dimensions of the four fundamental subspaces, as derived in the Fundamental Theorem of Linear Algebra (Theorem 7.5). We can, however, derive the orthonormal bases for the four fundamental subspaces directly from the SVD. Neither the SVD (Theorem 12.1) nor the spectral decomposition (Theorem 11.27), which we used in deriving the SVD, needed the Fundamental Theorem of Linear Algebra (Theorem 7.5). Therefore, the SVD can be used as a canonical form that reveals the orthonormal bases for the four fundamental subspaces (hence, their dimensions as well). The Fundamental Theorem of Linear Algebra follows easily. We present this as a theorem.

Theorem 12.2 SVD and the four fundamental subspaces. *Let $\boldsymbol{A} = \boldsymbol{U}\boldsymbol{D}\boldsymbol{V}'$ be the SVD of an $m \times n$ matrix as in Theorem 12.1.*

(i) The (left singular) column vectors of $\boldsymbol{U}_1 = [\boldsymbol{u}_1 : \boldsymbol{u}_2 : \ldots : \boldsymbol{u}_r]$ form an orthonormal basis for $\mathcal{C}(\boldsymbol{A})$.

(ii) *The (right singular) column vectors of* $V_2 = [v_{r+1} : v_{r+2} : \ldots : v_n]$ *form an orthonormal basis for* $\mathcal{N}(A) = [\mathcal{C}(A')]^{\perp}$.

(iii) *The (right singular) column vectors of* $V_1 = [v_1 : v_2 : \ldots : v_r]$ *form an orthonormal basis for* $\mathcal{R}(A) = \mathcal{C}(A')$.

(iv) *The (left singular) column vectors of* $U_2 = [u_{r+1} : u_{r+2} : \ldots : u_m]$ *form an orthonormal basis for* $\mathcal{N}(A') = [\mathcal{C}(A)]^{\perp}$.

Proof. We use the same notations and definitions as in Theorem 12.1.

Proof of (i): Since the columns of U_1 form an orthonormal set, they are an orthonormal basis for $\mathcal{C}(U_1)$. We will now prove that $\mathcal{C}(A) = \mathcal{C}(U_1)$. Recall that $A = U_1\Sigma V_1'$ and $U_1 = AV_1\Sigma^{-1}$. Therefore,

$$\mathcal{C}(A) = \mathcal{C}(U_1\Sigma V_1') \subseteq \mathcal{C}(U_1) = \mathcal{C}(AV_1\Sigma^{-1}) \subseteq \mathcal{C}(A) .$$

Therefore, the columns of U_1 form an orthonormal basis for $\mathcal{C}(A)$.

Proof of (ii): The columns of V_2 form an orthonormal basis for $\mathcal{C}(V_2)$. We will show that $\mathcal{C}(V_2) = \mathcal{N}(A)$. From the SVD $A = UDV'$ it follows that

$$[AV_1 : AV_2] = AV = UD = [U_1 : U_2]\begin{bmatrix} \Sigma & O \\ O & O \end{bmatrix} = [U_1\Sigma : O]$$
$$\Longrightarrow AV_2 = O \Longrightarrow \mathcal{C}(V_2) \subseteq \mathcal{N}(A) .$$

To prove the reverse inclusion, consider a vector $x \in \mathcal{N}(A)$. This vector is $n \times 1$, so it also belongs to $\Re^n$. The columns of V are an orthonormal basis for $\Re^n$, so there exists a vector α such that

$$x = V\alpha = [V_1 : V_2]\begin{bmatrix} \alpha_1 \\ \alpha_2 \end{bmatrix} = V_1\alpha_1 + V_2\alpha_2 ,$$

where α_1 and α_2 are $r \times 1$ and $(n-r) \times 1$ subvectors of a conformable partition of α. Since $AV_2 = O$, pre-multiplying both sides of the above by A yields

$$0 = Ax = AV_1\alpha_1 + AV_2\alpha_2 = AV_1\alpha_1 = U_1\Sigma V_1'V_1\alpha_1 = U_1\Sigma\alpha_1 .$$

Using the fact $U_1'U_1 = I$, the above leads to

$$0 = U_1'0 = U_1'U_1\Sigma\alpha_1 = \Sigma\alpha_1 \Longrightarrow \alpha_1 = \Sigma^{-1}0 = 0 .$$

Therefore, $x = V_2\alpha_2 \in \mathcal{C}(V_2)$ and we have proved that any vector in $\mathcal{N}(A)$ lies in $\mathcal{C}(V_2)$ as well. This establishes the reverse inclusion and the proof of (ii) is complete.

Proof of (iii): Taking transposes of both sides of $A = U_1\Sigma V_1'$, we obtain an SVD of the transpose of A: $A' = V_1\Sigma U_1'$. Applying (i) to the SVD of A', establishes that $\mathcal{C}(A') = \mathcal{C}(V_1)$. This proves that the columns of V_1 are an orthonormal basis for $\mathcal{C}(A')$, which is the same as $\mathcal{R}(A)$.

Proof of (iv): The columns of U_2 form an orthonormal basis for $\mathcal{C}(U_2)$. Applying (ii) to the SVD $A' = VD'U'$, it follows that $\mathcal{C}(U_2) = \mathcal{N}(A')$. Therefore, the columns of U_2 are an orthonormal basis for $\mathcal{N}(A')$. □

Theorem 12.2 establishes the Fundamental Theorem of Linear Algebra in a more concrete manner. Not only do we obtain the dimensions of the four subspaces, the SVD explicitly offers the orthonormal bases for the four subspaces. In fact, many texts, especially those with a more numerical focus, will introduce the SVD at the outset. The *rank* of an $m \times n$ matrix $\boldsymbol{A}$ can be defined as the number of nonzero (i.e., positive) singular values of $\boldsymbol{A}$. With this definition of rank, Theorem 12.2 reveals that the rank of $\boldsymbol{A}$ is precisely the number of vectors in a basis for the column space of $\boldsymbol{A}$. So, the SVD-based definition of rank agrees with the more classical definition we have encountered earlier. The SVD immediately reveals that the rank of $\boldsymbol{A}$ is equal to the rank of $\boldsymbol{A}'$ and it also establishes the well-known Rank-Nullity Theorem.

The SVD provides a more numerically stable way of computing the rank of a matrix than methods that reduce matrices to echelon forms. Moreover, it also provides numerically superior methods to obtain orthonormal bases for the fundamental subspaces. In practical linear algebra software, it is usually the SVD, rather than RREF's, that is used to compute the rank of a matrix.

12.3 SVD and linear systems

The SVD provides us with another mechanism to analyze systems of linear equations and, in particular, the least squares problem. For example, consider the possibly over-determined system $\boldsymbol{A}\boldsymbol{x} = \boldsymbol{b}$, where $\boldsymbol{A}$ is $m \times n$ with $m \geq n$ and $\boldsymbol{b}$ is $m \times 1$. Suppose that the rank of $\boldsymbol{A}$ is n and write the SVD of $\boldsymbol{A}$ as

$$\boldsymbol{A} = [\boldsymbol{U}_1 : \boldsymbol{U}_2] \begin{bmatrix} \boldsymbol{\Sigma} \\ \boldsymbol{O} \end{bmatrix} \boldsymbol{V}' = \boldsymbol{U}_1 \boldsymbol{\Sigma} \boldsymbol{V}' ,$$

where $\boldsymbol{U}_1$ is $m \times n$ with orthonormal columns, $\boldsymbol{\Sigma}$ is $n \times n$ and diagonal with positive diagonal entries, and $\boldsymbol{V}$ is an $n \times n$ orthogonal matrix. Then,

$$\boldsymbol{A}\boldsymbol{x} = \boldsymbol{b} \Longrightarrow \boldsymbol{U}_1'\boldsymbol{A}\boldsymbol{x} = \boldsymbol{U}_1'\boldsymbol{b} \Longrightarrow \boldsymbol{U}_1'\boldsymbol{U}_1\boldsymbol{\Sigma}\boldsymbol{V}'\boldsymbol{x} = \boldsymbol{b} \Longrightarrow \boldsymbol{\Sigma}\tilde{\boldsymbol{x}} = \tilde{\boldsymbol{b}} ,$$

where $\tilde{\boldsymbol{x}} = \boldsymbol{V}'\boldsymbol{x}$ and $\tilde{\boldsymbol{b}} = \boldsymbol{U}_1'\boldsymbol{b}$. The SVD has reduced the system $\boldsymbol{A}\boldsymbol{x} = \boldsymbol{b}$ into a diagonal system. Solve the diagonal system $\boldsymbol{\Sigma}\tilde{\boldsymbol{x}} = \tilde{\boldsymbol{b}}$ and then obtain $\boldsymbol{x} = \boldsymbol{V}\tilde{\boldsymbol{x}}$. Therefore,

$$\boldsymbol{x} = \boldsymbol{V}\tilde{\boldsymbol{x}} = \boldsymbol{V}\boldsymbol{\Sigma}^{-1}\tilde{\boldsymbol{b}} = \boldsymbol{V}\boldsymbol{\Sigma}^{-1}\boldsymbol{U}_1'\boldsymbol{b} . \tag{12.11}$$

The $\boldsymbol{x}$ in (12.11) satisfies

$$\boldsymbol{A}\boldsymbol{x} = \boldsymbol{A}\boldsymbol{V}\tilde{\boldsymbol{x}} = \boldsymbol{U}_1\boldsymbol{\Sigma}\boldsymbol{V}'\boldsymbol{V}\tilde{\boldsymbol{x}} = \boldsymbol{U}_1\boldsymbol{\Sigma}\tilde{\boldsymbol{x}} = \boldsymbol{U}_1\tilde{\boldsymbol{b}} = \boldsymbol{U}_1\boldsymbol{U}_1'\boldsymbol{b} .$$

When $m = n$ and $\rho(\boldsymbol{A}) = n$, then the system is nonsingular, $\boldsymbol{U}_1$ is $n \times n$ orthogonal and $\boldsymbol{A}\boldsymbol{x} = \boldsymbol{b}$. Hence, the $\boldsymbol{x}$ computed in (12.11) is the unique solution.

When $m > n$ and $\rho(\boldsymbol{A}) = n$, there are more equations than variables. The system is under-determined and may or may not be consistent. It will be consistent if and only if $\boldsymbol{b}$ belongs to the column space of $\boldsymbol{A}$. The $\boldsymbol{x}$ in (12.11) now yields $\boldsymbol{A}\boldsymbol{x} = \boldsymbol{U}_1\boldsymbol{U}_1'\boldsymbol{b}$, which implies that

$$\boldsymbol{A}'\boldsymbol{A}\boldsymbol{x} = \boldsymbol{A}'\boldsymbol{U}_1\boldsymbol{U}_1'\boldsymbol{b} = \boldsymbol{V}\boldsymbol{\Sigma}\boldsymbol{U}_1'\boldsymbol{U}_1\boldsymbol{U}_1'\boldsymbol{b} = \boldsymbol{V}\boldsymbol{\Sigma}(\boldsymbol{U}_1'\boldsymbol{U}_1)\boldsymbol{U}_1'\boldsymbol{b} = \boldsymbol{V}\boldsymbol{\Sigma}\boldsymbol{U}_1'\boldsymbol{b} = \boldsymbol{A}'\boldsymbol{b} .$$

So, $\boldsymbol{x}$ is a solution for the normal equations. Thus, solving an under-determined system using the SVD results in the least squares solution for $\boldsymbol{Ax} = \boldsymbol{b}$.

Let us now turn to an over-determined system $\boldsymbol{Ax} = \boldsymbol{b}$, where $\boldsymbol{A}$ has more columns than rows, i.e., $m < n$. There are more variables than equations. Assume that the rank of $\boldsymbol{A}$ is equal to m. Such a system is always consistent but the solution is not unique. The SVD of $\boldsymbol{A}$ is

$$\boldsymbol{A} = \boldsymbol{U}\left[\boldsymbol{\Sigma} : \boldsymbol{O}\right]\boldsymbol{V}' = \boldsymbol{U}\left[\boldsymbol{\Sigma} : \boldsymbol{O}\right]\begin{bmatrix}\boldsymbol{V}_1'\\ \boldsymbol{V}_2'\end{bmatrix} = \boldsymbol{U\Sigma V}_1' ,$$

where $\boldsymbol{U}$ and $\boldsymbol{V} = [\boldsymbol{V}_1 : \boldsymbol{V}_2]$ are $m \times m$ and $n \times n$ orthogonal matrices, respectively, $\boldsymbol{\Sigma}$ is $m \times m$ diagonal with positive entries along the diagonal and $\boldsymbol{O}$ is the $m \times (n-m)$ matrix of zeroes. Now,

$$\boldsymbol{Ax} = \boldsymbol{b} \Longrightarrow \boldsymbol{U\Sigma V}_1'\boldsymbol{x} = \boldsymbol{b} \Longrightarrow \boldsymbol{V}_1'\boldsymbol{x} = \boldsymbol{\Sigma}^{-1}\boldsymbol{U}'\boldsymbol{b} .$$

Caution: $\boldsymbol{V}_1\boldsymbol{V}_1'$ is *not* equal to the identity matrix. So, we cannot multiply both sides of the above equation to obtain $\boldsymbol{x} = \boldsymbol{V}_1\boldsymbol{\Sigma}^{-1}\boldsymbol{U}'\boldsymbol{b}$. However, $\boldsymbol{x}_p = \boldsymbol{V}_1\boldsymbol{\Sigma}^{-1}\boldsymbol{U}'\boldsymbol{b}$ is one particular solution for $\boldsymbol{Ax} = \boldsymbol{b}$ because

$$\boldsymbol{Ax}_p = \boldsymbol{A}(\boldsymbol{V}_1\boldsymbol{\Sigma}^{-1}\boldsymbol{U}'\boldsymbol{b}) = \boldsymbol{U\Sigma V}_1'\boldsymbol{V}_1\boldsymbol{\Sigma}^{-1}\boldsymbol{U}'\boldsymbol{b} = \boldsymbol{U\Sigma\Sigma}^{-1}\boldsymbol{U}'\boldsymbol{b} = \boldsymbol{b} ,$$

where we have used the facts that $\boldsymbol{V}_1$ has orthonormal columns, so $\boldsymbol{V}_1'\boldsymbol{V}_1 = \boldsymbol{I}_m$, and $\boldsymbol{UU}' = \boldsymbol{I}_m$ because $\boldsymbol{U}$ is orthogonal. That $\boldsymbol{x}_p$ is not a unique solution can be verified by adding any other vector from $\mathcal{N}(\boldsymbol{A})$ to $\boldsymbol{x}_p$. The columns of $\boldsymbol{V}_2$ form an orthonormal basis for the null space of $\boldsymbol{A}$. Since $\boldsymbol{AV}_2 = \boldsymbol{O}$, any vector of the form $\boldsymbol{x}_p + \boldsymbol{V}_2\boldsymbol{y}$, where $\boldsymbol{y}$ is any arbitrary vector in $\Re^{(n-m)}$, is a solution for $\boldsymbol{Ax} = \boldsymbol{b}$.

Finally, it is easy to compute a generalized inverse from the SVD, which can then be used to construct a solution for $\boldsymbol{Ax} = \boldsymbol{b}$ irrespective of the rank of $\boldsymbol{A}$. Suppose $\boldsymbol{A} = \boldsymbol{UDV}'$ is the SVD of an $m \times n$ matrix $\boldsymbol{A}$, where $\boldsymbol{U}$ and $\boldsymbol{V}$ are orthogonal matrices of order $m \times m$ and $n \times n$, respectively. Define

$$\boldsymbol{D}^+ = \begin{bmatrix}\boldsymbol{\Sigma}^{-1} & \boldsymbol{O}\\ \boldsymbol{O} & \boldsymbol{O}\end{bmatrix} .$$

Then $\boldsymbol{A}^+ = \boldsymbol{VD}^+\boldsymbol{U}'$ is the Moore-Penrose generalized inverse of $\boldsymbol{A}$. This is easily confirmed by verifying that $\boldsymbol{A}^+$ satisfies the properties of the Moore-Penrose inverse. Also,

$$\boldsymbol{AA}^+\boldsymbol{b} = \boldsymbol{UDV}'\boldsymbol{VD}^+\boldsymbol{U}'\boldsymbol{b} = \boldsymbol{UDD}^+\boldsymbol{U}'\boldsymbol{b} = \boldsymbol{UU}'\boldsymbol{b} = \boldsymbol{b} ,$$

which means that $\boldsymbol{x} = \boldsymbol{A}^+\boldsymbol{b}$ is a solution for $\boldsymbol{Ax} = \boldsymbol{b}$. This solution can also be expressed in outer-product form as

$$\boldsymbol{x} = \boldsymbol{A}^+\boldsymbol{b} = \boldsymbol{VD}^+\boldsymbol{U}'\boldsymbol{b} = \left(\sum_{i=1}^{r}\frac{\boldsymbol{v}_i\boldsymbol{u}_i'}{\sigma_i}\right)\boldsymbol{b} = \sum_{i=1}^{r}\frac{\boldsymbol{v}_i\boldsymbol{u}_i'}{\sigma_i}\boldsymbol{b} = \sum_{i=1}^{r}\frac{\boldsymbol{v}_i\boldsymbol{u}_i'\boldsymbol{b}}{\sigma_i} = \sum_{i=1}^{r}\frac{\boldsymbol{u}_i'\boldsymbol{b}}{\sigma_i}\boldsymbol{v}_i . \tag{12.12}$$

Therefore, the solution for $\boldsymbol{Ax} = \boldsymbol{b}$ is a linear combination of the right singular vectors associated with the r largest singular values of $\boldsymbol{A}$.

12.4 SVD, data compression and principal components

Among the numerous applications of the SVD in data sciences, one especially popular is *data compression.* Statisticians and data analysts today encounter massive amounts of data and are often faced with the task of finding the most significant features of the data. Put another way, given a table of mn numbers that have been arranged into an $m \times n$ matrix $\boldsymbol{A}$, we seek an approximation of $\boldsymbol{A}$ by a much smaller matrix using much less than those mn original entries.

The rank of a matrix is the number of linearly independent columns or rows. It is, therefore, a measure of redundancy. A matrix of low rank has a large amount of redundancy, while a nonsingular matrix has no redundancies. However, for matrices containing massive amounts of data, finding the rank using exact arithmetic is impossible. True linear dependencies, or lack thereof, in the data are often corrupted by *noise* or *measurement error.* In exact arithmetic, the number of nonzero singular values is the exact rank. A common approach to finding the rank of a matrix in exact arithmetic is to reduce it to a simpler form, generally row reduced echelon form, by elementary row operations. However, this method is unreliable when applied to floating point computations on computers.

The SVD provides a numerically reliable estimate of the *effective rank* or *numerical rank* of a matrix. In floating point arithmetic, the rank is the number of singular values that are greater in magnitude than the noise in the data. Numerical determination of rank is based upon a criterion for deciding when a singular value from the SVD should be treated as zero. This is a practical choice which depends on both the matrix and the application. For example, an estimate of the exact rank of the matrix is computed by counting all singular values that are larger than $\epsilon \times \sigma_{\max}$, where $\sigma_{\max}$ is the largest singular value and ϵ is numerical working precision. All smaller singular values are regarded as round-off artifacts and treated as zero.

The simplest forms of data compression are *row compression* and *column compression.* Let $\boldsymbol{A}$ be an $m \times n$ matrix with effective rank $r < \min\{m, n\}$. Consider the SVD $\boldsymbol{A} = \boldsymbol{U}\boldsymbol{D}\boldsymbol{V}'$ as in (12.1). Then,

$$\begin{bmatrix} \boldsymbol{U}_1' \\ \boldsymbol{U}_2' \end{bmatrix} \boldsymbol{A} = \boldsymbol{U}'\boldsymbol{A} = \boldsymbol{D}\boldsymbol{V}' = \begin{bmatrix} \boldsymbol{\Sigma}_{r\times r} & \boldsymbol{O} \\ \boldsymbol{O} & \boldsymbol{O} \end{bmatrix} \begin{bmatrix} \boldsymbol{V}_1' \\ \boldsymbol{V}_2' \end{bmatrix} = \begin{bmatrix} \boldsymbol{\Sigma}\boldsymbol{V}_1' \\ \boldsymbol{O} \end{bmatrix} .$$

The matrix $\boldsymbol{U}_1'\boldsymbol{A} = \boldsymbol{\Sigma}\boldsymbol{V}_1'$ is $r \times n$ with full effective row rank. So, there are no redundancies in its rows. Thus, $\boldsymbol{U}$ represents orthogonal transformations on the rows of $\boldsymbol{A}$ to obtain the row compressed form $\begin{bmatrix} \boldsymbol{\Sigma}\boldsymbol{V}_1' \\ \boldsymbol{O} \end{bmatrix}$. This row compression does not alter the null space of $\boldsymbol{A}$ because $\mathcal{N}(\boldsymbol{A}) = \mathcal{N}(\boldsymbol{U}'\boldsymbol{A}) = \mathcal{N}(\boldsymbol{\Sigma}\boldsymbol{V}_1')$. Therefore, in applications where we seek solutions for a rank deficient system $\boldsymbol{A}\boldsymbol{x} = \boldsymbol{0}$, the row compression does not result in loss of information. Column compression works by noting that

$$\boldsymbol{A}\boldsymbol{V} = \boldsymbol{U}\boldsymbol{D} = [\boldsymbol{U}_1 : \boldsymbol{U}_2] \begin{bmatrix} \boldsymbol{\Sigma}_{r\times r} & \boldsymbol{O} \\ \boldsymbol{O} & \boldsymbol{O} \end{bmatrix} = [\boldsymbol{U}_1\boldsymbol{\Sigma} : \boldsymbol{O}] .$$

The matrix $\boldsymbol{U}_1\boldsymbol{\Sigma}$ is $m \times r$ with full effective column rank. So, there is no loss of information in its columns and indeed $\mathcal{C}(\boldsymbol{A}) = \mathcal{C}(\boldsymbol{AV}) = \mathcal{C}(\boldsymbol{U}_1\boldsymbol{\Sigma})$.

In statistical applications, we often wish to analyze the linear model

$$\boldsymbol{b} = \boldsymbol{Ax} + \boldsymbol{\eta}\ , \quad \text{where } \boldsymbol{A} \text{ is } m \times n \text{ and } \boldsymbol{\eta} \text{ is random noise.}$$

What is usually sought is a least squares solution obtained by minimizing the sum of squares of the random noise components. This is equivalent to finding an $\boldsymbol{x}$ that minimizes the sum of squares $\|\boldsymbol{Ax} - \boldsymbol{b}\|^2$. Irrespective of the rank of $\boldsymbol{A}$, the solution to this problem is given by $\boldsymbol{x} = \boldsymbol{A}^+\boldsymbol{b}$ (see (12.12)). In situations where $\boldsymbol{A}$ has a very large number of columns, redundancies are bound to exist.

Suppose that the effective rank of $\boldsymbol{A}$ is $r < m$. Row compression using the SVD transforms the linear model to an under-determined system

$$\begin{bmatrix} \boldsymbol{U}_1' \\ \boldsymbol{U}_2' \end{bmatrix} \boldsymbol{b} = \boldsymbol{U}'\boldsymbol{b} = \boldsymbol{U}'\boldsymbol{Ax} + \boldsymbol{U}'\boldsymbol{\eta} = \begin{bmatrix} \boldsymbol{\Sigma V}_1' \\ \boldsymbol{O} \end{bmatrix} \boldsymbol{x} + \boldsymbol{U}'\boldsymbol{\eta} \Longrightarrow \boldsymbol{U}_1'\boldsymbol{b} = \boldsymbol{\Sigma V}_1'\boldsymbol{x} + \boldsymbol{U}_1'\boldsymbol{\eta}\ .$$

Since $\|\boldsymbol{U}'\boldsymbol{b} - \boldsymbol{U}'\boldsymbol{Ax}\|^2 = \|\boldsymbol{U}'(\boldsymbol{b} - \boldsymbol{Ax})\|^2 = (\boldsymbol{b} - \boldsymbol{Ax})'\boldsymbol{UU}'(\boldsymbol{b} - \boldsymbol{Ax}) = \|\boldsymbol{b} - \boldsymbol{Ax}\|^2$, the $\boldsymbol{x}$ solving the row compressed least squares problem will also solve the original problem. Analyzing row compressed linear systems are often referred to as ***compressive sensing*** and have found applications in different engineering applications, especially signal processing.

Column compression of the linear model leads to what is known as ***principal components regression*** in statistics. Assume that the effective rank of $\boldsymbol{A}$ is much less than the number of columns n. Column compression with the SVD leads to

$$\boldsymbol{b} = \boldsymbol{AV}(\boldsymbol{V}'\boldsymbol{x}) + \boldsymbol{\eta} = [\boldsymbol{U}_1\boldsymbol{\Sigma} : \boldsymbol{O}] \begin{bmatrix} \boldsymbol{V}_1'\boldsymbol{x} \\ \boldsymbol{V}_2'\boldsymbol{x} \end{bmatrix} + \boldsymbol{\eta} = \boldsymbol{U}_1\boldsymbol{\Sigma V}_1'\boldsymbol{x} + \boldsymbol{\eta} = \boldsymbol{AV}_1\boldsymbol{y} + \boldsymbol{\eta}\ ,$$

where $\boldsymbol{y} = \boldsymbol{V}_1'\boldsymbol{x}$ and $\boldsymbol{U}_1\boldsymbol{\Sigma} = \boldsymbol{AV}_1$ is $m \times r$ with full column rank. Dimension reduction is achieved because we are regressing on the r columns $\boldsymbol{Av}_1, \boldsymbol{Av}_2, \ldots, \boldsymbol{Av}_r$, instead of all the n columns of $\boldsymbol{A}$. The least squares solution in terms of $\boldsymbol{x}$ is still given by the outer product form in (12.12).

The right singular vectors, $\boldsymbol{v}_i$'s, provide the directions of dominant variability in a data matrix. To be precise, if $\boldsymbol{A}$ is a centered data matrix so that the mean of each column is zero, and $\boldsymbol{A} = \boldsymbol{UDV}'$ is an SVD of $\boldsymbol{A} \in \Re^{m \times n}$ with singular values $\sigma_1 \geq \sigma_2 \cdots$, then the vector $\boldsymbol{Av}_1 = \sigma_1\boldsymbol{u}_1$ has the largest sample variance among all linear combinations of the columns of $\boldsymbol{A}$. This variance is given by $\text{Var}(\boldsymbol{Av}_1) = \sigma_1^2/m$. The normalized vector $\boldsymbol{u}_1 = \boldsymbol{Av}_1/\sigma_1$ is called the ***first principal component*** of $\boldsymbol{A}$. Once the first principal component is found, we seek the vector with maximum sample variance among all vectors that are orthogonal to the first. This is achieved by finding the first principal component of the deflated matrix $\boldsymbol{A} - \sigma_1\boldsymbol{u}_1\boldsymbol{v}_1'$. The truncated form of the SVD reveals this to be along the direction of the right singular vector $\boldsymbol{v}_2$ of $\boldsymbol{A}$. Proceeding in this manner, we obtain an ordering of the variation in the data.

The outer-product form of the SVD makes this ordering of variation clear. Each term in the sum (12.12) is organized according to the dominant directions. Thus, the first term $(\boldsymbol{u}_1'\boldsymbol{b}/\sigma_1)\boldsymbol{v}_1$ is the solution component of the linear system $\boldsymbol{A}\boldsymbol{x} = \boldsymbol{b}$ along the dominating direction of the data matrix. The second term, $(\boldsymbol{u}_2'\boldsymbol{b}/\sigma_2)\boldsymbol{v}_2$ is the component along the second most dominant direction, and so on. The column compressed linear model, which regresses on the principal component directions $\boldsymbol{A}\boldsymbol{v}_1, \boldsymbol{A}\boldsymbol{v}_2, \ldots, \boldsymbol{A}\boldsymbol{v}_r$ is called ***principal component regression*** or *PCR* and is a widely deployed data mining technique when n is huge.

12.5 Computing the SVD

The singular values of an $m \times n$ matrix $\boldsymbol{A}$ are precisely the square of the eigenvalues of $\boldsymbol{A}'\boldsymbol{A}$ and $\boldsymbol{A}\boldsymbol{A}'$. A natural thought for computing the SVD would be to apply the symmetric eigenvalue algorithms to $\boldsymbol{A}'\boldsymbol{A}$ and $\boldsymbol{A}\boldsymbol{A}'$. This, however, is not ideal because computation of $\boldsymbol{A}'\boldsymbol{A}$ and $\boldsymbol{A}\boldsymbol{A}'$ can be ill-conditioned for dense $\boldsymbol{A}$ and may lead to loss of information. In practice, therefore, SVD's are computed using a two-step procedure by first reducing $\boldsymbol{A}$ to a bidiagonal matrix and then finding the SVD of this bidiagonal matrix.

Suppose $\boldsymbol{A}$ is $m \times n$ with $m \geq n$. In the first step, $\boldsymbol{A}$ is reduced to a bidiagonal matrix $\boldsymbol{B}$ using Householder reflections from the left and right. We construct orthogonal matrices $\boldsymbol{Q}'$ and $\boldsymbol{P}$ such that

$$\boldsymbol{Q}'\boldsymbol{A}\boldsymbol{P} = \begin{bmatrix} \boldsymbol{B} \\ \boldsymbol{O} \end{bmatrix} .$$

This is described in Section 8.8. Using orthogonal transformations ensures that the bidiagonal matrix $\boldsymbol{B}$ has the same singular values as $\boldsymbol{A}$. To be precise, suppose that $\boldsymbol{A}\boldsymbol{v} = \sigma\boldsymbol{u}$, where σ is a singular value of $\boldsymbol{A}$ with associated singular vectors $\boldsymbol{u}$ and $\boldsymbol{v}$. Let $\tilde{\boldsymbol{u}} = \boldsymbol{Q}\boldsymbol{u}$ and $\tilde{\boldsymbol{v}} = \boldsymbol{P}'\boldsymbol{v}$, Then,

$$\sigma\tilde{\boldsymbol{u}} = \sigma\boldsymbol{Q}\boldsymbol{u} = \boldsymbol{Q}\boldsymbol{A}\boldsymbol{v} = \begin{bmatrix} \boldsymbol{B} \\ \boldsymbol{O} \end{bmatrix} \boldsymbol{P}'\boldsymbol{v} = \begin{bmatrix} \boldsymbol{B} \\ \boldsymbol{O} \end{bmatrix} \tilde{\boldsymbol{v}} ,$$

which shows that σ is a singular value of $\boldsymbol{B}$ with associated singular vectors $\tilde{\boldsymbol{u}}$ and $\tilde{\boldsymbol{v}}$. If only singular values (and not the singular vectors) are required, then the Householder reflections cost $4mn^2 - 4n^3/3$ flops. When m is much larger than n computational benefits accrue by first reducing $\boldsymbol{A}$ to a triangular matrix with the $\boldsymbol{QR}$ decomposition and then use Householder reflections to further reduce the matrix to bidiagonal form. The total cost is $2mn^2 + 2n^3$ flops.

In the second step, a variant of the $\boldsymbol{QR}$ algorithm is used to compute the SVD of $\boldsymbol{B}$. One of two feasible strategies are pursued. The first is to construct $\boldsymbol{B}'\boldsymbol{B}$, which is easily verified to be tridiagonal and is better conditioned than $\boldsymbol{A}'\boldsymbol{A}$. We apply the tridiagonal $\boldsymbol{QR}$ algorithm which is modified so as not to explicitly form the matrix $\boldsymbol{B}'\boldsymbol{B}$. This yields the eigenvalues of $\boldsymbol{B}'\boldsymbol{B}$, which are the singular values of $\boldsymbol{B}$. Singular vectors, if desired any specific set of singular values, can be computed by applying inverse iterations to $\boldsymbol{B}\boldsymbol{B}'$ and $\boldsymbol{B}'\boldsymbol{B}$ for left and right singular vectors, respectively.

The second strategy makes use of the simple fact that for any matrix $\boldsymbol{B}$, the eigenvalues of

$$\tilde{\boldsymbol{B}} = \begin{bmatrix} \boldsymbol{O} & \boldsymbol{B} \\ \boldsymbol{B}' & \boldsymbol{O} \end{bmatrix}$$

are the singular values of $\boldsymbol{B}$. To see why, consider the partitioned form of the eigenvalue equations

$$\begin{bmatrix} \lambda\boldsymbol{I} & -\boldsymbol{B} \\ -\boldsymbol{B}' & \lambda\boldsymbol{I} \end{bmatrix} \begin{bmatrix} \boldsymbol{u} \\ \boldsymbol{v} \end{bmatrix} = \begin{bmatrix} \boldsymbol{u} \\ \boldsymbol{v} \end{bmatrix} \Longrightarrow \begin{cases} \boldsymbol{B}\boldsymbol{v} = \lambda\boldsymbol{u} \\ \boldsymbol{B}'\boldsymbol{u} = \lambda\boldsymbol{v} \end{cases}$$

and note that the above implies

$$\boldsymbol{B}'\boldsymbol{B}\boldsymbol{v} = \lambda\boldsymbol{B}'\boldsymbol{u} = \lambda^2\boldsymbol{v} \text{ and } \boldsymbol{B}\boldsymbol{B}'\boldsymbol{u} = \lambda\boldsymbol{B}\boldsymbol{v} = \lambda^2\boldsymbol{u} .$$

These equations reveal that λ is indeed a singular value of $\boldsymbol{B}$ with associated singular vectors $\boldsymbol{u}$ (left) and $\boldsymbol{v}$ (right). The above is true for any matrix $\boldsymbol{B}$. When $\boldsymbol{B}$ is bidiagonal, the added advantage is that $\tilde{\boldsymbol{B}}$ may be brought to a tridiagonal form by permuting its rows and columns, which means that $\tilde{\boldsymbol{B}}$ is orthogonally similar to a tridiagonal matrix. The tridiagonal $\boldsymbol{QR}$ algorithm can now be applied to obtain the singular values.

The second step is iterative and can, in theory, require an infinite number of steps. However, convergence up to machine precision is usually achieved in $O(n)$ iterations, each costing $O(n)$ flops for the bidiagonal matrix. Therefore, the first step is more expensive and the overall cost is $O(mn^2)$ flops.

Golub-Kahan-Lanczos bidiagonalization for large and sparse matrices

As discussed earlier, the first step in computing the SVD of an $m \times n$ matrix $\boldsymbol{A}$ is to compute unitary matrices $\boldsymbol{Q}$ and $\boldsymbol{P}$ such that $\boldsymbol{Q}'\boldsymbol{A}\boldsymbol{P}$ has an upper-bidiagonal form. To be precise, consider the case when $m \geq n$ (the case when $m < n$ can be explored with the transpose). We need to find orthogonal matrices $\boldsymbol{Q}$ and $\boldsymbol{P}$ of order $m \times m$ and $n \times n$, respectively, such that $\boldsymbol{Q}'\boldsymbol{A}\boldsymbol{P} = \begin{bmatrix} \boldsymbol{B} \\ \boldsymbol{O} \end{bmatrix}$, where $\boldsymbol{B}$ is $n \times n$ upper-bidiagonal.

For dense matrices, the orthogonal matrices $\boldsymbol{Q}$ and $\boldsymbol{P}$ are obtained by a series of Householder reflections alternately applied from the left and right. This was described in Section 8.8 and is known as the *Golub-Kahan bidiagonalization* algorithm. However, when $\boldsymbol{A}$ is large and sparse, Householder reflections tend to be wasteful because they introduce dense submatrices in intermediate steps. We mentioned this earlier when discussing reduction to Hessenberg forms for eigenvalue algorithms. Matters are not very different for SVD computations. Therefore when $\boldsymbol{A}$ is large and sparse, bidiagonalization is achieved using a recursive procedure known as the *Golub-Kahan-Lanczos* or *Lanczos-Golub-Kahan* method, which we describe below.

Observe that

$$\boldsymbol{Q}'\boldsymbol{A}\boldsymbol{P} = \begin{bmatrix} \boldsymbol{B} \\ \boldsymbol{O} \end{bmatrix} \Longrightarrow \boldsymbol{A}\boldsymbol{P} = \boldsymbol{Q}\begin{bmatrix} \boldsymbol{B} \\ \boldsymbol{O} \end{bmatrix} \text{ and } \boldsymbol{A}'\boldsymbol{Q} = \boldsymbol{P}\left[\boldsymbol{B}' : \boldsymbol{O}\right] .$$

Let $Q = [Q_1 : Q_2]$, where Q_1 has n columns. The above equations imply that $AP = Q_1 B$ and $A'Q_1 = PB'$. The Golub-Kahan-Lanczos algorithm is conceptually simple: solve the equations $AP = Q_1 B$ and $A'Q_1 = PB'$ for Q_1, P and B. Suppose we write $AP = Q_1 B$ as

$$A[p_1 : p_2 : p_3 : \ldots : p_n] = [q_1 : q_2 : q_3 : \ldots : q_n] \begin{bmatrix} \alpha_1 & \beta_1 & 0 & \ldots & 0 \\ 0 & \alpha_2 & \beta_2 & \cdots & 0 \\ 0 & 0 & \alpha_3 & \ldots & 0 \\ \vdots & \vdots & \vdots & \ddots & \vdots \\ 0 & 0 & 0 & \ldots & \alpha_n \end{bmatrix} \tag{12.13}$$

and $A'Q_1 = PB'$ as

$$A'[q_1 : q_2 : \ldots : q_n] = [p_1 : p_2 : \ldots : p_n] \begin{bmatrix} \alpha_1 & 0 & 0 & \ldots & 0 \\ \beta_1 & \alpha_2 & 0 & \cdots & 0 \\ 0 & \beta_2 & \alpha_3 & \cdots & 0 \\ \vdots & \vdots & \vdots & \ddots & \vdots \\ 0 & 0 & 0 & \ldots & \alpha_n \end{bmatrix} . \tag{12.14}$$

Let p_1 be any $n \times 1$ vector such that $\|p_1\| = 1$. Equating the first columns in $AP = Q_1 B$, we obtain

$$Ap_1 = \alpha_1 q_1 .$$

The condition $\|q_1\| = 1$ implies that $\alpha_1 = \|Ap_1\|$ and $q_1 = \frac{Ap_1}{\|Ap_1\|}$. Thus, we have determined α_1 and q_1. Next, we find β_1 and p_2 by equating the first column of $A'Q_1 = PB'$, which yields

$$A'q_1 = \alpha_1 p_1 + \beta_1 p_2 \Longrightarrow \beta_1 p_2 = A'q_1 - \alpha_1 p_1 .$$

We now let p_2 be the normalized vector in the direction of $A'q_1 - \alpha_1 p_1$ and obtain β_1 from the restriction that $\|p_2\| = 1$. To be precise, we set

$$v = A'q_1 - \alpha_1 p_1 \ , \ \beta_1 = \|v\| \text{ and } p_2 = \frac{v}{\|v\|} = \frac{A'q_1 - \alpha_1 p_1}{\|A'q_1 - \alpha_1 p_1\|} .$$

This shows that by alternating between equating the first columns in $AP = Q_1' B$ and $A'Q_1 = PB'$, we are able to determine the nonzero elements in the first row of B along with the first column of Q and the first two columns of P (note: the first column of P was initialized). This completes the first iteration.

Now consider a generic step. The first j steps have provided the nonzero elements in the first j rows of B along with first j columns of Q_1 and the first $j+1$ columns of P. Thus, we have determined scalars α_i and β_i for $i = 1, 2, \ldots, j$ and orthonormal sets $\{q_1, q_2, \ldots, q_j\}$ and $\{p_1, p_2, \ldots, p_{j+1}\}$. Equating the $j + 1$-th columns of $AP = Q_1 B$ gives us

$$Ap_{j+1} = \beta_j q_j + \alpha_{j+1} q_{j+1} \Longrightarrow \alpha_{j+1} q_{j+1} = Ap_{j+1} - \beta_j q_j .$$

This determines α_{j+1} and $\boldsymbol{q}_{j+1}$ as follows:

$$\alpha_{j+1} = \|\boldsymbol{A}\boldsymbol{p}_{j+1} - \beta_j \boldsymbol{q}_j\| \text{ and } \boldsymbol{q}_{j+1} = \frac{\boldsymbol{A}\boldsymbol{p}_{j+1} - \beta_j \boldsymbol{q}_j}{\boldsymbol{A}\boldsymbol{p}_{j+1} - \beta_j \boldsymbol{q}_j} .$$

Equating the $j+1$-th columns of $\boldsymbol{A}'\boldsymbol{Q}_1 = \boldsymbol{P}\boldsymbol{B}$ gives us

$$\boldsymbol{A}'\boldsymbol{q}_{j+1} = \alpha_{j+1}\boldsymbol{p}_{j+1} + \beta_{j+1}\boldsymbol{p}_{j+2} \Longrightarrow \beta_{j+1}\boldsymbol{p}_{j+2} = \boldsymbol{A}'\boldsymbol{q}_{j+1} - \alpha_{j+1}\boldsymbol{p}_{j+1} .$$

This determines β_{j+1} and $\boldsymbol{p}_{j+2}$ as follows:

$$\beta_{j+1} = \|\boldsymbol{A}'\boldsymbol{q}_{j+1} - \alpha_{j+1}\boldsymbol{p}_{j+1}\| \text{ and } \boldsymbol{p}_{j+2} = \frac{\boldsymbol{A}'\boldsymbol{q}_{j+1} - \alpha_{j+1}\boldsymbol{p}_{j+1}}{\|\boldsymbol{A}'\boldsymbol{q}_{j+1} - \alpha_{j+1}\boldsymbol{p}_{j+1}\|} .$$

We assemble the above into the *Golub-Kahan-Lanczos* algorithm:

1. Start with any normalized vector $\boldsymbol{p}_1$ such that $\|\boldsymbol{p}_1\| = 1$.
2. Set $\beta_0 = 0$ and construct the following sequence for $j = 1, 2, \ldots, n$:
   (a) Set $\boldsymbol{v} = \boldsymbol{A}\boldsymbol{p}_j - \beta_{j-1}\boldsymbol{q}_{j-1}$.
   (b) Set $\alpha_j = \|\boldsymbol{v}\|$ and $\boldsymbol{q}_j = \dfrac{\boldsymbol{v}}{\|\boldsymbol{v}\|}$.
   (c) Set $\boldsymbol{v} = \boldsymbol{A}'\boldsymbol{q}_j - \alpha_j\boldsymbol{p}_j$.
   (d) Set $\beta_j = \|\boldsymbol{v}\|$ and $\boldsymbol{p}_{j+1} = \dfrac{\boldsymbol{v}}{\|\boldsymbol{v}\|}$.

Collecting the computed quantities from the first t steps of the algorithm, yields the Golub-Kahan-Lanczos decompositions. This is obtained by equating the first t columns in (12.13) and in (12.14). This obtains

$$\boldsymbol{A}\boldsymbol{P}_t = \boldsymbol{Q}_t\boldsymbol{B}_t \text{ and } \boldsymbol{A}'\boldsymbol{Q}_t = \boldsymbol{P}_t\boldsymbol{B}_t' + \beta_t\boldsymbol{p}_{t+1}\boldsymbol{e}_t' , \tag{12.15}$$

where $\boldsymbol{Q}_t'\boldsymbol{Q}_t = \boldsymbol{I}_t = \boldsymbol{P}_t'\boldsymbol{P}_t$ and $\boldsymbol{B}_t$ is the $t \times t$ leading principal submatrix of $\boldsymbol{B}$. This is analogous to the Arnoldi decomposition we encountered in (11.21).

Equation 12.15 easily reveals that the columns of $\boldsymbol{P}_t$ and $\boldsymbol{Q}_t$ are orthonormal in each step of the Golub-Kahan-Lanczos algorithm. The first few steps can be easily verified. Once the normalized vector $\boldsymbol{p}_{t+1}$ is constructed, we find $\boldsymbol{P}_t'\boldsymbol{p}_{t+1} = \boldsymbol{0}$ because

$$\begin{aligned} \boldsymbol{P}_t'(\boldsymbol{A}'\boldsymbol{q}_t - \alpha_t\boldsymbol{p}_t) &= \boldsymbol{P}_t'\boldsymbol{A}'\boldsymbol{q}_t - \alpha_t\boldsymbol{P}_t'\boldsymbol{p}_t = (\boldsymbol{A}_t\boldsymbol{P}_t)'\boldsymbol{q}_t - \alpha_t\boldsymbol{e}_t \\ &= \boldsymbol{B}_t'\boldsymbol{Q}_t'\boldsymbol{q}_t - \alpha_t\boldsymbol{e}_t = \boldsymbol{B}_t'\boldsymbol{e}_t - \alpha_t\boldsymbol{e}_t = \alpha_t\boldsymbol{e}_t - \alpha_t\boldsymbol{e}_t = 0 . \end{aligned}$$

This confirms that $\boldsymbol{P}_{t+1} = [\boldsymbol{P}_t : \boldsymbol{p}_{t+1}]$ has orthonormal columns. Similarly, $\boldsymbol{Q}_t'\boldsymbol{q}_{t+1} = \boldsymbol{0}$ because

$$\begin{aligned} \boldsymbol{Q}_t'(\boldsymbol{A}\boldsymbol{p}_{t+1} - \beta_t\boldsymbol{q}_t) &= \boldsymbol{Q}_t'\boldsymbol{A}\boldsymbol{p}_{t+1} - \beta_t\boldsymbol{Q}_t'\boldsymbol{q}_t = (\boldsymbol{A}'\boldsymbol{Q}_t)'\boldsymbol{p}_{t+1} - \beta_t\boldsymbol{e}_t \\ &= (\boldsymbol{P}_t\boldsymbol{B}_t' + \beta_t\boldsymbol{p}_{t+1}\boldsymbol{e}_t')'\boldsymbol{p}_{t+1} - \beta_t\boldsymbol{e}_t \\ &= \boldsymbol{B}_t\boldsymbol{P}_t'\boldsymbol{p}_{t+1} + \beta_t\boldsymbol{p}_{t+1}'\boldsymbol{p}_{t+1}\boldsymbol{e}_t - \beta_t\boldsymbol{e}_t = \beta_t\boldsymbol{e}_t - \beta_t\boldsymbol{e}_t = \boldsymbol{0} , \end{aligned}$$

where we have used $\boldsymbol{P}_t'\boldsymbol{p}_{t+1} = \boldsymbol{0}$ and $\boldsymbol{p}_{t+1}'\boldsymbol{p}_{t+1} = 1$. Thus, $\boldsymbol{Q}_{t+1} = [\boldsymbol{Q}_t : \boldsymbol{q}_{t+1}]$ has orthonormal columns.

12.6 The Jordan Canonical Form

We now return to the problem of finding the simplest structures to which any square matrix can be reduced via similarity transformations. Schur's triangularization (Theorem 11.15) ensures that every square matrix with real or complex entries is unitarily similar to an upper-triangular matrix $\boldsymbol{T}$. It does not, however, say anything more about patterns in the nonzero part of $\boldsymbol{T}$. A natural question asks, then, if we can relax "*unitarily similar*" to "*similar*" and find an even simpler structure that can make many of the entries in the nonzero part of $\boldsymbol{T}$ to be zero. In other words, what is the sparsest triangular matrix to which a square matrix will be similar? The answer resides in the ***Jordan Canonical Form***, which we discuss below.

Throughout this section, $\boldsymbol{A}$ will be an $n \times n$ matrix.

Definition 12.2 *A* **Jordan chain** *of length m is a sequence of nonzero vectors $\boldsymbol{x}_1, \boldsymbol{x}_2, \ldots, \boldsymbol{x}_m$ in $\mathbb{C}^n$ such that*

$$\boldsymbol{A}\boldsymbol{x}_1 = \lambda\boldsymbol{x}_1, \text{ and } \boldsymbol{A}\boldsymbol{x}_i = \lambda\boldsymbol{x}_i + \boldsymbol{x}_{i-1} \text{ for } i = 1, 2, \ldots, m \tag{12.16}$$

for some eigenvalue λ of $\boldsymbol{A}$.

Observe that $\boldsymbol{x}_1$ is an eigenvector of $\boldsymbol{A}$ associated with the eigenvalue λ. Therefore, Jordan chains are necessarily associated with eigenvalues of $\boldsymbol{A}$. The vectors $\boldsymbol{x}_2, \boldsymbol{x}_3, \ldots, \boldsymbol{x}_m$ are called *generalized eigenvectors*, which we define below.

Definition 12.3 *A nonzero vector $\boldsymbol{x}$ is said to be* **generalized eigenvector** *associated with eigenvalue λ of $\boldsymbol{A}$ if*

$$(\boldsymbol{A} - \lambda\boldsymbol{I})^k\boldsymbol{x} = \boldsymbol{0} \text{ for some positive integer } k. \tag{12.17}$$

If $\boldsymbol{x}$ is an eigenvector associated with eigenvalue λ, then it satisfies (12.17) with $k = 1$. Therefore, an eigenvector of $\boldsymbol{A}$ is also a generalized eigenvector of $\boldsymbol{A}$. The converse is false: a generalized eigenvector is *not* necessarily an eigenvector.

Generalized eigenvectors are meaningful only when associated with an eigenvalue. If λ is not an eigenvalue, then $\boldsymbol{A} - \lambda\boldsymbol{I}$ is nonsingular and we can argue that

$$(\boldsymbol{A} - \lambda\boldsymbol{I})^k\boldsymbol{x} = \boldsymbol{0} \Longrightarrow (\boldsymbol{A} - \lambda\boldsymbol{I})[(\boldsymbol{A} - \lambda\boldsymbol{I})^{k-1}\boldsymbol{x}] = \boldsymbol{0} \Longrightarrow (\boldsymbol{A} - \lambda\boldsymbol{I})^{k-1}\boldsymbol{x} = \boldsymbol{0}$$
$$\Longrightarrow \ldots \Longrightarrow (\boldsymbol{A} - \lambda\boldsymbol{I})\boldsymbol{x} = \boldsymbol{0} \Longrightarrow \boldsymbol{x} = \boldsymbol{0}\,.$$

Therefore, $(\boldsymbol{A} - \lambda\boldsymbol{I})^k$ is also nonsingular and there is no nonzero $\boldsymbol{x}$ that will satisfy Definition 12.3. Definition 12.3 also implies that for a generalized eigenvector $\boldsymbol{x}$, there exists a *smallest* integer k such that $(\boldsymbol{A} - \lambda\boldsymbol{I})^k\boldsymbol{x} = \boldsymbol{0}$.

Definition 12.4 Index of a generalized eigenvector. *The* **index** *of a generalized eigenvector $\boldsymbol{x}$ associated with eigenvalue λ of $\boldsymbol{A}$ is the smallest integer k for which $(\boldsymbol{A} - \lambda\boldsymbol{I})^k\boldsymbol{x} = \boldsymbol{0}$.*

Every vector in a Jordan chain is a generalized eigenvector.

Lemma 12.1 *Let $\boldsymbol{w}_k$ be the k-th vector in the Jordan chain (12.16). Then $\boldsymbol{w}_k$ is a generalized eigenvector with index k.*

Proof. Clearly $\boldsymbol{w}_1$ in (12.16) is an eigenvector, so it is a generalized eigenvector with index 1. For the second vector in the chain, we have $(\boldsymbol{A} - \lambda\boldsymbol{I})\boldsymbol{w}_2 = \boldsymbol{w}_1 \neq \boldsymbol{0}$, which implies that $(\boldsymbol{A}-\lambda\boldsymbol{I})^2\boldsymbol{w}_2 = (\boldsymbol{A}-\lambda\boldsymbol{I})\boldsymbol{w}_1 = \boldsymbol{0}$. Therefore, $\boldsymbol{w}_2$ is a generalized eigenvector of index 2. In general,

$$(\boldsymbol{A} - \lambda\boldsymbol{I})\boldsymbol{w}_k = \boldsymbol{w}_{k-1} \Longrightarrow (\boldsymbol{A} - \lambda\boldsymbol{I})^2\boldsymbol{w}_k = (\boldsymbol{A} - \lambda\boldsymbol{I})\boldsymbol{w}_{k-1} = \boldsymbol{w}_{k-2}$$
$$\Longrightarrow \ldots \Longrightarrow (\boldsymbol{A} - \lambda\boldsymbol{I})^{k-1}\boldsymbol{w}_k = \boldsymbol{w}_1 \Longrightarrow (\boldsymbol{A} - \lambda\boldsymbol{I})^k\boldsymbol{w}_k = \boldsymbol{0} .$$

Therefore, $\boldsymbol{w}_k$ is a generalized eigenvector of index k. □

The Jordan chain (12.16) can be expressed in terms of the generalized eigenvector of highest index, $\boldsymbol{w}_m$, as

$$\mathcal{J}_\lambda(\boldsymbol{w}_m) = \{(\boldsymbol{A} - \lambda\boldsymbol{I})^{m-1}\boldsymbol{w}_m, (\boldsymbol{A} - \lambda\boldsymbol{I})^{m-2}\boldsymbol{w}_m, \ldots, (\boldsymbol{A} - \lambda\boldsymbol{I})\boldsymbol{w}_m, \boldsymbol{w}_m\} . \tag{12.18}$$

The last vector in a Jordan chain is the generalized eigenvector with highest index.

Example 12.4 Consider the following 3×3 matrix, known as a *Jordan block*,

$$\boldsymbol{A} = \begin{bmatrix} \lambda & 1 & 0 \\ 0 & \lambda & 1 \\ 0 & 0 & \lambda \end{bmatrix} .$$

Since $\boldsymbol{A}$ is upper-triangular, λ is the only eigenvalue of $\boldsymbol{A}$. It can be easily verified that $\boldsymbol{e}_1 = (1,0,0)'$ satisfies $\boldsymbol{A}\boldsymbol{e}_1 = \lambda\boldsymbol{e}_1$. Therefore, $\boldsymbol{e}_1$ is an eigenvector of $\boldsymbol{A}$. Direct multiplication also reveals $\boldsymbol{A}\boldsymbol{e}_2 = \lambda\boldsymbol{e}_2 + \boldsymbol{e}_1$ and $\boldsymbol{A}\boldsymbol{e}_3 = \lambda\boldsymbol{e}_3 + \boldsymbol{e}_2$. Therefore, $\boldsymbol{e}_1, \boldsymbol{e}_2, \boldsymbol{e}_3$ are members of a Jordan chain. Furthermore, $\boldsymbol{A} - \lambda\boldsymbol{I}$ is a 3×3 upper-triangular matrix with 0 along its diagonal. Therefore, $(\boldsymbol{A} - \lambda\boldsymbol{I})^3 = \boldsymbol{O}$. Therefore, every vector in $\Re^3$ is a generalized eigenvector of index at most 3. ■

In Example 12.4 the Jordan chain of $\boldsymbol{A}$ constitutes a basis for $\Re^3$. Matrices for which this is true are rather special and form the building blocks of what will be seen to be the simplest forms to which any square matrix is similar. We provide a formal name and definition to such a matrix.

Definition 12.5 *A* **Jordan block** *is an $n \times n$ matrix of the form*

$$\boldsymbol{J}_n(\lambda) = \begin{bmatrix} \lambda & 1 & 0 & \ldots & 0 \\ 0 & \lambda & 1 & \ldots & 0 \\ \vdots & \vdots & \vdots & \ddots & \vdots \\ 0 & 0 & 0 & \ldots & 1 \\ 0 & 0 & 0 & \ldots & \lambda \end{bmatrix} , \tag{12.19}$$

which has a real or complex number λ along the diagonal, 1's along the super-diagonal and 0's everywhere else.

In particular, a 1×1 Jordan block is a scalar. A Jordan block appears naturally from a Jordan chain as below.

Lemma 12.2 *Let $\boldsymbol{A}$ be an $n \times n$ matrix and let $\boldsymbol{P} = [\boldsymbol{p}_1 : \boldsymbol{p}_2 : \ldots : \boldsymbol{p}_m]$ be an $n \times m$ matrix ($m \leq n$) whose columns form a Jordan chain associated with an eigenvalue λ of $\boldsymbol{A}$. Then $\boldsymbol{AP} = \boldsymbol{PJ}_m(\lambda)$.*

Proof. The vectors $\boldsymbol{p}_1, \boldsymbol{p}_2, \ldots, \boldsymbol{p}_m$ form a Jordan chain associated with λ, as defined in Definition 12.2. Therefore,

$$\boldsymbol{AP} = \boldsymbol{A}[\boldsymbol{p}_1 : \boldsymbol{p}_2 : \ldots : \boldsymbol{p}_m] = [\lambda\boldsymbol{p}_1 : \lambda\boldsymbol{p}_2 + \boldsymbol{p}_1 : \ldots : \lambda\boldsymbol{p}_m + \boldsymbol{p}_{m-1}]$$

$$= [\boldsymbol{p}_1 : \boldsymbol{p}_2 : \ldots : \boldsymbol{p}_m] \begin{bmatrix} \lambda & 1 & 0 & \ldots & 0 \\ 0 & \lambda & 1 & \ldots & 0 \\ \vdots & \vdots & \vdots & \ddots & \vdots \\ 0 & 0 & 0 & \ldots & 1 \\ 0 & 0 & 0 & \ldots & \lambda \end{bmatrix} = \boldsymbol{PJ}_m(\lambda) ,$$

which completes the proof. □

Every matrix has at least one, possibly complex, eigenvector. Jordan blocks have the least possible number of eigenvectors. The following lemma clarifies.

Lemma 12.3 *The $n \times n$ Jordan block matrix $\boldsymbol{J}_n(\lambda)$ has a single eigenvalue λ. The standard basis vectors for $\Re^n$, $\{\boldsymbol{e}_1, \boldsymbol{e}_2, \ldots, \boldsymbol{e}_n\}$, form a Jordan chain, with $\boldsymbol{e}_1$ as the only independent eigenvector, for $\boldsymbol{J}_n(\lambda)$.*

Proof. Since $\boldsymbol{J}_n(\lambda)$ is upper-triangular, its diagonal elements are its eigenvalues (recall Example 11.3). Therefore, λ is the only eigenvalue of $\boldsymbol{J}_n(\lambda)$. Observe that

$$\boldsymbol{J}_n(\lambda) - \lambda\boldsymbol{I} = \begin{bmatrix} 0 & 1 & 0 & \ldots & 0 \\ 0 & 0 & 1 & \ldots & 0 \\ \vdots & \vdots & \vdots & \ddots & \vdots \\ 0 & 0 & 0 & 0 & 1 \\ 0 & 0 & 0 & 0 & 0 \end{bmatrix} = \begin{bmatrix} \boldsymbol{e}_2' \\ \boldsymbol{e}_3' \\ \vdots \\ \boldsymbol{e}_n' \\ \boldsymbol{0}' \end{bmatrix} .$$

It is easily verified that $(\boldsymbol{J}_n(\lambda) - \lambda\boldsymbol{I})\boldsymbol{e}_1 = \boldsymbol{0}$ and $(\boldsymbol{J}_n(\lambda) - \lambda\boldsymbol{I})\boldsymbol{e}_i = \boldsymbol{e}_{i-1}$ for $i = 2, 3, \ldots, n$. Therefore, the standard basis forms a Jordan chain. Also, $(\boldsymbol{J}_n(\lambda) - \lambda\boldsymbol{I})\boldsymbol{x} = \boldsymbol{0}$ if and only if $\boldsymbol{x}$ is orthogonal to each of the $\boldsymbol{e}_i$'s for $i = 2, 3, \ldots, n$. Therefore, any eigenvector of $\boldsymbol{J}_n(\lambda)$ must be in the span of $\{\boldsymbol{e}_1\}$, so $\boldsymbol{e}_1$ is the only independent eigenvector. □

If $\boldsymbol{A}$ is a general $n \times n$ matrix, it is not necessary that a *single* Jordan chain of $\boldsymbol{A}$ forms a basis for $\Re^n$ or $\mathbb{C}^n$.

Example 12.5 Let A be the 4×4 matrix

$$A = \begin{bmatrix} \lambda & 1 & 0 & 0 \\ 0 & \lambda & 1 & 0 \\ 0 & 0 & \lambda & 0 \\ 0 & 0 & 0 & \lambda \end{bmatrix}.$$

This is not a Jordan block because the $(3,4)$-th element, which is on the super-diagonal, is not 1. It is block-diagonal composed of two Jordan blocks associated with λ:

$$A = \begin{bmatrix} J_3(\lambda) & \mathbf{0} \\ \mathbf{0}' & J_1(\lambda) \end{bmatrix}.$$

Here, A is upper-triangular so λ is its only eigenvalue. Also, $J_1(\lambda)$ is 1×1, so it is simply the scalar λ. Observe that

$$A - \lambda I = \begin{bmatrix} 0 & 1 & 0 & 0 \\ 0 & 0 & 1 & 0 \\ 0 & 0 & 0 & 0 \\ 0 & 0 & 0 & 0 \end{bmatrix} = \begin{bmatrix} e_2' \\ e_3' \\ \mathbf{0}' \\ \mathbf{0}' \end{bmatrix},$$

where e_i' represents the ith row of I_4. The row-space of A is spanned by $\{e_2, e_3\}$. The null space of A, which is the orthocomplement of the row space, is, therefore, spanned by $\{e_1, e_4\}$. This implies that A has two linearly independent eigenvectors e_1 and e_4. It is easily verified that $(A - \lambda I)e_i = \mathbf{0}$ for $i = 1, 4$, and $(A - \lambda I)e_i = e_{i-1}$ for $i = 2, 3$. This produces *two* Jordan chains: one comprising three vectors, $\{e_1, e_2, e_3\}$, and the second comprising the single vector, $\{e_4\}$. Neither of these two chains, by themselves, constitute a basis for $\Re^4$. ■

Block diagonal matrices with Jordan blocks associated with a common eigenvalue along the diagonal (e.g., A in Example 12.5) is called a *Jordan segment.*

Definition 12.6 *A* **Jordan segment** *associated with an eigenvalue λ is a block-diagonal matrix of the form*

$$J(\lambda) = \begin{bmatrix} J_{n_1}(\lambda) & O & \dots & O \\ O & J_{n_2}(\lambda) & \dots & O \\ \vdots & \vdots & \ddots & O \\ O & O & \dots & J_{n_k}(\lambda) \end{bmatrix},$$

where each $J_{n_i}(\lambda)$ is an $n_i \times n_i$ Jordan block matrix.

In Example 12.5, neither of the two Jordan chains for the 4×4 matrix A constitutes a basis for $\Re^4$. Nevertheless, the standard basis for $\Re^4$ consists of vectors belonging to two Jordan chains with no common vectors. More generally, we can construct an $n \times n$ matrix with k different eigenvalues such that there exists a basis for $\Re^n$ or $\mathbb{C}^n$ consisting exclusively of vectors that belong to some Jordan chain of A. We call such a basis a *Jordan basis* for A and are important enough to merit their own definition.

Definition 12.7 *A basis for $\Re^n$ or $\mathbb{C}^n$ is called a* **Jordan basis** *for an $n \times n$ matrix if it is the union of one or more mutually exclusive Jordan chains (i.e., having no elements in common).*

For an $n \times n$ diagonalizable matrix, we know that there exists a set of n linearly independent eigenvectors that form a basis for $\mathbb{C}^n$ or $\Re^n$, as the case may be. This basis qualifies as a Jordan basis because each eigenvector belongs to a Jordan chain of length 1. Therefore, every diagonalizable matrix has a special Jordan basis comprising exclusively of eigenvectors. Generalized eigenvectors are not needed there. A natural question, then, is whether every square matrix has a Jordan basis consisting of generalized eigenvectors? The answer is yes and we will offer a formal statement and proof of this theorem soon. But first let us see what we can say about similarity of a square matrix that has a Jordan basis.

Theorem 12.3 *Let $\boldsymbol{A}$ be an $n \times n$ matrix. Let $\boldsymbol{P} = [\boldsymbol{p}_1 : \boldsymbol{p}_2 : \ldots : \boldsymbol{p}_n]$ be an $n \times n$ nonsingular matrix whose columns are a Jordan basis for $\boldsymbol{A}$. Then,*

$$\boldsymbol{P}^{-1}\boldsymbol{A}\boldsymbol{P} = \boldsymbol{J} = \begin{bmatrix} \boldsymbol{J}_{n_1}(\lambda_1) & \boldsymbol{O} & \ldots & \boldsymbol{O} \\ \boldsymbol{O} & \boldsymbol{J}_{n_2}(\lambda_2) & \cdots & \boldsymbol{O} \\ \vdots & \vdots & \ddots & \vdots \\ \boldsymbol{O} & \boldsymbol{O} & \ldots & \boldsymbol{J}_{n_k}(\lambda_k) \end{bmatrix}, \tag{12.20}$$

where λ_i's are the eigenvalues of $\boldsymbol{A}$ counting multiplicities (i.e., not necessarily distinct) and $\sum_{i=1}^{k} n_i = n$.

Proof. For clarity, assume that the Jordan basis for $\boldsymbol{A}$ have been placed as columns in $\boldsymbol{P}$ such that the first n_1 columns form a Jordan chain associated with eigenvalue λ_1, the next n_2 columns form Jordan chains associated with an eigenvalue λ_2 (which may or may not be equal to λ_1) and so on. Let $\boldsymbol{P}_1 = [\boldsymbol{p}_1 : \boldsymbol{p}_2 : \ldots : \boldsymbol{p}_{n_1}]$ be $n \times n_1$ whose columns form a Jordan chain associated with an eigenvalue λ_1. Then, Lemma 12.2 yields $\boldsymbol{A}\boldsymbol{P}_1 = \boldsymbol{J}_{n_1}(\lambda_1)$. Next, let $\boldsymbol{P}_2 = [\boldsymbol{p}_{n_1+1} : \boldsymbol{p}_{n_1+2} : \ldots : \boldsymbol{p}_{n_1+n_2}]$ be $n \times n_2$ with columns that are a Jordan chain associated with eigenvalue λ_2. Again, from Lemma 12.2 we obtain $\boldsymbol{A}\boldsymbol{P}_2 = \boldsymbol{P}_2\boldsymbol{J}_{n_2}(\lambda_2)$. Proceeding in this manner and writing $\boldsymbol{P} = [\boldsymbol{P}_1 : \boldsymbol{P}_2 : \ldots : \boldsymbol{P}_k]$, we obtain

$$\boldsymbol{A}\boldsymbol{P} = [\boldsymbol{P}_1\boldsymbol{J}_{n_1}(\lambda_1) : \boldsymbol{P}_2\boldsymbol{J}_{n_2}(\lambda_2) : \ldots : \boldsymbol{P}_k\boldsymbol{J}_{n_k}(\lambda_k)] = \boldsymbol{P}\boldsymbol{J},$$

where $\boldsymbol{J}$ is as in (12.20). Since $\boldsymbol{P}$ is nonsingular, we have $\boldsymbol{P}^{-1}\boldsymbol{A}\boldsymbol{P} = \boldsymbol{J}$. □

The matrix $\boldsymbol{J}$ is called a *Jordan matrix* and merits its own definition.

Definition 12.8 *A* **Jordan matrix** *is a square block-diagonal matrix*

$$\boldsymbol{J} = \begin{bmatrix} \boldsymbol{J}_{n_1}(\lambda_1) & \boldsymbol{O} & \ldots & \boldsymbol{O} \\ \boldsymbol{O} & \boldsymbol{J}_{n_2}(\lambda_2) & \cdots & \boldsymbol{O} \\ \vdots & \vdots & \ddots & \vdots \\ \boldsymbol{O} & \boldsymbol{O} & \ldots & \boldsymbol{J}_{n_k}(\lambda_k) \end{bmatrix},$$

where each $\boldsymbol{J}_{n_i}(\lambda_i)$ is an $n_i \times n_i$ Jordan block and the λ_i's are not necessarily distinct.

Note that a Jordan matrix can also be written in terms of *distinct* eigenvalues and the associated Jordan segments (recall Definition 12.6). Thus, if $\mu_1, \mu_2, \ldots, \mu_r$ represent the distinct values of $\lambda_1, \lambda_2, \ldots, \lambda_k$ in Definition 12.8, we can collect the Jordan blocks associated with each of the distinct eigenvalues into the Jordan segments and represent the Jordan matrix as

$$\begin{bmatrix} \boldsymbol{J}(\mu_1) & \boldsymbol{O} & \ldots & \boldsymbol{O} \\ \boldsymbol{O} & \boldsymbol{J}(\mu_2) & \cdots & \boldsymbol{O} \\ \vdots & \vdots & \ddots & \vdots \\ \boldsymbol{O} & \boldsymbol{O} & \ldots & \boldsymbol{J}(\mu_r) \end{bmatrix}, \tag{12.21}$$

where each $\boldsymbol{J}(\mu_i)$ is the Jordan segment associated with μ_i and has order equal to the sum of the order of Jordan blocks associated with μ_i.

Theorem 12.3 tells us that if there exists a Jordan basis for $\boldsymbol{A}$, then $\boldsymbol{A}$ is similar to a Jordan matrix. This is known as the Jordan Canonical Form (JCF) or Jordan Normal Form (JNF) of $\boldsymbol{A}$. Conversely, it is immediate from (12.20) that if $\boldsymbol{A}$ is indeed similar to $\boldsymbol{J}$, then there exists a Jordan basis for $\boldsymbol{A}$, namely the columns of $\boldsymbol{P}$.

What we now set out to prove is that every square matrix is similar to a Jordan matrix. By virtue of Theorem 12.3, it will suffice to prove that there exists a Jordan basis associated with every square matrix. We begin with the following simple lemma.

Lemma 12.4 *A Jordan basis for a matrix $\boldsymbol{A}$ is also a Jordan basis for for $\boldsymbol{B} = \boldsymbol{A} + \theta \boldsymbol{I}$ for any scalar θ.*

Proof. If μ is an eigenvalue of $\boldsymbol{A}$ and $\boldsymbol{x}$ is an associated eigenvector, then $\boldsymbol{A}\boldsymbol{x} = \mu\boldsymbol{x}$ and $\boldsymbol{B}\boldsymbol{x} = (\boldsymbol{A} + \theta\boldsymbol{I})\boldsymbol{x} = (\mu + \theta)\boldsymbol{x}$. So $\mu + \theta$ is an eigenvalue of $\boldsymbol{B}$ with $\boldsymbol{x}$ an associated eigenvector. Also, for any Jordan chain $\{\boldsymbol{u}_1, \boldsymbol{u}_2, \ldots, \boldsymbol{u}_k\}$ associated with eigenvalue μ for $\boldsymbol{A}$, it is immediate that

$$\boldsymbol{B}\boldsymbol{u}_1 = (\mu + \theta)\boldsymbol{u}_1 \text{ and } \boldsymbol{B}\boldsymbol{u}_i = (\mu + \theta)\boldsymbol{u}_i + \boldsymbol{u}_{i-1} \text{ for } i = 2, 3, \ldots, k.$$

Therefore, $\{\boldsymbol{u}_1, \boldsymbol{u}_2, \ldots, \boldsymbol{u}_k\}$ is a Jordan chain for $\boldsymbol{B}$ associated with eigenvalue $\mu + \theta$. Since a Jordan basis is the union of distinct Jordan chains, the vectors in a Jordan basis for $\boldsymbol{A}$ forms a Jordan basis for $\boldsymbol{B}$. □

We now prove the *Jordan Basis Theorem*, which states that every square matrix admits a Jordan basis. The proof is based upon Filippov (1971) and Ortega (1987).

Theorem 12.4 Jordan Basis Theorem. *Every $n \times n$ matrix with real or complex entries admits a Jordan basis for $\mathbb{C}^n$.*

Proof. We will proceed by induction on the size of the matrix. The case for 1×1 matrices is trivial. Suppose that the result holds for all square matrices of order $\leq$

$n-1$. Let $\boldsymbol{A}$ be an $n \times n$ matrix and let λ be an eigenvalue of $\boldsymbol{A}$. Therefore, 0 is an eigenvalue of $\boldsymbol{A}_\lambda = \boldsymbol{A} - \lambda \boldsymbol{I}$ so $\boldsymbol{A}_\lambda$ is singular and so $r = \rho(\boldsymbol{A}_\lambda) < n$. Also, Lemma 12.4 ensures that every Jordan basis for $\boldsymbol{A}_\lambda$ is also a Jordan basis for $\boldsymbol{A} = \boldsymbol{A}_\lambda + \lambda \boldsymbol{I}$. So, it will suffice to establish the existence of a Jordan basis for $\boldsymbol{A}_\lambda$. We proceed according to the following steps.

Step 1: *Using the induction hypothesis, find a Jordan basis for $\mathcal{C}(\boldsymbol{A}_\lambda)$.* Let $\boldsymbol{W}$ be an $n \times r$ matrix whose columns constitute a basis for $\mathcal{C}(\boldsymbol{A}_\lambda)$. Since $\mathcal{C}(\boldsymbol{A}_\lambda \boldsymbol{W}) \subset \mathcal{C}(\boldsymbol{A}_\lambda) = \mathcal{C}(\boldsymbol{W})$, each $\boldsymbol{A}_\lambda \boldsymbol{w}_i$ belongs to the column space of $\boldsymbol{W}$. Therefore, there exists an $r \times r$ matrix $\boldsymbol{B}$ such that $\boldsymbol{A}_\lambda \boldsymbol{W} = \boldsymbol{W}\boldsymbol{B}$. Since $\boldsymbol{B}$ is $r \times r$ and $r < n$, there exists a Jordan basis for $\boldsymbol{B}$. By Theorem 12.3, $\boldsymbol{B}$ is similar to a Jordan matrix, say $\boldsymbol{J_B}$. Let $\boldsymbol{S}$ be an $r \times r$ nonsingular matrix such that $\boldsymbol{B} = \boldsymbol{S}\boldsymbol{J_B}\boldsymbol{S}^{-1}$. Hence, we can conclude that

$$\boldsymbol{A}_\lambda \boldsymbol{W} = \boldsymbol{W}\boldsymbol{B} \Longrightarrow \boldsymbol{A}_\lambda \boldsymbol{W} = \boldsymbol{W}\boldsymbol{S}\boldsymbol{J_B}\boldsymbol{S}^{-1} \Longrightarrow \boldsymbol{A}_\lambda \boldsymbol{Q} = \boldsymbol{Q}\boldsymbol{J_B} \,,$$

where $\boldsymbol{Q} = \boldsymbol{W}\boldsymbol{S}$ is $n \times r$ and $\rho(\boldsymbol{Q}) = \rho(\boldsymbol{W}\boldsymbol{S}) = \rho(\boldsymbol{W}) = r$. Thus, $\boldsymbol{Q}$ is of full column rank and its columns are a Jordan basis for $\mathcal{C}(\boldsymbol{A}_\lambda)$.

Let $\boldsymbol{q}_1, \boldsymbol{q}_2, \ldots, \boldsymbol{q}_r$ denote the r columns of $\boldsymbol{Q}$ that form a Jordan basis for $\mathcal{C}(\boldsymbol{A}_\lambda)$. Assume that this basis contains k Jordan chains associated with the 0 eigenvalue of $\boldsymbol{A}_\lambda$; we will refer to these chains as *null Jordan chains*. The first vector of every null Jordan chain is an eigenvector associated with the eigenvalue 0. Also, the number of null Jordan chains, k, is equal to the number of linearly independent eigenvectors of $\boldsymbol{A}_\lambda$ associated with eigenvalue 0. Since, every nonzero vector in $\mathcal{N}(\boldsymbol{A}_\lambda) \cap \mathcal{C}(\boldsymbol{A}_\lambda)$ is an eigenvector associated with the eigenvalue 0, it follows that $k = \dim(\mathcal{N}(\boldsymbol{A}_\lambda) \cap \mathcal{C}(\boldsymbol{A}_\lambda))$. The subsequent steps extend this basis to a Jordan basis for $\mathbb{C}^n$.

Step 2: *Choose some special pre-images of vectors at the end of the null Jordan chains as part of the extension.* Let $\boldsymbol{q}_{j_1}, \boldsymbol{q}_{j_2}, \ldots, \boldsymbol{q}_{j_k}$ denote the vectors at the end of the k null Jordan chains. Since each $\boldsymbol{q}_{j_i}$ is a vector in $\mathcal{C}(\boldsymbol{A}_\lambda)$, there exist vectors $\boldsymbol{h}_i$ such that $\boldsymbol{A}_\lambda \boldsymbol{h}_i = \boldsymbol{q}_{j_i}$ for $i = 1, 2, \ldots, k$. Appending each $\boldsymbol{h}_i$ to the end of the corresponding null Jordan chain results in a collection of $r + k$ vectors arranged in non-overlapping Jordan chains.

We need an additional $n - r - k$ vectors in Jordan chains and prove the linear independence of the entire collection of n vectors to establish a basis for $\mathbb{C}^n$.

Step 3: *Choose remaining linearly independent vectors that extend the basis from $\mathcal{N}(\boldsymbol{A}_\lambda) \cap \mathcal{C}(\boldsymbol{A}_\lambda)$ to $\mathcal{N}(\boldsymbol{A}_\lambda)$.* Since $\mathcal{N}(\boldsymbol{A}_\lambda) \cap \mathcal{C}(\boldsymbol{A}_\lambda)$ is a subspace of $\mathcal{N}(\boldsymbol{A}_\lambda)$ and $\dim(\mathcal{N}(\boldsymbol{A}_\lambda)) = n - r$, we will be able to find $n - r - k$ linearly independent vectors in $\mathcal{N}(\boldsymbol{A}_\lambda)$ that are not in $\mathcal{N}(\boldsymbol{A}_\lambda) \cap \mathcal{C}(\boldsymbol{A}_\lambda)$. Let $\boldsymbol{z}_1, \boldsymbol{z}_2, \ldots, \boldsymbol{z}_{n-r-k}$ be such a set of vectors. Since $\boldsymbol{A}_\lambda \boldsymbol{z}_i = \boldsymbol{0}$ for $i = 1, 2, \ldots, n - r - k$, each $\boldsymbol{z}_i$ constitutes a Jordan chain of length one associated with eigenvalue 0 of $\boldsymbol{A}_\lambda$.

The collection of n vectors in $\mathbb{C}^n$

$$\{\boldsymbol{h}_1, \boldsymbol{h}_2, \ldots, \boldsymbol{h}_k, \boldsymbol{q}_1, \boldsymbol{q}_2, \ldots, \boldsymbol{q}_r, \boldsymbol{z}_1, \boldsymbol{z}_2, \ldots, \boldsymbol{z}_{n-r-k}\} \tag{12.22}$$

belong to some Jordan chain of $\boldsymbol{A}_\lambda$. To prove that they form a Jordan basis, it only remains to prove that they are linearly independent.

Step 4: *Establish the linear independence of the set in (12.22).* Consider any homogeneous linear relationship:

$$\sum_{i=1}^{k} \alpha_i \boldsymbol{h}_i + \sum_{i=1}^{r} \beta_i \boldsymbol{q}_i + \sum_{i=1}^{n-r-k} \gamma_i \boldsymbol{z}_i = \boldsymbol{0} . \tag{12.23}$$

Multiplying both sides of (12.23) by $\boldsymbol{A}_\lambda$ yields

$$\sum_{i=1}^{k} \alpha_i \boldsymbol{A}_\lambda \boldsymbol{h}_i + \sum_{i=1}^{r} \beta_i \boldsymbol{A}_\lambda \boldsymbol{q}_i = \boldsymbol{0} \tag{12.24}$$

since $\boldsymbol{A}_\lambda \boldsymbol{z}_i = \boldsymbol{0}$ for $i = 1, 2, \ldots, n-r-k$. By construction, each $\boldsymbol{A}_\lambda \boldsymbol{h}_i = \boldsymbol{q}_{j_i}$ for $i = 1, 2, \ldots, k$, where the $\boldsymbol{q}_{j_i}$'s are the end vectors in the null Jordan chains of $\boldsymbol{A}_\lambda$. Therefore, we can rewrite (12.24) as

$$\sum_{i=1}^{k} \alpha_i \boldsymbol{q}_{j_i} + \sum_{i=1}^{r} \beta_i \boldsymbol{A}_\lambda \boldsymbol{q}_i = \boldsymbol{0} . \tag{12.25}$$

Recall that the first vector in a Jordan chain is an eigenvector. Therefore,

$$\boldsymbol{A}_\lambda \boldsymbol{q}_i = \begin{cases} \boldsymbol{0} & \text{if } \boldsymbol{q}_i \text{ is the first vector in a null chain;} \\ \boldsymbol{q}_{i-1} & \text{if } \boldsymbol{q}_i \text{ is not the first vector in a null chain;} \\ \mu \boldsymbol{q}_i + \boldsymbol{q}_{i-1} & \text{if } \boldsymbol{q}_i \text{ belongs to a chain corresponding to } \mu; \end{cases} \tag{12.26}$$

where μ is some nonzero eigenvalue of $\boldsymbol{A}_\lambda$ and $\boldsymbol{q}_{i-1} = \boldsymbol{0}$ if $\boldsymbol{q}_i$ is an eigenvector (the first vector of the chain) associated with μ. In particular, for the end vectors in null Jordan chains, i.e., the $\boldsymbol{q}_{j_i}$'s, either $\boldsymbol{A}_\lambda \boldsymbol{q}_{j_i} = \boldsymbol{0}$ or $\boldsymbol{A}_\lambda \boldsymbol{q}_{j_i} = \boldsymbol{q}_{j_i - 1}$. It is clear from (12.26) that (12.25) is a linear combination of the $\boldsymbol{q}_i$'s, where the $\boldsymbol{q}_{j_i}$'s appear only in the first sum and not in the second sum in (12.25). Since the $\boldsymbol{q}_i$'s are all linearly independent, this proves that each $\alpha_i = 0$ for $i = 1, 2, \ldots, k$ and reduces (12.23) to

$$\sum_{i=1}^{r} \beta_i \boldsymbol{q}_i + \sum_{i=1}^{n-r-k} \gamma_i \boldsymbol{z}_i = \boldsymbol{0} .$$

If all the β_i's are zero, then all the γ_i's must also be zero because the $\boldsymbol{z}_i$'s are linearly independent. If at least one β_i is nonzero, then a vector in the span of $\boldsymbol{z}_i$'s belongs to the span of $\boldsymbol{q}_i$'s. But this is impossible because the $\boldsymbol{q}_i$'s are in $\mathcal{C}(\boldsymbol{A}_\lambda)$, while the $\boldsymbol{z}_i$'s, by construction, are in $\mathcal{N}(\boldsymbol{A}_\lambda)$ but are *not* in $\mathcal{C}(\boldsymbol{A}_\lambda)$. Therefore, $\beta_i = 0$ for $i = 1, 2, \ldots, r$ and $\gamma_i = 0$ for $i = 1, 2, \ldots, n-r-k$, which means that all the coefficients in (12.23) are zero.

Thus, the set in (12.22) is linearly independent and is a Jordan basis of $\mathbb{C}^n$. □

The following corollary is often called the ***Jordan Canonical Form (JCF) Theorem***.

Corollary 12.1 *Every $n \times n$ matrix is similar to a Jordan matrix.*

Proof. This is an immediate consequence of Theorems 12.3 and 12.4. □

As mentioned earlier, if $\boldsymbol{A} = \boldsymbol{PJP}^{-1}$ is the JCF of an $n \times n$ matrix $\boldsymbol{A}$, then $\boldsymbol{J}$ can be expressed as a block-diagonal matrix composed of Jordan segments along its diagonals, as in (12.21). Each Jordan segment is block-diagonal and composed of a certain number of Jordan blocks as described in Definition 12.6. The columns in $\boldsymbol{P}$ corresponding to a Jordan block constitute a Jordan chain associated with that eigenvalue. The columns in $\boldsymbol{P}$ corresponding to a Jordan segment constitutes the collection of all Jordan chains associated the eigenvalue of that Jordan segment.

Uniqueness: The matrix $\boldsymbol{J}$ in the JCF of $\boldsymbol{A}$ is structurally *unique* in that the number of Jordan segments in $\boldsymbol{J}$, the number of Jordan blocks in each segment, and the sizes of each Jordan block are uniquely determined by the entries in $\boldsymbol{A}$. Every matrix that is similar to $\boldsymbol{A}$ has the same structural Jordan form. In fact, two matrices are similar if and only if they have the same JCF. The matrix $\boldsymbol{P}$, or the Jordan basis, is not, however, unique. The Jordan structure contains all the necessary information regarding the eigenvalues:

- $\boldsymbol{J}$ has exactly one Jordan segment for each eigenvalue of $\boldsymbol{A}$. Therefore, the number of Jordan segments in $\boldsymbol{A}$ is the number of distinct eigenvalues of $\boldsymbol{A}$. The size of the Jordan segment gives us the algebraic multiplicity of that eigenvalue.
- Each Jordan segment, say $\boldsymbol{J}(\mu_i)$, consists of ν_i Jordan blocks, where ν_j is the dimension of the null space of $\boldsymbol{A} - \mu_i \boldsymbol{I}$. Therefore, the number of Jordan blocks in $\boldsymbol{J}(\mu_i)$, i.e., the number of Jordan blocks associated with eigenvalue μ_i, gives us the geometric multiplicity of μ_i.
- The size of the largest Jordan block in a Jordan segment $\boldsymbol{J}(\mu)$ gives us the index of the eigenvalue μ.
- It is also possible to show that the number of $k \times k$ Jordan blocks in the segment $\boldsymbol{J}(\mu)$ is given by

$$\rho((\boldsymbol{A} - \mu \boldsymbol{I})^{k+1}) - 2\rho((\boldsymbol{A} - \mu \boldsymbol{I})^{k}) + \rho((\boldsymbol{A} - \mu \boldsymbol{I})^{k-1}) \ ,$$

  where $\rho(\cdot)$ is the rank.

12.7 Implications of the Jordan Canonical Form

We conclude this chapter with a list of some important consequences of the JCF. We do not derive these in detail, leaving them to the reader. The machinery developed so far is adequate for deriving these properties.

- **Square matrices are similar to their transposes.** This is a remarkable result and one that is not seen easily without the JCF. Consider the $m \times m$ matrix

$$\boldsymbol{P} = \begin{bmatrix} 0 & 0 & \dots & 0 & 1 \\ 0 & 0 & \dots & 1 & 0 \\ \vdots & \vdots & \ddots & \vdots & \vdots \\ 0 & 1 & \dots & 0 & 0 \\ 1 & 0 & \dots & 0 & 0 \end{bmatrix} = [\boldsymbol{e}_m : \boldsymbol{e}_{m-1} : \dots : \boldsymbol{e}_1] \ .$$

Note that $P^{-1} = P$ (it is a reflector) and that $J_m(\lambda)' = P^{-1}J_m(\lambda)P$, where $J(\lambda)$ is an $m \times m$ Jordan block, as in Definition 12.5. Thus any Jordan block $J_m(\lambda)$ is similar to its transpose $J_m(\lambda)'$. If J is a Jordan matrix with Jordan blocks $J_{n_i}(\lambda_i)$ (see Definition 12.8) and P is the block-diagonal matrix with blocks $P_1, P_2, \ldots, P_k$ such that $J_{n_i}(\lambda_i)' = P_i^{-1}J_{n_i}(\lambda_i)P_i$ for $i = 1, 2, \ldots, k$, then it is easy to see that $J' = P^{-1}JP$. Therefore, a Jordan matrix and its transpose are similar. Finally, since any square matrix A is similar to a Jordan matrix, say $Q^{-1}AQ = J$, it follows that $J' = Q'A'(Q')^{-1}$, so J' is similar to A'. So A is similar to J, which is similar to J' and J' is similar to A'. It follows that A is similar to A'. More explicitly,

$$
\begin{aligned}
A = QJQ^{-1} = QPJ'P^{-1}Q^{-1} &= QPQ'A'(Q')^{-1}P^{-1}Q^{-1} \\
&= (QPQ')A'(QPQ')^{-1} .
\end{aligned}
$$

- **Invariant subspace decompositions.** If the JCF of a matrix A is diagonal, then that matrix has n linearly independent eigenvectors, say $p_1, p_2, \ldots, p_n$, each of which spans a one-dimensional invariant subspace of A, and

$$\mathbb{C}^n = Sp(p_1) \oplus Sp(p_2) \oplus \cdots \oplus Sp(p_n) .$$

This decomposes $\mathbb{C}^n$ as a direct sum of one-dimensional invariant subspaces of A. If the JCF is not diagonal, then the generalized eigenvectors do not generate one-dimensional invariant subspaces but the set of vectors in each Jordan chain produces a direct sum decomposition of $\mathbb{C}^n$ in terms of invariant subspaces. To be precise, let $P^{-1}AP = J$ with J written as a block-diagonal matrix with Jordan segments as blocks, as in (12.21). If $P = [P_1 : P_2 : \ldots : P_r]$ is a conformable partition of P so that the columns of P_i belong to Jordan chains associated with distinct eigenvalue μ_i, then it can be shown that

$$\mathbb{C}^n = \mathcal{C}(P_1) \oplus \mathcal{C}(P_2) \oplus \cdots \oplus \mathcal{C}(P_r) ,$$

where each $\mathcal{C}(P_i)$ is an invariant subspace for A. Furthermore, if the algebraic multiplicity of μ_i is m_i and the index of μ_i is p_i, then it can be shown that $\dim(\mathcal{N}[(A - \mu_i I)^{p_i}]) = m_i$ and that the columns of P_i form a basis for $\mathcal{N}[(A - \mu_i I)^{p_i}]$. So the above decomposition can also be expressed in terms of these null spaces as

$$\mathbb{C}^n = \mathcal{N}[(A - \mu_1 I)^{p_1}] \oplus \mathcal{N}[(A - \mu_2 I)^{p_2}] \oplus \cdots \oplus \mathcal{N}[(A - \mu_r I)^{p_r}] . \quad (12.27)$$

- **Real matrices with real eigenvalues.** If A is a real matrix and all its eigenvalues are real, then we may take all the eigenvectors and eigenvalues of A to be real. The sum of the span of the Jordan chains decompose $\Re^n$ as a direct sum of invariant subspaces.

- **Eigenvalues of $p(A)$.** We have seen earlier that if λ is an eigenvalue of A, then $p(\lambda)$ is an eigenvalue of $p(A)$. The JCF reveals that all of the eigenvalues of $p(A)$ are obtained in this fashion. Let $P^{-1}AP = J$ be the JCF of A. Note that

$$A^m = \left(PJP^{-1}\right)^m = PJ^mP^{-1} \text{ for } m = 1, 2, \ldots; ,$$

which implies that $p(\boldsymbol{A}) = \boldsymbol{P}p(\boldsymbol{J})\boldsymbol{P}^{-1}$, which is equal to

$$\boldsymbol{P}\begin{bmatrix} p(\boldsymbol{J}_{n_1}(\lambda_1)) & \boldsymbol{O} & \dots & \boldsymbol{O} \\ \boldsymbol{O} & p(\boldsymbol{J}_{n_2}(\lambda_2)) & \dots & \boldsymbol{O} \\ \vdots & \vdots & \ddots & \vdots \\ \boldsymbol{O} & \boldsymbol{O} & \dots & p(\boldsymbol{J}_{n_k}(\lambda_k)) \end{bmatrix}\boldsymbol{P}^{-1}.$$

Therefore, $p(\boldsymbol{A})$ is similar to $p(\boldsymbol{J})$. It is easily verified that $p(\boldsymbol{J})$ is triangular, which means that the eigenvalues of $p(\boldsymbol{A})$ are the diagonal elements of $p(\boldsymbol{J})$, namely the $p(\lambda_i)$'s.

- **The Cayley-Hamilton Theorem revisited.** The JCF helps us prove the Cayley-Hamilton Theorem. Let $p(t)$ be the characteristic polynomial of the $n \times n$ matrix $\boldsymbol{A}$. Using the JCF, $\boldsymbol{A} = \boldsymbol{P}\boldsymbol{J}\boldsymbol{P}^{-1}$, where $\boldsymbol{J}$ is the Jordan matrix as in (12.8), it is possible to write

$$\begin{aligned} p(\boldsymbol{A}) &= (\boldsymbol{A} - \lambda_1\boldsymbol{I})^{n_1}(\boldsymbol{A} - \lambda_2\boldsymbol{I})^{n_2}\cdots(\boldsymbol{A} - \lambda_{n_k}\boldsymbol{I})^{n_k} \\ &= \boldsymbol{P}(\boldsymbol{J} - \lambda_1\boldsymbol{I})^{n_1}(\boldsymbol{J} - \lambda_2\boldsymbol{I})^{n_2}\cdots(\boldsymbol{J} - \lambda_{n_k}\boldsymbol{I})^{n_k}\boldsymbol{P}^{-1}, \end{aligned}$$

where $\sum_{i=1}^{k} n_i = n$. Each $(\boldsymbol{J} - \lambda_i\boldsymbol{I})^{n_i}$ is of the form

$$\begin{bmatrix} (\boldsymbol{J}_{n_1}(\lambda_1) - \lambda_i)^{n_i} & \boldsymbol{O} & \dots & \boldsymbol{O} \\ \boldsymbol{O} & (\boldsymbol{J}_{n_2}(\lambda_2) - \lambda_i\boldsymbol{I})^{n_i} & \dots & \boldsymbol{O} \\ \vdots & \vdots & \ddots & \vdots \\ \boldsymbol{O} & \boldsymbol{O} & \dots & (\boldsymbol{J}_{n_k}(\lambda_k) - \lambda_i\boldsymbol{I})^{n_i} \end{bmatrix}.$$

Since each $\boldsymbol{J}_{n_i}(\lambda_i) - \lambda_i\boldsymbol{I}$ is upper-triangular with 0 along its diagonal, it follows that $(\boldsymbol{J}_{n_i}(\lambda_i) - \lambda_i\boldsymbol{I})^{n_i} = \boldsymbol{O}$ for $i = 1, 2, \dots, m$. This means that each of the block-diagonal matrices in the above expansion has at least one block as $\boldsymbol{O}$. And these zero blocks are situated in such a way that $\prod_{i=1}^{m}(\boldsymbol{J}_{n_i}(\lambda_i) - \lambda_i\boldsymbol{I})^{n_i} = \boldsymbol{O}$, which can be verified easily. Therefore, $p(\boldsymbol{A}) = \boldsymbol{O}$.

12.8 Exercises

1. Find the SVD of the following matrices:

$$\text{(a)} \begin{bmatrix} 1 & 0 \\ -1 & 1 \end{bmatrix}, \quad \text{(b)} \begin{bmatrix} 1 & 1 \\ 0 & 0 \end{bmatrix}, \quad \text{(c)} \begin{bmatrix} 2 & -1 & 1 \\ 1 & 1 & 2 \\ 0 & 1 & 1 \\ 1 & -2 & -1 \end{bmatrix}.$$

2. True or false: If δ is a singular value of $\boldsymbol{A}$, then δ^2 is a singular value of $\boldsymbol{A}^2$.
3. Let $\boldsymbol{A}$ be $m \times n$. Prove that $\boldsymbol{A}$ and $\boldsymbol{PAQ}$ have the same singular values, where $\boldsymbol{P}$ and $\boldsymbol{Q}$ are $m \times m$ and $n \times n$ orthogonal matrices, respectively.
4. True or false: The singular values of $\boldsymbol{A}$ and $\boldsymbol{A}'$ are the same.
5. If $\boldsymbol{A}$ is nonsingular, then prove that the singular values of $\boldsymbol{A}^{-1}$ are the reciprocals of the singular values of $\boldsymbol{A}$.

6. If $\boldsymbol{A}$ is nonsingular with singular values $\sigma_1, \sigma_2, \ldots, \sigma_n$, then prove that $\prod_{i=1}^{n} \sigma_i = \text{abs}(|\boldsymbol{A}|)$, where abs$(|\boldsymbol{A}|)$ is the absolute value of the determinant of $\boldsymbol{A}$.
7. True or false: The singular values of a real symmetric matrix are the same as its eigenvalues.
8. True or false: Similar matrices have the same set of singular values.
9. Recall the setting of Theorem 12.2. Prove that the orthogonal projectors onto the four fundamental subspaces are given by: (i) $\boldsymbol{P}_{\mathcal{C}(\boldsymbol{A})} = \boldsymbol{U}_1\boldsymbol{U}_1'$, (ii) $\boldsymbol{P}_{\mathcal{N}(\boldsymbol{A})} = \boldsymbol{V}_2\boldsymbol{V}_2'$, (iii) $\boldsymbol{P}_{\mathcal{C}(\boldsymbol{A}')} = \boldsymbol{V}_1\boldsymbol{V}_1'$, and (iv) $\boldsymbol{P}_{\mathcal{N}(\boldsymbol{A}')} = \boldsymbol{U}_2\boldsymbol{U}_2'$.
10. Let $\boldsymbol{A}$ be an $m \times n$ matrix such that $\mathcal{C}(\boldsymbol{A})$ and $\mathcal{N}(\boldsymbol{A})$ are complementary. If $\boldsymbol{A} = \boldsymbol{U}_1\boldsymbol{\Sigma}\boldsymbol{V}_1'$ as in Theorem 12.1, then show that $\boldsymbol{P} = \boldsymbol{U}_1(\boldsymbol{V}_1'\boldsymbol{U}_1)^{-1}\boldsymbol{V}_1'$ is the (oblique) projector onto $\mathcal{C}(\boldsymbol{A})$ along $\mathcal{N}(\boldsymbol{A})$. *Hint: Recall Theorem 6.8 or (7.16).*
11. Let $\boldsymbol{X}$ be an $n \times p$ matrix with rank p. Use the SVD to show that
$$\boldsymbol{X}^+ = (\boldsymbol{X}'\boldsymbol{X})^{-1}\boldsymbol{X}' \,.$$
12. Let $\boldsymbol{A}^+$ be the Moore-Penrose inverse of $\boldsymbol{A}$. Show that the orthogonal projectors onto the four fundamental subspaces are given by: (i) $\boldsymbol{P}_{\mathcal{C}(\boldsymbol{A})} = \boldsymbol{A}\boldsymbol{A}^+$, (ii) $\boldsymbol{P}_{\mathcal{N}(\boldsymbol{A})} = \boldsymbol{I} - \boldsymbol{A}^+\boldsymbol{A}$, (iii) $\boldsymbol{P}_{\mathcal{C}(\boldsymbol{A}')} = \boldsymbol{A}^+\boldsymbol{A}$, and (iv) $\boldsymbol{P}_{\mathcal{N}(\boldsymbol{A}')} = \boldsymbol{I} - \boldsymbol{A}\boldsymbol{A}^+$.
13. Consider the linear system $\boldsymbol{A}\boldsymbol{x} = \boldsymbol{b}$ and let $\boldsymbol{x}^+ = \boldsymbol{A}^+\boldsymbol{b}$. If $\mathcal{N}(\boldsymbol{A}) = \{\boldsymbol{0}\}$, then show that $\boldsymbol{x}^+$ is the least-squares solution, i.e., $\boldsymbol{A}'\boldsymbol{A}\boldsymbol{x}^+ = \boldsymbol{A}'\boldsymbol{b}$. If $\mathcal{N}(\boldsymbol{A}) \neq \{\boldsymbol{0}\}$, then show that $\boldsymbol{x}^+ \in \mathcal{R}(\boldsymbol{A})$ is the least squares solution with minimum Euclidean norm. If $\boldsymbol{b} \in \mathcal{C}(\boldsymbol{A})$ and $\mathcal{N}(\boldsymbol{A}) = \{\boldsymbol{0}\}$, then show that $\boldsymbol{x}^+ = \boldsymbol{A}^+\boldsymbol{b}$ is the unique solution to $\boldsymbol{A}\boldsymbol{x} = \boldsymbol{b}$.
14. Let $\boldsymbol{A}$ and $\boldsymbol{B}$ be $m \times n$ such that $\boldsymbol{A}\boldsymbol{A}' = \boldsymbol{B}\boldsymbol{B}'$. Show that there is an orthogonal matrix $\boldsymbol{Q}$ such that $\boldsymbol{A} = \boldsymbol{B}\boldsymbol{Q}$.
15. Find the Jordan blocks of $\begin{bmatrix} 2 & 1 \\ 0 & 2 \end{bmatrix}$ and $\begin{bmatrix} 1 & 0 & 0 & 0 \\ 0 & 2 & 1 & 0 \\ 0 & 0 & 3 & 0 \\ 0 & 0 & 0 & 4 \end{bmatrix}$.
16. True or false: If $\boldsymbol{A}$ is diagonalizable, then every generalized eigenvector is an ordinary eigenvector.
17. True or false: If $\boldsymbol{A}$ has JCF $\boldsymbol{J}$, then $\boldsymbol{A}^2$ has JCF $\boldsymbol{A}^2$.
18. True or false: The index of an $n \times n$ nilpotent matrix can never exceed n.
19. Explain the proof of Theorem 12.4 when $\mathcal{C}(\boldsymbol{A}_\lambda) \cap \mathcal{N}(\boldsymbol{A}_\lambda) = \{\boldsymbol{0}\}$.
20. Reconstruct the proof of Theorem 12.4 for any $n \times n$ nilpotent matrix.

CHAPTER 13

Quadratic Forms

13.1 Introduction

Linear systems of equations are the simplest types of algebraic equations. The simplest nonlinear equation is the quadratic equation

$$p(x) = ax^2 + bx + c = 0$$

in a single variable x and $a \neq 0$. Its solution often relies upon the popular algebraic technique known as "completing the square." Here, the first two terms are combined to form a perfect square so that

$$p(x) = a\left(x + \frac{b}{2a}\right)^2 + c - \frac{b^2}{4a} = 0 ,$$

which implies that

$$\left(x + \frac{b}{2a}\right)^2 = \frac{b^2 - 4ac}{4a^2} \Longrightarrow x = \frac{-b \pm \sqrt{b^2 - 4ac}}{2a} ,$$

where the last step follows by taking the square root of both sides and solving for x. This provides a complete analysis of the quadratic equation in a single variable. Not only do we obtain the solution of the quadratic equation, we can also get conditions when $p(x)$ is positive or negative. For example, $p(x) \geq 0$ whenever $a \geq 0$ and $c \geq b^2/4a$. This, in turn, reveals that $p(x)$ attains its minimum value when $x = -b/2a$ and this minimum value is $c - b^2/4a$. Our goal in this chapter is to explore and analyze quadratic functions in *several variables*.

If $\boldsymbol{a} = (a_1, a_2, \ldots, a_n)'$ is an $n \times 1$ column vector, then the function $f(\boldsymbol{x}) = \sum_{i=1} a_i x_i = \boldsymbol{a}'\boldsymbol{x}$ that assigns the value $\boldsymbol{a}'\boldsymbol{x}$ to each $\boldsymbol{x} = (x_1, x_2, \ldots, x_n)'$ is called a ***linear function*** or ***linear form*** in $\boldsymbol{x}$. Until now we have focused mostly upon linear functions of vectors, or collections thereof. For example, if $\boldsymbol{A}$ is an $m \times n$ matrix, then the matrix-vector product $\boldsymbol{A}\boldsymbol{x}$ is a collection of m linear forms in $\boldsymbol{x}$.

A scalar-valued (real-valued for us) function $f(\boldsymbol{x}, \boldsymbol{y})$ of two vectors, say $\boldsymbol{x} \in \Re^m$ and $\boldsymbol{y} \in \Re^n$, is said to be a ***bilinear function*** or ***bilinear form*** in $\boldsymbol{x}$ and $\boldsymbol{y}$ if $f(\boldsymbol{x}, \boldsymbol{y})$

can be expressed as

$$f(\boldsymbol{x}, \boldsymbol{y}) = \sum_{i,j=1}^{m,n} a_{ij} x_i y_j = \boldsymbol{x}' \boldsymbol{A} \boldsymbol{y} = \boldsymbol{y}' \boldsymbol{A}' \boldsymbol{x} \, , \tag{13.1}$$

for some $m \times n$ matrix $\boldsymbol{A} = \{a_{ij}\}$. The quantity $\boldsymbol{x}'\boldsymbol{A}\boldsymbol{y}$ is a scalar, so it is equal to its transpose. Therefore, $\boldsymbol{x}'\boldsymbol{A}\boldsymbol{y} = (\boldsymbol{x}'\boldsymbol{A}\boldsymbol{y})' = \boldsymbol{y}'\boldsymbol{A}'\boldsymbol{x}$, which explains the last equality in (13.1).

Note that $f(\boldsymbol{x}, \boldsymbol{y}) = \boldsymbol{x}'\boldsymbol{A}\boldsymbol{y}$ is a linear function of $\boldsymbol{y}$ for every fixed value of $\boldsymbol{x}$ because $\boldsymbol{x}'\boldsymbol{A}\boldsymbol{y} = \boldsymbol{u}'\boldsymbol{y}$, where $\boldsymbol{u}' = \boldsymbol{x}'\boldsymbol{A}$. It is also a linear function of $\boldsymbol{x}$ for every fixed value of $\boldsymbol{y}$ because $\boldsymbol{x}'\boldsymbol{A}\boldsymbol{y} = \boldsymbol{x}'\boldsymbol{v} = \boldsymbol{v}'\boldsymbol{x}$, where $\boldsymbol{v}' = \boldsymbol{y}'\boldsymbol{A}'$. Hence, the name bilinear.

In general, $\boldsymbol{x}$ and $\boldsymbol{y}$ in (13.1) do not need to have the same dimension (i.e., m need not be equal to n). However, if $\boldsymbol{x}$ and $\boldsymbol{y}$ have the same dimension, then $\boldsymbol{A}$ is a square matrix. Our subsequent interest in this book will be on a special type of bilinear forms, where not only will $\boldsymbol{x}$ and $\boldsymbol{y}$ have the same dimension, but they will be equal. The bilinear form now becomes a *quadratic form*, $f(\boldsymbol{x}) = \boldsymbol{x}'\boldsymbol{A}\boldsymbol{x}$.

Quadratic forms arise frequently in statistics and econometrics primarily because of their conspicuous presence in the exponent of the multivariate Gaussian or Normal distribution. The distributions of quadratic forms can often be derived in closed forms and they play an indispensable role in the statistical theory of linear regression. In the remainder of this chapter we focus upon properties of quadratic forms and especially upon the characteristics of the square matrix $\boldsymbol{A}$ that defines them.

13.2 Quadratic forms

We begin with a formal definition of a quadratic form.

Definition 13.1 *For an $n \times n$ matrix $\boldsymbol{A} = \{a_{ij}\}$ and an $n \times 1$ vector $\boldsymbol{x} = [x_1 : x_2 : \ldots : x_n]'$, a* **quadratic form** *$q_{\boldsymbol{A}}(\boldsymbol{x})$ is defined to be the real-valued function*

$$q_{\boldsymbol{A}}(\boldsymbol{x}) = \boldsymbol{x}'\boldsymbol{A}\boldsymbol{x} = \sum_{i,j=1}^{n} a_{ij} x_i x_j = \sum_{i=1}^{n} a_{ii} x_i^2 + \sum_{i=1}^{n} \sum_{j \neq i} a_{ij} x_i x_j \, . \tag{13.2}$$

The matrix $\boldsymbol{A}$ is referred to as the matrix associated with the quadratic form $\boldsymbol{x}'\boldsymbol{A}\boldsymbol{x}$.

It easily follows from (13.1) that $q_{\boldsymbol{A}}(\boldsymbol{x}) = q_{\boldsymbol{A}'}(\boldsymbol{x})$ for all $\boldsymbol{x} \in \Re^n$.

Quadratic forms are homogeneous polynomials of degree two in several variables. They can also be looked upon as high-dimensional generalizations of conics in geometry. Sometimes the matrix $\boldsymbol{A}$ is dropped from the notation and $q_{\boldsymbol{A}}(\boldsymbol{x})$ is simply written as $q(\boldsymbol{x})$. This is sometimes preferred because a quadratic form can yield the same value for more than one matrix.

Example 13.1 Quadratic forms in n-variables are often referred to as n-ary quadratic forms. A general unary quadratic form is of the form $q(x) = \alpha x^2$; the associated matrix is the 1×1 matrix (scalar) $\{\alpha\}$.

A general binary quadratic form is $q(x_1, x_2) = \alpha_1 x_1^2 + \alpha_2 x_2^2 + \alpha_{12} x_1 x_2$. The associated symmetric matrix is $\begin{bmatrix} \alpha_1 & \alpha_{12}/2 \\ \alpha_{12}/2 & \alpha_2 \end{bmatrix}$.

The following is an example of a quadratic form in three variables and an associated (symmetric) matrix:

$$q(x,y,z) = x^2 + 3y^2 - 2z^2 + 5xy + 3zx + 4yz = [x, y, z] \begin{bmatrix} 1 & 5/2 & 3/2 \\ 5/2 & 3 & 2 \\ 3/2 & 2 & -2 \end{bmatrix} \begin{bmatrix} x \\ y \\ z \end{bmatrix} .$$

Note that the associated matrix above is symmetric. But it is not unique. The matrix

$$\begin{bmatrix} 1 & 5 & 3 \\ 0 & 3 & 4 \\ 0 & 0 & -2 \end{bmatrix}$$

would also produce the same quadratic form.

Any homogeneous polynomial of degree two in several variables can be expressed in the form $\boldsymbol{x}'\boldsymbol{A}\boldsymbol{x}$ for some *symmetric* matrix $\boldsymbol{A}$. To see this, consider the general homogeneous polynomial of degree two in n variables $\boldsymbol{x} = (x_1, x_2, \ldots, x_n)'$:

$$f(x_1, x_2, \ldots, x_n) = \sum_{i=1}^{n} \alpha_i x_i^2 + \sum_{1 \le i < j \le n} \beta_{ij} x_i x_j .$$

Collecting the n-variables into $\boldsymbol{x} = (x_1, x_2, \ldots, x_n)'$, we see that the above polynomial can be expressed as $\boldsymbol{x}'\boldsymbol{A}\boldsymbol{x}$, where $\boldsymbol{A} = \{a_{ij}\}$ is the $n \times n$ symmetric matrix whose (i,j)-th elements are given by

$$a_{ij} = \begin{cases} \alpha_i & \text{if } \; i = j , \\ \frac{\beta_{ij}}{2} & \text{if } \; i < j , \\ \frac{\beta_{ij}}{2} & \text{if } \; i > j . \end{cases}$$

Since $a_{ij} = a_{ji} = \beta_{ij}/2$, the matrix $\boldsymbol{A}$ is symmetric. We present this fact as a theorem and offer a proof using the neat trick that the transpose of a real number is the same real number.

Theorem 13.1 *For the n-ary quadratic form $\boldsymbol{x}'\boldsymbol{B}\boldsymbol{x}$, where $\boldsymbol{B}$ is any $n \times n$ matrix, there is a symmetric matrix $\boldsymbol{A}$ such that $\boldsymbol{x}'\boldsymbol{B}\boldsymbol{x} = \boldsymbol{x}'\boldsymbol{A}\boldsymbol{x}$ for every $\boldsymbol{x} \in \Re^n$.*

Proof. Since $\boldsymbol{x}'\boldsymbol{B}\boldsymbol{x}$ is a real number, its transpose is equal to itself. So,

$$\boldsymbol{x}'\boldsymbol{B}\boldsymbol{x} = (\boldsymbol{x}'\boldsymbol{B}\boldsymbol{x})' = \boldsymbol{x}'\boldsymbol{B}'\boldsymbol{x} \Longrightarrow \boldsymbol{x}'\boldsymbol{B}\boldsymbol{x} = \frac{\boldsymbol{x}'\boldsymbol{B}\boldsymbol{x} + \boldsymbol{x}'\boldsymbol{B}'\boldsymbol{x}}{2} = \boldsymbol{x}'\left(\frac{\boldsymbol{B}+\boldsymbol{B}'}{2}\right)\boldsymbol{x}$$

$$= \boldsymbol{x}'\boldsymbol{A}\boldsymbol{x} , \quad \text{where } \boldsymbol{A} = \frac{\boldsymbol{B}+\boldsymbol{B}'}{2} \text{ is a symmetric matrix.}$$

□

The above result proves that the matrix of any quadratic form can always be forced to be symmetric. Using a symmetric matrix for the quadratic form can yield certain interesting properties that are not true when the matrix is not symmetric. The following is one such example.

Lemma 13.1 *Let $\boldsymbol{A} = \{a_{ij}\}$ be an $n \times n$ symmetric matrix. Then, $\boldsymbol{x}'\boldsymbol{A}\boldsymbol{x} = 0$ for every $\boldsymbol{x} \in \Re^n$ if and only if $\boldsymbol{A} = \boldsymbol{O}$ (the null matrix).*

Proof. Clearly, if $\boldsymbol{A} = \boldsymbol{O}$, then $\boldsymbol{x}'\boldsymbol{A}\boldsymbol{x} = 0$.

Now assume that $\boldsymbol{x}'\boldsymbol{A}\boldsymbol{x} = 0$ for every $\boldsymbol{x} \in \Re^n$. Taking $\boldsymbol{x} = \boldsymbol{e}_i$, we find that $a_{ii} = \boldsymbol{e}_i'\boldsymbol{A}\boldsymbol{e}_i = 0$. This confirms that every diagonal element of $\boldsymbol{A}$ is zero. Now comes a neat trick: taking $\boldsymbol{x} = \boldsymbol{e}_j + \boldsymbol{e}_k$, we obtain

$$\begin{aligned}
\boldsymbol{x}'\boldsymbol{A}\boldsymbol{x} = 0 &\Longrightarrow (\boldsymbol{e}_j + \boldsymbol{e}_k)'\boldsymbol{A}(\boldsymbol{e}_j + \boldsymbol{e}_k) = 0 \\
&\Longrightarrow \boldsymbol{e}_j'\boldsymbol{A}\boldsymbol{e}_j + \boldsymbol{e}_k'\boldsymbol{A}\boldsymbol{e}_j + \boldsymbol{e}_j'\boldsymbol{A}\boldsymbol{e}_k + \boldsymbol{e}_k'\boldsymbol{A}\boldsymbol{e}_k = 0 \\
&\Longrightarrow a_{jj} + a_{kj} + a_{jk} + a_{kk} = a_{kj} + a_{jk} = 0 \text{ because } a_{jj} = a_{kk} = 0 \\
&\Longrightarrow 2a_{jk} = 0 \text{ because } \boldsymbol{A} \text{ is symmetric } \Longrightarrow a_{jk} = 0\,.
\end{aligned}$$

This proves that each element of $\boldsymbol{A}$ is zero. □

A related question concerns the equality of two quadratic forms $\boldsymbol{x}'\boldsymbol{A}\boldsymbol{x}$ and $\boldsymbol{x}'\boldsymbol{B}\boldsymbol{x}$. If two quadratic forms are equal, then what can we say about the respective matrices? The following theorem holds the answer.

Theorem 13.2 *Let $\boldsymbol{x}'\boldsymbol{A}\boldsymbol{x}$ and $\boldsymbol{x}'\boldsymbol{B}\boldsymbol{x}$ be two quadratic forms, where $\boldsymbol{A}$ and $\boldsymbol{B}$ are any two $n \times n$ matrices.*

(i) $\boldsymbol{x}'\boldsymbol{A}\boldsymbol{x} = \boldsymbol{x}'\boldsymbol{B}\boldsymbol{x}$ for every $\boldsymbol{x} \in \Re^n$ if and only if $\boldsymbol{A} + \boldsymbol{A}' = \boldsymbol{B} + \boldsymbol{B}'$.

(ii) If one of the two matrices, say $\boldsymbol{A}$, is symmetric, then $\boldsymbol{x}'\boldsymbol{A}\boldsymbol{x} = \boldsymbol{x}'\boldsymbol{B}\boldsymbol{x}$ for every $\boldsymbol{x} \in \Re^n$ if and only if $\boldsymbol{A} = \dfrac{\boldsymbol{B} + \boldsymbol{B}'}{2}$.

Proof. **Proof of (i):** If $\boldsymbol{A} + \boldsymbol{A}' = \boldsymbol{B} + \boldsymbol{B}'$, then

$$\boldsymbol{x}'\boldsymbol{A}\boldsymbol{x} = \boldsymbol{x}'\left(\frac{\boldsymbol{A} + \boldsymbol{A}'}{2}\right)\boldsymbol{x} = \boldsymbol{x}'\left(\frac{\boldsymbol{B} + \boldsymbol{B}'}{2}\right)\boldsymbol{x} = \boldsymbol{x}'\boldsymbol{B}\boldsymbol{x}\,.$$

This proves the "if" part.

The proof of the "only if" part is a bit trickier and makes use of Lemma 13.1. If $\boldsymbol{x}'\boldsymbol{A}\boldsymbol{x} = \boldsymbol{x}'\boldsymbol{B}\boldsymbol{x}$ for every $\boldsymbol{x} \in \Re^n$, then

$$\begin{aligned}
\boldsymbol{x}'(\boldsymbol{A} + \boldsymbol{A}')\boldsymbol{x} &= \boldsymbol{x}'\boldsymbol{A}\boldsymbol{x} + \boldsymbol{x}'\boldsymbol{A}'\boldsymbol{x} = 2\boldsymbol{x}'\boldsymbol{A}\boldsymbol{x} = 2\boldsymbol{x}'\boldsymbol{B}\boldsymbol{x} \\
&= \boldsymbol{x}'\boldsymbol{B}\boldsymbol{x} + \boldsymbol{x}'\boldsymbol{B}'\boldsymbol{x} = \boldsymbol{x}'(\boldsymbol{B} + \boldsymbol{B}')\boldsymbol{x} \\
&\Longrightarrow \boldsymbol{x}'\boldsymbol{C}\boldsymbol{x} = 0 \text{ for every } \boldsymbol{x} \in \Re^n, \text{ where } \boldsymbol{C} = (\boldsymbol{A} + \boldsymbol{A}') - (\boldsymbol{B} + \boldsymbol{B}')\,.
\end{aligned}$$

The matrix $\boldsymbol{C}$ is symmetric, so $\boldsymbol{C} = \boldsymbol{O}$ from Lemma 13.1. This proves (i).

Proof of (ii): This follows easily from part (i). Suppose $\boldsymbol{A}$ is symmetric. Then, $\boldsymbol{A} = \boldsymbol{A}'$ and we find that the necessary and sufficient condition for $\boldsymbol{x}'\boldsymbol{A}\boldsymbol{x}$ and $\boldsymbol{x}'\boldsymbol{B}\boldsymbol{x}$ to be equal for every $\boldsymbol{x} \in \Re^n$ is

$$2\boldsymbol{A} = \boldsymbol{A} + \boldsymbol{A}' = \boldsymbol{B} + \boldsymbol{B}' \Longrightarrow \boldsymbol{A} = \frac{\boldsymbol{B} + \boldsymbol{B}'}{2} .$$

This completes the proof. □

Based upon the above results, we see that restricting $\boldsymbol{A}$ to a symmetric matrix yields an attractive uniqueness property in the following sense: if $\boldsymbol{x}'\boldsymbol{A}\boldsymbol{x} = \boldsymbol{x}'\boldsymbol{B}\boldsymbol{x}$ for every $\boldsymbol{x} \in \Re^n$ and $\boldsymbol{A}$ and $\boldsymbol{B}$ are both symmetric, then $\boldsymbol{A} = \boldsymbol{B}$. For these reasons, matrices in quadratic forms are often taken to be symmetric by default. Soon we will be restricting our attention to symmetric matrices. However, in the next section we outline certain properties for general, i.e., not necessarily symmetric, matrices in quadratic forms.

13.3 Matrices in quadratic forms

The range of a quadratic form can yield useful characterizations for the corresponding matrix. The most interesting case arises when the quadratic form is greater than zero whenever $\boldsymbol{x}$ is nonzero. We have a special name for such quadratic forms.

Definition 13.2 Positive definite quadratic form. *Let $\boldsymbol{A}$ be an $n \times n$ matrix. The quadratic form $q_{\boldsymbol{A}}(\boldsymbol{x}) = \boldsymbol{x}'\boldsymbol{A}\boldsymbol{x}$ is said to be* **positive definite** *if*

$$q_{\boldsymbol{A}}(\boldsymbol{x}) = \boldsymbol{x}'\boldsymbol{A}\boldsymbol{x} = \sum_{i=1}^{n}\sum_{j=1}^{n} a_{ij}x_i x_j > 0 \textit{ for all nonzero } \boldsymbol{x} \in \Re^n .$$

A symmetric matrix in a positive definite quadratic form is called a **positive definite matrix**.

We will later study positive definite matrices in greater detail. Here, we consider some implications of a positive definite quadratic form $\boldsymbol{x}'\boldsymbol{A}\boldsymbol{x}$ on $\boldsymbol{A}$ even if $\boldsymbol{A}$ is not necessarily symmetric.

A few observations are immediate. Because $q_{\boldsymbol{A}}(\boldsymbol{x}) = q_{\boldsymbol{A}'}(\boldsymbol{x})$, $q_{\boldsymbol{A}}(\boldsymbol{x})$ is positive definite if and only if $q_{\boldsymbol{A}'}(\boldsymbol{x})$ is. Theorems 13.1 and 13.2 ensure that $q_{\boldsymbol{A}}(\boldsymbol{x})$ is positive definite if and only if $q_{\boldsymbol{B}}(\boldsymbol{x})$ is, where $\boldsymbol{B}$ is the symmetric matrix $\boldsymbol{B} = \dfrac{\boldsymbol{A} + \boldsymbol{A}'}{2}$.

If $q_{\boldsymbol{A}}(\boldsymbol{x})$ and $q_{\boldsymbol{B}}(\boldsymbol{x})$ are positive definite, then so is $q_{\boldsymbol{A}+\boldsymbol{B}}(\boldsymbol{x})$ because

$$q_{\boldsymbol{A}+\boldsymbol{B}}(\boldsymbol{x}) = \boldsymbol{x}'(\boldsymbol{A}+\boldsymbol{B})\boldsymbol{x} = \boldsymbol{x}'\boldsymbol{A}\boldsymbol{x} + \boldsymbol{x}'\boldsymbol{B}\boldsymbol{x} = q_{\boldsymbol{A}}(\boldsymbol{x}) + q_{\boldsymbol{B}}(\boldsymbol{x}) > 0 \text{ for all } \boldsymbol{x} \neq \boldsymbol{0} .$$

If $\boldsymbol{D}$ is $n \times n$ diagonal with $d_{ii} > 0$, then $q_{\boldsymbol{D}}(\boldsymbol{x}) = \boldsymbol{x}'\boldsymbol{D}\boldsymbol{x} = \sum_{i=1}^{n} d_{ii}x_i^2 > 0$ for all nonzero $\boldsymbol{x} \in \Re^n$. The following lemma shows that if $q_{\boldsymbol{A}}(\boldsymbol{x})$ is positive definite, then the matrix $\boldsymbol{A}$ must have positive diagonal elements.

Lemma 13.2 *If $q_{\boldsymbol{A}}(\boldsymbol{x})$ is positive definite, then each diagonal element of $\boldsymbol{A}$ is positive.*

Proof. Since $q_{\boldsymbol{A}}(\boldsymbol{x}) = \boldsymbol{x}'\boldsymbol{A}\boldsymbol{x} > 0$ whenever $\boldsymbol{x} \neq \boldsymbol{0}$, choosing $\boldsymbol{x} = \boldsymbol{e}_i$ reveals that $\boldsymbol{e}_i'\boldsymbol{A}\boldsymbol{e}_i = a_{ii} > 0$. □

It is easy to construct matrices that have negative entries but yield positive definite quadratic forms. The following is an example with a lower-triangular matrix.

Example 13.2 Let $\boldsymbol{A} = \begin{bmatrix} 2 & 0 \\ -2 & 2 \end{bmatrix}$. If $\boldsymbol{x} = [x_1 : x_2]'$, then

$$q_{\boldsymbol{A}}(\boldsymbol{x}) = \boldsymbol{x}'\boldsymbol{A}\boldsymbol{x} = 2x_1^2 + 2x_2^2 - 2x_1x_2 = x_1^2 + x_2^2 + (x_1 - x_2)^2 > 0$$

for all nonzero $\boldsymbol{x} \in \Re^2$. ■

How do we confirm if a quadratic form is positive definite? We could "complete the square" as was done for the single variable quadratic equation at the beginning of this chapter. It is convenient to consider symmetric quadratic forms (Theorem 13.1). Let us illustrate with the following bivariate example.

Example 13.3 Consider the following quadratic form in $\boldsymbol{x} = [x_1 : x_2]'$:

$$q_{\boldsymbol{A}}(\boldsymbol{x}) = a_{11}x_1^2 + 2a_{12}x_1x_2 + a_{22}x_2^2 = \boldsymbol{x}'\boldsymbol{A}\boldsymbol{x}\ , \quad \text{where } \boldsymbol{A} = \begin{bmatrix} a_{11} & a_{12} \\ a_{12} & a_{22} \end{bmatrix}$$

and $a_{11} \neq 0$. Then,

$$\begin{aligned} q_{\boldsymbol{A}}(\boldsymbol{x}) &= a_{11}\left(x_1 + 2\frac{a_{12}}{a_{11}}x_2\right) + a_{22}x_2^2 = a_{11}\left(x_1 + \frac{a_{12}}{a_{11}}x_2\right)^2 - \frac{a_{12}^2}{a_{11}}x_2^2 + a_{22}x_2^2 \\ &= a_{11}y_1^2 + \left(a_{22} - \frac{a_{12}^2}{a_{11}}\right)y_2^2\ , \end{aligned}$$

where $y_1 = x_1 + (a_{12}/a_{11})x_2$ and $y_2 = x_2$. Completing the square has brought the quadratic form to a *diagonal form* in $\boldsymbol{y} = [y_1 : y_2]'$ with diagonal matrix $\boldsymbol{D}$,

$$q_{\boldsymbol{D}}(\boldsymbol{y}) = d_{11}y_1^2 + d_{22}y_2^2 = \boldsymbol{y}'\boldsymbol{D}\boldsymbol{y}\ , \quad \text{where } d_{11} = a_{11} \text{ and } d_{22} = a_{22} - \frac{a_{12}^2}{a_{11}}\ ,$$

using the change of variable

$$\boldsymbol{y} = \begin{bmatrix} y_1 \\ y_2 \end{bmatrix} = \begin{bmatrix} 1 & a_{12}/a_{11} \\ 0 & 1 \end{bmatrix} \begin{bmatrix} x_1 \\ x_2 \end{bmatrix} .$$

The diagonal form makes the conditions for $q_{\boldsymbol{A}}(\boldsymbol{x})$ to be positive definite very clear: d_{11} and d_{22} must *both* be positive or, equivalently, $a_{11} > 0$ and $a_{22} > a_{12}^2/a_{11}$. ■

Completing the square can be applied to general quadratic forms but becomes a bit more cumbersome with several variables. It is more efficient, both computationally and for gaining theoretical insight, to deduce the conditions of positive definiteness by considering the properties of the associated matrix. We will later see several necessary and sufficient conditions for positive definite matrices. The following theorem offers a useful necessary condition: if the quadratic form $q_{\boldsymbol{A}}(\boldsymbol{x})$ is positive definite, then $\boldsymbol{A}$ must be nonsingular.

Theorem 13.3 *If $q_{\boldsymbol{A}}(\boldsymbol{x})$ is positive definite, then $\boldsymbol{A}$ is nonsingular.*

Proof. Suppose that $\boldsymbol{A}$ is singular. In that case, we will be able to find a nonzero vector $\boldsymbol{u}$ such that $\boldsymbol{A}\boldsymbol{u} = \boldsymbol{0}$. This means that $q_{\boldsymbol{A}}(\boldsymbol{u}) = \boldsymbol{u}'\boldsymbol{A}\boldsymbol{u} = 0$ for some $\boldsymbol{u} \neq \boldsymbol{0}$, which contradicts the positive definiteness of $q_{\boldsymbol{A}}(\boldsymbol{x})$. □

Example 13.4 Nonsingular matrices need not produce positive definite quadratic forms. If $\boldsymbol{A} = \begin{bmatrix} -1 & 0 \\ 0 & -1 \end{bmatrix}$, then $q_{\boldsymbol{A}}(\boldsymbol{x}) \leq 0$ for all $\boldsymbol{x} \in \Re^2$. ■

Linear systems $\boldsymbol{A}\boldsymbol{x} = \boldsymbol{b}$, where $q_{\boldsymbol{A}}(\boldsymbol{x})$ is positive definite, occur frequently in statistics, economics, engineering and other applied sciences. Such systems are often called positive definite systems and they constitute an important problem in numerical linear algebra. Theorem 13.3 ensures that a linear system whose coefficient matrix is positive definite will always have a unique solution.

The following theorem reveals how we can "complete the square" in expressions involving the sum of a quadratic form and a linear form. The result is especially useful in studying Bayesian linear models in statistics.

Theorem 13.4 Completing the square. *If $\boldsymbol{A}$ is a symmetric $n \times n$ matrix such that $q_{\boldsymbol{A}}(\boldsymbol{x})$ is positive definite, $\boldsymbol{b}$ and $\boldsymbol{x}$ are $n \times 1$, and c is a scalar, then*

$$\boldsymbol{x}'\boldsymbol{A}\boldsymbol{x} - 2\boldsymbol{b}'\boldsymbol{x} + c = (\boldsymbol{x} - \boldsymbol{A}^{-1}\boldsymbol{b})'\boldsymbol{A}(\boldsymbol{x} - \boldsymbol{A}^{-1}\boldsymbol{b}) + c - \boldsymbol{b}'\boldsymbol{A}^{-1}\boldsymbol{b}\,.$$

Proof. Theorem 13.3 ensures that $\boldsymbol{A}^{-1}$ exists because $\boldsymbol{A}$ is positive definite. One way to derive the above relationship is to simply expand the right hand side and show that it is equal to the left hand side, keeping in mind that $\boldsymbol{A}$ and $\boldsymbol{A}^{-1}$ are both symmetric. This is easy and left to the reader.

An alternative proof is to use the method of "undetermined coefficients." Here, we want to find an $n \times 1$ vector $\boldsymbol{u}$ and a scalar d so that

$$\boldsymbol{x}'\boldsymbol{A}\boldsymbol{x} - 2\boldsymbol{b}'\boldsymbol{x} + c = (\boldsymbol{x} - \boldsymbol{u})'\boldsymbol{A}(\boldsymbol{x} - \boldsymbol{u}) + d = \boldsymbol{x}'\boldsymbol{A}\boldsymbol{x} - 2\boldsymbol{u}'\boldsymbol{A}\boldsymbol{x} + \boldsymbol{u}'\boldsymbol{A}\boldsymbol{u} + d\,.$$

This implies that $\boldsymbol{u}$ and d must satisfy $(\boldsymbol{b}' - \boldsymbol{u}'\boldsymbol{A})\boldsymbol{x} + c - \boldsymbol{u}'\boldsymbol{A}\boldsymbol{u} - d = 0$. Since this is true for all $\boldsymbol{x}$, we can set $\boldsymbol{x} = \boldsymbol{0}$ to obtain $d = c - \boldsymbol{u}'\boldsymbol{A}\boldsymbol{u}$. This means that $(\boldsymbol{b}' - \boldsymbol{u}'\boldsymbol{A})\boldsymbol{x} = 0$ for all $\boldsymbol{x}$, which implies that $\boldsymbol{u}'\boldsymbol{A} = \boldsymbol{b}'$ or, equivalently, that $\boldsymbol{A}\boldsymbol{u} = \boldsymbol{b}$. Hence, $\boldsymbol{u} = \boldsymbol{A}^{-1}\boldsymbol{b}$. Substituting this in the expression obtained for d, we obtain $d = c - \boldsymbol{b}'\boldsymbol{A}^{-1}\boldsymbol{b}$. This completes the proof. □

The above result finds several applications in Bayesian statistics and hierarchical linear models. One example involves simplifying expressions of the form

$$(\boldsymbol{x} - \boldsymbol{\mu})'\boldsymbol{V}(\boldsymbol{x} - \boldsymbol{\mu}) + (\boldsymbol{y} - \boldsymbol{W}\boldsymbol{x})'\boldsymbol{A}(\boldsymbol{y} - \boldsymbol{W}\boldsymbol{x})\,, \tag{13.3}$$

where $\boldsymbol{y}$ is $n \times 1$, $\boldsymbol{x}$ and $\boldsymbol{\mu}$ are $p \times 1$, $\boldsymbol{W}$ is $n \times p$, and $\boldsymbol{A}$ and $\boldsymbol{V}$ are positive definite of order $n \times n$ and $p \times p$, respectively. Expanding (13.3) yields

$$\begin{aligned} &\boldsymbol{x}'\boldsymbol{V}\boldsymbol{x} - 2\boldsymbol{\mu}'\boldsymbol{V}\boldsymbol{x} + \boldsymbol{\mu}'\boldsymbol{V}\boldsymbol{\mu} + \boldsymbol{y}'\boldsymbol{A}\boldsymbol{y} - 2\boldsymbol{y}'\boldsymbol{A}\boldsymbol{W}\boldsymbol{x} + \boldsymbol{x}'\boldsymbol{W}'\boldsymbol{A}\boldsymbol{W}\boldsymbol{x} \\ &= \boldsymbol{x}'(\boldsymbol{V} + \boldsymbol{W}'\boldsymbol{A}\boldsymbol{W})\boldsymbol{x} - 2(\boldsymbol{V}\boldsymbol{\mu} + \boldsymbol{W}'\boldsymbol{A}\boldsymbol{y})'\boldsymbol{x} + c\,, \end{aligned}$$

where $c = \boldsymbol{\mu}'\boldsymbol{V}\boldsymbol{\mu} + \boldsymbol{y}'\boldsymbol{A}\boldsymbol{y}$ does not depend upon $\boldsymbol{x}$. Applying Theorem 13.4, we find that (13.3) can be expressed as a quadratic form in $\boldsymbol{x}$ plus some constant as

$$(\boldsymbol{x} - \boldsymbol{B}^{-1}\boldsymbol{b})'\boldsymbol{B}(\boldsymbol{x} - \boldsymbol{B}^{-1}\boldsymbol{b})\,, \text{ where } \boldsymbol{B} = \boldsymbol{V} + \boldsymbol{W}'\boldsymbol{A}\boldsymbol{W} \text{ and } \boldsymbol{b} = \boldsymbol{V}\boldsymbol{\mu} + \boldsymbol{W}'\boldsymbol{A}\boldsymbol{y}\,.$$

It is useful to know that certain transformations do not destroy positive definiteness of quadratic forms.

Theorem 13.5 *Let $\boldsymbol{A}$ be an $n \times n$ matrix such that $q_{\boldsymbol{A}}(\boldsymbol{x})$ is positive definite. If $\boldsymbol{P}$ is an $n \times k$ matrix with full column rank, then $q_{\boldsymbol{B}}(\boldsymbol{u})$ is also positive definite, where $\boldsymbol{B} = \boldsymbol{P}'\boldsymbol{A}\boldsymbol{P}$ is $k \times k$.*

Proof. Suppose, if possible, that $q_{\boldsymbol{B}}(\boldsymbol{x})$ is not positive definite. This means that we can find some nonzero $\boldsymbol{u} \in \Re^k$ such that $q_{\boldsymbol{B}}(\boldsymbol{u}) \leq 0$. For this $\boldsymbol{u}$ we can argue that

$$q_{\boldsymbol{B}}(\boldsymbol{u}) = \boldsymbol{u}'\boldsymbol{B}\boldsymbol{u} \leq 0 \Longrightarrow \boldsymbol{u}'\boldsymbol{P}'\boldsymbol{A}\boldsymbol{P}\boldsymbol{u} \leq 0 \Longrightarrow \boldsymbol{x}'\boldsymbol{A}\boldsymbol{x} = q_{\boldsymbol{A}}(\boldsymbol{x}) \leq 0\,,$$

where $\boldsymbol{x} = \boldsymbol{P}\boldsymbol{u}$ is nonzero. This contradicts the positive definiteness of $q_{\boldsymbol{A}}(\boldsymbol{x})$, so $q_{\boldsymbol{B}}(\boldsymbol{u})$ must be positive definite. □

Remark: In the above result we used $\boldsymbol{x}$ to denote the argument for $q_{\boldsymbol{A}}$ and $\boldsymbol{u}$ for $q_{\boldsymbol{B}}$ because $\boldsymbol{x}$ and $\boldsymbol{u}$ reside in different spaces—$\boldsymbol{x} \in \Re^n$ and $\boldsymbol{u} \in \Re^k$. Note that when using quadratic forms the dimension of $\boldsymbol{x}$ is automatically determined by the dimension of the associated matrix. So, there should be no confusion if we write that $q_{\boldsymbol{A}}(\boldsymbol{x})$ and $q_{\boldsymbol{B}}(\boldsymbol{x})$ are both positive definite even when $\boldsymbol{A}$ and $\boldsymbol{B}$ are not of the same dimension. It will be implicit that the $\boldsymbol{x}$'s will reside in different spaces depending upon the dimensions of $\boldsymbol{A}$ and $\boldsymbol{B}$.

The following corollary finds wide applicability in algebraic derivations.

Corollary 13.1 *If $\boldsymbol{A}$ is an $n \times n$ matrix such that $q_{\boldsymbol{A}}(\boldsymbol{x})$ is positive definite and $\boldsymbol{P}$ is an $n \times n$ nonsingular matrix, then $q_{\boldsymbol{P}'\boldsymbol{A}\boldsymbol{P}}(\boldsymbol{x})$ is also positive definite.*

Proof. If $\boldsymbol{P}$ is nonsingular, it has full column rank and this result is an immediate consequence of Theorem 13.5 with $k = n$. □

Corollary 13.1 shows that a positive definite quadratic form remains positive definite if a nonsingular transformation is applied to the variables. Therefore, if $\boldsymbol{x}'\boldsymbol{A}\boldsymbol{x}$ is a positive definite quadratic form and we apply a change of variable from $\boldsymbol{x}$ to $\boldsymbol{y}$ using the nonsingular transformation $\boldsymbol{x} = \boldsymbol{P}\boldsymbol{y}$ (or, equivalently, $\boldsymbol{y} = \boldsymbol{P}^{-1}\boldsymbol{x}$), then the quadratic form $\boldsymbol{x}'\boldsymbol{A}\boldsymbol{x}$ changes to $\boldsymbol{y}'\boldsymbol{B}\boldsymbol{y}$, which is still positive definite.

We saw earlier that a positive definite matrix must have positive diagonal elements. Below we show that the implications are, in fact, more general.

Theorem 13.6 *Let $q_{\boldsymbol{A}}(\boldsymbol{x})$ be positive definite. If $\boldsymbol{B}$ is a principal submatrix of $\boldsymbol{A}$, then $q_{\boldsymbol{B}}(\boldsymbol{u})$ is positive definite.*

Proof. Suppose that $\boldsymbol{B}$ is the principal submatrix obtained from $\boldsymbol{A}$ by extracting the row and column numbers given by $1 \leq i_1 < i_2 < \cdots < i_k \leq n$. Form the $n \times k$ matrix $\boldsymbol{P} = [\boldsymbol{e}_{i_1} : \boldsymbol{e}_{i_2} : \ldots : \boldsymbol{e}_{i_k}]$ and note that $\boldsymbol{B} = \boldsymbol{P}'\boldsymbol{A}\boldsymbol{P}$. Also, $\boldsymbol{P}$ has full column rank, which means that $q_{\boldsymbol{B}}(\boldsymbol{u})$ is positive definite (see Theorem 13.5). □

The following result tells us that every matrix associated with a positive definite quadratic form has an $\boldsymbol{LDU}$ decomposition with positive pivots (recall Section 3.4). This means that a positive definite linear system can be solved without resorting to row interchanges or pivoting.

Theorem 13.7 *If $\boldsymbol{A}$ is an $n \times n$ matrix such that $q_{\boldsymbol{A}}(\boldsymbol{x})$ is positive definite, then $\boldsymbol{A} = \boldsymbol{LDU}$, where $\boldsymbol{L}$, $\boldsymbol{D}$ and $\boldsymbol{U}$ are $n \times n$ unit lower-triangular, diagonal and unit upper-triangular matrices, respectively. Furthermore, all the diagonal entries in $\boldsymbol{D}$ are positive.*

Proof. From Theorem 13.6, all the principal submatrices of $\boldsymbol{A}$ are nonsingular. In particular, all its leading principal submatrices are nonsingular. Theorem 3.1 ensures that $\boldsymbol{A}$ has an $\boldsymbol{LU}$ decomposition and, hence, an $\boldsymbol{LDU}$ decomposition (see Section 3.4). Therefore, $\boldsymbol{A} = \boldsymbol{LDU}$, where $\boldsymbol{L}$ and $\boldsymbol{U}$ are unit lower and unit upper-triangular and $\boldsymbol{D}$ is diagonal.

It remains to prove that the diagonal elements of $\boldsymbol{D}$ are positive. Observe that $\boldsymbol{DUL}^{-1'} = \boldsymbol{L}^{-1}\boldsymbol{AL}^{-1'}$ is nonsingular because of Corollary 13.3. Therefore, $\boldsymbol{DUL}^{-1'}$ has positive diagonal entries. Since $\boldsymbol{U}$ and $\boldsymbol{L}^{-1'}$ are both unit upper-triangular, so is their product: $\boldsymbol{UL}^{-1'}$. Therefore, the diagonal entries of $\boldsymbol{DUL}^{-1'}$ are the same as those of $\boldsymbol{D}$. □

Example 13.5 Consider the bivariate quadratic form in $\boldsymbol{x} = [x_1 : x_2]$:

$$q_{\boldsymbol{A}}(\boldsymbol{x}) = x_1^2 + 6x_1x_2 + 10x_2^2 = \boldsymbol{x}'\boldsymbol{A}\boldsymbol{x} \;, \quad \text{where } \boldsymbol{A} = \begin{bmatrix} 1 & 3 \\ 3 & 10 \end{bmatrix} .$$

Completing the square as in Example 13.3, reveals that

$$q_{\boldsymbol{A}}(\boldsymbol{x}) = (x_1 + 3x_2)^2 - 9x_2^2 + 10x_2^2 = (x_1 + 3x_2)^2 + x_2^2 \;,$$

which immediately reveals that $q_{\boldsymbol{A}}(\boldsymbol{x})$ is positive definite. This also reveals the $\boldsymbol{LDU}$ decomposition for the symmetric matrix $\boldsymbol{A}$. Let $\boldsymbol{y} = \boldsymbol{L}'\boldsymbol{x}$, where

$$\boldsymbol{y} = \begin{bmatrix} y_1 \\ y_2 \end{bmatrix} \text{ and } \boldsymbol{L}' = \begin{bmatrix} 1 & 3 \\ 0 & 1 \end{bmatrix} .$$

Then, $q_{\boldsymbol{A}}(\boldsymbol{x}) = \boldsymbol{y}'\boldsymbol{D}\boldsymbol{y} = \boldsymbol{x}'\boldsymbol{LDL}'\boldsymbol{x}$, where $\boldsymbol{D} = \begin{bmatrix} 1 & 0 \\ 0 & 1 \end{bmatrix}$. Therefore,

$$\boldsymbol{A} = \begin{bmatrix} 1 & 3 \\ 3 & 10 \end{bmatrix} = \begin{bmatrix} 1 & 3 \\ 0 & 1 \end{bmatrix} \begin{bmatrix} 1 & 0 \\ 0 & 1 \end{bmatrix} \begin{bmatrix} 1 & 0 \\ 3 & 1 \end{bmatrix} = \boldsymbol{LDU} \;,$$

where $U = L'$, is the desired decomposition. This could also have been computed by writing $A = LDL'$ and solving for the entries in L and D.

Theorem 13.7 ensures that even if A is an *unsymmetric* matrix in a positive definite quadratic form, it will have an LDU decomposition. Consider, for example, the matrix $A = \begin{bmatrix} 1 & 2 \\ 4 & 10 \end{bmatrix}$, which also leads to the same quadratic form $q_A(x)$ as above. Writing

$$A = \begin{bmatrix} 1 & 2 \\ 4 & 10 \end{bmatrix} = \begin{bmatrix} 1 & 0 \\ l & 1 \end{bmatrix} \begin{bmatrix} d_1 & 0 \\ 0 & d_2 \end{bmatrix} \begin{bmatrix} 1 & u \\ 0 & 1 \end{bmatrix} = \begin{bmatrix} d_1 & d_1 u \\ l d_1 & l d_1 u + d_2 \end{bmatrix}$$

and solving for d_1, l, u and d_2 (in this sequence), we obtain

$$d_1 = 1\,, \quad l = 4/d_1 = 4\,, \quad u = 2/d_1 = 2 \text{ and } d_2 = 10 - l d_1 u = 2\,.$$

Therefore,

$$A = \begin{bmatrix} 1 & 2 \\ 4 & 10 \end{bmatrix} = \begin{bmatrix} 1 & 0 \\ 4 & 1 \end{bmatrix} \begin{bmatrix} 1 & 0 \\ 0 & 2 \end{bmatrix} \begin{bmatrix} 1 & 2 \\ 0 & 1 \end{bmatrix} = LDU$$

is the desired LDU factorization for A. ■

Of course quadratic forms need not necessarily have positive range. In fact, based upon their ranges we can define the following classes of quadratic forms. We include positive definite quadratic forms once again for the sake of completion.

Definition 13.3 *Let $q_A(x) = x'Ax$ be an n-ary quadratic form.*

- *$q_A(x)$ is said to be* **nonnegative definite (or n.n.d.)** *if $x'Ax \geq 0$ for all $x \in \mathbb{R}^n$.*
- *$q_A(x)$ is said to be* **positive definite (p.d.)** *if $x'Ax > 0$ for all nonzero $x \in \mathbb{R}^n$.*
- *$q_A(x)$ is said to be* **non-positive definite (n.p.d.)** *if $x'Ax \leq 0$ for all $x \in \mathbb{R}^n$.*
- *$q_A(x)$ is said to be* **negative definite (n.d.)** *if $x'Ax < 0$ for all $x \in \mathbb{R}^n$.*
- *$q_A(x)$ is said to be* **indefinite (n.d.)** *if it is neither n.n.d nor n.p.d.*

Observe that every positive definite quadratic form is nonnegative definite but not the reverse. If $q_A(x) \geq 0$ for all $x \in \Re^n$ but there is some nonzero x for which $q_A(x) = 0$, then $q_A(x)$ is nonnegative definite but not positive definite. Such forms are sometimes referred to as *positive semi-definite* although there really is not much uniformity among authors in how they precisely define these different classes.

Example 13.6 Let $q_A(x)$ be a 2-ary quadratic form with $A = \begin{bmatrix} 1 & 1 \\ 1 & 1 \end{bmatrix}$. Then

$$q_A(x) = x'Ax = x_1^2 + x_2^2 + 2x_1x_2 = (x_1 + x_2)^2 \geq 0 \text{ for all } x \in \Re^2\,.$$

However $q_A(x) = 0$ whenever $x_1 = -x_2$, so $q_A(x)$ is nonnegative definite but not positive definite. ■

If $q_{\boldsymbol{A}}(\boldsymbol{x})$ is negative definite, then $q_{-\boldsymbol{A}}(\boldsymbol{x})$ is positive definite. Similarly, if $q_{\boldsymbol{A}}(\boldsymbol{x})$ is non-positive definite, then $q_{-\boldsymbol{A}}(\boldsymbol{x})$ is nonnegative definite. That is why for most purposes it suffices to restrict attention to only nonnegative definite and positive definite forms. In the next section we study matrices associated with such forms.

13.4 Positive and nonnegative definite matrices

When studying matrices associated with positive definite quadratic forms, it helps to restrict our attention to symmetric matrices. This is not only because of the "uniqueness" offered by symmetric matrices (see Theorem 13.2), but also because symmetry yields several interesting and useful results that are otherwise not true. Therefore, if $q_{\boldsymbol{A}}(\boldsymbol{x})$ is positive definite and $\boldsymbol{A}$ is symmetric, we call $\boldsymbol{A}$ a positive definite matrix. An analogous definition holds for nonnegative definite matrices. Here is a formal definition.

Definition 13.4 Positive definite matrix.

- *A symmetric $n \times n$ matrix $\boldsymbol{A}$ is said to be* **positive definite** *(shortened as p.d.) if $\boldsymbol{x}'\boldsymbol{A}\boldsymbol{x} > 0$ for every nonzero vector $\boldsymbol{x} \in \Re^n$.*
- *A symmetric $n \times n$ matrix $\boldsymbol{A}$ is said to be* **nonnegative definite** *(shortened as n.n.d.) if $\boldsymbol{x}'\boldsymbol{A}\boldsymbol{x} \geq 0$ for every $\boldsymbol{x} \in \Re^n$.*

When we say that a matrix is positive definite or nonnegative definite, it is implicit that the matrix is symmetric.

From now on, unless explicitly mentioned, we assume that $\boldsymbol{A}$ is symmetric whenever we refer to the quadratic form $\boldsymbol{x}'\boldsymbol{A}\boldsymbol{x}$.

Clearly, every positive definite matrix is also nonnegative definite but not the reverse. As we saw in Example 13.6, the matrix $\boldsymbol{A} = \begin{bmatrix} 1 & 1 \\ 1 & 1 \end{bmatrix}$ is nonnegative definite but not positive definite.

In Section 3.4 we defined an $n \times n$ symmetric matrix $\boldsymbol{A}$ to be positive definite if all its pivots were positive. Theorem 3.5 showed that such a matrix would always admit a ***Cholesky Decomposition*** $\boldsymbol{A} = \boldsymbol{T}\boldsymbol{T}'$, where $\boldsymbol{T}$ is $n \times n$ lower-triangular with positive diagonal elements. We will soon see that this is equivalent to Definition 13.4. But first, let us list some properties of positive definite matrices that follow from those seen in Section 13.3 for general (i.e., not necessarily symmetric) matrices associated with positive definite quadratic forms. We list them below.

1. If $\boldsymbol{A}$ and $\boldsymbol{B}$ are positive definite matrices, then so is their sum $\boldsymbol{A} + \boldsymbol{B}$.
2. If $\boldsymbol{A}$ is positive definite, then all its diagonal entries must be positive (Lemma 13.2).
3. Positive definite matrices are always nonsingular (see Theorem 13.3).

4. If A is $n \times n$ positive definite and P is $n \times k$ with full column rank, then the matrix $B = P'AP$ is also positive definite (Theorem 13.5). In particular, if P is $n \times n$ nonsingular, then $P'AP$ is positive definite if and only if A is positive definite (Corollary 13.1).
5. If A is positive definite, then all its principal submatrices are positive definite (Theorem 13.6).
6. If A is positive definite, then A has an LDL' decomposition, where D is a diagonal matrix with positive diagonal entries, and so A has a Cholesky decomposition (see Section 3.4). This follows from Theorem 13.7 when A is symmetric.

The last property in the above list is important enough to merit its own theorem. It proves that a matrix is positive definite if and only if it has a Cholesky decomposition. This proves the equivalence of definition in Section 3.4 using positive pivots and Definition 13.4.

Theorem 13.8 *A matrix A is positive definite if and only if $A = R'R$ for some nonsingular upper-triangular matrix R.*

Proof. If A is positive definite, then $q_A(x) > 0$ for all nonzero $x \in \Re^n$ and Theorem 13.7 ensures that A has an LDU decomposition with positive diagonal entries in D. Because A is symmetric, $U = L'$ and we can write $A = LDL' = R'R$, where $R = D^{1/2}L'$ and $D^{1/2}$ is the diagonal matrix whose diagonal entries are the square roots of the corresponding diagonal entries in D. Since L is unit lower-triangular, R is upper-triangular with positive diagonal elements and, therefore, is nonsingular. Thus, every positive definite matrix has a Cholesky decomposition.

Now suppose that $A = R'R$, where R is nonsingular. Clearly A is symmetric and

$$x'Ax = x'R'Rx = y'y = \sum_{i=1}^{n} y_i^2 \geq 0 ,$$

where $y = Rx$. Because R is nonsingular, $y = Rx \neq 0$ unless $x = 0$. This means that $x'Ax = y'y > 0$ whenever $x \neq 0$. Hence, A is positive definite. □

Example 13.7 Consider the symmetric matrix A in Example 13.5, where we computed $A = LDL'$. It is easy to see that

$$A = \begin{bmatrix} 1 & 3 \\ 3 & 10 \end{bmatrix} = \begin{bmatrix} 1 & 0 \\ 3 & 1 \end{bmatrix} \begin{bmatrix} 1 & 3 \\ 0 & 1 \end{bmatrix} = R'R\,, \text{ where } R = \begin{bmatrix} 1 & 3 \\ 0 & 1 \end{bmatrix}$$

is the Cholesky decomposition of A. ■

The Cholesky decomposition tells us that the determinant of a positive definite matrix must be positive.

Corollary 13.2 *If A is p.d., then $|A| > 0$.*

Proof. Since $\boldsymbol{A}$ is p.d., we have $\boldsymbol{A} = \boldsymbol{R}'\boldsymbol{R}$ for nonsingular upper-triangular $\boldsymbol{R}$, so $|\boldsymbol{A}| = |\boldsymbol{R}'\boldsymbol{R}| = |\boldsymbol{R}|^2 > 0$. □

While the Cholesky decomposition is a nice characterization for positive definite matrices, it does not extend to nonnegative definite matrices quite so easily. This is because a nonnegative definite matrix may be singular and, therefore, may have zero pivots, which precludes the $\boldsymbol{LDL}'$ decomposition. However, if we drop the requirement that the factorization be in terms of triangular matrices, we can get useful characterizations that subsume both nonnegative definite and positive definite matrices. Eigenvalues of nonnegative definite matrices can be helpful.

Since nonnegative definite matrices are symmetric, their eigenvalues are real numbers. In fact, their eigenvalues are nonnegative real numbers and this provides a useful characterization for both nonnegative definite and positive definite matrices.

Theorem 13.9 *The following statements are true for an $n \times n$ symmetric matrix $\boldsymbol{A}$:*

(i) $\boldsymbol{A}$ is n.n.d if and only if all its eigenvalues are nonnegative.

(ii) $\boldsymbol{A}$ is p.d. if and only if all its eigenvalues are positive.

Proof. **Proof of (i):** Since $\boldsymbol{A}$ is an $n \times n$ real symmetric matrix, it has a spectral decomposition. Therefore, there exists some orthogonal matrix $\boldsymbol{P}$ such that $\boldsymbol{A} = \boldsymbol{P\Lambda P}'$, where $\boldsymbol{\Lambda} = \text{diag}(\lambda_1, \lambda_2, \ldots, \lambda_n)$ and λ_i's are (real) eigenvalue of $\boldsymbol{A}$.

If each $\lambda_i \geq 0$, we can write

$$\boldsymbol{x}'\boldsymbol{A}\boldsymbol{x} = \boldsymbol{x}'\boldsymbol{P}'\boldsymbol{\Lambda}\boldsymbol{P}\boldsymbol{x} = \boldsymbol{y}'\boldsymbol{\Lambda}\boldsymbol{y} = \sum_{i=1}^{n} \lambda_i^2 y_i^2 \geq 0 ,$$

where $\boldsymbol{y} = \boldsymbol{P}\boldsymbol{x}$. This proves that $\boldsymbol{A}$ is nonnegative definite.

Now suppose $\boldsymbol{A}$ is nonnegative definite, so $\boldsymbol{x}'\boldsymbol{A}\boldsymbol{x} \geq 0$ for every $\boldsymbol{x} \in \Re^n$. Let λ be an eigenvalue of $\boldsymbol{A}$ and $\boldsymbol{u}$ the corresponding eigenvector. Then, $\boldsymbol{A}\boldsymbol{u} = \lambda\boldsymbol{u}$ so

$$\lambda = \frac{\boldsymbol{u}'\boldsymbol{A}\boldsymbol{u}}{\boldsymbol{u}'\boldsymbol{u}} \geq 0 ,$$

which completes the proof for (i).

Proof of (ii): The proof for positive definite matrices follows closely that of (i). If $\boldsymbol{A}$ has strictly positive eigenvalues, then the spectral decomposition $\boldsymbol{A} = \boldsymbol{P\Lambda P}'$ yields

$$\boldsymbol{x}'\boldsymbol{A}\boldsymbol{x} = \boldsymbol{x}'\boldsymbol{P}'\boldsymbol{\Lambda}\boldsymbol{P}\boldsymbol{x} = \boldsymbol{y}'\boldsymbol{\Lambda}\boldsymbol{y} = \sum_{i=1}^{n} \lambda_i^2 y_i^2 > 0 ,$$

where $\boldsymbol{y} = \boldsymbol{P}\boldsymbol{x}$. This proves that $\boldsymbol{A}$ is positive definite.

Conversely, suppose $\boldsymbol{A}$ is positive definite and λ is an eigenvalue of $\boldsymbol{A}$ with $\boldsymbol{u}$ as the corresponding eigenvector. If $\lambda \leq 0$ then

$$\boldsymbol{u}'\boldsymbol{A}\boldsymbol{u} = \boldsymbol{u}'(\boldsymbol{A}\boldsymbol{u}) = \boldsymbol{u}'(\lambda\boldsymbol{u}) = \lambda\boldsymbol{u}'\boldsymbol{u} = \lambda\|\boldsymbol{u}\|^2 \leq 0 ,$$

which contradicts the positive definiteness of $\boldsymbol{A}$. Therefore, every eigenvalue of $\boldsymbol{A}$ must be strictly positive. □

The above results help us derive several useful properties for p.d. matrices.

Corollary 13.3 *If $\boldsymbol{A}$ is positive definite, then $|\boldsymbol{A}| > 0$.*

Proof. Since the determinant of a matrix is the product of its eigenvalues, Theorem 13.9 implies that $|\boldsymbol{A}| > 0$ whenever $\boldsymbol{A}$ is positive definite. □

Here is another useful result.

Corollary 13.4 *If $\boldsymbol{A}$ is positive definite, then so is $\boldsymbol{A}^{-1}$.*

Proof. Recall that if $\boldsymbol{A}$ is positive definite, then it is nonsingular. So $\boldsymbol{A}^{-1}$ exists. The eigenvalues of $\boldsymbol{A}^{-1}$ are simply the reciprocal of the eigenvalues of $\boldsymbol{A}$, so the former are all positive whenever the latter are. Part (ii) of Theorem 13.9 tells us that $\boldsymbol{A}^{-1}$ is positive definite as well. □

It is worth remarking that Theorem 13.9 is true only for symmetric matrices. It is possible to find an unsymmetric matrix with positive eigenvalues that does not render a positive definite quadratic form. Here is an example.

Example 13.8 Consider the matrix $\boldsymbol{A} = \begin{bmatrix} 1 & -4 \\ 0 & 2 \end{bmatrix}$. Its eigenvalues are 1 and 2; both are positive. The corresponding quadratic form

$$q_{\boldsymbol{A}}(\boldsymbol{x}) = [x_1 : x_2] \begin{bmatrix} 1 & -4 \\ 0 & 2 \end{bmatrix} \begin{bmatrix} x_1 \\ x_2 \end{bmatrix} = x_1^2 - 4x_1x_2 + 2x_2^2$$

is not positive definite because $x_1 = x_2 = 1$ produces $q_{\boldsymbol{A}}(\boldsymbol{x}) = -1$. ■

The Cholesky decomposition does not exist for singular matrices. However, nonnegative definite matrices will always admit a factorization "similar" to the Cholesky if we drop the requirement that the factors be triangular or nonsingular.

Theorem 13.10 *The following statements are true for an $n \times n$ symmetric matrix $\boldsymbol{A}$:*

(i) $\boldsymbol{A}$ is n.n.d. if and only if there exists a matrix $\boldsymbol{B}$ such that $\boldsymbol{A} = \boldsymbol{B}'\boldsymbol{B}$.

(ii) $\boldsymbol{A}$ is p.d. if and only if $\boldsymbol{A} = \boldsymbol{B}'\boldsymbol{B}$ for some nonsingular (hence square) matrix $\boldsymbol{B}$.

Proof. **Proof of (i):** If $\boldsymbol{A} = \boldsymbol{B}'\boldsymbol{B}$ for any matrix $\boldsymbol{B}$, then $\boldsymbol{A}$ is nonnegative definite because

$$\boldsymbol{x}'\boldsymbol{A}\boldsymbol{x} = \boldsymbol{x}'\boldsymbol{B}'\boldsymbol{B}\boldsymbol{x} = \boldsymbol{y}'\boldsymbol{y} = \sum_{i=1}^{n} y_i^2 \geq 0\,, \quad \text{where } \boldsymbol{y} = \boldsymbol{B}\boldsymbol{x}\,.$$

This proves the "if" part.

Now suppose that A is nonnegative definite. By Theorem 13.9, all the eigenvalues of A are nonnegative. So, if $A = P'\Lambda P$ is the spectral decomposition of A, then we can define $\Lambda^{1/2} = \text{diag}(\sqrt{\lambda_1}, \sqrt{\lambda_2}, \ldots, \sqrt{\lambda_n})$ and write

$$A = P'\Lambda P = P'\Lambda^{1/2}\Lambda^{1/2}P = B'B\,, \text{ where } B = \Lambda^{1/2}P\,.$$

Note that B has the same order as A and is well-defined as long as the λ_i's are nonnegative (even if some of them are zero).

Proof of (ii): If $A = B'B$, then A is nonnegative definite (from (i)). If B is nonsingular, then B must be $n \times n$ and $\rho(B) = n$. Therefore, $\rho(A) = \rho(B'B) = \rho(B) = n$, so A is itself nonsingular. Because A is nonsingular and symmetric, all its eigenvalues must be nonzero. In addition, because A is nonnegative definite, none of its eigenvalues can be negative, so they are all positive. Therefore, A is positive definite. This proves the "if" part.

If A is positive definite, then the Cholesky decomposition (Theorem 13.8) immediately yields a choice for a nonsingular B. Alternatively, the spectral decomposition constructed in (i) for a positive definite matrix also yields a nonsingular B. $\Lambda^{1/2}$ is diagonal with positive diagonal entries and P' is nonsingular because it is orthogonal. Therefore, $B = \Lambda^{1/2}P'$ is the product of two nonsingular matrices, hence it is nonsingular. □

Theorem 13.10 implies that if A is n.n.d. but not p.d., then A must be singular. In fact, if A is $n \times n$ with $\rho(A) = r < n$, then the spectral decomposition can be used to find an $n \times n$ matrix B with $\rho(B) = r$ such that $A = B'B$. If $r = n$, then $\rho(B) = n$, which means that B is nonsingular and so is A. This is an alternative argument why a p.d. matrix must be nonsingular (recall Theorem 13.3). Also, if $A = B'B$ is p.d. then $A^{-1} = (B^{-1})(B^{-1})'$ which implies that A^{-1} is again positive definite.

Corollary 13.5 *Let A and B be two $n \times n$ matrices, where A is symmetric and B is n.n.d. Then all the eigenvalues of AB and BA are real numbers.*

Proof. Since B is positive definite, Theorem 13.10 ensures that there is a matrix R such that $B = R'R$. The nonzero eigenvalues of $AB = AR'R$ are the same as those of RAR' (by applying Theorem 11.3). The matrix $R'AR$ is symmetric because A is symmetric and, hence, all its eigenvalues are real (recall Theorem 11.25). Therefore, all the eigenvalues of AB are real. Since the nonzero eigenvalues of AB are the same as those of BA (Theorem 11.3), all the eigenvalues of BA are also real. □

Theorem 13.11 Square root of an n.n.d. matrix. *If A is an $n \times n$ p.d. (n.n.d.) matrix, then there exists an p.d. (n.n.d.) matrix B such that $A = B^2$.*

Proof. Let $A = P\Lambda P'$ be the spectral decomposition of A. All the eigenvalues of A are positive (nonnegative). Let $\Lambda^{1/2}$ be the diagonal matrix with

$\sqrt{\lambda_1}, \sqrt{\lambda_2}, \ldots, \sqrt{\lambda_n}$ along its diagonal, where $\sqrt{\lambda_i}$ is the positive (nonnegative) square root of λ_i. Let $\boldsymbol{B} = \boldsymbol{P}\boldsymbol{\Lambda}^{1/2}\boldsymbol{P}'$. Then,

$$\boldsymbol{B}^2 = (\boldsymbol{P}\boldsymbol{\Lambda}^{1/2}\boldsymbol{P}')(\boldsymbol{P}\boldsymbol{\Lambda}^{1/2}\boldsymbol{P}') = \boldsymbol{P}\boldsymbol{\Lambda}^{1/2}(\boldsymbol{P}'\boldsymbol{P})\boldsymbol{\Lambda}^{1/2}\boldsymbol{P}' = \boldsymbol{P}\boldsymbol{\Lambda}\boldsymbol{P}' = \boldsymbol{A}$$

because $\boldsymbol{P}'\boldsymbol{P} = \boldsymbol{I}$. Also, $\boldsymbol{B}$ is clearly p.d. (n.n.d.) when $\boldsymbol{A}$ is p.d. (n.n.d). □

The matrix $\boldsymbol{B}$ in Theorem 13.11 is sometimes referred to as the ***symmetric square root*** of a positive definite matrix. It can also be shown that while there exists several $\boldsymbol{B}$'s satisfying $\boldsymbol{A} = \boldsymbol{B}^2$, there is a ***unique*** p.d. (n.n.d) square root of a p.d. (n.n.d.) matrix $\boldsymbol{A}$. However, the nomenclature here is not universal and sometimes any matrix $\boldsymbol{B}$ (symmetric or not) satisfying $\boldsymbol{A} = \boldsymbol{B}'\boldsymbol{B}$ as in Theorem 13.10 is referred to as a square root of $\boldsymbol{A}$. When $\boldsymbol{B}$ is not symmetric, we call it the "unsymmetric" square root. For instance, when $\boldsymbol{B}'$ is the lower-triangular factor from the Cholesky decomposition, it may be referred to as the Cholesky square root. We will reserve the notation $\boldsymbol{A}^{1/2}$ to denote the unique p.d. or n.n.d. square root as in Theorem 13.11.

Corollary 13.6 *If $\boldsymbol{A}^{1/2}$ is the p.d. square root of a p.d. matrix $\boldsymbol{A}$, then $(\boldsymbol{A}^{-1})^{1/2} = (\boldsymbol{A}^{1/2})^{-1}$.*

Proof. Let $\boldsymbol{B} = \boldsymbol{A}^{1/2}$. Then,

$$\boldsymbol{A} = \boldsymbol{B}^2 = \boldsymbol{B}\boldsymbol{B} \Longrightarrow \boldsymbol{A}^{-1} = (\boldsymbol{B}\boldsymbol{B})^{-1} = \boldsymbol{B}^{-1}\boldsymbol{B}^{-1} = (\boldsymbol{B}^{-1})^2 ,$$

which means that $\boldsymbol{B}^{-1} = (\boldsymbol{A}^{1/2})^{-1}$ is a square root of $\boldsymbol{A}^{-1}$. □

If $\boldsymbol{A} = \boldsymbol{L}\boldsymbol{L}'$ is the Cholesky decomposition for a p.d. $\boldsymbol{A}$, then $\boldsymbol{A}^{-1} = (\boldsymbol{L}^{-1})'\boldsymbol{L}^{-1}$. This is still a factorization of $\boldsymbol{A}^{-1}$ as ensured by Theorem 13.10, where the first factor, $(\boldsymbol{L}^{-1})'$, is now upper-triangular and the second factor, $\boldsymbol{L}^{-1}$, is lower-triangular. Strictly speaking, this is not a Cholesky decomposition for $\boldsymbol{A}^{-1}$, where the first factor is lower-triangular and the second factor is upper-triangular. However, if we have obtained the Cholesky decomposition for $\boldsymbol{A}$, we can cheaply compute $\boldsymbol{A}^{-1} = (\boldsymbol{L}^{-1})'\boldsymbol{L}^{-1}$, which suffices for most computational purposes, and one does not need to apply Crout's algorithm on $\boldsymbol{A}^{-1}$ to obtain its Cholesky decomposition.

Another important result that follows from Theorem 13.10 is that orthogonal projectors are nonnegative definite matrices.

Corollary 13.7 *Orthogonal projectors are nonnegative definite.*

Proof. If $\boldsymbol{P}$ is an orthogonal projector, then $\boldsymbol{P}$ is symmetric and idempotent. Therefore $\boldsymbol{P} = \boldsymbol{P}^2 = \boldsymbol{P}'\boldsymbol{P}$, which implies that $\boldsymbol{P}$ is nonnegative definite. □

Remark: Projectors are singular (except for $\boldsymbol{I}$) so they are not positive definite.

Relationships concerning the definiteness of a matrix and its submatrices are useful. Suppose we apply Theorem 13.6 to a block-diagonal matrix $\boldsymbol{A}$. Since, each diagonal block is a principal submatrix of $\boldsymbol{A}$, it follows that whenever $\boldsymbol{A}$ is positive definite, so is each diagonal block. The following theorem relates positive definite matrices to their Schur's complements.

Theorem 13.12 *Let* $\boldsymbol{A} = \begin{bmatrix} \boldsymbol{A}_{11} & \boldsymbol{A}_{12} \\ \boldsymbol{A}'_{12} & \boldsymbol{A}_{22} \end{bmatrix}$ *be a symmetric matrix partitioned such that* $\boldsymbol{A}_{11}$ *and* $\boldsymbol{A}_{22}$ *are symmetric square matrices.*

(i) $\boldsymbol{A}$ *is p.d. if and only if* $\boldsymbol{A}_{11}$ *and* $\boldsymbol{A}_{22} - \boldsymbol{A}'_{12}\boldsymbol{A}_{11}^{-1}\boldsymbol{A}_{12}$ *are both p.d.*

(ii) $\boldsymbol{A}$ *is p.d. if and only if* $\boldsymbol{A}_{22}$ *and* $\boldsymbol{A}_{11} - \boldsymbol{A}_{12}\boldsymbol{A}_{22}^{-1}\boldsymbol{A}'_{12}$ *are both p.d.*

Proof. **Proof of (i):** We first prove the "only if" part. Suppose $\boldsymbol{A}$ is positive definite. Since $\boldsymbol{A}_{11}$ is a principal submatrix, $\boldsymbol{A}_{11}$ is also positive definite and, therefore, nonsingular. Now, reduce $\boldsymbol{A}$ to a block diagonal form using a nonsingular transformation $\boldsymbol{P}$ as below:

$$\boldsymbol{P}'\boldsymbol{A}\boldsymbol{P} = \begin{bmatrix} \boldsymbol{I} & \boldsymbol{O} \\ -\boldsymbol{A}'_{12}\boldsymbol{A}_{11}^{-1} & \boldsymbol{I} \end{bmatrix} \begin{bmatrix} \boldsymbol{A}_{11} & \boldsymbol{A}_{12} \\ \boldsymbol{A}'_{12} & \boldsymbol{A}_{22} \end{bmatrix} \begin{bmatrix} \boldsymbol{I} & -\boldsymbol{A}_{11}^{-1}\boldsymbol{A}_{12} \\ \boldsymbol{O} & \boldsymbol{I} \end{bmatrix}$$
$$= \begin{bmatrix} \boldsymbol{A}_{11} & \boldsymbol{O} \\ \boldsymbol{O} & \boldsymbol{A}_{22} - \boldsymbol{A}'_{12}\boldsymbol{A}_{11}^{-1}\boldsymbol{A}_{12} \end{bmatrix},$$

where $\boldsymbol{P} = \begin{bmatrix} \boldsymbol{I} & -\boldsymbol{A}_{11}^{-1}\boldsymbol{A}_{12} \\ \boldsymbol{O} & \boldsymbol{I} \end{bmatrix}$ is nonsingular. Therefore, $\boldsymbol{P}'\boldsymbol{A}\boldsymbol{P}$ is positive definite as well. Since each of its block diagonal submatrices are also principal submatrices of $\boldsymbol{A}$, they are positive definite. This proves the positive definiteness of $\boldsymbol{A}_{11}$ and $\boldsymbol{A}_{22} - \boldsymbol{A}'_{12}\boldsymbol{A}_{11}^{-1}\boldsymbol{A}_{12}$.

Now we prove the "if" part. Assume that $\boldsymbol{A}_{11}$ and $\boldsymbol{A}_{22} - \boldsymbol{A}'_{12}\boldsymbol{A}_{11}^{-1}\boldsymbol{A}_{12}$ are both positive definite and let $\boldsymbol{D}$ be the block-diagonal matrix with these two matrices as its diagonal blocks. Clearly $\boldsymbol{D}$ is positive definite. Note that $\boldsymbol{A} = (\boldsymbol{P}')^{-1}\boldsymbol{D}\boldsymbol{P}^{-1}$. Since $\boldsymbol{P}$ is nonsingular, we conclude that $\boldsymbol{A}$ is positive definite.

The proof of (ii) is similar and left to the reader. □

The above result ensures that if a matrix is positive definite, so are its Schur's complements. This is widely used in the theory of linear statistical models.

Nonnegative definiteness also has some implications for the column spaces of a submatrix. The following is an important example.

Theorem 13.13 *Let* $\boldsymbol{A}$ *be a nonnegative definite matrix partitioned as*

$$\boldsymbol{A} = \begin{bmatrix} \boldsymbol{A}_{11} & \boldsymbol{A}_{12} \\ \boldsymbol{A}'_{12} & \boldsymbol{A}_{22} \end{bmatrix}, \tag{13.4}$$

where $\boldsymbol{A}_{11}$ *and* $\boldsymbol{A}_{22}$ *are square. Then* $\mathcal{C}(\boldsymbol{A}_{12}) \subseteq \mathcal{C}(\boldsymbol{A}_{11})$.

Proof. Theorem 13.10 ensures that there is a matrix $\boldsymbol{B}$ such that $\boldsymbol{A} = \boldsymbol{B}'\boldsymbol{B}$. Let $\boldsymbol{B} = [\boldsymbol{B}_1 : \boldsymbol{B}_2]$ where $\boldsymbol{B}_1$ has the same number of columns as $\boldsymbol{A}_{11}$. Then,

$$\boldsymbol{A} = \begin{bmatrix} \boldsymbol{A}_{11} & \boldsymbol{A}_{12} \\ \boldsymbol{A}'_{12} & \boldsymbol{A}_{22} \end{bmatrix} = \boldsymbol{B}'\boldsymbol{B} = \begin{bmatrix} \boldsymbol{B}'_1 \\ \boldsymbol{B}'_2 \end{bmatrix} [\boldsymbol{B}_1 : \boldsymbol{B}_2] = \begin{bmatrix} \boldsymbol{B}'_1\boldsymbol{B}_1 & \boldsymbol{B}'_1\boldsymbol{B}_2 \\ \boldsymbol{B}'_2\boldsymbol{B}_1 & \boldsymbol{B}'_2\boldsymbol{B}_2 \end{bmatrix},$$

which shows that $\boldsymbol{A}_{11} = \boldsymbol{B}_1'\boldsymbol{B}_1$ and $\boldsymbol{A}_{12} = \boldsymbol{B}_1'\boldsymbol{B}_2$. Therefore,

$$\mathcal{C}(\boldsymbol{A}_{12}) \subseteq \mathcal{C}(\boldsymbol{B}_1') = \mathcal{C}(\boldsymbol{B}_1'\boldsymbol{B}_1) = \mathcal{C}(\boldsymbol{A}_{11}) ,$$

which completes the proof. □

Let $\boldsymbol{P}$ be a permutation matrix. Because $\boldsymbol{P}$ is non-singular, $\boldsymbol{P}'\boldsymbol{A}\boldsymbol{P}$ and $\boldsymbol{A}$ have the same definiteness category (n.n.d. or p.d.). Using the fact that any principal submatrix of $\boldsymbol{A}$ can be brought to the top left corner by applying the same permutation to the rows and columns, several properties of leading principal submatrices of n.n.d. and p.d. matrices can be extended to principal submatrices. The next result follows thus from the preceding theorem.

Corollary 13.8 *Let* $\boldsymbol{A} = \{a_{ij}\}$ *be an* $n \times n$ *n.n.d. matrix. Let* $\boldsymbol{x} = (a_{i_1j}, a_{i_2j}, \ldots, a_{i_kj})'$, *where* $1 \le i_1 < i_2 < \ldots < i_k \le n$ *and* $1 \le j \le n$, *be the* $k \times 1$ *vector formed by extracting a subset of* k *elements from the* j*-th column of* $\boldsymbol{A}$. *Then,* $\boldsymbol{x} \in \mathcal{C}(\boldsymbol{A}_{II})$, *where* $I = \{i_1, i_2, \ldots, i_k\}$.

Proof. Using permutations of rows and the corresponding columns, we can assume that $\boldsymbol{A}_{11} = \boldsymbol{A}_{II}$ in the n.n.d. matrix in Theorem 13.13. If $j \in I$, then $\boldsymbol{x}$ is a column of $\boldsymbol{A}_{II}$; so it clearly belongs to $\mathcal{C}(\boldsymbol{A}_{II})$. If $j \in I$, then $\boldsymbol{x} \in \mathcal{C}(\boldsymbol{A}_{12}) \subseteq \mathcal{C}(\boldsymbol{A}_{11}) = \mathcal{C}(\boldsymbol{A}_{II})$, as in Theorem 13.13. □

Recall that if $\boldsymbol{A}$ is p.d., then each of its diagonal elements must be positive. If, however, $\boldsymbol{A}$ is n.n.d, then one or more of its diagonal elements may be zero. In that case, we have the following result.

Corollary 13.9 *If* $\boldsymbol{A}$ *is n.n.d. and* $a_{ii} = 0$, *then* $a_{ij} = a_{ji} = 0$ *for all* $1 \le j \le n$.

Proof. Suppose $\boldsymbol{A}$ is n.n.d. and $a_{ii} = 0$. We can find a permutation matrix $\boldsymbol{P}$ such that the first row and first column of $\boldsymbol{B} = \boldsymbol{P}'\boldsymbol{A}\boldsymbol{P}$ are the i-th row and i-th column of $\boldsymbol{A}$. Note that this implies $b_{11} = a_{ii} = 0$. Applying Corollary 13.8 to $\boldsymbol{B}$ yields that $a_{ij} = a_{ji} = 0$ for all j. □

Theorem 13.14 *If* $\boldsymbol{A}$ *and* $\boldsymbol{C}$ *are n.n.d. matrices of the same order, then* $\boldsymbol{A} + \boldsymbol{C}$ *is n.n.d. and* $\mathcal{C}(\boldsymbol{A}+\boldsymbol{C}) = \mathcal{C}(\boldsymbol{A}) + \mathcal{C}(\boldsymbol{C})$. *If at least one of* $\boldsymbol{A}$ *and* $\boldsymbol{C}$ *is positive definite, then* $\boldsymbol{A} + \boldsymbol{C}$ *is positive definite.*

Proof. If $\boldsymbol{A}$ and $\boldsymbol{C}$ are both $n \times n$ and n.n.d, then

$$\boldsymbol{x}'(\boldsymbol{A}+\boldsymbol{C})\boldsymbol{x} = \boldsymbol{x}'\boldsymbol{A}\boldsymbol{x} + \boldsymbol{x}'\boldsymbol{C}\boldsymbol{x} \ge 0 \text{ for every } \boldsymbol{x} \in \Re^n .$$

Therefore, $\boldsymbol{A} + \boldsymbol{C}$ is n.n.d. and, clearly, if either $\boldsymbol{A}$ or $\boldsymbol{C}$ is p.d., then $\boldsymbol{A} + \boldsymbol{C}$ is too. Writing $\boldsymbol{A} = \boldsymbol{B}'\boldsymbol{B}$ and $\boldsymbol{C} = \boldsymbol{D}'\boldsymbol{D}$ for some $\boldsymbol{B}$ and $\boldsymbol{D}$, we obtain

$$\begin{aligned} \mathcal{C}(\boldsymbol{A}+\boldsymbol{C}) &= \mathcal{C}\left([\boldsymbol{B}' : \boldsymbol{D}'] \begin{bmatrix} \boldsymbol{B} \\ \boldsymbol{D} \end{bmatrix}\right) = \mathcal{C}(\boldsymbol{B}' : \boldsymbol{D}') = \mathcal{C}(\boldsymbol{B}') + \mathcal{C}(\boldsymbol{D}') \\ &= \mathcal{C}(\boldsymbol{A}) + \mathcal{C}(\boldsymbol{C}) , \end{aligned}$$

which completes the proof. □

13.5 Congruence and Sylvester's Law of Inertia

We say that a quadratic form $q_{\boldsymbol{D}}(\boldsymbol{x}) = \boldsymbol{x}'\boldsymbol{D}\boldsymbol{x}$ is in ***diagonal form*** whenever $\boldsymbol{D}$ is a diagonal matrix. This means that $\boldsymbol{x}'\boldsymbol{D}\boldsymbol{x} = \sum_{i=1}^{n} d_{ii}x_i^2$ and there are no cross-product terms. Quadratic forms in diagonal form are more transparent and are the simplest representations of a quadratic form. And they are easy to classify. For example, it is nonnegative definite if all the d_{ii}'s are nonnegative, positive definite if all they are strictly positive, non-positive definite if all the d_{ii}'s are less than or equal to zero, and negative definite if they are all strictly negative. Classification of all quadratic forms up to equivalence can thus be reduced to the case of diagonal forms.

The following theorem is, therefore, particularly appealing.

Theorem 13.15 *Any quadratic form $q_{\boldsymbol{A}}(\boldsymbol{x}) = \boldsymbol{x}'\boldsymbol{A}\boldsymbol{x}$, where $\boldsymbol{A}$ is symmetric, can be reduced to a diagonal form using an orthogonal transformation.*

Proof. Because $\boldsymbol{A}$ is symmetric, there is an orthogonal matrix $\boldsymbol{P}$ such that $\boldsymbol{PAP}' = \boldsymbol{\Lambda}$, where $\boldsymbol{\Lambda} = \text{diag}\{\lambda_1, \lambda_2, \ldots, \lambda_n\}$. Letting $\boldsymbol{x} = \boldsymbol{P}'\boldsymbol{y}$ (or, equivalently, $\boldsymbol{y} = \boldsymbol{Px}$ because $\boldsymbol{P}^{-1} = \boldsymbol{P}'$) we obtain

$$q_{\boldsymbol{A}}(\boldsymbol{x}) = \boldsymbol{x}'\boldsymbol{A}\boldsymbol{x} = \boldsymbol{y}'\boldsymbol{PAP}'\boldsymbol{y} = \boldsymbol{y}'\boldsymbol{\Lambda}\boldsymbol{y} = q_{\boldsymbol{\Lambda}}(\boldsymbol{y}) = \sum_{i=1}^{n} \lambda_i y_i^2 \,.$$

This completes the proof. □

The above theorem demonstrates how the nature of the quadratic form is determined by the eigenvalues of the symmetric matrix $\boldsymbol{A}$ associated with it. This is especially convenient because symmetry ensures that all the eigenvalues are real numbers and offers another reason why it is convenient to consider only symmetric matrices in quadratic forms.

Diagonalizing a quadratic form using the orthogonal transformation provided by the spectral decomposition has a nice geometric interpretation and generalizes the concept of bringing an ellipse (in two-dimensional analytical geometry) into "standard form." The orthogonal transformation rotates the standard coordinate system so that the graph of $\boldsymbol{x}'\boldsymbol{A}\boldsymbol{x} = c$, where c is a constant, is aligned with the new coordinate axes. If $\boldsymbol{A}$ is positive definite, then all of its eigenvalues are positive, which implies that $q_{\boldsymbol{A}}(\boldsymbol{x}) = \boldsymbol{x}'\boldsymbol{A}\boldsymbol{x} = c$ for any positive constant c is an ellipsoid centered at the origin. The coordinate system obtained by transforming $\boldsymbol{y} = \boldsymbol{Px}$ in Theorem 13.15 ensures that the principal axes of the ellipsoid are along the axes of the new coordinate system.

Note that diagonalization of quadratic may as well be achieved by nonsingular transformations that are not necessarily orthogonal. Recall Corollary 13.1 and the discussion following it. While we only considered a positive definite quadratic form there, a similar argument makes it clear that the quadratic forms $q_{\boldsymbol{A}}(\boldsymbol{x})$ and $q_{\boldsymbol{B}}(\boldsymbol{x})$ enjoy the same definiteness whenever $\boldsymbol{B} = \boldsymbol{P}'\boldsymbol{AP}$ for some nonsingular matrix $\boldsymbol{P}$. This matter can be elucidated further using *congruent matrices*.

Definition 13.5 Congruent matrices. *An $n \times n$ matrix $\boldsymbol{A}$ is said to be* **congruent** *to an $n \times n$ matrix $\boldsymbol{B}$ if there exists a nonsingular matrix $\boldsymbol{P}$ such that $\boldsymbol{B} = \boldsymbol{P}'\boldsymbol{A}\boldsymbol{P}$. We denote congruence by $\boldsymbol{A} \cong \boldsymbol{B}$.*

The following result shows that congruence defines an "equivalence" relation in the following sense.

Lemma 13.3 *Let $\boldsymbol{A}$, $\boldsymbol{B}$ and $\boldsymbol{C}$ be $n \times n$ matrices. Then:*

(i) Congruence is reflexive: $\boldsymbol{A} \cong \boldsymbol{A}$.
(ii) Congruence is symmetric: if $\boldsymbol{A} \cong \boldsymbol{B}$, then $\boldsymbol{B} \cong \boldsymbol{A}$.
(iii) Congruence is transitive: if $\boldsymbol{A} \cong \boldsymbol{B}$ and $\boldsymbol{B} \cong \boldsymbol{C}$, then $\boldsymbol{A} \cong \boldsymbol{C}$.

Proof. **Proof of (i):** This is because $\boldsymbol{A} = \boldsymbol{I}\boldsymbol{A}\boldsymbol{I}$.

Proof of (ii): Suppose $\boldsymbol{A} \cong \boldsymbol{B}$. Then, $\boldsymbol{B} = \boldsymbol{P}'\boldsymbol{A}\boldsymbol{P}$ for some nonsingular matrix $\boldsymbol{P}$. But this means that $\boldsymbol{A} = (\boldsymbol{P}^{-1})'\boldsymbol{B}\boldsymbol{P}^{-1}$, so clearly $\boldsymbol{B} \cong \boldsymbol{A}$.

Proof of (iii): Since $\boldsymbol{A} \cong \boldsymbol{B}$ and $\boldsymbol{B} \cong \boldsymbol{C}$, there exist $n \times n$ nonsingular matrices $\boldsymbol{P}$ and $\boldsymbol{Q}$ such that $\boldsymbol{B} = \boldsymbol{P}'\boldsymbol{A}\boldsymbol{P}$ and $\boldsymbol{C} = \boldsymbol{Q}'\boldsymbol{B}\boldsymbol{Q}$. This means that

$$\boldsymbol{C} = \boldsymbol{Q}'\boldsymbol{P}'\boldsymbol{A}\boldsymbol{P}\boldsymbol{Q} = (\boldsymbol{P}\boldsymbol{Q})'\boldsymbol{A}(\boldsymbol{P}\boldsymbol{Q}) \ , \quad \text{where } \boldsymbol{P}\boldsymbol{Q} \text{ is nonsingular.}$$

Therefore, $\boldsymbol{A} \cong \boldsymbol{C}$. □

Since nonsingular matrices can be obtained as a product of elementary matrices, it follows that $\boldsymbol{A}$ and $\boldsymbol{B}$ are congruent to each other if and only if one can be obtained from the other using a sequence of elementary row operations and the corresponding column operations. One example of congruence is produced by the $\boldsymbol{L}\boldsymbol{D}\boldsymbol{L}'$ transformation for a nonsingular symmetric matrix $\boldsymbol{A}$, where $\boldsymbol{L}$ is nonsingular unit lower-triangular and $\boldsymbol{D}$ is diagonal with the pivots of $\boldsymbol{A}$ as its diagonal entries (see Section 3.4). The transformation $\boldsymbol{y} = \boldsymbol{L}'\boldsymbol{x}$ reduces the quadratic form $\boldsymbol{x}'\boldsymbol{A}\boldsymbol{x}$ to the diagonal form $\boldsymbol{y}'\boldsymbol{D}\boldsymbol{y}$. If $\boldsymbol{A}$ is singular, however, then it will have zero pivots and arriving at an $\boldsymbol{L}\boldsymbol{D}\boldsymbol{L}'$ decomposition may become awkward. The spectral decomposition, though, still works equally well here.

The simplest form to which a real-symmetric matrix is congruent to is

$$\boldsymbol{E} = \begin{bmatrix} \boldsymbol{I}_P & \boldsymbol{O} & \boldsymbol{O} \\ \boldsymbol{O} & -\boldsymbol{I}_N & \boldsymbol{O} \\ \boldsymbol{O} & \boldsymbol{O} & \boldsymbol{O} \end{bmatrix} . \tag{13.5}$$

To see why this can be done, suppose $\boldsymbol{A}$ is an $n \times n$ real-symmetric matrix which has P positive eigenvalues (counting multiplicities), N negative eigenvalues (counting multiplicities) and $n - (P + N)$ zero eigenvalues. We can find an orthogonal matrix $\boldsymbol{P}$ such that $\boldsymbol{P}'\boldsymbol{A}\boldsymbol{P} = \boldsymbol{\Lambda}$, where $\boldsymbol{\Lambda}$ is diagonal with the first P entries along the diagonal being the positive eigenvalues, the next N entries along the diagonal being the negative eigenvalues and the remaining entries are all zero. Thus, we can write

$$\Lambda = \begin{bmatrix} D_1 & O & O \\ O & -D_2 & O \\ O & O & O \end{bmatrix} ,$$

where D_1 and D_2 are diagonal matrices of order $P \times P$ and $N \times N$, respectively. Notice that both D_1 and D_2 have positive diagonal elements—the entries in D_1 are the positive eigenvalues of A and those in D_2 are the absolute values of the negative eigenvalues of A. This means that we can legitimately define $D_1^{1/2}$ and $D_2^{1/2}$ as matrices whose entries along the diagonal are the square roots of the corresponding entries in D_1 and D_2, respectively. Therefore,

$$\Lambda = \begin{bmatrix} D_1^{1/2} & O & O \\ O & D_2^{1/2} & O \\ O & O & I \end{bmatrix} \begin{bmatrix} I_P & O & O \\ O & -I_N & O \\ O & O & O \end{bmatrix} \begin{bmatrix} D_1^{1/2} & O & O \\ O & D_2^{1/2} & O \\ O & O & I \end{bmatrix} .$$

And because their diagonal entries are positive, $D_1^{1/2}$ and $D_2^{1/2}$ have inverses, say $D_1^{-1/2}$ and $D_2^{-1/2}$, respectively. Letting $D^{-1/2} = \text{diag}\{D_1^{-1/2}, D_2^{-1/2}, I\}$, we obtain $D^{-1/2}P'APD^{-1/2} = E$, which shows that $A \cong E$ and E has the structure in (13.5).

There is an interesting result that connects congruence with the number of positive, negative and zero eigenvalues of a real-symmetric matrix. We first assign this "triplet" a name and then discuss the result.

Definition 13.6 The inertia of a real-symmetric matrix. *The* **inertia** *of a real-symmetric matrix is defined as the three numbers* (P, N, Z), *where*

- P *is the number of positive eigenvalues,*
- N *is the number of negative eigenvalues, and*
- Z *is the number of zero eigenvalues*

of the matrix. The quantity $P - N$ *is often called the* **signature** *of a real-symmetric matrix.*

Note: $P + N$ is the rank of the matrix.

The following result shows that inertia is invariant under congruence.

Theorem 13.16 *Two real symmetric matrices are congruent if and only if they have the same inertia.*

Proof. Let A and B be two $n \times n$ matrices with inertia (p_A, n_A, z_A) and (p_B, n_B, z_B), respectively. Note that $n = p_A + n_A + z_A = p_B + n_B + z_B$ because the multiplicities of the eigenvalues are counted in the inertia. The spectral decompositions of A and B yield $P_A A P_A' = \Lambda_A$ and $P_B B P_B' = \Lambda_B$, where Λ_B is an $n \times n$ diagonal matrix with the eigenvalues of $\boldsymbol{B}$ along its diagonal. Therefore, $\boldsymbol{A} \cong \boldsymbol{\Lambda}_A$ and $\boldsymbol{B} \cong \boldsymbol{\Lambda}_B$.

If $\boldsymbol{A} \cong \boldsymbol{B}$, then $\boldsymbol{\Lambda}_A \cong \boldsymbol{\Lambda}_B$ because congruence is transitive. Therefore, there exists a nonsingular matrix $\boldsymbol{Q}$ such that $\boldsymbol{\Lambda}_B = \boldsymbol{Q}'\boldsymbol{\Lambda}_A\boldsymbol{Q}$. Without loss of generality, we can assume that the positive diagonal eigenvalues of $\boldsymbol{A}$ occupy the first p_A positions on the diagonal in $\boldsymbol{\Lambda}_A$.

Suppose, if possible, $p_A > p_B$. We show below that this will lead to a contradiction. Partition $\boldsymbol{Q} = [\boldsymbol{Q}_1 : \boldsymbol{Q}_2]$, where $\boldsymbol{Q}_1$ has p_B columns (so $\boldsymbol{Q}_2$ has $n - p_B$ columns), and $\boldsymbol{\Lambda}_A = [\boldsymbol{\Lambda}_{A1} : \boldsymbol{\Lambda}_{A2}]$, where $\boldsymbol{\Lambda}_{A1}$ comprises the first p_A columns of $\boldsymbol{\Lambda}_A$. Thus,

$$\boldsymbol{\Lambda}_{A1} = \begin{bmatrix} \lambda_1 & 0 & \dots & 0 \\ 0 & \lambda_2 & \dots & 0 \\ \vdots & \vdots & \ddots & \vdots \\ 0 & 0 & \dots & \lambda_{p_A} \\ 0 & 0 & \dots & 0 \\ \vdots & \vdots & \ddots & \vdots \\ 0 & 0 & \dots & 0 \end{bmatrix},$$

where $\lambda_i > 0$ for $i = 1, 2, \dots, p_A$ are the positive eigenvalues of $\boldsymbol{A}$.

Observe that both $\mathcal{C}(\boldsymbol{Q}_2)$ and $\mathcal{C}(\boldsymbol{\Lambda}_{A1})$ are subspaces of $\Re^n$. Now consider the dimension of $\mathcal{C}(\boldsymbol{Q}_2) \cap \mathcal{C}(\boldsymbol{\Lambda}_{A1})$:

$$\begin{aligned} \dim(\mathcal{C}(\boldsymbol{Q}_2) \cap \mathcal{C}(\boldsymbol{\Lambda}_{A1})) &= \dim(\mathcal{C}(\boldsymbol{Q}_2)) + \dim(\mathcal{C}(\boldsymbol{Q}_2)) - \dim(\mathcal{C}(\boldsymbol{Q}_2) + \mathcal{C}(\boldsymbol{\Lambda}_{A1})) \\ &= (n - p_B) + p_A - \dim(\mathcal{C}(\boldsymbol{Q}_2) + \mathcal{C}(\boldsymbol{\Lambda}_{A1})) \\ &= (n - \dim(\mathcal{C}(\boldsymbol{Q}_2) + \mathcal{C}(\boldsymbol{\Lambda}_{A1}))) + (p_A - p_B) > 0 \, . \end{aligned}$$

So there is a nonzero $\boldsymbol{x} \in \mathcal{C}(\boldsymbol{Q}_2) \cap \mathcal{C}(\boldsymbol{\Lambda}_{A1})$. For such an $\boldsymbol{x}$, we can write

$$\boldsymbol{x} \in \mathcal{C}(\boldsymbol{Q}_2) \Longrightarrow \boldsymbol{x} = \boldsymbol{Q}_2\boldsymbol{u} \Longrightarrow \boldsymbol{x} = \boldsymbol{Q}\begin{bmatrix} \boldsymbol{0} \\ \boldsymbol{u} \end{bmatrix}$$

for some nonzero $\boldsymbol{u} \in \Re^{n-p_B}$. If $\boldsymbol{z} = \begin{bmatrix} \boldsymbol{0} \\ \boldsymbol{u} \end{bmatrix}$, note that the first p_A elements in $\boldsymbol{z}$ are all zero. Therefore,

$$\boldsymbol{x}'\boldsymbol{\Lambda}_A\boldsymbol{x} = \boldsymbol{z}'\boldsymbol{\Lambda}_A\boldsymbol{z} = [\boldsymbol{0} : \boldsymbol{u}']\,\boldsymbol{\Lambda}_A\begin{bmatrix} \boldsymbol{0} \\ \boldsymbol{u} \end{bmatrix} \leq 0 \qquad (13.6)$$

because all the diagonal elements in $\boldsymbol{\Lambda}_A$ except for the first p_A are zero.

It also follows for any nonzero $\boldsymbol{x} \in \mathcal{C}(\boldsymbol{Q}_2) \cap \mathcal{C}(\boldsymbol{\Lambda}_{A1})$ that $\boldsymbol{x} \in \mathcal{C}(\boldsymbol{\Lambda}_{A1})$, which means that the last $n - p_A$ elements of $\boldsymbol{x}$ are zero. This means that

$$\boldsymbol{x}'\boldsymbol{\Lambda}_A\boldsymbol{x} = \sum_{i=1}^{p_A} \lambda_i x_i^2 > 0 \, . \qquad (13.7)$$

Equations (13.6) and (13.7) are contradictory to each other. Therefore, we cannot have $p_A > p_B$. Hence, $p_A \leq p_B$. By symmetry, the reverse inequality will follow and we obtain $p_A = p_B$. By applying the above arguments to $-\boldsymbol{\Lambda}_A$ and $-\boldsymbol{\Lambda}_B$, we can show that $n_A = n_B$. This proves that any two congruent matrices will have the same inertia.

We leave the proof of the converse to the reader. □

The implications of the above theorem for quadratic forms is important. Suppose that a quadratic form $q_{\boldsymbol{A}}(\boldsymbol{x})$ has been reduced to two different diagonal quadratic forms $q_{\boldsymbol{D}_1}(\boldsymbol{x})$ and $q_{\boldsymbol{D}_2}(\boldsymbol{x})$ using nonsingular transformations. This means that $\boldsymbol{D}_1$ and $\boldsymbol{D}_2$ are congruent and, hence, have the same inertia. Therefore, while $\boldsymbol{D}_1$ and $\boldsymbol{D}_2$ may be two different diagonal matrices, the number of positive, negative and zero entries in $\boldsymbol{D}_1$ must be the same as those in $\boldsymbol{D}_2$.

We now have the tools to classify an n-ary quadratic form $\boldsymbol{x}'\boldsymbol{A}\boldsymbol{x}$ (or the corresponding $n \times n$ real-symmetric matrix $\boldsymbol{A}$) using only P and N.

1. $\boldsymbol{A}$ is positive definite if and only if $P = n$.
2. $\boldsymbol{A}$ is nonnegative definite if and only if $N = 0$.
3. $\boldsymbol{A}$ is negative definite if and only if $N = n$.
4. $\boldsymbol{A}$ is non-positive definite if and only if $P = 0$.
5. $\boldsymbol{A}$ is indefinite if and only if $P \geq 1$ and $N \geq 1$.

Observe that the first item in the above list immediately yields the following fact: $\boldsymbol{A}$ *is positive definite if and only if it is nonnegative definite and nonsingular.*

13.6 Nonnegative definite matrices and minors

The Cholesky decomposition ensures that the determinant of positive definite matrices is positive. And if the matrix is n.n.d. but not p.d., then the determinant will be zero. We next turn to some characterizations of nonnegative definite matrices in terms of minors.

Theorem 13.17 *If $\boldsymbol{A}$ is p.d., then all its principal minors are positive. Conversely, if all the* **leading** *principal minors of a real symmetric matrix $\boldsymbol{A}$ are positive, then $\boldsymbol{A}$ is positive definite.*

Proof. Suppose $\boldsymbol{A}$ is positive definite. Then $\boldsymbol{A} = \boldsymbol{B}'\boldsymbol{B}$ for some non-singular matrix $\boldsymbol{B}$, so $|\boldsymbol{A}| = |\boldsymbol{B}|^2 > 0$. Theorem 13.6 tells us that every principal submatrix of $\boldsymbol{A}$ is also positive definite. This means that all principal minors of $\boldsymbol{A}$ are positive.

Now suppose $\boldsymbol{A}$ is an $n \times n$ real-symmetric matrix all of whose leading principal minors are positive. We prove the converse by induction on n.

If $n = 1$ the result is immediate. Assume that the result is true for matrices of order $n - 1$. Let $\boldsymbol{A}_k$ denote the $k \times k$ order leading principal submatrix of $\boldsymbol{A}$ formed by the rows and columns of $\boldsymbol{A}$ indexed by $\{1, 2, \ldots, k\}$. By hypothesis, $a_{11} > 0$.

Using elementary Type-III row operations, sweep out the elements in the first column of $\boldsymbol{A}$ below a_{11}. Because $\boldsymbol{A}$ is symmetric, we can use exactly the same Type-III operations on the columns to annihilate the elements in the first row of $\boldsymbol{A}$. In other words, we can find a nonsingular matrix $\boldsymbol{G}$ such that

$$\boldsymbol{GAG}' = \begin{bmatrix} a_{11} & \boldsymbol{0}' \\ \boldsymbol{0} & \boldsymbol{B}_{22} \end{bmatrix} = \boldsymbol{B} .$$

Suppose $\boldsymbol{C} = \boldsymbol{B}_{22}$ is the matrix obtained from $\boldsymbol{B}$ by deleting the first row and the first column. The $k \times k$ leading principal minors of $\boldsymbol{A}$ and $\boldsymbol{B}$ are related by

$$|\boldsymbol{A}_k| = |\boldsymbol{B}_k| = a_{11}|\boldsymbol{C}_{k-1}| \, , \quad \text{for } k = 2, ..., n \, ,$$

because $\boldsymbol{B}_k$ is obtained from $\boldsymbol{A}_k$ by elementary Type-III operations. This proves that the leading principal minors of $\boldsymbol{C}$ are positive. Therefore, by induction hypothesis, $\boldsymbol{C}$ is positive definite because it is $(n-1) \times (n-1)$. This implies that $\boldsymbol{B} = \begin{bmatrix} a_{11} & \boldsymbol{0}' \\ \boldsymbol{0} & \boldsymbol{C} \end{bmatrix}$ is positive definite. Since $\boldsymbol{A}$ is congruent to $\boldsymbol{B}$, $\boldsymbol{A}$ is positive definite as well. □

The following is an analogue for nonnegative definite matrices. The proof is slightly different from that for p.d. matrices, so we provide the details.

Theorem 13.18 *A real symmetric matrix* $\boldsymbol{A}$ *is n.n.d. if and only if all principal minors of* $\boldsymbol{A}$ *are non-negative.*

Proof. If $\boldsymbol{A}$ is $n \times n$ and n.n.d., then $\boldsymbol{A} = \boldsymbol{B}'\boldsymbol{B}$ for some square matrix $\boldsymbol{B}$. This means that $|\boldsymbol{A}| = |\boldsymbol{B}|^2 \geq 0$ and we now include the possibility that $|\boldsymbol{B}| = 0$ because $\boldsymbol{B}$ will be singular if $\boldsymbol{A}$ is n.n.d. but not positive definite. It is easy to prove (on the same lines as Theorem 13.6) that every principal submatrix of $\boldsymbol{A}$ is also n.n.d., which means that the determinant of that principal submatrix is nonnegative. This proves the "only if" part.

We prove the "if" part by induction on the order of $\boldsymbol{A}$. The result is obviously true for 1×1 matrices, which serves as the base case for induction. Assume the result is true for all $(n-1) \times (n-1)$ matrices and suppose that $\boldsymbol{A}$ is $n \times n$ with all principal minors being nonnegative.

First consider the case where all the diagonal entries of $\boldsymbol{A}$ are zero. The 2×2 principal minor formed by the i-th and j-th rows and columns is given by $-a_{ij}^2$. Because $\boldsymbol{A}$ has all principal minors as nonnegative, it follows that $a_{ij} = 0$. This shows that $\boldsymbol{A} = \boldsymbol{O}$, so $\boldsymbol{A}$ is nonnegative definite ($\boldsymbol{x}'\boldsymbol{A}\boldsymbol{x} = 0$ for every $\boldsymbol{x}$).

Now suppose that $\boldsymbol{A}$ has at least one positive diagonal entry. That diagonal element can always be brought to the $(1, 1)$-th position in the matrix $\boldsymbol{P}'\boldsymbol{A}\boldsymbol{P}$, where $\boldsymbol{P}$ is the permutation matrix that permutes the appropriate rows and columns. Note that all principal minors of $\boldsymbol{P}'\boldsymbol{A}\boldsymbol{P}$ would still be non-negative and $\boldsymbol{P}'\boldsymbol{A}\boldsymbol{P}$ would have the same definiteness category as $\boldsymbol{A}$. So, without loss of generality, assume that $a_{11} > 0$.

We now make a_{i1} and a_{1i} zero for $i = 2, ..., n$ as in the proof of Theorem 13.17

and consider the matrices $\boldsymbol{B}$ and $\boldsymbol{C}$ defined there. Let $I = \{1, i_1, ..., i_k\}$, where $2 \leq i_1 < i_2 < \cdots < i_k \leq n$, and form the submatrices $\boldsymbol{A}_{II}$ and $\boldsymbol{B}_{II}$. Note that

$$|\boldsymbol{A}_{II}| = |\boldsymbol{B}_{II}| = a_{11}|\boldsymbol{C}_{JJ}| , \text{ where } J = \{i_1 - 1, i_2 - 1, \ldots, i_k - 1\} .$$

Hence all the principal minors of $\boldsymbol{C}$ are non-negative and, by the induction hypothesis, $\boldsymbol{C}$ is nonnegative definite. This means that $\boldsymbol{B} = \begin{bmatrix} a_{11} & \boldsymbol{0}' \\ \boldsymbol{0} & \boldsymbol{C} \end{bmatrix}$ is nonnegative definite. Since $\boldsymbol{A}$ is congruent to $\boldsymbol{B}$, $\boldsymbol{A}$ is nonnegative definite as well. □

We point out that one cannot weaken the requirement of *all principal minors* being nonnegative to *leading principal minors* being nonnegative for the "if" part of Theorem 13.18. In other words, there can exist a matrix that is not n.n.d. but all its leading principal minors are nonnegative. One such example is the matrix $\begin{bmatrix} 0 & 0 \\ 0 & -1 \end{bmatrix}$.

Theorem 13.18 can be used to deduce parallel conclusions for nonpositive definite matrices as well. Note that $\boldsymbol{A}$ is nonpositive definite if and only if $-\boldsymbol{A}$ is nonnegative definite. This means that $\boldsymbol{A}$ is nonpositive definite if and only if all the principal minors of $-\boldsymbol{A}$ are nonnegative. Using elementary properties of determinants, one can argue that $\boldsymbol{A}$ is nonpositive definite if and only if all principal minors of $\boldsymbol{A}$ with even order are nonnegative and all principal minors of $\boldsymbol{A}$ with odd order are nonpositive.

13.7 Some inequalities related to quadratic forms

Special properties of positive and nonnegative definite matrices lead to several important inequalities in matrix analysis. For example, the Cauchy-Schwarz inequality can be proved using n.n.d. matrices.

Theorem 13.19 Cauchy-Schwarz inequality revisited. *If $\boldsymbol{x}$ and $\boldsymbol{y}$ are two $n \times 1$ vectors, then*

$$(\boldsymbol{x}'\boldsymbol{y})^2 \leq (\boldsymbol{x}'\boldsymbol{x})(\boldsymbol{y}'\boldsymbol{y}) .$$

Equality holds if and only if $\boldsymbol{x}$ and $\boldsymbol{y}$ are collinear.

Proof. Theorem 13.10 ensures that the 2×2 matrix

$$\boldsymbol{A} = \begin{bmatrix} \boldsymbol{x}'\boldsymbol{x} & \boldsymbol{x}'\boldsymbol{y} \\ \boldsymbol{y}'\boldsymbol{x} & \boldsymbol{y}'\boldsymbol{y} \end{bmatrix} = \begin{bmatrix} \boldsymbol{x}' \\ \boldsymbol{y}' \end{bmatrix} [\boldsymbol{x} : \boldsymbol{y}]$$

is nonnegative definite because $\boldsymbol{A} = \boldsymbol{B}'\boldsymbol{B}$, where $\boldsymbol{B} = [\boldsymbol{x} : \boldsymbol{y}]$. Theorem 13.9 ensures that the eigenvalues of $\boldsymbol{A}$ are nonnegative, which implies $|\boldsymbol{A}| \geq 0$. Therefore,

$$0 \leq |\boldsymbol{A}| = (\boldsymbol{x}'\boldsymbol{x})(\boldsymbol{y}'\boldsymbol{y}) - (\boldsymbol{x}'\boldsymbol{y})^2$$

and the result follows. If $\boldsymbol{A}$ is positive definite, then $|\boldsymbol{A}| > 0$ and we have strict inequality. Therefore, equality holds if and only if $\boldsymbol{A}$ is singular, which means if and only if the rank of $\boldsymbol{A}$, which is equal to the rank of $\boldsymbol{B}$, is 1. This happens if and only if $\boldsymbol{x}$ and $\boldsymbol{y}$ are collinear. □

The Cauchy-Schwarz inequality leads to the following inequalities involving n.n.d. or p.d. matrices.

Theorem 13.20 *Let $\boldsymbol{x}$ and $\boldsymbol{y}$ be $n \times 1$ vectors and $\boldsymbol{A}$ an $n \times n$ matrix.*

(i) If $\boldsymbol{A}$ is n.n.d. then $(\boldsymbol{x}'\boldsymbol{A}\boldsymbol{y})^2 \leq (\boldsymbol{x}'\boldsymbol{A}\boldsymbol{x})(\boldsymbol{y}'\boldsymbol{A}\boldsymbol{y})$.

(ii) If $\boldsymbol{A}$ is p.d. then $(\boldsymbol{x}'\boldsymbol{y})^2 \leq (\boldsymbol{x}'\boldsymbol{A}\boldsymbol{x})(\boldsymbol{y}'\boldsymbol{A}^{-1}\boldsymbol{y})$.

Proof. **Proof of (i):** Since $\boldsymbol{A}$ is n.n.d. there exists an $n \times n$ matrix $\boldsymbol{B}$ such that $\boldsymbol{A} = \boldsymbol{B}'\boldsymbol{B}$. Let $\boldsymbol{u} = \boldsymbol{B}\boldsymbol{x}$ and $\boldsymbol{v} = \boldsymbol{B}\boldsymbol{y}$. Then $\boldsymbol{u}'\boldsymbol{v} = \boldsymbol{x}'\boldsymbol{B}'\boldsymbol{B}\boldsymbol{y} = \boldsymbol{x}'\boldsymbol{A}\boldsymbol{y}$, $\boldsymbol{u}'\boldsymbol{u} = \boldsymbol{x}'\boldsymbol{B}'\boldsymbol{B}\boldsymbol{x} = \boldsymbol{x}'\boldsymbol{A}\boldsymbol{x}$ and $\boldsymbol{v}'\boldsymbol{v} = \boldsymbol{y}'\boldsymbol{B}'\boldsymbol{B}\boldsymbol{y} = \boldsymbol{y}'\boldsymbol{A}\boldsymbol{y}$. Using the Cauchy-Schwarz inequality we conclude

$$(\boldsymbol{x}'\boldsymbol{A}\boldsymbol{y})^2 = (\boldsymbol{u}'\boldsymbol{v})^2 \leq (\boldsymbol{u}'\boldsymbol{u})^2(\boldsymbol{v}'\boldsymbol{v})^2 = (\boldsymbol{x}'\boldsymbol{A}\boldsymbol{x})^2(\boldsymbol{y}'\boldsymbol{A}\boldsymbol{y})^2 .$$

Proof of (ii): If $\boldsymbol{A}$ is positive definite and $\boldsymbol{A} = \boldsymbol{B}'\boldsymbol{B}$, then $\boldsymbol{A}^{-1}$ exists and $\boldsymbol{A}^{-1} = \boldsymbol{B}^{-1}(\boldsymbol{B}^{-1})'$. Let $\boldsymbol{u} = \boldsymbol{B}\boldsymbol{x}$ and $\boldsymbol{v} = (\boldsymbol{B}^{-1})'\boldsymbol{y}$. Then, $\boldsymbol{u}'\boldsymbol{v} = \boldsymbol{x}'\boldsymbol{B}'(\boldsymbol{B}')^{-1}\boldsymbol{y} = \boldsymbol{x}'\boldsymbol{y}$, $\boldsymbol{u}'\boldsymbol{u} = \boldsymbol{x}'\boldsymbol{B}'\boldsymbol{B}\boldsymbol{x} = \boldsymbol{x}'\boldsymbol{A}\boldsymbol{x}$ and $\boldsymbol{v}'\boldsymbol{v} = \boldsymbol{y}'\boldsymbol{B}^{-1}(\boldsymbol{B}^{-1})'\boldsymbol{y} = \boldsymbol{y}'\boldsymbol{A}^{-1}\boldsymbol{y}$ and we conclude

$$(\boldsymbol{x}'\boldsymbol{y})^2 = (\boldsymbol{u}'\boldsymbol{v})^2 \leq (\boldsymbol{u}'\boldsymbol{u})^2(\boldsymbol{v}'\boldsymbol{v})^2 = (\boldsymbol{x}'\boldsymbol{A}\boldsymbol{x})^2(\boldsymbol{y}'\boldsymbol{A}^{-1}\boldsymbol{y})^2 .$$

□

Corollary 13.10 *If $\boldsymbol{A} = \{a_{ij}\}$ is n.n.d., then*

$$|a_{ij}| \leq \max\{a_{11}, a_{22}, \ldots, a_{nn}\} .$$

Proof. Using (i) of Theorem 13.20 with $\boldsymbol{x} = \boldsymbol{e}_i$ and $\boldsymbol{y} = \boldsymbol{e}_j$, we obtain

$$a_{ij}^2 = (\boldsymbol{e}_i'\boldsymbol{A}\boldsymbol{e}_j)^2 \leq (\boldsymbol{e}_i'\boldsymbol{A}\boldsymbol{e}_i)(\boldsymbol{e}_j'\boldsymbol{A}\boldsymbol{A}_j) = a_{ii}a_{jj} \leq \max a_{ii}^2, a_{jj}^2$$

for $i, j = 1, 2, \ldots, n$. The result follows. □

This means that the maximum element in an n.n.d. matrix must lie along its diagonal.

Here is another interesting inequality involving the traces of n.n.d. matrices.

Theorem 13.21 *If $\boldsymbol{A}$ and $\boldsymbol{B}$ are both $n \times n$ nonnegative definite matrices, then*

$$0 \leq tr(\boldsymbol{A}\boldsymbol{B}) \leq tr(\boldsymbol{A})tr(\boldsymbol{B}) .$$

Proof. The strategy will be to first prove the result when $\boldsymbol{A}$ is diagonal and then extend it to the more general n.n.d. $\boldsymbol{A}$. First assume that $\boldsymbol{A}$ is diagonal with nonnegative entries a_{ii}. Since $\boldsymbol{B} = \{b_{ij}\}$ is n.n.d., we have $b_{ii} \geq 0$. Clearly, $b_{ii} \leq \sum_{j=1}^{n} b_{jj}$ because the b_{ii}'s are nonnegative. This means that if we replace each b_{ii} in the sum $\sum_{i=1}^{n} a_{ii}b_{ii}$ by $\sum_{j=1}^{n} b_{jj}$, we obtain the inequality

$$0 \leq \sum_{i=1}^{n} a_{ii}b_{ii} \leq \sum_{i=1}^{n} a_{ii}\left(\sum_{j=1}^{n} b_{jj}\right) = \left(\sum_{i=1}^{n} a_{ii}\right)\left(\sum_{j=1}^{n} b_{jj}\right) .$$

Since A is diagonal, we have $\sum_{i=1}^n a_{ii}b_{ii} = \text{tr}(AB)$. Therefore,

$$0 \leq \sum_{i=1}^{n} a_{ii}b_{ii} = \text{tr}(AB) \leq \left(\sum_{i=1}^{n} a_{ii}\right)\left(\sum_{j=1}^{n} b_{jj}\right) = \text{tr}(A)\text{tr}(B) .$$

This proves the result when A and B are n.n.d. but A is diagonal.

Now consider the case when A is not necessarily diagonal. Since A is n.n.d, it has a spectral decomposition $P'AP = \Lambda$, where P is orthogonal and Λ is diagonal with nonnegative numbers along its diagonal. Clearly, $P'BP$ is nonnegative definite. Also,

$$\text{tr}(\Lambda) = \text{tr}(P'AP) = \text{tr}(APP') = \text{tr}(A) \text{ and } \text{tr}(P'BP) = \text{tr}(BPP') = \text{tr}(B) .$$

Then, using the result for diagonal matrices, we obtain

$$\text{tr}(AB) = \text{tr}(P\Lambda P'B) = \text{tr}(\Lambda P'BP) \leq \text{tr}(\Lambda)\text{tr}(P'BP) = \text{tr}(A)\text{tr}(B) .$$

□

The determinant for n.n.d. matrices satisfy certain useful inequalities.

Theorem 13.22 *Let A be an n.n.d. matrix of order n, partitioned so that*

$$A = \begin{bmatrix} A_{11} & u \\ u' & a_{nn} \end{bmatrix} .$$

Then, $|A| \leq a_{nn}|A_{11}|$.

Proof. If A is n.n.d. but not p.d., then $|A| = 0$ and $|A| \leq a_{nn}A_{nn}$ follows from the preceding theorem. So consider the case when A is positive definite, which means that A_{11} is positive definite as well and, hence, nonsingular. From Theorem 10.10 we know that the determinant of A is

$$|A| = |A_{11}| \left(a_{nn} - u'A_{11}^{-1}u\right) .$$

Since A_{11} is p.d., so is A_{11}^{-1}, and hence $u'A_{11}^{-1}u \geq 0$. Therefore,

$$\frac{|A|}{|A_{11}|} = a_{nn} - u'A_{11}^{-1}u \leq a_{nn} ,$$

which proves that $|A| \leq a_{nn}|A_{11}|$. Further, equality holds if and only if $u'A_{11}^{-1}u = 0$, which happens if and only if $u = 0$. □

The above result is often restated in terms of cofactors. Since $|A_{11}| = A_{11}$, the preceding inequality can be restated as $|A| \leq a_{nn}A_{nn}$ whenever A is nonnegative definite. In fact, essentially the same argument proves the more general result that $|A| \leq a_{ii}A_{ii}$, where $i = 1, 2, \ldots, n$ because a_{ii} can be moved to the (n, n)-th position by permuting the rows and columns that do not alter the value of the determinant.

The following result says that the determinant of an n.n.d. matrix cannot exceed the product of its diagonal elements.

Theorem 13.23 *If $\boldsymbol{A}$ is $n \times n$ and n.n.d., then $|\boldsymbol{A}| \leq a_{11}a_{22}...a_{nn}$.*

Proof. The proof is on the same lines as Theorem 13.22. We use induction. The case for $n = 1$ is trivial and the result can be easily verified for 2×2 matrices as well. We leave this for the reader to verify.

Now suppose that the result is true for all matrices of order $n-1$ and let

$$\boldsymbol{A} = \begin{bmatrix} \boldsymbol{A}_{11} & \boldsymbol{u} \\ \boldsymbol{u}' & a_{nn} \end{bmatrix} .$$

The matrix $\boldsymbol{A}_{11}$ is of order $n-1$ so, by induction hypothesis, its determinant is less than or equal to the product of its diagonal elements. That is, $|\boldsymbol{A}_{11}| \leq a_{11}a_{22}\cdots a_{n-1,n-1}$. Theorem 13.22 tells us that $|\boldsymbol{A}| \leq a_{nn}|\boldsymbol{A}_{11}|$, which implies that $|\boldsymbol{A}| \leq |\boldsymbol{A}_{11}|a_{nn} \leq a_{11}a_{22}\cdots a_{n-1,n-1}a_{nn}$. □

If $\boldsymbol{A}$ is p.d., then $|\boldsymbol{A}| = a_{11}a_{22}\cdots a_{nn}$ if and only if $\boldsymbol{A}$ is diagonal. This is because equality holds in the above results if and only if it holds throughout the above string of inequalities, which implies that $\boldsymbol{u} = \boldsymbol{0}$.

The following inequality applies to any real $n \times n$ matrix (not necessarily n.n.d. or even symmetric).

Theorem 13.24 Hadamard's inequality. *For any real $n \times n$ matrix $\boldsymbol{B} = \{b_{ij}\}$,*

$$|\boldsymbol{B}|^2 \leq \prod_{j=1}^{n}(b_{1j}^2 + ... + b_{nj}^2) .$$

Proof. If $\boldsymbol{B} = \{b_{ij}\}$ is any square matrix, then $\boldsymbol{A} = \boldsymbol{B}'\boldsymbol{B}$ is n.n.d. The j-th diagonal element of $\boldsymbol{A}$ is given by

$$a_{jj} = \boldsymbol{b}_{*j}'\boldsymbol{b}_{*j} = \|\boldsymbol{b}_{*j}\| = b_{1j}^2 + ... + b_{nj}^2 .$$

Applying Theorem 13.23 to $\boldsymbol{A}$, we obtain

$$|\boldsymbol{B}|^2 = |\boldsymbol{A}| \leq a_{11}a_{22}\cdots a_{nn} = \prod_{j=1}^{n}(b_{1j}^2 + ... + b_{nj}^2) ,$$

which completes the proof. □

The next two inequalities concern the sum of two n.n.d. matrices.

Lemma 13.4 *If $\boldsymbol{C}$ is an $n \times n$ and n.n.d., then $|\boldsymbol{I} + \boldsymbol{C}| \geq 1 + |\boldsymbol{C}|$. The inequality is strict if $n \geq 2$ and $\boldsymbol{C}$ is p.d.*

Proof. Since $\boldsymbol{C}$ is n.n.d., all its eigenvalues are nonnegative. Let $\lambda_1, \lambda_2, \ldots, \lambda_n$ be the eigenvalues of $\boldsymbol{C}$. Then the eigenvalues of $\boldsymbol{I} + \boldsymbol{C}$ are $1 + \lambda_i$, $i = 1, 2, \ldots, n$.

Since the determinant is the product of eigenvalues, we have

$$|\boldsymbol{I}+\boldsymbol{C}| = \prod_{i=1}^{n}(1+\lambda_i) = 1 + \prod_{i=1}^{n}\lambda_i + \text{terms involving products of } \lambda_i\text{'s}$$
$$\geq 1 + \prod_{i=1}^{n}\lambda_i \,, \tag{13.8}$$

where the last inequality follows easily because the λ_i's are all nonnegative. Since $|\boldsymbol{C}| = \prod_{i=1}^{n}\lambda_i$, it follows that $|\boldsymbol{I}+\boldsymbol{C}| \geq 1 + |\boldsymbol{C}|$.

If $n = 1$, then we have determinants of scalars and we have equality. If $n \geq 2$ and $\boldsymbol{C}$ is positive definite, then all the λ_i's are strictly positive and we have a strict inequality in (13.8), which leads to $|\boldsymbol{I}+\boldsymbol{C}| > 1 + |\boldsymbol{C}|$. □

We use the above lemma to obtain the following important result.

Theorem 13.25 *If $\boldsymbol{A}$ and $\boldsymbol{B}$ are both $n \times n$ and n.n.d., then*

$$|\boldsymbol{A}+\boldsymbol{B}| \geq |\boldsymbol{A}| + |\boldsymbol{B}| \,.$$

Moreover, if $n \geq 2$ and $\boldsymbol{A}$ and $\boldsymbol{B}$ are p.d., then we obtain strict inequality above.

Proof. The proof is not difficult but involves some tricks using the square root of an n.n.d. matrix. Since $\boldsymbol{A}$ is n.n.d., Theorem 13.11 ensures that it has an n.n.d. square root matrix $\boldsymbol{A}^{1/2}$ such that $\boldsymbol{A} = \boldsymbol{A}^{1/2}\boldsymbol{A}^{1/2}$. Observe that

$$|\boldsymbol{A}+\boldsymbol{B}| = |\boldsymbol{A}^{1/2}\boldsymbol{A}^{1/2}+\boldsymbol{B}| = |\boldsymbol{A}^{1/2}(\boldsymbol{I}+\boldsymbol{A}^{-1/2}\boldsymbol{B}\boldsymbol{A}^{-1/2})\boldsymbol{A}^{1/2}|$$
$$= |\boldsymbol{A}^{1/2}||\boldsymbol{I}+\boldsymbol{A}^{-1/2}\boldsymbol{B}\boldsymbol{A}^{-1/2}||\boldsymbol{A}^{1/2}| = |\boldsymbol{A}||\boldsymbol{I}+\boldsymbol{A}^{-1/2}\boldsymbol{B}\boldsymbol{A}^{-1/2}|$$
$$= |\boldsymbol{A}||\boldsymbol{I}+\boldsymbol{C}| \,, \tag{13.9}$$

where $\boldsymbol{C} = \boldsymbol{A}^{-1/2}\boldsymbol{B}\boldsymbol{A}^{-1/2}$. Note that $\boldsymbol{C}$ is n.n.d. because $\boldsymbol{B}$ is. Lemma 13.4 we know that $|\boldsymbol{I}+\boldsymbol{C}| \geq 1 + |\boldsymbol{C}|$, which means that

$$|\boldsymbol{A}+\boldsymbol{B}| = |\boldsymbol{A}||\boldsymbol{I}+\boldsymbol{C}| \geq |\boldsymbol{A}|(1+|\boldsymbol{C}|) = |\boldsymbol{A}| + |\boldsymbol{A}||\boldsymbol{C}| = |\boldsymbol{A}| + |\boldsymbol{B}| \,, \tag{13.10}$$

where the last step follows from the fact that

$$|\boldsymbol{A}||\boldsymbol{C}| = |\boldsymbol{A}||\boldsymbol{A}^{-1/2}||\boldsymbol{B}||\boldsymbol{A}^{-1/2}| = |\boldsymbol{A}||\boldsymbol{A}^{-1/2}||\boldsymbol{A}^{-1/2}|\boldsymbol{B}| = |\boldsymbol{A}||\boldsymbol{A}^{-1}|\boldsymbol{B}| = |\boldsymbol{B}| \,.$$

This completes the proof when $\boldsymbol{A}$ and $\boldsymbol{B}$ are both nonnegative definite.

If $n = 1$, the $\boldsymbol{A}$ and $\boldsymbol{B}$ are nonnegative scalars and we have equality. If $\boldsymbol{A}$ and $\boldsymbol{B}$ are p.d. and $n \geq 2$, then so $\boldsymbol{C}$ too is p.d. and all its eigenvalues are strictly positive. In that case, the inequality in (13.10) is strict, which implies that $|\boldsymbol{A}+\boldsymbol{B}| > |\boldsymbol{A}| + |\boldsymbol{B}|$.
□

This inequality has a number of implications. Clearly, since n.n.d. matrices have nonnegative determinants,

$$|\boldsymbol{A}+\boldsymbol{B}| \geq |\boldsymbol{A}| \text{ and } |\boldsymbol{A}+\boldsymbol{B}| \geq |\boldsymbol{B}| \,, \tag{13.11}$$

whenever $\boldsymbol{A}$ and $\boldsymbol{B}$ are n.n.d. and of the same order. Here is another corollary.

Corollary 13.11 *If $\boldsymbol{A}$ and $\boldsymbol{B}$ are both n.n.d and $\boldsymbol{A}-\boldsymbol{B}$ is n.n.d. as well, then $|\boldsymbol{A}| \geq |\boldsymbol{B}|$.*

Proof. Using Theorem 13.25, we obtain

$$|\boldsymbol{A}| = |\boldsymbol{A}-\boldsymbol{B}+\boldsymbol{B}| \geq |\boldsymbol{A}-\boldsymbol{B}| + |\boldsymbol{B}| \geq |\boldsymbol{B}|$$

because $|\boldsymbol{A}-\boldsymbol{B}| \geq 0$. □

Theorem 13.25 helps us derive several other inequalities. The Schur's complement too plays an important role sometimes. We saw this in Theorem 13.22. Below is another inequality useful in statistics and econometrics.

Theorem 13.26 *Let $\boldsymbol{M} = \begin{bmatrix} \boldsymbol{A} & \boldsymbol{B} \\ \boldsymbol{B}' & \boldsymbol{D} \end{bmatrix}$ be nonnegative definite, where $\boldsymbol{A}$ is square. Then, $|\boldsymbol{M}| \leq |\boldsymbol{A}||\boldsymbol{D}|$. Equality holds if and only if either $\boldsymbol{A}$ is singular or $\boldsymbol{B} = \boldsymbol{O}$.*

Proof. Since $\boldsymbol{M}$ is nonnegative definite, all its principal submatrices are nonnegative definite (the nonnegative definite analogue of Theorem 13.6), so $\boldsymbol{A}$ and $\boldsymbol{D}$ are both nonnegative definite. If $\boldsymbol{A}$ is singular, then $|\boldsymbol{A}| = 0$, which means that $\boldsymbol{M}$ cannot be positive definite (Theorem 13.17) and so $|\boldsymbol{M}| = 0$. Therefore, equality holds.

Now suppose that $\boldsymbol{A}$ is positive definite. Theorem 10.10 tells us that $|\boldsymbol{M}| = |\boldsymbol{A}||\boldsymbol{D} - \boldsymbol{B}'\boldsymbol{A}^{-1}\boldsymbol{B}|$. Also,

$$|\boldsymbol{D}| = |\boldsymbol{D} - \boldsymbol{B}'\boldsymbol{A}^{-1}\boldsymbol{B} + \boldsymbol{B}'\boldsymbol{A}^{-1}\boldsymbol{B}| \geq |\boldsymbol{D} - \boldsymbol{B}'\boldsymbol{A}^{-1}\boldsymbol{B}|$$

because $\boldsymbol{B}'\boldsymbol{A}^{-1}\boldsymbol{B}$ is nonnegative definite. Therefore,

$$|\boldsymbol{M}| = |\boldsymbol{A}||\boldsymbol{D} - \boldsymbol{B}'\boldsymbol{A}^{-1}\boldsymbol{B}| \leq |\boldsymbol{A}||\boldsymbol{D}| .$$

Equality occurs if and only if the last inequality is an equality. This happens if and only if $\boldsymbol{B}'\boldsymbol{A}^{-1}\boldsymbol{B} = \boldsymbol{O}$, which happens if and only if $\boldsymbol{B} = \boldsymbol{O}$. □

The above inequality can be further refined when $\boldsymbol{A}$, $\boldsymbol{B}$ and $\boldsymbol{D}$ in Theorem 13.26 are all square and of the same size. Then, we can write

$$|\boldsymbol{D}| \geq |\boldsymbol{D} - \boldsymbol{B}'\boldsymbol{A}^{-1}\boldsymbol{B}| + |\boldsymbol{B}'\boldsymbol{A}^{-1}\boldsymbol{B}| = \frac{|\boldsymbol{M}|}{|\boldsymbol{A}|} + \frac{|\boldsymbol{B}|^2}{|\boldsymbol{A}|} ,$$

which implies that

$$|\boldsymbol{D}||\boldsymbol{A}| - |\boldsymbol{B}|^2 \geq |\boldsymbol{M}| \geq 0 \text{ and, in particular, } |\boldsymbol{B}|^2 \leq |\boldsymbol{A}||\boldsymbol{D}| . \tag{13.12}$$

The Schur's complement leads to several other inequalities involving determinants of positive definite matrices. Here is another, rather general, inequality that leads to several others.

Theorem 13.27 *Let $\boldsymbol{A}$, $\boldsymbol{B}$, $\boldsymbol{C}$, $\boldsymbol{D}$, $\boldsymbol{U}$ and $\boldsymbol{V}$ be square matrices of the same size. If $\boldsymbol{U}$ and $\boldsymbol{V}$ are positive definite, then*

$$|\boldsymbol{AUA}' + \boldsymbol{BVB}'| \cdot |\boldsymbol{CUC}' + \boldsymbol{DVD}'| \geq |\boldsymbol{AUC}' + \boldsymbol{BVD}'|^2 .$$

Proof. Consider the matrix

$$\begin{bmatrix} A & B \\ C & D \end{bmatrix} \begin{bmatrix} U & O \\ O & V \end{bmatrix} \begin{bmatrix} A' & C' \\ B' & D' \end{bmatrix} = \begin{bmatrix} AUA' + BVB' & AUC' + BVD' \\ CUA' + DVB' & CUC' + DVD' \end{bmatrix} .$$

The above matrix is positive definite because U and V are positive definite. This is clear from the factored form on the left hand side of the above identity. Applying the inequality in (13.12) to the matrix on the right hand side, we obtain

$$|AUA' + BVB'| \cdot |CUC' + DVD'| - |AUC' + BVD'|^2 \geq 0 ,$$

which establishes the desired inequality. □

This has several interesting special cases. If $A = D = I$, we obtain

$$|U + BVB'| \cdot |CUC' + V| \geq |UC' + BV|^2 . \tag{13.13}$$

Taking $U = V = I$ in (13.13) yields

$$|I + BB'| \cdot |I + CC'| \geq |C' + B|^2 . \tag{13.14}$$

Finally, if $U = V = I$ in Theorem 13.27, we obtain

$$|AA' + BB'| \cdot |CC' + DD'| \geq |AC' + BD'|^2 . \tag{13.15}$$

The following is a Cauchy-Schwarz type of inequality involving determinants. It applies to two general $m \times n$ matrices that need not even be square, let alone symmetric or positive definite.

Theorem 13.28 *If A and B are both $m \times n$ matrices, then*

$$|A'B|^2 \leq |A'A||B'B| .$$

Proof. First of all, observe that $A'B$, $A'A$ and $B'B$ are all $n \times n$, so their determinants are well-defined. If $A'B$ is singular, then $|A'B| = 0$ and the inequality holds because $A'A$ and $B'B$, being nonnegative definite, have nonnegative determinants.

Now assume that $A'B$ is nonsingular, which means that

$$n = \rho(A'B) \leq \min\{\rho(A), \rho(B)\} .$$

But this implies that both A and B must have rank n, i.e., they have full column rank, and, hence, $A'A$ and $B'B$ are positive definite. Now, make use of a useful trick. Decomposing $B'B$ in terms of the projectors of A yields

$$B'B = B'(P_A + I_m - P_A)B = B'P_AB + B'(I_m - P_A)B .$$

Also note that since P_A is symmetric and idempotent, it is n.n.d. and so are $B'P_AB$ and $B'(I_m - P_A)B$. Using Theorem 13.25, we see that

$$|B'B| \geq |B'P_AB| + |B'(I_m - P_A)B| \geq |B'P_AB| \geq |B'P_AB| ,$$

because $|B'(I_m - P_A)B| \geq 0$. Therefore,

$$|B'B| \geq |B'P_AB| = |B'A(A'A)^{-1}A'B| = |B'A||A'A|^{-1}|A'B| = \frac{|A'B|^2}{|A'A|} ,$$

which implies that $|B'B||A'A| \geq |A'B|^2$. □

Next, we obtain some results on the extreme values of quadratic forms. Let $A = P\Lambda P'$ be the spectral decomposition of A, where P is orthogonal and Λ is diagonal with eigenvalues of A along its diagonal. Then,

$$x'Ax = x'P\Lambda P'x = \lambda_1 y_1^2 + \lambda_2 y_2^2 + \cdots + \lambda_n y_n^2 , \tag{13.16}$$

where y_i's are the elements of $y = P'x$. If A has at least one positive eigenvalue, then it is clear from (13.16) that $x'Ax$ can be made bigger than any given finite number by suitably choosing the y_i's. This means that $\sup_x x'Ax = \infty$. If, on the other hand, all the eigenvalues of A are zero, then $x'Ax = 0$ for all x. Hence, $\sup_x x'Ax = 0$. Matters become more interesting when we restrict $\|x\| = 1$.

Theorem 13.29 Maxima and Minima of quadratic forms. *For any quadratic form $x'Ax$,*

$$\max_{\|x\|=1} x'Ax = \lambda_1 \text{ and } \min_{\|x\|=1} x'Ax = \lambda_n ,$$

where λ_1 and λ_n are the largest and the smallest eigenvalues of A.

Proof. Let $\lambda_1 \geq \lambda_2 \geq \cdots \lambda_n$ be the eigenvalues of A. Let $A = P\Lambda P'$ be the spectral decomposition so that Λ is diagonal with λ_i as its i-th diagonal element. Since P is orthogonal, we conclude $\|y\|^2 = x'PP'x = x'x = 1$, which means that $\|x\| = 1$ if and only if $\|y\| = 1$. Consider (13.16) and note that

$$\max_{\|x\|=1} x'Ax = \max_{\|y\|=1} y'\Lambda y = \max_{\sum_{i=1}^n y_i^2=1} \sum_{i=1}^n \lambda_i y_i^2 \leq \lambda_1 \sum_{i=1}^n y_i^2 = \lambda_1 .$$

Similarly, for the minimum of $x'Ax$, we note

$$\min_{\|x\|=1} x'Ax = \min_{\|y\|=1} y'\Lambda y = \min_{\sum_{i=1}^n y_i^2=1} \sum_{i=1}^n \lambda_i y_i^2 \geq \lambda_n \sum_{i=1}^n y_i^2 = \lambda_n .$$

Equality holds when x is an eigenvector of unit norm associated with λ_1 in the first case and with λ_n in the second case. □

Note that the restriction to unit norm in Theorem 13.29 is sometimes replaced by the equivalent conditions

$$\max_{x\neq 0} \frac{x'Ax}{x'x} = \lambda_1 \text{ and } \min_{x\neq 0} \frac{x'Ax}{x'x} = \lambda_n . \tag{13.17}$$

The next result is known as the *Courant-Fischer Theorem*, attributed to mathematicians Ernst Fischer, who provided the result for matrices in 1905, and Richard Courant, who extended the result for infinite-dimensional operators (beyond the scope of this book).

Theorem 13.30 Courant-Fischer Theorem. *Let A be a real symmetric matrix of order $n \geq 2$ with eigenvalues $\lambda_1 \geq \lambda_2 \geq \ldots \geq \lambda_n$ and let k be any integer between*

2 and n. Then,

$$\min_{\boldsymbol{x}\in\mathcal{N}(\boldsymbol{B}')} \frac{\boldsymbol{x}'\boldsymbol{A}\boldsymbol{x}}{\boldsymbol{x}'\boldsymbol{x}} \leq \lambda_k \leq \max_{\boldsymbol{x}\in\mathcal{N}(\boldsymbol{C}')} \frac{\boldsymbol{x}'\boldsymbol{A}\boldsymbol{x}}{\boldsymbol{x}'\boldsymbol{x}} ,$$

where $\boldsymbol{B}$ and $\boldsymbol{C}$ are any two $n \times (k-1)$ matrices.

Proof. Let $\boldsymbol{U}'\boldsymbol{A}\boldsymbol{U} = \boldsymbol{\Lambda}$ be the spectral decomposition, where $\boldsymbol{\Lambda}$ is diagonal with λ_i as its i-th diagonal entry and $\boldsymbol{U}$ is the orthogonal matrix whose columns $\boldsymbol{u}_1, \boldsymbol{u}_2, \ldots, \boldsymbol{u}_n$ are real orthonormal eigenvectors of $\boldsymbol{A}$ corresponding to $\lambda_1, \lambda_2, \ldots, \lambda_n$, respectively. Let $\boldsymbol{U}_k = [\boldsymbol{u}_1 : \boldsymbol{u}_2 : \ldots : \boldsymbol{u}_k]$ be the $n \times k$ matrix with the first k columns of $\boldsymbol{U}$ as its columns. Then $\boldsymbol{U}_k'\boldsymbol{U}_k = \boldsymbol{I}_k$. Now, $\boldsymbol{C}'\boldsymbol{U}_k$ is $(k-1)\times k$. It has fewer rows than columns, which means that there exists a non-null $k \times 1$ vector $\boldsymbol{w}$ such that $\boldsymbol{C}'\boldsymbol{U}_k\boldsymbol{w} = \boldsymbol{0}$. Writing $\boldsymbol{y} = \boldsymbol{U}_k\boldsymbol{w}$, we find that $\boldsymbol{C}'\boldsymbol{y} = \boldsymbol{C}'\boldsymbol{U}_k\boldsymbol{w} = \boldsymbol{0}$ and $\boldsymbol{y}'\boldsymbol{y} = \boldsymbol{w}'\boldsymbol{U}_k'\boldsymbol{U}_k\boldsymbol{w} = \boldsymbol{w}'\boldsymbol{w}$. Therefore,

$$\max_{\boldsymbol{C}'\boldsymbol{x}=\boldsymbol{0}} \frac{\boldsymbol{x}'\boldsymbol{A}\boldsymbol{x}}{\boldsymbol{x}'\boldsymbol{x}} \geq \frac{\boldsymbol{y}'\boldsymbol{A}\boldsymbol{y}}{\boldsymbol{y}'\boldsymbol{y}} = \frac{\boldsymbol{w}'\boldsymbol{U}_k'\boldsymbol{A}\boldsymbol{U}_k\boldsymbol{w}}{\boldsymbol{w}'\boldsymbol{w}} = \frac{\sum_{i=1}^k \lambda_i w_i^2}{\sum_{i=1}^k w_i^2} \geq \lambda_k \frac{\sum_{i=1}^k w_i^2}{\sum_{i=1}^k w_i^2} = \lambda_k .$$

This establishes the desired upper bound for λ_k. The lower bound is established using an analogous argument and is left to the reader. □

The proof of the Courant-Fischer Theorem also reveals when equality is attained. If we set $\boldsymbol{C} = [\boldsymbol{u}_1 : \boldsymbol{u}_2 : \ldots : \boldsymbol{u}_{k-1}]$, then it is easy to verify that $\max_{\boldsymbol{x}\in\mathcal{N}(\boldsymbol{C}')} \boldsymbol{x}'\boldsymbol{A}\boldsymbol{x}/(\boldsymbol{x}'\boldsymbol{x}) = \lambda_k$. The lower bound can also be attained with an appropriate choice of $\boldsymbol{B}$. In other words, for each $k = 2, 3, \ldots, n$ there exists $n\times(k-1)$ matrices $\boldsymbol{B}_k$ and $\boldsymbol{C}_k$ such that

$$\min_{\boldsymbol{x}\in\mathcal{N}(\boldsymbol{B}_k')} \frac{\boldsymbol{x}'\boldsymbol{A}\boldsymbol{x}}{\boldsymbol{x}'\boldsymbol{x}} = \lambda_k = \max_{\boldsymbol{x}\in\mathcal{N}(\boldsymbol{C}_k')} \frac{\boldsymbol{x}'\boldsymbol{A}\boldsymbol{x}}{\boldsymbol{x}'\boldsymbol{x}} . \tag{13.18}$$

The following theorem provides the maximum of the ratio of two quadratic forms when at least one of the matrices is positive definite.

Theorem 13.31 Ratio of quadratic forms. *Let $\boldsymbol{A}$ and $\boldsymbol{B}$ be $n \times n$ matrices and suppose that $\boldsymbol{B}$ is positive definite. Then,*

$$\max_{\boldsymbol{x}\neq\boldsymbol{0}} \frac{\boldsymbol{x}'\boldsymbol{A}\boldsymbol{x}}{\boldsymbol{x}'\boldsymbol{B}\boldsymbol{x}} = \lambda_1(\boldsymbol{B}^{-1}\boldsymbol{A}) ,$$

where $\lambda_1(\boldsymbol{B}^{-1}\boldsymbol{A})$ is the largest eigenvalue of $\boldsymbol{B}^{-1}\boldsymbol{A}$. Also, the maximum is attained at $\boldsymbol{x}_0$ if and only if $\boldsymbol{x}_0$ is an eigenvector of $\boldsymbol{B}^{-1}\boldsymbol{A}$ corresponding to $\lambda_1(\boldsymbol{B}^{-1}\boldsymbol{A})$.

Proof. Since $\boldsymbol{B}$ is positive definite, there is a nonsingular matrix $\boldsymbol{R}$ such that $\boldsymbol{B} = \boldsymbol{R}'\boldsymbol{R}$ (Theorem 13.10). Although $\boldsymbol{B}^{-1}\boldsymbol{A}$ is not symmetric, it has real eigenvalues. This is because $\boldsymbol{B}^{-1} = \boldsymbol{R}^{-1}(\boldsymbol{R}^{-1})'$ and the eigenvalues of $\boldsymbol{B}^{-1}\boldsymbol{A} = \boldsymbol{R}^{-1}(\boldsymbol{R}^{-1})'\boldsymbol{A}$ are the same as those of the symmetric matrix $(\boldsymbol{R}^{-1})'\boldsymbol{A}\boldsymbol{R}^{-1}$. If $\boldsymbol{y} = \boldsymbol{R}\boldsymbol{x}$, then,

$$\max_{\boldsymbol{x}\neq\boldsymbol{0}} \frac{\boldsymbol{x}'\boldsymbol{A}\boldsymbol{x}}{\boldsymbol{x}'\boldsymbol{B}\boldsymbol{x}} = \max_{\boldsymbol{y}\neq\boldsymbol{0}} \frac{\boldsymbol{y}'\left(\boldsymbol{R}^{-1}\right)'\boldsymbol{A}\boldsymbol{R}^{-1}\boldsymbol{y}}{\boldsymbol{y}'\boldsymbol{y}} = \lambda_1\left(\left(\boldsymbol{R}^{-1}\right)'\boldsymbol{A}\boldsymbol{R}^{-1}\right) = \lambda_1(\boldsymbol{B}^{-1}\boldsymbol{A}) .$$

Also, $\boldsymbol{x}_0'\boldsymbol{A}\boldsymbol{x}_0/\boldsymbol{x}_0'\boldsymbol{B}\boldsymbol{x}_0 = \lambda_1(\boldsymbol{B}^{-1}\boldsymbol{A})$ if and only if $\boldsymbol{R}\boldsymbol{x}_0$ is an eigenvector of $(\boldsymbol{R}^{-1})'\boldsymbol{A}\boldsymbol{R}^{-1}$ corresponding to $\lambda_1(\boldsymbol{B}^{-1}\boldsymbol{A})$, which is the same as saying that $\boldsymbol{x}_0$ is an eigenvector of $\boldsymbol{B}^{-1}\boldsymbol{A}$ corresponding to $\lambda_1(\boldsymbol{B}^{-1}\boldsymbol{A})$. □

We next determine the minimum value of a positive definite quadratic form in $\boldsymbol{x}$ when $\boldsymbol{x}$ is subject to linear constraints. This result is widely used in the study of the general linear hypotheses in statistics.

Theorem 13.32 *Let $\boldsymbol{x}'\boldsymbol{\Lambda}\boldsymbol{x}$ be a positive definite quadratic form and let $\boldsymbol{A}\boldsymbol{x} = \boldsymbol{b}$ be a consistent system. Then*

$$\min_{\boldsymbol{A}\boldsymbol{x}=\boldsymbol{b}} \boldsymbol{x}'\boldsymbol{\Lambda}\boldsymbol{x} = \boldsymbol{b}'\boldsymbol{G}\boldsymbol{b}\,,$$

where $\boldsymbol{G}$ is a generalized inverse of $\boldsymbol{A}\boldsymbol{\Lambda}^{-1}\boldsymbol{A}'$. Furthermore, the minimum is attained at the unique point $\boldsymbol{x}_0 = \boldsymbol{\Lambda}^{-1}\boldsymbol{A}'\boldsymbol{G}\boldsymbol{b}$.

Proof. It is enough to prove this result for the case $\boldsymbol{\Lambda} = \boldsymbol{I}$. The general case follows easily from the special case by writing $\boldsymbol{\Lambda} = \boldsymbol{R}'\boldsymbol{R}$ where $\boldsymbol{R}$ is non-singular.

Recall that $\boldsymbol{H} = \boldsymbol{A}'(\boldsymbol{A}\boldsymbol{A}')^-$ is a generalized inverse of $\boldsymbol{A}$ and define the flat

$$\mathcal{W} = \{\boldsymbol{x} : \boldsymbol{A}\boldsymbol{x} = \boldsymbol{b}\} = \boldsymbol{H}\boldsymbol{b} + \mathcal{N}(\boldsymbol{A})\,.$$

Note that $\boldsymbol{H}\boldsymbol{b} \in \mathcal{W}$ and $-\boldsymbol{H}\boldsymbol{b} \in \mathcal{C}(\boldsymbol{A}') = (\mathcal{N}(\boldsymbol{A}))^{\perp}$, which implies that $\boldsymbol{H}\boldsymbol{b}$ is the orthogonal projection of $\boldsymbol{0}$ onto $\mathcal{W}$. It now follows that $\min\{\boldsymbol{x}'\boldsymbol{x} : \boldsymbol{A}\boldsymbol{x} = \boldsymbol{b}\}$ is attained at $\boldsymbol{H}\boldsymbol{b}$ and at no other point.

Also observe that $(\boldsymbol{A}\boldsymbol{A}^T)^-$ can itself be taken to be the Moore-Penrose inverse. Therefore,

$$\boldsymbol{b}'\boldsymbol{H}'\boldsymbol{H}\boldsymbol{b} = \boldsymbol{b}'(\boldsymbol{A}\boldsymbol{A}')^-\boldsymbol{b}\,,$$

which proves the theorem for $\boldsymbol{\Lambda} = \boldsymbol{I}$. □

13.8 Simultaneous diagonalization and the generalized eigenvalue problem

It is of interest to study the simultaneous reduction of two or more quadratic forms to diagonal forms by the same non-singular transformation. We note that this is not always possible and it is easy to construct quadratic forms that cannot be diagonalized by the same non-singular transformation. The conditions when such diagonalizations are possible, therefore, are interesting.

Theorem 13.33 *If both $\boldsymbol{A}$ and $\boldsymbol{B}$ are symmetric and at least one is positive definite, then there exists a nonsingular matrix $\boldsymbol{T}$ such that $\boldsymbol{x} = \boldsymbol{T}\boldsymbol{y}$ transforms $\boldsymbol{x}'\boldsymbol{A}\boldsymbol{x}$ to a diagonal form $\boldsymbol{y}'\boldsymbol{D}\boldsymbol{y}$ and $\boldsymbol{x}'\boldsymbol{B}\boldsymbol{x}$ to $\boldsymbol{y}'\boldsymbol{y}$.*

Proof. Without any loss of generality we may take $\boldsymbol{B}$ to be positive definite. If $\boldsymbol{Q}'\boldsymbol{B}\boldsymbol{Q} = \boldsymbol{\Lambda}$ is the spectral decomposition of $\boldsymbol{B}$, then $\boldsymbol{M} = \boldsymbol{Q}\boldsymbol{\Lambda}^{-1/2}$ is a nonsingular matrix such that $\boldsymbol{M}'\boldsymbol{B}\boldsymbol{M} = \boldsymbol{I}$.

Now $M'AM$ is symmetric, so there exists an orthogonal matrix P such that $P'M'AMP$ is diagonal. Clearly, $P'M'BMP = P'P = I$, which proves that if $T = MP$, then

$$T'AT = D \,, \quad \text{where } D \text{ is diagonal and } T'BT = I \,.$$

Therefore, $x = Ty$ is the required transformation that diagonalizes both $x'Ax$ and $x'Bx$. □

Theorem 13.33 also has a variant for nonnegative matrices.

Theorem 13.34 *Let A and B be $n \times n$ real-symmetric matrices such that A is n.n.d. and $\mathcal{C}(B) \subseteq \mathcal{C}(A)$. Then there exists a non-singular matrix P such that $P'AP$ and $P'BP$ are both diagonal.*

Proof. Let $\rho(A) = r$. By Theorem 13.10, there exists an $r \times n$ matrix R such that $A = R'R$. Then $\mathcal{C}(B) \subseteq \mathcal{C}(A) = \mathcal{C}(R')$, which implies that $B = R'CR$ for some $r \times r$ real-symmetric matrix C. Since C is real symmetric, there exists an orthogonal matrix Q such that $C = Q'DQ$, where D is diagonal. So, we can write

$$A = R'Q'QR \text{ and } B = R'Q'DQR \,.$$

Since QR is an $r \times n$ matrix with rank r, there exists a non-singular matrix P such that $QRP = [I_r : O]$. Therefore,

$$P'AP = P'R'Q'QRP = \begin{bmatrix} I_r \\ O \end{bmatrix} [I_r : O] = \begin{bmatrix} I_r & O \\ O & O \end{bmatrix}$$

and

$$P'BP = P'R'Q'DQRP = \begin{bmatrix} I_r \\ O \end{bmatrix} D[I_r : O] = \begin{bmatrix} D & O \\ O & O \end{bmatrix} .$$

This completes the proof. □

If both the matrices are n.n.d., then they can be simultaneously diagonalized by a nonsingular transformation.

Theorem 13.35 *If $x'Ax$ and $x'Bx$ are both nonnegative definite, then they can be simultaneously reduced to diagonal forms by a non-singular transformation.*

Proof. Since A and B are both n.n.d., so is $A + B$ and $\mathcal{C}(B) \subseteq \mathcal{C}(A + B)$ (recall Theorem 13.14). Theorem 13.34 ensures that there exists a nonsingular matrix P such that $P'(A + B)P$ and $P'BP$ are both diagonal. Thus, $P'AP$ is diagonal. □

The problem of simultaneously reducing quadratic forms to diagonal forms is closely related to what is known as the ***generalized eigenvalue problem*** in matrix analysis. Here, one studies the existence and computation of ***generalized eigenvalues*** λ and ***generalized eigenvectors*** $x \neq 0$ satisfying

$$Ax = \lambda Bx \,, \tag{13.19}$$

where $\boldsymbol{A}$ and $\boldsymbol{B}$ are $n \times n$ matrices. If $\boldsymbol{B}$ is nonsingular, (13.19) reduces to the standard eigenvalue equation $\boldsymbol{B}^{-1}\boldsymbol{A}\boldsymbol{x} = \lambda\boldsymbol{x}$ and the generalized eigenvalues and eigenvectors are simply the standard eigenvalues and eigenvectors of $\boldsymbol{B}^{-1}\boldsymbol{A}$. In this case, there are n eigenvalues. On the other hand, if $\boldsymbol{B}$ is singular, then matters are somewhat more complicated and it is possible that there are an infinite number of generalized eigenvalues.

An important special case of (13.19) arises when $\boldsymbol{A}$ is real symmetric and $\boldsymbol{B}$ is positive definite. This is known as the ***symmetric generalized eigenvalue problem***. Since $\boldsymbol{B}$ is positive definite, it is nonsingular so (13.19) reduces to the standard eigenvalue problem $\boldsymbol{B}^{-1}\boldsymbol{A}\boldsymbol{x} = \lambda\boldsymbol{x}$. However, this is no longer a symmetric eigenvalue problem because $\boldsymbol{B}^{-1}\boldsymbol{A}$ is not necessarily symmetric. Nevertheless, since $\boldsymbol{B}^{-1}$ is also positive definite, Corollary 13.5 ensures that its eigenvalues are all real.

The following theorem connects the generalized eigenvalue problem with simultaneous diagonalization.

Theorem 13.36 *Let $\boldsymbol{A}$ and $\boldsymbol{B}$ be $n \times n$ matrices, where $\boldsymbol{A}$ is real symmetric and $\boldsymbol{B}$ is positive definite. Then the generalized eigenvalue problem $\boldsymbol{A}\boldsymbol{x} = \lambda\boldsymbol{B}\boldsymbol{x}$ has n real generalized eigenvalues, and there exist n generalized eigenvectors $\boldsymbol{x}_1, \boldsymbol{x}_2, \ldots, \boldsymbol{x}_n$ such that $\boldsymbol{x}_i'\boldsymbol{B}\boldsymbol{x}_j = \boldsymbol{0}$ for $i \neq j$. Also, there exists a nonsingular matrix $\boldsymbol{T}$ such that $\boldsymbol{T}'\boldsymbol{A}\boldsymbol{T} = \boldsymbol{D}$ and $\boldsymbol{T}'\boldsymbol{B}\boldsymbol{T} = \boldsymbol{I}$, where $\boldsymbol{D}$ is a diagonal matrix whose diagonal entries are the eigenvalues of $\boldsymbol{B}^{-1}\boldsymbol{A}$.*

Proof. The trick is to convert $\boldsymbol{A}\boldsymbol{x} = \lambda\boldsymbol{B}\boldsymbol{x}$ to a symmetric eigenvalue problem and borrow from established results. The positive definiteness of $\boldsymbol{B}$ allows us to do this. Theorem 13.10 ensures that $\boldsymbol{B} = \boldsymbol{R}'\boldsymbol{R}$ for some nonsingular matrix $\boldsymbol{R}$. Observe that

$$\begin{aligned}\boldsymbol{A}\boldsymbol{x} = \lambda\boldsymbol{B}\boldsymbol{x} &\Longrightarrow \boldsymbol{A}\boldsymbol{x} = \lambda\boldsymbol{R}'\boldsymbol{R}\boldsymbol{x} \Longrightarrow (\boldsymbol{R}')^{-1}\boldsymbol{A}\boldsymbol{x} = \lambda\boldsymbol{R}\boldsymbol{x} \\ &\Longrightarrow (\boldsymbol{R}^{-1})'\boldsymbol{A}\boldsymbol{R}^{-1}\boldsymbol{R}\boldsymbol{x} = \lambda\boldsymbol{R}\boldsymbol{x} \Longrightarrow \boldsymbol{C}\boldsymbol{z} = \lambda\boldsymbol{z}\ ,\end{aligned}$$

where $\boldsymbol{C} = (\boldsymbol{R}^{-1})'\boldsymbol{A}\boldsymbol{R}^{-1}$ is real symmetric and $\boldsymbol{z} = \boldsymbol{R}\boldsymbol{x}$ is an eigenvector associated with the eigenvalue λ of $\boldsymbol{C}$. Since $\boldsymbol{C}$ is real symmetric, it has n real eigenvalues $\lambda_1, \lambda_2, \ldots, \lambda_n$ and a set of orthonormal eigenvectors $\boldsymbol{z}_1, \boldsymbol{z}_2, \ldots, \boldsymbol{z}_n$ such that $\boldsymbol{C}\boldsymbol{z}_i = \lambda_i\boldsymbol{z}_i$ for $i = 1, 2, \ldots, n$ (the spectral decomposition ensures this). Observe that $\boldsymbol{x}_i = \boldsymbol{R}^{-1}\boldsymbol{z}_i$ are generalized eigenvectors and

$$\boldsymbol{x}_i'\boldsymbol{B}\boldsymbol{x}_j = \boldsymbol{x}_i'\boldsymbol{R}'\boldsymbol{R}\boldsymbol{x}_j = \boldsymbol{z}_i'\boldsymbol{z}_j = 0 \text{ if } i \neq j\ .$$

Let $\boldsymbol{P} = [\boldsymbol{z}_1 : \boldsymbol{z}_2 : \ldots : \boldsymbol{z}_n]$ be the $n \times n$ orthogonal matrix with $\boldsymbol{z}_i$'s as its columns. Then $\boldsymbol{P}'\boldsymbol{C}\boldsymbol{P} = \boldsymbol{D}$ is the spectral decomposition of $\boldsymbol{C}$, where $\boldsymbol{D}$ is a diagonal matrix with λ_i's along its diagonals. Letting $\boldsymbol{T} = \boldsymbol{R}^{-1}\boldsymbol{P}$, we see that

$$\begin{aligned}\boldsymbol{T}'\boldsymbol{A}\boldsymbol{T} &= \boldsymbol{P}'(\boldsymbol{R}^{-1})'\boldsymbol{A}\boldsymbol{R}^{-1}\boldsymbol{P} = \boldsymbol{P}'\boldsymbol{C}\boldsymbol{P} = \boldsymbol{D} \\ \text{and } \boldsymbol{T}'\boldsymbol{B}\boldsymbol{T} &= \boldsymbol{P}'(\boldsymbol{R}^{-1})'\boldsymbol{R}'\boldsymbol{R}\boldsymbol{R}^{-1}\boldsymbol{P} = \boldsymbol{I}\ .\end{aligned}$$

Finally, note that each λ_i satisfies $\boldsymbol{A} = \lambda_i\boldsymbol{B}\boldsymbol{x}_i$ and, hence, is an eigenvalue of $\boldsymbol{B}^{-1}\boldsymbol{A}$. □

The above result also has computational implications for the symmetric generalized eigenvalue problem. Rather than solving the unsymmetric standard eigenvalue problem $\boldsymbol{B}^{-1}\boldsymbol{A}\boldsymbol{x} = \lambda\boldsymbol{x}$, it is computationally more efficient to solve the equivalent symmetric problem $\boldsymbol{C}\boldsymbol{z} = \lambda\boldsymbol{z}$, where $\boldsymbol{C} = (\boldsymbol{R}^{-1})'\boldsymbol{A}\boldsymbol{R}^{-1}$, $\boldsymbol{z} = \boldsymbol{R}\boldsymbol{x}$ and $\boldsymbol{R}$ can, for numerical efficiency, be obtained from the Cholesky decomposition $\boldsymbol{B} = \boldsymbol{R}'\boldsymbol{R}$.

Simultaneous diagonalization has several applications and can be effectively used to derive some inequalities involving n.n.d. matrices. For example, we can offer a different proof for Theorem 13.25. Because $\boldsymbol{A}$ and $\boldsymbol{B}$ are both $n \times n$ and n.n.d., they can be simultaneously diagonalized using a nonsingular matrix $\boldsymbol{P}$. Suppose $\boldsymbol{P}'\boldsymbol{A}\boldsymbol{P} = \boldsymbol{D}_{\boldsymbol{A}}$ and $\boldsymbol{P}'\boldsymbol{B}\boldsymbol{P} = \boldsymbol{D}_{\boldsymbol{B}}$, where $\boldsymbol{D}_{\boldsymbol{A}}$ and $\boldsymbol{D}_{\boldsymbol{B}}$ are diagonal matrices with diagonal entries $\alpha_1, \alpha_2, \cdots, \alpha_n$ and $\beta_1, \beta_2, \cdots, \beta_n$ that are all nonnegative. Then $\boldsymbol{P}'(\boldsymbol{A} + \boldsymbol{B})\boldsymbol{P} = \boldsymbol{D}_{\boldsymbol{A}+\boldsymbol{B}}$, where $\boldsymbol{D}_{\boldsymbol{A}+\boldsymbol{B}}$ is diagonal with entries $\alpha_1 + \beta_1, \alpha_2 + \beta_2, \cdots, \alpha_n + \beta_n$. This implies

$$|\boldsymbol{A} + \boldsymbol{B}| = \frac{(\alpha_1 + \beta_1)\cdots(\alpha_n + \beta_n)}{|\boldsymbol{P}|^2} \geq \frac{\alpha_1 \cdots \alpha_n}{|\boldsymbol{P}|^2} + \frac{\beta_1 \cdots \beta_n}{|\boldsymbol{P}|^2} = |\boldsymbol{A}| + |\boldsymbol{B}| \,.$$

If $\boldsymbol{A}$ and $\boldsymbol{B}$ are both p.d., and $n \geq 2$, then α's and β's are strictly positive and strict inequality holds above. Below is another useful inequality that is easily proved using simultaneous diagonalization.

Theorem 13.37 *Let $\boldsymbol{A}$ and $\boldsymbol{B}$ be two positive definite matrices of the same order such that $\boldsymbol{A} - \boldsymbol{B}$ is nonnegative definite. Then $\boldsymbol{B}^{-1} - \boldsymbol{A}^{-1}$ is nonnegative definite.*

Proof. Theorem 13.33 ensures that there exists a non-singular matrix $\boldsymbol{P}$ such that $\boldsymbol{P}'\boldsymbol{A}\boldsymbol{P} = \boldsymbol{C}$ and $\boldsymbol{P}'\boldsymbol{B}\boldsymbol{P} = \boldsymbol{D}$ are both diagonal. Because $\boldsymbol{A} - \boldsymbol{B}$ is n.n.d, so is $\boldsymbol{C} - \boldsymbol{D} = \boldsymbol{P}'(\boldsymbol{A} - \boldsymbol{B})\boldsymbol{P}$, which means that $c_{ii} \geq d_{ii}$ for each i. Hence $1/d_{ii} \geq 1/c_{ii}$ and $\boldsymbol{D}^{-1} - \boldsymbol{C}^{-1}$ is nonnegative definite. This means that $\boldsymbol{B}^{-1} - \boldsymbol{A}^{-1} = \boldsymbol{P}(\boldsymbol{D}^{-1} - \boldsymbol{C}^{-1})\boldsymbol{P}'$ is nonnegative definite. □

Hitherto, we have discussed sufficient conditions for two quadratic forms to be simultaneously diagonalizable using the same nonsingular transformation. However, if we restrict ourselves to orthogonal transformations, the condition becomes more stringent and we can derive necessary and sufficient condition for quadratic forms to be simultaneously diagonalizable. We present this below for two matrices. Note that this result applies to any real symmetric matrix, not just p.d. or n.n.d matrices.

Theorem 13.38 *Let $\boldsymbol{A}$ and $\boldsymbol{B}$ be two $n \times n$ real symmetric matrices. There exists an $n \times n$ orthogonal matrix $\boldsymbol{P}$ such that $\boldsymbol{P}'\boldsymbol{A}\boldsymbol{P}$ and $\boldsymbol{P}'\boldsymbol{B}\boldsymbol{P}$ are each diagonal if and only if $\boldsymbol{A}\boldsymbol{B} = \boldsymbol{B}\boldsymbol{A}$.*

Proof. If $\boldsymbol{P}'\boldsymbol{A}\boldsymbol{P}$ and $\boldsymbol{P}'\boldsymbol{B}\boldsymbol{P}$ are both diagonal, then $\boldsymbol{P}'\boldsymbol{A}\boldsymbol{P}$ and $\boldsymbol{P}'\boldsymbol{B}\boldsymbol{P}$ commute. Since $\boldsymbol{P}$ is orthogonal we have $\boldsymbol{P}\boldsymbol{P}' = \boldsymbol{I}$ and

$$\begin{aligned} \boldsymbol{A}\boldsymbol{B} &= \boldsymbol{A}\boldsymbol{P}\boldsymbol{P}'\boldsymbol{B} = \boldsymbol{P}\boldsymbol{P}'\boldsymbol{A}\boldsymbol{P}\boldsymbol{P}'\boldsymbol{B}\boldsymbol{P}\boldsymbol{P}' = \boldsymbol{P}(\boldsymbol{P}'\boldsymbol{A}\boldsymbol{P})(\boldsymbol{P}'\boldsymbol{B}\boldsymbol{P})\boldsymbol{P}' \\ &= \boldsymbol{P}(\boldsymbol{P}'\boldsymbol{B}\boldsymbol{P})(\boldsymbol{P}'\boldsymbol{A}\boldsymbol{P})\boldsymbol{P}' = \boldsymbol{B}\boldsymbol{P}\boldsymbol{P}'\boldsymbol{A} = \boldsymbol{B}\boldsymbol{A} \,. \end{aligned}$$

This proves the "only if" part.

The "if" part is slightly more involved. Assume that $AB = BA$. Suppose that $\lambda_1, \lambda_2, \ldots, \lambda_k$ are the *distinct* eigenvalues of A with multiplicities $m_1, m_2, \ldots, m_k$, respectively, where $\sum_{i=1}^k m_i = n$. The spectral decomposition of A ensures that there is an orthogonal matrix Q such that

$$Q'AQ = \Lambda = \begin{bmatrix} \lambda_1 I_{m_1} & O & \ldots & O \\ O & \lambda_2 I_{m_2} & \ldots & O \\ \vdots & \vdots & \ddots & \vdots \\ O & O & \ldots & \lambda_k I_{m_k} \end{bmatrix}. \tag{13.20}$$

Consider the real symmetric matrix $C = Q'BQ$. This is also an $n \times n$ matrix and let us partition it conformably with (13.20) as

$$C = \begin{bmatrix} C_{11} & C_{12} & \ldots & C_{1k} \\ C_{21} & C_{22} & \ldots & C_{2k} \\ \vdots & \vdots & \ddots & \vdots \\ C_{k1} & C_{k2} & \ldots & C_{kk} \end{bmatrix},$$

where each C_{ij} is $m_i \times m_j$. Since $AB = BA$ and $QQ' = I$, we find

$$\begin{aligned} \Lambda C &= Q'AQQ'BQ = Q'ABQ = Q'BAQ = Q'B(QQ')AQ \\ &= Q'BQQ'BQ = C\Lambda \,. \end{aligned}$$

Writing $\Lambda C = C\Lambda$ in terms of the blocks in Λ and C, we obtain

$$\begin{bmatrix} \lambda_1 C_{11} & \lambda_1 C_{12} & \ldots & \lambda_1 C_{1k} \\ \lambda_2 C_{21} & \lambda_2 C_{22} & \ldots & \lambda_2 C_{2k} \\ \vdots & \vdots & \ddots & \vdots \\ \lambda_k C_{k1} & \lambda_k C_{k2} & \ldots & \lambda_k C_{kk} \end{bmatrix} = \begin{bmatrix} \lambda_1 C_{11} & \lambda_2 C_{12} & \ldots & \lambda_k C_{1k} \\ \lambda_1 C_{21} & \lambda_2 C_{22} & \ldots & \lambda_k C_{2k} \\ \vdots & \vdots & \ddots & \vdots \\ \lambda_1 C_{k1} & \lambda_1 C_{k2} & \ldots & \lambda_k C_{kk} \end{bmatrix}.$$

Equating the (i, j)-th blocks, we see that $\lambda_i C_{ij} = \lambda_j C_{ij}$ or that

$$(\lambda_i - \lambda_j) C_{ij} = O \,.$$

Since the λ_i's are distinct, we conclude that $C_{ij} = O$ whenever $i \neq j$. In other words, all the off-diagonal blocks in C are zero, so C is a block diagonal matrix.

At this point, we have that $Q'AQ$ is diagonal and $C = Q'BQ$ is block diagonal. Our next step will be to find another orthogonal matrix that will reduce C to a diagonal matrix and will not destroy the diagonal structure of $Q'AQ$. The key lies in the observation that each C_{ii} is real symmetric matrix because C is real symmetric. Therefore, each C_{ii} has a spectral decomposition $Z_i' C_{ii} Z_i = D_i$, where D_i is an $m_i \times m_i$ diagonal matrix with the eigenvalues of C_{ii} along its diagonal and Z_i is an $m_i \times m_i$ orthogonal matrix, so $Z_i' Z_i = I$ for $i = 1, 2, \ldots, k$.

Let Z be the block diagonal matrix with Z_i's as its diagonal blocks. Clearly, Z is

$n \times n$ and $\boldsymbol{Z}'\boldsymbol{Z} = \boldsymbol{I}_n$, so $\boldsymbol{Z}$ is orthogonal. Now form $\boldsymbol{P} = \boldsymbol{Q}\boldsymbol{Z}$ and observe that

$$\boldsymbol{P}'\boldsymbol{B}\boldsymbol{P} = \boldsymbol{Z}'\boldsymbol{C}\boldsymbol{Z} = \begin{bmatrix} \boldsymbol{D}_1 & \boldsymbol{O} & \dots & \boldsymbol{O} \\ \boldsymbol{O} & \boldsymbol{D}_2 & \dots & \boldsymbol{O} \\ \vdots & \vdots & \ddots & \vdots \\ \boldsymbol{O} & \boldsymbol{O} & \dots & \boldsymbol{D}_k \end{bmatrix},$$

which is diagonal. Also,

$$\boldsymbol{P}'\boldsymbol{A}\boldsymbol{P} = \boldsymbol{Z}'\boldsymbol{Q}'\boldsymbol{A}\boldsymbol{Q}\boldsymbol{Z} = \begin{bmatrix} \lambda_1\boldsymbol{Z}_1'\boldsymbol{Z}_1 & \boldsymbol{O} & \dots & \boldsymbol{O} \\ \boldsymbol{O} & \lambda_2\boldsymbol{Z}_2'\boldsymbol{Z}_2 & \dots & \boldsymbol{O} \\ \vdots & \vdots & \ddots & \vdots \\ \boldsymbol{O} & \boldsymbol{O} & \dots & \lambda_k\boldsymbol{Z}_k'\boldsymbol{Z}_k \end{bmatrix},$$

which is equal to $\boldsymbol{\Lambda}$ in (13.20) because $\boldsymbol{Z}_i'\boldsymbol{Z}_i = \boldsymbol{I}_{m_i}$. This proves that $\boldsymbol{P}'\boldsymbol{A}\boldsymbol{P}$ and $\boldsymbol{P}'\boldsymbol{B}\boldsymbol{P}$ are both diagonal. □

The above process can be repeated for a set of p matrices that commute pairwise. However, there is a neat alternative way to prove this result using the lemma below, which, in its own right, is quite a remarkable result.

Lemma 13.5 *If $\boldsymbol{A}$ is any $n \times n$ matrix and $\boldsymbol{x}$ is any $n \times 1$ nonzero vector, then $\boldsymbol{A}$ has an eigenvector of the form*

$$\boldsymbol{y} = c_0\boldsymbol{x} + c_1\boldsymbol{A}\boldsymbol{x} + c_2\boldsymbol{A}^2\boldsymbol{x} + \dots + c_k\boldsymbol{A}^k\boldsymbol{x}$$

for some nonnegative integer k.

Proof. If $\boldsymbol{x}$ is itself an eigenvector, then the above is true for $k = 0$ and $c_0 = 1$. In that case $\{\boldsymbol{x}, \boldsymbol{A}\boldsymbol{x}\}$ is linearly dependent. Now suppose that $\boldsymbol{x}$ is not an eigenvector of $\boldsymbol{A}$. The set $\{\boldsymbol{x}, \boldsymbol{A}\boldsymbol{x}, \boldsymbol{A}^2\boldsymbol{x}, \dots\}$ consisting of $\boldsymbol{x}$ and $\boldsymbol{A}^i\boldsymbol{x}$ for $i = 1, 2, \dots$ must eventually become linearly dependent for $i \geq n$ because no set comprising $n+1$ or more $n \times 1$ vectors can be linearly independent. Let k be the smallest positive integer such that $\{\boldsymbol{x}, \boldsymbol{A}\boldsymbol{x}, \dots, \boldsymbol{A}^k\boldsymbol{x}\}$ is linearly dependent. Suppose

$$c_0\boldsymbol{x} + c_1\boldsymbol{A}\boldsymbol{x} + c_2\boldsymbol{A}^2\boldsymbol{x} + \dots + c_k\boldsymbol{A}^k\boldsymbol{x} = \boldsymbol{0}, \tag{13.21}$$

where $c_k \neq 0$. Let $\lambda_1, \lambda_2, \dots, \lambda_k$ be the roots of the polynomial

$$f(t) = c_0 + c_1 t + c_2 t^2 + \dots + c_k t^k,$$

so that $f(t) = c_k(t - \lambda_1)(t - \lambda_2)\cdots(t - \lambda_k)$ and

$$\begin{aligned} f(\boldsymbol{A}) &= c_0 + c_1\boldsymbol{A} + c_2\boldsymbol{A}^2 + \dots + c_k\boldsymbol{A}^k \\ &= c_k(\boldsymbol{A} - \lambda_1\boldsymbol{I})(\boldsymbol{A} - \lambda_2\boldsymbol{I})\cdots(\boldsymbol{A} - \lambda_k\boldsymbol{I}). \end{aligned}$$

From (13.21) it follows that $f(\boldsymbol{A})\boldsymbol{x} = \boldsymbol{0}$. Consider the vector $\boldsymbol{y} = c_k(\boldsymbol{A} - \lambda_2\boldsymbol{I})\cdots(\boldsymbol{A} - \lambda_k\boldsymbol{I})\boldsymbol{x}$. Note that $\boldsymbol{y} \neq \boldsymbol{0}$—otherwise k would not be the smallest integer for which (13.21) is true. Also,

$$(\boldsymbol{A} - \lambda_1\boldsymbol{I})\boldsymbol{y} = c_k(\boldsymbol{A} - \lambda_1\boldsymbol{I})(\boldsymbol{A} - \lambda_2\boldsymbol{I})\cdots(\boldsymbol{A} - \lambda_k\boldsymbol{I})\boldsymbol{x} = f(\boldsymbol{A})\boldsymbol{x} = \boldsymbol{0},$$

which implies that $\boldsymbol{Ay} = \lambda_1 \boldsymbol{y}$ so $\boldsymbol{y}$ is an eigenvector of $\boldsymbol{A}$. $\square$

Note that the above result holds for any matrix. The polynomial $f(t)$ constructed above is, up to the constant c_k, the minimal polynomial of $\boldsymbol{A}$. Note that λ_1 is an eigenvalue of $\boldsymbol{A}$ and it can be argued, using symmetry, that each of the λ_i's is an eigenvalue of $\boldsymbol{A}$. If $\boldsymbol{A}$ is real and symmetric, then all its eigenvalues are real and $\boldsymbol{y}$ is real whenever $\boldsymbol{x}$ is real. We will use this result to prove the following extension of Theorem 13.38.

Theorem 13.39 *Let $\boldsymbol{A}_1, \boldsymbol{A}_2, \ldots, \boldsymbol{A}_p$ be a set of $n \times n$ real-symmetric matrices. Then, there exists an $n \times n$ orthogonal matrix $\boldsymbol{P}$ such that $\boldsymbol{P}'\boldsymbol{A}_i\boldsymbol{P}$ are each diagonal, for $i = 1, 2, \ldots, p$, if and only if $\boldsymbol{A}_1, \boldsymbol{A}_2, \cdots, \boldsymbol{A}_p$ commute pairwise.*

Proof. If each $\boldsymbol{P}'\boldsymbol{A}_i\boldsymbol{P}$ are diagonal, then $\boldsymbol{P}'\boldsymbol{A}_i\boldsymbol{P}$ and $\boldsymbol{P}'\boldsymbol{A}_j\boldsymbol{P}$. Since $\boldsymbol{P}$ is orthogonal, $\boldsymbol{PP}' = \boldsymbol{I}$ and we have

$$\begin{aligned} \boldsymbol{A}_i\boldsymbol{A}_j &= \boldsymbol{A}_i\boldsymbol{PP}'\boldsymbol{A}_j = \boldsymbol{PP}'\boldsymbol{A}_i\boldsymbol{PP}'\boldsymbol{A}_j\boldsymbol{PP}' = \boldsymbol{P}(\boldsymbol{P}'\boldsymbol{A}_i\boldsymbol{P})(\boldsymbol{P}'\boldsymbol{A}_j\boldsymbol{P})\boldsymbol{P}' \\ &= \boldsymbol{P}(\boldsymbol{P}'\boldsymbol{A}_j\boldsymbol{P})(\boldsymbol{P}'\boldsymbol{A}_i\boldsymbol{P})\boldsymbol{P}' = \boldsymbol{A}_j\boldsymbol{PP}'\boldsymbol{A}_i = \boldsymbol{A}_j\boldsymbol{A}_i \,. \end{aligned}$$

This proves the "only if" part.

We prove the "if" part by induction on n, the size of the matrix. The result is trivial for $n = 1$. So, assume it holds for all real-symmetric matrices of order $n - 1$ and let $\boldsymbol{A}_1, \boldsymbol{A}_2, \cdots, \boldsymbol{A}_p$ be a collection of $n \times n$ matrices such that $\boldsymbol{A}_i\boldsymbol{A}_j = \boldsymbol{A}_j\boldsymbol{A}_i$ for every pair $i \neq j$. Let $\boldsymbol{x}_1$ be a real eigenvector of $\boldsymbol{A}_1$ corresponding to λ_1 and let

$$\boldsymbol{x}_2 = c_0\boldsymbol{x}_1 + c_1\boldsymbol{A}_2\boldsymbol{x}_1 + c_2\boldsymbol{A}_2^2\boldsymbol{x}_1 + \cdots + c_k\boldsymbol{A}_2^k\boldsymbol{x}_1$$

be a real eigenvector of $\boldsymbol{A}_2$, as ensured by Lemma 13.5. Observe that

$$\begin{aligned} \boldsymbol{A}_1\boldsymbol{x}_2 &= \boldsymbol{A}_1\left(c_0\boldsymbol{x}_1 + c_1\boldsymbol{A}_2\boldsymbol{x}_1 + c_2\boldsymbol{A}_2^2\boldsymbol{x}_1 + \cdots + c_k\boldsymbol{A}_2^k\boldsymbol{x}_1\right) \\ &= c_0\boldsymbol{A}_1\boldsymbol{x}_1 + c_1\boldsymbol{A}_1\boldsymbol{A}_2\boldsymbol{x}_1 + c_2\boldsymbol{A}_1\boldsymbol{A}_2^2\boldsymbol{x}_1 + \cdots + c_k\boldsymbol{A}_1\boldsymbol{A}_2^k\boldsymbol{x}_1 \\ &= c_0\boldsymbol{A}_1\boldsymbol{x}_1 + c_1\boldsymbol{A}_2\boldsymbol{A}_1\boldsymbol{x}_1 + c_2\boldsymbol{A}_2^2\boldsymbol{A}_1\boldsymbol{x}_1 + \cdots + c_k\boldsymbol{A}_2^k\boldsymbol{A}_1\boldsymbol{x}_1 \\ &= c_0\lambda_1\boldsymbol{x}_1 + c_1\boldsymbol{A}_2\lambda_1\boldsymbol{x}_1 + c_2\boldsymbol{A}_2^2\lambda_1\boldsymbol{x}_1 + \cdots + c_k\boldsymbol{A}_2^k\lambda_1\boldsymbol{x}_1 \\ &= \lambda_1(c_0\boldsymbol{x}_1 + c_1\boldsymbol{A}_2\boldsymbol{x}_1 + c_2\boldsymbol{A}_2^2\boldsymbol{x}_1 + \cdots + c_k\boldsymbol{A}_2^k\boldsymbol{x}_1) = \lambda_1\boldsymbol{x}_2 \,. \end{aligned}$$

Therefore, $\boldsymbol{x}_2$ is a real eigenvector of $\boldsymbol{A}_1$ as well. That is, $\boldsymbol{x}_2$ is a real eigenvector of $\boldsymbol{A}_1$ and $\boldsymbol{A}_2$. Similarly, by letting $\boldsymbol{x}_3$ be a real eigenvector of $\boldsymbol{A}_3$ spanned by $\boldsymbol{x}_2, \boldsymbol{A}_3\boldsymbol{x}_2, \boldsymbol{A}_3^2\boldsymbol{x}_2, \ldots$, we can establish that $\boldsymbol{x}_3$ is a common eigenvector of $\boldsymbol{A}_1$, $\boldsymbol{A}_2$ and $\boldsymbol{A}_3$. Continuing this process, we eventually obtain a real vector $\boldsymbol{x}$ that is an eigenvector of $\boldsymbol{A}_1, \boldsymbol{A}_, \ldots, \boldsymbol{A}_p$.

Suppose we normalize this vector and define $\boldsymbol{q}_1 = \boldsymbol{x}/\|\boldsymbol{x}\|$ so that $\|\boldsymbol{q}_1\| = 1$. Let $\boldsymbol{Q} = [\boldsymbol{q}_1 : \boldsymbol{Q}_2]$ be an $n \times n$ orthogonal matrix with $\boldsymbol{q}_1$ as its first column and let λ_i be the eigenvalue of $\boldsymbol{A}_i$ associated with $\boldsymbol{x}$. Using the facts that $\boldsymbol{Aq}_i = \lambda_1\boldsymbol{q}_i$ and $\boldsymbol{Q}_2'\boldsymbol{q}_1 = \boldsymbol{0}$, we obtain

$$\boldsymbol{Q}'\boldsymbol{A}_i\boldsymbol{Q} = \begin{bmatrix} \boldsymbol{q}_1' \\ \boldsymbol{Q}_2' \end{bmatrix} \boldsymbol{A}_i[\boldsymbol{q}_1 : \boldsymbol{Q}_2] = \begin{bmatrix} \lambda_1 & \boldsymbol{q}_1'\boldsymbol{A}_i\boldsymbol{Q}_2 \\ \lambda_i\boldsymbol{Q}_2'\boldsymbol{q}_1 & \boldsymbol{Q}_2'\boldsymbol{A}\boldsymbol{Q}_2 \end{bmatrix} = \begin{bmatrix} \lambda_1 & \boldsymbol{0}' \\ \boldsymbol{0} & \boldsymbol{Q}_2'\boldsymbol{A}_i\boldsymbol{Q}_2 \end{bmatrix}$$

for $i = 1, 2, \ldots, p$. Observe that the $\boldsymbol{Q}'\boldsymbol{A}_i\boldsymbol{Q}$'s commute pairwise because the $\boldsymbol{A}_i$'s commute pairwise. This implies that the $\boldsymbol{Q}_2'\boldsymbol{A}_i\boldsymbol{Q}_2$'s also commute pairwise. Since each $\boldsymbol{Q}_2'\boldsymbol{A}_i\boldsymbol{Q}_2$ has size $(n-1) \times (n-1)$, the induction hypothesis ensures that there is an orthogonal matrix $\boldsymbol{Z}$ such that $\boldsymbol{Z}'(\boldsymbol{Q}_2'\boldsymbol{A}_i\boldsymbol{Q}_2)\boldsymbol{Z}$ is diagonal for each $i = 1, 2, \ldots, p$. Letting

$$\boldsymbol{P} = \boldsymbol{Q}\begin{bmatrix} 1 & \boldsymbol{0}' \\ \boldsymbol{0} & \boldsymbol{Z} \end{bmatrix},$$

we obtain, for each $i = 1, 2, \ldots, p$,

$$\boldsymbol{P}'\boldsymbol{A}_i\boldsymbol{P} = \begin{bmatrix} 1 & \boldsymbol{0}' \\ \boldsymbol{0} & \boldsymbol{Z}' \end{bmatrix}(\boldsymbol{Q}'\boldsymbol{A}_i\boldsymbol{Q})\begin{bmatrix} 1 & \boldsymbol{0}' \\ \boldsymbol{0} & \boldsymbol{Z} \end{bmatrix} = \begin{bmatrix} \lambda_i & \boldsymbol{0}' \\ \boldsymbol{0} & \boldsymbol{Z}'(\boldsymbol{Q}_2'\boldsymbol{A}_i\boldsymbol{Q}_2)\boldsymbol{Z} \end{bmatrix},$$

which is diagonal. □

13.9 Exercises

1. Express each of the following quadratic forms as $\boldsymbol{x}'\boldsymbol{A}\boldsymbol{x}$, where $\boldsymbol{A}$ is symmetric:
   (a) $(x_1 - x_2)^2$, where $\boldsymbol{x} \in \Re^2$;
   (b) $x_1^2 - 2x_2^2 + 3x_3^2 - 4x_1x_2 + 5x_1x_3 - 6x_2x_3$, where $\boldsymbol{x} \in \Re^3$;
   (c) x_2x_3, where $\boldsymbol{x} \in \Re^3$;
   (d) $x_1^2 + 2x_3^2 - x_1x_3 + x_1x_2$, where $\boldsymbol{x} \in \Re^3$;
   (e) $(x_1 + 2x_2 + 3x_3)^2$, where $\boldsymbol{x} \in \Re^3$.
2. Find $\boldsymbol{A}$ such that $(\boldsymbol{u}'\boldsymbol{x})^2 = \boldsymbol{x}'\boldsymbol{A}\boldsymbol{x}$, where $\boldsymbol{u}$ and $\boldsymbol{x}$ are $n \times 1$.
3. Let $\boldsymbol{x} = \{x_i\}$ be an $n \times 1$ vector. Consider the following expressions:
   (a) $n\bar{x}^2$, where $\bar{x} = \dfrac{1}{n}(x_1 + x_2 + \cdots + x_n)$;
   (b) $\sum_{i=1}^n (x_i - \bar{x})^2$.

   Express each of the above as $\boldsymbol{x}'\boldsymbol{A}\boldsymbol{x}$, where $\boldsymbol{A}$ is symmetric. Show that in each of the above cases, $\boldsymbol{A}$ is idempotent and they add up to $\boldsymbol{I}_n$.
4. Consider the quadratic form $q_{\boldsymbol{A}}(\boldsymbol{x}) = \boldsymbol{x}'\boldsymbol{A}\boldsymbol{x}$, where $\boldsymbol{x} \in \Re^3$ and

   $$\boldsymbol{A} = \begin{bmatrix} 1 & 1 & 0 \\ 1 & 2 & 1/2 \\ 0 & 1/2 & -1 \end{bmatrix}.$$

   Let $y_1 = x_1 - x_3$, $y_2 = x_2 - x_3$ and $y_3 = x_3$. Find the matrix $\boldsymbol{B}$ such that $\boldsymbol{y}'\boldsymbol{B}\boldsymbol{y} = \boldsymbol{x}'\boldsymbol{A}\boldsymbol{x}$.
5. Reduce each of the quadratic forms in Exercise 1 to diagonal form. Express $\boldsymbol{A} = \boldsymbol{L}\boldsymbol{D}\boldsymbol{L}'$ in each case.
6. Reduce $x_1x_2 + x_2x_3 + x_3x_1$ to a diagonal form.
7. If $\boldsymbol{A}$ and $\boldsymbol{B}$ are nonnegative definite, then prove that $\boldsymbol{C} = \begin{bmatrix} \boldsymbol{A} & \boldsymbol{O} \\ \boldsymbol{O} & \boldsymbol{B} \end{bmatrix}$ is nonnegative definite. Is $\boldsymbol{C}$ p.d. if $\boldsymbol{A}$ is p.d. and $\boldsymbol{B}$ is n.n.d?

8. If A is $n \times n$ and nonnegative definite and P is any $n \times k$ matrix, then prove that $P'AP$ is nonnegative definite.

9. Let A and B be nonnegative definite matrices of the same order. Prove that $A + B = O$ if and only if $A = B = O$.

10. True or false: If A and B are symmetric and $A^2 + B^2 = O$, then $A = B = O$.

11. Let $\mathbf{1}$ be the $n \times 1$ vector of 1's. Prove that $(1 - \rho)I_n + \rho\mathbf{1}\mathbf{1}'$ is positive definite if and only if $-1/(n-1) < \rho < 1$.

12. If A is real and symmetric, then prove that $\alpha I + A$ is positive definite for some real number α.

13. If A is n.n.d., then show that $x'Ax = 0$ if and only if $Ax = 0$.

14. If A is $n \times n$, then show that $x'Ax = 0$ if and only if $u'Ax = 0$ for every $u \in \Re^n$.

15. Prove that every orthogonal projector is a nonnegative definite matrix. Is there an orthogonal projector that is positive definite?

16. If A is p.d., then show that A^k is p.d. for every positive integer k.

17. If A is n.n.d. and p is any positive integer, then prove that there exists a unique n.n.d. matrix B such that $B^p = A$.

18. Let A be an $n \times n$ n.n.d. matrix and $\rho(A) = r$. If $p \geq r$, then prove that there exists an $n \times p$ matrix B such that $A = BB'$. If $p = r$, then explain why this is a rank factorization of A.

19. Let A and B be $n \times n$ n.n.d. matrices. Prove that the eigenvalues of AB are nonnegative.

20. True or false: If A is p.d. (with real entries) and B is real and symmetric, then all the eigenvalues of AB are real numbers.

21. Let $A = \begin{bmatrix} A_{11} & u \\ u' & \alpha \end{bmatrix}$, where A_{11} is positive definite. Prove the following:

    (a) If $\alpha - u'A_{11}^{-1}u > 0$, then A is positive definite.

    (b) If $\alpha - u'A_{11}^{-1}u = 0$, then A is nonnegative definite.

    (c) If $\alpha - u'A_{11}^{-1}u < 0$, then A is neither p.d. nor n.n.d.

22. If A is p.d., then show that $\begin{bmatrix} A & I \\ I & A^{-1} \end{bmatrix}$ is also p.d.

23. If A is $n \times n$ and p.d. and P is $n \times p$, then prove that (i) $\rho(P'AP) = \rho(P)$, and (ii) $\mathcal{R}(P'AP) = \mathcal{R}(P)$.

24. If A is $n \times n$ and n.n.d. but not p.d., then prove that there exists an $n \times n$ n.n.d. matrix B such that $\rho(A + B) = \rho(A) + \rho(B) = n$.

25. Let A be an $n \times n$ n.n.d. matrix and let B be $n \times p$. Prove the following:

    (a) $\mathcal{C}(A + BB') = \mathcal{C}(A : B) = \mathcal{C}(B : (I - P_B)A)$;

    (b) $A + BB'$ is positive definite if and only if $\rho((I - P_B)A) = \rho(I - P_B)$.

26. **Polar decomposition.** Let A be an $n \times n$ nonsingular matrix. Prove that there

exists an orthogonal matrix $\boldsymbol{Q}$ and a positive definite matrix $\boldsymbol{B}$, such that $\boldsymbol{A} = \boldsymbol{QB}$. *Hint: Use the SVD of* $\boldsymbol{A}$ *or the square root of* $\boldsymbol{A}'\boldsymbol{A}$.

27. Are $\boldsymbol{Q}$ and $\boldsymbol{B}$ unique in the polar decomposition $\boldsymbol{A} = \boldsymbol{QB}$?

28. Prove that a symmetric matrix is n.n.d. if and only if it has an n.n.d. generalized inverse.

29. If $\boldsymbol{B} = \boldsymbol{A}^{-1}$, where $\boldsymbol{A}$ is real symmetric, then prove that $\boldsymbol{A}$ and $\boldsymbol{B}$ have the same signature.

30. If $\boldsymbol{A}$ is real symmetric and $\boldsymbol{P}$ is of full column rank, then prove that $\boldsymbol{A}$ and $\boldsymbol{P}'\boldsymbol{AP}$ have the same rank and the same signature.

31. If $\boldsymbol{A}$ and $\boldsymbol{B}$ are symmetric matrices of the same order, then let $\boldsymbol{A} \succcurlyeq \boldsymbol{B}$ denote that $\boldsymbol{A} - \boldsymbol{B}$ is n.n.d. Prove the following:

    (a) If $\boldsymbol{A} \succcurlyeq \boldsymbol{B}$ and $\boldsymbol{B} \succcurlyeq \boldsymbol{A}$, then $\boldsymbol{A} = \boldsymbol{B}$.

    (b) If $\boldsymbol{A} \succcurlyeq \boldsymbol{B}$ and $\boldsymbol{B} \succcurlyeq \boldsymbol{C}$, then $\boldsymbol{A} \succcurlyeq \boldsymbol{C}$.

32. If $\boldsymbol{A}_1, \boldsymbol{A}_2, \ldots, \boldsymbol{A}_k$ be $n \times n$ symmetric matrices and $\boldsymbol{A}_1$ is p.d., then prove that there exists a nonsingular matrix $\boldsymbol{P}$ such that $\boldsymbol{P}'\boldsymbol{A}_i\boldsymbol{P}$ is diagonal for $i = 1, 2, \ldots, k$.

33. Let $\boldsymbol{A}$ and $\boldsymbol{B}$ be two $n \times n$ positive definite matrices with respective Cholesky decompositions $\boldsymbol{A} = \boldsymbol{L_A}\boldsymbol{L}_{\boldsymbol{A}}'$ and $\boldsymbol{B} = \boldsymbol{L_B}\boldsymbol{L}_{\boldsymbol{B}}'$. Consider the SVD $\boldsymbol{L}_{\boldsymbol{B}}^{-1}\boldsymbol{L_A} = \boldsymbol{UDV}'$, where $\boldsymbol{D}$ is $n \times n$ diagonal. If $\boldsymbol{Q} = (\boldsymbol{L}_{\boldsymbol{B}}')^{-1}\boldsymbol{U}$, then show that $\boldsymbol{Q}'\boldsymbol{AQ}$ and $\boldsymbol{Q}'\boldsymbol{BQ}$ are both diagonal.

34. Let $\boldsymbol{A}$ and $\boldsymbol{B}$ be real symmetric matrices. If $\boldsymbol{A} - \boldsymbol{B}$ is n.n.d., then show that $\lambda_i \geq \mu_i$ for $i = 1, 2, \ldots, n$, where $\lambda_1 \geq \lambda_2 \geq \cdots \geq \lambda_n$ and $\mu_1 \geq \mu_2 \geq \cdots \geq \mu_n$ are the eigenvalues of $\boldsymbol{A}$ and $\boldsymbol{B}$, respectively.

35. Consider the following inequality for real numbers $\alpha_1, \alpha_2, \ldots, \alpha_n$ and $\beta_1, \beta_2, \ldots, \beta_n$:

$$\left(\prod_{i=1}^{n} \alpha_i\right)^{1/n} + \left(\prod_{i=1}^{n} \beta_i\right)^{1/n} \leq \left(\prod_{i=1}^{n} (\alpha_i + \beta_i)\right)^{1/n}.$$

Using the above inequality, prove that

$$|\boldsymbol{A} + \boldsymbol{B}|^{1/n} \geq |\boldsymbol{A}|^{1/n} + |\boldsymbol{B}|^{1/n},$$

where $\boldsymbol{A}$ and $\boldsymbol{B}$ are both $n \times n$ positive definite matrices.

CHAPTER 14

The Kronecker Product and Related Operations

14.1 Bilinear interpolation and the Kronecker product

Partitioned matrices arising in statistics, econometrics and other linear systems often reveal a special structure that allow them to be expressed in terms of two (or more) matrices. Rather than writing down the partitioned matrices explicitly, they can be described as a special product, called the ***Kronecker product***, of matrices of smaller dimension. Not only do Kronecker products offer a more compact notation, this operation enjoys a number of attractive algebraic properties that help with manipulation, computation and also provide insight about the linear system.

Among the simplest examples of a linear system with a coefficient matrix expressible as a Kronecker product, is the one associated with *bilinear interpolation*. Suppose we know the value of a bivariate function $f(x, y)$ at four locations situated in the corners of a rectangle. Let (a_1, b_1), (a_2, b_1), (a_1, b_2) and (a_2, b_2) denote these four points and let $f_{ij} = f(a_i, b_j)$ be the corresponding values of the function at those locations. Based upon this information, we wish to interpolate at a location (x_0, y_0) inside the rectangle. Bilinear interpolation assumes that $f(x, y) = \beta_0 + \beta_1 x + \beta_2 y + \beta_3 xy$, computes the four coefficients using the values of $f(a_i, b_j)$ from the four locations and uses this function to interpolate at (x_0, y_0).

Substituting (a_i, b_j) for (x, y) and equating $f(a_i, b_j) = f_{ij}$ produces the following linear system:

$$\begin{bmatrix} 1 & a_1 & b_1 & a_1 b_1 \\ 1 & a_2 & b_1 & a_2 b_1 \\ 1 & a_1 & b_2 & a_1 b_2 \\ 1 & a_2 & b_2 & a_2 b_2 \end{bmatrix} \begin{bmatrix} \beta_0 \\ \beta_1 \\ \beta_2 \\ \beta_3 \end{bmatrix} = \begin{bmatrix} f_{11} \\ f_{12} \\ f_{21} \\ f_{22} \end{bmatrix} . \tag{14.1}$$

It can be verified that the above system is nonsingular when the (a_i, b_j)'s are situated in the corners of a rectangle (although there are other configurations that also yield nonsingular systems). Therefore, we can solve for $\boldsymbol{\beta} = (\beta_0, \beta_1, \beta_2, \beta_3)'$, which determines the function $f(x, y)$. The interpolated value at (x_0, y_0) is then $f(x_0, y_0)$.

The coefficient matrix in (14.1) can be written as

$$\begin{bmatrix} 1 \times \begin{bmatrix} 1 & a_1 \\ 1 & a_2 \end{bmatrix} & b_1 \times \begin{bmatrix} 1 & a_1 \\ 1 & a_2 \end{bmatrix} \\ 1 \times \begin{bmatrix} 1 & a_1 \\ 1 & a_2 \end{bmatrix} & b_2 \times \begin{bmatrix} 1 & a_1 \\ 1 & a_2 \end{bmatrix} \end{bmatrix} = \begin{bmatrix} \boldsymbol{A} & b_1\boldsymbol{A} \\ \boldsymbol{A} & b_2\boldsymbol{A} \end{bmatrix} , \text{ where } \boldsymbol{A} = \begin{bmatrix} 1 & a_1 \\ 1 & a_2 \end{bmatrix} . \tag{14.2}$$

More compactly, we write (14.2) as

$$\boldsymbol{B} \otimes \boldsymbol{A} , \text{ where } \boldsymbol{B} = \begin{bmatrix} 1 & b_1 \\ 1 & b_2 \end{bmatrix} \text{ and } \boldsymbol{A} = \begin{bmatrix} 1 & a_1 \\ 1 & a_2 \end{bmatrix} ,$$

where $\otimes$ is the operation that multiplies each entry in $\boldsymbol{B}$ with the entire matrix $\boldsymbol{A}$. This is called the Kronecker product. Notice that not all arrangements of the equations in (14.1) will lead to a Kronecker product representation. For example, if we switch the second and third equations in (14.1), we lose the Kronecker structure.

Are Kronecker products merely a way to represent some partitioned matrices? Or is there more to them? Treated as an algebraic operation, Kronecker products enjoy several attractive properties. In particular, when a large matrix can be partitioned and expressed as a Kronecker product of two (or more) matrices, then several properties of the large partitioned matrix can be derived from the smaller constituent matrices. As one example, if $\boldsymbol{A}$ and $\boldsymbol{B}$ are nonsingular, then so is the matrix $\boldsymbol{B} \otimes \boldsymbol{A}$ in (14.2). Not only that, the solution can be obtained by solving linear systems involving $\boldsymbol{A}$ and $\boldsymbol{B}$ instead of the larger $\boldsymbol{B} \otimes \boldsymbol{A}$. In fact, finding inverses, determinants, eigenvalues and carrying out many other matrix operations are much easier with Kronecker products. We derive such results in the subsequent sections.

14.2 Basic properties of Kronecker products

The general definition of the Kronecker product of two matrices is given below.

Definition 14.1 *Let $\boldsymbol{A}$ be an $m \times n$ matrix and $\boldsymbol{B}$ a $p \times q$ matrix. The* **Kronecker product** *or* **tensor product** *of $\boldsymbol{A}$ and $\boldsymbol{B}$, written as $\boldsymbol{A} \otimes \boldsymbol{B}$, is the $mp \times nq$ partitioned matrix whose (i,j)-th block is the $p \times q$ matrix $a_{ij}\boldsymbol{B}$ for $i = 1, 2, \ldots, m$ and $j = 1, 2, \ldots, n$. Thus,*

$$\boldsymbol{A} \otimes \boldsymbol{B} = \begin{bmatrix} a_{11}\boldsymbol{B} & a_{12}\boldsymbol{B} & \ldots & a_{1n}\boldsymbol{B} \\ a_{21}\boldsymbol{B} & a_{22}\boldsymbol{B} & \ldots & a_{2n}\boldsymbol{B} \\ \vdots & \vdots & & \vdots \\ a_{m1}\boldsymbol{B} & a_{m2}\boldsymbol{B} & \ldots & a_{mn}\boldsymbol{B} \end{bmatrix} .$$

The Kronecker product, as defined in Definition 14.1 is sometimes referred to as the *right Kronecker product.* This is the more popular definition, although some authors (Graybill, 2001) define the Kronecker product as the left Kronecker product, which, in our notation, is $\boldsymbol{B} \otimes \boldsymbol{A}$. For us, the Kronecker product will always mean the right

Kronecker product. Kronecker products are also referred to as *direct products* or *tensor products* in some fields.

Example 14.1 The Kronecker product of two 2×2 matrices produces a 4×4 matrix:

$$\begin{bmatrix} 1 & 2 \\ 3 & 4 \end{bmatrix} \otimes \begin{bmatrix} 5 & 6 \\ 7 & 8 \end{bmatrix} = \begin{bmatrix} 1 \times 5 & 1 \times 6 & 2 \times 5 & 2 \times 6 \\ 1 \times 7 & 1 \times 8 & 2 \times 7 & 2 \times 8 \\ 3 \times 5 & 3 \times 6 & 4 \times 5 & 4 \times 6 \\ 3 \times 7 & 3 \times 8 & 4 \times 7 & 4 \times 8 \end{bmatrix} = \begin{bmatrix} 5 & 6 & 10 & 12 \\ 7 & 8 & 14 & 16 \\ 15 & 18 & 20 & 24 \\ 21 & 24 & 28 & 32 \end{bmatrix}.$$

Here is an example of a Kronecker product of a 3×2 matrix and a 3×3 matrix:

$$\begin{bmatrix} 1 & 2 \\ 3 & 4 \\ 0 & 1 \end{bmatrix} \otimes \begin{bmatrix} 5 & 6 & 0 \\ 7 & 8 & 0 \\ 0 & 0 & 1 \end{bmatrix} = \begin{bmatrix} 5 & 6 & 0 & 10 & 12 & 0 \\ 7 & 8 & 0 & 14 & 16 & 0 \\ 0 & 0 & 1 & 0 & 0 & 2 \\ 15 & 18 & 0 & 20 & 24 & 0 \\ 21 & 24 & 0 & 28 & 32 & 0 \\ 0 & 0 & 3 & 0 & 0 & 4 \\ 0 & 0 & 0 & 5 & 6 & 0 \\ 0 & 0 & 0 & 7 & 8 & 0 \\ 0 & 0 & 0 & 0 & 0 & 1 \end{bmatrix}. \blacksquare$$

Some immediate consequences of Definition 14.1 are listed below.

- The Kronecker product of $\boldsymbol{A}$ and $\boldsymbol{B}$ differs from ordinary matrix multiplication in that it is defined for any two matrices $\boldsymbol{A}$ and $\boldsymbol{B}$. It does not require that the number of columns in $\boldsymbol{A}$ be equal to the number of rows in $\boldsymbol{B}$. Each element of $\boldsymbol{A} \otimes \boldsymbol{B}$ is the product of an element of $\boldsymbol{A}$ and an element of $\boldsymbol{B}$. If, for example, $\boldsymbol{A}$ is 2×2 and $\boldsymbol{B}$ is 3×2, then

$$\boldsymbol{A} \otimes \boldsymbol{B} = \begin{bmatrix} a_{11}b_{11} & a_{11}b_{12} & a_{12}b_{11} & a_{12}b_{12} \\ a_{11}b_{21} & a_{11}b_{22} & a_{12}b_{21} & a_{12}b_{22} \\ a_{11}b_{31} & a_{11}b_{32} & a_{12}b_{31} & a_{12}b_{32} \\ a_{21}b_{11} & a_{21}b_{12} & a_{22}b_{11} & a_{22}b_{12} \\ a_{21}b_{21} & a_{21}b_{22} & a_{21}b_{21} & a_{22}b_{22} \\ a_{21}b_{31} & a_{21}b_{32} & a_{22}b_{31} & a_{22}b_{32} \end{bmatrix}.$$

  In general, if $\boldsymbol{A}$ is $m \times n$ and $\boldsymbol{B}$ is $p \times q$, then $a_{ij}b_{kl}$ (i.e., the (k, l)-th entry in $a_{ij}\boldsymbol{B}$) appears at the intersection of row $p(i-1)+k$ and column $q(j-1)+l$ in $\boldsymbol{A} \otimes \boldsymbol{B}$.

- The dimensions of $\boldsymbol{A} \otimes \boldsymbol{B}$ and $\boldsymbol{B} \otimes \boldsymbol{A}$ are the same. If $\boldsymbol{A}$ is $m \times n$ and $\boldsymbol{B}$ is $p \times q$, then both $\boldsymbol{A} \otimes \boldsymbol{B}$ and $\boldsymbol{B} \otimes \boldsymbol{A}$ have mp rows and nq columns. However, Kronecker products are *not* commutative, i.e., $\boldsymbol{A} \otimes \boldsymbol{B} \neq \boldsymbol{B} \otimes \boldsymbol{A}$. Here is a simple example. Let $\boldsymbol{A}$ and $\boldsymbol{B}$ be the following 2×2 matrices:

$$\boldsymbol{A} = \begin{bmatrix} 1 & 0 \\ 0 & 1 \end{bmatrix} \text{ and } \boldsymbol{B} = \begin{bmatrix} 1 & 2 \\ 3 & 4 \end{bmatrix}.$$

Then,

$$\boldsymbol{A} \otimes \boldsymbol{B} = \begin{bmatrix} \boldsymbol{B} & 0\boldsymbol{B} \\ 0\boldsymbol{B} & \boldsymbol{B} \end{bmatrix} = \begin{bmatrix} 1 & 2 & 0 & 0 \\ 3 & 4 & 0 & 0 \\ 0 & 0 & 1 & 2 \\ 0 & 0 & 3 & 4 \end{bmatrix},$$

while

$$\boldsymbol{B} \otimes \boldsymbol{A} = \begin{bmatrix} 1\boldsymbol{A} & 2\boldsymbol{A} \\ 3\boldsymbol{A} & 4\boldsymbol{A} \end{bmatrix} = \begin{bmatrix} 1 & 0 & 2 & 0 \\ 0 & 1 & 0 & 2 \\ 3 & 0 & 4 & 0 \\ 0 & 3 & 0 & 4 \end{bmatrix}.$$

- If $\boldsymbol{a} = (a_1, a_2, \ldots, a_m)'$ is an $m \times 1$ vector (i.e., $n = 1$ in Definition 14.1), then the definition of the Kronecker product implies that

$$\boldsymbol{a} \otimes \boldsymbol{B} = \begin{bmatrix} a_1\boldsymbol{B} \\ a_2\boldsymbol{B} \\ \vdots \\ a_m\boldsymbol{B} \end{bmatrix}.$$

  In particular, if $\boldsymbol{b}$ is a $p \times 1$ vector $\boldsymbol{a} \otimes \boldsymbol{b}$ is the $mp \times 1$ column vector consisting of m subvectors, each with dimension $p \times 1$, where the i-th subvector is $a_i\boldsymbol{b}$. Also,

$$\boldsymbol{a} \otimes \boldsymbol{b}' = \begin{bmatrix} a_1\boldsymbol{b} \\ a_2\boldsymbol{b}' \\ \vdots \\ a_n\boldsymbol{b}' \end{bmatrix} = \boldsymbol{a}\boldsymbol{b}' = [b_1\boldsymbol{a} : b_2 : \boldsymbol{a} : \ldots : b_p\boldsymbol{a}] = \boldsymbol{b}' \otimes \boldsymbol{a}\,.$$

- If $\boldsymbol{a}' = (a_1, a_2, \ldots, a_n)$ is a $1 \times n$ row vector (i.e., $m = 1$ in Definition 14.1), then

$$\boldsymbol{a}' \otimes \boldsymbol{B} = \left[a_1\boldsymbol{B} : a_2\boldsymbol{B} : \ldots : a_n\boldsymbol{B}\right]$$

  is an $p \times nq$ partitioned matrix with one row of n blocks, each of size $p \times q$, where the i-th block is $a_i\boldsymbol{B}$.

- If $\boldsymbol{0}$ is a vector of zeroes and $\boldsymbol{O}$ is a matrix of zeroes, then

$$\boldsymbol{0} \otimes \boldsymbol{A} = \boldsymbol{A} \otimes \boldsymbol{0} = \boldsymbol{0} \text{ and } \boldsymbol{O} \otimes \boldsymbol{A} = \boldsymbol{A} \otimes \boldsymbol{O} = \boldsymbol{O}\,.$$

  Note that the size of $\boldsymbol{0}$ and $\boldsymbol{O}$ in the results will, in general, be different from those in the Kronecker product operations.

- If $\boldsymbol{D}$ is a diagonal matrix with diagonal entries d_i, then $\boldsymbol{D} \otimes \boldsymbol{A}$ is the block diagonal matrix with blocks $d_i\boldsymbol{A}$. In the special case when $\boldsymbol{D} = \boldsymbol{I}$, the identity matrix, we have

$$\boldsymbol{I} \otimes \boldsymbol{A} = \begin{bmatrix} A & O & \ldots & O \\ O & A & \ldots & O \\ \vdots & \vdots & \ddots & \vdots \\ O & O & \ldots & A \end{bmatrix}.$$

Also note that

$$A \otimes I = \begin{bmatrix} a_{11}I & a_{12}I & \dots & a_{1m}I \\ a_{21}I & a_{22}I & \dots & a_{2n}I \\ \vdots & \vdots & \ddots & \vdots \\ a_{m1}I & a_{m2}I & \dots & a_{mn}I \end{bmatrix} .$$

Clearly, $I \otimes A \neq A \otimes I$.

- When both D and A are diagonal, so is $D \otimes A$. More precisely, if D is $m \times m$ and A is $n \times n$, then $D \otimes A$ is $mn \times mn$ diagonal with $d_i a_{jj}$ as the $n(i-1)+j$-th diagonal entry.
- It is often convenient to derive properties of Kronecker products by focusing upon smaller matrices and vectors before moving to more general settings. For example, if $A = [a_{*1} : a_{*2}]$ is $m \times 2$ with only two columns, then it is easily seen that

$$A \otimes B = \begin{bmatrix} a_{*1} \otimes B : a_{*2} \otimes B \end{bmatrix} .$$

Analogously, if $A = \begin{bmatrix} a'_{1*} \\ a'_{2*} \end{bmatrix}$ is $2 \times n$, then it is easily verified that

$$A \otimes B = \begin{bmatrix} a'_{1*} \otimes B \\ a'_{2*} \otimes B \end{bmatrix} .$$

More generally, if we partition an $m \times n$ matrix A in terms of its column vectors, we easily see that

$$A \otimes B = \begin{bmatrix} a_{*1} \otimes B : a_{*2} \otimes B : \dots : a_{*n} \otimes B \end{bmatrix} .$$

If we partition A in terms of its row vectors, we obtain

$$A \otimes B = \begin{bmatrix} a'_{1*} \otimes B \\ a'_{2*} \otimes B \\ \vdots \\ a'_{n*} \otimes B \end{bmatrix} .$$

- The above results on row and column vectors imply that if we combine some of the rows and columns to form a partition of A, then $A \otimes B$ can be expressed in terms of the Kronecker product of the individual blocks in A and B. For example, it is easily verified that

$$\begin{bmatrix} A_1 : A_2 \end{bmatrix} \otimes B = \begin{bmatrix} A_1 \otimes B : A_2 \otimes B \end{bmatrix} \text{ and } \begin{bmatrix} A_1 \\ A_2 \end{bmatrix} \otimes B = \begin{bmatrix} A_1 \otimes B \\ A_2 \otimes B \end{bmatrix} .$$

If we further split the above blocks in A to obtain a 2×2 partition of A, we have

$$A \otimes B = \begin{bmatrix} A_{11} & A_{12} \\ A_{21} & A_{22} \end{bmatrix} \otimes B = \begin{bmatrix} A_{11} \otimes B & A_{12} \otimes B \\ A_{21} \otimes B & A_{22} \otimes B \end{bmatrix} .$$

Let us make this more explicit. Suppose A_{11} is $m_1 \times n_1$, A_{12} is $m_1 \times n_2$, A_{21} is

$m_2 \times n_1$ and $\boldsymbol{A}_{22}$ is $m_2 \times n_2$, where $m_1 + m_2 = m$ and $n_1 + n_2 = n$. Then,

$$\boldsymbol{A} \otimes \boldsymbol{B} = \left[\begin{array}{ccc|ccc} a_{11}\boldsymbol{B} & \dots & a_{1n_1}\boldsymbol{B} & a_{1,n_1+1}\boldsymbol{B} & \dots & a_{1n}\boldsymbol{B} \\ \vdots & \ddots & \vdots & \vdots & \ddots & \vdots \\ a_{m_1 1}\boldsymbol{B} & \dots & a_{m_1 n_1}\boldsymbol{B} & a_{m_1,n_1+1}\boldsymbol{B} & \dots & a_{m_1 n}\boldsymbol{B} \\ \hline a_{m_1+1,1}\boldsymbol{B} & \dots & a_{m_1+1,n_1}\boldsymbol{B} & a_{m_1+1,n_1+1}\boldsymbol{B} & \dots & a_{m_1+1,n}\boldsymbol{B} \\ \vdots & \ddots & \vdots & \vdots & \ddots & \vdots \\ a_{m1}\boldsymbol{B} & \dots & a_{mn_1}\boldsymbol{B} & a_{m,n_1+1}\boldsymbol{B} & \dots & a_{mn}\boldsymbol{B} \end{array}\right]$$
$$= \left[\begin{array}{c|c} \boldsymbol{A}_{11} \otimes \boldsymbol{B} & \boldsymbol{A}_{12} \otimes \boldsymbol{B} \\ \hline \boldsymbol{A}_{21} \otimes \boldsymbol{B} & \boldsymbol{A}_{22} \otimes \boldsymbol{B} \end{array}\right].$$

Proceeding even further, the above results can be used to easily verify the following property for general partitioned matrices:

$$\begin{bmatrix} \boldsymbol{A}_{11} & \boldsymbol{A}_{12} & \cdots & \boldsymbol{A}_{1c} \\ \boldsymbol{A}_{21} & \boldsymbol{A}_{22} & \dots & \boldsymbol{A}_{2c} \\ \vdots & \vdots & \ddots & \vdots \\ \boldsymbol{A}_{r1} & \boldsymbol{A}_{r2} & \dots & \boldsymbol{A}_{rc} \end{bmatrix} \otimes \boldsymbol{B} = \begin{bmatrix} \boldsymbol{A}_{11} \otimes \boldsymbol{B} & \boldsymbol{A}_{12} \otimes \boldsymbol{B} & \cdots & \boldsymbol{A}_{1c} \otimes \boldsymbol{B} \\ \boldsymbol{A}_{21} \otimes \boldsymbol{B} & \boldsymbol{A}_{22} \otimes \boldsymbol{B} & \dots & \boldsymbol{A}_{2c} \otimes \boldsymbol{B} \\ \vdots & \vdots & \ddots & \vdots \\ \boldsymbol{A}_{r1} \otimes \boldsymbol{B} & \boldsymbol{A}_{r2} \otimes \boldsymbol{B} & \dots & \boldsymbol{A}_{rc} \otimes \boldsymbol{B} \end{bmatrix}. \tag{14.3}$$

Thus, if $\boldsymbol{A}$ is partitioned into blocks, then $\boldsymbol{A} \otimes \boldsymbol{B}$ can also be partitioned with blocks, where each block is the Kronecker product of the corresponding block of $\boldsymbol{A}$ and the matrix $\boldsymbol{B}$.

For further developments, we collect some additional elementary algebraic properties in the next theorem.

Theorem 14.1 *Let $\boldsymbol{A}$, $\boldsymbol{B}$ and $\boldsymbol{C}$ be any three matrices.*

(i) $\alpha \otimes \boldsymbol{A} = \boldsymbol{A} \otimes \alpha = \alpha\boldsymbol{A}$, *for any scalar* α.

(ii) $(\alpha\boldsymbol{A}) \otimes (\beta\boldsymbol{B}) = \alpha\beta(\boldsymbol{A} \otimes \boldsymbol{B})$, *for any scalars* α *and* β.

(iii) $(\boldsymbol{A} \otimes \boldsymbol{B}) \otimes \boldsymbol{C} = \boldsymbol{A} \otimes (\boldsymbol{B} \otimes \boldsymbol{C})$.

(iv) $(\boldsymbol{A} + \boldsymbol{B}) \otimes \boldsymbol{C} = (\boldsymbol{A} \otimes \boldsymbol{C}) + (\boldsymbol{B} \otimes \boldsymbol{C})$, *if* $\boldsymbol{A}$ *and* $\boldsymbol{B}$ *are of same size.*

(v) $\boldsymbol{A} \otimes (\boldsymbol{B} + \boldsymbol{C}) = (\boldsymbol{A} \otimes \boldsymbol{B}) + (\boldsymbol{A} \otimes \boldsymbol{C})$, *if* $\boldsymbol{B}$ *and* $\boldsymbol{C}$ *are of same size.*

(vi) $(\boldsymbol{A} \otimes \boldsymbol{B})' = \boldsymbol{A}' \otimes \boldsymbol{B}'$.

Proof. Most of the above results are straightforward implications of Definition 14.1. We provide sketches of the proofs below.

Proof of (i): Treating α as a 1×1 matrix, we see from Definition 14.1 that $\alpha \otimes \boldsymbol{A}$, $\boldsymbol{A} \otimes \alpha$ and $\alpha\boldsymbol{A}$ all are matrices of the same size as $\boldsymbol{A}$. Furthermore, the (i, j)-th entry in $\alpha \otimes \boldsymbol{A}$ is equal to αa_{ij}, which is also the (i, j)-th entry in $\boldsymbol{A} \otimes \alpha$ and $\alpha\boldsymbol{A}$. This proves part (i).

Proof of (ii): It is easily seen that $(\alpha \boldsymbol{A}) \otimes (\beta \boldsymbol{B})$ and $\alpha\beta(\boldsymbol{A} \otimes \boldsymbol{B})$ are matrices of the same size. If we write $(\alpha \boldsymbol{A}) \otimes (\beta \boldsymbol{B})$ as a partitioned matrix with blocks that are scalar multiples of $\boldsymbol{B}$, we see that the (i, j)-th block is $\alpha\beta a_{ij}\boldsymbol{B}$, which is also the (i, j)-th block in $\alpha\beta(\boldsymbol{A} \otimes \boldsymbol{B})$. Therefore, $(\alpha \boldsymbol{A}) \otimes (\beta \boldsymbol{B})$ and $\alpha\beta(\boldsymbol{A} \otimes \boldsymbol{B})$ are equal.

Proof of (iii): Suppose that $\boldsymbol{A}$ is $m \times n$. Then we can write

$$
\begin{aligned}
(\boldsymbol{A} \otimes \boldsymbol{B}) \otimes \boldsymbol{C} &= \begin{bmatrix} a_{11}\boldsymbol{B} & a_{12}\boldsymbol{B} & \dots & a_{1n}\boldsymbol{B} \\ a_{21}\boldsymbol{B} & a_{22}\boldsymbol{B} & \dots & a_{2n}\boldsymbol{B} \\ \vdots & \vdots & \ddots & \vdots \\ a_{m1}\boldsymbol{B} & a_{m2}\boldsymbol{B} & \dots & a_{mn}\boldsymbol{B} \end{bmatrix} \otimes \boldsymbol{C} \\
&= \begin{bmatrix} (a_{11}\boldsymbol{B}) \otimes \boldsymbol{C} & (a_{12}\boldsymbol{B}) \otimes \boldsymbol{C} & \dots & (a_{1n}\boldsymbol{B}) \otimes \boldsymbol{C} \\ (a_{21}\boldsymbol{B}) \otimes \boldsymbol{C} & (a_{22}\boldsymbol{B}) \otimes \boldsymbol{C} & \dots & (a_{2n}\boldsymbol{B}) \otimes \boldsymbol{C} \\ \vdots & \vdots & \ddots & \vdots \\ (a_{m1}\boldsymbol{B}) \otimes \boldsymbol{C} & (a_{m2}\boldsymbol{B}) \otimes \boldsymbol{C} & \dots & (a_{mn}\boldsymbol{B}) \otimes \boldsymbol{C} \end{bmatrix} \\
&= \begin{bmatrix} a_{11}(\boldsymbol{B} \otimes \boldsymbol{C}) & a_{12}(\boldsymbol{B} \otimes \boldsymbol{C}) & \dots & a_{1n}(\boldsymbol{B} \otimes \boldsymbol{C}) \\ a_{21}(\boldsymbol{B} \otimes \boldsymbol{C}) & a_{22}(\boldsymbol{B} \otimes \boldsymbol{C}) & \dots & a_{2n}(\boldsymbol{B} \otimes \boldsymbol{C}) \\ \vdots & \vdots & \ddots & \vdots \\ a_{m1}(\boldsymbol{B} \otimes \boldsymbol{C}) & a_{m2}(\boldsymbol{B} \otimes \boldsymbol{C}) & \dots & a_{mn}(\boldsymbol{B} \otimes \boldsymbol{C}) \end{bmatrix} \\
&= \boldsymbol{A} \otimes (\boldsymbol{B} \otimes \boldsymbol{C}) \,,
\end{aligned}
$$

where the second equality follows from (14.3) and the third equality follows part (ii). This establishes part (iii).

Proof of (iv): Partition $(\boldsymbol{A} + \boldsymbol{B}) \otimes \boldsymbol{C}$ so that the (i, j) block is

$$(a_{ij} + b_{ij})\boldsymbol{C} = (a_{ij}\boldsymbol{C}) + (b_{ij}\boldsymbol{C}) \,.$$

The right hand side is equal to the sum of the (i, j)-th blocks in $\boldsymbol{A} \otimes \boldsymbol{C}$ and $\boldsymbol{B} \otimes \boldsymbol{C}$. Since $\boldsymbol{A}$ and $\boldsymbol{B}$ are of the same size, this sum is equal to the (i, j)-th block in $(\boldsymbol{A} + \boldsymbol{B}) \otimes \boldsymbol{C}$. This proves part (iv).

Proof of (v): Partition $\boldsymbol{A} \otimes (\boldsymbol{B} + \boldsymbol{C})$ so that the (i, j)-th block is

$$a_{ij}(\boldsymbol{B} + \boldsymbol{C}) = a_{ij}\boldsymbol{B} + a_{ij}\boldsymbol{C} \,,$$

which is equal to the sum of the (i, j)-th blocks in $\boldsymbol{A} \otimes \boldsymbol{B}$ and $\boldsymbol{A} \otimes \boldsymbol{C}$. Since $\boldsymbol{B}$ and $\boldsymbol{C}$ are of the same size, this is equal to the (i, j)-th block in $\boldsymbol{A} \otimes \boldsymbol{B} + \boldsymbol{A} \otimes \boldsymbol{C}$, thereby establishing part (v).

Proof of (vi): If $\boldsymbol{A}$ is $m \times n$, then we can write

$$
\begin{aligned}
(\boldsymbol{A} \otimes \boldsymbol{B})' &= \begin{bmatrix} a_{11}\boldsymbol{B} & a_{12}\boldsymbol{B} & \dots & a_{1n}\boldsymbol{B} \\ a_{21}\boldsymbol{B} & a_{22}\boldsymbol{B} & \dots & a_{2n}\boldsymbol{B} \\ \vdots & \vdots & \ddots & \vdots \\ a_{m1}\boldsymbol{B} & a_{m2}\boldsymbol{B} & \dots & a_{mn}\boldsymbol{B} \end{bmatrix}' = \begin{bmatrix} a_{11}\boldsymbol{B}' & a_{21}\boldsymbol{B}' & \dots & a_{m1}\boldsymbol{B}' \\ a_{12}\boldsymbol{B}' & a_{22}\boldsymbol{B}' & \dots & a_{m2}\boldsymbol{B}' \\ \vdots & \vdots & \ddots & \vdots \\ a_{1n}\boldsymbol{B}' & a_{2n}\boldsymbol{B}' & \dots & a_{mn}\boldsymbol{B}' \end{bmatrix} \\
&= \boldsymbol{A}' \otimes \boldsymbol{B}' \,,
\end{aligned}
$$

which completes the proof. □

Part (iii) is useful because it allows us to unambiguously define $\boldsymbol{A} \otimes \boldsymbol{B} \otimes \boldsymbol{C}$. Also note the effect of transposing Kronecker products. The result is $(\boldsymbol{A} \otimes \boldsymbol{B})' = \boldsymbol{A}' \otimes \boldsymbol{B}'$ and *not* $\boldsymbol{B}' \otimes \boldsymbol{A}'$. Thus, unlike for regular matrix multiplication, the order of the matrices is not reversed by transposition.

Part (vi) is true for conjugate transposes as well:

$$(\boldsymbol{A} \otimes \boldsymbol{B})^* = \begin{bmatrix} \bar{a}_{11}\boldsymbol{B}^* & \bar{a}_{21}\boldsymbol{B}^* & \dots & \bar{a}_{m1}\boldsymbol{B}^* \\ \bar{a}_{12}\boldsymbol{B}^* & \bar{a}_{22}\boldsymbol{B}^* & \dots & \bar{a}_{m2}\boldsymbol{B}^* \\ \vdots & \vdots & \ddots & \vdots \\ \bar{a}_{1n}\boldsymbol{B}^* & \bar{a}_{2n}\boldsymbol{B}^* & \dots & \bar{a}_{mn}\boldsymbol{B}^* \end{bmatrix} = \boldsymbol{A}^* \otimes \boldsymbol{B}^* . \tag{14.4}$$

Kronecker products help to simplify characteristics for larger matrices in terms of its components. One simple example concerns the trace of a matrix. The next result shows that the trace of the Kronecker product between two matrices is the product of their individual traces. This result is of course meaningful only when the matrices in question are square. The proof is simple and requires no additional machinery.

Theorem 14.2 *Let $\boldsymbol{A}$ and $\boldsymbol{B}$ be $m \times m$ and $p \times p$ matrices, respectively. Then*

$$tr(\boldsymbol{A} \otimes \boldsymbol{B}) = tr(\boldsymbol{A})tr(\boldsymbol{B}) = tr(\boldsymbol{B} \otimes \boldsymbol{A}) .$$

Proof. Note that

$$\begin{aligned} \operatorname{tr}(\boldsymbol{A} \otimes \boldsymbol{B}) &= \operatorname{tr} \begin{bmatrix} a_{11}\boldsymbol{B} & a_{12}\boldsymbol{B} & \dots & a_{1m}\boldsymbol{B} \\ a_{21}\boldsymbol{B} & a_{22}\boldsymbol{B} & \dots & a_{2m}\boldsymbol{B} \\ \vdots & \vdots & \ddots & \vdots \\ a_{m1}\boldsymbol{B} & a_{m2}\boldsymbol{B} & \dots & a_{mm}\boldsymbol{B} \end{bmatrix} \\ &= \operatorname{tr}(a_{11}\boldsymbol{B}) + \operatorname{tr}(a_{22}\boldsymbol{B}) + \cdots + \operatorname{tr}(a_{mm}\boldsymbol{B}) \\ &= a_{11}\operatorname{tr}(\boldsymbol{B}) + a_{22}\operatorname{tr}(\boldsymbol{B}) + \cdots + a_{mm}\operatorname{tr}(\boldsymbol{B}) \\ &= (a_{11} + a_{22} + \cdots + a_{mm})\operatorname{tr}(\boldsymbol{B}) = \operatorname{tr}(\boldsymbol{A})\operatorname{tr}(\boldsymbol{B}) . \end{aligned}$$

The above immediately implies that $\operatorname{tr}(\boldsymbol{B} \otimes \boldsymbol{A}) = \operatorname{tr}(\boldsymbol{A} \otimes \boldsymbol{B})$. □

The following result is extremely useful in algebraic manipulations with Kronecker products.

Theorem 14.3 The mixed product rule. *Let $\boldsymbol{A}$, $\boldsymbol{B}$, $\boldsymbol{C}$ and $\boldsymbol{D}$ be matrices of sizes such that the products $\boldsymbol{AC}$ and $\boldsymbol{BD}$ are well-defined. Then,*

$$(\boldsymbol{A} \otimes \boldsymbol{B})(\boldsymbol{C} \otimes \boldsymbol{D}) = (\boldsymbol{AC}) \otimes (\boldsymbol{BD}) .$$

Proof. Let $\boldsymbol{A}$, $\boldsymbol{B}$, $\boldsymbol{C}$ and $\boldsymbol{D}$ be matrices of sizes $m \times n$, $p \times q$, $n \times r$ and $q \times s$, respectively. Then, we can write the left hand side, $\boldsymbol{F} = (\boldsymbol{A} \otimes \boldsymbol{B})(\boldsymbol{C} \otimes \boldsymbol{D})$, as

$$\boldsymbol{F} = \begin{bmatrix} a_{11}\boldsymbol{B} & a_{12}\boldsymbol{B} & \dots & a_{1n}\boldsymbol{B} \\ a_{21}\boldsymbol{B} & a_{22}\boldsymbol{B} & \dots & a_{2n}\boldsymbol{B} \\ \vdots & \vdots & \ddots & \vdots \\ a_{m1}\boldsymbol{B} & a_{m2}\boldsymbol{B} & \dots & a_{mn}\boldsymbol{B} \end{bmatrix} \begin{bmatrix} c_{11}\boldsymbol{D} & c_{12}\boldsymbol{D} & \dots & c_{1r}\boldsymbol{D} \\ c_{21}\boldsymbol{D} & c_{22}\boldsymbol{D} & \dots & c_{2r}\boldsymbol{D} \\ \vdots & \vdots & \ddots & \vdots \\ c_{n1}\boldsymbol{D} & c_{n2}\boldsymbol{D} & \dots & c_{nr}\boldsymbol{D} \end{bmatrix}$$

$$= \begin{bmatrix} \boldsymbol{F}_{11} & \boldsymbol{F}_{12} & \dots & \boldsymbol{F}_{1r} \\ \boldsymbol{F}_{21} & \boldsymbol{F}_{22} & \dots & \boldsymbol{F}_{2r} \\ \vdots & \vdots & \ddots & \vdots \\ \boldsymbol{F}_{m1} & \boldsymbol{F}_{m2} & \dots & \boldsymbol{F}_{mr} \end{bmatrix},$$

where $\boldsymbol{F}_{ij} = (\sum_{k=1}^{n} a_{ik}c_{kj})\boldsymbol{BD} = (\boldsymbol{a}'_{i*}\boldsymbol{c}_{*j})(\boldsymbol{BD})$. Since $\boldsymbol{a}'_{i*}\boldsymbol{c}_{*j}$ is the (i,j)-th element of the product $\boldsymbol{AC}$, it follows that $\boldsymbol{F} = (\boldsymbol{AC}) \otimes (\boldsymbol{BD})$. □

The above *mixed product rule* is easily generalized for $2k$ matrices

$$(\boldsymbol{A}_1 \otimes \boldsymbol{A}_2)(\boldsymbol{A}_3 \otimes \boldsymbol{A}_4) \cdots (\boldsymbol{A}_{2k-1} \otimes \boldsymbol{A}_{2k}) = (\boldsymbol{A}_1\boldsymbol{A}_3 \cdots \boldsymbol{A}_{2k-1})(\boldsymbol{A}_2\boldsymbol{A}_4 \cdots \boldsymbol{A}_{2k}), \tag{14.5}$$

whenever the matrix products in (14.5) are well-defined.

The Kronecker product $(\boldsymbol{AC}) \otimes (\boldsymbol{BD})$ is often simply written as $\boldsymbol{AC} \otimes \boldsymbol{BD}$ with the understanding that the ordinary matrix multiplications are computed first and then the Kronecker product is computed. Also, a repeated application of the above result yields the following:

$$(\boldsymbol{A} \otimes \boldsymbol{B})(\boldsymbol{U} \otimes \boldsymbol{V})(\boldsymbol{C} \otimes \boldsymbol{D}) = (\boldsymbol{AU} \otimes \boldsymbol{BV})(\boldsymbol{C} \otimes \boldsymbol{D}) = \boldsymbol{AUC} \otimes \boldsymbol{BVD}. \tag{14.6}$$

There are several important implications of Theorem 14.3 and (14.6). For example, if $\boldsymbol{A}$ is $m \times n$ and $\boldsymbol{B}$ is $p \times q$, then Theorem **??** implies that

$$\boldsymbol{A} \otimes \boldsymbol{B} = (\boldsymbol{A} \otimes \boldsymbol{I}_p)(\boldsymbol{I}_n \otimes \boldsymbol{B}), \tag{14.7}$$

which shows that a Kronecker product can be always be expressed as a matrix product. We will see several others in subsequent sections.

14.3 Inverses, rank and nonsingularity of Kronecker products

As mentioned in Section 14.1, where we considered a simple linear system for bilinear interpolation, we can ascertain the nonsingularity of a linear system whose coefficient matrix is expressible as a Kronecker product by checking whether the individual matrices in the Kronecker product are themselves nonsingular. The advantage here is that these individual matrices are of much smaller size than the original system. The following theorem makes this clear.

Theorem 14.4 *The following statements are true:*

(i) If $\boldsymbol{A}$ and $\boldsymbol{B}$ are two nonsingular matrices, then $\boldsymbol{A} \otimes \boldsymbol{B}$ is also nonsingular and

$$(\boldsymbol{A} \otimes \boldsymbol{B})^{-1} = \boldsymbol{A}^{-1} \otimes \boldsymbol{B}^{-1} .$$

(ii) Let $\boldsymbol{A}$ be $m \times n$ and $\boldsymbol{B}$ be $p \times q$. If $\boldsymbol{G_A}$ and $\boldsymbol{G_B}$ are any two generalized inverses of $\boldsymbol{A}$ and $\boldsymbol{B}$, then $\boldsymbol{G_A} \otimes \boldsymbol{G_B}$ is a generalized inverse of $\boldsymbol{A} \otimes \boldsymbol{B}$.

(iii) $(\boldsymbol{A} \otimes \boldsymbol{B})^{+} = \boldsymbol{A}^{+} \otimes \boldsymbol{B}^{+}$.

Proof. **Proof of (i):** Suppose that $\boldsymbol{A}$ is $m \times m$ and $\boldsymbol{B}$ is $p \times p$. Using Theorem 14.3 we see that

$$(\boldsymbol{A} \otimes \boldsymbol{B})(\boldsymbol{A}^{-1} \otimes \boldsymbol{B}^{-1}) = (\boldsymbol{A}\boldsymbol{A}^{-1} \otimes \boldsymbol{B}\boldsymbol{B}^{-1}) = \boldsymbol{I}_m \otimes \boldsymbol{I}_p = \boldsymbol{I}_{mp} ,$$

which proves part (i).

Proof of (ii): Using (14.6) we obtain that

$$(\boldsymbol{A} \otimes \boldsymbol{B})(\boldsymbol{G_A} \otimes \boldsymbol{G_B})(\boldsymbol{A} \otimes \boldsymbol{B}) = \boldsymbol{A}\boldsymbol{G_A}\boldsymbol{A} \otimes \boldsymbol{B}\boldsymbol{G_B}\boldsymbol{B} = \boldsymbol{A} \otimes \boldsymbol{B} ,$$

which proves that $\boldsymbol{G_A} \otimes \boldsymbol{G_B}$ is a generalized inverse of $\boldsymbol{A} \otimes \boldsymbol{B}$ (recall Theorem 9.9).

Proof of (iii): This follows from a straightforward verification that $\boldsymbol{A}^{+} \otimes \boldsymbol{B}^{+}$ satisfies the four conditions in Definition 9.3. We leave the details as an exercise. □

Theorem 14.4, in conjunction with (14.6), leads to the following attractive property concerning similarity of Kronecker products.

Theorem 14.5 *Let $\boldsymbol{A}$ and $\boldsymbol{B}$ be two square matrices. If $\boldsymbol{A}$ is similar to $\boldsymbol{C}$ and $\boldsymbol{B}$ is similar to $\boldsymbol{D}$, then $\boldsymbol{A} \otimes \boldsymbol{B}$ is similar to $\boldsymbol{C} \otimes \boldsymbol{D}$.*

Proof. Since $\boldsymbol{A}$ and $\boldsymbol{B}$ are similar to $\boldsymbol{C}$ and $\boldsymbol{D}$, respectively, there exist nonsingular matrices $\boldsymbol{P}$ and $\boldsymbol{Q}$ such that $\boldsymbol{P}^{-1}\boldsymbol{A}\boldsymbol{P} = \boldsymbol{C}$ and $\boldsymbol{Q}^{-1}\boldsymbol{B}\boldsymbol{Q} = \boldsymbol{D}$. Then,

$$\begin{aligned}(\boldsymbol{C} \otimes \boldsymbol{D}) &= (\boldsymbol{P}^{-1}\boldsymbol{A}\boldsymbol{P}) \otimes (\boldsymbol{Q}^{-1}\boldsymbol{B}\boldsymbol{Q}) = (\boldsymbol{P}^{-1} \otimes \boldsymbol{Q}^{-1})(\boldsymbol{A} \otimes \boldsymbol{B})(\boldsymbol{P} \otimes \boldsymbol{Q}) \\ &= (\boldsymbol{P} \otimes \boldsymbol{Q})^{-1}(\boldsymbol{A} \otimes \boldsymbol{B})(\boldsymbol{P} \otimes \boldsymbol{Q}) ,\end{aligned}$$

which shows that $\boldsymbol{C} \otimes \boldsymbol{D}$ and $\boldsymbol{A} \otimes \boldsymbol{B}$ are similar matrices with $\boldsymbol{P} \otimes \boldsymbol{Q}$ being the associated change of basis matrix. □

In Theorem 14.2 we saw that the trace of the Kronecker product of two matrices is equal to the product of traces of the two matrices. We can now show that a similar result holds for the rank as well. In fact, we can make use of (9.2) to prove the following.

Theorem 14.6 *Let $\boldsymbol{A}$ and $\boldsymbol{B}$ be any two matrices. Then the rank of $\boldsymbol{A} \otimes \boldsymbol{B}$ is the product of the ranks of $\boldsymbol{A}$ and $\boldsymbol{B}$. That is,*

$$\rho(\boldsymbol{A} \otimes \boldsymbol{B}) = \rho(\boldsymbol{A}) \times \rho(\boldsymbol{B}) = \rho(\boldsymbol{B} \otimes \boldsymbol{A}) .$$

Proof. Let $\boldsymbol{G_A}$ and $\boldsymbol{G_B}$ be any two generalized inverses of $\boldsymbol{A}$ and $\boldsymbol{B}$, respectively.

Theorem 14.4 tells us that $\boldsymbol{G_A} \otimes \boldsymbol{G_B}$ is a generalized inverse of $\boldsymbol{A} \otimes \boldsymbol{B}$. Using (9.2) we can conclude that

$$\begin{aligned}\rho(\boldsymbol{A} \otimes \boldsymbol{B}) &= \rho[(\boldsymbol{A} \otimes \boldsymbol{B})(\boldsymbol{G_A} \otimes \boldsymbol{G_B})] = \operatorname{tr}[(\boldsymbol{A} \otimes \boldsymbol{B})(\boldsymbol{G_A} \otimes \boldsymbol{G_B})] \\ &= \operatorname{tr}[(\boldsymbol{AG_A}) \otimes (\boldsymbol{BG_B})] = \operatorname{tr}(\boldsymbol{AG_A})\operatorname{tr}(\boldsymbol{BG_B}) = \rho(\boldsymbol{AG_A})\rho(\boldsymbol{BG_B}) \\ &= \rho(\boldsymbol{A})\rho(\boldsymbol{B}) \ .\end{aligned}$$

In the chain of equalities above, the first and second equalities result from (9.2). The third follows from Theorem 14.3, the fourth from Theorem 14.2 and the last equality follows (again) from (9.2). □

It is worth noting that $\boldsymbol{A} \otimes \boldsymbol{B}$ can be square even if neither $\boldsymbol{A}$ nor $\boldsymbol{B}$ is square. If $\boldsymbol{A}$ is $m \times n$ and $\boldsymbol{B}$ is $p \times q$, then $\boldsymbol{A} \otimes \boldsymbol{B}$ is square whenever $mp = nq$. The following result explains precisely when a Kronecker product is nonsingular.

Theorem 14.7 *Let $\boldsymbol{A}$ be $m \times n$ and $\boldsymbol{B}$ be $p \times q$, where $mn = pq$. Then $\boldsymbol{A} \otimes \boldsymbol{B}$ is nonsingular if and only if both $\boldsymbol{A}$ and $\boldsymbol{B}$ are square and nonsingular.*

Proof. If $\boldsymbol{A}$ and $\boldsymbol{B}$ are nonsingular (hence square), then $\rho(\boldsymbol{A}) = m = n$ and $\rho(\boldsymbol{B}) = p = q$. Therefore, $\boldsymbol{A} \otimes \boldsymbol{B}$ is $mp \times mp$ and Theorem 14.6 tells us that $\rho(\boldsymbol{A} \otimes \boldsymbol{B}) = \rho(\boldsymbol{A}) \times \rho(\boldsymbol{B}) = mp$. Thus $\boldsymbol{A} \otimes \boldsymbol{B}$ is nonsingular.

Now suppose that $\boldsymbol{A} \otimes \boldsymbol{B}$ is nonsingular, and hence square, so that $mp = nq$. Therefore, $\rho(\boldsymbol{A} \otimes \boldsymbol{B}) = mp$. From Theorem 14.6 and a basic property of rank, we note that

$$mp = \rho(\boldsymbol{A} \otimes \boldsymbol{B}) = \rho(\boldsymbol{A}) \times \rho(\boldsymbol{B}) \leq \min\{m, n\} \times \min\{p.q\} \ .$$

From the above we conclude that $m \leq n$ and $p \leq q$. However, it is also true that $nq = \rho(\boldsymbol{A} \otimes \boldsymbol{B})$, which would imply that $nq \leq \min\{m, n\} \times \min\{p.q\}$ and, therefore, $n \leq m$ and $q \leq p$. This establishes $m = n$ and $p = q$, so $\boldsymbol{A}$ and $\boldsymbol{B}$ are both square. The facts that $\rho(\boldsymbol{A}) \leq m$, $\rho(\boldsymbol{B}) \leq p$ and $\rho(\boldsymbol{A})\rho(\boldsymbol{B}) = mp$ ensure that $\rho(\boldsymbol{A}) = m$ and $\rho(\boldsymbol{B}) = p$. Therefore, $\boldsymbol{A}$ and $\boldsymbol{B}$ are both nonsingular. □

14.4 Matrix factorizations for Kronecker products

Several standard matrix factorizations, including $\boldsymbol{LU}$, $\boldsymbol{QR}$, Schur's triangularization and other eigenvalue revealing decompositions, including the SVD and the Jordan, can be derived for $\boldsymbol{A} \otimes \boldsymbol{B}$ using the corresponding factorizations for $\boldsymbol{A}$ and $\boldsymbol{B}$. Theorem 14.3 and (14.6) are conspicuous in these derivations. We begin with a few useful results that show that Kronecker products of certain types of matrices produce a matrix of the same type.

Lemma 14.1 *If $\boldsymbol{A}$ and $\boldsymbol{B}$ are both upper (lower) triangular, then $\boldsymbol{A} \otimes \boldsymbol{B}$ is upper (lower) triangular.*

Proof. We will prove the result for upper-triangular matrices. The lower-triangular case will follow by taking transposes and using the fact that $(\boldsymbol{A} \otimes \boldsymbol{B})' = \boldsymbol{A}' \otimes \boldsymbol{B}'$ (part (vi) of Theorem 14.1).

Let $\boldsymbol{A}$ be $m \times m$ and $\boldsymbol{B}$ be $n \times n$. $\boldsymbol{A} = \{a_{ij}\}$ is upper-triangular implies that $a_{ij} = 0$ whenever $i > j$. Therefore,

$$\boldsymbol{A} \otimes \boldsymbol{B} = \begin{bmatrix} a_{11}\boldsymbol{B} & a_{12}\boldsymbol{B} & \dots & a_{1m}\boldsymbol{B} \\ \boldsymbol{O} & a_{22}\boldsymbol{B} & \dots & a_{2m}\boldsymbol{B} \\ \vdots & \vdots & \ddots & \vdots \\ \boldsymbol{O} & \boldsymbol{O} & \dots & a_{mm}\boldsymbol{B} \end{bmatrix}.$$

The above implies that all block matrices below the diagonal blocks in $\boldsymbol{A} \otimes \boldsymbol{B}$ are zero. Now examine the block matrices on the diagonal, or $a_{ii}\boldsymbol{B}$ for $i = 1, 2, \dots, m$. Each of these diagonal blocks is upper-triangular because $\boldsymbol{B}$ is upper-triangular, so all entries below the diagonal in $a_{ii}\boldsymbol{B}$ are zero. Since the diagonal entries of $\boldsymbol{A} \otimes \boldsymbol{B}$ are precisely the diagonal entries along its diagonal blocks, it follows that all entries below the diagonal in $\boldsymbol{A} \otimes \boldsymbol{B}$ are zero. Therefore, $\boldsymbol{A} \otimes \boldsymbol{B}$ is upper-triangular. □

If, in addition, $\boldsymbol{A}$ and $\boldsymbol{B}$ are both unit upper (lower) triangular, i.e., their diagonal elements are all 1, then it is easily verified that $\boldsymbol{A} \otimes \boldsymbol{B}$ is also unit upper (lower) triangular.

Lemma 14.2 *If $\boldsymbol{A}$ and $\boldsymbol{B}$ are both orthogonal matrices, then $\boldsymbol{A} \otimes \boldsymbol{B}$ is an orthogonal matrix.*

Proof. Let $\boldsymbol{A}$ and $\boldsymbol{B}$ are orthogonal matrices of order $m \times m$ and $n \times n$, respectively. This means that $\boldsymbol{A}'\boldsymbol{A} = \boldsymbol{A}\boldsymbol{A}' = \boldsymbol{I}_m$ and $\boldsymbol{B}'\boldsymbol{B} = \boldsymbol{B}\boldsymbol{B}' = \boldsymbol{I}_n$. Therefore,

$$\begin{aligned} (\boldsymbol{A} \otimes \boldsymbol{B})'(\boldsymbol{A} \otimes \boldsymbol{B}) &= (\boldsymbol{A}'\boldsymbol{A}) \otimes (\boldsymbol{B}'\boldsymbol{B}) = \boldsymbol{I}_m \otimes \boldsymbol{I}_n = \boldsymbol{I}_{mn} \\ \text{and } (\boldsymbol{A} \otimes \boldsymbol{B})(\boldsymbol{A} \otimes \boldsymbol{B})' &= (\boldsymbol{A}\boldsymbol{A}') \otimes (\boldsymbol{B}\boldsymbol{B}') = \boldsymbol{I}_m \otimes \boldsymbol{I}_n = \boldsymbol{I}_{mn} . \end{aligned}$$

If $\boldsymbol{A}$ and $\boldsymbol{B}$ are complex matrices, then we use conjugate transposes instead of transposes and (14.4) to obtain $(\boldsymbol{A} \otimes \boldsymbol{B})^*(\boldsymbol{A} \otimes \boldsymbol{B}) = \boldsymbol{I}_{mn} = (\boldsymbol{A} \otimes \boldsymbol{B})(\boldsymbol{A} \otimes \boldsymbol{B})^*$. □

Our next result shows that the Kronecker product of two permutation matrices is again a permutation matrix. The proof will use the definition of a permutation matrix as a square matrix that has exactly one 1 in every row and all other entries are zeroes.

Lemma 14.3 *If $\boldsymbol{A}$ and $\boldsymbol{B}$ are both permutation matrices, then $\boldsymbol{A} \otimes \boldsymbol{B}$ is a permutation matrix.*

Proof. If $\boldsymbol{A} = \{a_{ij}\}$ is an $m \times m$ permutation matrix, then for each i there exists exactly one j such that $a_{ij} = 1$ and all other elements in the i-th row are zero. This also means that for each j there exists exactly one i such that $a_{ij} = 1$ and all other entries in the j-th column are zero. A similar result is true for $\boldsymbol{B}$. Note that

$$\boldsymbol{A} \otimes \boldsymbol{B} = \begin{bmatrix} a_{11}\boldsymbol{B} & a_{12}\boldsymbol{B} & \dots & a_{1m}\boldsymbol{B} \\ a_{21}\boldsymbol{B} & a_{22}\boldsymbol{B} & \dots & a_{2m}\boldsymbol{B} \\ \vdots & \vdots & \ddots & \vdots \\ a_{m1}\boldsymbol{B} & a_{m2}\boldsymbol{B} & \dots & a_{mm}\boldsymbol{B} \end{bmatrix}.$$

Any generic column of $\boldsymbol{A} \otimes \boldsymbol{B}$ can be expressed as

$$\begin{bmatrix} a_{1r}b_{1t} \\ a_{1r}b_{2t} \\ \vdots \\ a_{1r}b_{nt} \\ \vdots \\ a_{mr}b_{1t} \\ a_{mr}b_{2t} \\ \vdots \\ a_{mr}b_{nt} \end{bmatrix},$$

where r and t are two integers. For example, the first column has $r = 1, t = 1$, the second column has $r = 1, t = 2$, the n-th column has $r = 1, t = n$, the $n+1$-th column has $r = 2, t = 1$ and so on.

Only one of the a_{ir}'s is equal to 1 and all others are zero. Similarly, only one of the b_{it}'s is equal to one and all others are zero. Hence, the above column of $\boldsymbol{A} \otimes \boldsymbol{B}$ has only one element equal to 1 and all others are zero. Since the above column was arbitrarily chosen, this holds true for all columns of $\boldsymbol{A} \otimes \boldsymbol{B}$. A similar argument reveals that every row of $\boldsymbol{A} \otimes \boldsymbol{B}$ will contain exactly one element equal to 1 and all others equal to 0. This proves that $\boldsymbol{A} \otimes \boldsymbol{B}$ is a permutation matrix. □

We use some of the above results to show how several popular matrix factorizations of Kronecker products follow from the corresponding factorizations of the constituent matrices.

Theorem 14.8 *The following statements are true:*

(i) If $\boldsymbol{P_A}\boldsymbol{A} = \boldsymbol{L_A}\boldsymbol{U_A}$ and $\boldsymbol{P_B}\boldsymbol{B} = \boldsymbol{L_B}\boldsymbol{U_B}$ are $\boldsymbol{LU}$ decompositions (with possible row interchanges) for nonsingular matrices $\boldsymbol{A}$ and $\boldsymbol{B}$, respectively, then

$$(\boldsymbol{P_A} \otimes \boldsymbol{P_B})(\boldsymbol{A} \otimes \boldsymbol{B}) = (\boldsymbol{L_A} \otimes \boldsymbol{L_B})(\boldsymbol{U_A} \otimes \boldsymbol{U_B}) \tag{14.8}$$

gives an $\boldsymbol{LU}$ decomposition for $\boldsymbol{A} \otimes \boldsymbol{B}$ with the corresponding row interchanges given by $\boldsymbol{P_A} \otimes \boldsymbol{P_B}$.

(ii) If $\boldsymbol{A} = \boldsymbol{L_A}\boldsymbol{L}'_{\boldsymbol{A}}$ and $\boldsymbol{B} = \boldsymbol{L_B}\boldsymbol{L}'_{\boldsymbol{B}}$ are Cholesky decompositions for nonsingular matrices $\boldsymbol{A}$ and $\boldsymbol{B}$, respectively, then

$$\boldsymbol{A} \otimes \boldsymbol{B} = (\boldsymbol{L_A} \otimes \boldsymbol{L_B})(\boldsymbol{L_A} \otimes \boldsymbol{L_B})' \tag{14.9}$$

gives the Cholesky decomposition for $\boldsymbol{A} \otimes \boldsymbol{B}$.

(iii) If $\boldsymbol{A} = \boldsymbol{Q_A R_A}$ *and* $\boldsymbol{B} = \boldsymbol{Q_B R_B}$ *are* $\boldsymbol{QR}$ *decompositions for square matrices* $\boldsymbol{A}$ *and* $\boldsymbol{B}$*, respectively, then*

$$\boldsymbol{A} \otimes \boldsymbol{B} = (\boldsymbol{Q_A} \otimes \boldsymbol{Q_B})(\boldsymbol{R_A} \otimes \boldsymbol{R_B}) \tag{14.10}$$

gives the $\boldsymbol{QR}$ *decomposition for* $\boldsymbol{A} \otimes \boldsymbol{B}$.

Proof. The mixed product rule of Kronecker products (Theorem 14.3) makes these derivations almost immediate.

Proof of (i): By repeated application of Theorem 14.3, we obtain

$$(\boldsymbol{P_A} \otimes \boldsymbol{P_B})(\boldsymbol{A} \otimes \boldsymbol{B}) = (\boldsymbol{P_A A}) \otimes (\boldsymbol{P_B B}) = (\boldsymbol{L_A U_A}) \otimes (\boldsymbol{L_B U_B})$$
$$= (\boldsymbol{L_A} \otimes \boldsymbol{L_B})(\boldsymbol{U_A} \otimes \boldsymbol{U_B}) .$$

Lemma 14.1 ensures that $(\boldsymbol{L_A} \otimes \boldsymbol{L_B})$ is lower-triangular and $\boldsymbol{U_A} \otimes \boldsymbol{U_B}$ is upper-triangular. Furthermore, since $\boldsymbol{L_A}$ and $\boldsymbol{L_B}$ are both unit lower-triangular, as is the convention for $\boldsymbol{LU}$ decompositions, so is $\boldsymbol{L_A} \otimes \boldsymbol{L_B}$. Finally, Lemma 14.3 ensures that $\boldsymbol{P_A} \otimes \boldsymbol{P_B}$ is also a permutation matrix, which implies that (14.8) is an $\boldsymbol{LU}$ decomposition for $\boldsymbol{A} \otimes \boldsymbol{B}$, with possible row interchanges determined by the permutation matrix $\boldsymbol{P_A} \otimes \boldsymbol{P_B}$.

Proof of (ii): Theorem 14.3 immediately yields

$$\boldsymbol{A} \otimes \boldsymbol{B} = (\boldsymbol{L_A L_A'}) \otimes (\boldsymbol{L_B L_B'}) = (\boldsymbol{L_A} \otimes \boldsymbol{L_B})(\boldsymbol{L_A} \otimes \boldsymbol{L_B})' .$$

Lemma 14.1 ensures that $\boldsymbol{L_A} \otimes \boldsymbol{L_B}$ is lower-triangular, which implies that (14.9) is the Cholesky decomposition for $\boldsymbol{A} \otimes \boldsymbol{B}$.

Proof of (iii): Theorem 14.3 immediately yields

$$\boldsymbol{A} \otimes \boldsymbol{B} = (\boldsymbol{Q_A R_A}) \otimes (\boldsymbol{Q_B R_B}) = (\boldsymbol{Q_A} \otimes \boldsymbol{Q_B})(\boldsymbol{R_A} \otimes \boldsymbol{R_B}) .$$

Lemma 14.1 ensures that $\boldsymbol{R_A} \otimes \boldsymbol{R_A}$ is upper-triangular and Lemma 14.2 ensures that $\boldsymbol{Q_A} \otimes \boldsymbol{Q_B}$ is orthogonal, which implies that (14.10) is the $\boldsymbol{QR}$ decomposition for $\boldsymbol{A} \otimes \boldsymbol{B}$. □

Part (ii) of Theorem 14.8 implies that if $\boldsymbol{A}$ and $\boldsymbol{B}$ are both positive definite, then so is $\boldsymbol{A} \otimes \boldsymbol{B}$ because a matrix is positive definite if and only if it has a Cholesky decomposition.

Theorem 14.5 revealed how Kronecker products preserve similarity. Theorem 11.14 tells us that every square matrix is similar to an upper-triangular matrix. Therefore, the Kronecker product of any two square matrices will be similar to the Kronecker product of two triangular matrices. Furthermore, if the similarity transformations for the constituent matrices are unitary (orthogonal), then so will be the similarity transformation for the Kronecker product, which leads to the Schur's triangular factorization for Kronecker products.

Theorem 14.9 *The following statements are true:*

(i) If $Q_A^* A Q_A = T_A$ and $Q_B^* B Q_B = T_B$ are Schur's triangular factorizations (recall Theorem 11.15) for square matrices A and B, then

$$(Q_A \otimes Q_B)^*(A \otimes B)(Q_A \otimes Q_B) = T_A \otimes T_B \tag{14.11}$$

gives a Schur's triangular factorization for $A \otimes B$.

(ii) If $P_A' A P_A = \Lambda_A$ and $P_B' B P = \Lambda_B$ are spectral decompositions (recall Theorem 11.27) for real symmetric matrices A and B, then

$$(P_A \otimes P_B)'(A \otimes B)(P_A \otimes P_B) = \Lambda_A \otimes \Lambda_B \tag{14.12}$$

gives the spectral decomposition for $A \otimes B$.

Proof. **Proof of (i):** Using the result on conjugate transposes in (14.4) and the mixed product rule in (14.6), we obtain

$$\begin{aligned}(Q_A \otimes Q_B)^*(A \otimes B)(Q_A \otimes Q_B) &= (Q_A^* \otimes Q_B^*)(A \otimes B)(Q_A \otimes Q_B) \\ &= (Q_A^* A Q_A) \otimes (Q_B^* B Q_B) = T_A \otimes T_B \,.\end{aligned}$$

Since $Q_A \otimes Q_B$ is orthogonal, the above gives the Schur's triangular factorization for $A \otimes B$.

Proof of (ii): In the special case where A and B are real and symmetric, Theorem 11.27 ensures that A and B are similar to diagonal matrices with real entries. The proof now follows from part (i) and by noting that the Kronecker product of two diagonal matrices is again a diagonal matrix. □

If $A = P_A J_A P_A^{-1}$ and $B = P_B J_B P_B^{-1}$ are two Jordan decompositions for square matrices A and B, then

$$A \otimes B = (P_A \otimes P_B)(J_A \otimes J_B)(P_A \otimes P_B)^{-1} \,. \tag{14.13}$$

Note, however, that $J_A \otimes J_B$ is not necessarily in Jordan form, so the above is not really a Jordan decomposition for $A \otimes B$. Further reduction is required to bring $J_A \otimes J_B$ to a Jordan form and $A \otimes B$ will be similar to the Jordan form of $J_A \otimes J_B$.

Finally, we conclude with a theorem on the SVD of Kronecker products.

Theorem 14.10 *Let $A = U_A D_A V_A'$ and $B = U_B D_B V_B'$ be singular value decompositions (as in Theorem 12.1) for an $m \times n$ matrix A and a $p \times q$ matrix B. Then*

$$A \otimes B = (U_A \otimes U_B)(D_A \otimes D_B)(V_A \otimes V_B)'$$

gives a singular value decomposition of $A \otimes B$ (up to a simple reordering of the diagonal elements of $D_A \otimes D_B$ and the respective singular vectors).

Proof. Using the mixed product rule in (14.6), we obtain

$$\begin{aligned}A \otimes B &= (U_A D_A V_A') \otimes (U_B D_B V_B') \\ &= (U_A \otimes U_B)(D_A \otimes D_B)(V_B' \otimes V_B') \\ &= (U_A \otimes U_B)(D_A \otimes D_B)(V_A \otimes V_B)' \,.\end{aligned}$$

Note that $\boldsymbol{U_A} \otimes \boldsymbol{U_B}$ and $\boldsymbol{V_A} \otimes \boldsymbol{V_B}$ are orthogonal (Lemma 14.2) and $\boldsymbol{D_A} \otimes \boldsymbol{D_B}$ is diagonal with diagonal entries $\sigma_{\boldsymbol{A},i}\sigma_{\boldsymbol{B},j}$, where $\sigma_{\boldsymbol{A},i}$'s and $\sigma_{\boldsymbol{B},j}$'s are the diagonal elements in $\boldsymbol{D_A}$ and $\boldsymbol{D_B}$, respectively. If we want to reorder the singular values (to be consistent with Theorem 12.1), we can do so by reordering the $\sigma_{\boldsymbol{A},i}\sigma_{\boldsymbol{B},j}$'s and the corresponding left and right singular vectors. □

Theorem 14.10 also reveals what we have already seen in Theorem 14.6. If $\boldsymbol{A}$ and $\boldsymbol{B}$ have ranks r_A and r_B, respectively, then there are $r_{\boldsymbol{A}}$ positive numbers in $\{\sigma_{\boldsymbol{A},i} : i = 1, 2. \ldots, \min\{m, n\}\}$ and $r_{\boldsymbol{B}}$ positive numbers in $\{\sigma_{\boldsymbol{B},j} : j = 1, 2, \ldots, \min\{p, q\}\}$. The nonzero singular values of $\boldsymbol{A} \otimes \boldsymbol{B}$ are the $r_{\boldsymbol{A}} r_{\boldsymbol{B}}$ positive numbers in the set $\{\sigma_{A,i}\sigma_{B,j} : i = 1, 2, \ldots, \min m, n;\ j = 1, 2, \ldots, \min\{p, q\}\}$. Zero is a singular value of $\boldsymbol{A} \otimes \boldsymbol{B}$ with multiplicity $\min\{mp, nq\} - r_{\boldsymbol{A}} r_{\boldsymbol{B}}$. It follows that the rank of $\boldsymbol{A} \otimes \boldsymbol{B}$ is $r_{\boldsymbol{A}} r_{\boldsymbol{B}}$.

14.5 Eigenvalues and determinant

Let us now turn to eigenvalues and eigenvectors of Kronecker products. Suppose that λ is an eigenvalue of $\boldsymbol{A}$ and μ is an eigenvalue of $\boldsymbol{B}$. Then, there exist eigenvectors $\boldsymbol{x}$ and $\boldsymbol{y}$ such that $\boldsymbol{Ax} = \lambda\boldsymbol{x}$ and $\boldsymbol{By} = \lambda\boldsymbol{y}$. We can now conclude that

$$(\boldsymbol{A} \otimes \boldsymbol{B})(\boldsymbol{x} \otimes \boldsymbol{y}) = (\boldsymbol{Ax}) \otimes (\boldsymbol{By}) = (\lambda\boldsymbol{x}) \otimes (\mu\boldsymbol{y}) = (\lambda\mu)(\boldsymbol{x} \otimes \boldsymbol{y}) ,$$

which means that $\lambda\mu$ is an eigenvalue of $\boldsymbol{A} \otimes \boldsymbol{B}$ with associated eigenvector $\boldsymbol{x} \otimes \boldsymbol{y}$. In fact, the following theorem shows that all the eigenvalues of $\boldsymbol{A} \otimes \boldsymbol{B}$ can be found by multiplying the eigenvalues of $\boldsymbol{A}$ and $\boldsymbol{B}$.

Theorem 14.11 *Let $\lambda_1, \ldots, \lambda_m$ be the eigenvalues (counting multiplicities) of an $m \times m$ matrix $\boldsymbol{A}$, and let $\mu_1, \ldots, \mu_p$ be the eigenvalues (counting multiplicities) of a $p \times p$ matrix $\boldsymbol{B}$. Then the set $\{\lambda_i\mu_j : i = 1, \ldots, m; j = 1, \ldots, p\}$ contains all the mp eigenvalues of $\boldsymbol{A} \otimes \boldsymbol{B}$.*

Proof. Based upon Schur's Triangularization Theorem (Theorem 11.15), we know that there exist unitary matrices $\boldsymbol{P}$ and $\boldsymbol{Q}$ such that

$$\boldsymbol{P}^*\boldsymbol{AP} = \boldsymbol{T_A} \ \text{ and } \ \boldsymbol{Q}^*\boldsymbol{BQ} = \boldsymbol{T_B} ,$$

where $\boldsymbol{T_A}$ and $\boldsymbol{T_B}$ are upper-triangular matrices whose diagonal entries contain the eigenvalues of $\boldsymbol{A}$ and $\boldsymbol{B}$, respectively. Theorem 14.9 tells us that

$$(\boldsymbol{P} \otimes \boldsymbol{Q})^*(\boldsymbol{A} \otimes \boldsymbol{B})(\boldsymbol{P} \otimes \boldsymbol{Q}) = \boldsymbol{T_A} \otimes \boldsymbol{T_B} . \tag{14.14}$$

Since similar matrices have the same set of eigenvalues (Theorem 11.6), $\boldsymbol{A} \otimes \boldsymbol{B}$ and $\boldsymbol{T_A} \otimes \boldsymbol{T_B}$ have the same set of eigenvalues. Since $\boldsymbol{T_A}$ and $\boldsymbol{T_B}$ are both upper-triangular, so is $\boldsymbol{T_A} \otimes \boldsymbol{T_B}$ and its diagonal elements, given by $\{\lambda_i\mu_j : i = 1, \ldots, m; j = 1, \ldots, p\}$, are its eigenvalues. □

Since the determinant of a matrix is the product of its eigenvalues, we have the following result.

Theorem 14.12 *If A is $m \times m$ and B is $p \times p$, then*

$$|A \otimes B| = |A|^p |B|^m = |B \otimes A| .$$

Proof. If $\lambda_1, \lambda_2, \ldots, \lambda_m$ are the eigenvalues of A and $\mu_1, \mu_2, \ldots, \mu_p$ are the eigenvalues of B, then (recall Theorem 11.5)

$$|A| = \prod_{i=1}^{m} \lambda_i \text{ and } |B| = \prod_{j=1}^{p} \mu_j .$$

From Theorem 14.11, we have

$$|A \otimes B| = \prod_{j=1}^{p}\prod_{i=1}^{m} \lambda_i \mu_j = \prod_{j=1}^{p}\left(\prod_{i=1}^{m} \mu_j \lambda_i\right) = \left(\prod_{j=1}^{p} \mu_j^m\right)\left(\prod_{i=1}^{m} \lambda_i\right) = \prod_{j=1}^{p}\left(\mu_j^m |A|\right)$$
$$= |A|^p \left(\prod_{j=1}^{p} \mu_j^m\right) = |A|^p \left(\prod_{j=1}^{p} \mu_j\right)^m = |A|^p |B|^m .$$

Applying the above result to $B \otimes A$ immediately reveals that $|B \otimes A|$ is again $|A|^p |B|^m$, thereby completing the proof. □

We have already seen that $A \otimes B$ is nonsingular if and only if A and B are both singular. The above result on determinants can also be used to arrive at this fact. $A \otimes B$ is nonsingular if and only if $|A \otimes B|$ is nonzero. This happens only if $|A|^p |B|^m$ is nonzero, which happens if and only if $|A|$ and B are both nonzero. Finally, $|A|$ and $|B|$ are both nonzero if and only if A and B are both nonsingular.

14.6 The vec and commutator operators

In matrix analysis it is often convenient to assemble the entries in an $m \times n$ matrix into an $mn \times 1$ vector. Such an operation on the matrix is called the *vectorization* of a matrix and denoted by the ***vec*** operator.

Definition 14.2 *The* **vectorization** *of a matrix converts a matrix into a column vector by stacking the columns of the matrix on top of one another. If $A = [a_1 : a_2 : \ldots : a_n]$ is an $m \times n$ matrix, then the vectorization operator is*

$$vec(A) = \begin{bmatrix} a_1 \\ a_2 \\ \vdots \\ a_n \end{bmatrix} ,$$

which is an $mn \times 1$ vector partitioned into n subvectors, each of length m and corresponding to a column of A.

Note that $\text{vec}(\boldsymbol{A}) = [a_{1,1}, \ldots, a_{m,1}, a_{1,2}, \ldots, a_{m,2}, \ldots, a_{1,n}, \ldots, a_{m,n}]'$, where $\boldsymbol{A} = \{a_{ij}\}$ is $m \times n$. Vectorization stacks the *columns and not the rows* on top of one another. This is purely by convention but it has implications. One needs to be careful that $\text{vec}(\boldsymbol{A})$ is not necessarily equal to $\text{vec}(\boldsymbol{A}')$. In fact, we should distinguish between $(\text{vec}(\boldsymbol{A}))'$ and $\text{vec}(\boldsymbol{A}')$. The former is the transpose of $\text{vec}(\boldsymbol{A})$, hence a row vector, while the latter vectorizes $\boldsymbol{A}'$ by stacking up the columns of $\boldsymbol{A}'$, i.e., the rows of $\boldsymbol{A}$, on top of one another into a column vector. Clearly, they are not the same.

Example 14.2 Vectorization. For a 2×2 matrix, vectorization yields the 4×1 vector:

$$\text{vec}\left(\begin{bmatrix} a & b \\ c & d \end{bmatrix}\right) = \begin{bmatrix} a \\ c \\ b \\ d \end{bmatrix} .$$

For a 3×2 matrix, we obtain a 6×1 vector:

$$\text{vec}\left(\begin{bmatrix} 1 & 2 \\ 3 & 4 \\ 0 & 1 \end{bmatrix}\right) = \begin{bmatrix} 1 \\ 3 \\ 0 \\ 2 \\ 4 \\ 1 \end{bmatrix} . \quad \blacksquare$$

A few other basic properties are presented in the next theorem.

Theorem 14.13 *If $\boldsymbol{a}$ and $\boldsymbol{b}$ are any two vectors (of possibly varying size), and $\boldsymbol{A}$ and $\boldsymbol{B}$ are two matrices of the same size, then the following statements are true:*

(i) $\text{vec}(\boldsymbol{a}) = \text{vec}(\boldsymbol{a}') = \boldsymbol{a}$.

(ii) $\text{vec}(\boldsymbol{a}\boldsymbol{b}') = \boldsymbol{b} \otimes \boldsymbol{a}$.

(iii) $\text{vec}(\alpha\boldsymbol{A} + \beta\boldsymbol{B}) = \alpha \cdot \text{vec}(\boldsymbol{A}) + \beta \cdot \text{vec}(\boldsymbol{B})$ where α and β are scalars.

Proof. All of these follow from Definition 14.2.

Proof of (i): The "proof" consists of some elementary observations. If $\boldsymbol{a}$ is $m \times 1$, it is a column vector with one "column," so $\text{vec}(\boldsymbol{a})$ will simply be itself. The row vector $\boldsymbol{a}'$ is $1 \times m$, so it has m columns but each column is a scalar. Therefore, stacking up the columns yields the column vector $\boldsymbol{a}$. Thus, $\text{vec}(\boldsymbol{a}) = \text{vec}(\boldsymbol{a}') = \boldsymbol{a}$.

Proof of (ii): If $\boldsymbol{a}$ is $m \times 1$ and $\boldsymbol{b}$ is $n \times 1$, then $\boldsymbol{a}\boldsymbol{b}'$ is $m \times n$ and

$$\text{vec}(\boldsymbol{a}\boldsymbol{b}') = \text{vec}([b_1\boldsymbol{a}, \ldots, b_n\boldsymbol{a}]) = \begin{bmatrix} b_1\boldsymbol{a} \\ b_2\boldsymbol{a} \\ \vdots \\ b_n\boldsymbol{a} \end{bmatrix} = \boldsymbol{b} \otimes \boldsymbol{a} .$$

Proof of (iii): Since $\boldsymbol{A}$ and $\boldsymbol{B}$ are of the same size, say $m \times n$, $\alpha\boldsymbol{A} + \beta\boldsymbol{B}$ is well-defined for all scalars α and β. Now,

$$\begin{aligned}\text{vec}(\alpha\boldsymbol{A} + \beta\boldsymbol{B}) &= \text{vec}\left(\alpha\left[\boldsymbol{a}_1 : \boldsymbol{a}_2 : \ldots \boldsymbol{a}_n\right]\right) + \text{vec}\left(\beta\left[\boldsymbol{b}_1 : \boldsymbol{b}_2 : \ldots \boldsymbol{b}_n\right]\right) \\ &= \text{vec}\left(\left[\alpha\boldsymbol{a}_1 + \beta\boldsymbol{b}_1 : \alpha\boldsymbol{a}_2 + \beta\boldsymbol{b}_2 : \ldots \alpha\boldsymbol{a}_n + \beta\boldsymbol{b}_n\right]\right) \\ &= \begin{bmatrix}\alpha\boldsymbol{a}_1 + \beta\boldsymbol{b}_1 \\ \alpha\boldsymbol{a}_2 + \beta\boldsymbol{b}_2 \\ \vdots \\ \alpha\boldsymbol{a}_n + \beta\boldsymbol{b}_n\end{bmatrix} = \alpha\begin{bmatrix}\boldsymbol{a}_1 \\ \boldsymbol{a}_2 \\ \vdots \\ \boldsymbol{a}_n\end{bmatrix} + \beta\begin{bmatrix}\boldsymbol{b}_1 \\ \boldsymbol{b}_2 \\ \vdots \\ \boldsymbol{b}_n\end{bmatrix} \\ &= \alpha\text{vec}(\boldsymbol{A}) + \beta\text{vec}(\boldsymbol{B}) .\end{aligned}$$

This completes the proof. □

The following theorem reveals a useful relationship between the vectorized product of three matrices and a matrix-vector product. We will see later that this has applications in solving linear systems involving Kronecker products.

Theorem 14.14 *If $\boldsymbol{A}$, $\boldsymbol{B}$ and $\boldsymbol{C}$ are matrices of sizes $m \times n$, $n \times p$, and $p \times q$, respectively, then*

$$vec(\boldsymbol{ABC}) = (\boldsymbol{C}' \otimes \boldsymbol{A})vec(\boldsymbol{B}) .$$

Proof. Let us write $\boldsymbol{B}$ as an outer-product

$$\boldsymbol{B} = \boldsymbol{BI}_p = [\boldsymbol{b}_1 : \boldsymbol{b}_2 : \ldots : \boldsymbol{b}_p]\begin{bmatrix}\boldsymbol{e}_1' \\ \boldsymbol{e}_2' \\ \vdots \\ \boldsymbol{e}_p'\end{bmatrix} = \sum_{i=1}^{p} \boldsymbol{b}_i\boldsymbol{e}_i' ,$$

where $\boldsymbol{b}_i$ and $\boldsymbol{e}_i$ are the i-th columns of $\boldsymbol{B}$ and $\boldsymbol{I}_p$, respectively. Using the linearity of the vec operator (part (iii) of Theorem 14.13), we obtain

$$\begin{aligned}\text{vec}(\boldsymbol{ABC}) &= \text{vec}\left\{\boldsymbol{A}\left(\sum_{i=1}^{p}\boldsymbol{b}_i\boldsymbol{e}_i'\right)\boldsymbol{C}\right\} = \text{vec}\left\{\sum_{i=1}^{p}(\boldsymbol{Ab}_i\boldsymbol{e}_i'\boldsymbol{C})\right\} \\ &= \sum_{i=1}^{p}\text{vec}(\boldsymbol{Ab}_i\boldsymbol{e}_i'\boldsymbol{C}) = \sum_{i=1}^{p}\text{vec}\{(\boldsymbol{Ab}_i)(\boldsymbol{C}'\boldsymbol{e}_i)'\} = \sum_{i=1}^{p}(\boldsymbol{C}'\boldsymbol{e}_i) \otimes (\boldsymbol{Ab}_i) \\ &= (\boldsymbol{C}' \otimes \boldsymbol{A})\sum_{i=1}^{p}(\boldsymbol{e}_i \otimes \boldsymbol{b}_i) , \end{aligned} \tag{14.15}$$

where the second to last equality follows from part (ii) of Theorem 14.13. Part (ii) of Theorem 14.13 also ensures that $\boldsymbol{e}_i \otimes \boldsymbol{b}_i = \text{vec}(\boldsymbol{b}_i\boldsymbol{e}_i')$, so

$$\sum_{i=1}^{p}(\boldsymbol{e}_i \otimes \boldsymbol{b}_i) = \sum_{i=1}^{p}\text{vec}(\boldsymbol{b}_i\boldsymbol{e}_i') = \text{vec}\left(\sum_{i=1}^{p}\boldsymbol{b}_i\boldsymbol{e}_i'\right) = \text{vec}(\boldsymbol{B}) .$$

Substituting this expression into what we obtained in (14.15) renders $\text{vec}(\boldsymbol{ABC}) = (\boldsymbol{C}' \otimes \boldsymbol{A})\text{vec}(\boldsymbol{B})$. □

The next theorem shows how the trace of the product of two matrices can be expressed using vectorization.

Theorem 14.15

(i) If $\boldsymbol{A}$ and $\boldsymbol{B}$ are both $m \times n$, then

$$tr(\boldsymbol{A}'\boldsymbol{B}) = (vec(\boldsymbol{A}))'\, vec(\boldsymbol{B})\ .$$

(ii) If the product $\boldsymbol{ABCD}$ is well-defined, then

$$tr(\boldsymbol{ABCD}) = (vec(\boldsymbol{D}'))'(\boldsymbol{C}' \otimes \boldsymbol{A})vec(\boldsymbol{B}) = (vec(\boldsymbol{D}))'(\boldsymbol{A} \otimes \boldsymbol{C}')vec(\boldsymbol{B}')\ .$$

Proof. **Proof of (i):** The trace of $\boldsymbol{A}'\boldsymbol{B}$ is the sum of the diagonal entries in $\boldsymbol{A}'\boldsymbol{B}$. The i-th diagonal entry for $\boldsymbol{A}'\boldsymbol{B}$ is the inner product of the i-th row of $\boldsymbol{A}'$ and the i-th column of $\boldsymbol{B}$. The i-th row of $\boldsymbol{A}'$ is the i-th column of $\boldsymbol{A}$ so the i-th diagonal element is $\boldsymbol{a}_i'\boldsymbol{b}_i$, where $\boldsymbol{a}_i$ and $\boldsymbol{b}_i$ are the i-th column of $\boldsymbol{A}$ and $\boldsymbol{B}$, respectively. This implies that

$$\mathrm{tr}(\boldsymbol{A}'\boldsymbol{B}) = \sum_{i=1}^{n} \boldsymbol{a}_i'\boldsymbol{b}_i = \begin{bmatrix}\boldsymbol{a}_1' : \boldsymbol{a}_2' : \ldots : \boldsymbol{a}_n'\end{bmatrix} \begin{bmatrix}\boldsymbol{b}_1 \\ \boldsymbol{b}_2 \\ \vdots \\ \boldsymbol{b}_n\end{bmatrix} = (\mathrm{vec}(\boldsymbol{A}))'\, \mathrm{vec}(\boldsymbol{B})\ .$$

Proof of (ii): We use (i) and Theorem 14.14. The first equality in (ii) is true because

$$\begin{aligned}\mathrm{tr}(\boldsymbol{ABCD}) &= \mathrm{tr}(\boldsymbol{D}(\boldsymbol{ABC})) = (\mathrm{vec}(\boldsymbol{D}'))'\mathrm{vec}(\boldsymbol{ABC}) \\ &= (\mathrm{vec}(\boldsymbol{D}'))'(\boldsymbol{C}' \otimes \boldsymbol{A})\mathrm{vec}(\boldsymbol{B})\ .\end{aligned}$$

The second equality in (ii) is derived as follows

$$\begin{aligned}\mathrm{tr}(\boldsymbol{ABCD}) &= \mathrm{tr}((\boldsymbol{ABCD})') = \mathrm{tr}(\boldsymbol{D}'\boldsymbol{C}'\boldsymbol{B}'\boldsymbol{A}') = (\mathrm{vec}(\boldsymbol{D}))'\mathrm{vec}(\boldsymbol{C}'\boldsymbol{B}'\boldsymbol{A}') \\ &= (\mathrm{vec}(\boldsymbol{D}))'(\boldsymbol{A} \otimes \boldsymbol{C}')\mathrm{vec}(\boldsymbol{B}')\ .\end{aligned}$$

□

We mentioned earlier that $\mathrm{vec}(\boldsymbol{A})$ and $\mathrm{vec}(\boldsymbol{A}')$ are not the same. The former stacks up the columns of $\boldsymbol{A}$, while the latter stacks up the rows of $\boldsymbol{A}$. Note, however, that both $\mathrm{vec}(\boldsymbol{A})$ and $\mathrm{vec}(\boldsymbol{A}')$ contain the entries of $\boldsymbol{A}$ but arranged differently. Clearly, one can be obtained from the other by permuting the elements. Therefore, there exists an $mn \times mn$ permutation matrix $\boldsymbol{K}_{mn}$ corresponding to an $m \times n$ matrix $\boldsymbol{A}$, such that

$$\mathrm{vec}(\boldsymbol{A}') = \boldsymbol{K}_{mn}\mathrm{vec}(\boldsymbol{A})\ . \tag{14.16}$$

This matrix is referred to as the ***vec-permutation*** matrix or a ***commutator*** matrix. The motivation behind the second name will become clear later. Observe that $\boldsymbol{K}_{mn}$ is a permutation matrix. It does not depend upon the numerical values of the entries in $\boldsymbol{A}$, only how they are arranged.

Example 14.3 Consider the 2×2 matrix $\boldsymbol{A} = \begin{bmatrix} a & b \\ c & d \end{bmatrix}$. Then,

$$\text{vec}(\boldsymbol{A}) = \begin{bmatrix} a \\ c \\ b \\ d \end{bmatrix} \quad \text{and} \quad \text{vec}(\boldsymbol{A}') = \begin{bmatrix} a \\ b \\ c \\ d \end{bmatrix} .$$

Note that $\text{vec}(\boldsymbol{A}')$ can be obtained from $\text{vec}(\boldsymbol{A})$ by permuting the second and third entries. This means that $\boldsymbol{K}_{22}$ is the permutation matrix that interchanges the second and third rows of $\text{vec}(\boldsymbol{A})$ and is given by

$$\boldsymbol{K}_{22} = \begin{bmatrix} 1 & 0 & 0 & 0 \\ 0 & 0 & 1 & 0 \\ 0 & 1 & 0 & 0 \\ 0 & 0 & 0 & 1 \end{bmatrix} .$$

It is easily verified that $\text{vec}(\boldsymbol{A}') = \boldsymbol{K}_{22}\text{vec}(\boldsymbol{A})$.

Consider $\boldsymbol{K}_{32}$, which corresponds to 3×2 matrices. If $\boldsymbol{A} = \begin{bmatrix} 1 & 2 \\ 3 & 4 \\ 5 & 6 \end{bmatrix}$, then

$$\text{vec}(\boldsymbol{A}) = \begin{bmatrix} 1 \\ 3 \\ 5 \\ 2 \\ 4 \\ 6 \end{bmatrix} \quad \text{and} \quad \text{vec}(\boldsymbol{A}') = \begin{bmatrix} 1 \\ 2 \\ 3 \\ 4 \\ 5 \\ 6 \end{bmatrix} .$$

Swapping the second and third rows of $\text{vec}(\boldsymbol{A})$, followed by interchanging the second and fourth rows of the resulting vector and concluding by a swap of fourth and fifth rows yields $\text{vec}(\boldsymbol{A}')$. The composition of these permutations results in

$$\boldsymbol{K}_{32} = \begin{bmatrix} 1 & 0 & 0 & 0 & 0 & 0 \\ 0 & 0 & 0 & 1 & 0 & 0 \\ 0 & 1 & 0 & 0 & 0 & 0 \\ 0 & 0 & 0 & 0 & 1 & 0 \\ 0 & 0 & 1 & 0 & 0 & 0 \\ 0 & 0 & 0 & 0 & 0 & 1 \end{bmatrix} .$$

As for general permutation matrices, the easiest way to construct commutators is to perform the corresponding permutations (in the same sequence) on the identity matrix. ■

Since commutators are permutation matrices, they are orthogonal matrices. In general $\boldsymbol{K}_{mn} \neq \boldsymbol{K}_{nm}$. When $m = n$, they are equal and some authors prefer to write $\boldsymbol{K}_n$ instead of $\boldsymbol{K}_{nn}$ or $\boldsymbol{K}_{n^2}$. It can also be verified that $\boldsymbol{K}_{mn}$ is symmetric when $m = n$ or when either m or n is equal to one.

The following lemma shows the relationship between $\boldsymbol{K}_{mn}$ and $\boldsymbol{K}_{nm}$ in general.

Lemma 14.4 *If K_{mn} is a commutator matrix, then $K'_{mn} = K_{mn}^{-1} = K_{nm}$.*

Proof. The first equality follows from the fact that K_{mn} is a permutation matrix and so is orthogonal. Therefore, $K_{mn}K'_{mn} = K'_{mn}K_{mn} = I_{mn}$. Let A be any $m \times n$ matrix. Then,

$$K_{nm}\text{vec}(A') = K_{nm}K_{mn}\text{vec}(A) .$$

Since the above holds for every matrix A, it holds for every vector vec(A) in $\Re^{mn}$. Therefore, $K_{nm}K_{mn} = I_{mn}$, which implies that $K_{mn}^{-1} = K_{nm}$. □

We now come to the "reason" why K_{mn} is referred to as the *commutator*. Recall that $A \otimes B$ is not necessarily equal to $A \otimes B$. However, the following theorem shows that one can be obtained from the other by applying commutators to their rows and columns. In this sense, the commutators make Kronecker products *commute*.

Theorem 14.16 *Let A be $m \times n$ and B be $p \times q$. Then,*

$$K_{pm}(A \otimes B)K_{nq} = B \otimes A .$$

Proof. A useful trick is to consider an arbitrary $n \times q$ matrix C and apply the transformation $(A \otimes B)K_{nq}$ to vec(C). Thus,

$$\begin{aligned}(A \otimes B)K_{nq}\text{vec}(C) &= (A \otimes B)\text{vec}(C') = \text{vec}(BC'A') \\ &= K_{mp}\text{vec}(ACB') = K_{mp}(B \otimes A)\text{vec}(C) .\end{aligned}$$

Since C is arbitrary, the above holds for every vector vec(C) in $\Re^{nq}$. Therefore,

$$(A \otimes B)K_{nq} = K_{mp}(B \otimes A) .$$

Multiplying both sides of the above by K_{pm} yields

$$K_{pm}(A \otimes B)K_{nq} = K_{pm}K_{mp}(B \otimes A) = B \otimes A ,$$

where the last equality follows from the fact that $K_{pm}K_{mp} = I_{mp}$ (Lemma 14.4). □

14.7 Linear systems involving Kronecker products

In statistical modeling we often encounter linear systems whose coefficient matrix is a Kronecker product. For example, consider the system of equations

$$(A \otimes B)x = y , \tag{14.17}$$

where A is $m \times n$, B is $p \times q$, x is $nq \times 1$ and y is $mp \times 1$.

Assume that A and B are known in (14.17). If A and B are large matrices, then the Kronecker product $A \otimes B$ will be massive and storing it may become infeasible. For a given $nq \times 1$ vector x, how then should we compute y in (14.17)? Can we exploit the block structure of the Kronecker product matrix to compute y without explicitly forming $A \otimes B$?

Let us partition the $nq \times 1$ vector $\boldsymbol{x}$ as follows

$$\boldsymbol{x} = \begin{bmatrix} \boldsymbol{x}_1 \\ \boldsymbol{x}_2 \\ \vdots \\ \boldsymbol{x}_n \end{bmatrix}, \text{ where each } \boldsymbol{x}_i \in \Re^q . \tag{14.18}$$

Then, we can rewrite (14.17) as

$$\begin{bmatrix} a_{11}\boldsymbol{B} & a_{12}\boldsymbol{B} & \dots & a_{1n}\boldsymbol{B} \\ a_{21}\boldsymbol{B} & a_{22}\boldsymbol{B} & \dots & a_{2n}\boldsymbol{B} \\ \vdots & \vdots & \ddots & \vdots \\ a_{m1}\boldsymbol{B} & a_{m2}\boldsymbol{B} & \dots & a_{mn}\boldsymbol{B} \end{bmatrix} \begin{bmatrix} \boldsymbol{x}_1 \\ \boldsymbol{x}_2 \\ \dots \\ \boldsymbol{x}_n \end{bmatrix} = \begin{bmatrix} \boldsymbol{y}_1 \\ \boldsymbol{y}_2 \\ \vdots \\ \boldsymbol{y}_m \end{bmatrix},$$

where each $\boldsymbol{y}_i$ is $p \times 1$. In fact, each $\boldsymbol{y}_i$ can be expressed as

$$\begin{aligned} \boldsymbol{y}_i &= a_{i1}\boldsymbol{B}\boldsymbol{x}_1 + a_{i2}\boldsymbol{B}\boldsymbol{x}_2 + \dots + a_{in}\boldsymbol{B}\boldsymbol{x}_n \\ &= \left[\boldsymbol{B}\boldsymbol{x}_1 : \boldsymbol{B}\boldsymbol{x}_2 : \dots : \boldsymbol{B}\boldsymbol{x}_n\right] \begin{bmatrix} a_{i1} \\ a_{i2} \\ \vdots \\ a_{in} \end{bmatrix} = \boldsymbol{B}\left[\boldsymbol{x}_1 : \boldsymbol{x}_2 : \dots : \boldsymbol{x}_n\right] \begin{bmatrix} a_{i1} \\ a_{i2} \\ \vdots \\ a_{in} \end{bmatrix} \\ &= \boldsymbol{B}\boldsymbol{X}\boldsymbol{a}_{i*} \text{ for } i = 1, 2, \dots, m , \end{aligned}$$

where $\boldsymbol{X} = [\boldsymbol{x}_1 : \boldsymbol{x}_2 : \dots : \boldsymbol{x}_n]$ is the $q \times n$ matrix with $\boldsymbol{x}_i$ as its columns, and $\boldsymbol{a}_{i*} = [a_{i1}, a_{i2}, \dots, a_{in}]'$ is the i-th row vector of $\boldsymbol{A}$ expressed as an $n \times 1$ column. If we form the $p \times m$ matrix $\boldsymbol{Y}$ with $\boldsymbol{y}_i$ as its columns, then we obtain the following relationship:

$$\boldsymbol{Y} = \left[\boldsymbol{y}_1 : \boldsymbol{y}_2 : \dots : \boldsymbol{y}_m\right] = \boldsymbol{B}\boldsymbol{X}\left[\boldsymbol{a}_{1*} : \boldsymbol{a}_{2*} : \dots : \boldsymbol{a}_{m*}\right] = \boldsymbol{B}\boldsymbol{X}\boldsymbol{A}' . \tag{14.19}$$

For the last equality above, note that $\boldsymbol{a}_{i*}$ is the i-th column of $\boldsymbol{A}'$ because $\boldsymbol{a}_{i*}'$ is the i-th row of $\boldsymbol{A}$.

Equation (14.19) provides us with a more effective way to compute $\boldsymbol{y} = (\boldsymbol{A} \otimes \boldsymbol{B})\boldsymbol{x}$, where $\boldsymbol{A}$ and $\boldsymbol{B}$ are known matrices of order $m \times n$ and $p \times q$, respectively, and $\boldsymbol{x}$ is a known $nq \times 1$ vector. We summarize this below.

- Partition $\boldsymbol{x}$ as in (14.18) and construct the $q \times n$ matrix $\boldsymbol{X} = [\boldsymbol{x}_1 : \boldsymbol{x}_2 : \dots : \boldsymbol{x}_n]$, where each $\boldsymbol{x}_i$ is $q \times 1$. Thus, $\boldsymbol{x} = \text{vec}(\boldsymbol{X})$.
- Form the $p \times m$ matrix $\boldsymbol{Y} = \boldsymbol{B}\boldsymbol{X}\boldsymbol{A}'$.
- Take the columns of $\boldsymbol{Y} = [\boldsymbol{y}_1 : \boldsymbol{y}_2 : \dots : \boldsymbol{y}_m]$ and stack them on top of one another to form the $mp \times 1$ column vector $\boldsymbol{y}$. In other words, form $\boldsymbol{y} = \text{vec}(\boldsymbol{Y})$.

Computing $\boldsymbol{y}$ as above requires floating point operations (flops) in the order of $npq + qnm$. If, on the other hand, we had explicitly constructed $\boldsymbol{A} \otimes \boldsymbol{B}$, then the cost to compute $\boldsymbol{y}$ would have been in the order of $mpnq$ flops. In addition to these computational savings, we obtain substantial savings in storage by avoiding the explicit construction of $\boldsymbol{A} \otimes \boldsymbol{B}$ and, instead, just storing the smaller matrices $\boldsymbol{A}$ and $\boldsymbol{B}$.

Let us now turn to solving linear systems such as (14.17). Now $\boldsymbol{y}$ is given and we want to find a solution $\boldsymbol{x}$. Suppose $\boldsymbol{A}$ and $\boldsymbol{B}$ are both nonsingular (hence square). Then,

$$\boldsymbol{x} = (\boldsymbol{A} \otimes \boldsymbol{B})^{-1}\boldsymbol{y} = (\boldsymbol{A}^{-1} \otimes \boldsymbol{B}^{-1})\boldsymbol{y} ,$$

which implies that we need only the inverses of the constituent matrices $\boldsymbol{A}$ and $\boldsymbol{B}$ and not the inverse of the bigger matrix $\boldsymbol{A} \otimes \boldsymbol{B}$. In practice, it may be inefficient to compute the inverses of $\boldsymbol{A}$ and $\boldsymbol{B}$ and a more effective way to obtain $\boldsymbol{x}$ is to solve for $\boldsymbol{X}$ in (14.19). Note that

$$(\boldsymbol{A} \otimes \boldsymbol{B})\boldsymbol{x} = \boldsymbol{y} \Longleftrightarrow \boldsymbol{B}\boldsymbol{X}\boldsymbol{A}' = \boldsymbol{Y} \Longleftrightarrow \boldsymbol{X} = \boldsymbol{B}^{-1}\boldsymbol{Y}(\boldsymbol{A}')^{-1} ,$$

where $\boldsymbol{x} = \text{vec}(\boldsymbol{X})$ and $\boldsymbol{y} = \text{vec}(\boldsymbol{Y})$. We do not explicitly compute the inverses and solve linear systems instead. The following steps show to solve for $\boldsymbol{x}$ in (14.17).

- Given $\boldsymbol{B}$ and $\boldsymbol{Y}$, solve the matrix equation $\boldsymbol{B}\boldsymbol{Z} = \boldsymbol{Y}$ for $\boldsymbol{Z}$.
- Given $\boldsymbol{Z}$ and $\boldsymbol{A}$, solve the matrix equation $\boldsymbol{A}\boldsymbol{X}' = \boldsymbol{Z}'$ for $\boldsymbol{X}$.
- Obtain $\boldsymbol{x} = \text{vec}(\boldsymbol{X})$.

Note that $\boldsymbol{Z} = \boldsymbol{B}^{-1}\boldsymbol{Y}$ and $\boldsymbol{X}' = \boldsymbol{A}^{-1}\boldsymbol{Z}'$. Therefore, $\boldsymbol{X} = \boldsymbol{Z}(\boldsymbol{A}')^{-1} = \boldsymbol{B}^{-1}\boldsymbol{Y}(\boldsymbol{A}')^{-1}$, which is indeed a solution for (14.19). If $\boldsymbol{A}$ is $n \times n$ and $\boldsymbol{B}$ is $p \times p$, then the cost of doing this with matrix factorization methods such as Gaussian elimination or $\boldsymbol{LU}$ is only in the order of $n^3 + p^3$ flops. On the other hand, if we explicitly form $\boldsymbol{A} \otimes \boldsymbol{B}$ and then solve the system, the cost will be in the order of n^3p^3 flops.

The following theorem provides general existence and uniqueness conditions for the solution of systems such as (14.19).

Theorem 14.17 *Consider the matrix equation*

$$\boldsymbol{A}\boldsymbol{X}\boldsymbol{B} = \boldsymbol{C} , \tag{14.20}$$

where $\boldsymbol{A}$, $\boldsymbol{B}$ and $\boldsymbol{C}$ are known matrices with sizes $m \times n$, $p \times q$ and $m \times q$, respectively, and $\boldsymbol{X}$ is an unknown $n \times p$ matrix.

(i) The system (14.20) has a solution $\boldsymbol{X}$ if and only if $\boldsymbol{A}\boldsymbol{A}^+\boldsymbol{C}\boldsymbol{B}^+\boldsymbol{B} = \boldsymbol{C}$.

(ii) The general solution is of the form

$$\boldsymbol{X} = \boldsymbol{A}^+\boldsymbol{C}\boldsymbol{B}^+ + \boldsymbol{Y} - \boldsymbol{A}^+\boldsymbol{A}\boldsymbol{Y}\boldsymbol{B}\boldsymbol{B}^+ , \tag{14.21}$$

where $\boldsymbol{Y}$ is any arbitrary $n \times p$ matrix.

(iii) The solution is unique if $\boldsymbol{B}\boldsymbol{B}^+ \otimes \boldsymbol{A}^+\boldsymbol{A} = \boldsymbol{I}$.

Proof. **Proof of (i):** Using Theorem 14.14, we can rewrite $\boldsymbol{A}\boldsymbol{X}\boldsymbol{B} = \boldsymbol{C}$ as the linear system

$$(\boldsymbol{B}' \otimes \boldsymbol{A})\text{vec}(\boldsymbol{X}) = \text{vec}(\boldsymbol{C}) . \tag{14.22}$$

From a basic property of generalized inverses, the above system has a solution if and only if

$$(\boldsymbol{B}' \otimes \boldsymbol{A})(\boldsymbol{B}' \otimes \boldsymbol{A})^+\text{vec}(\boldsymbol{C}) = \text{vec}(\boldsymbol{C}) .$$

Theorem 14.4 tells us that $(\boldsymbol{B}' \otimes \boldsymbol{A})^+ = (\boldsymbol{B}')^+ \otimes \boldsymbol{A}^+$. Using this and the fact that $(\boldsymbol{B}^+)' = (\boldsymbol{B}')^+$ (recall Lemma 9.4), the above condition can be rewritten as

$$\begin{aligned}\text{vec}(\boldsymbol{C}) &= (\boldsymbol{B}' \otimes \boldsymbol{A})((\boldsymbol{B}')^+ \otimes \boldsymbol{A}^+)\text{vec}(\boldsymbol{C}) = \left(\boldsymbol{B}'(\boldsymbol{B}')^+ \otimes \boldsymbol{A}\boldsymbol{A}^+\right)\text{vec}(\boldsymbol{C}) \\ &= \left(\boldsymbol{B}'(\boldsymbol{B}^+)' \otimes \boldsymbol{A}\boldsymbol{A}^+\right)\text{vec}(\boldsymbol{C}) = \left((\boldsymbol{B}^+\boldsymbol{B})' \otimes \boldsymbol{A}\boldsymbol{A}^+\right)\text{vec}(\boldsymbol{C}) \\ &= \text{vec}(\boldsymbol{A}\boldsymbol{A}^+\boldsymbol{C}\boldsymbol{B}^+\boldsymbol{B}) .\end{aligned}$$

Therefore, $\boldsymbol{AXB} = \boldsymbol{C}$ can be solved for $\boldsymbol{X}$ if and only if $\boldsymbol{C} = \boldsymbol{A}\boldsymbol{A}^+\boldsymbol{C}\boldsymbol{B}^+\boldsymbol{B}$.

Proof of (ii): Theorem 9.17 provides the general solution for (14.22) as

$$\text{vec}(\boldsymbol{X}) = (\boldsymbol{B}' \otimes \boldsymbol{A})^+\text{vec}(\boldsymbol{C}) + \left(\boldsymbol{I} - (\boldsymbol{B}' \otimes \boldsymbol{A})^+(\boldsymbol{B}' \otimes \boldsymbol{A})\right)\text{vec}(\boldsymbol{Y}) ,$$

where $\boldsymbol{Y}$ is an arbitrary $n \times p$ matrix. Using the fact that $(\boldsymbol{B}\boldsymbol{B}^+)' = \boldsymbol{B}\boldsymbol{B}^+$ (recall Definition 9.3), we can rewrite the above as

$$\begin{aligned}\text{vec}(\boldsymbol{X}) &= (\boldsymbol{B}' \otimes \boldsymbol{A})^+\text{vec}(\boldsymbol{C}) + \text{vec}(\boldsymbol{Y}) - \left((\boldsymbol{B}\boldsymbol{B}^+)' \otimes \boldsymbol{A}^+\boldsymbol{A}\right) \\ &= \left((\boldsymbol{B}^+)' \otimes \boldsymbol{A}^+\right)\text{vec}(\boldsymbol{C}) + \text{vec}(\boldsymbol{Y}) - \left(\boldsymbol{B}\boldsymbol{B}^+ \otimes \boldsymbol{A}^+\boldsymbol{A}\right)\text{vec}(\boldsymbol{Y}) \\ &= \text{vec}(\boldsymbol{A}^+\boldsymbol{C}\boldsymbol{B}^+) + \text{vec}(\boldsymbol{Y}) - \text{vec}(\boldsymbol{A}^+\boldsymbol{A}\boldsymbol{Y}\boldsymbol{B}\boldsymbol{B}^+) \\ &= \text{vec}\left(\boldsymbol{A}^+\boldsymbol{C}\boldsymbol{B}^+ + \boldsymbol{Y} - \boldsymbol{A}^+\boldsymbol{A}\boldsymbol{Y}\boldsymbol{B}\boldsymbol{B}^+\right) .\end{aligned}$$

Therefore, $\boldsymbol{X} = \boldsymbol{A}^+\boldsymbol{C}\boldsymbol{B}^+ + \boldsymbol{Y} - \boldsymbol{A}^+\boldsymbol{A}\boldsymbol{Y}\boldsymbol{B}\boldsymbol{B}^+$.

Proof of (iii): This follows from the expression for the general solution in (ii). □

For computing quadratic forms involving Kronecker products without explicitly storing or computing $\boldsymbol{A} \otimes \boldsymbol{B}$, the following theorem is sometimes useful.

Theorem 14.18 *Let $\boldsymbol{A} = \{a_{ij}\}$ and $\boldsymbol{B} = \{b_{ij}\}$ be $m \times m$ and $n \times n$ matrices, respectively, and let $\boldsymbol{x}$ be any $mn \times 1$ vector. Then, there exists an $n \times n$ matrix $\boldsymbol{C}$ and an $m \times m$ matrix $\boldsymbol{D}$ such that*

$$\boldsymbol{x}'(\boldsymbol{A} \otimes \boldsymbol{B})\boldsymbol{x} = tr(\boldsymbol{CB}) = tr(\boldsymbol{DA}) .$$

Proof. Partition the $mn \times 1$ vector $\boldsymbol{x}$ into m subvectors, each of length n, as $\boldsymbol{x}' = [\boldsymbol{x}_1' : \boldsymbol{x}_2' : \ldots : \boldsymbol{x}_m']$, where each $\boldsymbol{x}_i$ is $n \times 1$. Then,

$$\begin{aligned}\boldsymbol{x}'(\boldsymbol{A} \otimes \boldsymbol{B})\boldsymbol{x} &= \sum_{i=1}^m\sum_{j=1}^m a_{ij}\boldsymbol{x}_i'\boldsymbol{B}\boldsymbol{x}_j = \sum_{i=1}^m\sum_{j=1}^m a_{ij}\text{tr}(\boldsymbol{x}_j\boldsymbol{x}_i'\boldsymbol{B}) \\ &= \text{tr}\left(\sum_{i=1}^m\sum_{j=1}^m a_{ij}\boldsymbol{x}_j\boldsymbol{x}_i'\boldsymbol{B}\right) = \text{tr}(\boldsymbol{CB}) ,\ \text{with } \boldsymbol{C} = \sum_{i=1}^m\sum_{j=1}^m a_{ij}\boldsymbol{x}_j\boldsymbol{x}_i' .\end{aligned}$$

Clearly $\boldsymbol{C}$ is $n \times n$ since each $\boldsymbol{x}_j\boldsymbol{x}_i'$ is $n \times n$. This establishes the first equality.

The second equality can be obtained from the first using Theorem 14.16, which ensures that there exists a permutation (hence orthogonal) matrix $\boldsymbol{P}$, such that

$A \otimes B = P'(B \otimes A)P$. Setting $y = Px$ and partitioning, y into n subvectors, each of length m, so that $y' = [y_1' : y_2' : \ldots : y_n']$, where each y_i is $m \times 1$, we obtain

$$x'(A \otimes B)x = x'P'(B \otimes A)Px = \sum_{i=1}^{n}\sum_{j=1}^{n} b_{ij} y_i' A y_j = \sum_{i=1}^{n}\sum_{j=1}^{n} b_{ij} \text{tr}\left(y_j y_i' A\right)$$

$$= \text{tr}\left(\sum_{i=1}^{n}\sum_{j=1}^{n} b_{ij} y_j y_i' A\right) = \text{tr}\,(DA)\ , \quad \text{where } D = \sum_{i=1}^{n}\sum_{j=1}^{n} b_{ij} y_j y_i' \ .$$

□

The second equality in the above theorem can also be derived directly, without resorting to Theorem 14.16, as below:

$$x'(A \otimes B)x = x' \begin{bmatrix} a_{11}B & a_{12}B & \ldots & a_{1m}B \\ a_{21}B & a_{22}B & \ldots & a_{2m}B \\ \vdots & \vdots & \ddots & \vdots \\ a_{m1}B & a_{m2}B & \ldots & a_{mm}B \end{bmatrix} x = \sum_{i=1}^{m}\sum_{j=1}^{m} a_{ij} x_i' B x_j$$

$$= \text{tr}\,(DA)\ , \quad \text{where } D = \begin{bmatrix} x_1'Bx_1 & x_2'Bx_1 & \ldots & x_m'Bx_1 \\ x_1'Bx_2 & x_2'Bx_2 & \ldots & x_m'Bx_2 \\ \vdots & \vdots & \ddots & \vdots \\ x_1'Bx_m & \ldots & \ldots & x_m'Bx_m \end{bmatrix} .$$

It is easily verified that this D is equal to that obtained in Theorem 14.18.

14.8 Sylvester's equation and the Kronecker sum

Let A, B and C be given matrices of size $m \times m$, $n \times n$ and $m \times n$, respectively. The matrix equation

$$AX + XB = C\ , \tag{14.23}$$

in the unknown $m \times n$ matrix X is known as ***Sylvester's equation*** and plays a prominent role in theoretical and applied matrix analysis. Rewriting (14.23) by equating the columns, we obtain

$$Ax_i + Xb_i = c_i\ , \quad \text{for } i = 1, 2, \ldots, n\ ,$$

where x_i, b_i and c_i denote the i-th columns of X, B and C, respectively. Writing Xb_i as a linear combination of the columns of X, we further obtain

$$Ax_i + \sum_{j=1}^{n} b_{ji} x_j = c_i\ , \quad \text{for } i = 1, 2, \ldots, n\ .$$

The above system can now be collected into the following linear system in vec$(\boldsymbol{X})$:

$$\begin{bmatrix} \boldsymbol{A}+b_{11}\boldsymbol{I}_m & b_{21}\boldsymbol{I}_m & \dots & b_{n1}\boldsymbol{I}_m \\ b_{12}\boldsymbol{I}_m & \boldsymbol{A}+b_{22}\boldsymbol{I}_m & \dots & b_{n2}\boldsymbol{I}_m \\ \vdots & \vdots & \ddots & \vdots \\ b_{1n}\boldsymbol{I}_m & b_{2n}\boldsymbol{I}_m & \dots & \boldsymbol{A}+b_{nn}\boldsymbol{I}_m \end{bmatrix} \begin{bmatrix} \boldsymbol{x}_1 \\ \boldsymbol{x}_2 \\ \vdots \\ \boldsymbol{x}_n \end{bmatrix} = \begin{bmatrix} \boldsymbol{c}_1 \\ \boldsymbol{c}_2 \\ \vdots \\ \boldsymbol{c}_m \end{bmatrix} .$$

The above system can be compactly written as

$$[(\boldsymbol{I}_n \otimes \boldsymbol{A}) + (\boldsymbol{B}' \otimes \boldsymbol{I}_m)] \operatorname{vec}(\boldsymbol{X}) = \operatorname{vec}(\boldsymbol{C}) . \tag{14.24}$$

From (14.24) it is clear that Sylvester's equation has a unique solution if and only if $(\boldsymbol{I}_n \otimes \boldsymbol{A}) + (\boldsymbol{B}' \otimes \boldsymbol{I}_m)$ is nonsingular. This motivates the study of matrices having forms as the coefficient matrix in (14.24). The following definition is useful.

Definition 14.3 *Let $\boldsymbol{A}$ and $\boldsymbol{B}$ be two matrices of size $m \times m$ and $n \times n$, respectively. The* **Kronecker sum** *of $\boldsymbol{A}$ and $\boldsymbol{B}$, denoted by $\boldsymbol{A} \boxplus \boldsymbol{B}$, is*

$$\boldsymbol{A} \boxplus \boldsymbol{B} = (\boldsymbol{I}_n \otimes \boldsymbol{A}) + (\boldsymbol{B} \otimes \boldsymbol{I}_m) .$$

If $\boldsymbol{x}$ is an eigenvector associated with eigenvalue λ of $\boldsymbol{A}$ and if $\boldsymbol{y}$ is an eigenvector associated with eigenvalue μ of $\boldsymbol{B}$, then $\lambda + \mu$ is an eigenvalue of $\boldsymbol{A} \boxplus \boldsymbol{B}$ and $\boldsymbol{y} \otimes \boldsymbol{x}$ is an eigenvector corresponding to $\lambda + \mu$. This is because

$$\begin{aligned} [(\boldsymbol{I}_n \otimes \boldsymbol{A}) + (\boldsymbol{B} \otimes \boldsymbol{I}_m)] (\boldsymbol{y} \otimes \boldsymbol{x}) &= (\boldsymbol{I}_n \otimes \boldsymbol{A})(\boldsymbol{y} \otimes \boldsymbol{x}) + (\boldsymbol{B} \otimes \boldsymbol{I}_m)(\boldsymbol{y} \otimes \boldsymbol{x}) \\ &= (\boldsymbol{y} \otimes \boldsymbol{A}\boldsymbol{x}) + (\boldsymbol{B}\boldsymbol{y} \otimes \boldsymbol{x}) \\ &= (\boldsymbol{y} \otimes \lambda\boldsymbol{x}) + (\mu\boldsymbol{y} \otimes \boldsymbol{x}) \\ &= \lambda(\boldsymbol{y} \otimes \boldsymbol{x}) + \mu(\boldsymbol{y} \otimes \boldsymbol{x}) = (\lambda + \mu)(\boldsymbol{y} \otimes \boldsymbol{x}) . \end{aligned}$$

Therefore, if $\lambda_1, \lambda_2, \dots, \lambda_m$ are the m eigenvalues (counting multiplicities) of $\boldsymbol{A}$ and $\mu_1, \mu_2, \dots, \mu_n$ are the n eigenvalues (counting multiplicities) of $\boldsymbol{B}$, then the eigenvalues of $\boldsymbol{A} \boxplus \boldsymbol{B}$ are the mn numbers in the set $\{\lambda_i + \mu_j : i = 1, 2, \dots, m,\ j = 1, 2, \dots, n\}$.

Returning to Sylvester's equation, we see that the coefficient matrix $\boldsymbol{A} \boxplus \boldsymbol{B}'$ is nonsingular if and only if it has no zero eigenvalues, which means that $\lambda_i + \mu_j \neq 0$ for any eigenvalue λ_i of $\boldsymbol{A}$ and any eigenvalue μ_j of $\boldsymbol{B}$ (the eigenvalues of $\boldsymbol{B}$ and $\boldsymbol{B}'$ are the same, including their multiplicities). Therefore, Sylvester's equation has a unique solution if and only if $\boldsymbol{A}$ and $-\boldsymbol{B}$ have no eigenvalues in common.

In practice Sylvester's equation is rarely solved in the form presented in (14.24). We first transform (14.23) into one with triangular coefficient matrices by reducing $\boldsymbol{A}$ and $\boldsymbol{B}$ to their Schur's triangular form. Let $\boldsymbol{Q}_A'\boldsymbol{A}\boldsymbol{Q}_A = \boldsymbol{T}_A$ and $\boldsymbol{Q}_B'\boldsymbol{B}\boldsymbol{Q}_B = \boldsymbol{T}_B$ be the Schur triangular forms for $\boldsymbol{A}$ and $\boldsymbol{B}$. Then (14.23) can be written as

$$\begin{aligned} & \boldsymbol{Q}_A\boldsymbol{T}_A\boldsymbol{Q}_A'\boldsymbol{X} + \boldsymbol{X}\boldsymbol{Q}_B\boldsymbol{T}_B\boldsymbol{Q}_B' = \boldsymbol{C} \\ \Longrightarrow\ & \boldsymbol{T}_A(\boldsymbol{Q}_A'\boldsymbol{X}\boldsymbol{Q}_B) + (\boldsymbol{Q}_A'\boldsymbol{X}\boldsymbol{Q}_B)\boldsymbol{T}_B = \boldsymbol{Q}_A'\boldsymbol{C}\boldsymbol{Q}_B \\ \Longrightarrow\ & \boldsymbol{T}_A\boldsymbol{Z} - \boldsymbol{Z}\boldsymbol{T}_B = \boldsymbol{D} , \end{aligned}$$

where $Z = Q'_A X Q_B$ and $D = Q'_A C Q_B$. This is Sylvester's equation with triangular coefficient matrices. This matrix equation is solved for Z and then $X = Q_A Z Q'_B$ is obtained. Further details on Sylvester's equation and its applications can be found in Laub (2005).

14.9 The Hadamard product

The ***Hadamard product*** (also known as the ***Schur product***) is a binary operation that takes two matrices of the same dimensions, and produces another matrix whose (i, j)-th is the product of elements (i, j)-th elements of the original two matrices. Here we provide a brief outline of the Hadamard product. A more in-depth treatment may be found in Horn and Johnson (2013).

Definition 14.4 *Let $A = \{a_{ij}\}$ and $B = \{b_{ij}\}$ be $m \times n$ matrices. The* **Hadamard product** *of A and B is the $m \times n$ matrix $C = \{c_{ij}\}$ whose (ij)-th element is $c_{ij} = a_{ij}b_{ij}$. The Hadamard product is denoted by $C = A \odot B$.*

If A and B are both $m \times n$, then

$$A \odot B = \begin{bmatrix} a_{11}b_{11} & a_{12}b_{12} & \dots & a_{1n}b_{1n} \\ a_{21}b_{21} & a_{22}b_{22} & \dots & a_{2n}b_{2n} \\ \vdots & \vdots & \ddots & \vdots \\ a_{m1}b_{m1} & a_{m2}b_{m2} & \dots & a_{mn}b_{mn} \end{bmatrix}.$$

Example 14.4 Consider the following 2×2 matrices

$$A = \begin{bmatrix} 1 & 3 \\ -1 & 2 \end{bmatrix} \text{ and } B = \begin{bmatrix} 2 & -1 \\ -5 & 8 \end{bmatrix}.$$

Then, the Hadamard product of A and B is given by

$$A \odot B = \begin{bmatrix} 1 & 3 \\ -1 & 2 \end{bmatrix} \odot \begin{bmatrix} 2 & -1 \\ -5 & 8 \end{bmatrix} = \begin{bmatrix} 2 & -3 \\ 5 & 16 \end{bmatrix}.$$

The following properties of the Hadamard product are immediate consequences of Definition 14.4.

Theorem 14.19 *Let $A = \{a_{ij}\}$, $B = \{b_{ij}\}$ and $C = \{c_{ij}\}$ be $m \times n$ matrices.*

(i) $A \odot B = B \odot A$.

(ii) $(A \odot B) \odot C = A \odot (B \odot C)$.

(iii) $(A + B) \odot C = A \odot C + B \odot C$.

(iv) $A \odot (B + C) = A \odot B + A \odot C$.

(v) $\alpha(A \odot B) = (\alpha A) \odot B = A \odot (\alpha B)$.

(vi) $(A \odot B)' = A' \odot B' = B' \odot A'$.

Proof. **Proof of (i):** The (i,j)-th element of $\boldsymbol{A} \odot \boldsymbol{B}$ is $a_{ij}b_{ij} = b_{ij}a_{ij}$, which is the (i,j)-th element of $\boldsymbol{B} \odot \boldsymbol{A}$.

Proof of (ii): The (i,j)-th element of $(\boldsymbol{A} \odot \boldsymbol{B}) \odot \boldsymbol{C}$ is $(a_{ij}b_{ij})c_{ij} = a_{ij}(b_{ij}c_{ij})$, which is also the (i,j)-th element of $\boldsymbol{A} \odot (\boldsymbol{B} \odot \boldsymbol{C})$.

Proof of (iii): The (i,j)-th element of $(\boldsymbol{A} + \boldsymbol{B}) \odot \boldsymbol{C}$ can be expressed as

$$(a_{ij} + b_{ij})c_{ij} = a_{ij}c_{ij} + b_{ij}c_{ij} \ ,$$

which is the (i,j)-th element of $\boldsymbol{A} \odot \boldsymbol{C} + \boldsymbol{B} \odot \boldsymbol{C}$.

Proof of (iv): The (i,j)-th element of $\boldsymbol{A} \odot (\boldsymbol{B} + \boldsymbol{C})$ can be expressed as

$$a_{ij}(b_{ij} + c_{ij}) = a_{ij}b_{ij} + a_{ij}c_{ij} \ ,$$

which is the (i,j)-th element of $\boldsymbol{A} \odot \boldsymbol{B} + \boldsymbol{A} \odot \boldsymbol{C}$.

Proof of (v): The (i,j)-th elements of $\alpha(\boldsymbol{A} \odot \boldsymbol{B})$, $(\alpha\boldsymbol{A}) \odot \boldsymbol{B}$ and $\boldsymbol{A} \odot (\alpha\boldsymbol{B})$ are all equal to $\alpha a_{ij}b_{ij}$.

Proof of (vi): The (i,j)-th entry of $(\boldsymbol{A} \odot \boldsymbol{B})'$ is the (j,i)-th entry of $\boldsymbol{A} \odot \boldsymbol{B}$, which is $a_{ji}b_{ji}$. And $a_{ji}b_{ji}$ is also the (i,j)-th entry of $\boldsymbol{A}' \odot \boldsymbol{B}'$ because a_{ji} and b_{ji} are the (i,j)-th entries of $\boldsymbol{A}'$ and $\boldsymbol{B}'$, respectively. The last equality follows from (i). □

Hadamard products do *not* satisfy the mixed product rule (Theorem 14.3) that Kronecker products do. However, the following result involving Hadamard products of vectors is useful.

Theorem 14.20 Mixed product rule for vectors. *Let $\boldsymbol{a}$ and $\boldsymbol{b}$ be $m \times 1$, and $\boldsymbol{u}$ and $\boldsymbol{v}$ be $n \times 1$ vectors. Then,*

$$(\boldsymbol{a} \odot \boldsymbol{b})(\boldsymbol{u} \odot \boldsymbol{v})' = (\boldsymbol{a}\boldsymbol{u}') \odot (\boldsymbol{b}\boldsymbol{v}') \ .$$

Proof. The proof is simply a direct verification. If a_i, b_i, u_i and v_i denote the i-th elements of $\boldsymbol{a}$, $\boldsymbol{b}$, $\boldsymbol{u}$ and $\boldsymbol{v}$, respectively, then

$$\boldsymbol{a} \odot \boldsymbol{b} = \begin{bmatrix} a_1b_1 \\ a_2b_2 \\ \vdots \\ a_mb_m \end{bmatrix} \quad \text{and} \quad \boldsymbol{u} \odot \boldsymbol{v} = \begin{bmatrix} u_1v_1 \\ u_2v_2 \\ \vdots \\ u_nv_n \end{bmatrix}$$

are $m \times 1$ and $n \times 1$, respectively. Therefore, $(\boldsymbol{a} \odot \boldsymbol{b})(\boldsymbol{u} \odot \boldsymbol{v})'$ is the $m \times n$ matrix whose (i,j)-th element is $(a_i b_i)(u_j v_j) = (a_i u_j)(b_i v_j)$ and

$$(\boldsymbol{a} \odot \boldsymbol{b})(\boldsymbol{u} \odot \boldsymbol{v})' = \begin{bmatrix} (a_1u_1)(b_1v_1) & (a_1u_2)(b_1v_2) & \dots & (a_1u_n)(b_1v_n) \\ (a_2u_1)(b_2v_1) & (a_2u_2)(b_2v_2) & \dots & (a_2u_n)(b_2v_n) \\ \vdots & \vdots & \ddots & \vdots \\ (a_mu_1)(b_mv_1) & (a_mu_2)(b_mv_2) & \dots & (a_mu_n)(b_mv_n) \end{bmatrix}$$
$$= \left(\begin{bmatrix} a_1 \\ a_2 \\ \vdots \\ a_m \end{bmatrix} [u_1 : u_2 : \dots : u_n] \right) \odot \left(\begin{bmatrix} b_1 \\ b_2 \\ \vdots \\ b_m \end{bmatrix} [v_1 : v_2 : \dots : v_n] \right)$$
$$= (\boldsymbol{a}\boldsymbol{u}') \odot (\boldsymbol{b}\boldsymbol{v}') .$$

This proves the result. □

Note that $\boldsymbol{A} \odot \boldsymbol{B} = \boldsymbol{AB}$ if and only if both $\boldsymbol{A}$ and $\boldsymbol{B}$ are diagonal. In general, the Hadamard product can be related with ordinary matrix multiplication using diagonal matrices. For this, it will be convenient to introduce the $\texttt{diag}(\cdot)$ function, which acts on a square matrix to produce a vector consisting of its diagonal elements. To be precise, if $\boldsymbol{A} = \{a_{ij}\}$ is an $n \times n$ matrix, then

$$\texttt{diag}(\boldsymbol{A}) = \begin{bmatrix} a_{11} \\ a_{22} \\ \vdots \\ a_{nn} \end{bmatrix} .$$

Observe that $\texttt{diag}(\cdot)$ is linear in its argument since

$$\texttt{diag}(\alpha\boldsymbol{A} + \beta\boldsymbol{B}) = \begin{bmatrix} \alpha a_{11} + \beta b_{11} \\ \alpha a_{22} + \beta b_{22} \\ \vdots \\ \alpha a_{nn} + \beta b_{nn} \end{bmatrix} = \alpha \begin{bmatrix} a_{11} \\ a_{22} \\ \vdots \\ a_{nn} \end{bmatrix} + \beta \begin{bmatrix} b_{11} \\ b_{22} \\ \vdots \\ b_{nn} \end{bmatrix}$$
$$= \alpha \cdot \texttt{diag}(\boldsymbol{A}) + \beta \cdot \texttt{diag}(\boldsymbol{B}) . \tag{14.25}$$

The following theorem shows how the Hadamard product and the $\texttt{diag}$ function can be used to express the diagonal elements of a matrix $\boldsymbol{A}$ in certain cases.

Theorem 14.21 *Let $\boldsymbol{A}$ be an $n \times n$ matrix such that $\boldsymbol{A} = \boldsymbol{P\Lambda Q}$, where $\boldsymbol{P}$, $\boldsymbol{\Lambda}$ and $\boldsymbol{Q}$ are all $n \times n$ and $\boldsymbol{\Lambda}$ is diagonal. Then,*

$$\texttt{diag}(\boldsymbol{A}) = (\boldsymbol{P} \odot \boldsymbol{Q}')\, \texttt{diag}(\boldsymbol{\Lambda}) .$$

Proof. Let $\boldsymbol{p}_{*i}$ be the i-th column vector of $\boldsymbol{P}$, $\boldsymbol{q}'_{i*}$ be the i-th row vector of $\boldsymbol{Q}$ and

λ_i be the i-th diagonal entry of $\boldsymbol{\Lambda}$. Using the linearity of diag in (14.25), we obtain

$$\begin{aligned}
\texttt{diag}(\boldsymbol{A}) &= \texttt{diag}(\boldsymbol{P\Lambda Q}) = \texttt{diag}\left(\sum_{i=1}^{n} \lambda_i \boldsymbol{p}_{*i}\boldsymbol{q}'_{i*}\right) = \sum_{i=1}^{n} \lambda_i \texttt{diag}(\boldsymbol{p}_{*i}\boldsymbol{q}'_{i*}) \\
&= \sum_{i=1}^{n} \lambda_i \texttt{diag}\left(\begin{bmatrix} p_{1i}q_{i1} & p_{1i}q_{i2} & \dots & p_{1i}q_{in} \\ p_{2i}q_{i1} & p_{2i}q_{i2} & \dots & p_{2i}q_{in} \\ \vdots & \vdots & \ddots & \vdots \\ p_{ni}q_{i1} & p_{ni}q_{i2} & \dots & p_{ni}q_{in} \end{bmatrix}\right) \\
&= \sum_{i=1}^{n} \lambda_i \begin{bmatrix} p_{1i}q_{i1} \\ p_{2i}q_{i2} \\ \vdots \\ p_{ni}q_{in} \end{bmatrix} = \begin{bmatrix} p_{11}q_{11} & p_{12}q_{21} & \dots & p_{1n}q_{n1} \\ p_{21}q_{12} & p_{22}q_{22} & \dots & p_{2n}q_{n2} \\ \vdots & \vdots & \ddots & \vdots \\ p_{n1}q_{1n} & p_{n2}q_{2n} & \dots & p_{nn}q_{nn} \end{bmatrix} \begin{bmatrix} \lambda_1 \\ \lambda_2 \\ \vdots \\ \lambda_n \end{bmatrix} \\
&= (\boldsymbol{P} \odot \boldsymbol{Q}')\, \texttt{diag}(\boldsymbol{\Lambda}) \, .
\end{aligned}$$

□

Theorem 14.21 can be applied to several matrix decompositions. For example, if $\boldsymbol{A}$ is an $n \times n$ diagonalizable matrix with eigenvalues $\lambda_1, \lambda_2, \ldots, \lambda_n$, then there exists a nonsingular matrix $\boldsymbol{P}$ such that $\boldsymbol{A} = \boldsymbol{P\Lambda P}^{-1}$, where $\boldsymbol{\Lambda}$ is diagonal with the λ_i's as its diagonal entries. Theorem 14.21 reveals a relationship between the diagonal entries of $\boldsymbol{A}$ and its eigenvalues:

$$\texttt{diag}(\boldsymbol{A}) = \left[\boldsymbol{P} \odot (\boldsymbol{P}^{-1})'\right] \texttt{diag}(\boldsymbol{\Lambda}) \, . \tag{14.26}$$

Similarly, if $\boldsymbol{A} = \boldsymbol{UDV}'$ is the SVD of the square matrix $\boldsymbol{A}$ with real entries, then Theorem 14.21 relates the diagonal entries of $\boldsymbol{A}$ with its singular values:

$$\texttt{diag}(\boldsymbol{A}) = (\boldsymbol{U} \odot \boldsymbol{V})\texttt{diag}(\boldsymbol{D}) \, . \tag{14.27}$$

The $\texttt{diag}(\boldsymbol{A})$ function, as defined above, takes a square matrix as input and produces a vector consisting of the diagonal entries of $\boldsymbol{A}$. Another function that in some sense carries out the inverse operation can be defined as $\boldsymbol{D_x}$, which takes an $n \times 1$ vector $\boldsymbol{x}$ as input and produces an $n \times n$ diagonal matrix with the entries of $\boldsymbol{x}$ along its diagonal. To be precise,

$$\boldsymbol{D_x} = \begin{bmatrix} x_1 & 0 & \dots & 0 \\ 0 & x_2 & \dots & 0 \\ \vdots & \vdots & \ddots & \vdots \\ 0 & 0 & \dots & x_n \end{bmatrix}, \text{ where } \boldsymbol{x} = \begin{bmatrix} x_1 \\ x_2 \\ \vdots \\ x_n \end{bmatrix} .$$

This function is sometimes useful in expressing the entries of the vector $(\boldsymbol{A} \odot \boldsymbol{B})\boldsymbol{x}$. In fact, the i-th entry of $(\boldsymbol{A} \odot \boldsymbol{B})\boldsymbol{x}$ is precisely the (i,i)-th diagonal entry of $\boldsymbol{AD_xB}'$.

Theorem 14.22 *If $\boldsymbol{A}$ and $\boldsymbol{B}$ are both $m \times n$ and $\boldsymbol{x}$ is $n \times 1$, then*

$$\textit{diag}(\boldsymbol{AD_xB}') = (\boldsymbol{A} \odot \boldsymbol{B})\boldsymbol{x} \, .$$

Proof. Let $\boldsymbol{a}_i$ and $\boldsymbol{b}_i$ denote the i-th column vectors of $\boldsymbol{A}$ and $\boldsymbol{B}$, respectively, and note that $\boldsymbol{b}_i'$ is then the i-th row vector of $\boldsymbol{B}'$. Then,

$$\begin{aligned}
\texttt{diag}(\boldsymbol{A}\boldsymbol{D}_{\boldsymbol{x}}\boldsymbol{B}') &= \texttt{diag}\left(\sum_{i=1}^{n} x_i \boldsymbol{a}_i \boldsymbol{b}_i'\right) = \sum_{i=1}^{n} x_i \texttt{diag}(\boldsymbol{a}_i\boldsymbol{b}_i') \\
&= \sum_{i=1}^{n} x_i \texttt{diag}\left(\begin{bmatrix} a_{1i}b_{1i} & a_{1i}b_{2i} & \dots & a_{1i}b_{mi} \\ a_{2i}b_{1i} & a_{2i}b_{2i} & \dots & a_{2i}b_{mi} \\ \vdots & \vdots & \ddots & \vdots \\ a_{mi}b_{1i} & a_{mi}b_{2i} & \dots & a_{mi}b_{mi} \end{bmatrix}\right) \\
&= \sum_{i=1}^{n} x_i \begin{bmatrix} a_{1i}b_{1i} \\ a_{2i}b_{2i} \\ \vdots \\ a_{mi}b_{mi} \end{bmatrix} = \sum_{i=1}^{n} \boldsymbol{c}_i x_i = (\boldsymbol{A} \odot \boldsymbol{B})\boldsymbol{x} ,
\end{aligned}$$

where the last equality follows from noting that $\boldsymbol{c}_i = \left[a_{1i}b_{1i}, a_{2i}b_{2i}, \dots, a_{mi}b_{mi}\right]'$ is the i-th column of $\boldsymbol{A} \odot \boldsymbol{B}$. □

The above expression is useful in deriving alternate expressions for quadratic forms involving Hadamard products. If $\boldsymbol{A} = \{a_{ij}\}$ is $n \times n$ and $\boldsymbol{y}$ is $n \times 1$, then

$$\boldsymbol{y}'\texttt{diag}(\boldsymbol{A}) = \sum_{i=1}^{n} a_{ii} y_i = \text{tr}(\boldsymbol{D}_{\boldsymbol{y}}\boldsymbol{A}) . \tag{14.28}$$

Using Theorem 14.22 and (14.28), we obtain

$$\boldsymbol{y}'(\boldsymbol{A} \odot \boldsymbol{B})\boldsymbol{x} = \boldsymbol{y}'\texttt{diag}(\boldsymbol{A}\boldsymbol{D}_{\boldsymbol{x}}\boldsymbol{B}') = \text{tr}(\boldsymbol{D}_{\boldsymbol{y}}\boldsymbol{A}\boldsymbol{D}_{\boldsymbol{x}}\boldsymbol{B}') = \text{tr}(\boldsymbol{A}\boldsymbol{D}_{\boldsymbol{x}}\boldsymbol{B}'\boldsymbol{D}_{\boldsymbol{y}}) . \tag{14.29}$$

If the vectors are complex, then $\boldsymbol{y}^*(\boldsymbol{A} \odot \boldsymbol{B})\boldsymbol{x} = \text{tr}(\boldsymbol{D}_{\boldsymbol{y}}^*\boldsymbol{A}\boldsymbol{D}_{\boldsymbol{x}}\boldsymbol{B}')$.

We now turn to an important result: the Hadamard product of two positive definite matrices is positive definite.

Theorem 14.23 Positive definiteness of Hadamard products. *If $\boldsymbol{A}$ and $\boldsymbol{B}$ are positive definite matrices of the same order, then $\boldsymbol{A} \odot \boldsymbol{B}$ is positive definite.*

Proof. Using (14.29), we can write

$$\boldsymbol{x}'(\boldsymbol{A} \odot \boldsymbol{B})\boldsymbol{x} = \text{tr}(\boldsymbol{A}\boldsymbol{D}_{\boldsymbol{x}}\boldsymbol{B}'\boldsymbol{D}_{\boldsymbol{x}}) = \text{tr}(\boldsymbol{A}\boldsymbol{D}_{\boldsymbol{x}}\boldsymbol{B}\boldsymbol{D}_{\boldsymbol{x}}) ,$$

where the last equality is true because $\boldsymbol{B}$ is positive definite, hence symmetric. Let $\boldsymbol{A} = \boldsymbol{L}_{\boldsymbol{A}}\boldsymbol{L}_{\boldsymbol{A}}'$ and $\boldsymbol{B} = \boldsymbol{L}_{\boldsymbol{B}}\boldsymbol{L}_{\boldsymbol{B}}'$ be Cholesky decompositions. Then,

$$\begin{aligned}
\boldsymbol{x}'(\boldsymbol{A} \odot \boldsymbol{B})\boldsymbol{x} &= \text{tr}(\boldsymbol{A}\boldsymbol{D}_{\boldsymbol{x}}\boldsymbol{B}\boldsymbol{D}_{\boldsymbol{x}}) = \text{tr}(\boldsymbol{L}_{\boldsymbol{A}}\boldsymbol{L}_{\boldsymbol{A}}'\boldsymbol{D}_{\boldsymbol{x}}\boldsymbol{L}_{\boldsymbol{B}}\boldsymbol{L}_{\boldsymbol{B}}'\boldsymbol{D}_{\boldsymbol{x}}) \\
&= \text{tr}\left[\boldsymbol{L}_{\boldsymbol{A}}(\boldsymbol{L}_{\boldsymbol{A}}'\boldsymbol{D}_{\boldsymbol{x}}\boldsymbol{L}_{\boldsymbol{B}})\boldsymbol{L}_{\boldsymbol{B}}'\boldsymbol{D}_{\boldsymbol{x}}\right] = \text{tr}\left[(\boldsymbol{L}_{\boldsymbol{A}}'\boldsymbol{D}_{\boldsymbol{x}}\boldsymbol{L}_{\boldsymbol{B}})\boldsymbol{L}_{\boldsymbol{B}}'\boldsymbol{D}_{\boldsymbol{x}}\boldsymbol{L}_{\boldsymbol{A}}\right] \\
&= \text{tr}\left[(\boldsymbol{L}_{\boldsymbol{B}}'\boldsymbol{D}_{\boldsymbol{x}}\boldsymbol{L}_{\boldsymbol{A}})'\boldsymbol{L}_{\boldsymbol{B}}'\boldsymbol{D}_{\boldsymbol{x}}\boldsymbol{L}_{\boldsymbol{A}}\right] = \text{tr}(\boldsymbol{C}'\boldsymbol{C}) \geq 0 ,
\end{aligned}$$

where $\boldsymbol{C} = \boldsymbol{L}'_{\boldsymbol{B}}\boldsymbol{D}_{\boldsymbol{x}}\boldsymbol{L}_{\boldsymbol{A}}$. Also,

$$\operatorname{tr}(\boldsymbol{C}'\boldsymbol{C}) = 0 \Longrightarrow \boldsymbol{C} = \boldsymbol{O} \Longrightarrow \boldsymbol{D}_{\boldsymbol{x}} = \boldsymbol{O} \Longrightarrow \boldsymbol{x} = \boldsymbol{0}\,.$$

Therefore, $\boldsymbol{x}'(\boldsymbol{A} \odot \boldsymbol{B})\boldsymbol{x} > 0$ for every $\boldsymbol{x} \neq \boldsymbol{0}$, so $\boldsymbol{A} \odot \boldsymbol{B}$ is positive definite. □

We now turn to a relationship between Kronecker and Hadamard products. The Hadamard product between two $m \times n$ matrices is another $m \times n$ matrix. The Kronecker product between two $m \times n$ matrices is a much larger matrix of size $m^2 \times n^2$. In fact, the Hadamard product is a principal submatrix of the Kronecker product. Consider the following example.

Example 14.5 Consider the matrices $\boldsymbol{A}$ and $\boldsymbol{B}$ in Example 14.4. Then,

$$\boldsymbol{A} \otimes \boldsymbol{B} = \begin{bmatrix} 2 & -1 & 6 & -3 \\ -5 & 8 & -15 & 24 \\ -2 & 1 & 4 & -2 \\ 5 & -8 & -10 & 16 \end{bmatrix} \quad \text{and} \quad \boldsymbol{A} \odot \boldsymbol{B} = \begin{bmatrix} 2 & -3 \\ 5 & 16 \end{bmatrix}.$$

The entries at the intersection of the first and fourth rows and columns of $\boldsymbol{A} \otimes \boldsymbol{B}$ constitutes $\boldsymbol{A} \odot \boldsymbol{B}$. Can we find a matrix $\boldsymbol{P}$ such that $\boldsymbol{A} \odot \boldsymbol{B} = \boldsymbol{P}'(\boldsymbol{A} \otimes \boldsymbol{B})\boldsymbol{P}$? The matrix $\boldsymbol{P}$ will "select" or "extract" the appropriate elements from $\boldsymbol{A} \otimes \boldsymbol{B}$ and, therefore, is called a selection matrix. In this example, we find this to be $\boldsymbol{P} = \begin{bmatrix} 1 & 0 \\ 0 & 0 \\ 0 & 0 \\ 0 & 1 \end{bmatrix} = \begin{bmatrix} \boldsymbol{E}_{11} \\ \boldsymbol{E}_{22} \end{bmatrix}$, where $\boldsymbol{E}_{ij}$ denotes a 2×2 matrix whose (i, j)-th entry is 1 and all other entries are 0. ■

The above example is representative of the more general situation. Let $\boldsymbol{E}_{ij}^{(n)}$ denote an $n \times n$ matrix with 1 as its (i, j)-th entry and 0 everywhere else. We will only need $\boldsymbol{E}_{ii}^{(n)}$'s, which are $n \times n$ diagonal matrices with δ_{ij} as its j-th diagonal entry (δ_{ij} is the Kronecker delta, equaling 1 if $i = j$ and 0 otherwise). Define the ***selection matrix*** or ***extraction matrix***

$$\boldsymbol{P}'_n = [\boldsymbol{E}_{11}^{(n)} : \boldsymbol{E}_{22}^{(n)} : \ldots : \boldsymbol{E}_{nn}^{(n)}]\,.$$

Notice that the $\boldsymbol{E}_{ii}^{(n)}$'s are $n \times n$ and symmetric so $\boldsymbol{P}'_n$ is $n \times n^2$ and $\boldsymbol{P}_n$ is the $n^2 \times n$ matrix formed by stacking up the $\boldsymbol{E}_{ii}^{(n)}$'s above each other. With this definition of the selection matrix, we have the following.

Theorem 14.24 *If $\boldsymbol{A}$ and $\boldsymbol{B}$ are any two $m \times n$ matrices, then*

$$\boldsymbol{A} \odot \boldsymbol{B} = \boldsymbol{P}'_m(\boldsymbol{A} \otimes \boldsymbol{B})\boldsymbol{P}_n\,.$$

Proof. The proof is by direct computation:

$$P'_m(A \otimes B)P_n = [E_{11}^{(m)} : E_{22}^{(m)} : \ldots : E_{mm}^{(m)}] \begin{bmatrix} a_{11}B & a_{12}B & \ldots & a_{1n}B \\ a_{21}B & a_{22}B & \ldots & a_{2n}B \\ \vdots & \vdots & \ddots & \vdots \\ a_{m1}B & a_{m2}B & \ldots & a_{mn}B \end{bmatrix} \begin{bmatrix} E_{11}^{(n)} \\ E_{22}^{(n)} \\ \vdots \\ E_{nn}^{(n)} \end{bmatrix}$$

$$= \sum_{i=1}^{m}\sum_{j=1}^{n} a_{ij}E_{ii}^{(m)}BE_{jj}^{(n)} = \sum_{j=1}^{n}\left(\sum_{i=1}^{m} a_{ij}E_{ij}^{(m)}\right)BE_{jj}^{(n)}$$

$$= \sum_{j=1}^{n}\begin{bmatrix} a_{1j} & 0 & \ldots & 0 \\ 0 & a_{2j} & \ldots & 0 \\ \vdots & \vdots & \ddots & \vdots \\ 0 & 0 & \ldots & a_{mj} \end{bmatrix} BE_{jj}^{(n)}$$

$$= \sum_{j=1}^{n}\begin{bmatrix} a_{1j}b_{11} & a_{1j}b_{12} & \ldots & a_{1j}b_{1n} \\ a_{2j}b_{21} & a_{2j}b_{22} & \ldots & a_{2j}b_{2n} \\ \vdots & \vdots & \ddots & \vdots \\ a_{mj}b_{m1} & a_{mj}b_{m2} & \ldots & a_{mj}b_{mn} \end{bmatrix}\begin{bmatrix} \delta_{1j} & 0 & \ldots & 0 \\ 0 & \delta_{2j} & \ldots & 0 \\ \vdots & \vdots & \ddots & \vdots \\ 0 & 0 & \ldots & \delta_{nj} \end{bmatrix}$$

$$= \begin{bmatrix} \sum_{j=1}^{n} a_{1j}b_{11}\delta_{1j} & \sum_{j=1}^{n} a_{1j}b_{12}\delta_{2j} & \ldots & \sum_{j=1}^{n} a_{1j}b_{1n}\delta_{nj} \\ \sum_{j=1}^{n} a_{2j}b_{21}\delta_{1j} & \sum_{j=1}^{n} a_{2j}b_{22}\delta_{2j} & \ldots & \sum_{j=1}^{n} a_{2j}b_{2n}\delta_{nj} \\ \vdots & \vdots & \ddots & \vdots \\ \sum_{j=1}^{n} a_{mj}b_{m1}\delta_{1j} & \sum_{j=1}^{n} a_{mj}b_{m2}\delta_{2j} & \ldots & \sum_{j=1}^{n} a_{mj}b_{mn}\delta_{nj} \end{bmatrix}$$

$$= \begin{bmatrix} a_{11}b_{11} & a_{12}b_{12} & \ldots & a_{1n}b_{1n} \\ a_{21}b_{21} & a_{22}b_{22} & \ldots & a_{2n}b_{2n} \\ \vdots & \vdots & \ddots & \vdots \\ a_{m1}b_{m1} & a_{m2}b_{m2} & \ldots & a_{mn}b_{mn} \end{bmatrix} = A \odot B\,,$$

where $\delta_{ij} = 1$ if $i = j$ and 0 otherwise. $\square$

If A and B are both $n \times n$, then the above simplifies to $A \odot B = P'_n(A \otimes B)P_n$.

We conclude with some results involving determinants of Hadamard products. To facilitate the development, consider the following partitioned matrices

$$A = \begin{bmatrix} a_{11} & a'_{12} \\ a_{21} & A_{22} \end{bmatrix} \text{ and } B = \begin{bmatrix} b_{11} & b'_{12} \\ b_{21} & B_{22} \end{bmatrix},$$

where A and B are both $n \times n$, and A_{22} and B_{22} are both $(n-1) \times (n-1)$. It is easily verified that

$$A \odot B = \begin{bmatrix} a_{11}b_{11} & a'_{12} \odot b'_{12} \\ a_{21} \odot b_{21} & A_{22} \odot B_{22} \end{bmatrix}.$$

Schur's formula for determinants (Theorem 10.10) can be used to express $|A \odot B|$ in terms of the determinant of the Schur's complement of $a_{11}b_{11}$,

$$|A \odot B| = a_{11}b_{11}\left|A_{22} \odot B_{22} - \frac{(a_{21} \odot b_{21})(a_{12} \odot b_{12})'}{a_{11}b_{11}}\right|. \tag{14.30}$$

It will be useful to relate (14.30) with the corresponding Schur's complements of a_{11} in $\boldsymbol{A}$ and b_{11} in $\boldsymbol{B}$. Using (Theorem 14.20) and some further algebra yields

$$\begin{aligned}
\boldsymbol{A}_{22} \odot \boldsymbol{B}_{22} - \frac{(\boldsymbol{a}_{21} \odot \boldsymbol{b}_{21})(\boldsymbol{a}_{12} \odot \boldsymbol{b}_{12})'}{a_{11}b_{11}} &= \boldsymbol{A}_{22} \odot \boldsymbol{B}_{22} - \frac{(\boldsymbol{a}_{21}\boldsymbol{a}_{12}') \odot (\boldsymbol{b}_{21}\boldsymbol{b}_{12}')}{a_{11}b_{11}} \\
&= \boldsymbol{A}_{22} \odot \boldsymbol{B}_{22} - \left(\frac{\boldsymbol{a}_{21}\boldsymbol{a}_{12}'}{a_{11}}\right) \odot \left(\frac{\boldsymbol{b}_{21}\boldsymbol{b}_{12}'}{b_{11}}\right) \\
&= \boldsymbol{A}_{22} \odot \left(\boldsymbol{B}_{22} - \frac{\boldsymbol{b}_{21}\boldsymbol{b}_{12}'}{b_{11}} + \frac{\boldsymbol{b}_{21}\boldsymbol{b}_{12}'}{b_{11}}\right) - \left(\frac{\boldsymbol{a}_{21}\boldsymbol{a}_{12}'}{a_{11}}\right) \odot \left(\frac{\boldsymbol{b}_{21}\boldsymbol{b}_{12}'}{b_{11}}\right) \\
&= \boldsymbol{A}_{22} \odot \left(\boldsymbol{B}_{22} - \frac{\boldsymbol{b}_{21}\boldsymbol{b}_{12}'}{b_{11}}\right) + \boldsymbol{A}_{22} \odot \frac{\boldsymbol{b}_{21}\boldsymbol{b}_{12}'}{b_{11}} - \left(\frac{\boldsymbol{a}_{21}\boldsymbol{a}_{12}'}{a_{11}}\right) \odot \left(\frac{\boldsymbol{b}_{21}\boldsymbol{b}_{12}'}{b_{11}}\right) \\
&= \boldsymbol{A}_{22} \odot \left(\boldsymbol{B}_{22} - \frac{\boldsymbol{b}_{21}\boldsymbol{b}_{12}'}{b_{11}}\right) + \left(\boldsymbol{A}_{22} - \frac{\boldsymbol{a}_{21}\boldsymbol{a}_{12}'}{a_{11}}\right) \odot \left(\frac{\boldsymbol{b}_{21}\boldsymbol{b}_{12}'}{b_{11}}\right) .
\end{aligned}$$

The above implies that

$$\boldsymbol{A}_{22} \odot \boldsymbol{B}_{22} - \frac{(\boldsymbol{a}_{21} \odot \boldsymbol{b}_{21})(\boldsymbol{a}_{12} \odot \boldsymbol{b}_{12})'}{a_{11}b_{11}} = \boldsymbol{A}_{22} \odot \boldsymbol{T} + \boldsymbol{S} \odot \left(\frac{\boldsymbol{b}_{21}\boldsymbol{b}_{12}'}{b_{11}}\right) , \tag{14.31}$$

where $\boldsymbol{S} = \boldsymbol{A}_{22} - \dfrac{\boldsymbol{a}_{21}\boldsymbol{a}_{12}'}{a_{11}}$ and $\boldsymbol{T} = \boldsymbol{B}_{22} - \dfrac{\boldsymbol{b}_{21}\boldsymbol{b}_{12}'}{b_{11}}$ are the Schur's complements of a_{11} in $\boldsymbol{A}$ and b_{11} in $\boldsymbol{B}$, respectively.

In 1930, Sir Alexander Oppenheim proved a beautiful determinantal inequality for positive definite matrices. We present a proof by Markham (1986) that is straightforward and does not require any new machinery.

Theorem 14.25 Oppenheim's inequality. *If $\boldsymbol{A} = \{a_{ii}\}$ and $\boldsymbol{B} = \{b_{ij}\}$ are both $n \times n$ positive definite matrices, then*

$$|\boldsymbol{A} \odot \boldsymbol{B}| \geq \left(\prod_{i=1}^{n} a_{ii}\right) |\boldsymbol{B}| .$$

Proof. The proof is by induction on the size of the matrices. The case $n = 1$ is trivial (equality holds) and the case $n = 2$ is easy to derive. We leave this verification to the reader. Consider the following partition of the $n \times n$ matrices $\boldsymbol{A}$ and $\boldsymbol{B}$

$$\boldsymbol{A} = \begin{bmatrix} a_{11} & \boldsymbol{a}_{12}' \\ \boldsymbol{a}_{12} & \boldsymbol{A}_{22} \end{bmatrix} \text{ and } \boldsymbol{B} = \begin{bmatrix} b_{11} & \boldsymbol{b}_{12}' \\ \boldsymbol{b}_{12} & \boldsymbol{B}_{22} \end{bmatrix} ,$$

where $\boldsymbol{a}_{12}$ and $\boldsymbol{b}_{12}$ are $(n-1) \times 1$, while $\boldsymbol{A}_{22}$ and $\boldsymbol{B}_{22}$ are both $(n-1) \times (n-1)$ and symmetric. Define the following matrices:

$$\boldsymbol{W} = \frac{\boldsymbol{b}_{12}\boldsymbol{b}_{12}'}{b_{11}} , \quad \boldsymbol{S} = \boldsymbol{A}_{22} - \frac{\boldsymbol{a}_{12}\boldsymbol{a}_{12}'}{a_{11}} \text{ and } \boldsymbol{T} = \boldsymbol{B}_{22} - \boldsymbol{W} .$$

Since $\boldsymbol{A}$ and $\boldsymbol{B}$ are positive definite, so are the Schur's complements of a_{11} in $\boldsymbol{A}$ and b_{11} in $\boldsymbol{B}$, respectively (recall Theorem 13.12). Thus, $\boldsymbol{S}$ and $\boldsymbol{T}$ are positive definite. Being a principal submatrix of a positive definite matrix, $\boldsymbol{A}_{22}$ is also positive definite (recall Theorem 13.6). Theorem 13.10 ensures that $\boldsymbol{W}$ is nonnegative definite. Since the Hadamard product of two positive definite matrices is again positive definite (Theorem 14.23), the above facts imply that $\boldsymbol{A}_{22} \odot \boldsymbol{T}$ and $\boldsymbol{S} \odot \boldsymbol{W}$ are both nonnegative definite.

Therefore, using (14.30) and (14.31) for symmetric matrices $\boldsymbol{A}$ and $\boldsymbol{B}$, we obtain

$$\begin{aligned}|\boldsymbol{A} \odot \boldsymbol{B}| &= a_{11}b_{11} \left| \boldsymbol{A}_{22} \odot \boldsymbol{B}_{22} - \frac{(\boldsymbol{a}_{12} \odot \boldsymbol{b}_{12})(\boldsymbol{a}_{12} \odot \boldsymbol{b}_{12})'}{a_{11}b_{11}} \right| \\ &= a_{11}b_{11} \left| \boldsymbol{A}_{22} \odot \boldsymbol{T} + \boldsymbol{S} \odot \boldsymbol{W} \right| \geq a_{11}b_{11} |\boldsymbol{A}_{22} \odot \boldsymbol{T}| \\ &\geq a_{11}b_{11} \left(a_{22}a_{33} \cdots a_{nn} |\boldsymbol{T}| \right) = \left(\prod_{i=1}^{n} a_{ii} \right) (b_{11}|\boldsymbol{T}|) = \left(\prod_{i=1}^{n} a_{ii} \right) |\boldsymbol{B}| .\end{aligned}$$

In the above, the first "$\geq$" is ensured by Theorem 13.25, the induction hypothesis ensures the second "$\geq$" and the last equality follows from Schur's formula for determinants (Theorem 10.10). □

By symmetry, it follows that

$$|\boldsymbol{A} \odot \boldsymbol{B}| = |\boldsymbol{B} \odot \boldsymbol{A}| \geq \left(\prod_{i=1}^{n} b_{ii} \right) |\boldsymbol{A}| . \tag{14.32}$$

Hadamard's inequality, obtained independently in Theorem 13.24, is an immediate consequence of Oppenheim's inequality with $\boldsymbol{B} = \boldsymbol{I}$. In that case, each $b_{ii} = 1$ and

$$|\boldsymbol{A}| = (b_{11}b_{22} \cdots b_{nn})|\boldsymbol{A}| \leq |\boldsymbol{A} \odot \boldsymbol{I}| = a_{11}a_{22} \cdots a_{nn} . \tag{14.33}$$

We conclude this chapter with the following inequality, which is a consequence of Oppenheim's and Hadamard's inequality.

Theorem 14.26 *If $\boldsymbol{A}$ and $\boldsymbol{B}$ are $n \times n$ n.n.d. matrices, then* $|\boldsymbol{A} \odot \boldsymbol{B}| \geq |\boldsymbol{A}||\boldsymbol{B}|$.

Proof. Let a_{ij} be the elements of $\boldsymbol{A}$. Then,

$$|\boldsymbol{A} \odot \boldsymbol{B}| \geq (a_{11}a_{22} \cdots a_{nn}) |\boldsymbol{B}| \geq |\boldsymbol{A}||\boldsymbol{B}| = |\boldsymbol{AB}| ,$$

where the first and second "$\geq$" follow from Oppenheim's and Hadamard's inequalities, respectively. □

14.10 Exercises

1. If $\boldsymbol{x} \in \Re^m$ and $\boldsymbol{y} \in \Re^n$, then show that $\boldsymbol{x}' \otimes \boldsymbol{y} = \boldsymbol{y}\boldsymbol{y}'$.
2. If $\boldsymbol{A}$ is a square matrix, then show that $(\boldsymbol{I} \otimes \boldsymbol{A})^k = \boldsymbol{I} \otimes \boldsymbol{A}^k$ and $(\boldsymbol{A} \otimes \boldsymbol{I})^k = \boldsymbol{A}^k \otimes \boldsymbol{I}$.
3. True or false: If $\boldsymbol{A}$ and $\boldsymbol{B}$ are square matrices, then $\boldsymbol{I} \otimes \boldsymbol{A}$ and $\boldsymbol{B} \otimes \boldsymbol{I}$ commute.
4. If $\boldsymbol{A}$ and $\boldsymbol{B}$ are both idempotent, then show that $\boldsymbol{A} \otimes \boldsymbol{B}$ is idempotent.
5. If $\boldsymbol{P}_{\boldsymbol{X}}$ denotes the orthogonal projector onto $\mathcal{C}(\boldsymbol{X})$, then show that

$$\boldsymbol{P}_{\boldsymbol{A} \otimes \boldsymbol{B}} = \boldsymbol{P}_{\boldsymbol{A}} \otimes \boldsymbol{P}_{\boldsymbol{B}} ,$$

where $\boldsymbol{A}$ and $\boldsymbol{B}$ have full column rank.

6. If $\boldsymbol{A}$ is $n \times n$, then show that $\text{tr}(\boldsymbol{A}) = (\text{vec}(\boldsymbol{I}_n))'\text{vec}(\boldsymbol{B})$.
7. Let $\boldsymbol{A}$ be $m \times n$ and $\boldsymbol{B}$ be $m \times n$ and $p \times q$. Prove that

$$\boldsymbol{A} \otimes \boldsymbol{B} = (\boldsymbol{A} \otimes \boldsymbol{I}_p)(\boldsymbol{I}_n \otimes \boldsymbol{B}) = \boldsymbol{K}_{mp}(\boldsymbol{I}_p \otimes \boldsymbol{A})\boldsymbol{K}_{pn}(\boldsymbol{I}_n \otimes \boldsymbol{B}) .$$

Let $m = n$ and $p = q$. Using the above expression (and without using Theorem 14.12), prove that:

$$|\boldsymbol{A} \otimes \boldsymbol{B}| = |\boldsymbol{A}|^p |\boldsymbol{B}|^m .$$

8. **Tensor product surface interpolation.** Consider interpolation over a gridded data (x_i, y_j, f_{ij}) for $i = 1, 2, \ldots, m$, $j = 1, 2, \ldots, n$, where $a_1 = x_1 < x_2 < \cdots < x_m = b_2$ and $a_2 = y_1 < y_2 < \cdots < y_n = b_2$. For each i, j we treat $f_{ij} = f(x_i, y_j)$ as the value of an unknown function or surface $f(x, y)$. We define a ***tensor product surface***

$$g(x, y) = \sum_{i=1}^{m}\sum_{j=1}^{n} c_{ij}\phi_i(x)\psi_j(y) = \boldsymbol{\phi}(x)'\boldsymbol{C}\boldsymbol{\psi}(y) ,$$

where $\phi_i(x)$'s and $\psi_j(y)$'s are specified basis functions (such as splines), c_{ij}'s are unknown coefficients, $\boldsymbol{\phi}(x) = [\phi_1(x), \phi_2(x), \ldots, \phi_m(x)]'$ and $\boldsymbol{\psi}(y) = [\psi_1(y), \psi_2(y), \ldots, \psi_n(y)]'$. We wish to find the coefficients c_{ij} so that $g(x, y)$ will interpolate $f(x, y)$ over the grid of (x_i, y_j)'s. Define the following:

$$\boldsymbol{\Phi} = \begin{bmatrix} \boldsymbol{\phi}(x_1)' \\ \boldsymbol{\phi}(x_2)' \\ \vdots \\ \boldsymbol{\phi}(x_m)' \end{bmatrix} , \ \boldsymbol{\Psi} = \begin{bmatrix} \boldsymbol{\psi}(y_1)' \\ \boldsymbol{\psi}(y_2)' \\ \vdots \\ \boldsymbol{\psi}(y_n)' \end{bmatrix} , \boldsymbol{F} = \{f_{ij}\} \text{ and } \boldsymbol{C} = \{c_{ij}\} .$$

Show that the desired $g(x, y)$ is obtained by solving $\boldsymbol{\Phi}\boldsymbol{C}\boldsymbol{\Psi}' = \boldsymbol{F}$ for $\boldsymbol{C}$. Prove that this solution is unique if and only if $\boldsymbol{\Phi}$ and $\boldsymbol{\Psi}$ are nonsingular.

9. Show that $(\boldsymbol{A} \otimes \boldsymbol{B})^+ = \boldsymbol{A}^+ \otimes \boldsymbol{B}^+$.
10. Consider the matrix equation $\boldsymbol{AXB} = \boldsymbol{C}$ as in Theorem 14.17. Show that the system has a solution for all $\boldsymbol{C}$ if $\boldsymbol{A}$ has full row rank and $\boldsymbol{B}$ has full column rank.
11. Let $\boldsymbol{A}$ have full column rank and $\boldsymbol{B}$ have full row rank in the matrix equation $\boldsymbol{AXB} = \boldsymbol{C}$ as in Theorem 14.17. Suppose there exists a solution $\boldsymbol{X}_0$. Show that $\boldsymbol{X}_0$ is the unique solution.
12. If $\boldsymbol{A}$ and $\boldsymbol{B}$ are square matrices that are diagonalizable, show that

$$\exp(\boldsymbol{A} \boxplus \boldsymbol{B}) = \exp((\boldsymbol{I} \otimes \boldsymbol{A}) + (\boldsymbol{B} \otimes \boldsymbol{I})) = \exp(\boldsymbol{B}) \otimes \exp(\boldsymbol{A}) .$$

CHAPTER 15

Linear Iterative Systems, Norms and Convergence

15.1 Linear iterative systems and convergence of matrix powers

Mathematical models in statistics, economics, engineering and basic sciences often take the form of *linear iterative systems*,

$$\boldsymbol{x}_{k+1} = \boldsymbol{A}\boldsymbol{x}_k \ , \quad \text{for } k = 0, 1, 2, \ldots, m \ , \tag{15.1}$$

where $\boldsymbol{A}$ is an $n \times n$ coefficient matrix and each $\boldsymbol{x}_k$ is an $n \times 1$ vector. Systems such as (15.1) can be real or complex, meaning that $\boldsymbol{A}$ and the $\boldsymbol{x}_k$'s can have entries that are either real or complex numbers. At the 0-th iteration (i.e., the start) the initial "state" of the system $\boldsymbol{x}_0$ is usually known. Equation (15.1) implies that

$$\boldsymbol{x}_k = \boldsymbol{A}\boldsymbol{x}_{k-1} = \boldsymbol{A}(\boldsymbol{A}\boldsymbol{x}_{k-2}) = \boldsymbol{A}^2\boldsymbol{x}_{k-2} = \cdots = \boldsymbol{A}^{k-1}\boldsymbol{x}_1 = \boldsymbol{A}^{k-1}(\boldsymbol{A}\boldsymbol{x}_0) = \boldsymbol{A}^k\boldsymbol{x}_0 \ .$$

Therefore, if we know the initial state $\boldsymbol{x}_0$, then each iterate is determined simply by premultiplying $\boldsymbol{x}_0$ with the corresponding power of the coefficient matrix $\boldsymbol{A}$.

Of particular interest is the exploration of the long-run behavior of the system and, more specifically, whether every solution $\boldsymbol{x}_k$ of (15.1) converges to $\boldsymbol{0}$ as $k \to \infty$. If that is the case, then the system is called ***asymptotically stable***. That is, irrespective of what the initial state $\boldsymbol{x}_0$ is, the solution $\boldsymbol{A}^k\boldsymbol{x}_0$ converges to $\boldsymbol{0}$ as $k \to \infty$.

Theorem 15.1 *Every solution vector of the linear iterative system* $\boldsymbol{x}_k = \boldsymbol{A}\boldsymbol{x}_{k-1}$ *converges to* $\boldsymbol{0}$ *as* $k \to \infty$ *if and only if* $\boldsymbol{A}^k \to \boldsymbol{O}$ *as* $k \to \infty$.

Proof. Suppose $\lim_{k\to\infty} \boldsymbol{A}^k = \boldsymbol{O}$, meaning that that each entry of $\boldsymbol{A}^k$ converges to 0 as $k \to \infty$. Then, clearly the system is asymptotically stable because

$$\lim_{k\to\infty} \boldsymbol{x}_k = \lim_{k\to\infty} \boldsymbol{A}^k\boldsymbol{x}_0 = (\lim_{k\to\infty} \boldsymbol{A}^k)\boldsymbol{x}_0 = \boldsymbol{O}\boldsymbol{x}_0 = \boldsymbol{0} \ .$$

The converse is also true. If the linear iterative system in (15.1) is asymptotically stable, then $\lim_{k\to\infty} \boldsymbol{A}^k = \boldsymbol{O}$. Why is this true? Asymptotic stability ensures that $\boldsymbol{A}^k\boldsymbol{x}_0$ converges to $\boldsymbol{0}$ as $k \to \infty$ for *every* initial state $\boldsymbol{x}_0$. Consider (15.1) with the initial state $\boldsymbol{x}_0 = \boldsymbol{e}_j$, the j-th column of the identity matrix. Then, $\boldsymbol{x}_k = \boldsymbol{A}^k\boldsymbol{e}_j$, which is the j-th column of $\boldsymbol{A}^k$. Since $\boldsymbol{x}_k$ converges to $\boldsymbol{0}$ as $k \to \infty$, the j-th column of $\boldsymbol{A}^k$ converges to $\boldsymbol{0}$. Repeating this argument by taking $\boldsymbol{x}_0 = \boldsymbol{e}_j$ for $j = 1, 2, \ldots, n$ we see that every column of $\boldsymbol{A}^k$ converges to $\boldsymbol{0}$ and, hence, $\lim_{k\to\infty} \boldsymbol{A}^k = \boldsymbol{O}$. □

Suppose that λ is an eigenvalue of $\boldsymbol{A}$ and $\boldsymbol{v}$ is a corresponding eigenvector. Then,

$$\boldsymbol{A}(\lambda^k \boldsymbol{v}) = \lambda^k(\boldsymbol{A}\boldsymbol{v}) = \lambda^k(\lambda \boldsymbol{v}) = \lambda^{k+1}\boldsymbol{v} \,, \tag{15.2}$$

which shows that $\boldsymbol{x}_k = \lambda^k \boldsymbol{v}$ is a solution for (15.1). If $\lambda_1, \lambda_2, \ldots, \lambda_r$ are eigenvalues of $\boldsymbol{A}$ and $\boldsymbol{v}_1, \boldsymbol{v}_2, \ldots, \boldsymbol{v}_r$ are the corresponding eigenvectors, then any linear combination $\boldsymbol{x}_k = c_1\lambda_1^k \boldsymbol{v}_1 + c_2\lambda_2^k \boldsymbol{v}_k + \cdots + c_r\lambda_r^k \boldsymbol{v}_r$ also satisfies (15.1) because

$$\boldsymbol{A}\boldsymbol{x}_k = \boldsymbol{A}\left(\sum_{i=1}^{r} c_i\lambda_i^k \boldsymbol{v}_i\right) = \sum_{i=1}^{r} c_i\lambda_i^k \boldsymbol{A}\boldsymbol{v}_i = \sum_{i=1}^{r} c_i\lambda_i^{k+1}\boldsymbol{v}_i = \boldsymbol{x}_{k+1} \,.$$

In fact, when $\boldsymbol{A}$ is diagonalizable the general solution for (15.1) is of the above form.

When $\boldsymbol{A}$ is diagonalizable, the solution $\boldsymbol{x}_k = \boldsymbol{A}^k \boldsymbol{x}_0$ for (15.1) can be expressed conveniently in terms of the eigenvalues and eigenvectors of $\boldsymbol{A}$. If $\boldsymbol{A} = \boldsymbol{P}\boldsymbol{\Lambda}\boldsymbol{P}^{-1}$, then $\boldsymbol{A}^k = \boldsymbol{P}\boldsymbol{\Lambda}^k\boldsymbol{P}^{-1}$, where $\boldsymbol{\Lambda}$ is diagonal with the eigenvalues of $\boldsymbol{A}$ along its diagonal and so $\boldsymbol{\Lambda}^k$ is also diagonal with the eigenvalues raised to the power k along its diagonal. Let $\boldsymbol{v}_1, \boldsymbol{v}_2, \ldots, \boldsymbol{v}_n$ be the columns of $\boldsymbol{P}$. Therefore,

$$\boldsymbol{x}_k = \boldsymbol{A}^k\boldsymbol{x}_0 = \boldsymbol{P}\boldsymbol{\Lambda}^k\boldsymbol{P}^{-1}\boldsymbol{x}_0 = a_1\lambda_1^k\boldsymbol{v}_1 + a_2\lambda_2^k\boldsymbol{v}_2 + \cdots + a_n\lambda_n^k\boldsymbol{v}_n \,, \tag{15.3}$$

where a_i is the i-th entry in the $n \times 1$ vector $\boldsymbol{P}^{-1}\boldsymbol{x}_0$. In fact, $a_i = \boldsymbol{u}_i'\boldsymbol{x}_0$, where $\boldsymbol{u}_i'$ is the i-th row of $\boldsymbol{P}^{-1}$ which is also a left eigenvector of $\boldsymbol{A}$ (recall Definition 11.2). Therefore, the solution for (15.3) can be expressed in terms of the initial state of the system, and the eigenvalues, and the left and right eigenvectors of $\boldsymbol{A}$.

Theorem 15.1 demonstrates how the convergence of powers of a matrix characterize the long-run behavior of linear iterative systems. Finding limits of powers of matrices has other applications as well. For example, it provides a sufficient condition for matrices of the form $\boldsymbol{I} - \boldsymbol{A}$ to be nonsingular.

Theorem 15.2 *If $\lim_{n\to\infty} \boldsymbol{A}^n = \boldsymbol{O}$, then $\boldsymbol{I} - \boldsymbol{A}$ is nonsingular and*

$$(\boldsymbol{I} - \boldsymbol{A})^{-1} = \boldsymbol{I} + \boldsymbol{A} + \boldsymbol{A}^2 + \cdots + \boldsymbol{A}^k + \cdots = \sum_{k=0}^{\infty} \boldsymbol{A}^k \,,$$

where $\boldsymbol{A}^0$ is defined as $\boldsymbol{I}$.

Proof. Multiplying $(\boldsymbol{I} - \boldsymbol{A})$ by $\sum_{k=0}^{n} \boldsymbol{A}^k$ and then passing to the limit yields

$$\begin{aligned}(\boldsymbol{I} - \boldsymbol{A})(\boldsymbol{I} + \boldsymbol{A} + \boldsymbol{A}^2 + \cdots + \boldsymbol{A}^n) &= (\boldsymbol{I} - \boldsymbol{A}) + (\boldsymbol{A} - \boldsymbol{A}^2) + \cdots + (\boldsymbol{A}^n - \boldsymbol{A}^{n+1}) \\ &= \boldsymbol{I} - \boldsymbol{A}^{n+1} \to \boldsymbol{I} \text{ as } n \to \infty\end{aligned}$$

because $\lim_{n\to\infty} \boldsymbol{A}^n = \boldsymbol{O}$. Therefore, $(\boldsymbol{I} - \boldsymbol{A})^{-1} = \sum_{k=0}^{\infty} \boldsymbol{A}^k$. □

The above series expansion for $\boldsymbol{I} - \boldsymbol{A}$ is often called the ***Neumann series***, named after the German mathematician Carl Gottfried Neumann, and has applications in econometric and statistical modeling. It is widely used in numerical linear algebra to approximate the inverse of a sum of matrices. For example, we obtain $(\boldsymbol{I} - \boldsymbol{A})^{-1} \approx \boldsymbol{I} + \boldsymbol{A}$ as a first-order approximation. This can be generalized to find a first-order approximation for $(\boldsymbol{A} + \boldsymbol{B})^{-1}$ when $\boldsymbol{A}^{-1}$ exists and the entries of $\boldsymbol{B}$ are (roughly speaking) small enough relative to those of $\boldsymbol{A}$. Then, $\lim_{n\to\infty}(\boldsymbol{A}^{-1}\boldsymbol{B})^n = \boldsymbol{O}$ and

$$\begin{aligned}(\boldsymbol{A}+\boldsymbol{B})^{-1} &= \left[\boldsymbol{A}(\boldsymbol{I} - (-\boldsymbol{A}^{-1}\boldsymbol{B}))\right]^{-1} = \left(\boldsymbol{I} - (-\boldsymbol{A}^{-1}\boldsymbol{B})\right)^{-1}\boldsymbol{A}^{-1} \\ &\approx \left(\boldsymbol{I} + (-\boldsymbol{A}^{-1}\boldsymbol{B})\right)\boldsymbol{A}^{-1} = \boldsymbol{A}^{-1} - \boldsymbol{A}^{-1}\boldsymbol{B}\boldsymbol{A}^{-1}\,.\end{aligned}$$

This first-order approximation has important implications in understanding numerical stability of linear systems (see, e.g., Meyer, 2001). If the entries in $\boldsymbol{A}$ are "perturbed" by the entries in $\boldsymbol{B}$, say due to round-off errors in floating point arithmetic, then this error is propagated in $\boldsymbol{A}^{-1}$ by $\boldsymbol{A}^{-1}\boldsymbol{B}\boldsymbol{A}^{-1}$, which can be large when $\boldsymbol{A}^{-1}$ has large entries, even if the entries in $\boldsymbol{B}$ are small. Therefore, even small perturbations can cause numerical instabilities in matrices whose inverses have large entries.

Given the importance of matrices that satisfy $\lim_{n\to\infty}\boldsymbol{A}^n = \boldsymbol{O}$, some authors assign a special name to them: ***convergent matrices***, by which we mean that each entry of $\boldsymbol{A}^k$ converges to 0 as $n \to \infty$. However, testing whether each entry of $\boldsymbol{A}^n$ converges to 0 can become analytically cumbersome; for one, we will need to multiply $\boldsymbol{A}$ n times with itself to obtain $\boldsymbol{A}^n$. It will, therefore, be convenient if we can derive equivalent conditions for convergence using the properties of $\boldsymbol{A}$. To facilitate this analysis, we will introduce more general definitions of *norms* as a notion of distance.

15.2 Vector norms

A ***norm*** is a function that helps formalize the notion of ***length*** and distance in a vector space. We have already seen one example of length or norm of vectors in $\Re^n$ as defined in Definition 1.9. This is sometimes known as the standard ***Euclidean vector norm*** and it gives the ***Euclidean distance*** of a vector from the origin. Euclidean distance is not the only way to measure distance. For example, living in a city often entails navigating on a grid of blocks with streets. In such cases, it is more appropriate to measure distance "by blocks" and not how the crow flies. For example, if you travel three blocks north on Street A, then four blocks west on Street B and five blocks south on Street C, then you have traveled $|3| + |-4| + |-5| = 12$ blocks. This distance, often called the "Manhattan distance" because of how distances are measured in New York City, is formally known as the 1-norm. The following is a general definition of a norm motivated by our intuition of length in Euclidean geometry.

Definition 15.1 *A* **norm** *on a real or complex vector space $\mathcal{V}$ is a function that maps each vector $\boldsymbol{x} \in \mathcal{V}$ to a real number $\|\boldsymbol{x}\|$ such that the following axioms are satisfied:*

(i) Positivity: $\|\boldsymbol{x}\| \geq 0$ for every $\boldsymbol{x} \in \mathcal{V}$ and $\|\boldsymbol{x}\| = 0$ if and only if $\boldsymbol{x} = \boldsymbol{0}$.

(ii) Homogeneity: $\|\alpha\boldsymbol{x}\| = |\alpha|\|\boldsymbol{x}\|$ for every $\boldsymbol{x} \in \mathcal{V}$ and real or complex number α.

(iii) Triangle inequality: $\|\boldsymbol{x} + \boldsymbol{y}\| \leq \|\boldsymbol{x}\| + \|\boldsymbol{y}\|$ *for every* $\boldsymbol{x}, \boldsymbol{y} \in \mathcal{V}$.

Positivity agrees with our intuition that length must essentially be a positive quantity. Homogeneity ensures that multiplying a vector by a positive number alters its length (stretches or shrinks it) but does not change its direction. Finally, the triangle inequality ensures that the shortest distance between any two points is a straight line.

Example 15.1 The following are some examples of vector norms on $\mathcal{V} = \Re^n$. Let $\boldsymbol{x} \in \Re^n$ with elements x_i.

- The ***Euclidean norm*** computes length in the classical Pythagorean manner by taking the square root of the sum of squares of each element in $\boldsymbol{x}$:

$$\|\boldsymbol{x}\|_2 = (x_1^2 + x_2^2 + \cdots + x_n^2)^{1/2} = \left(\sum_{i=1}^{n} |x_i|^2\right)^{1/2} .$$

- The *Manhattan distance* or 1-***norm*** of $\boldsymbol{x}$ is the sum of the absolute of its entries:

$$\|\boldsymbol{x}\|_1 = \max\{|x_1|, |x_2|, \ldots, |x_n|\} = \sum_{i=1}^{n} |x_i| .$$

- The ***max-norm*** or the ∞-***norm*** of $\boldsymbol{x}$ is its maximal entry (in absolute value):

$$\|\boldsymbol{x}\|_\infty = \max\{|x_1|, |x_2|, \ldots, |x_n|\} = \max_{1 \leq i \leq n} |x_i| .$$

- The p-***norm*** is defined as

$$\|\boldsymbol{x}\|_p = (x_1^p + x_2^p + \cdots + x_n^p)^{1/p} = \left(\sum_{i=1}^{p} |x_i|^p\right)^{1/p} .$$

The above norms are valid for $\mathcal{V} = \mathbb{C}^n$ as well. We have already verified the axioms in Definition 15.1 for the Euclidean norm. The triangle inequality was a consequence of the Cauchy-Schwarz inequality (recall Corollary 7.1). The verifications for the 1-norm and *max-norm* are easy and left as an exercise. The triangle inequality for these norms can be derived using the elementary property of real numbers that $|u + v| \leq |u| + |v|$ for any two real or complex numbers u and v. Deriving the triangle inequality for the p-norm is more involved and not trivial. It is called the ***Minkowski's inequality*** and relies upon a more general version of the Cauchy-Schwarz inequality called Holder's inequality.

The Euclidean distance and the 1-norm are special cases of the p-norm with $p = 2$ and $p = 1$, respectively. More remarkably, the ∞-norm is a limiting case of the p-norm. To see why this is true, suppose that $\alpha = \max_i |x_i|$ is the maximal absolute value in $\boldsymbol{x}$ and assume that k entries in $\boldsymbol{x}$ are equal to α. Construct $\boldsymbol{y} \in \Re^n$ from the entries in $\boldsymbol{x}$ such that $y_1 = y_2 = \cdots = y_k = \alpha$ and $y_{k+1}, y_{k+2}, \ldots, y_n$ are equal to the remaining entries in $\boldsymbol{x}$. Since the elements of $\boldsymbol{y}$ are a permutation of those in $\boldsymbol{x}$, as $p \to \infty$

$$\|\boldsymbol{x}\|_p = \left(\sum_{i=1}^{n} |y_i|^p\right)^{1/p} = |y_1| \left(k + \left|\frac{y_{k+1}}{y_1}\right|^p + \cdots + \left|\frac{y_n}{y_1}\right|^p\right)^{1/p} \to |y_1| \,. \tag{15.4}$$

Since $y_1 = \alpha = \max_i |x_i| = \|\boldsymbol{x}\|_\infty$, we have $\lim_{p\to\infty} \|\boldsymbol{x}\|_p = \|\boldsymbol{x}\|_\infty$.

Vector norms allow us to analyze limiting behavior of elements in vector spaces by extending notions of convergence for real numbers to vectors. Therefore, we say that an infinite sequence of vectors $\{\boldsymbol{x}_1, \boldsymbol{x}_2, \ldots, \} \subset \Re^n$ converges to $\boldsymbol{x}$ if $\|\boldsymbol{x}_k - \boldsymbol{x}\| \to \boldsymbol{0}$. This would have been well-defined except one outstanding issue. The definition depends upon the specific norm we choose and has the potential difficulty that $\boldsymbol{x}_k \to \boldsymbol{x}$ with respect to one norm but not with another. However, this cannot happen in finite dimensional spaces because all norms, while not necessarily equal in value, are always "close" to one another in a certain sense. Let us make this precise.

Let $\|\cdot\|_a$ and $\|\cdot\|_b$ be any two norms on $\Re^n$. Then, there exist positive constants α and β such that

$$\alpha \leq \frac{\|\boldsymbol{x}\|_a}{\|\boldsymbol{x}\|_b} \leq \beta \,, \quad \forall \quad \text{nonzero vectors in } \Re^n \,. \tag{15.5}$$

We say that any two norms on $\Re^n$ are ***equivalent***. To see why (15.5) is true, we use some basic real analysis. Using the triangle inequality, one can establish that $f(\boldsymbol{x}) = \|\boldsymbol{x}\|$ is a *continuous* function from $\Re^n$ to $\Re^1$. Construct the unit sphere with respect to the first norm and define it by $S_a = \{\boldsymbol{x} \in \Re^n : \|\boldsymbol{x}\|_a = 1\}$. Clearly S_a is a closed and bounded set in $\Re^n$ and a basic result from real analysis ensures that any continuous function mapping S_a to $\Re$ will attain its maximum and minimum value at points in S_a. Therefore, if we restrict the continuous function $g(\boldsymbol{x}) = \|\boldsymbol{x}\|_b$ to the set S_a, then we have well-defined real numbers

$$\alpha = \min\{g(\boldsymbol{x}) : \boldsymbol{x} \in S_a\} \;\text{ and }\; \beta = \max\{g(\boldsymbol{x}) : \boldsymbol{x} \in S_a\} \,.$$

From the above definition, clearly α and β are positive finite numbers such that $\alpha \leq \beta$ with equality holding if and only if $\|\boldsymbol{x}\|_a = \|\boldsymbol{x}\|_b$. Consider any vector $\boldsymbol{u}$ such that $\|\boldsymbol{u}\|_a = 1$, i.e., $\boldsymbol{u}$ is a vector in S_a. Since $\alpha \leq g(\boldsymbol{x}) \leq \beta$ for every $\boldsymbol{x}$ in S_a, it follows that $\alpha \leq g(\boldsymbol{u}) \leq \beta$. Now consider any vector $\boldsymbol{x} \in \Re^n$ (not necessarily in S_a). Then, the vector $\boldsymbol{u} = \boldsymbol{x}/\|\boldsymbol{x}\|_a$ is again in S_a and, hence,

$$\alpha \leq g\left(\frac{\boldsymbol{x}}{\|\boldsymbol{x}\|_a}\right) \leq \beta \Longrightarrow \alpha \leq \left\|\frac{\boldsymbol{x}}{\|\boldsymbol{x}\|_a}\right\|_b \leq \beta \Longrightarrow \alpha \leq \frac{\|\boldsymbol{x}\|_b}{\|\boldsymbol{x}\|_a} \leq \beta \,,$$

where we have used the homogeneity axiom of $\|\cdot\|_b$ in the last step.

Example 15.2 Equivalence of Euclidean and ∞-norms. Consider the Euclidean norm and the ∞-norm for vectors in $\Re^n$. Observe that

$$\|\boldsymbol{x}\|_\infty^2 = \left(\max_{1\leq i\leq n} |x_i|\right)^2 = \max_{1\leq i\leq n} |x_i|^2 \leq \sum_{i=1}^{n} |x_i|^2 = \|\boldsymbol{x}\|_2^2 \,,$$

where the last step is true because $\sum_{i=1}^{n} |x_i|^2$ contains $\max_{1\leq i\leq n} |x_i|^2$ as one its terms. Also,

$$\|\boldsymbol{x}\|_2^2 = |x_1|^2 + |x_2|^2 + \cdots + |x_n|^2 \leq n\|\boldsymbol{x}\|_\infty^2$$

because $|x_i| \leq \|\boldsymbol{x}\|_\infty$ for each $i = 1, 2, \ldots, n$. The above two inequalities combine to produce

$$\frac{1}{\sqrt{n}}\|\boldsymbol{x}\|_2 \leq \|\boldsymbol{x}\|_\infty \leq \|\boldsymbol{x}_2\| ,$$

which is a special case of (15.5) with $\alpha = 1/\sqrt{n}$ and $\beta = 1$. ■

The equivalence of norms provides us with a well-defined notion of convergent sequences of vectors. Let us restrict ourselves to $\Re^n$ although these results are true for general finite dimensional vector spaces. As for real numbers, we say that the infinite sequence $\boldsymbol{x}_1, \boldsymbol{x}_2, \ldots$ of vectors in $\Re^n$ is convergent if there exists a vector $\boldsymbol{a}$ such that $\|\boldsymbol{x}_n - \boldsymbol{a}\| \to \boldsymbol{0}$ as $n \to \infty$. The equivalence of norms in (15.5) ensures that if this condition is true for one norm, it must be true for all norms (in finite dimensional vector spaces). Hence, we can ambiguously write $\boldsymbol{x} \to \boldsymbol{a}$. Also, $\boldsymbol{x}_n \to \boldsymbol{a}$ if and only if each entry in $\boldsymbol{x}_n$ converges to the corresponding element in $\boldsymbol{a}$, i.e., $x_{n,i} \to a_i$, where $x_{n,i}$ and a_i are the i-th entries in $\boldsymbol{x}$ and $\boldsymbol{a}$. Again, we do not need to prove this for every norm. It is easy to establish this for the standard Euclidean norm and (15.5) ensures that it will hold for any valid norm in $\Re^n$.

Example 15.3 Asymptotic stability of linear iterative systems. Let us revisit the asymptotic stability of the linear iterative system (15.1) and convergence of matrix powers using matrix norms. Suppose, for convenience, that the coefficient matrix $\boldsymbol{A}$ is diagonalizable. Consider an eigenvalue λ of $\boldsymbol{A}$ such that $|\lambda| < 1$, and let $\boldsymbol{v}$ be an eigenvector associated with λ. The solution $\boldsymbol{x}_k = \lambda \boldsymbol{v}$ satisfies

$$\|\boldsymbol{x}_k\|_\infty = \|\lambda^k \boldsymbol{v}\| = |\lambda|^k \|\boldsymbol{v}\| \to 0 \text{ as } k \to \infty$$

because $|\lambda| < 1$ and so $|\lambda|^k \to 0$ as $k \to \infty$. Therefore, $\boldsymbol{x}_k \to \boldsymbol{0}$ as $k \to \infty$. This observation leads to a simple sufficient condition for asymptotic stability: if all the eigenvalues of $\boldsymbol{A}$ have modulus strictly less than one, i.e., $|\lambda| < 1$, then the system is asymptotically stable. This follows immediately from (15.3) because each of the terms on the right hand side of (15.3) converges to $\boldsymbol{0}$. On the other hand, if any eigenvalue λ of $\boldsymbol{A}$ has modulus greater than or equal to one and $\boldsymbol{v}$ is an eigenvector associated with λ, then $\boldsymbol{x}_k = \lambda^k \boldsymbol{v}$ does not converge to $\boldsymbol{0}$ and so the system is not asymptotically stable. Hence, *a necessary and sufficient condition for a linear iterative system to be asymptotically stable is that every eigenvalue of $\boldsymbol{A}$ must have modulus less than one.*

Theorem 15.1 states that a linear iterative system is asymptotically stable if and only if $\boldsymbol{A}^k \to \boldsymbol{O}$ as $k \to \infty$. From the above argument, we conclude that a necessary and sufficient condition for $\lim_{k\to\infty} \boldsymbol{A}^k = \boldsymbol{O}$, where $\boldsymbol{A}$ is diagonalizable, is that all the eigenvalues of $\boldsymbol{A}$ must have modulus strictly less than one. ■

15.3 Spectral radius and matrix convergence

Example 15.3 reveals the importance of the magnitude of the eigenvalues of a matrix with respect to linear iterative systems and convergence of powers of matrices. We introduce the *spectral radius*.

Definition 15.2 *The* **spectral radius** *of a square matrix* $\boldsymbol{A}$ *is defined as the maximum of the absolute values of all its eigenvalues:*

$$\kappa(\boldsymbol{A}) = \max\{|\lambda_1|, |\lambda_2|, \ldots, |\lambda_n|\} .$$

The results in Example 15.3 are true for nondiagonalizable matrices as well. However, the derivations are a bit more involved and require the Jordan canonical form.

Theorem 15.3 *For any square matrix* $\boldsymbol{A}$ *with real or complex entries,* $\lim_{k\to\infty} \boldsymbol{A}^k = \boldsymbol{O}$ *if and only if* $\kappa(\boldsymbol{A}) < 1$.

Proof. Every square matrix is similar to a Jordan matrix (recall Section 12.6). If $\boldsymbol{P}^{-1}\boldsymbol{A}\boldsymbol{P} = \boldsymbol{J}$ is the JCF of $\boldsymbol{A}$, where $\boldsymbol{J}$ block diagonal with Jordan blocks (recall Definition 12.5) along the diagonal, then $\boldsymbol{A}^k = \boldsymbol{P}\boldsymbol{J}^k\boldsymbol{P}^{-1}$. The matrix $\boldsymbol{J}^k$ is again block diagonal with i-th block of the form

$$\boldsymbol{J}_{n_i}(\lambda_i)^k = \begin{bmatrix} \lambda_i^k & \binom{k}{1}\lambda_i^{k-1} & \binom{k}{2}\lambda_i^{k-2} & \cdots & \binom{k}{n_i-1}\lambda_i^{k-n_i+1} \\ 0 & \lambda_i^k & \binom{k}{1}\lambda_i^{k-1} & \cdots & \binom{k}{n_i-2}\lambda_i^{k-n_i+2} \\ \vdots & \vdots & \vdots & \vdots & \vdots \\ 0 & 0 & \cdots & \lambda_i^k & \binom{k}{1}\lambda_i^{k-1} \\ 0 & 0 & \cdots & 0 & \lambda_i^k \end{bmatrix} .$$

That is, $\boldsymbol{J}_{n_i}(\lambda_i)^k$ is upper-triangular with λ_i^k along the diagonal and the $(i, i+j)$-th element is $\binom{k}{j}\lambda^{k-j}$ for $j = 1, 2, \ldots, n_i - i$. If $\lim_{k\to\infty} \boldsymbol{A}^k = \boldsymbol{O}$, then $\boldsymbol{J}^k \to \boldsymbol{O}$ and $\boldsymbol{J}_{n_i}(\lambda_i)^k \to \boldsymbol{O}$ as $k \to \infty$. Therefore, each of its diagonal entries converge to 0. Thus, $\lambda_i^k \to 0$ as $k \to \infty$, which implies that $|\lambda_i| < 1$. This shows that every eigenvalue of $\boldsymbol{A}$ has modulus strictly less than to one.

Conversely, suppose any eigenvalue λ of $\boldsymbol{A}$ satisfies $|\lambda| < 1$. Then,

$$\left|\binom{k}{j}\lambda^{k-j}\right| = \frac{k(k-1)(k-2)\cdots(k-j+1)}{j!}|\lambda|^{k-j} \leq \frac{k^j}{j!}|\lambda|^{k-j} \to 0$$

as $k \to \infty$ because $k^j \to \infty$ with polynomial speed but $|\lambda|^{k-j} \to 0$ with exponential speed. One can also use other methods of calculus, e.g., L'Hospital's rule, to establish this limit. Thus, $\lim_{k\to\infty} \boldsymbol{A}^k = \boldsymbol{O}$ if and only if every eigenvalue of $\boldsymbol{A}$ has modulus strictly less than one, i.e., $\kappa(\boldsymbol{A}) < 1$. □

The spectral radius governs the convergence of the Neumann series (Theorem 15.2).

Theorem 15.4 *The Neumann series* $\sum_{k=0}^{\infty} \boldsymbol{A}^k$ *converges if and only if* $\kappa(\boldsymbol{A}) < 1$.

Proof. If $\kappa(\boldsymbol{A}) < 1$, Theorem 15.3 ensures that $\boldsymbol{A}^n \to \boldsymbol{O}$ as $n \to \infty$, which, by virtue of Theorem 15.2, implies that the Neumann series converges.

Next suppose that the Neumann series $\sum_{k=0}^{\infty} \boldsymbol{A}^k$ converges. This means that $\sum_{k=0}^{\infty} \boldsymbol{J}^k$ converges, where $\boldsymbol{A} = \boldsymbol{P}\boldsymbol{J}\boldsymbol{P}^{-1}$ is the JCF of $\boldsymbol{A}$. Thus, $\sum_{k=0}^{\infty} \boldsymbol{J}_{n_i}(\lambda_i)^k$ converges for every Jordan block $\boldsymbol{J}_{n_i}(\lambda_i)$. In particular, the sum of the diagonals will converge, i.e., $\sum_{k=0}^{\infty} \lambda_i^k$ converges for every eigenvalue λ_i. This is a geometric series of scalars and converges only if $|\lambda_i| < 1$. Therefore, $\kappa(\boldsymbol{A}) < 1$. □

Theorems 15.3 and 15.4 together establish the equivalence of the following three statements: (i) $\lim_{n\to\infty} \boldsymbol{A}^n = \boldsymbol{O}$, (ii) $\kappa(\boldsymbol{A}) < 1$, and (iii) the Neumann series $\sum_{k=0}^{\infty} \boldsymbol{A}^k$ converges. When any of these equivalent conditions holds, $\boldsymbol{I} - \boldsymbol{A}$ is nonsingular and $(\boldsymbol{I} - \boldsymbol{A})^{-1}$ is given by the Neumann series (Theorem 15.2).

Matrices with all entries as nonnegative real numbers arise in econometrics and statistics. The following result is useful.

Theorem 15.5 *Let $\boldsymbol{A} = \{a_{ij}\}$ be $n \times n$ with each $a_{ij} \geq 0$. Then, $\alpha\boldsymbol{I}_n - \boldsymbol{A}$ is nonsingular if and only if $\kappa(\boldsymbol{A}) < \alpha$ and every element of $(\alpha\boldsymbol{I}_n - \boldsymbol{A})^{-1}$ is nonnegative.*

Proof. If $\kappa(\boldsymbol{A}) < \alpha$, then clearly $\alpha > 0$ and $\kappa((1/\alpha)\boldsymbol{A}) < 1$. By Theorem 15.4, the Neumann series of $(1/\alpha)\boldsymbol{A}$ converges and, by virtue of Theorem 15.2, $(\alpha\boldsymbol{I} - \boldsymbol{A})$, which we can write as $\alpha(\boldsymbol{I} - (1/\alpha)\boldsymbol{A})$, is easily seen to be nonsingular. Furthermore,

$$(\alpha\boldsymbol{I} - \boldsymbol{A})^{-1} = \frac{1}{\alpha}\left(\boldsymbol{I} - \frac{1}{\alpha}\boldsymbol{A}\right)^{-1} = \frac{1}{\alpha}\sum_{k=0}^{\infty}\frac{\boldsymbol{A}^k}{\alpha^k}$$

and each element in $(\alpha\boldsymbol{I} - \boldsymbol{A})^{-1}$ is nonnegative because the elements in $(\boldsymbol{A}/\alpha)^k$ are nonnegative. This proves the "if" part.

To prove the other direction, suppose that $\alpha\boldsymbol{I}_n - \boldsymbol{A}$ is nonsingular and each of the elements of $(\alpha\boldsymbol{I}_n - \boldsymbol{A})^{-1}$ is nonnegative. For convenience, let $|\boldsymbol{x}|$ denote the vector whose elements are the absolute values of the corresponding entries in $\boldsymbol{x}$. That is, the modulus operator is applied to each entry in $\boldsymbol{x}$. Using the triangle inequality for real numbers, $|u + v| \leq |u| + |v|$, we obtain $|\boldsymbol{A}\boldsymbol{x}| \leq \boldsymbol{A}|\boldsymbol{x}|$, where "$\leq$" means element-wise inequality. Applying this to an eigen-pair $(\lambda, \boldsymbol{v})$ of $\boldsymbol{A}$, we obtain

$$\begin{aligned}
|\lambda||\boldsymbol{x}| = |\lambda\boldsymbol{x}| = |\boldsymbol{A}\boldsymbol{x}| \leq \boldsymbol{A}|\boldsymbol{x}| &\Longrightarrow (\alpha\boldsymbol{I} - \boldsymbol{A})|\boldsymbol{x}| \leq (\alpha - |\lambda|)|\boldsymbol{x}| \\
&\Longrightarrow \boldsymbol{0} \leq |\boldsymbol{x}| \leq (\alpha - |\lambda|)(\alpha\boldsymbol{I} - \boldsymbol{A})^{-1}|\boldsymbol{x}| \Longrightarrow \alpha - |\lambda| \geq 0\,,
\end{aligned}$$

where the last step follows because each of the elements of $(\alpha\boldsymbol{I}_n - \boldsymbol{A})^{-1}$ is nonnegative. Therefore, $|\lambda| \leq \alpha$ for any eigenvalue λ of $\boldsymbol{A}$. Hence, $\kappa(\boldsymbol{A}) < \alpha$. □

The spectral radius of a matrix governs the convergence of matrix powers and of linear iterative systems. However, as we have seen in Section 11.8, evaluating eigenvalues of matrices is computationally demanding and involves iterative algorithms. Therefore, computing the spectral radius to understand the behavior of linear iterative systems will be onerous. An alternative approach uses the concept of ***matrix norms***, which formalizes notions of the length of a matrix.

15.4 Matrix norms and the Gerschgorin circles

Let us restrict ourselves to matrices with real entries, although most of these results can be easily extended to complex matrices. How can we extend vector norms to matrix norms? What properties should a matrix norm possess? One place to start may be with a plausible definition of inner products for matrices. We would like this inner product to have properties analogous to the Euclidean inner product for vectors (recall Lemma 7.1). Then, the norm induced by this inner product would be a matrix norm.

Definition 15.3 Inner product for matrices. *Let $\Re^{m\times n}$ be the collection of all $m \times n$ matrices with real entries. An inner product on $\Re^{m\times n}$ is a function $\langle\cdot,\cdot\rangle$: $\Re^{m\times n} \times \Re^{m\times n} \to \Re$ that maps a pair of matrices in $\Re^{m\times n}$ to the real line $\Re$ and satisfies the following axioms:*

(i) Symmetry: $\langle \boldsymbol{A}, \boldsymbol{B}\rangle = \langle \boldsymbol{B}, \boldsymbol{A}\rangle$ for every $\boldsymbol{A}, \boldsymbol{B} \in \Re^{m\times n}$.

(ii) Bilinearity (linearity in both arguments):

$$\langle \alpha\boldsymbol{A}, \beta\boldsymbol{B} + \gamma\boldsymbol{C}\rangle = \alpha\beta\langle \boldsymbol{A}, \boldsymbol{B}\rangle + \alpha\gamma\langle \boldsymbol{A}, \boldsymbol{C}\rangle$$

for every $\boldsymbol{A}, \boldsymbol{B}, \boldsymbol{C} \in \Re^{m\times n}$. The symmetry in (i) and the linearity above implies linearity in the first argument: $\langle \alpha\boldsymbol{A} + \beta\boldsymbol{B}, \gamma\boldsymbol{C}\rangle = \alpha\gamma\langle \boldsymbol{A}, \boldsymbol{C}\rangle + \beta\gamma\langle \boldsymbol{B}, \boldsymbol{C}\rangle$.

(iii) Positivity: $\langle \boldsymbol{A}, \boldsymbol{A}\rangle > 0$ for every $\boldsymbol{A} \neq \boldsymbol{O} \in \Re^{m\times n}$ and $\langle \boldsymbol{O}, \boldsymbol{O}\rangle = 0$.

Any inner product for matrices, as defined in Definition 15.3, will induce a norm just as Euclidean inner products do. This norm can serve as a ***matrix norm***, i.e., $\|\boldsymbol{A}\| = \langle \boldsymbol{A}, \boldsymbol{A}\rangle$ for any $\boldsymbol{A}$ in $\Re^{m\times n}$. Analogous to Euclidean spaces, any inner product on $\Re^{m\times n}$ will satisfy the Cauchy-Schwarz inequality:

$$\langle \boldsymbol{A}, \boldsymbol{B}\rangle \leq \|\boldsymbol{A}\|\|\boldsymbol{B}\| \ . \tag{15.6}$$

This is proved analogous to the more familiar Cauchy-Schwarz inequality for Euclidean spaces (Theorem 7.1) but we sketch it here for the sake of completeness: For any real number α, we have

$$\begin{aligned} 0 &\leq \langle \boldsymbol{A} - \alpha\boldsymbol{B}, \boldsymbol{A} - \alpha\boldsymbol{B}\rangle = \langle \boldsymbol{A}, \boldsymbol{A}\rangle - 2\alpha\langle \boldsymbol{A}, \boldsymbol{B}\rangle + \alpha^2\langle \boldsymbol{B}, \boldsymbol{B}\rangle \\ &= \|\boldsymbol{A}\|^2 - 2\alpha\langle \boldsymbol{A}, \boldsymbol{B}\rangle + \alpha^2\|\boldsymbol{B}\|^2. \end{aligned}$$

If $\boldsymbol{B} = \boldsymbol{O}$, then we have equality in (15.6). Let $\boldsymbol{B} \neq \boldsymbol{0}$ and set $\alpha = \langle \boldsymbol{A}, \boldsymbol{B}\rangle/\|\boldsymbol{B}\|^2$ as a particular choice. Then the above expression becomes

$$0 \leq \|\boldsymbol{A}\|^2 - 2\frac{\langle \boldsymbol{B}, \boldsymbol{A}\rangle^2}{\|\boldsymbol{B}\|^2} + \frac{\langle \boldsymbol{A}, \boldsymbol{B}\rangle^2}{\|\boldsymbol{B}\|^2} = \|\boldsymbol{B}\|^2 - \frac{\langle \boldsymbol{A}, \boldsymbol{B}\rangle^2}{\|\boldsymbol{B}\|^2},$$

which proves (15.6). When $\boldsymbol{B} = \boldsymbol{O}$ or $\boldsymbol{A} = \boldsymbol{O}$ we have equality. Otherwise, equality follows if and only if $\langle \boldsymbol{A} - \alpha\boldsymbol{B}, \boldsymbol{A} - \alpha\boldsymbol{B}\rangle = 0$, which implies $\boldsymbol{A} = \alpha\boldsymbol{B}$.

One particular instance of an inner product is obtained using the trace function:

$$\langle \boldsymbol{A}, \boldsymbol{B}\rangle = \text{tr}(\boldsymbol{B}'\boldsymbol{A}) = \text{tr}(\boldsymbol{A}'\boldsymbol{B}) = \text{tr}(\boldsymbol{B}\boldsymbol{A}') = \text{tr}(\boldsymbol{A}\boldsymbol{B}') \ \text{ for any } \ \boldsymbol{A}, \boldsymbol{B} \in \Re^{m\times n} \ . \tag{15.7}$$

It is easy to verify that (15.7) satisfies all the conditions in Definition 15.3 and, therefore, also satisfies the Cauchy-Schwarz inequality

$$\text{tr}(A'B) = \langle A, B\rangle \le \|A\|\|B\| = \sqrt{\text{tr}(A'A)}\sqrt{\text{tr}(B'B)}\,. \tag{15.8}$$

The norm induced by (15.7) is called the *Frobenius norm.*

Definition 15.4 *The* **Frobenius norm** *of an $m \times n$ matrix $A = \{a_{ij}\}$ is defined by*

$$\|A\|_F = \sqrt{tr(A'A)} = \sqrt{\sum_{i=1}^{m} \|a_{i*}\|_2^2} = \sqrt{\sum_{j=1}^{n} \|a_{*j}\|_2^2} = \sqrt{\sum_{i=1}^{m}\sum_{j=1}^{n} |a_{ij}|^2}\,,$$

where a_{i*} and a_{*j} denote, respectively, the i-th row and j-th column of A.

Another way to look at the Frobenius norm is as follows. The linear space $\Re^{m\times n}$ is essentially the same as $\Re^{mn}$ because the entries in A can be stacked to form an $mn \times 1$ vector in $\Re^{mn}$. For example, one could stack up the columns of A one above the other. Likewise, any vector in $\Re^{mn}$ can be arranged in an $m\times n$ matrix. One could then employ any vector norm in $\Re^{mn}$ to define a matrix norm. Using the Euclidean norm this way yields the Frobenius norm.

Example 15.4 Consider the matrix

$$A = \begin{bmatrix} 3 & 1 \\ 2 & -4 \end{bmatrix}.$$

Extract its entries and place them in a 4×1 vector. The Euclidean norm of this vector produces the Frobenius norm

$$\|A\|_F = \left[3^2 + 1^2 + 2^2 + (-4)^2\right] = 30\,. \quad \blacksquare$$

What are some of the properties of the Frobenius norm that seem to be consistent with what we would expect a norm to be? In fact, the Frobenius norm meets the three axioms analogous to those in Definition 15.1. For example, the positivity condition is satisfied because (recall (1.18))

$$\|A\|_F^2 = \text{tr}(A'A) = \text{tr}(AA') > 0 \;\text{ for every }\; A \neq O \in \Re^{m\times n}$$

and $\|A\|_F = O$ if and only if $A = O$. Homogeneity is also immediate because

$$\|\alpha A\|_F = \sqrt{\text{tr}((\alpha A)'\alpha A)} = \sqrt{\alpha^2\text{tr}(A'A)} = |\alpha|\sqrt{\text{tr}(A'A)} = |\alpha|\|A\|_F\,.$$

It also satisfies the triangle inequality. Writing $\langle A, B\rangle_F = \text{tr}(A'B)$, we find

$$\begin{aligned}\|A+B\|_F^2 &= \langle A+B, A+B\rangle_F = \|A\|_F^2 + \|B\|_F^2 + 2\langle A, B\rangle_F \\ &\le \|A\|_F^2 + \|B\|_F^2 + 2\|A\|_F\|B\|_F = (\|A\|_F + \|B\|_F)^2\,,\end{aligned}$$

where the "$\le$" follows as a result of the Cauchy-Schwarz inequality in (15.6).

One operation that distinguishes the matrix space $\Re^{m\times n}$ from the vector space $\Re^{mn\times 1}$ is matrix multiplication. The Cauchy-Schwarz inequality ensures that

$$\|Ax\|_2^2 = \sum_{i=1}^{n} |a_{i*}'x|^2 \leq \sum_{i=1}^{m} \|a_{i*}\|^2\|x\|^2 = \left(\sum_{i=1}^{m} \|a_{i*}\|^2\right)\|x\|^2 = \|A\|_F^2\|x\|_2^2 ,$$

where a_{i*}' is the i-th row of A. Therefore, $\|Ax\|_2 \leq \|A\|_F\|x\|_2$.

Turning to the Frobenius norm of the matrix $C = AB$, where A and B are $m \times n$ and $n \times p$, respectively. Then, $c_{*j} = Ab_{*j}$, where c_{*j} and b_{*j} are the j-th columns of C and B, respectively. We obtain

$$\begin{aligned}\|AB\|_F^2 = \|C\|_F^2 &= \sum_{j=1}^{p} \|c_{*j}\|^2 = \sum_{j=1}^{p} \|Ab_{*j}\|^2 \\ &\leq \sum_{j=1}^{p} \|A\|_F^2\|b_{*j}\|^2 = \|A\|_F^2 \sum_{j=1}^{p} \|b_{*j}\|^2 = \|A\|_F^2\|B\|_F^2 ,\end{aligned} \tag{15.9}$$

which implies that $\|AB\|_F \leq \|A\|_F\|B\|_F$. This is often called the *submultiplicative* property.

The properties of the Frobenius norm can be regarded as typical of matrix norms arising from inner products on matrix spaces (see Definition 15.3). This means that any reasonable definition of matrix norms must include the submultiplicative property in (15.9). We now provide a formal definition of matrix norms.

Definition 15.5 General matrix norms. *A* **matrix norm** *is a function $\|\cdot\|$ that maps any matrix A in $\Re^{m\times n}$ or $\mathbb{C}^{m\times n}$ into a real number $\|A\|$ that satisfies the following axioms:*

(i) Positivity: $\|A\| \geq 0$ and $\|A\| = 0$ if and only if $A = O$.

(ii) Homogeneity: $\|\alpha A\| = |\alpha|\|A\|$ for every real or complex number α.

(iii) Triangle inequality: $\|A + B\| \leq \|A\| + \|B\|$.

(iv) Submultiplicativity: $\|AB\| \leq \|A\|\|B\|$ whenever AB is well-defined.

Every vector norm on $\Re^n$ on $\mathbb{C}^n$ *induces* a matrix norm. We prove this for $\Re^n$ only.

Theorem 15.6 *Let $\|\cdot\|$ be any vector norm on $\Re^n$ (see Definition 15.1) and let*

$$\|A\| = \max_{\{x\in\Re^n : \|x\|=1\}} \|Ax\| \text{ for every } A \in \Re^{m\times n} .$$

Then, $\|A\|$ is a legitimate matrix norm.

Proof. The sphere $S = \{x \in \Re^n : \|x\| = 1\}$ is a closed and bounded subset in $\Re^n$. This ensures that $\|A\| < \infty$. We now verify that $\|A\|$ satisfies the axioms in Definition 15.5.

Positivity: Clearly $\|A\| \geq 0$ and $\|O\| = 0$. Suppose that $\|A\| = 0$. Therefore,

$\|\boldsymbol{Au}\| = 0$ for every vector $\boldsymbol{u} \in S$ and, hence, $\boldsymbol{Au} = \boldsymbol{0}$ for every $\boldsymbol{u} \in S$. Consider any nonzero vector $\boldsymbol{v} \in \Re^n$ and let $\boldsymbol{u} = \boldsymbol{v}/\|\boldsymbol{v}\|$. Since $\boldsymbol{u} \in S$, we can conclude that

$$\boldsymbol{Av} = \boldsymbol{A}(\|\boldsymbol{v}\|\boldsymbol{u}) = \|\boldsymbol{v}\|\boldsymbol{Au} = \boldsymbol{0} \text{ for every nonzero } \boldsymbol{v} \in \Re^n .$$

This means that $\boldsymbol{A} = \boldsymbol{O}$ is the zero matrix.

Homogeneity: If α is any real number, then

$$\|\alpha\boldsymbol{A}\| = \max_{\boldsymbol{x}\in S} \|\alpha\boldsymbol{Au}\| = \max_{\boldsymbol{x}\in S} |\alpha|\|\boldsymbol{Au}\| = |\alpha| \max_{\boldsymbol{x}\in S} \|\boldsymbol{Au}\| = |\alpha|\|\boldsymbol{A}\| .$$

Triangle inequality: If $\boldsymbol{A}$ and $\boldsymbol{B}$ are two $m \times n$ matrices, then

$$\begin{aligned}\|\boldsymbol{A}+\boldsymbol{B}\| &= \max_{\boldsymbol{x}\in S} \|\boldsymbol{Ax}+\boldsymbol{Bx}\| \leq \max_{\boldsymbol{x}\in S} \{\|\boldsymbol{Ax}\| + \|\boldsymbol{Bx}\|\} \\ &\leq \max_{\boldsymbol{x}\in S} \|\boldsymbol{Ax}\| + \max_{\boldsymbol{x}\in S} \|\boldsymbol{Bx}\| = \|\boldsymbol{A}\| + \|\boldsymbol{B}\| ,\end{aligned}$$

where we have used the triangle inequality of the vector norm in conjunction with the fact that the maximum of the sum of two positive quantities cannot exceed the sum of their individual maxima.

Submultiplicativity: Let $\boldsymbol{v}$ be any vector in $\Re^n$ and let $\boldsymbol{u} = \boldsymbol{v}/\|\boldsymbol{v}\|$ be the unit vector obtained by normalizing $\boldsymbol{v}$. Since $\boldsymbol{u} \in S$, we note that $\|\boldsymbol{Au}\| \leq \|\boldsymbol{A}\|$ and we can conclude that

$$\|\boldsymbol{Av}\| = \|\boldsymbol{A}(\|\boldsymbol{v}\|\boldsymbol{u})\| = \|\boldsymbol{v}\|\|\boldsymbol{Au}\| \leq \|\boldsymbol{v}\|\|\boldsymbol{A}\| = \|\boldsymbol{A}\|\|\boldsymbol{v}\| . \tag{15.10}$$

We now apply (15.10) to conclude that

$$\begin{aligned}\|\boldsymbol{AB}\| &= \max_{\boldsymbol{x}\in S} \|\boldsymbol{ABx}\| = \max_{\boldsymbol{x}\in S} \|\boldsymbol{A}(\boldsymbol{Bx})\| \\ &\leq \max_{\boldsymbol{x}\in S} \{\|\boldsymbol{A}\|\|\boldsymbol{Bx}\|\} = \|\boldsymbol{A}\| \max_{\boldsymbol{x}\in S} \|\boldsymbol{Bx}\| = \|\boldsymbol{A}\|\|\boldsymbol{B}\| .\end{aligned}$$

This establishes the submultiplicative property and confirms the axioms in Definition 15.5. □

Remark: The matrix norm $\|\boldsymbol{A}\|$ that has been induced by a vector norm is the maximum extent to which $\boldsymbol{A}$ stretches any vector on the unit sphere. The property $\|\boldsymbol{Av}\| \leq \|\boldsymbol{A}\|\|\boldsymbol{v}\|$ derived in (15.10) is often referred to as the *compatibility* of a matrix norm with a vector norm.

Example 15.5 Matrix norm induced by the Euclidean vector norm. The Frobenius norm is obtained by stacking the elements of a matrix into a vector and then applying the Euclidean norm to it. Is the Frobenius norm induced by the Euclidean norm? Contrary to what one might intuit, the answer is no. In fact,

$$\|\boldsymbol{A}\|_2 = \max_{\|\boldsymbol{x}\|_2=1} \|\boldsymbol{Ax}\|_2 = \sqrt{\lambda_{\max}} , \tag{15.11}$$

where $\|\boldsymbol{A}\|_2$ denotes the matrix norm of $\boldsymbol{A}$ induced by the Euclidean vector norm, and λ_{max} is the largest eigenvalue of $\boldsymbol{A}'\boldsymbol{A}$. Equivalently, $\sqrt{\lambda_{max}}$ is the largest singular value of $\boldsymbol{A}$. The proof of (15.11) follows from Theorem 13.29.

Suppose we wish to find the induced norm $||\boldsymbol{A}||_2$ for the non-singular matrix

$$\boldsymbol{A} = \begin{bmatrix} 2 & 0 \\ -1 & 3 \end{bmatrix}.$$

The two eigenvalues are obtained as $\lambda_1 = 7+\sqrt{13}$ and $\lambda_2 = 7-\sqrt{13}$. Consequently,

$$||\boldsymbol{A}||_2 = \sqrt{\lambda_{\max}} = \sqrt{7+\sqrt{13}}\,. \quad \blacksquare$$

When $\boldsymbol{A}$ is nonsingular, matrix norms induced by vector norms satisfy the following relationship:

$$\min_{\boldsymbol{x}:||\boldsymbol{x}||=1} ||\boldsymbol{A}\boldsymbol{x}|| = \frac{1}{||\boldsymbol{A}^{-1}||} \quad \text{and} \quad ||\boldsymbol{A}^{-1}||_2 = \frac{1}{\min_{\boldsymbol{x}:||\boldsymbol{x}||=1} ||\boldsymbol{A}\boldsymbol{x}||_2} = \frac{1}{\sqrt{\lambda_{\min}}}, \tag{15.12}$$

where $||\boldsymbol{A}^{-1}||_2$ is the matrix norm of $\boldsymbol{A}^{-1}$ induced by the Euclidean vector norm and $\lambda_{\min}$ is the minimum eigenvalue. The proof is straightforward and left to the reader as an exercise. Figure 15.1 presents a visual depiction of $||\boldsymbol{A}||_2$ and $||\boldsymbol{A}^{-1}||_2$. The sphere on the left is mapped by $\boldsymbol{A}$ to the ellipsoid on the right. The value of $||\boldsymbol{A}||$ is the farthest distance from the center to any point on the surface of the ellipsoid, while $||\boldsymbol{A}^{-1}||_2$ is the shortest distance from the center to a point on the surface of the ellipsoid.

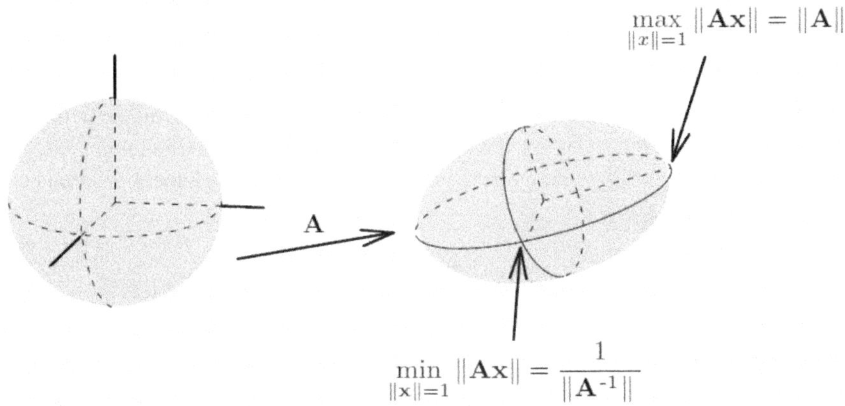

Figure 15.1 *The induced matrix 2-norm in* $\Re^3$.

We now describe the matrix norms induced by the vector norms $||\cdot||_1$ and $||\cdot||_\infty$.

Theorem 15.7 *If $\boldsymbol{A}$ is an $m \times n$ matrix, then*

$$\|\boldsymbol{A}\|_1 = \max_{\|\boldsymbol{x}\|_1=1} \|\boldsymbol{A}\boldsymbol{x}\|_1 = \max_{j=1,2,\ldots,n} \sum_{i=1}^{m} |a_{ij}|$$

$$\text{and } \|\boldsymbol{A}\|_\infty = \max_{\|\boldsymbol{x}\|_\infty=1} \|\boldsymbol{A}\boldsymbol{x}\|_\infty = \max_{i=1,2,\ldots,m} \sum_{j=1}^{n} |a_{ij}| .$$

Proof. We prove the result only for real matrices. If $\boldsymbol{a}'_{i*}$ is the i-th row of $\boldsymbol{A}$, then the triangle inequality for real numbers ensures that for every $\boldsymbol{x}$ with $\|\boldsymbol{x}\| = 1$,

$$\begin{aligned}
\|\boldsymbol{A}\boldsymbol{x}\|_1 &= \sum_{i=1}^{m} |\boldsymbol{a}'_{i*}\boldsymbol{x}| = \sum_{i=1}^{m} \left| \sum_{j=1}^{n} a_{ij}x_j \right| \\
&\leq \sum_{i=1}^{m}\sum_{j=1}^{n} |a_{ij}||x_j| = \sum_{j=1}^{n} \left(|x_j| \sum_{i=1}^{m} |a_{ij}| \right) \\
&\leq \left(\sum_{j=1}^{n} |x_j| \right) \left(\max_{j=1,2,\ldots,n} \sum_{i=1}^{m} |a_{ij}| \right) = \max_{j=1,2,\ldots,n} \sum_{i=1}^{m} |a_{ij}| .
\end{aligned}$$

Furthermore, suppose that $\boldsymbol{a}_{*k}$ is the column with largest absolute sum and let $\boldsymbol{x} = \boldsymbol{e}_k$. Note that $\|\boldsymbol{e}_k\|_1 = 1$ and $\|\boldsymbol{A}\boldsymbol{e}_k\|_1 = \|\boldsymbol{A}_{*k}\|_1 = \max_{j=1,2,\ldots,n} \sum_{i=1}^{m} |a_{ij}|$. This proves the first result.

Next, consider the ∞-norm. For every $\boldsymbol{x}$ with $\|\boldsymbol{x}\|_\infty = 1$ we obtain

$$\|\boldsymbol{A}\boldsymbol{x}\|_\infty = \max_{i=1,2,\ldots,m} \left| \sum_{j=1}^{n} a_{ij}x_j \right| \leq \max_{i=1,2,\ldots,m} \sum_{j=1}^{n} |a_{ij}||x_j| \leq \max_{i=1,2,\ldots,m} \sum_{j=1}^{n} |a_{ij}| . \tag{15.13}$$

This proves that $\|\boldsymbol{A}\|_\infty \leq \max_{i=1,2,\ldots,m} \sum_{j=1}^{n} |a_{ij}|$. Now suppose that the k-th row of $\boldsymbol{A}$ has the largest absolute row sum of all the rows. Let $\boldsymbol{x}$ be the vector with entries $x_j = 1$ if $a_{kj} \geq 0$ and $x_j = -1$ if $a_{kj} < 0$. Then, it is easily verified that $\|\boldsymbol{x}\|_\infty = 1$. Also, since $a_{kj}x_j = |a_{kj}|$ for every $j = 1, 2, \ldots, n$, the k-th element of $\boldsymbol{A}\boldsymbol{x}$ is equal to $\sum_{j=1}^{n} |a_{kj}|$. Therefore,

$$\|\boldsymbol{A}\|_\infty \geq \|\boldsymbol{A}\boldsymbol{x}\|_\infty \geq \sum_{j=1}^{n} |a_{kj}| = \max_{i=1,2,\ldots,m} \sum_{j=1}^{n} |a_{ij}|.$$

This, together with (15.13), implies that $\|\boldsymbol{A}\boldsymbol{x}\|_\infty = \max_{i=1,2,\ldots,m} \sum_{j=1}^{n} |a_{ij}|$. □

Recall that $\lim_{k\to\infty} \boldsymbol{A}^k = \boldsymbol{O}$ if and only if the spectral radius of $\boldsymbol{A}$ is less than one (Theorem 15.3). Suppose that λ is an eigenvalue of an $n \times n$ matrix $\boldsymbol{A}$ and $\boldsymbol{u}$ is an associated eigenvector associated with λ. Then,

$$\lambda\boldsymbol{u} = \boldsymbol{A}\boldsymbol{u} \Longrightarrow |\lambda|\|\boldsymbol{u}\| = \|\boldsymbol{A}\boldsymbol{u}\| \leq \|\boldsymbol{A}\|\|\boldsymbol{u}\| \Longrightarrow |\lambda| \leq \|\boldsymbol{A}\| . \tag{15.14}$$

Since $\kappa(\boldsymbol{A})$ is the maximum of the absolute values of the eigenvalues of $\boldsymbol{A}$, $\kappa(\boldsymbol{A}) \leq \|\boldsymbol{A}\|$. Therefore, if $\|\boldsymbol{A}\| < 1$, then the spectral radius is also strictly less than 1 and Theorem 15.3 ensures that $\lim_{k\to\infty} \boldsymbol{A}^k = \boldsymbol{O}$. Thus, $\|\boldsymbol{A}\| < 1$ is a sufficient condition for the matrix powers to converge to the zero matrix.

The converse, however, is not true. Consider, for example, the matrix $\boldsymbol{A} = \begin{bmatrix} 0 & 1 \\ 0 & 0 \end{bmatrix}$. Clearly $\boldsymbol{A}^k = \boldsymbol{O}$ for every $k \geq 2$. However, $\|\boldsymbol{A}\| = 1$ for most of the standard matrix norms. This reveals the somewhat unfortunate fact that matrix norms are not a conclusive tool to ascertain convergence. Indeed there exist matrices such that $\kappa(\boldsymbol{A}) < 1$ but $\|\boldsymbol{A}\| \geq 1$. The following, more precise, relationship between the spectral radius and the matrix norm shows that the limit of a specific function of $\|\boldsymbol{A}\|$ governs the convergence of powers of $\boldsymbol{A}$.

Theorem 15.8 *If $\boldsymbol{A}$ is a square matrix and $\|\boldsymbol{A}\|$ is any matrix norm, then*

$$\kappa(\boldsymbol{A}) = \lim_{k\to\infty} \|\boldsymbol{A}^k\|^{1/k} .$$

Proof. It is immediately seen that if λ is an eigenvalue of $\boldsymbol{A}$, then λ^k is an eigenvalue of $\boldsymbol{A}^k$. This implies that $(\kappa(\boldsymbol{A}))^k = \kappa(\boldsymbol{A}^k)$. Therefore,

$$(\kappa(\boldsymbol{A}))^k = \kappa(\boldsymbol{A}^k) \leq \|\boldsymbol{A}^k\| \Longrightarrow \kappa(\boldsymbol{A}) \leq \|\boldsymbol{A}^k\|^{1/k} ,$$

where the "$\leq$" follows from (15.14). Let $\epsilon > 0$ and define $\boldsymbol{B} = \dfrac{\boldsymbol{A}}{\kappa(\boldsymbol{A}) + \epsilon}$. Clearly, the spectral radius of $\boldsymbol{B}$ is less than one, i.e., $\kappa(\boldsymbol{B}) < 1$. Theorem 15.3 ensures that $\lim_{k\to\infty} \boldsymbol{B}^k = \boldsymbol{O}$ and, hence,

$$0 = \lim_{k\to\infty} \|\boldsymbol{B}^k\| = \lim_{k\to\infty} \frac{\|\boldsymbol{A}^k\|}{(\kappa(\boldsymbol{A}) + \epsilon)^k} .$$

Therefore, we can find a positive integer K_ϵ such that $\dfrac{\|\boldsymbol{A}^k\|}{(\kappa(\boldsymbol{A}) + \epsilon)^k} < 1$ whenever $k > K_\epsilon$ and for any such k we can conclude that

$$\kappa(\boldsymbol{A}) \leq \|\boldsymbol{A}^k\|^{1/k} < \frac{\|\boldsymbol{A}^k\|^{1/k}}{\kappa(\boldsymbol{A}) + \epsilon} \Longrightarrow \kappa(\boldsymbol{A}) \leq \|\boldsymbol{A}^k\|^{1/k} \leq \kappa(\boldsymbol{A}) + \epsilon .$$

Thus, for any arbitrary positive number $\epsilon > 0$, $\|\boldsymbol{A}^k\|^{1/k}$ is sandwiched between $\kappa(\boldsymbol{A})$ and $\kappa(\boldsymbol{A}) + \epsilon$ for sufficiently large k. Therefore, $\kappa(\boldsymbol{A}) = \lim_{k\to\infty} \|\boldsymbol{A}^k\|^{1/k}$. □

Gerschgorin circles

In many analytical situations and applications of matrix analysis, one requires only a bound on the eigenvalues of $\boldsymbol{A}$. We do not need to compute the precise eigenvalues. Instead, we seek an upper bound on the spectral radius $\kappa(\boldsymbol{A})$. Since $\kappa(\boldsymbol{A}) \leq \|\boldsymbol{A}\|$, every matrix norm supplies an upper bound on the eigenvalues. However, this may be a somewhat crude approximation. A better approximation is provided by a remarkable result attributed to the Soviet mathematician Semyon A. Gerschgorin, which says that the eigenvalues of a matrix are trapped within certain discs or circles centered around the diagonal entries of the matrix.

Definition 15.6 Gerschgorin circles. *Let $\boldsymbol{A} = \{a_{ij}\}$ be an $n \times n$ matrix with real or complex entries. The Gerschgorin circles associated with $\boldsymbol{A}$ are*

$$\mathcal{D}(a_{ii}, r_i) = \{z : |z - a_{ii}| < r_i\} \;, \quad \text{where } r_i = \sum_{\substack{j=1 \\ j \neq i}}^{n} |a_{ij}| \text{ for } i = 1, 2, \ldots, n \,.$$

Each Gerschgorin circle of $\boldsymbol{A}$ has a diagonal entry as center and the sum of the absolute values of the non-diagonal entries in that row as the radius. The **Gerschgorin domain** *$\mathcal{D}_{\boldsymbol{A}}$ is defined as the union of the n Gerschgorin circles $\cup_{i=1}^{n} \mathcal{D}(a_{ii}, r_i)$.*

Theorem 15.9 Gerschgorin Circle Theorem. *Every eigenvalue of an $n \times n$ matrix $\boldsymbol{A}$ lies in at least one of the Gerschgorin discs of $\boldsymbol{A}$.*

Proof. Let λ be an eigenvalue of $\boldsymbol{A}$ and $\boldsymbol{x}$ be an eigenvector associated with λ. Also assume that $\boldsymbol{x}$ has been normalized using the ∞-norm, i.e., $\|\boldsymbol{x}\|_\infty = 1$. This means that $|x_k| = 1$ and $|x_j| \leq 1$ for $j \neq k$. Since $\boldsymbol{A}\boldsymbol{x} = \lambda \boldsymbol{x}$, equating the k-th rows yields

$$\sum_{j=1}^{n} a_{kj} x_j = \lambda x_k \Longrightarrow \sum_{\substack{j=1 \\ j \neq k}} a_{kj} x_j = (\lambda x_k - a_{kk} x_k) = (\lambda - a_{kk}) x_k \,.$$

Since, $|(\lambda - a_{kk}) x_k| = |\lambda - a_{kk}||x_k| = |\lambda - a_{kk}|$, we can write

$$|\lambda - a_{kk}| = |(\lambda - a_{kk}) x_k| = \left| \sum_{\substack{j=1 \\ j \neq k}} a_{kj} x_j \right| \leq \sum_{\substack{j=1 \\ j \neq k}} |a_{kj}||x_j| \leq \sum_{\substack{j=1 \\ j \neq k}} |a_{kj}| = r_k \,.$$

Thus, λ lies in the Gerschgorin circle $\mathcal{D}(a_{kk}, r_k)$, where $r_k = \sum_{j \neq k}^{n} |a_{kj}|$ □

Corollary 15.1 *Every eigenvalue of an $n \times n$ matrix $\boldsymbol{A}$ lies in at least one of the Gerschgorin circles $D(a_{jj}, c_j)$, where the radius $c_j = \sum_{i \neq j}^{n} |a_{ij}|$ is the sum of the absolute values of the non-diagonal entries in the j-th column.*

Proof. The eigenvalues of $\boldsymbol{A}$ and $\boldsymbol{A}'$ are the same. Apply Theorem 15.9 to $\boldsymbol{A}'$. □

In fact, since the eigenvalues of $\boldsymbol{A}$ and $\boldsymbol{A}'$ are the same, the eigenvalues of $\boldsymbol{A}$ will lie in the intersection of the Gerschgorin domains $\mathcal{D}_{\boldsymbol{A}} \cap \mathcal{D}_{\boldsymbol{A}'}$. This refines the search for eigenvalues. Other leads can be obtained if the union of p Gerschgorin circles does not intersect any of the other $n - p$ circles. Then, there will be exactly p eigenvalues (counting multiplicities) in the union of the p circles. This result is a bit more difficult to establish and we refer the reader to Meyer (2001) or Horn and Johnson (2013).

The Gerschgorin Circle Theorem can be useful in ascertaining whether a matrix $\boldsymbol{A}$ is nonsingular by checking whether any of the Gerschgorin circles includes 0. If none of them do, then the matrix must be nonsingular because 0 cannot be an eigenvalue of $\boldsymbol{A}$. We illustrate with *strictly diagonally dominant* matrices.

Definition 15.7 Diagonally dominant matrices. *An $n \times n$ matrix $\boldsymbol{A} = \{a_{ij}\}$ is said to be* **diagonally dominant** *if the absolute value of each diagonal entry is not less than the sum of the absolute values of the non-diagonal entries in that row, i.e., if*

$$|a_{ii}| \geq \sum_{\substack{j=1 \\ j \neq i}}^{n} |a_{ij}| \ , \ \textit{for each} \ \ i = 1, 2, \ldots, n \ .$$

The matrix is called **strictly diagonally dominant** *if the above inequality is strict for each diagonal entry.*

Example 15.6 The matrix

$$\boldsymbol{A} = \begin{bmatrix} -4 & 1 & 2 \\ -2 & 3 & 0 \\ -2 & -2 & 5 \end{bmatrix}$$

is strictly diagonally dominant because $|-4| > |1| + |2|$, $|3| > |-2| + |0|$ and $|5| > |-2| + |-2|$. Another example of a strictly diagonally dominant matrix is the identity matrix. ■

We now see how the Gerschgorin Circle Theorem can be used to conclude that a diagonally dominant matrix is nonsingular.

Theorem 15.10 *A strictly diagonally dominant matrix is nonsingular.*

Proof. Let $\boldsymbol{A} = \{a_{ij}\}$ be an $n \times n$ strictly diagonally dominant matrix. Since $|0 - a_{ii}| > r_i$ for each $i = 1, 2, \ldots, n$, where $r_i = \sum_{j \neq i} |a_{ij}|$ is the radius of the i-th Gerschgorin circle, it follows that 0 is not contained in any of the Gerschgorin circles. Hence, 0 cannot be an eigenvalue and so $\boldsymbol{A}$ is nonsingular. □

The converse of this result is of course not true and there are numerous nonsingular matrices that are not diagonally dominant.

15.5 The singular value decomposition—revisited

Let us revisit the singular value decomposition or SVD (recall Section 12.1). We provide an alternate derivation of the SVD that does not depend upon eigenvalues and also show how it provides a useful low-rank approximation.

Theorem 15.11 The singular value decomposition (SVD). *Let $\boldsymbol{A} \in \Re^{m\times n}$, where $m \geq n$. There exist orthogonal matrices $\boldsymbol{U} \in \Re^{m\times m}$ and $\boldsymbol{V} \in \Re^{n\times n}$ such that*

$$\boldsymbol{A} = \boldsymbol{U}\begin{bmatrix} \boldsymbol{\Sigma} \\ \boldsymbol{O} \end{bmatrix}\boldsymbol{V}' , \text{ where } \boldsymbol{\Sigma} = \begin{bmatrix} \sigma_1 & 0 & \dots & 0 \\ 0 & \sigma_2 & \dots & 0 \\ \vdots & \vdots & \ddots & \vdots \\ 0 & 0 & \dots & \sigma_n \end{bmatrix} \in \Re^{n\times n} . \tag{15.15}$$

and each σ_i is a nonnegative real number.

Proof. Consider the maximization problem

$$\sup_{\boldsymbol{x}\in\Re^n : \|\boldsymbol{x}\|_2=1} \|\boldsymbol{A}\boldsymbol{x}\|_2 .$$

Every continuous function attains its supremum over a closed set. Since $S = \{\boldsymbol{x} \in \Re^n : \|\boldsymbol{x}\|_2 = 1\}$ is a closed set, there exists a vector $\boldsymbol{v}_1 \in S$ such that $\|\boldsymbol{A}\boldsymbol{v}_1\|_2 = \|\boldsymbol{A}\|_2$. If $\boldsymbol{u}_1$ is the unit vector (i.e., of length one) in the direction of $\boldsymbol{A}\boldsymbol{v}_1$, then we can write $\boldsymbol{A}\boldsymbol{v}_1 = \sigma_1\boldsymbol{u}_1$, where $\|\boldsymbol{u}_1\|_2 = 1$ and $\sigma_1 = \|\boldsymbol{A}\|_2$. Since $\boldsymbol{u}_1$ and $\boldsymbol{v}_1$ are both vectors having unit norm, the set $\{\boldsymbol{u}_1\}$ can be extended to an orthonormal basis in $\Re^m$ and the set $\{\boldsymbol{v}_1\}$ can be extended to an orthonormal basis in $\Re^m$. Placing the extension vectors as columns we obtain two matrices $\boldsymbol{U}_2$ and $\boldsymbol{V}_2$ such that

$$\boldsymbol{U}_1 = [\boldsymbol{u}_1 : \boldsymbol{U}_2] \in \Re^{m\times m} \text{ and } \boldsymbol{V}_1 = [\boldsymbol{v}_1 : \boldsymbol{V}_2] \in \Re^{n\times n}$$

are both orthogonal matrices (recall Lemma 8.2). Observe that

$$\boldsymbol{u}_1'\boldsymbol{A}\boldsymbol{v}_1 = \sigma_1\boldsymbol{u}_1'\boldsymbol{u}_1 = \sigma_1 \text{ and } \boldsymbol{U}_2'\boldsymbol{A}\boldsymbol{v}_1 = \sigma_1\boldsymbol{U}_2'\boldsymbol{u}_1 = \boldsymbol{0} .$$

Therefore,

$$\boldsymbol{U}_1'\boldsymbol{A}\boldsymbol{V}_1 = \begin{bmatrix} \boldsymbol{u}_1' \\ \boldsymbol{U}_2' \end{bmatrix}\boldsymbol{A}[\boldsymbol{v}_1 : \boldsymbol{V}_2] = \begin{bmatrix} \sigma_1 & \boldsymbol{u}_1'\boldsymbol{A}\boldsymbol{V}_2 \\ 0 & \boldsymbol{U}_2'\boldsymbol{A}\boldsymbol{V}_2 \end{bmatrix} = \begin{bmatrix} \sigma_1 & \boldsymbol{w}' \\ \boldsymbol{0} & \boldsymbol{B} \end{bmatrix} ,$$

where $\boldsymbol{w}' = \boldsymbol{u}_1'\boldsymbol{A}\boldsymbol{V}_2$ and $\boldsymbol{B} = \boldsymbol{U}_1'\boldsymbol{A}\boldsymbol{V}_2$. Writing $\boldsymbol{A}_1 := \boldsymbol{U}_1'\boldsymbol{A}\boldsymbol{V}_1$, we obtain

$$\frac{1}{\sigma_1^2 + \|\boldsymbol{w}\|^2}\left\|\boldsymbol{A}_1\begin{bmatrix} \sigma_1 \\ \boldsymbol{w} \end{bmatrix}\right\|_2^2 = \frac{1}{\sigma_1^2 + \|\boldsymbol{w}\|^2}\left\|\boldsymbol{A}_1\begin{bmatrix} \sigma_1^2 + \|\boldsymbol{w}\|^2 \\ \boldsymbol{B}\boldsymbol{w} \end{bmatrix}\right\|_2^2 \geq \sigma_1^2 + \|\boldsymbol{w}\|^2 .$$

This implies that $\boldsymbol{w} = \boldsymbol{0}$ because $\|\boldsymbol{A}_1\|_2^2 = \|\boldsymbol{U}_1'\boldsymbol{A}\boldsymbol{V}_1\|_2^2 = \sigma_1^2$. Hence,

$$\boldsymbol{A}_1 = \boldsymbol{U}_1'\boldsymbol{A}\boldsymbol{V}_1 = \begin{bmatrix} \sigma_1 & \boldsymbol{0}' \\ \boldsymbol{0} & \boldsymbol{B} \end{bmatrix} .$$

The proof is now completed by induction on the dimension of $\boldsymbol{A}$. The result is trivial for 1×1 matrices. Assume that every matrix whose size (number of rows or columns) is less than $\boldsymbol{A}$ has an SVD. In particular, the $(m-1)\times(n-1)$ matrix $\boldsymbol{B}$ has an SVD. Therefore, there are orthogonal matrices $\boldsymbol{U}_B \in \Re^{(m-1)\times(m-1)}$ and $\boldsymbol{V}_B \in$

$\Re^{(n-1)\times(n-1)}$ such that

$$B = U_B D_B V_B' , \text{ where } D_B = \begin{bmatrix} \Sigma_B \\ O \end{bmatrix} \text{ and } \Sigma_B = \begin{bmatrix} \sigma_2 & 0 & \dots & 0 \\ 0 & \sigma_3 & \dots & 0 \\ \vdots & \vdots & \ddots & \vdots \\ 0 & 0 & \dots & \sigma_n \end{bmatrix}$$

is an $(n-1) \times (n-1)$ diagonal matrix with each diagonal entry, σ_i, being a non-negative real number. Then, we have

$$A = U_1 \begin{bmatrix} \sigma_1 & 0' \\ 0 & B \end{bmatrix} V_1 = U_1 \begin{bmatrix} 1 & 0' \\ 0 & U_B \end{bmatrix} \begin{bmatrix} \sigma_1 & 0' \\ 0 & D_B \end{bmatrix} \begin{bmatrix} 1 & 0' \\ 0 & V_B' \end{bmatrix} V_1' .$$

Let us define

$$U = U_1 \begin{bmatrix} 1 & 0' \\ 0 & U_B \end{bmatrix} , \quad \Sigma = \begin{bmatrix} \sigma_1 & 0' \\ 0 & \Sigma_B \end{bmatrix} \text{ and } V = V_1 \begin{bmatrix} 1 & 0' \\ 0 & V_B' \end{bmatrix} .$$

It is easily verified that $A = U \begin{bmatrix} \Sigma \\ O \end{bmatrix} V'$ and that U and V are orthogonal matrices. Hence, the theorem is proved. □

The SVD presented above is no different from what was discussed in Section 12.1, except that we presented only the case when $m \geq n$, where m and n are the number of rows and columns in A. When $m < n$, the SVD is obtained by applying Theorem 15.11 to the transpose of A. This derivation employs very basic properties of norms and does not require defining eigenvalues. It allows us to introduce the SVD early, as is often done in texts on matrix computations, and use it to obtain basis vectors for the four fundamental subspaces of A and estimate the rank of a matrix. It can also be used to easily derive theoretical results such as the Rank-Nullity Theorem (Theorem 5.1) and the fundamental theorem of ranks (Theorem 5.2).

Low-rank approximations

The SVD also leads to certain *low-rank approximations* for matrices. Let A be an $m \times n$ matrix with rank r. The outer product form of the SVD (recall (12.9)), is

$$A = UDV' = \sum_{i=1}^{\min\{m,n\}} \sigma_i u_i v_i' = \sum_{i=1}^{r} \sigma_i u_i v_i' ,$$

where the singular vectors have been arranged so that $\sigma_1 \geq \sigma_2 \cdots \sigma_r > 0 = \sigma_{r+1} = \sigma_{r+2} = \cdots = \sigma_n$. While, in theory, the rank of A is r, many of the singular values can, in practice, be close to zero. It may be practically prudent to approximate A with a matrix of lower rank. An obvious way of doing this would be to simply truncate the SVD to $k \leq r$ terms. Therefore,

$$A = \sum_{i=1}^{r} \sigma_i u_i v_i' \approx \sum_{i=1}^{k} \sigma_i u_i v_i' ,$$

where $k \leq r$ and $\sigma_i \approx 0$ for $i = k+1, k+2, \ldots, r$. This truncated SVD is widely used in practical applications. It is used to remove additional noise in datasets, for compressing data and stabilizing the solution of problems that are ill-conditioned.

The truncated SVD provides an *optimal* solution to the problem of approximating a given matrix by one of lower rank. In fact, the truncated SVD is the matrix with rank k which is "closest" to $\boldsymbol{A}$, where "closeness" is determined by the Frobenius matrix norm $\|\boldsymbol{X}\|_F = \sqrt{\sum_{i=1}^{m}\sum_{j=1}^{n} x_{ij}^2}$. We will prove this.

Let $\Re^{m\times n}$ be the vector space of all $m \times n$ matrices and consider the inner product

$$\langle \boldsymbol{A}, \boldsymbol{B}\rangle = \text{tr}(\boldsymbol{A}'\boldsymbol{B}) = \sum_{j=1}^{n} \boldsymbol{a}_{*j}'\boldsymbol{b}_{*j} = \sum_{i=1}^{m}\sum_{j=1}^{n} a_{ij}b_{ij} \,. \tag{15.16}$$

The norm induced by the above inner product is the Frobenius norm: $\langle \boldsymbol{A}, \boldsymbol{A}\rangle = \text{tr}(\boldsymbol{A}'\boldsymbol{A}) = \|\boldsymbol{A}\|_F^2$. We now have the following lemma.

Lemma 15.1 *Let $\boldsymbol{A}$ be a matrix in $\Re^{m\times n}$ and let $\boldsymbol{A} = \boldsymbol{UDV}'$ be the SVD of $\boldsymbol{A}$. Then the rank-one matrices $\boldsymbol{B}_{ij} = \boldsymbol{u}_i\boldsymbol{v}_j'$ for $i = 1, 2, \ldots, m$ and $j = 1, 2, \ldots, n$ form an orthonormal basis for $\Re^{m\times n}$.*

Proof. It easily follows that $\langle \boldsymbol{B}_{ij}, \boldsymbol{B}_{kl}\rangle = \delta_{(i,k)(j,l)}$, where $\delta_{(i,k)(j,l)} = 1$ if $i = k$ *and* $j = l$ and zero otherwise, because

$$\langle \boldsymbol{B}_{ij}, \boldsymbol{B}_{kl}\rangle = \langle \boldsymbol{u}_i\boldsymbol{v}_j', \boldsymbol{u}_k\boldsymbol{v}_l'\rangle = \text{tr}(\boldsymbol{v}_j\boldsymbol{u}_i'\boldsymbol{u}_k\boldsymbol{v}_l') = (\boldsymbol{u}_i'\boldsymbol{u}_k)\text{tr}(\boldsymbol{v}_j\boldsymbol{v}_l') = (\boldsymbol{u}_i'\boldsymbol{u}_k)(\boldsymbol{v}_l'\boldsymbol{v}_j) \,.$$

This proves that the $\boldsymbol{B}_{ij}'$'s constitute an orthonormal set of rank-one matrices with respect to the inner product in (15.16). Since there are mn such matrices, they form an orthonormal basis. □

Theorem 15.12 *Let $\boldsymbol{A}$ be an $m \times n$ real matrix with rank $r > k$. Construct the truncated SVD: $\boldsymbol{A}_k = \sum_{i=1}^{k}\sigma_i\boldsymbol{u}_i\boldsymbol{v}_i'$, where $\sigma_1 \geq \sigma_2 \geq \cdots \geq \sigma_k > 0$ are the singular values of $\boldsymbol{A}$, $\boldsymbol{u}_i$'s and $\boldsymbol{v}_i$'s are the associated left and right singular vectors. Then,*

$$\|\boldsymbol{A} - \boldsymbol{A}_k\|_F = \min_{\rho(\boldsymbol{Z})=k} \|\boldsymbol{A} - \boldsymbol{Z}\|_F \,.$$

Proof. Lemma 15.1 tells us that any $\boldsymbol{Z} \in \Re^{m\times n}$ can be expressed as $\boldsymbol{Z} =$

$\sum_{i,j} \theta_{ij} \boldsymbol{B}_{ij}$, where $\boldsymbol{B}_{ij} = \boldsymbol{u}_i \boldsymbol{v}_j'$. Using the orthogonality of $\boldsymbol{B}_{ij}$'s, we obtain

$$\begin{aligned}
\|\boldsymbol{A} - \boldsymbol{Z}\|_F^2 &= \left\| \sum_{i=1}^{r} \sigma_i \boldsymbol{B}_{ii} - \sum_{i,j} \theta_{ij} \boldsymbol{B}_{ij} \right\|_F^2 \\
&= \left\| \sum_{i=1}^{r} (\sigma_i - \theta_{ii}) \boldsymbol{B}_{ii} - \sum_{i=r+1}^{\min\{m,n\}} \theta_{ii} \boldsymbol{B}_{ii} - \sum_{i \neq j} \theta_{ij} \boldsymbol{B}_{ij} \right\|_F^2 \\
&= \left\| \sum_{i=1}^{r} (\sigma_i - \theta_{ii}) \boldsymbol{B}_{ii} \right\|_F^2 + \left\| \sum_{i=r+1}^{\min\{m,n\}} \theta_{ii} \boldsymbol{B}_{ii} \right\|_F^2 + \left\| \sum_{i \neq j} \theta_{ij} \boldsymbol{B}_{ij} \right\|_F^2 \\
&= \sum_{i=1}^{r} (\sigma_i - \theta_{ii})^2 + \sum_{i=r+1}^{\min\{m,n\}} \theta_{ii}^2 + \sum_{i \neq j} \theta_{ij}^2 .
\end{aligned}$$

The first step toward minimizing $\|\boldsymbol{A} - \boldsymbol{Z}\|_F^2$ is to set $\theta_{ij} = 0$ whenever $i \neq j$ and $\theta_{ii} = 0$ for $i = r+1, r+2, \ldots, n$. This constrains the minimizing $\boldsymbol{Z}$ to be

$$\boldsymbol{Z} = \sum_{i=1}^{r} \theta_{ii} \boldsymbol{B}_{ii} \,, \quad \text{which implies that } \|\boldsymbol{A} - \boldsymbol{Z}\|_F^2 = \sum_{i=1}^{r} (\sigma_i - \theta_{ii})^2 \,.$$

A further constraint comes from $\rho(\boldsymbol{Z}) = k < r$; $\boldsymbol{Z}$ can have exactly k nonzero coefficients for the $\boldsymbol{B}_{ii}$'s. Since the σ_i's are arranged in descending order, the objective function is minimized by setting

$$\theta_{ii} = \sigma_i \ \text{ for } \ i = 1, 2, \ldots, k \ \text{ and } \ \theta_{ii} = 0 \ \text{ for } i = k+1, k+2, \ldots, r \,.$$

Therefore, $\boldsymbol{A}_k = \sum_{i=1}^{k} \sigma_i \boldsymbol{u}_i \boldsymbol{v}_i'$ is the best rank-k approximation. □

The above result, put differently, says that $\boldsymbol{A}_k = \boldsymbol{U}_k \boldsymbol{\Sigma}_k \boldsymbol{V}_k'$ is the best rank-k approximation for $\boldsymbol{A}$, where $\boldsymbol{\Sigma}_k$ is the $k \times k$ diagonal matrix with the k largest singular values of $\boldsymbol{A}$ along the diagonal and $\boldsymbol{U}_k$ and $\boldsymbol{V}_k$ are $m \times k$ and $n \times k$ matrices formed by the k corresponding left and right singular vectors, respectively.

15.6 Web page ranking and Markov chains

A rather pervasive modern application of linear algebra is in the development of Internet search engines. The emergence of Google in the late 1990s, brought about an explosion of interest in algorithms for ranking web pages. Google, it is fair to say, set itself apart with its vastly superior performance in ranking web pages. While with the other search engines of that era we often had to wade through screen after screen of irrelevant web pages, Google seemed to be effectively delivering the "right stuff." This highly successful algorithm, that continues to be fine tuned, is a beautiful application of linear algebra. Excellent accounts are provided by Langville and Meyer (2005; 2006) and also the book by Eldén (2007). We offer a very brief account here.

Assume that the web of interest consists of r pages, where each page is indexed by an integer $k \in \{1, 2, \ldots, r\}$. One way to model the web is to treat it as a relational *graph* consisting of n vertices corresponding to the pages and a collection of directed edges between the vertices. An edge originating at vertex i and pointing to vertex j indicates that there is a link from page i to page j. Depending upon the direction of the edge between two vertices, we distinguish between two types of links: the pages that have links *from* page i are called ***outlinks*** of page i, while the pages that link *to* page i are called the ***inlinks*** (or ***backlinks***) of page i. In other words, pages that have outlinks to i constitute the set of its inlinks.

Page ranking is achieved by assigning a score to any given web page, which reflects the importance of the web page. It is only reasonable that this importance score will depend upon the number of inlinks to that page. However, letting the score depend only upon the inlinks will be flawed. For example, a particular web page's score will become deceptively high if a number of highly irrelevant and unimportant pages link to it. To compensate for such anomalies, what one seeks is a score for page i that will be a weighted average of the ranks of web pages with outlinks to i. This will ensure that the rank of i is increased if a highly ranked page j is one of its inlinks. In this manner, an Internet search engine can be looked upon as elections in a "democracy," where pages "vote" for the importance of other pages by linking to them.

To see specifically how this is done, consider a "random walk" model on the web, where a surfer currently visiting page i is equally likely to choose any of the pages that are outlinks of i. Furthermore, we do assume that the surfer will not get stuck in page i and must choose one of the outlinks as the next visit. Therefore, if page i has r_i outlinks, then the probability of the surfer selecting a page j is $1/r_i$ if page j is linked from i and 0 otherwise. This motivates the definition of an $r \times r$ *transition matrix* $\boldsymbol{P} = \{p_{ij}\}$ such that p_{ij} is the probability that the surfer who is currently situated in page i will visit page j in the next "click." Therefore,

$$p_{ii} = 0 \text{ and } p_{ij} = \begin{cases} 1/r_i & \text{if the number of outlinks from page } i \text{ is } r_i \\ 0 & \text{otherwise.} \end{cases} \tag{15.17}$$

By convention, we do not assume that a page links to itself, so $p_{ii} = 0$ for $i = 1, 2, \ldots, r$. Also, we assume that the surfer cannot get "stuck" in a page without any outlinks. Thus, all pages should have a positive number of outlinks, i.e., $r_i > 0$ for each i. It is easy to see that the sum of the entries in each row of $\boldsymbol{P}$ is equal to one, i.e., $\boldsymbol{P1} = \boldsymbol{1}$. Such matrices are called ***row-stochastic***.

Clearly $\boldsymbol{1}$ is an eigenvector of $\boldsymbol{P}$ associated with eigenvalue 1. In fact, 1 happens to be the largest eigenvalue for any row-stochastic matrix, irrespective of whether its entries are negative or positive.

Theorem 15.13 *The maximum eigenvalue of a row-stochastic matrix is equal to* 1.

Proof. This is easily derived from the relationship between the spectral radius and the matrix norm. Since 1 is an eigenvalue of $\boldsymbol{P}$, the spectral radius $\kappa(\boldsymbol{P})$ cannot be smaller than one. Since, $\boldsymbol{P1} = \boldsymbol{1}$, it is clear that the row sums of $\boldsymbol{P}$ are all equal to one and, hence, $\|\boldsymbol{P}\|_\infty = 1$. Therefore,

$$1 \leq \kappa(\boldsymbol{P}) \leq \|\boldsymbol{P}\|_\infty = 1 \implies \kappa(\boldsymbol{P}) = 1 .$$

□

Readers familiar with probability will recognize $\boldsymbol{P}$ to be the transition probability matrix of a ***Markov chain***. Let X_t be a random variable representing the current state (web page) of the surfer. A Markov chain, for our purposes, is a discrete stochastic process that assumes that the future states the surfer visits depends only upon the current state and not any of the past states that the surfer has already visited. This is expressed as the conditional probability $P(X_{t+1} = j \,|\, X_t = i) = p_{ij}$, which forms the (i,j)-th entry in the transition matrix $\boldsymbol{P}$. We will assume that the Markov chain is *homogeneous*, i.e., the transition matrix does not depend upon t. Basic laws of probability ensure that $\sum_{j=1}^{r} p_{ij} = 1$, which means that the transition matrix $\boldsymbol{P}$ is row-stochastic with all entries nonnegative. It is not necessary that the diagonal elements of $\boldsymbol{P}$ are zero (as in (15.17)). That was the case only for a model of the web page. In general, $\boldsymbol{P}$ is a transition probability matrix if all its entries are nonnegative and its row sums are all equal to one, i.e., $\boldsymbol{P}\mathbf{1} = \mathbf{1}$. In fact, every row-stochastic matrix with nonnegative entries corresponds to a transition probability matrix of a Markov chain.

The transition matrix helps us update the probability of visiting a web page at each step. To be precise, let $\boldsymbol{\pi}(k) = [\pi_1(k), \pi_2(k), \ldots, \pi_r(k)]'$ be the $r \times 1$ *probability vector*, i.e., $\boldsymbol{\pi}(k)'\mathbf{1} = \sum_{i=1}^{r} \pi_i(k) = 1$, such that $\pi_i(k)$ is the probability of the surfer visiting web page i at time k. Using basic laws of probability, we see that the transition matrix maps the probabilities of the current state $\boldsymbol{\pi}(k)$ to the next state $\boldsymbol{\pi}(k+1)$ because

$$\pi_j(k+1) = \sum_{i=1}^{r} p_{ij}\pi_i(k) \;\text{ for }\; j = 1, 2, \ldots, r ,$$
$$\implies \boldsymbol{\pi}(k+1)' = \boldsymbol{\pi}(k)'\boldsymbol{P} , \tag{15.18}$$

which can also be written as the linear iterative system $\boldsymbol{\pi}(k+1) = \boldsymbol{P}'\boldsymbol{\pi}(k)$. The probability vector $\boldsymbol{\pi}(k)$ is called the k-th *state* vector. It is easy to see that $\boldsymbol{\pi}(k+1)$ is also a probability vector and the solution $\boldsymbol{\pi}(k)' = \boldsymbol{\pi}(0)'\boldsymbol{P}^k$ generates a sequence of probability vectors through successive iterations. The Markov chain attains *steady state* if the system in (15.18) converges.

Powers of the transition matrix are, therefore, important in understanding the convergence of Markov chains. The (i,j)-th entry in $\boldsymbol{P}^k$ gives the probability of moving from state i to state j in k steps. To explore convergence of Markov chains in full generality will be too much of a digression for this book. We will, however, explore transition matrices which, when raised to a certain power, have nonzero entries. Such transition matrices are called *regular*.

Definition 15.8 *A transition matrix $\boldsymbol{P}$ is said to be* **regular** *if for some positive integer k, all the entries in $\boldsymbol{P}^k$ have strictly positive entries.*

If $\boldsymbol{P}$ is a regular transition matrix, then $\lim_{n\to\infty} \boldsymbol{P}^n$ exists. To prove this requires a bit of work. We first derive this result for transition matrices all of whose entries are strictly positive, and subsequently extend it to regular transition matrices. Transition matrices with all entries strictly positive have particularly simple eigenspaces, as the following result shows.

Theorem 15.14 *Consider a Markov chain with $r > 1$ states having an $r \times r$ transition matrix $\boldsymbol{P} = \{p_{ij}\}$ all of whose entries are positive. Then, any vector $\boldsymbol{v}$ satisfying $\boldsymbol{P}\boldsymbol{v} = \boldsymbol{v}$ must be of the form $\boldsymbol{v} = \alpha\boldsymbol{1}$ for some scalar α.*

Proof. Since $\boldsymbol{P}$ is row-stochastic, $\boldsymbol{P}\boldsymbol{1} = \boldsymbol{1}$, which implies that $(1, \boldsymbol{1})$ is an eigenpair of $\boldsymbol{P}$. Suppose $\boldsymbol{P}\boldsymbol{v} = \boldsymbol{v}$. Assume, if possible, that $\boldsymbol{v}$ does not have all equal entries and let v_i be the minimum entry in $\boldsymbol{v}$. Since there is at least one entry in $\boldsymbol{v}$ that is greater that u_i, and that each $p_{ij} > 0$, we conclude

$$v_i = \boldsymbol{p}'_{i*}\boldsymbol{v} = \sum_{j=1}^{r} p_{ij} v_j > \sum_{j=1}^{r} p_{ij} v_i = v_i \ ,$$

where $\boldsymbol{p}'_{i*}$ is the i-th row of $\boldsymbol{P}$. Clearly the above leads to a contradiction. Therefore, all the entries in $\boldsymbol{v}$ must be equal, i.e., $v_1 = v_2 = \ldots = v_r = \alpha$ and, hence, $\boldsymbol{v} = \alpha\boldsymbol{1}$. □

Theorem 15.14 states that if $\boldsymbol{P}$ is a transition probability matrix with no zero entries, then the eigenspace of $\boldsymbol{P}$ corresponding to the eigenvalue 1 is spanned by $\boldsymbol{1}$. Hence, it has dimension one, i.e., $\dim(\mathcal{N}(\boldsymbol{P} - \lambda\boldsymbol{I})) = 1$. Furthermore, with $\alpha = 1/n$ in Theorem 15.14, the vector $\boldsymbol{v} = (1/n)\boldsymbol{1}$ is a probability eigenvector of $\boldsymbol{P}$.

Remarkably, the powers $\boldsymbol{P}^n$ approach a matrix each of whose rows belong to the one-dimensional eigenspace corresponding to the eigenvalue 1. We show this below. Our development closely follows Kemeny and Snell (1976) and Grinstead and Snell (1997).

Lemma 15.2 *Consider a Markov chain with $r > 1$ states having an $r \times r$ transition matrix $\boldsymbol{P} = \{p_{ij}\}$ all of whose entries are positive, i.e., $p_{ij} > 0$ for all i, j, with the minimum entry being d. Let $\boldsymbol{u}$ be a $r \times 1$ vector with non-negative entries, the largest of which is M_0 and the smallest of which is m_0. Also, let M_1 and m_1 be the maximum and minimum elements in $\boldsymbol{P}\boldsymbol{u}$. Then,*

$$M_1 \leq dm_0 + (1-d)M_0, \quad \textit{and} \quad m_1 \geq dM_0 + (1-d)m_0 \ ,$$

and hence

$$M_1 - m_1 \leq (1-2d)(M_0 - m_0).$$

Proof. Each entry in the vector $\boldsymbol{P}\boldsymbol{u}$ is a weighted average of the entries in $\boldsymbol{u}$. The largest weighted average that could be obtained in the present case would occur if all but one of the entries of $\boldsymbol{u}$ have value M_0 and one entry has value m_0, and this one small entry is weighted by the smallest possible weight, namely d. In this case, the weighted average would equal $dm_0 + (1-d)M_0$. Similarly, the smallest possible weighted average equals: $dM_0 + (1-d)m_0$. Thus,

$$M_1 - m_1 \leq (dm_0 + (1-d)M_0) - (dM_0 + (1-d)m_0) = (1-2d)(M_0 - m_0) ,$$

which completes the proof. □

The next lemma shows that the entries in $\boldsymbol{P}^n\boldsymbol{u}$ are the same.

Lemma 15.3 *In the setup of Lemma 15.2, let M_n and m_n be the maximum and minimum elements of $\boldsymbol{P}^n\boldsymbol{u}$, respectively. Then $\lim_{n\to\infty} M_n$ and $\lim_{n\to\infty} m_n$ both exist and are equal to the same number.*

Proof. The vector $\boldsymbol{P}^n\boldsymbol{u}$ is obtained from the vector $\boldsymbol{P}^{n-1}\boldsymbol{u}$ by multiplying on the left by the matrix $\boldsymbol{P}$. Hence each component of $\boldsymbol{P}^n\boldsymbol{u}$ is an average of the components of $\boldsymbol{P}^{n-1}\boldsymbol{u}$. Thus,

$$M_0 \geq M_1 \geq M_2 \cdots$$

and

$$m_0 \leq m_1 \leq m_2 \cdots .$$

Each of the above sequences is monotone and bounded:

$$m_0 \leq m_n \leq M_n \leq M_0 .$$

Hence each of these sequences will have a limit as $n \to \infty$.

Let $\lim_{n\to\infty} M_n = M$ and $\lim_{n\to\infty} m_n = m$. We know that $m \leq M$. We shall prove that $M - m = 0$, which will be the case if $M_n - m_n$ tends to 0. Recall that d is the smallest element of $\boldsymbol{P}$ and, since all entries of $\boldsymbol{P}$ are strictly positive, we have $d > 0$. By our lemma

$$M_n - m_n \leq (1-2d)(M_{n-1} - m_{n-1}),$$

from which it follows that

$$M_n - m_n \leq (1-2d)^n (M_0 - m_0).$$

Since $r \geq 2$, we must have $d \leq 1/2$, so $0 \leq 1 - 2d < 1$, so $M_n - m_n \to 0$ as $n \to \infty$. Since every component of $\boldsymbol{P}^n\boldsymbol{u}$ lies between m_n and M_n, each component must approach the same number $M = m$. □

We now show that $\lim_{n\to\infty} \boldsymbol{P}^n$ exists and is a well-defined transition matrix.

Theorem 15.15 *If $\boldsymbol{P} = \{p_{ij}\}$ is an $r \times r$ transition matrix all of whose entries are positive, then*

$$\lim_{n\to\infty} \boldsymbol{P}^n = \boldsymbol{W} = \boldsymbol{1}\boldsymbol{\alpha}'$$

where $\boldsymbol{W}$ is $r \times r$ with identical row vectors, $\boldsymbol{\alpha}'$, and strictly positive entries.

Proof. Lemma 15.3 implies that if we denote the common limit $M = m = \alpha$, then

$$\lim_{n\to\infty} \boldsymbol{P}^n\boldsymbol{u} = \alpha\boldsymbol{1} .$$

In particular, suppose we choose $\boldsymbol{u} = \boldsymbol{e}_j$ where $\boldsymbol{e}_j$ is the $r \times 1$ vector with its j-th component equal to 1 and all other components equaling 0. Then we obtain an α_j such that $\lim_{n\to\infty} \boldsymbol{P}^n \boldsymbol{e}_j = \alpha_j \mathbf{1}$. Repeating this for each $j = 1, \ldots, r$ we obtain

$$\lim_{n\to\infty} \boldsymbol{P}^n [\boldsymbol{e}_1 : \ldots : \boldsymbol{e}_r] = [\alpha_1 \mathbf{1} : \ldots : \alpha_r \mathbf{1}] = \mathbf{1}\boldsymbol{\alpha}' ,$$

which implies that $\lim_{n\to\infty} \boldsymbol{P}^n = \boldsymbol{W}$, where each row of $\boldsymbol{W}$ is $\boldsymbol{\alpha}' = [\alpha_1, \ldots, \alpha_r]$.

It remains to show that all entries in $\boldsymbol{W}$ are strictly positive. Note that $\boldsymbol{P}\boldsymbol{e}_j$ is the j-th column of $\boldsymbol{P}$, and this column has all entries strictly positive. The minimum component of the vector $\boldsymbol{P}\boldsymbol{u}$ was defined to be m_1, hence $m_1 > 0$. Since $m_1 \leq m$, we have $m > 0$. Note finally that this value of m is just the j-th component of $\boldsymbol{\alpha}$, so all components of $\boldsymbol{\alpha}$ are strictly positive. □

Observe that the matrix $\boldsymbol{W}$ is also row-stochastic because each $\boldsymbol{P}^n$ is row-stochastic. Hence, each row of $\boldsymbol{W}$ is a probability vector. Theorem 15.15 is also true when $\boldsymbol{P}$ is a regular transition matrix.

Theorem 15.16 *If $\boldsymbol{P} = \{p_{ij}\}$ is an $r \times r$ regular transition matrix, then*

$$\lim_{n\to\infty} \boldsymbol{P}^n = \boldsymbol{W} ,$$

where $\boldsymbol{W}$ is $r \times r$ with identical row vectors and strictly positive entries.

Proof. Since $\boldsymbol{P}$ is regular, there is a positive integer N such that all the entries of $\boldsymbol{P}^N$ are positive. We apply Theorem 15.15 to $\boldsymbol{P}^N$. Let d_N be the smallest entry in $\boldsymbol{P}^N$. From the inequality in Lemma 15.2, we obtain

$$M_{nN} - m_{nN} \leq (1 - 2d_N)^n (M_{0N} - m_{0N}) ,$$

here M_{nN} and m_{nN} are defined analogously as the maximum and minimum elements of $\boldsymbol{P}^{Nn}\boldsymbol{u}$. As earlier, this implies that $M_{nN} - m_{nN} \to 0$ as $n \to \infty$. Thus, the non-increasing sequence $M_n - m_n$, has a subsequence that converges to 0. From basic real analysis, we know that this must imply that the entire sequence must also tend to 0. The rest of the proof imitates Theorem 15.15. □

For the linear iterative system (15.18) with a regular transition matrix we obtain

$$\lim_{k\to\infty} \boldsymbol{\pi}(k)' = \lim_{k\to\infty} \boldsymbol{\pi}(0)'\boldsymbol{P}^k = \boldsymbol{\pi}(0)'\boldsymbol{W} , \text{ where } \lim_{k\to\infty} \boldsymbol{P}^k = \boldsymbol{W} . \tag{15.19}$$

Thus, Markov chains with regular transition matrices are guaranteed to converge to a steady state. This limit $\boldsymbol{W}\boldsymbol{\pi}(0)$ is again a probability vector (because each iteration produces a probability vector) and is called the ***stationary distribution*** of the Markov chain. More precisely, since $\boldsymbol{\pi}(0)$ is a probability vector, we obtain

$$\boldsymbol{\pi}(0)'\boldsymbol{W} = \boldsymbol{\pi}(0)'[\alpha_1\mathbf{1} : \ldots : \alpha_r\mathbf{1}] = [\alpha_1, \alpha_2, \ldots, \alpha_r] = \boldsymbol{\alpha}' ,$$

where $\boldsymbol{W}$ and $\boldsymbol{\alpha}$ are as in Theorem 15.15. Clearly $\boldsymbol{\alpha}$ is a probability vector because

$$\mathbf{1} = \lim_{n\to\infty} \boldsymbol{P}^n\mathbf{1} = \boldsymbol{W}\mathbf{1} = \mathbf{1}(\boldsymbol{\alpha}'\mathbf{1}) ,$$

which implies that $\sum_{i=1}^{r} \alpha_i = \boldsymbol{\alpha}'\mathbf{1} = 1$. Thus, $\boldsymbol{\alpha}'$ is precisely the steady-state or stationary distribution. In fact, the $\boldsymbol{\alpha}'$ remains "stationary" with respect to $\boldsymbol{P}$ in the following sense:

$$\boldsymbol{\alpha}'\boldsymbol{P} = (\boldsymbol{e}_1'\boldsymbol{W})\boldsymbol{P} = \boldsymbol{e}_1'(\lim_{n\to\infty} \boldsymbol{P}^n)\boldsymbol{P} = \boldsymbol{e}_1' \lim_{n\to\infty} \boldsymbol{P}^{n+1} = \boldsymbol{e}_1'\boldsymbol{W} = \boldsymbol{\alpha}' .$$

This leads to a more general definition of a stationary distribution.

Definition 15.9 *A probability vector $\boldsymbol{\pi}$ is said to be the* **stationary distribution** *of a Markov chain if it satisfies*

$$\boldsymbol{\pi}'\boldsymbol{P} = \boldsymbol{\pi}' \quad \textit{or, equivalently,} \quad \boldsymbol{P}'\boldsymbol{\pi} = \boldsymbol{\pi} ,$$

where $\boldsymbol{P}$ is the transition matrix of the Markov chain.

This definition applies to Markov chains with transition matrices that may or may not be regular. However, if $\boldsymbol{P}$ is a regular transition matrix and $\boldsymbol{\pi}$ is a stationary distribution in the sense of Definition 15.9, then $\boldsymbol{\pi}$ must coincide with $\boldsymbol{\alpha}$. This is because

$$\boldsymbol{\pi}' = \boldsymbol{\pi}'\boldsymbol{P} = \boldsymbol{\pi}'\boldsymbol{P}^n = \boldsymbol{\pi}' \lim_{n\to\infty} \boldsymbol{P}^n = \boldsymbol{\pi}'\boldsymbol{W} = \boldsymbol{\pi}'[\alpha_1\mathbf{1} : \ldots : \alpha_r\mathbf{1}] = \boldsymbol{\alpha}' ,$$

where we used the fact $\boldsymbol{\pi}'\mathbf{1} = 1$ in the last equality. Thus, Markov chains with regular transition matrices have a unique stationary distribution. The stationary distribution is a probability eigenvector of the transpose of $\boldsymbol{P}'$ associated with the eigenvalue 1.

Note that Definition 15.9 does not say anything about convergence of $\boldsymbol{P}^n$ to a steady state. It is all about the *existence* of the probability vector $\boldsymbol{\pi}$. Since $\boldsymbol{P}$ is row-stochastic, $(1, \mathbf{1})$ is an eigen-pair for $\boldsymbol{P}$. Any eigenvalue of a matrix is also an eigenvalue of its transpose. Therefore, 1 is an eigenvalue of $\boldsymbol{P}'$ and there corresponds an eigenvector $\boldsymbol{v}$ such that $\boldsymbol{P}'\boldsymbol{v} = \boldsymbol{v}$. It is tempting to identify this eigenvector as the stationary distribution, but $\boldsymbol{v}$ may not be a probability eigenvector. The key is question is whether every entry in $\boldsymbol{v}$ is nonnegative (recall that $\boldsymbol{v} \neq \mathbf{0}$ because $\boldsymbol{v}$ is an eigenvector). If so, then $\boldsymbol{\pi} = \boldsymbol{v}/\|\boldsymbol{v}\|$ will meet the requirement in Definition 15.9.

So, are we guaranteed to find an eigenvector of $\boldsymbol{P}'$ associated with eigenvalue 1 such that all the entries in $\boldsymbol{v}$ are nonnegative? The answer, quite remarkably, is yes and relies upon a well-known result attributed to the German mathematicians Oskar Perron and Ferdinand Georg Frobenius, known as the *Perron-Frobenius Theorem*. This theorem states that *any* square matrix $\boldsymbol{A}$ with nonnegative entries has its spectral radius $\kappa(\boldsymbol{A})$ as an eigenvalue and there exists an associated eigenvector with nonnegative entries. In particular, if $\boldsymbol{A}$ is nonnegative and row-stochastic (hence a transition matrix), this theorem assures us that there exists an eigenvector with nonnegative entries associated with the eigenvalue $\kappa(\boldsymbol{A}) = 1$. This ensures that *every* finite Markov chain has a stationary distribution. There is, however, no guarantee that this stationary distribution is *unique*. Nor does it ensure the convergence of a general Markov chain to its stationary distribution.

Let us return to the problem of ranking web pages. Ideally, we would like to define the so-called *PageRank* vector to be the *unique* stationary distribution for the

Markov chain associated with the web. Thus, the highest ranked page would be the one where the surfer has the highest probability of visiting once the chain has reached steady state. The second highest rank will be assigned to the page with the second highest probability, and so on. There is, however, some practical problems with this approach. From what we have seen, a regular transition matrix ensures a unique stationary distribution. This may be too restrictive an assumption for the web and, in general, the uniqueness of a probability eigenvector associated with the eigenvalue 1 of $\boldsymbol{P}'$ is not guaranteed. Fortunately, the Perron-Frobenius theory assures us of the existence of a unique probability eigenvector of $\boldsymbol{P}'$ associated with the eigenvalue 1 if the matrix is ***irreducible***.

Definition 15.10 Reducible matrix. *A square matrix $\boldsymbol{A}$ is said to be* **reducible** *if there exists a permutation matrix $\boldsymbol{M}$ such that*

$$\boldsymbol{M}'\boldsymbol{A}\boldsymbol{M} = \begin{bmatrix} \boldsymbol{X} & \boldsymbol{Y} \\ \boldsymbol{O} & \boldsymbol{Z} \end{bmatrix} ,$$

where $\boldsymbol{X}$ and $\boldsymbol{Z}$ are both square. Matrices that are not reducible are called **irreducible**.

A Markov chain is called *irreducible* if its transition matrix is not reducible. Assuming that the Markov chain for the web is irreducible means that there is a path leading from any page to any other page. Thus, the surfer has a positive probability of visiting any other page from the current location. Put another way, the surfer does not get trapped within a "subgraph" of the Internet. Given the size of the Internet, the assumption of irreducibility still seems like a stretch of imagination. Nevertheless, it is required to ensure a unique PageRank vector. An easy trick to impose irreducibility, even when $\boldsymbol{P}$ is not, is to create the convex combination

$$\tilde{\boldsymbol{P}} = \alpha\boldsymbol{P} + (1-\alpha)\frac{\boldsymbol{1}\boldsymbol{1}'}{r} ,$$

where α is some scalar between 0 and 1 and $\boldsymbol{P}$ is $r \times r$. It is easy to verify that $\tilde{\boldsymbol{P}}$ is another transition matrix. It has nonnegative entries and is row-stochastic because

$$\tilde{\boldsymbol{P}}(\alpha)\boldsymbol{1} = \left(\alpha\boldsymbol{P} + (1-\alpha)\frac{\boldsymbol{1}\boldsymbol{1}'}{r}\right)\boldsymbol{1} = \alpha\boldsymbol{P}\boldsymbol{1} + (1-\alpha)\frac{\boldsymbol{1}\boldsymbol{1}'}{r}\boldsymbol{1} = \alpha\boldsymbol{1} + (1-\alpha)\boldsymbol{1} = \boldsymbol{1} .$$

Therefore, $\tilde{\boldsymbol{P}}$ has 1 as its largest eigenvalue. It corresponds to a modified web, where a link has been added from every page to every other page, and the Markov chain posits that the surfer visiting a page will visit any other random page with probability $1-\alpha$. The matrix $\tilde{\boldsymbol{P}}$ is irreducible and the unique probability eigenvector associated with it is the PageRank vector. Furthermore, let $\{1, \lambda_2, \lambda_3, \ldots, \lambda_n r\}$ be the set of eigenvalues (counting multiplicities) of $\boldsymbol{P}$. Extend the vector $\boldsymbol{1}/\sqrt{r}$ to an orthonormal basis of $\Re^r$ and place the extension vectors as columns of the $r \times (r-1)$ matrix $\boldsymbol{Q}_1$. Then, $\boldsymbol{Q} = [\boldsymbol{1}/\sqrt{r} : \boldsymbol{Q}_1]$ is an orthogonal matrix and $\boldsymbol{Q}_1'\boldsymbol{1} = \boldsymbol{0}$, which implies

$$\begin{aligned} \boldsymbol{Q}'\boldsymbol{P}\boldsymbol{Q} &= \begin{bmatrix} \boldsymbol{1}'/\sqrt{r} \\ \boldsymbol{Q}_1' \end{bmatrix} \boldsymbol{P}[\boldsymbol{1}/\sqrt{r} : \boldsymbol{Q}_1] = \begin{bmatrix} 1 & \boldsymbol{1}'\boldsymbol{P}\boldsymbol{Q}_1/\sqrt{r} \\ \boldsymbol{Q}_1'\boldsymbol{1}/\sqrt{r} & \boldsymbol{Q}_1'\boldsymbol{P}\boldsymbol{Q}_1 \end{bmatrix} \\ &= \begin{bmatrix} 1 & \boldsymbol{u} \\ \boldsymbol{0} & \boldsymbol{W} \end{bmatrix}, \text{ where } \boldsymbol{u} = \boldsymbol{1}'\boldsymbol{P}\boldsymbol{Q}_1/\sqrt{r} \text{ and } \boldsymbol{W} = \boldsymbol{Q}_1'\boldsymbol{P}\boldsymbol{Q}_1 . \end{aligned}$$

The eigenvalues of $\boldsymbol{Q}'\boldsymbol{P}\boldsymbol{Q}$ are the same as those of $\boldsymbol{P}$ and their characteristic polynomial can be expressed as

$$|\lambda \boldsymbol{I}_r - \boldsymbol{P}| = |\lambda \boldsymbol{I}_r - \boldsymbol{Q}'\boldsymbol{P}\boldsymbol{Q}| = |\lambda \boldsymbol{I}_{r-1} - \boldsymbol{W}| .$$

Since 1 is an eigenvalue of $\boldsymbol{P}$, $\{\lambda_2, \lambda_3, \ldots, \lambda_r\}$ is the set of $r-1$ eigenvalues (counting multiplicities) of $\boldsymbol{W}$. Also,

$$\begin{aligned} \boldsymbol{Q}'\tilde{\boldsymbol{P}}\boldsymbol{Q} &= \alpha \boldsymbol{Q}'\boldsymbol{P}\boldsymbol{Q} + (1-\alpha) \begin{bmatrix} \boldsymbol{1}'/\sqrt{r} \\ \boldsymbol{Q}_1' \end{bmatrix} \frac{\boldsymbol{1}\boldsymbol{1}'}{r}[\boldsymbol{1}/\sqrt{r} : \boldsymbol{Q}_1] \\ &= \alpha \begin{bmatrix} 1 & \boldsymbol{u} \\ \boldsymbol{0} & \boldsymbol{W} \end{bmatrix} + (1-\alpha) \begin{bmatrix} 1 & \boldsymbol{0}' \\ \boldsymbol{0} & \boldsymbol{O} \end{bmatrix} = \begin{bmatrix} 1 & \alpha\boldsymbol{u} \\ \boldsymbol{0} & \alpha\boldsymbol{W} \end{bmatrix} . \end{aligned}$$

Therefore, $\{1, \alpha\lambda_2, \alpha\lambda_3, \ldots, \alpha\lambda_r\}$ is the set of eigenvalues (counting multiplicities) of $\tilde{\boldsymbol{P}}$. In many cases, including that of the Google matrix, the eigenvalue 1 of $\boldsymbol{P}$ has multiplicity greater than one. Then, the second largest eigenvalue (in magnitude) of $\tilde{\boldsymbol{P}}$ is α. This is not the only irreducible model for the web and several variants have been proposed. For a more detailed treatment, we refer the reader to Langville and Meyer (2006).

15.7 Iterative algorithms for solving linear equations

Let us return to the basic problem of solving a system of n linear equations in n unknowns. Let $\boldsymbol{A}\boldsymbol{x} = \boldsymbol{b}$ be a linear system that is square. Iterative methods for solving such linear systems provide an attractive alternative to direct methods such as Gaussian elimination and $\boldsymbol{LU}$ decompositions and are especially effective when n is large and $\boldsymbol{A}$ is sparse.

The underlying idea is to solve the linear system $\boldsymbol{A}\boldsymbol{x} = \boldsymbol{b}$ by transforming it to an iterative system of the form

$$\boldsymbol{x}_{k+1} = \boldsymbol{T}\boldsymbol{x}_k + \boldsymbol{u} , \quad \text{for } k = 0, 1, 2, \ldots \tag{15.20}$$

and $\boldsymbol{x}_0$ is a fixed starting point of the scheme. Successive iterations allow us to express the k-th iterate in terms of $\boldsymbol{x}_0$:

$$\begin{aligned} &\boldsymbol{x}_1 = \boldsymbol{T}\boldsymbol{x}_0 + \boldsymbol{u} ; \; \boldsymbol{x}_2 = \boldsymbol{T}^2\boldsymbol{x}_0 + (\boldsymbol{I} + \boldsymbol{T})\boldsymbol{u} ; \; \boldsymbol{x}_3 = \boldsymbol{T}^3\boldsymbol{x}_0 + (\boldsymbol{I} + \boldsymbol{T} + \boldsymbol{T}^2)\boldsymbol{u} ; \\ &\ldots \; \boldsymbol{x}_k = \boldsymbol{T}^k\boldsymbol{x}_0 + (\boldsymbol{I} + \boldsymbol{T} + \cdots + \boldsymbol{T}^{k-1})\boldsymbol{u} . \end{aligned} \tag{15.21}$$

A general scheme for arriving at (15.20) or (15.21) is based upon *splitting* the matrix $\boldsymbol{A}$ into two parts: $\boldsymbol{A} = \boldsymbol{M} + \boldsymbol{N}$, where $\boldsymbol{M}$ is nonsingular. Then,

$$\boldsymbol{A}\boldsymbol{x} = \boldsymbol{b} \Longrightarrow \boldsymbol{M}\boldsymbol{x} = -\boldsymbol{N}\boldsymbol{x} + \boldsymbol{b} \Longrightarrow \boldsymbol{x} = -\boldsymbol{M}^{-1}\boldsymbol{N}\boldsymbol{x} + \boldsymbol{M}^{-1}\boldsymbol{b} ,$$

which yields the linear iterative scheme $\boldsymbol{x}_{k+1} = \boldsymbol{T}\boldsymbol{x}_k + \boldsymbol{u}$ for $k = 1, 2, \ldots$, where $\boldsymbol{T} = -\boldsymbol{M}^{-1}\boldsymbol{N}$ and $\boldsymbol{u} = \boldsymbol{M}^{-1}\boldsymbol{b}$. If $\kappa(\boldsymbol{T}) < 1$, then $\boldsymbol{T}^k \to \boldsymbol{O}$ as $k \to \infty$ and $(\boldsymbol{I} - \boldsymbol{T})^{-1}$ exists. Therefore, $\boldsymbol{A}^{-1}$ also exists. Taking limits in (15.21), we obtain

$$\begin{aligned}\lim_{k\to\infty} \boldsymbol{x}_k &= \lim_{k\to\infty} \boldsymbol{T}^k\boldsymbol{x}_0 + \lim_{k\to\infty} (\boldsymbol{I} + \boldsymbol{T} + \cdots + \boldsymbol{T}^{k-1})\boldsymbol{u} = (\boldsymbol{I} - \boldsymbol{T})^{-1}\boldsymbol{u} \\ &= (\boldsymbol{I} - \boldsymbol{T})^{-1}\boldsymbol{M}^{-1}\boldsymbol{b} = [\boldsymbol{M}(\boldsymbol{I} - \boldsymbol{T})]^{-1}\boldsymbol{b} = (\boldsymbol{M} + \boldsymbol{N})^{-1}\boldsymbol{b} = \boldsymbol{A}^{-1}\boldsymbol{b}\,. \end{aligned} \tag{15.22}$$

This shows that if we are able to split $\boldsymbol{A} = \boldsymbol{M} + \boldsymbol{N}$ in a way that $\boldsymbol{M}^{-1}$ exists and the spectral radius of $\boldsymbol{M}^{-1}\boldsymbol{N}$ is strictly less than one, then the linear iterative system in (15.20) will converge to the unique solution $\boldsymbol{A}^{-1}\boldsymbol{b}$ of $\boldsymbol{A}\boldsymbol{x} = \boldsymbol{b}$ for every $\boldsymbol{x}_0$.

15.7.1 The Jacobi method

Consider the system of linear equations $\boldsymbol{A}\boldsymbol{x} = \boldsymbol{b}$, where $\boldsymbol{A}$ is $n \times n$ and nonsingular. Let us split $\boldsymbol{A}$ as

$$\boldsymbol{A} = \boldsymbol{L} + \boldsymbol{D} + \boldsymbol{U}\,, \tag{15.23}$$

where $\boldsymbol{D}$ is a diagonal matrix containing the diagonal elements of $\boldsymbol{A}$, $\boldsymbol{L}$ is strictly lower-triangular containing entries below the diagonal of $\boldsymbol{A}$, and $\boldsymbol{U}$ is strictly upper-triangular containing entries above the diagonal of $\boldsymbol{A}$. This decomposition is a simple additive decomposition and has nothing to do with the $\boldsymbol{LU}$ or $\boldsymbol{LDU}$ factorizations obtained from Gaussian elimination. For example, if

$$\boldsymbol{A} = \begin{bmatrix} a_{11} & a_{12} & a_{13} \\ a_{21} & a_{22} & a_{23} \\ a_{31} & a_{32} & a_{33} \end{bmatrix},$$

then $\boldsymbol{A} = \boldsymbol{L} + \boldsymbol{D} + \boldsymbol{U}$, where

$$\boldsymbol{L} = \begin{bmatrix} 0 & 0 & 0 \\ a_{21} & 0 & 0 \\ a_{31} & a_{32} & 0 \end{bmatrix}, \quad \boldsymbol{D} = \begin{bmatrix} a_{11} & 0 & 0 \\ 0 & a_{22} & 0 \\ 0 & 0 & a_{33} \end{bmatrix} \text{ and } \boldsymbol{U} = \begin{bmatrix} 0 & a_{12} & a_{13} \\ 0 & 0 & a_{23} \\ 0 & 0 & 0 \end{bmatrix}.$$

We now rewrite the system

$$\boldsymbol{A}\boldsymbol{x} = (\boldsymbol{L} + \boldsymbol{D} + \boldsymbol{U})\boldsymbol{x} = \boldsymbol{b} \text{ as } \boldsymbol{D}\boldsymbol{x} = -(\boldsymbol{L} + \boldsymbol{U})\boldsymbol{x} + \boldsymbol{b}\,.$$

This yields the ***Jacobi*** fixed point iterative scheme for $k = 1, 2, \ldots$:

$$\boldsymbol{x}_{k+1} = \boldsymbol{T}\boldsymbol{x}_k + \boldsymbol{u}\,, \text{ where } \boldsymbol{T} = -\boldsymbol{D}^{-1}(\boldsymbol{L} + \boldsymbol{U}) \text{ and } \boldsymbol{u} = \boldsymbol{D}^{-1}\boldsymbol{b} \tag{15.24}$$

assuming that all the diagonal entries in $\boldsymbol{A}$, and hence in $\boldsymbol{D}$, are nonzero so that $\boldsymbol{D}^{-1}$ is well-defined. The explicit element-wise form of (15.24) is

$$x_i(k+1) = \sum_{\substack{j=1 \\ j\neq i}}^{n} \left(-\frac{a_{ij}}{a_{ii}}\right) x_i(k) + \frac{b_i}{a_{ii}}\,, \text{ for } k = 1, 2, \ldots, \tag{15.25}$$

where a_{ij}'s are the entries in $\boldsymbol{A}$, b_i's are the entries in $\boldsymbol{b}$ and $x_i(k+1)$ denotes the i-th entry in $\boldsymbol{x}_{k+1}$. The element-wise iterative scheme in (15.25) reveals that the algorithm does not depend upon the sequence in which the equations are considered and can process equations independently (or in parallel).

Jacobi's method will converge to the unique solution of $\boldsymbol{A}\boldsymbol{x} = \boldsymbol{b}$ for all initial vectors $\boldsymbol{x}_0$ and for all right-hand sides $\boldsymbol{b}$ when $\boldsymbol{A}$ is *strictly diagonally dominant*. Indeed, if $\boldsymbol{A}$ is strictly diagonally dominant, then $|a_{ii}| > \sum_{j \neq i} |a_{ij}|$ for each i. This implies that the absolute row sums of $\boldsymbol{T}$ are also less than one because

$$\sum_{j=1}^{n} |t_{ij}| = \frac{1}{|a_{ii}|} \sum_{j=1}^{n} |a_{ij}| < 1$$

and, hence, $\|\boldsymbol{T}\|_\infty < 1$. Using the bound in (15.14), we obtain that $\kappa(\boldsymbol{T}) < 1$ and so, as discussed in (15.22), the Jacobi iterations in (15.24) will converge.

15.7.2 The Gauss-Seidel method

The ***Gauss-Seidel*** method emerges from a slight modification of the Jacobi. Again, we split $\boldsymbol{A}$ as in (15.23) but we now rewrite the system

$$\boldsymbol{A}\boldsymbol{x} = (\boldsymbol{L} + \boldsymbol{D} + \boldsymbol{U})\boldsymbol{x} = \boldsymbol{b} \,, \text{ as } (\boldsymbol{L} + \boldsymbol{D})\boldsymbol{x} = -\boldsymbol{U}\boldsymbol{x} + \boldsymbol{b} \,.$$

This yields the ***Gauss-Seidel*** fixed point iterative scheme for $k = 1, 2, \ldots$:

$$\boldsymbol{x}_{k+1} = \boldsymbol{T}\boldsymbol{x}_k + \boldsymbol{u} \,, \text{ where } \boldsymbol{T} = -(\boldsymbol{L} + \boldsymbol{D})^{-1}\boldsymbol{U} \text{ and } \boldsymbol{u} = (\boldsymbol{L} + \boldsymbol{D})^{-1}\boldsymbol{b} \tag{15.26}$$

assuming $(\boldsymbol{L} + \boldsymbol{D})$ is nonsingular. Writing (15.26) as $(\boldsymbol{L} + \boldsymbol{D})\boldsymbol{x}_{k+1} = -\boldsymbol{U}\boldsymbol{x}_k + \boldsymbol{b}$, we obtain the following explicit element-wise equation:

$$a_{i1}x_1(k+1) + a_{i2}x_2(k+1) + \cdots + a_{i,i}x_i(k+1) = b_i - a_{i,i+1}x_{i+1}(k) - \cdots - a_{in}x_n(k) \,.$$

This can be rewritten as

$$x_i(k+1) = \frac{b_i - \sum_{j=1}^{i-1} a_{ij}x_j(k+1) - \sum_{j=i+1}^{n} a_{ij}x_j(k)}{a_{ii}} \,, \tag{15.27}$$

for $k = 1, 2, \ldots,$. Thus, the Gauss-Seidel scheme can be looked upon as a refinement over the Jacobi method. It evaluates $x_i(k+1)$ by using the latest (and improved) values $x_1(k+1), x_2(k+1), \ldots, x_{i-1}(k+1)$ in the current iterate along with $x_{i+1}(k), x_{i+2}(k), \ldots, x_n(k)$ from the previous iterate. As was the case for the Jacobi method, if $\boldsymbol{A}$ is strictly diagonally dominant then Gauss-Seidel too is guaranteed to converge. The proof is a bit more involved (but not that difficult) than for the Jacobi method.

The Gauss-Seidel iteration scheme in (15.27) is particularly effective with serial computers because one can easily replace $x_i(k)$ with its updated value $x_i(k+1)$ and save on storage. On the other hand, the Jacobi method will require storage of the entire vector $\boldsymbol{x}_k$ from the previous iterate until $\boldsymbol{x}_{k+1}$ has been computed. However, the Jacobi method can be very effective in parallel or distributed computing environments are widely used for solving large and sparse systems today. With regard to convergence, since Gauss-Seidel makes use of the newer updates it often outperforms the Jacobi method, although this is not universally the case.

15.7.3 The Successive Over-Relaxation (SOR) method

Successive Over-Relaxation (SOR) attempts to improve the Gauss-Seidel method by introducing a nonzero scalar parameter α to be adjusted by the user. We decompose $\boldsymbol{A}$ as in (15.23) but write it as

$$\boldsymbol{A} = \boldsymbol{L} + \boldsymbol{D} + \boldsymbol{U} = (\boldsymbol{L} + \alpha\boldsymbol{D}) + ((1-\alpha)\boldsymbol{D} + \boldsymbol{U}) \; . \tag{15.28}$$

We rewrite the linear system $\boldsymbol{A}\boldsymbol{x} = [(\boldsymbol{L} + \alpha\boldsymbol{D}) + ((1-\alpha)\boldsymbol{D} + \boldsymbol{U})]\,\boldsymbol{x} = \boldsymbol{b}$ as

$$(\boldsymbol{L} + \alpha\boldsymbol{D})\,\boldsymbol{x} = -\left((1-\alpha)\boldsymbol{D} + \boldsymbol{U}\right)\boldsymbol{x} + \boldsymbol{b} \, ,$$

which leads to the following SOR iterative scheme for $k = 1, 2, \ldots,$:

$$(\boldsymbol{L} + \alpha\boldsymbol{D})\,\boldsymbol{x}_{k+1} = -\left((1-\alpha)\boldsymbol{D} + \boldsymbol{U}\right)\boldsymbol{x}_k + \boldsymbol{b} \, . \tag{15.29}$$

This, for numerical purposes, is often more conveniently expressed as

$$(\omega\boldsymbol{L} + \boldsymbol{D})\,\boldsymbol{x}_{k+1} = -\left((\omega-1)\boldsymbol{D} + \omega\boldsymbol{U}\right)\boldsymbol{x}_k + \omega\boldsymbol{b} \, , \tag{15.30}$$

where $\omega = 1/\alpha$. Observe that the SOR iterations reduce to the Gauss-Seidel iterations in (15.26) when $\omega = 1$. The element-wise version of the SOR is obtained from the i-th row of (15.30) as

$$x_i(k+1) = (1-\omega)x_i(k) + \omega\frac{b_i - \sum_{j=1}^{i-1} a_{ij}x_i(k+1) - \sum_{j=i+1}^{n} a_{ij}x_i(k)}{a_{ii}} \tag{15.31}$$

for $k = 1, 2, \ldots$. As in the Gauss-Seidel approach, the $x_i(k)$'s are updated sequentially for $i = 1, 2, \ldots, n$. Equation 15.31 reveals that the SOR method is, in fact, a weighted average of the previous iterate and the Gauss-Seidel iterate. When $\omega = 1$, we obtain the Gauss-Seidel iterate. If $\omega < 1$, we obtain an *under-relaxed* method, while $\omega > 1$ produces an *over-relaxed* method. Over-relaxation is the scheme that works well in most practical instances.

The spectral radius $\kappa(\boldsymbol{T}_\omega)$, where $\boldsymbol{T}_\omega = -(\omega\boldsymbol{L} + \boldsymbol{D})^{-1}((\omega-1)\boldsymbol{D} + \omega\boldsymbol{U})$, determines the rate of convergence of the SOR method. Numerical experiments reveal that clever choices of ω will produce substantially accelerated convergence (see, e.g., Olver and Shakiban, 2006).

15.7.4 The conjugate gradient method

We now consider the system $\boldsymbol{A}\boldsymbol{x} = \boldsymbol{b}$, where $\boldsymbol{A}$ is $n \times n$ and positive definite with real entries. Since positive definite matrices are nonsingular, the system has a unique solution. Positive definite systems appear frequently in statistics and econometrics and can be solved directly using the Cholesky decomposition of $\boldsymbol{A}$. This is effective but can be slow for large systems.

The conjugate gradient method transforms the problem of solving a linear system into one of optimization. The method is based upon the observation that the unique solution of $\boldsymbol{A}\boldsymbol{x} = \boldsymbol{b}$ also happens to be the minimizer of the quadratic function

$$f(\boldsymbol{x}) = \frac{1}{2}\boldsymbol{x}'\boldsymbol{A}\boldsymbol{x} - \boldsymbol{b}'\boldsymbol{x} \ . \tag{15.32}$$

This is a consequence of Theorem 13.4. Taking $c = 0$ in Theorem 13.4 yields

$$\boldsymbol{x}'\boldsymbol{A}\boldsymbol{x} - 2\boldsymbol{b}'\boldsymbol{x} = (\boldsymbol{x} - \boldsymbol{A}^{-1}\boldsymbol{b})'\boldsymbol{A}(\boldsymbol{x} - \boldsymbol{A}^{-1}\boldsymbol{b}) - \boldsymbol{b}'\boldsymbol{A}^{-1}\boldsymbol{b} < (\boldsymbol{x} - \boldsymbol{A}^{-1}\boldsymbol{b})'\boldsymbol{A}(\boldsymbol{x} - \boldsymbol{A}^{-1}\boldsymbol{b})$$

because $\boldsymbol{b}'\boldsymbol{A}^{-1}\boldsymbol{b} > 0$ whenever $\boldsymbol{b} \neq \boldsymbol{0}$ (recall that $\boldsymbol{A}^{-1}$ is also positive definite whenever $\boldsymbol{A}$ is). Alternatively, using calculus, we see that the gradient of $f(\boldsymbol{x})$ in (15.32) is $\nabla f(\boldsymbol{x}) = \boldsymbol{A}\boldsymbol{x} - \boldsymbol{b}$. The extrema of $f(\boldsymbol{x})$ is attained at the points where $\nabla f(\boldsymbol{x}) = \boldsymbol{0}$. Since $\boldsymbol{A}$ is positive definite, the solution of $\boldsymbol{A}\boldsymbol{x} = \boldsymbol{b}$ is the unique minimizer for $f(\boldsymbol{x})$. So, we can solve $\boldsymbol{A}\boldsymbol{x} = \boldsymbol{b}$ by minimizing $f(\boldsymbol{x})$.

The *method of steepest descent* is one approach to minimize $f(\boldsymbol{x})$. Here, we define the *residual* vector

$$\boldsymbol{r}(\boldsymbol{x}) = -\nabla f(\boldsymbol{x}) = \boldsymbol{b} - \boldsymbol{A}\boldsymbol{x} \ \text{ for any } \ \boldsymbol{x} \in \Re^n \ . \tag{15.33}$$

Note that $\boldsymbol{r}(\boldsymbol{x}) = \boldsymbol{0}$ if and only if $\boldsymbol{x}$ solves $\boldsymbol{A}\boldsymbol{x} = \boldsymbol{b}$. The residual $\boldsymbol{r}(\boldsymbol{x})$ provides a measure of how close $\boldsymbol{x}$ comes to solving the system and it also provides the direction of steepest change in $f(\boldsymbol{x})$. This suggests the following iterative system

$$\boldsymbol{x}_{k+1} = \boldsymbol{x}_k + \alpha_k \boldsymbol{r}_k \ , \ \text{ where } \ \boldsymbol{r}_k := \boldsymbol{r}(\boldsymbol{x}_k) = \boldsymbol{b} - \boldsymbol{A}\boldsymbol{x}_k \ . \tag{15.34}$$

The scalar α_k is chosen to minimize $f(\boldsymbol{x}_{k+1})$ (treated as a function of α_k). Note that

$$\begin{aligned} f(\boldsymbol{x}_{k+1}) &= \frac{1}{2}\boldsymbol{x}_{k+1}'\boldsymbol{A}\boldsymbol{x}_{k+1} - \boldsymbol{b}'\boldsymbol{x}_{k+1} \ , \ \text{ where } \boldsymbol{x}_{k+1} \text{ is as in (15.34)} \ , \\ &= \frac{1}{2}(\boldsymbol{r}_k'\boldsymbol{A}\boldsymbol{r}_k)\alpha_k^2 - [\boldsymbol{r}_k'(\boldsymbol{b} - \boldsymbol{A}\boldsymbol{x}_k)]\alpha_k + f(\boldsymbol{x}_k) \\ &= \frac{1}{2}(\boldsymbol{r}_k'\boldsymbol{A}\boldsymbol{r}_k)\alpha_k^2 - (\boldsymbol{r}_k'\boldsymbol{r}_k)\alpha_k + f(\boldsymbol{x}_k) \ . \end{aligned}$$

Since $f(\boldsymbol{x}_k)$ does not depend upon α_k, it is easily verified that the minimum value of $f(\boldsymbol{x}_{k+1})$ is attained at $\alpha_k = \dfrac{\boldsymbol{r}_k'\boldsymbol{r}_k}{\boldsymbol{r}_k'\boldsymbol{A}\boldsymbol{r}_k}$. While this choice of α_k ensures that successive residuals are orthogonal, i.e., $\boldsymbol{r}_k \perp \boldsymbol{r}_{k+1}$, the convergence of this steepest descent algorithm can be slow in practice, especially if the ratio $|\lambda_{\max}/\lambda_{min}|$ becomes large, where $\lambda_{\max}$ and $\lambda_{\min}$ are the maximum and minimum eigenvalues of $\boldsymbol{A}$.

A much more competitive algorithm is obtained if we modify the steepest descent iteration in (15.34) by replacing the $\boldsymbol{r}_k$'s with *conjugate vectors* with respect to $\boldsymbol{A}$. Two vectors, $\boldsymbol{q}_i$ and $\boldsymbol{q}_j$, are said to be conjugate with respect to $\boldsymbol{A}$ if $\boldsymbol{q}_i'\boldsymbol{A}\boldsymbol{q}_j = 0$. The conjugate gradient algorithm can be described in the following steps.

- We start with an initial vector $\boldsymbol{x}_0$. For convenience, we set $\boldsymbol{x}_0 = \boldsymbol{0}$. Thus, the initial residual vector is $\boldsymbol{r}_0 := \boldsymbol{b} - \boldsymbol{A}\boldsymbol{x}_0 = \boldsymbol{b}$.

- We take the first conjugate vector to be $\boldsymbol{q}_1 = \boldsymbol{r}_0$ and construct the first iterate

$$\boldsymbol{x}_1 = \boldsymbol{x}_0 + \alpha_1 \boldsymbol{q}_1 = \alpha_1 \boldsymbol{q}_1 \;, \quad \text{where } \boldsymbol{q}_1 = \boldsymbol{r}_0 \text{ and } \alpha_1 = \frac{\|\boldsymbol{r}_0\|^2}{\boldsymbol{q}_1'\boldsymbol{A}\boldsymbol{q}_1} \;.$$

Here, α_1 is chosen so that the corresponding residual

$$\boldsymbol{r}_1 = \boldsymbol{b} - \boldsymbol{A}\boldsymbol{x}_1 = \boldsymbol{r}_0 - \alpha_1 \boldsymbol{A}\boldsymbol{q}_1 = \boldsymbol{r}_0 - \alpha_1 \boldsymbol{A}\boldsymbol{r}_0$$

is orthogonal to $\boldsymbol{r}_0$. This yields

$$0 = \boldsymbol{r}_0'\boldsymbol{r}_1 = \boldsymbol{r}_0'(\boldsymbol{r}_0 - \alpha_1 \boldsymbol{A}\boldsymbol{r}_0) = \|\boldsymbol{r}_0\|^2 - \alpha_1 \boldsymbol{r}_0'\boldsymbol{A}\boldsymbol{r}_0$$
$$\Longrightarrow \alpha_1 = \frac{\|\boldsymbol{r}_0\|^2}{\boldsymbol{r}_0'\boldsymbol{A}\boldsymbol{r}_0} = \frac{\|\boldsymbol{r}_0\|^2}{\boldsymbol{q}_1'\boldsymbol{A}\boldsymbol{q}_1} \;.$$

This step is essentially the same as in the method of steepest descent.

- In the second step, we depart from the method of steepest descent by choosing the second direction as $\boldsymbol{q}_2 = \boldsymbol{r}_1 + \theta_1 \boldsymbol{q}_1$, where θ_1 is a scalar that will be chosen so that $\boldsymbol{q}_2'\boldsymbol{A}\boldsymbol{q}_1 = 0$. This yields

$$0 = \boldsymbol{q}_2'\boldsymbol{A}\boldsymbol{q}_1 = (\boldsymbol{r}_1 + \theta_1 \boldsymbol{q}_1)'\boldsymbol{A}\boldsymbol{q}_1 = \boldsymbol{r}_1'\boldsymbol{A}\boldsymbol{q}_1 + \theta_1 \boldsymbol{q}_1'\boldsymbol{A}\boldsymbol{q}_1 \Longrightarrow \theta_1 = -\frac{\boldsymbol{r}_1'\boldsymbol{A}\boldsymbol{q}_1}{\boldsymbol{q}_1'\boldsymbol{A}\boldsymbol{q}_1} \;.$$

A further simplification in the expression for θ_1 is possible. Note that

$$\boldsymbol{r}_1'\boldsymbol{A}\boldsymbol{q}_1 = \boldsymbol{r}_1'\left(\boldsymbol{r}_0 - \frac{1}{\alpha_1}\boldsymbol{r}_1\right) = \boldsymbol{r}_1'\boldsymbol{r}_0 - \frac{1}{\alpha_1}\|\boldsymbol{r}_1\|^2 = -\frac{\|\boldsymbol{r}_1\|^2}{\alpha_1}$$

because $\boldsymbol{r}_1'\boldsymbol{r}_0 = 0$. Therefore,

$$\theta_1 = -\frac{\boldsymbol{r}_1'\boldsymbol{A}\boldsymbol{q}_1}{\boldsymbol{q}_1'\boldsymbol{A}\boldsymbol{q}_1} = \frac{\|\boldsymbol{r}_1\|^2}{\alpha_1 \boldsymbol{q}_1'\boldsymbol{A}\boldsymbol{q}_1} = \frac{\|\boldsymbol{r}_1\|^2}{\|\boldsymbol{r}_0\|^2} \;.$$

Therefore, the second conjugate gradient direction is

$$\boldsymbol{q}_2 = \boldsymbol{r}_1 + \theta_1 \boldsymbol{q}_1 = \boldsymbol{r}_1 + \frac{\|\boldsymbol{r}_1\|^2}{\|\boldsymbol{r}_0\|^2}\boldsymbol{q}_1 \;.$$

The second step of the conjugate gradient iteration scheme will update $\boldsymbol{x}_2 = \boldsymbol{x}_1 + \alpha_2 \boldsymbol{q}_2$, where α_2 is chosen so that the residual vector

$$\boldsymbol{r}_2 = \boldsymbol{b} - \boldsymbol{A}\boldsymbol{x}_2 = \boldsymbol{b} - \boldsymbol{A}(\boldsymbol{x}_1 + \alpha_2 \boldsymbol{q}_2) = \boldsymbol{r}_1 - \alpha_2 \boldsymbol{A}\boldsymbol{q}_2$$

is orthogonal to $\boldsymbol{r}_1$. This yields

$$\begin{aligned} 0 &= \boldsymbol{r}_1'\boldsymbol{r}_2 = \boldsymbol{r}_1'(\boldsymbol{r}_1 - \alpha_2 \boldsymbol{A}\boldsymbol{q}_2) = \|\boldsymbol{r}_1\|^2 - \alpha_2 \boldsymbol{r}_1'\boldsymbol{A}\boldsymbol{q}_2 \\ &= \|\boldsymbol{r}_1\|^2 - \alpha_2(\boldsymbol{q}_2 - \theta_1 \boldsymbol{q}_1)'\boldsymbol{A}\boldsymbol{q}_2 = \|\boldsymbol{r}_1\|^2 - \alpha_2 \boldsymbol{q}_2'\boldsymbol{A}\boldsymbol{q}_2 + \alpha_2 \theta_1 \boldsymbol{q}_1'\boldsymbol{A}\boldsymbol{q}_2 \\ &= \|\boldsymbol{r}_1\|^2 - \alpha_2 \boldsymbol{q}_2'\boldsymbol{A}\boldsymbol{q}_2 \;\text{ because }\; \boldsymbol{q}_1'\boldsymbol{A}\boldsymbol{q}_2 = 0 \;. \end{aligned}$$

This implies that $\alpha_2 = \dfrac{\|\boldsymbol{r}_1\|^2}{\boldsymbol{q}_2'\boldsymbol{A}\boldsymbol{q}_2}$. The second iteration of the conjugate gradient algorithm can now be summarized as

$$\boldsymbol{x}_2 = \boldsymbol{x}_1 + \frac{\|\boldsymbol{r}_1\|^2}{\boldsymbol{q}_2'\boldsymbol{A}\boldsymbol{q}_2}\boldsymbol{q}_2 = \frac{\|\boldsymbol{r}_0\|^2}{\boldsymbol{r}_0'\boldsymbol{A}\boldsymbol{r}_0}\boldsymbol{q}_1 + \frac{\|\boldsymbol{r}_1\|^2}{\boldsymbol{q}_2'\boldsymbol{A}\boldsymbol{q}_2}\boldsymbol{q}_2 \;. \tag{15.35}$$

- Proceeding in this manner, we can describe a generic iteration of the conjugate gradient method as follows. Starting with $\boldsymbol{x}_0 = \boldsymbol{0}$ (for convenience), we compute the residual $\boldsymbol{r}_0 = \boldsymbol{b} - \boldsymbol{A}\boldsymbol{x}_0 = \boldsymbol{b}$ and set the first conjugate direction as $\boldsymbol{q}_1 = \boldsymbol{r}_0$. Then, we compute the following:

$$\boldsymbol{q}_{k+1} = \boldsymbol{r}_k + \frac{\|\boldsymbol{r}_k\|^2}{\|\boldsymbol{r}_{k-1}\|^2}\boldsymbol{q}_k \; ; \tag{15.36}$$

$$\boldsymbol{x}_{k+1} = \boldsymbol{x}_k + \frac{\|\boldsymbol{r}_k\|^2}{\boldsymbol{q}'_{k+1}\boldsymbol{A}\boldsymbol{q}_{k+1}}\boldsymbol{q}_{k+1} \; ; \tag{15.37}$$

$$\boldsymbol{r}_{k+1} = \boldsymbol{b} - \boldsymbol{A}\boldsymbol{x}_{k+1} = \boldsymbol{r}_k - \frac{\|\boldsymbol{r}_k\|^2}{\boldsymbol{q}'_{k+1}\boldsymbol{A}\boldsymbol{q}_{k+1}}\boldsymbol{A}\boldsymbol{q}_{k+1} \,. \tag{15.38}$$

  Each conjugate direction vector is of the form $\boldsymbol{q}_{k+1} = \boldsymbol{r}_k + \theta_k\boldsymbol{q}_k$, where $\theta_k = \|\boldsymbol{r}_k\|^2/\|\boldsymbol{r}_{k-1}\|^2$ ensures that $\boldsymbol{q}'_i\boldsymbol{A}\boldsymbol{q}_{k+1} = 0$ for $i = 1, 2, \ldots, k$. Each update of the solution approximation is of the form $\boldsymbol{x}_{k+1} = \boldsymbol{x}_k + \alpha_{k+1}\boldsymbol{q}_{k+1}$, where $\alpha_{k+1} = \|\boldsymbol{r}_k\|^2/\boldsymbol{q}'_{k+1}\boldsymbol{A}\boldsymbol{q}_{k+1}$ ensures that $\boldsymbol{r}'_k\boldsymbol{r}_{k+1} = 0$. Observe that the solution approximation $\boldsymbol{x}_{k+1}$ belongs to $Sp\{\boldsymbol{q}_1, \boldsymbol{q}_2, \ldots, \boldsymbol{q}_{k+1}\}$ because

$$\boldsymbol{x}_{k+1} = \boldsymbol{x}_k + \alpha_{k+1}\boldsymbol{q}_{k+1} = \alpha_1\boldsymbol{q}_1 + \alpha_2\boldsymbol{q}_2 + \cdots + \alpha_{k+1}\boldsymbol{q}_{k+1} \,.$$

The conjugate gradient method has some remarkable features. Unlike purely iterative methods such as the Jacobi, Gauss-Seidel and SOR, the conjugate gradient method is guaranteed to eventually terminate at the *exact* solution. This is because the n conjugate directions form an "orthogonal" basis of $\Re^n$, where "orthogonality" (sometimes called $\boldsymbol{A}$-orthogonality) means that $\boldsymbol{q}'_i\boldsymbol{A}\boldsymbol{q}_j = 0$. The residual vector $\boldsymbol{r}_n = \boldsymbol{b} - \boldsymbol{A}\boldsymbol{x}_n$, where $\boldsymbol{x}_n = \sum_{i=1}^{n}\alpha_i\boldsymbol{q}_i$, is, by construction, orthogonal to all the vectors in the conjugate basis $\{\boldsymbol{q}_1, \boldsymbol{q}_2, \ldots, \boldsymbol{q}_n\}$. Therefore $\boldsymbol{r}_n = \boldsymbol{0}$ and indeed $\boldsymbol{x}_n$ is the exact solution of $\boldsymbol{A}\boldsymbol{x} = \boldsymbol{b}$. Thus, the conjugate gradient method exactly solves a positive definite linear system without incurring the expenses involved in directly solving one. Most computations are rapidly performed using Euclidean dot products and one matrix-vector multiplication to evaluate $\boldsymbol{A}\boldsymbol{q}_k$. Furthermore, the method produces successive iterates that converge to the true solution. Sufficiently close approximations are often obtained in a few iterations, so the process can be terminated much earlier in practice.

15.8 Exercises

The problems below assume that the reader is familiar with elementary real analysis.

1. Find the general solution of $\boldsymbol{x}_{k+1} = (\alpha\boldsymbol{A} + \beta\boldsymbol{I})\boldsymbol{x}_k$, where α and β are fixed constants, in terms of the general solution of $\boldsymbol{x}_{k+1} = \boldsymbol{A}\boldsymbol{x}_k$.
2. True or false: If $\lim_{k\to\infty}\boldsymbol{A}^k = \boldsymbol{B}$, then $\boldsymbol{B}$ is idempotent.
3. Let $\boldsymbol{x} = \begin{bmatrix} 1 \\ 2 \\ 3 \\ 4 \end{bmatrix}$. Find $\|\boldsymbol{x}\|_1$, $\|\boldsymbol{x}\|_2$ and $\|\boldsymbol{x}\|_\infty$.

4. If $\|\boldsymbol{x} - \boldsymbol{y}\|_2 = \|\boldsymbol{x} + \boldsymbol{y}\|_2$, then find $\boldsymbol{x}'\boldsymbol{y}$, where $\boldsymbol{x}, \boldsymbol{y} \in \Re^n$.
5. If $\boldsymbol{x}_1, \boldsymbol{x}_2, \ldots, \boldsymbol{x}_n, \ldots$ is an infinite sequence of vectors in $\Re^n$, then prove that $\lim_{n\to\infty} \boldsymbol{x}_n = \boldsymbol{a}$ if and only if $\lim_{n\to\infty} \|\boldsymbol{x}_n - \boldsymbol{a}\| = 0$, where $\|\cdot\|$ is a vector norm.
6. True or false: $\|\boldsymbol{I}_n\| = 1$ for every induced matrix norm.
7. Show that $\|\boldsymbol{A}\|_F = \|\boldsymbol{A}'\|_F$ and $\|\boldsymbol{A}\|_2 = \|\boldsymbol{A}'\|_2$, where $\boldsymbol{A}$ is a real matrix.
8. If $\boldsymbol{A}$ is a real matrix, then show that $\left\| \begin{bmatrix} \boldsymbol{A} & \boldsymbol{O} \\ \boldsymbol{O} & \boldsymbol{B} \end{bmatrix} \right\|_2 = \max\{\|\boldsymbol{A}\|_2, \|\boldsymbol{B}\|_2\}$.
9. If $\boldsymbol{A}$ is $m \times n$, then show that $\|\boldsymbol{U}'\boldsymbol{A}\boldsymbol{V}\|_2 = \|\boldsymbol{A}\|_2$ when $\boldsymbol{U}\boldsymbol{U}' = \boldsymbol{I}_m$ and $\boldsymbol{V}'\boldsymbol{V} = \boldsymbol{I}_n$.
10. Find a matrix $\boldsymbol{A}$ such that $\|\boldsymbol{A}^2\| \neq \|\boldsymbol{A}\|^2$ under the 2- and ∞-norms.
11. Let $\boldsymbol{A} = \boldsymbol{P}\boldsymbol{J}\boldsymbol{P}^{-1}$ be the JCF of $\boldsymbol{A}$. Show that $\boldsymbol{A}^k = \boldsymbol{P}\boldsymbol{J}^k\boldsymbol{P}^{-1}$, where $\boldsymbol{J}^k$ is as described in Theorem 15.3.
12. True or false: $\kappa(\boldsymbol{A} + \boldsymbol{B}) = \kappa(\boldsymbol{A}) + \kappa(\boldsymbol{B})$.
13. True or false: Similar matrices have the same spectral radius.
14. If $\boldsymbol{A}$ is nonsingular, then prove that $\|\boldsymbol{A}^{-1}\|_2 = \dfrac{1}{\min_{\|\boldsymbol{x}\|=1} \|\boldsymbol{A}\boldsymbol{x}\|_2}$.
15. If $\sigma_1 \geq \sigma_2 \geq \cdots \sigma_r > 0$ are the positive singular values of $\boldsymbol{A}$, then prove that
$$\|\boldsymbol{A}\|_2 = \sigma_1 \text{ and } \|\boldsymbol{A}\|_F = \sqrt{\sigma_1^2 + \sigma_2^2 + \cdots + \sigma_r^2}\ .$$
16. Let $\boldsymbol{A}_1, \boldsymbol{A}_2, \ldots, \boldsymbol{A}_n, \ldots,$ be an infinite sequence of matrices of the same order. Prove that
$$\sum_{n=0}^{\infty} \|\boldsymbol{A}_n\|_\infty < \infty \Longrightarrow \sum_{n=0}^{\infty} \boldsymbol{A}_n = \boldsymbol{A}_*$$
for some matrix $\boldsymbol{A}_*$ with finite entries.
17. Show that $\exp(\boldsymbol{A}) = \sum_{k=0}^{\infty} \dfrac{\boldsymbol{A}^k}{k!}$ converges for any matrix $\boldsymbol{A}$ with finite real entries.
18. If $\boldsymbol{A} = \begin{bmatrix} \lambda & 1 \\ 0 & \lambda \end{bmatrix}$, prove that $\exp(\boldsymbol{A}) = \exp(\lambda) \begin{bmatrix} 1 & 1 \\ 0 & 1 \end{bmatrix}$.
19. Find the singular values of a symmetric idempotent matrix.
20. If $\boldsymbol{P}_{\boldsymbol{A}}$ is an orthogonal projector onto $\mathcal{C}(\boldsymbol{A})$, where $\boldsymbol{A}$ has full column rank, then find $\|\boldsymbol{P}_{\boldsymbol{A}}\|_2$.
21. Using the SVD, prove the fundamental theorem of ranks:
$$\rho(\boldsymbol{A}) = \rho(\boldsymbol{A}'\boldsymbol{A}) = \rho(\boldsymbol{A}\boldsymbol{A}') = \rho(\boldsymbol{A})\ .$$
22. If $\sigma_1 \geq \sigma_2 \geq \cdots \sigma_r > 0$ are the positive singular values of $\boldsymbol{A}$ and $\|\boldsymbol{B}\|_2 < \sigma_r$, then prove that $\rho(\boldsymbol{A} + \boldsymbol{B}) \geq \rho(\boldsymbol{A})$.
23. Let $\sigma_1 \geq \sigma_2 \geq \cdots \geq \sigma_n$ be the singular values of $\boldsymbol{A}$. Prove that $\lim_{k\to\infty} (\boldsymbol{A}'\boldsymbol{A})^k = \boldsymbol{O}$ if and only if $\sigma_1 < 1$.

24. True or false: If all the singular values of $\boldsymbol{A}$ are strictly less than one, then $\lim_{k\to\infty} \boldsymbol{A}^k = \boldsymbol{O}$.

25. If $\boldsymbol{A}$ is symmetric and strictly diagonally dominant with all positive diagonal entries, then show that $\boldsymbol{A}$ is positive definite.

26. Prove that if $\boldsymbol{P}_1$ and $\boldsymbol{P}_2$ are $n \times n$ transition matrices, then so is $\boldsymbol{P}_1\boldsymbol{P}_2$.

27. True or false: If $\boldsymbol{P}$ is a transition matrix, then so is $\boldsymbol{P}^{-1}$.

28. A doubly stochastic transition matrix has both its row sums and column sums equal to one. Find the stationary distribution of a Markov chain with a doubly stochastic transition matrix.

CHAPTER 16

Abstract Linear Algebra

Linear algebra, it turns out, is not just about manipulating matrices. It is about understanding *linearity*, which arises in contexts far beyond systems of linear equations involving real numbers. Functions on the real line that satisfy $f(x+y) = f(x)+f(y)$ and $f(\alpha x) = \alpha f(x)$ are called ***linear functions*** or ***linear mappings***. A very simple example of a linear function on $\Re$ is the straight line through the origin $f(x) = \alpha x$. But linearity is encountered in more abstract concepts as well. Consider, for instance, the task of taking the derivative of a function. If we treat this as an "operation" on any function $f(x)$, denoted by $D_x(f) = df(x)/dx$, then

$$D_x(f+g) = \frac{d}{dx}(f(x)+g(x)) = \frac{d}{dx}f(x) + \frac{d}{dx}g(x) = D_x(f+g) \text{ and} \tag{16.1}$$

$$D_x(\alpha f) = \frac{d}{dx}(\alpha f(x)) = \alpha \frac{d}{dx}f(x) = \alpha D_x(f) \, . \tag{16.2}$$

This shows that D_x behaves like a like a linear mapping. Integration, too, has similar properties. Linear algebra can be regarded as a branch of mathematics concerning linear mappings over vector spaces in as general a setting as possible. Such settings go well beyond the field of real and complex numbers. This chapter offers a brief and brisk overview of more abstract linear algebra. For a more detailed and rigorous treatment, we refer to Hoffman and Kunze (1971), Halmos (1993) and Axler (2004) among several other excellent texts. We will see that most of the results obtained for Euclidean spaces have analogues in more general vector spaces. We will state many of them but leave most of the proofs to the reader because they are easily constructed by imitating their counterparts in Euclidean spaces.

16.1 General vector spaces

Chapter 4 dealt with Euclidean vector spaces, where the "vectors" were points in $\Re^n$. These were n-tuples of real numbers. While we rarely alluded to complex numbers, much of the results in Chapter 4 were applicable to points in $\mathbb{C}^n$, i.e., n-tuples of complex numbers. To bring these different vector spaces under a common conceptual framework, so that the these concepts can be applied in broader contexts, we provide a more general treatment of vector spaces. We begin with the definition of a *field*.

Definition 16.1 *A* **field** *is a set $\mathcal{F}$ equipped with two operations:*

(i) Addition: *denoted by "+" that acts on two elements $x, y \in \mathcal{F}$ to produce another element $x + y \in \mathcal{F}$;*

(ii) Multiplication: *denoted by "$\cdot$" that acts on two elements $x, y \in \mathcal{F}$ to produce another element $x \cdot y \in \mathcal{F}$. The element $x \cdot y$ is often denoted simply as xy.*

A field is denoted by $(\mathcal{F}, +, \cdot)$. Addition and multiplication satisfy the following axioms:

[A1] Addition is associative: $x + (y + z) = (x + y) + z$ *for all* $x, y, z \in \mathcal{F}$.

[A2] Addition is commutative: $x + y = y + x$ *for all* $x, y \in \mathcal{F}$.

[A3] Additive identity: $0 \in \mathcal{F}$ *such that* $x + 0 = x$ *for all* $x \in \mathcal{F}$.

[A4] Additive inverse: *For every $x \in \mathcal{F}$, there exists an element in $-x \in \mathcal{F}$ such that $x + (-x) = 0$.*

[M1] Multiplication is associative: $x \cdot (y \cdot z) = (x \cdot y) \cdot z$ *for all* $x, y, z \in \mathcal{F}$.

[M2] Multiplication is commutative: $x \cdot y = y \cdot x$ *for all* $x, y \in \mathcal{F}$.

[M3] Multiplicative identity: *There exists an element in $\mathcal{F}$, denoted by* 1*, such that $x \cdot 1 = x$ for all $x \in \mathcal{F}$.*

[M4] Multiplicative inverse: *For every $x \in \mathcal{F}$, $x \neq 0$, there exists an element $x^{-1} \in \mathcal{F}$ such that $x \cdot x^{-1} = 1$.*

[D] Distributive: $x \cdot (y + z) = x \cdot y + y \cdot z$ *for all* $x, y, z \in \mathcal{F}$.

A field is a particular ***algebraic structure***. Readers familiar with *groups* will recognize that the set $(\mathcal{F}, +)$ forms an ***abelian group***. Axioms (A1)–(A4) ensures this. Without the commutativity (A2) it would be simply a ***group***. Axioms (M1)–(M4) imply that $(\mathcal{F} \setminus \{0\}, \cdot)$ is also an abelian group. Generally, when no confusion arises, we will write $x \cdot y$ as xy, without the $\cdot$.

Example 16.1 Examples of fields. Following are some examples of fields. The first two are most common.

1. $(\Re, +, \cdot)$ is a field with $+, \cdot$ being ordinary addition and multiplication, respectively.
2. $(\mathrm{C}, +, \cdot)$ with $+, \cdot$ with standard addition and multiplication of complex numbers.
3. The set of rational numbers consisting of numbers which can be written as fractions a/b, where a and $b \neq 0$ are integers.
4. Finite fields are fields with finite numbers. An important example is $(\mathcal{F}, +, \cdot)$, where $\mathcal{F} = \{1, 2, \ldots, p - 1\}$ with p being a prime number, and the operations of addition and multiplication are defined by performing the operation in the set of integers Z, dividing by p and taking the remainder.

Vector spaces can be defined over any field in the following manner.

Definition 16.2 *A* **vector space** *over a field $\mathcal{F}$ is the set $\mathcal{V}$ equipped with two operations:*

(i) Vector addition: *denoted by "+" adds two elements $\boldsymbol{x}, \boldsymbol{y} \in \mathcal{V}$ to produce another element $\boldsymbol{x} + \boldsymbol{y} \in \mathcal{V}$;*

(ii) Scalar multiplication: *denoted by "$\cdot$" multiplies a vector $\boldsymbol{x} \in \mathcal{V}$ with a scalar $\alpha \in \mathcal{F}$ to produce another vector $\alpha \cdot \boldsymbol{x} \in \mathcal{V}$. We usually omit the "$\cdot$" and simply write this vector as $\alpha \boldsymbol{x}$.*

The vector space is denoted as $(\mathcal{V}, \mathcal{F}, +, \cdot)$. The operations of vector addition and scalar multiplication satisfy the following axioms.

[A1] Vector addition is commutative: $\boldsymbol{x} + \boldsymbol{y} = \boldsymbol{y} + \boldsymbol{x}$ *for every* $\boldsymbol{x}, \boldsymbol{y} \in \mathcal{V}$.

[A2] Vector addition is associative: $(\boldsymbol{x}+\boldsymbol{y})+\boldsymbol{z} = \boldsymbol{x}+(\boldsymbol{y}+\boldsymbol{z})$ *for every* $\boldsymbol{x}, \boldsymbol{y}, \boldsymbol{z} \in \mathcal{V}$.

[A3] Additive identity: *There is an element* $\mathbf{0} \in \mathcal{V}$ *such that* $\boldsymbol{x} + \mathbf{0} = \boldsymbol{x}$ *for every* $\boldsymbol{x} \in \mathcal{V}$.

[A4] Additive inverse: *For every* $\boldsymbol{x} \in \mathcal{V}$, *there is an element* $(\boldsymbol{x}) \in \mathcal{V}$ *such that* $\boldsymbol{x} + (-\boldsymbol{x}) = \mathbf{0}$.

[M1] Scalar multiplication is associative: $a(b\boldsymbol{x}) = (ab)\boldsymbol{x}$ *for every* $a, b \in \mathcal{F}$ *and for every* $\boldsymbol{x} \in \mathcal{V}$.

[M2] First Distributive property: $(a + b)\boldsymbol{x} = a\boldsymbol{x} + b\boldsymbol{x}$ *and for every* $a, b \in \mathcal{F}$ *and for every* $\boldsymbol{x} \in \mathcal{V}$.

[M3] Second Distributive property: $a(\boldsymbol{x} + \boldsymbol{y}) = a\boldsymbol{x} + a\boldsymbol{y}$ *for every* $\boldsymbol{x}, \boldsymbol{y} \in \mathcal{V}$ *and every* $a \in \Re^1$.

[M4] Unit for scalar multiplication: $1\boldsymbol{x} = \boldsymbol{x}$ *for every* $\boldsymbol{x} \in \mathcal{V}$.

Note the use of the $\cdot$ operation in M1. Multiplication for the elements in the field and scalar multiplication are both at play here. When we write $a(b\boldsymbol{x})$, we use two scalar multiplications: first the vector $b\boldsymbol{x}$ is formed by multiplication between the scalar b and the vector $\boldsymbol{x}$ and then $a(b\boldsymbol{x})$ is formed by multiplying the scalar a with the vector $b\boldsymbol{x}$. Also, in M2 the "+" on the left hand side denotes addition for elements in the field, while that on the right hand side denotes vector addition.

Example 16.2 Some examples of vector spaces are listed below.

1. Let $\mathcal{V} = \Re^n$ and $\mathcal{F} = \Re$ with addition defined by $\boldsymbol{x}+\boldsymbol{y} = (x_1+y_1, \ldots, x_n+y_n)$, where $\boldsymbol{x} = (x_1, \ldots, x_n)$ and $\boldsymbol{y} = (y_1, \ldots, y_n)$, and scalar multiplication defined by $a\boldsymbol{x} = (ax_1, \ldots, ax_n)$. This is the familiar Euclidean space of n-tuples of real numbers.
2. The set $\Re^{m \times n}$ consisting of all of $m \times n$ real matrices is a vector space over $\mathcal{F} = \Re$ under the standard operations of addition and scalar multiplication for matrices as in Definitions 1.7 and 1.8.
3. The set $\mathbb{C}^{m \times n}$ consisting of all of $m \times n$ complex matrices forms a vector space over $\mathcal{F} = \mathbb{C}$ under the standard operations of addition and scalar multiplication for complex matrices.

4. **Vector spaces of functions**: The elements of $\mathcal{V}$ can be functions. Define function addition and scalar multiplication as

$$(f+g)(x) = f(x) + g(x) \text{ and } (af)(x) = af(x) . \tag{16.3}$$

Then the following are examples of vector spaces of functions.

   (a) The set of all polynomials with real coefficients.
   (b) The set of all polynomials with real coefficients and of degree $\leq n-1$. We write this as

$$\mathcal{P}_n = \left\{p(x) = a_0 + a_1 x + a_2 x^2 + \cdots + a_{n-1} x^{n-1} : a_i \in \Re\right\} . \tag{16.4}$$

   (c) The set of all real-valued continuous functions defined on $[0,1]$.
   (d) The set of real-valued functions that are differentiable on $[0,1]$.
   (e) Each of the above examples are special cases for an even more general construction. For *any* non-empty set X and any field $\mathcal{F}$, define $\mathcal{F}^X = \{f : X \to \mathcal{F}\}$ to be a space of functions with addition and scalar multiplication defined as in (16.3) for all $x \in X$, $f, g \in \mathcal{F}^X$ and $a \in \mathcal{F}$. Then F^X is a vector space of functions over the field $\mathcal{F}$. This vector space is denoted by $\Re^X$ when the field is chosen to be the real line.

The vector space axioms are easily verified for all the above examples and are left to the reader. ■

The most popular choices of $\mathcal{F}$ are the real line $\Re$ or the complex plane $\mathbb{C}$. The former are called ***real vector spaces*** and the latter are called ***complex vector spaces***. In what follows we will take $\mathcal{F} = \Re$ (and perhaps the complex plane in certain cases). However, we will not necessarily assume that the elements in $\mathcal{V}$ are points in $\Re^n$ or $\mathbb{C}^n$. We will denote vector spaces over the field of real or complex numbers simply by $(\mathcal{V}, +, \cdot)$. When dealing with abstract vector spaces, we must be careful about not taking apparently "obvious" statements for granted. For example, the following statements are true for all $\boldsymbol{x} \in \mathcal{V}$ and all $\alpha \in \mathcal{F}$ but need to be "proved" using the axioms of a vector space.

1. $0 \cdot \boldsymbol{x} = \boldsymbol{0}$ and $\alpha \cdot \boldsymbol{0} = \boldsymbol{0}$.
2. $\alpha \cdot \boldsymbol{x} = \boldsymbol{0} \Rightarrow \alpha = 0$ or $\boldsymbol{x} = \boldsymbol{0}$
   $(-1) \cdot \boldsymbol{x} = -\boldsymbol{x}$.

Consider, for example, how to establish the first statement. Let $\boldsymbol{y} = 0 \cdot \boldsymbol{x}$. Then,

$$\boldsymbol{y} + \boldsymbol{y} = 0 \cdot \boldsymbol{x} + 0 \cdot \boldsymbol{x} = (0+0) \cdot \boldsymbol{x} = 0 \cdot \boldsymbol{x} = \boldsymbol{y} ,$$

where the second equality follows from the distributive property. Adding $-\boldsymbol{y}$, the (vector) additive inverse of $\boldsymbol{y}$ to both sides of the above equation yields

$$\boldsymbol{y} + \boldsymbol{y} + (-\boldsymbol{y}) = \boldsymbol{y} + (-\boldsymbol{y}) = \boldsymbol{0} \Longrightarrow \boldsymbol{y} + \boldsymbol{0} = \boldsymbol{0} \Longrightarrow \boldsymbol{y} = \boldsymbol{0} .$$

Next, consider the statement $\alpha \cdot \boldsymbol{0} = \boldsymbol{0}$. If $\boldsymbol{w} = \alpha \boldsymbol{0}$, then

$$\boldsymbol{w} + \boldsymbol{w} = \alpha \cdot \boldsymbol{0} + \alpha \cdot \boldsymbol{0} = \alpha \cdot (\boldsymbol{0} + \boldsymbol{0}) = \alpha \cdot \boldsymbol{0} = \boldsymbol{w} ,$$

where we have used $\mathbf{0}+\mathbf{0}=\mathbf{0}$ because $\mathbf{0}$ is the additive identity, so when it is added to any element in $\mathcal{V}$ (including $\mathbf{0}$) it returns that element. *It is not because* $\mathbf{0}$ *is an* $n \times 1$ *vector of the real number zero (the origin in the Euclidean vector space), which we do not assume.* We leave the verification of the other two statements as exercises.

Let $(\mathcal{V}, +, \cdot)$ be a vector space. If $\mathcal{S} \subseteq \mathcal{V}$ is non-empty such that

$$\alpha \boldsymbol{x} + \boldsymbol{y} \in \mathcal{S} \text{ for all } \alpha \in \mathcal{F} \text{ and all } \boldsymbol{x}, \boldsymbol{y} \in \mathcal{S} ,$$

then $\mathcal{S}$ is called a ***subspace*** of $\mathcal{V}$. The argument in Theorem 4.1 can be reproduced to prove that every subspace contains the additive identity (A3) under vector addition and, indeed, subspaces are vector spaces in their own right. Furthermore, as we saw for Euclidean spaces in Chapter 4, the intersection and sum of subspaces are also subspaces; unions are not necessarily so.

The concepts of linear independence, basis and dimension are analogous to those for Euclidean spaces. A finite set $X = \{\boldsymbol{x}_1, \boldsymbol{x}_2, \ldots, \boldsymbol{x}_n\}$ of vectors is said to be ***linearly dependent*** if there exist scalars $a_i \in \mathcal{F}$, not all zero, such that $\sum_{i=1}^{n} a_i \boldsymbol{x}_i = \mathbf{0}$. If, on the other hand, $\sum_{i=1}^{n} a_i \boldsymbol{x}_i = \mathbf{0}$ implies $a_i = 0$ for all $i = 1, \ldots, n$, then we say X is ***linearly independent***. If X is a linearly dependent set, then there exists some $a_i \neq 0$ such that $\sum_{j=1}^{n} a_j \boldsymbol{x}_j = \mathbf{0}$. Therefore, we can write $\boldsymbol{x}_i$ as a linear combination of the remaining vectors:

$$\boldsymbol{x}_i = \sum_{j=1 \neq i}^{n} \left(-\frac{a_j}{a_i} \right) \boldsymbol{x}_j .$$

From the definition, it is trivially true that $X = \{\mathbf{0}\}$ is a linearly dependent set. Also, it is immediate from the definition that any set containing a linearly dependent subset is itself linearly dependent. A consequence of this observation is that no linearly independent set can contain the $\mathbf{0}$ vector.

Example 16.3 Let $X = \{a_1, a_2, \ldots, \}$, where $a_1, a_2, \ldots$ is a sequence of distinct real numbers. Consider the indicator functions $f_i(x) = 1(x = a_i)$, which is equal to 1 if $x = a_i$ and zero otherwise. These functions belong to the vector space $\Re^X$ (see the last item in Example 16.2). It is easy to verify that the sequence $\{f_1, f_2 \ldots, \}$ is linearly independent. ■

Let $X = \{\boldsymbol{x}_1, \boldsymbol{x}_2, \ldots\}$ be a countable (not necessarily finite) collection of vectors. Then, the ***linear span*** (or simply span) of X, denoted by $Sp(X)$, is defined as the collection of all linear combinations of the vectors in X:

$$Sp(X) = \{\boldsymbol{y} : \boldsymbol{y} = \sum_{i=1}^{k} a_i \boldsymbol{x}_i;\ a_i \in \mathcal{F};\ \boldsymbol{x}_i \in X;\ k \in \mathrm{N}\},$$

where $\mathrm{N} = \{1, 2, \ldots\}$. Theorem 4.4 can be imitated to prove that the span of a non-empty set of vectors is a subspace. It is easy to show that a set of vectors $A \subseteq \mathcal{V}$ is linearly dependent if and only if there exists an $\boldsymbol{x} \in A$ such that $\boldsymbol{x} \in Sp(A \setminus \{\boldsymbol{x}\})$. Also, if A is linearly independent and $\boldsymbol{y} \notin A$, then $A \cup \{\boldsymbol{y}\}$ is linearly dependent if and only if $\boldsymbol{y} \in Sp(A)$.

Spans of linearly independent sets yield ***coordinates*** of vectors. Let us suppose that $X = \{\boldsymbol{x}_1, \ldots, \boldsymbol{x}_n\}$ is a linearly independent set and suppose $\boldsymbol{y} \in Sp(X)$. Then, there exists scalars $\{\theta_i\}_{i=1}^n$ such that $\boldsymbol{y} = \sum_{i=1}^n \theta_i \boldsymbol{x}_i$. Lemma 4.5 can be imitated to prove that these θ_i's are *unique* in the sense that if $\{\alpha_1, \alpha_2, \ldots, \alpha_n\}$ are any other set of scalars such that $\boldsymbol{y} = \sum_{i=1}^n \alpha_i \boldsymbol{x}_i$, then $\theta_i = \alpha_i$. The θ_i's are well-defined and we call them the ***coordinates*** of $\boldsymbol{y}$ with respect to X.

A set X is called a ***basis*** of $\mathcal{V}$ if (a) X is a linearly independent set of vectors in $\mathcal{V}$, and (b) $Sp(X) = \mathcal{V}$. Most of the results concerning linear independence, spans and basis that were obtained in Chapter 4 for Euclidean vector spaces carries over easily to more abstract vector spaces. The following are especially relevant.

- The second proof of Theorem 4.19 can be easily imitated to prove the following: *If $\mathcal{S}$ is a subspace of $\mathcal{V}$ and $\mathcal{B} = \{\boldsymbol{b}_1, \boldsymbol{b}_2, \ldots, \boldsymbol{b}_l\}$ is a linearly independent subset of $\mathcal{S}$, then any spanning set $\mathcal{A} = \{\boldsymbol{a}_1, \boldsymbol{a}_2, \ldots, \boldsymbol{a}_k\}$ of $\mathcal{S}$ cannot have fewer elements than $\mathcal{B}$. In other words, we must have $k \geq l$.*
- The above result leads to the following analogue of Theorem 4.20: *Every basis of a subspace must have the same number of elements.*
- Theorem 4.21 can be imitated to prove for general finite dimensional vector spaces that *(i) every spanning set of $\mathcal{S}$ can be reduced to a basis, and (ii) every linearly independent subset in $\mathcal{S}$ can be extended to a basis of $\mathcal{S}$.*

The ***dimension*** of $\mathcal{V}$ is the number of elements in the basis of a finite dimensional vector space and is a concept invariant to our choice of a basis. We write this as $\dim(\mathcal{V})$. A vector space $\mathcal{V}$ is defined to be ***finite-dimensional*** if it has a finite number of elements in any basis. The Euclidean spaces are finite dimensional because $\Re^m$ has dimension m. An example of an infinite dimensional vector space is $\Re^X$ because, as we saw in Example 16.3, it contains an infinite collection of linearly independent functions as its basis.

The following facts are easily established on the same lines as in Chapter 4. If X is a basis of $\mathcal{V}$, there can be no linearly independent set in $\mathcal{V}$ that contains more elements than X. In this sense, we characterize the basis as a ***maximal linearly independent set***. It is also true that no set with fewer elements than X can be a spanning set of $\mathcal{V}$. Therefore, we say that a basis must also be a ***minimal spanning set*** of $\mathcal{V}$. From these, we can conclude that if $\mathcal{V}$ is a finite dimensional vector space with dimension m and X is a linearly independent set with m elements, then X must be a basis of $\mathcal{V}$.

We will assume, unless explicitly mentioned, that $\mathcal{V}$ is a finite-dimensional vector space. In many cases, we can associate each element of one vector space with an element in another vector space. Consider, for example, the elements of vector spaces $\Re^n$ and $\mathcal{P}_n$ over the field of real numbers, as defined in (16.4). The two spaces are clearly different. However, we can associate the element $(a_0, a_1, \ldots, a_{n-1})$ in $\Re^n$ with the element $\sum_{i=0}^{n-1} a_i x^i$ in $\mathcal{P}_n$. Not only is this a one-one correspondence, it also preserves addition and scalar multiplication in the following sense: if $\boldsymbol{a} = (a_0, a_1, \ldots, a_{n-1})$ and $\boldsymbol{b} = (b_0, b_1, \ldots, b_{n-1})$ are two points in $\Re^n$ and α is a scalar, then $\boldsymbol{a}+\boldsymbol{b}$ corresponds with the element $\sum_{i=0}^{n-1}(a_i+b_i)x^i$ in $\mathcal{P}_n$, and $\alpha\boldsymbol{a}$ corresponds to $\sum_{i=0}^{n-1}\alpha a_i x^i$ in $\mathcal{P}_n$. Therefore, while $\Re^n$ and $\mathcal{P}_n$ have different elements, they essentially have the same *structure* and share the same structural properties. We say that $\Re^n$ and $\mathcal{P}_n$ are ***isomorphic***. A formal definition of this concept follows.

Definition 16.3 *Two vector spaces $\mathcal{V}_1$ and $\mathcal{V}_2$ over the same field $\mathcal{F}$ are* **isomorphic** *if there is a map ψ from $\mathcal{V}_1$ to $\mathcal{V}_2$ such that:*

(i) $\psi(\boldsymbol{x})$ is **linear**, *which means that*

$$\psi(\boldsymbol{x}+\boldsymbol{y})=\psi(\boldsymbol{x})+\psi(\boldsymbol{y}) \text{ and } \psi(\alpha\boldsymbol{x})=\alpha\psi(\boldsymbol{x}) \text{ for all } \boldsymbol{x},\boldsymbol{y}\in\mathcal{V}_1 \text{ and } \alpha\in\mathcal{F}\,;$$

(ii) $\psi(\boldsymbol{x})$ is a one-to-one and onto function.

The map ψ is said to be an **isomorphism**.

If $\mathcal{V}_1$ and $\mathcal{V}_2$ are isomorphic, then they will have the same dimension because to any basis in one there is a corresponding basis in the other. In fact, the converse is true for finite-dimensional vector spaces: any two vector spaces over the same field that have the same dimension are isomorphic.

Theorem 16.1 *Two vector spaces $\mathcal{V}_1$ and $\mathcal{V}_2$ over the same field $\mathcal{F}$ are* **isomorphic** *if and only if they have the same dimension.*

Proof. Assume that $\mathcal{V}_1$ and have the same dimension. Fix bases $\{\boldsymbol{x}_1,\boldsymbol{x}_2,\ldots,\boldsymbol{x}_n\}$ and $\{\boldsymbol{y}_1,\boldsymbol{y}_2,\ldots,\boldsymbol{y}_n\}$ for $\mathcal{V}_1$ and $\mathcal{V}_2$, respectively. For any $\boldsymbol{x}\in\mathcal{V}_1$ such that $\boldsymbol{x}=\sum_{i=1}^n\alpha_i\boldsymbol{x}_i$, define the map

$$\psi(\boldsymbol{x})=\alpha_1\boldsymbol{y}_1+\alpha_2\boldsymbol{y}_2+\cdots+\alpha_n\boldsymbol{y}_n\;.$$

Clearly $\psi:\mathcal{V}_1\to\mathcal{V}_2$. It is also easy to check that it is linear in the sense of Definition 16.3. Take any two vectors in $\mathcal{V}_1$ and write them as $\boldsymbol{u}=\sum_{i=1}^n\alpha_i\boldsymbol{x}_i$ and $\boldsymbol{v}=\sum_{i=1}^n\beta_i\boldsymbol{x}_i$. Then, since $\boldsymbol{u}+\gamma\boldsymbol{v}=\sum_{i=1}^n(\alpha_i+\gamma\beta_i)\boldsymbol{x}_i$, we find

$$\psi(\boldsymbol{u}+\gamma\boldsymbol{v})=\sum_{i=1}^n(\alpha_i+\gamma\beta_i)\boldsymbol{y}_i=\sum_{i=1}^n\alpha_i\boldsymbol{y}_i+\gamma\sum_{i=1}^n\beta_i\boldsymbol{y}_i=\psi(\boldsymbol{u})+\gamma\psi(\boldsymbol{v})\;.$$

Since the vectors in $\mathcal{V}_1$ can be *uniquely* expressed as a linear combination of the $\boldsymbol{x}_i$'s, it follows that ψ is a well-defined one-to-one and onto map. This establishes the "if" part of the theorem.

To prove the "only if" part, we assume that $\mathcal{V}_1$ and $\mathcal{V}_2$ are isomorphic. Let $\psi:\mathcal{V}_1\to\mathcal{V}_2$ be an isomorphism and let $\{\boldsymbol{x}_1,\boldsymbol{x}_2,\ldots,\boldsymbol{x}_n\}$ be any basis for $\mathcal{V}_1$. The $\psi(\boldsymbol{x}_i)$'s are linearly independent because

$$\sum_{i=1}^n\alpha_i\psi(\boldsymbol{x}_i)=\boldsymbol{0}\Longrightarrow\psi\Big(\sum_{i=1}^n\alpha_i\boldsymbol{x}_i\Big)=\boldsymbol{0}\Longrightarrow\sum_{i=1}^n\alpha_i\boldsymbol{x}_i=\boldsymbol{0}$$
$$\Longrightarrow\alpha_1=\alpha_2=\cdots=\alpha_n=0\;,$$

where we have used the elementary fact that if $\psi(\boldsymbol{x}) = \boldsymbol{0}$, then $\boldsymbol{x} = \boldsymbol{0}$. Therefore, $\dim(\mathcal{V}_2) \geq \dim(\mathcal{V}_1)$. The reverse inequality can be proved analogously by considering an isomorphism from $\mathcal{V}_2$ to $\mathcal{V}_1$. $\square$

It is easy to see that isomorphism is a transitive relation, which means that if $\mathcal{V}_1$ and $\mathcal{V}_2$ are isomorphic, $\mathcal{V}_2$ and $\mathcal{V}_3$ are isomorphic, then $\mathcal{V}_1$ and $\mathcal{V}_3$ are isomorphic as well. An easy and important consequence of Theorem 16.1 is that every finite-dimensional vector space of dimension m is isomorphic to $\Re^m$. This means that pretty much all the results we have obtained for Euclidean spaces in Chapter 4 have exact analogues for subspaces of more general (finite-dimensional) vector spaces.

16.2 General inner products

We have already seen the important role played by inner products in Euclidean vector spaces, especially in extending the concept of orthogonality to n-dimensions. Inner products can be defined for more general vector spaces as below.

Definition 16.4 *An* **inner product** *in a real or complex vector space $\mathcal{V}$ is, respectively, a real or complex valued function defined on $\mathcal{V} \times \mathcal{V}$, denoted by $\langle \boldsymbol{u}, \boldsymbol{v} \rangle$ for $\boldsymbol{u}, \boldsymbol{v} \in \mathcal{V}$, that satisfies the following four axioms:*

1. *Positivity: $\langle \boldsymbol{u}, \boldsymbol{u} \rangle \geq 0 \;\; \forall \boldsymbol{u} \in \mathcal{V}$.*
2. *Definiteness: If $\langle \boldsymbol{v}, \boldsymbol{v} \rangle = 0$, then $\boldsymbol{v} = 0$ for all $\boldsymbol{v} \in \mathcal{V}$.*
3. *Conjugate Symmetry: $\langle \boldsymbol{u}, \boldsymbol{v} \rangle = \overline{\langle \boldsymbol{v}, \boldsymbol{u} \rangle}$ for all $\boldsymbol{u}, \boldsymbol{v} \in \mathcal{V}$.*
4. *Linearity: $\langle \alpha \boldsymbol{u} + \beta \boldsymbol{v}, \boldsymbol{w} \rangle = \alpha \langle \boldsymbol{u}, \boldsymbol{w} \rangle + \beta \langle \boldsymbol{v}, \boldsymbol{w} \rangle$ for all $\boldsymbol{u}, \boldsymbol{v}, \boldsymbol{w} \in \mathcal{V}$,*

where $\overline{\langle \boldsymbol{v}, \boldsymbol{u} \rangle}$ is the complex conjugate of $\langle \boldsymbol{v}, \boldsymbol{u} \rangle$. An **inner product space** *is a vector space with an inner product.*

Examples of inner product spaces include the $\Re^n$ and $\mathbb{C}^n$. For $\Re^n$, the symmetry condition is simply $\langle \boldsymbol{u}, \boldsymbol{v} \rangle = \langle \boldsymbol{v}, \boldsymbol{u} \rangle$. We have discussed inner products in $\Re^n$ in great detail in Chapter 7. Lemma 7.1 shows that the standard inner product, which was introduced way back in (1.5), satisfies the properties in Definition 16.4.

If our vector space is over the field of complex numbers, then

$$\begin{aligned} \langle \boldsymbol{u}, \beta_1 \boldsymbol{v}_1 + \beta_2 \boldsymbol{v}_2 \rangle &= \overline{\langle \beta_1 \boldsymbol{v}_1 + \beta_2 \boldsymbol{v}_2, \boldsymbol{u} \rangle} = \overline{\beta_1 \langle \boldsymbol{v}_1, \boldsymbol{u} \rangle + \beta_2 \langle \boldsymbol{v}_2, \boldsymbol{u} \rangle} \\ &= \overline{\beta_1 \langle \boldsymbol{v}_1, \boldsymbol{u} \rangle} + \overline{\beta_2 \langle \boldsymbol{v}_2, \boldsymbol{u} \rangle} = \overline{\beta}_1 \overline{\langle \boldsymbol{v}_1, \boldsymbol{u} \rangle} + \overline{\beta}_2 \overline{\langle \boldsymbol{v}_2, \boldsymbol{u} \rangle} \\ &= \overline{\beta}_1 \langle \boldsymbol{u}, \boldsymbol{v}_1 \rangle + \overline{\beta}_2 \langle \boldsymbol{u}, \boldsymbol{v}_2 \rangle \;. \end{aligned} \tag{16.5}$$

For $\mathbb{C}^n$, the standard inner product is defined as

$$\langle \boldsymbol{u}, \boldsymbol{v} \rangle = \boldsymbol{v}^* \boldsymbol{u} = \sum_{i=1}^{n} \bar{v}_i u_i \;, \tag{16.6}$$

where $\boldsymbol{v}^*$, the conjugate transpose of $\boldsymbol{v}$, is a row vector whose i-th entry $\bar{v}_i$ is the complex conjugate of v_i. Clearly, (16.6) satisfies the conditions in Definition 16.4.

The vector spaces $\Re^{m\times n}$ and $\mathbb{C}^{m\times n}$ consisting of $m \times n$ real and complex matrices is also an inner product space. A popular inner product on $\mathbb{C}^{m\times n}$ used in matrix analysis is

$$\langle \boldsymbol{A}, \boldsymbol{B} \rangle = \text{tr}(\boldsymbol{B}^* \boldsymbol{A}) \; . \tag{16.7}$$

It is easily verified that (16.7) satisfies the axioms in Definition 16.4. For $\Re^{m\times n}$, we have $\boldsymbol{B}^* = \boldsymbol{B}'$ in (16.7).

Several definitions, concepts and properties associated with Euclidean inner product spaces easily carry over to general inner product spaces. For example, note that $\langle \boldsymbol{u}, \boldsymbol{u} \rangle = \overline{\langle \boldsymbol{u}, \boldsymbol{u} \rangle}$ so $\langle \boldsymbol{u}, \boldsymbol{u} \rangle$ is real, irrespective of whether $\mathcal{F}$ is real or complex. Thus, there is the concept of a ***norm*** or length of a vector $\boldsymbol{u}$, which we define as $\|\boldsymbol{u}\| = \sqrt{\langle \boldsymbol{u}, \boldsymbol{u} \rangle}$. We say that two vectors $\boldsymbol{u}$ and $\boldsymbol{v}$ are ***orthogonal*** with respect to an inner product if $\langle \boldsymbol{u}, \boldsymbol{v} \rangle = 0$. A set is said to be an orthogonal set if any two vectors in it are orthogonal. Note that all the concepts related to orthogonality are with respect to a specified inner product. When the context is clear, we simply say the vectors are orthogonal and omit the redundant phrase "with respect to the inner product."

A set is said to be ***orthonormal*** if, in addition to any two vectors being orthogonal, each vector has unit norm, i.e., $\|\boldsymbol{u}\| = 1$. Thus, for any two vectors in an orthonormal set $\langle \boldsymbol{x}_i, \boldsymbol{x}_j \rangle = 1$ if $i = j$ and 0 whenever $i \neq j$. We call an orthonormal set ***complete*** if it is not contained in any larger orthonormal set. Orthogonal vectors are linearly independent. The Gram-Schmidt orthogonalization procedure, outlined for real Euclidean spaces in Section 7.4, can be applied to real or complex inner product spaces with any legitimate inner product. This implies that every nonzero finite-dimensional inner product space has an orthonormal basis. Also, an orthogonal set in a finite dimensional inner product space can be extended to an orthonormal basis (recall Theorem 7.3).

If $\boldsymbol{u}_1, \boldsymbol{u}_2, \ldots, \boldsymbol{u}_r$ is an orthonormal basis for a finite dimensional subspace $\mathcal{S}$ of an inner product space $\mathcal{V}$ and $\boldsymbol{v}$ is any vector in $\mathcal{V}$, then

$$\boldsymbol{w} = \langle \boldsymbol{v}, \boldsymbol{u}_1 \rangle \boldsymbol{u}_1 + \langle \boldsymbol{v}, \boldsymbol{u}_2 \rangle \boldsymbol{u}_2 + \cdots + \langle \boldsymbol{v}, \boldsymbol{u}_2 \rangle \boldsymbol{u}_2 \tag{16.8}$$

is the vector in $\mathcal{S}$ such that $\boldsymbol{v} - \boldsymbol{w}$ is orthogonal to *every* vector in $\mathcal{S}$. Let $\boldsymbol{x} \in \mathcal{S}$ so that $\boldsymbol{x} = \sum_{i=1}^{r} \zeta_i \boldsymbol{u}_i$. Then,

$$\langle \boldsymbol{v} - \boldsymbol{w}, \boldsymbol{x} \rangle = \left\langle \boldsymbol{v} - \boldsymbol{w}, \sum_{i=1}^{r} \zeta_i \boldsymbol{u}_i \right\rangle = \sum_{i=1}^{r} \bar{\zeta}_i \langle \boldsymbol{v} - \boldsymbol{w}, \boldsymbol{u}_i \rangle = 0$$

because it is easily verified from (16.8) that $\langle \boldsymbol{v} - \boldsymbol{w}, \boldsymbol{u}_i \rangle = 0$. The vector $\boldsymbol{w}$ is called the ***orthogonal projection*** of $\boldsymbol{v}$ onto $\mathcal{S}$.

Theorem 16.2 Bessel's inequality. *Let $X = \{\boldsymbol{x}_1, \ldots, \boldsymbol{x}_k\}$ be any finite orthonormal set of vectors in an inner product space. Then,*

$$\sum_{i=1}^{k} |\langle \boldsymbol{x}, \boldsymbol{x}_i \rangle|^2 \leq \|\boldsymbol{x}\|^2 .$$

Equality holds only when $\boldsymbol{x} \in Sp(X)$.

Proof. Let be the orthogonal projection of a vector $\boldsymbol{x}$ onto $Sp(X)$. Then, $\boldsymbol{w} = \sum_{i=1}^{r} \alpha_i \boldsymbol{x}_i$, where $\alpha_i = \langle \boldsymbol{x}, \boldsymbol{x}_i \rangle$. Therefore,

$$\begin{aligned}
0 \leq \|\boldsymbol{x} - \boldsymbol{w}\|^2 &= \langle \boldsymbol{x} - \boldsymbol{w}, \boldsymbol{x} - \boldsymbol{w} \rangle = \|\boldsymbol{x}\|^2 - \langle \boldsymbol{w}, \boldsymbol{x} \rangle - \langle \boldsymbol{x}, \boldsymbol{w} \rangle + \|\boldsymbol{w}\|^2 \\
&= \|\boldsymbol{x}\|^2 - \sum_{i=1}^{k} \alpha_i \langle \boldsymbol{x}_i, \boldsymbol{x} \rangle - \sum_{i=1}^{k} \bar{\alpha}_i \langle \boldsymbol{x}, \boldsymbol{x}_i \rangle + \sum_{i=1}^{n} \alpha_i \bar{\alpha}_i \\
&= \|\boldsymbol{x}\|^2 - \sum_{i=1}^{k} \alpha_i \overline{\alpha}_i - \sum_{i=1}^{k} \overline{\alpha}_i \alpha_i + \sum_{i=1}^{k} \alpha_i \overline{\alpha}_i \\
&= \|\boldsymbol{x}\|^2 - 2\sum_{i=1}^{k} \alpha_i \overline{\alpha}_i + \sum_{i=1}^{k} \alpha_i \overline{\alpha}_i = \|\boldsymbol{x}\|^2 - \sum_{i=1}^{k} |\alpha_i|^2 = \|\boldsymbol{x}\|^2 - \sum_{i=1}^{k} |\langle \boldsymbol{x}, \boldsymbol{x}_i \rangle|^2 ,
\end{aligned}$$

which yields Bessel's inequality. Equality holds only when $\boldsymbol{x} = \boldsymbol{w}$, which happens when $\boldsymbol{x} \in Sp(X)$. □

The *Cauchy-Schwarz inequality* can be derived easily from Bessel's inequality.

Theorem 16.3 Cauchy-Schwarz revisited. *If $\boldsymbol{x}$ and $\boldsymbol{y}$ are any two vectors in an inner product space $\mathcal{V}$, then*

$$|\langle \boldsymbol{x}, \boldsymbol{y} \rangle| \leq \|\boldsymbol{x}\| \|\boldsymbol{y}\| .$$

Equality holds only when $\boldsymbol{y}$ is linearly dependent on $\boldsymbol{x}$.

Proof. If $\boldsymbol{y} = \boldsymbol{0}$, then both sides vanish and the result is trivially true. If $\boldsymbol{y}$ is non-null, then the set $\{\boldsymbol{y}/\|\boldsymbol{y}\|\}$ is an orthonormal set with only one element. Applying Bessel's inequality to this set, we obtain

$$\left| \left\langle \boldsymbol{x}, \frac{\boldsymbol{y}}{\|\boldsymbol{y}\|} \right\rangle \right| \leq \|\boldsymbol{x}\| ,$$

from which the inequality follows immediately. The condition for equality, from Bessel's inequality, is that $\boldsymbol{y}$ be linearly dependent on $\boldsymbol{x}$. □

Two subspaces spaces $\mathcal{S}_1$ and $\mathcal{S}_2$ are orthogonal if *every* vector in $\mathcal{S}_1$ is orthogonal to *every* vector in $\mathcal{S}_2$. The *orthogonal complement* of a subspace $\mathcal{S} \subset \mathcal{V}$ is the set of all vectors in $\mathcal{V}$ that are orthogonal to $\mathcal{S}$ and is written as

$$\mathcal{S}^{\perp} = \{\boldsymbol{v} \in \mathcal{V} : \langle \boldsymbol{u}, \boldsymbol{v} \rangle = 0 \text{ for all } \boldsymbol{u} \in \mathcal{S}\} . \tag{16.9}$$

Note that $\mathcal{S}^\perp$ depends upon the specific choice of the inner product. It is easily checked that $\mathcal{S}^\perp$ is always a subspace and is *virtually disjoint* with $\mathcal{S}$ in the sense that $\mathcal{S} \cap \mathcal{S}^\perp = \{\mathbf{0}\}$, irrespective of the specific choice of the inner product. Several other properties of orthogonal complements developed in Chapter 7. For example, Lemmas 7.9 and 7.10 can be imitated to establish that if $\mathcal{S}$ and $\mathcal{T}$ are equal subspaces, then $\mathcal{S}^\perp = \mathcal{T}^\perp$.

If $\{\boldsymbol{u}_1, \boldsymbol{u}_2, \ldots, \boldsymbol{u}_r\}$ is a basis for a subspace $\mathcal{S} \subseteq \mathcal{V}$ of a finite dimensional inner product space $\mathcal{V}$ and if $\{\boldsymbol{v}_1, \boldsymbol{v}_2, \ldots, \boldsymbol{v}_k\}$ is a basis for $\mathcal{S}^\perp$, then it is easy to argue that $\{\boldsymbol{u}_1, \boldsymbol{u}_2, \ldots, \boldsymbol{u}_r, \boldsymbol{v}_1, \boldsymbol{v}_2, \ldots, \boldsymbol{v}_k\}$ is a basis for $\mathcal{V}$. This implies

$$\dim(\mathcal{S}) + \dim(\mathcal{S}^\perp) = \dim(\mathcal{V}) \tag{16.10}$$

and that $\mathcal{V}$ is the direct sum $\mathcal{V} = \mathcal{S} \oplus \mathcal{S}^\perp$. Therefore, every vector in $\boldsymbol{x} \in \mathcal{V}$ can be expressed *uniquely* as $\boldsymbol{x} = \boldsymbol{u} + \boldsymbol{v}$, where $\boldsymbol{u} \in \mathcal{S}$ and $\boldsymbol{v} \in \mathcal{S}^\perp$. All these facts are easily established for general finite dimensional vector spaces and left to the reader. The vector $\boldsymbol{u}$ in the unique decomposition of $\boldsymbol{x} \in \mathcal{V}$ is the orthogonal projection of $\boldsymbol{x}$ onto $\mathcal{S}$.

16.3 Linear transformations, adjoint and rank

We now introduce a special type of function that maps one vector space to another.

Definition 16.5 *A* **linear transformation** *or* **linear map** *or* **linear function** *is a map $A : \mathcal{V}_1 \rightarrow \mathcal{V}_2$, where $\mathcal{V}_1$ and $\mathcal{V}_2$ are vector spaces over a field $\mathcal{F}$, such that*

$$A(\boldsymbol{x} + \boldsymbol{y}) = A(\boldsymbol{x}) + A(\boldsymbol{y}) \;\; \textit{and} \;\; A(\alpha \boldsymbol{x}) = \alpha A(\boldsymbol{x})$$

for all $\boldsymbol{x}, \boldsymbol{y} \in \mathcal{V}_1$ and all scalars $\alpha \in \mathcal{F}$. The element $A(\boldsymbol{x}) \in \mathcal{V}_2$, often written simply as $A\boldsymbol{x}$, is called the **image** *of $\boldsymbol{x}$ under* **A**. *A linear transformation from a vector space to itself, i.e., when $\mathcal{V}_1 = \mathcal{V}_2 = \mathcal{V}$, is called a* **linear operator** *on $\mathcal{V}$.*

For checking the two linearity conditions in Definition 16.5, it is enough to check whether

$$A(\alpha \boldsymbol{x} + \beta \boldsymbol{y}) = \alpha A(\boldsymbol{x}) + \beta A(\boldsymbol{y}), \; \forall\, \boldsymbol{x}, \boldsymbol{y} \in \mathcal{V}_1 \text{ and } \forall\, \alpha, \beta \in \mathcal{F} . \tag{16.11}$$

In fact, we can take either $\alpha = 1$ or $\beta = 1$ (but not both) in the above check. The linearity condition has a nice geometric interpretation. Since adding two vectors amounts to constructing a parallelogram from the two vectors, A maps parallelograms to parallelograms. Since scalar multiplication amounts to stretching a vector by a factor of α, the image of a stretched vector under A is stretched by the same factor.

The above definition implies that $A(\mathbf{0}) = \mathbf{0}$ by taking $\alpha = 0$. Also, the linearity in linear transformation implies that

$$A(\alpha_1 \boldsymbol{x}_1 + \alpha_2 \boldsymbol{x}_2 + \cdots \alpha_k \boldsymbol{x}_k) = \alpha_1 A(\boldsymbol{x}_1) + \alpha_2 A(\boldsymbol{x}_2) + \cdots + \alpha_k A(\boldsymbol{x}_k)$$

for any finite set of vectors $\boldsymbol{x}_1, \boldsymbol{x}_2, \ldots, \boldsymbol{x}_k \in \mathcal{V}_1$ and scalars $\alpha_1, \alpha_2, \ldots, \alpha_k \in \mathcal{F}$. In other words, linear transformations preserve the operations of addition and scalar multiplication in a vector space.

Example 16.4 Some examples of linear transformations are provided below.

- The map $0 : \mathcal{V}_1 \to \mathcal{V}_2$ such that $0(\boldsymbol{x}) = \mathbf{0}$ is a trivial linear transformation. It maps any element in $\mathcal{V}_1$ to the $\mathbf{0}$ element in $\mathcal{V}_2$.
- The map $I : \mathcal{V} \to \mathcal{V}$ such that $I(\boldsymbol{x}) = \boldsymbol{x}$ for every $\boldsymbol{x} \in \mathcal{V}$ is a linear transformation from $\mathcal{V}$ onto itself. It is called the ***identity map***.
- If $\mathcal{V}_1 = \Re^n$, $\mathcal{V}_2 = \Re^m$ and $\boldsymbol{A}$ is any $m \times n$ matrix with real entries, then $f(\boldsymbol{x}) = \boldsymbol{A}\boldsymbol{x}$ defined by matrix multiplication is a linear transformation that maps $\boldsymbol{x} \in \Re^n$ to $\boldsymbol{A}\boldsymbol{x}$ in $\Re^m$. This means that any operation that can be described by the action of a matrix on a vector is a linear transformation. Examples include rotation, projection and reflection. It is not necessary that every linear transformation arises as the product of a matrix and a vector.
- If $\Re^{m \times n}$ is the vector space of $m \times n$ real matrices, then $\text{vec}(\boldsymbol{A}) : \Re^{m \times n} \to \Re^{mn}$, which stacks the columns of an $m \times n$ matrix $\boldsymbol{A}$ into an $mn \times 1$ vector is a linear transformation.
- Differentiation (see Eq.(16.1)) and integration are examples of linear transformations on functional vector spaces. As a more specific example, the differentiation is a linear transformation from $\mathcal{P}_n$ to $\mathcal{P}_{n-1}$ because

$$\frac{d}{dt}(\alpha_0 + \alpha_1 t + \cdots + \alpha_n t^n) = \alpha_1 + \alpha_2 t + \cdots + \alpha_n t^{n-1} \in \mathcal{P}_{n-1} .$$

Definition 16.6 *Let A and B be two linear transformations, both from $\mathcal{V}_1 \to \mathcal{V}_2$ under the same field $\mathcal{F}$. Then, their* **sum** *is defined as the transformation $A + B : \mathcal{V}_1 \to \mathcal{V}_3$ such that*

$$(A + B)(\boldsymbol{x}) = A_1(\boldsymbol{x}) + A_2(\boldsymbol{x}) \ \textit{for all} \ \boldsymbol{x} \in \mathcal{V}_1 .$$

The **scalar multiplication** *of a linear transformation results in the transformation αA defined as*

$$(\alpha A)(\boldsymbol{x}) = \alpha A(\boldsymbol{x}) \ \textit{for all} \ \boldsymbol{x} \in \mathcal{V}_1 \ \textit{and} \ \alpha \in \mathcal{F} .$$

Let $A : \mathcal{V}_1 \to \mathcal{V}_2$ and $B : \mathcal{V}_2 \to \mathcal{V}_3$, where $\mathcal{V}_1$, $\mathcal{V}_2$ and $\mathcal{V}_3$ are vector spaces under the same field $\mathcal{F}$. Then their **composition** *is defined as the map $B \odot A$ or simply $BA : \mathcal{V}_1 \to \mathcal{V}_3$ such that*

$$(BA)(\boldsymbol{x}) = B(A(\boldsymbol{x})) \ \textit{for all} \ \boldsymbol{x} \in \mathcal{V}_1 .$$

It is easily seen that sum and composition of functions (not necessarily linear) are ***associative***. If A, B and C are three functions from $\mathcal{V}_1 \to \mathcal{V}_3$, then

$$\begin{aligned} [(A + B) + C](\boldsymbol{x}) &= (A + B)(\boldsymbol{x}) + C(\boldsymbol{x}) = A(\boldsymbol{x}) + B(\boldsymbol{x}) + C(\boldsymbol{x}) \\ &= A(\boldsymbol{x}) + (B + C)(\boldsymbol{x}) = [A + (B + C)](\boldsymbol{x}) , \end{aligned}$$

which is unambiguously written as $(A+B+C)(\boldsymbol{x})$. If $A : \mathcal{V}_1 \to \mathcal{V}_2$, $B : (\mathcal{V}_2, \to \mathcal{V}_3$ and $C : \mathcal{V}_3 \to \mathcal{V}_4$, then

$$[C(BA)](\boldsymbol{x}) = C(BA(\boldsymbol{x})) = C(B(A(\boldsymbol{x})) = (CB)(A(\boldsymbol{x})) = [(CB)A](\boldsymbol{x}) ,$$

which is unambiguously written as $CBA(\boldsymbol{x})$.

Scalar multiplication, sums and compositions of linear transformations are linear.

Theorem 16.4 *Let A and B be linear transformations from $\mathcal{V}_1$ to $\mathcal{V}_2$ and let C be a linear transformation from $\mathcal{V}_2$ to $\mathcal{V}_3$, where $\mathcal{V}_1$, $\mathcal{V}_2$ and $\mathcal{V}_3$ are vector spaces over the same field $\mathcal{F}$.*

(i) αA is a linear transformation from $\mathcal{V}_1$ to $\mathcal{V}_2$, where α is any scalar in $\mathcal{F}$.

(ii) $A + B$ is a linear transformation from $\mathcal{V}_1$ to $\mathcal{V}_2$.

(iii) CA is a linear transformation from $\mathcal{V}_1$ to $\mathcal{V}_3$.

Proof. **Proof of (i)**: For scalar multiplication, we verify (16.11):

$$\begin{aligned}(\alpha A)(\beta_1 \boldsymbol{x} + \beta_2 \boldsymbol{y}) &= \alpha A(\beta_1 \boldsymbol{x} + \beta_2 \boldsymbol{y}) = \alpha(\beta_1 A(\boldsymbol{x}) + \beta_2 A(\boldsymbol{y}))\\ &= \alpha\beta_1 A(\boldsymbol{x}) + \alpha\beta_2 A(\boldsymbol{y}) = \beta_1 \alpha A(\boldsymbol{x}) + \beta_2 \alpha A(\boldsymbol{x})\\ &= \beta_1(\alpha A)(\boldsymbol{x}) + \beta_2(\alpha A)(\boldsymbol{y}) .\end{aligned}$$

Proof of (ii): For addition, we verify (16.11) using the linearity of A and B:

$$\begin{aligned}(A+B)(\alpha \boldsymbol{x} + \beta \boldsymbol{y}) &= A(\alpha \boldsymbol{x} + \beta \boldsymbol{y}) + B(\alpha \boldsymbol{x} + \beta \boldsymbol{y})\\ &= \alpha A(\boldsymbol{x}) + \beta A(\boldsymbol{y}) + \alpha B(\boldsymbol{x}) + \beta B(\boldsymbol{y})\\ &= \alpha(A(\boldsymbol{x}) + B(\boldsymbol{x})) + \beta(A(\boldsymbol{x}) + B(\boldsymbol{y}))\\ &= \alpha(A+B)(\boldsymbol{x}) + \beta(A+B)(\boldsymbol{y}) .\end{aligned}$$

Proof of (iii): For composition, we verify (16.11) by first using the linearity of A and then that of C:

$$\begin{aligned}CA(\alpha \boldsymbol{x} + \beta \boldsymbol{y}) &= C\left(A(\alpha \boldsymbol{x} + \beta \boldsymbol{y})\right) = C\left(\alpha A(\boldsymbol{x}) + \beta A(\boldsymbol{y})\right)\\ &= \alpha C(A(\boldsymbol{x})) + \beta C(A(\boldsymbol{y})) = \alpha CA(\boldsymbol{x}) + \beta CA(\boldsymbol{y}) .\end{aligned}$$

□

Let $\mathcal{L}(\mathcal{V}_1, \mathcal{V}_2)$ be the set of all linear functions $A : \mathcal{V}_1 \to \mathcal{V}_2$, where $\mathcal{V}_1$ and $\mathcal{V}_2$ are two vector spaces under the same field $\mathcal{F}$. Parts (i) and (ii) of Theorem 16.4 ensure that $\mathcal{L}(\mathcal{V}_1, \mathcal{V}_2)$ is itself a vector space under the operations of addition of two linear transformations and multiplication of a linear transformation by a scalar, as defined in Definition 16.6.

Real- or complex-valued linear transformations (i.e., $A : \mathcal{V} \to \mathbb{C}$) play important roles in linear algebra and analysis. They are called ***linear functionals***. More generally, linear functionals are linear transformations from $\mathcal{V} \to \mathcal{F}$, where $\mathcal{V}$ is a vector space over the field $\mathcal{F}$.

Theorem 16.5 *Let $\mathcal{V}$ be a finite-dimensional inner product space over the field of complex numbers and let $A : \mathcal{V} \to \mathbb{C}$ be a linear functional on $\mathcal{V}$. Then there exists a* **unique** *$y \in \mathcal{V}$ such that*

$$A(\boldsymbol{x}) = \langle \boldsymbol{x}, \boldsymbol{y} \rangle \ \forall\, \boldsymbol{x} \in \mathcal{V} .$$

In other words, every linear functional is given by an inner product.

Proof. Let $X = \{\boldsymbol{x}_1, \ldots, \boldsymbol{x}_n\}$ be an orthonormal basis of $\mathcal{V}$ and let $A(\boldsymbol{x}_i) = a_i$ for $i = 1, 2, \ldots, n$, where each $a_i \in \mathbb{C}$. Construct the vector

$$\boldsymbol{y} = \bar{a}_1 \boldsymbol{x}_1 + \bar{a}_2 \boldsymbol{x}_2 + \cdots + \bar{a}_n \boldsymbol{x}_n .$$

Taking the inner product of both sides with any $\boldsymbol{x}_i \in X$ yields

$$\begin{aligned} \langle \boldsymbol{x}_i, \boldsymbol{y} \rangle &= \langle \boldsymbol{x}_i, \bar{a}_1 \boldsymbol{x}_1 \rangle + \langle \boldsymbol{x}_i, \bar{a}_2 \boldsymbol{x}_2 \rangle + \cdots + \langle \boldsymbol{x}_i, \bar{a}_n \boldsymbol{x}_n \rangle \\ &= a_1 \langle \boldsymbol{x}_i, \boldsymbol{x}_1 \rangle + a_2 \langle \boldsymbol{x}_i, \boldsymbol{x}_2 \rangle + \cdots + a_n \langle \boldsymbol{x}_i, \boldsymbol{x}_n \rangle \\ &= a_i \langle \boldsymbol{x}_i, \boldsymbol{x}_i \rangle = a_i . \end{aligned}$$

Thus, $A(\boldsymbol{x}_i) = \langle \boldsymbol{x}_i, \boldsymbol{y} \rangle$ for every $\boldsymbol{x}_i \in X$. Now consider any vector $\boldsymbol{x} \in \mathcal{V}$. Express this vector in terms of the orthonormal basis X:

$$\boldsymbol{x} = \alpha_1 \boldsymbol{x}_1 + \alpha_2 \boldsymbol{x}_2 + \cdots + \alpha_n \boldsymbol{x}_n .$$

Therefore,

$$\begin{aligned} A(\boldsymbol{x}) &= A(\alpha_1 \boldsymbol{x}_1 + \alpha_2 \boldsymbol{x}_2 + \cdots + \alpha_n \boldsymbol{x}_n) = \alpha_1 A(\boldsymbol{x}_1) + \alpha_2 A(\boldsymbol{x}_2) + \cdots + \alpha_n A(\boldsymbol{x}_n) \\ &= \alpha_1 \langle \boldsymbol{x}_1, \boldsymbol{y} \rangle + \alpha_2 \langle \boldsymbol{x}_2, \boldsymbol{y} \rangle + \cdots + \alpha_n \langle \boldsymbol{x}_n, \boldsymbol{y} \rangle \\ &= \langle \alpha_1 \boldsymbol{x}_1 + \alpha_2 \boldsymbol{x}_2 + \cdots + \alpha_n \boldsymbol{x}_n, \boldsymbol{y} \rangle = \langle \boldsymbol{x}, \boldsymbol{y} \rangle . \end{aligned}$$

It remains to prove that $\boldsymbol{y}$ is unique. If $\boldsymbol{u}$ is another vector satisfying the above, then we must have $\langle \boldsymbol{x}, \boldsymbol{y} \rangle = \langle \boldsymbol{x}, \boldsymbol{u} \rangle$ for all $\boldsymbol{x}$, which implies $\langle \boldsymbol{x}, \boldsymbol{y} - \boldsymbol{u} \rangle = 0$ for all $\boldsymbol{x}$. Taking $\boldsymbol{x} = \boldsymbol{y} - \boldsymbol{u}$ we obtain $\|\boldsymbol{y} - \boldsymbol{u}\| = 0$ so $\boldsymbol{y} = \boldsymbol{u}$. □

Theorem 16.6 The adjoint of a linear transformation. *Let A be a linear transformation from $A : (\mathcal{V}_1, \mathcal{F}) \to (\mathcal{V}_2, \mathcal{F})$, where $\mathcal{V}_1$ and $\mathcal{V}_2$ are inner product spaces over $\mathcal{F}$. Then there exists a* unique *linear transformation $A^* : (\mathcal{V}_2, \mathcal{F}) \to (\mathcal{V}_1, \mathcal{F})$ such that*

$$\langle A\boldsymbol{x}, \boldsymbol{y} \rangle = \langle \boldsymbol{x}, A^* \boldsymbol{y} \rangle \ \forall\, \boldsymbol{x} \in \mathcal{V}_1 \ \text{and} \ \boldsymbol{y} \in \mathcal{V}_2 .$$

We refer to A^* as the **adjoint** *of A.*

Proof. Let $\boldsymbol{y} \in \mathcal{V}_2$. Define the linear functional $g(\boldsymbol{x}) = \langle A\boldsymbol{x}, \boldsymbol{y} \rangle$ from $\mathcal{V}_1 \to \mathcal{F}$. By Theorem 16.5, there exists a unique $\boldsymbol{u} \in \mathcal{V}_1$ such that $g(\boldsymbol{x}) = \langle \boldsymbol{x}, \boldsymbol{u} \rangle$ for all $\boldsymbol{x} \in \mathcal{V}_1$. Given any linear transformation A, this procedure maps a $\boldsymbol{y} \in \mathcal{V}_2$ to a $\boldsymbol{u} \in \mathcal{V}_1$. Define $A^* : \mathcal{V}_2 \to \mathcal{V}_1$ as this map. That is, $A^*(\boldsymbol{y}) = \boldsymbol{u}$. Therefore,

$$\langle A\boldsymbol{x}, \boldsymbol{y} \rangle = \langle \boldsymbol{x}, A^* \boldsymbol{y} \rangle \ \forall\, \boldsymbol{x} \in \mathcal{V}_1 \ \text{and} \ \boldsymbol{y} \in \mathcal{V}_2.$$

We first confirm that A^* is a well-defined function. Indeed, if $A^*(\boldsymbol{y}) = \boldsymbol{u}_1$ and $A^*(\boldsymbol{y}) = \boldsymbol{u}_2$, then $\langle A\boldsymbol{x}, \boldsymbol{y} \rangle = \langle \boldsymbol{x}, \boldsymbol{u}_1 \rangle = \langle \boldsymbol{x}, \boldsymbol{u}_2 \rangle \ \forall\, \boldsymbol{x} \in \mathcal{V}_1$. Hence, $\boldsymbol{u}_2 = \boldsymbol{u}_1$.

That A^* is linear follows by considering the following equation for any $\boldsymbol{x} \in \mathcal{V}_1$:

$$\begin{aligned}\langle \boldsymbol{x}, A^*(\alpha \boldsymbol{y}_1 + \beta \boldsymbol{y}_2)\rangle &= \langle A\boldsymbol{x}, \alpha \boldsymbol{y}_1 + \beta \boldsymbol{y}_2\rangle = \overline{\alpha}\langle A\boldsymbol{x}, \boldsymbol{y}_1\rangle + \overline{\beta}\langle A\boldsymbol{x}, \boldsymbol{y}_2\rangle \\ &= \overline{\alpha}\langle \boldsymbol{x}, A^*\boldsymbol{y}_1\rangle + \overline{\beta}\langle \boldsymbol{x}, A^*\boldsymbol{y}_2\rangle = \langle \boldsymbol{x}, \alpha A^*\boldsymbol{y}_1\rangle + \langle \boldsymbol{x}, \beta A^*\boldsymbol{y}_2\rangle \\ &= \langle \boldsymbol{x}, \alpha A^*\boldsymbol{y}_1 + \beta A^*\boldsymbol{y}_2\rangle.\end{aligned}$$

Since this holds for all $\boldsymbol{x} \in \mathcal{V}_1$, we conclude that $A^*(\alpha \boldsymbol{y}_1 + \beta \boldsymbol{y}_2) = \alpha A^*\boldsymbol{y}_1 + \beta A^*\boldsymbol{y}_2$.

Finally, suppose B is another linear transformation satisfying the inner product identity $\langle A\boldsymbol{x}, \boldsymbol{y}\rangle = \langle \boldsymbol{x}, B\boldsymbol{y}\rangle$ for all $\boldsymbol{x} \in \mathcal{V}_1$ and $\boldsymbol{y} \in \mathcal{V}_2$. Then, $\langle \boldsymbol{x}, A^*\boldsymbol{y} - B\boldsymbol{y}\rangle = 0 \;\forall\, \boldsymbol{x} \in \mathcal{V}_1$. Hence, $A^*(\boldsymbol{y}) = B(\boldsymbol{y})$ for all $\boldsymbol{y} \in \mathcal{V}_2$, which implies that $A^* = B$. The adjoint is indeed unique. □

The adjoint of the adjoint: If $A : \mathcal{V}_1 \to \mathcal{V}_2$, then $A^{**} = A$, where we write $(A^*)^*$ as A^{**}. First, observe that $A^* : \mathcal{V}_2 \to \mathcal{V}_1$ so $A^{**} : \mathcal{V}_2 \to \mathcal{V}_1$. Thus, A^{**} and A are maps between the same two vector spaces. The result now follows from below:

$$\langle \boldsymbol{x}, A^{**}\boldsymbol{y}\rangle = \langle A^*\boldsymbol{x}, \boldsymbol{y}\rangle = \overline{\langle \boldsymbol{y}, A^*\boldsymbol{x}\rangle} = \overline{\langle A\boldsymbol{y}, \boldsymbol{x}\rangle} = \langle \boldsymbol{x}, A\boldsymbol{y}\rangle \;\forall\, \boldsymbol{x} \in \mathcal{V}_1,\; \boldsymbol{y} \in \mathcal{V}_2\,. \tag{16.12}$$

The adjoint of products: If $A : \mathcal{V}_1 \to \mathcal{V}_2$ and $B : \mathcal{V}_2 \to \mathcal{V}_3$, then $(BA)^* = A^*B^*$ because

$$\langle \boldsymbol{x}, (BA)^*\boldsymbol{y}\rangle = \langle BA\boldsymbol{x}, \boldsymbol{y}\rangle = \langle A\boldsymbol{x}, B^*\boldsymbol{y}\rangle = \langle \boldsymbol{x}, A^*B^*\boldsymbol{y}\rangle \;\forall\, \boldsymbol{x} \in \mathcal{V}_1,\; \boldsymbol{y} \in \mathcal{V}_3\,. \tag{16.13}$$

Note that the above operations are legitimate because $(BA)^* : \mathcal{V}_3 \to \mathcal{V}_1$, $A^* : \mathcal{V}_2 \to \mathcal{V}_1$ and $B^* : \mathcal{V}_3 \to \mathcal{V}_2$.

16.4 The four fundamental subspaces—revisited

There are four fundamental subspaces associated with a linear transformation $A : \mathcal{V}_1 \to \mathcal{V}_2$. The first is often called the ***range*** or ***image*** and is defined as

$$\text{Im}(A) = \{A\boldsymbol{x} : \boldsymbol{x} \in \mathcal{V}_1\}\,. \tag{16.14}$$

The second is called the ***kernel*** or ***null space*** and is defined as

$$\text{Ker}(A) = \{\boldsymbol{x} \in \mathcal{V}_1 : A\boldsymbol{x} = \boldsymbol{0}\}. \tag{16.15}$$

Observe that these definitions are very similar to the column and null spaces of matrices. However, we do not call (16.14) the column space of A because there are no "columns" in a linear function. The third and fourth fundamental subspaces are namely $\text{Im}(A^*)$ and $\text{Ker}(A^*)$. It is obvious from (16.14) that $\text{Im}(BA) \subseteq \text{Im}(B)$ whenever the composition BA is well-defined. Similarly, (16.15) immediately reveals that $\text{Ker}(A) \subseteq \text{Ker}(BA)$.

The image and kernel of A^* are intimately related by the following *fundamental theorem of linear transformations*, which is analogous to our earlier result on matrices.

Theorem 16.7 Fundamental Theorem of Linear Transformations. *Let $\mathcal{V}_1$ and $\mathcal{V}_2$ be inner product spaces and let $A : \mathcal{V}_1 \to \mathcal{V}_2$ be a linear transformation. Then,*

$$Ker(A^*) = Im(A)^\perp \text{ and } Im(A^*) = Ker(A)^\perp .$$

Proof. The first relationship follows because

$$\begin{aligned} \boldsymbol{u} \in \text{Ker}(A^*) &\iff A^*\boldsymbol{u} = \boldsymbol{0} \iff \langle \boldsymbol{x}, A^*\boldsymbol{u} \rangle = 0 \; \forall \; \boldsymbol{x} \in \mathcal{V}_1 \\ &\iff \langle A\boldsymbol{x}, \boldsymbol{u} \rangle = 0 \; \forall \; \boldsymbol{x} \in \mathcal{V}_1 \iff \boldsymbol{u} \in \text{Im}(A)^\perp . \end{aligned}$$

Replacing A^* by A in the above immediately yields $\text{Ker}(A) = \text{Im}(A^*)^\perp$. Since both are subspaces, we have equality of their orthocomplements as well. Therefore, $\text{Im}(A^*) = \text{Ker}(A)^\perp$. □

Using (16.10) and Theorem 16.7, we conclude that

$$\dim(\text{Im}(A)) + \dim(\text{Ker}(A^*)) = m \text{ and } \dim(\text{Im}(A^*)) + \text{Ker}(A) = n . \quad (16.16)$$

The dimension of $\text{Im}(A)$ is called the ***rank*** of A and is denoted by $\rho(A)$, while the dimension of $\text{Ker}(A)$ is called the ***nullity*** of A and is denoted by $\nu(A)$. There is an analogue of the Rank-Nullity Theorem for linear transformations that relates these two dimensions in exactly the same way as for matrices.

Theorem 16.8 *Let $\mathcal{V}_1$ and $\mathcal{V}_2$ be inner product spaces and let $A : \mathcal{V}_1 \to \mathcal{V}_2$ be a linear transformation. Then,*

$$\dim(Im)(A) + \dim(Ker(A)) = n, \text{ where } n = \dim(\mathcal{V}_1).$$

Proof. Write $k = \dim(\text{Ker}(A))$ and let $X = \{\boldsymbol{x}_1, \boldsymbol{x}_2, \dots, \boldsymbol{x}_k\}$ be a basis of $\text{Ker}(A)$. Find extension vectors $\{\boldsymbol{x}_{k+1}, \boldsymbol{x}_{k+2}, \dots, \boldsymbol{x}_n\}$ such that $\{\boldsymbol{x}_1, \boldsymbol{x}_2, \dots, \boldsymbol{x}_n\}$ is a basis for $\mathcal{V}_1$. We claim $Y = \{A\boldsymbol{x}_{k+1}, A\boldsymbol{x}_{k+2}, \dots, A\boldsymbol{x}_n\}$ is a basis of $\text{Im}(A)$.

To prove that Y spans $\text{Im}(A)$, consider any vector $A\boldsymbol{x}$ with $\boldsymbol{x} \in \mathcal{V}_1$. Express $\boldsymbol{x}$ in terms of the basis X as $\boldsymbol{x} = \alpha_1\boldsymbol{x}_1 + \alpha_2\boldsymbol{x}_2 + \cdots + \alpha_n\boldsymbol{x}_n$. Applying the linear transformation A to both sides, yields

$$\begin{aligned} A(\boldsymbol{x}) &= A(\alpha_1\boldsymbol{x}_1 + \alpha_2\boldsymbol{x}_2 + \cdots + \alpha_n\boldsymbol{x}_n) = \alpha_1 A(\boldsymbol{x}_1) + \alpha_2 A(\boldsymbol{x}_2) + \cdots + \alpha_n A(\boldsymbol{x}_n) \\ &= \alpha_{k+1}A(\boldsymbol{x}_{k+1}) + \alpha_{k+2}A(\boldsymbol{x}_{k+2}) + \cdots + \alpha_n A(\boldsymbol{x}_n) \in Sp(Y) , \end{aligned}$$

where the last equality follows from the fact that $A(\boldsymbol{x}_i) = \boldsymbol{0}$ for $i = 1, 2, \dots, n$.

It remains to prove that Y is linearly independent. Consider the homogeneous system

$$\beta_1 A(\boldsymbol{x}_{k+1}) + \beta_2 A(\boldsymbol{x}_{k+2}) + \cdots + \beta_{n-k} A(\boldsymbol{x}_n) = \boldsymbol{0} .$$

We want to show that $\beta_i = 0$ for $i = 1, 2, \dots, n-k$. Note that the above implies

$$\boldsymbol{0} = A(\beta_1\boldsymbol{x}_{k+1} + \beta_2\boldsymbol{x}_{k+2} + \cdots + \beta_{n-k}\boldsymbol{x}_{k+n}) \Longrightarrow \boldsymbol{u} \in \text{Ker}(A) ,$$

where $\boldsymbol{u} = \beta_1\boldsymbol{x}_{k+1} + \beta_2\boldsymbol{x}_{k+2} + \cdots + \beta_{n-k}\boldsymbol{x}_{k+n}$. Since $\boldsymbol{u} \in \text{Ker}(A)$, we can express it in terms of the basis vectors in X. Therefore,

$$\beta_1\boldsymbol{x}_{k+1} + \beta_2\boldsymbol{x}_{k+2} + \cdots + \beta_{n-k}\boldsymbol{x}_{k+n} = \boldsymbol{u} = \theta_1\boldsymbol{x}_1 + \theta_2\boldsymbol{x}_2 + \cdots + \theta_k\boldsymbol{x}_k ,$$

which implies that

$$\theta_1 \boldsymbol{x}_1 + \theta_2 \boldsymbol{x}_2 + \cdots + \theta_k \boldsymbol{x}_k + (-\beta_1)\boldsymbol{x}_{k+1} + (-\beta_2)\boldsymbol{x}_{k+2} + \cdots + (-\beta_{n-k})\boldsymbol{x}_n = \mathbf{0} .$$

Since $\{\boldsymbol{x}_1, \boldsymbol{x}_2, \ldots, \boldsymbol{x}_n\}$ is a basis for $\mathcal{V}_1$, the vectors are linearly independent, so each of the β_i's (and θ_i's) are 0. This proves that Y is a linearly independent set.

Thus, Y is a basis for $\text{Im}(A)$ and it contains $n - k$ elements. So, $\dim(\text{Im}(A)) = n - \dim(\text{Ker}(A))$. $\square$

The rank of the adjoint: An important consequence of the Rank-Nullity Theorem above is that the rank of a linear transformation equals the rank of its adjoint:

$$\dim(\text{Im}(A)) = \dim(\text{Im}(A^*)) . \tag{16.17}$$

This follows immediately from (16.16) and Theorem 16.8. In fact, we have a more general result as an analogue of the fundamental theorem of ranks for matrices. This states that

$$\dim(\text{Im}(A)) = \dim(\text{Im}(A^*A)) = \dim(\text{Im}(AA^*)) = \dim(\text{Im}(A^*)). \tag{16.18}$$

This is also a consequence of Theorem 16.8. Observe that

$$\begin{aligned} \boldsymbol{x} \in \text{Ker}(A) &\Longrightarrow A\boldsymbol{x} = \mathbf{0} \Longrightarrow A^*A\boldsymbol{x} = \mathbf{0} \Longrightarrow \boldsymbol{x} \in \text{Ker}(A^*A) \\ &\Longrightarrow \langle \boldsymbol{x}, A^*A\boldsymbol{x} \rangle = 0 \Longrightarrow \langle A\boldsymbol{x}, A\boldsymbol{x} \rangle = 0 \Longrightarrow A\boldsymbol{x} = \mathbf{0} \Longrightarrow \boldsymbol{x} \in \text{Ker}(A) . \end{aligned}$$

Therefore, $\text{Ker}(A) = \text{Ker}(A^*A)$ and their dimensions are equal. Theorem 16.8 ensures that $\dim(\text{Im}(A)) = \dim(\text{Im}(A^*A))$. The other half of the result follows by interchanging the roles of A and A^* and we obtain $\dim(\text{Im}(A^*)) = \dim(\text{Im}(AA^*))$. The final result follows from (16.17).

One can also derive (16.17) using the first equality in (16.18), which ensures that $\dim(\text{Im}(A)) = \dim(\text{Im}(A^*A)) \leq \dim(\text{Im}(A^*))$. The "$\leq$" holds because $\text{Im}(A^*A) \subseteq \text{Im}(A^*)$. Applying this inequality to $B = A^*$, we conclude

$$\dim(\text{Im}(A)) \leq \dim(\text{Im}(A^*)) = \dim(\text{Im}(B)) \leq \dim(\text{Im}(B^*)) = \dim(\text{Im}(A)) ,$$

where we used $B^* = A^{**} = A$; see (16.12). Thus, $\dim(\text{Im}(A)) = \dim(\text{Im}(A^*))$.

16.5 Inverses of linear transformations

Let $A : \mathcal{V}_1 \to \mathcal{V}_2$ be a linear transformation satisfying the conditions:

(i) If $A\boldsymbol{x}_1 = A\boldsymbol{x}_2$, then $\boldsymbol{x}_1 = \boldsymbol{x}_2$;

(ii) For every $\boldsymbol{y} \in \mathcal{V}_2$, there exists (at least one) $\boldsymbol{x} \in \mathcal{V}_1$ such that $A\boldsymbol{x} = \boldsymbol{y}$.

The first condition says that A is a ***one-one*** map. The second says that A is ***onto***. If A is one-one, then

$$A\boldsymbol{u} = \mathbf{0} \Longrightarrow A\boldsymbol{u} = A(\mathbf{0}) \Longrightarrow \boldsymbol{u} = \mathbf{0} . \tag{16.19}$$

In other words, if A is one-one, then $\text{Ker}(A) = \{\mathbf{0}\}$. Conversely, if $\text{Ker}(A) = \{\mathbf{0}\}$, then

$$A\boldsymbol{x}_1 = A\boldsymbol{x}_2 \Longrightarrow A(\boldsymbol{x}_1 - \boldsymbol{x}_2) = \mathbf{0} \Longrightarrow \boldsymbol{x}_1 - \boldsymbol{x}_2 \in \text{Ker}(A) \Longrightarrow \boldsymbol{x}_1 = \boldsymbol{x}_2 \; .$$

Therefore the transformation A is one-one. Thus, A is one-one if and only if $\text{Ker}(A) = \{\mathbf{0}\}$. Clearly, A is onto if and only if $\text{Im}(A) = \mathcal{V}_2$.

Let us suppose that A is one-one and onto. Since A is onto, for each $\boldsymbol{y} \in \mathcal{V}_2$ there exists $\boldsymbol{x} \in \mathcal{V}_1$ such that $A\boldsymbol{x} = \boldsymbol{y}$. Since A is also one-one, this $\boldsymbol{x}$ is unique. Let $B : \mathcal{V}_2 \to \mathcal{V}_1$ map any $\boldsymbol{y} \in \mathcal{V}_2$ to the corresponding $\boldsymbol{x} \in \mathcal{V}_1$ such that $A\boldsymbol{x} = \boldsymbol{y}$. In other words, $B\boldsymbol{y} = \boldsymbol{x}$ if and only $A\boldsymbol{x} = \boldsymbol{y}$. Let $\boldsymbol{y}_1 = \boldsymbol{y}_2$ be any two elements in $\mathcal{V}_1$ and let $B\boldsymbol{y}_1 = \boldsymbol{x}_1$ and $B\boldsymbol{y}_2 = \boldsymbol{x}_2$. Then, $A\boldsymbol{x}_1 = \boldsymbol{y}_1 = \boldsymbol{y}_2 = A\boldsymbol{x}_2$, which implies that $\boldsymbol{x}_1 = \boldsymbol{x}_2$ because A is one-one. Thus, the map is well-defined. Furthermore, note that

$$AB(\boldsymbol{y}) = A(B\boldsymbol{y}) = A\boldsymbol{x} = \boldsymbol{y} \text{ and } BA(\boldsymbol{x}) = B(\boldsymbol{y}) = \boldsymbol{x} \; . \tag{16.20}$$

In other words, we have $AB(\boldsymbol{y}) = \boldsymbol{y}$ for every $\boldsymbol{y} \in \mathcal{V}_2$ and $BA(\boldsymbol{x}) = \boldsymbol{x}$ for every $\boldsymbol{x} \in \mathcal{V}_1$. We call B the ***inverse*** of A. This means that $AB = I_{\mathcal{V}_2}$ and $BA = I_{\mathcal{V}_1}$, where $I_{\mathcal{V}_2}(\boldsymbol{x}) = \boldsymbol{x}$ for every $\boldsymbol{x} \in \mathcal{V}_2$ and $I_{\mathcal{V}_1}(\boldsymbol{x}) = \boldsymbol{x}$ for every $\boldsymbol{x} \in \mathcal{V}_1$ are the identity maps in $\mathcal{V}_2$ and $\mathcal{V}_1$, respectively.

Definition 16.7 *A linear transformation $A : \mathcal{V}_1 \to \mathcal{V}_2$ is* **invertible** *if there exists a map $B : \mathcal{V}_2 \to \mathcal{V}_1$ such that*

$$AB = I_{\mathcal{V}_2} \text{ and } BA = I_{\mathcal{V}_1} \; ,$$

where $I_{\mathcal{V}_2}$ and $I_{\mathcal{V}_1}$ are the identity maps in $\mathcal{V}_2$ and $\mathcal{V}_1$, respectively. We call B the **inverse** *of A.*

Example 16.5 The identity map $I(\boldsymbol{x})$ is invertible and $I^{-1} = I$. On the other hand, the null map $0(\boldsymbol{x})$ is not invertible as it is neither one-one nor onto. ■

If $B : \mathcal{V}_2 \to \mathcal{V}_1$ and $C : \mathcal{V}_2 \to \mathcal{V}_1$ are both inverses of $A : \mathcal{V}_1 \to \mathcal{V}_2$, then

$$B = BI_{\mathcal{V}_2} = B(AC) = (BA)C = I_{\mathcal{V}_1}C = C \; .$$

Therefore, if A is invertible, then it has a ***unique*** inverse, which we denote by A^{-1}. Also, Definition 16.7 implies that $B^{-1} = (A^{-1})^{-1} = A$. The following result shows that A^{-1}, when it exists, is linear.

Lemma 16.1 *If $A : \mathcal{V}_1 \to \mathcal{V}_2$ is an invertible linear transformation, then its inverse is also linear.*

Proof. If $A\boldsymbol{x}_1 = \boldsymbol{y}_1$ and $A\boldsymbol{x}_2 = \boldsymbol{y}_2$, then $A(\alpha_1\boldsymbol{x}_1 + \alpha_2\boldsymbol{x}_2) = \alpha_1 A\boldsymbol{x}_1 + \alpha_2 A\boldsymbol{x}_2 = \alpha_1\boldsymbol{y}_1 + \alpha_2\boldsymbol{y}_2$. Therefore, $B(\alpha_1\boldsymbol{y}_1 + \alpha_2\boldsymbol{y}_2) = \alpha_1\boldsymbol{x}_1 + \alpha_2\boldsymbol{x}_2 = \alpha_1 B(\boldsymbol{y}_1) + \alpha_2 B(\boldsymbol{y}_2)$, showing that B is linear. □

We already saw how to construct the inverse B in (16.20) when A is one-one and onto. The next theorem shows the converse is also true.

Theorem 16.9 *A linear transformation is invertible if and only if it is one-one and onto.*

Proof. If $A : \mathcal{V}_1 \to \mathcal{V}_2$ is a one-one and onto linear transformation, we can construct a well-defined $B : \mathcal{V}_2 \to \mathcal{V}_1$ that maps each $\boldsymbol{y} \in \mathcal{V}_2$ to the element $\boldsymbol{x} \in \mathcal{V}_1$ such that $A\boldsymbol{x} = \boldsymbol{y}$. This B satisfies (16.20) and is the inverse. This proves the "if" part.

If $A : \mathcal{V}_1 \to \mathcal{V}_2$ is invertible and $A\boldsymbol{x}_1 = A\boldsymbol{x}_2$, then multiplying both sides by A^{-1} yields $\boldsymbol{x}_1 = \boldsymbol{x}_2$. This proves that A must be one-one. Also, if $\boldsymbol{y} \in \mathcal{V}_2$, then $\boldsymbol{y} = I_{\mathcal{V}_2}\boldsymbol{y} = (AA^{-1})\boldsymbol{y} = A(A^{-1}\boldsymbol{y})$, so $\boldsymbol{x} = A^{-1}\boldsymbol{y} \in \mathcal{V}_1$ is an element that is mapped by A to $\boldsymbol{y}$. Thus, $\boldsymbol{y} \in \operatorname{Im}(A)$ and A is onto. □

Let $A : \mathcal{V}_1 \to \mathcal{V}_2$ and $B : \mathcal{V}_2 \to \mathcal{V}_3$ be linear transformations. Assuming the inverses, $B^{-1} : \mathcal{V}_3 \to \mathcal{V}_2$ and $A^{-1} : \mathcal{V}_2 \to \mathcal{V}_1$, exist, note that

$$A^{-1}B^{-1}BA\boldsymbol{x} = A^{-1}(B^{-1}B)(A\boldsymbol{x}) = A^{-1}(A\boldsymbol{x}) = \boldsymbol{x} \text{ for all } \boldsymbol{x} \in \mathcal{V}_1.$$

It can, similarly, be proved that $ABB^{-1}A^{-1}\boldsymbol{x} = \boldsymbol{x}$ for all $\boldsymbol{x} \in \mathcal{V}_2$. Therefore,

$$(BA)^{-1} = A^{-1}B^{-1}. \tag{16.21}$$

The preceding results do not require that $\mathcal{V}_1$ and $\mathcal{V}_2$ be inner product spaces. They can be vector spaces over any common field $\mathcal{F}$. When $\mathcal{V}_1$ and $\mathcal{V}_2$ *are* inner product spaces, we can derive an expression for the inverse of the adjoint transformation. Taking $A = B^*$ in (16.21) and using the fact that $(BA)^* = A^*B^*$ (see (16.13)), we find that $(B^{-1})^*B^*\boldsymbol{x} = (BB^{-1})^*\boldsymbol{x} = I^*(\boldsymbol{x}) = \boldsymbol{x}$ for all $\boldsymbol{x} \in \mathcal{V}_1$. Thus,

$$(B^*)^{-1} = (B^{-1})^*. \tag{16.22}$$

In general, the conditions for linear transformations $A : \mathcal{V}_1 \to \mathcal{V}_2$ to be one-one are different for being onto. However, consider the special case when $\dim(\mathcal{V}_1) = \dim(\mathcal{V}_2) = n$. Recall that A is one-one if and only if $\operatorname{Ker}(A) = \{\boldsymbol{0}\}$. The Rank-Nullity Theorem ensures that $\dim(\operatorname{Im}(A)) = \dim(\mathcal{V}_1) = n$. Since $\operatorname{Im}(A) \subseteq \mathcal{V}_2$ and $\dim(\operatorname{Im}(A)) = n = \dim(\mathcal{V}_2)$, it follows that $\operatorname{Im}(A) = \mathcal{V}_2$. This is the definition of A being onto. Conversely, if A is onto then, by definition, $\operatorname{Im}(A) = \mathcal{V}_2$ and $\dim(\operatorname{Ker}(A)) = n - \dim(\operatorname{Im}(A)) = n - n = 0$, which implies that $\operatorname{Ker}(A) = \{\boldsymbol{0}\}$. Therefore, if $A : \mathcal{V}_1 \to \mathcal{V}_2$ is a linear transformation with $\dim(\mathcal{V}_1) = \dim(\mathcal{V}_2) = n$, then the following three statements are equivalent:

(i) A is one-one;

(ii) A is onto;

(iii) $\operatorname{Ker}(A) = \{\boldsymbol{0}\}$.

When $A : \mathcal{V} \to \mathcal{V}$ is invertible, we say that the composition $AA^{-1} = A^{-1}A = I$ where $I : \mathcal{V} \to \mathcal{V}$ is the ***identity*** map $I(\boldsymbol{x}) = \boldsymbol{x}$ for all $\boldsymbol{x} \in \mathcal{V}$.

16.6 Linear transformations and matrices

Consider a linear transformation $A : \mathcal{V}_1 \to \mathcal{V}_2$, where $\dim(\mathcal{V}_1) = n$ and $\dim(\mathcal{V}_2) = m$ are finite. Let $\mathcal{X} = \{\boldsymbol{x}_1, \ldots, \boldsymbol{x}_n\}$ be a basis for $\mathcal{V}_1$ and $\mathcal{Y} = \{\boldsymbol{y}_1, \ldots, \boldsymbol{y}_m\}$ be a basis for $\mathcal{V}_2$. We can express any $A\boldsymbol{x}_j$ in terms of the basis $\mathcal{Y}$:

$$A\boldsymbol{x}_j = a_{1j}\boldsymbol{y}_1 + a_{2j}\boldsymbol{y}_2 + \cdots + a_{mj}\boldsymbol{y}_m \text{ for } j = 1, \ldots, n \,. \tag{16.23}$$

The a_{ij}'s are the ***coordinates*** of the linear transformation A with respect to the bases $\mathcal{X}$ and $\mathcal{Y}$. These coordinates can be placed into an $m \times n$ matrix as below:

$$\boldsymbol{A} = \begin{bmatrix} a_{11} & a_{12} & \ldots & a_{1n} \\ a_{21} & a_{22} & \ldots & a_{2n} \\ \vdots & \vdots & \ddots & \vdots \\ a_{m1} & a_{m2} & \ldots & a_{mn} \end{bmatrix} \,. \tag{16.24}$$

Note that the coordinates of $A\boldsymbol{x}_j$ in (16.23) are placed as the j-th column in $\boldsymbol{A}$. This matrix $\boldsymbol{A}$ represents the linear map A with respect to the bases $\mathcal{X}$ and $\mathcal{Y}$. A stricter notation would require that we write this matrix as $[A]_{\mathcal{X}}^{\mathcal{Y}}$ to explicitly show the bases. However, often this will be implicit and we will simply write $\boldsymbol{A}$. By this convention, if $\boldsymbol{A}$ is a matrix of the linear transformation $A : \mathcal{V}_1 \to \mathcal{V}_2$, where $\dim(\mathcal{V}_1) = n$ and $\dim(\mathcal{V}_2) = m$, then $\boldsymbol{A}$ will have m rows and n columns.

Here are a few simple examples of finding the matrix of a linear map.

Example 16.6 Rotation. Let $A_\theta : \Re^2 \to \Re^2$ be the map that rotates any point in $\Re^2$ counter-clockwise by an angle θ about the origin. Let $\mathcal{X} = \mathcal{Y} = \{\boldsymbol{e}_1, \boldsymbol{e}_2\}$ be the standard canonical basis in $\Re^2$, where $\boldsymbol{e}_1 = (1,0)'$ and $\boldsymbol{e}_2 = (0,1)'$. Observe, from elementary trigonometry, that A_θ maps $\boldsymbol{e}_1$ to the vector $(\cos\theta, \sin(\theta))'$, while it maps $\boldsymbol{e}_2$ to $(-\sin\theta, \cos\theta)'$. Thus, we can write

$$A_\theta(\boldsymbol{e}_1) = \cos\theta\boldsymbol{e}_1 + \sin\theta\boldsymbol{e}_2 \text{ and } A_\theta(\boldsymbol{e}_2) = -\sin\theta\boldsymbol{e}_1 + \cos\theta\boldsymbol{e}_2 \,.$$

From the above, it is clear that A_θ is linear and its matrix with respect to $\mathcal{X}$ and $\mathcal{Y}$ is

$$\boldsymbol{A} = [A_\theta]_{\mathcal{X}}^{\mathcal{Y}} = \begin{bmatrix} \cos\theta & -\sin\theta \\ \sin\theta & \cos\theta \end{bmatrix} \,.$$

Any point $\boldsymbol{x} \in \Re^2$ is mapped to $\boldsymbol{A}\boldsymbol{x}$ as a result of A_θ. ■

Example 16.7 Reflection. Let $A_m : \Re^2 \to \Re^2$ be the map that reflects a point in the line $y = mx$. Let $\mathcal{X} = \mathcal{Y}$ be the standard canonical basis in $\Re^2$, as in Example 16.6. From elementary analytical geometry, we find that

$$A_m(\boldsymbol{e}_1) = \frac{1-m^2}{1+m^2}\boldsymbol{e}_1 + \frac{2m}{1+m^2}\boldsymbol{e}_2 \text{ and } A_m(\boldsymbol{e}_2) = \frac{2m}{1+m^2}\boldsymbol{e}_1 + \frac{m^2-1}{1+m^2}\boldsymbol{e}_2 \,.$$

From the above, it is clear that A_m is linear and its matrix with respect to $\mathcal{X}$ and $\mathcal{Y}$ is

$$\boldsymbol{A} = [A_\theta]_{\mathcal{X}}^{\mathcal{Y}} = \frac{1}{1+m^2}\begin{bmatrix} 1-m^2 & 2m \\ 2m & m^2-1 \end{bmatrix} \,.$$

Thus, the reflection of any point $\boldsymbol{x} \in \Re^2$ in the line $y = mx$ is given by $\boldsymbol{Ax}$. Interestingly, what is the matrix for reflection about the y-axis, i.e., the line $x = 0$? We leave this for the reader to figure out. ■

Example 16.8 The identity and null maps. The matrix of the identity map $I(\boldsymbol{x}) = \boldsymbol{x} \ \forall \ \boldsymbol{x} \in \Re^n$ is the $n \times n$ identity matrix $\boldsymbol{I}_n$. The matrix of the null map $0(\boldsymbol{x}) = \boldsymbol{0} \ \forall \ \boldsymbol{x} \in \Re^n$ is the null matrix with all entries as zero. ■

If $\boldsymbol{A}$ is an $m \times n$ real matrix, then $A(\boldsymbol{x}) = \boldsymbol{Ax}$ is a linear map from $\Re^n$ to $\Re^m$. Conversely, every linear transformation $A : \Re^n \to \Re^m$ can be expressed as a matrix-vector multiplication $A(\boldsymbol{x}) = \boldsymbol{Ax}$. This is demonstrated in the following example.

Example 16.9 Let $A : \Re^n \to \Re^m$ be a linear transformation. Let $\boldsymbol{e}_1, \boldsymbol{e}_2, \ldots, \boldsymbol{e}_n$ and $\tilde{\boldsymbol{e}}_1, \tilde{\boldsymbol{e}}_2, \ldots, \tilde{\boldsymbol{e}}_m$ be the standard canonical bases in $\Re^n$ and $\Re^m$, respectively. We write the latter as $\tilde{\boldsymbol{e}}_i$'s to avoid confusion with the former. Since each $A\boldsymbol{e}_j$ is a vector in $\Re^m$, we can write it as a linear combination of the $\tilde{\boldsymbol{e}}_i$'s, say

$$A\boldsymbol{e}_j = a_{1j}\tilde{\boldsymbol{e}}_1 + a_{2j}\tilde{\boldsymbol{e}}_2 + \cdots + a_{mj}\tilde{\boldsymbol{e}}_m = [\tilde{\boldsymbol{e}}_1 : \tilde{\boldsymbol{e}}_2 : \ldots : \tilde{\boldsymbol{e}}_m] \begin{bmatrix} a_{1j} \\ a_{2j} \\ \vdots \\ a_{mj} \end{bmatrix} = \boldsymbol{I}_{m\times m}\boldsymbol{a}_{*j}$$

$$= \boldsymbol{a}_{*j} \text{ where } \boldsymbol{a}_{*j} = \begin{bmatrix} a_{1j} \\ a_{2j} \\ \vdots \\ a_{mj} \end{bmatrix}, \text{ for } j = 1, 2, \ldots, n\,.$$

The matrix of A with respect to the standard bases in $\Re^n$ and $\Re^m$ is the $m \times n$ matrix

$$\boldsymbol{A} = [\boldsymbol{a}_{*1} : \boldsymbol{a}_{*2} : \ldots : \boldsymbol{a}_{*n}] = \begin{bmatrix} a_{11} & a_{12} & \cdots & a_{1n} \\ a_{21} & a_{22} & \cdots & a_{2n} \\ \vdots & \vdots & \ddots & \vdots \\ a_{m1} & a_{m2} & \cdots & a_{mn} \end{bmatrix}.$$

Let us now compute $A\boldsymbol{x}$ for a general vector $\boldsymbol{x} \in \Re^n$ in terms of these bases. Let $\boldsymbol{x}' = [x_1 : x_2 : \ldots : x_n]$. Then,

$$A(\boldsymbol{x}) = A(x_1\boldsymbol{e}_1 + x_2\boldsymbol{e}_2 + \cdots + x_n\boldsymbol{e}_n) = x_1A\boldsymbol{e}_1 + x_2A\boldsymbol{e}_2 + \cdots + x_nA\boldsymbol{e}_n$$

$$= x_1\boldsymbol{a}_{*1} + x_2\boldsymbol{a}_{*2} + \cdots + x_n\boldsymbol{a}_{*n} = [\boldsymbol{a}_{*1} : \boldsymbol{a}_{*2} : \ldots : \boldsymbol{a}_{*n}] \begin{bmatrix} x_1 \\ x_2 \\ \vdots \\ x_n \end{bmatrix} = \boldsymbol{Ax}\,.$$

Therefore, *every* linear transformation $A : \Re^n \to \Re^m$ is given by a matrix multiplication $A\boldsymbol{x} = \boldsymbol{Ax}$, where $\boldsymbol{A}$ is the $m \times n$ matrix representing A with respect to the standard bases in $\Re^n$ and $\Re^m$. ■

Sums: The standard definition of the sum of two matrices has a nice correspondence with the sum of two linear transformations. Let A and B be transformations from $\mathcal{V}_1 \to \mathcal{V}_2$ and let $C = A + B$. Let $\mathcal{X} = \{\boldsymbol{x}_1, \dots, \boldsymbol{x}_n\}$ be a basis for $\mathcal{V}_1$ and $\mathcal{Y} = \{\boldsymbol{y}_1, \dots, \boldsymbol{y}_m\}$ be a basis for $\mathcal{V}_2$. If $\boldsymbol{A} = \{a_{ij}\}$ and $\boldsymbol{B} = \{b_{ij}\}$ are the matrices $[A]_{\mathcal{X}}^{\mathcal{Y}}$ and $[B]_{\mathcal{X}}^{\mathcal{Y}}$, then

$$\begin{aligned} C\boldsymbol{x}_j &= (A+B)\boldsymbol{x}_j = A\boldsymbol{x}_j + B\boldsymbol{x}_j \\ &= a_{1j}\boldsymbol{y}_1 + a_{2j}\boldsymbol{y}_2 + \cdots + a_{mj}\boldsymbol{y}_m + b_{1j}\boldsymbol{y}_1 + b_{2j}\boldsymbol{y}_2 + \cdots + b_{mj}\boldsymbol{y}_m \\ &= c_{1j}\boldsymbol{y}_1 + c_{2j}\boldsymbol{y}_2 + \cdots + c_{mj}\boldsymbol{y}_j \ , \quad \text{where } c_{ij} = a_{ij} + b_{ij} \ , \end{aligned}$$

for each $j = 1, \dots, m$. This shows that

$$[C]_{\mathcal{X}}^{\mathcal{Y}} = \boldsymbol{C} = \{c_{ij}\} = \{a_{ij} + b_{ij}\} = \boldsymbol{A} + \boldsymbol{B} = [A]_{\mathcal{X}}^{\mathcal{Y}} + [B]_{\mathcal{X}}^{\mathcal{Y}} \ . \tag{16.25}$$

Products: Let $B : \mathcal{V}_1 \to \mathcal{V}_2$ and $A : \mathcal{V}_2 \to \mathcal{V}_3$ be two linear transformations and let $C = AB : \mathcal{V}_1 \to \mathcal{V}_3$ be the composition of A and B. Let $\mathcal{X} = \{\boldsymbol{x}_1, \boldsymbol{x}_2, \dots, \boldsymbol{x}_n\}$ be a basis for $\mathcal{V}_1$, $\mathcal{Y} = \{\boldsymbol{y}_1, \boldsymbol{y}_2, \dots, \boldsymbol{y}_p\}$ be a basis for $\mathcal{V}_2$ and $\mathcal{Z} = \{\boldsymbol{z}_1, \boldsymbol{z}_2, \dots, \boldsymbol{z}_m\}$ be a basis for $\mathcal{V}_3$. Consider the three matrices

$$\boldsymbol{A} = \{a_{ij}\} = [A]_{\mathcal{Y}}^{\mathcal{Z}} \ , \ \boldsymbol{B} = \{b_{ij}\} = [B]_{\mathcal{X}}^{\mathcal{Y}} \ \text{and} \ \boldsymbol{C} = \{c_{ij}\} = [C]_{\mathcal{X}}^{\mathcal{Z}} \ .$$

Observe that $\boldsymbol{A}$ is $m \times p$, $\boldsymbol{B}$ is $p \times n$ and $\boldsymbol{C}$ is $m \times n$. Thus, $\boldsymbol{C} = \boldsymbol{AB}$ is well-defined. Then, for each $j = 1, 2, \dots, n$,

$$\begin{aligned} C\boldsymbol{x}_j &= (AB)\boldsymbol{x}_j = A(B\boldsymbol{x}_j) = A(b_{1j}\boldsymbol{y}_1 + b_{2j}\boldsymbol{y}_2 + \cdots + b_{pj}\boldsymbol{y}_p) \\ &= b_{1j}A\boldsymbol{y}_1 + b_{2j}A\boldsymbol{y}_2 + \cdots + b_{pj}A\boldsymbol{y}_p \\ &= b_{1j}\left(\sum_{i=1}^m a_{i1}\boldsymbol{z}_i\right) + b_{2j}\left(\sum_{i=1}^m a_{i2}\boldsymbol{z}_i\right) + \cdots + b_{pj}\left(\sum_{i=1}^m a_{ip}\boldsymbol{z}_i\right) \\ &= \sum_{k=1}^p b_{kj}\sum_{i=1}^m a_{ik}\boldsymbol{z}_i = \sum_{i=1}^m\left(\sum_{k=1}^p a_{ik}b_{kj}\right)\boldsymbol{z}_i = \sum_{i=1}^m c_{ij}\boldsymbol{z}_i \ , \end{aligned}$$

where $c_{ij} = \sum_{k=1}^p a_{ik}b_{kj} = \boldsymbol{a}_{i*}'\boldsymbol{b}_{*j}$, where $\boldsymbol{a}_{i*}'$ and $\boldsymbol{b}_{*j}$ are the i-th row of $\boldsymbol{A}$ and j-th column of $\boldsymbol{B}$, respectively. Thus,

$$[AB]_{\mathcal{X}}^{\mathcal{Z}} = [C]_{\mathcal{X}}^{\mathcal{Z}} = \boldsymbol{C} = \boldsymbol{AB} = [A]_{\mathcal{Y}}^{\mathcal{Z}}[B]_{\mathcal{X}}^{\mathcal{Y}} \ . \tag{16.26}$$

It can also be shown that the scalar multiplication for matrices, $\alpha\boldsymbol{A}$, will correspond to the linear transformation αA, where $\boldsymbol{A}$ is the matrix of A with respect to two fixed bases. We leave the details to the reader. Several elementary properties of matrices follow immediately from the corresponding properties of linear transformations. For example, associativity of matrix multiplication: $\boldsymbol{ABC} = \boldsymbol{A}(\boldsymbol{BC}) = (\boldsymbol{AB})\boldsymbol{C}$ is a consequence of the associativity of composition of linear transformations.

The adjoint of a complex matrix $\boldsymbol{A} = \{a_{ij}\}$ is defined as $\boldsymbol{A}^* = \{\bar{a}_{ji}\}$. In other words, $\boldsymbol{A}^*$ is formed by transposing $\boldsymbol{A}$ and then taking the complex conjugate of each element. If $\boldsymbol{A}$ is a real matrix, like most of the matrices in this text, then the adjoint is equal to the transpose, i.e., $\boldsymbol{A}^* = \boldsymbol{A}' = \{a_{ji}\}$.

How does the adjoint of a matrix relate to the adjoint of a linear transformation? A natural guess is that if $\boldsymbol{A}$ is the matrix of a linear map A between two finite-dimensional inner product spaces, then $\boldsymbol{A}^*$ is the matrix of the adjoint transformation A^*. This is *almost* true, with the caveat that the matrix representations must be with respect to *orthogonal bases*. The following theorem states and proves this result.

Theorem 16.10 *Let $A : \mathcal{V}_1 \to \mathcal{V}_2$ be a linear transformation over finite-dimensional inner product spaces $\mathcal{V}_1$ and $\mathcal{V}_2$. If $\boldsymbol{A}$ is a matrix of A with respect to a specified pair of orthornormal bases for $\mathcal{V}_1$ and $\mathcal{V}_2$, then $\boldsymbol{A}^*$ is the matrix of $A^* : \mathcal{V}_2 \to \mathcal{V}_1$ with respect to the same orthonormal bases.*

Proof. Let $\mathcal{X} = \{\boldsymbol{x}_1, \boldsymbol{x}_2, \ldots, \boldsymbol{x}_n\}$ and $\mathcal{Y} = \{\boldsymbol{y}_1, \boldsymbol{y}_2, \ldots, \boldsymbol{y}_m\}$ be orthonormal bases for $\mathcal{V}_1$ and $\mathcal{V}_2$, respectively. Express each $A\boldsymbol{x}_i$ and $A^*\boldsymbol{y}_j$ as

$$A\boldsymbol{x}_i = a_{1i}\boldsymbol{y}_1 + a_{2i}\boldsymbol{y}_2 + \cdots + a_{mi}\boldsymbol{y}_m \text{ and } A^*\boldsymbol{y}_j = b_{1j}\boldsymbol{x}_1 + b_{2j}\boldsymbol{x}_2 + \cdots + b_{nj}\boldsymbol{x}_n \,.$$

Therefore, $[A]_{\mathcal{X}}^{\mathcal{Y}} = \boldsymbol{A} = \{a_{ij}\}$ and $[A^*]_{\mathcal{Y}}^{\mathcal{X}} = \boldsymbol{B} = \{b_{ij}\}$. We find that

$$\begin{aligned} b_{ij} &= \sum_{k=1}^{n} b_{kj}\langle \boldsymbol{x}_k, \boldsymbol{x}_i\rangle = \left\langle \sum_{k=1}^{n} b_{kj}\boldsymbol{x}_k, \boldsymbol{x}_i \right\rangle = \langle A^*\boldsymbol{y}_j, \boldsymbol{x}_i\rangle = \langle \boldsymbol{y}_j, A\boldsymbol{x}_i\rangle \\ &= \left\langle \boldsymbol{y}_j, \sum_{k=1}^{m} a_{ki}\boldsymbol{y}_k \right\rangle = \sum_{k=1}^{m} \bar{a}_{ki}\langle \boldsymbol{y}_j, \boldsymbol{y}_k\rangle = \bar{a}_{ji} \,, \end{aligned}$$

which implies that $[A^*]_{\mathcal{Y}}^{\mathcal{X}} = \boldsymbol{B} = \{b_{ij}\} = \{\bar{a}_{ji}\} = \boldsymbol{A}^*$. □

If A is a linear map from $\mathcal{V}_1$ to $\mathcal{V}_2$, then the adjoint A^* is a map from $\mathcal{V}_2$ to $\mathcal{V}_1$. If $\mathcal{X}$ and $\mathcal{Y}$ are two orthonormal bases in $\mathcal{V}_1$ and $\mathcal{V}_2$, respectively, so that $\boldsymbol{A} = [A]_{\mathcal{X}}^{\mathcal{Y}}$, then $\boldsymbol{A}^* = [A^*]_{\mathcal{Y}}^{\mathcal{X}}$. Observe that the $\mathcal{X}$ and $\mathcal{Y}$ flip positions in $[A]$ and $[A^*]$ to indicate the bases in the domain (subscript) and image (superscript) of A and A^*.

Consider the adjoint of an adjoint matrix. Fix any two orthonormal bases $\mathcal{X}$ and $\mathcal{Y}$ in the domain and image of A and let $\boldsymbol{A}$ be the matrix of A with respect to the bases. Theorem 16.10 ensures that $[A^*]_{\mathcal{Y}}^{\mathcal{X}} = \boldsymbol{A}^*$. Using (16.12), we find that

$$(\boldsymbol{A}^*)^* = [(A^*)^*]_{\mathcal{X}}^{\mathcal{Y}} = [A]_{\mathcal{X}}^{\mathcal{Y}} = \boldsymbol{A} \,.$$

Let $A : \mathcal{V}_1 \to \mathcal{V}_2$ and $B : \mathcal{V}_2 \to \mathcal{V}_3$ be linear transformations. Fixing orthonormal bases $\mathcal{X}$, $\mathcal{Y}$ and $\mathcal{Z}$ in $\mathcal{V}_1$, $\mathcal{V}_2$ and $\mathcal{V}_3$, respectively, let $\boldsymbol{A} = [A]_{\mathcal{X}}^{\mathcal{Y}}$ and $\boldsymbol{B} = [B]_{\mathcal{Y}}^{\mathcal{Z}}$ so that $\boldsymbol{BA} = [BA]_{\mathcal{X}}^{\mathcal{Z}}$. Now, using Theorem 16.10 and (16.13), we obtain

$$(\boldsymbol{BA})^* = [(BA)^*]_{\mathcal{Z}}^{\mathcal{X}} = [A^*B^*]_{\mathcal{Z}}^{\mathcal{X}} = [A^*]_{\mathcal{Y}}^{\mathcal{X}}[B^*]_{\mathcal{Z}}^{\mathcal{Y}} = \boldsymbol{A}^*\boldsymbol{B}^* \,.$$

These results were directly obtained for matrices in Chapter 1.

16.7 Change of bases, equivalence and similar matrices

Let $\mathcal{X} = \{\boldsymbol{x}_1, \ldots, \boldsymbol{x}_n\}$ and $\mathcal{Y} = \{\boldsymbol{y}_1, \ldots, \boldsymbol{y}_m\}$ be bases of $\mathcal{V}_1$ and $\mathcal{V}_2$, respectively, and suppose $\boldsymbol{A} = \{a_{ij}\}$ is the matrix of $A : \mathcal{V}_1 \to \mathcal{V}_2$ with respect to these two bases. Let $\boldsymbol{x} \in \mathcal{V}_1$ such that $\boldsymbol{x} = \alpha_1\boldsymbol{x}_1 + \alpha_2\boldsymbol{x}_2 + \cdots + \alpha_n\boldsymbol{x}_n$. Then,

$$A(\boldsymbol{x}) = A(\alpha_1\boldsymbol{x}_1 + \alpha_2\boldsymbol{x}_2 + \cdots + \alpha_n\boldsymbol{x}_n) = \alpha_1 A\boldsymbol{x}_1 + \alpha_2 A\boldsymbol{x}_2 + \cdots + \alpha_n A\boldsymbol{x}_n$$
$$= \sum_{j=1}^{n} \alpha_j \left(\sum_{i=1}^{m} a_{ij}\boldsymbol{y}_i \right) = \sum_{i=1}^{m} \left(\sum_{j=1}^{n} a_{ij}\alpha_j \right) \boldsymbol{y}_i = \sum_{i=1}^{m} (\boldsymbol{a}'_{i*}\boldsymbol{\alpha})\boldsymbol{y}_i ,$$

where $\boldsymbol{a}'_{i*}$ is the i-th row of $\boldsymbol{A}$ and $\boldsymbol{\alpha}$ is $n \times 1$ vector with α_j as the j-th element. The coordinates of $A\boldsymbol{x}$ are given by

$$\begin{bmatrix} \boldsymbol{a}'_{1*}\boldsymbol{\alpha} \\ \boldsymbol{a}'_{2*}\boldsymbol{\alpha} \\ \vdots \\ \boldsymbol{a}'_{m*}\boldsymbol{\alpha} \end{bmatrix} = \begin{bmatrix} \boldsymbol{a}'_{1*} \\ \boldsymbol{a}'_{2*} \\ \vdots \\ \boldsymbol{a}'_{m*} \end{bmatrix} \boldsymbol{\alpha} = \boldsymbol{A}\boldsymbol{\alpha} .$$

In other words, if $\boldsymbol{\alpha}$ is the coordinate vector of $\boldsymbol{x}$ with respect to $\mathcal{X}$ in $\mathcal{V}_1$, then $\boldsymbol{A}\boldsymbol{\alpha}$ is the coordinate vector of $A(\boldsymbol{x})$ with respect to $\mathcal{Y}$ in $\mathcal{V}_2$, where $\boldsymbol{A}$ is the matrix of A with respect to these two bases. We can write this more succinctly as

$$[A\boldsymbol{x}]_{\mathcal{X}}^{\mathcal{Y}} = [A]_{\mathcal{X}}^{\mathcal{Y}}[\boldsymbol{x}]_{\mathcal{X}}^{\mathcal{X}} = \boldsymbol{A}\boldsymbol{\alpha} .$$

Example 16.9 is a special case of this.

Let $\mathcal{V}$ be a vector space with $\dim(\mathcal{V}) = n$ and let $\mathcal{X} = \{\boldsymbol{x}_1, \ldots, \boldsymbol{x}_n\}$ and $\mathcal{Y} = \{\boldsymbol{y}_1, \ldots, \boldsymbol{y}_n\}$ be any two of its bases. Let $I : \mathcal{V} \to \mathcal{V}$ be the identity map $I(\boldsymbol{x}) = \boldsymbol{x}$ for all $\boldsymbol{x} \in \mathcal{V}$. Clearly, the matrix of I with respect to $\mathcal{X}$ and itself—-we simply write "with respect to $\mathcal{X}$" in such cases—i.e., $[I]_{\mathcal{X}}^{\mathcal{X}}$, is the identity matrix $\boldsymbol{I}$. What is the matrix of I with respect to $\mathcal{X}$ and $\mathcal{Y}$? We express the members of $\mathcal{X}$ mapped by I as a linear combination of the members in $\mathcal{Y}$. Suppose

$$\boldsymbol{x}_j = p_{1j}\boldsymbol{y}_1 + p_{2j}\boldsymbol{y}_2 + \cdots + p_{nj}\boldsymbol{y}_n \quad \text{for } j = 1, 2, \ldots, n . \tag{16.27}$$

The matrix of I with respect to $\mathcal{X}$ and $\mathcal{Y}$ is given by the $n \times n$ matrix

$$[I]_{\mathcal{X}}^{\mathcal{Y}} = \boldsymbol{P} = \begin{bmatrix} p_{11} & p_{12} & \cdots & p_{1n} \\ p_{21} & p_{22} & \cdots & p_{2n} \\ \vdots & \vdots & \ddots & \vdots \\ p_{n1} & p_{n2} & \cdots & p_{nn} \end{bmatrix} . \tag{16.28}$$

Similarly, we can express the members of $\mathcal{Y}$ mapped by I as a linear combination of the members in $\mathcal{X}$, say

$$\boldsymbol{y}_j = q_{1j}\boldsymbol{x}_1 + q_{2j}\boldsymbol{x}_2 + \cdots + q_{nj}\boldsymbol{x}_n \quad \text{for } j = 1, 2, \ldots, n . \tag{16.29}$$

The matrix of I with respect to $\mathcal{Y}$ and $\mathcal{X}$ is given by the $n \times n$ matrix

$$[I]_{\mathcal{Y}}^{\mathcal{X}} = \boldsymbol{Q} = \begin{bmatrix} q_{11} & q_{12} & \cdots & q_{1n} \\ q_{21} & q_{22} & \cdots & q_{2n} \\ \vdots & \vdots & \ddots & \vdots \\ q_{n1} & q_{n2} & \cdots & q_{nn} \end{bmatrix} . \tag{16.30}$$

The matrices $\boldsymbol{P} = [I]_{\mathcal{X}}^{\mathcal{Y}}$ and $\boldsymbol{Q} = [I]_{\mathcal{U}}^{\mathcal{W}}$ are often called ***transition matrices*** as they describe *transitions* from one basis to another within the same vector space.

We can now combine the above information to note that

$$\boldsymbol{x}_j = \sum_{i=1}^{n} p_{ij}\boldsymbol{y}_i = \sum_{i=1}^{n} p_{ij}\left(\sum_{k=1}^{n} q_{ki}\boldsymbol{x}_k\right) = \sum_{k=1}^{n}\left(\sum_{i=1}^{n} q_{ki}p_{ij}\right)\boldsymbol{x}_k .$$

This means that

$$\boldsymbol{q}'_{k*}\boldsymbol{p}_{*j} = \sum_{i=1}^{n} q_{ki}p_{ij} = \delta_{kj} = \begin{cases} 1 & \text{if } k = j \\ 0 & \text{if } k \neq j \end{cases} ,$$

where $\boldsymbol{q}'_{*k}$ is the k-th row of $\boldsymbol{Q}$ and $\boldsymbol{p}_{*j}$ is the j-th column of $\boldsymbol{P}$. It follows that $\boldsymbol{QP} = \boldsymbol{I}$. Also,

$$\boldsymbol{y}_j = \sum_{i=1}^{n} q_{ij}\boldsymbol{x}_i = \sum_{i=1}^{n} q_{ij}\left(\sum_{k=1}^{n} p_{ki}\boldsymbol{y}_k\right) = \sum_{k=1}^{n}\left(\sum_{i=1}^{n} p_{ki}q_{ij}\right)\boldsymbol{y}_k ,$$

which implies that

$$\boldsymbol{p}'_{k*}\boldsymbol{q}_{*j} = \sum_{i=1}^{n} p_{ki}q_{ij} = \delta_{kj} = \begin{cases} 1 & \text{if } k = j \\ 0 & \text{if } k \neq j \end{cases} ,$$

where $\boldsymbol{p}'_{*k}$ is the k-th row of $\boldsymbol{P}$ and $\boldsymbol{q}_{*j}$ is the j-th column of $\boldsymbol{Q}$. It follows that $\boldsymbol{PQ} = \boldsymbol{I}$. We have, therefore, established that

$$[I]_{\mathcal{Y}}^{\mathcal{X}}[I]_{\mathcal{X}}^{\mathcal{Y}} = \boldsymbol{QP} = \boldsymbol{I} = \boldsymbol{PQ} = [I]_{\mathcal{X}}^{\mathcal{Y}}[I]_{\mathcal{Y}}^{\mathcal{X}} . \tag{16.31}$$

Since $\boldsymbol{I} = [I]_{\mathcal{X}}^{\mathcal{X}} = [I]_{\mathcal{Y}}^{\mathcal{Y}}$, we could also write the above as

$$[I]_{\mathcal{Y}}^{\mathcal{X}}[I]_{\mathcal{X}}^{\mathcal{Y}} = [I]_{\mathcal{X}}^{\mathcal{X}} = [I]_{\mathcal{Y}}^{\mathcal{Y}} = [I]_{\mathcal{X}}^{\mathcal{Y}}[I]_{\mathcal{Y}}^{\mathcal{X}} .$$

In other words, $\left([I]_{\mathcal{X}}^{\mathcal{Y}}\right)^{-1} = [I]_{\mathcal{Y}}^{\mathcal{X}}$.

The above results become especially transparent in Euclidean spaces. The following example illustrates.

Example 16.10 Let $\mathcal{V} = \Re^n$ and let $\mathcal{X} = \{\boldsymbol{x}_1, \ldots, \boldsymbol{x}_n\}$ and $\mathcal{Y} = \{\boldsymbol{y}_1, \ldots, \boldsymbol{y}_n\}$ be any two of its bases. Construct the $n \times n$ matrices $\boldsymbol{X} = [\boldsymbol{x}_1 : \boldsymbol{x}_2 : \ldots : \boldsymbol{x}_n]$ and $\boldsymbol{Y} = [\boldsymbol{y}_1 : \boldsymbol{y}_2 : \ldots : \boldsymbol{y}_n]$. Clearly $\boldsymbol{X}$ and $\boldsymbol{Y}$ are nonsingular. Then, (16.27) and (16.29) imply that we solve $\boldsymbol{YP} = \boldsymbol{X}$ and $\boldsymbol{XQ} = \boldsymbol{Y}$ for $\boldsymbol{P}$ and $\boldsymbol{Q}$ to obtain the change of basis. Therefore, $\boldsymbol{QP} = \boldsymbol{X}^{-1}\boldsymbol{Y}\boldsymbol{Y}^{-1}\boldsymbol{X} = \boldsymbol{I}_n$ and $\boldsymbol{PQ} = \boldsymbol{Y}^{-1}\boldsymbol{X}\boldsymbol{X}^{-1}\boldsymbol{Y} = \boldsymbol{I}_n$. ■

Let $\boldsymbol{A} = [A]_{\mathcal{X}}^{\mathcal{Y}} = \{a_{ij}\}$ be the matrix of a linear transformation $A : \mathcal{V}_1 \to \mathcal{V}_2$ with respect to bases $\mathcal{X} = \{\boldsymbol{x}_1, \boldsymbol{x}_2, \ldots, \boldsymbol{x}_n\}$ of $\mathcal{V}_1$ and $\mathcal{Y} = \{\boldsymbol{y}_1, \boldsymbol{y}_2, \ldots, \boldsymbol{y}_m\}$ of $\mathcal{V}_2$. Let $\boldsymbol{B} = [A]_{\mathcal{U}}^{\mathcal{W}} = \{b_{ij}\}$ represent the same transformation A with respect to two other bases, $\mathcal{U} = \{\boldsymbol{u}_1, \boldsymbol{u}_2, \ldots, \boldsymbol{u}_n\}$ of $\mathcal{V}_1$ and $\mathcal{W} = \{\boldsymbol{w}_1, \boldsymbol{w}_2, \ldots, \boldsymbol{w}_m\}$ of $\mathcal{V}_2$. How are $\boldsymbol{A}$ and $\boldsymbol{B}$ related?

This question is easily answered using the following composition:

$$\boldsymbol{B} = [A]_{\mathcal{U}}^{\mathcal{W}} = [I]_{\mathcal{Y}}^{\mathcal{W}}[A]_{\mathcal{X}}^{\mathcal{Y}}[I]_{\mathcal{U}}^{\mathcal{X}} = \boldsymbol{QAP}^{-1} ,$$

where $\boldsymbol{P} = [I]_{\mathcal{X}}^{\mathcal{U}}$ is $n \times n$ and $\boldsymbol{Q} = [I]_{\mathcal{Y}}^{\mathcal{W}} = \{q_{ij}\}$ is $m \times m$. Both are invertible (i.e., $\boldsymbol{P}^{-1}$ and $\boldsymbol{Q}^{-1}$ exist) because I is always one-one and onto.

Let us see what goes on under the hood. For each $j = 1, 2, \ldots, n$, we can write

$$A\boldsymbol{x}_j = a_{1j}\boldsymbol{y}_1 + a_{2j}\boldsymbol{y}_2 + \cdots + a_{mj}\boldsymbol{y}_m \text{ and } A\boldsymbol{u}_j = b_{1j}\boldsymbol{w}_1 + b_{2j}\boldsymbol{w}_2 + \cdots + b_{mj}\boldsymbol{w}_m .$$

Since $\boldsymbol{P} = [I]_{\mathcal{X}}^{\mathcal{U}}$ and $\boldsymbol{Q} = [I]_{\mathcal{Y}}^{\mathcal{W}}$, we can similarly write

$$\boldsymbol{x}_j = p_{1j}\boldsymbol{u}_1 + p_{2j}\boldsymbol{u}_2 + \cdots + p_{nj}\boldsymbol{u}_n \text{ and } \boldsymbol{y}_i = q_{1j}\boldsymbol{w}_1 + q_{2j}\boldsymbol{w}_2 + \cdots + q_{mj}\boldsymbol{w}_m$$

for each $j = 1, 2, \ldots, n$ and $i = 1, 2, \ldots, m$. We will find the coordinates of $A\boldsymbol{x}_j$ with respect to $\mathcal{W}$ in two different ways that will evince the relationship between $\boldsymbol{A}$ and $\boldsymbol{B}$. First, observe that for each $\boldsymbol{x}_j \in \mathcal{X}$,

$$\begin{aligned} A(\boldsymbol{x}_j) &= \sum_{i=1}^{m} a_{ij}\boldsymbol{y}_i = \sum_{i=1}^{m} a_{ij}\left(q_{1j}\boldsymbol{w}_1 + q_{2j}\boldsymbol{w}_2 + \cdots + q_{mj}\boldsymbol{w}_m\right) \\ &= \sum_{i=1}^{m} a_{ij}\left(\sum_{k=1}^{m} q_{ki}\boldsymbol{w}_k\right) = \sum_{k=1}^{m}\left(\sum_{i=1}^{m} q_{ki}a_{ij}\right)\boldsymbol{w}_k = \sum_{k=1}^{m}\left(\boldsymbol{q}'_{k*}\boldsymbol{a}_{*j}\right)\boldsymbol{w}_k , \end{aligned} \tag{16.32}$$

where $\boldsymbol{q}'_{k*}$ is the k-th row of $\boldsymbol{Q}$ and $\boldsymbol{a}_{*j}$ is the j-th column of $\boldsymbol{A}$. Next, observe that

$$\begin{aligned} A(\boldsymbol{x}_j) &= A\left(p_{1j}\boldsymbol{u}_1 + p_{2j}\boldsymbol{u}_2 + \cdots + p_{nj}\boldsymbol{u}_n\right) = \sum_{i=1}^{n} p_{ij}A\boldsymbol{u}_i \\ &= \sum_{i=1}^{n} p_{ij}\left(\sum_{k=1}^{m} b_{ki}\boldsymbol{w}_k\right) = \sum_{k=1}^{m}\left(\sum_{i=1}^{n} b_{ki}p_{ij}\right)\boldsymbol{w}_k = \sum_{k=1}^{m}\left(\boldsymbol{b}'_{k*}\boldsymbol{p}_{*j}\right)\boldsymbol{w}_k , \end{aligned} \tag{16.33}$$

where $\boldsymbol{b}'_{k*}$ is the k-th row of $\boldsymbol{B}$ and $\boldsymbol{p}_{*j}$ is the j-th column of $\boldsymbol{P}$. Thus, (16.32) and (16.33) both reveal the coordinates of $A\boldsymbol{x}_j$ with respect to the basis $\mathcal{W}$. These coordinates must be equal to each other. Therefore, $\boldsymbol{b}'_{k*}\boldsymbol{p}_{*j} = \boldsymbol{q}'_{k*}\boldsymbol{a}_{*j}$ for $k = 1, 2, \ldots, m$ and $j = 1, 2, \ldots, n$, which means that $\boldsymbol{BP} = \boldsymbol{QA}$ or, equivalently, $\boldsymbol{B} = \boldsymbol{QAP}^{-1}$.

Matrices $\boldsymbol{B}$ and $\boldsymbol{A}$ are said to be ***equivalent*** if there exists nonsingular matrices $\boldsymbol{G}$ and $\boldsymbol{H}$ such that $\boldsymbol{B} = \boldsymbol{GAH}$. In other words, equivalent matrices are derivable from one another using invertible (or *non-singular*) transformations. In the special case where $\mathcal{V}_1 = \mathcal{V}_2 = \mathcal{V}$ and we take $\mathcal{Y} = \mathcal{X}$ and $\mathcal{W} = \mathcal{U}$, we end up with $\boldsymbol{B} = \boldsymbol{PAP}^{-1}$, whereupon $\boldsymbol{B}$ and $\boldsymbol{A}$ are said to be ***similar***.

Example 16.11 Let $A : \Re^n \to \Re^m$ be a linear transformation. Suppose we are given two bases $\mathcal{X}$ and $\mathcal{U}$ for $\Re^n$, and two bases $\mathcal{Y}$ and $\mathcal{W}$ for $\Re^m$. We construct $\boldsymbol{X} = [\boldsymbol{x}_1 : \boldsymbol{x}_2 : \ldots : \boldsymbol{x}_n]$ and $\boldsymbol{U} = [\boldsymbol{u}_1 : \boldsymbol{u}_2 : \ldots : \boldsymbol{u}_n]$ as $n \times n$ non-singular matrices with columns as the vectors in the bases $\mathcal{X}$ and $\mathcal{U}$, respectively. The transition matrix $[I]_{\mathcal{X}}^{\mathcal{U}} = \boldsymbol{P}$ is then related to $\boldsymbol{X}$ and $\boldsymbol{U}$ as $\boldsymbol{X} = \boldsymbol{UP}$, which we solve for $\boldsymbol{P}$. Similarly, for $\Re^m$ we form $m \times m$ non-singular matrices $\boldsymbol{Y} = [\boldsymbol{y}_1 : \boldsymbol{y}_2 : \ldots : \boldsymbol{y}_m]$ and $\boldsymbol{W} = [\boldsymbol{w}_1 : \boldsymbol{w}_2 : \ldots : \boldsymbol{w}_m]$, so that $\boldsymbol{Y} = \boldsymbol{WQ}$, which we solve for the transition matrix Q. If $A = [A]_{\mathcal{X}}^{\mathcal{Y}}$ and $B = [A]_{\mathcal{U}}^{\mathcal{W}}$, then clearly $AX = YA$ and $AU = WB$. Multiplying both sides by P reveals that $WBP = AUP = AX = YA$ and, since W is nonsingular, $BP = W^{-1}YA = QA$. This proves that A and B are similar. Given the matrix with respect to one pair of bases, say A, we can easily obtain B by solving $BP = QA$ or, equivalently, $P'B' = A'Q'$ for B'. ■

16.8 Hilbert spaces

The vector spaces that we have dealt with thus far have been finite dimensional and, primarily, subspaces of an n-dimensional Euclidean space, $\Re^n$, for some finite positive integer n. However, in modern statistical methods and machine learning we often encounter vector spaces that are not finite dimensional. Among the infinite dimensional spaces, *Hilbert spaces* are the closest relative to any finite dimensional space. They have many important features of a finite dimensional space such as the concept of an angle between two vectors, orthogonality, orthogonal complements, orthogonal projections and so on.

Before we define a Hilbert space, let us recall the definition of an inner product Definition 16.4. This definition does not require the vector space to be finite dimensional. If an inner product exists on $\mathcal{V}$, then we can define the notion of orthogonality (or *perpendicular* vectors) on $\mathcal{V}$ by defining two vectors u and v to be *orthogonal* if and only if $\langle u, v\rangle = 0$.

A norm, denoted by $\|\cdot\|$, is a measure of magnitude of a vector. If a norm exists, then $\forall u, v, w \in \mathcal{V}$ and $\alpha \in \Re$, it satisfies:

(i) $\|v\| \geq 0$ and $\|v\| = 0$ if and only if $v = 0$.

(ii) $\|\alpha v\| = |\alpha|\|v\|$.

(iii) $\|u + v\| \leq \|u + w\| + \|v + w\|$.

When an inner product exists on $\mathcal{V}$, a norm can be defined easily by the relation $\|v\|^2 = \langle v, v\rangle$ for all $v \in \mathcal{V}$. In that case we will say that the norm is induced by the inner product. The inner product and the norm are related by the Cauchy-Schwarz inequality (Theorem 16.3). When a norm exists, we can naturally define a *metric* or a distance between two vectors in $\mathcal{V}$ as $d(u, v) = \|u - v\|$. A metric induces a ***topology*** on the vector space $\mathcal{V}$ by defining open balls or neighborhoods around each vector. Thus, a vector space with an inner product is rich in algebraic and topological structures. In addition, if we require that the vector space is complete in the metric induced by the inner product, i.e., any *Cauchy sequence* of vectors will converge in $\mathcal{V}$, then we call such a vector space a *Hilbert space*. We reserve the symbol $\mathcal{H}$ specifically for a vector space that is also a Hilbert space.

Definition 16.8 *A vector space $\mathcal{H}$ with an inner product $\langle\cdot,\cdot\rangle$ is a* **Hilbert space** *if it is complete in the metric topology induced by the inner product.*

It is perhaps the concept of orthogonality on Hilbert spaces that makes them extremely useful in statistics and machine learning. We are already familiar with the definition and usefulness of *orthonormal bases* for finite dimensional vector spaces. Fortunately, Hilbert spaces also admit orthonormal bases, although in the infinite dimensional case the definition is slightly more general.

Definition 16.9 *A collection of vectors* $\{\boldsymbol{v}_\lambda : \lambda \in \Lambda\}$ *in* $\mathcal{H}$ *is an* **orthonormal basis** *if:*

(i) $\langle \boldsymbol{v}_\lambda, \boldsymbol{v}_{\tilde{\lambda}} \rangle = \delta_{\lambda\tilde{\lambda}}\ \forall \lambda, \tilde{\lambda} \in \Lambda.$

(ii) $Sp\{\boldsymbol{v}_\lambda : \lambda \in \Lambda\} = \mathcal{H}.$

Property (ii) means that any vector $\boldsymbol{v} \in \mathcal{H}$ can be expressed as the limit of a linear combination of vectors in the orthonormal basis. Since the point of this chapter is to show that many of the tools developed in the finite dimensional case continues to be useful in an infinite dimensional Hilbert space, and more so if Λ is countable, we will now focus on the case when the Hilbert space is ***separable***, i.e., it admits a countable orthonormal basis.

When the Hilbert space is separable, i.e., $\Lambda = \{\lambda_k,\ k \geq 1\}$ is countable, we will simply write an orthonormal basis as $\{\boldsymbol{v}_k,\ k \geq 1\}$. In a separable Hilbert space, any orthonormal basis constitutes a linearly independent collection of vectors, where linear independence means that any finite subcollection of $\{\boldsymbol{v}_k, k \geq 1\}$ will be linearly independent in the usual sense.

Theorem 16.11 *If* $\{\boldsymbol{v}_k,\ k \geq 1\}$ *is an orthonormal basis for a separable Hilbert space, then* $\{\boldsymbol{v}_k,\ k \geq 1\}$ *is a linearly independent collection of vectors.*

Proof. Let $\{\boldsymbol{v}_{k_1},\ \ldots, \boldsymbol{v}_{k_n}\}$ be a finite subcollection of $\{\boldsymbol{v}_k,\ k \geq 1\}$. Suppose $\sum_{i=1}^n \alpha_i \boldsymbol{v}_{k_i} = 0$ for a collection of scalars $\{\alpha_i,\ i = 1, \ldots, n\}$. Then for any j,

$$0 = \langle 0, \boldsymbol{v}_{k_j} \rangle = \left\langle \sum_{i=1}^n \alpha_i \boldsymbol{v}_{k_i}, \boldsymbol{v}_{k_j} \right\rangle = \sum_{i=1}^n \alpha_i \langle \boldsymbol{v}_{k_i}, \boldsymbol{v}_{k_j} \rangle = \sum_{i=1}^n \alpha_j \delta_{ij} = \alpha_j\ ,$$

where $\delta_{ij} = 1$ if $i = j$ and 0 if $i \neq j$. $\square$

A similar calculation reveals that any vector $\boldsymbol{v} \in \mathcal{H}$ has the unique representation $\boldsymbol{v} = \sum_{k=1}^\infty \langle \boldsymbol{v}, \boldsymbol{v}_k \rangle \boldsymbol{v}_k$. To draw the parallel with Euclidean spaces, consider the standard orthonormal basis $\{\boldsymbol{e}_1, \boldsymbol{e}_2, \ldots, \boldsymbol{e}_n\}$ in $\Re^n$ where $\boldsymbol{e}_i$ is the vector of zeros with a one in the ith place. Then any vector $\boldsymbol{x} \in \Re^n$ can be represented as $\boldsymbol{x} = x_1\boldsymbol{e}_1 + \cdots + x_n\boldsymbol{e}_n$ where $x_i = \langle \boldsymbol{x}, \boldsymbol{e}_i \rangle$ are the coordinates of $\boldsymbol{x}$ in the Cartesian orthonormal system $\{\boldsymbol{e}_1, \boldsymbol{e}_2, \ldots, \boldsymbol{e}_n\}$. Following the same terminology, a vector $\boldsymbol{v}$ in $\mathcal{H}$ is represented by its coordinates $\langle \boldsymbol{v}, \boldsymbol{v}_k \rangle$ in the orthonormal system $\{\boldsymbol{v}_k,\ k \geq 1\}$.

A ***functional*** on a Hilbert space $\mathcal{H}$ is a function $f : \mathcal{H} \longrightarrow \Re$. For two Hilbert spaces, $\mathcal{H}_1$ and $\mathcal{H}_2$ an ***operator*** from $\mathcal{H}_1$ to $\mathcal{H}_2$ is a function $T : \mathcal{H}_1 \longrightarrow \mathcal{H}_2$. When the operator is from $\mathcal{H}$ to itself, then the reason that we will simply call it an operator on $\mathcal{H}$ is clear from the context. The norm of a functional f is given by $\|f\| = \sup_{\|\boldsymbol{v}\|\leq 1} |f(\boldsymbol{v})|$, and that of an operator T from $\mathcal{H}_1$ to $\mathcal{H}_2$ is $\|T\| = \sup_{\|\boldsymbol{v}\|_1\leq 1} \|T(\boldsymbol{v})\|_2$, where $\|\cdot\|_1$ denotes the norm in $\mathcal{H}_1$ and $\|\cdot\|_2$ denotes that in $\mathcal{H}_2$. Functionals are operators from $\mathcal{H}$ to $\Re$ where $\Re$ is the Hilbert space with the usual scalar product as the inner product.

An operator is ***linear*** if $T(\alpha\boldsymbol{u}+\beta\boldsymbol{v}) = \alpha T(\boldsymbol{u})+\beta T(\boldsymbol{v})$ for all $\boldsymbol{u}, \boldsymbol{v} \in \mathcal{H}$ and scalars $\alpha, \beta \in \Re$. It is continuous at $\boldsymbol{v}$ if, for every sequence of vectors $\boldsymbol{v}_k$, $\lim_{k\to\infty} \|\boldsymbol{v} - \boldsymbol{v}_k\|_1 = 0$ implies $\lim_{k\to\infty} \|T(\boldsymbol{v}) - T(\boldsymbol{v}_k)\|_2 = 0$. An operator that is continuous at every $\boldsymbol{v} \in \mathcal{H}_1$ is said to be a ***continuous*** operator. For linear Hilbert space operators, continuity is equivalent to having finite operator norm, i.e., being a ***bounded linear operator***. Of special interest are the classes of continuous linear functionals and the class of continuous linear operators on a Hilbert space $\mathcal{H}$. The class of all linear functionals on $\mathcal{H}$ is called the ***dual*** or the ***dual space*** of $\mathcal{H}$. The dual is usually denoted by $\mathcal{H}^*$. A nice feature of Hilbert spaces is that continuous linear functionals and continuous linear operators behave almost like vectors and matrices.

For any continuous linear operator T on $\mathcal{H}$, the image and kernel of T are defined analogous to (16.14) and (16.15):

$$\text{Im}(T) = \{\boldsymbol{v} \in \mathcal{H} : \boldsymbol{v} = T\boldsymbol{u} \text{ for some } \boldsymbol{u} \in \mathcal{H}\} \text{ and } \text{Ker}(T) = \{\boldsymbol{v} \in \mathcal{H} : T\boldsymbol{v} = \boldsymbol{0}\} .$$

Theorem 16.12 *For any continuous linear operator on $\mathcal{H}$, Ker(T) is a closed linear subspace of $\mathcal{H}$.*

Proof. If $\boldsymbol{v}_n \in \text{Ker}(T)$ and $\lim_{n\to\infty} \boldsymbol{v}_n = \boldsymbol{v}$, then $T\boldsymbol{v} = \lim_{n\to\infty} T\boldsymbol{v}_n = \boldsymbol{0}$. Therefore, $\boldsymbol{v} \in \text{Ker}(T)$. $\square$

Orthogonality in Hilbert spaces

Orthogonality leads to a sensible discussion about the shortest distance between a vector $\boldsymbol{v}$ and closed subspace $\mathcal{V}$ of $\mathcal{H}$. We define the vector in $\mathcal{V}$ closest to $\boldsymbol{v}$ as the ***projection*** of v onto $\mathcal{V}$. To study projectors, it will be convenient to define orthogonal complements in Hilbert spaces.

Definition 16.10 Orthogonal complement in a Hilbert space. *For any subspace $\mathcal{V} \in \mathcal{H}$, define the orthogonal complement of $\mathcal{V}$ in $\mathcal{H}$ as*

$$\mathcal{V}^\perp = \{\boldsymbol{v} \in \mathcal{H} : \langle \boldsymbol{u}, \boldsymbol{v}\rangle = 0 \textit{ for each } \boldsymbol{u} \in \mathcal{V}\} .$$

It is easy to verify that $\mathcal{V}^\perp$ is a closed linear subspace of $\mathcal{H}$ and is left to the reader.

Theorem 16.13 Projection theorem. *Let $\mathcal{V}$ be a closed subspace of a Hilbert space $\mathcal{H}$ and let $\boldsymbol{v} \in \mathcal{H}$ be an arbitrary vector. Then, there exists a unique $\hat{\boldsymbol{v}} \in \mathcal{V}$ such that $\boldsymbol{v} - \hat{\boldsymbol{v}} \perp \boldsymbol{u} = 0$ for each $\boldsymbol{u} \in \mathcal{V}$.*

Proof. Imitating approaches for finite dimensional vector spaces (as in) would run into problems because we would require that if $\{\boldsymbol{v}_\lambda,\ \lambda \in \Lambda_\mathcal{V}\}$ is an orthonormal basis for $\mathcal{V}$, then it can be extended to an orthonormal basis $\{\boldsymbol{v}_\lambda,\ \lambda \in \Lambda\}$ of $\mathcal{H}$ with $\Lambda_\mathcal{V} \subset \Lambda$. We will, instead, provide a different proof which uses the fact that $\mathcal{V}$ is a closed subspace. For any $\boldsymbol{v} \in \mathcal{H}$, define the distance of $\boldsymbol{w}$ to $\mathcal{V}$ as

$$d(\boldsymbol{w}) := \inf\{\|\boldsymbol{u} - \boldsymbol{w}\| : \boldsymbol{u} \in \mathcal{V}\}\ .$$

Since $\mathcal{V}$ is a subspace, $d(\boldsymbol{w}) \leq \|\boldsymbol{w}\|$ and, therefore, finite. Also let $d_{\boldsymbol{v}} = d(\boldsymbol{v})$. Suppose $\boldsymbol{z} \in \mathcal{H}$. By definition, $d(\boldsymbol{w}) \leq \|\boldsymbol{u} - \boldsymbol{w}\|$ for any $\boldsymbol{u} \in \mathcal{V}$. Then by the triangle inequality $d(\boldsymbol{w}) \leq \|\boldsymbol{u} - \boldsymbol{w}\| \leq \|\boldsymbol{u} - \boldsymbol{z}\| + \|\boldsymbol{w} - \boldsymbol{z}\|$. By taking an infimum over $\boldsymbol{u}$, we have $d(\boldsymbol{w}) \leq d(\boldsymbol{z}) + \|\boldsymbol{w} - \boldsymbol{z}\|$. Similarly, $d(\boldsymbol{z}) \leq d(\boldsymbol{w}) + \|\boldsymbol{w} - \boldsymbol{z}\|$ and we can conclude that $|d(\boldsymbol{w}) - d(\boldsymbol{z})| \leq \|\boldsymbol{w} - \boldsymbol{z}\|$. Therefore, $d : \mathcal{H} \to \Re^+$ is continuous.

Since $\mathcal{V} \setminus \{\boldsymbol{v}\}$ is a closed set, there exists a sequence $\boldsymbol{u}_n \in \mathcal{V}$ such that $d_{\boldsymbol{v}} = \lim \|\boldsymbol{u}_n - \boldsymbol{v}\|$. Therefore,

$$\begin{aligned}\|\boldsymbol{u}_n - \boldsymbol{u}_m\|^2 + \|\boldsymbol{u}_n + \boldsymbol{u}_m\|^2 &= \|(\boldsymbol{u}_n - \boldsymbol{v}) - (\boldsymbol{u}_m - \boldsymbol{v})\|^2 \\ &\quad + \|(\boldsymbol{u}_n - \boldsymbol{v}) + (\boldsymbol{u}_m - \boldsymbol{v})\|^2 \\ &\leq 2(\|\boldsymbol{u}_n - \boldsymbol{v}\|^2 + \|\boldsymbol{u}_m + \boldsymbol{v}\|^2)\ ,\end{aligned}$$

which implies the following:

$$\begin{aligned}\|\boldsymbol{u}_n - \boldsymbol{u}_m\|^2 &\leq 2(\|\boldsymbol{u}_n - \boldsymbol{v}\|^2 + \|\boldsymbol{u}_m + \boldsymbol{v}\|^2) - \|(\boldsymbol{u}_n - \boldsymbol{v}) + (\boldsymbol{u}_m - \boldsymbol{v})\|^2 \\ &\leq 2(\|\boldsymbol{u}_n - \boldsymbol{v}\|^2 + \|\boldsymbol{u}_m + \boldsymbol{v}\|^2) - 4\|(\boldsymbol{u}_n + \boldsymbol{u}_m)/2 - \boldsymbol{v}\|^2 \\ &\leq 2(\|\boldsymbol{u}_n - \boldsymbol{v}\|^2 + \|\boldsymbol{u}_m + \boldsymbol{v}\|^2) - 4d_{\boldsymbol{v}}^2 \text{ because } (\boldsymbol{u}_m + \boldsymbol{u}_n)/2 \in \mathcal{V}\ .\end{aligned}$$

Letting $m, n \to \infty$ we get $\|\boldsymbol{u}_n - \boldsymbol{u}_m\|^2 \leq 4d_{\boldsymbol{v}}^2 - 4d_{\boldsymbol{v}}^2 = 0$. Hence, the sequence $\{\boldsymbol{u}_n\}$ is a Cauchy sequence. Since $\mathcal{V}$ is a Hilbert space in its own right, it is complete and, hence, there exists $\hat{\boldsymbol{v}} \in \mathcal{V}$ such that $d_{\boldsymbol{v}} = \|\hat{\boldsymbol{v}} - \boldsymbol{v}\|$.

Now consider any $\boldsymbol{u} \in \mathcal{V}$. If $\boldsymbol{u} = \boldsymbol{0}$, then $\langle \boldsymbol{v} - \hat{\boldsymbol{v}}, \boldsymbol{u}\rangle = 0$. Otherwise, normalize $\boldsymbol{u}$ so that $\|\boldsymbol{u}\| = 1$. Let $c = \langle \boldsymbol{v} - \hat{\boldsymbol{v}}, \boldsymbol{u}\rangle$. Then,

$$\begin{aligned}\|\boldsymbol{v} - (\hat{\boldsymbol{v}} + c\boldsymbol{u})\|^2 &= \|(\boldsymbol{v} - \hat{\boldsymbol{v}}) - c\boldsymbol{u}\|^2 \\ &= \|\boldsymbol{v} - \hat{\boldsymbol{v}}\|^2 + c^2\|\boldsymbol{u}\|^2 - 2c\langle \boldsymbol{v} - \hat{\boldsymbol{v}}, \boldsymbol{u}\rangle \\ &= \|\boldsymbol{v} - \hat{\boldsymbol{v}}\|^2 + c^2 - 2c^2 \\ &= \|\boldsymbol{v} - \hat{\boldsymbol{v}}\|^2 - 2c^2\ .\end{aligned}$$

But since both $\hat{\boldsymbol{v}}$ and $\boldsymbol{u}$ are in $\mathcal{V}$, so is $(\hat{\boldsymbol{v}} + c\boldsymbol{u})$. Hence, $\|\boldsymbol{v} - (\hat{\boldsymbol{v}} + c\boldsymbol{u})\|^2 = \|\boldsymbol{v} - \hat{\boldsymbol{v}}\|^2 - 2c^2$ violates the definition of $\hat{\boldsymbol{v}}$ as the point in $\mathcal{V}$ closest to $\boldsymbol{v}$ unless $c = 0$.

The proof that this projection is unique is left to the reader. □

Thus, given any vector $\boldsymbol{v} \in \mathcal{H}$ we can write it uniquely as $\boldsymbol{v} = \hat{\boldsymbol{v}} + \hat{\boldsymbol{v}}^\perp$ where $\boldsymbol{v} \in \mathcal{V}$ and $\hat{\boldsymbol{v}}^\perp \in \mathcal{V}^\perp$. Therefore, $\mathcal{H} = \mathcal{V} \oplus \mathcal{V}^\perp$. Also,

$$\|\boldsymbol{v}\|^2 = \|\hat{\boldsymbol{v}}\|^2 + \|\hat{\boldsymbol{v}}^\perp\|^2\ .$$

The operator that maps each v to the unique projection vector $\hat{v} \in \mathcal{V}$ (closest to v among all vectors in $\mathcal{V}$) is called the ***orthogonal projection operator*** or ***orthogonal projector***. We will denote the operator by $P_{\mathcal{V}}$ to explicitly show that it depends on the space, $\mathcal{V}$, where it is projecting every vector in $\mathcal{H}$. If the space $\mathcal{V}$ is clear from the context, then we will simply denote the orthogonal projection operator as P. Thus, $Pv = \hat{v}$. Similarly, for each v, let $P^{\perp}$ be the operator that maps v to the unique vector $\hat{v}^{\perp}$. Then, from the uniqueness of the decomposition $v = \hat{v} + \hat{v}^{\perp}$, we have $P^{\perp} = I - P$ where I is the identity operator. We will denote $\hat{v}^{\perp}$ by $(I - P)v$.

For any projection operator P, the space $\mathcal{V}$ is the image of P and the space $\mathcal{V}^{\perp}$ is the kernel of P. From the definition of $I - P$ we have

$$\text{Im}(P) = \text{Ker}(I - P) \text{ and } \text{Ker}(P) = \text{Im}(I - P) .$$

For any $c_1, c_2 \in \Re$ and $v_1, v_2 \in \mathcal{H}$, $\langle (c_1 v_1 + c_2 v_2) - (c_1 P v_1 + c_2 P v_2, u\rangle = c_1 \langle v - P v_1, u\rangle + c_2 \langle v_2 - P v_2, u\rangle = 0 + 0 = 0$. So, by the uniqueness of the projection theorem, we obtain $P(c_1 v_1 + c_2 v_2) = c_1 P v_1 + c_2 P v_2$. This establishes that P is a linear operator. Also, $\|Pv_1 - Pv_2\|^2 = \|v_1 - v_2\|^2 - \|(I-P)(v_1 - v_2)\|^2$. Thus, $\|Pv_1 - Pv_2\| \leq \|v_1 - v_2\|$. Therefore, P is a bounded (hence continuous) linear operator.

The following theorem (analogous to Theorem 16.5) is instrumental in further study of continuous linear operators on Hilbert spaces.

Theorem 16.14 Riesz representation theorem. *Suppose f is a continuous linear functional on a Hilbert space $\mathcal{H}$ endowed with the inner product $\langle \cdot, \cdot \rangle$. Then there exists a unique $v_f \in \mathcal{H}$ such that $f(v) = \langle v, v_f \rangle$ for all $v \in \mathcal{H}$ and $\|f\| = \|v_f\|$.*

Proof. **Existence:** By the projection theorem, $\mathcal{H} = \text{Ker}(f) \oplus \text{Ker}(f)^{\perp}$. If $\text{Ker}(f) = \mathcal{H}$, then choose $v_f = \mathbf{0}$. Otherwise, choose $w \neq \mathbf{0}$ such that $w \in \text{Ker}(f)^{\perp}$ and $\|w\| = 1$. For any $v \in \mathcal{H}$ choose $u = wf(v) - vf(w)$. Then $u \in \text{Ker}(f)$. Thus, $0 = \langle u, w\rangle = f(v) - \langle v, w\rangle f(w)$. Therefore, $f(v) = \langle v, v_f\rangle$ where $v_f = f(w)w$.

Uniqueness: If u_f is another vector such that for all $v \in \mathcal{H}$ we have $f(v) = \langle v, v_f\rangle = \langle v, u_f\rangle$. Then, using the linearity of the inner product $0 = \langle v, v_f - u_f\rangle$ for all $v \in \mathcal{H}$. In particular, choosing $v = v_f - u_f$ we get $\|v_f - u_f\|^2 = \langle v_f - u_f, v_f - u_f\rangle = 0$, which implies $v_f = u_f$.

Equality of norm: Note that $f(u) = \langle u, v_f\rangle$ is a continuous linear functional in u. The Cauchy-Schwarz inequality ensures that $|f(u)| \leq \|u\|\|v_f\|$. Hence, $\|f\| = \sup_{\|u\|\leq 1} |f(u)| \leq \|v_f\|$. If $v_f = \mathbf{0}$, then $f \equiv \mathbf{0}$ and $\|f\| = \|v_f\|$. Otherwise, choose $u = v_f/\|v_f\|$. Then, $|f(u)| = \langle v_f/\|v_f\|, v_f\rangle = \|v_f\|^2/\|v_f\| = \|v_f\|$. Therefore, $\|f\| = \|v_f\|$. □

The Riesz representation theorem ensures that well-defined adjoints exist in Hilbert spaces.

Theorem 16.15 Adjoints in Hilbert spaces. *Let $\mathcal{H}$ be a Hilbert space and let $T : \mathcal{H} \to \mathcal{H}$ be a continuous linear operator. Then, there exists a continuous linear* *operator $T^* : \mathcal{H} \to \mathcal{H}$, called the* **adjoint** *of T, such that for all $\boldsymbol{u}, \boldsymbol{v} \in \mathcal{H}$ we have $\langle \boldsymbol{u}, T\boldsymbol{v} \rangle = \langle T^*\boldsymbol{u}, \boldsymbol{v} \rangle$.*

Proof. For any $\boldsymbol{u} \in \mathcal{H}$, define an operator $T_{\boldsymbol{u}}$ by $T_{\boldsymbol{u}}\boldsymbol{v} = \langle \boldsymbol{u}, T\boldsymbol{v} \rangle$ for all $\boldsymbol{v} \in \mathcal{H}$. Using linearity and continuity of T and the inner product, it is immediate that $T_{\boldsymbol{u}}$ is a continuous linear operator. Then by the Riesz representation theorem, there exists a unique vector $\boldsymbol{u}^*$ such that $T_{\boldsymbol{u}}\boldsymbol{v} = \langle \boldsymbol{u}^*, \boldsymbol{v} \rangle$ for all $\boldsymbol{v} \in \mathcal{H}$. Define the operator T^* as $T^*\boldsymbol{u} = \boldsymbol{u}^*$. Again, using linearity of T and the inner product it is easy to check that T^* is a continuous linear operator. □

Definition 16.11 Self-adjoint operator. *Let $\mathcal{H}$ be a Hilbert space and let $T : \mathcal{H} \to \mathcal{H}$ be a continuous linear operator with adjoint T^*. Then T is called* **self-adjoint** *if $T = T^*$.*

For a real Hilbert space we will call T *symmetric* if it is self-adjoint. For general Hilbert spaces, symmetric continuous linear operators play the same role as symmetric matrices in finite dimensional spaces. Once this parallel is established, it is instructive to draw the parallel to real symmetric matrices whenever possible. The following theorem characterizes all orthogonal projection operators on a real Hilbert space.

Theorem 16.16 Projection operator. *A continuous linear operator P on a real Hilbert space $\mathcal{H}$ is an orthogonal projection operator to a closed linear subspace of $\mathcal{H}$ if and only if $P^2 = P$ and P is symmetric.*

Proof. If P is a projection operator, then for each $\boldsymbol{v} \in \mathcal{H}$ we have $P\boldsymbol{v} \in \mathcal{V}$ and hence $P^2\boldsymbol{v} = P(P\boldsymbol{v}) = P\boldsymbol{v}$. Thus, $P^2 = P$. Also, for any $\boldsymbol{u}, \boldsymbol{v} \in \mathcal{H}$, we have

$$0 = \langle \boldsymbol{v} - P\boldsymbol{v}, P\boldsymbol{u} \rangle - \langle P\boldsymbol{v}, \boldsymbol{u} - P\boldsymbol{u} \rangle = \langle \boldsymbol{v}, P\boldsymbol{u} \rangle - \langle P\boldsymbol{v}, \boldsymbol{u} \rangle \, .$$

Hence P is symmetric. Now assume $P^2 = P$ and P is symmetric. Then, for any $\boldsymbol{v} \in \mathcal{H}$ we can write $\boldsymbol{v} = P\boldsymbol{v} + (I - P)\boldsymbol{v}$. Here $P\boldsymbol{v} \in \text{Im}(P)$ and $\langle \boldsymbol{v} - P\boldsymbol{v}, P\boldsymbol{v} \rangle = \langle \boldsymbol{v}, P\boldsymbol{v} \rangle - \langle P\boldsymbol{v}, P\boldsymbol{v} \rangle = \langle \boldsymbol{v}, P\boldsymbol{v} \rangle - \langle \boldsymbol{v}, P^2\boldsymbol{v} \rangle = \langle \boldsymbol{v}, P\boldsymbol{v} \rangle - \langle \boldsymbol{v}, P\boldsymbol{v} \rangle = 0$. The result now follows from the uniqueness of the projection theorem. □

16.9 Exercises

1. Let $\boldsymbol{x}$ and $\boldsymbol{y}$ be any two vectors in a vector space $\mathcal{V}$ over a field $\mathcal{F}$. Show that

$$\mathcal{S} = \{\alpha\boldsymbol{x} + \beta\boldsymbol{y}` : \alpha, \beta \in \mathcal{F}\}$$

   is a subspace of $\mathcal{V}$.

2. Let $\mathcal{V}$ be a vector space over a field $\mathcal{F}$ and let $\boldsymbol{x}, \boldsymbol{y} \in \mathcal{V}$. The vector $\boldsymbol{x} + (-\boldsymbol{y})$ is denoted as $\boldsymbol{x} - \boldsymbol{y}$. Prove that
   (a) $\boldsymbol{x} - \boldsymbol{y}$ is the unique solution of $\boldsymbol{v} + \boldsymbol{x} = \boldsymbol{u}$;
   (b) $(\boldsymbol{x} - \boldsymbol{y}) + \boldsymbol{z} = (\boldsymbol{x} + \boldsymbol{z}) - \boldsymbol{y}$;

(c) $\alpha(\boldsymbol{x} - \boldsymbol{y}) = \alpha\boldsymbol{x} - \alpha\boldsymbol{y}$, where $\alpha \in \mathcal{F}$.

3. Let m and n be integers such that $0 \leq m \leq n$. Show that $\mathcal{P}_m$ is a subspace of $\mathcal{P}_n$.
4. Let $\mathbb{Q}$ be the field of rational real numbers. Show that $\mathbb{Q}$ forms a subspace of $\Re$ over the field $\mathbb{Q}$.
5. Let $\mathcal{V}$ be a vector space and let $A \subset \mathcal{V}$ be a finite set of vectors in $\mathcal{V}$. Prove that A is linearly dependent if and only if there exists an $\boldsymbol{x} \in A$ such that $\boldsymbol{x} \in Sp(A \backslash \{\boldsymbol{x}\})$.
6. Let $\mathcal{V}$ be a vector space and let $A \subset \mathcal{V}$ be a finite set of vectors in $\mathcal{V}$. If A is a linearly independent set and $\boldsymbol{y} \notin A$, then prove that $A \cup \{\boldsymbol{y}\}$ is linearly dependent if and only if $\boldsymbol{y} \in Sp(A)$.
7. Let A be a linearly independent subset of a subspace $\mathcal{S}$. If $\boldsymbol{x} \notin \mathcal{S}$, then prove that $A \cup \{\boldsymbol{x}\}$ is linearly independent.
8. Show that $\{\boldsymbol{x}_1, \boldsymbol{x}_2, \ldots, \boldsymbol{x}_k\}$ are linearly independent if and only if

$$\sum_{i=1}^{k} \alpha_i \boldsymbol{x}_i = \sum_{i=1}^{k} \beta_i \boldsymbol{x}_i \Longrightarrow \alpha_i = \beta_i \ , \quad \text{for } i = 1, 2, \ldots, k \ .$$

9. Prove that $1 + t + t^2$, $2 - 3t + 4t^2$ and $1 - 9t + 5t^2$ form a linearly independent set in $\mathcal{P}_3$.
10. If ψ is an isomorphism, then prove that $\psi(\boldsymbol{x}) = 0$ if and only if $\boldsymbol{x} = \boldsymbol{0}$.
11. If ψ is an isomorphism, then prove that $\psi(-\boldsymbol{x}) = -\psi(\boldsymbol{x})$.
12. Let $\psi : \mathcal{V}_1 \to \mathcal{V}_2$, where $\mathcal{V}_1$ and $\mathcal{V}_2$ are two vector spaces. If $\{\boldsymbol{x}_1, \boldsymbol{x}_2, \ldots, \boldsymbol{x}_k\}$ is a linearly independent set in $\mathcal{V}_1$, then prove that $\{\psi(\boldsymbol{x}_1), \psi(\boldsymbol{x}_2), \ldots, \psi(\boldsymbol{x}_k)\}$ is a linearly independent set in $\mathcal{V}_2$.
13. True or false: If $\psi : \mathcal{V}_1 \to \mathcal{V}_2$, where $\mathcal{V}_1$ and $\mathcal{V}_2$ are two vector spaces and $\mathcal{S} \subseteq \mathcal{V}_1$ is a subspace of $\mathcal{V}_1$, then $\psi(\mathcal{S}) = \{\psi(\boldsymbol{x}) : \boldsymbol{x} \in \mathcal{S}\}$ is a subspace of $\mathcal{V}_2$ with the same dimension.
14. Find a matrix representation for the linear transformation that rotates a point in $\Re^3$ clockwise by 60 degrees around the x-axis.
15. Find a matrix representation for the linear transformation that reflects a point in $\Re^3$ through the (x, y)-plane.
16. Let $\boldsymbol{u}$ be a fixed vector in $\Re^2$. When is $A : \Re^2 \to \Re^2$ such that $A(\boldsymbol{x}) = \boldsymbol{x} + \boldsymbol{u}$ a linear transformation?
17. True or false: If $\boldsymbol{A}$ is a fixed symmetric matrix in $\Re^{n \times n}$ and $\boldsymbol{x} \in \Re^n$, then the $Q(\boldsymbol{x}) = \boldsymbol{x}'\boldsymbol{A}\boldsymbol{x}$ is a linear function $Q : \Re^n \to \Re$.
18. Let $A : \mathcal{V}_1 \to \mathcal{V}_2$ be a linear transformation between vector spaces $\mathcal{V}_1$ and $\mathcal{V}_2$. If $A(\boldsymbol{x}_1), A(\boldsymbol{x}_2), \ldots, A(\boldsymbol{x}_k)$ are linearly independent vectors in $\mathcal{V}_2$, then prove that $\boldsymbol{x}_1, \boldsymbol{x}_2, \ldots, \boldsymbol{x}_k$ are linearly independent vectors in $\mathcal{V}_1$. *This says that the pre-images of linearly independent vectors are linearly independent.*
19. Let $\mathcal{V}$ be a vector space and let $\boldsymbol{u} \in \mathcal{V}$ be a fixed vector. Define the function $A : \Re \to \mathcal{V}$ so that $A(x) = x\boldsymbol{u}$ for every $x \in \Re$. Prove that $A(x)$ is a linear map. Prove that every linear map $A : \Re \to \mathcal{V}$ is of this form.

20. Let $\boldsymbol{u} \in \Re^3$ be a fixed vector. Define the map: $A(\boldsymbol{x}) = \boldsymbol{u} \times \boldsymbol{x}$, where "$\times$" is the usual cross product in $\Re^3$. Find the matrix representation of A.

21. Let $\mathcal{V}_1$ and $\mathcal{V}_2$ be two vector spaces with basis vectors $\{\boldsymbol{x}_1, \boldsymbol{x}_2, \ldots, \boldsymbol{x}_n\}$ and $\{\boldsymbol{y}_1, \boldsymbol{y}_2, \ldots, \boldsymbol{y}_n\}$, respectively. Can you find a linear function $A : \mathcal{V}_1 \rightarrow \mathcal{V}_2$ such that $A(\boldsymbol{x}_i) = \boldsymbol{y}_i$ for $i = 1, 2, \ldots, n$? Is this linear function *unique*?

22. Prove that (i) $(A + B)^* = A^* + B^*$ and (ii) $(\alpha A)^* = \alpha A^*$, where $\alpha \in \Re$.

23. Let $\mathcal{V}$ be an inner product space and let $A : \mathcal{V} \rightarrow \mathcal{V}$ be an invertible linear transformation. Prove that the following statements are equivalent:
    (a) $\langle A(\boldsymbol{x}), A(\boldsymbol{y}) \rangle = \langle \boldsymbol{x}, \boldsymbol{y} \rangle$ for all $\boldsymbol{x}, \boldsymbol{y} \in \mathcal{V}$;
    (b) $\|A(\boldsymbol{x})\| = \|\boldsymbol{x}\|$ for all $\boldsymbol{x} \in \mathcal{V}$;
    (c) $A^* = A^{-1}$.

24. **Dual space.** Let $\mathcal{V}$ be a vector space $\mathcal{V}$. The dual space of $\mathcal{V}$ is defined to be the set of all linear functions $A : \mathcal{V} \rightarrow \Re$. We denote the dual space as $\mathcal{V}^* = \mathcal{L}(\mathcal{V}_1, \Re)$. Show that $\mathcal{V}^*$ is a vector space. If $\mathcal{V} = \Re^n$, then prove that every linear function $A : \Re^n \rightarrow \Re$ is given by multiplication by a row vector or $1 \times n$ matrix. In other words, prove that $A(\boldsymbol{x}) = \boldsymbol{a}'\boldsymbol{x}$, where $\boldsymbol{a}'$ is a $1 \times n$ row vector.

25. **Dual basis.** Let $\boldsymbol{v}_1, \boldsymbol{v}_2, \ldots, \boldsymbol{v}_n$ be a basis of a vector space $\mathcal{V}$. Define linear functions $A_1, A_2, \ldots, A_n$ such that

$$A_i(\boldsymbol{v}_j) = \delta_{ij} = \begin{cases} 1, & \text{if } i = j \\ 0, & \text{if } i \neq j \end{cases} .$$

    (a) If $\boldsymbol{x} = \sum_{i=1}^n x_i \boldsymbol{v}_i$, then prove that $A_i(\boldsymbol{x}) = x_i$. That is, $A_i(\boldsymbol{x})$ gives the i-th coordinate of $\boldsymbol{x}$ when expressed in terms of $\boldsymbol{v}_1, \boldsymbol{v}_2, \ldots, \boldsymbol{v}_n$.
    (b) Prove that $A_1, A_2, \ldots, A_n$ forms a basis for the dual vector space of $\mathcal{V}$. This is called the ***dual basis*** of $\mathcal{V}^*$.
    (c) Suppose that $\mathcal{V} = \Re^n$. If $\boldsymbol{V} = [\boldsymbol{v}_1 : \boldsymbol{v}_2 : \ldots : \boldsymbol{v}_n]$, then show that the rows of $\boldsymbol{V}^{-1}$ can be identified as the dual basis of $\mathcal{V}^* = (\Re^n)^*$.

References

Abadir, K.M. and Magnus, J.R. (2005). *Matrix Algebra (Econometric Exercises), Vol. 1*. Cambridge, UK: Cambridge University Press.

Axler, S. (2004). *Linear Algebra Done Right* (Second edition). New York: Springer.

Bapat, B.R. (2012). *Linear Algebra and Linear Models* (Third edition). New York: Springer.

Ben-Israel, A. and Greville, T.N.E. (2003). *Generalized Inverses* (Second Edition). New York: Springer.

Christensen, R. (2011). *Plane Answers to Complex Questions: The Theory of Linear Models* (Fourth Edition). New York: Springer.

Eldén, L. (2007). *Matrix Methods in Data Mining and Pattern Recognition*. Philadelphia: Society for Industrial and Applied Mathematics (SIAM).

Faraway, J.J. (2005). *Linear Models with R*. Boca Raton, FL: Chapman and Hall/CRC Press.

Filippov, A.F. (1971). A short proof of the theorem on reduction of a matrix to Jordan form. *Moscow University Mathematics Bulletin*, **26**, 70–71.

Gentle, J.E. (2010), *Matrix Algebra: Theory, Computations, and Applications in Statistics*.

Golub, G.H. and Van Loan, C.F. (2013). *Matrix Computations* (Fourth edition). Baltimore: The Johns Hopkins University Press.

Graybill, F.A. (2001). *Matrices with Applications in Statistics* (Second Edition). Pacific Grove, CA: Duxbury Press.

Grinstead, C.N. and Snell, L.J. (1997). *Introduction to Probability*. Providence, RI: American Mathematical Society.

Halmos, P.R. (1993). *Finite-Dimensional Vector Spaces*. New York: Springer.

Harville, D.A. (1997), *Matrix Algebra from a Statistician's Perspective*, New York: Springer.

Healy, M. (2000). *Matrices for Statistics*. Oxford: Clarendon Press.

Henderson, H.V. and Searle, S.R. (1981), On deriving the inverse of a sum of matrices. *SIAM Review*, **23**, 53–60.

Hoffman, K. and Kunze, R. (1971). *Linear Algebra* (Second edition). New Jersey: Prentice Hall.

Horn, R.A. and Johnson, C.R. (2013). *Matrix Analysis* (Second Edition). Cambridge, UK: Cambridge University Press.

Householder, A.S. (2006). *The theory of matrices in numerical analysis*. New York: Dover Publications.

Kalman, D. (1996). A singularly valuable decomposition: The SVD of a matrix. *The College Mathematics Journal*, **27**, 2–23.

Kemeny, J.G. and Snell, L.J. (1976). *Finite Markov Chains*. New York: Springer-Verlag.

Langville, A.N. and Meyer, C.D. (2005). A survey of eigenvector methods of web information retrieval. *SIAM Review*, **47**, 135–161.

Langville, A.N. and Meyer, C.D. (2006). *Google's PageRank and Beyond: The Science of Search Engine Rankings*. Princeton, NJ: Princeton University Press.

Laub, A.J. (2005). *Matrix Analysis for Scientists and Engineers*. Philadelphia: Society for Industrial and Applied Mathematics (SIAM).

Mackiw, G. (1995). A note on the equality of the column and row rank of a matrix. *Mathematics Magazine*, **68**, 285–286.

Markham, T.L. (1986). Oppenheim's inequality for positive definite matrices. *The American Mathematical Monthly*, **93**, 642–644.

Meyer, C.D. (2001). *Matrix Analysis and Applied Linear Algebra*. Philadelphia: Society for Industrial and Applied Mathematics (SIAM).

Monahan, J.F. (2008). *A Primer on Linear Models*. Boca Raton, FL: Chapman and Hall/CRC Press.

Olver, P.J. and Shakiban, C. (2006). *Applied Linear Algebra*. Upper Saddle River, NJ: Pearson Prentice Hall.

Ortega, J.M. (1987). *Matrix Theory: A Second Course*. New York: Plenum Press.

Putanen, S., Styan, G.P.H. and Isotallo, J. (2011). *Matrix Tricks for Linear Statistical Models: Our Personal Top Twenty*. New York: Springer.

Rao, A.R. and Bhimasankaram, P. (2000). *Linear Algebra* (Second Edition). New Delhi, India: Hindustan Book Agency.

Rao, C.R. (1973). *Linear Statistical Inference and its Applications* (Second Edition). New York: John Wiley & Sons Inc.

Rao, C.R. and Mitra, S.K. (1972). *Generalized Inverse of Matrices and Its Applications*. New York: John Wiley & Sons Inc.

Schott, J.R. (2005). *Matrix Analysis for Statistics* (Second Edition). Hoboken, NJ: John Wiley & Sons.

Schur, I. (1917). Über Potenzreihen, die im Innern des Einheitskreises Beschränkt sind [I]. *Journal für die reine und angewandte mathematik*, **147**, 205–232.

Searle, S.R. (1982). *Matrix Algebra Useful for Statistics*. Hoboken, NJ: John Wiley & Sons.

Seber, G.A.F. and Lee, A.J. (2003). *Linear Regression Analysis.* Hoboken, NJ: John Wiley & Sons.

Stapleton, J. (1995). *Linear Statistical Models.* Hoboken, NJ: John Wiley & Sons.

Stewart, G.W. (1998). *Matrix Algorithms: Volume 1: Basic Decompositions.* Philadelphia: Society for Industrial and Applied Mathematics (SIAM).

Stewart, G.W. (2001). *Matrix Algorithms Volume 2: Eigensystems.* Philadelphia: Society for Industrial and Applied Mathematics (SIAM).

Strang, G. (1993). The fundamental theorem of linear algebra. *American Mathematical Monthly*, **100**, 848–855.

Strang, G. (2005). *Linear Algebra and Its Applications*, Fourth Edition. Cambridge, MA: Wellesley Cambridge Press.

Strang, G. (2009). *Introduction to Linear Algebra*, Fourth Edition. Cambridge, MA: Wellesley Cambridge Press.

Trefethen, L.N. and Bau III, D. (1997). *Numerical Linear Algebra.* Philadelphia: Society for Industrial and Applied Mathematics (SIAM).

Wilkinson, J.H. (1965). Convergence of the LR, QR and related algorithms. *The Computer Journal*, **8**, 77–84.

Index